THE ROUTLEDGE COMPANION TO ENVIRONMENTAL ETHICS

Written for a wide range of readers in environmental science, philosophy, and policy-oriented programs *The Routledge Companion to Environmental Ethics* is a landmark, comprehensive reference work in this interdisciplinary field. Not merely a review of theoretical approaches to the ethics of the environment, the *Companion* focuses on specific environmental problems and other concrete issues. Its 65 chapters, all appearing in print here for the first time, have been organized into the following eleven parts:

    I. Animals
   II. Land
  III. Water
   IV. Climate
    V. Energy and Extraction
   VI. Cities
  VII. Agriculture
 VIII. Environmental Transformation
   IX. Policy Frameworks and Response Measures
    X. Regulatory Tools
   XI. Advocacy and Activism

The volume not only explains the nuances of important core philosophical positions, but also cuts new pathways for the integration of important ethical and policy issues into environmental philosophy. It will be of immense help to undergraduate students and other readers coming up to the field for the first time, but also serve as a valuable resource for more advanced students as well as researchers who need a trusted resource that also offers fresh, policy-centered approaches.

Benjamin Hale is Associate Professor in the Departments of Philosophy and Environmental Studies at the University of Colorado, Boulder. His primary area of research is environmental and public health ethics, and he is the author of the book, *The Wild and the Wicked: On Nature and Human Nature* (2016).

Andrew Light is University Professor of Philosophy, Public Policy, and Atmospheric Sciences, and Director of the Institute for Philosophy and Public Policy at George Mason University. He is currently on leave, serving as Assistant Secretary of Energy for International Affairs at the

U.S. Department of Energy. He is the author of over 100 articles and book chapters on climate change, restoration ecology, and urban sustainability, and has authored, co-authored, and edited 19 books, including *Environmental Values* (Routledge, 2008), *Moral and Political Reasoning in Environmental Practice* (2003), *Technology and the Good Life?* (2000), and *Environmental Pragmatism* (Routledge, 1996). He was previously a Senior Fellow at the Center for American Progress and a Distinguished Senior Fellow at the World Resources Institute in Washington, D.C., and served as Senior Advisor and India Counselor to the U.S. Special Envoy for Climate Change in the U.S. Department of State.

Lydia A. Lawhon is Research Associate in the Environmental Studies Program at the University of Colorado, Boulder. Her research broadly investigates the drivers of practical conflicts between people and large carnivores and the political conflicts between people over large carnivore management.

ROUTLEDGE PHILOSOPHY COMPANIONS

Routledge Philosophy Companions offer thorough, high quality surveys and assessments of the major topics and periods in philosophy. Covering key problems, themes and thinkers, all entries are specially commissioned for each volume and written by leading scholars in the field. Clear, accessible and carefully edited and organised, *Routledge Philosophy Companions* are indispensable for anyone coming to a major topic or period in philosophy, as well as for the more advanced reader.

Also available:

THE ROUTLEDGE COMPANION TO THE FRANKFURT SCHOOL
Edited by Peter E. Gordon, Espen Hammer, Axel Honneth

THE ROUTLEDGE COMPANION TO FEMINIST PHILOSOPHY
Edited by Ann Garry, Serene J. Khader, and Alison Stone

THE ROUTLEDGE COMPANION TO PHILOSOPHY OF PSYCHOLOGY, SECOND EDITION
Edited by Sarah Robins, John Symons, and Paco Calvo

THE ROUTLEDGE COMPANION TO MEDIEVAL PHILOSOPHY
Edited by Richard Cross and JT Paasch

THE ROUTLEDGE COMPANION TO PHILOSOPHY OF PHYSICS
Edited by Eleanor Knox and Alastair Wilson

THE ROUTLEDGE COMPANION TO ENVIRONMENTAL ETHICS
Edited by Benjamin Hale, Andrew Light, and Lydia A. Lawhon

For more information about this series, please visit: https://www.routledge.com/Routledge-Philosophy-Companions/book-series/PHILCOMP

THE ROUTLEDGE COMPANION TO ENVIRONMENTAL ETHICS

Edited by Benjamin Hale, Andrew Light,
and Lydia A. Lawhon

Routledge
Taylor & Francis Group

NEW YORK AND LONDON

Cover image: ©Vibro1 via Getty Images

First published 2023
by Routledge
605 Third Avenue, New York, NY 10158

and by Routledge
4 Park Square, Milton Park, Abingdon, Oxon, OX14 4RN

Routledge is an imprint of the Taylor & Francis Group, an informa business

Library of Congress Cataloging-in-Publication Data
A catalog record for this title has been requested

ISBN: 978-1-138-78492-5 (hbk)
ISBN: 978-1-032-29119-2 (pbk)
ISBN: 978-1-315-76809-0 (ebk)

DOI: 10.4324/9781315768090

Typeset in Bembo
by codeMantra

CONTENTS

List of Figures   *xiii*
List of Tables   *xiv*
Notes on Contributors   *xv*

Introduction   1
Benjamin Hale and Andrew Light

PART I
Animals   **3**

1 Animal Cognition and Moral Status   5
   Robert C. Jones

2 Eating   20
   Dan Hooley and Nathan Nobis

3 Experimentation   31
   Larry Carbone

4 Companion Animals   40
   Clare Palmer and T. J. Kasperbauer

5 Species and Wildlife   51
   Lee Brann, Alexandra Laird, and Alexander Lee

6 Wild Animals   63
   Jeff Sebo

7 Hunting   72
   Nathan Kowalsky

Contents

PART II
Land   **85**

 8 Forests   87
   Lydia A. Lawhon

 9 Mountains: Rethinking Thinking Like a Mountain   96
   David Strong

10 Wilderness   110
   Michael Paul Nelson

11 National Parks   124
   Holmes Rolston, III

12 Landscape   133
   Allen Carlson

13 Property   143
   Markku Oksanen

PART III
Water   **155**

14 Water Quality and Availability   157
   Jeremy J. Schmidt

15 Wetlands   168
   J. Baird Callicott

16 Rivers and Watersheds   179
   Alan J. Rabideau and Kenneth E. Shockley

17 Ocean Policy   192
   Carl Safina

18 Fishing and Harvesting   205
   Mary Lyn Stoll

19 The Ethics of Marine Protected Areas   215
   Nathan J. Bennett and Kai M. A. Chan

PART IV
Climate   **231**

20 Moral Bases of Responses to Climate Change   233
    David R. Morrow

21 Climate Modeling   245
    Wendy S. Parker

22 Climate Change Mitigation   256
    Marcus Hedahl, Kyle Fruh, and Lindsay Whitlow

23 Climate Justice and Equity   271
    Steve Vanderheiden

24 Geoengineering   280
    Benjamin Hale and Michael Pellegrino

25 Skepticism and Denialism   293
    Jay Odenbaugh

PART V
Energy and Extraction   **315**

26 Fossil Fuels   317
    Kian Mintz-Woo

27 Mining   327
    Jessica M. Smith

28 Nuclear Power   338
    John Nolt

29 Hydropower   349
    Jonas Anshelm and Simon Haikola

30 Renewable Energy   359
    Anne Schwenkenbecher and Martin Brueckner

31 Natural Gas and Fracking   374
    Adam Briggle

Contents

32 Energy Poverty   387
    Robin Attfield

PART VI
Cities   **395**

33 Urban Sustainability   397
    Steven A. Moore and Meghan Kleon

34 Urban Parks and Open Space   412
    Roger Paden

35 Suburbs and Exurbs   422
    Robert Kirkman

36 Transportation   431
    Lisa Schweitzer

37 Waste and Consumption   442
    Jen Everett and Rich Cameron

PART VII
Agriculture   **457**

38 Food   459
    David M. Kaplan

39 Industrial Agriculture   470
    Paul B. Thompson

40 Biotechnology   481
    Dane Scott

41 Sustainable Agriculture   492
    Alastair Iles

42 Community Gardens   503
    Stephanie Ross

PART VIII
Environmental Transformation   **513**

43 Remediation   515
    Marion Hourdequin

44 Restoration   529
    Mark Woods

45 Assisted Migration and Reintroduction   546
    Ronald L. Sandler

46 Zoos and Conservation   554
    Ben A. Minteer, James P. Collins, and Aireona Bonnie Raschke

47 Rewilding   568
    Derek Turner

48 Novel Ecosystems   583
    Allen Thompson

PART IX
Policy Frameworks and Response Measures   **601**

49 Pollution and Polluter Pays   603
    Aaron Lercher

50 Constitutional Rights   613
    Kristian Skagen Ekeli

51 Libertarianism   623
    Matt Zwolinski

52 Prediction and Forecasting   636
    Arielle Tozier de la Poterie and Meaghan Daly

53 Disaster Response   652
    Bruce Jennings

PART X
Regulatory Tools   **665**

54 Command and Control   667
    Joshua Preiss

55 Economic Instruments   675
    Joseph E. Aldy

56 Cost-Benefit Analysis   685
    David Schmidtz

57 Risk Assessment   696
    Sven Ove Hansson

58 Precautionary Principles   707
    Kevin C. Elliott

59 Adaptive Management   717
    R. Bruce Hull

PART XI
Advocacy and Activism   **727**

60 Education   729
    Matt Ferkany

61 Everyday Aesthetics   741
    Yuriko Saito

62 Community Participation   752
    W.S.K. "Scott" Cameron

63 Environmental Justice   767
    Robert Melchior Figueroa

64 Environmental Civil Disobedience   783
    Jennifer Welchman

65 Lawbreaking and Ecoterrorism   794
    Ned Hettinger

Index   *807*

FIGURES

22.1 Major GHG concentrations over the past 800,000 years; 10,000 years; and 100 years (from Carter 2015, data sources: Solomon et al. 2007; Luthi et al. 2008 & Hartmann et al. 2013)   257

22.2 Global patterns of impacts in recent decades attributed to climate change, based on studies since the AR4. Impacts are shown at a range of geographic scales. Symbols indicate categories of attributed impacts, the relative contribution of climate change (major or minor) to the observed impact, and confidence in attribution (from Field et al. 2014)   258

25.1 Oreskes' 2004 Consensus Study   298

25.2 Doran and Zimmerman's 2009 Study   299

25.3 Gallup Global Warming Opinion Groups   305

25.4 Kahan et al. Scientific Literacy Study   306

25.5 Kahan et al. Scientific Literacy Study   306

33.1 Changing Neighborhoods: Photographs of Social Reform from 7 Chicago Settlement Houses. Courtesy, The University of Illinois at Chicago, JAMC 0000-0038-588   399

33.2 The Fenway and Simmons College, Boston, Mass. Designed by America's first landscape architect, Frederick Law Olmstead, "the Fens" is a historic example of an ecosociotechnical system. Photo from Detroit Publishing Co/US Library of Congress, LC-D4-73371   403

33.3 Le Corbusier's *Plan Voisin* for Paris, 1922. Photo by SiefkinDR, CC BY-SA 4.0   404

33.4 Bioswale Detail for Elmer Avenue Retrofit. Courtesy, Landscapeperformance.org   406

47.1 A wolf tree in Cockaponset State Forest, Connecticut, USA   570

57.1 Combinations of roles   701

TABLES

19.1 Ethical schools of thought and implications for MPA policy
(after Moore and Russell 2009)   216
19.2 Objectives of Thai MPAs (as cited in Christie and Ole-Moyoi 2011)   217
19.3 Summary of potential benefits of MPAs for humans and nature. Note that many
social benefits might feed back to enhanced ecological protection, and that many
ecological benefits can yield social benefits (e.g. as supporting ecosystem services)   222
19.4 Key ethical considerations in MPA governance, planning, and management   224
20.1 Options for responding to climate change (based on Heyward 2013)   236
25.1 The climate decision   297
47.1 Varieties of rewilding   569

CONTRIBUTORS

Joseph E. Aldy is Professor of the Practice of Public Policy at the Harvard Kennedy School, a University Fellow at Resources for the Future, a Faculty Research Fellow at the National Bureau of Economic Research, and a Senior Adviser at the Center for Strategic and International Studies.

Jonas Anshelm is Professor in the Department of Thematic Studies at Linköping University, Sweden. He received his PhD in technology and social change in 1990 from the university. Today, most of his teaching is conducted within the doctoral education program at Tema Technology. His research and teaching is devoted to the history of ideas, environmental controversies, and conflicts related to socio-technical systems. Since the early 1990s, he has been studying different aspects of energy and environmental politics, mainly using a discourse analytical approach.

Robin Attfield has served as a Professor of Philosophy at Cardiff University since 1991, although he was a Lecturer at this institution from January 1968. His research concerns ethics, the history of ideas, environmental philosophy and philosophy of religion. Attfield has served as a member of the Council of the Royal Institute of Philosophy, and as Chair of the Cardiff Branch of that Institute, and has also participated in a UNESCO Working Party on Environmental Ethics.

Nathan J. Bennett is Research Associate with the OceanCanada Partnership at the University of British Columbia (Canada), an independent consultant who works on human dimensions of ocean governance and marine conservation, and is the Chair of the People and the Ocean Specialist Group for the International Union for the Conservation of Nature (IUCN). Nathan Bennett graduated from the University of Victoria with a PhD in Geography in 2013. He is a recipient of a Trudeau Scholarship, Banting Postdoctoral Fellowship, Liber Ero Fellowship, Fulbright Visiting Scholar Award, and early career contributions award from the Society for Conservation Biology.

Lee Brann is a graduate of the Environmental Studies Master's Program at CU–Boulder and current graduate student in the Wildlife and Fisheries Biology Master's Program at Clemson University. His work centers on the interactions between conservation science, policy, and ethics.

Adam Briggle is Associate Professor in the Department of Philosophy and Religion at the University of North Texas. He received his PhD from the University of Colorado in 2006 in environmental studies. He is particularly interested in how society might better articulate and handle the philosophical issues that are posed by the rapid advancement of technoscience.

Martin Brueckner is Senior Lecturer in Sustainability at Murdoch University. He is a social ecologist with a background in management, economics and environmental policy. His research is focused on industry-community relations, the political economy of sustainable development as well as regional and community sustainability with a social justice emphasis.

J. Baird Callicott is University Distinguished Research Professor Emeritus and Regents Professor of Philosophy, ret. at the University of North Texas. He is co-Editor-in-Chief of the Encyclopedia of Environmental Ethics and Philosophy and author or editor of many books and author of dozens of journal articles, encyclopedia articles, and book chapters in environmental philosophy and ethics. Callicott has served the International Society for Environmental Ethics as President and Yale University as Bioethicist-in-Residence. He is the leading contemporary exponent of Aldo Leopold's land ethic and has elaborated an Earth ethic, *Thinking Like a Planet* (Oxford University Press 2013), in response to climate change. Most recently, he returned to his roots in classical scholarship with *Greek Natural Philosophy: The Presocratics and their importance for Environmental Philosophy* (Cognella 2018).

Rich Cameron is Professor in the Department of Philosophy at DePauw University, where he has taught since 2005. He received his MA and PhD from the Department of Philosophy at the University of Colorado, Boulder in 2000. His areas of interests include ancient Greek philosophy, metaphysics, climate change, and the economics of wealth and happiness.

W.S.K. "Scott" Cameron was a professor of philosophy at Loyola Marymount University for over 20 years before he died at age 54 after a long battle with cancer. He attended Queen's University in Ontario for his BA and MA in philosophy and went on to earn his doctorate in philosophy in 1989 from Fordham University. He was a valued member of the environmental ethics community and we were lucky to procure this essay before he passed.

Larry Carbone has spent 40 years working in laboratory animal care as an academic, as well as animal care staff, researcher, and veterinarian. He holds veterinary specialty board certifications in Animal Welfare and in Laboratory Animal Medicine, and a PhD in Science and Technology Studies. His research focuses on animal welfare science and policy, specifically, pain management for laboratory animals and how scientists write about animal pain. He is the author of *What Animals Want: Expertise and Advocacy in Laboratory Animal Welfare Policy* (Oxford University Press, 2004). He is currently an independent scholar and consultant based in San Francisco, California.

Allen Carlson is Professor of Philosophy (Emeritus) at the University of Alberta, Edmonton, Canada. He works in environmental philosophy and the aesthetics of nature and landscape. He has a number of publications, including *Aesthetics and the Environment: The Appreciation of Nature, Art and Architecture* (1999) and *Nature and Landscape: An Introduction to Environmental Aesthetics* (2008).

Kai M. A. Chan is a sustainability scientist whose work straddles social and natural systems with a focus on values, rewilding, and transformative change. He is a Professor and Canada Research Chair (*Rewilding and Social-Ecological Transformation*) at the University of British Columbia. Kai is also a member of the Royal Society of Canada's College of New Scholars, Artists and Scientists (2017), a Coordinating Lead Author of the IPBES Global Assessment, a Lead Editor for the new journal *People and Nature*, a member of Canada's Clean16 for 2020, and co-founder of CoSphere (a Community of Small-Planet Heroes).

James P. Collins is Virginia M. Ullman Professor of Natural History and the Environment in the School of Life Sciences at Arizona State University. His research group studies host-pathogen biology and its relationship to the decline of species, at times even to extinction. He also studies the scientific, ethical, and public policy issues surrounding the development and proposed environmental release of genetically modified organisms. He is co-author with Martha Crump of *Extinction in our Times: Global Amphibian Decline* (Oxford University Press) and co-editor with Ben Minteer and Jane Maienschein of *The Ark and Beyond: The Evolution of Zoo and Aquarium Conservation* (University of Chicago Press).

Meaghan Daly is Assistant Professor in Environmental Studies at the University of New England in Biddeford, Maine. Her research focuses on the human dimensions of climate change. She holds a PhD from the University of Colorado, Boulder.

Kristian Skagen Ekeli is Professor of Philosophy at the University of Stavanger, Norway. His research interests lie in political, legal and moral philosophy. Ekeli's articles have appeared in *Environmental Values*, *Inquiry*, *Journal of Political Philosophy*, *Law and Philosophy*, *Ratio Juris*, *Philosophy & Social Criticism*, and several other journals.

Kevin C. Elliott is Professor of Philosophy at Michigan State University, with joint appointments in Lyman Briggs College, the Department of Fisheries and Wildlife, and the Department of Philosophy. His scholarship lies at the intersection of the philosophy of science, research ethics, and environmental ethics. His work has been published in a wide variety of books and journal articles, including *Is a Little Pollution Good for You? Incorporating Societal Values in Environmental Research* (Oxford University Press, 2011) and *A Tapestry of Values: An Introduction to Values in Science* (Oxford University Press, 2017).

Jennifer Everett is Professor of Philosophy and Co-Director of the Environmental Fellows Program at DePauw University, which she joined in 2006 after teaching at the University of Alaska Anchorage and Carleton College. She received her PhD, MA, and an Interdisciplinary Certificate in Environmental Policy from the University of Colorado, Boulder. Today, her scholarly work focuses on environmental ethics, animal ethics, and interdisciplinary approaches to materiality.

Matt Ferkany is Associate Professor in the Department of Philosophy at Michigan State University. His scholarship focuses on normative ethical problems related to moral education, well-being, virtue, and environmental ethics and education. He received his PhD in Philosophy from the University of Wisconsin in 2007.

Robert Melchior Figueroa is Associate Professor of Environmental Justice and Philosophy in Oregon State University's School of History, Philosophy, and Religion. He received his PhD, MA, and MS at the University of Colorado at Boulder. His work in Environmental Justice Studies spans over three decades of conceptual development from introducing recognition justice and environmental identity, expanding restorative justice, and framing the social-ecology of moral terrains.

Kyle Fruh is Assistant Professor of Philosophy at Duke Kunshan University. He holds a PhD in Philosophy from Georgetown University. His research ranges over a variety of issues in applied ethics and ethical theory, with a particular focus on the phenomenon of moral heroism.

Simon Haikola is Senior Lecturer at Linköping University, Sweden. His work focuses on the interplay between environmental politics and society along different geographic and institutional scales. He received his PhD in technology and social change from Linköping University in 2012.

Benjamin Hale is Associate Professor in the Philosophy Department and the Environmental Studies Program at the University of Colorado Boulder. He is author of the book *The Wild and the Wicked: On Nature and Human Nature* (MIT Press: 2016), co-editor of the journal *Ethics, Policy & Environment*, and former officer (VP, 2013-16; President, 2016-2019) of the International Society for Environmental Ethics.

Sven Ove Hansson is Professor in philosophy at the Royal Institute of Technology (KTH), Stockholm. He is editor-in-chief of Theoria, member of the Royal Swedish Academy of Engineering Sciences and past president of the Society for Philosophy and Technology. He has published about 390 peer-reviewed papers on moral and political philosophy, decision theory, logic, and the philosophy of science and technology.

Marcus Hedahl is Associate Professor of Philosophy at the U.S. Naval Academy. He holds a BS in Physics from the University of Notre Dame and a PhD in Philosophy from Georgetown University. He previously served as a Dahrendorf Postdoctoral Research Fellow at the Grantham Research Institute on Climate Change and the Environment and as the Environmental Justice Fellow at the Kennedy Institute of Ethics.

Ned Hettinger is Professor of Philosophy Emeritus, College of Charleston. His research specialization is environmental philosophy, including environmental ethics and aesthetics.

Dan Hooley is an instructor at Capilano University. He completed his PhD in philosophy at the University of Toronto in 2018. His dissertation considered the political statute of non-human animals and such animals' place in our legal and political institutions.

Marion Hourdequin is Professor of Philosophy at Colorado College. She is the author of *Environmental Ethics: From Theory to Practice* (Bloomsbury) and co-editor, with David Havlick, of *Restoring Layered Landscapes* (Oxford). She is Vice-President/President Elect of the International Society for Environmental Ethics (ISEE) and serves as an Associate Editor for *Environmental Values*.

R. Bruce Hull is a Senior Fellow at Virginia Tech's Center for Leadership in Global Sustainability located in the Washington DC area. He writes, teaches and conducts professional development on leadership for sustainable development. He advises organizations, communities, and people adapting to the 2050 transitions. He has authored and edited numerous publications, including three books, *Leadership for Sustainability, Infinite Nature,* and *Restoring Nature* (http://cligs.vt.edu/).

Alastair Iles works on diversified farming systems, chemical policies, and the politics of making sustainability transitions at the Department of Environmental Science, Policy & Management, University of California - Berkeley. He served as a co-founding faculty director of the Berkeley Food Institute.

Bruce Jennings is Senior Fellow at the Center for Humans and Nature, Adjunct Associate Professor at Vanderbilt University, and Senior Advisor at The Hastings Center. He has written widely on ethics and public policy issues. His recent work includes *Ecological Governance: Toward a New Social Contract with the Earth* (West Virginia University Press, 2016), Emergency *Ethics: Public*

Health Preparedness and Response (Oxford University Press, 2016), and *Liberty and the Ecological Crisis: Freedom on a Finite Planet* (Routledge, 2019).

Robert C. Jones is Associate Professor of Philosophy at California State University, Dominguez Hills. He is the co-author of Chimpanzee Rights (Routledge 2018) and has published numerous articles and book chapters on animal ethics, animal cognition, and research ethics, and has given over forty talks on animal ethics.

David M. Kaplan is Professor and Chair of the Department of Philosophy and Religion at the University of North Texas. He is author of *Food Philosophy* (2020), editor of the *Encyclopedia of Food and Agricultural Ethics, 2ˢᵗ edition* (2019), and *The Philosophy of Food* (2012). He has published on food technologies, genetically-modified food, artificial ingredients, and junk food. He also manages the *Philosophy of Food Project* (http://www.food.unt.edu).

T. J. Kasperbauer is Postdoctoral Fellow at the Indiana University Center for Bioethics. He is the author of *Subhuman: The Moral Psychology of Human Attitudes to Animals* (Oxford University Press, 2018).

Robert Kirkman is Associate Professor in the School of Public Policy at the Georgia Institute of Technology in Atlanta. His focus is practical ethics, including the ethics and design in engineering and in the built environment.

Meghan Kleon is a writer and researcher living in Austin, Texas. She holds an MS in Sustainable Design from the University of Texas at Austin.

Nathan Kowalsky is Associate Professor of philosophy at St. Joseph's College, University of Alberta, Canada. Kowalsky's research focuses on environmental philosophy, environmental ethics, philosophy of religion, philosophy of culture, and philosophy of technology.

Alexandra Laird is a graduate student in the Philosophy Department at the University of Colorado, Boulder. Her research explores the ontology and values central to conservation biology.

Lydia A. Lawhon holds a PhD in environmental studies from the University of Colorado Boulder. She is a lecturer for the Masters of the Environment professional degree program at CU-Boulder. Her interests are in the political, social, and ethical dimensions of entrenched natural resource conflicts, particularly in the American West.

Alexander Lee is Assistant Professor of Philosophy at the Institute for Culture and Environment at Alaska Pacific University. His work explores moral obligations in the face of environmental change, focusing on wildlife conservation and land management policy in the American West.

Aaron Lercher holds an MLS as well as a PhD in philosophy, both received at the University of Buffalo, State University of New York. Today, he serves as the collection analysis librarian at Louisiana State University. His past research and work has focused on environmental rights, the right to liberty of ecological conscience, and information science. He is currently working in logic.

Andrew Light is University Professor of Philosophy, Public Policy, and Atmospheric Sciences, and Director of the Institute for Philosophy and Public Policy at George Mason University. He is currently on leave, serving as Assistant Secretary of Energy for International Affairs at the U.S.

Department of Energy. He is the author of over 100 articles and book chapters on climate change, restoration ecology, and urban sustainability, and has authored, co-authored, and edited 19 books, including Environmental *Values* (Routledge, 2008), *Moral and Political Reasoning in Environmental Practice* (MIT, 2003), *Technology and the Good Life?* (Chicago, 2000), and *Environmental Pragmatism* (Routledge, 1996). He previously was a Senior Fellow at the Center for American Progress and a Distinguished Senior Fellow at the World Resources Institute in Washington, D.C., and served as Senior Advisor and India Counselor to the U.S. Special Envoy for Climate Change in the U.S. Department of State.

Ben A. Minteer is Professor of environmental ethics and conservation and the Arizona Zoological Society Endowed Chair in the School of Life Sciences at Arizona State University. He has published many books, including most recently *The Fall of the Wild: Extinction, De-Extinction, and the Ethics of Conservation* (Columbia University Press) and (with Jane Maienschein and James P. Collins) *The Ark and Beyond: The Evolution of Zoo and Aquarium Conservation* (University of Chicago Press).

Kian Mintz-Woo is a permanent Lecturer in the philosophy department and Environmental Research Institute at University College Cork, Ireland. He is also a visiting scholar at the Equity and Justice group at the International Institute for Applied Systems Analysis, Austria. He was the recipient of the 2021 Andrew Light Award for Public Philosophy from the International Society for Environmental Ethics.

Steven A. Moore is a Bartlett Cocke Regents Professor Emeritus of Architecture and Planning at the University of Texas at Austin. His current book project is *Infrastructure: Changes in Maine's People, Nature and Technology.*

David R. Morrow serves as the Director of Research at the Institute for Carbon Removal Law and Policy, as well as the Forum for Climate Engineering Assessment, at American University. He is also a Research Fellow at George Mason University in the Institute for Philosophy and Public Policy.

Michael Paul Nelson holds the Ruth H. Spaniol Chair of Renewable Resources and is Professor of Environmental Ethics and Philosophy at Oregon State University. He serves as the Lead Principal Investigator for the HJ Andrews Experimental Forest Long-Term Ecological Research Program and is the Philosophy in Residence for the Isle Royale Wolf-Moose Project. He is the co-editor (with J. Baird Callicott) of *The Great New Wilderness Debate* and *The Wilderness Debate Rages On*, co-editor (with Kathleen Dean Moore) of *Moral Ground: Ethical Action for a Planet in Peril*, and co-author (with J. Baird Callicott) of *American Indian Environmental Ethics: An Ojibwa Case Study.*

Nathan Nobis is Professor of philosophy at Morehouse College in Atlanta, GA. He's the author of *Animals and Ethics 101* (Open Philosophy Press). His areas of research specialization include applied or practical ethics, ethical theory, and critical thinking.

John Nolt is Professor Emeritus in the Philosophy Department at the University of Tennessee and a Research Fellow of the Energy and Environment Program of the Howard H. Baker Jr. Center for Public Policy. His interests include environmental, intergenerational, and climate ethics; as well as formal axiology and logic. He strives to live sustainably and has been an environmental activist for many years.

Jay Odenbaugh is Professor of Philosophy at Lewis & Clark College. He received his Ph.D. from the University of Calgary. His areas of research are the philosophy of science and environmental philosophy. He is the author of Ecological Models (2019) from Cambridge University Press.

Markku Oksanen is Senior Lecturer in Philosophy at the University of Eastern Finland. He has published widely on green political theory and environmental ethics, including *Environmental Human Rights* (edited with Ashley Dodsworth and Selina O'Doherty, Routledge 2018) and *Rights of Nature* (edited with Daniel P. Corrigan, Routledge 2021).

Roger Paden is Associate Professor Emeritus in the Department of Philosophy at George Mason University. His areas of research include Environmental Ethics and Aesthetics, Philosophy of Architecture and Urban Planning, and Social and Political Philosophy. He is the author of *Architecture and Mysticism: Wittgenstein and the Meanings of the Palais Stonborough*, the co-editor of *The Aesthetics of Architecture*, and has published over 70 scholarly articles.

Clare Palmer is George T. and Gladys H. Abell Professor in Liberal Arts and Professor of Philosophy at Texas A&M University, USA. She is the author of *Animal Ethics in Context* (Columbia University Press, 2010) and, with Peter Sandøe and Sandra Corr, of *Companion Animal Ethics* (Wiley-Blackwell, 2015).

Wendy S. Parker is Professor of Philosophy at Virginia Tech. She received her PhD in History and Philosophy of Science from the University of Pittsburgh. Her research addresses issues in general philosophy of science – especially evidence, modeling, and explanation – as well as philosophical issues in meteorology and climate science.

Michael Pellegrino is a PhD student in Environmental Studies at the University of Colorado Boulder. He graduated from the University of Illinois at Urbana-Champaign with a BS in Mathematics and a Juris Doctor degree. He also graduated from Yale University with a Master of Environmental Management degree. Mike is a licensed attorney in Illinois and is admitted to practice law before the Supreme Court of the United States.

Joshua Preiss is Professor of Philosophy and Director of the Program in Philosophy, Politics, and Economics at Minnesota State University, Mankato. His research is in social and political philosophy, applied ethics, political economy, and the philosophy of economics. His recent book, *Just Work for All: The American Dream in the 21st Century* (Routledge 2021) concerns the future of work and the theory and practice of justice in a post-Covid economy.

Alan J. Rabideau is Professor of Civil, Structural, and Environmental Engineering at the University at Buffalo. He holds graduate degrees in Civil Engineering, Philosophy, and Environmental Sciences and Engineering. Current research interests include groundwater modeling and remediation, ecological restoration, and engineering ethics.

Aireona Bonnie Raschke is currently the Associate Director of Practice for the Center for Collaborative Conservation out of Colorado State University. She specializes in community-based conservation, and she previously led the Central Arizona Conservation Alliance in Phoenix, Arizona. She received her MA and PhD in Biology from Arizona State University.

Holmes Rolston, III is University Distinguished Professor of Philosophy Emeritus at Colorado State University in Fort Collins.

Stephanie Ross is Professor Emerita of Philosophy at the University of Missouri-St. Louis. She is the author of What Gardens Mean (Chicago 1998), Two Thumbs Up: How Critics Aid Appreciation (Chicago, 2020), and chapters and papers on a variety of topics in aesthetics.

Carl Safina is an ecologist and founder of The Safina Center. He is the inaugural holder of the Carl Safina Endowed Chair for Nature and Humanity at Stony Brook University.

Yuriko Saito is professor emerita of philosophy at the Rhode Island School of Design. She works in environmental aesthetics, everyday aesthetics, and Japanese aesthetics. Her *Everyday Aesthetics* (2007) and *Aesthetics of the Familiar: Everyday Life and World-Making* (2017) were both published by Oxford University. She is the editor of *Contemporary Aesthetics*.

Ronald Sandler serves as a professor of philosophy and Director of the Ethics Institute at Northeastern University. He received his PhD in philosophy from the University of Wisconsin-Madison in 2001. His primary research is centered on emerging technologies, ethical theory, and environmental ethics.

Jeremy J. Schmidt is Associate Professor at the Department of Geography, Durham University. He is the author of *Water: Abundance, Scarcity, and Security in the Age of Humanity* (2017). His research focuses on the social dimensions of energy, water, and land, as well as the social dimensions of human impacts on the Earth system. Prior to joining Durham University, Schmidt held post-doctoral positions in social anthropology at Dalhousie University and Harvard University.

David Schmidtz is editor in chief of *Social Philosophy & Policy,* and author of *Living Together: Reinventing Moral Science* (Oxford, 2022).

Anne Schwenkenbecher is Senior Lecturer in Philosophy at Murdoch University in Western Australia. Her PhD in Philosophy is from Humboldt University of Berlin. She researches and publishes extensively on collective action and responsibility, social epistemology, environmental philosophy, and political violence. Her latest monograph, *Getting Our Act Together: A Theory of Collective Moral Obligations* (Routledge, 2021), won the North American Society for Social Philosophy 2022 Book Award "as the best social philosophy book published in 2021."

Lisa Schweitzer is a professor at the University of Southern California, USC Sol Price School of Public Policy. She holds a PhD in urban planning from the University of California, Los Angeles. She specializes in social justice, environment, and transportation and writes about urbanism and ethics at www.lisaschweitzer.com.

Dane Scott holds a PhD in philosophy from Vanderbilt University, an MA in philosophical theology from the Graduate Theological Union, and a BS in soil science from the University of California Riverside. Today, he serves as the director of the Mansfield Ethics and Public Affairs Program at the University of Montana. He is also a professor of ethics in the College of Forestry and Conservation.

Jeff Sebo is Clinical Associate Professor of Environmental Studies, Affiliated Professor of Bioethics, Medical Ethics, Philosophy, and Law, and Director of the Animal Studies M.A. Program at New York University. He is author of *Saving Animals, Saving Ourselves* (Oxford University Press, 2022) and co-author of *Chimpanzee Rights* (Routledge, 2018) and *Food, Animals, and the Environment* (Routledge, 2018). He is also an executive committee member at the NYU Center for

Environmental and Animal Protection, a board member at Minding Animals International, and a senior research affiliate at the Legal Priorities Project.

Kenneth E. Shockley is Professor and Holmes Rolston III Chair in Environmental Ethics and Philosophy, Faculty Affiliate in the School of Global Environmental Sustainability, Faculty in the Graduate Degree Program in Ecology, and Director of the Mountain Campus Program in Environmental Humanities at Colorado State University. His work focuses on the climate ethics and policy, the expression of environmental values in public policy, collective responsibility, and ethical theory.

Jessica M. Smith is an anthropologist, focusing on the mining and energy industry. Her interests include social responsibility, engineers, labor, and gender. She is Director of Humanitarian Engineering Graduate Programs and Research and Professor in the Engineering, Design, and Society Department at the Colorado School of Mines.

Mary Lyn Stoll is associate professor of philosophy at the University of Southern Indiana. Her research specialization is business ethics with a focus on the intersection of business ethics with environmental ethics and media ethics. She is co-editor of *Stakeholder Ethics: Essential Readings in Ethical Leadership and Management* (Prometheus).

David Strong is Professor of Philosophy and Environmental Studies at Rocky Mountain College in Billings, Montana. He received his PhD in philosophy at the State University of New York at Stony Brook. He is author of *Crazy Mountains: Learning from Wilderness to Weigh Technology* and coeditor of *Technology and the Good Life?* His work focuses on centering things and the challenges raised for them by the modern technological transformation of our natural and built environments. He is also an avid backpacker and fly fisherman.

Allen Thompson is Professor of Philosophy at Oregon State University where he works on adaptation, hope, future generations, and virtue. He is editor in chief of the journal *Environmental Ethics*, co-editor of *Ethical Adaptation to Climate Change: Human Virtues of the Future* (MIT 2012) with Jeremy Bendik-Keymer and co-editor of *The Oxford Handbook of Environmental Ethics* (Oxford 2017) with Stephen M. Gardiner. From 2019 to 2022 he served as President of the International Society for Environmental Ethics.

Paul B. Thompson is a philosopher, professor, and W.K. Kellogg Chair in Agriculture, Food, and Community Ethics at Michigan State University. He received his PhD in philosophy at the State University of New York at Stony Brook. Today, he serves on the faculty at Michigan State University in the departments of philosophy, agricultural, food and resource economics, and community sustainability.

Arielle Tozier de la Poterie holds a PhD from the University of Colorado, Boulder. She currently works as an interdisciplinary analyst and social scientist for the German Red Cross and Integrated Risk Management Associates.

Derek Turner is Professor of Philosophy at Connecticut College, where he also serves as the Karla Heurich Harrison '28 Director of the Goodwin-Niering Center for the Environment. He is the author of *Making Prehistory: Historical Science and the Scientific Realism Debate* (2007); *Paleontology: A Philosophical Introduction* (2011); and *Paleoaesthetics and the Practice of Paleontology* (2019).

Steve Vanderheiden is Professor of Political Science and Environmental Studies at the University of Colorado at Boulder. He specializes in normative political theory and environmental politics.

Jennifer Welchman is a professor of philosophy at the University of Alberta (Canada). She has published articles and book chapters on environmental ethics and aesthetics. Her recent work has focused on issues in conservation biology.

Lindsay Whitlow is Professor of Biology at Seattle University. He holds a PhD in Ecology and Evolutionary Biology from the University of Michigan. His research focuses on the interactions between urban development and ecosystem health.

Mark Woods is the Chair of the Department of Philosophy at the University of San Diego. He received both his MA and PhD in philosophy at the University of Colorado, Boulder. He served as a postdoctoral fellow in environmental ethics at the Ethics Center at the University of South Florida. His scholarly work focuses on environmental philosophy and philosophical issues of war and peace. He is the author of *Rethinking Wilderness* (2017), and he is currently writing a book trilogy on environmental issues and armed conflicts.

Matt Zwolinski is Professor of Philosophy at the University of San Diego and director of USD's Center for Ethics, Economics, and Public Policy. He is the editor of *Arguing About Political Philosophy* and, with Benjamin Ferguson, the forthcoming *Routledge Companion to Libertarianism* and *Exploitation: Philosophy, Politics, and Economics*. With John Tomasi, he is currently writing *A Brief History of Libertarianism*, and with Miranda Fleischer, *Universal Basic Income: What Everyone Needs to Know*.

INTRODUCTION

Benjamin Hale and Andrew Light

Of the many environmental ethics readers available on library shelves, most organize around traditional distinctions in value theory and normative ethics – moral status, anthropocentrism, intrinsic value, duties to the distant, non-identity, and so on – typically with the objective of providing an environmental gloss on historically significant philosophical problems. In this volume instead, we've collected work centered on ethical questions that affect real policy issues. Our hope in doing so has been to build a volume that is at once sensitive to conceptual questions of interest to philosophers, but also more relevant to a broad set of academics in allied environmental fields. We additionally hope that it might serve as a stepping off point for philosophy students who are hoping to gain insight into the complicated policy terrain that may interest them.

We hope in this respect to offer a real alternative to the other volumes in environmental ethics. To pull this off, our approach has been to take standard conceptual topics in the environmental ethics corpus and offer them up to a finer grained practical analysis than is usually provided. We bring together ethicists and experts in 12 broad categories of environmental inquiry: Animals, Land, Water, Climate, Energy and Extraction, Cities, Agriculture, Environmental Transformation, Law and Policy, Regulatory Tools, Advocacy, and Activism. These categories track the sections in our table of contents. Each section, we hope, creates some potential for bringing philosophical work into the practical realm, while at the same time offering philosophers more detail on the practical questions that presumably inform their work. This gives the volume a shape that eschews general theories prevalent in environmental ethics and helps demonstrate more particularly the utility of philosophy for particular environmental questions.

That said, in keeping with the interests of our expected readership, the early sections of the compendium – starting with Animals and Lands – cover several traditional issues generally associated with the literature in environmental ethics. Topics and treatments in these sections tend toward the encyclopedic. Later sections of the compendium, however, have a much broader orientation, engaging topics rarely addressed from an ethical point of view. We've integrated work on water, climate, and energy to address some of the recent work coming out of the climate ethics and just transitions community. In even later sections, we've integrated work from folks working on topics that have mostly remained adjacent to the environmental ethics literature: cities, agriculture, transformation. And finally, near the end of the book, sensitive to the concern that environmental ethicists have only hitherto interpreted the world in various ways, but that the point is to change it, we look at mechanisms and devices that might work to this end. Policy frameworks, response measures, regulatory tools, advocacy, and activism all collectively make up the final portions of the book.

DOI: 10.4324/9781315768090-1

1

We have at least three reasons for designing the compendium in this way. First, in the world of environmental ethics and political theory, an understanding of the specifics of environmental policy has gained an increasingly important toehold. We seem to be leaving the days when the details of policy were a mere empty example to be attached at the end of a substantive discussion of values and ethics. As editors of the journal *Ethics, Policy and Environment*, we are seeing more environmental ethicists taking up practical policy matters, both in their research and in their teaching. In this respect, environmental ethics is poised to take a turn much like contemporary philosophy of science took only a few decades before. Namely, general philosophy of science has given way to a philosophy of particular sciences. There are now philosophers of physics, mathematics, biology, climate, geology, paleontology, and so on. This shift in the field is so dominant that many of the best PhD programs in the philosophy of science in the English-speaking world now require their students to earn an additional MA in an allied scientific field. Our volume anticipates, and may help to encourage, a similar transition in environmental ethics.

Second, there can be little doubt that many of the important issues that are currently under discussion in the policy realm could benefit from closer ethical interrogation. If we provide a volume that addresses, in a comprehensive and accessible way, the theoretical underpinnings of these discussions, we believe that many non-philosophers will be interested in the volume as well.

Finally, we believe that this volume, structured in this innovative way and paired with some of the work that the same authors have contributed to the journal that we co-edit, *Ethics, Policy & Environment*, can provide a basis for policy-makers and philosophers alike. Our goal throughout has been to develop a text that can bridge the divide between environmental ethics and environmental policy, with an eye toward providing advanced undergraduates support for further study. We have encouraged our authors to offer more than summary of theoretical approaches on a given issue, and instead to include more of a survey of the moral issues at stake in a discrete area of potential moral controversy.

Our hope is that the work offered here will find a home on the shelves of policy professionals, philosophers, and students. We have spoken at times of designing undergraduate courses around some of the essays, hoping there to inspire young students of environmental matters to think more deeply about some of the issues that matter to them. At the same time, we know that the essays are sophisticated and robust enough to provide graduate courses with important touchpoints and references that can guide further study. Some of the essays are some of the first focused philosophical essays on their respective topics. In this respect, the volume will also be of use to advanced graduate students and professionals as a reference work to get them started on future inquiries into particular policy questions. Beyond the relatively straightforward pedagogical goals of most compendia, our hope is that environmental managers and policy professionals will keep this book on their shelves for reference when thinking through the complexities of their given projects.

In this way, the *Routledge Companion to Environmental Ethics* will not only help readers of many backgrounds better understand the nuances of important core positions in environmental philosophy, but will also cut new pathways for integration of important ethical and policy issues into environmental philosophy.

PART I

Animals

1

ANIMAL COGNITION AND MORAL STATUS

Robert C. Jones

Introduction

As employed in the animal ethics literature, *sentience* refers to those conscious experiences with an attractive or aversive quality. These capacities include subjective sensations such as pain and pleasure, suffering, anxiety, and fear (Birch 2017). Sentience plays a central role in many theories of animal ethics. *Animal cognition* involves the study of animal minds, specifically, those mental capacities and processes that generate various flexible behaviors in animals. Animal cognition research investigates the cognitive mechanisms involved in abilities like learning, memory, perception, language, mindreading, and metacognition (Andrews 2020). Precisely what it means to say that animals have minds means different things to different people. Philosophers and researchers working on animal cognition include under that heading a variety of properties and capacities, from sentience to rationality, memory, and language. Loosely speaking, *moral status* is a measure of the moral value of a being (or entity) based on its welfare interests. These interests can used to determine both moral status and relative moral worth or ranking. For example, we can agree that human children have moral status, but what about kittens? Or rats? Or lizards? Or insects? Or trees? Or pebbles along a beach? Measuring the moral value of entities and deciding which ones have moral status and to what degree, occupy a substantial chunk of the work of applied ethicists.

Though few of us have taken the time to reflect upon the connection between sentience, cognition, and moral status, many people believe something like this: that the more intelligent a being is, the greater their moral status. Yet as common as such a belief may be, the justifications for the connection between the sentient and cognitive properties of animals and their moral status are often based on vague intuitions rather than well thought-out ethical principles. When pressed to say precisely which capacities connect to moral status, to what degree they connect to moral status, how they connect to moral status, and why those (but not other) capacities connect to moral status, most of us can do little more than scratch our heads.

This essay will explicate two philosophical views on the connection between sentience, cognition, and moral status. To that end, I will (a) clarify the notion of moral status, (b) explicate two of the prominent philosophical views arguing for increased moral status for nonhuman animals, highlighting the centrality of physiological and cognitive capacities to each theory, (c) discuss recent empirical findings regarding animal sentience and cognition, and (d) briefly look at one connection between animal sentience, cognition, moral status, and environmental ethics.

DOI: 10.4324/9781315768090-3

Moral Status

Key to understanding the concept of moral status is the notion of *interests*. Unlike rocks and teddy bears, human and nonhuman animals are experiential subjects, that is, there's a subjective "what it's like" *from the inside* to be a human or nonhuman animal. We animals are the kinds of things whose experiences matter to us, beings whose lives can go better or worse, *for us*. It is in this sense that humans and animals have interests, for example, interests in our own well-being. Having interests means two things. First, having interests means that a being can be wronged morally, for example, if its interests are unjustifiably harmed or violated and its well-being compromised. Second, interests create duties and obligations upon us agents to respect the interests of others.

To say that an entity has *moral status* is to say that its interests matter morally for its own sake, and not merely for the sake of the interests of others. This affords the being at least some of the protections afforded by moral norms (Beauchamp and Childress 2019). In the case of nonhuman animals, we can say that an animal has moral status when violating the animal's interests directly harms and wrongs the animal herself, not merely her "owner" or those who love or care about her and her interests.

Moral status need not be an all-or-nothing game. It makes sense to say that one entity has greater moral status than another; that, for example, a normal adult chimpanzee has greater moral status than a flea. Used in this sense, "moral status" specifies not only which entities belong to the moral community, but also the degree to which their interests count. These two senses of moral status reflect a distinction between what some philosophers call moral considerability and moral significance (Goodpaster 1978). On this view, an entity is said to be *morally considerable* just in case it is a bona fide member of the moral community in that it can be wronged in a morally relevant way. In this sense, the fact that a being is morally considerable means that we have moral obligations to that being. Saying that an entity is morally considerable is like saying that it's "in the club" of things whose interests we must consider. Other philosophers cut this distinction even more finely, distinguishing between which entities to consider morally, what considerations are relevant about them, and how much weight must we give to each of these considerations (Hale 2011).

Once a being is morally considerable, we may then need to adjudicate questions of relative moral value between beings. That is a question of moral significance. *Moral significance* speaks to the moral value of the members once admitted to the club (Gruen 2014). Just because two entities are "in the club", it doesn't follow that they are of equal moral value. Surely, all living human persons, chimpanzees, dogs, cats, deer, wolves, and birds are in the club. But does that imply that—all things being equal—a human person and a finch have equal moral worth? Or whether 500 gorillas or 5,000 wolves have less moral value than one human being? Though these are tricky questions, we can give a somewhat clear formulation of the concept of moral status as follows: an entity X has moral status just in case (a) moral agents have moral obligations to X, (b) X has basic welfare interests, and (c) the moral obligations owed to X are based on X's interests (DeGrazia 2008).

Humans, Persons, Agents, and Patients

When discussing moral status, it's important to distinguish the terms "human" from "person", and "moral agent" from "moral patient". Though colloquially the terms "human" and "person" are synonymous, philosophers make an important distinction between the two. The term *human* identifies a biologically descriptive feature of a being, namely, her membership in the species *Homo sapiens*. By contrast, *person* is often used to describe a normative feature of a being that confers moral status. Persons who possess the ability to make moral distinctions, deliberate, and freely choose (or fail to choose) to act morally are *moral agents*. Moral agents are subject to moral obligations and may be held morally accountable for their actions. In other words, moral agents may be held morally praiseworthy or blameworthy for their actions. However, those who lack the ability to morally deliberate and act on the results of those deliberations are termed *moral patients*. Moral patients lack the cognitive abilities required to do things like contemplate moral principles or deliberate about which actions are morally right or wrong and so cannot be held blameworthy for their actions, even in cases where a moral patient causes significant harm to another (Regan 1983).

Philosophically speaking, though neurotypical adult human beings are usually seen as paradigm persons, the question of whether infants, young children, severely cognitively other-abled adults, and permanently comatose humans are considered persons remains philosophically controversial (Andrews et al 2018). Legally, at the time of this writing, the U.S. courts recognize all and only human beings (and some corporations) as persons, whereas nonhuman animals are considered "things". However, what is uncontroversial is that those neuroatypical humans just described are not considered moral agents, but are considered paradigm cases of what are termed human *moral patients*. All but a small minority of philosophers hold that no animals are moral agents (Johnson 1983, Rowlands 2015). Most philosophers agree that nonhuman animals are never robust moral agents, but only moral patients. Given a wolf's or deer's cognitive capacities, it would be nonsensical to hold a wolf morally responsible for killing cattle or a deer for intensely over-foraging a particular habitat. Though moral patients such as human children and nonhuman animals cannot bear duties and responsibilities and thus not be held morally accountable for their actions, it does not then follow that moral patients lack interests or rights that moral agents are obligated to respect and consider. If anything, moral abominations like institutionalized eugenics programs suggest that it is those most vulnerable populations that demand moral vigilance on the part of moral agents.

Though there are quite a number of theories of moral status (Beauchamp and Childress 2019, Tannenbaum and Jaworska 2018), in the following section, I summarize just three.

The Anthropocentric Account

The *anthropocentric account* of moral status holds that merely being a member of the species *Homo sapiens* confers maximal moral status. The anthropocentric account claims that all and only human beings are bona fide members of the moral community. On the anthropocentric account, human beings comprise both a biological natural kind and a unique moral kind.

One virtue of the anthropocentric account is that it secures full moral status for all humans, including infants, the mildly and severely cognitively other-abled, and permanently comatose humans. While there are numerous and extensive problems with the anthropocentric account (Andrews et al 2018), I will address just a few.

According to the anthropocentric account, there exist some set of distinctively human properties that confer full moral status on all and only human beings. In order to justify this claim, the account must (a) identify which properties are *distinctively* human, (b) demonstrate that these properties are possessed universally by *all* humans, (c) provide an account of why *those* properties (but not others) are the morally relevant ones, and (d) explain why those properties are both necessary and sufficient to secure maximum moral status for all and only human beings. Here is where the problems begin.

There are a number of ways to read the anthropocentric account. One, the *species membership* version of the anthropocentric account clearly (though trivially) does indeed identify a property of being human that is both distinctively human and possessed universally by all humans, namely, that of *being human*. Yet, when pushed to justify why a contingent, arbitrary biological property like species membership is morally relevant, the anthropocentric account begs the question since the very question at the heart of the anthropocentric account is this: What is it about membership in the species *Homo sapiens* alone that uniquely confers full moral status? We all agree that morally relevant properties should not be arbitrary; skin or eye color are arbitrary and thus not valid characteristics that should determine one's moral status. So why is species membership not an arbitrary property? To state that species membership alone (in this case membership in the species *Homo sapiens*) is morally relevant tells us nothing whatsoever about *why* being human is morally relevant.

Being a human biological organism alone cannot be *necessary* for moral status. Surely, Spock's Vulcan father, Sarek, from the *Star Trek* series counts morally. Yet, on the species membership version of the anthropocentric account, he does not. Something has gone wrong. Contrary to the species membership version of the anthropocentric account, being a human biological organism alone cannot be necessary for moral status. Whether being a human biological organism is a sufficient condition for moral status is another story, yet even there the advocate of the anthropocentric account must answer the question, what makes membership in the species *Homo sapiens* alone sufficient for full moral status? Not a simple question to answer. Further, critics of the anthropocentric account charge that this kind of human exceptionalism reflects a deeper bias, that of *speciesism*, a bias in favor of the interests of humans over nonhumans based primarily on species membership that posits the moral superiority of all humans over all nonhumans.

In response, an advocate of the anthropocentric account may concede that it is not merely membership in the species *Homo sapiens*, but those distinctively human species-neurotypical cognitive abilities that do the moral work. These properties include things like intelligence, language, self-awareness, and rationality. This variety of the anthropocentric account—the *cognitive properties* version—claims that it is not merely membership in the species *Homo sapiens* that secures maximal moral status for all humans, but rather those uniquely human capacities that make human beings—and *only* human beings—*moral* beings.

Nevertheless, problems arise for even this modified interpretation of the anthropocentric account. Recall that in order for the anthropocentric account to secure full moral status for *all* humans, it must also demonstrate that the alleged distinctively human properties are possessed universally by all humans. However, if the moral divide between human and nonhuman animals rests on the possession of some set of uniquely human cognitive abilities possessed by all and only humans, then there will always exist some humans who lack these characteristics. That is, for any human cognitive capacity we select, there will always exist some human being who lacks this capacity, and some nonhuman animal who will possess this capacity, to varying degrees. The challenge that neuroatypical severely cognitively other-abled humans pose to the cognitive properties version of the anthropocentric account is significant. Simply put, the challenge for the speciesist is to account for the moral status of those human beings whose supposed morally relevant capacities fall short of whatever is held to justify the attribution of higher moral status to paradigm human beings than to (most or all) nonhuman animals (DeGrazia 2014).

Further, disability rights advocates rightly push back against both the cognitive properties version of the anthropocentric account and animal ethicists' use of neuroatypical severely cognitively other-abled humans to counter the view. Daniel Salomon argues that cases like these trade on a kind of neurotypicalism, the view that characteristics or properties that are species-typical are therefore cognitively normative. Salomon argues that neurotypicalism "privileges a form of cognitive processing characteristic of peoples who have a neurotypical (e.g. non-autistic) brain structure, while at least implicitly finding other forms of cognitive processing to be inferior, such as those natural to autists and nonhuman animals" (Salomon 2010:47). Sunaura Taylor argues persuasively that this kind of ableism is "intimately entangled with speciesism, and is deeply relevant to thinking through the ways nonhuman animals are judged, categorized, and exploited" (2017:58). That the most common and ubiquitous argument used to support the continued domination and oppression of nonhuman animals is that they lack any number of psycho-physical or cognitive processes that are species-typical to human beings "shows the extent to which speciesism uses ableist logics to function" (Taylor 2017:58). There exists an ugly and intimate connection between neurotypicalism, human supremacy, speciesism, and systemic animal oppression. Ableism (and thus, neurotypicalism) helps "construct the systems that render the lives and experiences of both nonhuman animals and disabled humans as less valuable and as discardable, which leads to a variety of oppressions that manifest differently" (Taylor 2017:58).

A common response to the challenge that neuroatypical severely cognitively other-abled humans pose to the cognitive properties version of the anthropocentric account is to modify the account to what I'll call the *species typicalism* version of the account (McMahan 2002). On the species typicalism version of the anthropocentric account, though some humans lack those cognitive abilities required for entrance into the moral club, they remain members of a species (*Homo sapiens*) whose *typical* members possess the requisite capacities, something that cannot be said for any member of a nonhuman animal species. Though a permanently and severely congenitally cognitively other-abled infant may forever lack the cognitive abilities required by the anthropocentric account, because they are a member of a species whose typical member possesses these abilities, they are then afforded full membership in the moral community.

The problem, however, is providing a non-question-begging account of how such "honorary" membership works. That is, in virtue of precisely what does the neuroatypical severely cognitively other-abled infant gain full membership into the moral community? Their honorary membership can't be due to their biological species membership. Were humans able someday through gene therapy to produce a chimpanzee who developed cognitive capacities comparable to those of a human person, surely this "Superchimp" would be entitled to the same moral status due human persons despite their being an atypical member of a species who typical members do not possess the kinds of cognitive capacities she now does (McMahan 2002). Conferring moral status on nonparadigm members of a species based on the capacities of paradigm members is unwarranted.

These and other challenges (Gruen 2011, McMahan 2002) have led philosophers to seek alternative grounds of moral status, grounds that reject speciesism and more robustly expand the bounds of moral status across the species boundary. Though a number of philosophers have offered further animal cognition-heavy accounts, including Rollin (1981), Sapontzis (1987), Rachels (1990), DeGrazia (1996), McMahan (2002), and Nussbaum (2006), in the next section, I explicate the two most influential systematic ethical theories built upon a set of core philosophical principles aimed at expanding the moral status of nonhuman animals and overthrowing the anthropocentric account.

The Sentientist Account

In contrast to the anthropocentric account, a number of philosophers have formulated non-speciesist, egalitarian accounts of moral status intended to expand the moral sphere to include nonhuman animals as robust members of the moral community. Easily, the most influential theory advocating increased moral status for nonhuman animals is that of Peter Singer. Singer's *Animal Liberation* (1975) remains the *locus classicus* of the contemporary animal liberation movement. Singer provides what can be described as a *sentientist account* of moral status. According to the sentientist account, the capacity to experience pain or pleasure is both necessary and sufficient for having morally considerable interests.

The central principle driving Singer's view is what he calls the *principle of equal consideration of interests* which requires that we give equal weight in our moral deliberations to the like interests of all those affected by our actions. However, since the possession of sentience is what grounds interests, and since humans are not the only sentient beings, to be consistent, we must extend the principle of equal consideration of interests to *all* sentient creatures. On Singer's utilitarian sentientist account, if a being suffers, there can be no moral justification for refusing to take that suffering into consideration. Privileging the interests of humans over nonhumans solely in virtue of species membership is yet another form of speciesism. Thus, no matter what the nature of the being, the principle of equality requires that its suffering be counted equally with the like suffering of any other being (Singer 1975).

Beyond sentience, Singer enumerates various cognitive capacities the possession of which may weigh moral status beyond mere sentience, where "the superior mental powers of normal adult humans make a difference: anticipation, more detailed memory, greater knowledge of what is happening, and so on" (Singer 2011:52). Thus, on the sentientist account, sentience and complex cognitive capacities bear on questions of interests, moral considerability, and moral significance.

The Rights Account

In *The Case for Animal Rights* (1983), Tom Regan rejects utilitarian arguments for animal liberation and provides instead a *rights-based account* of animal liberation. For Regan, what matters morally is the capacity to be the subject of experiences *that matter to oneself.* Possessing certain physiological, emotional, psychological, and cognitive capacities-over-and-above mere sentience makes one a *subject-of-a-life*:

> To be the subject of a life... involves more than merely being alive and more than merely being conscious.... [I]ndividuals are subjects-of-a-life if they have beliefs and desires; perception, memory, and a sense of the future, including their own future; an emotional life together with feelings of pleasure and pain; preference and welfare-interests; the ability to initiate action in pursuit of their desires and goals; a psychophysical identity over time; and an individual welfare in the sense that their experiential life fares well or ill for them, logically independent of their being the object of anyone else's interests.
>
> *(Regan 1983:243)*

This passage makes clear how crucial are facts about animal cognition to Regan's view of animal rights. He argues that since a great number of nonhuman animals are subjects-of-a-life whose individual value cannot be reduced to their utility to humans, it follows that animals possess what he calls *inherent value*, a type of moral worth independent of how animals are instrumentally valued by—or valuable to—humans. Importantly, on Regan's view, all who possess inherent value possess it equally. If individuals have equal inherent value, then any principle that declares what treatment is due them as a matter of justice must take their equal value into account. Possession of inherent value merits the respect due a subject-of-a-life. Inherent value, in combination with what Regan calls the Respect Principle—we are to treat those individuals who have inherent value in ways that respect their inherent value—confer upon nonhuman animals an absolute and (usually) inviolable *moral right* to the protection and maintenance of those conditions essential to their welfare and existence. Thus, for Regan, practices such as "wildlife management" that promote hunting or culling are morally wrong, irrespective of human or environmental demands, context, or culture (Gruen 2014).

Hopefully, the relationship between sentience, animal cognition, and moral status is clear. Let's now turn to what our best science can reveal about those aspects of sentience and animal cognition that bear on animals' moral status.

Sentience, Animal Cognition, and Moral Status

The query, which animals possess which morally relevant physiological and cognitive properties and to what extent? sits at the foundation of questions regarding the moral status of animals, their interests, rights, and treatment. Since a thorough discussion of the results of research into animal sentience and cognition is beyond the scope of this essay, I will provide only a summary of a few areas with implications for the moral status of animals. We start with research on animal pain and suffering, followed by a summary of one capacity that bears on the moral status of animals, namely, awareness.

Sentience

Common sense suggests that vertebrates experience pain and suffering, but what about invertebrates? Do scallops feel pain when shucked? Lobsters, when thrown into a pot of boiling water? Insects, when trampled? Here, intuitions break down. Since the answers to these questions bear on moral status, science can step in where commonsense intuitions falter.

Pain in humans is at least a two-step process. The first step involves the stimulation of special receptors called *nociceptors* that transmit injury detecting electrical impulses to the spinal cord, triggering an automatic reflex response (Tracey Jr 2017). At this first stage, there are no conscious, phenomenal aspects of the experience. In the second stage, the signal moves from the spinal cord to the neocortex at which point the phenomenal aspects of pain kick in and we experience the unpleasant sensation associated with tissue damage. Though researchers are clear about the mechanisms involved in the first stage, it is the second stage of the process—the affective aspect of pain—that remains somewhat mysterious. In addressing nonhuman animal pain, we can start with the following questions. First, which animal groups possess nociceptors (or exhibit a "nociceptive response"), and (how) do they respond to noxious stimuli, analgesics, and anesthetics? Do they exhibit pain-avoidance behavior? Do their responses to noxious stimuli involve tradeoffs between stimulus avoidance and other motivational requirements? We can further explore which organisms possess neural organs more complex than simple neural nets (e.g. organs such as ganglia, brain masses, or brains), and of these, which possess nociceptor-to-brain pathways (Elwood 2011). For the purposes of this essay, I assume—for sound scientific reasons—that all vertebrate species are sentient (Proctor 2012). However, the question of invertebrate sentience is by no means a settled issue.

Invertebrates

Since invertebrates are widely seen as evolutionarily less sophisticated than vertebrates and thus less likely to be sentient, the question of invertebrate pain remains open to debate (Carere and Mather 2019). While *Drosophila* have been found to possess transient receptor potential channels, structures central to mammalian pain (Neely et al 2011, Tobin and Bargmann 2004), one recent study suggests that insects may experience pain after injury (Khuong et al 2019). Other evidence of insect pain includes the discovery of opioid receptors in crickets (Dyakonova et al 1999), and nociceptive response in insect larvae (Neely et al 2011, Walters et al 2001). Though insects possess miniature brains, they exhibit sophisticated behaviors that seem to go beyond simple associative learning (Giurfa 2013). Chittka and Niven (2009) argue that it is the *neural circuitry* of insects' tiny brains—not brain region size—that facilitates highly differentiated motor repertoires, extensive social structures, and impressive cognitive feats. Barron and Klein (2016) argue persuasively from comparative functional neurobiology that the insect brain is capable of subjective experience. Whether such experience is sentient, or just conscious but insentient remains a matter of some discussion (DeGrazia 2020, Tye 2016). In their work on spider pain, Eisner and Camazine found that "[t]he sensing mechanism by which spiders detect injected harmful chemicals such as venoms…may be fundamentally similar to the one in humans that is coupled with the perception of pain" (1983:3382).

The question of crustacean pain is one that interests not only philosophers and scientists, but also the general public (Jones 2014, Wallace 2004). Physiologically, crustaceans possess nociceptors, ganglia, and nociceptor-to-ganglia pathways (Ross and Ross 2009). Though crustacean pain attribution is not widely accepted, some findings support crustacean sentience. For example, Puri and Faulkes (2015) found that crayfish appear to possess specialized nociceptors that respond quickly and strongly to high-temperature thermal stimuli. Barr et al (2008) observed that the behavioral pain response in prawns is inhibited by the application of benzocaine, a local anesthetic. Lozada et al (1988) showed that blocking the activation of opioid receptors in crabs reduced response to electric shock, and Elwood and Appel (2009) found that hermit crabs more willingly abandon their shells as the intensity of electrical shock increases. Barr et al conclude that such findings are "consistent with the idea that these crustaceans can experience pain" (2008:745). Though jellyfish and sea anemones exhibit a nociceptive response, since the nervous systems of these invertebrates take the form of neural nets, skepticism persists about pain in these creatures.

Nociceptors are found in a wide range of bilateria such as annelids, nematodes, and mollusks (Ross and Ross 2009, Smith and Lewin 2009, Tracey Jr 2017). Both marine and terrestrial worms are found to possess not only nociceptors but a ganglion and nociceptor-to-ganglia pathways (Ross and Ross 2009). Behaviorally, worms react to noxious stimuli, while anatomically, some terrestrial snails and earthworms are found to produce neurochemical opioids (Dalton and Widdowson 1989, Kavaliers and Ossenkopp 1991, Ross and Ross 2009).

Though bivalves appear to lack a sufficiently complex nervous system sufficient for sentience, researchers studying anesthesia in oysters found that magnesium chloride induced "anesthesia quickly, allowing rapid recovery with minimal stress and mortality" (Culloty and Mulcahy 1992:249), while both mussels and scallops show increased heart rate under predation threat (Kamenos et al 2006). Further, the presence of antibody-producing immunocytes and an increase in opiates in mussels and leeches subjected to trauma mirror the physiological changes seen in humans after coronary artery bypass surgery (Stefano et al 2002).

By contrast, cephalopod mollusks like cuttlefish, squid, and octopuses have a large and well-centralized nervous system condensed into a brain-like structure. Sentience and aspects of cognition like memory appear to have evolved not from a single evolutionary path but instead seem to have evolved in parallel with and distinct from vertebrates (Godfrey-Smith 2016). Octopuses possess a central nervous system that rivals in complexity that of phylogenetically distant vertebrates such as mice (Wollesen et al 2009). Researchers note that the stress response system present in these invertebrates is neuronally and endocrinologically similar to that of vertebrates (Mather and Anderson 2007, Stefano et al 2002). Octopuses are repelled by a natural stimulus (e.g. sea anemones), exhibit nociception when stung, and engage in tradeoffs when presented with such noxious stimuli (Mather 2008).

Though fish are obviously not invertebrates, the question of fish pain demands a brief discussion. Skepticism about pain in ray-finned fishes, specifically teleosts, persists. Since the economic costs of changing our commercial fishing practices would be great, there remain commercial interests in denying fish sentience (Brown 2015, 2016). Work by Sneddon and colleagues (2003, 2003a, b) suggests that fish experience pain in a manner similar to other vertebrates, and that fish perception and cognitive abilities match or exceed other vertebrates, exhibiting even psychological suffering in the form of fear and stress. The broad consensus from the scientific community is that fish are sentient (Brown 2016).

In some jurisdictions, scientifically based fish welfare guidelines have been implemented. For example, in Germany, all fish captured by anglers must be retained in accordance with animal welfare regulations. Policies in New South Wales, Australia recommend minimizing air exposure of caught fish, and using barbless hooks (Cooke and Sneddon 2007). In the UK, *Octopus vulgaris* has been added to the list of animals protected by the Animals (Scientific Procedures) Act of 1986, a group of protected animals that, before the octopus, contained solely vertebrates. In 2012, a group of prominent scientists released the "Cambridge Declaration on Consciousness". The declaration concludes that

> non-human animals have the neuroanatomical, neurochemical, and neurophysiological substrates of conscious states along with the capacity to exhibit intentional behaviors. Consequently, the weight of evidence indicates that humans are not unique in possessing the neurological substrates that generate consciousness. Non-human animals, including all mammals and birds, and many other creatures, including octopuses, also possess these neurological substrates.
>
> *(Low et al 2012)*

Given the complexity of the phenomenon of pain—that it requires not only nociception but neural complexity, perception, and some level of consciousness—conclusions with any degree of certainty regarding invertebrate pain seem far off. However, considering the growing body of physiological and behavioral evidence combined with the central role that sentience plays in theories of the moral status of nonhuman animals, the most prudent course of action may be to employ the precautionary principle with regard to treatment that may potentially cause pain and suffering to some invertebrates and vertebrates (Jones 2016).

Self-Awareness

What does it mean to say that an animal is self-aware and what might self-awareness have to do with ethics? As we've seen, for Regan, being the subject of a life—having things like a sense of self and one's future—confers inherent value and ultimately, rights. But there are many senses of the term "self-awareness" (DeGrazia 2009). Exploring them is beyond the bounds of this essay. The kind of self-awareness that I want to examine here is of two parts. The first is the ability to recognize or conceptualize oneself as an independent self, for example, understanding one's own image as an image of one's *self*. The second involves a type of self-awareness associated with memory called autonoetic (self-knowing) awareness. In this section, we'll look at the question of self-recognition; in the following section, we'll investigate autonoetic awareness as it constitutes what is called episodic memory.

The experimental paradigm related to self-awareness in nonhumans that has received the most attention is mirror self-recognition (MSR) (Gallup 1970). MSR intends to demonstrate whether subjects can recognize themselves in a mirror. "Passing" the MSR test involves touching a mark placed on the subject's forehead (or other part of its body not readily visible to itself) more frequently when there is a mirror available than when there is not.

Though many regard MSR as the "gold-standard" of self-awareness, some caution against seeing MSR as the only viable metric of self-awareness in nonhumans (Bekoff 2001, Rogers and Kaplan 2004), while others dispute the results of MSR altogether (Heyes 1994). Some note that self-awareness as measured in these kinds of tasks is limited to the visual modality and fail to take into account other modalities of self-representation (for example, auditory, tactile, and olfactory) as well as certain salient anatomical differences (Andrews 2011, Bekoff 2001, Rogers and Kaplan 2004). Others note that some species (for example, gorillas) exhibit gaze aversion which would contribute to their "failing" the test (Gallup 1994). Despite these challenges, MSR results remain central to questions of nonhuman animals' self-awareness.

Positive MSR results have been reported for a number of species, including chimpanzees, gorillas, orangutans, dolphins, Asian elephants, magpies, rhesus macaques (Andrews 2016), and orcas (Delfour and Marten 2001). Horowitz (2017) has demonstrated that dogs pass an "olfactory" MSR test, while evidence seems to indicate that perhaps even goldfish pass the MSR (Kohda et al 2018). When presented with a mirror, dolphin subjects use the mirror to examine the insides of their mouths, and even engage in sexual behavior much more often in front of the mirror. Upon being marked for the first and only time on the tongue, one dolphin engaged in a mouth opening-and-closing sequence never before observed (Reiss and Marino 2001). Though pigs have not passed the MSR test, research has shown that pigs can utilize information obtained from a mirror (Broom et al 2009).

We have been looking at just one aspect of self-awareness. Given that there are various morally relevant sorts of self-awareness, it would seem that many animals are self-aware (DeGrazia 2009).

Episodic Memory

Though memory can generally be characterized simply as the encoding, storage, and retrieval of information, of the various types of memory identified in the literature on memory, e.g. remembering facts (semantic memory) or remembering how to ride a bicycle (procedural memory), episodic memory is most relevant to the moral status of animals. *Episodic memory* involves the conscious recollection of unique personal experiences or episodes (Queensland Brain Institute 2016). Episodic memory in humans is believed to require *autonoetic* awareness, the ability to mentally situate oneself in a particular time and place accompanied by the one's perspectives and attendant emotions (Wheeler 2000). Evidence of conscious recollection of personal experience in most humans involves linguistic reports of vivid subjective experiences, so assessing episodic memory in nonhuman animals presents challenges. Since only the behavioral aspects of episodic memory can be observed in nonhuman animals, researchers refer to episodic memory in nonhumans as "episodic-*like*" memory.

That said, in a landmark study of episodic-like memory in nonhuman animals, Clayton and Dickinson (1998) demonstrated that Western scrub jays possess the ability to remember the location, time, and identity of various food items they cached, while later studies further demonstrated that scrub jays' memory for their caches incorporated an impressive degree of flexibility (Salwiczek et al 2008). Martin-Ordas et al (2010) noted episodic-like memory in chimpanzees, orangutans, and bonobos, while Jozet-Alves et al (2013) noted episodic-like memory in cuttlefish.

Similar results were obtained in a study of hummingbirds (Henderson et al 2006), while rats and mice exhibit episodic-like memory in their capacity for detailed representations of remembered events (Dere et al 2006, Panoz-Brown et al 2016). Episodic memory has been observed in bottle-nosed dolphins (Mercado et al 1998), while Hoffman et al (2009) and Basile and Hampton (2013) report episodic-like memory in rhesus macaques. There is even some evidence of episodic-like memory in bees (Pahl et al 2007). Not only has episodic memory been observed in the common domestic chicken, but hens demonstrate delayed gratification, self-control, and the ability to anticipate the future, all associated with the capacity for pre-emptive anxiety and the generation and frustration of expectations (Marino 2017). The evidence seems to support the inference that neither self-awareness nor episodic-like memory is unique to human beings, a conclusion that as we have seen carries significant moral relevance.

The ethical upshot of these results is that if folks like Singer and Regan are right, then a staggering number of nonhuman animals across countless species who currently possess little to no moral status are actually robust, bona fide members of the moral community. And if that's the case, then our treatment of them constitutes a moral atrocity.

Animal Liberation versus Environmentalism

Questions regarding the practical implications of the moral status of animals arise in varied human practices such as animal agriculture, biomedical research, entertainment (rodeos and circuses), and captivity (zoos and aquaria). However, in this section, I want to outline the deep philosophical divide between environmentalists and advocates of animal liberation.

As we have seen, the anthropocentric view of moral status holds that all and only humans have full moral status. Animal rights philosophers reject this view, holding that the primary units of moral concern are sentient, cognitively complex *individuals*. By contrast, environmentalists tend toward *holism*, where the focus of moral concern centers not on individuals, but on the integrity of ecological wholes such as ecosystems, the "land", and biotic communities.

While none of these collective entities is sentient or a subject-of-a-life, the preservation of these entities is a central concern for environmentalists. Further, environmental ethicists tend to see the goals of animal rights as anti-environmental (Varner 2011). As a consequence, the aims of animal rights advocates and environmentalists seem often to conflict (Brennan and Lo 2020). For example, when the stability of a particular ecosystem appears threatened by a so-called "invasive" species, environmentalists may call for a culling of the "invasive" animal population, while an animal liberationist may advocate for "compassionate conservation" (Bekoff 2013).

In his influential essay "Animal Liberation: A Triangular Affair" (1980), J. Baird Callicott outlines the central issue dividing animal liberationists from environmentalists. For holists, biotic communities are what have intrinsic value with the ecological whole being the ultimate measure of moral value. The value of individual organisms lies in their ecological function, and their well-being should be considered only inasmuch as they contribute to the ecological whole. Aldo Leopold's Maxim captures the moral heart of environmental holism: "A thing is right when it tends to preserve the integrity, stability, and beauty of the biotic community; It is wrong when it tends otherwise" (Leopold 1989:224–225). This tension between the primacy of individual sentient beings and environmental wholes remains the "most fundamental theoretical difference between environmental ethics and the ethics of animal liberation" (Callicott 1980:337).

Callicott sees environmental holism as life-affirming in its celebration of the value of ecosystems and biotic communities, whereas he sees animal liberation as "life-loathing" and "world-denying" and too narrowly individualistic (1980:333). When the interests of the wholes clash with the interests of the individuals that comprise it, the interests of individuals must be sacrificed. However, since environmentalists see human populations as, themselves, members of the biotic community, folks like Callicott are pressed to provide some non-speciesist principle that exempts humans from similar sacrifice (Callicott 1989, Varner 2001, 2011).

For example, environmental holists would condone "culling" rabbits to preserve a particular plant species—but they are reluctant to sacrifice human interests in similar situations. Yet, the most abundant species destroying biotic communities is *Homo sapiens*. If human individuals are just another element within the larger and more important biotic community as environmental holism would seem to imply, then environmental holists should call for us to "control" or "eliminate" some of these individuals for the sake of the larger whole, an implication Tom Regan labels "environmental fascism" (Regan 1983/2004:362).

But if humans cannot be sacrificed for the good of the whole, why can rabbits, deer, and wolves? Environmental holists reply by claiming that while the biotic community matters morally, it is not the only community that matters. We humans are part of various "nested" human communities, all of which have claims upon us; we are part of a tight-knit human community, but only a very loose human-rabbit community. Thus, our obligations to the biotic community may require the culling of rabbits, but may not require the culling of humans (Callicott 1999).

However, it would seem now that some relations within the biotic community carry more moral weight than others, namely, the relations between individual *human* members of a given biotic community. But the environmentalist reply will not suffice. It would seem now that some relations within the biotic community carry more moral weight than others, an implication derived not from Leopold's Maxim, but from the point of view of individual human members of a given biotic community. If our moral commitments to the biotic community are trumped by our obligations to the human community, and if other members of the biotic community are merely instrumentally valuable, then environmental holism collapses to just another anthropocentric view. And if that's the case, then animal liberation never really was a "triangular affair." Unacceptable implications of environmental holism such as these should give one pause before rejecting animal liberation as too individualistic. (Jones 2021) Folks like Callicott remain pressed to provide some non-speciesist principle that exempts humans from similar sacrifice (Callicott 1989, Varner 2001, 2011).

A Final Rumination

That Western scientific, philosophical, and cultural thought is speciesist is an indisputable fact. The denial of mind to animals has been a keystone to their exploitation and the exploitation of their habitats and the natural world. Eileen Crist argues persuasively that Western philosophy's denial and disparagement of animal minds is "causally implicated in the devastation of the biosphere", facilitating the "objectification of the natural world and its transformation into a domain of resources" (2013:45). Despite scientific consensus on animal consciousness and sentience, and despite legislative welfare regulations, billions of animals suffer unspeakable pain, suffering, and death, while their (and our) habitats, biotic communities, and ecosystems are poisoned, denigrated, and destroyed. Better science, near-certainty regarding animal consciousness, cognition, or sentience, or increased welfare legislation alone will not end the exploitation of the environment or the suffering that is visited upon billions of animals annually at the hand of speciesism and human supremacy. As John Sanbonmatsu argues, "[b]y telling ourselves that we have no 'choice' but to kill and to consume animals, thereby refusing responsibility for our participation in terror, we undermine our claims to being the kind of being that alone can exercise autonomous judgment" (2011:43). Our task then is to transcend our bad faith by untelling such stories about the supremacy of *Homo sapiens* and our domination of the natural world and its inhabitants.

References

Andrews K (2011) Beyond Anthropomorphism. In: Beauchamp TL and Frey RG (eds) *The Oxford Handbook of Animal Ethics*.

Andrews K (2011) Beyond Anthropomorphism: Attributing Psychological Properties to Animals. In: Beauchamp TL and Frey RG (eds) *The Oxford Handbook of Animal Ethics*, 469–494.

Andrews K (2016) Animal Cognition (Summer 2016). In: Zalta EN (ed) *The Stanford Encyclopedia of Philosophy*. Metaphysics Research Lab, Stanford University. Available at: https://plato.stanford.edu/archives/sum2016/entries/cognition-animal/ (accessed 13/05/19).

Andrews K (2020) *The animal mind: An introduction to the philosophy of animal cognition*. Routledge.

Andrews K, Comstock GL, Crozier GKD, Donaldson S, Fenton A, John TM, Johnson LSM, Jones RC, Kymlicka W and Meynell L (2018) *Chimpanzee Rights: The Philosophers' Brief*. Routledge.

Andrews, K. (2020). The animal mind: An introduction to the philosophy of animal cognition. Routledge.

Barr S, Laming PR, Dick JT and Elwood RW (2008) Nociception or pain in a decapod crustacean? *Animal Behaviour* 75(3): 745–751.

Barron AB and Klein C (2016) What insects can tell us about the origins of consciousness. *Proceedings of the National Academy of Sciences* 113(18): 4900–4908.

Basile BM and Hampton RR (2013) Monkeys recall and reproduce simple shapes from memory. *Current Biology* 23(20): 2078.

Beauchamp TL and Childress JF (2019) *Principles of Biomedical Ethics* (8th edition). Oxford University Press, USA.

Bekoff M (2001) Observations of scent-marking and discriminating self from others by a domestic dog (Canis familiaris): tales of displaced yellow snow. *Behavioural Processes* 55(2): 75–79.

Bekoff M (2013) *Ignoring Nature No More: The Case for Compassionate Conservation*. University of Chicago Press.

Birch J (2017) Animal sentience and the precautionary principle. *Animal Sentience: An Interdisciplinary Journal on Animal Feeling* 2(16): 1.

Brennan A and Lo Y-S (2020) Environmental Ethics (Summer 2020). In: Zalta EN (ed) *The Stanford Encyclopedia of Philosophy*. Metaphysics Research Lab, Stanford University. Available at: https://plato.stanford.edu/archives/sum2020/entries/ethics-environmental/ (accessed 28/05/20).

Broom DM, Sena H and Moynihan KL (2009) Pigs learn what a mirror image represents and use it to obtain information. *Animal Behaviour* 78(5): 1037–1041.

Brown C (2015) Fish intelligence, sentience and ethics. *Animal Cognition* 18(1): 1–17.

Brown C (2016) Fish pain: an inconvenient truth. *Animal Sentience: An Interdisciplinary Journal on Animal Feeling* 1(3): 32.

Callicott JB (1980) Animal liberation: A triangular affair. *Environmental Ethics* 2(4): 311–338.

Callicott JB (1989) *In Defense of the Land Ethic: Essays in Environmental Philosophy*. Suny Press.

Callicott JB (1999) Holistic environmental ethics and the problem of ecofascism. *Beyond the Land Ethic: More Essays in Environmental Philosophy*. SUNY Press, 59–76.

Carere C and Mather J (2019) *The Welfare of Invertebrate Animals*. Springer.

Chittka L and Niven J (2009) Are bigger brains better? *Current Biology* 19(21): R995–R1008.

Clayton NS and Dickinson A (1998) Episodic-like memory during cache recovery by scrub jays. *Nature* 395(6699): 272–274.

Cooke SJ and Sneddon LU (2007) Animal welfare perspectives on recreational angling. *Applied Animal Behaviour Science* 104(3–4): 176–198.

Crist E (2013) Ecocide and the extinction of animal minds. In: Bekoff M (ed) *Ignoring Nature No More: The Case for Compassionate Conservation*. Chicago: Chicago University Press, 45–61.

Culloty SC and Mulcahy MF (1992) An evaluation of anaesthetics for Ostrea edulis (L.). *Aquaculture* 107(2–3): 249–252.

Dalton LM and Widdowson PS (1989) The involvement of opioid peptides in stress-induced analgesia in the slug Arion ater. *Peptides* 10(1): 9–13.

DeGrazia D (1996) *Taking Animals Seriously: Mental Life and Moral Status*. Cambridge University Press.

DeGrazia D (2008) Moral status as a matter of degree? *The Southern Journal of Philosophy* 46(2): 181–198.

DeGrazia D (2009) *Self-awareness in Animals. The Philosophy of Animal Minds*. Cambridge: Cambridge University Press.

DeGrazia D (2014) On the moral status of infants and the cognitively disabled: a reply to Jaworska and Tannenbaum. *Ethics* 124(3): 543–556.

DeGrazia D (2020) Sentience and consciousness as bases for attributing interests and moral status: considering the evidence and speculating slightly beyond. In: Johnson L, Fenton A and Shriver A (eds) *Neuroethics and Nonhuman Animals*. Springer, 17–31.

Delfour F and Marten K (2001) Mirror image processing in three marine mammal species: killer whales (Orcinus orca), false killer whales (Pseudorca crassidens) and California sea lions (Zalophus californianus). *Behavioural Processes* 53(3): 181–190.

Dere E, Kart-Teke E, Huston JP and Silva MDS (2006) The case for episodic memory in animals. *Neuroscience & Biobehavioral Reviews* 30(8): 1206–1224.

Dyakonova VE, Schürmann F-W and Sakharov DA (1999) Effects of serotonergic and opioidergic drugs on escape behaviors and social status of male crickets. *Naturwissenschaften* 86(9): 435–437.

Eisner T and Camazine S (1983) Spider leg autotomy induced by prey venom injection: An adaptive response to "pain"? *Proceedings of the National Academy of Sciences* 80(11): 3382–3385.

Elwood RW (2011) Pain and suffering in invertebrates? *Ilar Journal* 52(2): 175–184.

Elwood RW and Appel M (2009) Pain experience in hermit crabs? *Animal Behaviour* 77(5): 1243–1246.

Gallup GG (1970) Chimpanzees: self-recognition. *Science* 167(3914): 86–87.

Gallup GG (1994) *Self-recognition: Research Strategies and Experimental Design*. Cambridge University Press.

Giurfa M (2013) Cognition with few neurons: higher-order learning in insects. *Trends in Neurosciences* 36(5): 285–294.

Godfrey-Smith P (2016) *Other Minds: The Octopus and the Evolution of Intelligent Life*. London: William Collins.

Goodpaster KE (1978) On being morally considerable. *The Journal of Philosophy* 75(6): 308–325.

Gruen L (2011) *Ethics and Animals: An Introduction*. Cambridge University Press.

Gruen L (2014) The moral status of animals. *The Stanford Encyclopedia of Philosophy*. Available at: http://plato.stanford.edu/archives/fall2014/entries/moral-animal/ (accessed 6/02/ 2014).

Hale B (2011) Moral considerability: deontological, not metaphysical. *Ethics & the Environment* 16(2): 37–62.

Henderson J, Hurly TA, Bateson M and Healy SD (2006) Timing in free-living rufous hummingbirds, Selasphorus rufus. *Current Biology* 16(5): 512–515.

Heyes CM (1994) Reflections on self-recognition in primates. *Animal Behaviour* 47(4): 909–919.

Hoffman ML, Beran MJ and Washburn DA (2009) Memory for "what", "where", and "when" information in rhesus monkeys (Macaca mulatta). *Journal of Experimental Psychology: Animal Behavior Processes* 35(2): 143.

Horowitz A (2017) Smelling themselves: dogs investigate their own odours longer when modified in an "olfactory mirror" test. *Behavioural Processes* 143: 17–24.

Johnson LE (1983) Can animals be moral agents? *Ethics and Animals* 4(2): 5.

Jones R (2014) The Lobster Considered. In: Bolger R and Korb S (eds) *Gesturing Toward Reality: David Foster Wallace and Philosophy*. Bloomsbury Academic, 85–102.

Jones R (2016) Fish sentience and the precautionary principle. *Animal Sentience* 1(3): 10.

Jones RC (2021) Animal ethics as a critique of animal agriculture, environmentalism, foodieism, locavorism, and clean meat. In: *Rethinking Food and Agriculture*. Kassam, A., & Kassam, L. (Eds.). Elsevier, 113–133.

Jozet-Alves C, Bertin M and Clayton NS (2013) Evidence of episodic-like memory in cuttlefish. *Current Biology* 23(23): R1033–R1035.

Kamenos NA, Calosi P and Moore PG (2006) Substratum-mediated heart rate responses of an invertebrate to predation threat. *Animal Behaviour* 71(4): 809–813.

Kavaliers M and Ossenkopp K-P (1991) Opioid systems and magnetic field effects in the land snail, Cepaea nemoralis. *The Biological Bulletin* 180(2): 301–309.

Khuong TM, Wang Q-P, Manion J, Oyston LJ, Lau M-T, Towler H, Lin YQ and Neely GG (2019) Nerve injury drives a heightened state of vigilance and neuropathic sensitization in Drosophila. *Science Advances* 5(7): eaaw4099.

Kohda M, Takashi H, Takeyama T, Awata S, Tanaka H, Asai J and Jordan A (2018) Cleaner wrasse pass the mark test. What are the implications for consciousness and self-awareness testing in animals? *BioRxiv* 397067.

Leopold A (1989) *A Sand County almanac, and sketches here and there*. Outdoor Essays & Reflections.

Low P, Panksepp J, Reiss D, Edelman D, Van Swinderen B and Koch C (2012) The Cambridge declaration on consciousness. In *Francis crick memorial conference*, Cambridge, England, 1–2.

Low, P., Panksepp, J., Reiss, D., Edelman, D., Van Swinderen, B., & Koch, C. (2012, July). The Cambridge declaration on consciousness. *Francis crick memorial conference*, Cambridge, England (pp. 1-2).

Lozada M, Romano A and Maldonado H (1988) Effect of morphine and naloxone on a defensive response of the crab Chasmagnathus granulatus. *Pharmacology Biochemistry and Behavior* 30(3): 635–640.

Marino L (2017) Thinking chickens: a review of cognition, emotion, and behavior in the domestic chicken. *Animal Cognition* 20(2): 127–147.

Martin-Ordas G, Haun D, Colmenares F and Call J (2010) Keeping track of time: evidence for episodic-like memory in great apes. *Animal Cognition* 13(2): 331–340.

Mather JA (2008) Cephalopod consciousness: behavioural evidence. *Consciousness and Cognition* 17(1): 37–48.

Mather JA and Anderson RC (2007) Ethics and invertebrates: a cephalopod perspective. *Diseases of Aquatic Organisms* 75(2): 119–129.

McMahan J (2002) *The Ethics of Killing: Problems at the Margins of Life*. Oxford University Press.

Mercado E, Murray SO, Uyeyama RK, Pack AA and Herman LM (1998) Memory for recent actions in the bottlenosed dolphin (Tursiops truncatus): repetition of arbitrary behaviors using an abstract rule. *Animal Learning & Behavior* 26(2): 210–218.

Neely GG, Keene AC, Duchek P, Chang EC, Wang Q-P, Aksoy YA, Rosenzweig M, Costigan M, Woolf CJ and Garrity PA (2011) TrpA1 regulates thermal nociception in Drosophila. *PloS One* 6(8):e24343.

Neely GG, Keene AC, Duchek P, Chang EC, Wang QP, Aksoy YA, Rosenzweig M, Costigan M, Woolf CJ, Garrity PA, Penninger JM. TrpA1 regulates thermal nociception in Drosophila. PLoS One. 2011;6(8):e24343. doi: 10.1371/journal.pone.0024343. Epub 2011 Aug 31. PMID: 21909389; PMCID: PMC3164203.

Nussbaum MC (2006) *Frontiers of Justice: Disability, Nationality, Species Membership*. Harvard University Press.

Pahl M, Zhu H, Pix W, Tautz J and Zhang S (2007) Circadian timed episodic-like memory–a bee knows what to do when, and also where. *Journal of Experimental Biology* 210(20): 3559–3567.

Panoz-Brown D, Corbin HE, Dalecki SJ, Gentry M, Brotheridge S, Sluka CM, Wu J-E and Crystal JD (2016) Rats remember items in context using episodic memory. *Current Biology* 26(20): 2821–2826.

Proctor H (2012) Animal sentience: where are we and where are we heading? *Animals* 2(4): 628–639.

Puri S and Faulkes Z (2015) Can crayfish take the heat? Procambarus clarkii show nociceptive behaviour to high temperature stimuli, but not low temperature or chemical stimuli. *Biology Open* 4(4): 441–448.

Queensland Brain Institute (2016) *Types of Memory*. Available at: https://qbi.uq.edu.au/brain-basics/memory/types-memory (accessed 26/05/20).

Rachels J (1990) *Created from animals: the Moral Implications of Darwinism*. Oxford University Press.

Regan T (1983) *The Case for Animal Rights*. University of California Press.

Reiss D and Marino L (2001) Mirror self-recognition in the bottlenose dolphin: a case of cognitive convergence. *Proceedings of the National Academy of Sciences* 98(10): 5937–5942.

Rogers LJ and Kaplan G (2004) All animals are not equal: The interface between scientific knowledge and legislation for animal rights. *Animal Rights: Current Debates and New Directions*. Sunstein, C. R., & Nussbaum, M. C. (Eds.). Oxford University Press. 175–201.

Rollin BE (1981) *Animal Rights & Human Morality*. Prometheus Books.

Ross LG and Ross B (2009) *Anaesthetic and Sedative Techniques for Aquatic Animals*. John Wiley & Sons.

Rowlands M (2015) *Can Animals be Moral?* Oxford University Press.

Salomon D (2010) From marginal cases to linked oppressions: reframing the conflict between the autistic pride and animal rights movements. *Journal for Critical Animal Studies* 8(1/2): 47–72.

Salwiczek LH, Dickinson A and Clayton NS (2008) *What do Animals Remember about their Past?* Elsevier.

Sanbonmatsu J (2011) The animal of bad faith: speciesism as an existential project. *Critical Theory and Animal Liberation* . Rowman & Littlefield, 29–45.

Sapontzis SF (1987) *Morals, Reason, and Animals*. Temple University Press.

Singer P (1975) *Animal Liberation: A New Ethics for our Treatment of Animals*. HarperCollins.

Singer P (2011) *Practical Ethics*. Cambridge University Press.

Smith EStJ and Lewin GR (2009) Nociceptors: a phylogenetic view. *Journal of Comparative Physiology A* 195(12): 1089–1106.

Sneddon LU (2003) The evidence for pain in fish: the use of morphine as an analgesic. *Applied Animal Behaviour Science* 83(2): 153–162.

Sneddon LU, Braithwaite VA and Gentle MJ (2003a) Novel object test: examining nociception and fear in the rainbow trout. *The Journal of Pain* 4(8): 431–440.

Sneddon LU, Braithwaite VA and Gentle MJ (2003b) Do fishes have nociceptors? Evidence for the evolution of a vertebrate sensory system. *Proceedings of the Royal Society of London. Series B: Biological Sciences* 270(1520): 1115–1121.

Stefano GB, Cadet P, Zhu W, Rialas CM, Mantione K, Benz D, Fuentes R, Casares F, Fricchione GL and Fulop Z (2002) The blueprint for stress can be found in invertebrates. *Neuroendocrinology Letters* 23(2): 85–93.

Tannenbaum J and Jaworska A (2018) The Grounds of Moral Status. *Stanford Encyclopedia of Philosophy*.

Taylor S (2017) *Beasts of Burden: Animal and Disability Liberation*. The New Press.

Tobin DM and Bargmann CI (2004) Invertebrate nociception: behaviors, neurons and molecules. *Journal of Neurobiology* 61(1): 161–174.

Tracey Jr WD (2017) Nociception. *Current Biology* 27(4): R129–R133.

Tye M (2016) *Tense Bees and Shell-shocked Crabs: Are Animals Conscious?* Oxford University Press.

Varner G (2001) Sentientism. *A Companion to Environmental Philosophy*. Blackwell Publishers Ltd 192–203.

Varner G (2011) Environmental ethics, hunting, and the place of animals. *The Oxford Handbook of Animal Ethics*. Beauchamp, T. L., & Frey, R. G. (Eds.). Oxford University Press.

Wallace DF (2004) Consider the Lobster. *Gourmet*, August, 2004.

Walters E, Illich P, Weeks J and Lewin M (2001) Defensive responses of larval Manduca sexta and their sensitization by noxious stimuli in the laboratory and field. *Journal of Experimental Biology* 204(3): 457–469.

Wheeler MA (2000) *Episodic Memory and Autonoetic Awareness*. Oxford University Press.

Wollesen T, Loesel R and Wanninger A (2009) Pygmy squids and giant brains: mapping the complex cephalopod CNS by phalloidin staining of vibratome sections and whole-mount preparations. *Journal of Neuroscience Methods* 179(1): 63–67.

See also Eating (Chapter 2), Experimentation (Chapter 3), Companion Animals (Chapter 4), Species and Wildlife (Chapter 5), Wild Animals (Chapter 6), Hunting (Chapter 6), Remediation (Chapter 43), Restoration (Chapter 44), Assisted Migration and Reintroduction (Chapter 45), Zoos and Conservation (Chapter 46), and Rewilding (Chapter 47).

Further Reading

Kristin Andrews' entry "Animal Cognition" in *The Stanford Encyclopedia of Philosophy* <http://plato.stanford.edu/archives/fall2014/entries/cognition-animal/> as well as her book *The Animal Mind: An Introduction to the Philosophy of Animal Cognition*, Routledge (2020), are excellent sources of the details of animal cognition. Lori Gruen's *Ethics and Animals: An Introduction*, Cambridge University Press (2011) is an excellent and comprehensive introduction to animal ethics.

2

EATING

Dan Hooley and Nathan Nobis

Introduction

Globally, approximately 70 billion land animals are raised and killed each year for human consumption. Farmed animals exist, of course, because human beings want to eat them. These animals' lives and existence, however, contribute to significant environmental damage. They must be fed and watered, and crops must be raised and transported to do this. This all requires massive amounts of water, land, fertilizer, and energy. These animals also produce huge quantities of manure and flatulence, which contributes significantly to air and water pollution and is a major driver of global climate change.

Human habits of eating animals, therefore, result in much environmental damage. Much of this could be avoided by simply eating plants, instead of animals who eat plants. Plant-based diets use far less water, land, fertilizer, and energy to produce compared to diets with animal products. And far fewer plants are needed to feed human beings directly.

While many human activities negatively affect the environment, many are very difficult to reduce or eliminate to a satisfactory level, and reducing the environmental impact of others can be quite financially costly for individuals. Not eating meat and other animal products, however, is not financially out of reach for most people. When it comes to efforts we can take to lessen our environmental impact, significantly reducing or abstaining from meat and other animal products is "low-hanging fruit." It's an action that dramatically helps the environment that, for most people, would not negatively affect their well-being: indeed, it may even enhance it.

Here, we consider how concern for the environment relates to our own eating habits, in particular, the consumption of animals and animal products. Based on broad concerns for the environment, we argue that there are strong moral reasons to radically reduce our consumption of animal-based foods, and that this reduction is a moral obligation. We concede that this conclusion is vague – it doesn't specify precisely how much meat consumption is allowable – but environmental concerns do clearly encourage and support raising far fewer animals and eating significantly less meat and animal products. Individuals living in the U.S., for example, where per-person yearly consumption of meat (excluding fish) easily tops 200 pounds per year, could eat a diet much closer to that of an individual in Indonesia (30 pounds per year) or, even better, India (eight pounds per year) (Food and Agriculture Organization of the United Nations 2017).

We will argue that these concerns alone, however, cannot ground a moral obligation for individuals to be strict vegetarians or vegans. Nevertheless, when concerns for the environment are combined with concerns for animals themselves, a powerful moral argument for veganism can be

DOI: 10.4324/9781315768090-4

made. We develop such an argument. Finally, we conclude with some brief thoughts about how non-human animals might fit into, and relate to, our concern for "the environment."

As a caveat, we acknowledge that, in some parts of the world, plants are very difficult to cultivate and so animals – typically, wild and free-roaming animals – are the only available food source for humans. Eating animals may be environmentally better in these circumstances, given the resources needed to cultivate very difficult land, or the environmental costs associated with importing all of one's food. These contexts, however, are increasingly rare in our urbanizing world and generally not the contexts for most readers of this collection, who likely get all or mostly all of their foods from supermarkets, restaurants, and other retailers. Our environmental argument applies to this agriculture context which is relevant for most readers.

Environmental Damage from Animal Agriculture

Animal agriculture, as it is normally practiced today, is an environmental disaster. Using animals for food is one of the most environmentally destructive technologies used by humans. Perhaps the most significant and pressing environmental damage caused by animal agriculture concerns global climate change. While the estimates of the precise contribution of animal agriculture to climate change vary, it is clear that animal agriculture is one of the biggest contributors to the warming of our planet. A 2006 report from the United Nations Food and Agriculture Organization, entitled *Livestock's Long Shadow*, estimated that 18% of greenhouse gases were attributable to animal agriculture, more than all of transportation combined (Steinfeld 2006). Goodland and Anhang (2009), however, have argued that this report seriously underestimates the contribution of animal agriculture to global climate change. They estimate that animal agriculture accounts for 32.6 billion tons of carbon dioxide per year or 51% of global greenhouse gas emissions.

Whatever the exact contribution, it is beyond dispute that animal agriculture is a major contributor to global climate change. Animal agriculture does this in a variety of ways. Much of this contribution comes from clearing land and forests to graze animals, feeding animals (which requires significantly more food, and energy-intensive inputs to produce this food, than if humans grew and ate plants directly), the life processes of farmed animals (including the waste they produce and flatulence), as well as all the energy needed to process and transport the "end products."

But animal agriculture also harms the environment in many other ways. Raising animals for food requires significantly more inputs (land, fertilizer, energy, and water) than would be required to only grow plants for human consumption. As a result, animal agriculture puts much more strain on finite resources, like land and water, than alternative methods of food production. Finally, because animals are produced in confinement in such large numbers, disposing of animal waste has become a significant environmental problem. Farmed animals produce more than three times the amount of waste produced by humans, and the excess waste and inappropriate land application of such large quantities of animal waste bring antibiotics (which animals are fed), hormones, pesticides, and heavy metals into our waterways, lakes, groundwater, soils, and airways (Pew Commission on Industrial Farm Animal Production 2008).

Eating and the Environment

Reasons to Reduce

The overwhelming environmental destruction caused by animal agriculture has implications for how we, as individuals, should approach eating.

Our starting point is the premise that individuals have some moral obligations to mitigate their impact on the environment, including global climate change. There are different ways one can work toward fulfilling this general obligation, some of which are financially easier than others. Unlike buying an electric car or outfitting our home with solar panels, the choice to eat a plant-based diet needn't be expensive and, in most cases, is unlikely to cost someone more than eating a diet heavy in animal products. For example, beans, lentils, and other legumes are a cheap and healthy way to get protein. Further, unlike other ways of reducing one's impact on the environment (such as riding public transit or using a bicycle to commute to work), eating a vegan diet is something nearly all of us *can* do, if we are willing to put in the effort. And we can do it in addition to whatever else we are doing to reduce our environmental impact. Nearly everyone living in developed countries has access to plant-based foods that they can purchase and consume. This is not true of many other ways that we can reduce our impact on climate change and the environment.

Choosing to consume a plant-based diet is an important way that we can reduce our own negative impact on the environment. In 2006, researchers at the University of Chicago found that someone who ditches a standard American diet, heavy in animal products, for a vegetarian diet, reduces their emissions as much as a person who trades in a standard car for a Toyota Prius (Eshel and Martin 2006). A recent study by researchers in the UK found similar results: in the UK, the carbon footprint of the average vegan was approximately 60% less than that of the average "heavy meat eater" (Scarborough et al 2014). And a systematic review of studies on the environmental impact of switching to a vegan diet showed that switching to a vegan diet brought the greatest reduction in emissions, with a median reduction in emissions of 51%, and reductions for those switching from a Western diet around 70% (Aleksandrowicz, Green, Joy, Smith, Haines 2016).

Taken together, these factors explain why we all have very strong moral reasons to alter our diet, and at least radically reduce the amount of meat and animal products we consume. This is something almost everyone can do, and usually relatively easily and inexpensively. While some effort is certainly required to switch to a mostly plant-based diet, and there are initial social costs, after this switch has been made, eating a mostly vegan diet can be done without a great deal of effort.

Environmentalism and Veganism

Clearly, there are moral reasons based on the environment to reduce the production and consumption of farmed animals. Some argue, however, that environmental concern justifies a moral obligation to not raise or consume *any* farmed animals or their products (like dairy and eggs) and, thus, that environmentalism necessitates veganism (Colb 2013). We disagree: environmental concerns, including our personal contributions to climate change, cannot, in themselves, ground a moral obligation to eat a strict vegan diet.

First, an adequate concern for the environment is consistent with some meat-eating, *if* such concern is consistent with other avoidable activities that contribute to some environmental degradation, like driving cars or flying in airplanes. Nearly all sports and recreational activities that involve energy consumption fall into this category. Raising some limited number of animals for food needn't be worse, environmentally, than some other environmentally unfriendly activities, so if the latter are morally acceptable on environmental grounds then so is eating some meat. Those who argue that environmental concern requires veganism seem to think that environmentalism always requires doing everything we can to eliminate negative impact on the environment: we think, however, that this demand is too much. Serious environmental concern is compatible with causing *some* negative environmental impact and this could allow for *some* limited animal agriculture and non-vegan eating.

Second, environmental arguments for veganism often overlook the climate impact that different foods have. While it is true that, *taken as a whole*, meat and animal products contribute significantly more to climate change than plants, the production of certain types of animals contributes much more than others, and in some circumstances, some plant foods appear to contribute more to climate change than some animal products. For example, ruminant animals (like lambs and cows) contribute significantly more emissions, per 1,000 calories produced, than other types of animals used for food (Tamar 2014). More surprisingly, some plants, like tomatoes and broccoli, may contribute more emissions, per 1,000 calories, than animal products like pork, chicken, milk, yogurt, cheese, and eggs (Tamar 2014). The emissions produced to sustain a plant-based diet are still significantly less than a diet that involves large portions of animal products like pork, chicken, milk, and eggs. But there certainly are some important exceptions. For example, a diet involving local, sustainably caught wild fish, in some circumstances, may contribute no more to climate change than a fully plant-based diet. So, while environmental concerns, by themselves, provide reasons to move away from animal products, these concerns don't always rule out all animal products.

Third, there appear to be effective, and presently inexpensive, ways individuals can pay to off-set the contribution their diet and lifestyle make to climate change. Instead of working to reduce their own emissions, individuals can pay organizations to implement projects to reduce emissions in other parts of the world. The organization *Giving What We Can* estimates that the cost for the average American to offset all of their yearly emissions is $105, if they give this to the most effective offsetting organizations (MacAskill 2015: 137–140). This is significant but within the reach of many individuals, who might choose to offset the climate contributions of their diet, rather than make lifestyle changes. If individuals can effectively offset their greenhouse gas emissions, then their diet would not contribute to at least one of the more pressing forms of environmental damage. This might not discharge *all* of their obligations, as there are other forms of environmental damage that animal agriculture contributes to, but it would discharge much of their obligations.

Taking into consideration these objections, we believe that environmental considerations do not require that anyone adopts a strict vegan or vegetarian diet.

An Objection: "Imperfect" Environmental Duties?

We want to suggest, however, that environmental concerns could, for some people, generate an obligation to at least *reduce* their consumption of meat and animal products *if* they do not offset their contributions to problems regarding climate change and other ways animal agriculture contributes to environmental damage.

One may ask how we can legitimately single out any one particular activity as one thing many are obligated to do to reduce their contributions to climate change. Even if we grant that there is a general obligation to reduce our negative environmental impact, some might deny that we are obligated to do any *particular* action to reduce our impact on the environment.

This position would suggest that our duties to the environment are, as Kant put it, "imperfect duties," roughly, general obligations that can be satisfied in a variety of ways. After all, there are many ways we can mitigate our negative impact on the environment. With respect to global climate change, people can often use public transit or a bicycle instead of a car, they can reduce their energy consumption in their homes, they can reduce long flights on airplanes, they can purchase carbon offsets, and so on. So even if we are willing to grant an individual obligation to reduce our negative, environmental impact, it is not clear why any *particular* way of reducing our environmental impact is obligatory. Some might say: as long as we are doing enough for the environment, we are fine: it doesn't matter how we choose to lessen our environmental impact.

To illustrate, imagine someone who considers her consumption of meat and animal products to be something that is central to her life and that gives her life great value and meaning. She is, however, an environmentalist, and when confronted by the facts about the way animal agriculture contributes so much to climate change, she is dismayed. Rather than deciding to forego or even reduce her consumption of these products, our meat-loving environmentalist decides that she will redouble her efforts to reduce her effect on the environment in other ways, to make up for her meat-eating ways: she will reduce or avoid altogether traveling on jets, she will make generous contribution to carbon offsets that more than cover her personal emissions, she decides against having a child, spends her summers planting trees, and commits only to using her bike and public transit for most of her transportation. Her objection is that she has done enough for the environment, so she need not reduce her meat intake.

We agree that this environmentalist will have done enough to meet her environmental obligations, such that *environmental reasons* do not ground an obligation to reduce her meat intake (although as we will see shortly, we think that other moral reasons ground an obligation not to purchase or consume any meat or animal products). Nevertheless, this is consistent with recognizing that other individuals, who choose not to reduce their environmental harm in other ways or who for other reasons cannot do so, may have an obligation to reduce their meat and animal product consumption. As we have already noted, choosing to abstain from meat and other animal products is one of the more effective and inexpensive ways by which individuals can reduce their contribution to climate change. As such, it is one way by which nearly everyone in developed countries can reduce their environmental impact, whereas other ways of doing so involve much greater financial costs that are difficult for some to meet.

There are strong reasons that support significantly reducing one's consumption of meat and other animal products. But these reasons do not always generate a moral obligation to reduce one's meat consumption. Whether or not they do will depend on the other efforts people make to reduce their environmental damage. However, as we will see shortly, the environmental impact of one's dietary choices is not the only morally relevant concern that confronts how we ought to eat.

Eating Animals

Evaluating the ethics of eating requires not only that we look at our diet's environmental impact, but also the ways our eating affects other animals. We must turn to how we treat the animals who end up on our plates, and our argument here is simple and straightforward. Raising and killing animals for food is wrong because of the ways these practices seriously harm other animals. If a practice causes serious harms to an individual or individuals, then it requires a moral justification, or else the practice is morally wrong. Serious harms require good reasons to justify them. We believe that attempts to justify the serious harms inflicted on animals raised for food do not succeed. Thus, the practice of raising and killing animals for food is wrong.

This argument depends on a simple, uncontroversial moral principle, that *it's wrong to cause serious harms unless there is a good reason to do so*. In addition to moral principles, our argument also depends on the facts about how animals are treated and some moral thinking about harms to animals, which we now briefly review.

Farmed Animal Facts

The vast majority of animals (99% in the U.S. and about 90% worldwide) that humans eat come from "factory farms," where the animals live in close confinement (Reese 2019; Witwicki 2019). The ways animals are harmed in these operations has been extensively documented (Halteman 2011; Singer 2002: 95–158).

Many of the ways that animals raised in factory farms are harmed stem from their extreme confinement. Egg-laying hens are confined in battery cages – wired cages, stacked on top of each other – where the birds lack the space to engage in natural behaviors, including basic things like walking on solid ground or spreading their wings. Sows (female pigs used for breeding) are confined for much of their lives in gestation crates, where they lack the space to even turn around. Like these animals, the vast majority of other farmed animals raised in the U.S. live in close confinement and this results in a variety of harms, including physical injuries, pain and suffering, disease, immobilization, boredom, psychological distress, and often death.

Animals on factory farms also experience painful body mutilations. The beaks of egg-laying hens are sliced or burnt off, pigs are castrated and have their tails cut off, and cows are branded, castrated, and dehorned. These mutilations cause animals severe pain – sometimes even chronic pain – and are all done without anesthetic.

The ways in which animals are harmed in factory farms are all standard industrial practices, not aberrations. In addition to these harms, animals raised for food are sometimes abused and injured by workers in other ways.

Animals raised on factory farms are also harmed in ways other than the pain and psychological harm that is inflicted upon them. These animals are harmed by being deprived of many of the goods crucial to their well-being. Think of a pig – a curious, intelligent, social creature – confined for its life in a small cage. This is not a good life for a pig. By failing to provide the space and resources needed for good lives, we seriously harm them: they are denied what they need for basic, natural, and social behaviors, and to live lives that are good *for them*.

"Factory Farming" versus So-Called "Humane" Farming

Many individuals feel that if farmed animals were given a genuinely good life, and then painlessly killed, there is nothing wrong with raising and killing them for food. If smaller farms could avoid harming animals in the ways we've noted above – not simply by not inflicting harms upon them, but also by providing them with the goods necessary for a flourishing life – then what is there to object to?

The first thing to note is that very few actual farms live up to this ideal. While some farms do give the animals they raise more space and better living conditions, the animals are often still seriously harmed. Many smaller farms still inflict painful body mutilations, such as castration, dehorning, and branding, on the animals. With this, many of these animals still face harms that come from transport to slaughter (such as abuse in handling, severe dehydration and hunger, and suffering from crowding as well as overheating or extreme cold). The biggest issue, however, is that animals raised for food are still harmed by an untimely death.

Even if animals enjoyed a good and flourishing life on an idyllic farm, we believe killing that animal for its meat seriously harms that animal, and thus requires a justification. Crucially, that death can seriously harm other animals is not simply a matter of whether or not the animal suffers or experiences pain in the process of being killed. Often, however, animals slaughtered for food in North America do experience a painful death. While the U.S. law mandates that cows and pigs be made unconscious before being killed, the rapid pace at which these animals are slaughtered means that many have their throat slit while still fully conscious (Pachirat 2011). This law, however, excludes birds, fish, and rabbits. Chickens and turkeys, for example, make up the vast majority of animals slaughtered in the U.S. (nearly 9 billion every year). They have their throats slit by a mechanical blade after being stunned in an electrified bath, and it is unknown what percentage of birds are rendered unconscious or merely immobilized prior to being killed (Shields and Raj 2010).

The Harm of Death

But even if these animals did not experience pain, killing them still seriously harms them. A painless death is not harmless death. For the vast majority of animals which humans kill for food, their lives are ended after a small fraction of their natural lifespan. Chickens raised for meat, to give just one example, are killed at about six weeks, while they can typically live between 8 and 12 years. Cutting their lives short seriously harms animals because the good lives they could have experienced are taken away from them.

Nearly all of us recognize this when it comes to our companion animals (and ourselves!). Imagine an individual who enjoys spending time with puppies, but does not wish to keep their dog past their puppy state, and so has each dog painlessly killed as they begin to age out of this stage of life (Lockwood 1979). Clearly, they have done something wrong. However, any adequate explanation of this wrong must appeal to the harm done to the puppy who is killed. The puppy has a very strong interest in continuing to live and is deprived of enjoyable future experiences. Doing this is wrong. However, there is no way to consistently maintain this view while denying it for farmed animals.

Death is a serious harm to other animals: it robs them of everything, their existence and the possibility of a valuable future. As a result, even when animals have lived good lives and are killed painlessly, ending their lives prematurely seriously harms them, and thus requires moral justification.

These Harms Aren't Justified

Animals raised for food – in both factory farms and less intensive farms – are seriously harmed. This should not be in doubt. These practices can only be justified, then, if these harms can be morally justified. However, there are no sufficient moral justifications that would justify these harms.

Two of the most common motivations for consuming meat and animal products – health and the pleasure one gets from eating these products – fail to justify the serious harms these practices inflict on farmed animals. Many individuals consume animal products because they believe that they are important to a healthy diet. However, it is now clear that humans can survive and flourish on a vegan, plant-based diet (Craig and Mangels 2009). Humans do not need to eat meat or other animal products to survive or even to live healthy lives.

With this, the pleasure humans get from eating meat and other animal products does not justify the serious harms we inflict on other animals. Many of us recognize this basic truth when it comes to practices unrelated to eating that inflict serious harms on other animals. We don't think that dog-fighting or cock-fighting are justified, even if it is the case that many humans get a great deal of pleasure from watching dogs or chickens fight. Why is this? Part of the answer, it seems, is that humans can engage in all sorts of leisure and recreational activities. We don't "need" to watch dogs or chickens fight to live an enjoyable or flourishing life. And choosing to do so means sacrificing animals' most basic interests – in not suffering and in continued existence – for pleasure. If we recognize this, however, it is hard to see how the same points do not also apply to animals humans raise and kill for food. It is true that many humans get pleasure from eating animal products, but it is unclear why, morally, this ought to matter. Humans can get pleasure in other ways, by eating plant-based foods, without causing serious harms to other animals. As a result, an appeal to the pleasure humans get from eating meat and other animal products fails to justify the serious harms we inflict on these animals.

Similar points apply to the "costs" that individuals might incur by giving up animal products. In addition to pleasure, many individuals continue to eat meat because it is what they are used to and because it is easier to continue to do what they have always done. However, the fact that giving up meat might bring some inconveniences does not justify participating in and contributing to these serious harms. Someone who painlessly kills their dog because it is no longer a puppy, and because taking care of this animal has become inconvenient, is clearly not justified in their action.

If these harms cannot be justified, then, we believe, humans have an obligation not to purchase or consume meat and other animal products (Hooley and Nobis 2016). Purchasing and consuming these products contribute to and financially support these practices, which cause serious harm to other animals. Once we recognize that these serious harms are not justified, we should withdraw our support from them.

Objections to Veganism

There are many critical responses to moral arguments for veganism, including from people who explicitly express concern for the environment, as well as others. Here, we briefly reply to a few common objections we often hear from our students and others:

We often hear that raising animals for food, and eating them, is *natural*, so it's not wrong. It is part of the natural order and is thus morally justified. But there's nothing at all "natural" about modern, mechanized industrial animal farming and slaughter. And just because some action is "natural," whatever that might mean (the claim that something is "natural" can have many different meanings, as discussions about sexual ethics show), this doesn't make it morally permissible. Acting violently or selfishly can be quite "natural," but is often wrong. Further, to claim that something is part of the natural order, in this context, only tells us that human beings have historically chosen to hunt, raise, and kill other animals. The mere fact that we have traditionally done something does not show that it is morally justified.

Others defend raising and killing animals by pointing to the fact that other animals eat other animals. Since we are also animals, they suggest that it cannot be wrong for us to join in, and eat animals. This, too, is a weak objection, and in some ways rather odd. Chickens, pig, and cows don't generally eat other animals. And unlike carnivorous animals, like lions, we don't *have* to eat meat. Further, unlike most animals, we can think about the consequences of our actions and choose to cause less harm when we can. Finally, just because animals do something doesn't make it OK for us to do it: e.g. some animals eat their babies, but it'd be profoundly wrong for us to do that. Animals' behavior is often not a good model for our own.

Related to the previous defense, some appeal to the fact that humans are omnivores, so it is not wrong to eat meat. But the claim that humans are omnivores can be interpreted as the claim that we *can* eat meat, or that we *should* eat meat. The first claim isn't controversial. Biologically, humans are capable of eating and digesting meat. However, the mere fact that humans can do this does not mean that we are morally justified in doing this. Defending this view requires reasons that would support raising and killing animals for food. The second interpretation assumes that it is okay to eat meat, without explaining why.

A different line of argument contends that animals are morally inferior to humans. They aren't rational, or very smart, or don't contribute to society in the ways humans generally do. And, because they lack the relevant capacity, it is okay to harm them and kill them for food. However, any appeal to human superiority that points to a specific capacity must deal with the fact that many human beings (and all of us at various points in our lives!) often lack the specific capacity in question, or possess it to various degrees. We all recognize that it's wrong to kill or eat human beings even if these humans aren't as smart or rational or whatever. If human beings who lack these advanced intellectual abilities, and even the potential for them, shouldn't be killed and eaten, then it is hard to see why it should be okay to do the same to other animals, with similar capacities, who just happen to be members of a different species.

But perhaps there is something inevitable about animal harm in the production of food. Some try to use this to justify the use and harms to animals in our present food system. If all farming methods cause animal death, then perhaps it is not wrong to eat meat.

This, too, is not plausible. Driving cars causes deaths, but we should still try to drive more safely and minimize deaths and injuries. Similar points apply to agriculture. People have tried to calculate how many animals are killed by different agricultural methods (these calculations are difficult and controversial) and have argued that, at least, current patterns of animal agriculture *certainly* don't minimize animal deaths or harms to the environment (Lamey 2007). While some animals are killed in the fields when producing grains and other vegetables, the evidence at this point suggests that significantly fewer animals would be killed if humans only ate plants. Humans should find ways to produce plants while further minimizing the numbers of animals killed in the field. However, presently eating a vegan diet is the best way to minimize animal suffering and death.

Lastly, sometimes here individuals suggest that their actions are inconsequential. If my decision to not eat animals is unlikely to lead to dramatic change, then how can it really make a difference? Perhaps it is morally okay to do what I want, including eating meat.

Unfortunately, few of us can change the entire world by our own efforts: what we do, as individuals, doesn't seem to make as much of a difference as we'd like to see it happen. This is especially true about the environment: one individual recycling, one individual using less energy, one individual taking the train instead of driving, and so on doesn't by itself fix the problems these actions are meant to address. But these actions, along with not eating meat, do make *some* difference (that world is different when we do them), and it often encourages others to make those differences also. And, unlike many other actions, we must eat, so we might as well eat in ways that are more likely to make a positive difference, and surely eating meat can't be that.

These are just a few common objections to arguments for veganism. Many more are discussed elsewhere (Hooley and Nobis 2016). Recall, however, that we argued above that environmental concerns do not, in themselves, require veganism, but that serious moral concern for animals themselves does. Thankfully, approaching the vegan ideal has major environmental benefits as well.

Conclusion: Animals and the Environment

We have argued that environmental concerns ground an individual's obligation to significantly reduce the amount of meat and animal products they purchase and consume. These concerns, however, cannot ground an obligation to eat a strict vegetarian or vegan diet. Nevertheless, we believe that most of us ought to eat a vegan diet. It is just that the grounding of this obligation stems from the ways raising and killing animals for food harms the animals themselves, not from broader environmental concerns.

Dividing our argument in this way provides a clearer sense on the basis of our moral obligations. However, it would be a mistake to read our argument and conclude from it that concerns about harms to animals and their well-being are entirely separate from "environmental concerns." While common, we think that this way of understanding "the environment" and how animals relate to it is problematic and needs revision.

Unfortunately, all too often concerns for the welfare and well-being of non-human animals are seen as distinct from, and sometimes competing with, concerns for "the environment." Perhaps nowhere is this better illustrated than in discussions about two oxen, Bill and Lou, at Green Mountain College in Vermont (Bidgwood 2012). Bill and Lou were oxen who worked for ten years plowing fields for Green Mountain College on the school's farm. After Lou sustained an injury and was unable to work, the college decided to slaughter the animals, in the name of sustainability. This produced a significant backlash and led to national attention. Despite the fact that animal sanctuaries offered to take care of Bill and Lou, Lou was killed by Green Mountain College (although not served in the cafeteria) and it appears that Bill was killed not that long after.

From the perspective of Green Mountain College, killing Lou and Bill was in the interest of environmental sustainability. The animals were viewed as resources: no longer able to work (since Bill refused to work without Lou), their bodies represented 2,000 pounds of meat that would otherwise "go to waste." Killing and using these animals, they thought, promoted the goal of benefiting the environment.

Killing Bill and Lou was morally indefensible for reasons we have already seen. But beyond this, the understanding of how animals, like Bill and Lou, relate to "the environment" that this action represents is rather odd and anthropocentric. Very few of us think, for example, that it would be a good idea to try and wipe out all of humanity to benefit the environment. Yet, this is despite the fact that humans are, by far, the most environmentally destructive species on earth! We don't think humans are simply resources who exist to benefit the environment. Rather, we recognize that humans are *inhabitants of* the environment and that our concern for the well-being of human beings explains, in part, our concern for broader environmental concerns.

Yet, when we recognize this, we ought to recognize that other animals, too, are inhabitants of the environment, sentient beings for whom the health of our shared environment matters. The environment is not for humans alone. And concern for the well-being of other non-human animals – and a recognition that we should work to avoid harming other animals whenever possible – should not be seen as competing with, or running against, concern for the environment. Animals, like us, are part of the environment that we care about, not resources that exist to benefit it.

In light of this, our moral argument for veganism shouldn't be seen as competing with, or as alien to, environmental concerns. Instead, concern for the well-being of other animals offers a way for us to imagine a much broader, and we believe more inspiring, conception of "the environment." For animals are residents of this earth just as much as other human beings. Our concern for the state of our shared environment, then, ought to include a concern for how this affects the lives and well-being of other animals, with whom we share this planet.

Related Topics

3. Experimentation
4. Companion Animals
6. Wild Animals
7. Hunting
20. Moral Bases of Responses to Climate Change
37. Waste and Consumption
38. Food

References

Aleksandrowicz, L., Green, R., Joy, E., Smith, P., and Haines, A. (2016) 'The impacts of dietary change on greenhouse gas emissions, land use, water use, and health: a systematic review', *PLoS ONE*, 11(11), p. e0165797. Available at: https://journals.plos.org/plosone/article?id=10.1371/journal.pone.0165797

Bidgwood, J. (2012) 'A casualty amid battle to save college oxen', *New York Times*, 13 November, p. A12.

Colb, S. (2013) *Mind if I order the cheeseburger? And other questions people ask vegans*. Brooklyn, NY: Lantern Books.

Craig, W.J., and Mangels, A.R. (2009) 'Position of the American Dietetic Association: vegetarian diets', *Journal of the American Dietetic Association*, 109(7), pp. 1266–1282.

Eshel, G., and Martin, P.A. (2006) 'Diet, energy, and global warming', *Earth Interactions*, 10(9), pp. 1–17.

Goodland, R., and Anhang, J. (2009) 'Livestock and climate change', *World Watch*, 22(6), pp. 10–19.

Halteman, M. (2011) 'Varieties of harm to animals in industrial agriculture', *Journal of Animal Ethics*, 1(2), pp. 122–131.

Hooley, D., and Nobis, N. (2016) 'A moral argument for veganism' in Chignell, A., Cuneo, T. and Halteman, M. (eds.) *Philosophy comes to dinner: arguments about the ethics of eating.* New York: Routledge, pp. 92–108.

Lamey, A. (2007) 'Food fight! Davis versus Regan on the ethics of eating Beef', *Journal of Social Philosophy*, 38(2), pp. 331–348.

Lockwood, M. (1979) 'Singer on killing and the preference for life', *Inquiry*, 22, pp. 157–170.

MacAskill, W. (2015) *Doing good better: effective altruism and how you can make a difference.* New York: Gotham Books.

Pachirat, T. (2011) *Every twelve seconds: industrial slaughter and the politics of sight.* New Haven, CT: Yale University Press.

Pew Commission on Industrial Animal Farm Production (2008) 'Putting meat on the table: industrial farm animal production in America', pp. 1–28. Available at: https://www.pewtrusts.org/en/research-and-analysis/reports/0001/01/01/putting-meat-on-the-table

Reese, J. (2019) 'US factory farming estimate', *Sentience Institute* [online]. Available at: https://www.sentienceinstitute.org/us-factory-farming-estimates (Accessed: 2 May 2019)

Scarborough, P. et al. (2014) 'Dietary greenhouse gas emissions of meat-eaters, fish-eaters, vegetarians and vegans in the UK'. *Climate Change*, 125(2), pp. 179–192.

Shields, S., and Raj, A. (2010) 'A critical review of electrical water-bath stun systems for poultry slaughter and recent developments in alternative technologies', *Journal of Applied Animal Welfare Science*, 13(4), pp. 281–299.

Singer, P. (2002) *Animal liberation.* New York: Harper Collins.

Steinfeld, H. (2006). *Livestock's long shadow: environmental issues and options.* Rome, Food and Agriculture Organization of the United Nations.

Tamar, H. (2014) 'Vegetarian or omnivore: the environmental implications of diet', *Washington Post*, March 10 [online]. Available at: http://www.washingtonpost.com/lifestyle/food/vegetarian-or-omnivore-the-environmental-implications-of-diet/2014/03/10/648fdbe8-a495-11e3-a5fa-55f0c77bf39c_story.html

Witwicki, K. (2019) 'Global farmed & factory farmed animals estimates', *Sentience Institute* [online]. Available at: https://www.sentienceinstitute.org/global-animal-farming-estimates (Accessed: 2 May 2019)

3

EXPERIMENTATION

Larry Carbone

Animals command a unique place in environmental ethics, whether they are "out there" in the natural world or close by in our human world.

But what is unique about animals that gives them special ethical consideration? Consider a squirrel living in an old growth redwood tree when the tree is felled. The destruction of the tree is a great loss: not only is that individual tree destroyed, so may be an important tree species, with disruption too to the tree's ecosystem. The squirrel, however, is a small cog in the ecosystem, one member of an abundant species. And yet, the harm that animal experiences is far different from what the tree, the species, the ecosystem suffer, even if she survives her injuries and lives to raise a next generation. Hers is a harm to which we would apply the word *suffering*. We would say with confidence that this is not a metaphor, this is not *like* suffering as humans experience it; this *is* suffering. For many people, this animal suffering has a moral weight that is qualitatively different from the harm inflicted in killing the last-ever redwood. It is a type of harm that has not typically been the concern of environmental ethicists.

This chapter is about animal suffering in the field and in one highly unnatural human-made environment: the modern biomedical research laboratory. Can animal suffering in experiments be justified? If so, on what grounds? If it can be justified, what limits, if any, should be placed? Does it matter who the animals are? Does it matter what the research project is? I will review current standards of care in treatment of laboratory animals, and suggest some changes in practice that could move animal research toward greater respect and concern for animal suffering.

Let us start with who and what the animals are. Scientists conduct experiments on all sorts of animals. There are the dogs, cats, rabbits, and monkeys that antivivisectionists – activists who would ban animal experimentation – showcase. There are the mice whose numbers far outweigh those larger animals. In addition, there are fish and frogs and nematode worms and amoebae and octopi and dozens of our other relatives in the kingdom *Animalia*. And some of those animals are highly unnatural creations: fish or mice with some of their own genes removed, or with genes borrowed from sea urchins or plants or people. Moreover, though much animal experimentation is performed in laboratories, companion animals in homes, animals on farms and in zoos, and free-living wild animals can all find themselves as experimental subjects.

For the animals' health, and to best standardize them as experimental subjects, most live in highly unnatural vivaria, breathing filtered air, eating sterilized foods, spared from temperature fluctuations, predators or infections. Like an agricultural CAFOS for very small animals, these laboratories themselves present environmental concerns. To meet government standards, they must hold temperatures and humidity within tight limits, at a great cost in energy use, use large

DOI: 10.4324/9781315768090-5

amounts of water, and must somehow contain, dispose of, or decontaminate the various toxins, infectious agents, genetically modified organisms, and radioactive materials that scientists use in their research (Chosewood and Wilson 2009, Coward 2015Honiotes, 2011, Institute for Laboratory Animal Research 2011).

Are all of these animals of equal moral concern? For many animal ethicists, *sentience*, more or less the capacity for suffering, is the primary concern (see the essay by Robert Jones in this volume) (Bentham 1789, Carbone 2011, Jones 2013). And though we may one day learn that we are wrong about this, the type of suffering about which we are concerned is the emotional pain and distress we know through our own lived human experiences, and as far as science has so far discovered, that requires a nervous system and brain of some sophistication. Suffering of moral concern may come from pain, from fear, from hunger, from social isolation, and a variety of other sources, but it seems to be unique to the animal kingdom, and it seems to be present in greater or lesser degree in different species. It does not appear to be present in plants, or in single-celled animals such as amoebae. It may be present to some degree in other animals: in insects and fish and other invertebrates (Magee and Elwood 2013). It is clearly present in mammals and birds, and it may be most present among the smartest of the mammals, the toothed whales, and of course, our primate relatives and ourselves (Brown 2015, Ferdowsian and Beauchamp 2013, Jones 2013).

To justify animal research in medicine, two conditions must be met. If either fails, then research, at least, those research projects that harm animals for human ends, cannot be justified. I call those two conditions the *speciesism* condition and the *utility* condition. Animals must be enough different from us to allow us to harm them in ways we would not harm our own species (*speciesism*) but enough like us that facts learned in animal experiments have *utility* in reliably yielding useful knowledge about human biology. Neither the speciesism concern nor the utility concern is sufficient alone to justify animal use; it is necessary that both conditions be met.

Is it possible that animals could be enough like humans biologically to serve as useful models of human biology, but enough different from humans in morally important ways that we could have permission to harm them? Notice that both of these questions require knowing something about animals, as well as clarifying moral values and staking out philosophical positions.

First, consider the *speciesism* condition. If we would not harm other people for a particular research project (or to eat them, or to restore an ecosystem, or for any other human purpose), then harming animals must only be justifiable if animals differ enough from people in morally important ways that good people would do unto them what we would not do unto ourselves.

We may privilege our species over all other animals through two routes: either we must show that we are far apart from them biologically in some morally important way, or we must find some non-biologically based justification. For the latter, some might feel that species membership alone is sufficient, and that no matter how mentally deficient a comatose person is, or how brilliant a chimpanzee, the person's interests always win out. We may not be better than the nonhuman animals, or more sentient and sensitive, but we are *us* and they are *other*, and that alone is sufficient justification for many people. Privileging one group, whether there are morally relevant real differences or not, is variously called racism, sexism, ethnocentrism, heterosexism, etc., when the groups are all human. For nonhumans, Ryder, Singer, and others use the word *speciesism* and reject species-bias as immoral (Ryder 1985, Singer 1990).

Anyone who would dismiss the sharp moral line between human and nonhuman animals faces the question of where, and whether, to redraw the line. Instead of simply moving the sharp line from human-nonhuman to animal-non-animal or vertebrate-non-vertebrate, Singer and others have suggested an alternative principle that would allow for gradient of treatment. Singer's proposal is to treat similar interests similarly, regardless of species. If a dog, a fish, a mouse, and a person all suffered similarly from the same painful stimulus, they would share an equal interest in avoiding that pain, and we would have the same obligation not to cause that pain to any of the four of them. This raises the biological question of whether we do all suffer equally from the same painful stimuli, and there is some evidence that we may not, that the human neocortex allows a level of emotional processing (and with that, the ability both to diminish the suffering or to increase the suffering, that arises from various pains and stressors) that just might make our interest in avoiding suffering greater than a mouse's, fish's, or dog's (Nuffield Council on Bioethics 2005). We might then recognize a hierarchy of moral concern, with most animals ranking of more concern than most plants, mammals being of higher concern than non-mammals, and maybe, just maybe, humans of higher concern than nonhumans, not because we are "us" but because our rich mental lives put us at the highest risk of suffering and with the highest potential for joy and fulfillment.

In practice, current laws, policies, guidelines, and practices do not treat human and nonhuman interests similarly. Before reviewing how law and practice grant limited respect to animal interests, we turn to the question of the usefulness of animal research.

No one can justify harming animals in laboratories unless animal experiments actually produce important and useful knowledge. If drug-safety testing does not predict safety for human patients, how can we justify harming animals for useless information? If animal experiments tell us that a drug will work on people, but it does not, can we justify animal research? We cannot.

Whether animal research produces useful knowledge for human medicine is a tricky claim to assess, and neither lists of apparent successes nor lists of apparent failures offer proof. As a veterinarian, I know from experience that medical knowledge from one species often does apply to my patient of a different species. But the bar for animal experimentation needs to be higher: it must be generally true that animal studies systematically produce knowledge that is likely to apply to people, with mechanisms for catching experimental results that would be false. LaFollette, Greek, and others have highlighted this challenge: they recognize myriad physical and emotional similarities between humans and nonhumans, but show ways in which our tools for knowing which similarities and which animal data are likely to extrapolate to people are lacking, both theoretically and empirically (Carbone 2012b, Greek and Greek 2004, LaFollette and Shanks 1996).

People differ both on how much animals' lesser cognitive capacities justify harming them for our ends, and on their assessment of how vital the information from animal experiments is. They also bring different values to the debate, disagreeing, for example, on how "necessary" medical progress is in the first place. A utilitarian justification of research requires defending the proposition that certain harm to animals presently in laboratories is outweighed by the potential benefits to humans in the future. The polar positions that would allow zero animal experimentation, or, however, complete and unlimited permission to experiment on animals are not the only possibilities. The positions embodied in most countries' laws and policies that some research on animals may be permissible with certain conditions and limitations. But how to decide what to allow? How to weigh the certain present harm to animals currently in laboratories against the potential benefits for future humans?

In considering harm to animals, the sentientist approach is to think primarily about the harm of suffering. As mentioned above, pain is not the only source of human or other animal suffering. In fact, overthinking pain leads us, mistakenly I believe, to think of suffering as exclusively bodily. But as DeGrazia and others have argued, physical insults to the body only cause morally significant suffering if they are processed by the mind as unpleasant (DeGrazia 1996). We know from our own experience that emotional harms, even without any physical component to them, can be excruciating, and there is evidence that for some nonhuman animals, psychological harms can cause great suffering. In fact, so sure are scientists that animals can suffer psychological harm that they have developed standardized methods for causing animal versions of Post-Traumatic Stress Syndrome, learned helplessness, depression, and anxiety, in hopes of understanding these conditions and developing treatments for them.

Animals may receive other harms without emotional suffering. Rollin considers any violation of an animal's basic nature, or *telos*, a harm, whether the animal is suffering or not (consider a caged tiger, even one who has adapted well to her environs) (Rollin 1995). A controversial question in animal ethics is whether killing animals (humanely, in ways that do not cause suffering) is a harm that must be justified (AVMA Panel on Euthanasia 2013). Thus, an animal may not be actively suffering for us to say s/he is being harmed, if we deny him or her the expanded opportunities for happiness and fulfillment the species is capable of enjoying.

Animal welfare scholars also debate whether the harms we must consider are limited to the physical and psychological insults relative to some neutral state of welfare, or do we owe more to animals in laboratories (and other places where humans use them as means to human ends) (Carbone 2004, Tannenbaum and Bennett 2015)? Are there kinds of positive states – of joy, happiness, intellectual stimulation, social harmony – above and beyond the theoretical neutral that we owe to these animals, and should failure to enrich their short, captive lives also be considered a harm we inflict? That is, when we cage animals and plan experiments on them, do we have merely a duty of *nonmaleficence* (i.e. a duty not to harm), or beyond that, a duty of *beneficence* (i.e. a duty to provide good things to them)?

Are all nonhuman animals the "others" whose harms must be justified? Possibly, the particular harms we call *suffering* seem to be limited to those animals, ourselves included, who have the neural structures and functions that can sustain, perhaps primitively, conscious emotions such as joy and sorrow and fear. In a sentientist ethic, those emotions need not be coupled with the high degree of self-awareness that humans and possibly very few other animals have. They need not include a sense of a self who persists over time. It may be sufficient to say that if an animal is capable of responding to pleasant or unpleasant events at an emotional level, that animal is capable of suffering. This may be quite widespread in the animal kingdom – even hermit crabs seem able to remember events (with the memories likely etched with strong emotional reactions) and avoid situations likely to cause those events to happen again (Magee and Elwood 2013).

Applied Ethics in Animal Experimentation

Laws and public policy are illuminating reflections of societal consensus on animal experimentation. Most countries have similar laws that regulate laboratory animal use. In most, a scientist must adhere to certain norms, and some sort of government entity or in-house ethics committee must approve an animal research project before it begins. The laws share certain features: they are speciesist in the Ryder/Singer sense; they do not treat similar interests similarly regardless of species; they focus on sentience and suffering rather than other harms (such as death); and they focus more on nonmaleficence than beneficence.

In the United States, one law (the Health Research Extension Act of 1985) covers use of any vertebrate animal, but only in government-funded research. The other law, the Animal Welfare Act (first passed in 1966), covers laboratory animals regardless of funding or location, but is restricted in species-coverage to a tiny minority of the animals in laboratories. While other laws, such as the Endangered Species Act, and some individual states' laws, have some impact on animal research, the AWA and the HREA are the most detailed and important in the United States.

All laboratory animal laws are speciesist in various ways. Foremost, they reflect a belief that using nonhumans (including hurting and killing them) is permissible, to meet various human interests in developing new knowledge. Almost no animal harm is too great, no human interest too trivial, to prohibit animal use. The Animal Welfare Act requires an annual report of animals on experiments in which their pain or distress is left untreated, if that would interfere with the scientific data collection. The Act requires the animal committee to review scientists' plans to conduct painful studies, and requires consultation with a veterinarian. A committee will review the degree of harm to be inflicted, the plans to reduce that harm, and the scientist's justification for causing harm, but is still free to approve the harmful and painful experiments (Carbone 2011).

Just as they draw a sharp line separating human from nonhuman, the laws also draw a sharp line between covered and non-covered animal species. Differences in degree of sentience inform these laws, but inconsistently. The American HREA covers all and only vertebrate animals (animals with backbones, comprising fish, amphibians, reptiles, mammals, and birds). Backbones are a rough but not always accurate proxy for sentience, and it is likely that an adult octopus (an invertebrate) is far more sentient than a hatchling (vertebrate) zebrafish (Mason 2011, Mather 2008). Indeed, the United Kingdom Animals Act has long included all vertebrate animals, as well as invertebrate octopi, under its aegis, in its attempt to provide for the welfare of all sentient animals. From a biological and ethical perspective, the AWA's exclusions are downright bizarre, with no biological basis for their most salient exclusion: laboratory-bred rats and mice, easily at least 97% of the mammals in laboratories. Under the AWA, hamsters, gerbils, and wild-caught mice are covered (legally, under this law, they are defined as *animals*), but their laboratory-bred cousins and offspring are not. The exclusion reduces the costs of enforcement for the government and the cost of compliance for laboratories, and also maps onto whatever public perceptions exist that hamsters are cute pet animals, while mice are vermin unworthy of consideration.

While laws draw sharp lines between species they cover and species they exclude, they also establish, though without clear definitions, species-based gradients of concern. The laws embrace the principle that researchers should replace animals with non-animals (bacteria, cells-in-culture, computer simulations) and also replace more-sentient animals with less-sentient whenever possible (Institute for Laboratory Animal Research 2011). Under this principle, flies are first choice before frogs, and frogs are a better choice than dogs or monkeys (and the moral principle dovetails with the lower cost, easier upkeep, and better public relations of choosing mice over monkeys). Assigning levels of sentience within a class (for example, among the mammals) is challenging, requiring careful selection of what criteria should count as evidence of sentience, and then careful research to find it. Despite this, research guidelines do occasionally describe rodents as less sentient than other mammals, with no evidence provided for that claim (Carter and Shieh 2015) or make reference to a "phylogenetic scale" in which mice are "lower" than other mammals, but again, without reference to biological support for such an assignment (Bennett et al. 1994, Orlans et al. 1998)

In 1959, Russell and Burch with the Universities Federation for Animal Welfare published a framework for reducing suffering (their term was *inhumanity*) in laboratory animals (Russell and Burch 1959). Their "3Rs" framework, as modified through the years in policies and laws, includes *replacement* of sentient animals with nonsentient tissue culture, nonsentient animals, and nonliving systems (including computer systems). Scientists may *reduce* harmful animal use in various ways, including use of statistical analyses that allow fewer subject animals to generate statistically significant data. Finally, *refinement* of experiments includes using the most humane available methods of restraint, the most humane housing, and the use of anesthetics and painkillers. Replacement, reduction, and refinement are sometimes referred to collectively as *Alternatives*, though that unfortunate word choice can lead people to think exclusively of replacement.

The laws concern themselves mostly with efforts to follow Russell and Burch and reduce harm to laboratory animals, specifically the harms of pain and distress. Laws require that institutional animal care and use committees (IACUCs) or in-house animals ethics committees must be established to review scientists' proposals to use animals, and these committees must include veterinarians as members. The committees review planned experiments, including the scientist's efforts to refine any potentially painful experiments, and the veterinarian's recommendations of painkilling anesthetics and analgesics.

Both American laws accept the premise that some experiments will cause animals pain or distress, and that painkillers may sometimes be withheld. Examples of untreated pain in laboratory animals include causing painful arthritis or distressing psychological conditions and leaving some comparison control groups untreated. The purpose is to best see the effects of a novel arthritis or anxiety medicine, but the result includes unalleviated animal suffering. Product safety testing is also often conducted without painkillers that might affect measuring just how toxic high exposure doses might be. The AWA requires annual reports of how many animals are in experiments in which anesthetics, analgesics, or tranquilizers are withheld. Of those species covered, roughly 7% per year are in such experiments. However, mice and rats are commonly used for safety testing, and for studies of pain, inflammation, cancer, and psychiatric diseases, and there are no public data available in the United States on their numbers. Regulatory guidance on adequately evaluating the need to withhold pain medication is scant. Is it sufficient for a scientist to state that a certain class of drugs affects immune function and therefore theoretically might affect a particular experiment in immunology? Or must s/he produce evidence that the same experiment, conducted side-by-side with and without pain management, produces different results and evidence that the pain drugs, not the pain itself, are affecting the animals' biology in a way that confounds the scientific data (Carbone 2012a, Kaldewaij 2008)?

As with the use of animals for food, the prevalent ethic is that pain and suffering are harms to animals that must be minimized when possible, but that death itself is not a harm. This "pain counts; death doesn't" ethic is essential to any justification of animals for food or for science. Not only is there no proscription against killing animals, killing (euthanasia) is in some ways the ultimate in pain relief. Thus, scientists may permissibly induce serious cancers in animals, cancers whose pain would be beyond a painkiller's ability to quell, so long as they commit to killing the animals once severe pain begins. Scientists must identify "humane endpoints" for their studies, the time points or defined criteria (weight loss, obvious signs of pain, etc.) at which they will remove an animal (usually by killing) from the experiment. The defense of this ethic is the belief that animals have so little self-consciousness and so little sense of the future that (painlessly) taking away their futures does not matter to them. Only a fraction of the animals killed in laboratories are in pain or distress; most are killed as healthy young or young adult animals. Killing these animals does indeed take away all prospect of future goods in their lives, and many argue that this is indeed a harm to them that should be avoided (AVMA Panel on Euthanasia 2013, Cigman 1989, Kaldewaij 2008, Regan 1989).

In 1996, an advisory board for the National Aeronautics and Space Administration (NASA)'s animal research activities elaborated the "Sundowner Principles," including a duty of nonmaleficence similar to the "do no harm" duty of human medical ethics:

"Vertebrate animals are sentient. This principle entails that the minimization of distress, pain and suffering is a moral imperative. Unless the contrary is established, investigators should consider that procedures that cause pain or distress in humans may cause pain or distress in other sentient animals" (National Air and Space Administration 1996).

But why stop at a duty of nonmaleficence? If we are to inflict confinement, pain, suffering, diseases on laboratory animals, might there be an obligation of beneficence, of striving to provide positive good things in their lives, even more than we might owe to companion animals or others we do not actively harm? Within current laws and policies, there are provisions that might indeed be early stages in articulating a duty of beneficence. Provisions in the AWA and other rules and regulations call for environmental enrichment to promote the psychological well-being of animals, especially primates. From a nonmaleficence perspective, these may serve solely to prevent outright emotional distress in animals, assuring that animals do not slip from a theoretical welfare-neutral state toward mental ill-health. Alternatively, there might, in fact, be efforts to move beyond a neutral not-suffering state toward some recognition of animal happiness, and an obligation to provide for that. Once, standards allowed for small barren solitary cages for laboratory animals. In recent years, with each edition of the *Guide for the Care and Use of Laboratory Animals*, the main standards the National Institutes of Health (NIH) require its grantees to follow, we find this expansion of attention to animals' social and emotional lives. Current guidelines call for efforts to enhance the psychological well-being of nonhuman primates, for housing social animals in compatible groups, for providing enriched environments full of cognitive challenges and chances to forage, burrow, build nests, and explore (Institute for Laboratory Animal Research 2011).

Not all animal experimentation is in the service of human medical progress. Scientists use animals for research in agriculture, veterinary medicine, and wildlife biology, and all of these can cause animal suffering and death. Wildlife biologists who trap, identify, and re-release wild animals must account for the possibilities of trapping injuries and animal fear and distress. Marking, identification, and sampling methods can be harmful, such as removing a toe or taking a blood or tissue sample. Anesthesia may make the technique less painful, but raises the choices of re-releasing animals before they are able to safely run or fly away or holding them extra hours in confinement they will find stressful. Collars and other tracking and identification devices may be heavy or in other ways may interfere with the animal's ability to move in the environment, defend against predators and aggression or obtain food. Animals may be "collected," that is, captured and killed for tissues. Others are captured and released, but most methods for capture, identifying, sedation, and DNA or other sample collection carry at least some risk of pain, distress, or lowered fitness. Biologists' presence may disrupt animals' lives. Relocation programs may stress and injure animals (Cooke et al. 2013, Farnsworth and Rosovsky 1993, Fedigan 2010, Green and Bradshaw 2004).

As with medical research, the animal subject in field studies rarely is the beneficiary of the research. Sometimes, human health drives animal field studies, as when scientists capture and kill fishes and other animals for sampling them as sentinels for environmental toxins. In other cases, scientists conduct their research on members of a species to benefit that species, or to learn general, basic knowledge about ecology and evolution. Efforts to help a species may sometimes justify harming an animal, but that should not be confused with helping that animal subject.

Scientists use the 3Rs framework to guide their efforts to make their animal experiments as humane as possible, with oversight from animal-use and animal ethics committees, veterinarians, and government regulators. Presently lacking, however, is a robust system to determine if an animal experiment, even one that causes great animal suffering, should be permitted in the first place. Agencies and foundations that fund animal research distribute finite dollars to what they judge to be the best applications. Animal care and use committees have no such finite cap: they can approve 100% of the proposals they review, fine-tuning the animal welfare details, but with no natural or regulatory cap on how many proposals they approve. Thus, an expectation that scientists only use, and only harm, animals when absolutely necessary for important research, remains a judgment call that is almost exclusively in the hands of the individual scientists.

Summary

Despite the highly unnatural genetic manipulations and their controlled human-made environment, laboratory animals retain the potential for suffering and happiness that we call *sentience*. Scientists can only justify their animal experiments if those experiments produce useful knowledge *and* if we have moral justification to harm animals for human ends. Both arms of this justification are controversial. In practice, scientists frame their plans to use animals around the 3Rs of replacement, reduction, and refinement, focused primarily on minimizing animal suffering and making some effort to enrich the lives of their captive subjects. In the 3Rs framework, killing animals is not a harm, *per se*, and is even a method for ending animal pain and suffering. This sentientist/ speciesist ethic, in which humans pay some heed to animal pain and suffering (but not death), but not the heed we pay to human suffering, is consistent with the dominant ethics of meat-eating, animal experimentation, and other human uses of nonhuman animals.

Related Topics

2. Eating
4. Companion Animals
6. Wild Animals

References

AVMA Panel on Euthanasia (2013) *AVMA guidelines for the euthanasia of animals: 2013 edition* [Online]. AVMA. Available at: http://www.avma.org/KB/Policies/Documents/euthanasia.pdf (Accessed 20 July 2013)

Bennett, B. T., Brown, M. J. & Schofield, J. C. (eds.) (1994) *Essentials for animal research: a primer for research personnel*. Beltsville, MD: National Agriculture Library.

Bentham, J. (1789) Excerpt from Chapter XVII, section 1. *The Principles of Morals and Legislation.*

Brown, C. (2015) 'Fish intelligence, sentience and ethics', *Animal Cognition*, 18, pp. 1–17

Carbone, L. (2004) *What animals want: expertise and advocacy in laboratory animal welfare policy.* New York: Oxford.

Carbone L. (2011) 'Pain in laboratory animals: the ethical and regulatory imperatives', *PLoS One*, 6, e21578.

Carbone, L. (2012a) 'Pain management standards', *Guide for the Care and Use of Laboratory Animals, Eighth Edition – Journal of the American Association for Laboratory Animal Science*, 51, pp. 1–5.

Carbone, L. (2012b) 'The utility of basic animal research', in *Animal research ethics: evolving views and practices, Hastings Center Report*, Special Report 42, pp. S12–S15.

Carter, M. & Shieh, J. C. (2015) *Guide to research techniques in neuroscience*, 2nd edition. San Diego, CA: Academic Press.

Chosewood, L.C. & Wilson, D. (eds.) (2009) *Biosafety in microbiological and biomedical laboratories*, 5th edition. Washington, DC: Department of Health and Human Services.

Cigman, R. (1989). 'Why death does not harm animals', in Regan, T. & Singer, P. (eds.) *Animal rights and human obligations*, 2nd edition. Englewood Cliffs, NJ: Prentice Hall.

Cooke, S. J., Nguyen, V. M., Murchie, K. J., Thiem, J. D., Donaldson, M. R., Hinch, S. G., Brown, R. S. & Fisk, A. (2013) 'To tag or not to tag: animal welfare, conservation, and stakeholder considerations in fish tracking studies that use electronic tags', *Journal of International Wildlife Law & Policy*, 16, p. 16.

Coward, C. (2015) 'How good IAQ can help save energy in labs', *ASHRAE Journal*, 57, pp. 12–24.

Degrazia, D. (1996) *Taking animals seriously.* Cambridge: Cambridge University Press.

Farnsworth, E. J. & Rosovsky, J. (1993) 'The ethics of ecological field experimentation', *Conservation Biology*, 7, pp. 463–472.

Fedigan, L. M. (2010) 'Ethical issues faced by field primatologists: asking the relevant questions', *American Journal of Primatology*, 72, pp. 754–71.

Ferdowsian, H. & Beauchamp, T. L. (2013) 'Animal experimentation,' in LaFollette, H., Brock, G., Deigh, J., Holroyd, J., Star, D. & Stroud, S. (eds.) *The international encyclopedia of ethics*. Blackwell Publishing Ltd.

Greek, J. S. & Greek, C. R. (2004) *What will we do if we don't experiment on animals?* Victoria, Canada: Trafford Publishing.

Green, J. J. & Bradshaw, C. J. A. (2004) 'The "capacity to reason" in conservation biology and policy: the southern elephant seal branding controversy', *Journal for Nature Conservation*, 12, pp. 25–39.

Institute for Laboratory Animal Research, National Research Council (2011) *Guide for the care and use of laboratory animals*, 8th edition. Washington, DC: National Academies Press.

Jones, R. C. (2013) Science, sentience, and animal welfare. *Biology & Philosophy*, 28, pp. 1–30.

Kaldewaij, F. (2008) 'Animals and the harm of death', in Armstrong, S. J. & Botzler, R. G. (eds.) *The animal ethics reader*, 2nd edition. London: Routledge.

Lafollette, H. & Shanks, N. 1996. *Brute Science: dilemmas of animal experimentation.* London: Routledge.

Magee, B. & Elwood, R. W. (2013) 'Shock avoidance by discrimination learning in the shore crab (Carcinus maenas) is consistent with a key criterion for pain', *Journal of Experimental Biology*, 216 (3), pp. 353–358.

Mason, G. (2011) 'Invertebrate welfare: where is the real evidence for conscious affective states?', *Trends in Ecology and Evolution*, 5, pp. 212–213.

Mather, J. A. (2008) 'Cephalopod consciousness: behavioural evidence', *Consciousness and Cognition*, 17, pp. 37–48.

National Air and Space Administration (1996) NASA principles for the ethical care and use of animals, p. 25. NPR 8910.1B-Appendix A. 28 May. Available at http://nodis3.gsfc.nasa.gov/displayDir.cfm?t=NP-Dandc=8910ands=1B (Accessed 10 May 2010).

Nuffield Council on Bioethics (2005) *The ethics of research involving animals*. London: Nuffield Council on Bioethics.

Orlans, F. B., Beauchamp, T. L., Dresser, R., Morton, D. B. & Gluck, J. P. (1998) *The human use of animals: case studies in ethical choice*. New York: Oxford University Press.

Regan, T. (1989) 'Why death does harm animals', in Regan, T. & Singer, P. (eds.) *Animal rights and human obligations*, 2nd edition. Englewood Cliffs: Prentice Hall.

Rollin, B. E. (1995) *The Frankenstein syndrome: ethical and social issues in the genetic engineering of animals*. Cambridge: Cambridge University Press.

Russell, W. M. S. & Burch, R. L. (1959) *The principles of humane experimental technique*. London: Methuen & Co. Ltd.

Ryder, R. D. (1985) 'Speciesism in the laboratory', in Singer, P. (ed.) *In defense of animals*. New York: Basil Blackwell, Inc.

Singer, P. (1990) *Animal liberation*. New York: Avon Books.

Tannenbaum, J. & Bennett, B. T. (2015) 'Russell and Burch's 3Rs then and now: the need for clarity in definition and purpose', *Journal of the American Association for Laboratory Animal Science*, 54, pp. 120–32.

4

COMPANION ANIMALS

Clare Palmer and T. J. Kasperbauer

Introduction

It's estimated that there are around 70 million dogs, and more cats, in the USA. In 2017–2018, 38% of US households owned at least one dog, and 25% at least one cat (AVMA 2018), and ownership figures are only slightly lower in the European Union. In most Western countries, the number of households keeping dogs and cats has been steadily growing for decades. On the face of it, this is surprising. After all, keeping dogs and cats in the home can be both expensive and inconvenient. Yet, obviously, for many people, the gains from living with animals are so significant that they outweigh the costs (Serpell & Paul 2011). Surveys repeatedly show that living with animals normally gives their owners pleasure; well over half of owners say that they perceive dogs and cats to be "members of the family"; and an even higher percentage describe their dogs and cats as companions (AMVA 2012).

While it may be the case that companion animals are normally good for their owners, keeping them, it's often argued, nonetheless raises ethical concerns. We will consider two kinds of ethical concerns here. The first relates to the creation and keeping of animals as human companions at all. "Companionship" is usually taken to be a two-way relationship, one of positive, non-coercive interaction between two parties. However, it's recently been argued that the term "companionship" brushes over the darker side of keeping animals in our homes, and that the practice of breeding and keeping companion animals might instead be seen as perpetuating a relationship based on human domination and non-human vulnerability—a traditional concern within environmental ethics. This concern is magnified by the fact that in many Western nations, companion animals are, like other elements of the non-human world, technically human property, lacking legal standing and representation. After explaining how we are using the term "companion animals," we will discuss such arguments, maintaining that animal companionship need not, in principle, be understood as coercive and exploitative, and outlining two recently proposed positive frameworks for living ethically with animal companions.

While keeping companion animals may not be *intrinsically* ethically problematic, it inevitably raises ethical questions, dilemmas, and conflicts in *practice*. The second group of ethical concerns we will discuss here relates to the broader impacts of companion animals on the environment. We will focus in particular on concerns about resource use, pollution, and predation. While these are traditional concerns of environmental ethics, they have only recently begun to be explored in the context of companion animals.

DOI: 10.4324/9781315768090-6

Key Terminology: Animal Companions

The most straightforward way of explaining how we will be using "companion animals" is to relate it to the widely used term "pet." Varner (2002) describes a "pet" as a being that is affectionately regarded by its owner (so not a pest); lives in or close to the home; is mobile (so not a plant); lives a life different in kind from its owner's (so not a human); possesses its own interests—that is, its life can go better or worse for it, and it has a welfare or a good of its own; and it depends on its owner in significant ways to help fulfill its interests.

This characterization includes most animals voluntarily kept by people in their homes: mammals, birds, fish, reptiles, amphibians, and insects. We take "companion animals" to be a subset of this broader group of pets, and understand companionship as a kind of interactive bond, where humans and animals recognize and are responsive to one another, and seek one another's company for comfort, consolation, play, and so on. This clearly describes the relationships many people have with their dogs and cats; dogs and cats are, as it were, paradigm animal companions, though sometimes people have similar relations with birds, rabbits, and other small mammals. Insects, fish, reptiles, and amphibians, though also voluntarily kept in the home, are rarely companions in this sense (and, some may argue, because of the kinds of beings they are, they cannot be). So, our discussion of companion animals here will focus on dogs and cats, as our most common non-human companions, including their impacts on the environment.

Is It Wrong to Keep Companion Animals?

From some ethical perspectives, the practice of keeping animal companions is *intrinsically* ethically problematic. As noted above, this worry is exacerbated by the fact that, in most places, companion animals are legally just property. We will begin by evaluating the claim that keeping companion animals is intrinsically wrong, and then consider arguments concerning companion animals as property, and conclude by briefly outlining two recent proposals for living ethically with companion animals.

The most well-known argument against breeding and keeping animals as companions has been developed by Gary Francione. Francione (2012) argues that animals bred to be companions "exist forever in a netherworld of vulnerability, dependent on us for everything, and at risk of harm from an environment that they do not really understand. We have bred them to be compliant and servile, or to have characteristics that are actually harmful to them but are pleasing to us." Breeding and keeping animals as companions seem to be understood here as a form of human domination of nature, where humans create non-humans that are essentially—and so inescapably—servile, and are unable to flourish without human support.

Francione argues that creating this built-in dependence is wrong, so much so that we should stop breeding such animals. It's worth emphasizing that the moral concern here is with the state of created dependence itself; it's not directly related to animals' subjective experiences, preferences, or desires. The concern persists "however well we treat our nonhuman companions" (Francione & Garner 2010: 79).

This is an interestingly different worry from the ethical concerns more commonly encountered in animal and environmental ethics. In the case of agricultural and laboratory animals, and wild animals in captivity, ethical concerns usually focus on animals' negative mental states (such as fear and suffering), and on the inability of animals, especially ones that are confined, to fully express natural, species-specific behaviors. For Francione, though, the objection concerns the state of dependence *itself*; we just should not create animals that *have* such essentially dependent natures, however happy and free to express natural behaviors they are.

But why should created dependence, in itself, be ethically problematic? (If it is, this may be part of a much bigger problem in environmental ethics, given how much responses to global environmental change rely on highly interventionist conservation strategies that may well generate dependence on humans.) One obvious concern might be that the vulnerability created by dependence necessarily leads to exploitation. A second concern may be that there's a sense in which it's just better not to be dependent, and instead to be self-sufficient and autonomous. But both of these concerns are problematic. While it's true that dependence does increase vulnerability, vulnerability does not *necessarily* lead to exploitation. After all, children are highly dependent on their parents, but we don't think that parent–child relationships are *necessarily* exploitative. Certainly, companion animals are vulnerable to those with whom they live, and that vulnerability can lead to welfare issues if their owners fail to properly meet their needs, but it's not clear why this must define the relationship.

So is the problem with the condition of dependence itself? It's not clear why this would be intrinsically problematic either. After all, all humans are highly dependent on others for some of the time (when infants, when elderly, when sick); some humans are highly dependent on others all of the time; and all of us are at least somewhat dependent on other humans in our daily lives. Should we try, then, to eliminate dependence between humans? Or to aim to create a society that doesn't contain dependent adult humans? As Donaldson and Kymlicka (2011: 83) argue, dependency is not, in itself, "inherently undignified or unnatural"; and there's clearly a risk that condemning dependency will have "pernicious consequences" for humans as well as animals.

The more obvious conclusion, in the case of companion animals, is that their created dependence gives them a special relationship to human beings—a relationship that's different from the one humans normally have with wild animals, for instance. Palmer (2010) argues that while, other things being equal, we have duties not to *harm* all sentient animals, we may have different responsibilities to care for animals in different contexts. Since we have made companion animals significantly dependent on us, and therefore vulnerable to our treatment, we have special responsibilities to care for them, responsibilities that we don't generally have to wild animals (unless they have also been made vulnerable and dependent due to some human activity).

A further concern—also expressed by Francione—is that companion animals are, almost everywhere, merely the legal property of their owners; and that as long as this is the case, keeping them is morally problematic. As property, it's argued, animals cannot have independent legal standing, which means that they are entitled to few legal protections (though many places do have anti-cruelty laws, which provide some restraint on what can legally be done to them). Korsgaard (2013: 629) notes: "Persons are the subjects of both rights and obligations, including the right to own property, while objects of property, being by their very nature for the use of persons, have no rights at all."

The worry that non-humans are not legal persons, and cannot therefore have independent legal standing, representation, or rights, has been an enduring concern of environmental ethics. Christopher Stone (1972) argued that natural objects such as streams or mountains should have some kind of legal standing, "rather like the standing of legal incompetents – human beings who have become vegetable" (p. 17). This would allow trustees, or guardians, to be appointed on their behalf. The property status of companion animals is a particularly acute instance of such concerns about property within environmental ethics, for three reasons. First, since companion animals are sentient, it can more obviously be claimed that they, unlike mountains or streams, have "interests," and that a change in their legal status may protect them from suffering. Second, since companion animals are commonly regarded as family members, the idea of buying and selling them seems incongruous, in a way that's less true of mountains and streams. And third, mere property status means that if someone harms or kills a companion animal, the owner is only likely to succeed in suing for the animal's fair market value, rather than (for instance) for damages for emotional harm, however important the animal was to them. For all these reasons, there's currently lively debate about the legal status of animal companions. Alternative legal regimes in which companion animals are at least not *solely* property are possible (and may to some degree already exist; for instance, in Norway and France, companion animals do have some kind of independent legal standing as "sentient beings").

However, not all ethicists, even those who take animals' independent moral status seriously, are troubled by the idea of companion animals being property. Cochrane (2012), for instance, argues that because animals don't understand what "being property" is, unlike humans, they don't have an interest in not being owned or in not being human companions:

> Some practices that are objectionable when done to humans are not objectionable when done to animals: keeping an animal as a pet is quite unlike keeping a human as a slave; using animals to undertake certain types of work is quite unlike coercing humans to labor; buying and selling animals is quite unlike trading human beings; and so on.
>
> *(Cochrane 2012: 11).*

On this view, animals aren't harmed by states such as "being property," of which they, in contrast to humans, can't be aware. However, Cochrane is clear that animals can still be harmed by states of which they *are* aware, such as pain and frustration; accepting his position does not necessarily lead to opposition to increased legal protections for companion animals.

So, neither created dependence nor property status seem to be, in themselves, insuperable objections to keeping companion animals. But both these concerns do highlight the importance of thinking carefully about appropriate ethical frameworks for living with animals as companions.

Ethical Frameworks for Keeping Companion Animals

Companion animals fit awkwardly into existing ethical, political, and legal frameworks, both those designed for humans and for animals. On the one hand, frameworks designed for people often presuppose certain capacities that non-humans are thought to lack, such as the ability to reflect on actions and to make moral decisions. On the other hand, frameworks designed for animals generally focus on animals with whom we have very different relationships from companionship—in particular farm and laboratory animals, kept for production and use. And unlike wild animals, companion animals are not independent and self-supporting, so we need to think about what's owed to them in ways that move beyond just non-intervention. This complex situation has led to proposals for new frameworks through which to think about positive ways of living with animal companions. Here, we will briefly outline two of these: Favre's idea of "living property" and Donaldson and Kymlicka's account of companion animal citizens.

Companion Animals as Living Property

As we've already noted, in most countries, companion animals are legal property. However, in general, people treat their animal companions as much more than mere property. Research suggests that owners are highly attached to their companions, and invest significant resources in taking care of them. Companion animals now support a massive industry—estimated to be worth $95.7 billion in the USA alone in 2019—producing tailor-made feed, litter, and accessories, and supplying services such as boarding, veterinary care, and grooming (APPA 2020).

For these reasons, it's sometimes argued that companion animals are more like family members than property (Milligan 2009, 2010)—a view reflected, as pointed out earlier in this chapter, by the responses of US dog and cat owners in surveys. Perhaps the most natural comparison within the family is between companion animals and human children. Burgess-Jackson (1998), for instance, argues that people have similar duties to the animals they take into their homes as they do their own children, because they have (normally) *chosen* to bring vulnerable, sentient beings into relations of dependence with them.

Yet, this family framework fits very awkwardly with the idea of companion animals as property, since, as noted above, we cannot own, buy, or sell our family members. Favre (2010) proposes that we can think of companion animals as both family members and property by considering animals to fall into a new category of what he calls "living property." Living property is still property; it can be kept, owned, and used. However, it also has its own interests, and should therefore be assigned some legal rights—for instance, access to sufficient space, protection from harm, and appropriate care.

The owners of animal companions, Favre suggests, should have similar legal responsibilities to meet their animals' basic needs as parents do for their own children. This means that "the rights of owners will have to be limited to some degree to accommodate some of the interests that their property asserts against them"; that those who don't own companion animals will still have some duties toward other people's animal companions; and that companion animals will have some rights themselves (Favre 2010: 1053). As Smith (2012: 85) notes, in this proposal, Favre takes advantage of a distinction between legal title and equitable title. Someone with legal title has control of the property, but someone with equitable title should benefit from it; the titleholder (in this case, the human owner) has a legal duty to take the interests of the beneficiary (the owned animal companion) into account.

Favre's framework, in contrast to Francione's, is built on ethical foundations that allow for the possibility that "positive human communities can include animals that are owned and used by humans" (Favre 2010: 1023). Both ownership and use, though, are constrained by the interests of the animals concerned. Companion animals have a distinct relationship with those who own them, and this relationship generates special responsibilities of provision and protection, some aspects of which should be governed by law.

Companion Animals as Citizens

Donaldson and Kymlicka (2011) propose, alternatively, that we understand companion animals within the framework of citizenship theory. Here, companion animals are best thought of as *citizens*, while wild animals have territorial *sovereignty*, and "liminal" animals (such as feral animals and those wild animals that live in urban areas) should be thought of as *denizens*. Donaldson and Kymlicka argue that since companion animals (and domesticated animals more generally) have been brought into our society, and don't have other possible forms of existence, we should include them in our social and political arrangements "on fair terms" (Donaldson & Kymlicka 2011: 101). Citizenship, they argue, is the appropriate social and political framework for companion animals: they should be granted residency and their interests should count in determining the public good.

While Donaldson and Kymlicka certainly agree that companion animals' interests should be represented in the law and that owners have special responsibilities to their companions, their claims are more ambitious and controversial than Favre's. Rather than drawing primarily on the family as a model, Donaldson and Kymlicka turn instead to disability theories of citizenship.

Typically, citizenship is thought to require *reflective* agency, where agents are expected not only to comply with laws and social norms but to be able to understand and deliberate about them. Donaldson and Kymlicka deny that this conception of citizenship is adequate in the case of either humans or animals. Rather, what is crucial to citizenship is that agents have a "subjective good"— that is, that they can have good or bad experiences such as pain, frustration, pleasure, and excitement—and are able to communicate that good. They argue that, "Domesticated animals may not *reflect* on the good, but they *have* a good—interests, preferences, desires—and an ability to act, or communicate, in order to achieve their good" (p. 112). Since animals have goods—for instance, in the case of dogs, the desire for space to run and play—the key goal is to find a way in which these goods can meaningfully enter the political process. This is where Donaldson and Kymlicka make the crucial link to human disabilities. As with some disabled human beings, non-human agents require intermediaries to contribute to social and political decision-making (a process that, drawing on disability theory, Donaldson and Kymlicka call "dependent agency"). And as beings with subjective goods, they should be provided with support so that their needs and interests can be better understood and incorporated into democratic decision-making processes.

Donaldson and Kymlicka's framework, then, looks very different from both Francione's and Favre's, and entails the recognition that companion animals have certain positive rights as co-citizens, as well as basic rights not to be harmed. Their citizenship framework has implications for a wide range of ethical issues raised by companion animals: selective breeding, acquisition, feeding, training, neutering, convenience surgeries such as declawing, exercise, medical care, relinquishment, and euthanasia (see Donaldson and Kymlicka 2011 for more discussion).

However, keeping companion animals raises ethical issues that go beyond the moral significance of the animals themselves, the special obligations we (as individuals or as a society) may have toward them, and the ethical and legal frameworks through which they are understood. Companion animals also have much broader and (it's often argued) negative social and environmental impacts in terms of resource use, pollution, and impacts on wildlife. These impacts raise questions about how and whether we should keep animal companions—not because the companion animal relationship is intrinsically wrong, but because in practice companion animals have negative environmental impacts that we should attempt to reduce or eliminate.

Environmental Impacts of Companion Animals

In 2014, the *Guardian* published a story entitled "Are pets bad for the environment?" which made a series of claims about the negative impacts of companion animals on resource use and the environment. According to the author, Erik Assaourdian of the WorldWatch Institute, "Two German Shepherds use more resources just for their annual food needs than the average Bangladeshi uses each year in total." Here, we'll consider some key claims of this kind, focusing first on worries about companion animals and resources, in particular food resources; second, on pollution and waste from feces; and third, worries about the impacts of companion animal predation on wildlife.

Food Consumption

As the number of companion animals has grown, so, obviously, has their demand for food. As cats are obligate carnivores, and dog diets frequently contain high quantities of meat, companion animal meat and fish consumption has become an area of particular concern. This is both because companion animals might compete for food humans could eat, thus reducing total human food supply and/or accessibility to food for some human populations, and on the grounds that companion animal food may have significant negative impacts on the environment. (There are also concerns, of course, about the welfare of the animals used to produce cat and dog food, but we'll put these on one side here.)

Most cats and dogs in Western countries eat commercially produced food. The ingredients usually include animal products, water, binders and thickeners, preservatives, grains or starches, animal meals (see below), and species-specific nutrients (such as taurine in the case of cats). In *Time to eat the dog? The real guide to sustainable living*, Brenda Vale and Robert Vale (2009) tried to calculate the environmental cost of feeding pets of different sizes in terms of their "eco footprints"—the amount of land that would be needed to support them. They concluded, for instance, that a big dog like a German shepherd would need the equivalent of 1.1 acres of land a year to supply it with food, while a cat had a lower (but still fairly substantial) eco-footprint of 0.15 hectares. In a different set of calculations, Okin (2017) argues that in the US, dogs and cats annually consume as much dietary energy as approximately 62 million Americans, 1/5 of the US population.

However, as the Vales note in passing, these calculations may be somewhat misleading, as few of the meat products in commercial cat and dog food come from animals kept for the purpose of feeding companion animals. Commercial dog and cat food generally contains secondary meat from the human food chain, including remains from mechanical deboning (Nestle & Nesheim 2010); animal by-products, such as brains and beaks, which are not widely eaten in Western countries; and rendered animal meals. Rendered animal meals may include 4D meat (classified as being from animals that were dead, dying, diseased, or disabled at the time they were inspected), used cooking oil, and expired meat from retail sources, which are ground, heated, sterilized, and dehydrated to kill viruses and bacteria. Rendered protein meal provides 5–40% of the protein and fats in most commercial cat and dog foods (Aldrich 2013).

Many of these ingredients are not intended for human consumption, and few of them are normally part of the human food supply. So, they don't seem to raise direct concerns about competition or price pressure for human food resources, and they also raise fewer environmental concerns than if the meat was from animals kept for this purpose. In fact, The National Renderers Association (n.d.) in the USA claims that rendering performs an important environmental function by "recycling animals and inedible materials" into usable commodities—11 billion pounds of reclaimed animal fat annually.

However, this isn't the whole story. Fish-based dog and cat foods include not only fish by-products but whole fish, especially anchovies, herring, mackerel, and sardines (Nestle & Nesheim 2010). These pelagic fish are an important protein source in many developing countries, especially in Sub-Saharan Africa, and are also widely used in aquaculture. So, their use in animal feed may directly reduce availability to local fishing communities who rely on small pelagic fish for protein, and may also, indirectly, be a factor constraining the growth of globally farmed salmon production (DeSilva & Turchini 2008). In addition, while small- and medium-sized pelagic fish such as anchovy and mackerel have generally been regarded as having more sustainable populations than other large ocean fish (such as tuna), concerns have been expressed about overfishing, especially since the populations of these pelagic fish are both highly variable and unstable (Freon et al. 2005).

The growth in the market for premium dog and cat food also raises questions here (Okin 2017). Some owners don't want to feed their companions rendered meals, animal by-products, and other substances that they would not eat themselves, especially if they suspect that the ingredients threaten their animal companions' health. So instead they choose premium commercial dog and cat food, or make their own food, using meat and fish of a quality they would be willing to eat themselves.

How far this "premium" demand, and the use of whole fish, impacts on human food resources and the environment is very difficult to judge. In absolute terms, there is "enough food in the world today for everyone to have the nourishment necessary for a healthy and productive life" (WFP 2014). But there are still significant limitations in *access* to food, and even those with sufficient access to food can't always afford it. There is also a problem with wasting food, in production, transportation, and homes; over-feeding cats and dogs is an additional form of waste. Moreover, human populations are growing, and climate change may impact on future food security (Wheeler & von Braun 2013); keeping cattle for meat products increases greenhouse gas emissions, and so adds to climate-related problems. So even if there is no *absolute* shortage of food, some individuals and communities, now and in the future, will not have enough; and dog and cat food, especially when premium or hand-made, is likely to have an impact on global food supply and/or prices. Cat and dog food that includes high-quality meat rather than rendered meals may also exact higher environmental as well as human costs, although the environmental (but not the human) costs may be somewhat offset if the animals used for meat are kept organically.

Excretion

One area in which cats and dogs present significant environmental problems is in terms of excretion, though these concerns vary to some degree between cats and dogs.

Indoor cats present a problem because they need litter trays, and the production of most litters has negative environmental impacts. Traditionally, and still popularly, cat litter is made from strip-mined and processed bentonite clay or fuller's earth. Reclamation and re-vegetation of these strip-mined sites have been difficult to achieve (Schuman 1999). Silica for silica gel litter may also be mined in damaging ways; and even more recent "environmentally friendly" litters made from reclaimed sawdust, corn, wheat, and recycled paper require processing and transportation that have environmental costs.

Disposal of both cat and dog waste is a yet more significant problem. Cat feces in particular are a potential source of zoonotic disease. Most problematic is *Toxoplasma gondii*, which can infect humans, and may well also have played a role in fatal outbreaks of toxoplasmosis in threatened southern sea otters off California (Conrad et al. 2005). When cats defecate outdoors, toxoplasmosis remains in the soil and can be washed out in rainwater; so outdoor disposal of feces is not recommended. Toxoplasmosis is also not destroyed by normal sewage treatment. This means that both flushing and composting of indoor cat feces are environmentally problematic. The alternative is to seal it in plastic and send it to landfill. But this is, obviously, also environmentally problematic, and prevents even biodegradable litter from decomposing as intended. So, cat feces appear to pose an environmental choice between the potential spreading of zoonotic disease or the creation of a significant waste burden.

Dog feces does not carry *Toxoplasma gondii* (although it does carry worms) and for this reason can, in principle, be flushed or digested in a small dog waste disposal system. But in practice, many owners don't dispose of dog feces safely—or pick it up at all—creating both a health hazard and a public nuisance.

Cat and dog feces thus present some risks to human health, and in some cases also to the health of wildlife populations. These concerns alone, though, given that there are so many other vectors of disease, don't seem significant enough to suggest that we should no longer keep companion animals; in addition, the challenges posed by companion animal excretion may be amenable to technological solutions. In contrast, the final issue we'll consider here—predation—seems to be more environmentally significant, and much less amenable to technological solutions.

Companion Animal Predation

The most significant environmental problem widely associated with companion animals is predation on wildlife. There's considerable evidence to suggest that dogs can be effective predators; for instance, they harass and hunt rabbits, squirrels, deer, kit foxes, and wild turkeys in the USA (Young et al. 2011) and beach-nesting birds and koalas in Australia (Lunney et al. 2007; Williams et al. 2009). However, there are far fewer free-roaming owned or feral dogs than cats in Western industrialized societies, and cats are a much bigger predation concern. So, we will focus here on companion cats with outdoor access, an issue that has become highly controversial—even prompting the publication in 2016 of a book called *Cat Wars: The Devastating Consequences of a Cuddly Killer.*

Chasing, pouncing, and hunting behavior is normal for cats. One study in Georgia, USA, using "kitty cams" found that 44% of owned cats with outdoor access actively engaged in hunting

(Loyd et al. 2013). Cats' predatory behavior kills significant numbers of small mammals (in particular mice and rabbits), birds, and lizards (see, for instance, Woods, McDonald & Harris 2003) and causes alarm and disturbance to other wildlife.

The main ethical concerns raised by cat predation are, first, the suffering and death of individual wild animals, and, second, the potential impact on environmental values caused by cat predation in wild ecosystems and on threatened species. Ethical perspectives that place particular value on subjective animal experience are likely to view cat predation as significantly problematic because of the fear, pain, and suffering it causes (on some views, death may also be of concern, independently of suffering). From these perspectives, it's the negative experiences of prey animals, not the value of the species to which they belong that is directly ethically relevant; what matters is that cats increase suffering and death in the world. (It's sometimes argued, though, that cats mostly take the "doomed surplus" [Smith 2009]—wildlife that would not survive anyway—and thereby don't, in practice, actually increase total suffering.)

From more holistic perspectives in environmental ethics—ones that emphasize the value of species and of healthy, functioning ecosystems—cat predation may also be perceived as ethically problematic. While a cat catching a common grackle, for instance, may not be of much concern, stalking an endangered sandpiper would be highly problematic. Where cats have been introduced to oceanic islands with endemic and vulnerable populations, predation has had major effects (Nogales et al. 2004); this is also true where cats are located close to endangered species of ground-nesting birds or rodents. It's less clear how to assess cats' ecological effects elsewhere. Suburban environments, for instance, can be key stopping places for migratory bird populations vulnerable to cats (Jessup 2004; Lepczyk et al. 2010; Longcore, Rich, & Sullivan 2009). However, many of these places have already undergone major ecological changes to create the suburbs in the first place (Dickman 2009).

Cat predation may, thus, appear problematic for different reasons. But there are also potentially significant ethical concerns in keeping cats away from wildlife. If subjective experience matters ethically, as most people agree, the subjective experiences of cats—not just their prey—must matter as well. But the only effective way to keep owned cats away from wildlife is to confine them indoors, and this is likely to have significant experiential welfare impacts, at least in terms of depriving cats of positive experiences they otherwise would have had (see Palmer and Sandøe [2014] and Abbate [2019] for further discussion). And there are even fewer options for cats who lack homes and are unlikely to find them. The only way to separate unowned cats, especially feral cats, from wildlife is likely by killing them. This obviously raises a further set of complex ethical issues.

The predatory activity of cats—especially where threatened species are involved—raises extremely difficult ethical questions. It brings together—and into conflict—ethical concerns about subjective animal welfare, the value of animal lives, and more holistic environmental values in a context where many people have very deeply held attachments and commitments. Heated debate about this issue is unlikely to be resolved very soon.

In Conclusion

The widespread practice of keeping animals as companions generates a range of ethical questions and concerns, only some of which we have been able to tackle here. One main concern is that keeping animals as companions is fundamentally unethical, because it is based on a relationship of dependence and vulnerability. Another important concern is that companion animals have a negative impact on resource availability and the environment. Yet despite these very different ethical concerns, the practice of keeping companion animals is growing and globalizing: between 2002 and 2012, there was an estimated 34% average combined increase in dog and cat ownership in Russia, Mexico, the Philippines, China, and Brazil (USDA 2013). These significant increases are likely to exacerbate existing ethical concerns. And while citizenship proposals such as Donaldson and Kymlicka's may help address ethical worries about animal domination and animal property, they don't obviously ease (and may exacerbate) environmental concerns. Given all these conflicting factors, we predict that companion animals will be a growing area of research—and of concern—in environmental ethics in the near future.

Acknowledgments

We are grateful to Peter Sandøe and Sandra Corr for allowing us to draw on joint research carried out with Clare Palmer for *Companion Animal Ethics* (Wiley Blackwell, 2015).

Related Topics

1. Animal Cognition and Moral Status

References

Abbate, C. E. (2020) "A Defense of Free-Roaming Cats from a Hedonist Account of Animal Wellbeing," *Acta Analytica* 35: 439–461

Aldrich, G. (2013) *Rendered ingredients in pet food*. Presentation. Available from: https://d10k7k7mywg42z.cloudfront.net/assets/511a48238ad7ca3a92000ce9/Greg_Aldrich___Rendered_Ingredients_in_Pet_Foods.pdf

APPA (2020) "American Pet Spending Reaches Record Breaking High: 95.7 Billion." Available from: https://www.americanpetproducts.org/press_releasedetail.asp?id=205

Assaourdian, E. (2014) "Are Pets Bad for the Environment?" *The Guardian*, 1 May. Available from: http://www.theguardian.com/sustainable-business/reduce-pets-sustainable-future-cats-dogs

AVMA (2012) "U.S. Pet Ownership Statistics." Available from: https://www.avma.org/KB/Resources/Statistics/Pages/Market-research-statistics-US-pet-ownership.aspx

AVMA (2018) "AMVA Pet Ownership and Demographics Sourcebook 2017–2018: Executive Summary." Available from: https://www.avma.org/sites/default/files/resources/AVMA-Pet-Demographics-Executive-Summary.pdf

Burgess-Jackson, K. (1998) "Doing Right by our Animal Companions," *The Journal of Ethics* 2: 159–185.

Cochrane, A. (2012) *Animal Rights without Liberation: Applied Ethics and Human Obligations*, New York: Columbia University Press.

Conrad, P. A., Miller, M. A., Kreuder, C., James, E. R., Mazet, J., Dabritz, H. … & Grigg, M. E. (2005) "Transmission of Toxoplasma: Clues From the Study of Sea Otters as Sentinels of *Toxoplasma gondii* Flow into the Marine Environment," *International Journal of Parasitology* 35: 1155–1168.

DeSilva, S. S. & Turchini, G. M. (2008) "Towards Understanding the Impacts of the Pet Food Industry on World Fish and Seafood Supplies," *Journal of Agricultural and Environmental Ethics* 21: 459–467

Dickman, C. R. (2009) "Housecats as Predators in the Australian Environment: Impacts and Management," *Human-Wildlife Conflicts* 3: 41–48.

Donaldson, S. & Kymlicka, W. (2011) *Zoopolis*, Oxford: Oxford University Press.

Favre, D. (2010) "Living Property: A New Status for Animals within the Legal System," *Marquette Law Review* 93: 1021–1071.

Francione, G. (2012, July 31) *"Pets": The Inherent Problems of Domestication*, Animal Rights: The Abolitionist Approach. Available from: http://www.abolitionistapproach.com/pets-the-inherent-problems-of-domestication.

Francione, G. & Garner, R. (2010) *Animal Rights Debate: Abolition or Regulation?* New York: Columbia University Press.

Freon, P., Cury, P., Shannon, L. & Roy, C. (2005) "Sustainable Exploitation of Small Pelagic Fish Stocks Challenged by Environmental and Ecosystem Changes: A Review," *Bulletin of Marine Science* 76: 385–462

Jessup, D. A. (2004) "The Welfare of Feral Cats and Wildlife," *Journal of the American Veterinary Medical Association* 225: 1377–1383.

Korsgaard, C. M. (2013) "Kantian Ethics, Animals, and the Law," *Oxford Journal of Legal Studies* 33: 629–648.

Lepczyk, C. A., Dauphine, N., Bird, D. M., Conant, S., Cooer, R. J., Duffy, D. C., … & Temple, S. A. (2010) "What Conservation Biologists Can Do to Counter Trap-Neuter-Return: Response to Longcore et al.," *Conservation Biology* 24: 627–629.

Longcore, T., Rich, C. & Sullivan, L. M. (2009) "Critical Assessment of Claims Regarding Management of Feral Cats by Trap-Neuter-Return," *Conservation Biology* 23: 887–894.

Loyd, K., Hernandez, S., Carroll, J., Abernathy, K. & Marshall, G. (2013) "Quantifying Free-Roaming Domestic Cat Predation Using Animal Borne Video Cameras," *Biological Conservation* 160: 183–189.

Lunney, D., Gresser, S., O'Neill, L. E., Matthews, A. & Rhodes, J. (2007) "The Impact of Fire and Dogs on Koalas at Port Stephens, New South Wales, Using Population Viability Analysis," *Pacific Conservation Biology* 13: 189–201.

Marra, P. & Santella, C. (2016) *Cat Wars: The Devastating Consequences of a Cuddly Killer*, Princeton, NJ: Princeton University Press.

Milligan, T. (2009) "Dependent Companions," *Journal of Applied Philosophy* 26: 402–413.

Milligan, T. (2010) *Beyond Animal Rights*, New York: Continuum.

National Renderers Association (n.d.) *The Environmental Impact*. Available from: http://www.nationalrenderers.org/environmental/

Nestle, M. & Nesheim, M. C. (2010) *Feed Your Pet Right*, New York: Free Press.

Nogales, M., Martin, A., Tershy, B.R., Donlan, C.J., Veitch, R., Puerta, N., … & Alonso, J. (2004) "A Review of Feral Cat Eradication on Islands," *Conservation Biology* 18: 310–319.

Okin, G.S. (2017) "Environmental Impacts of Food Consumption by Dogs and Cats." *PLoS ONE* 12: e0181301.

Palmer, C. (2010) *Animal Ethics in Context*, New York: Columbia University Press.

Palmer, C. & Sandøe, P. (2014) "For their Own Good: Captive Cats and Routine Confinement," in L. Gruen (ed.) *Ethics of Captivity*, Oxford: Oxford University Press.

Sandøe, P., Corr, S., & Palmer, C. (2015). *Companion Animal Ethics*. Oxford: Wiley Blackwell.

Schuman, G. E. (1999) "Reclamation of Abandoned Bentonite Mined Lands," in H. Wong (ed.) *Remediation and Management of Degraded Lands*, Boca Raton, FL: CRC Press.

Serpell, J. & Paul, E. S. (2011) "Pets in the Family: An Evolutionary Perspective," in C. A. Salmon & T. K. Shackelford (eds.) *The Oxford Handbook of Evolutionary Family Psychology*, Oxford: Oxford University Press.

Smith, R. (2009) "Cats: Warm Furry Friends or Natural Born Killers?" Expert comment. University of Huddersfield. Available from: http://www.hud.ac.uk/sas/comment/display/?articleid=rhs090909

Smith, K. K. (2012) *Governing Animals*, New York: Oxford University Press.

Stone, C. (1972) *Should Trees Have Standing?: Law, Morality and the Environment*, New York: Oxford University Press.

USDA (2013) *Processed Product Spotlight: Pet Food*. Available from: http://www.fas.usda.gov/data/processed-product-spotlight-pet-food.

Vale, R. & Vale, B. (2009) *Time to Eat the Dog? The Real Guide to Sustainable Living*, London: Thames and Hudson.

Varner, G. (2002) "Pets, Companion Animals and Domesticated Partners," in D. Benatar (ed.) *Ethics for Everyday*, New York: McGraw Hill.

WFP (2014) *Hunger: Frequently Asked Questions (FAQs)*. Available from: http://www.wfp.org/hunger/faqs.

Wheeler, T. & von Braun, J. (2013) "Climate Change Impacts on Global Food Security," *Science* 34(1): 508–513.

Williams, K., Weston, M., Henry, S. & McGuire, G. (2009) "Birds and Beaches, Dogs and Leashes: Dog Owners' Sense of Obligation to Leash Dogs in Victoria Australia," *Human Dimensions of Wildlife* 14: 89–101.

Woods, M., McDonald, R.A. & Harris, S. (2003) "Predation of Wildlife by Domestic Cats *Felis catus* in Great Britain," *Mammal Review* 33: 174–188.

Young, J. K., Olson, K. A., Reading, R. P., Amgalanbaatar, S. & Berger, J. (2011) "Is Wildlife Going to the Dogs? Impacts of Feral and Free-roaming Dogs on Wildlife Populations," *BioScience* 61: 125–132.

5

SPECIES AND WILDLIFE

Lee Brann, Alexandra Laird, and Alexander Lee

Discerning our moral responsibilities toward wild animals is a critical component of modern environmental thought, environmental policy, and environmental controversy. Orientations toward both the wise use and preservation of nature play a prominent role in wildlife management values (Gamborg, Palmer, & Sandoe 2012). Ethical debates often pivot around what animals are worth, how we should protect them, how we understand them, and how wildlife fits into nature conservation. For instance, the first National Wildlife Refuge, established in 1903, set aside Pelican Island "for the protection of native birds" (Executive Order No. 1,014, 1903). Similarly, with the passing of the Endangered Species Act in 1973, the U.S. Government asserts that species are of "esthetic, ecological, educational, historical, recreational, and scientific value to the Nation and its people" (ESA, 16 U.S.C. § 1532 2(a)(3)). Protecting nature *for* wildlife and defending the value *of* wildlife are central to these hallmark conservation efforts.

Concerns within wildlife ethics often hinge on four common kinds of arguments:

1   A longstanding presumptive argument has motivated much of the wildlife conservation in North America: we shouldn't mess with wild animals *because of their wildness*.
2   Rooted in the Judeo-Christian tradition and many cultural norms around the world is an argument that we owe something to wildlife *as stewards*.
3   The Natural Law tradition and common ontological perspectives about nature have been used to argue that wildlife are necessary components or serve essential functions of something like ecological integrity – arguing, in other words, that there is a right answer to the question of *what goes where*.
4   Finally, aspects of the debate surrounding natural kinds have sought to clarify moral responsibility by delineating the *categories* into which wildlife fits.

Each of these debates plays out against a backdrop of broader concerns over commensurability, considerability (scope), scale, and applying ethical theory in a rapidly changing world.

Furthermore, novel challenges of wildlife conservation in the Anthropocene might suggest that more traditional epistemic, ontological, and aesthetic concerns no longer matter – after all, if unsullied, untrammeled nature is nowhere to be found (that is, if no part of nature is free from human influence), then we can just hang up our hats and quit worrying about whether to protect or conserve it (McKibben 1989). However, we see that novel wildlife conservation debates give rise to an even greater need for ethical engagement. Environmental ethics must evolve beyond traditional questions to provide continued guidance on contemporary debates. This evolution will

DOI: 10.4324/9781315768090-7

be less historical and less reductive, on the one hand, and more relational, on the other hand (as has long been a cornerstone of conservation biology; Soule 1985).

Chapter Road Map

In this chapter, we break down the traditional ethical concerns regarding wild animals and offer a look at wildlife debates in light of novel environmental and conservation challenges. Authors have argued that animal welfare, utility, stewardship, respect, and social agreement are five underlying ethical approaches to wildlife management (Gamborg, Palmer & Sandoe 2012). We proceed by first expanding upon these approaches to discuss concerns over scale and the nature of obligations regarding wildlife. We then go on to investigate common ethical debates that overlap with epistemic questions (what we know about wildlife), ontological questions (what wildlife is), and aesthetic questions (what wildness, charisma, and beauty in wildlife are). Finally, we discuss a few responses to novel conservation challenges that express the need for an integrated wildlife ethic.

Traditional Concerns

Wildlife Ethics vs Animal Ethics

While it is tempting to conflate the worlds of wildlife ethics and animal ethics – which may appear congruous – we should identify the areas where these disciplines diverge. Iconic voices of both discourses – in the mode of, say, Leopold for wildlife ethics and Singer for animal ethics – do suggest overlap in foundational perspectives (Leopold 1949; Singer 1975; Sagoff 1984). Both seek to expand moral scope beyond the human realm and ask what level of initiative humans must take to meet moral obligations. Beyond these shared concerns, however, wildlife ethics begins to follow its own course.

Animal ethics consists primarily of a set of debates about whether and why all non-human animals, as individuals, should be afforded the same, or similar, appreciation, rights, care, treatment, and respect afforded to humans – the right to life perhaps being the foremost consideration. Concern for the welfare and interests of individual animals derives primarily from those individuals' capacity to suffer, their *telos* (purpose), or their autonomy (see, for example, Singer 1975; Regan 1980; Taylor 1981).

Wildlife ethics is far less occupied with individual welfare. The values, obligations, and rights that receive notice in wildlife ethics are generally assigned to the collective. Of course, defining the collective remains a challenge for ethicists in the wildlife arena. The abstract nature of collective entities that wildlife ethics considers – the biotic community, ecosystem, species, and species populations – in turn, calls for a different, perhaps more nuanced, ethical orientation (Sagoff 1984).

Individualism vs Holism

Individualism is the view that only individual beings have interests; therefore, only they can be said to be moral patients. On such a view, we don't owe moral consideration to beings or systems that cannot be said to have interests of their own. Popular variants of utilitarianism and biocentrism offer individualistic conceptions of animal ethics that are often applied to wildlife in many conservation contexts. Peter Singer, for instance, argues for a principle of equal consideration of interests, whereby we should weigh the interests of all affected parties in our decision making equally, regardless of their species. Interests on Singer's view are restricted to the individual, because he argues for utilitarian moral consideration based on sentience (Singer 1975). Paul Taylor argues for a similar principle of egalitarianism, whereby all organisms have equal moral worth, but deviates from Singer's sentientism by arguing that all individual organisms are "teleological centers of life." On Taylor's view, any organism, by virtue of being alive, can be made better or worse off, and therefore has welfare interests as an individual (Taylor 1981).

Individualistic approaches present frameworks for thinking about harming, disturbing, transplanting, and helping individual wild animals. If a conservation measure could be said to harm wildlife, managers may be left to compare the burden of that harm on the individuals with the benefit to the ecosystem or species. If instead, individual obligations to wild animals – for instance, to leave them be – are not compatible with a conservation effort, then managers may be faced with teasing apart duties concerning individuals with those concerning populations or ecosystems.

The holistic tradition holds that obligations may be directed at, and values may be located in, the biotic community, ecosystems, or species. For example, the integrity, stability, and beauty of the biotic community could be said to be the measure of right and wrong actions, with the welfare of individual creatures as subordinate (i.e. Leopold's "Land Ethic," Leopold 1949). The Land Ethic and many methodological approaches to environmental ethics like Deep Ecology, ecofeminism, or primitivism take a holistic stance – arguing that our responsibilities do not lay only with individual animals of a certain kind, but within our relationships to ecological wholes (Leopold 1949; Naess 1988; Shrader-Frechette 1996). Some foundations for this perspective can be found in evolutionary biology, for instance, in Darwin's observations on ties of kinship (Darwin 1859). Such ethics are readily apparent in debates concerning endangered species, ecological restoration, and large-scale preservation of ecosystems.

Axiological vs Deontological Approaches

Much ink has been spilt defending wildlife as valuable in its own right. Commonly, the value of wildlife is assumed – that is to say, wildlife is seen as valuable because it is wild. Wildness on such a view offers a possible type of value (Cookson 2011; Rolston 2012), but the source of value differs according to the theorist. Similar views concerning ecological integrity may find value in certain ecological composition, or the functions of certain species. Orientations of this kind present the conservation of wild animals in mostly axiological or value-focused terms.

Alternative views look instead at obligations regarding the natural world. Deontological and virtue-based approaches to ethics typically look at what is owed to nature with respect to wild animals and what responsibilities of stewardship or care we may hold regarding wildlife (see, for example, Norton 1995; Palmer 2006). For example, do humans have dominion over nature, or perhaps a responsibility to be good shepherds of nature?

Regardless of scale or ethical approach, wildlife conservation as a distinct conservation pursuit in practice presumes that wild animals should be protected *because of their wildness*. This presumptive argument may rest on the purported rights of wild autonomous things, the unique values of "wildness," species or ecosystem-level relationships, or something to do with individual animals. This ethical debate, as such, defends the presumptive argument in a number of ways.

Today, we grapple with the applications of this diverse discourse in many wide-ranging and novel wildlife problems. When, for example, is it ethically permissible to sacrifice individuals to bolster the holistic good of the biotic community? This is the fundamental question surrounding debates on hunting for conservation. Furthermore, is there even a biologically coherent notion of community with sufficient substance on which to ground our ethical stance? If there are "whole" communities, at what temporal or spatial scales are they situated? What organisms are necessary for ecosystem integrity? Are some species more valuable than others? Do we have a responsibility not just to protect, but also to resurrect lost systems through practices like rewilding or de-extinction?

Ultimately, these traditional ethical concerns play out through conservation measures and policies. The aforementioned ethical considerations in many ways guide the way wildlife conservation unfolds in practice. Epistemic, ontological, and aesthetic debates highlight distinct and important concerns for wildlife within the theoretical landscape.

Common Debates

Epistemic Debates: What Goes Where?

Central to teasing out the ethics of wildlife conservation has been a quest to answer the question, "what goes where?" After all, a Rhino in the Savannah doesn't spur the same conservation interest as one in a zoo. The "what" refers to the entities that we tend to include within the broad definition of wildlife – species, species populations, population segments, and so on. The "where" draws our attention to geographic place and the conditions and characteristics of a given environment at a certain time where certain wildlife should reside. Concepts like historic habitat range, suitable habitat, ecosystem type, and ecosystem health enter the ethical discourse on wildlife as it attempts to justify approaches to wildlife conservation (Carroll et al. 2010).

Asking "what goes where?" often assumes that by gaining clearer knowledge of the historical presence, abundance, behavior, and role of wildlife at certain locales, we might gain a clearer and more precise inventory of the values and obligations worthy of consideration. This, in turn, can better guide wildlife conservation initiatives and policies that aim to preserve value or uphold obligations, or so the story goes.

The challenge for any such justification is to avoid a version of the so-called "naturalistic fallacy": the claim that a thing is good because of its natural properties (Moore 1903). Merely describing a state of nature does not capture the reasons why certain goals ought to be set. Similarly, conservation often "appeals to nature" in setting goals: it is tempting to think a thing is right because it is natural without establishing further normative foundation. Attempts to avoid these old philosophical adversaries do so by rooting value to history, ecological community, aesthetics, and the extrinsic goods associated with historical continuity (see, for instance, Leopold 1949; Elliot 1982; Rolston 1985).

The value potentially associated with wildlife is manifold. But the values commonly voiced in the ethics discourse to date reveal a strong preoccupation with the "where" question. Natural historic value – a hallmark of traditional biocentric or ecocentric wildlife ethics – recognizes the value of species when they continue to reside in their historic geographic domains. Values related to ecosystem composition and ecological integrity – two fundamental pursuits of conservation – similarly rest on native species remaining in place. For example, obligations associated with "endangered" status and protections within endangered species policy (like the U.S. Endangered Species Act) often center on the extent to which a species occupies a historic range.

Today, however, environmental upheaval and habitat change have reduced the opportunities we have to preserve place-based and historical value. Aligning wildlife ethics with our historic knowledge of natural environments is complicated by the rapidly shifting conditions of those very environments. Perhaps most notably, climate change has begun to reconfigure our conceptual approach to things like habitat range, suitable habitat, ecological function, and ecological integrity. Some have aimed to avoid the need for any ecological baselines in light of these novel concerns by shifting discourse away from pinning down particular historic values and toward a discourse on reasons for pursuing given conservation goals more broadly (Lee, Hermans, & Hale 2014).

As we are forced to revisit our factual understanding of the natural world in the face of novel environmental challenges, we must also reevaluate the place-based and historical ethical orientations that were derived from this increasingly challenged understanding.

Ontological Debates: Do Species Matter?

Other debates suggest that wildlife ethics are centered on the moral considerability and status of species. Some proponents of holism argue that species are not only intrinsically valuable, but are often *more* valuable than individuals, since the loss of a species is a loss of the speciating process that makes the emergence of individuals possible (Rolston 1975, 1989).

Species-centered arguments are everywhere – environmental laws and conservation actions often frame conservation success in relation to species-level effects, despite the ambiguity of the term "species" (ESA, 16 U.S.C. § 1531; Ferraro 2003; Murdoch et al. 2007). The "species problem" is a longstanding one: biologists use over 20 different definitions for "species" (Hey 2001) and philosophers have long disagreed on the ontological status of a species (see, for example, Dupre 1993; Boyd 1999a; Hull 1999; Hey 2001; Franklin 2007).

So, are species natural kinds? To say that a kind is *natural* is to say that it picks out groupings that accurately reflect the structure of the natural world – natural kinds exist independently of human beings and would continue to exist even if there had never been anyone to think about them. "Water" and "gold" are typical illustrations. Other categories are non-natural in the sense that they reflect the concerns and actions of human beings (think "art," "trash," etc.).

There seems to be a real and important difference between thinking that taxonomic species classifications (e.g. Homo sapiens, Tyrannosaurus rex, Larix occidentalis, and Erethizon dorsatum) refer to groupings with real existence in an objective world, independent of science, and thinking that species words are a mere tool for our own classificatory purposes. If we reject the objective reality of species, it becomes much more challenging to defend the moral standing of species beyond the aggregate standing of the individuals that make it up.

The other half of the species problem is that there is no agreed upon convention for delimiting species, so the same group of organisms might be classified in many different ways. For example, the biological species concept (BSC) defines a species as a group of organisms that can successfully interbreed and produce offspring, while the phylogenetic species concept (PSC) defines species as all and only those members descended from a common set of ancestors. These are just two among many prominent concepts in the biological literature, and this has serious practical implications. Our estimations of earth's biodiversity and the allocation of conservation resources rest in large part on the concept of species that is used, which often results in shifting and indefinite policies (Stamos 2003).

The case of the red wolf is a clear example of this problem. The red wolf (*Canis rufus*) used to be widespread in the southeastern and south-central United States. By the 1970s, they were on the brink of extinction due to hunting and habitat destruction. The US Fish and Wildlife Service managed to establish a captive breeding population, and these efforts were originally hailed as a successful application of the Endangered Species Act. However, new genetic evidence brought the species status of the red wolf into question: some have argued that the red wolf does not constitute a species based on any of the prevailing concepts. For example, it does not satisfy the criteria of the BSC since it has been known to breed extensively with gray wolves (*Canis lupus*) and coyotes (*Canis latrans*) (Stamos 2003). In spite of this, is the red wolf worth protecting? There are many species that *do* satisfy various species concept requirements and are also in need of protection.

Different species concepts produce different outcomes when we attempt to calculate biodiversity. For example, according to Mayr and Short (1970), from the viewpoint of the polytypic BSC, there are 607 species of breeding birds in North America, but according to Cracraft's species concept (1983), there are 922. So however you lump or split individuals into species categories makes a considerable practical difference from the viewpoint of endangered species laws. Perhaps a uniform biological convention would help avoid conflict, but it is not clear how such a convention would appropriately reflect deeper ontology.

Aesthetic Debates: What Is Wildness?

The aesthetic dimensions of wildlife conservation provide yet another area of debate, particularly concerning the nature of wildness and charisma. Wildlife can be beautiful; and interactions with wildlife are often governed by different norms than those concerning domestic populations. This could be seen to suggest that "wildness" is why wildlife garner special concern: perhaps we ought not to interfere with wild animals because of their wildness.

Some contend that wildness is a state of being – an attribute, quality, or location of certain things (Ridder 2007; Rolston 2012) – and others argue that it is a relationship – a lack of control (free, not artificial, or free from human influence) (Simonsen 1981; Cookson 2011). A common concern asks to what extent naturalness and wildness are related. If they are, how then might we maintain wildness if it requires the absence of interference (a non-artificial "natural" state). We discuss this debate further in our discussion of the Anthropocene.

A symmetrical debate concerns the prioritization in conservation of charismatic stuff. Charismatic megafauna often garner greater concern as mascots for conservation (Brambilla et al. 2013). Such mascots include the Panda logo for the World Wildlife Foundation or polar bear imagery in the climate discourse. Many, if not most, of our large coordinated conservation efforts prioritize alluring animals (for example, people tend to donate more when conservation is tied to a "charismatic" animal, even if offered a choice between a charismatic species and an endangered one; Colléony et al. 2017). As scientific concern over widespread marine mammal decline began in the 1960s, subsequent conservation efforts concerning marine mammals gained serious public support through more emotional channels. From *Flipper* to *Free Willy* to Shamu to *Whale Wars*, cetacean concern in the public has helped to motivate legislative change and fueled countless programs to "save the whales." Few outside of ecological circles, however, find similar resonance in calls to save the snails – although several species of snail are in fact endangered.

Some argue that aesthetics or charisma are facts of the matter, *prima facie* sources of value (Rolston 2002). Perhaps charismatic animals evoke certain emotions, take on a certain social relationship, are beautiful, useful, or simply more readily apparent. Teasing apart the pragmatic reasons, aesthetic reasons, and arbitrary reasons is a longstanding challenge of prioritization.

Contemporary Debates

Decline and Change of Common Wildlife

A precipitous population decline in several species of common wildlife – specifically historically abundant species yet to garner an endangered status – demands focus and consideration in wildlife ethics. A 2016 study by the World Wildlife Fund noted an average 58% drop in the overall population of vertebrate species since 1970 (Oerlemans 2016). While ethical debate thus far has been predominantly concerned with rare species prone to outright extinction, the dramatic loss of common wildlife may call for a recalibration of certain population thresholds that have so far guided the ethical conversation.

Perhaps the cardinal question that arises here is exactly when wildlife can be said to exhibit or bestow value. Just as importantly, when do our obligations to wildlife kick in, so to speak? Does wildlife become valuable only when it is on the verge of vanishing from the earth entirely? Are we obligated merely to preserve vestigial populations of once-abundant species? An ethics of wildlife conservation that considers these questions may very well identify value and obligation far ahead of what has been recognized by the ethics of rare species.

Along with the waning abundance of species in their historic domains, we are also witnessing a remarkable change in behavior, territorial occupation, and fitness of some species in seemingly unnatural circumstances (Lodge & Shrader-Frechette 2003). Some species, it could be said, are doing better as a result of human alterations to the natural world than they would have otherwise. The racoons that envelop urban and suburban environments, the coyotes that have expanded into diverse eastern territories, and the global introduction of invasive species serve as prime examples of this phenomenon (Marris 2011; Flores 2016). Some might view the thriving condition of these species as a token of wildness, which is among the goals of the old conservation vision. But is this new form of wildness a substitute for the kind of wildness embodied by native species thriving in their natural homes? Emerging debate between a new conservation vision that seeks to support and develop novel ecology and the old conservation vision that looks to historical goals highlights the ethical division within wildlife ethics in answering related questions (Doak et al. 2014).

Extinction

Most of the globe now faces species extinction and biodiversity loss far beyond expected background levels. Records of earth's past life indicate a constant process of introduction and extinction of species, but evidence suggests that we are losing species diversity at a rate not seen since the die off at the end of the Cretacious period 65 million years ago (Barnosky et al. 2011; Laurance et al. 2014; Ceballos et al. 2015). This alarming loss is fundamentally due to anthropogenic climate change. Since the 1960s, we have codified a responsibility to prevent species extinction (ESA, 16 U.S.C. § 1531). What then do we make of this contradiction? Are there moral differences between extinction via natural selection (characterized in Darwin's Principle of Extinction) and more recent cases of human-caused extinction?

The standard responses to the question of if and why species matter claim that we have obligations to threatened species because they have either extrinsic or intrinsic values – species matter because they contribute to some larger good, e.g. the ecosystem, or they matter in their own right. Other arguments claim that our obligations are not to the species level at all. Lily-Marlene Russow argues that a special obligation to endangered species is inconsistent with a condemnation of speciesism. If we want to remain committed to the position that simply being a member of a certain species is not itself a morally relevant fact, then our obligation to protect animals has to rest on the value of individuals. She argues that individual animals have varying degrees of aesthetic value (beauty, rarity, historical significance, etc.), and that we have a moral obligation to protect and ensure the continued existence of things of aesthetic value (Russow 1981).

At which point should we consider a species extinct? To some extent, the species is lost long before the last member dies – it is lost once we no longer have a viable population. In that case, is the loss of the last member of a species any more significant than the loss of the second to last member, or the last member before endangerment of the population, or even the loss of a random individual?

The Anthropocene

Decline of common wildlife and extinction are both contemporary challenges stemming from broader human impacts on the environment. Since the mid-1990s, some have claimed that humanity has become the dominant force shaping the global environment (Crutzen 2006). The idea is that we have entered a new geologic time period, distinct from the past (the "Holocene"), because anthropogenic climate change, resource use, and landscape change have driven global environmental shifts. Several critical debates have resulted from this view.

Some herald the Anthropocene as the "end of nature," since the idea implies that humanity, not nature, is now responsible for how the world unfolds (McKibben 1989). In such a world shaped by people, then, can we say any wildlife is really "wild" anymore? If so, it appears we have to look beyond historical views of naturalness to ground our idea of wildness (Lee, Amir, & Hale forth-coming). Conservation in the Anthropocene forces managers and policy-makers to investigate our norms and standards in light of novelty. One response to this new notion is to give up histori-cal approaches to goal setting, and take on responsibility for novelty as gardeners of our ecological communities and set goals for what we want out of nature (Marris 2011). Historic ranges, native species, and ecological integrity all struggle as concepts to offer a path toward meaningful conser-vation targets if the world as we know it is in some sense an artifact of humanity.

Conservation Responses

Even as wildlife ethics begins to reexamine the traditional paradigm, there remains broad adherence to the conservation movement's early vision. While this vision becomes harder to attain, given the new environmental realities, one response has been to call for even greater vigilance in preserving the place-based, natural, and historic value of wildlife. Perhaps only by redoubling our conservation efforts and raising conservation standards can we keep these hopes alive. Another set of responses, however, has been to use new technology, adaptive policy, and engagement in deliberate heavy-handed manage-ment. Wildlife ethics should engage both types of contemporary conservation responses.

The Half-Earth Project, a program inaugurated by E.O. Wilson aiming to place half of the planet under protection, illustrates a renewed commitment to conservation (Wilson 2016). Ef-forts by the US National Park System to reestablish historically abundant megafauna also reflect the ambition presumably required for restoring historic value (Ripple & Beschta 2003; Ripple et al. 2011). A more aggressive practice in wildlife ethics might also involve redirecting the focus of conservation from the species level to the population level, in other words protecting species long before they dwindle to tiny fragments of their historic abundance. Ethics and policy debates related to ESA implementation have become increasingly focused on this very point (Vucetich, Nelson & Phillips 2006; Waples et al. 2015).

Many in the conservation community, however, concede that these kinds of measures cannot always be implemented by poorly funded conservation programs facing constant political opposi-tion. In fact, managing to protect anything at all is a herculean task. The question then becomes what wildlife conservation goals do we prioritize and how do we prioritize them? This ubiquitous conundrum – prioritization – opens another ethical can of worms. But it is nevertheless an ines-capable part of deliberation among those who must take action on behalf of wildlife.

Decision makers grappling with prioritization apply any number of criteria to navigate this problem. One strategy is to direct sparse resources toward saving wildlife facing the most clear, scientifically supported, and certain threats. Alternatively, decision makers could endeavor to save only those species that show the greatest potential to be saved, thus getting the most bang for our conservation buck. Or, the overall degree of risk to a species' viability – scientifically verified or not – could get priority consideration.

Other innovative technological and novel responses to conservation challenges are gaining ground. Wildlife ethics will increasingly need to consider such approaches. A relatively new strat-egy attempting to deal with declining biodiversity is the gene drive. Engineered gene drives are a type of genetic modification designed to spread a chosen trait through a population faster than the traditional Mendelian rate. One approach uses CRISPR technology: CRISPR components are incorporated into a "drive allele" that cuts a specific site on a chromosome and replaces the wild-type DNA sequences with a modified part (including a copy of the driver).

Gene drives could potentially provide a fundamental tool in the fight against diseases, invasive predators, and pests. Future efforts to eradicate invasive species are likely to involve gene drives as a form of species-specific pest control: for example, they could help New Zealand in their quest to eliminate introduced mammalian predators by 2050 (Russell et al. 2015). Gene drives can spread disadvantageous traits, like sterility, so we might be able to use them to wipe out disease-carrying pests, like mosquitoes (Kyrou et al. 2018). Gene drives might also be able to increase genetic diversity in small populations (Wisely et al. 2015) and provide other beneficial traits to populations in response to acute threats (see, for example, Rohwer 2018). Despite possible benefits, genetically engineering populations involve intentionally expanding anthropogenic change, and might simply increase, rather than undo, human impact (Sandler 2019). The ethical landscape of this new technology is far from clear (Callies 2019).

Deep de-extinction offers another management-heavy response. Deep de-extinction refers to efforts to revive lost species. Cloning, genetic editing, and synthetic biology appear to offer a path to saving species that have gone extinct even in the distant past. Such efforts rely on cutting-edge technology to not just avert conservation disaster, but to revert historic loss. For example, we may one day soon see the Passenger Pigeon come back to life (Sherkow 2013). Rewilding involves returning historic species (or their proxies) to landscapes where they once roamed, offering a possible means to put wildness back into landscapes through wildlife restoration, reintroduction, or substitution (Donlan et al. 2006). For example, African cheetah's could take on the role of the lost American cheetah and provide an ecological proxy in the American savannah.

Possible arguments in favor of de-extinction and rewilding include restoring value, creating value, justice, and a last resort of conservation success (Sandler 2017). Such rationales center on the possibility of creating whatever it is that is worthwhile in wildlife. If historical composition, wildness, or some level of biodiversity has value, then de-extinction or rewilding could offer a path to that value. If the loss of species involved injustice, then de-extinction may offer a path to justice. If conservation requires "saving" wild animals, then de-extinction offers a possible last resort when all else has already failed. Common ethical concerns facing such efforts highlight problems of unnaturalness, ecological health, animal welfare, and hubris (Sandler 2014).

What if we can no longer preserve or reconstruct the ecosystems or species of the past? We are often unable to remove all non-native species from an ecosystem, and even if we could, some of those species have begun to play critical roles: for example, over one third of California's butterfly species now rely on non-native plants for some or all of their nectar sources (Graves & Shapiro 2003). Since many attempts at restoration fail to protect or revive the historical character of an ecosystem, some argue that the traditional conservation approach is no longer as effective, and needs to be replaced with new approaches based on the pervasiveness and inevitability of large-scale human-caused change (Kareiva 2010, 2014).

Novel ecosystems are ecosystems that come about via anthropogenic activities like species introductions, species extinctions, and land-use changes. Species occur in combinations and abundances that differ from what has occurred previously, by virtue of human influence. If an ecosystem is novel, no amount of management action is likely to return it to its historical state – there's been too much change. Some ecologists argue that novel ecosystems have a different set of management requirements and should not be treated as restoration priorities (Hobbs et al. 2006; Hobbs, Higgs, & Harris 2009). Generally, conservation aims to prevent, reduce, or reverse biotic and abiotic changes. A different strategy might include maximizing *beneficial* change – though what that entails clearly depends on how "benefit" is defined. "To keep every cog and wheel is the first precaution of intelligent tinkering" (Leopold 1949:190). We have to decide which management goals are appropriate: when should we focus on conserving the original parts (species) of an ecosystem, and when should we focus on maintaining overall ecosystem function and processes regardless of how far the makeup strays from a historical composition? As Emma Marris puts it, "Once you admit that you can't put things back the way they were, you often find yourself having to choose between goals that all sound pretty good…" (Marris 2011:153).

The traditional divisions between animal ethics and wildlife ethics show the need for a multifaceted approach to conservation. Animals cannot be fully divorced from wildlife ethics, but concerns regarding wildlife do not reduce only to concerns regarding animals (see, for instance, Fellenz 2010). Individualistic and holistic approaches offer orientations for understanding ethics operating on different scales. Furthermore, the value of wildlife and our obligations regarding wildlife act as a normative backdrop for conservation policy. These traditional questions are not a discourse on a path to resolution, but rather needed foci in an integrated wildlife ethic.

Animal rights and welfare, utility, stewardship, respect, and social agreement are all ethical orientations, dimensions, reasons, and perspectives that we need to guide wildlife management (Gamborg, Palmer, & Sandoe 2012). Traditional ethical concerns and common debates revolve around the differences between these perspectives. Where utility demands the inclusion of externalities, conservation becomes more comprehensive. Where respect for nature establishes a duty to prevent extinction, conservation becomes more comprehensive. Those differences, however, do not only cleave up possible philosophical truths, but also divide perspectives that together paint a more complete picture of wildlife conservation. Each facet offers a tool. Novel conservation responses are seeing these tools put to use.

Climate change and the uncertain futures that come with it present novel challenges like the baseline problem, substitution problem, and anthropogenic large-scale wildlife loss. An integrated ethic that addresses scale, pragmatic constraints, and a comprehensive understanding of moral obligations to wildlife can face these challenges. Novel policy and management solutions are already grappling with novel conservation problems with an eye for normative clarity and ethical engagement.

References

Barnosky, A. D., Matzke, N., Tomiya, S., Wogan, G. O. U., Swartz, B., Quental, T. B., Marshall, C., McGuire, J. L., Lindsey, E. L., Maguire, K. C., Mersey, B., and Ferrer, E. A. (2011) "Has the earth's sixth mass extinction already arrived?" *Nature* 471 51–57.

Boyd, R. (1999a) "Homeostasis, species, and higher taxa," in R. Wilson (ed.) *Species: New Interdisciplinary Essays,* Cambridge, MA: MIT Press

Brambilla, M., Gustin, M., and Celada, C. (2013) "Species appeal predicts conservation status," *Biological Conservation* 160 209–213.

Callies, D. E. (2019) "The ethical landscape of gene drive research," *Bioethics* 33 (9) 1091–1097.

Carroll, C., Vucetich, J. A., Nelson, M. P., Rohlf, D. J., and Phillips, M. K. (2010) "Geography and recovery under the US Endangered Species Act," *Conservation Biology* 24 (2) 395–403.

Ceballos, G., Ehrlich, P. R., Barnosky, A. D., García, A., Pringle, R. M., and Palmer, T. M. (2015) "Accelerated modern human-induced species losses: entering the sixth mass extinction," *Science Advances* 1 (5) e1400253.

Colléony, A., Clayton, S., Couvet, D., Saint Jalme, M., and Prévot, A. (2017) "Human preferences for species conservation: animal charisma trumps endangered status," *Biological Conservation* 206 263–269.

Cookson, L. J. (2011) "A definition for wildness," *Ecopsychology* 3 (3) 187–193.

Cracraft, J. (1983) "Species concepts and speciation analysis," in R. F. Johnston (ed.) *Current Ornithology*, vol. 1, Boston: Springer.

Crutzen, P. J. (2006) "The 'Anthropocene'" in *Earth System Science in the Anthropocene*, Springer, Berlin: Springer 13–18. https://link.springer.com/chapter/10.1007/3-540-26590-2_3

Darwin, C. (1859) *On the Origin of Species*, London: John Murray.

Doak, D. F., Bakker, V. J., Goldstein, B. E., and Hale, B. (2014) "What is the future of conservation?" *Trends in Ecology & Evolution* 29 (2) 77–81.

Donlan, C. J., Berger, J., Bock, C. E., Bock, J. H., Burney, D. A., Estes, J. A., Foreman, D., Martin, P. S., Roemer, G. W., Smith, F. A., and Soulé, M. E. (2006) "Pleistocene rewilding: An optimistic agenda for twenty-first century conservation," *The American Naturalist* 168 (5) 660–681.

Dupré, J. (1993) *The Disorder of Things: Metaphysical Foundations of the Disunity of Science*, Cambridge: Harvard University Press.

Elliot, R. (1982) "Faking nature," *Inquiry* 28 (1) 81–93.

ESA, Endangered Species Act of 1973, 16 U.S.C. § 1531–1544 (1973).

Exec. Order No. 1,014, 3 C. F. R. (1903).

Fellenz, M. R. (2010) *The Moral Menagerie: Philosophy and Animal Rights*, Chicago: University of Illinois Press.

Ferraro, P. J. (2003) "Assigning priority to environmental policy interventions in a heterogenous world" *Journal of Policy Analysis and Management* 22 (1) 27–43.

Flores, D. (2016) *Coyote America: A Natural and Supernatural History*, New York: Basic Books.

Franklin, L. (2007) "Bacteria, sex, and systematics," *Philosophy of Science* 74 69–95.

Gamborg, C., Palmer, C., and Sandoe, P. (2012) "Ethics of wildlife management and conservation: what should we try to protect?" *Nature Education Knowledge* 3 (10) 8.

Graves, S., and Shaprio, A. (2003) "Exotics as host plants of the California Butterfly Fauna," *Biological Conservation* 110 (3) 413–433.

Hey, J. (2001) "The mind of the species problem," *Trends in Ecology and Evolution* 16 326–329.

Hobbs, R., Arico, S., Aronson, J., Baron, J. S., Bridgewater, P., Cramer, V. A., Epstein, P. R., Ewel, J. J., Klink, C. A., Lugo, A. E., and Norton, D. (2006) "Novel ecosystems: theoretical and management aspects of the new ecological world order," *Global Ecology and Biogeography* 15 1–7.

Hobbs, R., Higgs, E., and Harris, J. (2009) "Novel ecosystems: implications for conservation and restoration," *Trends in Ecology and Evolution* 24 (11) 599–605.

Hull, D. L. (1999) "On the plurality of species: questioning the party line," in R. Wilson (ed) *Species: New Interdisciplinary Essays,* Cambridge, MA: MIT Press.

Kareiva, P. (2010) "Conservation science: Trade-in to trade-up," *Nature* 466 (7304) 322.

——— (2014) "New conservation: setting the record straight and finding common ground," *Conservation Biology* 28 (3) 634–636.

Kitcher, P. (1984) "Species," *Philosophy of Science* 51 308–333.

Kyrou, K., Hammond, A. M., Galizi, R., Kranjc, N., Burt, A., Beaghton, A. K., Nolan, T., and Crisanti, A. (2018) "A CRISPR-Cas9 gene drive targeting *doublesex* causes complete population suppression in caged *Anopheles gambiae* mosquitoes," *Nature Biotechnology* 36 1062–1066.

Laurance, W. F., Sayer, J., and Cassman, K. G. (2014) "Agricultural expansion and its impacts on tropical nature," *Trends in Ecology and Evolution,* 29 (2) 107–116.Lee, A., Amir, A., and Hale, B. (forthcoming) "Wildness without naturalness," *Ethics, Policy, and Environment.*

Lee, A., Hermans, A. P., and Hale, B. (2014) "Restoration, obligation, and the baseline problem," *Environmental ethics* 36 (2) 171–186.

Leopold, A. (1949) *A Sand County Almanac: and Sketches Here and There*, New York: Oxford University Press.

Lodge, D. M., and Shrader-Frechette, K. (2003) "Nonindigenous species: ecological explanation, environmental ethics, and public policy," *Conservation Biology* 17 (1) 31–37.

Marris, E. (2011) *Rambunctious Garden: Saving Nature in a Post-Wild World*, New York: Bloomsbury Publishing USA.

Mayr, E., and Short, L. L. (1970) *Species Taxa of North American Birds: A Contribution to Comparative Systematics,* Cambridge, MA: Nuttall Ornithological Club.

McKibben, B. (1989) *The End of Nature*, New York: Random House.

Moore, G. E. (1903) *Principia Ethica,* Cambridge: Cambridge University Press.

Murdoch, W., Polasky, S., Wilson, K. A., Possingham, H. P., Kareiva, P., and Shaw, R. (2007) "Maximizing return on investment in conservation," *Biological Conservation* 139 375–388

Naess, A. (1988) "Deep ecology and ultimate premises," *Ecologist* 18 128–131.

Norton, B. (1995) "Caring for nature: A broader look at animal stewardship," in B. Norton, M. Hutchins, T. Maple, and E. Stevens (eds) *Ethics on the Ark: Zoos, Animal Welfare and Wildlife Conservation*, Washington, DC: Smithsonian Institution Press.

Oerlemans, N., ed. (2016) *Living Planet Report 2016. Risk and Resilience in a New Era.* Gland, Switzerland: WWF International.

Palmer, C. (2006) "Stewardship: A case study in environmental ethics," in R. Berry (ed) *Environmental Stewardship: Critical Perspectives. Past and Present.* London: T&T Clark.

Regan, T. (1980) "Animal rights, human wrongs," *Environmental Ethics* 2 (2) 99–120.

Ridder, B. (2007) "The naturalness versus wildness debate: ambiguity, inconsistency, and unattainable objectivity," *Restoration Ecology* 15 (1) 8–12.

Ripple, W. J., and Beschta, R. L. (2003) "Wolf reintroduction, predation risk, and cottonwood recovery in yellowstone National Park," *Forest Ecology and Management* 184 299–313.

Ripple, W. J., Painter, L. E., Beschta, R. L., and Gates, C. C. (2011) "Wolves, Elk, Bison, and secondary trophic cascades in Yellowstone National Park," *The Open Ecology Journal* 3 (1) 31–37.

Rohwer, Y. (2018) "A duty to cognitively enhance animals," *Environmental Values* 27 (2) 137–158.

Rolston, H. (1975) "Is there an ecological ethic?" *Ethics* 85 (2) 93–109.

—— (1985) "Duties to endangered species," *BioScience* 35 (11) 718–726.

—— (1989) *Environmental Ethics: Duties and Values in the Natural World*, Philadelphia: Temple University Press.

—— (2002) "From beauty to duty: aesthetics of nature and environmental ethics," in A. Berleant (ed.) *Environment and the Arts: Perspectives on Environmental Aesthetics,* Hampshire, UK: Ashgate Publishing.

—— (2012) *A New Environmental Ethics: The Next Millennium for Life on Earth*, New York: Routledge.

Russell, J. C., Innes, J. G., Brown, P. H., and Byrom, A. E. (2015) "Predator-free New Zealand: conservation country," *BioScience* 65 (5) 520–525.

Russow, L. (1981) "Why do species matter?" *Environmental Ethics* 3 (2) 101–112.

Sagoff, M. (1984) "Animal liberation and environmental ethics: bad marriage, quick divorce," *Osgoode Hall Law Journal* 22 297–307.

Sandler, R. (2014) "The ethics of reviving long extinct species," *Conservation Biology* 28 (2) 354–360.

—— (2017) "De-extinction: costs, benefits and ethics," *Nature Ecology & Evolution* 1 (4) 0105.

—— (2019) "The ethics of genetic engineering and gene drives in conservation," *Conservation Biology*, August 9, 2019.

Sherkow, J. S., and Greely, H. T. (2013) "What if extinction is not forever?" *Science* 340 (6128) 32–33.

Shrader-Frechette, K. (1996) "Individualism, holism, and environmental ethics," *Ethics and the Environment* 1 (1) 55–69.

Simonsen, K. H. (1981) "The value of wildness," *Environmental Ethics* 3 (3) 259–263.

Singer, P. (1975) *Animal Liberation: A New Ethics for Our Treatment of Animals*, New York: Harper Collins.

Soulé, M. E. (1985) "What is conservation biology?" *BioScience* 35 (11) 727–734.

Stamos, D. N., (2003) *The Species Problem: Biological Species, Ontology, and the Metaphysics of Biology*, Lanham, MD: Lexington Books.

Taylor, P. (1981) "Respect for life," *Environmental Ethics* 3 197–218.

Vucetich, J. A., Nelson, M. P., and Phillips, M. K. (2006) "The normative dimension and legal meaning of endangered and recovery in the US Endangered Species Act," *Conservation Biology* 20 (5) 1383–1390.

Waples, R. S., Adams, P. B., Bohnsack, J. A., and Taylor, B. L. (2015) "When is a species at risk in 'all or a significant portion of its range'?" *Endangered Species Research* 27 (2) 189–192.

Wilson, E. O. (2016) *Half-Earth: Our Planet's Fight for Life*, London: Liveright Publishing.

Wisely, S., Ryder, O., Santymire, R., Engelhardt, J., and Novak, B. (2015) "A road map for 21st century genetic restoration: gene pool enrichment of the black-footed ferret," *Journal of Heredity* 106 581–592.

6

WILD ANIMALS

Jeff Sebo

Introduction

Wild animals occupy a strange place in ethical theory. There are many more wild animals than humans or domesticated animals, and many wild animals are much worse off than many humans and domesticated animals. Yet, very few academics and advocates work on wild animal ethics or advocacy. This is true even within animal and environmental ethics and advocacy. For the most part, animal ethicists and advocates focus on domesticated animals, such as companion animals and farmed animals. And, for the most part, environmental ethicists and advocates focus on species and ecosystems. As a result, both groups have tended to neglect wild animals, whose needs are importantly different from the needs of domesticated animals, as well as from the needs of species and ecosystems of which wild animals are part.

Fortunately, this trend is starting to change. Increasingly, people are starting to recognize that wild animals have moral standing, and that our treatment of wild animals matters morally. For consequentialists (who think that morality is about consequences), this tends to lead to the idea that we should improve wild animal welfare where possible. Wild animal suffering is a massive and neglected problem. Thus, reducing wild animal suffering should be a top moral priority for consequentialists, all else equal. However, the main question for consequentialists is whether wild animal suffering is *tractable*. Do we know and care enough about wild animals, and about interactions between animals, species, and ecosystems, to be able to intervene in wild animal suffering in effective, achievable, and sustainable ways?

For non-consequentialists (who think that morality is about more than consequences), this recognition of wild animal moral standing tends to lead in a different direction. Instead of thinking that we should improve wild animal welfare when possible, non-consequentialists tend to think that we should leave wild animals alone, out of respect for wild animal autonomy. However, the main question for non-consequentialists is whether the source, scale, or complexity of wild animal suffering changes this duty at all. For example, when human activity is the source of wild animal suffering, do we have a duty to intervene? And, when wild animals are suffering a lot, and we have the power to prevent that suffering without violating rights, cultivating vice, or cultivating oppression, do we have a duty to intervene?

I will start this chapter with a discussion of wild animal welfare. Which wild animals have welfare, how much do they have, and what do they enjoy and prefer? I will then discuss wild animals in the Anthropocene. How are humans impacting wild animals, both directly (through, say, destruction of natural habitats) and indirectly (through, say, human-caused climate change)? Finally,

DOI: 10.4324/9781315768090-8

I will consider the ethical implications for consequentialists and non-consequentialists. Throughout, I will try to motivate the idea that, even if consequentialists and non-consequentialists accept different approaches to wild animal ethics in theory, they should accept a similar approach in practice, according to which we should attempt to reduce human-caused wild animal suffering provided that we can do so respectfully and effectively.

Wild Animal Welfare

Wild animals are more than parts of a whole, like drops of water or grains of sand. They are thinking, feeling individuals, and what they need individually can differ from what they need collectively. Thus, the first step in developing wild animal ethics is learning more about what wild animals are like as individuals. For example, we need to ask: Which wild animals have welfare? How much welfare do they have? What makes things good or bad for them? What makes *life* good or bad for them? These questions are partly scientific and partly philosophical, since they require asking about what animal lives are like as well as about how to evaluate these lives. We need to make progress on all these questions to know how we should treat wild animals. With that in mind, consider what we currently know about some of these questions.

First, which wild animals have welfare? That is, which wild animals have lives that can be good or bad for them? Many people agree that, if animals have both sentience (i.e. the ability to feel pleasure and pain) and agency (i.e. the ability to set and pursue goals), then they have welfare. So, which animals have these features? We know at least this much. All vertebrates (i.e. animals with a central nervous system) have a strong chance of having both sentience and agency, given our evidence. Meanwhile, many invertebrates, such as mollusks, have at least a moderate chance of having both features, and other invertebrates, such as arthropods, have at least a weak chance of having both features, given our evidence. However, since we can never experience other minds directly, we might never know for sure which animals have these features (Sebo 2018).

Second, how much welfare do wild animals have, if they have any at all? Plausibly, some animals have a higher capacity for welfare than others, in that they can have more complex experiences and motivations. For example, we might think that humans have a higher capacity for welfare than mice and that mice have a higher capacity for welfare than ants. How can we tell how much capacity for welfare a particular kind of animal has in practice? Some people think that we should use physical proxies, such as neuron counts, to answer this question. For example, if humans have 40 times as many neurons as mice, then, we might think, humans have 40 times as much welfare as mice at any given time. But of course, these estimates are only as reliable as these proxies, and these proxies might not be very reliable (Budolfson and Spears 2020).

Third, what makes things good or bad for wild animals? To answer this question, we need to know what causes wild animals to experience pleasure and pain, satisfies and frustrates their desires, and, perhaps, affects their lives in other ways. For example, we know that many wild animals have interest not only in pleasure and the absence of pain, but also in family, friendship, mental and physical stimulation, and more. As a result, we might think that, in order to flourish in life, many wild animals require access to many of these goods (which is part of why captivity can be bad for animals). With that said, there are millions of species and enormous variation within and across species. So, while we know that many wild animals have complex needs, it would take decades, if not longer, to learn what those needs are (Bekoff 2008).

Fourth, do wild animals have lives worth living, on average or in total? Many people assume that the answer is yes. However, this is far from clear. Granted, many wild animals have lives that are full of pleasure and desire-satisfaction. But many also have lives that are full of pain and desire-frustration, as a result of hunger, thirst, illness, injury, predation, and other natural threats. Indeed, the vast majority of wild animals die within days of being born, because many species employ a reproductive strategy that involves having hundreds or thousands of babies and leaving them to fend for themselves (Horta 2010). So, as much as we want to think that wild animals have good lives in nature, we need to take seriously the possibility that many wild animals would be better off in captivity, and that some might even be better off not existing at all.

When asking and answering these questions, it is important to be mindful of the possibility of bias. On the one hand, excessive anthropomorphism – the tendency to see nonhumans as similar to humans – might lead us to underestimate how good nature can be for nonhumans. On the other hand, excessive anthropodenial – the tendency to see nonhumans as different from humans – might lead us to overestimate how good nature can be for nonhumans. Meanwhile, a refusal to ask these questions at all can lead us to maintain the status quo, where we either neglect wild animals or evaluate their lives intuitively rather than critically. As long as we are careful and humble when evaluating the lives of wild animals, it seems likely that the results of these efforts will be at the very least be better than this speciesist status quo (Andrews 2020).

As I have been emphasizing, we are still early in our understanding of wild animal welfare. However, we can draw two general conclusions now. First, there are *a lot* of wild animals with *a lot* of welfare in the world. There are billions of humans and domesticated animals at any given time. Meanwhile, there are quintillions of wild animals at any given time. Even if we assumed for the sake of argument that only, say, 0.1% of wild animals have welfare at all, and even if we also assumed for the sake of argument that these wild animals have only, say, 0.1% as much welfare as humans and domesticated animals on average, wild animals would still have more welfare in total. Plausibly, the sheer number of wild animals and amount of wild animal welfare in the world is enough to make wild animal ethics vitally important (Tomasik 2015).

Second, there are likely no universal answers about whether wild animals have lives worth living, and about what kind of environment is best for wild animals. Some wild animals might have lives worth living and others might not. And, some might flourish more in relatively captive environments (such as sanctuaries), whereas others might flourish more in relatively wild environments (such as reserves). Plausibly, many wild animals will flourish more in such "middle ground" environments than in fully captive or wild environments (such as a zoo or aquarium, on the one hand, and a state of nature without any human support at all, on the other hand), though even this is unclear. So, we can safely say that wild animal ethics is complex. The problems that wild animals face are simple and clear. But the solutions, if there are any, are anything but.

The Anthropocene

We are now living in the Anthropocene, a geological epoch where human activity is the dominant force on the planet. This complicates wild animal ethics in several ways. How is human activity impacting wild animals, individually and collectively? How will attempts to mitigate and adapt to human-caused environmental change impact wild animals, individually and collectively? How, if at all, can we effectively promote wild animal welfare as part of these efforts? More generally, how does the Anthropocene change the nature of wild animals, and of our relationships with wild animals? For example, in a world reshaped by human activity, are any animals fully wild, or are all animals at least partly captive and domesticated? Similarly, are any harms fully natural, or are all harms at least partly human-caused?

We know that human activity is killing countless wild animals directly. For example, we are killing trillions of animals per year through deforestation, development, and industrial fishing and agriculture. We are also killing millions of animals per year through the wildlife trade and collisions with buildings and vehicles. We know that human activity is killing countless animals per year indirectly as well. For example, human-caused climate change will result in melting ice caps, rising sea levels, an increase in ocean acidification, flooding coastal regions, an increase in the frequency and intensity of extreme weather events such as fires and floods, and an increase in conflict over land, water, and food. As a result, it will not only introduce new threats for animals but also amplify existing threats for wild animals, such as resource scarcity (Sebo 2022).

This activity is already causing extinction rates to rise. Historically, about one species per year went extinct. Currently, about 10,000 species per year are going extinct. By the end of the century, about 100,000 species per year could be going extinct, and we could have lost anywhere from one quarter to one half of all species. Of course, this biodiversity loss is harmful in many ways. Not only does it represent the loss of countless inherently valuable nonhuman lives, but it also disrupts natural cycles on which many human and nonhuman animals depend, and it disrupts natural systems that have profound aesthetic, cultural, and historical values for human and non-human communities. In light of these impacts, many people believe that preserving biodiversity should be a top moral priority for humans and nonhumans alike (De Vos et al 2014).

However, when we consider wild animal welfare, our assessments of human-caused environmental changes become more complex. First, for each population that decreases, another population might increase to take its place. Thus, less biodiversity might not mean fewer animals. Second, individuals and species can have different needs. For example, if wild animals have bad lives, then increased populations can benefit the species but harm the individuals (and vice versa for decreased populations). Moreover, even if wild animals have good lives, increased populations can have positive welfare impacts *in total* but negative welfare impacts *on average* (and vice versa for decreased populations). Thus, assessing our impacts on nature requires considering and evaluating our impacts on individuals, species, and ecosystems holistically.

At present, we know very little about how human activity is impacting individual wild animals. We can make guesses. For example, it might be that animals who are lower on the food chain will do better than animals who are higher on the food chain. It might be that r-strategists (animals with relatively small bodies and short lifespans) will do better than K-strategists (animals with relatively large bodies and long lifespans). It might be that adaptive generalists (animals who can survive in a wider range of environments) will do better than niche specialists (animals who can survive in a narrower range of environments). And so on. But these are only guesses. And even if these guesses are correct, we would need to know more about the welfare of these animals in order to know if these changes are good or bad for animals overall.

We also know very little about how our attempts to mitigate and adapt to human-caused environmental changes will impact individual wild animals. In particular, whether particular mitigation or adaptation strategies are good or bad for wild animals depends on the details of these strategies, as well as on whether the relevant environmental changes are good or bad for wild animals in the first place. If we consider the needs of humans and nonhumans holistically, we might be able to make at least some choices that are mutually beneficial. For example, if we install animal-friendly windows on cars and buildings and animal overpasses on roads and tracks, then we can reduce collisions with wild animals, thereby benefiting many. But even these changes can have unpredictable long-term effects, and many other choices will involve clear trade-offs.

Relatedly, we know very little about how to effectively improve wild animal welfare in general. Nature is staggeringly complex. For example, suppose that you save an animal from dying. Is it possible that you did more harm than good for this animal by subjecting them to an even worse death later on? Is it possible that you did more harm than good in general by turning other animals into a meal or depriving them of a meal? And of course, if the effects of saving *one* animal are uncertain, then the effects of saving *many* animals are even more uncertain. For example, suppose that we attempt to reduce predation through population control (e.g. providing contraceptives to carnivores) or genetic control (e.g. turning carnivores into herbivores). In this case, it would be *incredibly* difficult to predict and control the trophic cascades that might result.

More fundamentally, the Anthropocene might change the nature of wild animals, and of our relationships with them. Thirty years ago, Bill McKibben famously declared the end of nature (McKibben 1989). We might likewise declare the end of wildness. When human activity impacts everything, we might think that no animals are fully wild. Instead, we might think that all animals are at least partly captive (insofar as we impact their behavior) and at least partly domesticated (insofar as we impact their evolution). Similarly, when human activity impacts everything, we might think that no harms are fully natural. Instead, we might think that all harms, including harms associated with hunger, thirst, illness, injury, and predation, are at least partly human-caused. This raises challenging questions about complicity that we must now consider.

Consequentialism and Wild Animals

With that in mind, consider how consequentialists might think about wild animal ethics. Consequentialism holds that morality is entirely a matter of consequences. For example, according to utilitarianism (the kind of consequentialism that, for simplicity, I will focus on here), we are morally required to maximize aggregate well-being in the world. Some utilitarians interpret well-being in terms of pleasure and the absence of pain, and so they think that we should maximize net pleasure in the world. Others interpret well-being in terms of desire-satisfaction and the absence of desire-frustration, and so they think that we should maximize net desire-satisfaction in the world. While there are important differences between these interpretations, we can ignore them for present purposes. (See Singer 1975 for a characteristically utilitarian approach to animal ethics.)

At least in theory, consequentialism is highly demanding. It holds that we should maximize well-being not only for, say, our family, our nation, or our generation, but also for all sentient beings from now until the end of time. Thus, for example, consequentialism implies that you morally ought to devote yourself to the projects that allow you to do the most good possible whether or not you personally care about them. At least in theory, consequentialism is also not at all restrictive about how we maximize well-being. It holds that we should maximize well-being by any means necessary. Thus, for example, suppose that you have to choose between killing one person and letting five people die. Consequentialism implies that you morally ought to kill one person instead of letting five people die, all else equal.

Given these features of consequentialism, many people see consequentialism as requiring that we attempt to improve wild animal welfare. As we have seen, wild animal suffering is an important, neglected, and at least potentially tractable problem. Wild animals experience a lot of suffering and they receive very little support. Thus, consequentialism implies that we morally ought to intervene in wild animal suffering, assuming that we can do so effectively, whether or not human activity is responsible. It also implies that we morally ought to *prioritize* intervening in wild animal suffering over many other good causes. After all, if we can do more good by supporting wild animal welfare than by supporting other good causes, then we have a moral duty to do so, all else equal (Johannsen 2020).

Many people also see consequentialism as requiring that we attempt to improve wild animal welfare *by any means necessary*. For example, suppose that we can reduce wild animal suffering by killing carnivores to reduce suffering associated with predation, as well as by killing herbivores to reduce suffering associated with overpopulation and resource scarcity. In this case, consequentialism would imply that we morally ought to kill these animals, all else equal. Moreover, suppose that wild animals experience more suffering than happiness, and that nothing that we can do can change that fact. In that case, consequentialism might imply that we morally ought to, say, pave much more of nature than we already are, so that we can reduce the number of wild animals who have to suffer in future generations (John and Sebo 2020).

However, consequentialism is more complex in practice. First, many consequentialists believe that, even if we have a moral duty to maximize well-being *in theory*, it would be a mistake to always attempt to maximize well-being *in practice*. For example, you might find that you need to spend at least *some* of your spare time and money on self-care, in order to make altruism sustainable for you. If so, then consequentialism implies that you are not only morally permitted but morally required to spend at least some of your spare time and money on self-care. Granted, you might still need to spend more of your time and money on altruism than many of us do. Still, nobody is morally required to do the impossible. Thus, nobody is required to maintain a level of altruism that, given our motivational limitations, is not realistically possible.

Second, many consequentialists believe that, even if we have a moral duty to attempt to maximize well-being by any means necessary *in theory*, it would be a mistake to always do so *in practice*. First, even if killing one to save five is net positive in the short term, it might not be net positive in the long run, because of indirect effects that are difficult to predict or control. For example, the act of killing someone might make you less empathetic, which might cause you to harm people unnecessarily in the future. It might also make *observers* less empathetic, which might cause *them* to harm people unnecessarily in the future (Gruen 2014). Thus, given how valuable norms of respect and compassion are, consequentialism sets a relatively high bar for performing actions that might erode these norms in practice, both for the agent and for observers.

Given these complications, consequentialism has more nuanced implications for wild animal ethics than many people assume. First, we need to consider how much support for wild animals is achievable and sustainable in practice. For example, even if it would be ideal in theory for us to allocate 99% of social benefits to wild animals, we might find that we need to substantially limit our support for wild animals in order to make support for them possible in practice. If so, then consequentialism implies that we are not only morally permitted but morally required to substantially limit our support for wild animals, as we do with our support for other nations and future generations. Granted, we should support wild animals much more than we are. But we do not need to aspire to a level of support that would be clearly impossible.

Similarly, we need to consider not only the direct but also the indirect impacts of interventions such as killing wild animals for the greater good. Is it possible that these interventions might do more harm than good, by bringing about trophic cascades that are difficult to control, or by eroding norms of respect and compassion for wild animals? Given how complex these social and ecological dynamics are, we might not be able to answer these questions with much confidence. In that case, we might think that we should err on the side of caution by focusing on *non-invasive* support for wild animals (while being willing to take invasive action too, in cases where doing so is clearly necessary). This approach would balance the benefits of supporting wild animals with the benefits of promoting norms of respect and compassion for wild animals in practice.

Non-consequentialism and Wild Animals

Now consider how non-consequentialists might think about wild animal ethics. Non-consequentialism holds that morality is about more than consequences. For example, according to rights theory (a standard kind of non-consequentialism, and the kind that I will focus on here), we are not morally required to pursue any particular goal, much less the most good possible. Instead, we are morally permitted to set and pursue our own goals in life. However, we are morally required to set and pursue goals in ways that allow others to do the same. Thus, for example, you might not have a duty to devote your life to helping others. But you do have a duty not to harm others unnecessarily. And if you do harm others unnecessarily, then you should attempt to reduce or repair that harm. (See Regan 1983 for a characteristically rights theoretic approach to animal ethics.)

At least in theory, non-consequentialism tends to be less demanding than consequentialism. While consequentialist theories require us to do the most good possible, non-consequentialist theories permit us to pursue our own goals in life, provided that we follow rules, respect rights, cultivate virtues, cultivate relationships of care, and so on along the way. Similarly, at least in theory, non-consequentialist theories tend to be more restrictive than consequentialist theories. Even if our goals are altruistic, we are not morally permitted to pursue those goals by any means necessary. Instead, we are morally permitted to pursue them only in morally permissible ways. Thus, for instance, if you have to kill one to save five, most non-consequentialists would say that you are morally required to let the five die rather than kill the one.

Given these features of non-consequentialism, many people deny that non-consequentialism requires us to attempt to improve wild animal welfare. First, we are not morally required to intervene in wild animal suffering, since we are not required to intervene in suffering in general. Surely, we might be morally required to help others *sometimes*. But even if we are, we are morally permitted to choose when, whom, and how to help (within certain limits that I will mention in a moment). Thus, for example, if you *want* to donate to charities that promote wild animal welfare, then you can do that, provided that these charities respect the rights of human and nonhuman animals. But if you would rather spend your money by, say, collecting comic books and donating to your favorite human rights causes, then you can do that as well.

Many people also deny that non-consequentialism permits (much less requires) us to attempt to improve wild animal welfare *by any means necessary*. For example, even if killing some wild animals can reduce suffering associated with predation or overpopulation, it might still be morally wrong according to non-consequentialism. After all, we might think, killing these wild animals treats them merely as a means to further ends, and is therefore morally wrong. Similarly, even if wild animals experience more suffering than happiness overall, and nothing we can do can change that, paving nature might still be morally wrong according to non-consequentialism. After all, we might think, even if paving nature reduces wild animal suffering, it still interferes with wild animal autonomy, and is therefore morally wrong.

However, like consequentialism, non-consequentialism is more complex in practice, for at least two reasons. First, many non-consequentialists believe that we morally ought to attempt to reduce and repair harms that we cause. On this view, while we might not be required to address problems that we have nothing to do with, we *are* morally required to address problems that we have something to do with. For example, you might not be required to save people from drowning in ponds in general (though it would, of course, be nice for you to do so). But if you push someone into a pond (intentionally or accidentally), then you *are* required to save them. In this case, saving them is not a matter of your *preventing* something bad from happening. It is instead a matter of your *causing* something bad to happen, and then attempting to reduce or repair this harm.

Second, many non-consequentialists believe that there can be at least *some* cases where, say, harming or killing someone is morally permissible. For example, while we might not be permitted to kill one as a *means* to saving five, we might be permitted to kill one as an *unavoidable side effect* of saving five. We might also be morally permitted to kill someone in *self-defense*, *other-defense*, or *for their own sake*. Moreover, many non-consequentialists accept that there can be "harm thresholds" above which we can be permitted to kill people for the greater good. For instance, while we might not be permitted to kill one to save *five*, we might be permitted to kill one to save, say, 1,000, 100,000, or 1 million. We might hope that such cases are rare. But if and when they occur, we can make choices that would normally be wrong (Kagan 2019).

Given these complications, non-consequentialism has more nuanced implications for wild animal ethics than many people assume as well. First, we need to consider how much the Anthropocene changes our moral relationship with wild animals. For example, when deforestation harms wild animals directly (through violence) and indirectly (through habitat destruction), are we at all responsible for these harms? Similarly, when human-caused climate change contributes to fires and floods, and then fires and floods harm wild animals directly (by causing them to burn or drown) or indirectly (by destroying natural habitats), are we at all responsible for these harms? Insofar as we are, we might have a moral duty to help increasingly many wild animals after all, since helping them would be a matter of reducing or repairing increasingly systematic human-caused harms.

Similarly, we need to consider how the scale and complexity of wild animal suffering change our moral relationship with wild animals. For example, if we are already interfering with wild animal autonomy, then the question is not *whether* to interfere but rather *how* to interfere as respectfully as possible. Moreover, given how many wild animals there are and how complex their interactions are, we might find that we regularly need to kill some animals as an unavoidable side effect of saving others; kill animals in self-defense, other-defense, or for their own sake; or kill animals in order to save, say, 1,000, 100,000, or 1 million times as many. As much as we might like to think of such cases as exceptional, that might not be true in this area of ethics. In wild animal ethics, these normally "exceptional" tragic cases might, in fact, be the rule.

Conclusion

If this is right, then ethicists might be able to agree on wild animal ethics more than we might have thought. On the one hand, consequentialists might think that we should intervene in wild animal suffering by any means necessary in theory. But they might accept a less demanding, and more restricted, interventionist view in practice, given our cognitive and motivational limitations. On the other hand, non-consequentialists might think that we should *not* intervene in wild animal suffering in theory. But they might also accept a more demanding, and less restricted, interventionist view in practice, given the source, scale, and complexity of some wild animal suffering. If so, then perhaps we can at least agree on the moral importance of reducing human-caused wild animal suffering, provided that we can do so respectfully and effectively.

If we accept roughly this view, a good first step in treating wild animals better would be to reduce the activities that harm them, and to increase support for research and advocacy around wild animal welfare. This will give us the knowledge and political will that we need to act effectively when the time comes. We can also pursue moderate interventions that benefit humans and nonhumans alike. For example, we can vaccinate wild animals to reduce the spread of zoonotic diseases. And we can install animal-friendly windows on buildings and vehicles, and animal overpasses and underpasses on transportation systems. These interventions, if implemented thoughtfully, would help many wild animals in the short term, and might also build momentum around the idea of helping wild animals even more in the long run.

It is hard to say what our next steps should be, since it all depends on answers to questions that we should currently treat as open. But, as suggested above, we can make some general predictions. For example, I think that we will need to extend political standing to wild animals (Donaldson and Kymlicka 2013). I also think that we will need to allocate a substantial percentage of social benefits to wild animals (Sebo 2022). Relatedly, I think that we will need to develop many more spaces for wild animals, ranging from relatively captive spaces like sanctuaries to relatively wild spaces like reserves. And, I think that we will need to build many more accommodations for wild animals in our own communities, to the point that we stop regarding them as human communities and start regarding them as multi-species communities.

When we look even farther into the future, we might see wild animal ethics as important for a further reason. In the far future, we might share the world not only with wild animals but also with new kinds of beings, such as artificial intelligences. In the same way that there are many more wild animals than humans and domesticated animals, there might be many more artificial beings than biological beings. And, in the same way that wild animals are much more diverse than humans and domesticated animals, artificial beings might be much more diverse than biological beings. Given this possibility, wild animal ethics is not only an invitation to expand our moral and political circle now. It is also an invitation to expand our moral and political imaginations, so that we can be prepared to expand our moral and political circle even more in the future.

References

Andrews, Kristin (2020). *The Animal Mind*, 2nd edition. New York: Routledge.

Bekoff, Mark (2008). *The Emotional Lives of Animals: A Leading Scientist Explores Animal Joy, Sorrow, and Empathy — and Why They Matter*. Novato: New World Library.

Budolfson, Mark and Dean Spears (2020). "Public Policy, Consequentialism, the Environment, and Nonhuman Animals," in Douglas Portmore, ed., *The Oxford Handbook of Consequentialism*. New York: Oxford University Press.

De Vos, Jurriaan M., Lucas N. Joppa, John L. Gittleman, Patrick R. Stephens, and Stuart L. Pimm (2014). "Estimating the Normal Background Rate of Species Extinction," *Conservation Biology* 29(2): 452–462.

Donaldson, Sue and Will Kymlicka (2013). *Zoopolis: A Political Theory of Animal Rights*. Oxford: Oxford University Press.

Gruen, Lori (2014). "Dignity, Captivity, and the Ethics of Sight," in Lori Gruen, ed., *The Ethics of Captivity*. New York: Oxford University Press.

Horta, Oscar (2010). "Debunking the Idyllic View of Natural Processes: Population Dynamics and Suffering in the Wild," *Telos: Revista Iberoamericana de Estudios Utilitaristas* 17(1): 73–90.

Johannsen, Kyle (2020). *Wild Animal Ethics: The Moral and Political Problem of Wild Animal Suffering*. New York: Routledge.

John, Tyler and Jeff Sebo (2020). "Consequentialism and Nonhuman Animals," in Douglas Portmore, ed., *The Oxford Handbook of Consequentialism*. New York: Oxford University Press.

Kagan, Shelly (2019). *How to Count Animals, More or Less*. New York: Oxford University Press.

McKibben, Bill (1989). *The End of Nature*. New York: Random House Trade Paperbacks.

Regan, Tom (1983). *The Case for Animal Rights*. Berkeley: The University of California Press.

Sebo, Jeff (2018). "The Moral Problem of Other Minds," *The Harvard Review of Philosophy* 25: 51–70.

Sebo, Jeff (2022). *Saving Animals, Saving Ourselves: Why animals matter for pandemics, climate change, and other catastrophes*. New York: Oxford University Press.

Singer, Peter (1975). *Animal Liberation*. New York: HarperCollins.

Tomasik, Brian (2015). "The Importance of Wild Animal Suffering," *Relations* 3(2): 133–152.

7

HUNTING

Nathan Kowalsky

And what is it to say goodbye to the swift pony and then the hunt? The end of living and the beginning of survival.

— *attributed to Chief Seattle*

The success of the North American Wildlife Conservation Model is attributed to hunting which was regulated to conserve game animal populations and has resulted in healthier herds than what existed a century ago (Heffelfinger et al. 2013). However, many wildlife management agencies are concerned by declining numbers of hunters, which makes it more difficult to keep herd numbers within the landscape's carrying capacity and minimize conflict with agriculture and dense human settlement. At the same time, hunting is similar to climate change in that it is one of the few environmental policy issues guaranteed to provoke strong opinions from just about anybody. Surely there are better tools for managing wildlife populations in the 21st century than the recreational killing of wild animals to eat their flesh or gain a trophy (Batavia et al. 2019)! Such practices are widely condemned by critics as barbaric blood sports out of place in any enlightened and progressive society.

However, one segment of the population that is seeing growth in hunting is, perhaps surprisingly, the gourmet food and especially local food movements (Cerulli 2012). These so-called "foodies" and "locavores" seek to both increase dietary quality by relying less on industrial agriculture and decrease environmental impact by pursuing locally and seasonally available foods (Pollan 2006). Seen in this way, hunting can be a way of being "mindful" of dietary consequences, confronting those consequences with "eyes open," while actually constituting "the path of least harm" (Cerulli 2010: 46; cf. Cahoone 2009). Rather than being morally odious, hunting might actually assist contemporary urban dwellers in establishing "a different relationship with the land, a sense of belonging, a kind of communion" (Cerulli 2010: 51).

Marc Bekoff has very harsh words for those who articulate this vision of ethical hunting, calling them "born-again meat-eating zealots." He claims that simply stalking animals "causes immense suffering," and that the "family and friends" of the hunted animal suffer from its loss as well. Bekoff adds, "Even if there were no suffering what gives [the hunter] the right to kill an animal for a thoroughly unnecessary meal?" Hunting involves

DOI: 10.4324/9781315768090-9

human exceptionalism and speciesism in which we conveniently place ourselves above other animals in importance, and in which animals are valued only for what they can do for us (instrumental or use value) but not for their own inherent or intrinsic value.

(Bekoff 2012)

Rather than being a valuable tool for ecosystem management which contributes positively to both personal responsibility and an embodied relationship with the land, hunting is excoriated because it inflicts death and pain onto individual animals, violates their right to life for trivial or unnecessary reasons, and embodies an anthropocentric refusal to recognize their intrinsic value.

Hunting illustrates, therefore, many of the theoretical debates that characterized the first 30 years of environmental ethics as an academic discipline. From animal liberation, nonanthropocentrism, and intrinsic natural value, through to nature-cultural dualism, natural disvalues, and ecofeminism, environmental ethics' various schools of thought underlie and shape real-world policy decisions about hunting. One of the tasks of environmental ethics is to assist in practical environmental decision making by uncovering what is at stake morally in a given policy issue. This chapter will proceed, therefore, by charting a path through the moral and socio-political issues raised by hunting. The first half will consider hunting in light of ethical theory, first with respect to sustainable anthropocentrism, second with respect to animal protectionism, and third with respect to ecological nonanthropocentrism. The second half of the chapter will explore how these aforementioned ethical issues are informed to a large extent by broader philosophical issues concerning the role of humanity within nature, landscape hermeneutics, the philosophies of sport, technology and biology, and the historical-cultural context of contemporary hunting. This chapter will not aim to decisively resolve these issues, but rather suggest what must be considered and pursued further by both researchers and policy makers who are concerned with the issue of hunting.

Sustainability, Conservation Management, and Anthropocentrism

There is no question that hunters value their prey animals instrumentally, which means that the animal is valued for the benefits it provides people: e.g. meat, hides, or trophies. As with clean water or air, it is important to maintain healthy populations of wild animals if their instrumental goods are to be reliably accessed by future generations who may also value them. Therefore, many, if not most, of the defenses of hunting claim it as ethical when it sustainably harvests wild animals and by extension, conserves the natural resource they comprise. For example, tags for hunting mule deer in Alberta (where I hunt) are limited and distributed by lottery to ensure that herd numbers will remain high enough to allow hunting to continue into the indefinite future. Conservation is also extended to the habitats that are necessary for prey species to flourish, which indirectly benefits many non-prey species that also depend on these habitats. For example, the non-profit waterfowl conservation organization Ducks Unlimited champions wetland habitat conservation so that there will be healthy populations of huntable waterfowl, but other species also benefit from this, from native plants to trumpeter swans. Furthermore, hunting is defended because of the instrumentally valuable economic benefits it brings to a variety of stakeholders, including hunting guides, outfitters, hoteliers, and any agencies or landowners that may be permitted to collect fees from hunters.

One of the benefits of exclusively evaluating hunting in terms of instrumental value is that this method appears to be relatively theory-neutral; everybody can agree that hunting should not drive prey species or their habitats to collapse. If hunting can be adequately assessed in terms of "important management principles" (Norton 1995: 344), there would be no need for "exotic appeals to hitherto unnoticed inherent values in nature" which are highly controversial and unlikely to influence policy reform (Norton 1995: 356). For example, while Robert Loftin (1984) and Gary Varner (1995) both find the killing and eating of wild animals to be morally wrong, they think that banning hunting would be a greater evil, as the ecosystems upon which wild animals depend for their welfare would be harmed by depriving environmental management agencies the financial support provided by hunting licenses and their most effective tool for managing species that would otherwise exceed the habitat's carrying capacity (cf. Dickman et al. 2019). Thus, pragmatists think that it should be possible to reach convergence on certain hunting policies regardless of whether one is opposed in principle to the procurement of wild animal bodies.

For Varner, however, this convergence is temporary. Given current technological and economic means, hunting may be the most efficient management tool to maintain both the general welfare of wild megafauna and the ecosystems on which they depend. However, if alternative funding models and artificial contraception methods become more viable, hunting will no longer be necessary (Varner 1995). But *should* such alternative methods be pursued? Technologically and economically useful or pragmatic means are always in service of value-laden ends. If hunting is a morally acceptable policy, then researching effective forms of wild animal contraception would be superfluous. If hunting is morally unacceptable, then such research is imperative. However, if the technological transformation of wild nonhuman reproductive processes is morally unacceptable, then such research is unacceptable. Because efficiency can only determine the most optimal way of achieving a given end, and not whether an end is morally acceptable (Ellul 1980), it cannot provide guidance on these questions. Therefore, beneath the pragmatic question of "what is the most effective tool of wildlife management?" is a deeper philosophical question of how humans *ought* to relate to wild nonhuman animals.

Because policy decisions about hunting are never theory-neutral, an *exclusive* focus on sustainable resource conservation or financial benefit can be morally problematic because the species and ecosystems at issue are managed to satisfy exclusively human goals (i.e. sufficient resources for humans to use later) rather than *also* taking into account the intrinsic value animals or natural systems may (or may not) have in their own right. This is why human-centered sustainability and conservation ethics are accused of being *anthropocentric* (Batavia et al. 2019). Moral anthropocentrism is perhaps the most controversial topic in environmental ethics, and is notoriously ill-defined. It is best understood as shorthand for "human chauvinism" (Midgley 1994; Routley 1973), or the idea that we humans are the only (or the most) valuable species in the universe. Such anthropocentrism has been identified as the primary cause of contemporary environmental problems (White 1967). Anthropocentrists do not think that they must comport themselves morally toward animals or nature directly; the only ethical consideration within an anthropocentric framework is how something might benefit or harm human beings, who exclusively possess intrinsic value. Therefore, an anthropocentrist would conserve wetland habitats or waterfowl only to the extent that human beings valued them; if there was another way to achieve human flourishing without protecting the environment, then destroying it would be a morally viable option.

Therefore, the environmental ethics of hunting cannot be reduced simply to a question of maintaining wildlife populations and ecosystem services sufficient for satisfying human needs or wants. Hunting must also be considered in light of the possibility that wild animals and their ecologies have intrinsic value, or value which is not simply reducible to what humans consider valuable for themselves. Can hunting be morally *nonanthropocentric*?

Extensionism, Pluralism, and Nonanthropocentrism

Ethical nonanthropocentrism denies the moral primacy of human beings and asserts the direct moral considerability of a wide variety of nonhumans in a variety of often-conflicting ways. As we shall see, this variety is of particular importance for hunting. One version of nonanthropocentrism is *moral extensionism*, which attempts to negate anthropocentrism by extending the sphere of human moral considerability to beings psychologically similar to humans, primarily sentient animals. As Jeremy Bentham famously said, "The question is not, 'Can they reason?' nor 'Can they talk?' but 'Can they suffer?'" (1823: 236). If humans are directly morally considerable on account of their ability to experience pleasure and suffer pain, then nonhumans who share an analogous ability should also be directly morally considerable and treated "humanely." This extension of humane sensibilities to animals developed into explicitly utilitarian animal welfare ethics (Singer 1975) as well as deontological animal rights ethics (Regan 1983), the latter of which focused on animals who can take an interest in the continuation of their lives rather than merely possessing the capacity for suffering. Thus understood, nonanthropocentric ethics consists in either beneficially contributing to animal welfare by minimizing especially unnecessary suffering, or by respecting at least "higher" animals' right to life.

All else being equal, hunting runs afoul of both forms of humane ethics. In terms of welfare, hunters rarely if ever kill their prey painlessly (although they do try to minimize it) and, if the animal is merely wounded, its suffering can be great and prolonged. Moreover, in the context of the industrialized world, there are almost no cases where the vital interests of the hunter's survival cannot be met by nonlethal means. Therefore, hunting almost always unnecessarily and painfully violates the right to life of another animal. So even if hunting sustainably conserves wildlife populations, it is not necessarily moral. If hunting is necessary for the management of wildlife populations, it is only the lesser of two evils and, as noted above, should be replaced as soon as equally effective nonpainful and nonlethal methods become available.

However, humane ethics extend a particularly individualistic sort of interhuman ethics onto nonhuman animals that may not be adequate for morally appreciating the various aspects of the environment itself. Indeed, it is a little known fact that humane ethics have been at loggerheads with environmental ethics almost since its inception as an academic sub-discipline (Hargrove 1992). In his magisterial study of Singer's *Animal Liberation* (1975) and Christopher Stone's *Should Trees Have Standing?* (1974), John Rodman (1977) pointed out that nature, rather than respecting the right to life and freedom from harm, systematically and uncaringly violates those rights (cf. Causey 1989). Relying in part on Rodman, J. Baird Callicott honed in on the utilitarian definition of evil as suffering and boldly declared

> in all soberness I see nothing wrong with pain.... A living mammal which experienced no pain would be one which had a lethal dysfunction of the nervous system. The idea that pain is evil and ought to be minimized or eliminated is as primitive a notion as that of a tyrant who puts to death messengers bearing bad news on the supposition that thus his well-being and security is improved.... If nature as a whole is good, then pain and death are also good.
>
> *(1980: 333)*

Callicott was subsequently excoriated for his anti-individualistic and non-utilitarian nonanthropocentrism, and (in)famously accused of "environmental fascism" (Regan 1983: 362).

Regardless of the vitriol, environmental or ecological nonanthropocentrism is far more tolerant of certain types of suffering and death than humane extensionist nonanthropocentrism, and thus more open to the possibility of ethical hunting. Holmes Rolston, III recognizes that from the perspective of the individual prey animal, being hunted is unquestionably value-negative. Nevertheless, he argues that it is a mistake to *remain* within the individual perspective and not consider higher levels of evaluation. Failing to look at animals in ecological relation constitutes moral myopia; treating the local, short-range perspectives of particular animals as if they were long-range or ultimate is "an overextension or aberration of the self-impulse" (Rolston 1992: 257), which is what human chauvinism is. He argues that "[j]udgements from the perspective of any prey [are] made from a human perspective" (1992: 253), specifically that of individualistic interhuman morality and sympathy with physiologically anthropomorphic animals, and that "the human experience of suffering must not be projected indiscriminately onto the animal world" (1992: 272).

Therefore, in contradistinction to humane ethics, environmental ethics tend toward nonanthropocentric recognitions of a multiplicity of values in nature. For Rolston, there is a species-specific variety of value systems in nature of which interhuman ethics is but one among many:

> the appropriate evaluative category is not nature's moral goodness, for there are no moral agents in nature. The appropriate category is one or more kinds of nonmoral goodness, better called nature's value. Such value is not to be mapped by projection from culture, much less from human moral systems within culture.
>
> *(1992: 252)*

Similarly, there may be a plurality of ethics for the treatment of animals depending on the community context they inhabit (Callicott 1989). The ethical way to treat a pet dog, for instance, will not necessarily be the same as the way we ought to treat a sled dog or a wild canid. Pluralistic nonanthropocentrism opens up the possibility that hunting may be an ethical way to treat certain wild animals, although it remains to be seen how that might be so.

Participatory Affirmation and Intrinsic Value

Aldo Leopold's land ethic famously states, "A thing is right when it tends to preserve the integrity, stability, and beauty of the biotic community. It is wrong when it tends otherwise" (1949: 224–225). Environmental ethics may positively value hunting (as did Leopold himself) if it fits into the biotic community (List 1997). One constitutive aspect of the biotic community's overarching system of value is carnivorous *predation*, an ecosystem process fundamental to "the world we have [which], in its general character, is the only world logically and empirically possible under the natural givens on Earth..." (Rolston 1992: 275). As such, Rolston suggests that predation and the value-laden ecological systems of which it is an inseparable part *ought to be*. Chaone Mallory isn't so sure, pointing out that "[human] males exhibiting the most egregious behavior are referred to as 'sexual predators.'" So even if hunting is a form of predation that fits into the biotic community, that doesn't mean predation is a good thing: "the problem with such a means of...'communing' with nature...is that it does so at the expense of another creature, a creature who undeniably has an interest in remaining alive" (2001: 79, 65). Nevertheless, Rolston argues that forms of hunting which provide experiential, embodied, or practical knowledge of predation constitute a physical symbol of the positive value of this ecosystemic process (he uses the religious term "sacrament" as shorthand for this; cf. 1988: 91). In Ned Hettinger's (1994) terms, hunting constitutes *participatory affirmation* of predation. Hunting, therefore, is a tacit avowal of the goodness of predation in the biotic community.

This environmental defense of hunting is a variation of the so-called predation problem posed to vegetarians (which says that because nonhuman carnivores naturally eat meat, it is acceptable for humans to eat meat too), for here it seems that hunting is acceptable for humans because nonhuman predators hunt too. Critics object, saying that nature is not a good model for human behavior – "is" does not equal "ought" – and that Social Darwinism is proof of this flaw in moral reasoning (cf. Kretz 2010). But Rolston (1992) is careful to point out that Social Darwinism categorically mistakes interhuman relations for interspecies relations, whereas hunting, as an interspecific relation, commits no such category mistake. What Rolston calls *following nature* in an "axiological" rather than "imitative ethical" sense means to analogously rather than literally mimic general forms of environmentally valuable natural behavior (Rolston 1988: 38–41). Moreover, I would add that participating in predation is also an appropriate species-specific behavior for humans because unlike coprophagy or brood parasitism, predation is a part of the human evolutionary inheritance, an inseparable part of our natural history, and thus something not likely to cause personal or social harm to humans when practiced by them.

With these refinements in place, environmental ethics can reverse the onus on anti-hunting: if predation is an inextricable part of ecosystemic value but hunting by humans is morally censured by humane ethics, then anti-hunting appears *opposed* to environmental ethics. Humane ethicists will need to show how censuring hunting by humans does not entail repudiating predation by nonhumans and thus entail an anti-environmental "hatred of nature" (Hettinger 1994: 20). Singer responds to the predation problem by arguing that, unlike humans, animals are not moral agents and thus cannot be held responsible for their actions – as much as "we may regret that this is the way the world is" (1975: 237). Jennifer Everett believes that it is natural for humans to be so concerned about the suffering of prey animals that humans will naturally feel "foreign to the world" and "alien to wild nature." Anti-hunters thus "respect [their own] nature" when they "lament" nature red in tooth and claw (2001: 61). Other critics suggest that humans should participate in less lethal natural processes, like berry picking (Moriarty & Woods 1997), or with less lethal methods, like wildlife photography (Luke 1997). But these alternatives to hunting still betray dissatisfaction with predation because they attempt to avoid "bloodshed in nature" by refusing to "wade right in and *add* to it" (Luke 1997: 41). A case can be made, therefore, that anti-hunting does in fact struggle to appreciate "what is actually the case in the [natural] world as compared to what we might wish to be the case" (Frasz 2009: 223).

Although hunting may sacramentally participate in intrinsically valuable ecosystemic processes, it can still appear disrespectful of the intrinsic value of individual animals within the biotic community. To value something intrinsically is to value it for its own sake, not simply for the instrumental benefits it may provide the valuer. That is, the valuer recognizes in the thing a value which is independent of whether that thing is also valuable *for* something else. The paradigmatic conception of intrinsic value comes from Immanuel Kant, who argued that rational beings (i.e. humans) are ends-in-themselves, and should not therefore be treated *only* as means to an end. Rolston argues that nature itself and all of its parts have their own intrinsic value, not just human beings, and that this intrinsic value exists objectively, even if some human beings do not always recognize it as being there.

Like Kant, however, Rolston recognizes that intrinsic value is not exclusive of instrumental valuing. Just because a human being is an end-in-itself does not mean that he or she cannot *also* be used as a means to an end. When we use other people as means, say in romantic or economic relationships, we must *simultaneously* respect their intrinsic value, dignity, or worth as human beings. Similarly, the intrinsic value of nature is not incompatible with the instrumental use of nature. Rolston argues that what one organism instrumentally values in another is what the other organism intrinsically values for itself. Ecology is, in fact, an economy, a system of value "transfer" (Rolston 1988: 222). Environmental ethics, then, requires that our instrumental use of nature also respects the intrinsic value of nature, which is ultimately reflected in how our actions "follow nature" or align with the "systemic value" of nature as a whole.

Therefore, while animal rights holds that higher animals have a right to life which, all else being equal, demands they not be killed, the objective intrinsic value of nature suggests that hunting is morally permissible when the use it makes of animals both sustainably conserves the instrumental value of the individual animal, its species, and the ecosystem of which it is a part, *and* respects the intrinsic value of the same by affirmatively participating in natural processes. Indeed, if wasting instrumental values should be seen as a form of disrespect, then the sustainable use of wild animals should be seen as a form of respect (Gunn 2001), thus connecting instrumental and intrinsic valuing in the same act. For example, a hunter who tracks a wounded animal to kill and retrieve it conserves the instrumental value of the animal's body and honors the death of the animal, both of which contribute to the hunter self-reflectively seeing her or his own self as respectfully participating in the overarching valuable natural process of predation. Environmental ethics which affirm the intrinsic value of animals in natural systems are not therefore in principle opposed to hunting, but are opposed to forms of hunting which do not conserve the instrumental value of nature, or exhibit a *merely* instrumental evaluation of the animal and its place within the larger ecosystem. When hunting follows nature axiologically by sacramentally participating in the systemic value of predation and conserving the instrumental value of the animal's body and species, hunting thereby respects the intrinsic value of both the animal itself and the biotic community of which it is a part.

Nature, Culture, Science, and Policy

Philosophical reasoning about hunting can only go so far until it becomes clear that many, if not all, of our moral intuitions are derived from and shaped by pre-reflective cultural contexts that we rarely, if ever, submit to critical scrutiny. The arguments above may fall on deaf ears if hunting clashes with other presuppositions about respecting natural value. Hunting seems obviously disrespectful of nature because it does not "leave nature alone" or preserve it from human impacts, on the assumption that human culture is an unnatural contaminant which automatically makes "the wilderness" less wild. However, when hunting is seen as "necessary for survival," it is considered acceptable, especially when it forms a part of a minority or Indigenous culture. Hunting in mainstream dominant cultures (e.g. the Euro-settler cultures of North America), meanwhile, is considered unacceptable especially when it appears as mere aggression for "sport" or fun using advanced technologies that cannot but dominate the animal by obliterating any chance it has of escaping. It is difficult to accept the notion that animals can be both killed for use and respected when they are seen as the pristine "other" against which is arrayed all the crushing power of industry and empire (cf. Kheel 1996). Environmental ethics must contend, therefore, not only with issues of moral principle, but more fundamentally with the metaphysical relationship between human culture and nonhuman nature. This entails an engagement with philosophies of biology, technology, and culture, and environmental philosophy more broadly.

While nature conservation entails sustainable use, nature *preservation* sees value in setting aside areas where humans do not use nature at all, at least not in a literally consumptive sense, but can appreciate nature aesthetically or in other ways that leave little or no trace, like low-impact camping. As the U.S.'s *Wilderness Act* suggests, a nature preserve is "an area where the earth and its community of life are untrammeled by man [sic], where man himself [sic] is a visitor and does not remain" (1964: sec. 2.c). But should nature be defined as that which is independent of human involvement and, inversely, should human culture be defined as other-than-nature and thus as an inevitably artificial and foreign threat to any "pure" nature? E. O. Wilson calls our species "ecologically abnormal" (1992: 272), as if the human spirit doesn't really belong down here on Earth – but perhaps has a destiny out among the stars. In environmental ethics, this view has been described as nature-culture dualism, whereas the other extreme, nature-culture monism, holds that human beings are so much a part of nature that anything they do is natural: the city of Chicago is just as much a part of nature as the Great Barrier Reef (Callicott 1992). Paul Veatch Moriarty and Mark Woods (1997) argue that because Rolston advocates nature-culture dualism, the fact that most hunting in mainstream Western society is a thoroughgoing cultural activity means that it cannot actually participate in nature and is thus deprived of its moral justification.

The relation between human culture and nonhuman nature also underlies policy decisions about hunting and wildlife biology. It may be that hunting adversely affects the constitution of certain animal populations, either directly or indirectly. For example, unlike most nonhuman predators, most humans who hunt do not pursue sick or old animals but rather those in their breeding prime and most likely to survive harsh winters (Kretz 2010). Moreover, even the need to use hunters to manage wildlife populations may itself be a consequence of the intentional eradication of nonhuman predators for the purpose of increasing the availability of game animals (Wenz 1983). Some of these concerns may be addressed empirically; for example, in "heavily hunted populations of white-tailed and mule deer, there is no indication that a low number of bucks negatively affects reproductive rates or overall population robustness" (Heffelfinger et al. 2013: 402). But underlying every empirical answer to a management question is a values paradigm which science itself is not equipped to address (Kuhn 1970). Philosophical questions arise like "what *ought* the sex and age ratios of a population to be?" or "what *should* the baseline for assessing population numbers be?" Similar questions abound in the literature on environmental restoration (Chapter 44). More fundamentally, how should humans be related to the rest of nature? Should population baselines be determined in terms of human absence and non-contact, or is it acceptable for nonhuman populations to show the effects of hunting by humans? If humans are and ought to be a part of nature, it seems that wildlife biology should accept human presence in principle and model species and ecosystem health accordingly.

This is a complex matter, however, because there is a wide variety of ways for humans to interact with the rest of nature, from the low-impact wilderness dwelling of hunter-gatherers to modern civilization's sixth great extinction event which continues unabated into the so-called Anthropocene epoch. Even if, in principle, humans belong in nature such that other species cannot be expected to be unaffected by their presence, hunting regulations may still need to be changed to ensure certain ecological goals within particular cultural and environmental contexts. Inversely, hunting policies may be implemented to minimize impacts on human culture, such as controlling white-tailed deer numbers with the aim of reducing agricultural crop damage or automobile collisions. Accordingly, management goals may be set to optimize a "Social Carrying Capacity" (Heffelfinger et al. 2013: 401), so that hunting becomes a technological solution which maintains and perpetuates the cultural status quo (King 2010). Indeed, hunting may be restricted to protect certain species that are threatened by other social arrangements that are more difficult to restrain, such as the oil and gas industry's effects on woodland caribou in northern Alberta (Environment Canada 2012).

Therefore, the question remains as to whether the societal status quo ought to be perpetuated, or whether policies should be introduced which require social change. Not all socio-political arrangements are equally laudable in terms of environmental values. What may seem like a pragmatic or politically unproblematic "solutions-based" approach to scientific questions about biology cannot ignore the fact that all "problems" are socially constructed. Any policy about hunting will inevitably entrench one arrangement of human culture while disadvantaging another. In most cases, however, the cultural dimensions of moral debates about hunting remain unstated, unrecognized, and unevaluated.

Agriculture, Necessity, and Sport

One place where culture does enter the hunting debate is when "hunting out of necessity" or traditional subsistence hunting is morally tolerated, whereas "hunting for pleasure" (sport, recreation, trophies, etc.) is not. Subsistence hunting constitutes a form of culture and is always culturally mediated (e.g. by gender, religion, tradition, and trade), making the moral relevance of its distinction from culturally mediated sport hunting problematic (List 2004). Even the very necessity of subsistence hunting is socially constructed: hunting for food is only made unnecessary by agricultural methods of protein production (animal or vegetable-based) in environments where that is possible. For "necessity" to have moral importance, critics of hunting would have to say that agriculture is *morally superior* to foraging. That is, whenever it is possible to transition a culture from foraging (where hunting is "necessary") to agricultural subsistence (where hunting is "unnecessary"), that transition *ought* to take place. To exculpate traditional cultures for hunting because they do so out of necessity assumes that such cultures only hunt because they have no other options, as if hunting is a horrific occupation that nobody would pursue if they had a choice. While some anti-hunting philosophers speak unflatteringly of our Paleolithic ancestors' (putative) tendencies for "bashing other human beings in the skull" (Pluhar 1991: 124) or "physical aggressiveness and sexual dominance" (Everett 2001: 59), seeing subsistence hunting as justified only by necessity is civilized prejudice pure and simple, viewing subsistence hunting cultures as uncultured, bloodthirsty, and savage. It negates the profoundly cultural aspect of such hunting while replicating the "vanishing Indian" trope, whereby these peoples tragically miss the boat of moral and cultural progress associated with the supplanting of foraging with domestication and tillage, ending up on the "wrong side of history" (Kowalsky 2014c).

Anti-hunting, therefore, even when making allowances for marginalized cultures, embodies a socially constructed prejudice in favor of agricultural civilization, rarely with any consideration of the environmental and social toll exacted by historical and contemporary agricultural practices (Scott 2017). It also uncritically assumes that hunting is an unrewarding occupation that only barely permits survival as opposed to the (putative) human flourishing unleashed by agriculture. Why, then, would anyone pursue it for recreational enjoyment as do sport and trophy hunters, if not for a perverse love of death (King 1991)? Environmental ethics answers that recreation is not necessarily a trivial good incapable of contributing to the justification of hunting. Rolston asks "why do humans enjoy nature even when we no longer need it for economic or life-supportive reasons" (1988: 7)? His answer is that recreation *re-creates* in the sense of re-enacting the process of predation (i.e. sacramental or participatory affirmation of that process), but also of re-charging ourselves by drawing succor from a source of profound goodness (i.e. the rejuvenation that comes from participation in natural processes like predation itself). Concerning trophy hunting specifically, Alistair Gunn rightly condemns hunters who "kill purely for the sake of acquiring prestigious evidence that they have killed an animal" (2001: 75), but when incorporated into the broader context of re-creating "the way the world is made" (Rolston 1988: 91) by conserving and respecting the instrumental and intrinsic values of nature, trophies can contribute significantly to the memorialization of the particularity of the individual animal (Marvin 2010), a non-trivial cultural good.

Wilderness, Technology, and Fair Chase

Because subsistence hunters no more hunt out of sheer necessity than recreational hunters hunt for the sheer hell of it (Cahoone 2009; List 2013), there may be no philosophical purpose in distinguishing between sport and subsistence hunting except to unmask the unrecognized cultural prejudices of anti-hunting. Even the idea that human culture is out of place in pristine nature may be the *effect* of the displacement of hunting by agriculture, rather than being an independent framework with which to assess hunting. As T. R. Kover shows, while hunting-based cultures spend much of their

> daily life...in a wild non-human environment which bears little witness to human control or importance,...agriculture is implicitly based on a deliberate and purposeful spatial separation between a human and a wild, non-human, natural domain and on bringing this domestic area under human control.
>
> *(2009: 237, 239)*

Yet in environmental hermeneutics, agricultural landscapes are often considered the happy medium between the exclusively cultural landscapes of the city and putatively uncultured wilderness landscapes, because through the process of cultivation, taming, and domestication, they harmoniously hybridize both culture and raw nature (Drenthen 2009a). But this perspective assumes that wild landscapes are not suitable for human dwelling, highlighting that a "sense of alienation is presupposed in the concept of wilderness" (Drenthen 2009b: 314).

Materially, the domestic order of the agricultural landscape must contend with incursions from the wild: weeds, vermin, and predators will invariably invade and threaten crops or livestock. From the agricultural perspective, hunting may be valued positively if it helps control animal pests, but negatively if it mimics too fully the death-dealing nature of the wild which steals the life-giving property (herds and crops) accumulated and stored by agrarian culture. This negative assessment of hunting can be seen, for example, in European folklore like Prokofiev's *Peter and the Wolf* (Kowalsky 2014b), and in philosophical classics like J.S. Mill's "On Nature" (1874), where "brutal and degraded...Patagonians or Esquimaux" hunter-gatherers are seen as savage humans who did not learn how to conquer wild nature like European civilization did. With this kind of contextual background, it is unsurprising that anti-hunting's attempt to exculpate subsistence hunters turns out to be self-defeating.

Once the ahistorical approach to the relationship between humanity and the rest of nature is overcome, it becomes clear that some kinds of culture may be more dualistic with relation to nature than others. But hunting's sacramental participation in wild natural processes is not likely to be seen as a positive thing, given agrarian cosmology's association of raw nature with death and disorder. For example, Roger King (1991) hopes that humans can participate in nature in non-violent, non-consumptive, and non-predatory ways, thus excluding hunting. But when hunting is presumed to be immoral because, as a cultural behavior, it is a violent, bloody, or harmful way to participate in nature, the problem actually seems to be that hunting connects humans *too closely* to the natural world, rather than separating us from it (Kover 2010). Hunting offers a different landscape hermeneutic and nature-culture metaphysic than agriculture, such that while at present human civilization appears to be fundamentally at odds with the natural world, hunting as a type of culture may "cultivate" the notion that humans can and should reconcile themselves to the transcendent moral order that is wild nature.

In fact, sport hunting may encourage this incorporation into a transcendent moral order by virtue of the way its own cultural dimensions enjoin the use of technology to subvert agrarian dualism. A "sporting" hunter is expected to follow the ethic of "fair chase," which has roots in the European distinction between aristocratic gentlemen and unentitled poachers. It requires sport hunters to give prey a fair chance at exercising their evasive capacities. Brian Luke (1997) points out that this leveling of the playing field, as it were, amounts to extending a right to life to prey animals, thus making hunting immoral on its own terms. But José Ortega y Gasset insists that hunting is defined by a leveling that does not efface the difference between species. The essential definition of hunting, for Ortega, is "a contest or confrontation between two systems of instincts. But for it to occur it is necessary that those instincts – not only the hunter's, but also the prey's – function freely" (1995: 64). So while fairness to the prey cannot be understood in terms of moral reciprocity whereby the animal might in some bizarre manner consent to being hunted (Vitali 2017), it is better understood in terms of the hunter's respect of species difference which transcends both the desire of the prey to stay alive and the power of the hunter to dominate.

If the prey's instincts are to function freely, then "[human] reason [must] try to preserve the distance that existed between [the hunter and the hunted] and, where possible, to improve it in favor of the animal" (Ortega 1995: 116). The human brain and hand have come together to create all manner of technological marvels, but these can sometimes eliminate the ability of the animal to exercise its own particular excellences at avoiding capture. Thus, Theodore Vitali argues that fair chase "requires, in turn, conscious efforts to handicap oneself in the hunting experience" (2010: 29), to not allow the distance between human and nonhuman (both physical and metaphysical) to be eliminated by technological gadgets like range finders, scent killers, or motorized vehicles. As Aristotelian ethics are relative to the constitution of each person, much technological handicapping will depend on the skill level of the individual hunter (e.g. whether or not one hunts with a rifle or a traditional bow), but certain technologies are too powerful or invasive (e.g. aircraft, automatic firearms, silencers, spotlights, and night vision goggles), and will thus be regulated or prohibited by law (Kowalsky 2014a). Such policies exist, therefore, because certain technologies unleash certain powers that *increase* the dualistic separation between human culture and the rest of nature, whereas the fair chase ethic of species difference advocates the limitation of technology to achieve a more holistic integration of the two (Clark 2017).

Conclusion

According to Ortega (1995), there are four enigmas at the heart of the ethics of hunting: the enigma of the animal, the enigma of death, the enigma of the morally ambiguous universe, and the enigma of the human place within it. We have seen how hunting forces us to inquire after the nature of animals, the nature of ourselves, the nature of nature, and how these might be integrated together. These are the ethical questions which haunt the shadows cast by conservation biology, wildlife management, and hunting policy. It is beyond the scope of this paper to resolve these enigmas (as if such resolution were even, finally, possible), but I have attempted to scry a trail through the thicket of issues they raise. Hunting is an embodied practice by which human beings can navigate the tension of being human in the natural world as it appears to us to be ecologically, rather than as agrarian civilization may want to envision it. It may also be beyond the scope of policy to resolve these enigmas, but policy is not thereby absolved of its inextricable entanglement with value and theory. This author would like to see increased collaboration between environmental philosophers and policy practitioners, for without engaging the philosophical issues at stake, policy will be more rather than less controversial by eliding over the issues that matter most to the general public. Inversely, theory and value are ultimately what lead to policy reform. Hunting will not emerge unscathed from its encounter with environmental ethics.

Related Topics

2. Eating
5. Species and Wildlife
6. Wild Animals
10. Wilderness
18. Fishing and Harvesting
38. Food
44. Restoration

References

Batavia, C., Nelson, M., Darimont, C., Paquet, P., Ripple, W. & Wallach, A. (2019) 'The elephant (head) in the room: a critical look at trophy hunting', *Conservation Letters*, 12(1), e12565.

Bekoff, M. (2012) 'Killing other animals for food does not make us human: there's nothing spiritual about killing innocent animals for unneeded meals [Blog]'. *Psychology Today, animal emotions blog.*

Bentham, J. (1823) *An introduction to the principles of morals & legislation*. London: W. Pickering.

Cahoone, L. (2009) 'Hunting as a moral good', *Environmental Values*, 18, pp. 67–89.

Callicott, J. B. (1980) 'Animal liberation: a triangular affair', *Environmental Ethics*, 2(4), pp. 311–338.

Callicott, J. B. (1989) 'Animal liberation & environmental ethics: back together again', in *In defense of the land ethic: essays in environmental philosophy*. Albany: State University of New York.

Callicott, J. B. (1990) 'The case against moral pluralism', *Environmental Ethics*, 12(2), pp. 99–12.

Callicott, J. B. (1992) 'La Nature est morte, vive la nature!' *Hastings Center Report*, 22(5), pp. 16–23.

Causey, A. (1989) 'On the morality of hunting', *Environmental Ethics*, 11(4), pp. 327–343.

Cerulli, T. (2010) 'Hunting like a vegetarian: same ethics, different flavors', in Kowalsky, N. (ed.) *Hunting – philosophy for everyone: in search of the wild life*. Malden: Wiley-Blackwell.

Cerulli, T. (2012) *The mindful carnivore: a vegetarian's hunt for sustenance*. New York: Pegasus.

Clark, G. (2017) '(Not) war by other means: why hunting is not war', in Hill, B. and White, J. (eds.) *God, Nimrod, and the world: exploring Christian perspectives on sport hunting*. Macon: Mercer University.

Dickman, A., Johnson, P., 't Sas-Rolfes, M., Di Minin, E., Loveridge, A., Good, C., Sibanda, L., Feber, R., Harrington, L., Mbizah, M., Cotterill, A., Burnham, D. & Macdonald, D. (2019) 'Is there an elephant in the room? a response to Batavia et al.', *Conservation Letters*, 12(1), e12603.

Drenthen, M. (2009a) 'Ecological restoration & place attachment: emplacing non-places?' *Environmental Values,* 18(3), pp. 285–312.

Drenthen, M. (2009b) 'Fatal attraction: wilderness in contemporary film', *Environmental Ethics*, 31(3), pp. 297–315.

Ellul, J. (1980) *The technological system*. Joachim Neugroschel (trans.). New York: Continuum.

Environment Canada (2012) *Recovery strategy for the woodland caribou (rangifer tarandus caribou), boreal population, in Canada*. Species at risk act recovery strategy series. Ottawa: Environment Canada.

Everett, J. (2001) 'Environmental ethics, animal welfarism, & the problem of predation: a Bambi lover's respect for nature', *Ethics & the Environment,* 6(1), pp. 42–67.

Frasz, G. (2009) 'The howl of the predator', *Environmental Ethics*, 31(2), pp. 223–224.

Gunn, A. (2001) 'Environmental ethics & trophy hunting', *Ethics & the Environment*, 6(1), pp. 68–95.

Hargrove, E. (1992) *The animal rights, environmental ethics debate: the environmental perspective*. Albany: State University of New York.

Heffelfinger, J., Geist, V. & Wisharts, W. (2013) 'The role of hunting in North American wildlife conservation', *International Journal of Environmental Studies*, 70(3), pp. 399–413.

Hettinger, N. (1994) 'Valuing predation in rolston's environmental ethics: Bambi lovers versus tree huggers', *Environmental Ethics*, 16(1), pp. 3–20.

Kheel, M. (1996) 'The killing game: an ecofeminist critique of hunting', *Journal of the Philosophy of Sport*, 23, pp. 30–44.

King, R. (1991) 'Environmental ethics & the case for hunting', *Environmental Ethics*, 13(1), pp. 59–85.

King, R. (2010) 'Hunting: a return to nature?' in Kowalsky, N. (ed.) *Hunting – philosophy for everyone: in search of the wild life*. Malden: Wiley-Blackwell.

Kover, T. R. (2009) 'The domestic order & its feral threat: the intellectual heritage of the neolithic landscape', in S. Bergmann, P. M. Scott, M. Jansdotter Samuelsson, & H. Bedford-Strohm (eds.) *Nature, space & the sacred: transdisciplinary perspectives*. Farnham: Ashgate.

Kover, T. R. (2010) 'Flesh, death, & tofu: hunters, vegetarians, & carnal knowledge', in Kowalsky, N. (ed.) *Hunting – philosophy for everyone: in search of the wild life*. Malden: Wiley-Blackwell.

Kowalsky, N. (2014a) 'Radical Albertans? Hunting as the subversion of heroic enlightenment', in R. Boschman & M. Trono (eds.) *Found in Alberta: environmental themes for the Anthropocene*. Waterloo, ON: Wilfrid Laurier University.

Kowalsky, N. (2014b) 'The hero, the wolf, & the hybrid: overcoming the overcoming of uncultured landscapes', in Drenthen, M. & Keulartz, J. (eds.) *Old World & New World perspectives on environmental philosophy*. New York: Springer.

Kowalsky, N. (2014c) 'Between relativism & romanticism: traditional ecological knowledge as social critique', in M. Vassányi (ed.) *Indigenous perspectives of North America*. Newcastle upon Tyne: Cambridge Scholars.

Kretz, L. (2010) 'A shot in the dark" the dubious prospects of environmental hunting', in Kowalsky, N. (ed.) *Hunting – philosophy for everyone: in search of the wild life*. Malden: Wiley-Blackwell.

Kuhn, T. (1970) *The structure of scientific revolutions*. 2nd edition. Chicago: University of Chicago.

Leopold, A. (1949) *A sand county almanac & sketches here & there*. Oxford: Oxford University.

List, C. (1997) 'Is hunting a right thing?' *Environmental Ethics,* 19(4), pp. 405–416.

List, C. (2004) 'On the moral distinctiveness of sport hunting', *Environmental Ethics,* 26(2), pp. 155–169.

List, C. (2013) *Hunting, fishing, & environmental virtue: reconnecting sportsmanship & conservation*. Corvalis: Oregon State University.

Loftin, R. (1984) 'The morality of hunting', *Environmental Ethics*, 6(3), pp. 241–250.

Luke, B. (1997) 'A critical analysis of hunters' ethics', *Environmental Ethics*, 19(1), pp. 25–44.

Mallory, C. (2001) 'Acts of objectification & the repudiation of dominance: Leopold, ecofeminism, & the ecological narrative', *Ethics & the Environment*, 6(2), pp. 59–89.

Marvin, G. (2010) 'Living with dead animals? Trophies as souvenirs of the hunt', in Kowalsky, N. (ed.) *Hunting – philosophy for everyone: in search of the wild life*. Malden: Wiley-Blackwell.

Midgley, M. (1994) 'The end of anthropocentrism?' in Attfield, R. and Belsey, A. (eds.) *Philosophy & the natural environment*. Royal Institute of Philosophy Supplement 36, pp. 103–112.

Mill, J. S. (1874) 'On nature', in *Nature, the utility of religion, & theism*. London: Longmans, Green, Reader, & Dyer.

Moriarty, P. V. & Woods, M. (1997) 'Hunting ≠ Predation', *Environmental Ethics*, 19(4), pp. 391–404.

Norton, B. (1995) 'Why I am not a nonanthropocentrist: Callicott & the failure of monistic inherentism', *Environmental Ethics*, 17(4), pp. 341–358.

Ortega y Gasset, J. (1995) *Meditations on hunting*. Howard B. Westcott (trans.). Belgrade, MT: Wilderness Adventure.

Pluhar, E. (1991) 'The Joy of killing', *Between the Species*, 7(3), pp. 121–128.

Pollan, M. (2006) *The omnivore's dilemma: a natural history of four meals*. New York: Penguin.

Regan, T. (1983) *The case for animal rights*. Berkeley: University of California.

Rodman, J. (1977) 'The liberation of nature?', *Inquiry*, 20, pp. 83–131.

Rolston, H. (1988) *Environmental ethics: duties to & values in the natural world*. Philadelphia: Temple University.

Rolston, H. (1992) 'Disvalues in nature', *The Monist*, 75(2), pp. 250–278.

Routley, R. (1973) 'Is there a need for a new, an environmental ethic?' *Proceedings of the XV World Congress of Philosophy* 1, Varna, pp. 205–210.

Scott, J. (2017) *Against the grain: a deep history of the earliest states*. New Haven and London: Yale University.

Singer, P. (1975) *Animal liberation: a new ethics for our treatment of animals*. New York: Avon.

Stone, C. (1974) *Should trees have standing? Toward legal rights for natural objects*. San Francisco: W. Kaufmann.

Varner, G. (1995) 'Can animal rights activists be environmentalists', in Marietta, D. and Embree, L. (eds.) *Environmental philosophy & environmental activism*. Lanham: Rowman & Littlefield.

Vitali, T. (2010) 'But they can't shoot back: what makes fair chase fair?' in Kowalsky, N. (ed.) *Hunting – philosophy for everyone: in search of the wild life*. Malden: Wiley-Blackwell.

Vitali, T. (2017) 'Killing what you love: a reflection on fair chase hunting', in Hill, B. and White, J. (eds.) *God, Nimrod, and the world: exploring Christian perspectives on sport hunting*. Macon: Mercer University.

Wenz, P. (1983) 'Ecology, morality, & hunting', in Miller, H. and Williams, W. (eds.) *Ethics & animals*. Clifton, NJ: Humana.

White, L. (1967) 'The historical roots of our ecologic crisis', *Science*, 155(3767), pp. 1203–1207.

Wilderness Act (1964) Public Law 88–577 (16 U.S. C. 1131–1136) 88th Congress, Second Session, 3 September.

Wilson, E. O. (1992) *The diversity of life*. Cambridge, MA: Harvard University.

Further Reading

Franklin, A. (2001) 'Neo-Darwinian leisures, the body & nature: hunting & angling in modernity', *Body & Society*, 7(4), pp. 57–76. (A sociological analysis of hunting as a leisure activity which offers an alternative vision of environmentalism.)

Kowalsky, N. & Kover, T. R. (eds.) (2017) Special issue on hunting and religion, *Studies in Religion/Sciences Religieuses*, 46(4), pp. 487–600. (An exploration of various religious aspects of hunting.)

Scruton, R. (1998) *On hunting*. London: Yellow Jersey. (A short philosophy of hunting in the European context.)

Stange, M. Zeiss (1997) *Woman the hunter*. Boston: Beacon. (A feminist defense of hunting.)

Zencey, E. (1987) 'On hunting', *The North American Review*, 272(2), pp. 59–63. (A cultural analysis of North American sport hunting.)

PART II

Land

8

FORESTS

Lydia A. Lawhon

On March 26, 2012, the Lower North Fork Fire ignited in southern Colorado. When firefighters brought it under control, it had burned only 4,140 acres, a relatively small fire in terms of geographic area, but had destroyed 23 homes and killed three people. The number of fatalities earned the fire the dubious distinction of the deadliest fire in Colorado's history. It was sparked by embers from a prescribed burn, or intentionally set wildland fire, which escaped from its planned perimeter. The purpose of the prescribed burn was to restore forest health and protect nearby communities from the impacts of a catastrophic wildfire.

Only two and a half months later, the High Park Fire started outside of Fort Collins, CO, and subsequently burned nearly 90,000 acres and 259 homes. One person died in the blaze. At this time, the High Park Fire was considered the most destructive fire in Colorado's history – only to have that designation eclipsed just two weeks later by the Waldo Canyon Fire near Colorado Springs, which destroyed more than 350 homes and killed two people.

The fire season of 2012 was historic. But a year later, the Black Forest Fire outside of Colorado Springs destroyed nearly 500 homes, with two lives lost (El Paso County, Colorado, Sheriff's Office 2014). As of the publication of this essay, the dubious designation of "most destructive wildfire in Colorado" remains with the Black Forest Fire.

The threat of wildfires to forests illuminates many of the reasons we care about forests – for their aesthetics, biodiversity, sources of fresh water and clean air, economic potential, recreational opportunities, and cultural and spiritual meaning (Price et al 2011) – as when a wildfire destroys a forest, we feel an acute loss. Nonetheless, wildfire is a critical ecosystem process for many forested ecosystems across the globe. In many places, but particularly in the American West, that process has been altered by a century of fire suppression policy (discussed below) and growth in housing developments in wildfire-prone areas, while also exacerbated by climate change (Radeloff et al 2018).

In the past several decades, the number of wildfires, as well as the area burned by wildfire, has increased across the western United States (Dennison et al 2014). Fire seasons are also getting longer (Westerling et al 2006; Westerling 2016). Wildfires in and of themselves tax financial and human resources, elevate risks to life and property, and present planning and management challenges for public agencies. However, they are a microcosm of larger questions surrounding the fate of forests. In this chapter, we discuss the problems surrounding forest management, particularly in the western United States, highlighting important ethical perspectives to consider in designing forest management and policy, including those surrounding wildfire as well as other management challenges, such as timber and recreation. Though the focus and examples are in the western United States, the questions posed could be adapted and applied in other forest contexts as well.

DOI: 10.4324/9781315768090-11

US Historical Context

Forests have captured the hearts of poets and philosophers for centuries, and were one of the core foci of the environmental movement for several decades. Congress passed the initial legislation supporting the creation of the US Forest Service (USFS) in 1897, with the Forest Service Organic Administration Act. This act laid out the purpose of the then-called Forest Reserves for timber production as well as forest and watershed protection. In 1905, the USFS, the managing federal agency for National Forests, was created and replaced the Forest Reserve system, with oversight transitioning from the Department of the Interior to the Department of Agriculture (Transfer Act of 1905).

In 1960, the Multiple Use Sustained Yield Act (MUSYA) codified the multiple-use mandate of the USFS and ensured that the USFS manage its lands for timber, range, wildlife, recreation, and water equally, with no use perceived to be greater or preferred over any other. The "sustained yield" clause was included to ensure the protection of natural resources in the long-term, avoiding practices that would negatively impact future production of resources, namely timber harvest. MUYSA states, "It is the policy of Congress that the national forests are established and shall be administered for outdoor recreation, range, timber, watershed, and wildlife and fish purposes" (Section 1). It goes on to define multiple use as "management of all the various renewable surface resources of the national forests so that they are utilized in the combination that will best meet the needs of the American people" (Section 4(a)). Accordingly, the mission of the USFS is "Caring for the land and serving people." The National Forest Management Act of 1976 was a process-oriented law that required National Forests and Grasslands to undergo comprehensive planning processes as well as reports on the status of renewable resources, such as timber. NFMA has been implemented through various planning rules, with the most recent promulgated in 2012.

Of course, timber has been a primary component of both public and private lands management in the United States over the past century. With the advent of the environmental movement in the 1970s, however, the public, through the advocacy work of environmental groups, increased scrutiny of timber extraction on federal forest lands, often because of its impacts on endangered species. The "timber wars" of the 1990s over the spotted owl in the Pacific Northwest were a key example of a landmark environmental battle between timber interests and environmental advocacy organizations working on behalf of the Endangered Species Act. Though timber sales from national forest lands dropped significantly toward the end of the 20th century, timber still remains an important component of management given the multiple-use mandate of USFS. Legal conflicts over timber extraction decisions and the associated potential impacts to endangered species still persist, though the level of public scrutiny and attention reached its zenith with the aforementioned timber wars in the 1990s. Nonetheless, there are notable and ongoing high-profile cases, such as ongoing lawsuits over timber management in the Tongass National Forest of southeast Alaska (see, for example, Nie 2008).

Recreation is another important driver of management decisions on publicly managed forests, federal, state, and local. Private forest landowners in some regions of the United States also manage their woodlots for recreational purposes as well, such as hiking or winter recreation. For the purposes of this essay, however, the focus will be on recreational demands on the National Forest system. Associated with demographic trends increasing the numbers of people living in or near national forests, demands for recreational opportunities strain existing infrastructure capacity and cause resource degradation, including soil erosion, disturbance to wildlife and vegetation, and overuse of facilities such as restrooms and parking areas (Marion et al 2016). Again, the multiple-use mandate requires that land managers strive to accommodate the varied uses of forests, which often entails building additional trails and infrastructure, as well as increasing the resources devoted to maintaining and enhancing public recreation opportunities. The increasing need for managers to focus on recreational management highlights the importance of forested lands to the public – from locals to visitors from near and afar. In the midst of the COVID-19 pandemic, Derks, Giessen, and Winkel found that the use of forests for recreation increased, and noted that in the 18th century, forests were once seen as critical infrastructure for utilitarian/economic uses; today, they are still critical infrastructure, but in support of citizens' mental and physical health (2020).

The inherent tension in managing forests is illustrated through the policy codified in the multiple-use mandate. People use forests for their economic value in extracting timber, minerals, and recreational development, as described above. They also reap spiritual benefits from visiting forests, replenishing their sense of beauty and nourishment (illustrated by the increasing popularity of the Japanese practice of shinrin-yoku, or "forest bathing"). Forests embody conservation values as they protect wildlife and ecosystems. Forest lands remain culturally, spiritually, and economically important to indigenous and native peoples, despite their forced and catastrophic removal from them over the 19th and 20th centuries. And increasingly, people rely on forests for their recreational and athletic pursuits. To the latter point, the value of time in nature and outdoor recreation is seen as an asset to public health (Wolf et al 2020; see also Razani et al 2020, on the value of public lands for supporting public health during the COVID-19 pandemic). From a global perspective, forests are watershed catchments, securing soil through root systems and mitigating soil erosion for the benefit of water quality. Forests also serve as carbon sinks, whereby they can sequester and store carbon (McKinley et al 2011). These diverse and valid values are all threatened by demographic shifts, climate change, and policy decisions, such as the fire suppression policy of the previous century.

The Land Ethic and Forest Management

It has been argued that forest managers have a responsibility to ensure the health and functionality of forests for ours and future generations, and for the species and ecological systems which depend on forests for their viability. For example, the "Grey Towers Protocol," drafted as a result of a conference convened by the Pinchot Institute for Conservation in the mid-1990s, developed a number of principles for land stewardship to be offered for consideration to land managers. Noting the value of scientific information in addressing pressing land management challenges, the report also argued that "in order for professional natural resource managers to exert leadership in the care of natural systems, they must first demonstrate that they truly care for the resources they manage. As Aldo Leopold put it, 'We can be ethical only in relation to something we can see, feel, understand, love, or otherwise have faith in'" (Sample 1995, p. viii). This idea was captured in the idea of a "stewardship ethic," which is closely tied to Leopold's land ethic, in that a steward has the power, along with the responsibility, to manage a particular piece of land (Sample 1995). The Grey Towers Protocol articulated the following principles for a stewardship ethic for land managers:

1   Land stewardship must be more than good "scientific management"; it must be a moral imperative.
2   Management activities must be within the physical and biological capabilities of the land, based upon comprehensive, up-to-date resource information and a thorough scientific understanding of the ecosystem's functioning and response.
3   The intent of management, as well as monitoring and reporting, should be making progress toward desired future resource conditions, not on achieving specific near-term resource output targets.
4   Stewardship means passing the land and resources, including intact, functioning forest ecosystems, to the next generation in a better condition than they were found (Sample 1995, ix).

These principles provide guidance for managers who are working to implement forest restoration projects and address increasing demands on forests for a diversity of public uses. Callicott (1989), elaborating on Leopold's land ethic, embraces the idea of a holistic (rather than individualistic) approach to environmental ethics, looking at a system as the sum of the individual parts. The objects of moral concern, therefore, can be entire ecosystems or species. By this argument, the stewardship ethic should encompass more than just the land; rather, it should include the processes, species, and ecosystems supported by it.

However, humans are the primary cause of degradation to forested ecosystems, whether through impacts resulting from recreational overuse, mismanagement of resource extraction, or wildfires, via unattended campfires, sparks thrown from power lines or vehicles, or cigarettes, among others. (In fact, researchers have found that human-caused wildfire accounts for a significant proportion (greater than 80%) of wildfire ignitions in the United States, while also impacting the geographic extent of wildfire impacts and lengthening the fire season (Fusco et al 2017).) Wilderness ethics in practice have been manifested in the "Leave No Trace" principles, which advocate for responsible enjoyment of the outdoors by people by minimizing human impacts (Leave No Trace n.d.). Managers will continue to face the challenge of weighing stewardship ethics and the multiple-use mandate, which legally confirms their obligations to provide opportunities within forests for public benefit, such as campgrounds, recreation areas, picnic areas, and trails, as well as timber extraction and protection of ecosystems and wildlife.

Ethical Considerations for Forest Management in the 21st Century

The multiple-use mandate challenges managers with the responsibility of seeking balance between land stewardship and opportunities for human endeavor. Acutely revealed in the controversies surrounding the old-growth forest battles of the Pacific Northwest in the late 1990s, when loggers and advocates for the endangered spotted owl vowed to die on their swords to protect their values, the question of "whose Nature" still reverberates today (Proctor 1996). In other words, for which values should land managers orient their decisions? For economic livelihoods? Endangered species safeguards? Cultural and aesthetic treasures? In some regards, wildfire mitigation is a common-ground issue behind which many decisions can be justified and for which public support can be garnered. However, there are a number of ethical questions that should be considered when making decisions about how lands should be used and where to invest mitigation resources, while also striving to accommodate competing values over forest management.

Leopold's central advocation was to extend a code of ethics to the land, making us responsible for ensuring its health and integrity – essentially, ensuring its stewardship into the future. However, the land ethic doesn't necessarily confer rights to nature, so the question remains as to whether forests have standing, or rights. Stone argues that, given our existing legal framework, natural objects do not have clear legal rights: "They have no standing in their own right; their unique damages do not count in determining outcome; and they are not the beneficiaries of awards" (or reparations to repair damage) (Stone 1972: 463). Conservation, even as mandated by law (e.g. Endangered Species Act), supports natural objects such as forests inasmuch as they are useful to or enjoyable for people. Stone further argues that, akin to a person who appoints power of attorney to another to act on their behalf should they not be able to do so for themselves, forests and other natural objects should similarly have "guardians" or "trustees" to advocate for their specific interests (Stone 1972).

Though laws such as the Endangered Species Act or the Wilderness Act have the goal of protecting natural objects, they do not explicitly appoint an actor to uphold their rights. This authority was provided to conservation organizations to legally challenge actions that threatened natural objects or places, primarily for public lands (though the ESA provides protection to endangered or threatened species that live on private lands as well). The role of environmental and conservation groups in forest management cannot be understated; their actions on behalf of forests (and other protected areas) have halted management decisions that violated environmental statutes and ensured the protection of important ecosystems. Furthermore, natural objects have rights to reparations for damages caused by development or negligence. In this vein, forests that have been degraded by various policies are deserving of attention and resources to restore them, or make them "whole." Procedurally, the National Environmental Policy Act (NEPA) affords a mechanism to articulate the rights of the environment and assess how human actions may impact them (Stone 1972).

Restoration Baselines

Stone's argument for standing for natural objects, such as forests, is premised on a somewhat static understanding of ecosystem processes, whereby we have a clear understanding of what "restored" or "whole" looks like. However, practically, this question is nuanced. For example, Native Americans often used fires to create the open forests and prairies which early Europeans encountered, illustrating that wildfire was human-caused prior to European settlement (see, for example, Pyne 1982; Cronon 2003). Therefore, in the case of wildfire, given the mandate to "restore" forests and reverse a century of policy incongruent with ecosystem processes, managers must determine, and justify, the appropriate fire regime to strive for in implementing restoration goals. Is it 100–150 years ago, prior to the advent of real impacts of human-caused climate change? Should it be prior to any European settlement? Should it be the regime that would have been in place with no human alteration, and only "natural" ignitions such as lightning?

This question can be applied and analyzed in the context of any forest management beyond only wildfire, to issues such as insect outbreaks, previously logged areas, and places impacted by intensive recreation or overuse. Managers are responsible for using their expertise to define what a "healthy" forest looks like, while also determining what their obligations are to try to achieve this state. A healthy forest, moreover, is not static – this determination is an ongoing question that forest managers continue to revisit over time based on changing environmental and ecological conditions.

Restoration Reparations

From a rights-based perspective, Stone argues that the perpetrator responsible for degrading a natural object should pay reparations into a "trust fund" that is overseen by the resource's guardian, and allocated to cover projects and research related to amending the wrongs experienced by the forest (Stone 1972). However, some impacts are difficult to tie to a specific perpetrator. For example, in areas that have experienced resource degradation from intensive recreation, such as trail erosion and expansion, trash, and human and dog waste, the question remains as to whom is responsible for paying for restoration. The nature of the restoration activities may be more clear in this case, but it is nearly impossible to pinpoint Stone's "perpetrator." Should all forest users be required to pay a fee that supports ongoing maintenance for a particular area? Users fees are commonplace in some particularly impacted regions; however, fees can also be seen as a barrier to participation in forest-related activities for those who cannot afford them. The impacts and ethical dimensions associated with limiting access or imposing fees should be closely examined in situations where forest degradation cannot be directly pinpointed to one particular resource user.

Restoration Scale

The scale at which restoration activities should be targeted is also important to consider. In a utilitarian sense, one might argue that the scale at which a project is initiated should provide the greatest good for the greatest number (and be cost-effective). An expensive project that only limits one small community situated in a wooded area may not be as desirable as a project of the same cost that also would benefit or protect water supply for an urban area downstream.

Forest landscape restoration has been popularized in recent years as an aspirational goal for land managers around the globe, both public and private, with the goals of "regaining ecological functionality and enhancing human well-being across deforested or degraded landscapes…it is restoring a whole landscape to meet present and future needs and to offer multiple benefits and land uses over time" (IUCN 2019, para. 1). Both technical (i.e. methods of restoration) and governance challenges arise in considering restoration at the landscape scale; resolving governance challenges may be more critical in contributing to the overall success of a project at this scale. Mansourian (2017) identifies questions to consider in governance, which are linked to others posited in this section as well: who are the stakeholders and decision-makers? How might stakeholders benefit or lose from particular decisions? What is the nature of the institutions involved in landscape-level restoration? The question of the scale of restoration cannot be resolved in isolation; it necessarily weaves in other ethical considerations as well.

There are several examples of collaboratives that focus on landscape-level forest restoration in the western United States, primarily to address the threat of catastrophic wildfire. The Front Range Roundtable brings together a diverse set of stakeholders from public agencies, municipal governments, non-profit organizations, academia, private interests and industry, and user groups to address the threat of wildfire risk and promote forest restoration in ten counties along the east side of the Colorado Rockies. Members of the collaborative share a common goal of wildfire risk reduction and a core set of values, including designing resilient communities, restoring and maintaining healthy landscapes, utilizing collaborative strategies, accounting for the reality of economic and political factors, and engaging directly with communities (Front Range Roundtable, n.d.). The Four Forest Restoration Initiative (4FRI) in northern Arizona aims to implement forest landscape restoration efforts to reduce wildfire fuels and improve overall ecosystem health, while also providing economic opportunities via jobs and contracts for restoration work. 4FRI's efforts are supported at the state level by the development of the Statewide Strategy for Restoring Arizona's Forests and the subsequent convening of the Arizona Forest Health Council (US Forest Service Four Forest Restoration Initiative n.d.).

Responsibilities

Given the checkerboard landscape of forest ownership in the western United States, particularly in the wildland-urban interface, the question arises as to who bears the responsibility of forest restoration efforts. Presumably, the landowner holds some part of this responsibility – whether an individual, community, state, or federal land management agency (e.g. U.S. Forest Service, National Park Service). However, the factors of resource availability and crisis management direct priorities, often resulting in reactive (e.g. actively fighting a wildfire) rather than proactive (e.g. taking measures to reduce wildfire risk) management. Proactive mitigation necessarily should be a collaborative effort among all landowners, as well as other critical stakeholders who depend on the forest resource. Water utilities, for example, depend on healthy forests for clean and abundant water supplies, and thus have a vested interest in working with landowners in their watersheds to ensure the health of the forest. Private corporations, such as breweries on the Front Range of Colorado, also depend on clean and abundant water. Outdoor recreation interests – whether the public through special-interest advocacy groups or the industry more broadly – also depend on aesthetic and intact forests. For all of these interests, it can be argued that they have an obligation to invest in protecting forested lands through the support of forest restoration projects.

Questions of responsibility may be constrained by the availability of resources. Large-scale forest restoration is expensive and demands technical expertise and trained personnel to successfully implement. State and federal land management agencies may have the expertise, but lack the substantial financial resources needed to carry out the project. This is where collaborative efforts, such as those discussed above, may be necessary to leverage funds and bring in other partners to achieve forest restoration goals. There are also social constraints. Proactive management is often hampered by nearby residents' values and understandings of the purpose of a forest management treatment as well. For example, prescribed burning is widely accepted by land managers as a cost-effective and efficient treatment for addressing wildfire risk and restoring forests. Nonetheless, often communities and neighboring landowners fear the escape of a prescribed burn and may not understand its benefits. It is nearly impossible to address all public perceptions of forest management, including those of future users.

The questions of baselines, reparations, scale, and responsibilities provide lenses through which we can start to unpack the influence of values and the legal and ethical obligations we have to manage and restore forests. There are no straightforward answers, but engaging with stakeholders and making these underlying and often unarticulated dimensions explicit can help to arrive at durable solutions that protect forest integrity (or the rights of forests in and of themselves) as well as the values we hold for forested places.

Equity Considerations for Forest and Wildfire Management

With changing demographics and priorities for forests, managers grapple with questions over equitable resource allocation. For example, federal and state funding may be disproportionately directed toward communities that have more local capacity for addressing wildfire mitigation and forest health management, as well as infrastructure development for other forest uses, such as recreation, leaving rural or under-privileged communities without similar resources. In turn, these communities may be more likely to have their nearby forests developed (or developed further) for more intensive, extractive forest uses.

To that end, questions about directing resources arise. Should a restoration project be implemented where the benefits to the surrounding human communities are greatest? Or, should it take place where the benefits to wildlife and forest health are significant, recognizing secondary advantages to nearby communities? Or, should it be in an area that has encountered substantial environmental degradation due to human activities, and the stewardship ethic guiding the actions of land managers directs that it should be restored? There may be justice concerns that arise out of these discussions, as well. A place of iconic ecological and cultural values (e.g. a National Park) may draw resources toward it, and by extension protect the neighboring gateway communities which are typically (though not always) higher income areas. In contrast, a small, remote rural community may not be as *politically* desirable for a large-scale restoration project. However, the overall impacts of a disaster, such as a wildfire, or an extractive economic development project may well be more catastrophic than they would be in a place where resources are ample.

Once the decision over where to focus proactive actions for forest protection has been made, if a project is successfully carried out in one target area, but then a severe wildfire destroys a community and ecosystems elsewhere, does the agency bear responsibility for addressing some of the harms to the impacted community and ecological resources because of their lack of action? This line of reasoning does not guide policy currently, but if there is a shift to more proactive risk mitigation instead of reactive responses to a disaster-in-progress, these types of questions could arise. Similarly, if an extractive project, such as timber harvest or oil and gas development, is approved by a land management agency, does the land management agency also bear some element of responsibility for the negative outcome?

Conclusion

With forest uses evolving, managers are challenged with balancing competing values among a diversity of stakeholders – from those who see the forest for the intrinsic value of ecosystems, including trees and wildlife habitat, to those who see the forest for its economic assets in the form of timber or biomass, or those that see the forest for its recreational opportunities. These values held by groups or individuals within each of these "categories" of stakeholders are not uniform either; nor do they necessarily hold steady over time. Aldo Leopold articulated as early as 1913 the various uses which forest lands served (primarily referring to federally managed forests, and not private lands): "timber, water, forage, farm, recreative, game, fish, and esthetic resources" (Leopold as cited in Flader 1998: 1). With changing climate, demographics, and permutations of these uses, as well as the potential for forests to mitigate and safeguard us in the battle to arrest climate change, forest managers benefit from tools to ascertain the ethical dimensions of their management decisions.

Works Cited

The Associated Press. (2018, June 12). Wildfire danger and drought forces San Juan national forest to close. *Colorado Public Radio News.*

Batavia, C., & Nelson, M. P. (2018). Translating climate change policy into forest management practice in a multiple-use context: the role of ethics. *Climatic Change, 148*(1–2), 81–94. http://doi.org/10.1007/s10584-018-2186-2

Callicott, J. B. (1989). *In Defense of the Land Ethic.* Albany: State University of New York Press.

Cronon, W. (2003). *Changes in the Land: Indians, Colonists, and the Ecology of New England.* New York: Hill & Wang.

Dennison, P. E., Brewer, S. C., Arnold, J. D., & Moritz, M. A. (2014). Large wildfire trends in the western United States, 1984–2011. *Geophysical Research Letters* (April), 2928–2933. https://doi.org/10.1002/2014GL061184

Derks, J., Giessen, L., & Winkel, G. (2020). COVID-19-induced visitor boom reveals the importance of forests as critical infrastructure. *Forest Policy and Economics, 118*(April), 102253. https://doi.org/10.1016/j.forpol.2020.102253

Donovan, G. H., & Brown, T. C. (2007). Be careful what you wish for: the legacy of Smokey Bear. *Frontiers in Ecology and the Environment, 5*(2), 73–79. https://doi.org/10.1890/1540-9295(2007)5[73:BCWY-WF]2.0.CO;2Flader, S. (1998). Aldo Leopold's Legacy to Forestry. *Forest History Today.* Retrieved from https://foresthistory.org/periodicals/1998/

Front Range Roundtable. (n.d.). Home. https://frontrangeroundtable.org/ May 13, 2019.

Fusco, E. J., Mahood, A. L., Balch, J. K., Bradley, B. A., Abatzoglou, J. T., & Nagy, R. C. (2017). Human-started wildfires expand the fire niche across the United States. *Proceedings of the National Academy of Sciences, 114*(11), 2946–2951. https://doi.org/10.1073/pnas.1617394114

Husari, S. J., & Mckelvey, K. S. (1996). Fire-management policies and programs. In *Sierra Nevada Ecosystem Project: Final Report to Congress* (Vol. II: Assess). Davis: University of California, Centers for Water & Wildland Resources.

IUCN. (2019). Forest landscape restoration. https://www.iucn.org/theme/forests/our-work/forest-landscape-restoration. Accessed May 12, 2019.

Jolly, W. M., Cochrane, M. a., Freeborn, P. H., Holden, Z. a., Brown, T. J., Williamson, G. J., & Bowman, D. M. J. S. (2015). Climate-induced variations in global wildfire danger from 1979 to 2013. *Nature Communications, 6*(May), 7537. https://doi.org/10.1038/ncomms8537

Leave No Trace Center for Outdoor Ethics. (n.d.). About. June 30, 2020. https://lnt.org/about/.

Mansourian, S. (2017). Governance and forest landscape restoration: a framework to support decision-making. *Journal for Nature Conservation, 37*, 21–30. https://doi.org/10.1016/j.jnc.2017.02.010

Marion, J. L., Leung, Y. F., Eagleston, H., & Burroughs, K. (2016). A review and synthesis of recreation ecology research findings on visitor impacts to wilderness and protected natural areas. *Journal of Forestry, 114*(3), 352–362. https://doi.org/10.5849/jof.15-498

McKinley, D. C., Ryan, M. G., Birdsey, R. A. et al. (2011). A synthesis of current knowledge on forests and carbon storage in the United States. *Ecological Applications, 21*, 1902–1924

Meldrum, J. R., Brenkert-Smith, H., Champ, P. A., Falk, L., Wilson, P., & Barth, C. M. (2018). Wildland–urban interface residents' relationships with wildfire: variation within and across communities. *Society & Natural Resources*, *31*(10), 1132–1148. https://doi.org/10.1080/08941920.2018.1456592

Multiple-Use and Sustained Yield Act of 1960, Public Law 86–517, 86th Congress (June 12, 1960)

Nie, M. (2008). *The Governance of Western Public Lands: Mapping Its Present and Future*. Lawrence: University of Kansas Press.

Price, M. F., Gratzer, G., Duguma, L. A., Kohler, T., Maselli, D., & Romeo, R. (2011). *Mountain Forests in a Changing World Mountain Forests in a Changing World Realizing Values, Addressing Challenges*. *FAO/MPS and SDC, Rome* (Vol. 29). https://doi.org/10.13140/2.1.2386.5283

Proctor, J. D. (1996). Whose nature? The contested moral terrain of ancient forests. In W. Cronon (Ed.), *Uncommon Ground* (pp. 269–297). New York: W.W. Norton & Company.

Pyne, S. J. (1982). *Fire in America; A Cultural History of Wildland and Rural Fire*. Princeton, NJ: Princeton University Press.

Radeloff, V. C., Hammer, R. B., Stewart, S. I., Fried, J. S., Holcomb, S. S., & McKeefry, J. F. (2005). The wildland-urban interface in the United States. *Ecological Applications*, *15*(3), 799–805. https://doi.org/10.1890/04-1413

Radeloff, V. C., Helmers, D. P., Kramer, H. A., Mockrin, M. H., Alexandre, P. M., Bar-Massada, A., … Stewart, S. I. (2018). Rapid growth of the US wildland-urban interface raises wildfire risk. *Proceedings of the National Academy of Sciences*, 201718850. https://doi.org/10.1073/pnas.1718850115

Razani, N., Radhakrishna, R., & Chan, C. (2020). Public lands are essential to public health during a pandemic. *Pediatrics*, *146*(2), e20201271. https://doi.org/10.1542/peds.2020-1271

Stone, C. (1972). Should trees have standing? – Toward legal rights for natural objects. *Southern California Law Review*, *45*, 450–501.

USDA Forest Service Four Forest Restoration Initiative. (2019). History of 4FRI. https://www.fs.usda.gov/main/4fri/about-history. May 12.

Veblen, T. T., Kitzberger, T., & Donnegan, J. (2000). Climatic and human influences on fire regimes in ponderosa pine forests in the Colorado front range. *Ecological Applications*, *10*(4), 1178–1195. https://doi.org/10.1890/1051-0761(2000)010[1178:CAHIOF]2.0.CO;2

Westerling, A. (2016). Increasing western US forest wildfire activity: sensitivity to changes in the timing of spring. *Philosophical Transactions of the Royal Society B-Biological Sciences*, *371*, 20150178 https://doi.org/10.1098/rstb.2015.0178

Westerling, A. L., Hidalgo, H. G., Cayan, D. R., & Swetnam, T. W. (2006). Warming and earlier spring increase Western U.S. forest wildfire activity. *Science*, *313*(5789), 940–943. https://doi.org/10.1126/science.1128834

Wolf, K. L., Derrien, M. M., & Kruger, L. E. (2020). Nature, outdoor experiences, and human health. In: S. Selin, L. K. Cerveny, D. J. Blahna, A. B. Miller (Eds.), *Igniting Research for Outdoor Recreation: Linking Science, Policy, and Action* (pp. 85–100). Gen. Tech. Rep. PNW-GTR-987. Portland, OR: U.S. Department of Agriculture, Forest Service, Pacific Northwest Research Station. https://www.fs.usda.gov/treesearch/pubs/60068

9

MOUNTAINS

Rethinking Thinking Like a Mountain

David Strong

We cannot lower the mountain,
therefore we must elevate ourselves.

—*Todd Skinner*

Today we move mountains out of the way rather than going around them.

Divided Highways: The Interstates and the Transformation of
American Life, a documentary (1997)

It would be no small advantage if every college were thus located at the base of a mountain.

—*Thoreau, speaking of Williams College*

Mountains are exhilarating and consoling; we are angered with assaults to them and profoundly saddened with their loss. For many of us in the environmental movement, particular, named mountains are our grounding tie, the source of our concern. For Bill McKibben, such a mountain is Crane Mountain in the Adirondacks of upstate New York:

> What I've done, in my daily life and my political work and my writing, I've done because of these woods, these very woods we're walking through. They captured my imagination and taught me, in my twenties, that the suburban life I'd grown up in was not as engaging as life out here. I fell in love with these hemlocks, these steep slopes, these patches of rock, these streams lit by leaf filtered sun.... Where I lived, the woods had already been preserved from simple destruction... But even the New York State Constitution can't stop... rising temperatures.... So that's where I spent my life, working (with little obvious effect) on global problems like climate change
>
> (133)

Particular mountains can be our partners in the dance of being. That is, we fall in love with them, becoming intimate with them. Over time, they can change the course of our life, transforming us in the process. Particular mountains can animate our lives, making all the difference in the world for us. Moreover, and more subtly, these particular mountains, for their part, come to life, stand out freshly and meaningfully, as never before, in this heartfelt discovery of them. As partners, they too are animated, as it were, as they are lived. Our intimacy with mountains illuminates a

DOI: 10.4324/9781315768090-12

kind mountainous relationship, what I call an "interanimating symmetry," between ourselves and these special things. Mountains therefore become elevated, that is, brought into relief, when disclosed in our close relationship with them. This point was brought home memorably to me when, seeing in the far distance the colorful ropes and clothing of people climbing a massive boulder of Pipestone Pass near Butte, Montana, I glimpsed for the first time this way that rock of the pass can speak to us. I also feel indebted to Cézanne when I'm stirred by the Pipestone landscape. Humans, and not only humans, are revealers.

No doubt, some clarification is in order. For my part, mountains, especially mountains of the Yellowstone watershed, have been central. If one were to track my footsteps in life, all my comings and goings, most of the paths would crisscross at my hometown, Livingston, Montana. Partly, this has to do with it being my hometown, but if this town had been in the Midwest, I doubt it would have this kind of pull for me, given that mountains and their rivers are in my blood. Partly, this has to do with my parents and siblings who still live nearby, but this is often a side course, for it is not family conversations that bring me back so often, were it not for the mountains and rivers we share. I fish the river, hike the trails, and ski the trails and local downhill ski area. I made friends with those who do the same. I tube the river with my family. We take in the entertainment the community offers and often do our shopping in Livingston, for the stores have enough selection without having to use a car that much. All of this activity takes place beneath the graceful symmetry of a forested, perfectly shaped cone peak, Livingston Peak or Mount Baldy, as most locals call it. This peak is always there, whether on Main Street or on the banks of the Yellowstone, whether driving from the east where I now live, Billings, or from the west, or from the north, or the south. Beneath its crown is a crossing place that gathers more of my life than any place on Earth.

Livingston Peak is not so close to town that it blocks the morning sun in winter, nor is it so far away that you have to look for it. It looks like the very tallest of all surrounding peaks, but it isn't. It looks like a symmetrical cone, but actually it's the end of a long ridge. No matter. It's a real mountain, a real peak, a real beauty. Beneath it, I see all the times I've hiked and played as a kid. I was taken to the foothills by my grandpa Howard, to the old Stumbo homestead, his in-laws from his sister's marriage. Fording the river horseback in his accustomed place, Leroy Stumbo one evening drowned, apparently, although questions remain as to the exact circumstances. Up Suce Creek, far past the Busbys, in a shack beside a dugout cave lived a miner friend of my grandfather. My father was frustrated by the hours that whiled away when my grandfather and Elmer Lee "sat around that old, potbellied stove, warming themselves and exchanging stories." A mysterious great, great uncle with perhaps a somewhat shady past (he may have been to prison), Art Bloodgood, used to spend all his time in the hills beneath Baldy and knew the place better than anyone. Somehow I think of Faulkner's Mississippi County when I think of this place. Alas, there is no Faulkner, at least to my knowledge, to lift it up and make it sing.

Yes, Livingston Peak is ecologically powerful. Behind it lies the rest of the Absaroka Range that runs 60 miles south along Montana's Paradise Valley to Yellowstone Park, then borders Yellowstone's North and East boundaries, forming the splendid array of wild peaks of the caldera that tourists see across Yellowstone Lake. The Yellowstone River itself begins on the range's Yaunts Peak, some 30 miles south of Yellowstone Park in the most inaccessible Wilderness in the lower continental US. About the same distance east of Yaunts Peak lies Franc's Peak, the Absaroka Range's highest. At 13,153 feet, Franc's peak is among the tallest peaks in Wyoming and taller than any peak in Montana. The time I scrambled up the peak we spooked a grizzly that skylined over the summit ahead of us. It had probably been feeding on cutworm moths flying there from the prairies of states of the Midwest. The Absarokas join the Wind River Range at Togwotee Pass, and the Winds, headwaters of the Green River, extend almost to Utah and Colorado at South Pass, important to the wagon trains on the Oregon Trail. When I see Livingston Peak, I think of Lander, Wyoming, at the other end of these ranges, a driving distance of over 350 miles. That's a lot of country to fill one's perception! I'd like to walk its full extent in my lifetime. In segments, I have walked or ridden horseback on trail crew most of the extent of the Absaroka Range from Livingston to the Thoroughfare River, just south of Yellowstone Park. I don't know anyone who has traversed the full extent of both the Absaroka and Wind ranges. This is a landscape of grizzlies, wolverines, bighorn sheep, bison, great herds of elk, and now wolves. It is a raw and fierce landscape; yet often, it possesses a warm, gentle, sweet, and serene beauty. Ecologically, it's rich; phenomenologically, resonant and deep.

Livingston Peak points upward like a raised finger or a spire. There it stands calmly, unmoved even when Livingston's great wind blasts a visible cloud of snow swirling far out east from the summit. The great wind, the calm mountain, the ever-flowing river. Simple things. Simple enough things. Final things. Thoreau may have his rooster crows that awaken him; my rooster is Livingston Peak. It reminds me of all of that life can be. It measures my life and the lives of my fellow citizens. It makes me feel what an honor and privilege it is to be a human walking the face of this beautiful planet. And it makes me feel anguish for all that has been lost on Earth, gone for want of thought.

What calls for thought here in our relationship with mountains? In our heedlessness, we are failing these mountains. We have failed to attend to the potential significance of mountains and to guard them. Thus, we need to heed and mind them again and for the first time. In a profound way, protection of mountains is the fundamental environmental and cultural issue of our time. We must learn to think like a mountain. What I mean by thinking like a mountain is somewhat different than Leopold intended to mean by the phrase, although this different sense of what I mean, as we shall see, can also be retrieved from Leopold's writings. By thinking like a mountain, I mean we come to understand, cherish, and guard our relationship, the interanimating symmetry between ourselves and mountains. Mountains get us to care about the quality of our relationship with them. The new challenge of our time is not to dominate mountains but to live by them.

The Transformation of the Earth: The Promising Turn and Ironic Twists of Ecological Conscience

It's been over a half-century since Aldo Leopold first published (1948) his *A Sand County Almanac* [*SCA*] containing the famous essay "Thinking like a Mountain" (Leopold 1968: 129–133). Since then, much has been accomplished that accords with the ecological conscience that Leopold sought to articulate and instill in his fellow citizens. In 1964, The Wilderness Act was passed, followed soon by the establishment of the EPA and the passage of other significant environmental legislation, including NEPA and the Endangered Species Act. The Earth Day was founded in 1970. The field of environmental ethics was initiated, numerous environmental journals were established, and the scope broadened from America to become international and global. In December 2013, the Western Governors' Association unveiled a multistate wildlife habitat mapping project that the governors hope will reduce conflict between wildlife and development by helping industry avoid development in crucial wildlife habitat (Sonner 2013). These and a host of other auspicious events have led Holmes Rolston to dub this period since the 1960s as having undergone "the environmental turn" (Rolston 2012: 2) when the natural world emerged as a focus of concern. "Environmental concern, according to Paul Hawken, is 'the largest movement in the world,' considering the number and force of environmental organizations around the globe" (Rolston 2012: 2).

While Rolston is indeed correct on one level about this turn, many sense that we are failing the deeper challenges before us, understanding the really "largest movement in the world," if viewed from the mountaintop. Having expended most of his book arguing for his vision of the environmental ethics of this environmental turn, Rolston devotes a portion of his final chapter to some disturbing findings. Considering graphs that show worldwide energy use, total GDP, damming of rivers, fertilizer consumption, and motor vehicle numbers, we see anything but an environmental turn for the better. Rather, from 1950 onward, all the trends indicated on the graphs skyrocket. To put it mildly, "the kind of development we've seen over the past 50 years is not sustainable" (Rolston 2012: 207). We don't seem to be able to say "enough" to endless growth and overconsumption. Bill McKibben puts his sense of the matter in terms of "history." Considering a future that conforms to the pattern what's gone on before and is for the most part continuing in the present, history

> at the moment looks as though it should go this way: dairy farms fail or consolidate; farmland turns into second homes or retirement homes or just home-homes, as Burlington [Vermont] sprawls south and north. Interfering with history is hard, because its momentum is so strong: the march of the big-box stores, the decline of the number of farmers, the demographic tides of our population.
>
> *(McKibben 2005: 57)*

Or take a current example of an environmental down-turn especially dear to Leopold: native prairieland.

More than 1.2 million acres of grassland have been lost since the federal government [in 2007] required that gasoline be blended with increasing amounts of ethanol, an Associated Press analysis of satellite data found. Plots that were wild grass or pastureland seven years ago are now corn and soybean fields (Brokaw and Gillum 2013).

A more accurate updated figure is 1.6 million acres of native prairie which was converted to cropland (Lark et al 2015). The key underlying this transformation of native prairie is the attraction of affluence. After the federal Renewable Fuel Standard policy of 2007, "the ethanol mandate," 44% of the nation's corn crop was used for fuel, about twice the percentage seen in 2006. Corn prices shot up, more than doubling, for now "farmers can make about $500 an acre planting corn" (Brokaw and Gillum 2013). Without adding any acreage, even a struggling corn farmer in 2006 was in a good position for more than doubling his income on corn in 2012. And by converting another 160 acres of prairie to corn each year as many are doing, a farmer could then make, by my calculations, some additional $80,000 on top of this already-increased income the first year and leap to an additional $80,000 to this newly increased income each year. Even discounting the kind of marginal farmland being developed, it's a good formula for getting rich at the cost of native prairie. But corn farmer Robert Malsam is "not about to apologize for ripping up prairieland to plant corn." "Farmers that own the land are free to farm their land to the extent they can make money on it or whatever purpose they need," says Jeff Loutt, CEO of a national ethanol refinery operation (Brokaw and Gillum 2013). And surely, it would seem unfair to deny to farmers what most of us wish for ourselves.

We can approach the transformation of the Earth in terms of its rate, scope, and magnitude. To lose 1.6 million acres of virgin grassland, "equivalent in area to the state of Delaware" (Lark et al 2015) and akin to the less than 6.9 million acres of the remaining unprotected and mountainous roadless wildland in Montana, in just six years, is an exceedingly rapid rate and extensive scope of transformation of the American landscape, of the Earth. It also shows a high degree of magnitude, for the use of herbicides to kill the grassland, the plowing of the land three times, the heavy use of inorganic fertilizers and pesticides, and the further use of herbicides to control "weeds," and destroys soils, plants, wildlife habitat, and ecosystems. All this conversion fits well within the terms of the analysis spelled out well by Leopold before the "environmental turn," only now this transformation, with more demand and more powerful technology and techniques, has intensified exponentially. It also fits well within what Bill McKibben calls history, just the way we expect to lose wildland, native prairie, open space, and so on to developments such as sprawl. "In fact, some 3,000 acres of productive farmland are lost to development each day in this country" according to the EPA ("Land-Use Overview"). The same kind of transformation is poised to set upon our mountainous regions. In central Appalachia, over 500 mountains have been flattened by Mountaintop Removal Mining where, instead of removing the coal from the mountain, the mountain is removed from the coal. "Mountaintop removal and the dumping of wastes and debris into adjacent valleys is the greatest earth-moving activity in the United States" (no date, Peck). This highly mechanized surface mining does not produce many jobs, but a high school graduate from a local community can step into a truck driving job with a salary starting at $65,000. Logging of what just a few decades ago was de facto wildland now mars the foothills just outside the Absaroka-Beartooth Wilderness boundary of Livingston Peak. Fracking for oil and gas is having a major impact on the Wind River Range near Pinedale, Wyoming, and is now targeting areas north of the Absaroka-Beartooth Wilderness. Montana's US Senator Steve Daines advocates deregulating environmental constraints and releasing nearly all of Montana's unprotected wildland for logging and other resource extraction, establishing in their place "Forest Reserve Revenue Areas." It would not take long for much of this wildland to be transformed, if it became legally possible to extract these resources.

Surely, we need to rethink the ethanol mandate, but my point is a larger one, connecting affluence with the transformation of the Earth. Yes, we can call on a plurality of solutions to a plurality of problems, e.g. the solutions in this case are multiple for the protection of native prairie (Lark 2020), at the very least a policy modification that better protects native grasslands, but somehow we sense that something bigger needs to change if these solutions are not to become mere stop gap measures. After all, the ethanol mandate itself was intended to ameliorate global warming through reducing greenhouse gas production from fossil fuels (which in this case backfired). More deeply, we sense that a pattern like what McKibben calls history – in reality "the largest movement in the world" – needs to be protested and interfered with before we can be sure we are really making an effective difference.

My contention is this: this view from the mountaintop shows us that the environmental turn without a corresponding technological turn will prove inconsequential. To be consequential, an environmental ethic must be more than an environmental ethic alone. A new relationship with nature requires a new relationship with technology. Likewise, a new relationship with technology requires a new relationship with nature. The resulting new ethic and aesthetic I call an environmental ethic within the philosophy of technology. We can locate the inchoate promptings of this new ethic even in Leopold's *SCA*. Just imagine how his ethic may have changed were Leopold to have had a powerful and sophisticated theory of technology at his disposal.

Revising "Thinking Like a Mountain": Living by the Mountains

The opening of the *SCA* reads, "There are some who can live without wild things, and some who cannot. These essays are the delights and dilemmas of one who cannot" (Leopold 1968: vii). How many people can be counted on to love nature with Leopold's kind of enthusiasm? Leopold asserts that many of his fellow citizens, the majority, are not in this position, initially at least. But we may still wonder whether it is the task of the *SCA* to get *all or most* people to love nature in such a way that they can now see that they, too, cannot live without wild things.

Indeed, from the standpoint of his explicitly developed environmental ethic, the land ethic, this may *seem* to be what Leopold is arguing, for the principles of the land ethic are universal. Leopold believes that from the beginning of history, humans have had a conception of the land that allowed and encouraged them to view nature as something that we can dispose of as we please. Such vision is radically anthropocentric, elevating humans to tyrants with the rest of nature beneath, serving in the role of mere utility. The land ethic challenges this vision by extending ethics to the land, requiring the land to be loved and respected. In J. Baird Callicott's interpretation of the land ethic (Callicott 1987), ecology plays the key role in calling forth the land ethic because ecology proves that we are not outside of nature and above it; rather, humans are plain members among its other nonhuman members. For the reason that they are now seen as members of a community to which we all belong, nonhuman members claim our respect. These are, of course, the animals, plants, and other organisms—and also, the soil, air, and water—that make up ecosystems. For Leopold, ecology humbles us and redirects our respect toward the larger community it shows us. Thinking ecologically and nonanthropocentrically like a mountain (metaphorically the land pyramid), we learn to respect the community and its members morally. He also argues for a land aesthetic in which ecology and evolution build breadth and depth into our perception of nature and natural things by supplying new and sophisticated background knowledge. Hence, natural things attain aesthetic value where before they had mere utility or were viewed as useless waste. Finally for Leopold, this land ethic and land aesthetic attempts to be universally valid. Regardless of whether we are Leopold or anyone else: "A thing is right when it tends to preserve the integrity, stability and beauty of the biotic community. It is wrong when it tends otherwise" (1968: 224–225). This land ethic should be binding for an epoch when people are informed by ecology and evolution. Once so informed, the land ethic, ecological conscience, should enable us to think nonanthropocentrically and restrain us from treating nature as ours to dispose of as we please. It promises to stop us from abusing the land (which the previous section finds it has not consequentially).

Yet, this land ethic interpretation of *SCA's* opening lines that we began with does not really hold up. If the billions of people on Earth love wild places to the degree that Leopold does when he says that he cannot live without them, their sheer numbers would overrun what little wild nature has been preserved. More to my point, we know from experience that Leopold's sentences ring of personal testimony, and most people will never be able to say that they cannot live without wild things and mean it. The land ethic does not require that level of concern for the land from each and every one of us; rather, Leopold is identifying himself for his readers as a lover of wild things just as another might identify herself as a lover of guitar, horses, farming, or cabinet-making. Leopold cannot expect all of us to share completely or be hardwired for his personal enthusiasm for wild things.

Given this personal testimonial reading of his opening lines, we are now poised to reframe Leopold's work by approaching it from the standpoint of an environmental ethic and aesthetic *within the philosophy of technology*. The second and third paragraphs of the Foreward will serve our purposes here:

> Like winds and sunsets, wild things were taken for granted until progress began to do away with them. Now we face the question whether a still higher "standard of living" is worth its cost in things natural, wild, and free. For us of the minority, the opportunity to see geese is more important than television, and the chance to find a pasque flower is a right as inalienable as free speech.

These wild things, I admit, had little human value until mechanization assured us of a good breakfast, and until science disclosed the drama of where they come from and how they live. The whole conflict thus boils down to a question of degree. We of the minority see a law of diminishing returns in progress; our opponents do not (vii).

Aside from the role of ecological and evolutionary disclosures ("the drama of where they come from and how they live") discussed previously, here Leopold believes that the importance of wild things shines brighter because they are disappearing. Technology, in doing away with them, assists paradoxically in a new apprehension of the importance of things, as Heidegger, too, remarks about the close proximity between the "danger" of extreme technology and the "saving powers of things," as he addresses what he calls "the turning" or reform of technology (1977: 42–49, 28–35). Those who cannot live without wild things see with clearer eyes that they cannot. Moreover, since technology assures us of a good breakfast, it emancipates us from an adversarial relationship with nature and makes possible a new relationship of love and respect for nature. No one can argue that Leopold is merely anti-technology.

Nevertheless, Leopold, in this same passage, is critical of technology for he sees a conflict between a high and rising standard of living and the continued existence of the wild things, between, in his words, "we of the minority" and "our opponents." This is a recurring theme in *SCA*. A case in point:

> The good soil that enabled the Flambeau [river valley] to grow the best cork pine for Paul Bunyan [and hence become logged] likewise enabled Rusk County, during recent decades, to sprout a dairy industry. These dairy farmers wanted cheaper electric power than that offered by local power companies, hence they organized a co-operative REA and in 1947 applied for a power dam, which, when built, would clip off the lower reaches of a fifty-mile stretch in process of restoration as canoe-water.
>
> *(116)*

To understand these reasons for "developing" the river as economic reasons only, as Leopold seems to, keeps the real issue submerged. We need to push these kinds of economic reasons to their ultimate ends. Affluence and increased affluence are the keys, the *means*, which unlock more products of technology, the *ends*. As we pursue headlong and thoughtlessly the promise of technology to liberate and enrich our lives, we continue to expand this framework with more and more commodities, a wider variety of them and the very latest models (Borgmann 1984; Strong 1995). Not only dairy farmers, corn farmers, and coal truck drivers, but most people are spellbound by this promise or vision. Television (and now the Internet and smartphones) itself is a model of this vision for it delivers the glamorous goods of the Earth to wherever we are whenever we want them in a quick and easy manner. And so the Flambeau River, the foothills of Livingston Peak, the mountaintops of West Virginia, and nearly everything else on Earth get swept into this framework and are made to serve this goal of greater consumption. In the past 300 years, and especially in the last 100 years, the most massive human-caused transformation of the Earth has taken place. For the first time, humans have become a major geologic force on the face of the planet. So, from a philosophy of technology standpoint, the opposition is between those thoughtlessly spellbound by the promising vision of technology and its consumer happiness ("our opponents") and those who take their bearings elsewhere ("we of the minority").

In Leopold's view, this expansion of the framework of technology, what McKibben sees as history, is not worth its cost to these wild things, mountains. Of the great meadow in his other mountain essay, "On Top," he writes: "Despite several opportunities to do so, I have never returned to White Mountain. I prefer not to see what tourists, roads, sawmills, and logging railroads have done for it, or to it" (Leopold 1968: 128). This testimony issues from his tie to the mountains, his tie to wild things that he cannot live without, not so much his ecological conscience. Leopold here, on my view, finds himself really most deeply opposed to the ontology of technology, the above ever-expanding framework that moves even mountains into its frame. And here is my point: to meet his concerns with wild things, it will not be enough to address only the value of the wild things, as he does. *He will also have to challenge and reform this expansion of technology;* otherwise, his concerns for wild things, even if they can be shown to have intrinsic value, will be subverted by technology, and his land ethic will prove inconsequential. In other words, he needs a theory of technology that is just as powerful as his theory of the ecological conscience.

Yet, even the "intrinsic value" he finds in biota and ecosystems needs to be reconsidered, as we will do later. His characterization of the deepest issue needs to shift in other respects as well, for Leopold does not understand clearly the relationship between modern technology and the impoverished, resource-view of nature. Land as mere resource, strictly speaking, is a function of our relationship to our overpowering technology and not, as Leopold thinks, the outcome of people uninformed by ecology, such as the ancient Hebrews. The chief factor here is to see modern technology as our road to happiness. When that happens, technology becomes a mode of being or way of life. Here, technology, because it serves nothing better than what it can procure, is taken to be an absolute *good in its own right* (an entertainment center, metaphorically) in contrast to technology being a *relative good*, an instrumental good, serving some other and more important end. If technology is taken to be good in its own right, if the goods of technology are taken to be good in their own right, then everything else becomes only a "relative" good, dependent upon how they serve technology as good in its own right. How important are "geese" to the service of "television"? Nature becomes a relative good only insofar as it serves consumption in one way or another. Thus, nature becomes degraded to resource. Hence, our relationship to technology as an absolute good in its own right calls forth an impoverishing relationship to nature, and this resource-view is something true of our epoch only.

On my view, breaking this thoughtless social agreement with our attraction to the promise and framework of technology is key. Hence, we need something more and other than Leopold's land ethic's new constraints, what Leopold calls "the key-log" (224). If we take technology as good in its own right, as a final good, in the same old way as we have been, then we may sense the need to constrain its rapidly growing powers to manipulate nature. But such constraints to protect nature, such as native prairie, forests, and mountaintops, will often be at a cost to what this vision assumes is the good life. The economy and the kind of assumed-but-never-talked-about good life the economy is the means to are inevitably at odds with the quality of the environment, especially if we are to protect the environment in any way more meaningful than nature as a relative good. In that kind of conflict, technology almost never loses. Rather than the land ethic's new constraints, we need a new relationship with technology. *If we do not take up with technology as an absolute good in its own right, then nature will not be relativized in relation to technology.* Nature will not be disclosed as a mere resource. A new relationship to technology evokes a new, freer relationship to nature – a liberated environmental ethic and aesthetic.

Wild things we cannot live without, including mountains, can show us this new relationship to technology. Wild things, taken as Leopold's personally takes them when he says he can't live without them, are good in their own right, a final good. Likewise, Leopold finds that farmers of the Sand Counties were "baited" with powerful economic (that is, on my interpretation, technological) reasons to resettle elsewhere in the 1930s; yet, they "did not want to go" (102). The reasons they stayed, he believes, were those very wild things he finds he cannot live without: the attraction of jackpines and a kinship with pasque flowers, sandhill cranes, and woodcocks. These wild things, we might say, manifest *staying powers*, counterbalancing counterforces that enable these farmers to be resolute and stay put in the face of the pull of what would be the too-much-technology of "our opponents." Speaking of the woodcocks' evening sky dance, he concludes:

> The drama of the sky dance is enacted nightly on hundreds of farms, the owners of which sigh for entertainment, but harbor the illusion that it is to be sought in theaters. They live *on* the land, but not *by* the land.
>
> *(1948: 34, emphasis mine)*

Wild things as good in their own right do not need to be made over into an entertainment commodity – an amusement for the entertainment center or the Internet or a videogame.

Leopold's point, applicable to the earlier corn farmers, is of particular significance for the mountainous regions of the American West. Rural counties here have generally been decreasing in population. However, counties near Wilderness areas (and all of these presently are mountainous areas) have seen an increase in population. People seem to want to live near the mountains. Some of them are like the stickers of Sand County, or like McKibben whose desire to live not only spatially close to the mountains but by the mountains has generally meant a "lower standard of living" (133). However, many of these homes are not in established towns, such as Bozeman, but rather they hug the Wilderness areas with homes big enough to be a Junior High, bespeaking conspicuous consumption. Such mountain-choking developments do much harm – their ecological and carbon footprints, habitat loss, compromises to natural beauty, and so on – to the very thing they profess to love. And these technological cocoons, ATVs (Wuerthner 2007), and associated airports are filled with enough tempting distractions to overwhelm any substantive but inchoate tie to the mountains that the families may have. Following Leopold's line of logic, they live spatially *close to* the mountains but not *by* the mountains. (The same can be said of communities writ large; they do not identify themselves as "Yellowstone Ecosystem Towns," unless that means living *off* the Ecosystem.)

The steep slopes and deep powder of mountains can test and show the masterful abilities of skiers, giving skiers something to live for in the winter, when skiing is a focal practice (Borgmann 1995). Some skiers do live simply by skiing, by the mountain, but most seem diverted by displays of affluence, the glamour, and glitz of it. The famous resorts are likened to more outdoor shopping malls, attracting nearly as many non-skiers as skiers, and status and celebrity have become important than the physical activity (Coleman 2004). Skiing equipment itself often compensates for lack of skill. "You can ski the moguls at 60 as though you were 40." A resort like Montana's Big Sky, a development that began in the 1960s when I was a kid, significantly compromises The Greater Yellowstone Ecosystem, separating off from the rest the Spanish Peaks range to the north.

When we really take up with mountains as good in their own right, as "centering things" (Borgmann 1992), technology quietly becomes displaced and balanced as a relative good. Metaphorically, technology, used appropriately, is good insofar as it gets us to the trailhead, but when it becomes the answering machine to the call of the wild, it separates us from those more important wild things, often destroying them. A new and vital relationship to nature, living *by* the mountains, calls for this new relationship to technology. When mountains animate us in this way, we are really *celebrating* nature rather than respecting it solely altruistically, especially if respect means keeping away from it. With a new relationship with technology, mountains are acknowledged as something more and other than a resource in the service of consumption (or, alternatively, as bearing intrinsic value from the standpoint of an objective and impersonal land pyramid).

To generalize, technology, on my account, taken up appropriately is not an absolute good to be maximized and universalized, that is, the seemingly unstoppable pattern of progress to include more and more within its framework of commodification; rather, technology, used appropriately, is a relative good, needing to be related to centering things like mountains that are other and better than itself. These more and other mountains that matter teach us the relative worth of technology and, through our comprehension of the cost of its excesses, teach us to *limit* technology.

At the core of conventional modern technology is a kind of asymmetry between humans and the world that is radically anthropocentric: our overpowering modern technology flattens and puts everything under our control and at our disposal (at least attempts do so), generating a kind of heedlessness (Strong 1995: 63–73). Without limitations, this asymmetry and hyper-anthropocentric flattening – the heedless seeing, heedless thinking, and heedless power that see only coal beneath mountaintops – will continue to expand and displace nearly all other relationships to our surroundings. This flattening also extends to ourselves, as we will see. To overcome flattening anthropocentrism, technology requires a counterbalance. If we live by them and not just on them, these animating wild things, mountains in particular, provide this counterbalance.

Meeting Mountains on Their Terms: Preserving the Interanimating Symmetry

Stepping back, we can see that, once the issue is reframed, the deeper task is to challenge the ontology of technology. Given that is the task, it is essential that Leopold (or anyone concerned with nature) understand explicitly (1) how the framework of technology hangs together, as we've seen above, (2) why exactly it follows, in Leopold's words, "a law of diminishing returns," and (3) what it will take to turn and counterbalance technology. For the second point, Borgmann's notion of the device in relation to the promise of technology would help Leopold to articulate what he mostly only senses, however keenly. Borgmann shows that modern devices, saturating our surroundings, evoke a life of consumption, where "to consume is to use up an isolated entity without preparation, resonance, or consequence" (1984: 53). Humans are not simply hardwired to want more and more insatiably and mal-adaptively because of our Pleistocene heredity, as Rolston and most economists believe (Rolston 2012: 197–198). Rather, we consume so much because we derive so little from it (Strong 1995: 87–94). Consumption as a way of life, Borgmann argues, leads to disengagement, disburdenment, disconnection, disembodiment, diversion, distraction, disorientation, loneliness, and other consequential disabilities (i.e. for human intelligence). In short, consumption encourages a passive, isolated, and unfulfilling way of life. Hubert Dreyfus and Sean Dorrance Kelly make a similar point memorably in terms of skill. The basic goal of technology, they explain, is to make everything accessible to everyone, improving our lives by making hard things easier, so that even a child could do them. The danger is

> not with individual gadgets, but the understanding of ourselves and of what we can aspire to that the technological way of life encourages. To aspire to a life that requires no skill to live it well is to embrace the flattened world of contemporary nihilism.
>
> *(2011: 214)*

As technology flattens mountains and prairies, it also flattens our own lives.

These de-skilled, fingertip control devices embody paradigmatically the fundamental transformation in the human relationship with their surroundings. With fingertip ease, technological devices overpower. This overpowering character of modern technology creates the *asymmetry* between humans and everything else suggested above. And it is this overpowering asymmetry that really is so troubling, and that mountains can teach us to reform.

In reference to the third point above, we need to understand what it will take to turn or reform technology. In other words, what will it take for us to outgrow technology as a way of life? One step is taken by recognizing the diminishing and negative returns of overconsumption. Another step is taken when people grasp the importance of wild things, mountains, as final things and see simultaneously how these things are threatened by technology displacing them. Mountains can lose their tops, as it were, or we can become too insulated and detached from them, or they can become simply crowded out of our too busy lives.

To address these points, stories and works of art generally become more primary than ecology from the standpoint of an environmental ethic within the philosophy of technology, although I would not want to overplay this point. Stories disclose the overlooked potential significance of mountains. Stories show us *animating nature*, the way the natural world in its own right calls for human participation and generates a higher quality of life. In contrast to "Thinking like a Mountain," where ecology is used to show how the mountain hangs together, in "On Top," Leopold's other mountain essay, Leopold does not point out ecological relationships to disclose the place and yet we are left with a poignant sense of loss, since, as we've seen, Leopold did not want to witness what roads had "done to it." What appeals to us about this place comes about through revealing stories – stories about humans in the place. It is a place whose remoteness and ruggedness make it inaccessible to all but those mounted on horses (or, in our time, I suppose, those willing to carry a backpack). The White Mountain's high plateau challenges young riders to be the first on top in the spring. Once there, the presence of the place is manifest sharply by its many moods. "The dullest rider, as well as his horse, felt these moods to the marrow of his bones" (125). The place could "invite" you to get down and roll; it made you aware that it was a rare day in "so rich a solitude." The great meadow on top, filled with "an infinity of bays and coves, points and stringers, peninsulas and parks" made each cove a new discovery and elicited from one the sense "that if anyone had been here before, he must of necessity have sung a song, or written a poem" (126). Surely, anyone out walking who has suddenly found him or herself singing, perhaps in response to a setting sun, can identify with something of this experience.

Leopold imagines an old founding rancher whose heart in his late years thrilled only to his bank account, his increased standard of living, and, in our terms, the prosperity of technology. Yet, an aspen with his initials shows that in his long forgotten youth, he, too, once felt spellbound by the glory of the mountain spring (127). Stories make us aware of the strength of the pulls and attractions of the place – of what animates human enthusiasm, activity, and poetry – and make us understand how diversions, roads, and automobiles threaten to displace, weaken, and destroy these powers. I do want to hike New Hampshire's White Mountains, but seeing bumper stickers bragging, "This Car Has Been To the Top of Mount Washington," does not make me or others want to include a hike to its top. The nearby Beartooth Highway may be "the most beautiful highway in the world," but we must remember that the superlative after all is being applied to a highway. Few would want another highway over the unroaded nearby Lake Plateau of the Beartooths. My essay here is the outcome of the story that began when a favorite trail in the Crazy Mountains was replaced by a road. We would be better off if more of us felt this kind of "road rage." Stories get us to comprehend animating nature juxtaposed to too-much technology, whether roads, subdivisions, broadcast towers, phone addiction, helicopter tourism in Grand Canyon, or merely a kid munching junk food while glued to a screen in the backseat of a van on the family vacation in Yellowstone.

To learn of mountains as good in their own right, to be called by them, requires this kind of lyrical or storied disclosure of them. These stories are experiential disclosures, often describing the times when the beauty or sublimity of the place draws us in, making our relationship with the place personal, intimate, and endearing. As Jack Turner lounges after summiting Grand Teton, something as a guide he's done countless times before, he suddenly hears white pelicans faintly clacking overhead, at the limit of visibility, at least a mile above the point on Earth he has struggled to reach. It is a moment of grace for him (1996: 69–80). These occasions are not, emphatically, disclosures that derive from a theoretical understanding of the matter; the connection between the self and the natural world is more intimate and direct than to allow for the intervention of a theory. Though there is self-transcendence even on the part of the imagined youthful founding rancher, it is not strictly speaking an ecologically informed transcendence, that is, one beyond an isolated individual who realizes that he or she is only a part of the greater ecological community, as Deep Ecologists put it. Nevertheless, it is transcendence, a sense of being, which the rancher drifted away from in his later years. The transcendence here is one found through intensive engagement, involvement, and participation with wild things that Henry Bugbee vividly depicts and studies so brilliantly in *The Inward Morning* (1999). In short, we learn another way to be.

Mountains show themselves as valuable in their own right when people respond to the mountains invitations to walk, hike, run, fish, climb, ski, or photograph them, but people in their response to these invitations show something of themselves as well. "On Top" shows who is and who is not equal to the top of mountain. Only people on horseback or backpack are capable of traversing its remoteness and are in a position to enjoy its intimate entertainments. Its tremendous lightning storms are capable of raising fear in any mortal, so we are told. Dreyfus and Kelly remind us that only people with their highly developed skills become equal to many of the kinds of things that make us special as human beings and keep our lives vital. If we want to be equal to mountains, we are required to stay or get in shape and actually get out there and take in the terrain with our lungs and legs; yet often getting out there in that way, just as getting to the top of the mountain, is really not enough. We need to maintain and cultivate our relationships with them over time, just as Thoreau did with Walden Pond and the woods surrounding Concord. Mountains are playing with us as much as we are playing with them when we are being with them on their own terms. Adequate accounts of just how they invite play, calm, joy, and solace do not come easily to many practitioners. Beyond getting out, really getting there, and getting something said, the mountain we have come to love may require restoration work and ripping roads. But this too, as the work of Eric Higgs has shown, can, if taken up with as a focal practice, restore us as well by restoring our ruptured bonds with the mountain (Higgs 2003).

Just as we come to grow and realize ourselves in meeting mountains on their own terms, mountains, too, simultaneously come to be realized in a way that would not be disclosed without the loving presence of humans, without your personal presence. Their staying powers, the fundamental ways they attract us, are exposed this way. Thus, the disclosure of mountains in this intimate sense is really a co-disclosure; mountains and ourselves are co-disclosed and co-animated through our engagement with them. If I may push the envelope here, we ourselves become simultaneously ecstatic as mountains stand forth and shine.

Unlike the deadening and destroying asymmetry of technological domination, there exists an interanimating symmetry between humans and wild things, such as Leopold's White Mountain, and this interanimating symmetry or correlational coexistence is being displaced by technology. This is the *mountainous* relationship we need to preserve for the sake of ourselves as well as the natural world, because technological commodification threatens to remove and flatten it.

Real, untamed mountains, freed from the threat of technological domination, have that lesson to teach us. Without untamed mountains, without these wild things, neither a disclosure of them or of us will take place. Mountains do not guarantee, of course, that this relationship will develop between particular humans and themselves, but their commodification will guarantee that it will not. In his early 20s, Turner tells us that he couldn't care less about Wilderness preservation.

> I simply liked climbing big walls and spires and exploring remote places, preferably before anyone else did. Like most rock climbers, I didn't even like to hike. I didn't know the name of a single wildflower.... (Years later, when I read Schopenhauer, I recognize myself in those days: 'in the mind of a man who is filled with his own aims, the world appears as a beautiful landscape appears on the plan of the battlefield').
>
> *(1996: 7)*

But the story of his life bears witness to how this narrow relationship with mountains can grow into something far wider and more meaningful, given a chance. The worth of our lives is at stake with mountains. By liberating these things (when we do not center our lives on technology), we liberate ourselves.

The old challenge mountains presented for technology was to bring them under control through tunneling under them, damming their streams, removing their tops, putting trams to their tops, or, more recently, making them easily available on the Nature Channel and Google maps. However, if we live by mountains, not merely close to them, they present a new and different challenge to the overall framework of technology. The new challenge is to interfere with this history – to reform technology, making room for mountains in their full presence and guarding our relationship with them. Mountains challenge us to meet them on their own terms, through engagement rather than domination. Here, in contrast to passive consumption, they serve as a guidepost for a new way to live our technology, a more thoughtful and mature way to be with mountains and a technology that serves them and the way of life they sponsor, where we learn to simplify our lives, leave behind most technological devices at the trailhead, and reburden ourselves again in order to enjoy a more vital life with these things.

Living by mountains, unless understood metaphorically, cannot be the only kind of "thing" that challenges the expanding framework of technology, for such a challenge would fail to concern most people. We've already seen how "wild things," not just actual mountains, are things that Leopold lives by, and we've seen how living by the land as a focal practice of farming, something spelled out much more explicitly in the works of Wendell Berry, differs with living merely on the land and otherwise trying to make a killing. McKibben's works also show us a "range" of possibilities for individuals and communities living by things rather than the consumption of commodities (14). Borgmann's philosophy of technology is helpful here, once again, for it shows that there exists a plurality of genuine things, both natural and cultural as well as urban and communal, that people may find, upon thoughtful consideration, that they cannot live without. These centering or "focal things" (Borgmann 1984) share with wild things a similar, if figurative, mountain of interanimating symmetry, for violin and violinist are required by each other. And they share a similar fate, e.g. the loss of habitat for musicians as symphonies around the country fold and live music is squeezed out by electronics. Once they understand what they have in common, a coalition of these minorities of practitioners will be in a better position to protest successfully the place of technology and interfere with history than would lovers of literal wild mountains alone (Strong 1995: 205–216) (or waiting for the silver bullet of some universal principle to gain widespread acceptance). But that would mean that these practitioners of focal things all understand – through a variety of mediums – what "the largest movement in the world" really is, how it threatens their particular focal things, and what effective steps in common, protecting things and guarding our relationship with them, can be taken to avoid such a fate. We need to learn to live by the mountains.

Related Topics

8. Forests
10. Wilderness
11. National Parks
12. Landscape
15. Wetlands
16. Rivers and Watersheds

References

Borgmann, A. (1995) 'The nature of reality and the reality of nature', in Soule, M. and Lease, G. (eds.) *Reinventing nature: responses to postmodern deconstruction*. Washington, DC: Island Press, 31–45.

Borgmann, A. (1992) *Crossing the postmodern divide*. Chicago: University of Chicago Press.

Borgmann, A. (1984) *Technology and the character of contemporary life: a philosophical inquiry*. Chicago: University of Chicago Press.

Brokaw, C. and Gillum, J. (2013) Prairies vanish in the US push for green energy. Retrieved from: http://bigstory.ap.org/article/prairies-vanish-us-push-green-energy

Bugbee, H. G. Jr. (1999) *The inward morning*. Athens: University of Georgia Press.

Callicott, J. B. (1987) *Companion to a sand county Almanac*. Madison: University of Wisconsin Press.

Coleman, A. G. (2004) *Skiing style: sport and culture in the rockies*. Lawrence: University of Kansas Press.

Dreyfus, H. and Kelly, S.D. (2011) *All things shining: reading the western classics to find meaning in a secular age*. New York: Free Press.

EPA (n.d.) Land-use overuse. Retrieved from: http://www.epa.gov/agriculture/ag101/landuse.html

Heidegger, M. (1977) *The question concerning technology and other essays*. William Lovitt, NY: Harper and Row.

Higgs, E. (2003) *Nature by design: people, natural process, and ecological restoration*. Cambridge, MA: The MIT Press.

Lark, T.J. (2020) 'Protecting our prairies: research and policy actions for conserving', *Land Use Policy* 97, 104727. Retrieved from: https://reader.elsevier.com/reader/sd/pii/S0264837717310372?token=D-46918FE1E5FE5A2147B899EC120BDC04F9F0EEE7BF247AC8A0376D1DB97A8F0F64BAADF-669C8F8E6B23842A8A46DBA2

Lark, T.J., Salmon, J.M. and Gibbs, H.K. (2015) 'Cropland expansion outpaces agricultural and biofuel policies in the United States', *Environmental Research Letters* 10, 044003. Retrieved from: https://doi.org/10.1088/1748-9326/10/4/044003.

Leopold, A. (1968) *A sand county almanac and sketches here and there.* New York: Oxford University Press.

McKibben, B. (2005) *Wandering home: a long walk across America's most hopeful landscape, Vermont's Champlain Valley and New York's Adirondacks.* New York: Crown Journeys.

Peck, R. (n.d.) 'Appalachian heartbreak', *National Resource Defense Council.* Retrieved from: https://www.nrdc.org/land/appalachian/files/appalachian.pdf

Rolston, H. III (2012) *A new environmental ethics: the next millennium for life on earth.* New York: Routledge.

Sonner, S. (2013) 'Western Governors' Association unveils multi-state wildlife habitat mapping project', *Huffington Post*, 12 December [online]. Retrieved from: http.//www.huffingtonpost.com/2013/12/12/western-governors-association-tool_n_4431914.html

Strong, D. (1995) *Crazy mountains: learning from wilderness to weigh technology.* Albany: State University of New York Press.

Turner, Jack (1996) *The abstract wild.* Tucson: University of Arizona Press.

Wuerthner, George, ed. (2007) *Thrillcraft: the environmental consequences of motorized recreation.* San Francisco: Foundation for Deep Ecology.

Further Reading

Abbey, E. (1991) *Desert solitaire: a season in the wilderness.* New York: Ballantine Books.

Abram, D. (1996) *The spell of the sensuous: perception and language in a more-than-human world.* New York: Pantheon Books.

Louv, Richard (2008) *Last child in the woods: saving our children from nature-deficit disorder.* Chapel Hill, NC: Algonquin Books.

Mitchell, Richard G. Jr. (1987) *Mountain experience.* Chicago: University of Chicago Press.

Snyder, Gary (2010) *The practice of the wild.* New York: North Point Press.

10

WILDERNESS

Michael Paul Nelson

2020 Introduction

I wrote the original version of this essay more than a quarter of a century ago, in 1993, for a wilderness anthology that J. Baird Callicott and I published in 1998.[1] As I review the essay for reprinting in this book, it is August of 2020. The world is in the throes of a hugely mishandled (at least in my country) and deadly global pandemic, an ever-looming climate crisis, an episode of massive biodiversity loss, and a long-overdue and painful social and racial reckoning. We sit, it seems, at the dark center of a Venn Diagram from hell,[2] deeply uncertain about even the immediate future. I confess that it feels weird, uncomfortable, inappropriately opulent, and possibly tone-deaf to write about arguments for, and critiques of, wilderness at this moment in time.

Some of the arguments in this essay also touch – sometimes indirectly, and sometimes awkwardly – on these prodigious current issues. I can imagine someone suggesting that in this time of upheaval and uncertainty, we need wilderness more than ever, urging that wilderness and areas thusly designated are a small gesture or symbol of respect for the more-than-human world that we are so desperately lacking. The argument would suggest that the pandemic, climate crisis, and biodiversity loss are entwined manifestations of human hubris and the utterly horrible relationship with nature embodied by the dominant Western worldview. Wilderness, as the epitome of the natural, is one of the few elements of humility in a world(view) gone mad, and we must fight fiercely for its preservation if we are to survive. I'm not sure what to think about this argument, especially when I weigh it against the need for social and racial justice, and acknowledge that certainly wilderness recreation and nature-based recreation more generally are largely pursued by those who are white, wealthy, and male.[3] Or when I consider the history of protected areas whose establishment was created by the removal of people of color native to that place.[4] I don't know whether wilderness is yet another quaint and outdated (and, at times in the past, blatantly racist) concept demonstrating our failed worldview, or whether it is a salve for our current woes – I really don't. But as you read, consider, and discuss this essay, I invite you to wrestle with these and other challenges presented by the concept of wilderness.

———————

DOI: 10.4324/9781315768090-13

The original version of this chapter first appeared in J. Baird Callicott and Michael P. Nelson, eds. *The Great New Wilderness Debate.* **(Athens, GA: University of Georgia Press, 1998):154–98.**

For more than 100 years, people have spoken powerfully in defense of those places we call wilderness. In this essay, I summarize, review, and critique their arguments.

It is important to review such arguments for the sake of historical interest, and to observe how our received view of wilderness[5] is manifest in such arguments. The rationales we employ on behalf of anything, including wilderness preservation, reflect our attitudes and values, and, in turn, our environmental policies. As Roderick Nash observes, "So it is that attitudes and values can shape a nation's environment just as do bulldozers and chain saws."[6] If, has been suggested, we need to rethink our classical concept of wilderness – and therefore our current policies with regard to those places taken to be wilderness – a review of where we came from can surely aid us in such an undertaking.

There are tensions in such a project. First, most of the wilderness preservation arguments below take the existence of wilderness for granted. The received view of wilderness, however, has been subject to intense debate since the early 1990s. Second, all of the following arguments for wilderness preservation are biased in certain ways; perhaps the most important is that the received view of wilderness and attendent arguments for wilderness preservation have a distinct Australian-American bias. Arguably, the presence of designated wilderness areas, or at least some recent memory of or belief in a once-pristine landscape, is required to hold a received view of wilderness. Many parts of the world lack a wilderness narrative that would allow them to cognitively align with the received view. As a result, the received view of, and arguments on behalf of, wilderness eminates largely from a distinct cultural perspective: namely, Protestant Christian, colonial, and post-colonial cultures in particular.

In the amalgamation of wilderness preservation arguments that follow, there is a general attempt to move from narrowly instrumental, egocentric, and anthropocentric values to broader social, biocentric, and even intrinsic values attributed to putative wilderness.

1. The Natural Resources Argument. Many of those places referred to as wilderness contain a variety of natural resources we can utilize to render our future more secure. This is perhaps the most narrowly anthropocentric, instrumental, and simple-minded wilderness preservation argument one could advance.

Upon reflection, however, this argument seems at least paradoxical, if not thoroughly self-contradictory. In theory at least, designated wilderness areas are places where the extraction of resources is strictly prohibited. Can we really extract resources from an area and still call it wilderness? Perhaps, if we could harvest on a very small, sustainable, and "non-trammeling" scale. For some, however, even small-scale natural resource exploitation runs contrary to the received view of wilderness as pristine. Moreover, resource use is a matter of degree and "small scale" is a relative term. Such ambiguities leave this argument for wilderness preservation paradoxical and troubling. Less problematic, perhaps, is the argument that we could envision wilderness areas as resource reserves to be exploited only by future generations. If we are saving vast resource-rich areas, not exploiting them, we are preserving them. And if those future generations used those presently preserved wilderness areas for resource extraction, then they would, at that point, no longer be wilderness areas. Resource reserves, it would seem, can only be wilderness areas in so far as their use is potential and not actual.

2. The Hunting Argument. One of the earliest wilderness preservation arguments asserted that the so-called wilderness areas were worthy of protection because some of them provided terrific venues for hunting, supplying the natural resource of wild game.

This special case of wilderness preservation as a sort of big, fierce, almost tribal proving grounds was (is) especially championed (or ridiculed) because of its identification with toxic masculinity.

Nowhere was this association of wilderness preservation, hunting, and masculinity more vehemently expressed than in the thoughts and writings of Theodore Roosevelt. Referring to modern Americans as overcivilized, slothful, and flabby, Roosevelt calls upon Americans to regain and develop those "fundamental frontier values," to lead a "life of strenuous endeavor," and to revel in the "savage virtues."[7] For Roosevelt, wilderness hunting was the means to accomplish this end:

> Every believer in manliness and therefore in manly sport, … every man who appreciates the majesty and beauty of wilderness and of wild life, should strike hands with the farsighted men who wish to preserve our material resources, in an effort to keep our forests and our game beasts, game-birds, and game-fish--indeed, all the living creatures.[8]

There seems two issues with this argument. First, this argument pertains only to some putative wilderness areas and not others: namely, it applies to those places containing animals that humans desire to hunt, but not to those largely devoid of big game. Second, a century later, the big-game-hunting argument for wilderness preservation might be viewed as an embarrassment to some wilderness advocates; it is tantamount to arguing for the establishment of zoos as places where people could go to taunt animals. More recently, many might ban hunting from wilderness areas as the epitome of an intrusive, exploitative, and destructive use of wild places.

3. The Pharmacopoeia Argument. Another special case of wilderness resource extraction is medicine. According to this argument, the areas of the Earth commonly referred to or thought of as wilderness contain and support the most species on Earth. Since around 80% of the world's medicines are derived from life forms,[9] these "wilderness" areas contain a great source of medicinal natural resources. Donella Meadows called this the "Madagascar periwinkle argument," referring to the celebrated rosy periwinkle (*Catharanthus roseus*) plant of Madagascar from which were derived the drugs vincristine and vinblastine, used in the treatment of leukemia.[10]

This argument seems unpersuasive if constructed in terms of the *proven* medicinal uses of wild species, since many medically useful species can be cultivated in plantations and laboratories, or their active ingredients can be isolated and synthesized. The argument is most forceful in reference to potential, and yet unknown, medicinal uses of wild species.

This argument deserves comment as well. First, only some designated wilderness areas in North America and Australia are species-rich rainforests and old-growth forests. Hence, the argument does little to support their preservation. Second, it might be noted that the people most knowledgeable about the medicinal uses of rainforest species are the local indigenous inhabitants. At times, these inhabitants were, in fact, removed from the land to create the so-called wilderness, since an area inhabited and used by humans is not, by some definitions, wilderness.

4. The Service Argument. In addition to natural resources, wilderness areas provide innumerable and invaluable services as well. Wetlands protect important river headways, and unbroken forests and oceans remove carbon dioxide from the atmosphere and replenish its oxygen – all valuable to human well-being.

While we might realize wilderness as a sufficient condition for the performance of these services, is it a necessary one? These services are not unique to uninhabited or uncultivated places; they are performed by non-wilderness ecosystems as well. Nevertheless, there are certain ecological services which can only be performed in large tracts of relatively untouched land. Some designated wilderness areas provide nurseries for species such as salmon, such as the Salmon-Huckleberry Wilderness in Oregon. Conservation biologists tell us that certain species, like grizzly bears, require large tracts of unbroken land to persist; and the Earth's oceans help to moderate global temperatures. While certain non-wilderness areas do perform some of these same services, they might not perform them as efficiently and parsimoniously. And some wilderness areas might be irreplaceable sources of important services.

Potentiality comes into play in this argument as well. Since we cannot be entirely confident of all the unique and crucial services provided by wilderness, for their *potential* services we ought to preserve them.

5. The Life-Support Argument. Holmes Rolston explains that there exists "a parallel between the good of the system and that of the individual." Further, we depend upon the healthy functioning of ecosystems: "So far as [we] are entwined with ecosystems, our choices... need to be within the capacities of biological systems, paying some attention to ecosystem value."[11] This prudential argument was made by thinkers like Rachel Carson, Barry Commoner, and Anne and Paul Ehrlich. The Ehrlichs likened species eradication to the popping of an airplane's rivets, with rivet-popping eventually leading to the demise of spaceship Earth.[12]

If this argument is to be persuasive, the preservation of wilderness must be linked with the preservation of species, and the preservation of species must be linked with human survival. These links seem questionable. Are species rivets? Is Earth a spaceship? Also at issue is taking survival in a literal sense. Even if rivets, we might survive some rivet-popping, even if only in diminished numbers as impoverished beings in an ecologically impoverished environment. In sum, the persuasiveness of such an argument is dependent upon a number of other truths.

6. The Physical Therapy Argument. One might argue that wilderness area-related activities are wonderful ways to enhance and even remedy our physical health. Primitivists, like Wilderness Society co-founder Robert Marshall, claim that the more closely we are associated with nature, the more physically healthy we will be. If wilderness is the purest representative form of nature, we would be healthier if we took our physical exercise there, it might be argued.

This may appear to be quite a weak argument for wilderness preservation since people in many parts of the world are physically fit despite having no access to designated wilderness areas. Exercising in would-be wilderness is at best only a sufficient, not a necessary, condition for physical well-being. Wilderness advocates might still try to argue that these proffered wilderness areas are the *best* source and measure of physical health.

It seems that there are actually two separate arguments in his classification: one is that designated wilderness areas provide us with a source and measure of physical health and the other is that these places serve as a great place to engage in certain sports (which may also aid our physical health), the latter we might call the Arena argument.

7. The Arena Argument. Even more elementary than supplying a source and measure of physical fitness, wilderness preservation is sometimes urged on the grounds that many designated wilderness areas provide us with superb and incomparable locales for athletic and recreational pursuits. Aldo Leopold saw wilderness areas as

> a means of perpetuating, in sport form, the...primitive skills in pioneering travel and subsistence...a series of sanctuaries for the primitive arts of wilderness travel.[13]

But we might ask why we need designated wilderness areas to do these things. These days, we often can maintain our fitness by running on the indoor treadmill, swimming laps, and going to spinning classes at the local gym. With regard to the "primitive" forms of exercise, I can paddle my canoe in a dam-created reservoir, hike in an industrial monoculture forest, and climb on a climbing-wall. The wilderness advocate may respond by claiming that "artificial" places are pale substitutes for the "real" thing; that designated wilderness areas provide the *best* locales for these sorts of activities.

The persuasiveness of this argument depends on untangling a number of issues. Are wilderness areas really the *best* locales for physical fitness? By what standard? Are they even necessary locales for physical fitness? Likely not. Should we preserve the so-called best means to a worthy end when there are many perfectly adequate means to the same end?

8. The Mental Therapy Argument. Perhaps we can defend the so-called wilderness on behalf of its actual and potential *psychological* health benefits.

Reflecting on the increasing public desire to visit America's National Parks, John Muir cites psychological dysfunction as a major cause. Muir refers to citified Americans as "tired, nerve-shaken, overcivilized," "half-insane," "choked with care like clocks full of dust," and bursting with "rust and disease," "sins and cobweb cares."[14] For Muir, visiting designated wilderness solves these problems; these places are "fountains of life."[15] Others (from Sigurd Olson to Wallace Stegner to Sigmund Freud) have argued that civilization represses, frustrates, and often breeds unhappiness and discontent in humans which can best be alleviated by periodic escape to what they took to be wilderness.

None of the proponents of the mental health argument for wilderness preservation explain exactly *how* the existence of designated wilderness areas contributes to sanity. Nor do they explain why there are so many seemingly psychologically fit people who have never visited a designated wilderness area, while there are "wilderness junkies" who are arguably not mentally healthy. However, even if these areas of alleged wilderness are only sufficient conditions for mental well-being, one might still wish to argue that they are both the *best* means of assuring mental health, and far more socially and individually cost-effective than therapists, 12-step programs, and prisons (many of the same counter-arguments about the "best" that we saw just above in the last paragraph of "The Arena Argument" apply here).

9. The Art Gallery Argument. Many search out putative wilderness for the aesthetic experience of beauty and sublimity. "Wild" places, it is argued, are like gigantic art galleries, and we should preserve them for that reason.

In fact, some have argued that aesthetic experience, of the sort so-called wilderness offers, can border on the religious or mystical. William Wordsworth wrote that experiencing the beauty of what his vision of wilderness was produces "a motion and a spirit, that impels...and rolls through all things."[16] Some even argue that designated wilderness areas are places where the very meaning of aesthetic quality can be ascertained and that, therefore, beauty itself is dependent upon such places. Muir, for example, boldly asserts that "None of Nature's landscapes are ugly so long as they are wild."[17]

According to this argument, these places are both necessary and sufficient conditions for a true sense of beauty. Hence, if the loss of this beauty is to be avoided, the preservation of wilderness areas is mandated.

10. The Inspiration Argument. For some, wilderness areas are important to maintain because they provide inspiration for the artistically and intellectually inclined, thereby helping to shape culture. Painter Thomas Cole, writer Edward Abbey, musician John Denver, and poet Robinson Jeffers all found their muse in what they took to be wilderness.

Some even assert that wilderness serves to inspire those in the intellectual arts as well. Some environmental philosophers, for example, regard what they take to be wilderness experience as a contemplative catalyst or cognitive genesis for the really big questions of philosophy: What is nature? What are humans in relationship to nature? How then ought we to live? The point is not that wilderness areas are the *only* muse for art; rather, they are excellent and unique ones, and it would be tragic to lose any such inspirational kindling.

11. The Cathedral Argument. For some, putative wilderness is a site for spiritual, mystical, or religious encounters. They are places to experience mystery, moral regeneration, spiritual revival, meaning, oneness, unity, wonder, awe, inspiration, or a sense of harmony with the rest of creation – all essentially religious experiences. Hence, we should no more destroy wilderness areas than we should turn the Sistine Chapel into a grain silo.

For Muir, "wilderness" was the highest manifestation of nature and so was a "window opening into heaven, a mirror reflecting the Creator," and all parts of it were seen as "sparks of the Divine

Soul."[18] He called wilderness desecrators "these temple destroyers, devotees of ravaging commercialism," suggesting that we might just "as well dam for water-tanks the people's cathedrals and churches, for no holier temple has ever been consecrated by the heart of man."[19] Transcendentalists, such as Emerson, Thoreau, and William Cullen Bryant, went so far as to claim that one could only genuinely understand moral and aesthetic truths in what they took to be a wilderness setting.

This might be quite powerful political argument for legally preserving designated wilderness areas. In the United States, for example, if the sorts of experiences and activities just listed are presumed to be essentially religious experiences and activities, then designated wilderness areas can be said to serve a religious function. Hence, designated American wilderness areas could be defended on the Constitutional grounds of freedom to worship as one so chooses.

12. The Laboratory Argument. Some (mostly wildlife, marine, and conservation biologists) argue that the preservation of designated areas of wilderness is important because it provides scientists with an unprecedented location and raw materials for certain kinds of scientific inquiry. Their scientific study is important not only for the sake of knowledge itself but also more instrumentally because a society can use this knowledge to form a better understanding of itself, the world around it, and hence it's proper role in that world.

Admittedly, there is a potential paradox involved with this argument – as there is with all wilderness preservation arguments that entail the use areas of putative wilderness by humans. If we use wilderness areas as laboratories (or gymnasiums, cathedrals, resource pools, etc.) in too dense or intrusive a fashion, they cease to be wilderness. Moreover, who determines, and how do they determine that point of transition?

13. The Storage Silo Argument. Taking the Laboratory argument one step further, it is often asserted by conservation biologists, among others, that many supposed terrestrial and aquatic wilderness areas are worth saving because they contain vast amounts of biodiversity, or what E.O. Wilson refers to as the "diversity of life." Beyond the argument that humans have no moral right to muck with the evolutionary and ecological workings of *all* ecosystems of a given type, maintaining these genetic reservoirs intact is instrumentally important because they function as a great safety device – holding a large portion of the world's accumulated evolutionary and ecological wisdom as they do. Hence, some wilderness areas can serve as places where various forms of biodiversity can be stored for a time when they might be needed for genetic engineering, agricultural rejuvenation, or some other crucial purpose. Hence, biotically rich and untrammeled wilderness areas store the information that can help us to better manage and rebuild our natural environment. Certain wilderness areas, some believe, are, therefore, the key to life on Earth.

It is foolish to knowingly extirpate natural processes: "To keep every cog and wheel is the first precaution of intelligent tinkering," as Leopold puts it.[20] As Holmes Rolston analogizes, "destroying wildlands is like burning unread books."[21] Though the preservation of biodiversity is arguably both wise and moral, a challenge with this argument is arguably the unclear correlation between designated wilderness areas and biodiversity.

14. The Standard of Land Health Argument. Aldo Leopold's arguments on behalf of wilderness preservation shifted and expanded as his thought progressed. For the later Leopold, wilderness areas serve as a measure "of what the land was, what it is, and what it ought to be," providing us with both an actual ecological control sample of healthy land and a normative measure of what we ought to strive toward. Areas of "wilderness" can serve, according to Warwick Fox, "to warn us of more general kinds of deterioration in the quality or quantity of the free 'goods and services' that are provided *by* our 'life support system'," like the proverbial canary in the coal mine.[22]

We would need a lot of wilderness to accomplish this end. In order to serve as a measure of land health, it is argued, designated wilderness areas must be large and varied, for there are many distinct types of biotic communities. Leopold tells us – in a passage remarkably in line with contemporary conservation biology – that "each biotic province needs its own wilderness for comparative studies of used and unused land… In short all available wild areas, large or small, are likely to have value as norms for land science."[23]

Although this appears to be a strong argument for preserving designated wilderness areas, it also appears, at least in part, to buy into the ontologically impossible notion that there are places totally untouched and unaffected by human actions. In order for "wilderness" to serve as a standard of land health, we must recognize that "wilderness" needs to be conceived of as a relative and tenuous concept.

15. The Classroom Argument. Those of us who regularly or only occasionally visit designated wilderness areas are aware that these locales often function as a sort of classroom for valuable lessons. They are unparalleled placed to learn about the workings of nature, to learn tangible skills such as navigation and survival, and to attain a "sense of place" or feeling for a particular geographic region and it's features. Wilderness experiences can help us put elements of our lives in perspective, or put our priorities in order. They can teach us proper values and a sense of valuation, and they can provide a sense of humility and help build our characters. Periodic trips to designated wilderness areas force us to recognize our proper place as stewards, not masters, of the land, endow us with long-sighted and ecological vision, train us to make better public-policy decisions, furnish us with a sense of individual responsibility, promote our self-confidence and self-image, teach us how to cooperate successfully with others, how to assess and take wise and appropriate risks, instill within us a reverence for all life and a proper sense of beauty. Groups such as Outward Bound and the Girl and Boy Scouts of America, for example, utilize designated wilderness areas as classrooms for just these reasons.

Two critical notes. First, does this classroom argument for the preservation of designated wilderness areas hold true for all of these areas? And second, while it may be true that the presence of designated wilderness areas is a sufficient condition for providing these valuable lessons, is it a necessary one? Or, is the argument that these are induplicable locales for environmental, life, and ethical wisdom? This argument appears to be weakened if either it does not hold true for all designated wilderness areas or there are other ways or places to learn these lessons.

16. The Ontogeny Argument. *Homo sapiens* and their communities, like all species and their respective communities, are deeply entrenched in nature and nature's processes. In other words, we fit our context. One might go on from this simple premise to argue that since what is thought of as wilderness is the paradigmatic form of nature and its processes, that this wilderness is and continues to be the source of our evolution. Hence, putative wilderness in its many forms not only ought to but must be preserved for continued human evolution.

"The love of wilderness…," writes Edward Abbey, "is…an expression of loyalty to the earth, the earth which bore us and sustains us, the only home we shall ever know, the only paradise we ever need--if only we had the eyes to see." Abbey asserts that the destruction of these would be a sin, then, against our origins, or the "true original sin."[24] Similarly, Walt Whitman believed that those who remain closer their evolutionary context become better people: "Now I see the secret of the making of the best persons./ It is to grow in the open air, and to eat and sleep with the earth."[25]

There is as irony in this argument. If "wilderness" is a place where humans are but visitors who do not remain or an area beyond what is cultivated or inhabited continuously by humans, how in the world could it be our literal ancestral home? If we have evicted ourselves from our ancestral paradise home irrevocably, how can wilderness be the locus or future of our human evolution?

17. The Cultural Diversity Argument. In one of his most familiar works, "Walking," Thoreau writes:

> Our ancestors were savages. The story of Romulus and Remus being suckled by a wolf is not a meaningless fable. The founders of every state which has risen to eminence have drawn

their nourishment and vigor from similar wild sources… In such soil grew Homer and Confucius and the rest, and out of such a wilderness comes the Reformer eating locusts and wild honey.[26]

So it is argued that, just as human beings generically derive from a context, specific cultures are derived from and are dependent upon a certain ontogenetic[27] context. And the wide variety of cultural variation or diversity stems from the fact that there have been a wide variety of natural ecosystems. A wide variety of designated wilderness areas ought to be preserved, it is thought, because they function as the foundation for the world's myriad cultures.

There are problems with this argument. If the diverse cultures remain in the wilderness that gave them birth, then the places they are in are not, by definition, wilderness. If various areas of the world are *designated* wilderness areas and the cultures they spawned are expelled from them (if these cultures still exist), as the Ik were from the Kidepo in Uganda and the !Kung were from Etosha National Park in Namibia, then the cultures are exterminated in order to preserve the wildernesses that spawned them. To argue that areas of wilderness ought to be preserved as museum pieces in honor of a part of the environment which helped shape each unique culture commits the fallacy of appeal to tradition. Slavery helped to form much of the American Deep South (and much of the rest of America, for that matter), and various acts of violence, wars, and systems of patriarchy have helped to mold world cultures. Should remnant enclaves of these institutions likewise be preserved? Should we designate one or two counties in Mississippi as Old South cultural reserves in which the plantation system, including slavery, is preserved?

18. The National Character Argument. In the United States, some view designated wilderness areas as monuments symbolically enshrining national values – places where the eagle flies, the buffalo roams, and the deer and the antelope play. Wallace Stegner calls the wilderness idea "something that has helped form our character and that has certainly shaped our history as a people…part of the geography of hope."[28]

There is yet another inherent paradox involved with the above three ontogeny-type arguments, and especially with the National Character argument. If wilderness preservation is a good thing because it is a representation of Euro-American ontogeny, then wilderness destruction seems to be a good thing as well. That part of our Euro-American ontogeny most recently evolved is the tendency to work to destroy, or at least severely alter and interrupt, what colonizing Euro-Americans took to be wilderness. So, we might argue that Euro-Americans and Euro-American wilderness colonization and destruction are part of Euro-American national character. The root problem is to suggest that something is good and worthy of perpetuation merely because it is part of our heritage. Slavery, violence, racism, and sexism are parts of our heritage too, but that is no reason to maintain as treasured national institutions. These are all versions of the Genetic Fallacy: the fallacious argument suggesting that someone, some institution, or some practice is good (or bad) because of its origins.

19. The Self-Realization Argument. One of the fundamental tenets of Deep Ecology is the notion of Self-realization. Deep Ecologists assert that in order truly and appropriately to perceive and understand the world, our place in it, and our duties to it, we must first dismiss the assumed but inaccurate bifurcation between self and nature, and instead grasp the depth of the relational reality of the totality. In addition to the general character- and self-image-building mentioned above, wilderness preservation becomes crucial for Deep Ecologists because designated wilderness areas are, for them, necessary components in this process of Self-realization – a sort of asylum of reorientation where this relational self ideally can take form. No designated wilderness areas; no Self-realization; no Deep Ecology; no proper view of self and world; no appropriate treatment of self and world; no continued self-existence.

This argument appears to fall prey to the recurring problem of confusing necessary and sufficient conditions. Merely because designated wilderness areas provide an arguably *sufficient* condition for Self-realization, it remains to be proven that they are a *necessary* condition for such Self-realization.

20. The Disease Sequestration Argument. Since most of all species host viruses, we can assume that there are at least as many viruses as there are living species, if not some multiple of that number. A successful virus either does not, or "learns" not to, kill its host so as to have a residence for its continued existence. However, as human population continues to surge and human invasion into places such as tropical habitats becomes more prevalent, humans cross never before traversed ecological and spatial boundaries, encountering new species. As humans intrude, put pressure upon, or destroy supposed wilderness, viruses will adapt and jump hosts, taking humans as their new hosts. Known viruses such as Guanarito, Marburg, Q fever, and Monkeypox are part of a long list of viruses which have jumped to human hosts with deleterious results. Ebola Zaire, a lethal airborne filovirus emerging in 55 African villages in 1976 subjecting nine out of ten of its victims to hideous deaths within days of infection, is an example of an encountered virus even more lethal than HIV.

Since these disease-causing agents are for the most part sequestered in many of the Earth's remaining wild areas, our intrusion into these areas is a serious issue. So, not only do many of the wild areas of the Earth serve as a disaster hedge by acting as a buffer, but much ostensible wilderness left intact is argued to protect us from potential viral and bacterial decimation.

However, this argument appears to be less an argument for preserving designated wilderness areas and more an argument for not intruding into those areas where these viruses thrive; and less an argument for preserving the Bob Marshall Wilderness Area and more an argument for staying the hell out of some very specific places. Moreover, there are elements of paradox and irony here: these lairs of deadly viruses are not exactly those places that we wish to *visit* for recreation, inspiration, religious awe, etc.

21. The Salvation of Freedom Argument. Edward Abbey defends the preservation of what he imagines to be wilderness for what he calls "political reasons": claiming we need designated and *de facto* wilderness areas, whether or not we ever set foot in them, to serve as potential sanctuaries from oppressive governmental structures. As Abbey writes, "We may need it [wilderness] someday not only as a refuge from excessive industrialism but also as a refuge from authoritarian government, from political oppression...as bases for guerrilla warfare against tyranny."

In order to attempt to alleviate the obvious charge that this is only a paranoid survivalist-type fantasy, Abbey cites as a historical fact that the worst of the world's tyrannies have occurred in those countries with the most industry and the least of what Abbey thinks of as wilderness. Centralized oppressive domination flourished in Germany, Hungary, and the Dominican Republic, according to Abbey, because "an urbanized environment gives the advantage to the power with the technological equipment." However, more rural insurrections, such as those in Cuba, Algeria, the American Colonies, and Vietnam, have favored the revolutionaries since there remained in those countries "mountains, desert, and jungle hinterlands," or areas of would-be wilderness.[29]

Abbey's evidence is quite shaky since the counter examples of the Soviet Union, with its vast expanses of "wilderness," and Cuba, since Castro, throw a monkey wrench into his political theory. Moreover, Michael Zimmerman has chronicled how the Nazi Germans advocated nature preservation more ardently than any other Europeans.[30]

Obviously, not all designated wilderness areas would provide us with base camps for guerrilla warfare against tyrannical governments, and one has to allow the possibility at least that there might be some non-wilderness areas which would. This argument seems, then, to be more an argument for areas from which to oppose tyranny rather than necessarily an argument for the preservation of designated wilderness areas. Moreover, the fact that something would provide us with the means to oppose tyrannical government is *not* sufficient justification for its preservation. Private atomic-bomb factories in each of our basements would put us in a good position to oppose tyrannical governments too, but clearly (or hopefully) Abbey would not have advocated *that*.

22. The Mythopoetic Argument. Some contemporary thinkers argue that wilderness preservation is critical for mythopoetic reasons, or as the optimum location for viewing the history of myth and absolutely crucial for the building of the myth of the future. Putative wilderness areas function as an essential source of meaning, vital for the future of humanity.

The "Men's Movement" of the 1990s is an example of a mythopoetic use of putative wilderness. In their search for a new way of understanding or realizing their manhood, those involved in the movement would gather in "wild places" because they sought to recover the roots of their maleness as hunters and companions of animals originally thought to be found in "wilderness." Robert Bly (the central figure in the "mythopoetic" branch of the movement) promoted the ideal of the "wildman." Men should strive, according to Bly, to be whole, healthy, and energized by realizing the "wildman" deep inside their psyches. Male mythologizing, accordingly, requires areas of would-be wilderness, and for those in the Men's Movement and others, without this "wilderness" a sense of the history of, standard for, or place for the future of myth-building is lost.

This argument appears to rest on some unproved premises: that "wilderness" is the sole or best source of future mythologizing, that our future is impoverished without this source of this kind of myth, and that areas other than designated wilderness areas cannot serve as adequate or even more appropriate mythological sources.

23. The Necessity Argument. As we have seen above, certain wilderness advocates believe in the truism that, historically speaking, the civilized world would not have evolved without the prior existence of wilderness. As Leopold claimed, "wilderness is the raw material out of which man has hammered the artifact called civilization."[31] One might argue that "wilderness" is *necessary* in a more philosophical sense, claiming that an idea or concept of wilderness is logically and metaphysically necessary (in other words, a sense of necessity that is not historically dependent) for the existence and complete understanding of the concepts of culture and civilization.

There can be no finish without a start; no good without evil; no yin without yang; and (according to folk-singer Arlo Guthrie) "no light without a dark to stick it in." Similarly then, wilderness might be said to be logically (wherever "civilization" has meaning, "wilderness" does also) and metaphysically (where one thing exists, the other must also) necessary for a complete and proper understanding of civilization. The move, then, to rid the world of designated wilderness areas, is tantamount to the attempt to deny and dismantle a necessary component for a complete understanding of the world.

Although some may contend that this is more of an explanation than a justification for wilderness, and that it does not require the preservation of, at the most, a very small bit of wilderness (and at the least only the concept of wilderness and hence no areas of wilderness at all), others might argue that to truly understand concepts such as culture, civilization, freedom, primitive, development, and perhaps others, we need physical wildness (such as that found in designated wilderness areas) to serve as a model or foil of contrast.

24. The Defense of Democracy Argument. Enemies of wilderness preservation and environmentalism in general are fond of charging wilderness advocates and environmentalists with the sin of elitism. Environmentalists, they allege, are selfish people who want to set aside vast stretches of land that only the physically fit and economically able can experience. Wilderness preservation, then, only benefits the elite few and therefore does not serve the general welfare – the greatest good for the greatest number.

We might also turn this argument on its ear and claim that precisely because wilderness preservation shows respect for the needs of a minority, wilderness preservation is, therefore, indicative of good democracy. The existence of things like opera houses, softball diamonds, art galleries, public swimming pools, and designated wilderness areas – all places used by only a minority – is a display of respect for minority rights, or as Nash says, "the fact that these things can exist is a tribute to nations that cherish and defend minority interests as part of their political ideology."[32] Concerning such an argument, Leopold wrote:

> There are those who decry wilderness sports as 'undemocratic' because the recreational carrying capacity of a wilderness is small, as compared with a golf links or a tourist camp. The basic error in such argument is that it applies the philosophy of mass-production to what is intended to counteract mass-production… Mechanized recreation already has seized nine-tenths of the woods and mountains; a decent respect of minorities should dedicate the other tenth to wilderness.[33]

One might also argue that designating areas as wilderness does not limit, but rather opens, access to more people than would privatization or land development (especially if we consider future generations of humans who would also require access since these wilderness areas are public access lands.

25. The Social Bonding Argument. One might argue that many designated wilderness areas serve as a valuable mechanisms in the critical process of social bonding. Those who collectively recreate in designated wilderness areas might attest to the benefits coming from social interaction in such a setting.

Because, and to the extent that, we are social animals and our continued survival depends on effective social interaction, we ought to value social bonding. Exposure to "wilderness" is thought to intensify experiences and provide a vector for high level and successful interpersonal cohesion. Therefore, as an important social mechanism, designated wilderness areas are assumed to be valuable and worthy of preservation.

While designated wilderness areas might be sufficient and effective ways to facilitate social bonding, they are not clearly necessary conditions for this end. Social bonding can and does take place in classrooms, theaters, offices, salons, and saloons. Conversely, we can just as easily imagine the cohesiveness of a group being destroyed by the pressures of wilderness travel. And furthermore, the use of designated wilderness areas by groups of people might be seen as jeopardizing the "wilderness" integrity of a place.

26. The Animal Welfare Argument. One might claim that, like us, wild animals also depend upon their respective home environments for their existence. Since they should be considered members of the community of beings deserving moral consideration, we owe it to those animals not to destroy their homes. Along similar lines, it was popular in the 1970s to argue that the "wild" things on Earth had the right to go about their business unmolested or unharmed by *Homo sapiens*, that we humans had no right to interfere with their freedom and dignity. Since designated wilderness areas are the places where this freedom was most fully realized, these areas should be preserved as "reservoirs of freedom for biota" or "harbors of ecological freedom."

It must be recognized that most of the animals that animal welfare ethicists are concerned with do not depend on designated wilderness areas for their habitat needs: chickens and cows live on farms, deer live in rural and exurban areas, cats and dogs often live in human homes. Indeed, most animals, both wild and domestic, do not require "wilderness" at all. This argument, then, would, at best, only apply to wilderness-dependent animals, such as grizzly bears, wolverines, and others with significant habitat needs.

27. The Gaia Hypothesis Argument. Expanding on the above argument, we might apply non-anthropocentric moral consideration to yet another sort of living organism, namely, the Earth or Gaia. Scientists, such as James Lovelock, defend what they refer to as the Gaia Hypothesis which postulates that the Earth itself, as a self-correcting system, is alive or is tantamount to a living organism. Like any living thing, certain of its parts are imperative to its proper functioning and viability. With regard to the Earth, or Gaia, certain wild ecosystems are arguably vital to its prospering, likened to the internal organs of a multicelled organism. Without a liver a human cannot live; without wilderness Earth perhaps cannot either. If planetary homeostasis is to continue, and Gaia is to live, areas of "wilderness" must be preserved. In sum, if we owe moral consideration to living beings, and if the Earth itself is alive, then the Earth deserves moral consideration. Disrupting Gaia's vital organs, like putative wilderness, becomes immoral.

The objections to the above two arguments are too complex to cover here. In sum, these arguments rest on the premise(s) that animals or Gaia deserves moral consideration, which corresponds to actions of preservation on behalf of humans. Clearly, one would have to justify this argument prior to make the larger argument stick. Moreover, with regard to the Gaia argument, one would have to provide good reason to believe that designated wilderness areas really are crucial to the life of Gaia.

28. The Future Generations Argument. A common defense of wilderness preservation revolves around the supposed moral obligations humans have to future generations of humans. One might maintain that, among other debts owed, current humans ought to pass the world on to future generations as we inherited it, with as many designated wilderness areas intact as possible. Destroying putative wilderness areas, when taking into account these future generations, would then become a matter of injustice and unfairness.

The truth is we really do not know for certain what future generations will want or need: perhaps they will not want or need designated wilderness areas. However, one might argue that this lack of knowledge is a reason to save designated wilderness areas, not ravage them. Since we do not know what future generations will want or need, we should keep their options open. Wilderness preserved provides options for future generations. They can keep and use their wilderness areas or they can develop them. Interestingly, however, all future generations might arguably be forced to preserve designated wilderness areas because they would be locked into the same "logic of wilderness preservation for future generations" argument that compelled us to save wilderness for them. This logical compulsion might then propel wilderness preservation indefinitely.

29. The Unknown and Indirect Benefits Argument. For some, we ought to err on the side of caution and preserve and designate more wilderness areas since most of the benefits emanating from these areas are thought to be indirect or unknown. The potential for goods and services may be unlimited, and the possible harms unknown. E.O. Wilson maintains that, "The wildlands are like a magic well: the more that is drawn from them in knowledge and benefits, the more there will be to draw."[34] But if we destroy designated wilderness areas, we would then apparently destroy tremendous amounts of information and potential benefits along with them.

For certain wilderness proponents, the promise of wilderness preservation lies ahead, in the future. Aldo Leopold resonated this sentiment when he suggested that we "should be aware of the fact that the richest values of wilderness lies not in the days of Daniel Boone, nor even in the present, but rather, in the future."[35] Hence, because we have a responsibility to follow a wise course of action, and since this would include conserving potential benefits to all of humanity, we apparently then have a responsibility to keep designated wilderness areas in existence.

30. The Intrinsic Value Argument. Some wilderness boosters claim that simply knowing there exists designated wilderness areas, regardless of whether or not they ever get to experience such areas, is reason enough for them to want to preserve wilderness. For these people, "wilderness" is valuable just because it exists, because it is. Designated wilderness areas, in this sense, have value in and of themselves – value regardless of, or in addition to, their value as a means to some other end – like clean water, recreation, or medicine. "Wilderness," then, is said to possess intrinsic value.

If we can justify the intrinsic value of designated wilderness areas, then wilderness preservation immediately becomes a moral issue, since designated wilderness areas would gain considerable ethical clout. This condition changes the argument about wilderness preservation quite a bit. Now, those who would destroy designated wilderness areas would have the burden-of-proof; they would have to demonstrate that something of great societal value would be lost if a wilderness area stood in its way. They would have the task of showing that something possessing intrinsic value should be sacrificed for the sake of something of merely instrumental value. "Wilderness is innocent 'til proven guilty," David Brower once quipped, "and they're going to have a tough time proving it guilty."[36]

Notes

1 J. Baird Callicott and Michael P. Nelson, eds. *The Great New Wilderness Debate* (Athens, GA: University of Georgia Press, 1998), p. 154–198.

2 Kathleen Dean Moore and Michael Paul Nelson, "The Venn Diagram from Hell," *Philosopher's Magazine*. 89(2), pp. 84–90.

3 https://outdoorindustry.org/resource/2019-outdoor-participation-report/, accessed August 12, 2020.

4 See, for example, Mark David Spence, *Dispossessing the Wilderness: Indian Removal and the Making of the National Parks* (London: Oxford University Press, 2000), and Philip Burnham, *Indian Country, God's Country: Native Americans and the National Parks* (New York: Island Press, 2000).

5 From J. Baird Callicott: "The idea of wilderness we have inherited—received—from its framers, going back now at least several centuries, but shaped most fully during the first half of the 20th century." The received wilderness idea is eloquently conveyed in the definition of "wilderness" in the oft-quoted Wilderness Act of 1964:
A wilderness, in contrast to those areas where man and his own works dominate the landscape, is hereby recognized as an area where the earth and its community of life are untrammeled by man, where man himself is a visitor who does not remain.
(Public Law 88-577)
In Callicott, "Contemporary Criticisms of the Received Wilderness Idea," in Cole, David N., McCool, Stephen F., Freimund, Wayne A., O'Loughlin, Jennifer, comps. 2000. *Wilderness Science in a Time of Change Conference— Volume 1: Changing Perspectives and Future Directions*, 1999 May 23–27; Missoula, MT. Proceedings RMRS-P-15-VOL-1. Ogden, UT: U.S. Department of Agriculture, Forest Service, Rocky Mountain Research Station.

6 Roderick Nash, "The Value of Wilderness," *Environmental Review* 3 (1977), p. 25.

7 Theodore Roosevelt, "The Pioneer Spirit and American Problems," in *The Works of Theodore Roosevelt* (New York: Charles Scribner's Sons, 1923), vol. 18, p. 23.

8 Roosevelt, "Wilderness Reserves; The Yellowstone Park," in *Works*, vol. 3, pp. 267–68.

9 N.R. Farnsworth, "Screening Plants for New Medicines," in E.O. Wilson, ed., *Biodiversity* (Washington, DC: National Academy Press, 1988), pp. 83–97.

10 Donella Meadows, "Biodiversity: The Key to Saving Life on Earth," *Land Stewardship Letter*, Summer 1990.

11 Holmes Rolston III, "Valuing Wildlands," *Environmental Ethics* 7 (1985), pp. 26–27.

12 George Sessions, "Ecosystem, Wilderness, and Global Ecosystem Protection," in Max Oeschlaeger, ed., *The Wilderness Condition: Essays on Environment and Civilization* (Washington, DC: Island Press, 1992), p. 99. Paul and Anne Ehrlich, *Extinction: The Causes and Consequences of the Disappearance of Species* (New York: Random House, 1981), pp. xi–xiv, 77–100, and Paul Ehrlich, "The Loss of Biodiversity: Causes and Consequences," in E.O. Wilson, ed., *Biodiversity*, pp. 21–27, both cited in Sessions.

13 Leopold, *A Sand County Almanac: With Essays on Conservation from Round River* (New York: Ballantine Books, 1966), pp. 269–272 (his emphasis).

14 John Muir, *Our National Parks* (Boston, MA: Houghton Mifflin and Co., 1901), chapt. 1.

15 Ibid., p. 1.

16 William Wordsworth, "Lines Composed a Few Miles Above Tintern Abbey" (1978).

17 Muir, *Our National Parks*, p. 4.

18 Muir, quoted in Roderick Nash, "The Value of Wilderness," in Vance Morton, ed., *For the Conservation of the Earth* (Golden, CO: Fulcrum, Inc., 1988), p. 23.

19 Muir, *The Yosemite* (San Francisco: Sierra Club Books, 1988), p. 176–177.

20 Leopold, *A Sand County Almanac*, p. 190.
21 Rolston, "Valuing Wildlands," p. 28.
22 Warwick Fox, *Toward a Transpersonal Ecology: Developing New Foundations for Environmentalism* (Boston, MA: Shambhala, 1990), p. 158. (his emphasis)
23 Leopold, *A Sand County Almanac*, pp. 274 and 276, respectively. (my emphasis)
24 Edward Abbey, *Desert Solitaire: A Season in the Wilderness* (New York: Avon Books, 1968), p. 190.
25 Walt Whitman, *Leaves of Grass* (Ithaca, NY: Cornell University Press, 1961), p. 319. (originally published in 1860)
26 Thoreau, "Walking," pp. 185 and 191, respectively.
27 "Ontogeny" refers to the origination and development of an organism.
28 Stegner, "The Wilderness Idea," pp. 97 and 102, respectively.
29 Abbey, *Desert Solitaire*, pp. 148–151.
30 Michael Zimmerman, "The Threat of Ecofascism," *Social Theory and Practice* 21 (1995), pp. 207–238.
31 Leopold, *A Sand County Almanac*, p. 264.
32 Nash, "The Value of Wilderness," p. 24.
33 Leopold, *A Sand County Almanac*, pp. 271–272.
34 E.O. Wilson, *The Diversity of Life* (Cambridge: Harvard University Press, 1992), p. 282.
35 Leopold, "Wilderness Values," in *1941 Yearbook of the Parks and Recreational Services* (Washington, DC: National Park Service, 1941), p. 28.
36 Brower, quoted from *Wild by Law*, a film by Lawrence Hott and Dianne Garey, Florentine Films, 1990.

11

NATIONAL PARKS

Holmes Rolston, III

National/Natural Parks

In the United States, "national parks" are mostly "natural parks," though there are national historical parks, such as Gettysburg. Fifty-nine national/natural parks are operated by the National Park
Service. There are also national monuments, wildlife reserves, and wilderness areas (sometimes
located inside parks), where many of the concerns discussed here also arise. There are, of course,
national forests and Bureau of Land Management lands, which may be considered "working landscapes," subject to differing policy and ethical standards.

Elsewhere in the world, there are national parks, sometimes similar to the American model,
but often differing greatly in their blending of natural history and cultural landscapes. In Europe,
there are some 500 protected areas in 39 countries. Around the world, there are over 100,000
protected areas, managed with varying goals and efficiency. One agency that monitors such conservation areas is the International Union for Conservation of Nature (IUCN). American parks
are much visited by internationals, one of their favorite U.S. destinations. By some accounts, all
national/natural parks ought to be thought of as world heritage sites (and a dozen U.S. parks have
been so designated).

Americans parks are "national" as well as "natural" parks, established by an act of the U.S.
Congress, and so they are created both by nature and by Congress, or perhaps it would be better
to say *created* by nature and *designated* by Congress. Certainly, some areas are chosen to be set aside
as parks. Criteria include natural beauty, unique geological features, wildlife, ecosystems, and
recreational opportunities. This has triggered a debate about appropriate motivations – whether
the focus is the nation and its people, or respect for nature, biodiversity, wildlife, ecosystems – the
community of non-human life as well.

The National Park Service has, since its founding, had dual and sometimes conflicting goals:
both to conserve nature as scenic resources for recreation for people and to conserve nature unimpaired for future generations (*The Organic Act of 1916*, 16 USC, sec. 1–4 et seq). Unsurprisingly,
legislation has incorporated various motives. Yellowstone Park was described, in the enacting
legislation, as a "pleasuring ground for the benefit and enjoyment of the people" (16 USC, Chapter 1, Subchapter V, Yellowstone National Park, sec. 21, March 1, 1872). By later accounts, Glen
Cole says, "The primary purpose of Yellowstone National Park is to preserve natural ecosystems
and opportunities for visitors to see and appreciate scenery and native plant and animal life as it
occurred in primitive America" (Cole 1969). Yellowstone ought to be a biotic whole, a natural
community, and humans ought to learn to take pleasure in it.

DOI: 10.4324/9781315768090-14

David Graber explains what he calls "the dilemma of wilderness in national parks":

> Wilderness has taken on connotations, and mythology, that specifically reflect latter-twentieth-century values of a distinctive Anglo-American bent. It now functions to provide solitude and counterpoint to technological society in a landscape that is managed to reveal as few traces of the passage of other humans as possible…. This wilderness is a social construct.

To think of these natural parks as being primitive or elemental nature is a myth of the urbane, mostly urban mind, the city folk who vacation there, enjoying a couple weeks of counterpoint, going "back to nature" in their campers and motor homes (Graber 1995, pp. 123–124).

Park interpreters agree with preserving nature for visitors to appreciate, though many will hesitate at going back to "primitive" nature. On a few occasions, the national parks have thought of creating "a reasonable illusion of primitive America" (Leopold et al 1963, p. 4). This may be appropriate in certain places, such as the pioneer homesteads in Cades Cove in Great Smoky Mountains National Park. But this is ill-conceived as a general national/natural parks policy. Parks are not museums of past nature (with some exceptions such as Dinosaur National Monument). We do not, or ought not, visit parks to go back to something. Parks are for preserving ongoing, continuing natural processes, not for deep-freezing in time vignettes of a pre-Columbian America. Park conservationists more often speak of "wild" nature, or "natural" processes, of nature "untrammeled" by humans, "spontaneous" nature, of values "intrinsic" to nature "in itself" or "on its own."

Sometimes, this challenge is taken further: "'Pristine nature' does not exist." Indeed, there has been none for the millennia humans have been agents on the planet (Kareiva and Marvier 2012, p. 965). So we can't save what isn't there. Such wild nature is a myth: "We create parks that are no less human constructions than Disneyland" (Kareiva, Lalasz, and Marvier 2011, p. 31). Wilderness advocates, however, may wonder if anyone who makes such a claim has ever done a backcountry trek in Yellowstone or the Bob Marshall Wilderness.

Park interpreters do wish to conserve traces of ancient native Americans in parks, but they mostly see the socially constructed nature issue as academic arcane wrangling, metaphysics or epistemology, with little effect undercutting their "realist" conviction that in national/natural parks we really do encounter what is "out there," independently of humans, in the spontaneously given, created world of evolutionary and ecological natural history. This knowledge is aided as much as obscured by what scientific naturalists have described and discovered. Yes, the trees were not green until I arrived, nor will they be after I am gone. But yes, more certainly, those trees were there (and photosynthesizing) before I came and will continue to do so after I am gone. We do know natural kinds, breeding populations, birth, death, wolves killing and eating elk, seasons passing, species lines ongoing in a life-death-life-death-life-death lineage, evolutionary natural history, mosquitoes biting, lightning starting fires, and geysers erupting. People go to parks to find nature, and they can do so successfully.

People in Parks/People vs. Parks

Parks have many people, about 300 million per year. The most-visited, the Great Smoky Mountains, has had over 9 million visitors in recent years, followed by the Grand Canyon with over 4 million. Parks are at risk of being "loved to death." People benefit from national parks. So park policy is to get people into parks, typically to provide campsites, perhaps also lodges, which people desire – but policy is often reluctant to enlarge these, attempting to pack the people into sacrifice areas, keeping as much wild as possible. Parks provide roads, trails, but confine backpackers to designated sites, and restrict their use – limited numbers of sites, away from streams and lakes. Officials may also restrict off-trail travel, especially in sensitive areas, such as alpine zones.

People in parks should appreciate and respect what is naturally there, rather than expect to be entertained. From about 1871 onward, the Park Service at Yosemite would build an enormous fire on the lip of Glacier Point at dusk. "Indian Love Call" was played, and the fire pushed over the cliff to the "ahs!" of spectators (Yosemite National Park 1941). In the early 1960s, the firefall was stopped as an inappropriate activity. In 1881, the Yosemite Park Service cut a tunnel through a giant sequoia, the Wawona tree. The tunnel was 8-feet wide, 10-feet high, and 27-feet long. People from all over the world drove first stagecoaches and later automobiles through it. The tree blew over in storms in 1969, perhaps weakened by the tunnel. Facing requests to cut another tunnel sequoia, the Park Service refused. A drive-through sequoia would mutilate for whimsy a majestic living thing. Visitors should rather learn why sequoias attain such age and size, or why this relict species has survived geological changes that destroyed its relatives. This change of ethic can be interpreted either or both as promoting a more appropriate human experience and as respect for value in the sequoia tree itself, having a good of its own.

In Yellowstone, park officials long put soap into geysers to break the surface tension and time the eruptions conveniently for tourists. Old Faithful was improved with colored floodlights and background music between eruptions. But no more, no floodlights at all at night. Enjoy it as it is, not enhanced for tourist pleasures. In Petrified Forest National Park, there is a famous log, also called "Old Faithful," on the Giant Logs trail, 35-feet long and weighing 44 tons. The log was hit by lightning in 1962, which broke off portions of the log. Park resource managers worked hard to repair the damage, so that hikers could see the huge fossil as it was. Today, the Park Service has changed that policy (ethic?), believing, as the Trail Guide says, that "it is best to let nature take its course." The park is not so much a museum as a place for ongoing natural processes.

Almost nobody can enjoy Turner Avenue in Mammoth Cave National Park. Discovered in 1954, these are rooms laden with gypsum crystals spun as fine threads, a rare formation known as "angel hair." So fragile are these needles that humans passing through and disturbing the air destroy the hairthin filaments. Since 1957, this part of the cave has been closed, never visited by tourists and only on exceptional occasions by mineralogists. A nonbiotic work of nature is here protected at the cost of depriving humans of access to it. This park policy is partly for humanistic reasons (to preserve angel hair for scientific research). But it also involves an appreciation of angel hair as a project of spontaneous nature. Angel hair counts morally in the sense that something naturally created results in a value that lays a claim on human behavior.

People are regularly told what they cannot do in parks. Members of the Hopi tribe, native Americans in Arizona, wish to engage in a ceremony that requires killing golden eagles. The eagle is captured as a chick, kept well, even reverenced, for months, then ritually suffocated, sending the spirit of the eagle to fly to the world of their Hopi ancestors, informing the ancestors of what the Hopis need in today's world, and many Hopis are poor. The ancestors engage powers that ensure that these needs are met. The eagle chicks are taken from Hopi-sacred lands, but these are now often in national parks and monuments. Although the Hopis received permission from the US Fish and Wildlife to take up to 40 eaglets, they were first refused by National Park Service officials, on grounds of wildlife conservation. In particular, they were refused admission to Wupatki National Monument, outside Flagstaff, Arizona (Shaffer 1999; Stevens and Velushi 1999).

The tribe protested and the Park Service reversed their decision, allowing them to capture eaglets and sacrifice them in their ritual (Revkin 2000). They also concede that such permission might be expanded to other areas. Subsequently, in 2012, the U.S. Fish and Wildlife Service gave a permit to the Northern Arapaho Tribe in Wyoming allowing them to kill two bald eagles. The Park Service cites preserving cultural riches as well as natural resources as part of their mandate. Most environmentalists find the decision more politically correct than morally correct. Native

Americans have a rich relationship with the natural world; they ought to be able to find other enabling rituals that do not require killing the eagles. More bluntly put, we analyze and critique our heritages and need not save every superstition among our cultural riches, especially if this requires cruelty to animals. The U.S. citizens have engaged in many behaviors in national parks of which they are now ashamed, such as shooting the wolves in Yellowstone. We shot over a hundred times as many wolves over the turn of the twentieth century as the Native Americans wish to sacrifice eagles.

Devil's Tower National Monument in Wyoming is a huge igneous magna intrusion into the sedimentary rocks of the Great Plains, 867-feet high – phonolite porphyry, similar to granite but without quartz. The Lakota tribe respects the Tower as sacred; it is a focal point of an annual June run of 500 miles in five days, when they gain spiritual renewal. Lakotas call it the Lodge of the Bear, and believe that it was created to save some children fleeing an angry bear. They would prefer that others not visit the Tower, at least during their ceremonies. Rock climbers, about 4,000 each year, also wish to scale the tower, and in 1995 the Park Service requested climbers not to climb in June and tried to ban professional guides in that month. Lakotas did not want others "climbing their church." The guides have protested the formal ban, but for the most part, climbers and other visitors defer voluntarily to the Native Americans in June.

Parks may be threatened as much by what people do outside the park and which comes in as by human behaviors within the park – for example, toxics such as acid rain, or nitrogen pollution from agricultural fertilizers. Hemlock woolly adelgids (*Adelges tsugae*), an invasive exotic species rather like aphids, introduced through horticulture, have eliminated hemlock trees in Shenandoah National Park (as well as most of the Eastern United States). A threat to all parks is climate change. Park managers often have little control over such outside threats, though they do what they can. The Obama administration, through then-Interior Secretary, Ken Salazar, ruled that a million acres around the Grand Canyon were off-limits to uranium mining for 20 years (the longest time possible under the current law). This was to protect the park from contamination endangering people and wildlife and also with concern about protecting Colorado River water, which provides drinking water for millions of people (Barringer 2012).

Many of these could be called "management issues." Park decisions may favor "hands off" rather than "hands on" management, at least of the wildlife and their systems; they may think more of managing people than managing nature. Some of these decisions are made without asking much about any moral element – for example, fishing regulations often have to be worked out in junction with state fish and wildlife agencies. Looked at more deeply, however, there are often background moral dimensions. Rocky Mountain National Park permits catch and release of greenback cutthroat trout, an endangered species. This gives anglers excitement, but does it harass the trout, or risk introducing them to pathogens and diseases? Perhaps respect for the species should trump anglers' pleasures.

About 40% of Yellowstone waters were barren of fish, largely because downstream waterfalls were too high for fish to jump. Early Yellowstone managers thought it wise to introduce fish for the pleasures of anglers, and from 1889 onward stocked millions of rainbow, brook, brown, and lake trout in many waters there. But, the Park in 1936 set a policy that the remaining barren waters should be left barren. The term "barren" is really a misnomer, because the waters are not barren of the "lower" forms of life. The tiger salamander (*Ambystoma tigrinum*) flourishes there in a niche fish might otherwise occupy; the phantom midge (*Chaoborus*) becomes dominant in the aquatic ecosystem, and even dragonflies and diving beetles play different roles. One ought to respect and appreciate an anomalous kind of richness in the barren waters. The charismatic species do not have all the grace of life, nor is the biodiversity all in the higher forms (Varley and Schullery 1983).

So people can and ought to enjoy their national parks, but the ideal is only as nature takes its course. This presents challenges not only for managing people but equally for managing wildlife.

Wildlife: "Let Nature Take Its Course"

One February morning in 1983, a bison fell through the ice into the Yellowstone River, and, struggling to escape, succeeded only in enlarging the hole. Toward dusk, a party of snowmobilers looped a rope around the animal's horns and, pulling, nearly saved it, but not quite. It grew dark and the rescuers abandoned their attempt. Temperatures fell to twenty below that night; in the morning, the bison was dead. The ice refroze around the dead bison. Coyotes and ravens ate the exposed part of the carcass. After the spring thaw, a grizzly bear was seen feeding on the rest, a bit of rope still attached to the horns.

The snowmobilers were disobeying park authorities who had ordered them not to rescue it. The snowmobilers were troubled by the callous attitude. A drowning human would have been saved at once; so would a drowning horse. The Bible commends getting an ox out of a ditch, even if this means breaking the Sabbath. The poor thing was freezing to death. A park ranger replied that the incident was natural and the bison should be left to its fate (Robbins 1984).

A snowmobiler protested, "If you're not going to help it, then why don't you put it out of its misery?" But mercy-killing too was contrary to the park ethic, which was, in effect: "Let it suffer!" That seems so inhumane, contrary to everything we are taught about being kind, doing to others as we would have them do to us, or respecting the right to life. Isn't it cruel to let nature take its course?

The snowmobilers thought so. One contacted radio commentator Paul Harvey, who made three national broadcasts attacking park service indifference: "The reason Jesus came to earth was to keep nature from taking its course." Was the Yellowstone ethic too callous, inhumane? This ethic seems rather to have concluded that a simple extension of compassion from human ethics or humane society ethics to wildlife is too nondiscriminating. To treat wild animals with compassion learned in culture does not appreciate their wildness. But some protest that we are carrying this let-nature-take-its-course ethic to extremes.

One April morning in Glacier National Park, a wolverine attacked a deer in deep snow but did not finish the attack, possibly interrupted by two workmen who saw the event from a distance, a rare sighting of an endangered species. The injured deer struggled out onto the ice of Lake McDonald, but, hamstrung, could move no further. Many visitors saw it; a photograph appeared in the local newspaper. Park officials declined to end its suffering. Possibly, the wolverine would return. So the lame deer suffered throughout the day, the night, and died the following morning (*Hungry Horse News* 1989). Can this be the right ethics for a wild animal, so inhumane and indifferent? Or has ethics here somehow gone wild in the bad sense, blinded by a philosophy of false respect for cruel nature? Park officials can sometimes be compassionate. The same spring that the lame deer was left to its fate, a bear was injured when hit by a truck, and Glacier Park officials mercy-killed the bear.

One Christmas Day in Theodore Roosevelt National Park in North Dakota, park visitors found two bucks with entangled antlers. One buck had already died and coyotes had eaten the hind parts, also nipping the rear of the live buck, emaciated from the ordeal. Taking compassion, the visitors sought the park ranger on duty, who did return to help them photograph the unusual event, but explained en route the park ethic. Wild animals should be left to their fates; humans should not interfere. The would-be rescuers seemed to agree. But that night they sneaked back to saw off the antlers of the dead buck. The freed buck trotted away; the rescuers left, with coyotes howling nearby, thwarted from their kill. "We're glad we had the opportunity to save his life," one of them said, although he faced a park citation and a $50–$100 fine. After all, the saving was on Jesus' birthday (*Dakota Country* 1988; Powell 1988).

The bighorn sheep of Yellowstone caught pinkeye (conjunctivitis) in the winter of 1981–1982. On craggy slopes, partial blindness can be fatal. A sheep misses a jump, feeds poorly, and is soon injured and starving in result. More than 300 bighorns, over 60% of the herd, perished (Meagher 1984; Thorne 1987). Wildlife veterinarians wanted to treat the disease, as they would have in any domestic herd, but the Yellowstone ethicists left them to suffer, seemingly not respecting their life. Their decision was that the disease was natural, and should be left to run its course. Had they no mercy? Was this inhumane?

But perhaps mercy and humanity are not the criteria for decision here. The ethic of compassion must be set in a bigger picture of animal welfare, recognizing the function of pain in the wild. The Yellowstone ethicists knew that, while intrinsic pain is a bad thing whether in humans or in sheep, pain in ecosystems is instrumental pain, through which the sheep are naturally selected for a more satisfactory adaptive fit. Pain in a medically skilled culture is pointless, once the alarm to health is sounded, but pain operates functionally in bighorns in their niche, even after it becomes no longer in the interests of the pained individual. Having interfered in the interests of the blinded sheep would have weakened the species. Simply to ask whether they suffer is not enough. We must ask whether they suffer with a beneficial effect on the wild population.

We treat our children who catch pinkeye. We put them to bed, draw the curtains, and physicians prescribe eyedrops with sodium sulfacetamide. The Chlamydia microbes are destroyed and the children back outside playing in a few days. But they are not genetically any different, nor will the next generation be different. When the grandchildren catch pinkeye, they will get eyedrops too. But that is an ethic for culture, where humans interrupt and relax natural selection. The welfare of the sheep still lies under the rigors of natural selection. As a result of the park ethic, only those sheep that were genetically more fit and able to cope with the disease survived; and this coping is now coded in the survivors. What we *ought* to do depends on what *is*. *The nature differs significantly from the culture, even when similar suffering is present in both.*

One spring, a sow grizzly and her three cubs walked across the ice of Yellowstone Lake to Frank Island, two miles from shore. They stayed several days to feast on two elk carcasses, when the ice bridge melted. Soon afterward, they were starving on an island too small to support them. The stranded bears were left to starve – if nature took its course. The mother could swim to the mainland, but she was not going to without her cubs. This time, park authorities rescued the mother and her cubs (Ozment 1984). The relevant difference was a consideration for an endangered species in an ecosystem, much interrupted by humans who have too long persecuted the grizzlies. A breeding mother and three cubs was a significant portion of the breeding population. The bears were not saved lest they suffer, but lest the species be imperiled.

It might seem now that, inconsistently, we refuse to let nature take its course. The Yellowstone ethicists let the bison drown, callous to its suffering; they let the starving bighorns die. But this time the Yellowstone ethicists promptly rescued the grizzlies and released them on the mainland, in order to protect an endangered species. They were not rescuing individual bears so much as saving the species. They thought that humans had already and elsewhere imperiled the grizzly, and that they ought to save this form of life.

Duties to wildlife are not simply at the level of individuals; they are also to species. Nor are they simply at the level of species; they are to these species in their ecosystems. Sometimes, that means, as with the sow grizzly and her cubs, that we rescue individual animals in trouble, where they are the last tokens of a type.

Recreation, Creation, Re-creation

Living and dying is the business of life; and now, in contrast to the daily business workplace, in national/natural parks, one is immediately confronted with life persisting in the midst of its perpetual perishing. True, one is on vacation; one doesn't want to be too somber. But the seasons are evident: spring with its flowering, fall with its dieback. I am at leisure, but the struggle out there is perennial; eating and being eaten, survival through adapted fit. Wild nature is a vast scene of sprouting, budding, flowering, fruiting, passing away, passing life on. Birth, death, re-birth, life forever regenerated – that is the law, the nature of life.

Recreation re-creates, and rejuvenates people when they are worn from work. But national/natural parks are much more, and more fortunately so. These grander parks in the greater outdoors provide dimensions of depth belied by their vacational and recreational settings. Parks are places where we can simultaneously get "away from it all," away from the workaday week, from the labors of town, factory, farm, and "back to it all," encountering elemental nature. One is immersed in natural history. No one sees the Grand Canyon aright unless helped by geologists. No one sees the Everglades without benefit of ecologists and botanists. Park visitors want to be naturalists as well as recreators. They are birdwatchers, fern enthusiasts, butterfly lovers, mineralogists; they want a scientific appreciation of what they see. They expect ranger-naturalists leading hikes and campfire programs in the evenings.

This is recreation, if you like, but this recreation is set within a search for natural history. Some of the recreation is relaxation; deeper down the re-creation is restoration of perspective on the nature of life, re-encounter with the creation. National parks recreation preserves human life by re-creating it. This enlarges our sense of duration, antiquity, awe, vastness, wonder. An immense stream of life has flowed over this continent Americans so lately inhabit, flowed there before us and still flows in these parks we visit, reminding us of our surrounding biosphere. When a tourist at Yellowstone learns that anaerobic bacteria still present in those steaming pools exist in an optimal thermal habitat that survives little changed from the time when life evolved in an oxygen-free atmosphere, and that further studies might furnish clues to the origins of life on Earth (Brock 1967), his or her recreation has touched creation.

America the nation state is but the tip of an iceberg; beneath the American society resident here lie the vast depths of the continental history. Scientific studies show that adults who take walks for exercise and restoration get more benefits when they walk in natural areas rather than in town (Berman et al 2008). Parks cure children (and adults) from "nature-deficit disorder" (Louv 2005). Citizens are better citizens who love their landscape – the purple mountains' majesties, prairies, and plains from sea to shining sea. Parks elicit cosmic questions, differently from town. The great outdoors works on a recreationist's soul, as well as on muscles and body. Parks generate experiences of "a motion and spirit that impels... and rolls through all things" (Wordsworth 1798).

We figure out who we are and where we are. The last part of that question must be answered in culture, for human life is "by nature" cultural. But the first part, the fundamental ground, is answered in nature. "By nature," too, we are embodied creatures, residents on landscapes, earthlings, placed in a more inclusive, more comprehensive community of life and life support. In that sense, natural parks protect a full answer to the question of human identity. Parks keep us from being entirely de-natured.

National parks enlarge our human identity in a singular way. In the defense of life on Earth since time immemorial, organisms have set up territorial boundaries. They defend their places, their resources, or else they cannot survive and reproduce. Humans can and ought to do this too. But now there is something new, never seen on Earth before during its billions of years of evolving species. Humans have set up boundaries in national parks, in wilderness areas, in biodiversity reserves; and we set ourselves apart by consciously setting apart bounded areas for others. Part of the planet is governed by culture; we also decide to set aside part to be governed by nature.

Roger DiSilvestro puts it this way:

> This is something truly new under the sun, and every protected wild place is a monument to humanity's uniqueness.... We not only can do, but we can choose not to do. Thus, what is unique about the boundaries we place around parks and other sanctuaries is that these

boundaries are created to protect a region from our own actions.... No longer can we think of ourselves as masters of the natural world. Rather, we are partners with it.

(DiSilvestro 1993, p. xiv–xv)

So parks recreation re-creates us by recognizing creation, of which we are both part, and which is apart from us. Yes, we are "stand outs" in the world; and equally there the world "stands out" over against us. In this dialectic, we rise to our fullest humanity and simultaneously get put in our place.

Now, paradoxically, a national park is not an American place at all, if by that we mean to Americanize it. What Americans wish to preserve there is places we humans only visit and where we resolve to leave Earth and its community of life on its own. In these places, almost enclaves in our national state, we need another Emancipation Proclamation. These are realms where Americans value freedom so highly that they give even wild nature the freedom to run itself. Life goes on – protected in the parks but on its own, wild and free. Americans do not want to be imperialists who trammel the wilds. And the most surprising benefit is that each new generation learns that, yes, culture transcends nature, but nature past, present, and future, once and future nature is the ground of culture. Humans emerge from nature, but nature is a womb that humans never entirely leave. These parks are presence and symbol of the timeless natural givens that support everything else. Americans, having been outdoors in their natural/national parks, return re-created with the signature of time and eternity.

References

Barringer, Felicity, 2012. "U.S. to Block New Uranium Mines Near Grand Canyon," *New York Times*, June 7, p. A16.

Berman, Marc G., John Jonides, and Stephen Kaplan, 2008. "The Cognitive Benefits of Interacting with Nature," *Psychological Science* 19: 1207–1212.

Brock, T. D., 1967. "Life at High Temperatures," *Science* 158: 1012–1019.

Cole, Glen F., 1969. *Elk and the Yellowstone Ecosystem*, Yellowstone Library, Yellowstone National Park, February.

Dakota Country, 1988. "Locked to the Death!" *Dakota Country*, March, pp. 12–13.

DiSilvestro, Roger L., 1993. *Reclaiming the Last Wild Places: A New Agenda for Biodiversity*. New York: John Wiley & Sons.

Graber, David M., 1995. "Resolute Biocentrism: The Dilemma of Wilderness in National Parks." Pages 123–135 in Michael E. Soulé and Gary Lease, eds., *Reinventing Nature? Responses to Postmodern Deconstruction*. Washington, DC: Island Press.

Hungry Horse News, 1989. "Nature Plays Cruel with Lame Deer," *Hungry Horse News*, Columbia Falls, MT, April 5: 1.

Kareiva, Peter, Robert Lalasz, and Michelle Marvier, 2011. "Conservation in the Anthropocene: Beyond Solitude and Fragility," *Breakthrough Journal*, No. 2, Fall: 29–37.

Kareiva, Peter, and Michelle Marvier, 2012. "What Is Conservation Science?" *BioScience* 62: 962–969.

Leopold, A. S., S. A. Cain, C. M. Cottam, I. N. Gabrielson, and T. L. Kimball, 1963. *Wildlife Management in the National Parks, Report of the Advisory Board on Wildlife Management*, March 4. Washington: U. S. Government Printing Office.

Louv, Richard, 2005. *The Last Child in the Woods: Saving Our Children from Nature-Deficit Disorder*. Chapel Hill, NC: Algonquin Books of Chapel Hill.

Meagher, Mary, 1984. Personal communication from Mary Meagher, Yellowstone Park Research Biologist, Yellowstone National Park, Wyoming.

Ozment, Pat, 1984. "Case Incident Record # 843601," Yellowstone National Park, filed August 18.

Powell, Robert D., 1988. Theodore Roosevelt National Park, "Case Incident Record # 870474," filed January 4.

Revkin, Andrew C., 2000. "U.S. Plan Would Sacrifice Baby Eagles to Hopi Ritual," *New York Times*, October 29, Sec. 1: 14.

Robbins, Jim, 1984. "Do Not Feed the Bears?" *Natural History* 93, no. 1 (January): 12–21.

Shafer, Mark, 1999. "Wupatki Won't Let Hopis Gather Golden Eagles," *Arizona Republic* (Phoenix) 31 July: A1–A2.

Stevens, Jan and Lukas Velushi, 1999. "Hopi Eaglet Ceremonies Thwarted," *Arizona Daily Sun* (Flagstaff) 29 July: 1, 11.

Thorne, Tom, 1987. "Born Looking for a Place to Die," *Wyoming Wildlife* 51, no. 3 (March): 10–19.

Varley, John and Paul Schullery, 1983. *Freshwater Wilderness: Yellowstone Fishes and their World.* Yellowstone National Park, WY: Yellowstone Library and Museum Association.

Wordsworth, W., 1798. *Lines Composed a Few Miles above Tintern Abbey.*

Yosemite National Park, 1941. *The Firefall--Explanation and History.* Yosemite National Park, CA: Yosemite Park and Curry Co.

12

LANDSCAPE

Allen Carlson

Landscape and Environmental Ethics

What are landscapes? Landscapes are not land. The land is what we stand on, live on. Yet as we experience the land, we create landscapes from the land and from that which constitutes it, from prairies and grassland, lakes and rivers, forests and mountains. Consider the following: you stand before a range of mountains. You are struck by the scenic prospect as you admire the ragged shapes of the rocky cliffs and the subtle colors of the forested slopes. You wonder at the towering, snow-capped peaks as you contemplate their complex nature and their ancient age. You reflect on the long history of this place and on what has happened here, perhaps imagining what you might encounter were you to explore its valleys and canyons. In this way, the experience of the land is transformed into the appreciation of a landscape. Landscapes are constructions of the human mind, composed by human thought and imagination (Santayana 1961). They are the creations of a mental process by which a spectator deems certain aspects of the land to be salient and thereby gives coherence, order, and unity to the resultant experience. Thus, landscapes, although constructed from the land, are yet one step removed from it, one step abstracted from the land.

So what is the relationship between landscapes and environmental ethics (Rolston 2002)? Aldo Leopold spoke of a land ethic, and from his thought and that of others, an environmental ethic has grown, in light of which it is not uncommon to accept an ethical responsibility for the land, even to think of having a moral regard and respect for it (Leopold 1966). But Leopold did not speak of a landscape ethic. Landscapes, as one step removed from land, are not immediate, direct objects of moral regard or respect. Rather, they are objects of aesthetic appreciation. In fact, aside from works of art, landscapes are the paradigmatic objects of aesthetic appreciation. Thus, the question of the relationship between landscapes and environmental ethics presupposes the more general question of the relationship between aesthetics and ethics. To address this question, it is initially important to recognize the powerful connection between our aesthetic appreciation of things and our attitudes and actions regarding them.

This connection between aesthetic appreciation and attitudes and actions is most evident in our treatment of art. When works of art, such as Leonardo's *Mona Lisa* (1503–1506), Handel's *Messiah* (1741), or Picasso's *Guernica* (1937), are judged to have an exceptional aesthetic value, they are preserved with great care and at great cost. And, in general, it is thought wrong to do otherwise. This connection is also clear in other domains of human experience. Contrary to the idea that our aesthetic appreciation of everyday events and activities is unimportant or trivial, it rather has considerable power in our day-to-day lives, which is demonstrated by the ways in which it is used

DOI: 10.4324/9781315768090-15

to achieve political, social, and commercial ends (Saito 2007). For example, Riefenstahl's film *Triumph of the Will* (1935) both depicts and itself exemplifies how aesthetic appeal was employed for social and political purposes in pre-World War II Nazi Germany. In a similar way, the persuasive power of the aesthetic is obvious in any modern political campaign. Perhaps the most familiar and striking use of aesthetics in shaping everyday attitudes and actions is in the contemporary consumer world where it helps sell everything from personal grooming products to shiny stainless steel kitchen appliances.

Nonetheless, other than works of art and perhaps even more so than matters of everyday life, it is landscapes over which aesthetic appreciation exercises its greatest power. Aesthetic appreciation influences attitudes and actions toward landscapes in that those that are judged aesthetically exceptional are thought to be, as great works of art, worthy of preservation. And it is considered a great loss when they are destroyed. Thus, aesthetic appreciation has played a role in the preservation of landscapes, especially in North America, that is hard to overestimate. For example, one prominent thinker in the field of environmental ethics claims that aesthetic appreciation "has made a terrific difference to American conservation policy and management" in that one of "the main reasons that we have set aside certain natural areas as national, state, and county parks is because they are considered beautiful," since the "kinds of country we consider to be exceptionally beautiful makes a huge difference when we come to decide which places to save, which to restore or enhance, and which to allocate to other uses" (Callicott 1994: 169–170).

Thus, in many areas of human experience—art, everyday life, and landscape—aesthetic appreciation greatly affects what we think and do. However, to confirm a significant relationship between aesthetics and ethics, it is necessary to determine if the attitudes and actions that follow from aesthetic appreciation are moral in nature. In general, this determination depends upon what things we hold to be proper objects of moral regard. For example, if we take other sentient creatures to have intrinsic value, then the issue of whether they should be preserved is a moral matter. If we do not, it may not be ethical, but only, for instance, prudential. Thus, since great art is typically deemed to have intrinsic value, the obligation to preserve it is a moral obligation. By contrast, many of the everyday products of commercial enterprise, such as the above-mentioned kitchen appliances, are not regarded as having intrinsic value, and thus our actions toward them are not considered ethically significant, even when these actions are largely guided by aesthetic appreciation.

What then about the relationship between landscapes and environmental ethics? The problem is that, as noted, although in the wake of the development of environmental ethics it is common to have moral regard for the land, landscapes, as one step removed from land, are not direct objects of moral regard. Here again, the comparison with art is instructive. Works of art, like landscapes, are one step removed from their source, in the case of art, the artist. However, even though one step removed, works of art are yet deemed to have intrinsic value when they are essentially tied to their human creators, for humans are a source of intrinsic value. In a similar fashion, landscapes can be deemed to have intrinsic value when they are essentially tied to land. In this case, landscapes, as land itself, have intrinsic value and are proper objects of moral regard—and their preservation an ethical responsibility. This is the foundation of the relationship between landscapes and environmental ethics.

However, there is a significant qualification concerning the intrinsic value of both art and landscapes. Concerning art, it is important to recognize that not all works of art have intrinsic value (unless by definition, in which case art without intrinsic value is deemed pseudo art). If art is trivialized by, for example, commercialization or gimmickry, it can be severed from human creativity and thus lose, or fail to attain, intrinsic value. Similarly, if a landscape is not solidly and fully grounded in the source of its intrinsic value, the land, then it too can lose, or fail to obtain, both intrinsic value and the status of being an object of moral regard. Thus, the way in which landscapes are created from the land by the human mind, how they are composed by human thought and imagination, is the key to the relationship between landscapes and environmental ethics. It is also the subject matter of various theories of landscape appreciation. As will become evident, some of these theories tend to weaken the link between landscapes and their potential source of intrinsic value, the land itself, and thus to undercut the relationship between landscapes and environmental ethics, while others tend to do the opposite.

Landscape and the Picturesque

In the Western world, the first major treatment of landscape appreciation is the theory of the picturesque. The idea of the picturesque emerged in the eighteenth century, when Western aesthetic thought reached an unprecedented level of development in the work of thinkers such as Archibald Alison, Edmund Burke, and Immanuel Kant. The central concept of this development was the notion of disinterestedness, which was held to be the criterion of the aesthetic. Proper aesthetic experience was to be disinterested and this meant that it was divorced from practical, personal, and utilitarian interests, such as, for example, the potential agricultural or resource production of the land. In this way, landscapes were abstracted from the land itself and could be aesthetically appreciated as things in themselves. These newly freed landscapes were initially appreciated as beautiful, sublime, or picturesque (Hipple 1957).

These three distinct categories of aesthetic appreciation were meant to apply to different kinds of landscapes: the beautiful to the pastoral and domesticated and the sublime to the intense, terrifying, and frightening. The picturesque was for landscapes in the middle ground between the beautiful and the sublime and soon became the most influential of the three. In part, this was due to the popularity of landscape paintings by artists such as Salvator Rosa and Claude Lorrain, as well as the idea that landscapes were more appealing when they resembled works of art. For example, a traditional device for picturesque landscape appreciation was the Claude-Glass—a small, tinted, convex mirror named after Claude Lorrain. It was designed for viewing a landscape as it would appear if it were a landscape painting, with the appropriate distance, color, and perspective. Thus, in picturesque appreciation, the world is divided into landscape scenes, with their subject matter and composition dictated by the arts, especially landscape painting. Landscape appreciation became the appreciation of artistically composed scenery, scenic views, and prospects (Hussey 1927; Andrews 1989).

Picturesque landscape appreciation was popularized in the late eighteenth century by William Gilpin, Uvedale Price, and Richard Payne Knight. At that time, it became the aesthetic ideal of English tourists, who pursued picturesque scenery in the Lake District, the Scottish Highlands, and the Alps. In North America, the idea of the picturesque inspired the writings of Ralph Waldo Emerson and Henry David Thoreau and the landscape paintings of artists such as Thomas Cole and Frederic Church (Conron 2000). Throughout the nineteenth and into the twentieth century, the picturesque tradition continued to influence landscape appreciation, especially that of the growing numbers of tourists who wanted to see and appreciate the scenic views pictured in travel brochures and on picture postcards. Consequently, picturesque appreciation has been an extremely significant factor in the preservation of landscapes that conform to its dictates. It is no exaggeration to say that it was the major force in the creation of many of the great parks and preserves of both Europe and North America (Callicott 1994).

Nonetheless, in spite of the enormous influence of picturesque landscape appreciation on landscape preservation, the relationship between it and environmental ethics is not strong. This is because the way in which picturesque landscapes are composed by human thought and imagination weakens the link between landscapes and their potential source of intrinsic value, the land itself. Under the spell of the picturesque, landscapes are preserved not because of intrinsic value derived from the land, but because they can be seen as something other than the land from which they are constructed. They can be seen as if they are works of art (Carlson 1979). In sum, picturesque landscape appreciation does not essentially tie landscapes to the land; indeed, it tends to sever the connection between the two. Thus, although it was, as a matter of fact, a powerful force in the preservation of many landscapes, it does not make their preservation an ethical matter.

Landscape Formalism

While landscape appreciation was still largely under the influence of the picturesque, the world of art was moving away from the traditional depictions that had long dominated painting. At the beginning of the twentieth century, in the hands of artists such as Paul Cézanne and Henri Matisse, art increasingly emphasized strong lines, dominant shapes, and bold colors. This movement was popularized in the English-speaking world as formalism by British art critic Clive Bell. Bell, focusing on visual art, argued that what he called "significant form," a moving combination of lines, shapes, and colors, was the essential property of art (Bell 1958). Bell's view exemplifies pure formalism, making significant form both necessary and sufficient for a work of art. Some contemporary thinkers support a more moderate interpretation of formalism, holding only that attention to formal significance is necessary in aesthetic appreciation, but not sufficient for it (Zangwill 2001).

The early advocates of formalism focused mainly on art, but, in line with formalist theory, aesthetic appreciation in general can be construed as the contemplation of the formal structure of any object of appreciation. Thus, formalists can find aesthetic value in landscapes, but do so primarily when they are experienced as some artists experience them, seeing landscapes as pure formal combinations of lines, shapes, and colors. However, the early formalists favored their own contemporary artists, such as Bell's favorite, Cezanne, and not those admired by the picturesque tradition. Throughout the first part of the twentieth century, various artists developed the formalist approach to landscape appreciation. For instance, it is evident in Georgia O'Keefe's fluid renderings of mountain landscapes of the American Southwest and in the vivid compositions of Canadian landscapes by members of the Group of Seven, such as Lawren Harris and Franklin Carmichael. Formalism also inspired landscape photographers early in the twentieth century, such as Edward Weston and Ansel Adams, and, more recently, has influenced Edward Burtynsky, whose works stunningly capture the lines, shapes, and colors of land devastated by industry, manufacturing, and resource extraction (Burtynsky *et al* 2009).

Notwithstanding the excellence of many formalist-influenced works of art, nor formalism's considerable impact on landscape appreciation, landscape formalism, like the picturesque tradition, does not secure a strong relationship between landscapes and environmental ethics (Saito 1998). Although it does not, like the picturesque ideal, restrict appreciation to classically scenic landscapes, as Burtynsky's landscape photographs clearly demonstrate, formalist landscape appreciation yet has stronger allegiances to artistic traditions than to the land. Moreover, the problem with formalism is not simply that it yields what is primarily an artistic way of appreciating landscapes, but also that formalist appreciation reduces landscapes to formal compositions. It flattens them to two-dimensional surfaces. The way formalist landscapes are composed by human thought and imagination connects landscapes to their potential source of intrinsic value, the land itself, by what is only an inconsequential aspect of the land—not simply its mere appearance, but its mere formal appearance. In this sense, formalist landscape appreciation involves a rather shallow kind of aesthetic appreciation (Parsons and Carlson 2004). Like picturesque appreciation, landscape formalism does not essentially tie landscapes to the land and make their preservation an ethical responsibility.

The Postmodern Landscape

At the opposite extreme from both picturesque landscape appreciation and formalism is a position that might be called postmodern landscape appreciation. While the picturesque tradition restricts aesthetic appreciation to a limited set of scenic landscapes and formalism reduces it to a minimum, the postmodern alternative embraces just about anything as a component of such appreciation (Heyd 2001). The label "postmodern landscape appreciation" seems appropriate for this point of view because of the comparison sometimes made between landscapes and literary texts as well as a common postmodern position on the reading of such texts (Carlson 2000). This is the position that in a text we rightly find not just the meaning its author intended, but also any of various meanings that the text may have acquired in one way or another or that we may find in it for one reason or another. Moreover, none of these possible alternative meanings has priority; no reading of a text is privileged. On postmodern landscape appreciation, whatever the human mind and imagination brings to the task of composing a landscape from the land is seemingly as important a component of its appreciation as any other. Just as no reading of a text is privileged, neither is any appreciation of a landscape.

In one sense, postmodern appreciation is an obvious reaction to both the picturesque tradition and formalism, since, as noted, each imposes certain restrictions on landscape appreciation. The human mind and especially the human imagination often rebel against such straightjacketing of appreciation, holding that deeper and more profound thought and imagination are required for adequate aesthetic appreciation. However, as with some rebellions, this reaction frequently runs toward the extreme. Thus, imagination, rather than thought, typically plays the greater role in postmodern landscape appreciation. For example, while human thought may contemplate a forest landscape and focus on the different kinds of trees, human imagination might find in the gnarled bark of the trunks the semblance of the faces of wrinkled old men and in the twisted branches the likeness of their aged and crooked limbs. Thus, the forest can seem an enchanting landscape of mystery and perhaps of danger, one that can give rise to whimsical tales of excitement and adventure. And, on the postmodern position, such imagination-fueled appreciation is seemingly as appropriate as any other.

Concerning the relationship between landscape appreciation and environmental ethics and the possibility of composing landscapes such that they, as the land itself, possess intrinsic value, thereby making their preservation an ethical obligation, the postmodern view faces certain difficulties. The most obvious problem is that since the postmodern position accepts as relevant to landscape construction anything from, on the one hand, thoughtful contemplation of the land itself to, on the other, the most fanciful flights of the imagination, there is no guarantee that postmodern appreciation will secure the intrinsic value of a landscape by essentially tying it to the land from which it is composed, since even if the former normally contributes to this outcome, the latter frequently does not (Eaton 1998). Moreover, in the latter case, as evident from the above example, aesthetic appreciation can be a very personal and subjective matter. And it is generally accepted that if beauty is, as is said, only in the eye of the beholder, then no general obligations, moral or otherwise, follow from it. If simply one individual finds something aesthetically appreciable, from that alone, it is difficult to argue that others should also find it so and think that it should be treated in a certain way, such as being preserved (Thompson 1995).

Landscape Cognitivism

As noted, the potential resources for composing landscapes range from contemplation of the land itself to flights of the imagination. Postmodern landscape appreciation often tends toward the latter. However, it is also possible to focus primarily on the former, arguing that the necessary resource for landscape construction is thoughtful and knowledgeable contemplation of the land. This alternative is pursued by landscape cognitivism, which, although it does not, like the picturesque tradition and landscape formalism, model the appreciation of landscapes on that of works of art, nonetheless takes a clue from the aesthetic appreciation of art. In art appreciation, cognitivism generally supports the view that knowledge about the origin and nature of the work in question is necessary for its appropriate appreciation. For example, to appropriately appreciate a sonata, it is important to have some understanding of Western musical traditions as well as Western classical music in particular and especially the sonata form. Likewise, to appreciate an abstract expressionist painting, it is necessary to possess some knowledge of painting, of Western painting traditions, of mid-twentieth-century North American painting, and so forth.

Landscape cognitivism defends an analogous view about the appreciation of landscapes. It holds that to appropriately aesthetically appreciate a landscape, one must have knowledge of the land from which it is composed. Given any landscape, the land in question can be vastly different in its origin and nature. It may be prairies, grasslands, forests, mountains, lakes, deserts, wetlands, or badlands. The picturesque tradition seemingly holds that many of the resultant landscapes are simply beneath aesthetic appreciation, while formalism reduces all to compositions of lines, shapes, and colors. By contrast, the cognitive account argues that all can be aesthetically appreciated for what they in fact are. To do so, the appreciator must focus thoughtful contemplation on the cognitive resources relevant to the landscape in question, which is knowledge about the particular origin and nature of the land from which it is composed. This is knowledge given by, in the case of natural landscapes, sciences such as geology, biology, and ecology (Carlson 1979, 2000). Consequently, for natural landscapes, this position is sometimes called scientific cognitivism.

Landscape cognitivism has three main advantages over some other positions. First, it does not, as suggested, limit the scope of the aesthetic appreciation of landscapes. It is sometimes aligned with the view known as Positive Aesthetics, the position that all natural landscapes, in so far as they are not ruined by human intrusions, have primarily positive aesthetic value (Carlson 1984). Second, and more relevant to the central issue here, landscape cognitivism speaks directly to the relationship between landscape appreciation and environmental ethics. Given that landscapes have intrinsic value when essentially tied to land, in which case they, as the land itself, are proper objects of moral obligation, landscape cognitivism, in requiring that aesthetic appreciation of landscapes be informed by knowledge of the land from which they are composed, helps forge this crucial link between landscapes and land. Third, since this link is established by factual knowledge of the land, landscape cognitivism is not open to the charge that it yields a personal and subjective aesthetic experience from which no general obligations follow. Rather, it provides a solid, objective foundation from which it can be argued that if a particular landscape is aesthetically appreciable, it should generally be found so and thus recognized as worthy of preservation (Carlson 1981; Hettinger 2005; Parsons 2006; Carlson 2010).

Landscape Cultural Cognitivism

Although landscape scientific cognitivism has several advantages over some other accounts of landscape appreciation, it has a major limitation. Since it concerns primarily natural landscapes and even those that are completely free from human intrusions, it, by itself, is relevant to comparatively few landscapes, since most are not pristine, but rather, to various degrees, human or, as they are sometimes called, cultural landscapes. Even land once thought to be predominantly untouched by human contact is increasingly shown to be more or less influenced by human activity. Early explorers and settlers often found it convenient to regard land they newly occupied as "terra nullius" or "land belonging to no one" and from this claim grew the myth of vast expanses of virgin, pristine land. But such land hardly exists, other than in places exceedingly inhospitable for human habitation. The resultant landscapes are in general not simply influenced by humans, but are by and large produced by them. They are the landscapes of human habitation: houses, towns, cities; human production: farms, factories, mines; human recreation: roads, parks, resorts.

When aesthetically appreciated and preserved, cultural landscapes are sometimes called heritage landscapes. As with natural landscapes, the appreciation and preservation of such landscapes require recognition of their intrinsic value, and that, in turn, requires knowledge of the origin and nature of the land from which they are constructed. However, in the case of cultural landscapes, vital knowledge is provided not just by natural science, but even more so by disciplines such as anthropology, economics, politics, and especially history (Carlson 2001, 2009). For example, many American parks and monuments are preserved not simply because they contain natural wonders, but also because they are home to human wonders. Consider Mesa Verde with its spectacular landscapes of cliffs and canyons that shelter equally spectacular cliff dwellings of the Ancestral Puebloan people known as the Anasazi. To aesthetically appreciate such landscapes, we must know both about the formation of the natural landscape and about the culture that first prospered but eventually perished in this rugged environment. Or consider a somewhat different kind of heritage landscape: the seemingly idyllic hills of the Gettysburg Battlefield. Only with knowledge of what happened on this land and how it changed the course of American history does one fully appreciate the particular landscape and understand its preservation. In such cases, in addition to intrinsic value derived from the land itself, perhaps these landscapes have what may be called value added, due to the human contributions to them—the toil and accomplishment as well as the suffering and loss. In this way, the intrinsic value of cultural landscapes is not only like that of natural landscapes but also somewhat like that of art.

Some cultural landscapes not typically recognized as heritage landscapes can nonetheless elicit various levels of aesthetic experience and may be, for a variety of reasons, worthy of preservation. For example, consider the vast agricultural landscapes of the North American Midwest. Without knowledge of what they are and how they came to be, without an understanding of the origin and nature of the land itself, they are frequently thought difficult to aesthetically appreciate and not worthy of preservation other than for what the land can produce. Yet, they can be landscapes of great beauty and were all lost; it would be not just an economic disaster but an aesthetic one as well (Carlson 1985). Other landscapes are more perplexing. Consider the landscapes depicted in Burtynsky's photographs of land devastated by industry, manufacturing, and resource extraction. Or consider cities, such as Detroit, where industrial decline and social strife have left appalling cityscapes of abandoned factories and deserted streets lined with burned-out buildings. Or consider the horrific landscapes in Germany and Poland where the ruins of World War II concentration camps still stand on the land. Maybe examples of such frightful landscapes should be preserved, perhaps some of their intrinsic value lies in reminding us of what can happen when things go terribly wrong. But to appreciate them we may need more than just knowledge of what they are; we may have to appeal to the sister notion of the picturesque, the sublime, by which eighteenth-century thinkers found it possible to aesthetically appreciate intense, terrifying, and frightening natural landscapes. Perhaps we need the sublime for the aesthetic appreciation of frightful cultural landscapes (Maskit 2007).

Landscape cultural cognitivism, together with scientific cognitivism, explains how cultural landscapes come to possess their intrinsic value, value that arises both from the land of which they are constructed by human thought and imagination and from the human contributions that have physically influenced and shaped that land. However, there are cultural landscapes that have not been shaped or even marked by humans. They are the products of little or no substantial human action, but only of human belief. And in many instances, such landscapes are considered the most important heritage landscapes. They are often sacred, ceremonial places, places of adoration, veneration, and worship. Yet, they are frequently not made so by what has happened on the land, but only by what is believed to have happened. They are the landscapes of folklore, myth, and religion. Mountains and similar formations seem to be particularly appropriate for this kind of landscape-making. For example, consider Devils Tower, Mount Taranaki, the Black Hills, Uluru (Ayers Rock), the San Francisco Peaks, Mount Sinai, Ship Rock, and Mount Olympus. Each is the focal point of a landscape created by the beliefs of certain people, culture, or religion and as such is considered to have great intrinsic value and to be fully worthy of preservation (Carlson 2000, 2009).

Such landscapes make especially clear a pressing issue concerning landscape preservation. Even more so than other heritage landscapes, cultural landscapes that are primarily the products of belief rather than of action are not simply aesthetically appreciated but also the objects of other strong emotional reactions, such as awe, wonder, and reverence. The intensity of such reactions to these landscapes, together with the fact that the beliefs involved are typically not universally held, frequently leads to controversy concerning the intrinsic value and the preservation of such landscapes. Since there are no objectively recognized links between these landscapes and the land from which they are composed nor any value-adding human physical changes to that land, which may be seen as contributing to intrinsic value, it is all too common for those who do not share the cultural belief system that has generated the landscapes to call into question the intrinsic value and the preservation-worthiness that are claimed for them. Thus, these landscapes are especially vulnerable to forces and interests that want to exploit them for other purposes, such as resource extraction or recreational use. Many of the mountain-centered landscapes mentioned above, for instance, that of the San Francisco Peaks, are persistently subject to these kinds of dangers. As with other landscapes, both natural and cultural, these landscapes and their preservation are continually threatened by special interests.

Landscape and Ethics

At the outset, this chapter introduced the question of the relationship between landscapes and environmental ethics. As noted, the question arises because landscapes are first and foremost objects of aesthetic appreciation and one step removed from the land to which contemporary environmental ethics sometimes grants intrinsic value. Thus, to possess intrinsic value, landscapes must be essentially tied to land. In this case, they, as land itself, can be proper objects of moral regard and their preservation deemed an ethical responsibility. This fundamental link to the land can be established by an account of the aesthetic appreciation of landscapes that underwrites the intrinsic value of landscapes by requiring that landscape aesthetic appreciation meets certain conditions. At least some theories of landscape aesthetic experience satisfy these conditions. However, given the nature of cultural landscapes, it seems that in many cases, human activity can add intrinsic value to landscapes over and above the value that is established by the tie to the land itself. The upshot is that the important relationship is not simply between landscapes and environmental ethics, but also between landscapes and ethics more generally. This is especially evident concerning heritage landscapes that are the objects of both aesthetic appreciation and intense emotional reactions, such as veneration and reverence. Such cases clearly demonstrate that the preservation of landscapes is not just a question of our regard for the land and thus a matter of environmental ethics, but also a question of our regard for our fellow humans and thus a matter of ethics. But this should not be surprising; environmental ethics is a branch of ethics.

Conclusion

In this chapter, the picturesque tradition, landscape formalism, scientific cognitivism, landscape cultural cognitivism, and the postmodern point of view are discussed as if they are sharply separate, conflicting, and, in some cases, even mutually exclusive accounts of landscape appreciation. In one sense, this way of presenting these different positions is appropriate in that they do not all directly and equally address the issues concerning the intrinsic value of landscapes and the relationship between landscapes and environmental ethics. However, concerning day-to-day landscape appreciation, they need not be treated as in conflict, but rather can be seen as emphasizing different dimensions of an overall aesthetic experience. Consider the example of the mountain range introduced at the beginning of this chapter. You might first attend to it because it strikes you as a classically scenic view, and then find yourself tracing the lines and contours of the scene and admiring the shades and hues of its colors. This initial appreciation may give way to contemplating the massive granite outcroppings and reflecting on the fact that they have been exposed by erosion, dissected by fractures, and sculpted by glaciers over millions of years. Moreover, knowing that this is the Bitterroot Range, which is cut by Lemhi Pass where Louis and Clark first crossed the Continental Divide in 1805, can add depth to the encounter. And imagining someday hiking over the pass and walking in the footsteps of the Corps of Discovery may give a personal dimension to the whole experience. In short, these different positions highlight different aspects of landscape appreciation. Some, such as the picturesque tradition and formalism, emphasize that which characteristically first opens an appreciator to aesthetic experience, while others, such as the cognitive accounts of the appreciation of the natural and cultural landscapes, focus on how the experience is deepened and how the intrinsic value of landscapes is secured. Last but not least, stressing the musings of the imagination shows how the whole can come to feel singular and special. In this way, each of the different accounts makes a contribution to understanding the aesthetic experience of landscapes and its role in their appreciation and preservation.

Related Topics

8. Forests
9. Mountains: Rethinking Thinking Like a Mountain
10. Wilderness
11. National Parks
15. Wetlands
16. Rivers and Watersheds
34. Urban Parks and Open Space
61. Everyday Aesthetics
63. Environmental Justice

References

Andrews, M. (1989) *The Search for the Picturesque*, Stanford, CA: Stanford University Press.

Bell, C. (1958) *Art* [1913], New York: G. P. Putnam's Sons.

Burtynsky, E. with M. Mitchell, W. E. Rees, and P. Roth (2009) *Burtynsky Oil*, Göttingen: Steidle/Corcoran.

Callicott, J. B. (1994) "The Land Aesthetic," in C. K. Chapple (ed.) *Ecological Prospects: Scientific, Religious, and Aesthetic Perspectives*, Albany: State University of New York Press.

Carlson, A. (1979) "Appreciation and the Natural Environment," *Journal of Aesthetics and Art Criticism* 37: 267–276.

——— (1981) "Nature, Aesthetic Judgment, and Objectivity," *Journal of Aesthetics and Art Criticism* 40: 15–27.

——— (1984) "Nature and Positive Aesthetics," *Environmental Ethics* 6: 5–34.

——— (1985) "On Appreciating Agricultural Landscapes," *Journal of Aesthetics and Art Criticism* 43: 301–312.

——— (2000) *Aesthetics and the Environment: The Appreciation of Nature, Art and Architecture*, London: Routledge.

——— (2001) "On Aesthetically Appreciating Human Environments," *Philosophy and Geography* 4: 9–24.

——— (2009) *Nature and Landscape: An Introduction to Environmental Aesthetics*, New York: Columbia University Press.

——— (2010) "Contemporary Environmental Aesthetics and the Requirements of Environmentalism," *Environmental Values* 19: 289–314.

Conron, J. (2000) *American Picturesque*, University Park: Pennsylvania State University Press.

Eaton, M. M. (1998) "Fact and Fiction in the Aesthetic Appreciation of Nature," *Journal of Aesthetics and Art Criticism* 56: 149–156.

Hettinger, N. (2005) "Allen Carlson's Environmental Aesthetics and the Protection of the Environment," *Environmental Ethics* 27: 57–76.

Heyd, T. (2001) "Aesthetic Appreciation and the Many Stories about Nature," *British Journal of Aesthetics* 41: 125–137.

Hipple Jr., W. J. (1957) *The Beautiful, the Sublime and the Picturesque in Eighteenth-Century British Aesthetic Theory*, Carbondale: Southern Illinois University Press.

Hussey, C. (1927) *The Picturesque: Studies in a Point of View*, London: Putnam.

Leopold, A. (1966) *A Sand County Almanac with Essays on Conservation from Round River* [1949/1953], New York: Oxford University Press.

Maskit, J. (2007) "'Line of Wreckage': Towards a Postindustrial Environmental Aesthetics," *Ethics, Place and Environment: A Journal of Philosophy and Geography* 10: 323–337.

Parsons, G. (2006) "Freedom and Objectivity in the Aesthetic Appreciation of Nature," *British Journal of Aesthetics* 46: 17–37.

Parsons, G. and A. Carlson (2004) "New Formalism and the Aesthetic Appreciation of Nature," *Journal of Aesthetics and Art Criticism* 62: 363–376.

Rolston III, H. (2002) "From Beauty to Duty: Aesthetics of Nature and Environmental Ethics," in A. Berleant (ed.) *Environment and the Arts: Perspectives on Environmental Aesthetics*, Aldershot: Ashgate.

Saito, Y. (1998) "The Aesthetics of Unscenic Nature," *Journal of Aesthetics and Art Criticism* 56: 101–111.

——— (2007) *Everyday Aesthetics*, Oxford: Oxford University Press.

Santayana, G. (1961) *The Sense of Beauty: Being the Outline of Aesthetic Theory* [1896], New York: Collier.

Thompson, J. (1995) "Aesthetics and the Value of Nature," *Environmental Ethics* 17: 291–305.

Zangwill, N. (2001) "Formal Natural Beauty," *Proceedings of the Aristotelian Society* 101: 209–224.

Further Readings

Berleant, A. and A. Carlson (eds) (2007) *The Aesthetics of Human Environments*, Peterborough, Canada: Broadview Press. (A collection of contemporary essays on the aesthetic appreciation of cultural landscapes.)

Brady, E. (2003) *Aesthetics of the Natural Environment*, Edinburgh: University of Edinburgh Press. (A major study of the aesthetics of landscapes with attention to the relationship between aesthetic appreciation and preservation.)

Carlson, A. and A. Berleant (eds) (2004) *The Aesthetics of Natural Environments*, Peterborough, Canada: Broadview Press. (A collection of contemporary essays on the aesthetic appreciation of natural landscapes, including several of the pieces cited in this chapter.)

Carlson, A. and S. Lintott, (eds) (2008) *Nature, Aesthetics, and Environmentalism: From Beauty to Duty*, New York: Columbia University Press. (A collection of historical and contemporary essays on the relationship between landscapes, aesthetic appreciation, and environmental ethics, including a number of the pieces cited in this chapter.)

Nassauer, J. I. (ed.) (1997) *Placing Nature: Culture and Landscape Ecology*, Washington, DC: Island Press. (A collection of contemporary essays on the appreciation of cultural landscapes by scholars from several different disciplines.)

13

PROPERTY

Markku Oksanen

Introduction

Robinson Crusoe, the classic tale by Daniel Defoe about the shipwrecked man, has inspired both environmental philosophers and scholars of property. The economist Harold Demsetz (1967: 347) has argued that "in the world of Robinson Crusoe property rights play no role". Some environmental philosophers, most notably Richard Routley (1973) and Mary Midgley (1985: 148), have pondered whether in the world of Crusoe there are any environmental duties, primarily, whether or not Crusoe ought to refrain from destroying the biota of the island. The cases arise because Crusoe lives in isolation, outside human communities and the rights and responsibilities accruing from common life. Does it make sense to say that the island Crusoe inhabits is *his* (private) property? If it makes sense, does it follow that he could do whatever he likes with and to his property? Or if it does not make sense, could he, nonetheless, wipe out all life on the island?

As the example above illustrates, the scopes of environmental ethics and the philosophy of property overlap as both fields explore the rights, duties, and responsibilities concerning the bio-physical environment and the resources and services it provides for humans. A common thought, therefore, is that "*all* approaches to environmental protection ultimately are property-based" (Cole 2002: ix) and, conversely, that all approaches to environmental protection and its moral basis use, interpret, and modify the idea of ownership. Also, it is widely accepted that the fate of human communities (and the planet's resource base as well) partly depends on property arrangements (Baland & Platteau 1996; Diamond 2005).

The two concepts of property and ownership designate a specific social institution composed of rules. These rules govern, most importantly, access to, and use of, various resources and, at the same time, they make these resources property objects. Some of these rules are rigid and formal (written); others are more flexible and based on age-old customs or folklore without having taken the strict shape of statutes. As a result, even in Western democracies, the property concepts and practices are manifold. It is difficult to say what could *not* be property objects, but usually these have been limited to land, natural resources, buildings, machines, moveable (personal) objects, and intellectual property (including commercially useful inventions and artistic creations). In the modern Western context, to call land "real property" is not controversial; the prominence of land is also exemplified by the location of this chapter in the section on "Land" in the book at hand. For the legal scholar Joseph William Singer (2000: 86), "a clearly defined plot of land with a single house where the unique owner lives" is a stereotypical image of property. Arguably, the status of land as property has diminished during the 20th century, largely due to growth of immaterial

DOI: 10.4324/9781315768090-16

property. However, the interest in land has recently got livelier from the global perspective which has resulted in a debate about land-grabbing, or the acquisition of agricultural land by foreign investors for agricultural, forestry, and conservation purposes (Pearce 2012).

For liberals, conservatives, and often their socialist critics, the core idea of private ownership is the following: to own is to be in position where one can solely use the resources on one's property as one sees fit and exclude others from its use (Schmidtz 2009: 184). This is, however, a sketchy idea that applies to a limited number of trivial cases and in the solitary world of Robinson Crusoe, even if there, since there might be no property on his island anyway.

This chapter consists of three interrelated sections in which this core idea is analyzed. The first section sheds light on the element of exclusion in property. It considers what kind of system of ownership an ecologically decent society has and whether the basic or sole type of ownership is private, common, state, or a mixture. The second section addresses the content of ownership, that is, the question of what an owner is entitled to do with or to his or her property. The question of the content is to some extent independent of the question of the type of ownership since the possible limits may bind individuals and communities as well as states. The third section combines the previous sections at a more concrete level. It addresses the subject of ownership by asking whether a reasonable environmental ethic requires owners to be of a certain character and how that character is related to the type of ownership. By limiting the discussion to these topics, many important issues are not analyzed here. Most notably, these include the issue of "original" ownership or appropriation, the typologies of justificatory arguments, and the question of justice in the distribution of property.

Types

Ownership is one of the great philosophical topics dating back to Plato's *The State*. Although Plato's writings were quite nuanced and do not promote a crude form of communism where the state is the sole owner of all resources, Aristotle in *Politics* defended individual households and private property against Plato by simplifying his views (Garnsey 2007: 27; Pierson 2013: 25). This maneuver has been repeated in subsequent philosophical debates that have often focused on comparing the merits of state and private property and related concepts in an ideal society (Macpherson 1978: 2). For example, the significant codification of Roman law, Justinian's Institutes (530–533 CE), begins with a dichotomy according to which things are either "private wealth or not" (Book Two), but a precise catalogue follows immediately: things can be "everybody's by the law of nature; the state's; a corporation's; or nobody's. But most things belong to individual's" (*Justinian's Institutes*: Book Two 2.1). So, it identifies four or five types of property in a single society. Throughout human history, property arrangements have been and still are heterogeneous, normatively sophisticated, and contingent on the nature of the resource, its location, and the specific cultural and political features of human societies – so much so that some scholars consider property as secondary to these particular issues (see Grey 1980: 81; Waldron [2012] on Rawls; Pierson [2013: 33–34] on Aristotle). The same applies to legal theorists for whom ownership consists of "bundle of rights" or "incidents" (Honoré 1987 [1961]) that has led to the thesis about the "disintegration of property" (Grey 1980). According to the thesis, the general concept of property has lost its utility for legal and political thought and has been replaced with specific rights, responsibilities, and obligations.

To put it differently, a great deal of philosophical discussion about ownership has focused on the nature and justification of monistic and universalistic arrangements according to which one mode of ownership should prevail in all societies and over all resources. These ideas are based on the utopian visions of the good society, the conceptions of natural rights and duties of self-preservation, the natural order of life in which even non-human animals have property (i.e. territories), or the necessary elements emerging from social action, such as the market. Until the emergence of environmental concern in the 1960s, the discussion on property was not, in effect, ecologically informed. Since then, in many fields of academic research, ownership has been regarded as an *institution* that is created, maintained, and adjusted by humans for specific purposes in specific time and place; so, ownership comes in several types (Cole 2002; Dales 1968: 61; Ostrom 1990). As a starting point, I assume it is best to keep an open mind and look for reasons for the overall heterogeneity of property arrangements, or "institutional diversity" (Becker & Ostrom 1995), with an emphasis on the sustainable use of resources and on the adaptability of human communities. By claiming so, I do not want to advance uncritically the idea that all existing heterogeneity helps communities to sustain. Clearly, some arrangements work better than others and some communities are more adaptable than others and ambiguity may be harmful.

This gives a rise to a problem about the normative foundation for exercising the social and ecological criticism of property arrangements. In the ecological assessment of ownership, several specific questions arise: How do property arrangements work in terms of intended outcomes and accidental consequences? What are morally considerable beings and how do arrangements influence these beings? And finally, how just are arrangements? To provide a meticulous answer to these questions, a comprehensive normative study of various property arrangements should be conducted taking into account both theoretical and empirical elements. In other words, alternative arrangements should not be ranked either on the ideological or on the empirical basis alone.

Are all resources covered under some property-rights regime? It depends on the definitions of property and resource (i.e. potential property object). The standard answer is, however, negative, since some resources (e.g. high-sea fisheries) are identified as open-access or non-property; the competing answer is that "humanity as a whole owns the earth" (Risse 2012: 111). The assessment of the significance of this divide goes beyond the scope of this chapter. In many cases, however, the distinction between non-property and property is not obvious anymore. In earlier times, genetic resources, the high seas, and the air were non-property because no one could hold a proper title to them. Still, the air that all humans breathe is non-property to any flesh-and-blood individual user, although in the European Union (EU), for the coal-using utility corporations, the atmosphere is no longer non-property. To protect the atmosphere from greenhouse gas emissions, the EU has implemented a system of emission trading (EU ETS, European Union Emissions Trading System). Although ETS involves market instruments and "something more sophisticated in the way of property rights" (Dales 1968: 65), and although some consider this as privatization, or establish private property regarding the atmosphere, the created system is not a full-blooded or a "mature" system of private property. Therefore, it is not clear whether these rights to emit carbon are rights of ownership, although for some scholars, property is rather *rights* than *things* (Coase 1960: 44; Dales 1968: 58). No doubt, emission rights are transferable licenses to some commercial activity over a certain period of time. Furthermore, the property characteristics of emission rights are distinctive: they are property objects for firms but exclusive only to competitive actors within a certain geographic area. The exclusion does not concern others, such as human individuals who retain the right to breathe and to emit greenhouse gases.

Land as a resource differs from the air; land is a basis where individuals or human groups can, or at least try to, form territories and exercise "sole and despotic dominion" (Blackstone 1966 [1765–1769]: 2) against other individuals or human groups. Since humans live on all habitable continents and islands, there is no and there has not been vacant land for original appropriation for ages. The colonialist talk about *Terra Nullius*, a land empty of effective and hard-working human populations with a legal and moral title to it (Boucher 2010), has no foundations. In brief, before the emergence of "high-tech property" (like the emission rights), land has been the most fundamental resource for all humans and therefore it deserves its status as the paradigm type of property.

In addition to non-property or open access, three pure types of ownership are commonly distinguished: state property, private property, and (regulated) common property (Bromley 1991:

31). In effect, the pure types are theoretical constructions, or ideal types, that fit into the real world uneasily. The atmosphere is a subject of all pure types: in creating emissions trading system, states transformed the non-property status of the atmosphere to the regulated state-controlled resource. The next step was to auction off rights of use to various collective users (mainly firms); they become collective owners of the rights to the resource use. Should nature conservation trusts have bought emission rights for non-use purposes, the atmosphere would belong to several different entities and be non-property for individuals for some forms of use. Resources in nature are inseparable from each other and this fact makes them subjects of many regimes simultaneously. The conclusion is significant in two ways. First, there are implications for the classification of ownership – in many cases, the pure types do not and cannot exist. This is the most challenging aspect, since institutional vagueness and uncertainty are widely considered as the greatest obstacle to rational, long-term resource management. As one group of scholars puts it, "In fact, most environmental problems can be seen as problems of incomplete, inconsistent, or unenforced property rights. Without a solution to the property rights problem, the environmental problem will remain" (Hanna *et al.* 1995: 17; cf. Dales 1968: 63–64). Second, right-libertarians and many liberals require that the right to exclude is and should be at the kernel of property rights in each and every case. This requirement is not, however, a realistic or possible way to purify property relations. It cannot, for example, be implemented into the atmosphere. And it is not welcome in regard to land in some cultures, as I will next argue.

Although the idea of equivocality and complexity regarding land ownership might be "almost unthinkable" in countries with "strong laws of trespass" (Campion & Stephenson 2014: 54), there are many exceptions in Western jurisdictions. More strongly, the right to exclude has not been a universal component of private property rights. In 2000, the British Parliament approved a law that expanded citizens' "right to roam" at the cost of landowners' right to exclude. It designates specific "access lands" in the countryside to which the walkers may enter (Anderson 2007). Outdoor activities are popular in the UK, and while there is very little land in public ownership, the legal reform gained enough support after a decades-long campaign (see Ramblers 2014; Shoard 1999) to pass. In the USA, however, the federal government is the largest owner of land with 33% (Cole 2002: 20) and public pressure to increase access to natural areas has been lower (Anderson 2007). However, the British reform is comparable to practice in which business premises are privately owned, but the owners (or the premise holders) are not legally free to choose their customers and thus to exclude members of the public on the basis of their ethnicity or race. Singer (2000: 41–43) considers this arrangement as an example of both how private property becomes public property in the sense that the owner is not permitted to disallow access and how property types regarding a single kind of resource – in this case, land – intermingle.

The British reform draws on the Nordic customary law that secures access to private land without explicit and revocable permission of the owner as well as to some of its resources and services. Despite the right-libertarian resistance to it since it transforms some private services and resources into public services and resources, the idea is important from an environmental ethical point of view. Anderson (2007: 413–417) summarizes the rationale for this reform: free movement, enjoyment of nature, the enhancement of mental and physical health, connection to history and culture, and therefore a sense of community. Furthermore, it allows environmentalists – both amateur naturalists and civil servants – to monitor the state of the natural environment and to make public reports about it (Oksanen & Kumpula 2013). On the flipside, the same right also allows mining industry and other developers to make surveys on private lands, thus potentially opening land to exploitation.

Thus far, there have not been any cases in the European Court of Human Rights (ECHR) to test whether the Nordic customary law of public access to private land is in accord with the right to property (when regarded as a human right), but it is unlikely that ECHR would defy a longstanding and legally recognized customary law like this should a case arise. Moreover, common access to nature can also be considered as a human right, equal to, if not weightier than, the owner's right to exclude. This arrangement also enables people to develop their attitudes to nature, since access to forests, wetlands, and elsewhere is not tied to ownership. Anybody can go virtually anywhere at any time. However, because it is impossible to restrict access, there is a risk of "overconsumption" of recreational values in nature, particularly in sensitive areas or at sensitive times, or disturbing its economic use. As to the former, the problem is global and common to many of the world's most popular protected areas; however, the more possibilities to access to private property, the wider array of places can be visited. The latter risk has not been widely realized since people know about the rules of conduct on private property in the same way as they know about etiquette in public houses and these rules are sanctioned (see also Campion & Stephenson 2013).

The right of public access in the Nordic system is restricted: in principle, it allows no hunting, no commercial fishing, no tree-falling, and no permanent camping. Moreover, it does not allow people to go to agricultural lands or other businesses, nor to near to the homes and houses in the yard. However, an indication of the strength of tradition is that there is still no equivalent to the word "trespassing" in Finnish language. The same note can be made of other languages; *Oxford English Dictionary* says that "the legal application of the words seems specially English".

Singer (2000: 137) makes an important proposal according to which legal rights are to be defined "with reference to common understandings and considered judgments about which expectations are reasonable". As applied to the norm of the right to roam, in the US context, the claims of free access to private land are not conventional expectations, whereas in the Nordic context, they are and thus denying them is unreasonable and, in fact, unlawful. In the UK, the situation has remained partly contentious considering the historic public footpaths that provided legitimate easement for landowners for centuries and the growing demand for the opening up of the countryside that resulted in the aforementioned reform.

In sum, a particular plot of land can be and is the subject of many property arrangements simultaneously; thus, property types regarding it are mixed, often at the cost of the owner's right to exclude. Instead of arguing in the language of universalism in favor of one arrangement and rules of one kind, it is reasonable not to reject historically tested institutions out of hand.

Contents

The institution of ownership is a *social* institution that consists of socially enforced rules. According to the standard characterization, the rules of ownership dictate the relations between various subjects with respect to the objects of property. James Grunebaum (1987: 4) claims that a definition of ownership as a relation between persons regarding things "goes back at least to Hobbes, is found in Locke, and is explicitly used by Marx" (see also Bromley 1991: 15; Demsetz 1967; Singer 2000: 134).

The relational nature of ownership might be harmless from an environmentalist perspective since it, before anything, regulates human interactions and organizations. Many non-anthropocentrists have found the idea of nature as property disturbing because property involves more than the exclusion of others; it involves substantial rights to use the thing and it signifies a wrong attitude to nature. This controversy is clearly manifest in the novel ideas of rights of nature and legal personhood of natural objects that involve the idea of nature as owning itself, but these ideas are reducible to regulate relationships between humans, too (Sanders 2018). Nevertheless, Aldo Leopold's chapter "The Land Ethic" in *A Sand County Almanac* begins with the claim that historically "the land" has been considered as an object of property and thus its use is shaped solely by economic interests. Leopold questions the decency of this attitude, and by doing so he paves the way for incompatibility between two attitudes: nature as property and nature as subject of moral concern (cf. Meyer 2009: 101–102). He does not, however, renounce the institution of land ownership altogether; instead, he takes steps toward a non-anthropocentric theory of property in which the right to use is not unrestricted but qualified (see Breen 2001; Freyfogle 1998: 136–140; Goodin 1990; Meyer 2015; Oksanen 1998; cf. Wienhues 2020). Leopold is reluctant to accept the uncontrolled expansion of governmental power through the legal enforcement of the land ethic and asks, "At what point will governmental conservation, like the mastodon, become handicapped by its own dimensions?" (Leopold 1989: 213). As an alternative, Leopold puts his trust in the ecological and ethical cultivation of the landowners and hopes that they will denounce the role of the conqueror and become plain members of the land-community (Leopold 1989: 204). In other words, he proposes voluntary conservation by the landowners.

There are also other reasons for not advocating incompatibilism. For instance, even though ownership would entitle the owner to exploit his property, there is nothing to guarantee that the resources outside the property boundaries would not be exploited as well. Contrary to the incompatibilist ideas, the actual environmental policies have taken this as their point of departure and expanded the range of ownership rather than diminished it. In the name of explication of property rights, they have been extended to oceans and marine resources, genetic resources, and the atmosphere; in earlier times, these were regarded as non-property or unregulated commons. This change is often deemed reasonable because it clarifies and imposes certain relations among humans, not because it provides free reign for the owners to do whatever they like to do. Therefore, established property relations can be separate from environmental ethical rules, and when environmental ethical rules constrain human behavior, they constrain owners and non-owners equally. And finally, many theorists have argued that the institution of ownership is conceptually a necessary part of society and we could not get rid of it even if we wanted to (Grunebaum 1987; Honoré 1987: 162); however, the details, or the content, of ownership are, of course, the matters needing crafting. Therefore, the core issue is about the content of ownership and about the acceptability of the public regulation of land use. In some cases, the ownership of a thing entitles the owner to destroy the object of property; in other cases, it does not but rather requires care (Oksanen 1998: 133–134).

Ownership is ecologically a peculiar institution in the respect that it both allows many activities that may result in the loss of natural values and it disallows other activities because of their harmful impact on others. Owner may also make conservationist decisions. Unlike the incompatibility thesis claims, its environmental record is not black-and-white. For the sake of argument, let us adopt a conception of private property according to which the owners are at liberty to do whatever they like with their properties as long as they do not harm other people. This delimits the scope of potential activity but to what extent? Whether or not environmentalism really contradicts property, we need to address a number of issues, such as what constitutes harm, on what kind of procedure the recognition of harm relies, who the harmed parties (or stakeholders) are, and how conflicts between parties are to be settled.

Historically, the legislation on nuisance or the neighborhood infringements dates back to the Roman law: "Use of one's property may not harm another" (in Latin, "sic tuo utere, ut alienum non laedas"). Elizabeth Brubaker (1995) has studied several cases in the Canadian legal history in which individual landowners have tried to defend themselves against polluting industrial firms. A common denominator in the court-room decisions of these cases has been that in the name of public interest, the individual right not to be harmed has been overlooked: many industrial activities have enjoyed legal protection, she maintains, on unjustified grounds. If the sacrosanct principles of the English common law – which prohibit trespassing, harming, and polluting waters – had not been overlooked, the state of the environment would be much better. In other words, the strengthening of property rights serves for the cause of conservation.

I want to comment on Brubaker in two ways. First, the scope of stakeholdership is narrow; a broader scope can be found in the famous harm principle by J.S. Mill (1910: 73): one is free to do what one likes to as long as one does not harm others. Potential stakeholders vary from non-owning countrymen to foreigners, future generations, and non-human animals. Some conceptions are plainly individualistic, requiring a designation of two persons and a causal, harmful relationship between them (e.g. an oil spill from the A's refinery to B's land harms B); other ideas are of a more collectivist nature in which both the harming party and the victim may lack proper identity (e.g. greenhouse gas emissions). Both cases have specific problems and therefore, to apply the harm principle is far from plain and simple. A second comment on Brubaker is that her solution might require a stringent interpretation of harm and the right to exclude; in that respect, it would be in apparent conflict with traditional public access rights, analyzed in the previous section. This conflict is not necessary, however, since public access rights do not include a right to spoil and the landowner would have a case against polluters.

In sum, there are both intrinsic and relational restrictions on the use of property. Intrinsic restrictions are those that are part of the conception of property and how it can be used irrespective of others, and relational restrictions stem from the fact that the use of property often affects other people and their property. Therefore, one may not do whatever she likes with or to her property.

Subjects

In an encyclopedia entry "Private property", David Schmidtz (2009: 186) appraises John Locke:

> Centuries before the analytical tools of game theory, ecology, and contemporary economics were available, Locke noted that the option of fencing the commons and creating private property enables people to make land more productive than land is when left in the wild.

Schmidtz here formulates the classical liberal view, according to which the private ownership of land is superior to other types of ownership because it outcompetes them in the pursuit of economic growth.

From an environmental and ethical perspective, we can ask – granting the soundness of Schmidtz's reasoning– which option is better: productive land or a less productive if not entirely wild land? It is obvious that different property arrangements bring about different outcomes, and they do not only speak in favor of private property. Let's consider three European examples.

a   While the communist Poland was seriously polluted by the extensive use of coal, a little over quarter of its land remained virtually wild and provided sanctuary for, most notably, the European bison which had become extinct elsewhere (Cole 2002: 105–107). So, communism's environmental record is ambiguous as long as this simplistic account holds: the bison survived the monarchic era and the transition of Poland from communism to capitalism and so its conservation may be independent of property arrangements.

b   The populations of big predators, such as the wolf and the brown bear, are kept in strict control in Finland; their hunting was allowed, or even encouraged by paying bounty, until 1975. Since then, hunting has been strictly controlled outside the reindeer-herding area in the north. The biologists in Finland unanimously believed that these predators could not have survived in the capitalist Finland during the 20th century if there were no incoming animals from the Soviet Union (Siivonen 1972). Again, the Finnish hunters and particularly the reindeer herdsmen in Lapland thought to maximize the production of reindeer and the stocks of game animals (e.g. the elk and the northern hare) by keeping the populations of the noxious animals as low as possible or by exterminating them. On the Soviet side of the border, the local people lacked such a motive; of course, some hunting took place before the Soviet collapse and there was some attempt to systematize hunting and exterminate pest animals (see Weiner 1988: 253), but their attempts were not very efficient.

c   In *Deforesting the Earth*, Michael Williams (2006: 391) claims that in the first half of 20th century "throughout the European Soviet Union there were extensive areas of relatively unproductive and uneconomical forest land" and the forest management was in stark contrast to that of the USA. He continues that in the post-Soviet era, "the Russian forest is an underused and wasting assets". To concretize: in 2013, an international group of scientists compiled a high-resolution map on global forest-cover and its changes (Hansen *et al.* 2013). The map of the year 2000 illustrates how models of ownership have affected the use of forest: the border between Finland and Russia/the Soviet Union is striking. The Finnish forest-owners, mostly households, aimed to maximize the economic value of the forests and to obtain stable returns from them, whereas the Soviet forests were more "on the loose" and tended to evolve toward wilderness character under state management (see Pulla *et al.* 2013: 8). As of today, the forest-cover in Finland has remained unchanged, whereas in the Russian side of the border, forests are disappearing quickly (Hansen et al. 2013; Potapov et al. 2012). Both in the Soviet Union and Russia, the forests are state property; the new element is the emergence of commercial logging by foreign firms through logging concessions and land leasing. All in all, the Finnish model is good at producing fiber, the present Russian state-based model has resulted in deforestation in some regions, and the demolished Soviet model was negligent; so there are two second-best models and one third-best and no ideal model that delivers both the wood and thick, biodiverse forests. This is, of course, a highly simplified account of the phenomena, and there are many other elements to be taken into account in a comprehensive account.

Private owners' tendency to pursue efficiency has unintended and often undesired outcomes, though. This was also recognized by the classical liberal thinker Adam Smith. He was of the opinion that lands for public pleasure and recreation should belong to the crown (Cole 2002: 23). D. H. Thoreau who echoes Smith explains the rationale:

> I think that each town should have a park, or rather a primitive forest, of five hundred or a thousand acres, either in one body or several – where a stick should never be cut for fuel – nor for the navy, nor to make wagons, but stand and decay for higher uses – a common possession forever, for instruction and recreation.
>
> *(Thoreau 1998 [1861]: 45)*

First, it seems implicit to Thoreau that private owners would use the resources located on the wilderness tract for their own goals, which are somewhat lower than those Thoreau professes. Second, and more important, when a place is common, the access to it is open to all people willing to use it "for instruction and recreation"; apparently, the right to use area for malicious purposes is denied even if it would bring about more prosperity.

Later on, John Muir (1998 [1901]: 60) conformed to the idea:

> Only what belongs to all alike is reserved, and every acre that is left should be held together under the federal government as a basis for a general policy of administration for the public good. The people will not always be deceived by selfish opposition, whether from and mining corporations or from sheepmen and prospectors, however cunningly brought forward underneath fables and gold.

Smith, Thoreau, and Muir all assume that private owners are incapable of providing certain kinds of nature experience to those willing to seek them. Still, the case is not closed, and defenders of private natural parks might say that if there are paying customers for wilderness experience, the market mechanism will produce it. This, of course, requires that there are no other more profitable forms of land use or that the owner simply does not change her mind about how to manage the land. Most countries do not rely on the market in this matter and conceptualize the best part of their nature as *national* parks.

From the viewpoint of the traditional Lockean ideology, if the land is spoiled or is not used productively, it returns to the commons where it is again able to be acquired by anyone who mixes it with one's labor. This clearly makes the private ownership of wilderness impossible since continuous possession requires continuous modification through labor, whereas wilderness is, by definition, left outside modification, unlabored or wild (Oksanen 1997). There are two problems in applying this proviso directly to the present world. First, there may not be countries in which negligence toward land property would open up possibilities for other to acquire it. Actually, in many countries, payment scheme programs and voluntary environmental governance arrangements have been instituted to support conservation on private lands. Second, wilderness can be considered as an immaterial good or ecosystem service where the production of it may belong to the individuals and firms. In many countries, there are arrangements of public/private partnerships. For example, a trust may acquire a piece of land and convert it to public ownership (Cole 2002: 28). In other cases, public land is converted to trusts making it a "semiprivate" or "semipublic". Furthermore, one of the emerging issues is conservation-justified land-grabbing: wealthy westerners acquire inexpensive land from less wealthy countries and then decide to not to use their acquisitions for production (Humes 2009; Pearce 2012: 264–275). It is clear that as far as this is legally permitted and economically possible, it dismisses the importance of democratic decision making in environmental matters and it may be regarded as unjust by the local people whose access to the resources might thus be limited. Furthermore, it may increase pressure on the unsustainable use of land somewhere else and thus expose government-driven conservation efforts to danger.

Conclusion

The case of Robinson Crusoe simplifies too much. Both environmental and property issues are social issues. In the end, it may not help us much whether Crusoe may destroy the island he inhabits or whether it is *his* property; the world we live in is more complicated. Therefore, policies on natural resources, land, and other environmental resources have been manifold and contradictory. In some cases, transferable and exclusive rights to some resource and to some action have been created and the range of property objects has expanded so as to include marine resources, genetic resources, and the atmosphere. In other cases, environmental regulation has resulted in precise terms on how resources can and ought to be used. Neither the right to exclude nor the right to use freely are full rights. By this, I mean that there are exceptions and qualifications so much so that these two "incidents" characterize property only in utopian dreams.

I have defended the customary public access to private lands and claimed that there can be systems of ownership lacking the power to exclude others. I have also defended institutional diversity, or mixed property regimes, against the claim that it is ecologically better to have a single type of ownership universally imposed. Moreover, I have also argued that ownership does not entitle the owner to do whatever she likes with and to her property; there are restrictions that are inherent to the property system. A society without the institution of ownership that governs the use of natural resources is logically inconceivable. A society without any proper concern for the physical conditions of human living in its institutional arrangements seems, if not inconceivable, at least intolerable: people of such a society cannot flourish and live well. The conflict between the two domains of rules has to be settled somehow but this process is constantly on-going within and between societies.

Related Topics

10. Wilderness
11. National Parks

Acknowledgments

The author would like to express his gratitude to Anne Kumpula, Minna Pappila, and Marcel Wissenburg for their helpful comments. This project has been supported by an Academy of Finland grant no. 130322.

References

Anderson, J. L. (2007) "Britain's Right to Roam: Redefining the Landowner's Bundle of Sticks," *The Georgetown International Environmental Law Review*, 19: 375–426.
Baland, J.-M. and J.-P. Platteau (1996) *Halting Degradation of Natural Resources. Is there a Role for Rural Communities?* Oxford: FAO and Oxford University Press.
Becker, C.D. and E. Ostrom (1995) "Human Ecology and Resource Sustainability: The Importance of Institutional Diversity," *Annual Review of Ecology and Systematics*, 26: 113–133.
Blackstone, Sir W. (1966 [1766–1769]) *Commentaries of the Laws of England*. Book 2, London: Dawson of Pall Mall.
Boucher, D. (2010) "The Law of Nations and the Doctrine of *Terra Nullius*," in P. Schöder and O. Asbach (eds.) *War, the State and International Law in Seventeenth-Century Europe*, Farnham: Ashgate, pp. 63–82.
Breen, S.D. (2001) "Ecocentrism, Weighted Interests and Property Theory," *Environmental Politics*, 10: 36–51.
Bromley, D.W. (1991) *Environment and Economy. Property Rights and Public Policy*, Oxford: Blackwell.
Brubaker, E. (1995) *Property Rights in the Defence of Nature*, Toronto: Earthscan.Campion, R. and J. Stephenson (2014) "Recreation on Private Property: Landowner Attitudes Towards Allemansrätt," *Journal of Policy Research in Tourism, Leisure and Events*, 6: 52–65. doi: 10.1080/19407963.2013.800873
Coase, R. (1960) "The Problem of Social Cost," *Journal of Law and Economics*, 3: 1–44.
Cole, D.H. (2002) *Pollution and Property: Comparing Ownership Institutions for Environmental Protection*, Cambridge: Cambridge University Press.
Dales J.H. (1968) *Pollution, Property and Prices. An Essay in Policy-Making and Economics*, Toronto: University of Toronto Press.
Demsetz, H. (1967) "Toward a Theory of Property Rights," *American Economic Review*, 57: 347–359.
Diamond, J. (2005) *Collapse. How Societies Choose to Fail or Survive*, London: Allen Lane.
Freyfogle, E.T. (1998) *Bounded People, Boundless Lands. Envisioning a New Land Ethic*, Washington, DC: Island Press.
Garnsey, P. (2007) *Thinking about Property. From Antiquity to the Age of Revolution*, Cambridge: Cambridge University Press.
Goodin, R.E. (1990) "Property Rights and Preservationist Duties," *Inquiry*, 33: 401–432.
Grey, T.C. (1980) "The Disintegration of Property," in J.R. Pennock and J.W. Chapman (eds.) *Nomos XXII. Property*, New York: New York University Press, pp. 69–85.
Grunebaum, J.O. (1987) *Private Ownership*, London and New York: Routledge & Kegan Paul.
Hanna, S., C. Folke and K.-G. Mäler. (1995) "Property Rights and Environmental Resources," in S. Hanna and M. Munasinghe (eds.) *Property Rights and the Environment. Social and Ecological Issues*, Stockholm: The Beijer Institute; and Washington: The World Bank, pp. 15–29.
Hansen, M.C. *et al.* (2013) "High-Resolution Global Maps of 21st-Century Forest Cover Change," *Science*, 342: 850–853.
Honoré, T. (1987) "Ownership," in *Making Law Bind. Essays Legal and Philosophical*, Oxford: Clarendon Press, pp. 161–192. Reprinted and revised from *Oxford Essays in Jurisprudence*, A.G. Guest (ed.) 1961.

Humes, E. (2009) *Eco Barons. The New Heroes of Environmental Activism*, New York: Ecco.

Justinian's Institutes. (1993) Trans. P. Birks and G. McLeod. Ithaca: Cornell University Press.

Leopold, A. (1989 [1949]) *A Sand County Almanac and Sketches Here and There*, New York: Oxford University Press.

Macpherson, C.B. (1978) "The Meaning of Property," in C.B. Macpherson (ed.) *Property. Mainstream and Critical Positions*, Oxford: Blackwell, pp. 1–14.

Meyer, J.M. (2009) "The Concept of Private Property and the Limits of the Environmental Imagination," *Political Theory*, 37: 99–127.

———— (2015) *Engaging the Everyday*, Cambridge, MA: MIT Press.

Midgley, M. (1985) *Evolution as a Religion*, London: Methuen.

Mill, J.S. (1910) *Liberty*, Everyman's Library. London: Dent.

Muir, J. (1998 [1901]) "Selections from 'On National Parks," in J.B. Callicott and M.P. Nelson (eds.) *The Great New Wilderness Debate*, Athens, GA: University of Georgia Press, pp. 48–61.

Oksanen, M. (1997) "The Lockean Provisos and the Privatisation of Nature," in T. Hayward and J. O'Neill (eds.) *Justice, Property and the Environment: Social and Legal Perspectives*, Aldershot: Ashgate, pp. 97–113.

———— (1998) "Environmental Ethics and Concepts of Private Ownership," in D.G. Dallmeyer and A.F. Ike (eds.) *Environmental Ethics and the Global Marketplace*, Athens: University of Georgia Press, 114–139.

Oksanen, M. and A. Kumpula. (2013) "Transparency in Conservation. Rare Species, Secret Files, and Democracy," *Environmental Politics*, 22: 975–991.

Ostrom, E. (1990) *Governing the Commons. The Evolution of Institutions for Collective Action*, Cambridge: Cambridge University Press.

Pearce, F. (2012) *The Landgrabbers. The New Fight over Who Owns the Earth*, London: Eden Project Books.

Pierson, C. (2013) *Just Property. A History in the Latin West*, vol. 1, Oxford: Oxford University Press.

Potapov, P. *et al.* (2012) "Forest Cover Change within the Russian European North after the Breakdown of Soviet Union (1990–2005)," *International Journal of Forestry Research*, vol. 2012, Article ID 729614.

Pulla, Pamela *et al.* (2013) *Mapping the distribution of forest ownership in Europe*, European Forest Institute. EFI Technical Report 83. http://www.efi.int/files/attachments/publications/efi_tr_88_2013.pdf (date of access 14 January, 2014).

Ramblers. (2014) "Winning the Right to Roam," http://www.ramblers.org.uk/what-we-do/what-we-have-done/past-campaigns/right-to-roam-crow.aspx (date of access 10 January, 2014).

Risse, M. (2012) *On Global Justice*, Princeton: Princeton University Press.

Routley, R. (1973) "Is There a Need for a New, an Environmental, Ethic?" *Proceedings of the XV World Congress of Philosophy*, No. 1, Varna, pp. 205–210.

Sanders, K. (2018) "'Beyond Human Ownership'? Property, Power and Legal Personality for Nature in Aotearoa New Zealand," *Journal of Environmental Law*, 30: 207–234. doi: 10.1093/jel/eqx029

Schmidtz, D. (2009) "Private Property," in J.B. Callicott and R. Frodeman (eds.) *Encyclopedia of Environmental Ethics and Philosophy*, Detroit: Macmillan Reference USA, pp. 184–187.

Shoard, M. (1999) *A Right to Roam. Should We Open up Britain's Countryside?* Oxford: Oxford University Press.

Siivonen, L. (ed.) (1972) *Suomen nisäkkäät*, vol. 2, [*Finnish Mammals*, vol. 2 – original in Finnish]. Helsinki: Otava.

Singer, J.W. (2000) *Entitlement. The Paradoxes of Property*, New Haven: Yale University Press.

Thoreau, H.D. (1998 [1861]). "Selections from 'Huckleberries'," in J.B. Callicott and M.P. Nelson (eds.) *The Great New Wilderness Debate*, Athens: University of Georgia Press, pp. 31–47.

Waldron, J. (2012) "Property and Ownership," *The Stanford Encyclopedia of Philosophy* (Spring 2012 Edition), Edward N. Zalta (ed.) http://plato.stanford.edu/archives/spr2012/entries/property/.

Weiner, D.R. (1988) "The Changing Face of Soviet Conservation," in D. Worster (ed.) *The Ends of the Earth. Perspectives on Modern Environmental History*, Cambridge: Cambridge University Press, pp. 252–274.

Wienhues, A. (2020) *Ecological Justice and the Extinction Crisis*, Bristol: Bristol University Press.

Williams, M. (2006) *Deforesting the Earth. From Prehistory to Global Crisis. An Abridgment*, Chicago: University of Chicago Press.

PART III

Water

14

WATER QUALITY AND AVAILABILITY

Jeremy J. Schmidt

Common sense once held that running water purified itself. That idea, like the notion that streams are nature's sewers, was wrong. Yet, these types of ideas have often found their way into customary ways of sharing water and even into formalized legal traditions that affect water quality and availability (Benidickson 2007). Interestingly, responses to these faulty ideas frequently reveal a disjuncture between theories about why we need more carefully crafted water policies and the history of actual decisions. In theory, we might expect that, as pressure on water increases, policies would become more restrictive. Yet, history reveals the opposite: water policies often become less restrictive in precisely these circumstances (Rose 1990). Why is this? What norms, or rules of right conduct, support such a response to pressures on water? Further, what do these norms imply for moral goods or harms that result from decisions that affect who gets water, when, how much, and of what quality?

This chapter examines ethical aspects of policies affecting water quality and availability. It begins by introducing key ideas that inform a growing literature on water ethics, a field that examines the ethical aspects of water management and policy while also examining what the practical demands of water management and policy imply for environmental ethics (see Schmidt and Peppard 2014). This chapter then considers how existing rights and practices for making water available reflect different ways of reasoning about the ends and aims of water use decisions. From there, it examines how decisions about water quantity intersect with considerations of water quality and the measurements and practices designed to help ensure the health of water for different purposes – from maintaining safe drinking water to ensuring sufficient oxygen content in rivers for fish and other aquatic species. One conclusion that will be drawn is that water policies often create ethical dilemmas in attempts to solve others. That is, policies that may be harmful for some individuals, groups, other species, or the environment may nevertheless be good for others. They may even create ethical obligations as the latter become accustomed to patterns of water use. As a consequence, understanding the ethical aspects of policies affecting water availability and quality requires close attention to the cultural norms and practices that different policies foster or proscribe, and not only an examination of the principles used to justify policy.

Good Water, Now

Many domains of public policy affect water, such as agriculture, energy, forestry, urban planning, and public health. One important interconnection that water policy has with other policy domains is that water quality is intractably linked to water availability. Which is to say, it is not

DOI: 10.4324/9781315768090-18

sufficient to simply have water available if it is not useable. Many water injustices, such as those regarding lead poisoning in Flint, Michigan, are of this type (Pauli 2019). The same holds true in the other direction. Good water needs to be available at certain times and places because water requirements are often time-sensitive and vary according to specific needs: bodily maintenance, crop production, sanitation, ecosystem function, and so on. Further, there is often no substitute for water. In short, we need good water, now.

Thinking about these interlinked domains requires thinking broadly about policy. In this regard, Mark Sagoff (2004, p. 14) has argued that the point of democratic, political deliberation is to "…figure out how to classify a problem and then on what principle society should respond to problems of that kind." Often, however, water problems are of several kinds. For instance, pollution from synthetic estrogen in waterways – a consequence of pharmaceutical forms of birth control – affects the abundance and biomass of fish (Kidd et al. 2014). Addressing this form of pollution is not just an ecological problem. It is also one of respecting the rights of women to reproductive choice. What is required in such cases is an appreciation of the role of judgment in public policy. It is not simply a matter of applying general principles to specific problems, a model ethicists refer to as applied ethics but which is increasingly seen as inadequate because it does not account for how people actually reason in complex situations (Hoffmaster and Hooker 2009). Judgment, by contrast, is a skill that we learn through practice with others (Arendt 1982).

Despite the fact that water challenges are perennial, Michael Nelson (2003) argues that water has persisted in a "metaphysical blindspot" in environmental ethics. Part of this was due, in Nelson's view, to an imagination and subsequent ordering of nature that treated complex social and ecological relationships in discrete, atomistic terms. One implication of Nelson's arguments is that an atomistic view of water presents problems both physically and socially – both materially and morally – because so many relationships among biological life forms and social forms of life are interdependent with water. For instance, water is critical to biogeochemical processes at scales from individual cells to global nutrient balances and atmospheric climate interactions that are often interlinked and not easily or accurately separated into discrete categories (Berner and Berner 1987). In other words, water is the "bloodstream of the biosphere" (Ripl 2003). Likewise, anthropologists argue that water is a "total social fact" because actions that affect water in one area of social life frequently affect many other domains – religious, economic, political, and institutional (Orlove and Caton 2010). For example, in 1992, water experts met in Dublin to draft recommendations for the 1992 Earth Summit in Rio de Janeiro. One of the Dublin Principles (1992) stated that water had an economic value in all of its competing uses. As this was applied to international development policies in the 1990s, many people objected to its total implications for other domains of social life, from religion to human rights (Sultana and Loftus 2012).

Once we recognize water policy and management challenges as combinations of many kinds of problems, it becomes evident that the ethical implications of policies are also entangled with longer histories of social norms, institutions, laws, and policies that have co-evolved alongside the complex social and ecological relationships that affect hydrological systems (Whiteley et al. 2008). These contexts mean that issues of equity are tied to specific places and that ethical considerations should be understood through gaining insight into existing practices (Ingram et al. 1986). This often also requires philosophical investigation because "the rules and regulations governing the use of water stem from legal water doctrines which themselves stem from the philosophies of law, equity and justice" (Tisdell 2003, p. 415). This again reaffirms that water ethics are not primarily about applying general or abstract principles to specific cases, but of judging which principles are relevant to, and applicable within, which contexts (Schmidt and Shrubsole 2013). There are now several compendiums and reviews on examining the interplay between principles and practice in water ethics (Brown and Schmidt 2010, Ziegler and Groenfeldt 2017).

Before considering ethical aspects of water availability and quality, it is important to remark on why water management and policy cannot be reduced to scientific or technical matters alone. There are many philosophical reasons to reject the divide between facts and values, not the least of which is that values permeate all areas of life, including scientific practices (Putnam 2002). In water policy, however, it has long been held that there are established matters of fact, such as the water's molecular makeup as H_2O. Yet, even this claim is not free of value judgments. For instance, water is not only H_2O. In order for water to have the characteristics it does (i.e. its pH level), H_2O must mix with other molecules, like H_3O and OH^-. Importantly, there is no "correct" microstructure to these mixtures (van Brakel 2000, VandeWall 2007). Some philosophers of science even argue that older and long-rejected views of water as an element (as opposed to a compound of hydrogen and oxygen) have important insights (Chang 2012). As a result, we cannot rely on science alone to fully answer even basic questions like what is water? This is because even basic facts about water are only made available within broader judgments about the nature of science itself. For example, competing views about stream restoration in the United States turn not only on different policy objectives, but on different scientific practices and commitments (Lave 2012). Since there are competing understandings of science, policy decisions must at the very least judge among them and those judgments are not value free. Further, linking science to policy is made more difficult due to human impacts on the Earth system. For instance, experts recently rejected the idea that the hydrological cycle varies from year to year within an overall envelope of stability. This was a key axiom of hydrological science but human perturbations to the climate have rendered it untenable (Milly et al. 2008). One consequence is that policy norms built on the idea that there is a "naturally" stable hydrological cycle must be rethought to the extent that they rely on this axiom (Schmidt 2013).

Water Availability

Although it is not universal, a common historical feature of many water policies is that water use decisions are fundamentally about the good of the community. In early 20th-century American water policy, the twin ideas of resource conservation and multi-purpose river basin planning were formed around a synonymy between "the People" and the nation-state, which formed the basis for treating water as a public good (Schmidt 2017a). In the U.S. case, as in many other nationalist projects, maintaining control over water became a defining feature of political identity (Blackbourn 2006, Solomon 2010, Wittfogel 1957). These projects, however, were not confined within state borders. In fact, the massive increase in water availability during the 20th century through large dam-building programs that took place in the name of "international development" was designed to promote Western ideals of liberal democracy (Conca 2006, Sneddon 2015). Of course, who or what is included in the "community" has important consequences for political participation, moral consideration, and norms regarding how and when water is made available (Chamberlain 2008, Shaw and Francis 2008). Indigenous communities in North America, for example, have been marginalized in water policy decisions because their status as part of the political and moral community – either their own or that of the state – has been denied under settler colonial rule (Curley 2019, Espeland 1998, Hoover 2017, LaDuke 1999, Phare 2009). These exclusionary practices often intersect with dimensions of oppression based on gender, class, or race (Gaard 2001, Green 2020, von Schnitzler 2016, Waldron 2018).

A broad trend after WWII was that states gradually shifted from policies oriented to the "community" and toward polices based on meeting individual preferences (Norton 2005). In environmental ethics, this is frequently identified as a shift from bureaucratic utilitarianism to individual utilitarianism (Brown 2008, Feldman 1995, Norton 2005). Both versions seek the utilitarian aim of achieving the greatest good for the greatest number, but they do so through different means of government intervention versus individual transactions. But this shift can also be seen as a shift in ethical reasoning, not just the means of governing. Under bureaucratic utilitarianism, for instance, the good of the "community" was the object of concern. Since the "community" was virtually synonymous with the state, the goal of water policy and management was to make water available in ways that would strengthen and support the state. In western North America in the late 19th century, for example, miners and agricultural settlers were worried that capitalists would seize upon the value of water and accumulate a monopoly through the speculative acquisition of rights (Schorr 2005). In response, they developed a doctrine known as prior appropriation, which gives priority to those first to use water over those who come later, with rights to water specifically designed to not function like property rights. In some versions of prior appropriation, such as in Western Canada, water rights were tied to land ownership (Percy 1977); this legal tie, known as appurtenance, functioned as a way of keeping water in communities, "as a community resource" (Sax 1994, p. 14). When water laws formalized these practices, they created rights and obligations with respect to the community as a whole, which came to rely on certain patterns of upstream-downstream relationships. Yet since the initial system of rights was designed with a certain community in mind – miners and agricultural settlers – it produced the types of exclusions identified earlier for those who did not count as part of the "community," namely, Indigenous peoples (Schmidt 2017b).

To continue with the example, the general rule for granting prior appropriation rights was based on a utilitarian formula where individual water rights should, on balance, benefit the community (Feldman 1995). Throughout the 20th century, the "greatest good" was sought by increasing water supply through dams and other infrastructures that characterize the "supply-side" era of North American water management. It was a simple solution: generating more goods was possible whenever there was more water available to generate those goods. But when new sources of increasing supply ran out, a dilemma arose: on the one hand, it was increasingly recognized that even if new sources of water supply were found, that path was not ecologically sustainable in the long run. On the other, if an absolute limit to water was incorporated into water policies, how could rights be reformed to protect existing obligations while not denying those presently without rights fair opportunity to obtain them? This dilemma belied a thorny ethical issue. If the supply side path of 20th-century water policy was acknowledged to not be sustainable, then the government could no longer claim to be discharging its duty to steward water in the public trust (Glenn 2010, Ingram and Oggins 1992, Sax 1969, Wilkinson 1989). But if the constitution of the community – the public – *for whom* the government was charged with stewarding water was opened up on ecological grounds, then this would also require revisiting policy norms that impacted everybody relying on shared waters, including Indigenous peoples in North America. And even though Indigenous peoples in the United States (though never in Canada) had been recognized as having prior rights to water since a 1908 Supreme Court decision in *Winters v. United States*, these rights were often not respected in material or cultural terms (Curley 2019, Wilkinson 2010).

In retrospect, this dilemma was solved by not revisiting the constitution of the "community" but by adopting new forms of policy rationale based on the satisfaction of individual preferences. Part of this shift included assessing how new forms of transferring water rights may allow individuals to express their preferences via market mechanisms (see Natural Resources Council 1992). Alongside these arguments for economic rationality, several economists also retold the history of water rights, such as those of prior appropriation, as forerunners to market-oriented environmental policy (Anderson and Leal 2001). These historical claims do not, as we saw above, reflect or even approximate the original intent of water doctrines designed to stop water from being treated like private property. A second aspect of the shift to individual preference satisfaction was a growing emphasis on stakeholder participation in planning forums through decentralized policy planning and consensus-based decision-making (see generally Sabatier 2005). This reflected ideas about deliberative democracy and the need to have fair procedures for policy formation if decisions were to be normatively legitimate (Norton 2005, Priscoli 2004). Unfortunately, this "procedural turn" was often informal, and not matched with formal authority to make regulatory decisions, or with all parties included in the design of procedures or institutions, and this meant that those who started out in positions of power often retained an advantage (Schmidt 2014).

The shift from bureaucratic to individual utilitarianism, and from state-led policies to procedural models of governing water, fits with a distinction that the philosopher Derek Parfit (2011, p. 45) makes between object-given versus subject-given forms of ethical reasoning. In an "object-given" view, our reasons for acting in one way and not another "are provided by the facts that make certain outcomes worth producing or preventing, or make certain things worth doing for their own sake." Under state-led models, for instance, pursuing the good of the community (often conflated with a particular view of the state itself) was an intrinsically desirable end. The alternate is a "subject-given" view where our reasons depend on how our preferences or desires affect our actions; or, more generally, on facts about us. In procedural water policies, whether through fair market transactions or stakeholder collaboration, reasons depend on how preferences or desires are pursued. The differences between these two styles of reasoning parallel the shifts in water policy that are based on facts about the water community (object-given) versus those based on enabling individual preference satisfaction (subject-given). When we move outside of the North American examples that have been used so far, these differences can help to understand ethical aspects of policy in some other contexts as well.

Colonial experiences, and later post-colonial experiences under the guise of "international development," often paralleled shifts between object-given versus subject-given forms of ethical reasoning. Under British colonial rule, for instance, sovereign Indigenous communities in New Zealand were denied rights to water as part of broader political and moral exclusions (Strang 2014). One reason for this concerns sovereignty, which in the imaginary of modern constitutional states arises through the self-constitution of individuals in a social contract (Taylor 2004). As such, states created through colonial rule and European settlement assumed that their own political communities had a sovereign right to water that others did not. In many Indigenous cosmologies, including those through which they entered into international treaties, sovereignty is based in alternate forms of relations between individuals and groups (Asch 2014, Simpson 2014). These challenges are not limited to British colonialism. The privatization of water rights in Chile in the 1980s, for instance, was based on faith in the ability of markets to satisfy individual preferences in a way that would increase the overall benefits derived from water. These reforms in Chile produced, at best, mixed results (see Bauer 2004). As later research would show, the places where water markets have been successful – like Spain and Peru – were places where they were designed to reflect communal norms regarding fairness. That is, the market was designed using principles that ensured facts about community relations and institutions, not individual preferences, provided the rationale (Trawick 2010).

During the massive urbanization of many societies in the late 20th century, an important aspect of contemporary water challenges became the debate over public versus private water delivery. But it is also important to note that "public water" is a concept formed alongside certain forms of international development, particularly the finance and governance models needed to make water available through modern infrastructure (Bakker 2013). In theory, public water is a great idea. But it can also have a more covert dimension because the imagination of the "public" is cultural. As a result, for many non-Western people, becoming part of the "public" recognized by the state is a deeply unsatisfying option because it may require relinquishing cultural forms of life, their own systems of governance, and their own water use practices (Boelens et al. 2010). By contrast, many marginalized groups, such as those in informal settlements in Mumbai, actively pursue recognition as part of the "public" in order to secure the provision of water services from the state (Anand 2017). So, the ethics of "public" water run in multiple directions that require attention to the histories, contexts, and political contests affecting decisions.

Water Quality

In many jurisdictions, the broadly shared and deeply worrisome prospect of degraded water quality has led either to the establishment of minimum standards for pollutant loads or to the prohibition of certain types of pollutants altogether. Some of the first whole-ecosystem experiments at the Experimental Lakes Area in Canada and the Hubbard Brook Experimental Forest in the United States targeted issues of water quality, notably acid rain and phosphate pollution in the North American Great Lakes. These experiments had direct impacts on policy – such as the banning of certain types of detergents – by establishing object-given reasons for policy, namely, the facts provided by the intrinsic good of having lakes that were not inundated with unserviceable pollutant loads.

Despite some of the gains resulting from pollution standards – including environmental laws like the *Clean Water Act* in the United States (see Lazarus 2004) – significant water quality concerns persist in many jurisdictions. These arise through a combination of factors, such as the growing number of pollutants entering waterways from both point and non-point sources, the complex and unforeseen interactions of different pollutants, the increased human appropriation of freshwater resources, human changes to land uses that affect the timing and magnitude of run-off, and disruptions to precipitation regimes – timing, quantity, temperature – owing to climate change. One result of these converging concerns has been an attempt to approach the science of water quality from a perspective that treats social and ecological variables together. That is, in ethical terms, to move out of the "metaphysical blindspot" that treats complex relationships and processes in discrete terms. Treating social and ecological concerns together, however, raises questions of how to compare different goods that water provides, such as those for economies, biodiversity, public health, and recreation, to name a few (Acreman 2001).

The difficulty of comparing different kinds of goods is not unique to water policy. Yet, water policy does have its own challenges, particularly those related to how different scientific categories are used to gather together social and ecological considerations. For instance, "environmental flows" or "instream flows" are categories designed to enable policy makers to identify the water required for healthy ecosystems – frequently measured through a combination of variables like oxygen content in water for fish, whether enough water is available to maintain riparian habitat, and so on (see Annear et al. 2002, Richter et al. 2012). Determining instream flow needs that would maintain adequate water quality for aquatic ecosystems is also an ethical issue because humans and non-humans are both dependent on water (Postel and Richter 2003). In addition, there are numerous practical challenges. One is the limited data available about water quality for many waterways. Another is that different measurement techniques make it difficult to compare data. In addition, measurements often begin after certain forms of pollution are recognized as harmful, which means that there is often no historical baseline data for waterways before certain types of pollution began (Acreman et al. 2008, Richter et al. 2003).

But the selection of what to measure and the methods for assessing water quality are not just technical issues. Rather, the selection of which ecological indicators to measure is a social and political process (see Bouleau et al. 2009). During fieldwork in the Canadian province of Alberta, for instance, I was puzzled as to why certain types of ecological measurements were being used and others were not. As it turned out, a significant part of the reason was that, under Alberta's newly created water governance format, the terms of financial agreements stipulated that no new studies could be commissioned when determining the state of a given river basin. As a result, decision-makers relied on existing data, most of which was collected for other purposes, such as recreational fishing. My findings in Alberta are similar to other research that has found that contests over who participates in choosing ecological indicators and the social processes and institutional arrangements used to achieve agreement are deeply political (Fernandez 2014).

One increasingly prevalent way to establish water quality goals is through the concept of eco-system services, where an economic value is placed on water by asking the question: what would it cost to do what a healthy ecosystem does for free? For instance, if we remove a wetland that is key to maintaining water quality and replace it with urban development, how much would it cost to use technology to maintain the same level of water quality? In this view, managing water quality is about trade-offs where the ethical obligation is to balance ecosystem health with the services they provide for human welfare (Falkenmark and Folke 2010). In the example above, we must trade off the value of a wetland with the value of urban development. This view faces at least three challenges in addition to those above regarding the selection and measurement of criteria. First, it requires that ecosystems be characterized in ways that allow different services to be "traded-off" against each other to best enhance human welfare. This requires that we take a view where all values are comparable with each other – that they are commensurable – and do not differ fun-damentally. But whether values are commensurate depends on broader social considerations. For instance, in Chile, payments for ecosystem services do not fit well with, and actually undermine, the cultural practices and prevailing norms that Indigenous groups use to value water (Boelens et al. 2014). Second, "ecosystem services" are anthropocentric because entire ecosystems are valued only with respect to the services they provide for humans. We are not, for instance, valuing the services that a healthy ecosystem provides for aquatic species except insofar as those species are relevant to human welfare. Anthropocentrism (or variants of it) is a defensible view, but it requires adequate arguments and not just stipulation (see Norton 1984). Third, valuing ecosystem services in ways that fit with economic models requires that we use models of ecology that fit with mod-els of economics (Norgaard 2010). This last challenge drives straight at the issue of the scientific practices of different private or public actors and the assumptions made about both what science is for, and how best to understand complex hydrological processes (Lave 2012). For instance, constraining science to models that fit with economic assumptions leaves out the numerous other ecological models and forms of knowledge that do not see nature in terms that fit economic ways of valuing natural goods.

Complexity and Common Water

This chapter started off with a disjuncture. In theory, water policies should become more restric-tive as water comes under more pressure. Yet, actual decisions reveal a historical trend toward less restrictive policies. How did this come about? And how has this disjuncture been defended in normative terms? One explanation is that there has been a change in the style of ethical reasoning used in water policies from object-given to subject-given rationale. In that case, and without de-bating the merits of either, we can see how policies have combined ethical reasoning with political decision-making in subtle yet identifiable ways. The moral theory underlying these policies has been broadly utilitarian, but the connection of that theory to different policies has changed along with shifting political challenges. But a shift in style is not a shift in substance. Many pressing ethical challenges in water policy remain regarding both the challenges of selecting water quality indicators and of ensuring that the "community" sharing water is conceived of in ways that can remediate historical and contemporary harms. In short, my argument is that this disjuncture per-sists because of a refusal to incorporate others into the community just in case they challenge the existing imagination of moral and material orders of water.

A second consideration the foregoing arguments leave us with is how we should reconnect politics and ethics. One touchstone I have been suggesting is that the concept of the public – or the community – presents an ongoing site of ethical and political contest. It is here that the water ethics literature is concerned with how alternate ways of imagining the political community, including the spaces, relationships, and sources of authority, it operates in reference to. Multiple constituencies share water, particularly in those places where colonialism persists and where a "strange multiplicity" besets constitutional matters among Indigenous peoples and states (Tully 1995). These constituencies have legitimate political claims to water, and their water uses will affect others downstream. As a result, reconsidering the public "common sense" in ways that respect alternate forms and principles of decision-making will require forums in which different ways of reaching judgment are practiced and respected. In order to address historical inequities regarding the denial of the rights of these communities to water, a key moment of interruption will need to occur that revisits the notion of "community" in water policy.

There are many potential models for the "community" but none, I would argue, that can be reached through philosophical reflection alone. Rather, the complex relationships and systems supported by water require both cultural and philosophical dimensions of water policies to be considered. This type of approach would seek appraisals of the ways that history, ecology, culture, and law have informed different iterations and reforms of water policy in particular places (Rademacher 2011, Rodriguez 2007). Social sciences have already shed light on key policy questions, like what the conditions are for effectively understanding the place of economic instruments in water policy decisions (Ballestero 2019, Trawick 2003, 2010). They have also helped to detail the ways that different spheres of cultural, religious, and legal values shape the political and moral norms that affect issues of gender equity (Mehta 2005, Zwarteveen and Meinzen-Dick 2001). The suggestion here, then, is that we connect the contextual richness of these types of studies with the styles of ethical reasoning that have been used to simplify complex systems and relationships into actionable policy items. This would allow us to see how the choice of a particular form of ethical reasoning in water policy is itself a choice made by those with standing in particular moral and political communities.

Related Topics

2. Eating
9. Mountains: Rethinking Thinking Like a Mountain
13. Property
15. Wetlands
16. Rivers and Watersheds
17. Ocean Policy
19. The Ethics of Marine Protected Areas
29. Hydropower
38. Food
39. Industrial Agriculture
41. Sustainable Agriculture
49. Pollution and Polluter Pays

References

Acreman, M. (2001) 'Ethical aspects of water and ecosystems', *Water Policy* (3), pp. 257–265.
Acreman, M. et al. (2008) 'Developing environmental standards for abstractions from UK rivers to implement the EU water framework directive', *Hydrological Sciences* (53), pp. 1105–1120.
Anand, N. (2017) *Hydraulic city: water and the infrastructures of citizenship in Mumbai.* Durham, NC: Duke University Press.
Anderson, T. & Leal, D. (2001) *Free market environmentalism,* Revised Edition. New York: Palgrave.
Annear, T.C. et al. (2002) *Instream flows for riverine resource stewardship.* Cheyenne, WY: Instream Flow Council.
Arendt, H. (1982) *Lectures on Kant's political philosophy.* Chicago, IL: The University of Chicago Press.

Asch, M. (2014) *On being here to stay: treaties and aboriginal rights in Canada*. Toronto: University of Toronto Press.

Bakker, K. (2013) 'Constructing 'public' water: the World Bank, urban water supply, and the biopolitics of development', *Environment and Planning D: Society and Space* (31), pp. 280–300.

Ballestero, A. (2019) *A future history of water*. Durham, NC: Duke University Press.

Bauer, C. (2004) *Siren song: Chilean water law as a model for international reform*. Washington, DC: Resources for the Future.

Benidickson, J. (2007) *The culture of flushing: a social and legal history of sewage*. Vancouver: UBC Press.

Berner, E.K. & Berner, R.A. (1987) *The global water cycle: geochemistry and environment*. Englewood Cliffs, NJ: Prentice-Hall.

Blackbourn, D. (2006) *The conquest of nature: water, landscape, and the making of modern Germany*. New York: W.W. Norton & Company.

Boelens, R., Getches, D. & Guerva-Gill, A. (Eds.) (2010) *Out of the mainstream: water rights, politics and identity*. London: Earthscan.

Boelens, R., Hoogesteger, J. & Rodriguez de Francisco, J.C. (2014) 'Commoditizing water territories: the clash between Andean water rights cultures and payment for environmental services policies', *Capitalism Nature Socialism* (25), pp. 84–102.

Bouleau, G. et al. (2009) 'How ecological indicators construction reveals social changes – the case of lakes and rivers in France', *Ecological Indicators* (9), pp. 1198–1205.

Brown, P.G. (2008) *The commonwealth of life: economics for a flourishing earth*. Montreal: Blackrose Books.

Brown, P.G. & Schmidt, J.J. (Eds.) (2010) *Water ethics: foundational readings for students and professionals*. Washington, DC: Island Press.

Chamberlain, G. (2008) *Troubled waters: religion, ethics, and the global water crisis*. Lanham: Rowman and Littlefield Publishers, Inc.

Chang, H. (2012) *Is water H_2O? Evidence, pluralism and realism*. Dordrecht: Springer Verlag.

Conca, K. (2006) *Governing water: contentious transnational politics and global institution building*. Cambridge: MIT Press.

Curley, A. (2019) '"Our Winters' rights": challenging colonial water laws', *Global Environmental Politics* (19), pp. 57–76.

Dublin Statement (1992) *The Dublin statement on water and sustainable development* [online]. Available at: http://www.wmo.ch/pages/prog/hwrp/documents/english/icwedece.html.

Espeland, W.N. (1998) *The struggle for water: politics, rationality, and identity in the American Southwest*. Chicago, IL: University of Chicago Press.Falkenmark, M. & Folke, C. (2010) 'Ecohydrosolidarity: a new ethics for stewardship of value-adding rainfall', in Brown, P.G. & Schmidt, J.J. (Eds) *Water ethics: foundational readings for students and professionals*. Washington, DC: Island Press, pp. 247–264.

Feldman, D. (1995) *Water resources management: in search of an environmental ethic*. Baltimore: Johns Hopkins University Press.

Fernandez, S. (2014) 'Much ado about minimum flows: unpacking indicators to reveal water politics', *Geoforum* (57), pp. 258–271.

Gaard, G. (2001) 'Women, water, energy: an ecofeminist approach', *Organization & Environment* (14), pp. 157–172.

Glenn, J.M. (2010) 'Crown ownership of water in situ in common law Canada: public trusts, classical trusts and fiduciary duties', *Les Cahiers de Droit* (3–4), pp. 493–520.

Green, L. (2020) *Rock | water | life: ecology & humanities for a decolonial South Africa*. Durham, NC: Duke University Press.

Hoffmaster, B. & Hooker, C. (2009) 'How experience confronts ethics', *Bioethics* (23), pp. 214–225.

Hoover, E. (2017) *The river is in us: fighting toxics in a Mohawk community*. Minneapolis: University of Minnesota Press.

Ingram, H. & Oggins, C. (1992) 'The public trust doctrine and community values in water', *Natural Resources Journal* (32), pp. 515–537.

Ingram, H., Scaff, L.A. & Silko, L. (1986) 'Replacing confusion with equity: alternatives for water policy in the Colorado River Basin', in Weatherford, G.D. & Brown, F.L. (Eds.) *New courses for the Colorado River: major issues for the next century*. Albuquerque: University of New Mexico Press, pp. 177–200.

Kidd, K.A. et al. (2014) 'Direct and indirect responses of a freshwater food web to a potent synthetic oestrogen', *Philosophical Transactions of the Royal Society B* (369), pp. 1–11.

LaDuke, W. (1999) *All our relations: native struggles for land and life*. Cambridge, MA: South End Press.

Lave, R. (2012) *Fields and streams: stream restoration, neoliberalism, and the future of environmental science*. Athens: University of Georgia Press.

Lazarus, R.J. (2004) *The making of environmental law.* Chicago, IL: The University of Chicago Press.

Mehta, L. (2005) *The politics and poetics of water: naturalizing scarcity in Western India.* New Delhi: Orient Longman Private Limited.

Milly, P.C.D. et al. (2008) 'Stationarity is dead: whither water management?' *Science* (319), pp. 573–574.

Natural Research Council. (1992) *Water transfers in the west: efficiency, equity, and the environment.* Washington, DC: National Academy Press.

Nelson, M.P. (2003) 'Earth, air, water...ethics', *Transactions: Scholarly Journal of the Wisconsin Academy of Sciences, Arts and Letters* (90), pp. 164–173.

Norgaard, R.B. (2010) 'Ecosystem services: from eye-opening metaphor to complexity blinder', *Ecological Economics* (69), pp. 1219–1227.

Norton, B.G. (1984) 'Environmental ethics and weak anthropocentrism', *Environmental Ethics* (6), pp. 131–148.

Norton, B.G. (2005) *Sustainability: a philosophy for adaptive ecosystem management.* Chicago, IL: University of Chicago Press.

Orlove, B. & Caton, S. (2010) 'Water sustainability: anthropological approaches and prospects', *Annual Review of Anthropology* (39), pp. 401–415.

Parfit, D. (2011) *On what matters* (2 volumes). New York: Oxford University Press.

Pauli, B. (2019) *Flint fights back: environmental justice and democracy in the Flint water crisis.* Cambridge: MIT Press.

Percy, D. (1977) 'Water rights in Alberta', *Alberta Law Review* (15), pp. 142–165.

Phare, M.S. (2009) *Denying the source: the crisis of First Nations water rights.* Surrey: Rocky Mountain Books.

Postel, S. & Richter, B. (2003) *Rivers for life: managing water for people and nature.* Washington, DC: Island Press.

Priscoli, J.D. (2004) 'What is public participation in water resources management and why is it important?' *Water International* (29), pp. 221–227.

Putnam, H. (2002) *The collapse of the fact/value distinction and other essays.* Cambridge: Harvard University Press.

Rademacher, A. (2011) *Reigning the river: urban ecologies and political transformation in Kathmandu.* Durham, NC: Duke University Press.

Richter, B. et al. (2003) 'Ecologically sustainable management: managing river flows for ecological integrity', *Ecological Applications* (13), pp. 206–224.

Richter, B. et al. (2012) 'A presumptive standard for environmental flow protection', *River Research and Applications* (28), pp. 1312–1321.

Ripl, W. (2003) 'Water: the bloodstream of the biosphere', *Philosophical Transactions of the Royal Society of London B* (358), pp. 1921–1934.

Rodriguez, S. (2007) *Acequia: water sharing, sanctity, and place.* Santa Fe: School for Advanced Research Press.

Rose, C.M. (1990) 'Energy and efficiency in the realignment of common-law water rights', *Journal of Legal Studies* (19), pp. 261–296.

Sabatier, P.A., Focht, W., Lubell, M., Trachtenberg, Z., Vedlitz, A. & Matlock, M. (Eds.) (2005) *Swimming upstream: collaborative approaches to watershed management.* Cambridge: The MIT Press.

Sagoff, M. (2004) *Price, principle, and the environment.* New York: Cambridge University Press.

Sax, J.L. (1969) 'The public trust doctrine in natural resources law: effective judicial intervention', *Michigan Law Review* (68), pp. 471–566.

Sax, J.L. (1994) 'Understanding transfers: community rights and the privatization of water', *West-Northwest Journal of Environmental Law and Policy* (1), pp. 13–16.

Schmidt, J.J. (2013) 'Integrating water management in the Anthropocene', *Society and Natural Resources* (26), pp. 105–112.

Schmidt, J.J. (2014) 'Water management and the procedural turn: norms and transitions in Alberta', *Water Resources Management* (28), pp. 1127–1141.

Schmidt, J.J. (2017a) *Water: abundance, scarcity, and security in the age of humanity.* New York: New York University Press.

Schmidt, J.J. (2017b) 'Water policy in Alberta: settler-colonialism, community, and capital', *Journal of the Southwest* (59), pp. 204–226.

Schmidt, J.J. & Peppard, C.Z. (2014) 'Water ethics on a human dominated planet: rationality, context and values in global governance', *WIREs Water* (1), pp. 533–547.

Schmidt, J.J. & Shrubsole, D. (2013) 'Modern water ethics: implications for shared governance', *Environmental Values* (22), pp. 359–379.

Schorr, D. (2005) 'Appropriation as agrarianism: distributive justice in the creation of property rights', *Ecology Law Quarterly* (32), pp. 3–71.

Shaw, S. & Francis, A. (eds.) (2008) *Deep blue: critical reflections on nature, religion and water.* London: Equinox.

Simpson, A. (2014) *Mohawk interruptus: political life across the borders of settler states.* Durham, NC: Duke University Press.

Sneddon, C. (2015) *Concrete revolution: large dams, Cold Water geopolitics, and the U.S. Bureau of Reclamation.* Chicago, IL: University of Chicago Press.

Solomon, S. (2010) *Water: the epic struggle for wealth, power, and civilization.* New York: HarperCollins Publishers.Strang, V. (2014) 'The Taniwha and the Crown: defending water rights in Aotearoa/New Zealand', *WIREs Water* (1), pp. 121–131.

Sultana, F. & Loftus, A. (Eds.) (2012) *The right to water: politics, governance and social struggles.* London: Routledge.

Taylor, C. (2004) *Modern social imaginaries.* Durham, NC: Duke University Press.

Tisdell, J.G. (2003) 'Equity and social justice in water doctrines', *Social Justice Research* (16), pp. 401–416.

Trawick, P.B. (2003) *The struggle for water in Peru: comedy and tragedy in the Andean commons.* Stanford, CA: Stanford University Press.

Trawick, P. (2010) 'Encounters with the moral economy of water: general principles for successfully managing the commons', in Brown, P.G. & Schmidt, J.J. (Eds.) *Water ethics: foundational readings for students and professionals.* Washington, DC: Island Press, pp. 155–161.

Tully, J. (1995) *Strange multiplicity: constitutionalism in an age of diversity.* Cambridge: Cambridge University Press.

van Brakel, J. (2000) *Philosophy of chemistry: between the manifest and the scientific image.* Leuven: Leuven University Press.

VandeWall, H. (2007) 'Why water is not H_2O, and other critiques of essentialist ontology from the philosophy of chemistry', *Philosophy of Science* (74), pp. 906–919.

von Schnitzler, A. (2016) *Democracy's infrastructure: techno-politics and protest after Apartheid.* Princeton, NJ: Princeton University Press.

Waldron, I. (2018) *There's something in the water: environmental racism in Indigenous and Black communities.* Halifax: Fernwood Publishing Ltd.

Whiteley, J.M., Ingram, H. & Perry, R.W. (Eds.) (2008) *Water, place & equity.* Cambridge: MIT Press.

Wilkinson, C.F. (1989) 'Aldo Leopold and western water law: thinking perpendicular to the prior appropriation doctrine', *Land and Water Law Review* (24), pp. 1–38.

Wilkinson, C.F. (2010) 'Indian water rights in conflict with state water rights: the case of the Pyramid Lake Paiute Tribe in Nevada, US', in Boelens, R., Gretches, D., & Guerva-Gill, A. (Eds.) *Out of the mainstream: water rights, politics and identity.* London: Earthscan, pp. 213–222.Wittfogel, K.A. (1957) *Oriental despotism: a comparative study of total power.* New Haven, CT: Yale University Press.

Ziegler, R. & Groenfeldt, D. (Eds.) (2017) *Global water ethics: towards a global ethics charter.* London: Routledge.

Zwarteveen, M. & Meinzen-Dick, R. (2001) 'Gender and property rights in the commons: examples of water rights in South Asia', *Agriculture and Human Values* (18), pp. 11–25.

Recommended Reading

Boelens, R., Getches, D. & Guerva-Gill, A. (Eds.) (2010) *Out of the mainstream: water rights, politics and identity* London: Earthscan.

Brown, P.G. & Schmidt, J.J. (Eds.) (2010) *Water ethics: foundational readings for students and professionals.* Washington, DC: Island Press.

Peppard, C.Z. (2014) *Just water: theology, ethics and the global water crisis.* Maryknoll, NY: Orbis Books.

Richter, B. (2014) *Chasing water: a guide for moving from scarcity to sustainability.* Washington, DC: Island Press.

Whiteley, J.M., Ingram, H. & Perry, R.W. (Eds.) (2008) *Water, place & equity.* Cambridge, MA: MIT Press.

15

WETLANDS

J. Baird Callicott

Wetlands and the Western Imaginary

Wetlands have not fared well in the Western imaginary. Throughout much of Western intellectual history, they were associated with danger, disease, and death. In Greek mythology, the second of Heracles's 12 labors was killing the Hydra, a venomous, many-headed reptilian monster who rose up out of the swamp adjacent to Lake Lerna and terrorized nearby villagers with poisonous vapors (Pseudo-Apollodorus 1921). In *Beowulf* (unknown 1977), the grotesque Grendel and his evil mother—outcast descendants of the biblical Cain—made their lair in the middle of a dark, forbidding fen. (Grendel jealously raids Heorot, the warm and convivial meade-hall of King Hroogar, wreaking murder and mayhem, but eventually meets his end at the hands of the title character.) A pop-culture variation on this ancient theme is the 1982 movie "Swamp Thing"—now become a cult classic. A generic name for a number of diseases, including malaria, is "swamp fever." The Hydra's poisonous vapors may have a basis in the phenomenon of "marsh gas"—a volatile product of anaerobic organic decay, mostly consisting of methane.

And think of the way wetlands function as metaphors. "Referee a paper for the journal *Wetlands Bulletin?*" "Sorry, I'm *swamp*ed; I'm *bog*ged down with work on an NSF grant proposal on biocomplexity-in-the-environment with the deadline looming; and the complexity of ecosystems just *bog*gles my mind." The word "fen," more frequently used in British than in American English, derives from the Old English "fenn"—meaning "mud" or "mire." And one might also say, "I want to avoid getting *mire*d in office politics— the issue of merit distribution is a perennial quag*mire*—that's why I work at home as much as possible." In the Anglophone discourse of agriculture, swamps, bogs, fens, and marshes are denominated "waste lands" in need of draining for productive use.

Renowned American analytic philosopher Donald Davidson (1987) imagined himself struck by lighting and killed *while hiking in a swamp.* (But who actually hikes in a swamp?) He imagined a second lightning strike to reassemble ambient atoms into a precise replica of himself, down to the last neuron of his brain, the question being would the replica be identical to Davidson in mind as well as body? (No definitive answer to this question has been achieved, nor does one matter in the context of this essay—if it matters at all.) Davidson's name for this imagined replica is "Swampman." Would hundreds of parasitic philosophy papers on this so-called "thought experiment" have been forthcoming if Davidson had named his replica "Prairieman?" Probably not. Swampman gives the doppleganger a foul air (pun intended) of spookiness and mystery, just the

DOI: 10.4324/9781315768090-19

sort of thing to attract philosophers more cognitively than emotionally self-aware like flies to … honey (shall we say).

Part of the moral rehabilitation of swamps, bogs, fens, and marshes has been to give them a new name. We recently bestowed the pleasant name "rainforests" on what we used to call "jungles"; likewise, in polite company, we now call the merger of water and soil "wetlands." Well aware of the prevailing attitude toward wetlands, Henry David Thoreau, John Muir, and Aldo Leopold—the three giants on whose shoulders we contemporary environmental philosophers stand—made a point of trying to shift that attitude to something more positive, each in his own unique way. Thoreau initiated the moral rehabilitation of wetlands, Muir advanced it, and Leopold completed it.

Thoreau on Wetlands

In a late essay, "Walking," which was published just weeks after his death in *The Atlantic Monthly*, Thoreau has a lot to say about swamps and bogs. His several discussions of them, curiously, reflect a common opinion; the first seems to come in the course of a nightmarish dream sequence:

> I saw the fences half consumed, their ends lost in the middle of the prairie, and some worldly miser with a surveyor looking after his bounds, while heaven had taken place around him, and he did not see the angels going to and fro, but was looking for an old post-hole in the midst of paradise. I looked again and saw him standing in the middle of a boggy, stygian fen, surrounded by devils, and he had found his bounds without a doubt, three little stones, where a stake had been driven, and looking nearer, I saw that the Prince of Darkness was his surveyor.
>
> *(Thoreau 1862: 660)*

Thoreau himself, I should note, worked as a surveyor. In any case, the conventional opinion of both prairies and wetlands is perfectly reflected here. Prairie: heaven and angels; swamp: devils and the Prince of Darkness. The adjective "stygian" refers to the river Styx—in Homer, the waterway into the underworld. Apparently forgetting this little bit of swampy *danse macabre*, Thoreau seems to reverse himself when, pursuing variations on his main good-wild-versus-lame-tame theme in "Walking," he turns again to the subject of wetlands.

> Hope and future for me are not in lawns and cultivated fields, not in towns and cities, but in the impervious and quaking swamps. When, formerly, I have analyzed my partiality for some farm which I had contemplated purchasing, I have frequently found that I was attracted solely by a few square rods of impermeable and unfathomable bog—a natural sink in one corner of it. That was the jewel which dazzled me. I derive more of my subsistence from the swamps which surround my native town than from the cultivated gardens in the village. There are no richer parterres to my eyes than the dense beds of dwarf andromeda… which cover these tender places on the earth's surface. Botany cannot go farther than tell me the names of the shrubs which grow there—the high blueberry, panicled andromeda, lamb-kill, azalea, and rhodora—all standing in the quaking sphagnum… . Yes, though you may think me perverse, if it were proposed to me to dwell in the neighborhood of the most beautiful garden that ever human art contrived, or else of a dismal swamp, I should certainly decide for the swamp. How vain, then, have been all your labors, citizens, for me! My spirits infallibly rise in proportion to the outward dreariness… . When I would recreate myself, I seek the darkest wood, the thickest and most interminable, and, to the citizen, the most dismal swamp. I enter the swamp as a sacred place—a *sanctum sanctorum*. There is the strength, the marrow of Nature.
>
> *(Thoreau 1862: 666)*

Notice that for Thoreau, no less than for the citizen (that is, it would seem, the denizen of the city), the actual quality of wetlands is the same. They are dismal, dreary, lonely. And the beauty that Thoreau mentions finding in them is quite pedestrian—a matter of the plants they harbor, which he compares favorably to more colorful garden flowers. Thoreau's preference for wetlands in comparison to other places seems to amount, upon analysis, to two things. First, wetlands are—in the middle landscape of his rural Massachusetts—*refugia* of wildness. In that quality, however, they do not seem to exceed the darkest wood, as Thoreau tacitly admits. So the other thing settles the question, In which biotic community is it best to recreate and in the neighborhood of which to dwell?: Thoreau's perversity, as he calls it; his contrarian proclivity. It might be possible to find another person in Concord who loves perfectly wild woods—say Ralph Waldo Emerson— but only Thoreau, our hero (and his own as well), could possibly love a wetland, so universally shunned were they still in the mid-nineteenth century.

In "Walking," Thoreau's final word of any consequence on wetlands confirms my analysis:

> I was surveying for a man the other day a single straight line one hundred and thirty-two rods long, through a swamp, at whose entrance might have been written the words which Dante read over the entrance to the infernal regions— "Leave all hope, ye that enter"—that is, of ever getting out again; where at one time I saw my employer actually up to his neck and swimming for his life in his property, though it was still winter. He had another similar swamp which I could not survey at all, because it was completely under water, and nevertheless, with regard to a third swamp, which I did survey from a distance, he remarked to me, true to his instincts, that he would not part with it for any consideration, on account of the mud which it contained. And that man intends to put a girdling ditch round the whole in the course of forty months, and so redeem it by the magic of his spade. I refer to him only as the type of a class [or token of a type as analytic philosophers would say].
>
> *(Thoreau 1862: 667)*

In "Walking," wetlands are for Thoreau little different than they were for the author of *Beowulf.* But Thoreau proclaims himself to be perverse, from the conventional citizen's point of view. Thoreau prizes them for the very qualities—which are not in dispute—that make them almost universally reviled among his contemporaries.

Muir on Wetlands

In the late summer of 1867, five years after the posthumous publication of "Walking," John Muir set out on his thousand-mile walk from the Ohio River through Kentucky, Tennessee, the Carolinas, Georgia, and Florida to the Gulf of Mexico. He kept a journal en route, which, much later in life, he revised. Muir died in 1914. *A Thousand Mile Walk to the Gulf* was published, also posthumously, in 1916. In that journal, for the first time to my knowledge in English letters, Muir averred that rights belonged to nonhuman natural entities and to nature as a whole, as well as to human beings.

Walking across Georgia toward Savannah, Muir encountered extensive cypress swamps. Here is his description of the swamp-loving plant from which such wetlands take their name:

> This remarkable tree, called cypress, is a taxiodium, grows large and high, and is remarkable for its flat crown. The whole forest seems level on the top, as if each tree had grown up against a ceiling, or had been rolled while growing. The taxiodium is the only level-topped tree I have seen. The branches, though spreading, are careful not to pass each other, and stop suddenly on reaching the general level, as if they had grown up against a ceiling.
>
> *(Muir 1916: 57)*

Strangely, Muir does not mention what to most casual observers is the cypress's most remarkable feature—its knees. He marveled at Spanish moss and was mightily impressed by live oaks. In his two chapters on the state, he mentions many encounters with Georgia wetlands, but only briefly to characterize their flora or to note that in them it was difficult to find a dry place to sleep.

Inspired by Alexander von Humboldt's botanical adventures in South America, Muir's goal was to experience tropical vegetation. And that is just what he believed he had found in north Florida. While walking east to west along a railroad right-of-way from Fernandina (where his steamboat from Savannah had landed him) to Gainesville, and then from Gainesville on to Cedar Key, he would slog off several hundred yards into the swamps on either side of the rail bed attracted by a new, captivating plant. "Who actually hikes in swamps?"—I asked before. Donald Davidson only in his arm-chair imagination, only John Muir in actuality. And all the while, Muir was terrified of the fauna, particularly of alligators:

> I struggled hard and kept my course, leaving the general direction only when drawn aside by a plant of extraordinary promise, that I wanted for a specimen... . In wading, I never attempted to keep my clothes dry, because the water was too deep, and the necessary care would consume too much time. Had the water that I was forced to wade been transparent it would have lost much of its difficulty. But as it was, I constantly expected to plant my feet on an alligator, and therefore proceeded with strained caution. The opacity of the water caused uneasiness also on account of my inability to determine its depth. In many places I was compelled to turn back, after wading forty or fifty yards, and try again a score of times before I was able to get across a single lagoon.
>
> *(Muir 1916: 119–20)*

At this time, Muir was coming down with malaria, which made his struggle across north Florida all the harder. Still, he could write:

> Many good people believe that alligators were created by the Devil, thus accounting for their all-consuming appetite and ugliness. But doubtless these creatures are happy and fill the place assigned to them by the great Creator of us all. Fierce and cruel they appear to us, but beautiful in the eyes of God. They, also, are his children, for He hears their cries, cares for them tenderly, and provides their daily bread.
>
> The antipathies existing in the Lord's great animal family must be wisely planned, like balanced repulsion and attraction in the mineral kingdom. How narrow we selfish creatures are in our sympathies! How blind to the rights of all the rest of creation! With what dismal irreverence we speak of our fellow mortals! Though alligators, snakes, etc. naturally repel us, they are not mysterious evils. They dwell happily in these flowery wilds, are part of God's family, unfallen, undepraved, and cared for with the same species of tenderness and love as is bestowed on angels in heaven and saints on earth.
>
> I think that most of the antipathies which haunt and terrify us are morbid productions of ignorance and weakness. I have better thoughts of those alligators now that I have seen them at home. Honorable representatives of the great saurians of an older creation, may you long enjoy your lilies and your rushes, and be blessed now and then with a mouthful of terror-stricken man by way of dainty!.
>
> *(Muir 1916: 98–99)*

For Muir as for Thoreau, the beauty of wetlands was a matter of plants. Immediately after this remarkable dissertation on alligators, without any literary transition whatsoever even attempted, Muir writes, "Found a beautiful lycopodium today, and many grasses in the dry sunlit places called 'barrens,' 'hummocks,' 'savannas,' etc. Ferns also are abundant" (Muir 1916: 99). He waxes poetically about magnolias (especially), palmettos, live oaks, various grasses, sedges, and flowering shrubs. As we see, he would go to any lengths—or rather depths—wading off his path through black Florida swamp water to collect a specimen. And for Muir, as for Thoreau, love of wetlands was born of a contrarian impulse, a kind of perversity, and an attitude toward humanity that bordered on misanthropy. We have seen how contemptuous is Thoreau of the quotidian "citizen." Muir, for his part, is delighted by the thought of a gator eating a human being (so long, it seems, as he himself is not the terror-stricken man consumed by way of dainty). And he concludes his Florida swamp chapters in *Walk* with this coda:

> But more than aught else mankind requires burning, as being in great part wicked, and if that transmundane furnace can be so applied and regulated as to smelt and purify us into conformity with the rest of the terrestrial creation, then the tophetization of the entire erratic genus *Homo* were a consummation devoutly to be prayed for. But, glad to leave these ecclesiastical fires and blunders, I joyfully return to the immortal truth and immortal beauty of Nature.
>
> *(Muir 1916: 141–42)*

It seems to me, once again, that Thoreau and Muir liked wetlands precisely because no one else liked them. And therefore no one else might be found in them. They were wild, untrammeled, unpenetrated by loathsome *Homo sapiens*, among Nature's several and secret strongholds. Also common to Thoreau and Muir is a biblical frame of reference for portraying wetlands. Wetlands oscillate in their imaginations between heaven and hell (another biblical name for which is Tophet), the divine and the diabolical. From the anthropocentric point of view, they are hellish and diabolical; from a more disinterested point of view, they are heavenly and divine.

The more or less shared wetland ethic of both Thoreau and Muir is biologically informed, to be sure, but more just by descriptive botany than by evolutionary biology and ecology. Both, that is, write of little except the charm of the hydrophilic flora when singing the praises of bogs and swamps. How could it have been otherwise? Although Thoreau (1887) is now widely acknowledged to have discovered ecological succession in forest communities, there was no distinct science of ecology in the mid-nineteenth century, and Thoreau died only three years after the publication of Charles Darwin's *Origin*. And the mid-nineteenth century is also when Muir was making his trek across the watery reaches of southeast Georgia and north Florida and documenting it in a journal. So even though the year in which *Walk* was published was the same year in which Frederick Clements's second major treatise, *Plant Succession: an Analysis of the Development of Vegetation*, was also published, *Walk* is hardly more informed by ecology than is "Walking." Less than a decade after the publication of Darwin's paradigm-shattering tome, Muir blends traces of the new Darwinian origin myth with the old Mosaic one. Recall that Muir refers to the alligators that haunted his every swampy step as "the great saurians of an older creation" and, in the midst of his biblically based manifesto on the rights of nature, he points out that "whole kingdoms of creatures enjoyed existence and returned to dust, ere man appeared to claim them" (Muir 1916: 140). There is also a hint of ecological intuitions in Muir's remarks about "wisely planned" "antipathies," like "balanced attraction and repulsion in the mineral kingdom"—in other words equilibria in predator-prey population dynamics.

Aldo Leopold on Wetlands

Aldo Leopold's environmental philosophy and the emergence of ecology perfectly coincided. And by the time Leopold was starting to think about the human-nature relationship in ethical terms, the general contours of the theory of evolution had become normal science. Although Thoreau and Muir were no less ardent champions of wetlands than was Leopold, of the three, only Leopold could manage to disentangle his experience of wetlands from biblical associations and articulate a (wet)land ethic that was fully informed by evolutionary biology and ecology. Indeed, the unifying theme of *A Sand County Almanac*, relentlessly prosecuted, is the exposition and promulgation of an evolutionary-ecological worldview and its axiological and normative implications. Leopold says as much in the "Foreword":

> [O]ur bigger-and-better society is now like a hypochondriac, so obsessed with its own eco-nomic health as to have lost the capacity to remain healthy... . Conservation is getting no-where because it is incompatible with our Abrahamic [i.e., biblical] concept of land. We abuse land because we regard it as a commodity belonging to us. When we see land as a community to which we belong, we may begin to use it with love and respect... . That land is a commu-nity is the basic concept of ecology, but that land is to be loved and respected is an extension of ethics.
>
> *(Leopold 1949: vii–viii)*

In the Foreword, Leopold (Leopold 1949: vii) also says that "Part II, 'Sketches Here and There,' recounts some of the episodes in my life that taught me gradually and sometimes painfully, that the company is out of step." This is a military metaphor—"company" standing for society—and would resonate with readers in the late 1940s who had just emerged from the War Years earlier in the decade. But it is also an allusion to Thoreau (1854: 350) who, in *Walden*, famously writes, "If a man does not keep pace with his companions, perhaps it is because he hears a different drummer. Let him step to the music which he hears, however measured or far away." Thoreau was content to hear a different drummer and to be out of step with the company. But not Leopold (1949: vii) who goes on to say, "I suppose it may be said that these essays tell the company how it may get back in step." Grand and hubristic as it may seem, Leopold's project in *A Sand County Almanac* is nothing short of worldview remediation. The prevailing worldview, he thought, was a toxic mix of biblical dominionism and economism. The worldview that he exposes and promulgates in *Sand County*, and draws out its axiological and normative implications, is the evolutionary-ecological worldview.

Leopold opens Part II of his masterpiece with "Marshland Elegy." Its main burden is the evolu-tionary aspect of the evolutionary-ecological worldview. He begins with a visual simile making a "bank of fog" covering "the wide morass" to be "[l]ike the white ghost of a glacier" immediately evoking deep, evolutionary time (Leopold 1949: 95). (The word "morass" is yet another word for a wetland, but its metaphorical meaning—"a situation that traps, confuses, or impedes; an over-whelming or confusing mass or mixture"—has eclipsed its primary meaning.) Leopold (1949: 95) then compares the distinctive bugling of cranes, first heard from a great distance, to a "tinkling of little bells," next to a "baying of some sweet-throated hound," and finally to a "clear blast of hunting horns." Drawing this poetic evocation of sight and sound to a close, Leopold (1949: 95) ends with a simple sentence: "A new day has begun on the crane marsh." And then he turns to the cognitive message of the essay:

> A sense of time lies thick and heavy on such a place. Yearly since the ice age it has awakened each spring to the clangor of cranes. The peat layers that comprise the bog are laid down in the basin of an ancient lake. The cranes stand, as it were, on the sodden pages of their own history. These peats are the compressed remains of the mosses that clogged the pools, of the tamaracks that spread over the mosses, of the cranes that bugled over the tamaracks since the retreat of the ice sheet. An endless caravan of generations has built of its own bones this bridge into the future, this habitat where the oncoming host may live and breed and die.
>
> *(Leopold 1949: 96)*

Leopold first refers to the place as a "morass," then as a "bog," and finally as a "marsh"—and will never refer to it as a "swamp" because that more specifically indicates a very shallow expanse of still water covering the soil, while "marsh" and "bog" are terms for spongy surfaces largely covered with mosses or grasses, often punctuated with pools. (Nor in the whole book will he ever refer to such places as "wetlands.") Even though the very names "bog" and "morass" are metaphorically associated with impediment and disorder, there is hardly a trace of the ambivalence found in the wetland writings of Thoreau and Muir. Rather, the overall impression of the crane marsh, which Leopold conveys, is one of wild beauty and profound scientific interest. We are not in the wetland with Leopold, as we are with Thoreau and Muir; instead, we are aurally and visually attentive, disinterested observers of it.

"To what end?" Leopold (1949: 96) then asks, implicitly confronting one of the obstacles to general acceptance of an evolutionary worldview. There is no end, purpose, or goal to evolutionary processes; no final cause; no *telos*. For many people, that would make human life—indeed all life; even life itself—meaningless: like a tale told by an idiot, full of sound and fury, signifying nothing. In answer, Leopold simply says, "Out on the bog a crane, gulping some luckless frog, springs his ungainly hulk into the air and flails the morning sun with mighty wings. The tamaracks re-echo with his bugled certitude. He seems to know" (Leopold 1949: 96). Aristotle (1926) distinguished between immanent and transcendent ends, the former being superior to the latter. Evolutionary processes have no transcendent ends, ends beyond themselves. But life—by every living being, one and all—may be lived as an end in itself, an immanent end. Certainly, that's what this crane does; and thus that's what he seems to know; but in any case, that's what Leopold is trying to teach his reader. We are not here for any other purpose than to live as fully and robustly as any other animal—as most visibly and audibly manifested by cranes. Leopold goes on, first about cranes and then about their wetlands:

> [O]ur appreciation of the crane grows with the slow unraveling of earthly history. His tribe, we now know, stems out of the remote Eocene. The other members of the fauna in which he originated are long since entombed within the hills. When we hear his call we hear no mere bird. We hear the trumpet in the orchestra of evolution. He is the symbol of our untamable past, of that incredible sweep of millennia which underlies and conditions the daily affairs of birds and men.
>
> And so they live and have their being—these cranes—not in the constricted present, but in the wider reaches of evolutionary time. Their annual return is the ticking of the geologic clock. Upon the place of their return they confer a peculiar distinction. Amid the endless mediocrity of the commonplace, a crane marsh holds a paleontological patent of nobility, won in the march of eons, and revocable only by shotgun. The sadness discernible in some marshes arises, perhaps, from their once having harbored cranes. Now they stand humbled, adrift in history.
>
> *(Leopold 1949: 96–97)*

After reading that, who could not but fall in love first with cranes, then with crane marshes, and eventually with all wetlands? Leopold devotes another *Sand County* essay to wetlands, "Clandeboye," set in Manitoba, Canada. "Clandeboye" closes Part II "Sketches Here and There," just as "Marshland Elegy" opens it—so disquisitions on wetlands frame *Sand County*'s central and pivotal section. In the essay, Leopold (1949: 158) acknowledges that others find the Clandeboye marsh distinguished only because "it is merely lonelier to look upon and stickier to navigate, than other boggy places." Of course, Leopold is of another opinion; taking his cue from several species of birds, he finds it distinguished for a reason similar to the distinction he finds in the Horicon Marsh, setting of the elegy—the sense of deep evolutionary history that it evokes.

And just as "Marshland Elegy" features an especially iconic species, the sandhill crane, so "Cladeboye" shines a spotlight on the elusive western grebe. But this time, we accompany not an onlooker and listener from afar; rather, we are taken right into the mire with our guide. Hoping to get an up-close-and-personal view of the grebe and its summer brood, Leopold (1949: 160) "buried [him]self, prone in the muck of a muskrat house. [And] while [his] clothes absorbed local color, [his] eyes absorbed the lore of the marsh."

From the descriptions of species after species of birds and mammals, we can imagine that Leopold (1949: 160) was bogged down in the morass for quite a long while. Indeed, he tells us that he was there so long that he "was starting to doze in the sun." But he snapped to "when there emerged from the open pool a wild red eye, glaring from the head of a bird" (Leopold 1949: 160). (The western grebe feeds on fish for which it dives beneath the surface; and its red eye, like that of the loon, is adapted for underwater vision.) The male of a mated pair, it seems, looked around, as if by periscope, before fully surfacing. Leopold (1949: 160–161) goes on: "Finding all quiet, the silver body emerged: big as a goose, with the lines of a slim torpedo. Before I was aware of when or whence, a second grebe was there, and on her broad back rode two pearly young, neatly enclosed in a corral of humped up wings. All rounded a bend before I could catch my breath." And Leopold concludes this final essay of Part II with an ironic, almost bitter hint of his philosophy of conservation—not wise use for the greatest good of the greatest number (of humans), not preservation of pristine nature, but the harmonious co-existence of humans and nature:

> The marshlands that once sprawled over the prairie from the Illinois to the Athabasca are shrinking northward. Man cannot live by marsh alone, therefore he must needs live marshless. Progress cannot abide that farmland and marshland, wild and tame, exist in mutual toleration and harmony.... So with dredge and dyke, tile and torch, we sucked the cornbelt dry, and now the wheat belt... Some day my marsh, dyked and pumped, will lie forgotten under the wheat, just as today and yesterday will lie forgotten under the years.
>
> *(Leopold 1949: 162)*

The Value and Valuation of Wetlands

As noted, wetlands are routinely referred to as "waste lands" and generally regarded as having little instrumental value. As both Thoreau and Leopold lament, they are often drained to make the land good for something because otherwise it's believed to be good for nothing. One might then fairly say that Thoreau, Muir, and Leopold value wetlands intrinsically and hope that the power of their pens can, in Leopold's words, get the company in step with that. Leopold (1949: 223) actually articulates the concept of intrinsic value, although when he wrote, the term was not current: "It is inconceivable to me that an ethical relation to land can exist without love, respect, and admiration for land, and a high regard for its value. By value, I of course mean something far broader than mere economic value; I mean value in the philosophical sense."

Presently, however, ecologists, conservation biologists, and environmental economists are making the case that wetlands *are* instrumentally—and therefore economically—valuable. If all wetlands are not good for the goods they supply—although some are, as I indicate shortly—then they are good for the services they render. The Millennium Ecosystem Assessment, sponsored by the United Nations and published between 2003 and 2005, funneled all the benefits that ecosystems provide for us humans into the concept of "ecosystem services," parsed into four categories: provisioning services (food, fiber, timber, and such), regulating services (water purification, flood control, pollination, microclimate regulation, and such), supporting services (soil building, oxygenating the atmosphere, nutrient cycling, macroclimate regulation, and such), and cultural services (recreation, aesthetic experience, tribal identity, education, and such) (Alcamo et al. 2003). And environmental economists have developed methods to express the value of all these services in a monetary metric (Hanley et al. 2013). That makes it possible to compare the instrumental value of an undrained wetland to what the value, if drained, of its tillable soil would be, producing say corn or wheat (in other words, rendering a single provisioning service). Economic valuation allows us to see if making the trade-off would be worth it—that is, would yield a net benefit—or not. We know the economic value of a drained wetland producing corn or wheat because such commodities have a market price. But many of the ecosystem services rendered by wetlands are not traded in markets and thus have no revealed price. Environmental economists have therefore devised indirect means of finding price equivalents for them—"shadow prices" they are called.

So, what sort of ecosystem services do wetlands render? That depends in part on the specific type of wetland.

Cypress swamps, such as the Okefenokee Swamp in Georgia, render many services, including water purification, hydrology regulation, and nutrient cycling, but their most valuable service is cultural—more particularly, they yield aesthetic experience and outdoor recreation. People camp near, paddle canoes, and fish in cypress swamps and enjoy the presence of the abundant wildlife that they harbor. Economists measure the value of such services—shadow-price—by means of the travel-cost method: how much money people spend on gasoline, lodging, food, equipment, and forgone wages to get to and enjoy such wetlands.

Tidal marshes also render similar cultural services, but their greatest ecosystem services fall in the provisioning and regulating categories. As to provisioning services, tidal marshes are seafood nurseries supplying juvenile finfish, crabs, and shrimp abundant insects and algae on which to feed. And their thick and stiff grasses, such as cord grass, shelter these small fry from herons and other avian predators. Tidal marsh vegetation also traps sediments and thus stabilizes shorelines and counters coastal erosion. As well, tidal marshes provide protective buffers between the sea and the dry land, absorbing storm waters and diminishing the energy of storm surges. Visible only to the biogeochemist, tidal marsh plants take up nitrogen and phosphorous runoff from farm fields in their watersheds and thus absorb the nutrient load of the waters that pass through them. That helps prevent offshore algal blooms and the "dead zones" that such blooms create when the die-off depletes the seawater of oxygen, leached away by the process of decay. Marsh plants also store carbon and thus help to regulate greenhouse gases. The value of these services too can be quantified. The provisioning services can be quantified by means of the market value of the seafood that would not have survived without the resources and protection tidal marshes provide them. The regulating services can be quantified by comparing the cost of hurricane destruction in areas where tidal marshes have been degraded with the lower costs in areas where they have been preserved or restored.

One may feel uneasy about economists shadow-pricing the instrumental value of wetlands—now having learned from Thoreau, Muir, and especially Leopold to value them intrinsically. Wouldn't that be tantamount to the commodification of nature? Is nothing natural sacred? Isn't Leopold's grand scheme to propound an alternative evolutionary-ecological worldview to biblical dominionism married to modern economism? Isn't valuing wetlands in economic terms an implicit endorsement of the very worldview that Leopold is trying to counter and replace?

Intrinsic and instrumental values are not mutually exclusive. Many things have both kinds of value. Indeed, we humans are prime examples of "things" having both kinds of value. Employees, for instance, are instrumentally valued by their employers. And their instrumental value is expressed in a monetary metric: the price of their labor or other services—everything from flipping burgers, to teaching school, to providing legal advice, to dunking basketballs, to running a Fortune 500 corporation. In some small businesses, employers value their employees intrinsically as well as instrumentally and will sacrifice profits during economic downturns to keep from laying some of them off. The kindness of strangers, however, is not entirely reliable. Therefore, the intrinsic value of human employees is protected by labor laws, such as the Fair Labor and Standards Act, the Occupational Safety and Health Act, and the Family and Medical Leave Act in the US (and other developed countries have similar legal protections for employees). The intrinsic value of human beings is a foundational principle of civil society, but so is the instrumental value of human beings—and that instrumental value is expressed in a monetary metric. The actual market value of various human services is neither perfect nor fair, but it is hard to imagine a modern society in which a market for human services (provisioning, regulating, supporting, and cultural) does not exist. Perhaps no two people intrinsically value one another more palpably than do spouses, but marriage is based on mutual instrumental value (whether actually priced or not) as well as mutual intrinsic value.

Leopold (1949: 225) himself recognized that intrinsically valuing nature does not preclude also instrumentally valuing it: "It of course goes without saying that economic feasibility limits the tether of what can or cannot be done for land. It always has and it always will. The fallacy the economic determinists have tied around our collective neck, and which we now need to cast off, is the belief that economics determines all land-use." There was a time in the US and other developed countries when there was a totally free market in human labor and other services, which led to human indignity and degradation—mining disasters, black and brown lung disease, stunted growth of children forced by circumstances to work in factories and sweat shops. There are now comparable laws protecting nature that implicitly recognize the intrinsic as well as the instrumental value of some aspects of nature—the Wilderness Act of 1964 and the Endangered Species Act of 1973, outstanding among them (Callicott 2006; Callicott and Grove-Fanning 2009). There are no *comparable* national laws protecting wetlands per se, certainly none protecting them irrespective of their instrumental value to humans. And the declared purposes of state laws protecting wetlands, such as those of Florida, are instrumental—the services they render are typically enumerated in a preamble (Florida Forest Stewardship 2006). The cultural heritage of the intrinsic value of wetlands bequeathed to us in the works of Thoreau, Muir, and Leopold, in tandem with the instrumental value of wetlands recognized in the more recent concepts of ecosystem services and economic valuation thereof, combine to make for a robust wetland environmental ethic.

US Wetland Policy

Over the past two centuries, more than half of the wetlands in the US (excluding Alaska) have been drained or filled (EPA Wetlands Protection and Recreation 2018). The US has pursued a policy of no net loss of wetlands, beginning with the George H. W. Bush Administration in 1989 (Wetlands Policy and Action Plan 1994). That policy was continued during the administrations of subsequent presidents Bill Clinton, George W. Bush, and Barak Obama. In order to accommodate economic development, the no-net-loss policy incentivizes mitigative wetland restoration and such schemes as tradable conservation credits. The principal policy instrument is the Clean Water Act of 1972, which requires permits issued by the US Army Corps of Engineers for both dredging and filling navigable waters of the US (Federal Water Pollution and Control Act and Amendments 1972). What constitutes navigable waters and the connection (literally) of shallow and still wetland waters to navigable streams and lakes has been the subject of litigation (Rapanos v. United States 547 U. S. 2006). The Trump Administration, however, managed to thoroughly undermine wetlands protection and restoration in the US along with other provisions of the Clean Water Act of 1972 and Amendments by narrowly defining "Waters of the US" (Trump Erodes Water Protections: 6 Things to Know 2020). The Consortium of Aquatic Science Societies has importuned the Biden Administration to undo the wetland mischief of the Trump Administration (Anonymous 2021). On 6 August 2021 the Biden Adminstration began the process of returning the definition of the Waters of the US to the status quo ante Trump. Public comment on the reversal ended on 8 February 2022 and at the time of this writing (17 March 2022), the process is proceeding apace.

Acknowledgments

Research for this chapter was supported in part by the National Socio-Environmental Synthesis Center (SESYNC) (NSF award # DBI-1052875); I benefited from discussions about stream and wetland restoration with members of the SESYNC working group on Ecological Restoration and Ecosystem Services.

References

Alcamo, J., N. J. Ash, C. D. Butler, J. B. Callicott, D. Capistrano, S. R. Carpenter, et al. (2003) *Ecosystems and Human Well-being: A Framework for Assessment.* Washington, DC: Island Press.

Anonymous (2021) "Aquatic Science Societies Call on Biden Administration to Restore Science-based WOTUS Rule." American Fisheries Society, May 2021. https://fisheries.org/2021/03/cass-wotus-letter-to-biden/, accessed May 7, 2021

Aristotle (1926) *The Nichomachean Ethics,* H. Rackman (trans.). London: William Heineman Ltd.

Callicott, J. B. (2006) "Chapter 4: Explicit and Implicit Values" in J. M. Scott, D. D. Goble, F. W. Davis (eds.) *The Endangered Species Act at Thirty: Conserving Biodiversity in Human-Dominated Landscapes, Volume 2.* Washington: Island Press, 36–48.

Callicott, J. B. and W. Grove-Fanning (2009) "Should Endangered Species Have Standing? Toward Legal Rights for Listed Species," *Social Philosophy and Policy* 26: 317–352.

Davidson, D. (1987) "Knowing One's Own Mind," *Proceedings and Addresses of the American Philosophical Association* 60: 441–458.

EPA Wetlands Protection and Restoration (2018) https://www.epa.gov/wetlands, accessed May 17, 2020.

Federal Water Pollution Control Act and Amendments (1972) https://en.wikipedia.org/wiki/Clean_Water_Act, accessed May 17, 2020.

Florida Forest Stewardship (2006) "Wetlands Regulations" http://www.sfrc.ufl.edu/Extension/florida_forestry_information/planning_and_assistance/wetlands_regulations.html#why, accessed May 17, 2020.

Hanley, N., J. Shogren, and B. White (2013) *Introduction to Environmental Economics.* New York: Oxford University Press.

Leopold, Aldo (1949) *A Sand County Almanac and Sketches Here and There.* New York: Oxford University Press.

Muir, J. (1916) *A Thousand Mile Walk to the Gulf,* W. F Badé (ed.) Boston: Houghton Mifflin Company.

Pseudo-Apollodorus. (1921) *Library* vol. 2, J. G. Frazer (trans.) London: William Heinemann Ltd.

Rapanos v. United States 547 U. S. 715 (2006) https://en.wikipedia.org/wiki/Rapanos_v_United_States, accessed Sept. 12, 2015.

Thoreau, H. D. (1854) *Walden or Life in the Woods.* Boston: Ticknor & Fields

Thoreau, H. D. (1862) "Walking," *The Atlantic Monthly* 9: 657–74.

Thoreau, H. D. (1887) *The Succession of Forest Trees and Wild Apples with a Biographical Sketch by Ralph Waldo Emerson.* Boston: Houghton Mifflin Company.

Trump Erodes Water Protections: 6 Things to Know (2020) https://www.politico.com/news/2020/01/23/trump-epa-curbs-water-protections-102779, accessed May 17, 2020

Unknown (1912) *Beowulf,* C. H. Tinker (trans.). New York: Newsom and Company.

Wetlands Policy and Action Plan (1994) http://www.fws.gov/policy/660fw1.html, accessed Sept. 12, 2015.

16

RIVERS AND WATERSHEDS

Alan J. Rabideau and Kenneth E. Shockley

Introduction

Rivers and their associated watersheds comprise one of the most visible, dynamic, and societally important types of ecosystems. Although rivers are often described using terms such as *natural*, *wild*, and *untamed*, in reality most rivers have been extensively modified by human societies and are heavily managed to supply water, food, and energy, support navigation, and limit the negative effects of flooding, particularly in urban settings. As scientific research has provided a more complete understanding of the ecological consequences of these modifications, contemporary practice has shifted toward projects that seek to retain the societal benefits of river systems in a sustainable manner while restoring attributes critical to the function of healthy rivers and watersheds. A particularly noteworthy example is the ongoing Kissimmee River Restoration Project, which will eventually restore more than 40 square miles of river floodplain, including 20,000 acres of wetlands and 44 miles of river channel (Koebel and Bousquin, 2014). However, as one of the most technically challenging and costly forms of environmental management, publicly funded river projects have also generated controversy, as expressed in the rapidly growing body of literature that spans the scientific disciplines of ecology, physical geography, hydrology, and civil engineering.

In this essay, we review the historical causes of river degradation and discuss challenges and prospects for developing ethically sound approaches to river restoration (RR). RR, and the state of our rivers more generally, constitutes not only a practical concern worthy of ethical reflection and reformation, but also a lens through which we can see a wider range of philosophically and ethically significant environmental issues. As reviewed by Norton (2005), O'Neil et al. (2008), and others, the focus of environmental ethics discourse has evolved from apparent conflicts between economic and ecological principles to more pragmatic approaches that integrate and balance other approaches to valuation. We highlight a similar trajectory in the field of RR. In the remainder of this essay, we describe the historical practices that have stimulated the need for restoration projects, review some of most commonly referenced ethical principles, and highlight two examples – dam removal and compensatory mitigation (CM) – that illustrate the need to engage the place-based history of the affected communities in a more holistic, pragmatic, and participatory manner.

DOI: 10.4324/9781315768090-20

Alan J. Rabideau and Kenneth E. Shockley

River and Watershed Management: From Degradation to Restoration

Rivers have played a central role in the development of the United States (U.S.). Prior to European settlement, many Native American communities relied on river ecosystems for sustenance, with long-standing cultural practices and traditions built around the fish species native to particular locations. Although many Indigenous communities have been decimated by the modification of rivers for dam construction and other purposes, the goal of restoring native fish populations such as salmon remains an important focus of modern river management, for cultural reasons as well as the socio-economic benefits associated with recreational fishing.

Following European settlement, river systems played an important role in the westward expansion of the U.S. and the development of the modern industrial economy (e.g. as summarized by Doyle, 2012, 2018). Although the relationships between human communities and adjoining rivers are expressed in a variety of context-sensitive and place-based forms, intensive modification and management of rivers has been a common feature in the development of the modern economy. For example, many communities in the Northeast U. S. were formed around the hydropower provided by damming small rivers, which supported the development of a manufacturing-based economy. In the arid West, rivers provided water for irrigation and drinking, generating much political controversy as increasing populations encountered the inherent limits of regional water cycles.

Of the many ways that humans modified river systems during the early 20th century, the cumulative impact of dam construction has been particularly notable; recently, the American Society of Civil Engineers has identified 90,580 U.S. dams, with the vast majority aging and in poor condition (ASCE, 2017). During approximately the same period of intensive dam construction, a related practice of river management was also pervasive: the *channelization* of rivers to reduce flooding, which was typically accomplished by the highly intrusive processes of dredging and earthwork to straighten and deepen stream channels, eventually affecting and compromising hundreds of thousands of miles of American rivers (Wohl et al., 2005).

In addition to the deliberate alteration of river systems through dam construction or channel modification, expanding human settlements have altered river systems by withdrawing large amounts of water for drinking, irrigation, or industrial use. Postel and Richter (2003) have presented a thorough analysis of the ecological devastation imposed on American rivers through excessive water withdrawal, particularly in the arid Southwest, exemplified by the extreme case of the Colorado and similar rivers whose entire flows are depleted before reaching their historical discharge points. Beyond the removal of water, changing land use patterns alter the timing of how water moves though the watershed, with increased land coverage by impervious surfaces responsible for more rapid conveyance of storm water in stream systems, thereby reducing the residence time of water to nourish the riverine ecosystems and, in many cases, promoting stream bank erosion during periods of increased flow.

The historical practices of dam construction, channel modification, and flow diversion were traditionally justified in economic terms because of the straightforward connection with marketed commodities such as irrigation water, energy production, and enhanced commerce from improved navigation. However, these benefits are frequently accompanied by substantial losses in other forms of value, including indirect economic benefits such as waste assimilation and recreational use, a variety of ecological functions such as habitat provision and nutrient cycling, and non-instrumental human values such as aesthetic pleasure, spiritual inspiration, and cultural identity. At the time of this writing, more than half of U.S. rivers have been characterized as exhibiting a "poor" biological condition, a comprehensive indicator of severe impairment (USEPA, 2016).

One response to the anthropogenic degradation of river systems has been the proliferation of RR projects designed to address specific impairments and/or improve targeted aspects of river function. Engineered RR measures can involve highly invasive activity, such as dam removal or earthwork to realign stream channels or emplace large rock structures, and/or involve low-tech measures such as replanting native vegetation or emplacing pools, cover stones, or woody debris. Recently, an interdisciplinary team of scholars conservatively estimated that over 1 billion dollars is spent on RR and related actions in the U.S. each year (Bernhardt et al., 2007), with many projects conducted at the scale of a river "reach" (short segment). As the magnitude of public investment has increased, scrutiny and criticism of RR projects have also intensified, questioning the success of particular projects as well as the scientific and professional foundations of the discipline (Lave, 2009, Kondolf et al., 2007, Bennett et al., 2011, Buchanan et al., 2012, Palmer et al., 2007, Palmer et al., 2014b). In approaching this topic from a philosophical perspective, we focus primarily on questions about how *value* has been and should be attributed to restored rivers.

Valuing Rivers and Watersheds

For many years, a central goal of many of those working in environmental philosophy has been the identification of a single principle that can be applied to resolve competing claims about value. The struggle to find such a principle is still evident in the various approaches to managing rivers and watersheds, including the use of economic valuation tools such as cost-benefit analysis or ecosystem services, programs that emphasize the ecological/biological aspects of river health/integrity, and claims related to "water rights" attributed to particular communities or cultures. It is hard to deny that each of these approaches captures an important principle that might be used to resolve competing value claims for particular RR projects. However, looking across the field as a whole, it is clear that despite some common features of particular classes of problems (e.g. dam removal), each river has unique characteristics, both in its particular function and relationship with the local ecosystem, and in the different ways that human communities define, interact with, and attribute value to the river and its watershed. Here, we follow O'Neill, Holland, and Light (2008) in recognizing that the resolution of conflicts related to rivers requires *pluralism* in considering the many ways of valuing and prioritizing various attributes of river systems. Key to this pluralism is recognizing the distinct contexts of particular RR projects, including the relevant history and local expression of riverine ecology.

In keeping with the pluralistic, context-sensitive approach, we emphasize forms of value arising from the human community that engages with a river ecosystem. For example, while the language of a "healthy" river seems to refer to objective qualities of the river itself, we contend that the *values* attributed to, and taken to be constitutive of a so-called healthy river are in fact projected from the *communities valuing* that river. While this perspective may appear to privilege certain viewpoints (e.g. economic valuation) and disadvantage others (e.g. non-instrumental valuation of the broader ecosystem), it is both necessary to capture the full spectrum of relevant considerations and helpful in framing the trade-offs necessary to implement RR projects under typical fiscal and practical constraints.

Economic Value of River Restoration

For many practitioners, RR projects are considered primarily in terms of the **economic value** gained by restoring a degraded system, including direct benefits such as flood/erosion control, enhancement of property value, improved transportation, and/or the use of extracted water for drinking, agriculture, industry, and/or recreation. Additional nonmarket value is often associated with "ecosystem services" that include waste assimilation, nutrient processing, and/or species habitat, which are commonly characterized using indirect methods such as willingness-to-pay surveys and/or contingent valuation techniques (Gilvear, Spray, and Casas-Mulet, 2013, Palmer et al., 2014a). While economic valuations of these sorts may serve useful in justifying the sometimes large cost of RR projects, monetization often fails to capture the diversity of values expressed by stakeholders and practitioners. For example, the cost of urban stream restoration projects can significantly exceed the quantifiable benefits associated with flood control, but projects frequently receive strong community support based on their expressed aesthetic and other values not directly expressed through the economics of flood control (Kenney et al., 2012). In general, it seems clear that a wide range of noneconomic values should be acknowledged in evaluating RR projects (Drenthen, 2011); we see this mirrored in concerns over ecological restoration (ER) more generally (Higgs, 2003, Clewell and Aronson, 2013).

As with many expressions of environmental management, an incomplete understanding of the relevant values can also affect the scientific understanding of the river system. For example, when economic values are the primary motivation for an RR project, the analysis and design are often directed at the *physical* attributes of the river channel, which are particularly relevant for transportation, flood control, and erosion. Such an emphasis on the *form* of a river channel is also consistent with a professional RR culture shaped by civil engineers and physical geographers, who routinely analyze channel geometry in terms of metrics such as *stability* that can be reasonably measured and represented in theoretical/computational models (Lave, 2009).

As discussed below, the goal of promoting a physically stable river channel often dominates small-scale restoration projects. In addition to emphasizing the straightforward connection between stability and the economic value of flood/erosion control, some practitioners have also asserted that the *physical* properties of a river system strongly influence its *biological* character. For example, physical reconfiguration of a stream channel to promote stability can also improve the exchange of water with the surrounding floodplain, a process believed to have substantial ecological benefits (USEPA, 2015). Furthermore, many channel modification projects use "natural" bank stabilization techniques (e.g. plants and woody debris rather than rocks) and physical measures believed to affect habitat diversity (creating localized pools and shaded areas within the stream, removing nonnative vegetation, etc.). However, the assumed relationship between the physical form of a channel and its biological function has been criticized as the unsubstantiated "field of dreams" hypothesis by scientists who have questioned the ecological effectiveness of many RR projects (Palmer, Ambrose, and Poff, 1997).

Ecological Value of River Restoration

Although economic valuation remains a common frame of reference, an increasing number of RR projects, particularly large-scale or rural riverine systems, explicitly emphasize *ecological* considerations as a central concern. The evolution from a focus on channel modification has been welcomed by river scientists (Palmer et al., 2014b), but the process of identifying a generally applicable set of ecologic criteria remains in its early stages. To capture these concerns, evaluations of complex riverine systems are often considered using the normative terms of the ecological **health** or **integrity** of the river and/or its supporting watershed. Such "thick" concepts are powerful because they contain both descriptive and evaluative content that track common intuitions, but establishing coherent and action-guiding definitions is fraught with difficulty (see Clewell and Aronson, 2013, ch 2; for the difficulty of integrating descriptive and evaluative content more generally in environmental contexts, see Shockley, 2012). Indeed, appeals to the *integrity* of rivers and their supporting watersheds echo the language of Aldo Leopold's famous claim that "[a] thing is right when it tends to preserve the integrity, stability and beauty of the biotic community" (Leopold, 1949). Such appeals to straightforwardly normative considerations when we address ER more generally, or RR in our more particular case, require that we accept the descriptive characterization of the "health" or "integrity" of the river as at best partial.

In the context of U.S. environmental management, the concept of a healthy river is often linked to the requirements of the Clean Water Act to restore and maintain the *physical, chemical, and biological integrity* of waterways. In practice, the Environmental Protection Agency (EPA) and other government agencies typically evaluate rivers using "snapshot" (single-time) characterization of broad suite of indicators that include biological (e.g. macroinvertebrates), chemical (e.g. nutrients, salinity, acidification), and physical (e.g. sediment, vegetation, fish habitat, and human disturbance of the riparian region) metrics. However, in some cases, evaluations are framed entirely in terms of the physical degradation of river systems, a practice that has been criticized as overly narrow (Bronner et al., 2013; Palmer et al., 2014b; Lave, 2009; Lake, Bond, and Reich, 2007).

An alternative approach, frequently applied in large-scale RR projects, is to characterize the degradation/restoration of a riverine system in terms of its *flow* characteristics, which are typically expressed over larger spatial and temporal scales and can be correlated with ecosystem properties (Arthington and Pusey, 2003). A critical tool in this approach is the *hydrograph*, which represents the temporal pattern of water flowing through a river (volume per time) as a function of the local precipitation, river channel geometry, and watershed drainage patterns. As detailed by Postel and Richter (2003), returning a river system to an approximation of its pre-disturbance "natural hydrograph" is often considered an effective way to achieve the objective of health or integrity. Flow-referenced improvements are also commonly pursued by restoring "connectivity" between the river channel and its floodplain (USEPA, 2015).

Flow-based indicators of ecological health/integrity are context-sensitive, conceptually straightforward to implement, and most relevant when future precipitation and runoff patterns are expected to approximate those that have shaped the evolution of the historic riverine ecosystem. However, application of these concepts is less straightforward when changes in land use and/or climate have irrevocably altered the historic hydrograph. For these situations, scientists have increasingly focused on river *function* and *processes* (e.g. biological productivity, waste removal, and sediment transport) as a measure of health or integrity (Fischenich, 2006, Beechie et al., 2010), although there are recognized difficulties with translating these concepts into measurable and easy-to-understand indicators. Of course, function-based treatments of the health or integrity of a riverine system, without the normative characterization referenced above, remain partial.

The challenges in applying concepts such as "health" and "integrity" reflect both *scientific* issues – the uncertain relationships between measurable parameters, the target function/process, and connection with overall ecologic health/integrity – as well as the *ethical* concern regarding with which processes/functions should be privileged in a particular local context. That is, a focus on physical/chemical/biological indicators sidesteps the issue of *why* particular communities place value on healthy river ecosystems, which is a matter of human history and culture as well as the river's ecological condition. Respecting the variety and complexity in how stakeholder communities value restored rivers requires a normative discourse that can accommodate difficult trade-offs and ethical pluralism (Clewell and Aronson, 2013, Drenthen, 2011, Welchman, 2016).

Restoration, History, and Culture

Characterizing river management projects as "restoration" implicitly draws attention to *history* as a central concern and invokes a range of normative considerations. In practice, historical conditions often guide RR project goals. For example, the most comprehensive RR design manual (NRCS, 2007) includes a "historical approach" as one of several alternatives, using aerial photographs, maps, and other past records to reestablish the configuration of a stream channel prior to its impairment. This approach is most clearly relevant when a river suffered a specific disturbance event, such as dam construction or channelization. In these simple cases, removal of the disturbance could lead to a return to a key aspect of former conditions, such as a natural hydrograph or the prevalence of a key species; for example, many projects in the Pacific Northwest emphasize restoration of native salmon habitat, which has strong cultural and economic significance (Apostol and Sinclair, 2006).

Although references to *historical* concerns are common in RR, it is important to recognize that natural and social scientists often consider *history* in very different ways. While the historical configuration of a river channel may provide scientific insight regarding the physical and ecological responses of a *river* to physical and climatic influences, the socio-economic and cultural histories of the affected *human communities* may be equally important in prioritizing competing values and resolving trade-offs. In defining *historical fidelity* as a central pillar of ER, philosopher Eric Higgs (Higgs, 2003) highlighted two primary aspects: (1) *nostalgia* for a past that is in some way better than current conditions, and (2) *narrative continuity* as an important human value. Both ideas reflect the common and intuitive notion that continuous memories enrich and provide meaning to communities and should therefore be preserved when it is feasible to do so (O'Neill, Holland, and Light, 2008). In some cases, both human and ecological histories might favor a particular action, such as promoting the return of a native or historically significant species. However, more complex proposals such as dam removal (discussed in more detail below) might raise conflicts between human- and ecology-centered notions of history.

Concerns regarding the overreliance on historical approaches are increasingly common within the RR literature (Brierley, 2008; Dufour and Piegay, 2009) and can be loosely grouped into three categories. First, the dynamic nature of rivers and the pervasiveness of land use and climatic changes suggest that, in most cases, recovery of a "historical" river configuration may not be achievable (Higgs, 2003). Second, selection of a specific target condition from the past is arbitrary and difficult to justify on its own terms, a common concern known as the "baseline" problem within the ER literature (Hale, Lee, and Hermans, 2014, Hale, Hermans, and Lee, 2013). Third, other aspects of the river system – economic value or ecosystem services, presence of a specific species, channel stability, etc. – may simply outweigh concerns related to nostalgia or narrative continuity. However, attention to the past need not be limited to a "snapshot" recreation of a specific historical form. For example, McCool (2010) discussed how the concepts of "rewilding" (Soule and Noss, 1998), "regardening," and "renaturing" have been applied to capture some elements of a historical condition in setting RR goals for RR projects while avoiding the "return to a snapshot" approach.

In addition to historical concerns, RR projects also engage other forms of human value that include aesthetic preferences and/or a desire for "natural" rivers, the educational benefits of riverine parks and nature preserves, the spiritual inspiration associated with the power and dynamism of flowing water, and/or cultural practices of Indigenous communities. Because these values are often particular to a community and place, they are frequently lumped together (possibly with historical considerations) and referred to as **stakeholder** or **social** concerns (Palmer et al., 2005). Despite their highly contextual nature, such concerns often play a significant, even dominant, role in shaping RR goals and metrics (Drenthen, 2011), echoing the complex interweaving of values in other types of ER projects (Stevens, 1996, Jordan, 2003, Clewell and Aronson, 2013, Higgs, 2003, Holland and O'Neill, 2003).

Resolving Conflicts

A review of the recent RR literature indicates a considerable diversity of opinion, with many scholars contending that because the considerable public investment in RR projects has generated few success stories, both the ends and means of RR projects should be reconsidered. In considering these claims, it is helpful to distinguish among three sorts of disagreements: (1) those that are scientific in nature (e.g. the proper understanding of the relationship between physical and ecological characteristics of a river), (2) those that involve conflicting values (e.g. economic value of flood protection vs ecological health/integrity of the watershed), and (3) those that relate to the

RR professional culture (e.g. formulation of achievable goals, manner of post-project evaluation, and degree of public participation). In some cases, these disagreements are further complicated by inconsistent use of terminology. For example, by labeling a project as *ecological restoration*, the management of a river is expected to engage its history in some meaningful fashion and/or provide functional ecological improvement. Conversely, by framing river management in terms of *water ethics* or *water rights*, attention is focused on the use of water as a resource for human communities and the associated issues of distributive justice.

Clearly, a successful environmental management project of any type should be directed at important human and environmental values, grounded in clearly formulated and measurable goals, supported by resources sufficient to achieve project objectives in a cost-effective manner, and evaluated over appropriate spatial and temporal scales to assess success and address learning objectives. However, the resolution of conflicts among, for example, economic, ecological, and other stakeholder values is not straightforward, even when decoupled from the other sorts of professional, technical, and scientific disagreements that frequently accompany RR projects.

The most direct approach – development of a set of general principles to guide RR projects – has proven elusive for practitioners. For example, following an extensive survey of RR practices (Bernhardt et al., 2007), ecologist Margaret Palmer and colleagues proposed five criteria for "ecologically successful stream restoration" (Palmer et al., 2005). Two of the standards related to matters of project implementation, but three of the criteria related to short- and long-term ecological condition and resilience. While the criteria did not specify *which* ecological attributes should be emphasized, the recommendations implied that only projects with ecological goals should be termed "restoration." Although such a priority was not explicitly justified, Palmer et al.'s criteria could be viewed as privileging ecological concerns over economic and/or stakeholder considerations, at least for the type of projects termed "restoration." While such a restriction may be a sensible response to the historical emphasis on physical channel manipulation (without sufficient attention to ecological concerns), it sidesteps the reality that many RR projects are motivated by human-centered economic concerns, with project scales and/or budgets that may not support substantial ecological engagement. Furthermore, even if ecological criteria are deemed essential to support the term "restoration," such metrics are challenged by considerable scientific uncertainty regarding causal relationships between restorative techniques and stream health/integrity.

Recognizing the difficulty in establishing a general set of principles for defining RR goals, an alternative, more pragmatic strategy would emphasize a *deliberative* process that engages scientists, regulators, stakeholders, and local communities in establishing project objectives and practices that are responsive to local values and concerns. Such an approach is consistent with the approach to *discursive democracy* advanced by Dryzek (1997, 2000), the concept of *adaptive management* as formalized by Norton (2005), and the *historical* approach advocated by O'Neil et al (2008). While an inclusive deliberative process could potentially navigate conflicts between competing values, implementation is often difficult in practice, particularly for small-scale projects that may not have the resources to identify, engage, and adaptively consult the relevant stakeholders. For example, such approaches may suffer from requiring more labor or local research than is typically available. Nonetheless, some elements of the deliberative process (e.g. stakeholder and community consultation) are evident in evolving RR guidance, such as the *Conservation Planning Process* recommended by regulatory agencies (NRCS, 2007). In general, the contextual nature of RR projects makes the development and application of general guidelines exceedingly difficult.

The Importance of *Scale* in River Restoration

As exemplified by Norton's concept of adaptive management (2005), the expression of a particular set of local values is often shaped by the *scale* of the system that is being considered for restoration.

As reviewed by Bernhardt et al. (2007), a large proportion of RR projects are conducted at the scale of a river *reach* (a short uninterrupted segment). Reach-scale projects are often initiated because of public concerns related to erosion and/or flood control; the combination of these factors frequently results in a focus on earthmoving activity to alter the physical configuration of the local river channel. For much of the past two decades, the concern with physical stability has dominated the practice of reach-scale restoration, which has become strongly associated with the *Natural Channel Design* (NCD) approach, a technique that has been endorsed by several state regulatory agencies and often considered synonymous with "restoration," despite criticism by both academics and practitioners (Simon et al., 2005; Simon et al., 2007; Lave, 2009; Hillman and Brierley, 2005).

The debates surrounding reach-scale restoration, including NCD methods, encompass all types of disagreement noted previously: scientific, ethical, and professional. For example, some scientists have criticized the techniques used in NCD, which emphasizes a form-based replication of a "reference reach" rather than more rigorous physics-based mathematical models that quantify and balance the forces believed to be acting on the river (Shields and Copeland, 2006, Simon et al., 2007). From a values perspective, the common emphasis on the physical stability of the reach seems to implicitly deemphasize overall ecological health and function, as the most commonly employed technologies (physical measures) and associated metrics (e.g. stability indices) may not promote ecological improvement. From a professional standpoint, the equation of "restoration" with a particular set of physical techniques, as promoted by some regulatory agencies, could inhibit the expression of more participatory and context-sensitive approaches to river management (Lave, 2009).

The response of practitioners to these criticisms has included some proposals to broaden the scope of reach-scale projects to more explicitly address ecological concerns, but many have defended the utility of reach-scale channel modifications and the use of more descriptive terminology such as *river repair* for this type of project (Brierley, 2008). As discussed elsewhere, emphasizing the term "restoration" can sometimes introduce baggage that confuses the resolution of trade-offs and development of appropriate site-specific goals and metrics. However, further complicating these concerns is the lack of a robust scientific understanding of the relationship between the physical attributes of river channels and ecological health and function, particularly when considered at the reach scale. In a recent review of the conceptual basis for RR, Palmer et al. (2014b) documented how many reach-scale projects have failed to achieve "restoration" because the underlying disturbance is expressed at much larger spatial scale, typically the larger drainage area or *watershed*.

In contrast to the thousands of small-scale RR projects, a number of very extensive large-scale restoration projects have been launched that include an entire or substantial portions of a river watershed, as summarized by Doyle and Drew (2008). Examples include the Kissimmee River (part of the Everglades system), the Platte River, the upper Mississippi River, and the multiple river systems that contribute to the California Bay and Chesapeake Bay Deltas. Each of these restoration projects reflects a multiple-decade commitment by consortia of federal, state, and/or local governments and nongovernmental organizations, with anticipated expenditures in the tens of billions of dollars (for one instructive case study, see Clewell and Aronson, 2013, 185–189).

Restoration projects conducted at the watershed or ecosystem scale are less likely to suffer from the incompletely formulated goals or inadequate monitoring programs that often plague small projects. However, the challenges of aggregating the concerns of multiple stakeholders can be formidable, further complicated by the misalignment between watershed and political boundaries. For the case studies referenced above, it is evident that intensive negotiations were required to resolve trade-offs among the multiple types of value engaged by the project. In all cases, benefits of the projects included a mixture of ecological, economic, and cultural considerations, with some, but not all, formally included in benefit-cost calculations. Also, in each case, establishment of the target conditions involved some attention to prior ecological conditions, expressed in relation to a pre-development hydrograph, a particular mixture of species, or both. However, none of the projects would be characterized as a fully realized "restoration" to specific historical reference condition. For projects of such large scales, the combined effects of climate change, invasive species, and the demands of expanded human settlements would effectively preclude such an approach.

The Challenge of Dam Removal

One of the most prevalent and complex contemporary examples of watershed-scale RR is the practice of *dam removal* to address aging infrastructure and/or concerns about ecological impacts. While the exact number of dams in the U.S. is unknown, the American Society of Civil Engineers recently designated 14,000 dams as "hazards" and over 4,000 as "deficient" (ASCE, 2013). The nonprofit organization, American Rivers (2020), estimates that approximately 1,702 dams have been removed over the past century, with 90 dams in 2019 (American Rivers, 2020). Many of these projects are very expensive and highly contested. For example, plans to remove four dams on the Klamath River, at a cost of approximately $1 billion, have been the subject of ongoing debate between advocates for retaining the substantial hydropower capacity provided by the dams and the ecologists and local communities who wish to promote the recovery of endangered salmon species; see *DamNation* (2014) for a popularized examination of this issue with respect to the removal of the Elwa Dams. Versions of this controversy are playing out in communities across the U.S. as aging dams are increasingly considered for removal.

Dam removal projects illuminate many of the challenges with the ethical management and restoration of rivers. First, most dams provide clear and measurable *economic* value to local communities through energy generation, flood protection, and/or water management. However, a return to the pre-development flow regime will usually provide *ecological* benefits in terms of species habitat and floodplain connectivity, and may also provide alternative economic benefits such as tourism. In addition, the presence of a dam was often integral to the development of a local human community whose history has generated a social/cultural attachment to the dam. Indeed, the historical preservation of dam structures has emerged as a relatively new "cultural" value that must be considered as part of the complex trade-offs in removal projects (American Rivers, 2008). Further complicating many projects is the fact that removing a large dam can generate *new* environmental impacts such as the release of large amounts of sediment that have accumulated over decades of operation (Bednarek, 2001).

In considering the complex problem of dam removal, we see once again the contextual nature of river valuation. No single perspective – economic, ecological, historical, and cultural – can be seen to clearly dominate in all cases. Despite these challenges, however, many projects have been successful in finding compromise solutions among local communities and stakeholder parties with seemingly disparate interests (e.g. the 2014 Klamath Restoration Agreement, USDOI, 2012). The key to these successes, and so to the development of an ethical approach to the management and restoration of rivers, resides not in a single set of "master" principles, but in deliberative processes that identify and promote, in an adaptive fashion, shared community values.

Compensatory Mitigation

Another example of complex trade-offs in RR valuation is the controversial practice of CM, a form of environmental management grounded in the requirement of the Clean Water Act for permitted projects to mitigate impacts to aquatic resources. Following a decades-long history of application to wetlands systems (NRC, 2001), CM has been increasingly applied to river systems, either by "restoring" a nonadjacent segment of a river that been adversely impacted by a project, or by implementing improvements to a completely separate river system. A related practice is *mitigation banking*, in which an RR project is implemented to provide "credits" that are sold to mitigate future development projects that result in unavoidable impacts to a stream or river. CM practices have been strongly criticized on both philosophical and scientific grounds.

Responding to the proliferation of ER and CM projects in the early 1980s, philosophers Eric Katz (Katz, 1992) and Robert Elliott (Elliot, 1982, Elliot, 1997) challenged the so-called "restoration thesis" as implicitly accepting destructive environmental practices on the mistaken view that lost value of undisturbed nature can be adequately replaced by recreating, precisely, the pre-disturbance conditions of the system. Defenders of ER have emphasized the positive benefits of restoration and challenged the implicit dualism that draws a sharp distinction between "pristine" and human-modified natural systems, but all parties have acknowledged that some "malicious" restoration projects have been used to excuse unacceptable environmental destruction (Light, 2005). Such concerns have been recently echoed in the RR literature, in which the concept of a "false image" has been applied to projects where a "restored" river system does not provide the type or magnitude of value that is purported to motivate the public investment (Cockerill and Anderson, 2014).

Many scientists have argued that current RR practices do not, in the vast majority of cases, provide the sort of value claimed in mitigation, in part due to the regulatory practice of attaching mitigation credits on the basis of the *length* of the restored stream rather than a demonstrated improvement in physical, chemical, or biological attributes and functionality (Bronner et al., 2013, Doyle and Shields, 2012). While the practice of CM could be improved by redefining stream "equivalence" in terms of *function* rather than *length*, such an approach does not completely address the contextual nature of environmental values associated with stream systems. For example, a function-based CM metric could be completely inadequate for addressing the rich collection of local social and cultural values, such as where an Indigenous community has lost a culturally significant and potentially irreplaceable stream.

A particularly problematic application of CM occurs when it is coupled with the practice of *mountaintop mining* of coal deposits, which typically includes the disposal of the removed earthen material into an adjoining valley, resulting in damage to the (usually present) stream system. Because of the requirements of the Clean Water Act, valley-fill activities that are unavoidable must be accompanied by CM, which is usually directed at a site that is different from the affected stream. Scientific investigations into the practices surrounding Appalachian coal mining impacts have consistently demonstrated that permitted mitigation projects have not met the objectives of replacing lost or degraded streams (Palmer et al., 2014b). At worst, the inclusion of stream restoration as a component of surface mining could be construed as "malicious" restoration, enabling a practice that is destructive and indefensible on moral grounds. At best, characterizing practices as "restoration" rather than as partial mitigation of extensive ecological damage redirects our attention from the substantial trade-offs at stake in pursuing economically viable but environmentally destructive forms of energy.

Despite these concerns, the availability of the market-generated financial resources associated with CM could, in some cases, *increase* both the value generated and the potential for long-term success of RR projects. For example, the recently implemented public/private Northern Virginia Stream Restoration Bank (NVRRB) includes clearly articulated ecological goals, a design strategy that considers the riparian system in its entirety rather than a limited number of reaches, an emphasis on plant species native to the region, a long-term (ten-year) monitoring program, resources and institutional structures for long-term stewardship of the restored river, and robust and adaptive public participation (Wetland Studies and Solutions, 2006). Although in many respects the NVRRB could be regarded as a model RR project, its ultimate success and moral justification will depend on the resolution of a challenge facing all CM projects: defining the appropriate use of compensatory "credits" in approving or rejecting the geographically separate projects that could destroy or damage sensitive river ecosystems. Interested readers are encouraged to track the progress of this evolving project at https://www.wetlands.com/nvsrb.

The Path Forward

Although many people think of rivers in terms of their beauty, wildness, and "natural" character, the progress of civilization has increasingly intertwined human settlements with the river systems upon which they depend. In the contemporary world, the vast majority of river systems have been extensively modified by human communities, with some rivers bearing little resemblance to their pre-settlement condition. The ecological, economic, and cultural significance of these modifications are substantial. Accordingly, RR provides fertile ground for the philosophical and ethical explorations of concepts such as "restoration," "wildness," and "natural," and a helpful lens for reflecting on the evolving relationship between the human and the other-than-human world.

Because of the observed degradation in river health and integrity resulting from decades of modifications, the paradigm for river management has shifted from exploitation for direct or indirect economic benefit, to "restoration" and "repair" to address a broader suite of values that includes social and cultural considerations as well as the traditionally emphasized ecological and economic concerns. Furthermore, the pursuit of ethical RR requires a pragmatic, adaptive response to changes in how these values are conceived and balanced by the affected communities. As we have described above, the complexities associated with the ethics of RR also point to the importance of maintaining a pluralistic approach to the value of rivers and riverine systems. A number of encouraging examples of contemporary projects that reflect such pluralist and adaptive approaches are described by Postel (2017). However, despite many success stories, the future of RR faces a number of practical challenges, including but not limited to (1) the establishment of stronger scientific and philosophical bases for characterizing, measuring, and balancing the noneconomic values associated with river systems, (2) the broadening of the concept and practices of "restoration" to more reliably promote ecologically healthy and resilient river ecosystems, and (3) the development of political and regulatory structures that engage and integrate the multiple forms of value associated with river systems in a manner that is both participatory and sensitive to the local place-based context.

References

American Rivers (2008) "Dam Removal and Historic Preservation: Reconciling Dual Objectives," Washington, DC. Available: www.americanrivers.org.

American Rivers (2020) "Map of U.S. Dams Removed since 1912," Raw Dataset—ARDamRemovalList_figshare_Feb2020. Figshare. Available: https://doi.org/10.6084/m9.figshare.5234068, Retrieved: 07/29/20.

American Society of Civil Engineers (ASCE) (2013) "2017 Infrastructure Report Card," Reston. Available: https://www.infrastructurereportcard.org/cat-item/dams/.

Apostol, D., Sinclair, M. (eds.) (2006) *Restoring the Pacific Northwest: The Art and Science of Ecological Restoration in Cascadia*, Washington, DC: Island Press.

Arthington, A. H., Pusey, B. J. (2003) "Flow Restoration and Protection in Australian Rivers," *River Research and Applications*, 19, 377–395.

Bednarek, T. A. (2001) "Undamming Rivers: A Review of the Ecological Impacts of Dam Removal," *Environmental Management*, 27, 803–814.

Beechie, T. J., Sear, D. A., Olden, J. D., Pess, G. R., Buffington, J. M., Moir, H., Roni, P., Pollock, M. M (2010) "Process-Based Principles for Restoring River Ecosystems," *BioScience*, 60, 209–222.

Bennett, S. J., Simon, A., Castro, J. M., Atkinson, J. F., Bronner, C. E., Blersch, S. S., Rabideau, A. J. (2011) "The Evolving Science of Stream Restoration," In: Simon, A., Bennett, S. J., Castro, J. M. (eds.) *Stream Restoration in Dynamic Fluvial Systems: Scientific Approaches, Analyses, and Tools*, Geophysical Monograph 194, pp. 1–8. Washington, DC: American Geophysical Union.

Bernhardt, E. S., Sudduth, E. B., Palmer, M. A., Allan, J. D., Meyer, J. L., Alexander, G., Follastad-Shah, J., Hassett, B., Jenkinson, R., Lave, R., Rumps, J., Pagano, L. (2007) "Restoring Rivers One Reach at a Time: Results From a Survey of US River Restoration Practitioners," *Restoration Ecology*, 15, 483–493.

Brierley, G. J. (2008) *River Futures: An Integrative Scientific Approach to River Repair*. Washington, DC: Island Press.

Bronner, C. E., Bartlett, A. M., Whiteway, S. L., Lambert, D. C., Bennett, S. J., Rabideau, A. J. (2013) "An Assessment of U.S. Compensatory Stream Mitigation Policy: Necessary Changes to Protect Ecosystem Functions and Services," *Journal of the American Water Resources Association*, 49, 449–462.

Buchanan, B. P., Walter, M. T., Nagle, G. N., Schneider, R. L. (2012) "Monitoring and Assessment of a River Restoration Project in Central New York," *River Research and Applications*, 28, 216–233.

Clewell, A. F., Aronson, J. (2013) *Ecological Restoration – Principles, Values and Structure of an Emerging Profession*. Washington, DC: Island Press.

Cockerill, K., Anderson, W. P. J. (2014) "Creating False Images: Stream Restoration in an Urban Setting," *Journal of the American Water Resources Association*, 50, 468–482.

DamNation (2014) [video] United States: Stoecker Ecological and Felt Soul Media.

Doyle, M., Drew, C. A. (eds.) (2008) *Large-Scale Ecosystem Restoration: Five Case Studies From the United States*. Washington, DC: Island Press.

Doyle, M. W. (2012) "America's Rivers and the American Experiment," *Journal of the American Water Resources Association*, 48, 820–837.

Doyle, M. W. (2018) *The Source: How Rivers Made America and America Remade Its Rivers*. New York: W. W. Norton.

Doyle, M. W., Shields, F. D. (2012) "Compensatory Mitigation for Streams Under the Clean Water Act: Reassessing Science and Redirecting Policy," *Journal of the American Water Resources Association*, 48, 494–509.

Drenthen, M. (2011) "Ways to Embrace a River. On the Need to Articulate a New River Ethic," In: De Groot, W., Warner, J. (eds.) *The Social Side of River Management*. New York: Nova Publishers.

Dryzek, J. S. (1997) *The Politics of the Earth: Environmental Discourses*. Oxford: Oxford University Press.

Dryzek, J. S. (2000) *Deliberative Democracy and Beyond: Liberals, Critics and Contestations*. Oxford: Oxford University Press.

Dufour, S., Piegay, H. (2009) "From the Myth of a Lost Paradise to Targeted River Restoration: Forget Natural References and Focus on Human Benefits," *River Research and Applications*, 25, 568–581.

Elliot, R. (1982) "Faking Nature," *Inquiry*, 25, 81–93.

Elliot, R. (1997) *Faking Nature*. London: Routledge.

Fischenich, C. (2006) "Functional Objectives for Stream Restoration," *Ecosystem Management and Restoration Research Program Technical Notes Collection*, Vicksburg, MS: U.S. Army Engineer Research and Development Center.

Gilvear, D. J., Spray, C. J., Casas-Mulet, R. (2013) "River Rehabilitation for the Delivery of Multiple Ecosystem Services at the River Network Scale," *Journal of Environmental Management*, 126, 30–43.

Hale, B., Lee, A., Hermans, A. (2014) "Clowning Around with Conservation: Adaptation, Reparation and the New Substitution Problem," *Environmental Values*, 23, 181–198.

Hale, B., Pérou Hermans, A., Lee, A. (2013) "Climate Adaptation, Moral Reparation, and the Baseline Problem," In: Moser, S. C., Boykoff, M. T. (eds.) *Successful Adaptation to Climate Change: Linking Science and Policy in a Rapidly Changing World*. Oxfordshire: Routledge.Higgs, E. (2003) *Nature By Design*. Cambridge: The MIT Press.

Hillman, M., Brierley, G. (2005) "A Critical Review on Catchment-scale Stream Rehabilitation Programmes," *Progress in Physical Geography*, 29, 50–76.

Holland, A., O'Neill, J. (2003) "Yew Trees, Butterflies, Rotting Roots, and Washing Lines: The Importance of Narrative," In: Light, A., De-Shalit (eds.) *Moral and Political Reasoning in Environmental Practice*. Cambridge: MIT Press.

Jordan, W (2003) *The Sunflower Forest: Ecological Restoration and the New Communion with Nature*. Berkley: University of California Press.

Katz, E (1992) "The Big Lie: Human Restoration of Nature," *Research in Philosophy and Technology*, 12, 231–241.

Kenney, M. A., Wilcock, P. R., Hobbs, B. F., Flores, N. E., Martinez, D. C. (2012) "Is Urban Stream Restoration Worth It?" *Journal of the American Water Resources Association*, 48, 603–615.

Koebel, J. W., Bousquin, S. G. (2014) "The Kissimmee River Restoration Project and Evaluation Program, Florida, U.S.A." *Restoration Ecology*, 22, 345–352.

Kondolf, G. M., Anderson, S., Lave, R., Pagano, L., Merenlender, A., Bernhardt, E. S. (2007) "Two Decades of River Restoration in California: What Can We Learn?" *Restoration Ecology*, 15, 516–523.

Lake, P. S., Bond, N., Reich, P. (2007) "Linking Ecological Theory with Stream Restoration," *Freshwater Biology*, 52, 597–615.

Lave, R. (2009) "The Controversy Over Natural Channel Design: Substantive Explanations and Potential Avenues for Resolution," *Journal of the American Water Resources Association*, 45, 1519–1532.

Leopold, A. (1949) *A Sand County Almanac, and Sketches Here and There*. New York: Oxford University Press.

Light, A. (2005) "Restoration, Autonomy and Domination," In: Heyd, T. (ed.) *Recognizing the Autonomy of Nature: Theory and Practice*. New York: Columbia University Press.

McCool, D. (2010) "Implementing River Restoration Projects," In: Hall, M. (ed.) *Restoration and History: The Search for a Usable Environmental Past*. New York: Routledge.

National Research Council (2001) *Envisioning the Agenda for Water Resources Research in the 21st Century*. Washington, DC: National Academy Press.

National Resources Conservation Service (NRCS) (2007) *Stream Restoration Design, Part 654 National Engineering Handbook*. Washington, DC: United States Department of Agriculture.

Norton, B. (2005) *Sustainability: A Philosophy of Adaptive Ecosystem Management*. Chicago: University of Chicago Press.

O'Neill, J., Holland, A., Light, A. (2008) *Environmental Values*. New York: Routledge.

Palmer, M., Allan, J. D., Meyer, J., Bernhardt, E. S. (2007) "River Restoration in the Twenty-first Century: Data and Experimental Knowledge to Inform Future Efforts," *Restoration Ecology*, 15, 472–481.

Palmer, M. A., Ambrose, R. F., Poff, N. L. (1997) "Ecological Theory and Community Restoration Ecology," *Restoration Ecology*, 5, 291–300.

Palmer, M. A., Bernhardt, E. S., Allan, J. D., Lake, P. S., Alexander, G., Brooks, S., Carr, J., Clayton, S., Dahm, C. N., Shah, J. F., Galat, D. L., Loss, S. G., Goodwin, P., Hart, D. D., Hassett, B., Jenkinson, R., Kondolf, G. M., Lave, R., Meyer, J. L., O'donnell, T. K., Pagano, L., Sudduth, E. (2005) "Standards for Ecologically Successful River Restoration," *Journal of Applied Ecology*, 42, 208–217.

Palmer, M. A., Filoso, S., Fanelli, R. M. (2014a) "From Ecosystems to Ecosystem Services: Stream Restoration as Ecological Engineering," *Ecological Engineering*, 65, 62–70.

Palmer, M. A., Hondula, K. L. (2014) "Restoration as Mitigation: Analysis of Stream Mitigation for Coal Mining Impacts in Southern Appalachia," *Environmental Science and Technology*, 48, 10552–10560.

Palmer, M. A., Hondula, K. L., Koch, B. J. (2014b) Ecological Restoration of Streams and Rivers: Shifting Strategies and Shifting Goals," *Annual Review of Ecology, Evolution, and Systematics*, 45, 247–269.

Postel, S. (2017) *Replenish*, Washington, DC: Island Press.

Postel, S., Richter, B. (2003) *Rivers for Life*. Washington, DC: Island Press.

Shields, F. D., and Copeland, R. R. (2006) "Empirical and Analytical Approaches for Stream Channel Design," *Proceedings of the Eighth Federal Interagency Sedimentation Conference*, 2–6 April, Reno.

Shockley, K. (2012) "Thinning the Thicket: Thick Concepts, Context, and Evaluative Frameworks," *Environmental Ethics*, 34, 227.

Simon, A., Doyle, M., Kondolf, M., Shields, F. D. Jr., Rhoads, B., Grant, G., Fitzpatrick, F., Juracek, K., Mcphillips, M., Macbroom, J. (2005) "How Well do the Rosgen Classification and Associated 'Natural Channel Design' Methods Integrate and Quantify Fluvial Processes and Channel Response?" In: Walton, R. (ed.), *Proceedings of World Water and Environmental Resources Congress*, May 15–19, Anchorage.

Simon, A., Doyle, M., Kondolf, M., Shields, F. D., Rhoads, B., Mcphillips, M. (2007) "Critical Evaluation of How the Rosgen Classification and Associated 'Natural Channel Design' Methods Fail to Integrate and Quantify Fluvial Processes and Channel Response," *Journal of the American Water Resources Association*, 43, 1–15.

Soule, M., Noss, R. (1998) "Rewilding and Biodiversity: Complementary Goals for Continental Conservation," *Wild Earth*, Fall 1998, 19–28.

Stevens, W. K. (1996) *Miracle Under the Oaks: The Revival of Nature in America*. New York: Gallery Books.

United States Department of Interior (USDOI) (2012) *Klamath Dam Removal Overview Report of the Secretary of the Interior: An Assessment of Science and Technical Information*. Washington, DC.

United States Environmental Protection Agency (USEPA) (2015) *Connectivity of Streams & Wetlands to Downstream Waters: A Review & Synthesis of the Scientific Evidence*. Washington, DC: USEPA Office of Research and Development.

United States Environmental Protection Agency (USEPA) (2016) *National Rivers and Streams Assessment 2008–2009: A Collaborative Survey*. Washington, DC: USEPA Office of Water and Office of Research and Development.

Welchman, J. (2016) "Environmental Versus Natural Heritage Stewardship: Nova Scotia's Annapolis River and the Canadian Heritage River System," In: Hourdequin, M., Havlick, G. (eds.) *Restoring Layered Landscapes: History, Ecology, and Culture*. New York: Oxford Press.

Wetland Studies and Solutions (2006) "Northern Virgina Stream Restoration Bank Concept Plan," Gainesville, VA. Available: http://reston.wetlandstudies.info/Resources/Concept/NVSRB_ConceptPlan.pdf.

Wohl, E., Angermeier, P. L., Bledsoe, B., Kondolf, G. M., Macdonnell, L., Merritt, D. M., Palmer, M. A., Poff, N. L., Tarboton, D. (2005) "River Restoration," *Water Resources Research*, 41, W10301, 1–12.

17

OCEAN POLICY

Carl Safina

Introduction

Think of Earth's life-supporting exterior as a system of five interacting components: the atmosphere of gases surrounding our planet; the hydrosphere of liquid water; the cryosphere of frozen-water glaciers and ice-caps; dry land; and the biosphere of all living things. Past changes to these required geological or cosmic factors such as volcanic activities and asteroid strikes. Now, though, humans, too, affect all components of our life-supporting systems.

Most basic to these human factors is the sheer number of us, because the overall effect of us is the product of whatever humans do multiplied by how many humans do it. In the seas and along coasts today, the greatest effects come from fishing, habitat alteration, fossil fuel burning, pollution by fertilizers, pesticides, toxic metals, plastics, and a wide human-originated chemical palette. With these have come new problems, including spreading dead zones, greatly increased rates of harmful and toxic algal blooms, new diseases ravaging corals and sponges, the decline of seagrass meadows, the global near-disappearance of large oyster reefs—and the main game changer, the destabilizing effects of carbon. The list goes deep.

A Harsh but Perhaps Not Unfair Observation about Limits to Ocean Policy

We do not occupy the sea with human habitation as we occupy land. Because we go to sea to do things but do not reside, the ocean is a commons, giving rise to national and international "ocean policy." (There is no analogous international land policy.) Therefore, there are three things to consider: What we say about what we do in the ocean, and what we actually do (and the gap between them). And perhaps most important, what we humans with our overrated abilities and our underrated limits of foresight and will actually could do. There is progress, yes, but we are creating bigger problems than solutions because our values are off. Our morals are not scaled to reality.

One of the main issues in ocean governance is the lack of governance itself. Few countries in all of Asia, Africa, Central and South America, and much of Europe have the resources, laws, enforcement capacity, or plain interest to reasonably control fishing or pollution, not to mention respond to the global levelers: climate warming and ocean acidification (which I'll discuss more below). Western Europe is improving. Canada is worsening, in its rush to bring oil development to its coasts and its destructive salmon farming. The United States, New Zealand, Australia, and the Falkland Islands are doing a decent job taming fishing in their claimed waters (out to 200 nautical miles). But all nations on Earth still have significant problems and ongoing political battles over recovery versus more exploitation.

DOI: 10.4324/9781315768090-21

The idea that fisheries could be managed in perpetuity by a policy of "maximum sustainable yield" was developed in the United Kingdom by Michael Graham based on his "great law of fishing: 'Fisheries that are unlimited become unprofitable.'" It was a good start and in the few places where it has been earnestly applied (i.e. Alaska), it has generally worked, but quantifying the maximum sustainable yield requires data that many fisheries lack. Where the technical capacity exists, sustainable yield has usually been approached with more cynicism than earnest effort, and the political and policy emphasis has usually fallen squarely on "maximum" while generally ignoring "sustainable." The concept's flaws include its ignoring of food webs; fish we catch, e.g. herring, are needed as food by other fish whose catch we *also* want to maximize, e.g. tuna. And political exigencies frequently cause fisheries managers to simply set catch limits far higher than the maximum that scientists calculate and recommend (Safina 2012). In most of the world, lack of technical and enforcement capacity, lack of will, and the diffuse nature of small-scale fishing make such a top-down approach practically impossible.

And wholly outside of national policy, illegal fishing takes between 11 and 26 million tons of sea life, with total catches off West Africa estimated at 40% higher than reported legal catches (Agnew et al. 2009). Further, fishing and the seafood supply chain entail human rights abuses, including substantial slave labor, especially in Southeast Asia. In 2015, for example, authorities in Indonesia freed over 500 Burmese and Thai fish-working men from cages after an Associated Press story exposed their plight (Tobia 2015), and there have been several recent reports of slave labor on fishing boats from Thailand and elsewhere.

Most of the open ocean remains ungoverned even on paper. The ocean is not enclosed in a rule of law. The Law of the Sea has never been ratified by the United States, a major ocean player. And as with all ocean treaties, there is no real enforcement mechanism, so the Law of the Sea remains generally ineffective. In 1995, the United Nations Fish Stocks Agreement came into being (United Nations 2010). Like all such U.N. agreements, it urges and encourages ratifying nations to implement certain standards. The 77 countries that are now parties to the agreement have, by their participation, "signaled"—the U.N.'s word—that they are responsible fishing nations. Signals are important and the agreement is an improvement over the void that preceded it. It is a good model and creates some peer pressure among nations and is a basis for discussion. I worked to support it while it was being drafted and debated, and I still support it in principle.

But agreements in principle do not by themselves maintain or restore abundance. Eighteen years after the U.N. agreement was finalized on paper, a global assessment of shark exploitation (for example) determined that fishing kills about 100 million sharks annually, and that fishing mortality "exceeds the average rebound rate for many shark populations... and explains the ongoing declines in most populations for which data exist... Global total shark mortality, therefore, needs to be reduced drastically" (Worm et al. 2013). When I was in a graduate school, scientists writing in peer-reviewed journals did not use adverbs such as "drastically."

Into the general breach have come various multi-national fishing treaties, some of the newer ones patterned on the U.N. agreement. They, too, are better than nothing. But they, too, are generally ineffective at agreeing to sustain or allow recovery of exploited fish populations. Many of the tuna populations for which fishing is supposedly controlled by treaty, for instance, are depleted (Safina 2001; Myers and Worm 2003; Pikitch 2012). None are clearly recovering. The Atlantic bluefin tuna is the poster child for half a century of cynical management (Safina 1993, 2008, 2010a). National delegations to fisheries negotiations are generally rewarded *not* for how well they cooperate to limit catch and sustain yields, but rather for how much more fish quota they get for their nation to catch. To the extent there is cooperation, the cooperation has often been for countries to ignore scientific recommendations and award themselves higher catch quotas rather than agreeing to limits that would benefit all long-term. Some countries agree, then cheat, as Japan has been convicted of doing (Safina 2010b). The Pacific Halibut Commission, formed by a bilateral agreement of Canada and the United States, is the only international fishing commission that seems to work. If I am correct about that, perhaps that is because of the simplicity of two adjacent players with capable enforcement abilities.

The nations of the world began extending their claims over the waters they control in the 1950s, and 200 nautical miles has become the world standard. But even in national waters, not much more than fishing can be controlled, even in principle. And as I alluded, most countries lack both capacity and political interest to manage fisheries. This has led some conservation groups to seek solutions not in traditional Western fisheries management but rather resort to international wildlife treaties such as the Convention on International Trade in Endangered Species (CITES, which has made some progress recently in at least monitoring trade of certain sharks); or to focus on establishing local rights to manage, protect, and establish reserves in local fishing areas (Safina and Jenks 2012).

Progress is needed on all fronts, using all approaches. Protected areas alone can't address all the problems. Many fish populations migrate across boundaries. And global changes, such as warming and declining ocean pH, plastic debris, and chemical pollutants, know no bounds. If ever there was a worldwide tragedy of the commons, the ocean has a strong claim to the title of most spectacularly tragic, most broken global commons. And that brings us back to population, where all roads lead. In Garrett Hardin's original Tragedy of the Commons paper in *Science*, his abstract, in total, was this one-liner: "The population problem has no technical solution; it requires a fundamental extension in morality."

To be fair, I could emphasize hopeful signs of progress as a guide to what works: things like monitoring and enforcement in wealthier nations, or local empowerment and a few spectacularly successful protected areas and cooperatives such as Cabo Pulmo and Punta Abre Ojos in Mexico, or Palau's move toward creating a nationwide protected ocean area. (Though the United States has a system of National Marine "Sanctuaries," fishing is allowed in them.) But because this volume is about global issues and ethics combined, I've chosen not to sprinkle sugar on the massive problems that are setting the global pace. Because our species collectively demonstrates little will and thus insufficient mental capacity for solving the big global issues or changing overall trends in the deterioration of the natural world, I'd rather not whitewash the big problems with uplifting anecdotes of minor successes. Our accelerating global problems are outdistancing and out-spending our relatively feeble efforts toward solutions because, fundamentally, the dysfunctional relationship between humans and the rest of the world reflects deep defects in both our comprehension and our valuation of the world. We can feel confident that they are defects because they continue to cause widespread deterioration of Earth's life support systems.

So in the rest of this chapter, I will focus on the real changes occurring in the seas. The search for a durable scale of values, by which humans could exist without obliterating other living things or robbing future generations of a rich and beautiful natural endowment, should perhaps be foremost among humanity's priorities. Obviously, it isn't.

I am not a fan of ornate formal philosophizing on ethics and morality. This late in the day in our relationship with the world, I harbor scant patience for philosophical and economic arguments that question whether we really owe anything to future generations, or to other living things, thus implying that we can give ourselves an E-Z Pass for our recklessness. We know the right answer. Arguments that discount our responsibilities to future generations merely dress short-sighted greed as a "moral" conclusion. They are not only hideous, but wrong. All people would prefer not to suffer, and creating scarcity and biological impoverishment creates suffering now and in the future. Morality is mainly about how to treat people who are not close to us. It matters little whether they live across town or across time. People prefer being the recipients of compassionate acts. The simply stated and widely understood Golden Rule, "do unto others as you'd like them to do unto you," is adequate to guide us to a better world; better by virtue of reducing suffering, as most would prefer. In practice, I don't think we need fancier—or flakier—arguments questioning or affirming our moral and ethical responsibilities. It's really just a choice about what kind of people we want to be, destructive or constructive, apathetic or compassionate, small- or big-picture thinkers, short- or long-term planners, and what kind of world we want to contribute to or detract from. Problems arise largely because these considerations, based on values, really are a matter of preference; who do we wish to be?

Philosophy notwithstanding, the harsh reality for our current topic is that the global ocean game-changers of today and the foreseeable future (and beyond) include climate warming, ocean acidification, plastic pollution, fertilizer and toxic runoff, and human population growth. These are genies entirely out of each bottle, problems much bigger than any response that could be contained within the term "ocean policy."

Climate Changes

Climate change has always occurred, but today the scale and rate of change are driven by human activities (IPCC 2007). From the start of the Industrial Revolution around 1800, fossil fuel burning (mainly coal, petroleum, and natural gas) has been the major new source of carbon dioxide to our atmosphere. (In addition, coal burning is the major source of mercury that contaminates seafood; mothers beware.) Carbon dioxide traps heat analogous to the way insulation in your attic traps heat. More insulation traps more heat. Because another oft-used analogy is the way the glass ceiling of a greenhouse traps heat, the heat-trapping is called the "greenhouse effect," and gases that trap and hold atmospheric heat are called greenhouse gases.

The concentration of CO_2 in our atmosphere has so far risen from 280 parts per million (ppm) in 1750 (Feely et al. 2004) to about 420 ppm today (NOAA 2022) and increasing.

Methane, which on a unit-by-unit basis is a more potent heat-trapping gas than is carbon dioxide, is released by human activities such as farming. The warming of the arctic by carbon dioxide is melting upper levels of the ground previously considered "permanently" frozen and known as "permafrost." This is also triggering the release of methane through the rotting of dead vegetation that had been trapped in ice for millennia (NOAA 2015).

About one-third of CO_2 emitted into the air is absorbed by our oceans, where it changes ocean pH (Feely et al. 2004). Thus, the two main physical effects of the rise greenhouse gas concentrations are change to the planetary heat balance through warming of the air and ocean, and acidification of ocean water. These changes are well underway and measured.

These physical effects bring a raft of changes among living things. All life will be affected. Some species will flourish; others will become extinct. Global food production will be greatly affected through changes to on-land agriculture and ocean fish abundance and distribution (Easterling et al. 2007).

Ocean Warming

In the oceans, warming waters are causing large-scale redistribution of plankton and many fish species (Nye et al. 2009; Cheung et al. 2010). The top 400 meters of the world ocean averages about half a degree Celsius warmer than 135 years ago. This seemingly small change represents a tremendous amount of trapped heat. Warming is already causing poleward shifts in ranges of various fish species. The geographic range of warmer-water species will expand, while the range of colder-water species will shrink (Barange and Perry 2009). Even small changes in water temperature can greatly influence a fish species. Growth rates respond to water temperature. Abnormally, high water temperatures can even stop fish feeding (Drinkwater 2005).

Although predictions vary between models, atmospheric CO_2 concentration will continue to rise and by 2100 will range from 730 to 1,020 ppm (Meehl et al. 2007). Sea surface temperatures will continue to increase and will be one to three degrees Celsius warmer on average by 2100 (Meehl et al. 2007).

Ocean warming predictions vary greatly between regions (Roemmich et al. 2012). Arctic Ocean surface water may be seven degrees Celsius warmer by the end of this century (Meehl et al. 2007). In some regions, warming will also reduce the vertical movement of deep water toward the surface (upwelling), reducing nutrient supply to the upper sunlit waters, thus reducing plankton abundance and therefore fish abundance (Barange and Perry 2009).

Most tropical reef-building corals grow best between 26 and 28 degrees Celsius, with some notable exceptions in both directions (Sammarco and Strychar 2009). Anyone who has seen coral pieces in gift shops knows that dead coral is white. Living corals get their beige and greenish hues are photosynthesizing dinoflagellates called zooxanthellae ("zokes" for short), living within each coral polyp. These make sugars that corals rely on for food. For reasons not well understood, high water temperature causes zooxanthellae to abandon corals. This leaves corals white, called "bleached." Bleached coral can survive six to eight weeks without their zooxanthellae. Prolonged heat spikes cause corals to starve and die.

Bleaching is new. It was recorded three times in the century between 1876 and 1979. Then, in the 13 years between 1980 and 1993, scientists recorded 60 bleaching events worldwide (Sammarco and Strychar 2009). Along with rapidly escalating frequency, the areal extent of bleaching has vastly increased. In 2002, extensive bleaching occurred on more than half of the world's barrier reefs, killing most corals in those places (Sammarco and Strychar 2009).

Coral reefs are not just riotous pageants of life that attract tourists. They are habitats creating abundance and diversity for thousands of species of fishes and many other forms of sea life. Reductions in coral will diminish tourism, disrupt local fish-dependent economic activities around tropical coasts and islands, and reduce food for people. Large-scale coral death would be a biological catastrophe.

The hope is for fast evolutionary adaptation and spread by corals already adapted to extremes of heat and pH. But as you'd expect, those adapted to extremes live only in the rare areas of extreme conditions. Even if they began spreading around the tropics immediately, there is no realistic chance that they could replace today's coral reefs and support dependent human and non-human living communities in any timescale meaningful for people alive now, or their foreseeable descendants. The likely timescale of any such adjustment is at thousands of years. In the past, response and adjustment of coral reefs to massive atmospheric CO_2 changes took millions of years (Hoegh-Guldberg 1999; Hoegh-Guldberg et al. 2007; Safina 2011).

Sea Level

Rising ocean temperature is causing sea levels to rise through expansion of water and melting of land-based ice. During the 20th century, sea levels rose 17 centimeters. The rate of rise has increased dramatically in the last 20 years, now averaging 3 millimeters per year, nearly double the 20th-century average. Predictions of additional sea level rise by 2100 range from about 20 to 50 centimeters (Bindoff et al. 2007). A rising world ocean means that we are in an era when essentially all the world's coasts are subject to erosion, increased flooding, and inundation of coastal wetlands. Beaches and wetlands pinched between the sea and human development cannot easily shift up and inland. Fisheries in or near habitats most prone to sea level rises, like mangrove forests and sea grass beds, may be severely degraded (FAO 2009).

Most reef-building corals live in shallow water where sunlight is adequate for photosynthesis by the aforementioned symbiotic zooxanthellae that make sugars that nourish them. To survive sea level rise, corals must grow upward at least as fast as sea level is rising. The maximum rate of reef growth is about 6 millimeters annually. Current sea level rise is about 3 millimeters annually (GBRMPA 2009). But reduction in ocean carbonate (see acidification, below) is slowing coral growth; increasing storm intensity is worsening damage to reefs; and overfishing of herbivorous fishes is allowing seaweeds to overgrow corals, stunting growth and greatly inhibiting reproduction. This is one example of a cascade of effects caused by various human pressures, globally degrading one of Earth's major habitat types.

Storms

Ocean and atmospheric warming combined increase storm severity, thus the frequency of major storms. Warm oceanic water gives hurricanes (cyclones) the energy that develops strong rotating winds and heavy rain. Rising water temperatures directly result in more numerous and stronger storms. The 0.5 degrees Celsius increase in sea surface temperature in the Atlantic Ocean resulted in a 40% increase in the frequency and strength of hurricanes making landfall along the U.S. East Coast during recent decades (Saunders and Lea 2008). On the Great Barrier Reef where no Category 5 cyclones (i.e. most damaging) occurred during the previous 35 years, there were three such storms during a recent five-year span (GBRMPA 2009). These very large cyclones damaged many coral reefs directly through the power of wind-generated waves and by sediment runoff that smothered marine life following heavy rains.

Water temperatures are predicted to increase by 1–3 degrees Celsius on world average by 2100 (Meehl et al. 2007). Large storms will cause considerable damage to coral reefs and other coastal habitats of crucial importance for generating fish. The Arctic is warming considerably faster than elsewhere, partly because when reflective Arctic sea ice melts, it is replaced by dark water that absorbs more heat. It appears likely that the Arctic Ocean will be largely ice-free in summer by 2050. This may make the sea more productive for some species moving north, but will likely cause deep declines in ice-adapted animals, including many species of seals and arctic whales, polar bears, and walrus.

Acidification of the Ocean

The increased carbon dioxide is not just warming the air and sea. About 22 million tons of CO_2 dissolves in our oceans every day (Petkewich 2009). When carbon dioxide (CO_2) enters the sea, it combines with water (H_2O) to form carbonic acid (H_2CO_3), which then releases one hydrogen ion (H+), leaving a molecule of bicarbonate (HCO_3^-). Acidity is a measure of hydrogen ions, so this process is called "ocean acidification." The more hydrogen ions, the more acid. (By convention, the more acid, the lower the pH; a pH of 1 is a very strong acid.) Compared to before the Industrial Revolution, the upper ocean already has about 30% more hydrogen ions.

The problem is not simply the pH. The problem is that those hydrogen ions easily bind with carbonate ions (CO3-2). Because the reaction uses up carbonate ions, a dearth of carbonate is experienced by animals that require calcium carbonate for making their skeletons and shells. Acidity proceeds unevenly in various ocean regions. On the West Coast of the United States, acidity has already made seawater fatal to oyster larvae.

This is a profoundly changing chemistry for our oceans. Even if global CO_2 emissions are stabilized—highly unlikely—ocean pH would not to return to pre-industrial levels for thousands of years (The Royal Society 2005).

The shell-building creatures most directly affected include reef-building corals that create fish habitat. They also include clams and oysters and scallops, snails, and various plankton species that help support the ocean food web (Anthony et al. 2008; Feng et al. 2008; Talmage and Gobler

2010). Carbonate scarcity slows their growth, making them more fragile, and sometimes, fatally deformed.

Carbonate concentrations in the upper few hundred feet of the ocean have already declined about 10%, on average, compared to seawater just before steam-engine times. Shellfish grown in water chemistry conditions that mimic pre-industrial levels grow faster, and have thicker, stronger shells, than those grown in today's conditions (Talmage and Gobler 2010). Ocean acidification can reduce these organisms' ability to grow, survive, reproduce, and photosynthesize (Albright et al. 2010; Doney et al. 2009; Hoegh-Guldberg et al. 2007; Kurihara 2008). Some effects of ocean acidification on fish are simply bizarre, such as causing some coral reef fish to swim toward the smell of their predators (Munday et al. 2010).

Perhaps most fundamentally for coral reefs is the question of whether most reef-building corals will be able to reproduce. Sadly, the answer seems clear. Most corals hatch as swimming larvae that must find and settle on substrates conducive to their growth and survival. The substrates most conducive—which coral larvae find by smell—are thin crusts produced by a genus of so-called coralline algae. Even when they find the right substrate, competition for space by many kinds of organisms is fierce and survival of young corals is low. The carbonate scarcity predicted by 2100 would—and likely will—greatly reduce growth and abundance of the coralline algae most important for coral larvae settlement, severely impairing the ability of larvae to settle successfully and survive (Doropoulos et al. 2012).

Fishing

While the changes to temperature and chemistry are the most systemic, the major way that humans have changed the ocean to date is by simply removing living things from the sea. We got too good at it. Over 80% of the world's fisheries are fully or overexploited (FAO 2010), meaning that there are few if any major populations available for greatly increased pressure. Effects of climate change on fish and fisheries are added onto the backdrop of widespread depletion due to overfishing.

Globally, fisheries catch more than 90 million tons of life from oceans and lakes. Seafood is a significant source of protein and nutrients for billions of people, particularly in developing countries. Fisheries employ 45 million people directly and 180 million indirectly, and generate roughly US$95 billion annually. Changes in fish productivity thus have wide-ranging consequences (FAO 2010).

Interactions between fishing and warming will be complicated for hundreds of species. Just one example: growth rates of Atlantic Cod plateau at water temperatures between 6 and 10 degrees Celsius and then steadily decrease as temperature rises further. This growth-temperature relationship varies between cod populations however (Clark et al. 2003). It gets complex. Bottom-water temperatures account for 90% of differences in growth rates between cod populations, and larger fish grow fastest at lower temperatures (Drinkwater 2005). Differences in genetics, food, and population density can result in large differences between neighboring populations. Rising water temperatures will likely benefit the northernmost cod populations of Canada and Greenland and prompt northern range extension. Northward range extension has already been observed for North Sea cod as a series of warm winters and strong southerly winds have pushed cod eggs and larvae northward (Rindorf and Lewy 2006). Meanwhile, populations in the United States and the United Kingdom that have been fished for centuries may disappear by 2100 due to overfishing and warming combined (Drinkwater 2005).

Climate change will affect the lives and livelihoods of fisherfolk and communities, thru (1) changing ocean productivity (declines in formerly productive areas near human population centers), (2) altered fish distribution patterns, (3) greater storm damage to infrastructure, and (4) consequent reduced food availability and affordability. Decreased fish catches could lead to increased malnutrition in already impoverished fish-dependent areas, possibly increasing conflict between fishing groups and nations (Badjeck et al. 2010).

Warming and Acidification's One-Two Punch: Three Examples

Many people think of overfishing as a past problem, and climate change and ocean acidification as hypothetical future events. But some ocean regions and species are still experiencing intense overfishing *and* many are already experiencing great changes from warming and acidification.

Although each species is uniquely affected, several general patterns exist. Fish embryos and shellfish larvae, the most sensitive life history stages, are most at risk. Long-lived species will be slower to recover from stress, though in most of the world, reductions in stress are only hypothetical. (An exception is in the U.S. waters, where laws outlawing overfishing and mandating recoveries—first embodied in the sweeping reforms of the Sustainable Fisheries Act of 1996 and later legal affirmations—coupled with monitoring and enforcement capacity that does not exist in most of the world are resulting in fairly broad recoveries of many coastal fish species.)

All marine ecosystems will be affected to some degree by climate change and ocean acidification (Beaugrand et al. 2008). The following examples highlight some of the major dynamics underway and likely to accelerate in the near future.

1   The Gulf of Maine in the northeastern United States is heavily reliant for its productivity on a species of copepod the size of a grain of rice, called Calanus finmarchicus. In its trillions, this tiny creature might be thought of as the rice of the sea. It is vital for the entire food chain, its stored lipids a crucial energy source for everything from herring to right whales. The Gulf of Maine is the southern range of this creature, so it is likely that as waters warm the population will move north and away from the region, greatly compounding problems caused by devastating overfishing, making recoveries of such crucial fishes as cod slower and less probable. How this plays out remains to be seen (Runge et al. 2014).

2   The Pacific Northwest of the United States and Canada. As mentioned, Pacific Northwest shellfish are already on the front lines of ocean acidification. Because of that region's very active upwelling system, more acidic waters are welling up from the depths of the Pacific Ocean and flowing inshore to the bays and estuaries of Oregon, Washington, and British Columbia. Oyster growers who'd experienced near-total die-off of larvae now must chemically buffer water flowing into their hatcheries. But this says nothing of the changes affecting wild shellfish populations, which may be doomed (Grossman 2011). The deep waters absorbed their carbon dioxide some three decades ago; it appears certain that the situation is going to get far worse.

3   Australia's Great Barrier Reef. The world's largest reef, visible from space, is managed as a marine park zoned for multiple use, including reserves where fishing is prohibited. But map boundaries and rulebooks cannot prevent high coral mortality caused by warming sea surface temperatures and acidification (Selig and Bruno 2010; Selig et al. 2012).

Spreading Dead Zones

Synthetic fertilizers have since the 1960s doubled the global nitrogen flow to living systems, washing down rivers and, since the 1970s, creating hundreds of oxygen-starved seafloor "dead zones" where most animals cannot survive. The fertilizers washed off farms and down waterways spark dense blooms of plankton, whose death and bacterial decay depletes the oxygen of waters near the seafloor. The first extensively documented dead zone at the mouth of the Mississippi River, discovered in the 1970s, now covers up to 22,000 square kilometers. There are now about 170 seafloor dead zones documented worldwide, mostly on the U.S. East Coast, Northern Europe, and Southeast Asia but in many other places too (Boesch 2008; Nellemann et al. 2008).

Harmful and Toxic Algal Blooms

Increasing runoff of fertilizer, as mentioned, is sparking more frequent dense blooms of algae. Some algae are toxic, so some blooms are harmful. Harmful algal blooms can cause mass die-offs of fish and other marine organisms. Some cause human fatalities. Consider the names of a few of the "shellfish poisoning" illnesses that can be caused by eating shellfish contaminated by harmful algae: amnesic, diarrhetic, neurotoxic, and paralytic shellfish poisoning. Some can be fatal to humans (Backer and McGillicuddy 2006).

As an example of how the problems we're causing can interact and worsen each other, nutrient pollution and acidification promote growth and increased toxicity of the harmful red tide species *Alexandrium fundyense* (NOAA 2014). Ocean acidification makes harmful algal blooms worse.

New Diseases Ravaging Corals and Sponges

Since the early 1980s, newly seen coral diseases have ravaged coral reefs, particularly throughout the Caribbean where the formerly most abundant species, staghorn and elkhorn corals, have virtually disappeared. Warming and the stress of bleaching allow corals to be colonized by fungi and bacteria, which appear to be key parts of the mortality syndrome (Burge et al. 2014; Tracy et al. 2015).

The Decline of Seagrass Meadows and Oyster Reefs

Seagrass meadows, perhaps the most underappreciated major marine ecosystem in the world, are crucial to various fishes, sea turtles, marine mammals, and local fishing peoples. Nutrient and sediment overload, mainly, are causing their global decline (Cullen-Unsworth et al. 2014; Short et al. 2014). The same forces, plus extreme overexploitation, have caused the worldwide near-disappearance of wild oyster reefs (Ermgassen et al. 2012).

Plasticization of the Sea

People now in their 60s and older came into the world while the ocean contained essentially no plastic. Wide commercial production did not begin until the 1950s. By 2015, the ocean contained an estimated 5.25 trillion plastic particles weighing 268,940 tons (Eriksen et al. 2014). One way to understand such incomprehensible numbers is to go to any beach. In some of the remotest places, such as the Northwest Hawaiian Islands in the center of the North Pacific, Madagascar, Alaska—plastic pretty much everywhere (Safina 2002). During a one-day visit to Alaska's Katmai National Park, I helped haul four tons of washed-up plastic from roughly two miles of shoreline (Howard 2013a, b). It's everywhere. Plastic tangles marine animals, and clogs their gut with potentially toxic indigestible matter or chokes them when mistaken for food, as when turtles mistake plastic bags for their normal jellyfish fare.

Concluding Thoughts

This is only a cursory survey of some of the world's most pressing ocean problems. I have not touched on massive problems of seabird die-offs due to declining plankton, or sea turtle decline—and some recoveries—due to different regional patterns of nesting beach protection, beach erosion, human consumption, and incidental catch (Safina 2002, 2006, 2014). The United Nations Convention on Biodiversity aimed to conserve the planet's diversity of living things, but its own assessment says, "biodiversity is in decline at all levels and geographical scales," a situation, "likely to continue for the foreseeable future" (Secretariat 2006). I don't believe that "ocean policy" can fix many of the worst ocean problems, because most of them originate on land. Meanwhile, there is work to do. Two things are needed: clean energy and a sharp reversal of our numbers.

If we could free ourselves from the tyranny of the fossil fuel industry, we could harness the energy that actually runs the whole planet: the power of the sun, the strength of the wind, the force of the tides, and the heat of the Earth. In the process, we would likely shrivel the worst petro-dictatorships. In the longer term, we could stabilize warming and ocean pH. We as individuals are less likely to see benefits from "doing the right thing." But isn't all of ethics about performing actions that do not benefit us directly, or foregoing some that would if others get hurt? Isn't that why we do not steal?

In the United States, especially, disinformation backed largely by ideologues deeply invested in fossil fuels creates the widespread impression that clean energy is the boutique endeavor of a cottage industry, not serious, not scalable, and not practical. But in 2013, Denmark generated 34% of its electricity from the wind. Portugal, Spain, and Ireland now get roughly 20% of their electricity from wind. On windy days, Ireland gets half its electricity that way. And in January of 2014, wind supplied, Lester Brown and co-authors write, "a whopping 62 percent of Denmark's electricity." Wind is gaining on nuclear power to become Spain's leading electricity source. In China, electricity from wind has surpassed that from nuclear plants. For several days in August 2014 in the United Kingdom, electricity generated from wind eclipsed that from coal. In South Australia, wind farms now supply more electricity than do coal plants. On September 30, 2014 in South Australia, power generated from wind and sunlight exceeded electricity demand. For 170 million Chinese households, rooftop solar water heaters provide hot water. In the United States, you'd never know that Iowa and South Dakota are already generating a quarter of their electricity from wind. Texas—which now gets nearly 10% of its electricity from wind—is building massive wind facilities and long-distance transmission lines to take wind-generated power to buyers in Louisiana and Mississippi (Brown et al. 2015). So clean energy is scalable. We get the jobs and the future that we plan for.

About the fundamental dilemma—our numbers—we have a great stroke of true luck: the one thing that appears to work to relieve the pressure-cooker of human population expansion also works to improve the widest human injustice of all. That thing is gender equity. Investing in women and girls has been called "the breakthrough strategy" and "a miracle solution" for achieving all the UN's millennium development goals (Lomoy 2010). In societies where women gain mere equal opportunity and a fair shot, birth rates harmonize to about two children per woman, low enough to stabilize population. The biggest slices of pie get cut at the least crowded tables. It's as true for the whole planet.

Acknowledgment

I thank Alan Duckworth for help researching and drafting earlier climate portions.

References

Agnew DJ, Pearce J, Pramod G, Peatman T, Watson R, Beddington JR, et al. (2009) "Estimating the Worldwide Extent of Illegal Fishing," *PlosONE* 4-E4570. doi:10.1371/journal.pone.0004570

Albright R, Mason B, Miller M, Langdon C. (2010) "Ocean Acidification Compromises Recruitment Success of the Threatened Caribbean Coral Acropora Palmata," *Proceedings of the National Academy of Sciences of the United States of America* 107:20400–20404.

Anthony KRN, Kline DI, Diaz-Pulldo G, Dove S, Hoegh-Guldberg O. (2008) "Ocean Acidification Causes Bleaching and Productivity Loss in Coral Reef Builders," *PNAS* 105:17422–17446.

Backer LC and Mcgillicuddy DJ Jr. (2006) "Harmful Algal Blooms," *Oceanography* 19:94–106.

Badjeck MC, Allison EH, Halls AS, Dulvy NK. (2010) "Impacts of Climate Variability and Change on Fishery-Based Livelihoods," *Marine Policy* 34:375–383.

Barange M, Perry RI. (2009) "Physical and Ecological Impacts of Climate Change Relevant to Marine and Inland Capture Fisheries and Aquaculture," in K Cochrane, C De Young, D Soto, T Bahri (eds.) *Climate Change Implications for Fisheries and Aquaculture: Overview of Current Scientific Knowledge.* FAO Fisheries and Aquaculture Technical Paper. No. 530. Rome, FAO, pp. 7–106.

Beaugrand G, Edwards M, Brander K, Luczak C, Ibanez F. (2008) "Causes and Projections of Abrupt Climate-Driven Ecosystem Shifts in the North Atlantic," *Ecology Letters* 11:1157–1168.

Bindoff NL, et al. (2007) "Observations: Oceanic Climate Change and Sea Level," in S Solomon, D Qin, M Manning, Z Chen, M Marquis, KB Avery, M Tignor, HL Miller (eds.) *Climate Change 2007: The Physical Science Basis. Contribution of Working Group I to the Fourth Assessment Report of the Intergovernmental Panel on Climate Change.* Cambridge and New York: Cambridge University Press.

Boesch DF. (2008) "Global Warming and Coastal Dead Zones," *National Wetlands Newsletter* 30:11–21.

Brown L et al. (2015) *The Great Transition,* W. W. Norton & Company.

Burge CA, Eakin CM, Friedman CS, et al. (2014) "Climate Change Influences on Marine Infectious Diseases: Implications for Management and Society," *Annual Review of Marine Science* 6:249–277.

Cheung WWL, Lam VWY, Sarmiento JL, Kearney K, Zeller D, Pauly D. (2010) "Large-Scale Redistribution of Maximum Fisheries Catch Potential, in the Global Ocean Under Climate Change," *Global Change Biology* 16:24–35.

Clark RA, Fox CJ, Livermore M. (2003) "North Sea Cod and Climate Change – Modeling the Effects of Temperature on Population Dynamics," *Global Change Biology* 9:1669–1680.

Cullen-Unsworth L, Nordlund LM, Paddock J, et al. (2014) "Seagrass Meadows Globally as a Coupled Social-Ecological System: Implications for Human Wellbeing," *Marine Pollution Bulletin* 83:387–397.

Doney SC, Fabry VJ, Feely RA, Kleypas JA. (2009) "Ocean Acidification: The Other CO2 Problem," *Annual Review of Marine Science* 1:169–192.

Doropoulos C, Ward S, Diaz-Pulido G, Hoegh-Guldberg O, Mumby PJ. (2012) "Ocean Acidification Reduces Coral Recruitment by Disrupting Intimate Larval-Algal Settlement Interactions," *Ecology Letters* 15:338–346.

Drinkwater KF. (2005) "The Response of Atlantic Cod (GadusMorhua) to Future Climate Change," *ICES Journal of Marine Science* 62:1327–1337.

Easterling WE, Aggarwal PK, Batima P, Brander KM, Erda L, Howden SM, Kirilenko A, Morton J, Soussana J-F, Schmidhuber J, Tubiello FN. (2007) "Food, Fibre and Forest Products," in: ML Parry, OF Canziani, JP Palutikof, PJ Van Der Linden, CE Hanson (eds.) *Climate Change 2007: Impacts, Adaptation and Vulnerability. Contribution of Working Group II to the Fourth Assessment Report of the Intergovernmental Panel on Climate Change.* Cambridge: Cambridge University Press, pp. 273–313.

Eriksen M, Lebreton LCM, Carson HS, Thiel M, Moore CJ, Borerro JC, et al. (2014) "Plastic Pollution in the World's Oceans: More Than 5 Trillion Plastic Pieces Weighing Over 250,000 Tons Afloat at Sea," *PlosONE* 9(12):E111913. doi:10.1371/journal.pone.0111913.

Ermgassen PS, Zu E, Spalding MD, Blake B, et al. (2012) "Historical Ecology with Real Numbers: Past and Present Extent and Biomass of an Imperilled Estuarine Habitat," *Proceedings of the Royal Society B-Biological Sciences* 279:3393–3400.

FAO. (2009) *Climate Change Implications for Fisheries and Aquaculture: Overview of Current Scientific Knowledge,* Food and Agriculture Organization of the United Nations, Fisheries and Aquaculture Department: Rome.

FAO. (2010) *The State of World Fisheries And Aquaculture 2010,* Food and Agriculture Organization of the United Nations: Rome.

Feely RA, Sabine CL, Lee K, Berelson W, Kleypas J, Fabry VJ, Millero FJ. (2004) "Impact of Anthropogenic CO2 on the Caco3 System in the Oceans," *Science* 305:362–366.

Feng Y, Warner ME, Zhang Y, Sun J, Fu F-X, Rose JM, Hutchins DA. (2008) "Interactive Effects of Increased Pco2, Temperature and Irradiance on the Marine Coccolithophore Emiliania Huxleyi (Prymnesiophyceae)," *European Journal of Phycology* 43:87–98.

GBRMPA. (2009) *Great Barrier Reef Outlook Report 2009,* Great Barrier Reef Marine Park Authority, Townsville, Queensland.

Grossman E. (2011) "Northwest Oyster Die-Offs Show Ocean Acidification has Arrived," *Yale Environment 360.*

Hardin G. (1968) "The Tragedy of the Commons," *Science* 162:1243–1248. doi:10.1126/science.162.3859.1243.

Hoegh-Guldberg O. (1999) "Climate Change, Coral Bleaching and the Future of the Worlds Coral Reefs," *Marine and Freshwater Research* 50:839–866.

Hoegh-Guldberg O, Mumby PJ, Hooten AJ, Steneck RS, Greenfield P, Gomez E, Harvell CD, Sale PF, Edwards AJ, Caldeira K, Knowlton N, Eakin CM, Iglesias-Prieto R, Muthiga N, Bradbury RH, Dubi A, Hatziolos ME. (2007) "Coral Reefs under Rapid Climate Change and Ocean Acidification," *Science* 318:1737–1742.

Howard BC. (2013a) "Gyre Expedition Probes Impact of Plastic Pollution on Remote Beaches," *National Geographic Ocean Views.* http://bit.ly/1gjk6vj.

Howard BC. (2013b) "Filmmakers Document the 'Weirdness' of Marine Garbage on the Gyre Expedition," *National Geographic Ocean Views.* http://bit.ly/1hnsi5h.

IPCC. (2007) "Climate Change 2007: The Physical Science Basis. Contribution of Working Group I to the Fourth Assessment Report of the Intergovernmental Panel on Climate Change," in S Solomon, D Qin, M Manning, Z Chen, M Marquis, KB Averyt, M Tignor, HL Miller (eds.) Cambridge and New York: Cambridge University Press. Online faqs: http://oceanservice.noaa.gov/education/pd/climate/factsheets/howhuman.pdf.

Kurihara H. (2008) "Effects of CO2-Driven Ocean Acidification on the Early Developmental Stages of Invertebrates," *Marine Ecology Progress Series* 373:275–284.

Lomoy J. (2010) "Investing in Women and Girls – the Breakthrough Strategy for Achieving the MDGs," *Organization for Economic Co-Operation and Development.* https://www.oecd.org/dac/gender-development/45704694.pdf.

Meehl GA, Stocker TF, Collins WD, Friedlingstein P, Gaye AT, Gregory JM, Kitoh A, Knutti R, Murphy JM, Noda A, Raper SCB, Watterson IG, Weaver AJ, Zhao ZC. (2007) "Global Climate Projections," in S Solomon, D Qin, M Manning, Z Chen, M Marquis, KB Averyt, M Tignor, HL Miller (eds.) *Climate Change 2007: The Physical Science Basis. Contribution of Working Group I to the Fourth Assessment Report of the Intergovernmental Panel on Climate Change.* Cambridge: Cambridge University Press.

Munday PL, Dixson DL, Mccormick MI, Meekan M, Ferrari MCO, Chivers DP. (2010) "Replenishment of Fish Populations is Threatened by Ocean Acidification," *Proceedings of the National Academy of Sciences of the United States of America* 107:12930–12934.

Myers RA, Worm B. (2003) "Rapid Worldwide Depletion of Predatory Fish Communities," *Nature* 423:280–283.

Nellemann C et al. (2008) "In Dead Water – Merging of Climate Change with Pollution, Over-Harvest, and Infestations in the World's Fishing Grounds," *United Nations Environment Programme*, Norway: GRID-Arendal.

NOAA. (2022) "Ocean Acidification Promotes Disruptive and Harmful Algal Blooms on our Coasts," http://coastalscience.noaa.gov/news/climate/ocean-acidification-promotes-disruptive-and-harmful-algal-blooms-on-our-coasts/.

NOAA. (2015) "Trends in Atmospheric Carbon Dioxide," http://www.esrl.noaa.gov/gmd/ccgg/trends/global.html.

Nye JA, Link JS, Hare JA, Overholtz WJ. (2009) "Changing Spatial Distribution of Fish Stocks in Relation to Climate and Population Size on the Northeast United States Continental Shelf," *Marine Ecology Progress Series* 393:111–129.

Petkewich R. (2009) "Off-Balance Ocean: Acidification from Absorbing Atmospheric CO2 Is Changing the Ocean's Chemistry," *Chemical & Engineering News* 87:56–58.

Pikitch E. (2012) "The Risks of Overfishing," *Science* 26:474–475.

Rindorf A and Lewy P. (2006) "Warm, Windy Winters Drive Cod North and Homing of Spawners Keeps them there," *Journal of Applied Ecology* 43:445–453.

Roemmich D, Gould WJ, Gilson J. (2012) '135 Years of Global Ocean Warming between the Challenger Expedition and the Argo Programme', Nature Climate Change.

Runge JA, et al. (2015) "Persistence of Calanus Finmarchicus in the Western Gulf of Maine During Recent Extreme Warming," *Journal of Plankton Research* 37(1):221–232. doi:10.1093/plankt/fbu098.

Safina C. (1993) "Bluefin Tuna, in the West Atlantic Negligent Management, and the Making of an Endangered Species," *Conservation Biology* 7:229–234.

Safina C. (2001) "Tuna Conservation," in BA Block, D Stevens (eds.) *Tuna Ecological Physiology and Evolution.* Cambridge, MA, USA: Academic Press, pp. 413–459.

Safina C. (2002) *Eye of the Albatross*, New York: Henry Holt Co.

Safina C. (2006) *Voyage of the Turtle*, New York: Henry Holt Co.

Safina C. (2010a) "After Two Decades of Delay, A Chance To Save Bluefin Tuna," *Yale E360*, 5 March. http://bit.ly/bllce1.

Safina, C. (2011) *The View From Lazy Point,* New York: Henry Holt Co.

Safina C. (2010b) "Japan Aargh," *Carlsafina.Org.* http://carlsafina.org/2010/02/22/japan-aargh/.

Safina C. (2012) "Maxing Out Our Take," *Science* 335:169–170.

Safina C. (2014) "How Climate Change Is Sinking Seabirds," *Audubon*, September–October.

Safina C and Jenks B. (2012) "How to Catch Fish and Save Fisheries," *New York Times*, 19 October.

Safina C and Klinger D. (2008) "Collapse of Bluefin Tuna in the Western Atlantic," *Conservation Biology* 22:243–246.

Sammarco PW, Strychar KB. (2009) "Effects of Climate Change/Global Warming on Coral Reefs: Adaptation/Exaptation in Corals, Evolution in Zooxanthellae, and Biogeographic Shifts," *Environmental Bioindicators* 4:9–45

Saunders MA, Lea AS. (2008) "Large Contribution of Sea Surface Warming to Recent Increase in Atlantic Hurricane Activity," *Nature* 451:557–561.

Secretariat of the Convention on Biological Diversity. (2006) *Global Biodiversity Outlook 2.* Montreal, 81 + Vii pp.

Selig ER, Bruno JF. (2010) "A Global Analysis of the Effectiveness of Marine Protected Areas in Preventing Coral Loss," *PlosOne* 5. Doi:10.1371/Journal.Pone.0009278.

Selig ER, Casey KS, Bruno JF. (2012) "Temperature-Driven Coral Decline: The Role of Marine Protected Areas," *Global Change Biology* 18:1561–1570.

Short FT, Coles R, Fortes MD, et al. (2014) "Monitoring in the Western Pacific Region Shows Evidence of Seagrass Decline in Line with Global Trends," *Marine Pollution Bulletin* 83:408–416.

Talmage SC, Gobler CJ. (2010) "Effects of Past, Present, and Future Ocean Carbon Dioxide Concentrations on the Growth and Survival of Larval Shellfish," *Proceedings of the National Academy of Sciences of the United States of America* 107:17246–17251.

The Royal Society. (2005) "Ocean Acidification Due to Increasing Atmospheric Carbon Dioxide," *The Royal Society, Policy Document* 12/05.

Tobia PJ. (2015) "You Probably Benefitted from Slave Labor Today," *PBS.Org.* https://www.pbs.org/newshour/world/slave-made-goods-now-available-freezer-section.

Tracy AM, Koren O, Douglas N, et al. (2015) "Persistent Shifts in Caribbean Coral Microbiota Are Linked to the 2010 Warm Thermal Anomaly," *Environmental Microbiology Reports* 7:471–479.

Worm B, et al. (2013) "Global Catches, Exploitation Rates, and Rebuilding Options for Sharks," *Marine Policy* 40:194–204.

United Nations. (2010) "Fish Stocks Agreement: Overview of What the Agreement Says and its Impact," http://www.un.org/depts/los/convention_agreements/reviewconf/fishstocks_en_b.pdf

18
FISHING AND HARVESTING

Mary Lyn Stoll

In many ways, the current state of global fisheries illustrates the dangers of pursuing relatively unregulated global economic development.[1] After depleting fish stocks at home, trawlers move on to decimate the seas of poor nations, their activities on the high seas often unregulated or subject to rules that mean little in the face of corrupt or ill-equipped agencies of enforcement. Just as companies may set up their charters from tiny post boxes on remote island nations to avoid facing penalties at home, so too fishing ships will fly under flags of convenience. If it becomes inconvenient to follow the stricter requirements of one government, a simple email or fax allows one to change the flag and the rules of engagement in minutes (Clover 2006: 137). Nearly all of the problems that most concern the critics of global economic structures come into play: undermining the power of state governments to protect their environments and their citizens, an economic system that robs poor people and future generations to serve the seemingly exorbitant desires of a wealthy global elite, and technological developments that ultimately undermine long-term economic and environmental stability (Steger 2013). The United Nations' Food and Agriculture Organization (FAO) reports that although 90% of global wild fish stocks were at biologically sustainable levels in 1974, this percentage decreased to just 65.8% by 2017 (FAO 2020: 47). In short, a collective race to the bottom has ensued wherein there simply will be no viable fish populations left in the seas within another 30 years if significant changes in fisheries management are not put into place (Worm et al. 2006, 2009). The prisoner's dilemma applies in full force. While it is rational for each individual fisherman or fisherwoman to fish as much as is possible, the collective decision to do so undermines the long-term viability of what could otherwise be a sustainable industry.

I begin by discussing the ways in which the global fishing industry creates an array of seemingly intractable problems and explain how its problems are often thought to be typical of how business works under globalization. I then examine whether or not creative institutional measures could avert total collapse of ocean ecosystems and the economies dependent upon the health and vitality of those ecosystems. While the public often assumes that changes must be driven by activists and governments, I argue that business must also be a part of any fully viable solution to sustaining aquatic ecosystems. Even if governments lack the ability or political will to adequately regulate and manage marine ecosystems, companies can still better meet their moral duties to poor people and the environment. Companies ought to begin by investing in efficiency for sustainability and by working with certification organizations to empower customers to support responsible fishing. Companies can also invest in the future of the industry and the planet by setting aside reserves, and by taking a more creative approach to the role of fish in the marketplace on a number of fronts in order to meet their moral duties of environmental stewardship.

DOI: 10.4324/9781315768090-22

The State of the World's Fisheries: An Industry Built for Disaster

The United Nations' FAO reports that 90% of global fish stocks are fully exploited or overfished (at least among fish stocks for which data are available) (Kituyi and Thompson 2018). It should be noted that the United Nations' FAO reports that world tonnage of fish goes up yearly, but the FAO is an agricultural organization whose primary goal is often agricultural rather than being focused primarily upon environmental sustainability (FAO 2020: 8). It should also be noted that the FAO relies on self-reporting by the fishing industry and has a track record of taking Chinese reports on fishing at face value, which few regard as accurate. There are several scientific studies which provide evidence that global fish stocks and harvests are in fact much worse off. Watson and Pauly published a report in *Nature* which showed that the world's catch went up yearly since 1950 until it tapered off in the 1980s and began to decline. They argue that there is actually a decline of about 770,000 tons in fish caught each year (Pauly et al. 2002). The Intergovernmental Science-Policy Platform on Biodiversity and Ecosystem Services (IPBES), a United Nations organization more focused on biodiversity, reports that "(b)iodiversity – the diversity within species, between species and of ecosystems – is declining faster than at any time in human history" (Diaz et al. 2019: 10). With respect to marine species, 35% of marine species have been rapidly declining since 1970. Overfishing has had the single largest negative impact upon ocean ecosystems in the last 50 years, even greater than pollution or climate change. 66% of the ocean has experienced increasing cumulative impacts due to human activity (*ibid.*). According to a study by Boris Worm from Dalhousie University in Canada, 29% of fish populations in 2005 were fully collapsed. Thus, if anything, global fish populations may be in a worse state than FAO numbers would indicate. Lest one might be tempted to believe that this is a recent crisis, note that the same study finds that 65% of fish stocks that have been exploited since the 1950s have collapsed. If we continue on our current path, scientists warn that the populations of all fish and shellfish will collapse by 2048 (Roberts 2007: 330). While more recent research suggests that there may be more hope for a few select fisheries if sustainable fisheries practices are put into place, the problem is that these measures are the exception rather than the rule. Those exceptional cases of well-managed fisheries also tend to be off the coasts of richer nations like the United States, leaving poor nations who are most dependent upon fish in especially dire straits. Moreover, as with cod off the coast in Eastern Canada, some species may have already reached a tipping point beyond which they cannot fully recover even with the best fisheries management plan (Worm et al. 2009). This is not a case of just a few exotic species going extinct as a single tract of rainforest is razed. Rather, the entire population of the world's oceans may well be unable to support commercial fishing of any kind in a mere 30 years unless we change our fishing industries substantially and drastically. The economic damage alone is astounding. But once one considers the damage done to the world's ecosystems and poor communities who rely upon fish to survive, the incipient harm is appalling.

All of this is without fully factoring in further harms caused by plastic pollution and climate change. IPBES reports that marine plastic pollution has gone up tenfold since 1980 (Diaz et al. 2019). An estimated 4.4–12.7 million metric tons of plastic find their way into the oceans annually (Borelle et al. 2017). By 2050, if sweeping changes are not made, there will be more plastic than fish in the ocean (Ellen MacArthur Foundation 2016). This is problematic because plastics can alter gene expression, cells, and tissues, affect population size and community structure, and impair reproduction and development in addition to altering species function, dispersion, and assembly in aquatic species (Borelle et al. 2017). To make matters worse, one study found that the single largest contributor to plastics pollution in the sea was actually the fishing industry (Laville 2019). Abandoned nets and gear impact aquatic ecosystems long after the fishing boats have left. Climate change will also further negatively impact aquatic ecosystems. A 2020 study by Dalke et al. found that 60% of fish species may be unable to survive in their current locations as a warming climate interrupts breeding and fish lifecycles (Dahlke et al. 2020). So at a time when the oceans need a rest, in the face of radical change and heightened risk, the world's fisheries just keep taking more and more.

One might wonder how it could ever make sense for an industry to engineer its own demise. Each business member of an industry, after all, is generally thought to have at least a *prima facie* teleological moral obligation toward the continuance of that industry. The problem is actually the result of a multitude of factors. First, it is important to remember that fishing essentially runs according to a hunter gatherer economic system. Thus, the tragedy of the commons applies in full force. Imagine a trip to pick wild blackberries. You would likely take all you could, thinking that by the time the next berry lovers ambled by new berries would have ripened for the taking. Problems arise, however, when too many gatherers are after too few blackberries. Since no one owns the berries, no one in particular is likely to keep you from taking too many from the commons. Fishermen have traditionally taken a similar approach as did our hypothetical wild blackberry picker. When specific species of fish did become economically nonviable, the skipper simply moved on to other species or new hunting grounds assuming that the fish had moved elsewhere. Now there simply are no more places to which the fish can flee. The fishing industry has gone global. While the nations of West Africa, for example, may not themselves have the investment capital to overfish their waters, the member nations of the European Union, Japan, South Korea, and Taiwan are more than willing to pay them bargain prices for the right to fish the waters of poorer nations now that their own national waters are depleted (Roberts 2007: 328).

Even the deep seas, once thought to be of no commercial value, are now being trawled. As the fish species of shallower waters go into decline, deep sea fishing suddenly becomes more attractive. Deep sea trawling, the marine equivalent of clearcutting, occurs when fishing crews drag nets across the seamounts where fish gather to feed and spawn. Complex ecosystems formed over decades or even centuries on layers of coral reefs are dredged up along with fish (Roberts 2007: 295). The process alters ecosystems greatly as deep sea trawlers can reach over a mile below the surface (Roberts 2007: 299). Deep sea trawling has already reduced fish populations in the Atlantic to just 20% of their 1970s levels (Clover 2006: 94). One study of coral bycatch on boats searching for orange roughy found that a full metric ton of coral was dragged up to the surface with every 2.25 tons of orange roughy. Harms done to the deep seas are actually far worse since the fish who live below 3,300 feet are more likely to live longer lives, reproduce in smaller numbers later in life, and live in aquatic habitats that simply support fewer fish by their very nature (Clover 2006: 89). It is a bit shocking to think that your dinner may have outlived you several times over, but orange roughy probably live 150 years and take 20 years to reach maturity. Even when orange roughy does mature, it lays tens of thousands of eggs as opposed to the millions that cod would lay. Deep sea fish also tend to have deciduous scales or no scales at all, which make them far more vulnerable to being harmed even if they manage to escape trawling nets. Nets lost in the deep seas also continue to do harm long after their owners have fled. Because the wave currents in the deep seas are so weak, fishing nets stay suspended in the water dooming countless fish to meaningless deaths (Clover 2006: 91–92). While the vulnerability of deep sea species is much higher, the regulations governing what would count as a sustainable yield are difficult to determine because scientists know so little about them. The global fishing industry destroys whole ecosystems faster than scientists can study them. This is especially worrisome since each year bottom trawlers clear an area the size of the entire nation of France. At this rate, it would take the industry only 16 years to destroy every inch of deep sea habitat with bottom trawling if there was no overlap in trawling locations (Roberts 2007: 330). To make matters worse, there is evidence that the takings from trawling have been greatly underestimated. UN FAO data shows that trawlers caught 14 million tons of fish between 1950 and 2015. But researchers found that 25 million tons of fish that were taken in that period were never reported. Overall, 42% of fish extracted by trawling were not reported to the UN FAO (Victorero et al. 2018).

Fisheries have also been able to keep takings high despite their being fewer fish in the sea by using improved technology. Sonar developed after World War II proved invaluable to fishermen. With Doppler radar, GPS, and sea bed mapping technology, ships can sail straight to the fish (Clover 2006: 77–83). Malcolm Clark, a fisheries scientist from New Zealand, notes that, "Our understanding of how to exploit the resource has moved much faster than our ability to manage it" (Clover 2006: 85).

Finally, there is one more reason the fishing industry is hurtling toward its own collapse. When scientists determine what a sustainable catch should be, the industry balks. They complain that they can see that there are still plenty of fish in the sea and that the scientists are imposing real hardships on human beings because of hypothetical risks to mere fish. Fishermen vote. Fish don't. So governments will often ignore the warnings of scientists. It also doesn't help matters that most fisheries ministers have positions designed to encourage the development of the industry understood in the short terms of election cycles, rather than looking toward long-term environmental and economic sustainability.

Fisheries Collapse: The Posterchild for How Globalization Undermines Social Justice

Proponents of the global economy argue that globalization brings increased global access to the best technology and the ability to increase efficiency by producing goods and services where they can be procured most cheaply. Both of these measures should help poor communities to participate in global economic growth. But it is precisely these two ostensive benefits of globalization that are in many ways responsible for the collapse of global fisheries. As access to fishing technology improved and ships extended their reach into relatively unfished waters, the global catch rose to unsustainable levels. By fishing in the few areas which had once been sheltered from the impacts of extensive fishing, the industry unwittingly undermined its own future while simultaneously depriving many poor people of fish that serve as a dietary staple. And the cycle repeats itself. While improvements in technology in other sectors often lead to prices going down, as, for instance, in the production of chicken, beef, and pork, investments in technology for catching fish just make prices go up (Clover 2006: 39). FAO reports that fish prices hit a record high in 2018 and are expected to increase (FAO 2020). This is because technology is used to increase the amount of the product available in other areas of agriculture. But in fishing, technology simply helps one to overfish wild stocks faster, resulting in more and more fish that are becoming increasingly rare being brought to market.

Poor communities are especially hard hit by the crash of fish stocks. According to the United Nations FAO, fish account for over one third of animal protein consumption in just 34 nations. Of those 34 countries, 18 are classified as Low Income Food Deficit Countries, and five are considered Small Island Developing Nations (UNFAO 2020). Unlike fishing trawlers which can simply take off for another part of the sea, poor families have no such recourse. Jamaica, for instance, is now so overfished that Jamaicans have developed a recipe for fish tea, made by boiling a juvenile fish, and then removing fins, bones, and scales so that only the protein packed broth remains. Under better fish stock conditions, these tiny young fish would never be harvested (Roberts 2007: 235). Kiribati in the Pacific Ocean, for instance, gets just 5% of the landed value of tuna caught in its waters by the Japanese. Not only are the Kiribati stripped of their resources at bargain prices, the locals are forced to turn to hunting wild animals of the forest to get enough food to eat. This strips the forests and brings people into closer contact with chimpanzees which may carry HIV and Ebola (Roberts 2007: 329). And yet while the stocks of fisheries off the coast of Africa have gone down 50% since 1945 when industrial exploitation began, the EU is still spending $227 million a year on access to distant waters in Kiribati and elsewhere (Clover 2006: 44, 47).

A lack of strong global governance in international waters exacerbates the problem as it is often unclear who, if anyone, is responsible for ensuring that international treaties on fishing limits are enforced. The practice of flying under flags of convenience prevents the efforts of any one nation from being entirely effective. To circumvent local fisheries' management regulations, one need only change flags to a nation that does not want to or simply cannot afford to police one's activities. Flags of convenience are problematic on several counts. They encourage fishing operations not only to duck taxes and conservation requirements, but to neglect labor, safety, and training concerns as well. Moreover, according to the United Nations Convention on the Law of the Sea, a ship is subject only to the jurisdiction of the flag state if it is no more than 200 miles from shore and therefore in national waters (Clover 2006: 147). So, a Canadian fisherwoman may watch cod swim past her only to be caught by Portuguese ships just a few miles outside Canadian national waters. The Canadian government ensures that the Canadian would face stiff fines if she fishes for the depleted cod within Canadian waters. But local officials generally cannot board or inspect ships outside their waters. Even when they do report an illegal activity, the ship's home nation may well refuse to impose any fines. In one especially egregious case, Canadians inspected the Portuguese ship Solstico in 2003 finding that 65% of the fish on the ship were under a moratorium and they had satellite documentation that the ship was fishing in banned locations. But when Solstico came back to port in Portugal, the Portuguese found no infractions (Clover 2006: 171–172). The lack of a global initiative to keep fishing sustainable is a further problem. Fish don't just stay put. Even when nations like the United States established fishing limits in their own waters, other nations simply increase their fish kill to compensate. So an endangered fish might swim past American ships only to be killed later in Japanese waters and then sent back to United States' markets. This takes away the motivation for those in the fishing industry to follow any rules since someone else will simply catch the fish and make the money if the virtuous skipper refuses to do so.

How Can the Fish Business Ever Be Sustainable?

Given how dire the situation is for the world's fish and how systematically corrupt and ill designed the global fishing industry and its regulators are, it may seem impossible to imagine that there could ever be a sustainable business approach to fish. But it is precisely because the global fishing industry is so deeply characteristic of the challenges that global economic expansion entails that it is so important that the world finds a way to behave more sustainably. If we cannot save the fish and thereby the people and ecosystems that are supported by them, we will likely have little luck facing the slew of problems that globalization and industrialization more generally have left us. Anyone who enjoys the benefits of the global economy should be morally compelled to own up to its flaws and find a way to do better.

When it comes to industry finding ways to improve sustainability in its operations, the solutions run from the mundane to measures that seem drastically counterintuitive to many in business. The most obvious place to start is with efficiency, but efficiency understood in terms of the triple bottom line. Efficiency with an eye toward environmental sustainability is always a significant and easy first step when dealing with the destruction of the environment. Fishing ships can begin by reducing the amount of bycatch. FAO estimates that 35% of wild caught fish are wasted or dumped (FAO 2020). That means that over 29 million tons of fish are pulled from the sea each year and never even taken back to port. This does not even count the aquatic life damaged by gear that never makes it onboard (Clover 2006: 73). By investing in more selective gear and better practices, a great deal of damage can be averted. Deep sea trawling, given how utterly destructive it is, should be abandoned altogether. For fishing communities that are not under the control of large corporate-funded ships, a multitude of solutions can be found simply by devoting resources to education and finding more sustainable fishing methods appropriate to indigenous populations of both humans and fish. Since poor populations in many places would not have overfished waters if not for the technologically boosted European ships off their shores, it is incumbent upon those who profit from fishing in the seas of poorer nations to recognize a collective duty to invest in community measures that ensure future fishing both for the poorer natives and only when feasible for their richer visitors.

In fact, there have already been a number of efforts to find ways in which the fishing industry and the environment can peacefully co-exist. Unilever, a fish buyer which owns Birdseye and Knorr, actually developed an alliance with the World Wildlife Federation (WWF) to certify fisheries. This organization later came to be known as the Marine Stewardship Council (MSC). The MSC is now independent of both WWF and Unilever, but its blue eco-symbol appears on fish from hundreds of certified fisheries. MSC reports that 14% of all wild caught fish are from fisheries that they have been working with (MSC 2020). Although Unilever did not meet its 1996 pledge to source all of its fish sustainably by 2005, they were able to ensure that 56% of their fish came from sources which Unilever believed to be sustainable, 49% of which was MSC-certified. The MSC requires that companies score 80% or better on three key principles: (1) the company cannot exploit the fish population, (2) takings must be restricted enough to maintain the fish population, and (3) the company must respect local fisheries requirements (Clover 2006: 287–288). Whole Foods, Target, Walmart, and others use the MSC as a guideline in stocking their wild caught and frozen fish (Clover 2006: 296). Although the MSC is a step in the right direction, the MSC rewards effort even if the company is not yet at fully sustainable best practice. Some worry that the MSC should not certify fish that are caught in areas where regulatory enforcement is lax (Clover 2006: 288). Consumers and businesses will have to decide for themselves if a commitment to improvement is sufficient or whether tighter standards such as those provided by the Monterey Bay Aquarium's Seafood Watch, the Marine Conservation Society's Good Fish Project, the Blue Ocean Institute, or Friend of the Sea better fit their own values and expectations. Even though Unilever's consumers may not have been clamoring for sustainability certifications, responsible companies can use their advertising budgets to let consumers know that their products, unlike those of their competitors, are sustainably sourced. Advertising which disseminates knowledge of how the fish got to the consumers' plate can help to create and fuel demand for sustainable business practices.

One thing these certifications do well is to recognize that merely farming fish is not always a perfect or even adequate solution. Like intensive animal agriculture on land, intensive aquaculture in the seas is not without its problems. Both leave us facing the problem of what to do with mass quantities of waste concentrated in a single location. Domesticated stocks which escape to interbreed with wild stocks will also likely undermine the quality of wild stocks when they are already vulnerable. Domesticated animals also must be fed a steady dose of antibiotics and anti-parasite drugs because the conditions of their lives are so unhealthy. This raises issues for wild stocks that may be negatively impacted, for humans left to deal with pollution, and, as with land issues, there are issues of whether or not it is fair to the fish to keep them in conditions so detrimental to their health. Finally, it is also important to bear in mind that fish farming is generally only profitable with carnivorous fish. The feed for carnivorous fish usually comes from the seas, which simply encourages the industry to overfish every other species left that wasn't palatable to people. Moreover, there are health issues associated with concentrating toxins as they move up the food chain into carnivorous fish (Clover 2006: 253–269).

In terms of setting limits to catch, it is also not the best solution. Limiting days at sea, for instance, only works for a brief period of time until technology catches up. Ultimately, limiting days at sea encourages less responsible fishing since skippers will likely quickly catch all they can and be less likely to take only the highest quality fish. Iceland's Individual Transferrable Quota system is somewhat better since it is based upon takings, but it is too problematic insofar as very poor nations would still be tempted to sell their rights too quickly. Community ownership is likely necessary to ensure that the poorest of the poor can have some hope of managing for their communal long-term best interests. Clover, incidentally, also endorses this position although he is perhaps more optimistic about ITQs and the abilities of private property to work for rather than against sustainability (Clover 2006: 250). While those who have ITQs are more likely to police illegal takings, they are also more likely to feel entitled to catch whatever they want whenever they want. But fish are wild creatures who no more belong to the fishing industry than to any other individual human. They are our global heritage and are of value in terms of biodiversity in ways that mere property ownership would never guarantee. I argue that it is better to think of ITQs, not as private property, which has its own checkered past when it comes to environmental issues, but rather in terms of a percentage of allowable takings where the sustainable taking is set by best science in conjunction with the precautionary principle, not an actual share in the fish population, whatever that population may be.

Just as sweatshop labor practices in the garment industry cannot be mitigated until buyers take responsibility for their subcontractors, so too the fishing industry will likely never become sustainable until fish buyers demand it and companies find ways to market sustainable fish-sourcing practices effectively. Currently, while McDonald's fish fillet actually meets MSC standards, the fish served at five star restaurants often does not (Clover 2006: 281, 186). Culinary innovators need to take the lead in seeing the value added to a dining experience that can leave one with a clear conscience and not just a cleared plate. Rather than renaming fish or refusing to acknowledge which species is actually being served, chefs have a moral duty to let diners know what they are really getting. Companies from all sectors can ensure that their catered events only serve sustainably sourced fish. Food processors must do the same. Endangered yellowfin tuna is routinely packaged in "dolphin safe" cans of tuna (Clover 2006: 208). By not making it clear exactly which species are in the can, and instead leaving the information vague by demarcating only between light and dark tuna, the industry encourages unsustainable fishing. While many would blanch at the idea of eating an endangered rhino, many of us regularly eat endangered fish species unwittingly. Knowledge and transparency will be key to morally appropriate corporate practice. The entire network of corporate power that helps to prop up unsustainable fishing industries can be brought to bear upon the fishing ships and processors to ensure transparency and sustainability. Until purchasers, both individuals and corporations, demand it, the value of sustainably sourced fish will not be able to win out in the marketplace.

Not all solutions to the ills of the global fishing industry are quite so obvious as simply relying on more efficient ways to fish or on transparency. Paradoxically for those in industry, the fishing industry can benefit itself and other industries simply by *not* fishing. While it seems antithetical to standard business production goals, the industry must learn to value not producing – to value the work it is not doing – if the industry is to remain viable. Only by giving the seas a rest can the seas continue to do the work of supporting fish in the long term. Much like an athlete who can only reach top performance levels if she gives herself adequate rest, the fishing industry must learn to give the seas a rest in order to remain profitable in the long run. Luckily, marine-protected areas have been growing at a rate of approximately 8% annually since 1960, although many are still ill funded and/or ill managed (Worm 2018).

Marine sanctuaries, places where fishing is forbidden, can be economically beneficial on a number of counts. First, there are the obvious tourism benefits. Part of saving the fish entails that people become better able to understand them. Just as we have learned that people objectify animals trapped in cages at zoos and see them as mere objects detained for their enjoyment, so too we need to realize that aquariums will likely be inadequate for the global change in attitudes that will be necessary if their populations are ever to recover from the harms already done. People can better learn to appreciate fish and the importance of harvesting them sustainably (if they are to be harvested at all) by viewing them as the wild creatures they are. In a marine preserve, snorklers and passengers of glass-bottomed boats can marvel at the abundance of fish below the water's surface. We should not underestimate the financial opportunities the seas afford beyond the dinner table. Goat Island Marine Reserve in Leigh, New Zealand, for instance, was established for purely scientific purposes, but the busy tourist-filled beaches are proof that it has become much more (Clover 2006: 253). Visitors to marine reserves will also see how much larger fish can be when they are healthy. Hopefully, greater knowledge of fish will encourage consumers to demand more sustainable fishing practices if they do choose to eat fish. Not only does Goat Island increase tourism, it provides valuable knowledge about what a truly sustainable yield might look like. Rather than using already depleted stocks as a model, scientists can see what aquatic habitats look like with minimal human interference. If scientists can provide a better idea of what a sustainable yield would be, the fishing industry will be more likely to have a future too. Marine reserves also provide the safe harbor fish need to replenish stocks. On the borders of Goat Island Marine Reserve, for instance, the lobster catch is booming (Clover 2006: 255). While it is important to value biodiversity for its own sake, making that a practical reality will often require finding ways to make biodiversity economically desirable. Marine reserves can do just that if only they are given the chance. Companies in the fishing industry that want to secure their own financial futures would do well to collectively support reserves both monetarily and with their advertising dollars (although importantly with no strings attached).

Some companies object that far from having a moral and fiduciary duty to invest in reserves, they are actually entitled to a government payout for leaving marine reserves unfished (Clover 2006: 265). This argument, however, holds little moral weight. The fishing industry is already the primary long-term beneficiary of the reserve. It is unfair to expect government to pay the industry off for not fishing in national waters that belong to the public or in international waters, which while they do not belong to any particular nation state, do not belong to industry either. It is especially problematic, moreover, to subsidize any further an industry which uses subsidies primarily to drive the industry itself into extinction by producing far too many fishing ships which then results in fisheries collapse. Finally, since it is in many ways the industry itself which has caused the problem, it is preposterous to insist that the industry should be paid to desist. In the case of national waters, this would be akin to expecting a landlord to pay off a tenant who willfully destroyed the property, while the landlord fixed the damage for him, rather than charging the tenant for the damages and insisting that he remove himself from the premises indefinitely. In terms of how many reserves would be needed, Dr. Bill Balantine from the Goat Island Mare Reserve suggest that we would need to set aside 10% as reserves to achieve educational and research objectives, 20% for species conservation, 30% to ensure sustainable fishing, and 50% if the seas are to withstand the negative effects of intensive industrial fishing practices (Clover 2006: 262–263).

But reserves are not the only creative innovation that could help save the world's fish. Provided that they were well regulated and enforced, economic measures could be taken to better value unfished areas. The North Atlantic Salmon Fund, for example, asked owners of river fishing in Europe and North America to come together to buy out the commercial fishing capacity. The buyouts resulted in the River Tweed having its best fishing year since the 1960s. By buying out the commercial fisheries, the North Atlantic Salmon Fund got Northeast drift nets removed which helped the fish population to grow (Clover 2006: 243). While the latter option may only work in cases where fishing is regulated well enough to make legal rights to the catch enforceable, it could be yet another creative way to help fisheries avoid collapse.

In sum, the state of the world's fish is dire and we simply cannot afford to ignore it. The challenge of sustainable fishing should not be underestimated, but neither should it be dismissed as impossible. There are a number of measures businesses can and ought to take even in a context of lax government enforcement and scientific uncertainty. Neither businesses nor consumers are entitled to shirk their moral duties to fish by claiming that improvement is impossible until governments institute global reforms. By investing in efficiency for sustainability, by working with certification organizations to empower customers to support responsible fishing, by investing in the future of the industry and the planet by setting aside reserves, and by taking a more creative approach to the role of fish in the marketplace on a number of fronts, companies can do a great deal to meet their moral duties of environmental stewardship while still meeting their other fiduciary duties. In the long run, this is the only way the fishing industry could ever meet its fiduciary duties much less its duties of stewardship. Precisely because the problems with the global fishing industry are a sort of hyper case for the moral issues raised by globalized industries everywhere, it is especially important that business can find ways to rise up to meet the task at hand. The future of the industry and of the ecosystems which are our lifeblood depends upon it.

Related Topics

2. Eating
5. Species and Wildlife
6. Wild Animals
17. Ocean Policy
19. The Ethics of Marine Protected Areas
63. Environmental Justice

Note

1 An earlier version of this chapter originally appeared in *Between the Species* in 2011 with the title "Fishing for a Sustainable Future" 13 (9), Article 6. The chapter has been updated and modified substantially since first publication.

References

Borelle, S.B. et al. (2017) "Why We Need an International Agreement on Marine Plastic Pollution," *Proceedings of the National Academy of Sciences of the United States*, 114 (38) 9994–9997, viewed 2 April 2019, <https://doi.org/10.1073/pnas.1714450114>.

Clover, C. (2006) *The End of the Line: How Overfishing Is Changing the World and What We Eat*, New York: The New Press.

Dahlke, F. et al. (2020) "Thermal Bottlenecks in the Life Cycle Define Climate Vulnerability of Fish," *Science*, July, viewed 3 July 2020, <https://science.sciencemag.org/content/369/6499/65>.

Diaz, S., et al., eds. (2019) *Report of the Plenary of the Intergovernmental Science-Policy Platform on Biodiversity and Ecosystem Services on the Work of its Seventh Session: Summary for Policymakers*, Bonn, Germany: IPBES Secretariat, viewed 7 July 2020, <https://ipbes.net/sites/default/files/ipbes_7_10_add.1_en_1.pdf>.

EllenMacArthurFoundation.(2016) "NewPlasticsEconomyReportOffersBlueprinttoDesignaCircularFuture forPlastics," TheEllenMacArthurFoundation,viewed25June2020 <https://www.ellenmacarthurfoundation.org/news/new-plastics-economy-report-offers-blueprint-to-design-a-circular-future-for-plastics>.

Food and Agriculture Organization of the United Nations (FAO). (2020) *The State of World Fisheries and Aquaculture: Sustainability in Action*, Rome, Italy: Food and Agriculture Organization, March, viewed 6 July 2020, <http://www.fao.org/3/ca9231en/CA9231EN.pdf>.

Kituyi, M. and P. Thomson. (2018) "90% of Fish Stocks Are Used Up – Fisheries Subsidies Must Stop," *United Nations Conference on Trade and Development*, 13 July, viewed 2 July 2020, <https://unctad.org/en/pages/newsdetails.aspx?OriginalVersionID=1812>.

Laville, S. (2019) "Dumped Fishing Gear Is Biggest Plastic Polluter in Ocean, Finds Report," *The Guardian*, 5 November 2019, viewed 7 July 2020, <https://www.theguardian.com/environment/2019/nov/06/dumped-fishing-gear-is-biggest-plastic-polluter-in-ocean-finds-report>.

MSC.ORG. (2006) "USA Leads the Way with Eco-friendly Fish," *Marine Stewardship Council*, 6 December, viewed 29 August 2009, <http://www.msc.org/newsroom/msc-news/archive-2006/usa-leads-the-way-with-eco-friendly-fish/?searchterm=best%20market>.

MSC.ORG. (2020) "Our Collective Impact," MSC.org, viewed 7 July 2020, <https://www.msc.org/what-we-are-doing/our-collective-impact>.

Pauly, D., et al. (2002) "Towards Sustainability in World Fisheries," *Nature* 418, August, viewed 6 July 2020, <http://depts.washington.edu/donbevan/2003/speakers/punt1.pdf>.

Roberts, C. (2007) *The Unnatural History of the Sea*. Washington, DC: Island Press/Shearwater Books.

Steger, M. B. (2013) *Globalization: A Very Short Introduction*. New York: Oxford University Press.

Victorero, L., et al. (2018) "Out of Sight, But Within Reach: A Global History of Bottom-Trawled Deep-Sea Fisheries From >400 m Depth," *Frontiers in Marine Science*, 11 April 2018, viewed 7 July 2020, <https://www.frontiersin.org/articles/10.3389/fmars.2018.00098/full>.

Worm, B. (2018) "Ecological Change in the Oceans and the Role of Fisheries," in International Ocean Institute-Canada (ed.), *The Future of Ocean Governance and Capacity Development*, Canada: Brill, pp. 232–232, viewed 7 July 2020, <https://www.researchgate.net/publication/333344153_Ecological_Change_in_the_Oceans_and_the_Role_of_Fisheries>.

Worm, B., et al. (2006) "Impacts of Biodiversity Loss on Ocean Ecosystem Services," *Science* 314 (5800), pp. 787–790.

Worm, B., et al. (2009) "Rebuilding Global Fisheries," *Science* 325 (5940), pp. 578–585.

Further Reading

Engel Jr., M. (2020) "Fishy Reasoning and the Ethics of Eating," *Between the Species* 23 (1), pp. 52–103, viewed 6 July 2020, <https://digitalcommons.calpoly.edu/bts/vol23/iss1/3>. (An ethical argument against fishing and eating fish given fish suffering and intelligence.)

Hilborn, R. and U. Hilborn. (2019) *A Sustainable Future for Global Fisheries*. Oxford: Oxford University Press. (A book by a fisheries scientist and his partner with a more optimistic view of both the current state of global fish stocks and the potential for sustainable management. It should be noted that although the authors' work has won several awards, they have also been criticized for not fully disclosing industry financial support.)

Roberts, C. (2013) *The Ocean of Life: The Fate of Man and the Sea*. New York: Penguin. (A more general account of the human relationship to oceans both historically and at present, especially as it relates to climate, pollution, invasive species, and overfishing.)

19

THE ETHICS OF MARINE PROTECTED AREAS

Nathan J. Bennett and Kai M. A. Chan

In the end, decisions concerning preservation and use of biodiversity will turn on our values and ways of moral reasoning...

(E.O. Wilson 1993, The diversity of life, 37)

Introduction

In a reframing of the words of the poet W.H. Auden, National Geographic explorer in residence Sylvia Earle suggests that "Thousands have lived without love, not one without [the ocean]" (Earle 2010). Humans simply cannot survive without the services produced by the ocean. The ocean provides oxygen, medicine, and food while regulating the climate and currents. Humans rely on the oceans for materials and physical sustenance but also for knowledge and recreation, for emotional and spiritual renewal, and intellectual and artistic inspiration. Seascapes and the ocean's depths have been a place of magic, mystery, and meaning for humanity throughout time. However, for decades, there have been clear signs that the health and productivity of the ocean have been declining and that the assets that make our lives possible are being drawn down (Allsopp et al. 2009; Holland & Pugh 2010; Longhurst 2010; Sumaila & Pauly 2011). For example, many fisheries have declined to mere fractions of historic levels, often forcing commercial fisheries into deeper and more remote areas. Larger fish and apex predators, such as sharks, have all but disappeared from their historic ecological roles. Important habitats, such as coral reefs, mangroves, and seagrass, are being destroyed and degraded. Large "dead zones", areas that are overloaded with nutrients, oxygen-deficient, and cannot support life, have appeared in the oceans. Other significant threats to the marine environment include climate change, invasive species and disease, pollution, ocean acidification, and coastal development.

There have been plenty of calls for a broader "ocean ethic" or "sea ethic" that would lead to changes in governance structures and management of the ocean and its resources (Miller & Kirk 1992; Dallmeyer 2003; Moore & Russell 2009). Along with fisheries regulations (e.g. catch quotas) and anti-pollution measures (e.g. International Convention for the Prevention of Pollution from Ships), marine-protected areas (MPAs) are one marine management tool and international policy initiative that is an extension of local, national, and global ethics of concern and stewardship for the marine environment. According to the International Union for the Conservation of Nature (IUCN), protected areas (PAs; terrestrial or marine) are "A clearly defined geographical space, recognized, dedicated and managed, through legal or other effective means, to achieve the long-term conservation of nature with associated ecosystem

DOI: 10.4324/9781315768090-23

services and cultural values" (Dudley 2008, pp.8–9). Importantly, this definition places central emphasis on "conservation of nature" or ecocentric values and secondary emphasis on anthropocentric values through "associated ecosystem services and cultural values". However, this definition of PA also creates a tension or constructive ambiguity about whether the central purpose of conservation is for nature (ecocentric values) or for humans (anthropocentric values). The definition of MPA provided by the IUCN – "Any area of the intertidal or subtidal terrain, together with its overlying water and associated flora, fauna, historical and cultural features, which has been reserved by law or other effective means to protect part or all of the enclosed environment" (Kelleher & Kenchington 1992, p.7) – is also somewhat ambiguous about the ultimate purpose of MPAs and what marine environmental values they are intended to protect. Yet, international commitments were made through the Convention on Biological Diversity to increase the areal coverage of MPAs to 10% of ecosystems and oceans by 2020 (CBD 2010) and according to MPA Atlas, there are more than 15,000 MPAs (http://www.mpatlas.org) covering 5.3% of the world's oceans (Marine Conservation Institute 2020).

MPAs, like all environmental policy instruments, are the extension of an ethic – or a sense of what constitutes right or good treatment of both the environment and/or humans. Far from neutral, MPAs are social constructions that are imbued with different societal values and can be seen as good or bad depending on the questions that are asked and the ethical lens that is applied. Most agree that the marine environment needs to be protected, but "why", "for whom", and "how" are questions of considerable debate and discussion. Marine environmental policies have been based largely on collective understandings of the end values or benefits that society hopes to achieve (i.e. consequentialist or utilitarian ethics), partly on our understanding of right or just processes or outcomes (i.e. deontological ethics), and less so on reflections on what it means to be a good or virtuous person or policy (i.e. virtue ethics) (Moore & Russell 2009). See Table 19.1 for a conceptual clarification on these three predominant ethical schools of thought, which we have interpreted liberally. We also assume here, but wish to state explicitly, that all three schools of thought and related concepts and questions can serve as guides for actions taken toward both human and non-human organisms. This chapter will explore the ethical seascape of MPA policy and practice through reflecting on the following three questions: (1) What desired values or potential benefits are assigned to the creation of MPAs?; (2) are the processes and outcomes of MPAs fair and just for ecological and human communities?; and (3) what constitutes right or good practice in the planning and management of MPAs?

Table 19.1 Ethical schools of thought and implications for MPA policy (after Moore and Russell 2009)

Ethical Theory	Locus of Morality	Basis of Judgment	Questions for MPA Policy
Consequentialist or utilitarian ethics	Consequences or outcomes	Maximization of desired outcomes and minimization of undesired outcomes	What are the assigned values or desired outcomes? How can the outcomes be defined and measured?
Deontological ethics	Process, policy and outcomes	Justice, equity, rights, duties	Are the process and outcomes fair and just? Are interests and rights respected?
Virtue ethics	Character or intentions	Virtues, characteristics, integrity	Does the policy embody appropriate virtues?

Utility: Of What Benefit Are Marine-Protected Areas

The list of ecological and social values assigned to MPAs – often referred to as potential benefits in conservation and MPA literatures – has grown steadily over time and been subject to both convergence and continued debate. We will build toward a summary of the potential "benefits for nature" and "benefits for humans" that have been eschewed by proponents of MPAs later in the chapter. In the following text, we examine how the ecological and social benefits of MPAs have been articulated by advocates and scientists, as well as in policy documents since the concept has gained traction internationally. We suggest that several factors have caused the list of assigned values to grow and change along with corresponding shifts in MPA policy and practice. In particular, new ecological understandings and concepts (e.g. outcomes associated with closed areas; concepts of ecosystems; and biodiversity), changing environmental contexts (e.g. climate change) and societal values (e.g. toward the ocean), and awareness of the social consequences of MPA processes and outcomes (e.g. justice, equity, and rights) have led to a constant rearticulation of the ecological and social values assigned to MPAs.

From the early 20th century onward, Western conceptualizations of MPAs have mirrored the visions and wilderness values being applied to terrestrial conservation initiatives (Sloan 2002). Terrestrial PAs, such as Yellowstone National Park, were created as "wilderness" areas that should be devoid of human populations (Dearden & Bennett 2015). This preservationist ethic, which focused on the aesthetic, emotional, and spiritual values of "wild" nature, was articulated by the likes of John Muir (Wolf 2009). Globally, PAs often took on these Western elitist anthropocentric values as their principle mandates – focusing primarily on the protection of nature for nonconsumptive values, including knowledge (science and education) and the enjoyment (tourism and recreation) of current and future generations. Similar to their terrestrial counterparts, the wilderness ideal pervaded MPA thinking – leading to the creation of reserves where natural values were in the service of (some) current and future humans – for aesthetics, for knowledge, and for recreation. This popular early vision of MPAs is well characterized by a line from *The Forest and the Sea* which states, "There are plans for the new National Park in the Virgin Island to include reef areas where... stringent restrictions will be enforced.... There should be places reserved for the fascinating pastime of fish-watching" (Bates 1960, p.76). Similarly, in Thailand, decisions regarding MPA locations were made in an ad hoc fashion with primary consideration given to protecting beautiful beaches and islands, along with surrounding waters as well as coral reefs, from fishing and development for the benefit of non-consumptive uses and users (World Heritage Nomination Document 2010). Current objectives of Thai MPAs still retain similar objectives (Table 19.2).

The potential ecological benefits associated with MPAs have changed over time as a result of (1) ongoing scientific learning about their ecological potential, (2) novel and popular ecological concepts and management paradigms, and (3) emerging environmental problems. From early in the 20th century, MPAs were promoted as "fish nurseries" and fishery-replenishment zones. Modern Western awareness of this benefit of MPAs came, in part, from the observation that fisheries stocks had recovered in closed areas of the North Sea after World War II (Claudet 2011, Holm 2012, Kittinger 2013). In the 1950s, the US National Park Service started to promote

Table 19.2 Objectives of Thai MPAs (as cited in Christie and Ole-Moyoi 2011)

To preserve and maintain the ecosystem integrity, biodiversity, and scenic beauty for use by the present and future generations without compromising them;

To provide the general public as a ground for education and research;

To provide the general public the opportunities for nature tourism and recreation, which are compatible with the park ecosystem and its carrying capacity.

"marine sanctuaries" that were both "wilderness" areas for non-consumptive values and "nursery grounds" for fish (Wallis 1959). The first broad international endorsement for MPAs came at the 1962 World Parks Congress in Seattle where Resolution 15 stated "…preservation of unspoiled marine habitat is needed for ethical and aesthetic reasons, for the protection of rare species, for the replenishment of stocks and valuable food species and for the provision of undisturbed areas for scientific research" (as cited in Sloan 2002, p.297). Here, the primary ecological benefits of MPAs are habitats, rare species, and food species. Also during the 1960s, scientists realized that Glacier Bay National Monument (created in 1925 and notably the first MPA in the world) in Alaska was a feeding area for several species of whales. Soon thereafter, Mexico was the first to site an MPA specifically for cetaceans. Laguna Ojo de Liebre was established in 1972 to protect gray whale mating and calving grounds (Hoyt 2011).

By 1987, the Resolution of the 17th Assembly of IUCN advocated for the creation of a global system of MPAs to protect examples of marine and estuarine systems for viability and genetic diversity, rare or endangered species and critical habitats for these species, and areas of importance to economically important species (Kelleher & Kenchington 1992, Appendix 2). It was not until 1992 and 1995, with the publication of *Guidelines for Establishing Marine Protected Areas* and *A Global Representative System of Marine Protected Areas*, respectively, that holistic management paradigms such as ecosystem-based management and new concepts such as sustainability and biodiversity also entered the lexicon of international MPA policy (Kelleher & Kenchington 1992; Kelleher et al. 1995). These policy documents and the ecological values contained therein have largely set the direction for MPA planning globally since. Increasingly, MPAs have also been seen as a precautionary tool to mitigate against threats from a multitude of outside stressors and to support adaptation to climate change (Dudley et al. 2009). Throughout these and other policy documents, the focus appears to be primarily on humans as the central beneficiaries of values from natural resources. As Salm and Clark (2000, p.2) state, "we believe that the principal goal of all MPAs is conservation of resources so they yield the greatest benefit to present generations without losing their potential to meet the needs and ambitions of future generations". It is notable that this thinking and indeed wording mirror the notion of sustainable development as championed in the Brundtland report, *Our Common Future* (WCED 1987).

Articulating the diversity of environmental values of MPAs is one thing; determining "which areas will protect these values?", "what size of area is the most effective?", and "how much is enough to protect these values?" has been a significant and ongoing challenge. The desire to maximize different ecological values has required more advanced science and more complicated planning processes than the ad hoc approach taken to achieve earlier "wilderness" values. Holistic scientific concepts such as "ecosystem" have also required MPA system planners and managers to value the system in its entirety rather than just component parts – such as isolated areas, critical habitats, rare or valuable species, or individual human impacts. Only integrated management regimes and networks of connected MPAs, it is argued, will protect ecosystem composition, structure, functioning, resilience, and the component parts (Roberts et al. 2001; Halpern et al. 2010; Rice & Houston 2011). Protecting biodiversity through MPAs requires safeguarding the full spectrum of ecosystems, species, and genetics – including representative examples of coral reefs, mangroves, seagrass and other habitats, critical habitats, breeding areas and migration bottlenecks, areas of endemism, areas of high productivity, particularly vulnerable species and habitats (Kelleher et al. 1995).

Ensuring representation and ecological connectivity has led to the planning of MPA networks at national, regional, and global scales. Creating regional and global systems plans has entailed advancing the science of MPAs, articulating definitions and measurements for new concepts such as biodiversity and ecosystem services, and developing new tools (e.g. Marxan) for mapping ecological and socio-economic values, making trade-offs and situating MPAs to achieve conservation values. Marxan and Marxan with Zones, open source computer software for spatial conservation planning developed by the University of Queensland (Marxan 2013), could be seen as an extension of a utilitarian or deontological ethic depending on how it is applied. It can enable efficient achievement of conservation and/or fisheries benefits while minimizing overall socio-economic costs, or it can enable the consideration of equitable distribution of benefits and costs.

Equity and Rights: Are Marine-Protected Areas Just?

Our paper thus far suggests that the potential benefits of MPAs are many – education, recreation, knowledge, aesthetic, spiritual, and ecological – and that the values assigned to MPAs have been primarily articulated through anthropocentric and utilitarian ethics. But this is not the whole story. A utilitarian accounting of the overall societal benefits or a cost-benefit analysis might simply suggest that wilderness and no-take MPAs provide the most benefit and ultimately happiness for the most people (through protecting marine resources and associated benefits to humans) and, therefore, that MPAs are a worthwhile environmental policy initiative. However, utilitarian ethics and practice do not account for the just and equitable distribution of benefits across scales, times, and groups, for rights to a fair process, or for the rights of nature or non-human organisms.

Wells (1992) suggested that the net benefits of PAs are more likely to accrue to people who are geographically distant, while local populations are the ones who bear the costs of conservation. This is an issue of scale. Indeed, there have been numerous and longstanding critiques of the negative impacts of terrestrial conservation on local communities and indigenous people (West & Brechin 1991; West et al. 2006; Dowie 2009). MPAs have also been the subject of critical examination. Previous studies have shown that MPAs can result in loss of livelihoods, increased poverty, decreased food security, forced migration, loss of tenure, increased conflict, and exacerbated vulnerabilities for local populations (see review in Bennett and Dearden 2014). Wilderness MPAs, created for knowledge and recreational values while excluding local fishers, might put primacy on the values of those with power and money more than those without (Faasen & Watts 2007). For example, in Cayos Cochinos MPAs in Honduras, local populations were excluded from areas that they historically used, while tourists could access these same areas (Brondo & Woods 2007). Some groups tend to be more impacted than others during the creation of MPAs and, if tourism is developed, the benefits often accrue disproportionately to certain groups, such as men or elites (Oracion et al. 2005; Walker & Robinson 2009). Finally, the top-down approach to creating MPAs has often meant that local voices and opinions have been excluded from MPA planning and implementation as well as ongoing governance and management processes (Jones 2009; Singleton 2009; Bavinck & Vivekanandan 2011; Sowman & Sunde 2018).

Though the benefits of MPAs for local resource users and communities may be just as numerous and as likely (Leisher et al. 2007; Mascia & Claus 2009; Bennett & Dearden 2014; Gill et al. 2019), the potential negative impacts of conservation on local populations raise serious challenges to deontological ethics – which encourages us to question whether the "means justify the ends". Other concerns to address include: Who should benefit from conservation?; are conservation outcomes defensible when local people are negatively impacted in the process?; when is it right that one group benefits if other groups experience losses?; and who has the right and responsibility to plan, establish, and manage MPAs?. One might argue that ethical consistency precludes the application of policy instruments whose negative social or economic side effects for local populations are as harmful and as plausible as the potential ecological benefits.

The aforementioned critiques and questions regarding social and environmental justice, equity, and the rights of local people have resulted in several important shifts in MPA and broader conservation thinking, policy, and practice. The first change is that the MPA mandate has been broadened to include ensuring that socio-economic benefits accrue to local populations (Pomeroy et al. 2004; Cinner et al. 2005; White et al. 2006; ICSF 2009; Samonte et al. 2010). This shift has led to the development of numerous methods for monitoring social and economic impacts (Bunce et al. 2000; Christie 2004; Gjertsen 2005; Leisher et al. 2007; Schreckenberg et al. 2010; Ban et al. 2019). Second, there has been increasing attention to alternative and community-based models of MPAs – such as traditional closures and Locally Managed Marine Areas (Govan et al. 2009; Orbach & Karrer 2010). In many contexts, these non-Western models of MPAs have had a long history of application and effectiveness but have often been lost as a result of top-down and centralized fisheries management and governance regimes (Johannes 1978, 1981; McClenachan & Kittinger 2013). Third, international conservation organizations and policies have increasingly recognized the importance of considering local rights and tenure, local participation in planning, governance and management, protection and inclusion of indigenous knowledge and culture, and sharing of benefits through alternative livelihood programs and compensation schemes (Borrini-Feyerabend et al. 2004; Borrini-Feyerabend et al. 2007; Cattermoul et al. 2008; CBD 2010; Bennett et al. 2017). Moreover, the clear international policy response has been that the benefits of and responsibility for conservation should be shared in a just and equitable manner. This stance is taken up in the Convention on Biological Diversity's Aichi Target 11 which states that terrestrial and marine PAs need to be "effectively and equitably managed" both with and for the benefit of all people (CBD 2010).

Alongside these policy shifts, there has been an increasing focus on the incorporation of multiple social and ecological values into MPA planning (Murray 2005; Ban et al. 2012; Rojas-Nazar et al. 2012) and widespread support for the creation of multiple use MPAs with a sustainable development mandate (Agardy 1993; Noel & Weigel 2007; Diegues 2008). These approaches are emblematic of a relativist approach to applying environmental ethics – the implications of a relativist view are that all values of all people need to be considered in policy decisions or that MPAs should be "everything to everyone". However, the consideration of people's rights and socio-economic needs first and foremost may also undermine the effectiveness of MPAs at producing marine environmental conservation outcomes – whether for anthropocentric or ecocentric purposes. After a talk at the Third International Marine Protected Areas Congress in Marseille, France, an audience member asked a panelist: "Everyone participated and you considered all opinions but is the MPA actually protecting anything?" That is, should positive environmental outcomes of MPAs be a necessary outcome or one subject to great compromise via social inclusion?

Engaging with an environmental deontological ethic also obliges us to consider the rights of humans to environmental benefits as well as the intrinsic rights of nature. Creating MPAs that do not protect environmental services or not creating MPAs might also cause harm to local populations. For example, local people require ample fish for livelihoods and subsistence, and communities rely on healthy mangroves for protection from erosion and storms. Both environmental outcomes may require conservation actions to be taken. Importantly, "no-take" MPAs that do not allow fishing have been shown to be the most effective at protecting important environmental values (Lester & Halpern 2008; Lester et al. 2009); however, when communities are present and resources lacking, MPAs that are designed to meet community goals may be more effective (McClanahan et al. 2006).

The rights and just treatment of other non-human organisms must also be considered (Chan & Satterfield 2013). To date, there has been limited direct engagement with questions regarding the rights of nature in MPA mandates and policy such as: Do species and ecosystems have moral standing? or do fish have as much of a right to live as we do?; what should we aim to protect? or are whales more important to protect than nudibranchs?; do we have an obligation to return populations and ecosystems to historic levels?; and what consideration should be given to invasive species? If both nature and humans are considered to have rights, this brings us also to the difficult question of whether and how moral obligations to nature and local populations can be traded off against one another. In some contexts, the "win-win" conservation and development ideal may not be possible and difficult trade-offs may be required to balance anthropocentric and ecocentric values.

Two other challenging but critical questions are: What is ultimately good for nature? and who should speak for nature? In the first point, it is easier to identify what benefits organisms, less so for species, and somewhat harder for ecosystems. Persuading policy makers and stakeholders that we have moral responsibility to each of these levels of biological organization, and therefore that they deserve legal consideration beyond their utility for people, is a demanding task (Chan 2011). It may also be essential that we agree on who has the responsibility and who has the right to speak for nature – governments, local communities, indigenous groups, stakeholders, industry, the general public, scientists, or NGOs? At different scales, MPA network planning, siting, implementation, and management processes formalize who gets a voice and, ultimately, whether and how *nature* is articulated and valued.

Integrity: Do Marine-Protected Areas Embody Appropriate Virtues?

Engaging with virtues-based ethics demands that we interrogate the integrity of MPA policy and perhaps even the virtues that conservation organizations or practitioners themselves embody. There are numerous virtues that might enable us to make ecologically and socially ethical choices, including reverence, patience, respect, humility, reflexivity, fairness, reason, moderation, justice, caution, commitment, awareness, and flexibility. Perhaps the most direct engagement with virtues is contained in the literature on co-management and "good" governance for PAs (Borrini-Feyerabend et al. 2007; Lockwood 2010) – which is slowly finding its way into MPA policy documents. Similar to virtues ethics (and also deontological ethics), literatures on good governance suggest that there are normative commitments about what is right and wrong that should guide policies and actions. For example, often suggested norms of so-called "good" governance include participatory, lawful, effective, efficient, accountable, transparent, responsive, and equitable. Increasingly national MPA policy documents contain terminology that echoes these norms. For example, the *National Framework for Canada's Network of Marine Protected Areas* contains guiding principles which endorse coherency, respect, openness, transparency, and inclusiveness (Government of Canada 2011). In a recently published paper titled "An appeal for a code of conduct for marine conservation", Bennett et al. (2017) reviewed and proposed a much broader set of principles that ought to guide the creation of MPAs.

How Can Ethics Guide Marine-Protected Area Policy and Practice

Applied environmental ethics seeks to create prescriptions for environmental issues and management problems. However, there are several clear barriers to making environmental ethics practicable for MPAs. First, in MPA policy and practice, environmental protection is often considered merely as a means to an anthropocentric end. Over time, the assigned values (or potential social and ecological benefits) associated with MPAs have multiplied and converged toward a long list that is summarized in Table 19.3. Briefly, the potential benefits of MPAs to humans include direct fisheries and non-fisheries benefits and indirect management, knowledge/education, and cultural benefits. The benefits of MPAs to nature include process, ecosystem, population, and species benefits (Angulo-Valdés & Hatcher 2010). Yet, the ecological benefits of MPAs are mostly framed simply as resources for the use of current and future generations. For example, the rationale to protect coral reefs is often for enjoyment, shoreline protection, genetic resources, and fisheries benefits for humans. However, non-human nature (individuals, species, and ecosystems) also has a non-instrumental value (Jax et al. 2013) – i.e. worth that does not stem from benefit to humans – and we have a moral responsibility to non-human organisms that *may* be sentient or conscious (Chan 2011).

Table 19.3 Summary of potential benefits of MPAs for humans and nature. Note that many social benefits might feed back to enhanced ecological protection, and that many ecological benefits can yield social benefits (e.g. as supporting ecosystem services)

Potential Benefits for Humans (Assigned Social Values)	Potential Benefits for Nature (Assigned Ecological Values)
• Support protection and management of cultural and historical heritage	• Maintain and restore ecosystem structure, functions and integrity
• Maintain places for human recreation and enjoyment	• Maintain trophic structure and food webs
• Safeguard ecosystems for scientific research and knowledge creation	• Protect of vulnerable habitats
• Support education and environmental awareness	• Protect of biodiversity – genetic, species, ecosystem
• Provide spiritual fulfillment and intellectual inspiration	• Conserve rare, endemic or long-lived species
• Improve fisheries yields and increase food security	• Protection of biogeographic and geologic features
• Support local and equitable social and economic development (alternative livelihoods, poverty, health, infrastructure)	• Conserve representative, unique and vulnerable habitats
• Increase social capital and social resilience	• Protect spawning areas and fish harvest refugia
• Promote or reinvigorate rights, tenure and access for local populations	• Maintain key areas for critical life stages (reproductive, nursery, feeding) and migration bottlenecks
• Improve participation in governance processes and structures	• Sustain and restore populations
• Support sustainable development and use of resources	• Ensure MPAs are designed and managed as a network
• Avoid more destructive forms of development	• Help return ecosystems to historic levels
• Safeguard traditional uses and values	• Provide examples of controlled habitats for ecological restoration
• Provide places for educating visitors about heritage, history, and culture	• Maintain the resilience of habitats and ecosystems to climate change
• Maintain nature-based adaptation against climate change	• Support larval export and spillover of adult species
• Reduce conflicts between user groups	• Restore biomass, density, richness, and size of species
• Safeguard benefits for future generations	• Protect from coastal erosion
	• Allow for transformation and sequestration of pollutants

A practicable consequentialist or utilitarian ethic for MPAs could benefit from a holistic consideration of humans and non-humans simultaneously but differentially. For example, humans are an integral part of complex and interconnected resource and food webs (Moore & Russell 2009). Though not superior from a purely ecological standpoint, humans are unique in terms of their ability to fundamentally alter natural systems and processes, to develop culture and histories, and to express values and make choices about their actions. Thus, anthropocentric and intrinsic ecocentric values might lead to more efficacious outcomes if considered in tandem since neither set of values is replaceable and, furthermore, they are mutually constitutive. Adopting a holistic "ecological ethic" may better account for ecocentric values and better protect human objectives in the long run (Curry 2011).

Second, in environmental ethics, consequentialist, deontological, and virtue-based ethics are often treated as competitors. We contend that it may be necessary to attend to all three types of ethics simultaneously during environmental decision-making. Historically, MPA policy and practice has engaged thoroughly in utilitarian or consequentialist ethics through defining and re-defining the social and ecological benefits associated with MPAs. MPA practitioners and policy have also engaged with questions of justice and rights – but only to a limited extent through focusing on equity and rights for local communities and certain groups. Engaging with deontological ethics suggests that attention must be given to the rights of local populations and to the just distribution of benefits across time (past, present, and future generations) and space (proximity to MPA) and to different groups (gender, class, race, etc.). For some, this engagement with human rights is threatening to the conservation mandate of MPAs; to others, it is fundamental to the long-term success of conservation. Consideration should also be given to the rights of non-humans, including the component parts (individuals, species, habitats) as well as entire ecosystems since the parts support the whole and vice versa. The health of living things requires attention to non-living elements (water, climate, minerals, etc.). Yet, discussions of the rights of nature or non-human organisms have been largely avoided in MPA thinking. As discussed above, questions of virtue and integrity have not emerged to the same extent in MPA practice and policy discussions – but can be seen in recent engagement with norms and principles of "good" governance.

The concept of relational values ("preferences, principles, and virtues associated with relationships, both interpersonal and as articulated by policies and social norms") might help to resolve some of the tensions between different schools of thought (Chan et al. 2016). Many instrumental values and the perception of intrinsic values stem from relationships, where the objects of valuation (e.g. coral reefs, a particular coral reef) are not substitutable, even in principle. In such contexts, relational values are present (Muraca 2011; Jax et al. 2013; Chan et al. 2018; Himes & Muraca 2018), and recognizing these may help to enliven and contextualize intrinsic and instrumental values. As Chan et al. (2016) argue, sacred sites are not sacred due to the instrumental benefits of their protection or for reasons independent of human valuation, but rather because of longstanding relationships between people and a place. In this context, benefits need not be seen as purely instrumental, and a good life (through the idea of eudaimonia) can be seen as more than simply the satisfaction of preferences but also living in accordance with moral principles (Kaltenborn et al. 2017; Knippenberg et al. 2018). In recognition of this promise of relational values to provide concepts and language for relating human and ecological concerns, the UN's Intergovernmental Science-Policy Platform on Biodiversity and Ecosystem Services (IPBES) has identified relational values alongside intrinsic and instrumental values as central considerations in the pursuit of human and non-human thriving (Díaz et al. 2015; Pascual et al. 2017; Chan et al. 2019).

Third, environmental ethics may not be able to provide us with clear decision rules for questions such as "What percentage of the ocean should be protected in MPAs?" Ethical considerations will likely not be specifically prescriptive – i.e. they do not tell us directly how many MPAs to create, how to plan and manage MPAs ethically, or who should bear the costs and responsibilities for MPAs (Balmford & Whitten 2003; Balmford et al. 2004). However, attending to questions of values, rights, and integrity regarding humans and non-humans can provide us with clear constraints, thus avoiding decisions that are inappropriate or unacceptable. Doing so might help us to formulate best practices in MPA governance, planning, siting, implementations and ongoing management. Several key considerations – phrased as questions – to be taken into account at each of these stages are outlined in Table 19.4. Moreover, successfully achieving balanced and beneficial social and ecological outcomes from MPAs is and always will be challenging: it requires good governance to safeguard rights and ensure a due process, thorough planning and effective management to realize desired ecological objectives, and compensatory mechanisms or alternative livelihoods to ensure equitable outcomes for local people. Best practice documents and monitoring and evaluation guides for each of these topics can be found elsewhere (Kelleher & Kenchington 1992; Salm et al. 2000; Pomeroy et al. 2004; Sobel & Dahlgren 2004; Cattermoul et al. 2008; Lockwood 2010; Bennett & Dearden 2012).

Table 19.4 Key ethical considerations in MPA governance, planning, and management

Governance

- Are governance institutions and processes participatory, inclusive, lawful, efficient, accountable, transparent, adaptive, and equitable?
- Are the rights (and/or interests) of nature and rights to nature taken into account in law and policy?
- What levels of biological organization are taken into account?
- Do governors and organizations embody appropriate virtues in their design, structure, and practice?

Setting targets

- Are conservation targets adequate to protect marine environmental values?
- Do conservation targets align with societal values for marine conservation?
- Given the risks associated with insufficient conservation targets, do targets reflect a precautionary approach and the kind of stewardship that a given jurisdiction aspires to provide?

Network Planning

- Does the MPA network protect representative samples of habitats and biodiversity in connected networks of MPAs?
- How can environmental and societal benefits be maximized and social costs be minimized?
- How do we make trade-offs between social and ecological values and current and future benefit?
- How do we respect the needs of local resource users while recognizing the ocean as a common good?

Implementation

- Is management capacity and financing sufficient to support effective management?
- Are equity and rights taken into account through creating appropriate livelihood interventions and compensation schemes and considering local views, governance structures, knowledge and tenure?
- What model of MPA – e.g. community-based vs national, no-take refugia vs multiple use MPA – is most appropriate to the social context?
- How can implementation balance procedural (means) and substantive justice (ends)?
- What rights or moral norms are implicitly codified in the implementation process?

Ongoing Management

- Who should be responsible for managing the area?
- Does management consider desired social and ecological values?
- Are management actions effectively producing ecological outcomes?
- Are precautions being taken to manage and mitigate against threats in the broader seascape?
- Are mechanisms in place to ensure social and economic benefits are equitably distributed?
- Do managers embody appropriate virtues?

Conclusion

In this chapter, we have explored the implications of three ethical schools of thought – consequentialist or utilitarian, deontological, and virtues-based ethics – for MPA policy and practice. Historically, MPA advocates and policy documents have (1) drawn significantly on consequentialist or utilitarian ethics through continually rearticulating the benefits of MPAs to humans and nature, (2) engaged with deontological ethics in a limited fashion with a focus on equity and rights for local resource users and communities, and (3) largely avoided engaging with virtues-based ethics except through emerging discussions of "good" governance. We suggest that future MPA policy and practice could draw on all three ethical schools of thought to avoid inappropriate actions and undesired outcomes and to formulate best practices in formulating governance, setting targets, network planning, implementation, and ongoing management. We contend also that for MPAs to be considered "good", advocates and managers need to be equally attentive to both human and non-human components of these linked social-ecological systems – in other words to adopt a holistic ecological ethic. Yet, there will always be difficult ethical questions and decisions to make regarding the conservation of the oceans and the application of MPAs in different contexts. Continued engagement with these ethical schools of thought and the questions they engender can allow MPA advocates, policy makers and practitioners to refine intentions, clarify policies, ameliorate actions, and improve the outcomes of MPAs.

Additional Reading

- Claudet, J., 2011. *Marine Protected Areas: A Multidisciplinary Approach*, Oxford: Cambridge University Press.
- Dallmeyer, D.G., 2003. *Values at Sea: Ethics for the Marine Environment*, Athens: University of Georgia Press.
- Earle, S., 2010. *The World is Blue: How Our Fate and the Ocean's are One*, Washington, DC: National Geographic.
- Roberts, C., 2007. *The Unnatural History of the Sea*, Washington, DC: Island Press.
- Salm, R.V., Clark, J.R., & Siirila, E., 2000. *Marine and Coastal Protected Areas: A Guide for Planners and Managers*, Gland, Switzerland: IUCN.

Related Topics

10. Wilderness
11. National Parks
17. Ocean Policy
47. Rewilding
56. Cost-Benefit Analysis
58. Precautionary Principles
59. Adaptive Management
62. Community Participation
63. Environmental Justice

References

Agardy, T., 1993. Accommodating ecotourism in multiple use planning of coastal and marine protected areas. *Ocean & Coastal Management*, 20(3), 219–239.

Allsopp, M. et al., 2009. *State of the World's Oceans*, New York: Springer.

Angulo-Valdés, J.A., & Hatcher, B.G., 2010. A new typology of benefits derived from marine protected areas. *Marine Policy*, 34(3), 635–644.

Balmford, A. et al., 2004. The worldwide costs of marine protected areas. *Proceedings of the National Academy of Sciences of the United States of America*, 101(26), 9694–9697.

Balmford, A., & Whitten, T., 2003. Who should pay for tropical conservation, and how could the costs be met? *Oryx*, 37(02), 238–250.

Ban, N.C. et al., 2012. Recasting shortfalls of marine protected areas as opportunities through adaptive management. *Aquatic Conservation: Marine and Freshwater Ecosystems*, 22(2), 262–271.

Ban, N.C., Gurney, G.G., Marshall, N.A., Whitney, C.K., Mills, M., Gelcich, S., Bennett, N.J., Meehan, M.C., Butler, C., Ban, S., Tran, T.C., Cox, M.E., & Breslow, S.J., 2019. Well-being outcomes of marine protected areas. *Nature Sustainability*, 2(6), 524. Available at: https://doi.org/10.1038/s41893-019-0306-2

Bates, M., 1960. *The Forest and the Sea: A Look at the Economy of Nature and the Ecology of Man*, New York: Vintage Books.

Bavinck, M., & Vivekanandan, V., 2011. Conservation, conflict and the governance of fisher wellbeing: Analysis of the establishment of the Gulf of Mannar National Park and Biosphere Reserve. *Environmental Management*, 47(4), 593–602.

Bennett, N., & Dearden, P., 2012. *From Outcomes to Inputs: What Is Required to Achieve the Ecological and Socio-Economic potential of Marine Protected Areas?* Victoria, BC: Marine Protected Areas Research Group, University of Victoria. Available at: http://dspace.library.uvic.ca:8080/handle/1828/4511 [Accessed May 1, 2013].

Bennett, N.J., & Dearden, P., 2014. Why local people do not support conservation: Community perceptions of marine protected area livelihood impacts, governance and management in Thailand. *Marine Policy*, 44, 107–116.

Bennett, N.J., Teh, L., Ota, Y., Christie, P., Ayers, A., Day, J.C., Franks, P., Gill, D., Gruby, R.L., Kittinger, J.N., Koehn, J.Z., Lewis, N., Parks, J., Vierros, M., Whitty, T.S., Wilhelm, A., Wright, K., Aburto, J.A., Finkbeiner, E.M., … Satterfield, T., 2017. An appeal for a code of conduct for marine conservation. *Marine Policy*, 81, 411–418. Available at: https://doi.org/10.1016/j.marpol.2017.03.035

Borrini-Feyerabend, G. et al., 2007. *Sharing Power: Learning-By-Doing in Co-Management of Natural Resources throughout the World*, London: Earthscan.

Borrini-Feyerabend, G., Kothari, A., & Oviedo, G., 2004. *Indigenous and Local Communities and Protected Areas: Towards Equity and Enhanced Conservation: Guidance on Policy and Practice for Co-managed Protected Areas and Community Conserved Areas*, Gland, Switzerland: IUCN.

Brondo, K.V., & Woods, L., 2007. Garifuna land rights and ecotourism as economic development in Honduras' Cayos Cochinos Marine Protected Area. *Ecological and Environmental Anthropology*, 3(1), 2–18.

Bunce, L. et al., 2000. *Socioeconomic Manual for Coral Reef Management*, Townsville, Australia: Global Coral Reef Monitoring Network, Australian Inst. Marine Science. Available at: http://www.reefbase.org/resource_center/publication/main.aspx?refid=10903 [Accessed October 30, 2010].

Cattermoul, B., Townsley, P., & Campbell, J., 2008. *Sustainable Livelihoods Enhancement and Diversification: A Manual for Practitioners*, Gland, Switzerland: IUCN/CORDIO/ICRAN. Available at: http://www.eldis.org/go/topics/dossiers/livelihoods-connect&id=44789&type=Document [Accessed November 4, 2009].

CBD, 2010. Aichi biodiversity targets. *Convention on Biological Diversity*. Available at: http://www.cbd.int/sp/targets [Accessed March 23, 2013].

Chan, K.M.A., 2011. Ethical extensionism under uncertainty of sentience: Duties to non-human organisms without drawing a line. *Environmental Values*, 20(3), 323–346.

Chan, K.M.A., Agard, J., Liu, J., Aguiar, A.P.D.D., Armenteras, D., Boedhihartono, A.K., Cheung, W.W.L., Hashimoto, S., Pedraza, G.C.H., Hickler, T., Jetzkowitz, J., Kok, M., Murray-Hudson, M., O'Farrell, P., Satterfield, T., Saysel, A.K., Seppelt, R., Strassburg, B., Xue, D., Selomane, O., Balint, L., Mohamed, A., Anderson, P., Barrington-Leigh, C., Beckmann, M., Boyd, D.R., Driscoll, J., Eyster, H., Fetzer, I., Gould, R.K., Gregr, E., Latawiec, A., Lazarova, T., Leclere, D., Muraca, B., Naidoo, R., Olmsted, P., Palomo, I., Singh, G., Sumaila, R., & Tubenchlak, F., 2019. Pathways towards a sustainable future. *Global Assessment Report of the Intergovernmental Science-Policy Platform on Biodiversity and Ecosystem Services*. E. S. Brondízio, J. Settle, S. Díaz and H. Ngo.

Chan, K.M.A., Balvanera, P., Benessaiah, K., Chapman, M., Díaz, S., Gómez-Baggethun, E., Gould, R.K., Hannahs, N., Jax, K., Klain, S.C., Luck, G., Martín-López, B., Muraca, B., Norton, B., Ott, K., Pascual, U., Satterfield, T., Tadaki, M., Taggart, J., & Turner, N.J., 2016. Why protect nature? Rethinking values and the environment. *PNAS*, 113(6), 1462–1465. Available at: http://www.pnas.org/content/113/6/1462.full

Chan, K.M.A., Gould, R.K., & Pascual, U., 2018. Editorial overview: Relational values: What are they, and what's the fuss about? *Current Opinion in Environmental Sustainability*, 35, A1–A7. Available at: http://www.sciencedirect.com/science/article/pii/S1877343518301222

Chan, K.M.A., & Satterfield, T., 2013. Justice, equity and biodiversity. *Encyclopedia of Biodiversity*, 4, 434–441.

Christie, P., 2004. Marine protected areas as biological successes and social failures in Southeast Asia. *American Fisheries Society Symposium*, 42, 155–164.

Christie, P., & Ole-Moyoi, L.K., 2011. *Status of Marine Protected Areas and Fish Refugia in the Bay of Bengal Large Marine Ecosystem*, Phuket, Thailand: BOBLME.

Cinner, J.E., Marnane, M.J., & Mcclanahan, T.R., 2005. Conservation and community benefits from traditional coral reef management at Ahus Island, Papua New Guinea. *Conservation Biology*, 19(6), 1714–1723.

Claudet, J., 2011. Introduction. In J. Claudet, ed. *Marine Protected Areas: A Multidisciplinary Approach*. Cambridge: Cambridge University Press, 1–9.

Curry, P., 2011. *Ecological Ethics*. Cambridge, UK: Polity.

Dallmeyer, D.G., 2003. *Values at Sea: Ethics for the Marine Environment*. Athens: University of Georgia Press.

Dearden, P., & Bennett, N.J., 2015. The role of aboriginal people in protected areas. In P. Dearden & R. Rollins, eds. *Parks and Protected Areas in Canada*. Don Mills, ON: Oxford University Press.

Díaz, S., Demissew, S., Joly, C., Lonsdale, W., Larigauderie, A., Adhikari, J.R., Arico, S., Báldi, A., Bartuska, A., Baste, I.A., Bilgin, A., Brondizio, E., Chan, K.M.A., Figueroa, V.E., Duraiappah, A., Fischer, M., Hill, R., Koetz, T., Leadley, P., Lyver, P., Mace, G.M., Martin-Lopez, B., Okumura, M., Pacheco, D., Pascual, U., Pérez, E.S., Reyers, B., Roth, E., Saito, O., Scholes, R.J., Sharma, N., Tallis, H., Thaman, R., Watson, R., Yahara, T., Hamid, Z.A., Akosim, C., Al-Hafedh, Y., Al-lahverdiyev, R., Amankwah, E., Asah, S.T., Asfaw, Z., Bartus, G., Brooks, L.A., Caillaux, J., Dalle, G., Darnaedi, D., Driver, A., Erpul, G., Escobar-Eyzaguirre, P., Failler, P., M. Fouda, A.M., Fu, B., Gundimeda, H., Hashimoto, S., Homer, F., Lavorel, S., Lichtenstein, G., Mala, W.A., Mandivenyi, W., Matczak, P., Mbizvo, C., Mehrdadi, M., Metzger, J.P., Mikissa, J.B., Moller, H., Mooney, H.A., Mumby, P., Nagendra, H., Nesshover, C., Oteng-Yeboah, A.A., Pataki, G., Roué, M., Rubis, J., Schultz, M., Smith, P., Sumaila, R., Takeuchi, K., Thomas, S., Verma, M., Yeo-Chang, Y., & Zlatanova, D., 2015. The IPBES conceptual framework - connecting nature and people. *Current Opinion in Environmental Sustainability*, 14(June), 1–16. Available at: http://www.sciencedirect.com/science/article/pii/S187734351400116X

Diegues, A.C., 2008. *Marine Protected Areas and Artisinal Fisheries in Brazil*, Chennai: International Collective in Support of Fishworkers. Available at: http://aquacomm.fcla.edu/1565/ [Accessed July 1, 2010].

Dowie, M., 2009. *Conservation Refugees: The Hundred-Year Conflict Between Global Conservation and Native Peoples*, Cambridge, MA: MIT Press.

Dudley, N., 2008. *Guidelines for Applying Protected Area Management Categories*, Gland, Switzerland: IUCN. Available at: http://www.iucn.org/dbtw-wpd/edocs/PAPS-016.pdf [Accessed August 17, 2012].

Dudley, N. et al., 2009. *Natural Solutions: Protected Areas Helping People Cope with Climate Change*, New York and Gland: IUCN-WCPA, TNC, UNDP, WCS, the World Bank, and WWF.

Earle, S., 2010. *The World is Blue: How Our Fate and the Ocean's Are One*, Washington, DC: National Geographic.

Faasen, H., & Watts, S., 2007. Local community reaction to the 'no-take' policy on fishing in the Tsitsikamma National Park, South Africa. *Ecological Economics*, 64(1), 36–46.

Gill, D.A., Cheng, S.H., Glew, L., Aigner, E., Bennett, N.J., & Mascia, M.B., 2019. Social synergies, tradeoffs, and equity in marine conservation impacts. *Annual Review of Environment and Resources*, 44(1), 347–372. Available at: https://doi.org/10.1146/annurev-environ-110718-032344

Gjertsen, H., 2005. Can habitat protection lead to improvements in human well-being? Evidence from marine protected areas in the Philippines. *World Development*, 33(2), 199–217.

Govan, H. et al., 2009. *Status and Potential of Locally-Managed Marine Areas in the South Pacific: Meeting Nature Conservation and Sustainable Livelihood Targets through Wide-Spread Implementation of LMMAs*, SPREP/WWF/WorldFish-Reefbase/CRISP.

Government of Canada, 2011. *National Framework for Canada's Network of Marine Protected Areas*, Ottawa, ON: Fisheries and Oceans Canada. Available at: http://www.dfo-mpo.gc.ca/oceans/publications/dmpaf-eczpm/framework-cadre2011-eng.asp [Accessed August 23, 2012].

Halpern, B.S., Lester, S.E., & McLeod, K.L., 2010. Placing marine protected areas onto the ecosystem-based management seascape. *Proceedings of the National Academy of Sciences*, 107(43), 18312–18317.

Himes, A., & Muraca, B., 2018. Relational values: The key to pluralistic valuation of ecosystem services. *Current Opinion in Environmental Sustainability*, 35, 1–7. Available at: http://www.sciencedirect.com/science/article/pii/S1877343517302634

Holland, G., & Pugh, D., 2010. *Troubled Waters: Ocean Science and Governance*, Cambridge: Cambridge University Press.

Holm, P., 2012. World War II and the "Great Acceleration" of North Atlantic Fisheries. *Global Environment*, 5(10), 66–91. Available at: https://doi.org/10.3197/ge.2012.051005

Hoyt, E., 2011. *Marine Protected Areas for Whales, Dolphins and Porpoises: A World Handbook for Cetacean Habitat Conservation and Planning*, New York: Earthscan.

ICSF, 2009. *Social Dimensions of Marine Protected Areas Implementation in India: Do Fishing Communities Benefit?*, Chennai: International Collective in Support of Fishworkers. Available at: http://aquacomm.fcla.edu/2029/ [Accessed July 15, 2010].

Jax, K. et al., 2013. Ecosystem services and ethics. *Ecological Economics*, 93, 260–268.

Johannes, R.E., 1978. Traditional marine conservation methods in Oceania and their demise. *Annual Review of Ecology and Systematics*, 9(1), 349–364.

Johannes, R.E., 1981. *Words of the Lagoon: Fishing and Marine Lore in the Palau District of Micronesia*. Berkeley: University of California Press.

Jones, P.J.S., 2009. Equity, justice and power issues raised by no-take marine protected area proposals. *Marine Policy*, 33(5), 759–765.

Kaltenborn, B.P., Linnell, J.D.C., Gómez-Baggethun, E., Lindhjem, H., Thomassen, J., & Chan, K.M., 2017. Ecosystem services and cultural values as building blocks for 'the good life'. A case study in the community of Røst, Lofoten Islands, Norway. *Ecological Economics*, 140, 166–176. Available at: http://www.sciencedirect.com/science/article/pii/S0921800916304463

Kelleher, G., Bleakley, C., & Wells, S., 1995. *A Global Representative System of Marine Protected Areas*. Washington, DC: Great Barrier Reef Marine Park Authority, World Bank, IUCN.

Kelleher, G., & Kenchington, R.A., 1992. *Guidelines for Establishing Marine Protected Areas*, IUCN.

Kittinger, J.N., 2013. Human dimensions of small-scale and traditional fisheries in the Asia-Pacific region. *Pacific Science*, 67(3), 315–325. Available at: https://doi.org/10.2984/67.3.1

Knippenberg, L., de Groot, W.T., van den Born, R.J.G., Knights, P., & Muraca, B., 2018. Relational value, partnership, eudaimonia: A review. *Current Opinion in Environmental Sustainability*, 35, 39–45. Available at: http://www.sciencedirect.com/science/article/pii/S1877343517302518

Leisher, C., van Beukering, P., & Scherl, L., 2007. *Nature's Investment Bank: How Marine Protected Areas Contribute to Poverty Reduction*. Gland: The Nature Conservancy/WWF International.

Lester, S.E. et al., 2009. Biological effects within no-take marine reserves: A global synthesis. *Marine Ecology Progress Series*, 384, 33–49.

Lester, S.E., & Halpern, B.S., 2008. Biological responses in marine no-take reserves versus partially protected areas. *Marine Ecology Progress Series*, 367, 49–56.

Lockwood, M., 2010. Good governance for terrestrial protected areas: A framework, principles and performance outcomes. *Journal of Environmental Management*, 91(3), 754–766.

Longhurst, A., 2010. *Mismanagement of Marine Fisheries*. Cambridge: Cambridge University Press.

Marine Conservation Institute, 2020. MPAtlas. Seattle, WA. Available at: www.mpatlas.org

Marxan, 2013. Marxan - informing conservation decisions globally. *The University of Queensland, Australia*. Available at: http://www.uq.edu.au/marxan/index.html?p=1.1.1 [Accessed January 18, 2014].

Mascia, M.B., & Claus, C.A., 2009. A property rights approach to understanding human displacement from protected areas: The case of marine protected areas. *Conservation Biology*, 23(1), 16–23.

McClanahan, T.R. et al., 2006. A comparison of marine protected areas and alternative approaches to coral-reef management. *Current Biology*, 16(14), 1408–1413.

McClenachan, L., & Kittinger, J.N., 2013. Multicentury trends and the sustainability of coral reef fisheries in Hawai'i and Florida. *Fish and Fisheries*, 14(3), 239–255.

Miller, M.L., & Kirk, J., 1992. Marine environmental ethics. *Ocean & Coastal Management*, 17(3–4), 237–251.

Moore, K.D., & Russell, R., 2009. Toward a new ethic for the oceans. In K. McLeod & H. Leslie, eds. *Ecosystem-Based Management for the Oceans*. Washington, DC: Island Press, 325–340.

Muraca, B., 2011. The map of moral significance: A new axiological matrix for environmental ethics. *Environmental Values*, 20(3), 375–396. Available at: http://dx.doi.org/10.3197/096327111X13077055166063

Murray, G.D., 2005. Multifaceted measures of success in two Mexican marine protected areas. *Society & Natural Resources: An International Journal*, 18(10), 889–905.

Noel, J.F., & Weigel, J.Y., 2007. Marine protected areas: From conservation to sustainable development. *International Journal of Sustainable Development*, 10(3), 233–250.

Oracion, E.G., Miller, M.L., & Christie, P., 2005. Marine protected areas for whom? Fisheries, tourism, and solidarity in a Philippine community. *Ocean & Coastal Management*, 48(3–6), 393–410.

Orbach, M., & Karrer, L.B., 2010. *Marine Managed Areas: What, Why, and Where*, Conservation International.

Pascual, U., Balvanera, P., Díaz, S., Pataki, G., Roth, E., Stenseke, M., Watson, R.T., Başak Dessane, E., Islar, M., Kelemen, E., Maris, V., Quaas, M., Subramanian, S.M., Wittmer, H., Adlan, A., Ahn, S., Al-Hafedh, Y.S., Amankwah, E., Asah, S.T., Berry, P., Bilgin, A., Breslow, S.J., Bullock, C., Cáceres, D., Daly-Hassen, H., Figueroa, E., Golden, C.D., Gómez-Baggethun, E., González-Jiménez, D., Houdet, J., Keune, H., Kumar, R., Ma, K., May, P.H., Mead, A., O'Farrell, P., Pandit, R., Pengue, W., Pichis-Madruga, R., Popa, F., Preston, S., Pacheco-Balanza, D., Saarikoski, H., Strassburg, B.B., van den Belt, M., Verma, M., Wickson, F. & Yagi, N., 2017. Valuing nature's contributions to people: The IPBES approach. *Current Opinion in Environmental Sustainability* 26–27, 7–16. Available at: http://www.sciencedirect.com/science/article/pii/S1877343517300040

Pomeroy, R.S., Parks, J.E., & Watson, L.M., 2004. *How Is your MPA Doing?: A Guidebook of Natural and Social Indicators for evaluating Marine Protected Area Management Effectiveness*, Gland, Switzerland: IUCN/WWF.

Rice, J., & Houston, K., 2011. Representativity and networks of Marine Protected Areas. *Aquatic Conservation: Marine and Freshwater Ecosystems*, 21(7), 649–657.

Roberts, C.M. et al., 2001. Designing marine reserve networks: Why small, isolated protected areas are not enough. *Conservation in Practice*, 2(3), 10–17.

Rojas-Nazar, Ú. et al., 2012. Combining information from benthic community analysis and social studies to establish no-take zones within a multiple uses marine protected area. *Aquatic Conservation: Marine and Freshwater Ecosystems*, 22(1), 74–86.

Salm, R.V., Clark, J.R., & Siirila, E., 2000. *Marine and Coastal Protected Areas: A Guide for Planners and Managers*, Gland, Switzerland: IUCN.

Samonte, G., Karrer, L.B., & Orbach, M., 2010. *People and Oceans: Managing Marine Areas for Human Well-Being*, Arlington, VA: Conservation International.

Schreckenberg, K. et al., 2010. *Social Assessment of Conservation Initiatives: A Review of Rapid Methodologies*, London: IIED.

Singleton, S., 2009. Native people and planning for marine protected areas: How "Stakeholder" processes fail to address conflicts in complex, real-world environments. *Coastal Management*, 37(5), 421–440.

Sloan, N.A., 2002. History and application of the wilderness concept in marine conservation. *Conservation Biology*, 16(2), 294–305.

Sobel, J.A., & Dahlgren, C., 2004. *Marine Reserves: A Guide to Science, Design, and Use*, Washington, DC: Island Press.

Sowman, M., & Sunde, J., 2018. Social impacts of marine protected areas in South Africa on coastal fishing communities. *Ocean & Coastal Management*, 157, 168–179. Available at: https://doi.org/10.1016/j.ocecoaman.2018.02.013

Sumaila, U.R., & Pauly, D., 2011. The "March of Folly" in global fisheries. In J.B.C. Jackson, K.E. Alexander, & E. Sala, eds. *Shifting Baselines*. Island Press/Center for Resource Economics, 21–32. Available at: http://www.springerlink.com/content/p2436262334722qt/abstract/ [Accessed July 6, 2012].

Walker, B.L.E., & Robinson, M.A., 2009. Economic development, marine protected areas and gendered access to fishing resources in a Polynesian lagoon. *Gender, Place & Culture: A Journal of Feminist Geography*, 16(4), 467–484.

Wallis, O.L., 1959. Research and interpretation of marine areas of the U.S. National Park Service. In *Proceedings of the Gulf and Caribbean Fisheries Institute*. Gulf and Caribbean Fisheries Institute, 134–138. Available at: http://aquaticcommons.org/13296/ [Accessed January 18, 2014].

WCED, 1987. *Our Common Future*, Oxford; New York: World Commission on Environment and Development/Oxford University Press.

Wells, M., 1992. Biodiversity conservation, affluence and poverty: Mismatched costs and benefits and efforts to remedy them. *Ambio*, 21(3), 237–243.

West, P., Igoe, J., & Brockington, D., 2006. Parks and peoples: The social impact of protected areas. *Annual Review of Anthropology*, 35(1), 251–277.

West, P.C., & Brechin, S.R., 1991. *Resident Peoples and National Parks: Social Dilemmas and Strategies in International Conservation*, Tucson, AZ: University of Arizona Press.

White, A., Aliño, P.M., & Meneses, A., 2006. *Creating and Managing Marine Protected Areas in the Philippines*, Cebu City: Fisheries Improved for Sustainable Harvest Project, Coastal Conservation and Education Foundation and University of the Philippines Marine Science Institute.

Wolf, C., 2009. Values at sea: Ethics for the marine environment. In D.G. Dallmeyer, ed. *Ecosystem-Based Management for the Oceans*. Athens: University of Georgia Press, 19–32.

World Heritage Nomination Document, 2010. *Nomination Document of the Andaman Bioregion of Thailand for UNESCO World Heritage Status*, Bangkok: Ministry of Natural Resources and Environment.

PART IV

Climate

20

MORAL BASES OF RESPONSES TO CLIMATE CHANGE

David R. Morrow

Half a millennium of cod fishing in the Grand Banks came to an end in 1992, when the Canadian government declared a moratorium to protect what remained of a once famous fishery (MacDowell 2012: 22, 291). The collapse of the cod population there reflected a classic tragedy of the commons (Hardin 1968). In the fishing free-for-all off Newfoundland, fishermen had an incentive to catch as much as they could, even though the combined effects of their catch would eventually destroy the fishery—and with it, each person's livelihood and way of life. For almost five centuries, the fishery absorbed humans' intrusions without too much damage. But in the twentieth century, technological innovations pushed the fishery beyond its limits, and the cod population collapsed.

Climate change involves a somewhat similar tragedy of the commons. Our everyday activities lead to the emission of greenhouse gases, such as carbon dioxide (CO_2) and methane. Even though the combined effects of our emissions are beginning to disrupt the world's climate—and with it, the lives and livelihoods of billions of people and countless plants and non-human animals, each person, each country, and each generation has an incentive to continue emitting those greenhouse gases (Gardiner 2011). Because of a cascade of technological innovations and population and economic growth in the last few centuries, our emissions are now pushing the planet beyond its limits.

In at least one way, our predicament is actually worse than that of the cod fishers. Fish reproduce fairly quickly. If the fishing industry had restrained itself, they could have continued fishing the Grand Banks indefinitely. Once emitted, however, CO_2 stays in the atmosphere for a long time. David Archer summarizes the atmospheric lifetime of CO_2 as "300 years, plus 25% that lasts forever [for policy purposes]" (Archer 2005: 5). A bit more precisely, for every ton of CO_2 emitted in a given year, roughly 60% remains after 20 years, roughly 40% remains after a century, and roughly 25% remains after 1,000 years (Joos et al. 2013). Roughly 10% remains at 10,000 years and 7% at 100,000 years (Archer 2005). This means that on the relevant timescale, our emissions will continuously accumulate in the atmosphere. Thus, if we are to prevent "dangerous anthropogenic interference in the climate system," as the United Nations Framework Convention on Climate Change requires, humanity's net CO_2 emissions must eventually go to zero. Practically speaking, then, we must regard the Earth system's capacity to safely absorb CO_2 emissions as a limited and rapidly dwindling resource, not a renewable one like cod.

This chapter examines the moral reasons for tackling this challenge and others associated with climate change. It also surveys the various options for responding and considers how different reasons for responding to climate change would influence what combination of options is best and how they ought to be pursued. While there are also thorny questions about individuals' obligations to reduce their own carbon footprints, this chapter does not address those questions (Chapter 62).

DOI: 10.4324/9781315768090-25

Why Should We Respond to Climate Change?

There are both moral and prudential reasons for responding to climate change.

The most obvious source of our collective obligation to address climate change is the enormous suffering that climate change could impose on current and future generations. The potential causes of this suffering are well rehearsed in the climate ethics literature and well documented by the Intergovernmental Panel on Climate Change (IPCC). For instance, sea-level rise and increased storm surges will flood some communities and inundate others, potentially forcing tens or hundreds of millions of people from their homes or homelands; heavy precipitation and droughts will increase in frequency and intensity, flooding some places and parching others; people will face more, longer, and hotter heatwaves and extreme temperatures; and warming and acidifying ocean waters will damage coral reefs and threaten entire marine ecosystems, as well as the people who depend on them. Some of these changes will arrive sooner than others, while others, such as sea-level rise, will only be fully realized over the course of several millennia (Field et al. 2014).

Different ways of responding to climate change could reduce these changes or the suffering they would cause. The costs of an efficient climate policy would be easily manageable in aggregate. William Nordhaus, a pioneering climate economist, argues that under optimistic assumptions about international participation and efficient policy design, limiting warming to 2℃ would cost somewhere in the neighborhood of "1.5 percent of world income, or about one year's growth in average income" (Nordhaus 2013: 178). Under less optimistic assumptions, however, staying below 2℃ may be prohibitively expensive. Specifically, meeting a much less ambitious target of 3.5℃ warming could cost 3% of global income, and meeting a target of 3.25℃ warming could cost up to 8% of global income (Nordhaus 2013: 176–181). Looking at a wide range of studies, the IPCC estimates that with well-designed policies, the investment required to reduce climate change significantly would slow the growth of global consumption by somewhere between 0.05 and 0.2% over the rest of this century, leaving the world between 3 and 11% poorer in 2100 than it would otherwise be. Like Nordhaus, the IPCC warns that these costs would rise under poorly designed or delayed policies (Clarke et al. 2014: 424–461). Notably, none of these costs incorporate the significant co-benefits of addressing climate change, such as improved air quality. These figures are speculative, of course, but the general lesson is that the world could greatly reduce climate change at manageable costs *if* it promptly builds good institutions and chooses good policies, but that the costs rise sharply as the quality of cooperation and policy declines.

Given that it is technically and economically feasible to prevent many of these harmful changes outright, the threats to people living now (and, depending on one's conception of self-interest, to the immediate descendants of people now living) provide a straightforward prudential justification for addressing climate change (Gardiner & Weisibach 2016: 170–200).

Many of the short- and medium-term hazards would threaten the world's most vulnerable people and that the long-term hazards would threaten future generations. This fact strengthens the moral reasons to reduce the severity of climate change and prepare people to weather the changes that cannot be avoided. The most straightforward argument appeals to consequentialist views of morality—that is, views according to which morality is fundamentally about making the world a better place. By reducing human suffering (among other effects), good climate policies will bring about a much better world than would exist without such policies. For the same or similar reasons, virtue ethics, care ethics, and most other ethical theories also entail a collective obligation to address climate change (Jamieson 2007; Hourdequin 2010; Krakoff 2011; Nolt 2011a; Suikkanen 2014).

If we focus not just on consequences but on obligations and rights—that is, if we adopt a deontological perspective—the picture is slightly more complicated. On the one hand, the widely recognized duties to respect persons and to avoid doing harm generate a collective obligation to mitigate. The obligation to prevent harms is especially pressing, from a deontological perspective, because in causing that harm and violating those rights, those of us responsible for most of humanity's greenhouse gas emissions are committing a grave injustice. Furthermore, imposing such severe harms and risks arguably violates the human rights of those affected, including their rights to life, health, and subsistence (Caney 2010).

Ethicists have debated, however, whether and to what extent this obligation is attenuated by the "non-identity problem." The non-identity problem, popularized by Derek Parfit, concerns the effects of our policies on people in the future whose very existence depends on our policy choices. Some people owe their very existence to some prior policy—for example, because the policy caused their parents to meet. Even if that policy causes them to suffer, it does not make them worse off than they otherwise would have been because without the policy, they would not have existed at all. Since harming people involves making them worse off than they otherwise would have been, the policy does not technically harm them (Parfit 1984).

Ambitious climate policies' social effects would ripple through global society over the course of several generations. Almost everyone in a (distant) future generation would owe their existence to those policies; and so those policies could not technically count as harming them, no matter how much havoc climate change had caused. Most commentators hold that the non-identity problem does not undermine the obligation to mitigate, though they disagree about exactly how to understand and resolve the problem in the case of climate change (FitzPatrick 2007; Davidson 2008; Attfield 2011; Gardiner 2011; Nolt 2011b; Broome 2012; Hartzell-Nichols 2013b; Kelleher 2015; Berkey 2017). Regardless of how that controversy is resolved, unmitigated climate change would harm a large number of people who already exist or whose future existence does not depend on choices about mitigation policy.

Many people see a further moral reason for responding to climate change, beyond the risks and injustices that human-caused climate change imposes on human beings: the damages that it is doing and will do to the rest of the world, including non-human animals, plants, and ecosystems (Jamieson 1992; Sandler 2014). If the non-human world is of direct moral concern, then those damages provide additional reasons to mitigate climate change, beyond the harms that humans would suffer because of them. The strength of these additional reasons is contested in the case of climate change (Palmer 2011) and, at any rate, they rest on controversial claims about the moral value of the non-human world. But the harm to the non-human world is still of concern from a purely human-centered perspective, and so provides an important reason to respond to climate change.

Straddling all of these concerns is an appeal to precaution. The precautionary approach in climate ethics emphasizes that in addition to all of the effects discussed above, climate change carries an unacceptably high chance of catastrophic outcomes, which humanity is morally obligated to try to avoid. The kinds of harms and injustices underlying this logic include mass extinctions, such as have happened only a handful of times in the history of the planet (and always because of natural causes); unexpectedly large amounts of warming, ushering in conditions not seen for tens of millions of years at unprecedented speeds; tipping points in various Earth systems that might trigger large, irreversible changes; and so on. Climate science cannot yet rule out these possibilities. Some philosophers argue that, in light of these risks, continuing with business as usual would be unconscionably reckless (Gardiner 2006; McKinnon 2009; Munthe 2011; Hartzell-Nichols 2013a). Even those who take a more traditional risk management approach, as opposed to a precautionary approach, warn that the threat of such low-probability, high-damage outcomes ought to influence climate policy (Wagner & Weitzman 2015: 48–79; Nordhaus 2013: 216–217).

In summary, the reasons for responding to climate change rest on the fact that climate change would cause significant suffering; the claim that the infliction of such suffering would constitute a grave injustice and would violate many people's human rights, including the world's most vulnerable people; the fact that climate change would significantly harm the non-human world; and the claim that the risk of catastrophe renders a failure to respond unconscionably reckless.

How Should We Respond to Climate Change?

What are our options for responding to this predicament?

The single most important response is to reduce our greenhouse gas emissions. Unless we do so, any other response will eventually become futile. The process of reducing our net greenhouse gas emissions is known as *mitigation* (Chapter 22), because it will mitigate human-caused climate change—that is, make it less severe. Mitigation is typically divided into emissions reductions, sometimes called emissions abatement, and sink enhancement, which involves accelerating the natural processes that remove greenhouse gases from the atmosphere.

Another important kind of response is *adaptation*, which involves improving human and non-human systems' capacity to withstand climatic changes. Adaptation ought to be seen as a supplement to mitigation rather than an alternative. Given that our past emissions have already set climate change in motion, and that humanity is set to emit still more over the coming decades, some amount of adaptation is already required. But given the severity of the risks created by un-mitigated climate change, adaptation is not a plausible standalone solution to the problem.

Other responses may also play some role in a portfolio of climate responses. The large-scale removal of greenhouse gases, when applied specifically to CO_2, is called *carbon dioxide removal* (CDR). The technologies for doing this are known as negative emissions technologies. A more controversial possible response is known as solar radiation management or *solar geoengineering*, which would involve reflecting a fraction of incoming sunlight back into space before it can warm the planet. A final kind of response is *rectification*, which includes attempts to redress climate-induced harms through monetary payments for "loss and damage" or other kinds of reparations. These, too, are best seen as supplements to mitigation rather than replacements.

The various options for responding to climate change are depicted in Table 20.1. A complete climate policy portfolio will combine some or all of these options, but it will have mitigation at its core.

Mitigation

In terms of mitigation, the basic task before us is clear: Morality demands that humanity reduce its net emissions of carbon dioxide to almost nothing. That is, morality requires that, eventually, humanity emit no more CO_2 each year than is absorbed by human activities such as reforestation or CDR. Given the limits and costs of sink enhancements and CDR, this means reducing CO_2 emissions to a mere fraction of their current level.

Humanity's collective emissions depend on the global population, per capita consumption, and the carbon intensity of consumption (i.e. how much carbon is emitted for each unit of consumption). Slowing or reversing population growth would help (Cafaro 2011, 2012). So would slowing or reversing the growth of consumption, though there are strong moral reasons not to slow the growth of consumption among the global poor. But so long as there are people consuming goods and services, the only hope for adequate mitigation is to decarbonize the world economy.

Table 20.1 Options for responding to climate change (based on Heyward 2013)

Mitigation	CDR	Adaptation	Solar Geoengineering	Rectification
Lessen the severity of climate change by reducing net GHG emissions	Remove CO_2 from the atmosphere	Adjust to climate change to reduce harm to human and natural systems	Reflect sunlight into space to cool the planet	Compensate people for loss and damage from climate change

Filling in the details of the obligation to decarbonize requires answering two questions: How quickly should we eliminate our greenhouse gas emissions? And how should the burdens (and co-benefits) of those emission reductions be distributed, both within and across generations? Because different means of eliminating emissions impose different costs on different people, these questions are intertwined. Analytically, though, we can take them, in turn.

Two approaches dominate the debate over how quickly to eliminate our emissions: a "risk management" approach and an "optimal pathway" approach.

The "risk management" approach sets some upper limit on the amount of risk we are (or ought to be) willing to bear. This is usually defined in terms of a maximum acceptable probability of crossing some threshold of "dangerous" climate change—though what counts as "dangerous" arguably depends on context, including the benefits derived from the putatively dangerous activity (Moellendorf 2014). Capping that acceptable risk implies a cap on cumulative emissions, which implies an "emissions budget" or "carbon budget" representing the amount of CO_2 humanity could collectively emit before reaching that cap. Emissions beyond that cap would be justifiable only if humanity could later remove them through negative emissions technologies, as assumed in many of the scenarios considered by the IPCC.

For instance, suppose that we ought to tolerate no more than a 33% chance of global average temperature rising by 2°C or more, relative to the preindustrial average. This roughly corresponds to the stated goal of the Paris Agreement under the United Nations Framework Convention on Climate Change. Depending on the details of exactly how one specifies the threshold, how one incorporates the effects of other greenhouse gases, and so on, this implies a carbon budget of somewhere between 590 and 1,240 billion tons of CO_2 from 2015 onward. That compares to just over 2,000 billion tons of CO_2 emitted since the beginning of the industrial age (Rogelj et al. 2016). (Note that budgets and emissions figures are often given in billions of tons of *carbon*, as opposed to carbon dioxide. Measured in terms of carbon, humanity has collectively emitted roughly 550 billion tons of carbon and can only emit 160–335 billion more tons to retain a one-third chance of keeping below the 2°C target.) To put that in context, as of 2015, humanity emitted approximately 40 billion tons of CO_2 annually, which would exhaust that carbon budget in a few decades.

Settling on a particular size for the carbon budget, however, does not determine the morally optimal rate at which to consume it. For a fairly small carbon budget, however, that question may be irrelevant: given the social and political difficulties in cutting emissions, we are likely to exhaust any plausible carbon budget before we completely decarbonize the economy. Thus, decarbonizing as quickly as is socially and politically feasible is likely to get us as close as possible to the morally optimal path as we can manage.

The dominant theoretical alternative to capping cumulative emissions is to mitigate at the economically optimal rate, meaning that in any given year, humanity reduces its emissions to the point where the marginal cost of abatement equals the marginal social cost of carbon. The marginal cost of abatement is the cost of eliminating or avoiding one more ton of emissions, assuming that humanity always chooses the lowest-cost option for eliminating that next ton. In the near term, the marginal costs of emissions abatement would be low—and sometimes even negative—as people pick low-hanging fruit such as increasing energy efficiency, replacing extremely dirty fuels or facilities that harm public health (e.g. through air pollution from coal-fired power plants), and so on. On the other side of the equation, the social cost of a ton of carbon is the sum of the costs imposed by a ton of carbon emissions, stretching indefinitely into the future. These costs include everything from higher utility bills during hotter summers to increased suffering from tropical diseases to the disappearance of island nations beneath rising seas. (Calculating and combining these costs are, of course, philosophically and technically difficult.) The *marginal* social cost of carbon, relative to any given level of emissions, is the sum of the costs created by emitting *one more* ton of carbon. At the point when the marginal cost of emissions abatement equals the marginal social cost of carbon, the costs of eliminating the next ton of emissions would exceed the costs created by that ton of emissions.

Just how aggressive the optimal mitigation pathway would be depends on seemingly technical details in the calculation of the social cost of carbon. Of particular interest from an ethical perspective, it depends on how to compare costs and benefits that accrue now with those that accrue in the future. Since the costs and benefits of climate policy stretch into the distant future, such technical details have a significant effect on economically optimal mitigation pathways. Economists use a conceptual tool known as the social discount rate to compare costs borne today with costs borne in the future. (To "discount" some future harm is to be willing to pay less money today to avoid that harm, much as a store is willing to take less money to part with discounted merchandise. It is to attach less value to the harm, much as the store is attaching less value to the merchandise.) If we discount future costs fairly steeply—if, say, we count 1 million dollars of damage in 2100 as equivalent to 10,000 dollars of damage today—then the marginal social cost of carbon is currently quite low. Hence, we ought (now) to pursue only inexpensive options for reducing our emissions. If we discount future costs at a very low rate, that same million dollars of damage in 2100 might be equivalent to half a million dollars of damage today, yielding a much higher marginal social cost of carbon and thereby justifying very aggressive mitigation. Prominent economists have drawn very different conclusions about the optimal mitigation pathway, largely because of differences in comparing costs and benefits across time (Stern 2007; Nordhaus 2008). The morality of discounting, especially in the context of climate change, has inspired vigorous debate (Cowen & Parfit 1992; Groom et al. 2005; Stern 2007; Nordhaus 2008; Caney 2009; Gardiner 2011; Broome 2012; Kelleher 2012; Moellendorf 2014).

Different moral reasons for mitigation tend to favor different approaches to designing mitigation policy. An exclusively consequentialist focus on human suffering, ecosystem destruction, reduced consumption, or other ills tends to favor the "optimal pathway" approach of balancing marginal abatement costs with the marginal social cost of carbon. Precaution could play a role here, as well, by adjusting the calculation of the social cost of carbon to better account for low-probability, high-damage climate catastrophes (Wagner & Weitzman 2015). (Nordhaus, long critical of ambitious near-term mitigation policies, has somewhat warmed to them after incorporating such possibilities into his analysis [Nordhaus 2013: 212–217].) Theoretically, this approach would maximize human welfare over time—or at least, it would maximize whatever is being measured in calculating the social cost of carbon, which would undoubtedly omit crucial elements of human welfare and morally important impacts on the non-human world. In practice, taking the optimal pathway approach generally means implementing a tax on carbon emissions, with the tax pegged to the social cost of carbon, since this explicitly ties mitigation to an estimate of the negative consequences of emissions.

Placing greater weight on the injustice that climate change could inflict tends to favor a risk management approach, as does focusing on the vices or lack of care manifested by greenhouse gas emissions. This is partly because the optimal pathway approach allows one person's gains to offset another's suffering, increased per capita consumption to offset the destruction of ecosystems and traditional cultures, and so on. If we regard such suffering and destruction not just as a downside to our activities but as a wrong which we are inflicting on others through our activities, then compensating for them by producing benefits for other persons appears morally dubious. Thus, the question is how much risk we are willing to accept, given the difficulties and trade-offs involved in avoiding them. In practice, taking the risk management approach generally means implementing a cap-and-trade system, which sets a supposedly definite cap on total carbon emissions; implementing various targets for different sectors of the economy, such as renewable energy standards for the electricity generation sector, fuel efficiency standards for automobiles, and so on; or various "command-and-control" policies specifying which technologies may or may not be used, calibrated to help meet specific mitigation targets. With enough confidence in the cost of abatement, a risk management approach could also be pursued through a tax on carbon emissions.

Similar considerations apply to questions about how to allocate the burdens of mitigation policy within and between generations. Climate ethicists have devoted a great deal of energy to these questions, the details of which are considered elsewhere in this volume (Hedahl & Fruh 2018) and in the philosophical literature (Moellendorf 2015). It is worth noting, however, that one's position on those questions is likely to be heavily influenced by whether one sees climate policy primarily as a matter of minimizing aggregate harm and distributing burdens appropriately or primarily as a matter of avoiding injustice, rights violations, vice, lack of care, or other non-consequentialist moral failings (Caney 2014). One's position is also likely to be heavily influenced by, among other things, whether one situates mitigation policy in the context of broader distributive questions (Caney 2012), as well as how one thinks about feasibility constraints (Gardiner & Weisbach 2016).

Adaptation

Adaptation consists of adjustments in human or natural systems to observed or expected climate change. This typically involves adjustments designed to reduce or prevent harm from climate change. It can also involve adjustments designed to take advantage of opportunities created by climate change, although this sort of adaptation receives less attention than the protective sort. Examples of protective adaptation include building sea walls to better withstand rising seas or increased storm surges; improving water storage or expanding irrigation to better withstand variation in rainfall; changing the kinds of crops or trees planted in a region; building coastal buildings on stilts; migrating away from areas that will face more severe droughts or floods; and developing social and legal institutions that better equip people to cope with the effects of climate change. Adaptation also extends to deliberate efforts to help non-human systems adjust to climate change, too, such as moving species into new territories through "assisted migration" or "assisted colonization" (Albrecht et al. 2013; Palmer & Larson 2014).

One way to understand the concept of adaptation and its relationship to mitigation is in terms of the IPCC's distinction between *hazards* and *impacts* (Field et al. 2014: 39). Climate change imperils the planet's inhabitants in various ways. The IPCC refers to these new or intensified perils and the risks thereof as "hazards." More intense heatwaves or hurricanes, for instance, are hazards. When those hazards harm (or benefit) human and natural systems, those harms (or benefits) are called "impacts." While mitigation aims to reduce the number and intensity of hazards, adaptation enables human and natural systems to better avoid or withstand the hazards created by climate change, thereby reducing the impacts of climate change on those systems.

Adaptation can reduce harmful impacts in two ways: by reducing a system's *exposure* to hazards or by reducing its *sensitivity* to hazards. Moving away from the coast, for instance, can reduce one's exposure to hurricanes, whereas building sturdier homes can reduce a costal community's sensitivity to them. Both approaches are often said to reduce a system's *vulnerability* to climate change, although other authors use the term "vulnerability" more narrowly to refer to a system's sensitivity to hazards.

The primary reason for adaptation, as with mitigation, is to reduce loss and suffering, either as an end in itself or as a way of avoiding injustice and human rights violations. With the goal of reducing harm in mind, adaptation can arguably serve as a partial substitute for mitigation (Tol 2005; Brooks 2012), at least with respect to some hazards in some contexts. Given the difficulty of adapting to the changes that would occur without mitigation, however, adaptation alone will not suffice (Jamieson 2005). Furthermore, there are practical and conceptual limits to adaptation as a way of reducing harm and injustice. Indigenous communities provide striking examples of these limits. Many indigenous communities value traditions and lifeways that are closely tied to particular places, species, ecosystems, and climates (Whyte 2013). While they may be able to adapt to climate change in the sense of developing new ways of ensuring their own survival and livelihoods, doing so may require altering or abandoning cherished traditions, values, and places. This constitutes an important harm that is *caused* by adaptation, rather than reduced by it (Heyward 2017). Such limits apply not only to indigenous communities, of course, but to any community whose traditions and values are closely tied to specific aspects of their local environment, such as the fishing communities near the Grand Banks.

When it comes to deciding how and where to adapt to climate change and who should pay for it, many of the same questions arise as in mitigation policy, with one major complication. Adaptation is inherently local. So, in addition to deciding who should pay for adaptation and how it should be done, policymakers must decide *where* it should be done and so who benefits from it. The general consensus among scholars studying the ethics of adaptation is that adaptation ought to prioritize the needs of the most vulnerable. This consensus holds whether we look at the issue as one of reducing suffering or addressing injustice, since the most vulnerable would suffer the most serious harms and rights violations despite having contributed little to causing them.

Precaution can also motivate adaptation, especially anticipatory adaptation aimed at preventing serious impacts from potential hazards. Consider, for instance, a region that depends on melting mountain snowpack for fresh water. Unless the region took proactive measures to secure another source of fresh water, grave problems would arise if climate change reduced the snowpack too drastically for too long (Hartzell-Nichols 2011: 689). Given how long it would take to complete such a project, a precautionary approach to climate change provides an additional reason for adaptation.

Geoengineering

Geoengineering offers another option for responding to climate change—or better yet, a pair of options, since it is often useful to treat the two kinds of geoengineering separately. Since CDR removes CO_2 from the atmosphere, whereas solar geoengineering merely suppresses the effects of increased greenhouse gas concentrations, the two types of geoengineering could contribute to our response to climate change in very different ways. The moral reasons for pursuing one or the other are therefore quite different, even though some of the moral questions they raise overlap. (See Chapter 24.)

There are several major proposed technologies for implementing CDR, ranging from reforestation to highly engineered systems to capture CO_2 directly from the air (National Research Council 2015a). Current international climate policy implicitly relies on large-scale deployment of these technologies, since widely endorsed goals, such as limiting global warming to well below 2℃, will be extremely difficult to achieve without them, especially given current emissions trajectories (Fuss et al. 2014; Anderson & Peters 2016). The moral reasons for pursuing one or more of these technologies echo the reasons for mitigation: because they reduce the atmospheric concentration of greenhouse gases, they reduce the severity of climate change. Furthermore, because they enable emitters to "clean up" their pollution, they offer a powerful tool for preventing further injustice.

There are also several prominent proposed technologies for solar geoengineering, such as injecting aerosols into the upper atmosphere to reflect sunlight back into space (National Research Council 2015b). While solar geoengineering has the power to cool the planet significantly, it compensates for greenhouse warming imperfectly, and so cannot restore the preindustrial climate. The difference in climate between a high-greenhouse-gas world cooled with solar geoengineering and the preindustrial world grows more significant and more dangerous as greenhouse gas concentrations rise. Thus, solar geoengineering could be, at best, a supplement for mitigation, not a replacement for it. The main moral reason to consider it, given its shortcomings and risks, is that, if it were used wisely and in combination with emissions reductions and a portfolio of other climate policies, solar geoengineering might be able to significantly reduce overall climate risk.

Rectification

Even with a smart, ambitious portfolio of climate policies, it is too late to prevent all suffering and loss from climate change. Indeed, some people have already felt its effects. This leaves just one last response: rectification or, as it is often called, compensation for loss and damage. Rectification can serve two broad aims, corresponding to the consequentialist and deontological aims for other kinds of responses to climate change: the alleviation of suffering, on the one hand, and corrective justice, on the other. That is, while rectification necessarily aims at compensating people for their losses, it could also aim at ensuring that those who are responsible for causing those losses make amends for their actions. Given the conceptual and technical difficulties in attributing particular losses to anthropogenic climate change—as opposed to, say, natural climate variability, voluntary choices or socioeconomic conditions that increased a community's vulnerability to climate change, and so on—as well as the conceptual, technical, and political difficulties associated with assigning moral responsibility for climate change, a corrective justice approach to rectification would be exceptionally difficult to implement. Arguably, then, the primary motivation for rectification ought to be compensating people fairly for their losses (Wallimann-Helmer 2015). Difficult questions arise, of course, in deciding what counts as a "fair remedy" for loss and damage, especially in cases involving loss of life and the loss of irreplaceable places or cultural artifacts (de Shalit 2011; Wallimann-Helmer 2015).

Conclusion

The fishing communities of Newfoundland mismanaged their environment, and they paid dearly for it. But nearly a quarter century after the Canadian government banned cod fishing in the Grand Banks, the cod are coming back (Rose & Rowe 2015). Perhaps someday soon, the fishing boats will return, too, and old men who once earned their living on the sea will teach their grandchildren to fish for cod themselves, but with more respect for nature.

Dealing with the global mismanagement of Earth's climate will be much harder. It will require a permanent transformation of the energy sector, the industrial sector, transportation, buildings and urban planning, agriculture, and land use, all aimed at reducing our net greenhouse gas emissions to zero. We will need to supplement these mitigation efforts with adaptation and rectification and perhaps some forms of geoengineering. Along the way, publics and policymakers will need to make a series of value judgments about climate policies with the goal of reducing suffering and loss, avoiding and correcting injustice, protecting rights, protecting the non-human world, and exercising appropriate precaution. To succeed, we, too, will have to equip our grandchildren to respect Earth's boundaries—both for its sake and for theirs.

Related Topics

22. Climate Change Mitigation
23. Climate Justice and Equity
24. Geoengineering
41. Sustainable Agriculture
43. Remediation
45. Assisted Migration and Reintroduction

48. Novel Ecosystems
49. Pollution and Polluter Pays
54. Command and Control
55. Economic Instruments
56. Cost-Benefit Analysis
57. Risk Assessment
58. Precautionary Principles

References

Albrecht, G.A. et al. (2013) 'The ethics of assisted colonization in the age of anthropogenic climate change', *Journal of Agricultural and Environmental Ethics*, 26(4), pp. 827–845.

Anderson, K. & Peters, G. (2016) 'The trouble with negative emissions', *Science*, 354(6309), pp. 182–183. Available at: http://science.sciencemag.org/content/354/6309/182.

Archer, D. (2005) 'Fate of fossil fuel CO2 in geologic time', *Journal of Geophysical Research C: Oceans*, 110(9), pp. 1–6.

Attfield, R. (2011) 'Nolt, future harm and future quality of life', *Ethics, Policy & Environment*, 14(1), pp. 11–13.

Berkey, B. (2017) 'Human rights, harm, and climate change mitigation', *Canadian Journal of Philosophy*, 47(2–3), pp. 416–435. Available at: https://www.tandfonline.com/doi/full/10.1080/00455091.2016.1268465.

Brooks, T. (2012) 'Climate change and negative duties', *Politics*, 32(1), pp. 1–9. Available at: http://doi.wiley.com/10.1111/j.1467-9256.2011.01419.x.

Broome, J. (2012) *Climate matters: ethics in a warming world*. New York: W. W. Norton.

Cafaro, P. (2011) 'Beyond business as usual: alternative wedges to avoid catastrophic climate change and create sustainable societies', in Arnold, D. G. (ed.) *The ethics of global climate change*. New York: Cambridge University Press, pp. 192–215.

Cafaro, P. (2012) 'Climate ethics and population policy', *Wiley Interdisciplinary Reviews: Climate Change*, 3(1), pp. 45–61.

Caney, S. (2009) 'Climate change and the future: discounting for time, wealth, and risk', *Journal of Social Philosophy*, 40(2), pp. 163–186. Available at: http://doi.wiley.com/10.1111/j.1467-9833.2009.01445.x.

Caney, S. (2010) 'Climate change, human rights and moral thresholds', in Humphreys, S. (ed.) *Human rights and climate change*. New York: Cambridge University Press, pp. 69–90.

Caney, S. (2012) 'Just emissions', *Philosophy & Public Affairs*, 40(4), pp. 255–300.

Caney, S. (2014) 'Two kinds of climate justice: avoiding harm and sharing burdens', *Journal of Political Philosophy*, 22(2), pp. 125–149.

Clarke, L. et al. (2014) 'Assessing transformation pathways', in *Climate change 2014: mitigation of climate change, contribution of Working Group III to the Fifth Assessment Report of the Intergovernmental Panel on Climate Change*. Cambridge: Cambridge University Press.

Cowen, T. & Parfit, D. (1992) 'Against the social discount rate', in Laslett, P. and Fishkin, J. (eds.) *Justice between age groups and generations*. New Haven, CT: Yale University Press, pp. 144–161.

Davidson, M.D. (2008) 'Wrongful harm to future generations: the case of climate change', *Environmental Values*, 17(4), pp. 471–488.

Field, C.B. et al. (2014) 'Technical summary', in Field, C.B. et al. (eds.) *Climate change 2014: impacts, adaptation, and vulnerability. Part A: global and sectoral aspects. Contribution of Working Group II to the Fifth Assessment Report of the Intergovernmental Panel on Climate Change*. Cambridge, UK and New York, NY: Cambridge University Press.

FitzPatrick, W.J. (2007) 'Climate change and the rights of future generations: social justice beyond mutual advantage', *Environmental Ethics*, 29, pp. 369–388.

Fuss, S. et al. (2014) 'Betting on negative emissions', *Nature Climate Change*, 4(10), pp. 850–853. Available at: http://dx.doi.org/10.1038/nclimate2392%5Cn10.1038/nclimate2392.

Gardiner, S.M. (2006) 'A core precautionary principle', *Journal of Political Philosophy*, 14(1), pp. 33–60.

Gardiner, S.M. (2011) *A perfect moral storm: the ethical tragedy of climate change*. New York: Oxford University Press.

Gardiner, S.M. & Weisbach, D.A. (2016) *Debating climate ethics*. New York: Oxford University Press.

Groom, B. et al. (2005) 'Declining discount rates: the long and the short of it', *Environmental and Resource Economics*, 32(4), pp. 445–493.

Hale, B. (2018) 'Geoengineering', in Hale, B. and Light, A. (eds.) *Routledge companion to environmental ethics.* Abingdon: Routledge.

Hardin, G. (1968) 'The tragedy of the commons', *Science*, 162, pp. 1243–1248.

Hartzell-Nichols, L. (2013a) 'From "the" precautionary principle to precautionary principles', *Ethics, Policy & Environment*, 16(3), pp. 308–320. Available at: http://www.tandfonline.com/doi/abs/10.1080/2155008 5.2013.844569 [Accessed February 24, 2014].

Hartzell-Nichols, L. (2013b) 'How is climate change harmful?', *Ethics & the Environment*, 17(2), pp. 97–110.

Hartzell-Nichols, L. (2011) 'Responsibility for meeting the costs of adaptation', *Wiley Interdisciplinary Reviews: Climate Change*, 2(5), pp. 687–700.

Hedahl, M. & Fruh, K. (2018) 'Mitigation', in Hale, B. and Light, A. (eds.) *Routledge companion to environmental ethics.* Abingdon: Routledge.

Heyward, C. (2017) 'Ethics and climate adaptation', in Gardiner, S. M. and Thompson, A. (eds.) *The Oxford handbook of environmental ethics.* New York: Oxford University Press, pp. 474–486.

Hourdequin, M. (2010) 'Climate, collective action and individual ethical obligations', *Environmental Values*, 19(4), pp. 443–464.

Jamieson, D. (1992) 'Ethics, public policy, and global warming', *Science, Technology & Human Values*, 17(2), pp. 139–153.

Jamieson, D. (2005) 'Adaptation, mitigation, and justice', in Sinnott-Armstrong, W. and Howarth, R. (eds.) *Perspectives on climate change.* Amsterdam: Elsevier, pp. 221–253.

Jamieson, D. (2007) 'When utilitarians should be virtue theorists', *Utilitas*, 19(2), p. 160.

Joos, F. et al. (2013) 'Carbon dioxide and climate impulse response functions for the computation of greenhouse gas metrics: a multi-model analysis', *Atmospheric Chemistry and Physics*, 13(5), pp. 2793–2825.

Kelleher, J.P. (2012) 'Energy policy and the social discount rate', *Ethics, Policy & Environment*, 15(1), pp. 45–50. Available at: http://www.tandfonline.com/doi/abs/10.1080/21550085.2012.672684.

Kelleher, J.P. (2015) Is there a sacrifice-free solution to climate change?', *Ethics, Policy & Environment*, 18(1), pp. 68–78. Available at: http://www.tandfonline.com/doi/full/10.1080/21550085.2015.1016959.

Krakoff, S. (2011) 'Parenting the planet', in Arnold, D.G. (ed.) *The ethics of global climate change.* New York: Cambridge University Press, pp. 145–169.

MacDowell, L.S. (2012) *An environmental history of Canada.* Vancouver, BC: University of British Columbia Press.

McKinnon, C. (2009) 'Runaway climate change: a justice-based case for precautions', *Journal of Social Philosophy*, 40(2), pp. 187–203.

Moellendorf, D. (2014) *The moral challenge of dangerous climate change: values, poverty, and policy.* New York: Cambridge University Press.

Moellendorf, D. (2015) 'Climate change justice', *Philosophy Compass*, 10(3), pp. 173–186.

Munthe, C. (2011) *The price of precaution and the ethics of risk.* Dordrecht: Springer.

National Research Council (2015a) *Climate intervention: carbon dioxide removal and reliable sequestration.* Washington, DC: National Academies Press.

National Research Council (2015b) *Climate intervention: reflecting sunlight to cool earth.* Washington, DC: National Academies Press.

Nolt, J. (2011a) 'Greenhouse gas emissions and the domination of posterity', in Arnold, D. G. (ed.) *The ethics of global climate change.* Cambridge: Cambridge University Press.

Nolt, J. (2011b) 'How harmful are the average American's greenhouse gas emissions?', *Ethics, Policy & Environment*, 14(1), pp. 3–10. Available at: http://www.tandfonline.com/doi/abs/10.1080/21550085.2011.56 1584 [Accessed March 19, 2012].

Nordhaus, W. (2008) *A question of balance: weighing the options on global warming policies.* New Haven, CT: Yale University Press.

Nordhaus, W. (2013) *The climate casino: risk, uncertainty, and economics for a warming world.* New Haven, CT: Yale University Press.

Palmer, C. (2011) 'Does nature matter? The place of the nonhuman in the ethics of climate change', in Arnold, D.G. (ed.) *The ethics of global climate change.* New York: Cambridge University Press, pp. 272–291.

Palmer, C. & Larson, B.M.H. (2014) 'Should we move the Whitebark Pine? Assisted migration, ethics and global environmental change', *Environmental Values*, 23(6), pp. 641–662.

Parfit, D. (1984) *Reasons and persons.* Oxford: Oxford University Press.

Rogelj, J. et al. (2016) 'Differences between carbon budget estimates unravelled', *Nature Climate Change*, 6(3), pp. 245–252. Available at: http://www.nature.com/doifinder/10.1038/nclimate2868.

Rose, G.A. & Rowe, S. (2015) 'Northern cod comeback', *Canadian Journal of Fisheries and Aquatic Sciences*, 72(12), pp. 1789–1798.

Sandler, R. (2014) 'The ethics of climate change mitigation', in Di Paola, M. and Pellegrino, G. (eds.) *Canned heat: ethics and politics of global climate change.* New Delhi and Abingdon: Routledge, pp. 61–79.

de Shalit, A. (2011) 'Climate change refugees, compensation, and rectification', *The Monist*, 94(3), pp. 310–328.

Stern, N. (2007) *The economics of climate change: the Stern review.* Cambridge: Cambridge University Press.

Suikkanen, J. (2014) 'Contractualism and climate change', in Di Paola, M. and Pellegrino, G. (eds.) *Canned heat: ethics and politics of global climate change.* New Delhi and Abingdon: Routledge, pp. 115–128.

Tol, R.S.J. (2005) 'Adaptation and mitigation: trade-offs in substance and methods', *Environmental Science and Policy*, 8(6), pp. 572–578.

Wagner, G. & Weitzman, M.L. (2015) *Climate shock: the economic consequences of a hotter planet.* Princeton, NJ: Princeton University Press.

Wallimann-Helmer, I. (2015) 'Justice for climate loss and damage', *Climatic Change*, 133(3), pp. 469–480.

Whyte, K.P. (2013) 'Justice forward: tribes, climate adaptation and responsibility', *Climatic Change*, 120(3), pp. 517–530.

21

CLIMATE MODELING

Wendy S. Parker

Introduction

In recent decades, three main questions have been at the heart of research into global climate change: Has earth's climate changed since pre-industrial times? If so, are human activities – especially activities that release greenhouse gases to the atmosphere – the primary cause? What would climate be like in the future under different greenhouse gas emission scenarios? These are questions of climate change *detection*, *attribution,* and *projection*, respectively.

Scientific investigation has answered the detection and attribution questions in the affirmative: warming of earth's climate is "unequivocal" and, at least for the second half of the 20th century, it is judged to be "extremely likely" (i.e. probability ≥0.95) that anthropogenic emissions of greenhouse gases were the dominant cause (IPCC 2013: Sec.D.3). The question of what climate would be like in the future under different emission scenarios is more difficult to answer, especially when it comes to details at regional and local scales. There is good reason to think, however, that if greenhouse gas emissions are not curbed considerably, significant additional warming of the global climate will occur, with serious negative consequences for humans and other living things.

Climate models – mathematical and computational representations of earth's climate system – are integral to the study of global climate change. As explained below, they play important roles not just when it comes to projecting future climate change, but also in the processes of detection and attribution. Moreover, they support other modeling endeavors. For instance, simple climate models are often embedded in integrated assessment models (IAMs) that are used to estimate the costs and benefits of climate policies. Likewise, results from climate models serve as an input to impact models, which investigate how changes in climate will affect agriculture, the spread of disease, water supplies, human migration, and so on.

Over the last decade, climate modeling has attracted significant attention from philosophers of science. Most of this attention has focused on epistemic aspects of climate modeling: How can we tell which results from climate models are trustworthy? Is agreement among climate model projections epistemically significant? If climate models can't tell us exactly what will happen, can they tell us rough probabilities for different outcomes? It is only very recently that philosophers have begun to engage with ethical and value dimensions of climate modeling practices. In many cases, these ethical and value dimensions are intertwined with epistemic ones, as discussed below (see also Tuana 2010 and her subsequent work on coupled epistemic-ethical issues).

This chapter provides a non-technical introduction to climate modeling and calls attention to some of the ways in which ethical and value considerations intersect with the development and use

DOI: 10.4324/9781315768090-26

of climate models. The next section introduces several types of climate model. The third section outlines a number of important uses to which climate models are put. The fourth section focuses on climate model development, explaining how ethical and social values could influence this development both in the setting of research priorities and in the management of inductive risk. The fifth section is concerned with ethical and value considerations in the estimation and communication of uncertainties associated with climate modeling results, especially results that are intended to inform policy decisions. The final section offers some concluding remarks.

The Climate Modeling Hierarchy

Earth's climate system is understood to encompass its atmosphere, oceans, land surface, ice, and biosphere. Climate models are mathematical and computational representations of earth's climate system. They come in a range of forms, from the very simple to the very complex. This collection of models is often referred to as the *climate modeling hierarchy*.

At the simple end of the hierarchy are *energy balance models* (EBMs). These are mathematical models that represent the overall flow of energy in and out of the climate system without representing different components of the system – atmosphere, oceans, land surface, etc. – and without representing earth's geography. Results from the simplest EBMs can be calculated by hand on the back of an envelope. *Earth system models of intermediate complexity* (EMICs) are computational models that represent some or all of the main components of the climate system as well as earth's geography, but in a relatively simple and coarse way. *General circulation models* (also known as *global climate models* or GCMs) explicitly represent a wider range of processes in the atmosphere, ocean, land surface, and cryosphere components of the climate system, and in comparatively greater detail. GCMs can be very computationally intensive, typically using substantial time on a supercomputer to complete a single multi-decadal simulation. More complex still are *earth system models* (ESMs), close cousins of GCMs that in addition represent some biospheric processes. *Regional climate models* (RCMs) represent many of the same processes as GCMs/ESMs but focus on particular geographical regions.

In general, these climate models are physical rather than statistical models: their core equations are built upon physical theory and are not merely curve-fits to past observations. In the atmosphere component of a GCM, for example, the "dynamical core" consists of fluid dynamical equations, equations reflecting conservation principles (e.g. conservation of mass) and basic thermodynamic equations. The computer is programmed to estimate solutions to this set of equations at each of thousands of points on a three-dimensional spatial grid, where grid points are associated with volumes of atmosphere; the computer solves the equations repeatedly, stepping forward in short time steps, to produce a simulation of the evolution of temperature, pressure, and other variables in those volumes. The length of the model's time step and the spacing of the grid points – typically 50–100 km in the horizontal for the atmosphere components of today's GCMs – determine the resolution of the model, i.e. the scale of phenomena that can be explicitly simulated, and are limited by available computing power.

Unfortunately, a number of very important climate system processes, such as the formation of clouds and precipitation, occur on spatial scales smaller than those resolved by today's climate models. Rather than ignoring these important sub-grid processes – clouds, for instance, affect the transfer of radiation through the atmosphere, which, in turn, can impact near-surface temperatures significantly – scientists attempt to *parameterize* them: to represent them as a function of grid-scale variables, in a simplified way. For example, early parameterizations of clouds in GCMs assumed that the cloudiness of a given volume of atmosphere was a function of the relative humidity at an associated grid point. In reality, while relative humidity are cloudiness are related, there is not a one-to-one relationship; the extent of cloudiness often depends on the details of air motions within the volume, the presence of tiny particles that can serve as nuclei for cloud droplets, and other factors. In many cases, parameterizations are informed by physical understanding but also incorporate empirical parameters that are tuned to observations.

It is far from obvious how best to parameterize sub-grid processes. This is one reason why there is not just one GCM/ESM used to answer questions about climate change, but dozens of them, developed at different modeling centers around the world. These models differ to some extent in the way they represent sub-grid processes, in the numerical values that they assign to model parameters, in their spatial resolution, and in other ways, reflecting alternative reasonable choices in model construction in light of existing knowledge, uncertainties, modeling goals, and pragmatic constraints. Similarly, there are also multiple EMICs, RCMs, etc., that differ in various ways.

Uses of Climate Models

Climate models, both individually and collectively, are used for numerous purposes in the study of climate change. For instance, they are used for basic research that aims to advance understanding of feedbacks in the climate system. The focus in this section is on the use of climate models to help address key questions about the threat of climate change that society faces today: in detection, attribution, and projection studies; in support of impacts studies; and in analyses of climate policies.

Detection studies investigate whether observed changes in climatic conditions might be due to *internal variability* in the climate system – oscillations and fluctuations that would occur even in the absence of any *forcing* to the system; the El Niño – Southern Oscillation (ENSO) phenomenon is a good example. Forcing occurs when processes change the radiative balance of the climate system, i.e. the flow of energy into and out of the system; examples of such processes (or forcing "factors") include changing solar output, rising greenhouse gas concentrations, and changing aerosol concentrations. Internal variability cannot be estimated from observations alone in a straightforward way, because an expansive global observational network has existed only for a century or so, a period during which the climate system in fact has undergone forcing. But climate models, especially GCMs and ESMs, can be used to simulate the climate system over many centuries under the assumption that known forcing factors remain constant at pre-industrial levels. It is in part because the variability seen in these simulations – the model-estimated internal variability – is much smaller than the changes in conditions observed since pre-industrial times that climate scientists have concluded that earth's climate is undergoing genuine change on global and regional scales (IPCC 2013: Table SPM.1 & Sec. D.3).

Similarly, in support of attribution studies, scientists use GCMs/ESMs to estimate the expected contributions of different forcing factors to climate change (Hegerl and Zwiers 2011; Bindoff et al. 2013). These individual contributions are not apparent in recent observations, since the latter often reflect the action of several forcing factors acting at once. With GCMs/ESMs, however, scientists can run simulations in which only one factor is changing in accordance with historical estimates, while the others are assumed constant. This provides an estimate of the pattern of response (or *fingerprint*) for an individual forcing factor. These fingerprints are used in a regression-style methodology (details omitted here) to see whether, in combination, they can account for observed changes in temperature and other variables; the methodology thereby also estimates the fraction of an observed change that is attributable to rising anthropogenic greenhouse gas emissions. On the basis of fingerprint studies and other evidence, the IPCC 5th Assessment Report (AR5) concluded that it is "extremely likely" (i.e. probability ≥0.95) that human influence has been the dominant cause of observed global warming since the mid-20th century. It also concluded that various other observed changes, including some changes on regional scales, were "likely" (i.e. probability ≥0.66) – or in some cases "very likely" (i.e. probability ≥0.90) – to have a substantial anthropogenic component (IPCC 2013: Sec.D.3).

GCMs/ESMs and other climate models are perhaps best known for their use in generating quantitative projections of future climate change. Such projections inform international policy discussions where the nominal aim is to act in ways that keep global warming below "dangerous" levels, often interpreted to be two degrees Celsius (2°C) warming over pre-industrial levels. In part on the basis of projections from climate models, the IPCC AR5 concluded that it is "likely" (i.e. probability ≥0.66) that global warming would exceed 2°C by 2100 under moderately high and high emission scenarios, and "more likely than not" (i.e. probability >0.50) that it would exceed 2°C by 2100 for a moderate emission scenario; only in a scenario involving substantial and sustained reductions in greenhouse gas emissions was the temperature change expected to remain below the 2°C threshold during the 21st century (IPCC 2013: Sec.E.1). Projections of future climate change from GCMs/ESMs are also used as boundary conditions for RCMs, to produce more detailed regional projections of changes in conditions. These regional results, in turn, are sometimes used as an input to impact models – models that simulate how crop yields or water supply or the spread of disease would change in a region in response to projected changes in climate.

While GCMs/ESMs are preferred over simpler models for many purposes, their computational expense can be prohibitive. For instance, one approach to exploring uncertainties about future climate change – discussed further below – involves running a climate model many times with different values for uncertain parameters on each run; this is computationally demanding unless the model is relatively simple. Another important use of simple climate models is in studying the costs and benefits of climate policies. Climate models on their own do not provide information about such costs and benefits, but IAMs that include climate models as one component sometimes are used for this purpose. IAMs themselves take various forms but are intended to represent both natural and social/human systems: a simple climate model might be coupled to an energy-economic model, a demographic model, etc. Assumptions about discounting of future benefits and costs strongly influence the results of IAM studies and raise both ethical and epistemic questions (see Chapter 20) (Broome 2008; Frisch 2013).

Values in Climate Model Development

Climate scientists build and deploy climate models in order to learn about the climate system. In this sense, climate model development is fundamentally an epistemic endeavor. This does not mean, however, that non-epistemic factors – including pragmatic constraints and social and ethical values – do not exert any influence in climate model development. On the contrary, pragmatic constraints, especially limited computing power, are a fact of life in climate modeling and shape a host of decisions in model development, including decisions about model resolution. That social and ethical values would exert an influence in climate model development might seem more worrisome, raising concerns about bias. But as explained below, such influence need not be problematic.

Kitcher (2001) reminds us that science does not seek just any truths; it seeks significant truths. A truth can be significant for epistemic reasons – for example, because it unifies diverse bodies of knowledge – but also for non-epistemic reasons, e.g. because of its connection with harms or benefits. Answering questions of climate change detection, attribution, and projection is a high priority precisely because of potential harms: the worry is that anthropogenic emissions of greenhouse gases are changing earth's climate in ways that will bring devastating harms to humans and to other living things. In this way, non-epistemic factors – the negative value attached to these harms and the associated desire to avoid or reduce them – fundamentally shape the epistemic goals of climate modeling. This influence of non-epistemic values parallels that in many other areas of science. For example, the negative value that people attach to disease is one motivation for medical research that seeks cures for diseases.

Just as one might argue on ethical grounds that certain neglected diseases ought to receive more attention from medical researchers than they currently do, one also might argue that certain questions about climate change ought to receive more attention from climate scientists than they currently do. Consider, for example, regional climate modeling studies. Many of these studies have focused on the developed world, where modeling resources are concentrated (an important exception is the CORDEX program – see e.g. Gutowski et al. 2016). Yet, many harmful impacts of climate change are expected to be more severe in the developing world. Moreover, thus far, the emissions responsible for climate change have come primarily from the developed world. This suggests that a justice-based argument might be made for increased efforts to model regional-scale climate change in the developing world, especially in support of studies that aim to identify robust strategies for adaptation (Dessai et al. 2009). Indeed, it may be that ethical considerations like these should play much more of a role in shaping epistemic priorities in climate modeling than they currently do.

Re-thinking such priorities could feed back into the process of climate model development. After all, which variables are included in a model, which processes are represented with more or less fidelity, which aspects of the model are a priority for improvement, and so on, depend on the goals of the modeling study, which, in turn, can vary depending on whose interests motivate the study in the first place (see also Biddle and Winsberg 2009; Winsberg 2012; Heaphy 2015; Intemann 2015; Parker and Winsberg 2018; Winsberg 2018). The process of building a GCM/ESM or RCM from scratch, however, is an arduous one. And changing an existing model to try to make it better suited to addressing a particular question will sometimes degrade its performance in other important respects, because the changes upset a previously achieved balance among sources of error in the model. Nevertheless, it is worth considering to what extent today's – or tomorrow's – climate models, especially RCMs, could be tailored to address the epistemic and practical priorities of particularly vulnerable populations.

The foregoing discussion points to some ways that non-epistemic values can influence decisions in climate model development indirectly, by shaping epistemic priorities. A prominent line of argument in philosophy of science, however, suggests that non-epistemic values sometimes should have a more direct role to play in model development. Why should social and ethical values have anything to do with, say, the numerical value assigned to a model parameter representing the fall speed of ice crystals in clouds? The argument relates to *inductive risk* – i.e. the risk of error in inferring a conclusion – and the consequences of such error (Hempel 1965; Douglas 2009).

Suppose there is substantial uncertainty about the numerical value that should be assigned to a parameter in a climate model; any value in a relatively wide range might plausibly turn out to be the one that gives the most accurate results for some high-priority variable, such as average summer rainfall in a particular region during a future period. Suppose further that assigning a value near the high end of the plausible range can be expected to result in higher rainfall predictions in that region, while assigning a value near the low end can be expected to result in lower rainfall predictions. In this way, the parameter setting affects the balance of inductive risk: choosing a higher value increases the risk that, if the model's rainfall predictions are in error, they will be too high, while choosing a lower value increases the risk that the predictions will be too low. According to the argument from inductive risk, if overestimating rainfall would be <u>worse</u> than underestimating it – because, let us suppose, the people in the region will prepare for the conditions that the model predicts, and they can easily cope with additional rain but would face serious problems if rainfall were to be less than expected – then this provides a reason for assigning a value near the low end of the plausible range for the parameter. In this way, the argument goes, non-epistemic values, in particular the negative value attached to the harms that can be expected to befall people in the region if rainfall is overestimated, could legitimately shape methodological choices made in climate model development (Parker and Lusk 2019).

It is important to recognize the limits of this argument. It does not allow that scientists can make methodological choices to obtain whatever result they desire; the goal is always to produce accurate results. Moreover, the argument concerns situations in which there is substantial uncertainty about which of several plausible choices in model development will lead to the most accurate results, and choices remain constrained by what is plausible on scientific grounds. In the example above, for instance, it would not be permissible to choose a parameter value outside of the plausible range. (But see Elliott and McKaughan 2014 for an example of how non-epistemic considerations might trump epistemic ones in some cases.) Proponents of the inductive risk argument also emphasize that, when such value-laden methodological choices are made, this should be clearly reported along with the results of the study; transparency about value influence is paramount (Douglas 2009; Kloprogge et al. 2011; Elliott and Resnik 2014).

Nevertheless, the view that non-epistemic values should enter, or even unavoidably enter, the scientific process in this way remains controversial (Betz 2013; Brown 2013; Hudson 2015). It does not appear to be widely accepted in the climate modeling community. When faced with uncertain methodological choices, the preferred strategy in climate modeling instead has been to try several different options (e.g. several different parameter values) to get a sense of the range of modeling results that are plausible. This will be discussed further in the next section. Note, however, that trying several options is not always feasible. Human and computational resources may be limited enough that, for a particular uncertain methodological choice, only one option can be tried in a given study. In this situation, in practice, the option that is used is likely to be a "default" option – one that is used by other modeling groups, or that is close to a measured value (which might itself have large associated uncertainty), or that has been found to work well in previous versions of the model, etc. Yet, the default option is not necessarily that which will give the most accurate results for the variables that are of greatest interest in the study at hand.

It is also important to recognize that, if non-epistemic values are to influence uncertain methodological choices in climate modeling in ways like those suggested by the argument from inductive risk, then some key questions need answering. First, whose values should have an influence? If a study is undertaken for the benefit of a particular group of stakeholders – say, a population in a particular region – then it seems that the values of that group of stakeholders should have an influence (Parker and Lusk 2019). But stakeholders often have conflicting values, raising additional questions about how those conflicts should be handled. Second, are some values illegitimate? Suppose, for example, that an energy company attaches negative value to regulation, which is more likely to occur if large changes in climate are expected; if the company hires climate scientists to perform modeling studies on its behalf, can the company legitimately direct the scientists to choose values for uncertain parameters in such a way that it is more likely that modeling results, if they turn out to be in error, will underestimate rather than overestimate changes in climate? Intemann (2015) provides a preliminary response to these questions, contending that non-epistemic value judgments in climate modeling (and beyond) should promote democratically endorsed epistemological and social aims of research (see also Schroeder 2021). What constitutes a democratically endorsed aim in practice, however, is not entirely clear. Moreover, this response may be too permissive, insofar as it seems to allow for pernicious or unjust aims, so long as they are democratically endorsed. Steel (2016) suggests that implementing the argument from inductive risk at a structural level within a scientific domain – e.g. in setting overarching evidential standards – may help to mitigate some concerns about whose and which values should exert an influence.

Values and the Limits of Climate Knowledge

Accurately characterizing and communicating both what is known about climate change and the extent to which uncertainties remain is important as governments, industrial firms, and communities increasingly seek information that will help them plan for future climate change and its impacts. In addition to the periodic assessments of the IPCC, which aim primarily to support international policymaking, there are activities undertaken by national government agencies, state and local climatology offices, academic researchers, and private consulting firms that aim to develop and translate scientific information about past, present, and future climate into formats and tools that are useful for particular groups of stakeholders. The latter activities recently have become known as *climate services* (Vaughan and Dessai 2014).

Climate services products often are intended to inform important real-world decisions, such as decisions related to water management, farming practices, disease control, etc. Users typically are seeking information about climate at regional and local scales, and often the information sought relates to future climate conditions and impacts. Unfortunately, it is particularly difficult to make confident, precise projections of future climate change on these scales. One reason is that internal variability at regional and local scales is typically much larger than at the global scale (Deser et al. 2014). Another reason is that the methods commonly used to produce projections at these scales – regional climate modeling and/or statistical downscaling – face significant scientific challenges. Regional climate modeling requires information about (future) conditions at the boundaries of the modeled region. Statistical downscaling usually involves the shaky assumption that correlations between GCM/ESM-scale conditions and local conditions seen in the past will continue in the future. When it comes to some questions asked of climate services, including some questions about future changes in regional climate, the "best available" methods and tools may be clearly inadequate or of largely unknown quality. In these situations, while it is possible to run climate models and generate results, it might be better to simply admit to stakeholders that the information they seek cannot yet be provided. At the very least, large uncertainties associated with results ought to be clearly communicated.

These recommendations accord with those of Adams et al. (2015), who begin to outline an ethical framework for climate services. They start from the assumption that "ethical climate service products and practices should contribute to human security at both individual and collective scales, and minimize negative consequences from climate impacts" (Adams et al. 2015: 3) and identify four guiding values for climate services: *integrity, transparency, humility,* and *collaboration*. They urge practitioners to work closely with users to understand the contexts in which decisions and actions will unfold. They also emphasize the importance of clearly communicating the limitations of climate services, in part by providing detailed descriptions of remaining uncertainties. Indeed, they stress that "honesty about one's ignorance is central to integrity" in this context (Adams et al. 2015: 3). The stakes are high, after all, since a failure to communicate uncertainties associated with climate services products may lead users to make maladaptive decisions, resulting in significant, otherwise-avoidable harms.

Gauging the extent of current uncertainty, however, is not a trivial matter. Ensemble climate modeling is a prominent approach (Parker 2013). It involves producing multiple climate simulations that address the same research question, such as how much the average near-surface temperature in a region would rise by 2100 under a particular emission scenario. The simulations are produced by running different climate models (a *multi-model ensemble* (MME)), by running a single climate model multiple times with different values assigned to one or more parameters on each run (a *perturbed-physics* or *perturbed-parameter ensemble* (PPE)), and/or by running a single climate model multiple times with different initial conditions on each run (an *initial condition ensemble* (ICE)). In this way, uncertainty about how to represent the climate system – about the equations, parameter values, or initial conditions that should be used when asking a particular question about the climate system – is translated into uncertainty about the answer to that question. Ensemble modeling studies illustrate the response to uncertain methodological choices mentioned in the previous section: the different simulations that comprise a given MME, PPE, or ICE reflect alternative choices in climate model development. Rather than appealing to non-epistemic values to select one option from the set of reasonable choices – one set of equations, or parameter values or initial conditions – multiple options are tried.

Results from today's ensemble studies nevertheless do not provide a complete picture of current uncertainty. Studies producing MMEs typically employ models that happen to be available, rather than a set of models that is designed to systematically or comprehensively sample uncertainty about how to represent the climate system (Tebaldi and Knutti 2007). This was the case, for instance, with the Coupled Model Inter-comparison Project 5 (CMIP5), an MME study that served as a key resource for the IPCC's 5AR (Taylor et al. 2012); its projections, produced with a few dozen state-of-the-art climate models developed at different modeling centers around the world, are more like a set of best-guess estimates than a set that can be expected to fully span current uncertainty. Likewise, studies producing PPEs often sample only a small number of alternative values for a subset of uncertain model parameters, because a more comprehensive exploration would require more computing time than is available. Studies producing MMEs and PPEs often also incorporate ICEs, but these too typically involve just a few alternative sets of initial conditions. And of course ICEs on their own do not probe uncertainty about modeling equations or parameter values.

The fact that today's ensemble studies are limited in these and other ways needs to be factored in when their results inform conclusions about the extent of current uncertainty. In the 5AR, for example, IPCC scientists concluded that, for emission scenarios investigated in CMIP5, it was only "likely" (i.e. probability $\geq 66\%$) that the global temperature change would fall within the 5–95% range projected by CMIP5 models. If the CMIP5 results were interpreted as sampling "true" uncertainty in light of current information, this range would have been deemed "very likely" (i.e. probability $\geq 90\%$). The downgrading of confidence from "very likely" to "likely" reflects the scientists' judgment of how the limitations of the CMIP5 ensemble (e.g. that some physical processes are represented too simply or not at all in the models, that parameter uncertainty was not thoroughly explored, etc.) should affect the interpretation of its results. As this example illustrates, a careful assessment of the extent of current uncertainty will typically require more than just performing an ensemble modeling study; reflection on other available information (e.g. theory, observations, other modeling studies, and information about model limitations), as well as expert judgment, will be required (Petersen 2012; Parker and Risbey 2015). There is a value-related challenge to this process: social values involved in setting epistemic priorities in climate model development can make a difference to uncertainty estimates generated with those models, in ways that sometimes are hard to unpack and take account of (Biddle and Winsberg 2009; Winsberg 2018: Ch.9; Parker and Winsberg (2018)).

Considerations of inductive risk also can re-surface in uncertainty assessment: there can be uncertainty about the extent of current uncertainty. For instance, after reflecting on available information, it may be unclear whether a future change in climate should be considered "likely" (i.e. probability $\geq 66\%$) or merely "more likely than not" (probability $> 50\%$) under a given emission scenario. The inductive risk arguments developed by philosophers like Douglas (2009) suggest that, here too, scientists ought to consider the consequences of error: if the consequences can be expected to be much worse if they erroneously report that the change is only "more likely than not" than if they erroneously report that it is "likely", then this is a reason in favor of reporting "likely". More generally, Douglas (2009) argues that scientists have a moral responsibility to take special care in reaching and reporting conclusions, including conclusions about the extent of uncertainty, when they can foresee that significant negative consequences might well occur if their conclusions are in error; this responsibility is simply an instance of the responsibility that all moral agents have to consider the consequences of their actions and to avoid acting negligently.

As noted earlier, however, the idea that non-epistemic values in practice must or should play a role in this way remains somewhat controversial. For instance, Betz (2013) argues that it is always possible to hedge conclusions so that the uncertainty associated with them is negligible. In the example above, this might mean reporting that the change in climate is "*at least* more likely than not" or, depending on the details of the case, perhaps just that such a change cannot be ruled out. But hedging may not always be an option in practice, nor always the best strategy. In practice, scientists are sometimes required to choose from a limited set of options when reporting conclusions (Steele 2012). For example, they may be asked to report – yes or no – whether the probability of an event occurring is greater than 0.05. In addition, oftentimes what decision makers want to know is how confident scientists are in a hypothesis of interest (e.g. that X causes Y), not just that scientists are very confident in a different, weaker hypothesis (e.g. that it is possible that X causes Y) (see also Intemann 2015).

Concluding Remarks

This chapter identified several ways in which ethical and value considerations intersect with the development and use of climate models. It is worth briefly reviewing those intersections here.

First, ethical and value considerations can affect climate modeling practices indirectly, by *shaping epistemic priorities*; such priorities can influence which models are built and how they are built – which processes are represented with greater/lesser fidelity, which comparisons with observations are made, and so on. In connection with this, it was suggested that a justice-based argument might be given for prioritizing modeling and other studies that can meaningfully inform adaptation decision for particularly vulnerable populations in the developing world.

A somewhat more controversial idea is that ethical and value considerations ought to influence even fine-grained methodological decisions in climate modeling, in order to reduce the risk of errors that would have particularly negative consequences. Opponents contend that allowing values to play such a role in the "internal" workings of science is undesirable and/or unnecessary – several different methodological options could be tried instead, for instance, to see how this affects results. This strategy of trying multiple options is not always feasible in practice, however, and may not always best serve the needs of users of climate information products. Nevertheless, it is a strategy commonly pursued in the form of ensemble climate modeling studies, which probe the extent of uncertainty about future climate change.

Taking care in assessing uncertainties and in formulating conclusions is a particularly important component of an emerging *ethics of climate services*. Climate services products are intended to inform real-world decisions, and so failing to take sufficient care in gauging and communicating uncertainties may lead users to make maladaptive decisions and may open climate services practitioners to charges of negligence. Taking sufficient care will often require more than just reporting a spread of modeling results; it will require expert judgment that takes account of recognized limitations of today's models – and value influences in their development – as well as relevant observational data and background theory.

These are just some of the ways in which ethical and value considerations intersect with climate modeling practices. There is much scope for further work, both at a general theoretical level and in the context of particular climate modeling applications and studies. Such work has the potential to help make climate modeling practices *better*, insofar as they become more responsive to the epistemic and practical needs of communities facing serious threats from climate change.

Further Reading

Eric Winsberg's (2018) *Philosophy and Climate Science* (Cambridge University Press) surveys a wide range of philosophical issues related to climate modeling. For an introduction to the science of climate change (and more), see Houghton, J. (2015) *Global Warming: The Complete Briefing* (Cambridge University Press). For a more advanced treatment, see the IPCC's perioid Assessment Reports, especially the *Technical Summary* for each report, available at: www.ipcc.ch. For more in-depth discussions and review articles on a wide range climate-related topics, check out the journal *WIREs Climate Change*.

Related Topics

20. Moral Bases of Responses to Climate Change
24. Geoengineering
52. Prediction and Forecasting
59. Adaptive Management

References

Adams, P. et al. (2015) "Call for an ethical framework for climate services," *WMO Bulletin* 64(2). Available online at http://www.wmo.int/bulletin/en/content/call-ethical-framework-climate-services.

Betz, G. (2013) "In defense of the value-free ideal," *European Journal for Philosophy of Science* 3(2): 207–220.

Biddle, J. and E. Winsberg (2009) "Value judgments and the estimation of uncertainty in climate modeling." In P. D. Magnus & J. Busch (eds.) *New Waves in the Philosophy of Science*. New York: Palgrave MacMillan.

Bindoff, N. L. et al. (2013) "Detection and attribution of climate change: from global to regional." In T.F. Stocker et al. (eds.) *Climate Change 2013: The Physical Science Basis*. New York: Cambridge University Press.

Broome, J. (2008) "The ethics of climate change," *Scientific American*, June 2008: 69–73.

Brown, M. (2013) "Values in science beyond underdetermination and inductive risk," *Philosophy of Science* 80(5): 829–839.

Deser, C. et al. (2014) "Projecting north American climate over the next 50 years: uncertainty due to internal variability," *Journal of Climate* 27: 2771–2296.

Dessai, S. et al. (2009) "Climate prediction: a limit to adaptation?" In W.N. Adger, I. Lorenzoni and K. L.O'Brien (eds.) *Adapting to Climate Change: Thresholds, Values, Governance*. Cambridge: Cambridge University Press.

Douglas, H. (2009) *Science, Policy, and the Value-Free Ideal*. Pittsburgh: Pittsburgh University Press.

Elliott, K.C. and D. McKaughan (2014) "Non-epistemic values and the multiple goals of science," *Philosophy of Science* 81(1): 1–24.

Elliott, K.C. and D. B. Resnik (2014) "Science, policy and the transparency of values," *Environmental Health Perspectives* 122(7): 647–650.

Frisch, M. (2013) "Modeling climate policies: a critical look at integrated assessment modeling," *Philosophy and Technology* 26(2): 117–137.

Gutowski, W. J. et al. (2016) "WCRP Coordinated Regional Downscaling EXperiment (CORDEX): a diagnostic MIP to CMIP6," *Geoscientific Model Development* 9(11): 4087–4095.

Heaphy, L. J. (2015) "The role of climate models in adaptation decision-making: the case of the UK Climate Projections 2009," *European Journal for Philosophy of Science* 5(2): 233–257.

Hegerl, G. and F. Zwiers (2011) "Use of models in detection and attribution of climate change," *WIREs Climate Change* 2: 570–591.

Hempel, C. G. (1965) "Science and human values," in Hempel, C. G. *Aspects of Scientific Explanation and other Essays in the Philosophy of Science*. New York: The Free Press.

Hudson, R. (2015) "Why we should not reject the value-free ideal," *Perspectives on Science* 24(2): 167–191.

Intemann, K. (2015) "Distinguishing between legitimate and illegitimate values in climate modeling," *European Journal for Philosophy of Science* 5(2): 217–232.

IPCC (2013) "Summary for policymakers," in Stocker T.F. et al. (eds.) *Climate Change 2013: The Physical Science Basis*. New York: Cambridge University Press.

Kitcher, P. (2001) *Science, Truth and Democracy*. Oxford: Oxford University Press.

Kloprogge, P., J. P. van der Sluijs, and A. C. Petersen (2011). "A method for the analysis of assumptions in model-based environmental assessments," *Environmental Modelling and Software* 26(3): 289–301.

Parker, W. S. (2013) "Ensemble modeling, uncertainty and robust predictions," *Wiley Interdisciplinary Reviews (WIREs) Climate Change* 4: 213–223.

Parker, W. S. and G. Lusk (2019) "Incorporating user values into climate services," *Bulletin of the American Meteorological Society* 100(9): 1643–1650

Parker, W. S. and J. S. Risbey (2015) "False precision, surprise and improved uncertainty assessment," *Philosophical Transactions of the Royal Society A* 373: 20140453.

Parker, W. S. and E. Winsberg (2018) "Values and evidence: how models make a difference," *European Journal for Philosophy of Science* 8(1): 125–142

Petersen, A. (2012) *Simulating Nature: A Philosophical Study of Computer-Simulation Uncertainties and Their Role in Climate Science and Policy Advice, Second Edition.* Boca Raton, FL: Chapman & Hall/CRC.

Schroeder, S. Andrew (2021) "Democratic value: a better foundation for public trust in science," *British Journal for the Philosophy of Science* 72(2): 545–562.

Steel, D. (2016) "Climate change and second-order uncertainty: defending a generalized, normative, and structural argument from inductive risk," *Perspectives on Science* 24(6): 696–721.

Steele, K. (2012) "The scientist qua policy advisor makes value judgments," *Philosophy of Science* 79: 893–904.

Taylor, K. E., R. J. Stouffer, and G. A. Meehl (2012) "An overview of CMIP5 and the experiment design," *Bulletin of the American Meteorological Society* 90: 485–498.

Tebaldi, C. and R. Knutti (2007) "The use of the multi-model ensemble in probabilistic climate projections," *Philosophical Transactions of the Royal Society A* 365: 2053–2075.

Tuana, N. (2010) "Leading with ethics, aiming for policy," *Synthese* 177(3): 471–492.

Vaughan, C. and S. Dessai (2014) "Climate services for society: origins, institutional arrangements, and design elements for an evaluation framework," *WIREs Climate Change* 5: 587–603.

Winsberg, E. (2012) "Values and uncertainties in the predictions of global climate models," *Kennedy Institute of Ethics Journal* 22(2): 111–137.

Winsberg, E. (2018) *Philosophy and Climate Science.* Cambridge: Cambridge University Press.

22

CLIMATE CHANGE MITIGATION

Marcus Hedahl, Kyle Fruh, and Lindsay Whitlow

If our atmosphere were a blanket into which we have been stuffing more and more insulation, we might try to adapt to our new surroundings, inching toward the edges in search of cooler air. Or, we might consider re-engineering our environment, opening a window so that the room gets colder and our super-insulated blanket ends up producing the appropriate amount of protection. Perhaps the most obvious solution in this analogy is mitigation: Finding some way to return the blanket to an appropriate level of insulation. In fact, early discussions of climate change focused almost exclusively on mitigation, more specifically, on *abatement*, the drawing down of our heretofore prolific greenhouse gas (GHG) emissions, and on *deconcentration*, reducing the concentration of GHGs already in the atmosphere (e.g. by enhancing carbon sinks (Jamieson 2014: Ch. 7)).

Given the current status of the climate crisis, however, preventing climate change altogether is no longer a possibility. There is simply no way to avoid serious, climate-related harms. We have already seen atmospheric carbon dioxide rise above 400 ppm and a temperature increase greater than 0.7 degrees centigrade since the beginning of the 20th century (Stocker et al. 2013). Even with an immediate and radical reduction in our emissions, an even greater temperature rise is almost certain. This dire state of affairs has made many recognize the need to adapt (Baer 2010), and even to examine limited geoengineering efforts (Gardiner 2010; Fruh & Hedahl 2019). Nonetheless, while mitigation, adaptation, and geoengineering are not mutually exclusive options, mitigation remains an indispensable part of any morally appropriate response to climate change.

Scientific and Economic Reasons for Mitigation

At a basic level, the planet's temperature depends upon how much heat from the sun is held in the atmosphere rather than escaping back into space. The chemical properties of GHGs make them a primary driver of this radiative forcing: The more GHGs are in the air, the warmer the atmosphere will be. Atmospheric concentrations of carbon dioxide (CO_2), methane (CH_4), and nitrous oxide (N_2O)—the major anthropogenic GHGs—have all arrived at levels unprecedented in over 800,000 years (see Figure 22.1). Perhaps even more troubling, since the industrial revolution, the rate at which those levels have been increasing is similarly unprecedented (Ritchie and Roser 2020).

Climate change has already had a significant impact on our planet (Field et al. 2014). The last 30 years have been warmer than any since 1850 and they are very likely the warmest in the last 1,400 years (Stocker et al. 2013). As temperature has risen, Arctic Sea ice has declined (Renner et al. 2014) and mountain glaciers have receded (WGMS 2011). Warmer air temperatures have

DOI: 10.4324/9781315768090-27

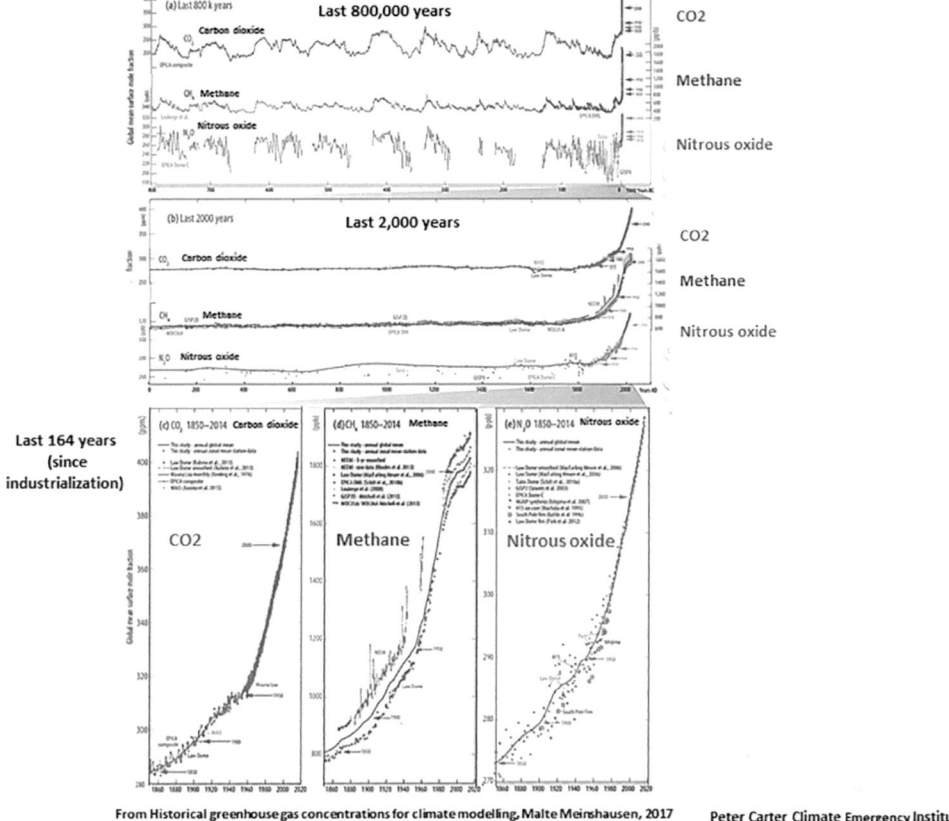

Atmospheric CO2, methane and nitrous oxide concentrations from 800,000 years ago, 2000 years ago & from industrialization until today 2017

From Historical greenhouse gas concentrations for climate modelling, Malte Meinshausen, 2017   Peter Carter Climate Emergency Institute

Figure 22.1 Major GHG concentrations over the past 800,000 years; 10,000 years; and 100 years (from Carter 2015, data sources: Solomon et al. 2007; Luthi et al. 2008 & Hartmann et al. 2013)

caused increases in ocean temperatures (Abraham et al. 2013), resulting in higher sea levels (Kuhlbrodt & Gregory 2012) as well as an increase in both the frequency and strength of hurricanes and typhoons (Mei et al. 2015). Earlier spring temperatures have disrupted critical ecosystem services on which human society depends (i.e. clean air and water, and crop pollination) (Staudinger et al. 2012). These impacts have been even more dramatic for those already vulnerable (Field et al. 2014). Water resources have become scarce and more highly variable (Haddeland et al. 2014), with associated negative consequences for crop yields (Lobell et al. 2011) as well as food security overall (Shindell et al. 2012). At the end of the last decade, in the Arctic winter, when temperatures should be plunging and sea ice should be expanding rapidly, temperatures were soaring, and sea ice was actually shrinking (Yulsman 2016). The collective scientific evidence is overwhelming: Climate change has already radically altered the ecosystems on which human health depends (Pecl et al. 2017).

The scientific evidence is also clear that the future impacts of climate change will be even more dire (see Figure 22.2). Heat waves, droughts, floods, and storms are likely to become more frequent or more severe (Coumou & Rahmstorf 2012). Precipitation will likely increase overall, but

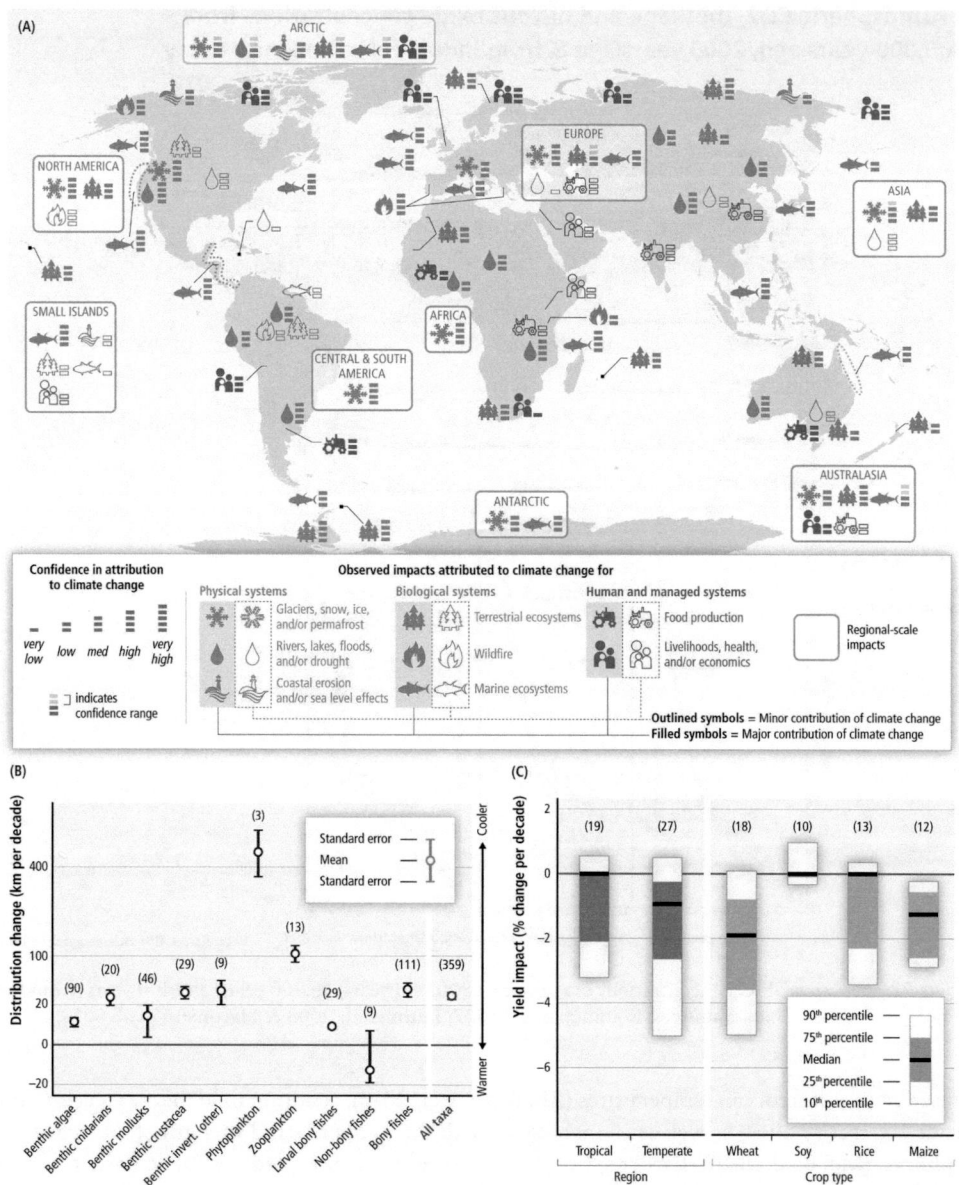

Figure 22.2 Global patterns of impacts in recent decades attributed to climate change, based on studies since the AR4. Impacts are shown at a range of geographic scales. Symbols indicate categories of attributed impacts, the relative contribution of climate change (major or minor) to the observed impact, and confidence in attribution (from Field et al. 2014)

there will be sharp regional variations, with some areas that now receive regular rainfall becoming arid (O'Gorman 2012). Coastal communities face risks of losing land to sea level rise (Kates et al. 2012; Masson-Delmotte et al. 2018) as well as loss of livelihoods through declines in fisheries catches (Cinner et al. 2012). Global health concerns will be exacerbated with climate changes, as factors that affect exposure, transmission, and resilience are heightened (McMichael 2013). Regional changes will often be variable, extreme, and much more difficult to predict (Aldous et al. 2011; Field et al. 2014).

Given the extent of these impacts, the scientific evidence highlights an urgent need to mitigate (Masson-Delmotte et al. 2018). The longer decisions are delayed, the fewer potential options will be available (Rogelj et al. 2013). Rich nations may be able to cope with some of these changes without enormous loss of life: They are often able to store food against the possibility of drought, to move people away from flooded areas, to fight the spread of disease-carrying insects, and to adapt to a rapidly changing environment (Lesnikowski et al. 2016). However, the cost of adaptation—even relatively limited adaptation—will be extreme (Dietz and Stern 2015). If the most conservative scientific estimates are accurate, unabated climate change would cost the world more than 5% of our total Gross Domestic Product (GDP) each year; if some of the more severe possibilities become realized, that cost could balloon past 20% (Stern 2007). While each average ton of CO_2 we emit causes *at least* (and likely significantly more than) $85 worth of damages, those emissions can be cut at a cost of *less than* (and likely significantly less than) $25 per ton (Stern 2007). The science and the economics are clear: Given the current climate crisis, mitigation is by far the most rational course of action.

Moral Reasons for Mitigation

The previous section shows that it is best to adopt mitigation strategies in response to climate change. But it's less clear that it is in the interest of any particular nation or individual to pursue mitigation, not least because mitigation doesn't come for free. A childless septuagenarian who wants to travel the world in private jets won't be made better off by mitigation-related restrictions on low-carrier flights. A country whose only resource of note is fossil fuel won't have its interests served by expansive programs to phase out oil dependence (Saudi Arabia, for example, has demanded compensation for any international agreement that would encourage shifting away from oil (Jamieson 2010: 269)). Even though climate change mitigation is *generally prudent*, even when it's not, an impressive array of reasons supports the conclusion that climate change mitigation remains *morally necessary*.

We can start to see those reasons by recognizing that when an especially damaging natural disaster destroys homes and takes lives, there is often a great outpouring of emotional and material support in the aftermath. Offering such support is morally good and, in many cases, failing to do so would constitute a moral failing. The rest of the global community has some duty to aid those who, through sheer accident, have fallen into desperation. We cannot, in good conscience, merely stand by while many die and suffer needlessly, especially when offering aid comes with relatively little cost. But, if there are good moral reasons to help those in need after a disaster, those same reasons also speak at least as strongly in favor of avoiding the hardship altogether. We have duties of beneficence to shield those who suffer from climate change's damaging effects and to assist them in the wake of climate-related harms (cf. Baer 2010: 252).

Yet, climate change does not merely place duties of beneficence upon us, because climate change is not a fully natural disaster. For those harmed by climate change, it is not the bare fact of desperate need that calls out for moral redress. Climate change does not merely present the rest of us with an opportunity to be a Good Samaritan. We are, each of us, more robber than Samaritan: It is our own emissions that are causing grievously wrongful harms (Fruh and Hedahl 2013). Once we recognize that mitigating climate change involves *refraining from harming* others rather than *being beneficent toward* others, we identify another, even more stringent and forceful moral reason for mitigation (Broome 2012: 55–59). In today's tightly interconnected world, the venerable requirement to not harm others places a number of new and significant obligations of justice upon us.

Consider relief efforts in Vanuatu after the destruction of Cyclone Pam in early 2015. There is an important moral difference between those efforts and the relief efforts in Haiti after the earthquake of 2010. We have good reasons to believe that Cyclone Pam would not have been as destructive without the warmer ocean temperatures created by GHG emissions. There is, obviously, a pernicious epistemic problem here due to the complexity of the climate system and our tangled, shared history of GHG emissions (Otto et al. 2017). One thing is nonetheless clear: Regardless of how much of the damage could be ascribed to climate change, the people of Vanuatu are not responsible for it. They have contributed almost nothing to climate change: Recent data ranks Vanuatu near the bottom for both per capita and total CO_2 emissions; they produce 0.000014 as much CO_2 as China (Boden et al. 2014). The failure to provide needed aid would therefore not just be cold-hearted, nor would the failure to take reasonable measures to prevent similar disasters be merely reckless; both failures would be unjust.

One of the distinguishing features of duties of justice is that they often correspond to the rights of others (Broome 2012: 52). When rising sea levels and more frequent, high-intensity storms cause deaths, displacement, and famine in poor, low-lying communities, climate change threatens both the so-called first-generation rights to life, liberty, and security and the so-called second-generation rights to food, shelter, and healthcare (Caney 2010a; Shue 2010b). When Tuvalu's 11,000 citizens are forced to leave their uninhabitable, flooding country, for instance, it would be an outrage to deny them refugee status (even though there is presently no UN provision for climate refugees (McAdam 2012)). In fact, we will almost certainly owe them much more than refugee status or piecemeal migration plans, for those harmed by anthropogenic climate change have a clear right not to be, and when they are, some form of restitution is required (Nine 2010; Dietrich & Wündisch 2015; Angell 2017; Buxton 2019).

Yet, these contentions about rights are difficult to reconcile with a prominent theoretical analysis of moral claim rights, an analysis that holds that a moral right exists if and only if a corresponding *directed obligation* exists (O'Neil 2005; Darwall 2012). Consider, for example, your right not to be assaulted. In this case, the content of your moral claim right is the same as the content of the corresponding directed obligation each and every one of us owes to you. You have a claim against each of us that we not assault you, and each of us, in turn, owes it *to you* to refrain from doing so. A directed obligation thus stands in contrast to a general duty that constrains behavior but is not owed to anyone in particular. While it could be the case, for example, that each of us ought to be more charitable, in standard circumstances at least these moral requirements are general rather than directed: No particular agent has a claim against anyone else's charity. Problematically, the traditional analysis of claim rights that insists on corresponding directed obligations as a necessary condition for the existence of a moral claim right must exclude nearly all ascriptions of rights associated with climate change. As things stand, for those most vulnerable to the most adverse effects of climate change, there simply is not yet any specifically designated body that their rights claims can be claims against.

Perhaps unsurprisingly, many take this limitation to provide sufficient grounds to reject a strict correlation between rights and directed obligations (Raz 1986; Sen 2004; Ashford 2006). Yet while this move away from a strict correlativity requirement between rights and directed obligations gives us the ability to call claims to goods and services rights and gives us the ability to call claims against being harmed through climate change rights, it risks losing the important inter-relational aspect of that designation. For now, those unjustly harmed by climate change cannot demand as their due any specific action from anyone in particular.

One solution to this tension is to contend that moral claim rights do require corresponding directed obligations, but that requirement is normative rather than descriptive. A moral right ought to engender directed obligations on the part of others, but it need not currently do so in order to be properly analyzed as a right (Hedahl 2014). If the world is aligned such that some moral rights do not correspond to another agent's or collective's directed duty, in some cases at least, we should regard that fact as a further normative failure (Shue 2010b). In the wake of Cyclone Pam, facing food shortages amid near-total destruction, the people of Vanuatu not only called out for aid provisions, but also appealed for the establishment of an institution to whom damage claims related to climate change could be addressed (Bartlett 2015). The fact that there is presently no such institution deepens rather than diminishes the sense in which rights are not being honored and climate victims are being wronged.

Considering climate change as an issue of rights and justice is powerful in part because it requires those harmed by climate change to be provided for and, to whatever extent possible, made whole. Yet, it would be bizarre to think that restitution, on its own, could be an adequate redress for harms due to climate change. The harms of climate change are ongoing, and if the perpetrators' contributions to climate change are also ongoing, restitution of even a very ambitious kind cannot hope to be adequate. As Simon Caney (2010a) has argued, "it is obviously impermissible for one person to assault someone else with a view to giving them a large benefit in order somehow to cancel out the harm" (172). However generous perpetrators of climate change are in facilitating adaptations for victims or in compensating them for past harms, the moral story of their actions is not complete without some form of mitigation. Justice demands discontinuing and forswearing, rather than merely compensating for, offending behavior.

The moral case for mitigation becomes even stronger, given that we know that the impacts of climate change will not be regionally uniform. Many developing countries are most vulnerable to the most adverse consequences (Stocker et al. 2013). Yet, the disparity in climate change burdens is not reducible to accidents of geography. It is not merely an unfortunate coincidence that those most likely to suffer the worst effects of climate change are those who are least likely to have contributed to the problem (Singer 2002: 31). The very same underlying social structures that lead to some bearing a greater causal responsibility for climate change lead to those same actors being in a better position to adapt to the hardships climate change imposes (McCormick 2016). In other words, anthropogenic climate change significantly magnifies existing distributional injustices. Individuals and nations therefore also have a moral duty to mitigate because to not do so would be to exacerbate existing unjustifiable inequalities (Shue 2010a; Francis 2015).

Differences in vulnerability become even more salient when we consider intergenerational justice (Barry 1997; Broome 2012). One purpose of the language of justice and rights is to provide normative protections for the vulnerable, and there is perhaps no population more vulnerable than the future generations of our planet (Shue 2010c). In a world without mitigation, future generations face sharply foreclosed options in carving out lives that are worth living. They face aggravated risks of catastrophe, increasingly volatile and unpredictable weather, and daunting challenges with respect to the maintenance of basic infrastructure and food supply (Shue 2010c). To unilaterally impose this state of affairs on future generations is unjust. However successful and sweeping adaptation measures might be, mitigation remains morally indispensable because of the irremediable vulnerability of future generations.

Finally, there are also important moral reasons to mitigate that are independent of human interests. Climate change could threaten as many as one quarter of the world's species in this century alone (Thomas et al. 2004). Plants and ecosystems are endangered, particularly those in Arctic and Alpine ecosystems (Parmesan 2006). Earlier spring temperatures have enabled more invasive species to thrive (Hellmann et al. 2008) and triggered earlier movements of migrating species (Both et al. 2006), both of which have decoupled ecological relationships with critical plant resources (Bellard et al. 2012). In addition, higher atmospheric concentrations of CO_2 have increased the level of carbonic acid in ocean water (Kroeker et al. 2013), reducing the levels of plankton that are the foundation of ocean food webs (Harley et al. 2006). Climate change also threatens the very possibility of wilderness by placing the maladroit influence of human beings squarely into each and every ecological relationship (McKibben 2006; Meyer 2006; Monbiot 2014). For many, these non-anthropocentric reasons provide further moral justification for the claim that mitigation (and especially abatement) must be part of any morally appropriate response to climate change.

Collective Duties to Mitigate

The previous section argued that climate change mitigation is a moral requirement. As former UN Human Rights Commissioner Mary Robinson notes, recognizing that imperative is not without consequence: Viewing climate change as a question of justice and injustice implies "a radical agenda—a transformative agenda" (2014: 19). Too often, collectives involved in international climate negotiations treat that undertaking as a technical issue—similar to a trade accord—in which participants merely seek to maximally advance their nation's own best interests. It has led some nations to proclaim, for example, that for now at least, "We've done our part" (as Poland did in the lead up to the 2013 Conference of Parties in Warsaw (Szulecki 2013)). That mindset is understandable, but it is fundamentally flawed: What collectives are obligated to do will be determined, in large part, by what we bind ourselves to do in procedurally fair negotiations (Richardson 2008). To believe that what constitutes a fair share of collective climate burdens already exists and is merely waiting to be discovered and executed not only betrays a misunderstanding of climate action as a zero-sum distribution of benefits and burdens, but also leads to an international paralysis that is devastating for the planet (cf. Light 2011). In other words, it fails to recognize the nature of our collective duties to mitigate.

These collective duties to mitigate are important for several reasons. Collectives (like states and corporations) are best positioned to bring about the most powerful changes in emissions patterns (cf. Broome 2012: 65). Additionally, it is largely as collectives that we have created the problem of climate change: Through our various institutions of governance and decision-making, we have adopted policies that have created the situation in which we find ourselves (Shue 2010a). Consider, for example, the problem of stranded assets. The amount of carbon our generation can safely emit (the amount we can emit while avoiding the most devastating consequences of climate change) is several times less than the amount of carbon contained within current fossil fuel assets, assets already accounted for in the economic assessments of various companies and countries (Stern 2011). Reasonable climate action will therefore unavoidably require some significant percentage of those assets (perhaps as much as 80%) to stay in the ground (Leaton et al. 2013). This is a collective problem: It was not created merely by individual desires for energy, but by collectively forged policies at the local, national, and international levels. And it requires a collective solution: Any effort to strand assets with economic value will require more than individual efficiency gains and individual reductions in consumption. To fail to notice our collective duties is to fail to take seriously both the history and the potential of collective action.

Furthermore, it is largely as collectives that we in the developed world continue to derive benefits from past emission-intensive activity (e.g. previous industrialization (Baer 2010)). So, while every state bears collective duties to mitigate climate change, those duties will require more sacrifice for some states than for others, in part because different states have contributed differently to the injustices of climate change. In particular, the history of GHG emissions requires that developed nations bear a greater burden than developing nations do. After all, "if someone has produced a harm … then they should rectify that situation. They, as the causers, are responsible for the ill-effects" (Caney 2010b: 126). This principle holds for our actions as individuals, as well as for our collective actions as nation states.

According to some theorists, however, developed nations can only be fully responsible for their emissions from the very recent past (since the last few decades of the 20th century), because only recently could they have been reasonably expected to know that their emitting behavior was harming others (Caney 2010b). In ethical theory, there is a nearly universally recognized limitation on the possibility of full responsibility if an agent was excusably ignorant of the consequences of her actions. We do not blame those who could not have known about the harms they were causing, nor, in many cases, do we hold that they can be required to compensate for the total costs of all the harms they caused *regardless* of what the costs of that compensation turn out to be (Bell 2013). In the case of climate change, for the vast majority of the 19th and 20th centuries, no state could have been reasonably expected to appreciate the effects of its GHG emissions, and so, on this line of reasoning, no state can be held fully responsible for those historic emissions (Caney 2010b: 134).

Yet, developed nations are still benefiting from their historical emissions. The infrastructure built in the 19th and early 20th centuries continues to provide advantages for those nations today. The fact that they continue to benefit from their historical emissions-generating activities justifies imposing a form of limited liability in which the burdens created by historical activity can be as great as but not greater than the benefits derived from those activities (Bell 2013: 400). While the membership in these collectives has changed, the collectives themselves have often endured. The United States is still the United States. If it wants to enjoy collective property protections created by the actions of its past members (e.g. if it wants to be able to rebuff Russia's contention that the sale of Alaska is no longer binding because no current members of Russia benefited from its sale), then it has to be willing to accept some moral consequences for its past actions. The duties faced by a collective as a collective must account in some way for its own past emissions.

Individual Duties to Mitigate

Some have argued that the duties of climate change fall *only* on large collectives, such as nation states and corporations (Sinnott-Armstrong 2005). According to such theorists, while nations can bind their citizens with laws and regulations, until such laws or regulations go into effect, individual agents do not have any moral duties to compensate victims of climate change or diminish their own GHG emissions. On this problematic line of reasoning, to say that collectives have moral duties to mitigate climate change is not yet to say that we, individually, have any such duties. Understanding the problems with this position allows us not only to recover the common sense notion that we all have individual duties of mitigation, but also to understand more clearly the variety of forms those duties take: To compensate for harms imposed, to attenuate our participation in injustice, and to agitate for institutional change.

While it's true that the causal contribution of any given individual lacks the dramatic scale of the causal impact of collective actors, such a criterion cannot be the only one that matters regarding duties to avoid acts of injustice. Individuals arguably have a general duty not to be complicit in injustices (Kutz 2000). If climate change results in injustices, then engaging in GHG-emitting activities at the very least raises concerns of complicity in those injustices. Yet to call these misdeeds merely complicitous is to understate the role we play in producing the harms of climate change. Each of us makes active, causal contributions to the GHG load in the atmosphere; we are not merely standing by and allowing the injustice to continue when we could do otherwise. Each of us is instead *participating* in creating the moral problems of climate change—even if our individual activities prove to be materially inconsequential. If we have some duty to attenuate our complicity in injustices, we have an even stronger duty to attenuate our participation in injustices (Fruh & Hedahl 2013).

Moreover, to believe that there are collective climate change duties only—that individuals must wait for collectives to convert those duties into their individual correlates in order to be duty bound—risks misrepresenting how representative collectives function. Even if all of an agent's individual activity were completely carbon neutral, she maintains some individual responsibility for the actions and status of the collective and institutional structures of which she is a part (May 1992; List and Pettit 2011). In short, each of us is performing actions that allow us to accrue benefits and pass on externalities that constitute rights-violating harms. We find ourselves in a morally fractured world, without collective agreement that would constrain us. The current system lurches forward, as do the aggregate and devastating consequences of our actions and our inactions. In this world, we must either allow that there is not any individual wrong in continuing to derive benefit from actions in a system delivering grave harms to others, or we must accept that each of us has some individual duty to stop or at least attenuate offending behavior and lobby for reform. Individual duties to mitigate are therefore a necessary consequence of any justification for climate change duties grounded in rights and justice.

In order to specify the content of these individual duties, we can begin by recognizing that one noteworthy strategy for individual GHG mitigation is sink enhancement, widely available in the form of carbon offsets. It is quite possible to reach a zero-carbon footprint by ensuring that the amount of GHGs you emit equals the amount of GHGs you cause to be taken out of the atmosphere (e.g. by paying for trees to be planted). John Broome (2012: 91) argues that this sort of carbon neutrality is precisely what is morally required. Theoretically, if everyone carried on emitting at present rates, but also invested in sufficient offsets to reach GHG neutrality, there would be nothing wrong with even very high levels of emissions.

Despite the moral significance of offsets, there are three important limitations to this approach. First, sink enhancement has limits: It's not clear what total amount of GHG emission can be effectively offset (Duncan et al. 2011). Second, sequestering carbon (through reforestation, for example) is a far less stable arrangement than leaving it in the geological formations from which it is originally extracted (Jamieson 2014: 205). Third, some types of offsets seem ill-suited to meet our full complement of climate change duties. For instance, "preventative" offsets work by ensuring that carbon that otherwise would have been emitted is not. One may, for example, look to achieve carbon neutrality by funding more efficient energy use in the developing world, thereby preventing more emissions than one creates (Broome 2012: 87). Putting your own resources into promoting abatement efforts of others is indisputably good, but it's not clear that it's a plausible means of completely canceling your own participation in causing climate harms. As a means of exhaustively discharging one's duties to mitigate climate change, helping *others* to reduce their participation sits uneasily in the logic of justice that lies at the heart of climate ethics.

So, while we ought to regard carbon offsets as an underutilized means for satisfying part of our duties with respect to attenuating our participation in climate injustice, they cannot stand alone. We must augment them with personal abatement; we have duties to cut back on our own GHG-intensive activity in order to minimize our participation in injustice. The duty will require more sacrifice for some than others, in large part because not everyone participates in the injustices of climate change equally. The average American is, for example, a notoriously profligate GHG emitter, and so our duty to not participate in injustice will be more invasive than the same duty borne by an average Ethiopian, despite the fact that it is, at bottom, the same duty (Singer 2002: 35).

Since the duty to attenuate one's participation in injustice attaches to emissions, rather than to any particular activity as such, individual moral agents will have some latitude in determining how to fulfill it. For some, cutting back on travel might be required, particularly air travel: For each passenger, two round-trip journeys from New York to San Francisco create a larger carbon footprint than the average global citizen produces in an entire year (Rosenthal 2013). For others, the duty to mitigate might (also) involve eating significantly less meat, particularly red meat: For most of us living in more affluent nations, changing to a vegetarian diet would have a far greater carbon impact than switching to a more fuel-efficient vehicle (Eshel et al. 2014). Finally, for some, this duty may even require considering having fewer biological children, perhaps by having a smaller family or perhaps adopting rather than procreating: The carbon impact of creating a child is, for nearly all of us, greater than the sum of all the "lifestyle" decisions we make during our adult lives (Rieder 2014).

Important as these duties to avoid participation in climate change are, they must not over-shadow our duties to alter our political and communal orders (Shrader-Frechette 2007). The legitimate demands of those wronged by anthropogenic climate change follow a familiar legal and social pattern: Claims of injustice are often advanced as demands against the *status quo*, as a fundamental challenge to an existing system of entrenched social practices, rather than merely a complaint about a specific act of exploitation or an individual instance of unjust treatment. These calls for justice are systemic; they call out for revolution or at the very least radical transformation of our shared social order (cf. Light 2011).

The moral duties of individual citizens to agitate for institutional change at the highest level of government are surely important and might be fulfilled in any number of ways: The global climate marches in the fall of 2014, aimed at influencing the UN, are a good example (cf. Broome 2012: 73–74). Each of us also belongs, however, to different social organizations and institutions, and each of us has different opportunities to push our various collective efforts to-ward change. Many of us involved in college communities, for example, may be better advised to focus our political efforts on the growing divestment movement rather than on lobbying Congress or Parliament directly (Gelles 2015). While divestment of university endowments from fossil fuel companies is unlikely to radically alter the industry's internal finances, it can undermine the industry's significant political power, an important requirement for meaningful structural change (Ansar et al. 2013).

Regardless of how we meet these individual duties to agitate for institutional change, we can only begin to do so by recognizing that combatting climate change will require dramatic political action at all levels. The solutions required to solve the climate crises are very likely not what they were 20, 10, or even 5 years ago (Hedahl & Rieder 2017). We need an Apollo Program, a Manhattan Project, not simply an expansion of New England's cap and trade market or mere compliance with the Paris Climate Accords (McKibben 2016; Rogelj et al. 2016).

Conclusion

Each person reading these words is almost assuredly an order of magnitude above sustainable GHG emissions (Vaughan 2009). This is, we have argued, incompatible with our climate change duties: Collective duties to mitigate our emissions, individual duties to attenuate our participation in injustice, and individual duties to agitate for political change. Faced with this harsh moral reality, denying climate change altogether ("Look at this snowball"), minimiz-ing its significance ("We have other priorities"), and becoming complacent to our continued inaction ("What difference does it make?") are all rather unsurprising responses. We are well equipped with psychological mechanisms to save ourselves from burdensome solutions by rationalizing away problems (Campbell & Kay 2014). Acknowledging the claims of justice is encumbering, and there is even the risk that the pursuit of climate justice could delay mean-ingful climate action (Broome 2012: 47).

Yet, turning away from these powerful moral reasons in the face of our moral misdeeds raises the specter of what Stephen Gardiner (2013) has called "moral corruption." Avoiding that cor-ruption may prove to be no small task. As Martin Luther King Jr. recognized, if reform can only be brought about through more sweeping changes in the underlying structures of society, then reformers must become revolutionaries (King 1967: 8). Climate justice presents us with a similar challenge: In the pivotal few years that lie ahead—possibly our last best chance to avoid the most devastating effects of climate change—we must aspire to a moral integrity that embraces both our past wrongs and our current, rather significant duties to mitigate (Masson-Delmotte et al. 2018). Doing so may be the only way to bring about the radical alterations of the *status quo* likely neces-sary to adequately respond to the climate crisis.

Related Topics

10. Wilderness
23. Climate Justice and Equity
24. Geoengineering
30. Renewable Energy
49. Pollution and Polluter Pays

References

Abraham, J., Baringer, M., Bindoff, N., Boyer, T., Cheng, L., Church, J., Conroy, J., Domingues, C., Fasullo, J., Gilson, J., Goni, G., Good, S., Gorman, J., Gouretski, V., Ishii, M., Johnson, G., Kizu, S., Lyman, J., Macdonald, A. Minkowycz, W., Moffitt, S., Palmer, M., Piola, A., Reseghetti, F., Schuckmann, K., Trenberth, K., Velicogna, I., & Willis, J. (2013) "A Review of Global Ocean Temperature Observations: Implications for Ocean Heat Content Estimates and Climate Change," *Reviews of Geophysics*, 51, 450–483.

Aldous, A., Fitzsimons, J., Richter, B., & Bach, L. (2011) "Droughts, Floods and Freshwater Ecosystems: Evaluating Climate Change Impacts and Developing Adaptation Strategies," *Marine and Freshwater Research*, 62 (3), 223–231.

Angell, K. (2017) "New Territorial Rights for Sinking States," *European Journal of Political Theory*, 1–21.

Ansar, A., Caldecott, B., & Tilbury, J. (2013) *Stranded Assets and the Fossil Fuel Divestment Campaign: What Does Divestment Mean for the Valuation of Fossil Fuel Assets?* Oxford: Smith School of Enterprise and the Environment, University of Oxford.

Ashford, E. (2006) "Global Ethics: The Inadequacy of Our Traditional Conceptions of the Duties Imposed by Human Rights," *Canadian Journal of Law and Jurisprudence*, 19, 217–235.

Baer, P. (2010) "Adaptation to Climate Change: Who Pays Whom?" in S. Gardiner, S. Caney, D. Jamieson, & H. Shue (eds.) *Climate Ethics: Essential Readings*, Oxford: Oxford University Press.

Barry, B. (1997) "Sustainability and Intergenerational Justice," *Theoria*, 45, 43–65.

Bartlett, C. (2015) "Adviser: Next 3 Months Crucial to Preventing Starvation on Vanuatu," *All Things Considered*, Mar. 20 (M. Block, Interviewer).

Bell, D. (2013) "Global Climate Justice, Historic Emissions, and Excusable Ignorance," *The Monist*, 94 (3), 391–411.

Bellard, C., Bertelsmeier, C., Leadley, P., Thuiller, W., & Courchamp, F. (2012) "Impacts of Climate Change on the Future of Biodiversity," *Ecology Letters*, 15 (4), 365–377.

Boden, T., Andres, B., & Marland, G. (2014) *World's Countries Ranked by 2010 Total Fossil-fuel CO_2 Emissions*, Carbon Dioxide Information Analysis Center, Oak Ridge National Laboratory, Aug. 22. http://cdiac.ornl.gov/trends/emis/top2010.tot [Accessed 30 May 2015].

Both, C., Bouwhuis, S., Lessells, C., & Visser, M. (2006) "Climate Change and Population Declines in a Long-distance Migratory Bird," *Nature*, 441 (7089), 81–83.

Broome, J. (2012) *Climate Matters*, New York: W.W. Norton & Company.

Buxton, R. (2019) "Reparative Justice for Climate Refugees," *Philosophy*, 94, 193–219.

Campbell, T., & Kay, A. (2014) "Solution Aversion: On the Relation Between Ideology and Motivated Disbelief," *Journal of Personality and Social Psychology*, 107 (5), 809–824.

Caney, S. (2010a) "Climate Change, Human Rights, and Moral Thresholds," in S. Gardiner, S. Caney, D. Jamieson, & H. Shue (eds.) *Climate Ethics: Essential Readings*, Oxford: Oxford University Press.

Caney, S. (2010b) "Cosmopolitan Justice, Responsibility, and Global Climate Change," in S. Gardiner, S. Caney, D. Jamieson, & H. Shue (eds.) *Climate Ethics: Essential Readings*, Oxford: Oxford University Press.

Carter, P. (2015) "Today's 2015 Global Climate Change Medical Emergency Response," *Climate Change Emergency Medical Response.org*. http://www.climate-change-emergency-medical-response.org [Accessed 28 June 2015].

Ciais, P., Sabine, C., Bala, G., Bopp, L., Brovkin, V., Canadell, J., Chhabra, A., DeFries, R., Galloway, R., Heimann, M., Jones, C., Le Quéré, C., Myneni, R., Piao, S., & Thornton, P. (2013) "Carbon and Other Biogeochemical Cycles," in T. Stocker, D. Qin, G. Plattner, M. Tignor, S. Allen, J. Boschung, A. Nauels, Y. Xia, V. Bex, & P. Midgley (eds.) *Climate Change 2013: The Physical Science Basis. Contribution of Working Group I to the Fifth Assessment Report of the Intergovernmental Panel on Climate Change*, Cambridge: Cambridge University Press.

Cinner, J., McClanahan, T., Graham, N., Daw, T., Maina, J., Stead, S., Wamukota, A., Brown, K., & Bodin, Ö. (2012) "Vulnerability of Coastal Communities to Key Impacts of Climate Change on Coral Reef Fisheries," *Global Environmental Change*, 22 (1), 12–20.

Coumou, D., & Rahmstorf, S. (2012) "A Decade of Weather Extremes," *Nature Climate Change*, 2 (7), 491–496.

Darwall, S. (2012) "Bi-polar Obligation," in R. Shafer-Landau (ed.) *Oxford Studies in Metaethics*, Oxford: Oxford University Press.

Dietrich, F., & Wündisch, J. (2015) "Territory Lost – Climate Change and the Violation of Self-Determination Rights," *Moral Philosophy and Politics*, 2 (1), 83–105.

Dietz, S., & Stern, N. (2015) "Endogenous Growth, Convexity of Damages and Climate Risk: Nordhaus' Framework Supports Deep Cuts in Carbon Emissions," *The Economic Journal*, 125, 574–620.

Duncan, D. & Morrisey, E. (2011), *The Concept of Geologic Carbon Sequestration*, Reston, VA: U.S. Geological Survey, Energy Resources Program. https://pubs.usgs.gov/fs/2010/3122/pdf/FS2010-3122.pdf. [Accessed 23 March 2022.]

Edenhofer, O., Pichs-Madruga, R., Sokona, Y., Farahani, E., Kadner, S., Seyboth, K., Adler, A., Baum, I., Brunner, S., Eickemeier, P., Kriemann, B., Savolainen, J., Schlömer, S., von Stechow, C., Zwickel, T., & Minx, J. (eds.) (2014) *Climate Change 2014: Mitigation of Climate Change. Contribution of Working Group III to the Fifth Assessment Report of the Intergovernmental Panel on Climate Change*, Cambridge: Cambridge University Press.

Eshel, G., Shepon, A., Makov, T., & Milo, R. (2014) "Land, Irrigation Water, Greenhouse Gas, and Reactive Nitrogen Burdens of Meat, Eggs, and Dairy Production in the United States," *Physical Sciences - Sustainability Science*, 111 (33), 11996–12001.

Field, C., Barros, V., Mach, K., Mastrandrea, M., van Aalst, M., Adger, W., Arent, D., Barnett, J., Betts, R., Bilir, T., Birkmann, J., Carmin, J., Chadee, D., Challinor, A., Chatterjee, M., Cramer, W., Davidson, D., Estrada, Y., Gattuso, J., Hijioka, Y., Hoegh-Guldberg, O., Huang, H., Insarov, G., Jones, R., Kovats, R., Lankao, P., Larsen, J., Losada, I., Marengo, J., McLean, R., Mearns, L., Mechler, R., Morton, J.F, Niang, I., Oki, T., Olwoch, J., Opondo, M., Poloczanska, E., Pörtner, H., Redsteer, M., Reisinger, A., Revi, A., Schmidt, D., Shaw, M., Solecki, W., Stone, D., Stone, J., Strzepek, K., Suarez, A., Tschakert, P., Valentini, R., Vicuña, S., Villamizar, A., Vincent, K., Warren, R., White, L., Wilbanks, T., Wong, P., & Yohe, G. (2014) "Technical Summary," in C. Field, V. Barros, D. Dokken, K. Mach, M. Mastrandrea, T. Bilir, M. Chatterjee, K. Ebi, Y. Estrada, R. Genova, B. Girma, E. Kissel, A. Levy, S. MacCracken, P. Mastrandrea, & L. White (eds.) *Climate Change 2014: Impacts, Adaptation, and Vulnerability. Part A: Global and Sectoral Aspects. Contribution of Working Group II to the Fifth Assessment Report of the Intergovernmental Panel on Climate Change.*

Francis. (2015) *Laudato Si* [Encyclical Letter on Care for our Common Home]. http://w2.vatican.va/content/francesco/en/encyclicals/documents/papa-francesco_20150524_enciclica-laudato-si.html [Accessed 19 June 2015.]

Fruh, K., & Hedahl, M. (2013) "Coping with Climate Change: What Justice Demands of Mormons, Surfers, and the Rest of Us," *Ethics, Policy & Environment*, 6 (3), 1–24.

Fruh, K., & Hedahl, M. (2019) "Climate Change Is Unjust War: Geoengineering and the Rising Tides of War," *The Southern Journal of Philosophy*, 57 (3), 378–401.

Gardiner, S. (2010) "Is 'Arming the Future' with Geoengineering Really the Lesser Evil? Some Doubts about the Ethics of Intentionally Manipulating the Climate System," in S. Gardiner, S. Caney, D. Jamieson, & H. Shue (eds.) *Climate Ethics: Essential Readings*, Oxford: Oxford University Press.

Gardiner, S. (2013) *A Perfect Moral Storm*, Oxford: Oxford University Press.

Gelles, D. (2015) "Fossil Fuel Divestment Movement Harnesses the Power of Shame," *New York Times*, June 13.

Haddeland, I., Heinke, J., Biemans, H., Eisner, S., Flörke, M., Hanasaki, N., Konzmann, M., Ludwig, F., Masaki, Y., Schewe, J., Stacke, T., Zachary, D., Tessler, Z., Wada, Y., & Wisser, D. (2014) "Global Water Resources Affected by Human Interventions and Climate Change," *Proceedings of the National Academy of Sciences*, 111 (9), 3251–3256.

Harley, C., Hughes, A., Hultgren, K., Miner, B., Sorte, C., Thornber, C., Rodriguez, L., Tomanek, L., & Williams, S. (2006) "The Impacts of Climate Change in Coastal Marine Systems," *Ecology Letters*, 9 (2), 228–241.

Hartmann, D., Tank, A., Rusticucci, M., Alexander, L., Brönnimann, S., Charabi, Y., Dentener, F., Dlugokencky, E., Easterling, D., Kaplan, A., Soden, B., Thorne, P., Wild, M., & Zhai, P. (2013) "Observations: Atmosphere and Surface," in T. Stocker, D. Qin, G. Plattner, M. Tignor, S. Allen, J. Boschung, A. Nauels, Y. Xia, V. Bex, & P.M. Midgley (eds.) *Climate Change 2013: The Physical Science Basis. Contribution of Working Group I to the Fifth Assessment Report of the Intergovernmental Panel on Climate Change*, Cambridge: Cambridge University Press.

Hedahl, M. (2014) "Directional Climate Justice: The Normative Relationship Between Moral Claim Rights and Directed Obligations," in A. Grear & C. Gearty (eds.) *Choosing a Future: The Social and Legal Aspects of Climate Change*, Northampton: Edward Elgar Press.

Hedahl, M., & Rieder, T. (2017) "Don't Feed the Trolls: Bold Climate Action in a New, Golden Age of Denialism," *Kennedy Institute of Ethics Journal: Special Edition on 2016 Election*, July 2017.

Hellmann, J., Byers, J., Bierwagen, B., & Dukes, J. (2008) "Five Potential Consequences of Climate Change for Invasive Species," *Conservation Biology*, 22 (3), 534–543.

Jamieson, D. (2010) "Adaptation, Mitigation, and Justice," in S. Gardiner, S. Caney, D. Jamieson, & H. Shue (eds.) *Climate Ethics: Essential Readings*, Oxford: Oxford University Press.

Jamieson, D. (2014) *Reason in a Dark Time*, Oxford: Oxford University Press.

Kates, R., Travis, W., & Wilbanks, T. (2012) "Transformational Adaptation when Incremental Adaptations to Climate Change Are Insufficient," *Proceedings of the National Academy of Sciences*, 109 (19), 7156–7161.

King, Jr. M. (1967) "SCLC Retreat Speech," Frogmore, South Carolina, May.

Kroeker, K., Kordas, R., Crim, R., Hendriks, I., Ramajo, L., Singh, G., & Gattuso, J. (2013) "Impacts of Ocean Acidification on Marine Organisms: Quantifying Sensitivities and Interaction with Warming," *Global Change Biology*, 19 (6), 1884–1896.

Kuhlbrodt, T., & Gregory, J. (2012) "Ocean Heat Uptake and its Consequences for the Magnitude of Sea Level Rise and Climate Change," *Geophysical Research Letters*, 39 (18), 1–6.

Kutz, C. (2000) *Complicity*, Cambridge: Cambridge University Press.

Leaton, J., Ranger, N., Ward, B., Sussams, L., & Brown, M. (2013) *Unburnable Carbon: Wasted Capital and Stranded Assets*, Carbon Tracker Initiative & Grantham Research Institute for Climate Change and the Environment. http://www.carbontracker.org/wp-content/uploads/2014/09/Unburnable-Carbon-2-Web-Version.pdf [Accessed 6 September 2015].

Lesnikowski, A., Ford, J., Biesbroek, R. et al. (2016) "National-Level Progress on Adaptation," *Nature Climate Change*, 6, 261–264.

Light, A. (2011) "Climate Ethics for Climate Action," in D. Schmidtz & E. Willott (eds.) *Environmental Ethics: What Really Matters? What Really Works?* Oxford: Oxford University Press.

List, C., & Pettit, P. (2011) *Group Agency: The Possibility, Design, and Status of Corporate Agents*, Oxford: Oxford University Press.

Lobell, D., Schlenker, W., & Costa-Roberts, J. (2011) "Climate Trends and Global Crop Production Since 1980," *Science*, 333 (6042), 616–620.

Lüthi, D., et al. (2008) "EPICA Dome C Ice Core 800KYr Carbon Dioxide Data. IGBP PAGES/World Data Center for Paleoclimatology," *Data Contribution Series # 2008-055*, NOAA/NCDC Paleoclimatology Program, Boulder.

Masson-Delmotte, V., Zhai, P., Pörtner, H.-O., Roberts, D., Skea, J., Shukla, P.R., Pirani, A., Moufouma-Okia, W., Péan, C., Pidcock, R., Connors, S., Matthews, J.B.R., Chen, Y., Zhou, X., Gomis, M.I., Lonnoy, E., Maycock, T., Tignor, M., & Waterfield, T. (2018) *Global Warming of 1.5°C: An IPCC Special Report on the Impacts of Global Warming of 1.5°C Above Pre-industrial Levels and Related Global Greenhouse Gas Emission Pathways, in the Context of Strengthening the Global Response to the Threat of Climate Change, Sustainable Development, and Efforts to Eradicate Poverty*.

May, L. (1992) *Sharing Responsibility*, Chicago: University of Chicago Press.

McAdam, J. (2012) *Climate Change, Forced Migration, and International Law*, Oxford: Oxford University Press.

McCormick, S. (2016) "Assessing Climate Change Vulnerability in Urban America: Stakeholder-Driven Approaches," *Climate Change*, 138, 397–410.

McKibben, B. (2006) *The End of Nature*, New York: Random House.

McKibben, B. (2016) "Recalculating the Climate Math," *New Republic*, Sep. 22. https://newrepublic.com/article/136987/recalculating-climate-math [Accessed 21 February 2017].

McKinley, D., Ryan, M., Birdsey, R., Giardina, C., Harmon, M., Heath, L., Houghton, R., Jackson, R., Morrison, J., Murray, B., Pataki, D., & Skog, K. (2011) "A Synthesis of Current Knowledge on Forests and Carbon Storage in the United States," *Ecological Applications*, 21, 1902–1924.

McMichael, A. J. (2013) "Globalization, Climate Change, and Human Health," *New England Journal of Medicine*, 368 (14), 1335–1343.

Mei, W., Xie, S., Primeau, F., McWilliams, J., & Pasquero, C. (2015) "Northwestern Pacific Typhoon Intensity Controlled by Changes in Ocean Temperatures," *Science Advances*, 1 (4), e1500014.

Meyer, S. (2006) *The End of the Wild*, Cambridge: The MIT Press.

Monbiot, G. (2014) *Feral*, Chicago: The University of Chicago Press.

Nine, C. (2010) "Ecological Refugees, State Borders, and the Lockean Proviso," *Journal of Applied Philosophy*, 27 (4), 359–375.

O'Gorman, P. (2012) "Sensitivity of Tropical Precipitation Extremes to Climate Change," *Nature Geoscience*, 5 (10), 697–700.

O'Neill, O. (2005) "The Dark Side of Human Rights," *International Affairs*, 81 (2), 427–439.

Otto, F. E. L., Skeie R. B., Fuglestvedt, J. S., Bernsten, T., & Allen, M. R., (2017) "Assigning Historic Responsibility for Extreme Weather Events," *Nature Climate Change*, 7, 757–759.

Parmesan, C. (2006) "Ecological and Evolutionary Responses to Recent Climate Change," *Annual Review of Ecology, Evolution, and Systematics*, 37 (1), 637–669.

Pecl, G. T. et al. (2017) "Biodiversity Redistribution Under Climate Change: Impacts on Ecosystems and Human Well-being," *Science*, 355 (6332), 1–9.

Raz, J. (1986) *The Morality of Freedom*, Oxford: Oxford University Press.

Renner, A., Gerland, S., Haas, C., Spreen, G., Beckers, J., Hansen, E., Nicolaus, M., & Goodwin, H. (2014) "Evidence of Arctic Sea Ice Thinning from Direct Observations," *Geophysical Research Letters*, 41, 5029–5036

Richardson, H. (2008) "Our Call: The Constitutive Importance of the People's Judgment," *Journal of Moral Philosophy*, 5 (1), 3–29.

Rieder, T. (2014) "Procreation, Adoption, and the Contours of Obligation" *Journal of Applied Philosophy*, doi: 10.1111/japp.12099.

Ritchie, H., & Roser, M. (2020) "CO_2 and Greenhouse Gas Emissions". Published online at OurWorldInData.org. https://ourworldindata.org/co2-and-other-greenhouse-gas-emissions

Robinson, M. (2013) "An Interview with Mary Robinson, President of the Mary Robinson Foundation – Climate Justice (C. Gearty, Interviewer)," in A. Grear & C. Gearty (eds.) *Choosing a Future: The Social and Legal Aspects of Climate Change*, Northampton: Edward Elgar Press.

Rogelj, J., McCollum, D, Reisinger, A., Meinshausen, M., & Riahi, K. (2013) "Probabilistic Cost Estimates for Climate Change Mitigation," *Nature*, 493 (7430), 79–83.

Rogelj, J., Michel den Elzen, M., Hoehne, N., et al. (2016) "Paris Agreement Climate Proposals Need a Boost to Keep Warming Well Below 2°C," *Nature*, 534, 631–639.

Rosenthal, E. (2013) "Your Biggest Carbon Sin May Be Air Travel," *The New York Times*, Jan. 26.

Sen, A. (2004) "Elements of a Theory of Human Rights," *Philosophy and Public Affairs*, 34 (2), 315–356.

Shindell, D., Kuylenstierna, J., Vignati, E., van Dingenen, R., Amann, M., Klimont, Z., Anenberg, S., Muller, N., Janssens-Maenhout, G., Raes, F., Schwartz, J., Faluvegi, G., Pozzoli, L., Kupiainen, K., Höglund-Isaksson, L., Emberson, L., Streets, D., Ramanathan, V., Hicks, K., Oanh, N., Milly, G., Williams, M., Demkine, V., & Fowler, D. (2012) "Simultaneously Mitigating Near-term Climate Change and Improving Human Health and Food Security," *Science*, 335 (6065), 183–189.

Shrader-Frechette, K. (2007) *Taking Action, Saving Lives: Our Duties to Protect Environmental and Public Health*, New York: Oxford University Press.

Shue, H. (2010a) "Global Environment and International Inequality," in S. Gardiner, S. Caney, D. Jamieson, & H. Shue (eds.) *Climate Ethics: Essential Readings*, Oxford: Oxford University Press.

Shue, H. (2010b) "Subsistence Emissions and Luxury Emissions," in S. Gardiner, S. Caney, D. Jamieson, & H. Shue (eds.) *Climate Ethics: Essential Readings*, Oxford: Oxford University Press.

Shue, H. (2010c) "Deadly Delays, Saving Opportunities: Creating a More Dangerous World?" in S. Gardiner, S. Caney, D. Jamieson, & H. Shue (eds.) *Climate Ethics: Essential Readings*, Oxford: Oxford University Press.

Singer, P. (2002) *One World*, New Haven: Yale University Press.

Sinnott-Armstrong, W. (2005) "It's Not My Fault: Global Warming and Individual Moral Obligations," in W. Sinnott-Armstrong & R. Howarth (eds.) *Perspectives on Climate Change: Science, Economics, Politics, Ethics*, Amsterdam: Elsevier Press.

Solomon, S., Qin, D., Manning, M., Chen, Z., Marquis, M., Averyt, K., Tignor, M., & Miller, H. (eds.) (2007) *Climate Change 2007: The Physical Science Basis. Contribution of Working Group I to the Fifth Assessment Report of the Intergovernmental Panel on Climate Change*, Cambridge: Cambridge University Press.

Staudinger, M., Grimm, N., Staudt, A., Carter, S., Stuart III, F., Kareiva, P., Ruckelshaus, M., & Stein, B. (2012) *Impacts of Climate Change on Biodiversity, Ecosystems, and Ecosystem Services: Technical Input to the 2013 National Climate Assessment*, United States Global Change Research Program.

Stern, N. (2007) *The Economics of Climate Change: The Stern Review*, Cambridge: Cambridge University Press.

Stern, N. (2011) "A Profound Contradiction at the Heart of Climate Change Policy," *Financial Times*, Dec. 8.

Stocker, T., Qin, D., Plattner, G., Tignor, M., Allen, S., Boschung, J., Nauels, A., Xia, Y., Bex, V., & Midgley, P. (eds.) (2013) *Climate Change 2013: The Physical Science Basis. Contribution of Working Group I to the Fifth Assessment Report of the Intergovernmental Panel on Climate Change*, Cambridge: Cambridge University Press.

Szulecki, K. (2013) "Poland Climate Summit," *Kultura Liberalna*, Nov. 12.

Thomas, C., Cameron, A., Green, R., Bakkenes, M., Beaumont, L., Collingham, Y., Erasmus, B., de Siqueira, M., Grainger, A., Hannah, L., Hughes, L., Huntley, B., van Jaarsveld, M., Midgley, G., Miles, M., Ortega-Huerta, M., Peterson, A., Phillips, O., & Williams, S. (2004) "Extinction Risk from Climate Change," *Nature*, 427, 145–148.

Vaughan, A. (2009) "Carbon Emissions per Person, by Country," *The Guardian*, Sep. 2.

WGMS (2011) "Glacier Mass Balance Bulletin No. 11 (2008–2009)," M. Zemp, S. Nussbaumer, I. Gärtner-Roer, M. Hoelzle, F. Paul, & W. Haeberli, (eds.), *ICSU(WDS)/IUGG(IACS)/UNEP/UNESCO/WMO*, Zurich: World Glacier Monitoring Service.

Yulsman, T. (2016) "Something Really Crazy Is Happening in the Arctic," *Discover*, Nov. 20. http://blogs.discovermagazine.com/imageo/2016/11/20/something-really-crazy-is-happening-in-the-arctic/#.WHPOcVs7RGZ [Accessed 9 January 2017].

23

CLIMATE JUSTICE AND EQUITY

Steve Vanderheiden

Anthropogenic climate change has aptly been cast as an injustice, in terms of its causes and effects, with justice serving as an objective for and set of constraints upon its remedy. The 1992 United Nations Framework Convention on Climate Change (UNFCCC) identifies it a "common concern of humankind," calling upon signatory parties to "protect the climate system for the benefit of present and future generations of humankind" (Principle 1) in recognition of its spatial and intertemporal dimensions. Caused by the combination of fossil fuel combustion that releases heat-trapping greenhouse gases from long-term storage in minerals like oil or coal and land use changes that diminish the capacities of terrestrial sinks like forests to assimilate these gases without increasing their atmospheric concentrations, climate change results from several quintessential activities associated with affluence. While those most responsible for causing climate change are among the world's most advantaged, those most vulnerable to its adverse effects in the near future are among the world's poorest, who are also the least responsible for causing the problem. The same is true of future generations, which are vulnerable to polluting actions of the present generation while contributing nothing toward their predicament. In its patterns of cause and effect, climate change can therefore be viewed as a massive negative externality that is created by the affluent and as a byproduct of their affluence, to be imposed upon the global poor, thereby exacerbating already wide inequalities among the two groups, and future generations.

Besides the injustice of global environmental externalities that disproportionately benefit the affluent while disproportionately harming the poor, remedies to climate change must take into account this inequity in the way that they assign burdens and create benefits. Insofar as the problem of climate change lies partly in the fact that it creates or exacerbates existing injustices, its remedy entails eliminating or reducing these. Benefits should accrue primarily to the least well off, as is likely with most mitigation action in that the poor are most vulnerable to harm from climate change and so would benefit most by reducing climate impacts, and the costs or remedial burdens associated with mitigation (or efforts to reduce the causes of climate change) should primarily be borne by the affluent. In recognition of this equity imperative at the core of climate justice, the UNFCCC declared that the international assignment of remedial burdens should be made "on the basis of equity and in accordance with their common but differentiated responsibilities and respective capabilities" (or CBDR, for short) of nation-state parties (Principle 1). In the short run, this principle was interpreted to mean that the group of developed countries most responsible for causing climate change (the Annex I state parties, including the United States, the European Union, Australia, Canada, and Japan) should "take the lead" by accepting the first round of greenhouse emissions caps under the 1997 Kyoto Protocol, roughly differentiating more from

DOI: 10.4324/9781315768090-28

less responsible parties. Over the long run, however, the CBDR principle may require parties to take on remedial burdens in proportion to their "differentiated responsibilities" and "respective capabilities" to decarbonize.

Since the UNFCCC's declaration, competing proposals for how to share the remedial burdens of mitigation have sought to most plausibly interpret these principles and to apply them to relevant differences among nation-states. Climate justice principles have been developed and applied to the resource-sharing and/or burden-sharing conflicts at the core of climate change mitigation and adaptation, with scholars and activists alike urging attention to variables by which either each state's responsibilities or capabilities might be differentiated, with resulting indices and formulae invoked on behalf of various substantive policy outcomes. While much of the foci of climate justice analyses have concerned national responsibility and capacity, with scholars assigning greater remedial burdens to states that have higher per capita emissions and incomes, relatively little attention has been paid to the role of equity within climate justice imperatives. As an ideal or principle to assist in unpacking differentiated responsibility and capacity, as well as one that has critical force on its own and in combination with other normative ideals, equity serves as the linchpin of climate justice, linking together its various elements and serving as the foundation for its critique against unmitigated climate change and guiding basis for its remedy.

Equity in Remedial Liability for Climate-Related Harm

The most prominent aspect of scholarly climate justice analyses is concerned with the assignment of remedial responsibility (or *liability*, to distinguish duties to remedy from causes), which Miller describes in terms of being "picked out, either individually or along with others, as having a responsibility towards the deprived or suffering party that is not shared equally among all agents" (2001, 454). As he argues of most cases in which remedial liability is at issue, the "overriding interest" in assigning remedial burdens is "to identify an agent who can remedy the deprivation or suffering that concerns us" (471), where the injustice of failing to provide a remedy would constitute a greater evil than the failure to properly assign its liability to the responsible parties according to the proper principles. In some cases, the urgency of providing an adequate remedy might require that those proximate to would-be victims who are able to come to their aid do so upon the basis of capacity alone, even if a more leisurely assessment of the respective responsibilities of various parties might identify another remedial agent. In the context of climate change, remedial liability may take on this level of urgency in some contexts of adaptation, were imperiled parties stand to suffer great harm unless protected by others against climate-related threats to which they are vulnerable, including agricultural crop losses, damage from storms and floods, and heat-related illness or death.. Here, proximate and capable rescuers are liable to assist as first responders, but the costs of such emergency assistance must later be assigned to responsible parties in accordance with climate justice principle of remedial liability.

In his account, which he develops for problems of global poverty rather than climate change, Miller identifies an agent's moral responsibility for contributing to the harm in question as the strongest criterion for assigning remedial responsibility and thus liability, but notes that other criteria sometimes apply where moral responsibility cannot be attributed, including mere causal responsibility, the capacity to assist, and special ties of community with victims. Others have developed hybrid principles for redressing climate-related harm, with Caney (2005, 2010) and Baer et al. (2009) defending remedial liability formulae that combine measures of responsibility in terms of national emissions and capacity in terms of per capita income. Like the UNFCCC's invocation of multiple justice principles in equity, responsibility, and capacity, scholars interpreting climate justice imperatives occasionally combine multiple principles within an assignment of national burdens in rectification of the injustice of climate change. In doing so, they make equity the central ideal that interacts with subsidiary principles in responsibility or capacity, focusing primarily upon the latter. Understanding equity as requiring outcome equality among equals but differentiation in accordance with morally relevant differences, responsibility and capacity identify two such differences, prescribing greater remedial burdens for those more responsible for climate-related harm or more able to reduce their contributions to that harm or otherwise mitigate its effects.

Equity can also function as a justice principle on its own, as with egalitarian principles of distributive justice that require *prima facie* equality of outcomes or equal resource shares for those associated with certain kinds of cooperative schemes. As I have argued (Vanderheiden 2008a), equity on its own can refer to such principles as applied to the resource-sharing problem associated with climate change mitigation efforts. In order to limit global emissions, state parties must be assigned caps upon national emissions, which can be viewed as entitlements to a crucial ecological service. Because terrestrial carbon sinks can absorb finite quantities of emissions without raising atmospheric concentrations of carbon dioxide, and insofar as some level of global emissions beyond this sustainable level can be accommodated while remaining within a specified concentration or without exceeding a given temperature target, that quantity becomes the global emissions budget from which various equitable emissions budgets are then assigned to nation-state parties. Since national carbon budgets entail a permission to use some specified amount of emissions absorptive capacity, the assignment of budgets involves the deliberate sharing of resources or ecological services. In determining the just shares to be assigned to each state party through their national emissions budgets, equity principles apply. Like egalitarian principles of distributive justice more generally, these may not entail equal per capita resource or emissions shares, but departures from equality should be justified on the basis of relevant differences between parties. Insofar as equity implies a default position of equality modified by adjustments to this baseline in proportion to the differences among parties of some relevant facts or features, equity in the context of mitigation could be taken to require principles of egalitarian distributive justice be applied to this allocation (Hayward 2006, Vanderheiden 2009).

Differentiated national responsibility and capabilities likewise involve equity principles, in that both entail departures from default equality in proportion to differences in responsibility for climate change (however, this is interpreted) or capacity to reduce emissions. The former is generally understood to involve some function of national emissions, with those advocating a historical responsibility principle assigning remedial liability in proportion to each nation's full historical emissions, going back to early industrialization (Shue 1999). This view, based in a standard of strict liability, derives remedial liability from causal responsibility, since greenhouse gases remain in the atmosphere for centuries after their initial emission. Other views ground remedial liability in fault-based standards, often dismissing emissions prior to some baseline year (e.g. 1990, during which the first assessment report of the Intergovernmental Panel on Climate Change was released) on grounds of excusable ignorance (Bell 2011), or distinguish between unavoidable "survival emissions" or avoidable and therefore culpable "luxury emissions" (Shue 1993), assigning remedial liability only in proportion to the former. According to the latter distinction, countries like India with per capita emissions barely sufficient to meet basic needs cannot be faulted for doing what is necessary for survival and are thus exempt from remedial liability, whereas affluent countries with emissions well beyond this threshold are held liable for their greenhouse pollution associated with luxury consumption. Insofar as fault is required of moral responsibility, epistemic constraints are treated in such views as precluding assignments of liability for harm that could not have been reasonably anticipated to result from some actions, and basic resource entitlements are presumed to excuse harm that results from other actions that are necessary for meeting the basic needs of otherwise liable parties.

The role of equity in assigning remedial responsibility for climate change also turns upon whether the remedial obligations are viewed as falling into one category that encompasses both mitigation and adaptation burdens or requires distinct categories and principles for each. Those seeking to allocate only one set of remedial burdens among state parties for both mitigation and adaptation often treat differentiated responsibility (sometimes in combination with capabilities) as that relevant difference to which equity applies, assigning greater remedial burdens to those states with greater responsibility (understood as a function of national emissions) or capability (typically understood in terms of per capita national income). Here, a single remedial principle or formula defines the overall burden that states are required to bear on behalf of both categories of remedial actions, with the balance of efforts toward each either left unspecified or determined on the basis of other factors. Others identify two distinct remedial burdens in mitigation and adaptation, with distinct but related principles and formulae attached to each. According to this view, mitigation involves a resource-sharing assignment based on equitable national emissions shares and forward-looking distributive justice principles, which provide the basis for national mitigation obligations. From equitable per capita entitlements to natural resources or ecological services, some states may be assigned greater mitigation burdens insofar as their higher current per capita emissions requires greater abatement costs to comply with those per capita targets, but the entitlements themselves would be assigned on the basis of equity. By contrast, according to this view, the burden-sharing problem of adaptation would be informed by CBDR principles that turn on backward-looking assessments of national responsibility, through which equity functions as described above. Whether on its own or in conjunction with differentiated responsibilities or capabilities, equity is widely viewed as providing the appropriate basis for climate justice.

Regardless of whether an interpretation of climate justice imperatives turns only upon backward-looking assessments of responsibility or whether it also includes egalitarian principles for forward-looking allocations of resource shares, the objective involves defining an equitable assignment of remedial burdens. Responsibility-based views presume a prima facie equality of burdens but take past contributions to climate change in the form of emissions as relevant to the differentiation of remedial liability. Approaches based on egalitarian resource justice presume prima facie equality in access to critical ecological services like emissions absorptive capacity, sometimes modifying these based on variability in resource access needed to reach a threshold of well-being, but entail variable compliance costs with equal per capita emissions entitlements on the basis of variations in current emissions (with abatement costs falling most heavily upon those with the highest current emissions), where differentiated responsibility is not taken to be relevant to the assignment of national carbon budgets. As shall be explored further below, alternative bases for assigning remedial liability to those found in CBDR principles likewise seek to express a conception of what equity requires in the context of climate change, reinforcing the centrality of the ideal within climate justice by adding an additional formulation to the conversation.

Equity, Responsibility, and CBDR Alternatives

In addition to these two applications of equity principles in the formulation of remedial burdens associated with climate change, scholars have identified other justice principles that could potentially serve to inform burden-sharing arrangements. Instead of identifying past or present greenhouse pollution levels as justifying differentiated remedial burdens, alternative approaches look to benefits derived from past exploitation of ecological services, especially where this exploitation has been unequal in its demand for resources and its consequent effect upon patterns of global inequality. Since the affluent states in the global North have on average emitted far more greenhouse gases than the poor states of the global South, focus upon benefits derived from rather than harm caused by this historical pollution offers an alternative normative basis for understanding the demands of equity, and the justice principles that it prescribes.

Because climate change results from billions of point sources that each makes minute contributions toward increased atmospheric concentrations of greenhouse gases, with no single emission or even any person's total annual emissions having any discernable effect upon global climate, the links between the actions in question and their consequences are difficult to draw (Sinnott-Armstrong 2005). Each time we exhale, we emit carbon dioxide, but no harm results from such marginal emissions. Rather, harm results from aggregate levels of greenhouse pollution, not from the actions of particular persons. Further dissociating climate-related harm from any specific source of greenhouse pollution is the probabilistic nature of dangerous weather events. Most such events, like storms or droughts, are thought to increase in intensity and frequency as the result of climate change, so for any particular weather event, one might estimate the probability that anthropogenic climate change was among its causes. But even in such cases, the event often could have resulted without human drivers and the precise contribution of climate change cannot be estimated with the level of certainty necessary for establishing culpability. We can know with sufficient certainty that humans cause climate change and that climate change is harmful, but for any particular harm for which climate change is posited as a contributory factor, the link between aggregate human causes and the harmful effect cannot be known with certainty. Links between the greenhouse pollution of any person or moderately sized group of persons and particular instances of harm is therefore less direct than in typical instances of culpable harm.

Injunctions against harming, upon which responsibility-based principles like those found in common conceptions of CBDR are often premised, have as a result been called into question as valid bases for remedial obligations. To the extent that assessments of remedial liability turn upon a view of moral responsibility that depends in part upon causal responsibility, as accounts of moral responsibility typically do, then weak links in the causal chain between the allegedly harmful acts and the threat or injury toward which remedial efforts are to be directed undermine those assessments. The polluter-pays principle, for example, assigns remedial responsibility for pollution-related harm on the basis of the polluter's contribution toward the harm (Page 2008). Within climate politics, responsibility is typically understood in terms of greenhouse emissions, which are assumed to cause climate-related harm in proportion to their exceeding some per capita baseline (Heyward 2007). Severing the causal link between identifiable polluting acts and harm that now requires that a remedy be provided and its costs assigned to responsible parties detaches that harm from those parties. While the difficulties in establishing culpability for climate-related harm are sometimes overstated, so also is the direct and linear link between emissions and harm, despite the nonlinear relationships involved in most specific climatic impacts and the many other intervening variables that affect any harm that is suffered.

Given the fragmented agency and indirect causality involved in linking climate-related harm for which some sort of remedy needs to be provided and the specific actions or actors that are responsible for it (or shares of overall remedial responsibility that each actors bears), some argue for alternative principles upon which to rest remedial imperatives. Some, for example, propose a beneficiary-pays principle (BPP), which has the additional advantage of being able to account for the emissions of previous generations without invoking contentious formulations of vicarious responsibility through which persons are morally responsible for the actions of their ancestors by virtue of a merely inherited culpability (Page 2012). Many (if not all) residents of developed countries have benefitted by many of the development activities and processes that have also been responsible for those countries having relatively high historical emissions, so a remedial liability principle based on BPP could assign greater remedial burdens to more affluent states without directly linking the emissions of those states to the harm in question, since it would merely need to establish those historical emissions records along with a link between historical emissions patterns and present affluence. Such a move merely swaps one empirically dubious causal claim for another, however, if intended to base differentiated remedial liability only upon the extent of present affluence that is directly linked to historical carbon emissions. The role that equity plays within applications of responsibility-based principles to burden-sharing arrangements in climate change mitigation or adaptation can nonetheless also be seen in beneficiary-based approaches. As before, justice requires that those with quantitatively greater links to past greenhouse emissions—whether as causes of harm or historical origins of current advantages—are obligated to pay or do more now in consequence.

A variation of the BPP seeks to rest remedial mitigation or adaptation obligations upon unjust enrichment, through which an additional equity-based analysis is added to that already associated with differentiated benefitting. Under this view, merely benefitting from some past acts committed by others provides insufficient grounds for being assigned remedial obligations now if more robust grounds for assigning remedial burdens can be identified. Partly in reply to historical entitlement views such as those developed by Nozick (1974), through which current holdings are justified insofar as the result from a chain of voluntary transactions regardless of how much inherited inequality these entail, the idea of unjust enrichment impugns only those benefits that result from unjust past appropriations or transfers. According to this view, those benefitting by unjustly transferred goods like stolen land have no current entitlement to it and must therefore disgorge those benefits, even if they acquired it from the previous owner justly (Butt 2007). Likewise, those benefitting from an unjust past appropriation—as, for example, with claims upon the atmospheric sink that failed to leave "enough, and as good" of this ecological service for others—would not now be entitled to the current benefits that accrued thereby. Here, a Nozickian entitlement theory is supplemented by an equity-based distributive principle in order to arrive at another principle by which the world's affluent might legitimately be required to bear disproportionate remedial costs in responding to climate change, without linking their historical emissions to current or future harm (Vanderheiden 2009). Here again, one complicated causal chain is swapped for another, as differentiated remedial burdens assigned in proportion to differentiated unjust enrichment would require the demonstration not only of inequitable past exploitation of a common ecological service and currently unequal outcomes between developing and developed societies, but also a demonstration that the former led to the latter.

Equity in Participation and Development Opportunity

Apart from the above-noted inequities in the causes and expected impacts of climate change, and the role of equity in generation justice principles for the assigning remedial liability for climate-related harm, equity plays a key role in two other posited climate justice imperatives. Through the civil society discourses by which climate justice demands are articulated and heard, and in the policy-making processes in which these are assimilated into the design of multilateral treaty frameworks and regulatory instruments, claims to greater equity in participation constitute substantive and instrumental demands for reform of dominant procedural models. As Schlosberg (2007) notes, social movements have defined environmental justice not only in terms of equity in the distribution of hazards or access to ecological goods and services, but also in terms of equal rights to participate in problem definition and policy formation processes and for recognition of diversity within the movement. In the Conferences of the Parties (COPs) to the UNFCCC, at which representatives of states negotiate the terms of major and minor components of mitigation and adaptation policies and implementation measures, non-state actors are effectively excluded altogether and control over the agenda is largely relegated to affluent states and major emitters. Although formally relying upon a consensus-based process designed to give all parties some voice by allowing for a universal veto among delegates to negotiating sessions, in practice the interests of the G77 developing states have been marginalized, despite the fact that many states within this negotiating block are among the world's most vulnerable to climate change.

Instantiating equity within international policy-making processes like COPs involves several theoretical and practical difficulties, and can likely only be roughly achieved rather than fully realized. Its demands for allowing departures from equality only upon the basis of relevant differences among parties or cases suggest avenues for potential procedural reforms but also introduce conflicts based on interested and contestable claims about what are and are not morally relevant differences. In multilateral treaty negotiations like those involved in the COP process, states are accorded equal voice and votes, despite arguably relevant differences in population size between giant countries like China and India and small ones like Tuvalu and Nauru. However, small island states like Tuvalu and Nauru are among the most vulnerable to climate change, which might be viewed as morally relevant under an all-affected principle of representation but which does not get incorporated into the COP system. Differences among states in economic and political power are not reflected in the assignment of unequal formal powers in the system, but informally manifest as an effective veto threat on the part of the United States, given its status as the biggest developed country emitter and global hegemon, and in disproportionate agenda setting power by the G8 industrialized countries, which command over 40% of global economic output and emit nearly half of all greenhouse gases. Absent the development of some alternative global governance structure that could avoid the Westphalian one-state, one-vote principle and could more effectively represent the interests of currently marginalized non-state and sub-state actors, gains in participating equity can at most hope to advance rather than fully embrace the ideal.

Incorporating a principle of political equality into international policy-making processes would be complicated by several problems, including the wide diversity among state parties to the UNFCCC in population size, level of development, economic and political power, and vulnerability to climate change, and the inherent limitations of global governance institutions that take states as primary agents and allow little room for representation of interests other than those of the nation-state itself. However, even within a system that concentrates power within an empowered space dominated by delegates of states, more equitable access to the COP's agenda setting and rulemaking functions by less powerful states, along with mechanisms for translating the rich climate justice discourses from the civil society actions that are regularly staged outside of the meetings into the meetings themselves (Stevenson & Dryzek 2014) could improve upon participative equity by making the process more deliberative. Allowing for the representation of non-state actors like indigenous peoples or other minority groups within such sessions could further enhance the voice of such groups within negotiating sessions, and allowing such groups some control over working group agendas could highlight issues and articulate interests that are not currently heard within empowered spaces. Aside from the substantive demands voiced by civil society climate justice movements that seek to influence the direction of COP negotiations, their demands for greater equity in participation can be viewed as of instrumental value to a more responsive process that better protects the interests of affected parties as well as being valuable in itself, as an expression of democratic equality within the climate regime.

Developing country parties have articulated a second climate justice demand that can also be expressed in terms of equity, charging the climate regime with achieving mitigation targets that are sufficient to avoid the inequitable imposition of harm upon the world's disadvantaged while doing so in a way that allows the world's poor people the opportunity to realize the goals of political, economic, and human development. These goals, which include poverty alleviation along with the more positive components of human development like the promotion of literacy and increased life expectancy, are widely viewed as requiring sufficient growth in the carbon footprints of the world's poorest nations and peoples to allow for the increased material welfare and improved life prospects that are conventionally associated with global justice claims, and are typically articulated in terms of development rights (Vanderheiden 2008b). Such rights appear under Principle 2 of the UNFCCC as calling for developing states not to be assigned "a disproportionate or abnormal burden," in Principle 4 calling for a right to "sustainable develop-ment" and recognizing the need for "economic development" as "essential for adopting measures to address climate change," and in Principle 5 calling for "sustainable economic growth and development in all Parties." Elaborating upon the CBDR principle that was viewed as requiring developed country parties to "take the lead" by accepting the first round of binding emissions caps under the 2007 Kyoto Protocol, these references to development interests in terms of rights call for sufficient greenhouse emissions allowances within developing countries to accomplish their development-related objectives. Overly restrictive limits on greenhouse emissions within devel-oping countries under the Protocol, as in an extension of its formula of calling for 5% reductions from 1990 baseline emissions by 2012, would have restricted development opportunities within those non-Annex 1 developing countries that would under such restrictions been prohibited from industrializing beyond their 1990 baseline.

Accommodating development rights along with the magnitude of aggregate emissions reduc-tions that many see as necessary for avoiding the worst impacts of climate change would entail even greater mitigation burdens for developed countries than are apparent in calls to reduce global emissions by 80% by 2050—a commonly cited target based on what is seen as necessary for lim-iting global temperature increased to 2 degree Celsius this century—since this overall decrease would need to occur while potentially allowing emissions growth within some developing coun-tries. Given their wide current disparity in per capita emissions, the combination of an aggressive mitigation program with the recognition of development rights would require a "contraction and convergence" scenario whereby carbon abatement outcomes in excess of that 80% overall target would allow for limited per capita emissions growth within some currently poor countries, sig-nificantly contracting emissions within developed countries and allowing for a convergence upon roughly equal per capita emissions between the global North and South. Here again, equity in na-tional per capita resource shares is viewed as necessary for realizing global equity in development goals and thus life prospects. While the costs for achieving these two equity objectives would not be equally borne by all nations or persons, the CBDR principles used to inform assignments of remedial liability would be based in equity by reflecting relevant differences among parties in its differentiated responsibility.

Conclusions

Climate justice analyses that identify climate change and an injustice and invoke justice principles in remedial objectives and constraints rely upon various and sometimes conflicting conceptions of justice. From forward-looking egalitarian principles of distributive justice that focus upon al-locating a shared resource or ecological service to backward-looking principles that seek to hold polluters responsible for the consequences of their polluting acts, the agents, facts, and normative bases for each prescribed remedy can involve quite different calculations and distinct bases for as-signing the unique remedial burdens associated with each. Viewing climate justice diagnoses and prescriptions as attempts to articulate the imperative to equitable share in the maintenance of the planet's climate system, however, links these various approaches together through their respective efforts to theorize the demands of equity and highlights the various ways in which the ideal can plausibly be interpreted. In examining implications for international climate policy that is issued from various interpretations, scholarly work in climate justice has thus contributed to the practical political project of instantiating equity within global institutions. This objective includes not only substantive equity outcomes like equal burden-sharing in climate change mitigation and adapta-tion, along with more equitable development opportunities for those residing in poor states, but also procedural equity commitments to more inclusive and deliberative decision making within climate policy development meetings.

References

Baer, P., Athanasiou, T., Kartha, S. and Kemp-Benedict, E. (2009) "Greenhouse Development Rights: A Proposal for a Fair Global Climate Treaty," *Ethics, Place, and Environment* 12(3): 267–281.

Bell, D. (2011) "Global Climate Justice, Historic Emissions, and Excusable Ignorance," *The Monist* 94(3): 391–411.

Butt, D. (2007) "On Benefitting from Injustice," *Canadian Journal of Philosophy* 37(1): 129–152.

Caney, S. (2005) "Cosmopolitan Justice, Responsibility, and Global Climate Change," *Leiden Journal of International Law* 18(4): 747–775.

Caney, S. (2010) "Climate Change and the Duties of the Advantaged," *Critical Review of International Social and Political Philosophy* 13(1): 203–228.

Hayward, T. (2006) "Global Justice and the Distribution of Natural Resources," *Political Studies* 54(2): 349–369.

Heyward, M. (2007) "Equity and International Climate Change Negotiations: A Matter of Perspective," *Climate Policy* 7(6): 518–534.

Miller, D. (2001) "Distributing Responsibilities," *Journal of Political Philosophy* 9(4): 453–471.

Nozick, R. (1974) *Anarchy, State and Utopia*, Malden, MA: Basic Books.

Page, E.A. (2008) "Distributing the Burdens of Climate Change," *Environmental Politics* 17(4): 556–575.

Page, E.A. (2012) "Give it Up for Climate Change: A Defense of the Beneficiary Pays Principle," *International Theory* 4(2): 300–330.

Schlosberg, D. (2007) *Defining Environmental Justice: Theories, Movements, and Nature*, New York: Oxford University Press.

Shue, H. (1993) "Subsistence Emissions and Luxury Emissions," *Law & Policy* 15(1): 39–60.

Shue, H. (1999) "Global Environment and International Equality," *International Affairs* 75(3): 531–545.

Sinnott-Armstrong, W. (2005) "It's Not *My* Fault: Global Warming and Individual Moral Obligations," *Advances in the Economics of Environmental Research* 5: 293–315

Stevenson, H. and Dryzek, J.S. (2014) *Democratizing Global Climate Governance*, New York: Cambridge University Press.

Vanderheiden, S. (2008a) *Atmospheric Justice: A Political Theory of Climate Change*, New York: Oxford University Press.

Vanderheiden, S. (2008b) "Climate Change, Environmental Rights, and Emissions Shares," in S. Vanderheiden (ed.) *Political Theory and Global Climate Change*, Cambridge, MA: The MIT Press.

Vanderheiden, S. (2009) "Allocating Ecological Space," *Journal of Social Philosophy* 40(2): 257–275.

24

GEOENGINEERING

Benjamin Hale and Michael Pellegrino

In early 2019, a team of scientists at Harvard University undertook what would be the most significant field test of a new technology aimed to lower global temperature, aspiring to find a quick technical fix that might reverse the devastating advance of climate change. The proposed Stratospheric Controlled Perturbation Experiment, or SCoPEx, involved releasing a weather balloon high into the atmosphere and triggering it to deploy a layer of aerosols "roughly one kilometer long and one hundred meters in diameter." Instrumentation could then be used to observe "changes in aerosol density, atmospheric chemistry, and light scattering" (Tollefson 2018; Keutsch Research Group 2019). If successful, the experiment promised to pave the way for future, larger-scale stratospheric aerosol injection (SAI) efforts, which could, in turn, establish SAI as a potentially desirable technique for combating climate warming. For a variety of reasons partly associated with concerns articulated below, the SCoPEx project has, at least as of this writing, been stuck in a holding pattern.

Harvard scientist David Keith, one of the more vocal proponents of climate engineering projects such as these (often called "geoengineering"), not only heads up the SCoPEx project, but also serves as the founder and director of a second commercial enterprise, aptly named "Carbon Engineering," that aims to produce and sell technologies built to directly capture carbon from the air. Unlike the atmospheric brightening effort proposed by SCoPEx, Carbon Engineering utilizes cooling towers containing a liquid hydroxide solution to capture CO_2 from the atmosphere and convert it into carbonate. If successful, Carbon Engineering's technology would function very differently than SCoPEx's, essentially changing the composition of the air instead of modifying the solar radiation that warms the planet. Radically different though they are, both technologies nevertheless fall in the same rough category of "geoengineering" technologies.

Weather control has long been the dream of farmers and war mongers (Fleming 2010). In this respect, geoengineering is nothing new. But the move from localized, one-off rainmaking efforts to global, temporally sustained climate shifts, introduces a whole new level of controversy. Can these two projects – different in so many technical ways – be described as essentially the same sort of technology? If so, should they be evaluated as morally equivalent? And if not, what accounts for the conceptual and moral difference?

This chapter is about geoengineering, sometimes also called "climate engineering," which is roughly describable as any intervention technology that aims to turn back the clock on climate change, and according to many definitions, to do so deliberately.

DOI: 10.4324/9781315768090-29

Some Brief Context

As the world has grown more concerned about climate change, policy makers and scientists have offered up a multitude of responses to anticipated climatic shifts. Some have suggested that the global community will need to find an appropriate agreement that reduces average individual footprint, and one, ideally, that is enforceable and monitorable. Others have suggested that we should accept our fate and learn to adapt to the problem. Still others, like David Keith and his Carbon Engineering team, have proposed that we need to engage in a global-scale clean-up effort, turning back the clock on emissions and activities that have gotten us into this predicament in the first place. These responses break down into three rough categories: those that propose mitigation, those that propose adaptation, and those that propose remediation.

Earlier chapters in this volume by Marcus Hedahl, Clare Heyward, and David Morrow, among others, explain that *mitigation* proposals are primarily oriented around reducing the drivers of climate change. For many years, mitigation proposals dominated the climate discussion. *Adaptation* proposals entered the discussion somewhat later, but when first introduced were widely viewed as abandoning earlier attempts to mitigate. The tension dominant throughout this discourse, in rough contour, was that adaptation efforts essentially amount to acquiescence. To advocate for adaptation at a time when mitigation still seemed a viable, if difficult, possibility appeared to many as giving up.

However, as climate projections have grown more dire and status quo emissions have "locked in" temperature shifts of 3–5°C for some regions of the planet (Global Linkages, UN Environment Programme 2019), the global community has come to accept that some amount of adaptation may be necessary regardless of mitigation efforts. More contemporary debates, in other words, have viewed mitigation and adaptation as two approaches that must work synergistically rather than antagonistically. Lurking in the background of this dispute between adaptation and mitigation has been a third dark-horse alternative – *remediation* – which over the years has taken on a variety of guises. Essentially, these proposals suggest that we should clean up the messes that we've made. Among the variety of remediation approaches for global carbon emissions, many geoengineering proposals fall in the remediation category and function neither as mitigation proposals nor adaptations proposals, but instead focus on the atmospheric mess and instead seek a technical solution to clean it up.

Technical Distinctions

Within the geoengineering literature, there are two key technical categories of geoengineering technology: solar radiation management (SRM) and carbon dioxide removal (CDR) (Caldeira et al. 2013). The first of these, SRM, seeks to avoid catastrophic climate change by managing the permeation and penetration of sunlight into our atmosphere, essentially preventing or reducing predicted warming. The idea here is that changing the albedo (the reflective capacity or "whiteness") of the earth will lead to global average temperature reduction. Technologies in the SRM category range from the mundane to the fantastic, and include proposals such as painting white the rooftops of houses and buildings, injecting small reflective particulates into the stratosphere, or suspending giant mirrors in orbit around the earth in order to deflect sunlight before it ever penetrates the atmosphere (Akbari et al. 2009).

Many people believe that of the various SRM proposals, stratospheric sulfate injection (SSI) is the most viable option. Not only is it relatively inexpensive and straightforward to deploy, but it stands the best chance of doing the work that needs to be done to limit damages from climate change (Shepherd et al. 2009). The proposed SCoPEx project at Harvard University, for example, is a prime early case of stratospheric injection technology (Tollefson 2018). SSI would involve jettisoning aerosolized sulfate particles into the atmosphere and suspending them into the strato-sphere long enough to reflect a sufficient quantity of solar radiation to result in a decrease in global temperatures (Caldeira et al. 2013). As injected particles would eventually precipitate out of the at-mosphere, this type of system would require constant upkeep (Vaughan and Lenton 2011). In this respect, SSI poses the risk that, if deployed, it may suddenly cease to be maintained. Some ethicists have worried that a technical breakdown would lead to a rapid reversal of progress, conceivably exacerbating a climate catastrophe. Though the technical challenges of such technologies seem surmountable, political feasibility is a somewhat more challenging hurdle, as SSI would require rewriting, or at least revisiting, existing environmental laws and redesigning existing international decision-making processes governing, say, NOx and SOx emissions. Nevertheless, advocates re-main optimistic that if the technology can be demonstrated as effective, the public will eventually accept the necessity of its deployment.

CDR proposals differ from SRM proposals in that they aim not to reduce the *effects* or *impacts* of climate change directly (by modifying average global temperature or the albedo of the planet), but instead aim to address the *primary driver* of these effects (atmospheric carbon concentrations) through mechanisms that extract carbon from the atmosphere (Caldeira et al. 2013). CDR pro-posals, like their SRM counterparts, range from the mundane to the fantastic (Barrett 2008). More mundane proposals include massive reforestation and afforestation efforts, where more fan-tastic proposals include wide-reaching ocean fertilization efforts (Caldeira et al. 2013). Alternative proposals include building sophisticated carbon scrubbing and direct air capture technologies, like the so-called "artificial trees" envisioned by the Canada-based company Carbon Engineering, and distributing them across the globe (Vaughan and Lenton 2011).

Each carbon removal technology relies on a slightly different technical mechanism to achieve its objectives. *Ocean fertilization* proposals aim to add nutrients to the oceans in order to spark a mid-ocean algal bloom that could, in principle, increase the oceanic absorption rate of carbon and store or sink that carbon back down onto the ocean floor. By its nature, this is a point-source technology that is perhaps most successfully deployed in a discrete region of the planet. *Distributed carbon scrubbing* methods, however, employ a large number of widely disseminated devices – artificial trees or roof-mounted scrubbers – that selectively absorb and trap CO_2 as air flows by. Some of these technologies include *direct air capture* methods, which pull streams of air into a device, utilizing chemical reactions to separate CO_2 from other atmospheric gases. The CO_2 collected in many of these methods is most often sequestered deep underground in geologic features that help prevent its escape. Others eschew technology altogether and rely on global reforestation or afforestation efforts.

Inasmuch as each CDR technologies rely on slightly different technical mechanisms with dif-ferent distributional contours, so too is each saddled with its own set of ethical complications. Where reforestation and afforestation efforts are widely viewed to be the least ethically problem-atic, they also, by dint of their scope and scale, are extremely unlikely to be deployed successfully. It is very hard to regrow devastated forests or to install a global network of artificial trees, for instance. Far less expensive and intensive CDR efforts, like ocean fertilization, are appealing due to cost and relative ease of deployment, but suffer from ecological risks and political complica-tions. Point-source deployment from a single state or non-state actor could result in unanticipated ecological catastrophe or even political destabilization. There is a real chance, for instance, that a successful ocean fertilization effort could result in runaway ocean hypoxia (Russel et al. 2012), essentially saving terrestrial systems at the expense of aquatic systems (Hale and Dilling 2011). Carbon capture technologies, while perhaps the most techno-optimistic in attempting to seques-ter carbon in a safe space, nevertheless require using large sections of the planet in order to store that carbon, raising justice considerations.

But there are wider problems with thinking about geoengineering in these narrow technical terms (Jamieson 2013). First, clumping geoengineering technologies into two technical categories

– either SRM or CDR – doesn't neatly capture the moral dimensions of the geoengineering proposals in question (Hale 2013; Jamieson 2013; Preston 2013). As an example, SRM technologies are designed to manage the earth's temperature by altering solar radiation. Implicit in this solution is the idea that the sole or primary problem of climate change is increasing global average temperature (Gardiner 2010). Questions about desertification, species loss, environmental justice, all are presumed to stem from the rise in temperature. Other important environmental problems related to emissions, like increased atmospheric pollution levels or acidification of ocean systems, are easily overlooked. Moreover, the SRM category appears to cluster a range of unrelated technologies under one heading (Jamieson 2013). Where technically ejecting sulfate aerosols into the stratosphere might produce the same temperature outcome as encouraging homeowners to install white roofs or to paint their driveways white, the technologies will require different deployment strategies and, in turn, engender different considerations regarding those affected.

CDR solutions suffer from similar complications. The primary function of CDR technologies is to tweak the primary driver of climate change – atmospheric carbon. Where the SRM proposals tend to emphasize temperature as the problem in need of correction, CDR proposals focus primarily on carbon as the central problem of climate change (Gardiner 2011). By limiting the characterization of the climate problem primarily to atmospheric carbon concentrations, other important drivers of climate change or other factors with substantial environmental impacts may be ignored, including greenhouse gases other than CO_2 and harmful land use practices (Irvine et al. 2016).

Dividing the categories in this way also neglects to address other potential technologies that may assist in the intentional driving of the climate (Jamieson 1996). For example, if atmospheric moisture were one concern with climate change – and it is – one might wonder whether global cloud seeding technologies should be included in this mix (Trenberth 1998). But this problem, and its corresponding technological solution, do not fall neatly within the two–part division that commonly characterizes the climate engineering discussion (Jamieson 2013).

Finally, these divisions, focused as they are on technical solutions to physical problems, tend to mask matters of authorization and legitimacy. It would be easy to think that because polluting industries and individual energy consumers have been inadvertently engaged in a decades-long process of altering the atmosphere through emissions in the transportation, energy, and agriculture sectors that this alone is enough to authorize the use of a geoengineering strategy. (In other words, it would be easy to believe that there is little moral question about the permissibility of geoengineering because we have already messed with the planetary thermostat.) Such an argument is reinforced by supposing that the primary distinction between different sorts of geoengineering technologies rests in the technology itself. It is, however, questionable whether it makes sense to consider reforestation in the same category as ocean iron fertilization or the planting of artificial trees, when one seems oriented toward restoring the world to a prior state and the others are more oriented toward moving the world to a new state (Hale 2013).

Alternative Conceptions of Geoengineering

There are other ways to conceptualize geoengineering that do not revolve around the technical and physical details of each technology. Many argue that what distinguishes geoengineering technologies from, say, everyday climate emissions, is that geoengineering technologies are fundamentally *intentional* (Jamieson 1996). Without some account of the intentional nature of geoengineering, even unintended shifts in climate, like climate change itself, appear to be describable as geoengineering efforts (see above). Though such intentionalistic accounts are attractive, and may at least serve as a straightforward means of differentiating geoengineering technologies from other forms of environmental modification, the intentional nature of geoengineering alone does not neatly capture many of the moral challenges relevant to different approaches to geoengineering.

Painting roofs white or planting trees – both of which are intentional actions aimed to shift the climate – ought not to be viewed in the same moral light as pumping sulfate aerosols into the atmosphere.

A different way of conceptualizing the distinctions among geoengineering technologies might employ a strategy like that employed with nuclear or medical technologies (Hale 2013). The *Dual Use Criterion* – emergent in part of attempts during the Cold War to regulate nuclear and chemical warfare – distinguishes between benevolent and malevolent uses of technologies under the presumption that it is the uses to which technologies are being put that determines their ethical valence (Harris 2016). Beneficial uses of nuclear technologies include, for instance, the provision of cheap and accessible energy, where malevolent uses are those in which the same technology might be used in war. In a geoengineering context, beneficial uses may include those technologies, for example, that remove CO_2 from the atmosphere in a manner that reduces the impacts of warming without increasing global inequality, where malevolent uses may include the deployment of technologies to sabotage weather patterns of a rival nation or group, or perhaps that could be used to promote national interest.

One criticism of using such a criterion is that it begs many of the most important ethical questions. If we already have answers to which *uses* are benevolent and which are malevolent, we are left in a position in which we still need further clarification on what we mean by "benevolent" or "malevolent." Are we concerned about harms, or rights, or sovereignty? Further arguments must be employed to demonstrate why certain uses are benevolent, while others are malevolent. In turn, clarifying the conception of geoengineering – that is, narrowing the definition and more clearly defining what is distinctive about geoengineering in particular – can help us better understand the ethical implications of its deployment.

A slightly alternative approach to differentiating geoengineering technologies seeks a middle ground between the above two approaches and suggests distinguishing between geo-steering technologies and geo-remediation technologies (Hale 2013). Hale argues that it is better to conceive of geoengineering as an attempt to steer the climate for a specific reason (or for a set of reasons) that we can attribute to a relevant actor (or set of actors). By taking heed of justificatory intent, this approach enables us to get past a complication introduced from overly reductionistic characterizations of the problem of climate change. If a geoengineering technology aims primarily to reduce harms caused by many billions of emitters, then a range of harm-reducing engineering projects may present themselves as possible solutions. However, depending on how the harms are construed – as shifting average temperature of the planet; as threats to humans, wildlife, and ecosystems; as economic losses due to increased climatic activity – the "solutions" may look very different. Where characterizations of climate harms as primarily regarding the shifting temperature of the planet may require controlling the thermostat of the earth, threats to humans, wildlife, and ecosystems may instead admit of somewhat more "adaptive" interventions. Equally so, if damages are understood primarily in economic terms, there may be yet different proposed solutions, including compensating victims or redistributing wealth. With the division between geo-steering technologies and geo-remediation technologies in view, one can avoid the complications associated with intention and dual use, but also drill down into the reasons that *justify* the environmental alteration.

The conception of geoengineering technologies one employs is important because this conception can shape how one understands the benefits, costs, and risks (and, consequently, the potential solutions) associated with those technologies.

Benefits, Costs, and Risks

Of course, none of these technologies would be considered in the first place were the threat of climate change not real and significant. On first assessment, the geoengineering question would seem to involve a relatively straightforward matter of how to weigh benefits and costs. But if conceptualizing risk and harm in the context of climate change raises challenging epistemic problems, as evidenced by the prolonged and contentious Stern-Nordhaus debates (Leonhardt 2007), it is all the more challenging when uncertain geoengineering proposals are factored in.

According to economist William Nordhaus's DICE model (used to estimate the long-term costs related to climate change), damages from climate change could be as high as 3.8% of the global GDP by the year 2100, even if the world's nations maintain the status quo (Nordhaus 2018). Should all nations engage in nearly optimal efforts to address climate change, Nordhaus still estimates that climate change damages would likely fall no lower than 2.8% of the global GDP (Nordhaus 2018). It is tempting to position geoengineering proposals in the context of projected outcomes like those above, since these technologies present the possibility of offsetting the costs and damages associated with climate change by generating overall benefits. But where harms are difficult to assess in the comparatively more straightforward case of climate change, they are compounded in the case of geoengineering technologies, as harm calculations must factor in not only risk and costs of climate change, but also the risk and cost of the technology itself (Svoboda 2012).

Where SRM proposals promise to reduce the temperature on the planet and thereby (a) decrease or reverse the melting of ice sheets on both land and sea, (b) decrease or reverse sea level rise, (c) increase primary productivity of plants, and increase the amount of CO_2 held in organic plant matter, most proposals are likely to be expensive. Wake Smith and Gernot Wagner estimate that stratospheric aerosol deployment could cost approximately 2.25 billion USD per year for the first 15 years of implementation (Smith and Wagner 2018). Further, such technology would involve designing a new kind of aircraft to deploy the aerosols effectively, and these planes would need to make at least 4,000 flights per year. Both factors would drastically increase the potential costs of such a program. Other, more conservative estimates put the cost of sulfate aerosol injection around 1 billion USD per year (Robock et al 2009; Moriyama et al 2017). So that is the first problem: that the costs and benefits are hard to establish.

Additionally, as before, it is not clear which harms we want to be preventing or benefits we want to be conferring – economic, environmental, or welfare, for instance. Take ocean iron fertilization as a second example. It is possible that the oceans could absorb carbon without problems, creating healthier marine ecosystems, more robust fish stocks, and more opportunities to feed human beings. But it is also possible that oceans could become hypoxic, killing marine flora and fauna alike, causing the collapse of multiple essential ocean systems. So that is a second problem: many of these costs and benefits involve seemingly incommensurable tradeoffs, and given the range of outcomes, it is not clear what threshold of risk we ought to tolerate.

If the debate about the nature of the harms is not resolved, it is reasonable to worry that whatever geoengineering technology is deployed may generate an incommensurable conflict even between categories of harms. For example, even if SRM were to lower global temperatures successfully, it could simultaneously cause oceans to become more acidic or deplete stratospheric ozone. So here we have yet a third problem: by focusing on only one characterization of environmental harm, the resulting state of affairs might easily replace old problems with new ones. The children's folk song "The Old Lady Who Swallowed a Fly" can serve as a cautionary argument against narrow technocratic solutions to climate change, as there is always some chance the outcome will be good but will bring about new states of the climate that are anticipated but nevertheless problematic (Preston 2017).

A fourth worry might be that in failing to adequately understand the harms engendered by climate change, geoengineering technologies simply fail to address the root cause of the problem (Gordijn and Ten Have 2012; Hale 2012). The so-called "band-aid objections" suggest that geoengineering might permit populations to continue to do damage at status quo levels while addressing the symptoms of the injury but not the root cause (Hale 2012).

A fifth worry is more epistemic in nature and is related to a point we made at the outset of this section: many of these direct and indirect harms are difficult to anticipate, or are uncertain (Preston 2017). Given the complexity of earth's systems, and given the large scope of many geoengineering proposals, geoengineering could create unintended or unforeseen global harms, even when a given technology successfully ameliorates a key climate problem. Once this technology is deployed, we are likely to discover that we did not properly understand the system we intended to repair or did not do our calculations correctly. Unanticipated harms raise policy questions about how we should make decisions under uncertain conditions. These harms could also be of such a size and scope that it would be unlikely they could be quickly addressed once they arise. In other words, there may be no going back once geoengineering is deployed.

Adding to the uncertainties of deploying geoengineering technologies, many new outcomes and social practices may be brought on by the deployment of the geoengineering technology itself. This sixth problem is not of mere uncertainty regarding unknown outcomes, but of *indeterminacy* (Hardin 2003). Indeterminate outcomes are additionally vexing because they are emergent out of whatever new state of the world a global-scale geoengineering solution invites. Because such harms are a consequence of actions not yet taken, they are not only uncertain, but are also unknowable and, in virtue of this, unmodelable. For example, given that geoengineering changes the global climate picture, including the behaviors of all actors who are contributing to climate change, it is entirely possible that the market for fossil fuels will change, which, in turn, will disrupt geopolitical arrangements. Indeterminacies of complex systems (Ladyman 2012), reflexive systems (Beck et al. 1994), and wicked systems (Lazarus 2009) make the epistemic challenge particularly vexing (Hardin 2003).

A seventh benefit-related concern regards the so-called "moral hazard." In its most elemental form, the worry is that if a reliable technology can bail the planet out of its predicament, individual actors might not then feel any need to change their behavior, or more distressingly, even increase their polluting behaviors. The idea here is that as society creates mechanisms to insure against downside risks, individual actors will change their exposure to risk. In the literature on geoengineering and the moral hazard, however, the term "moral hazard" is used loosely to describe many different, and sometimes conflicting, phenomena (Hale 2012). Straightening out conflicts in invocations of the term may help provide clarity on how to approach the question of how to respond.

Some of these concerns about downside risks could be responded to by invoking well-known Doctrine of Double Effect, which authorizes actions that have morally bad side effects if their good effects are large enough (Morrow 2014a, 2014b). To advance such a position, however, one would obviously have to find a way to weigh the urgency of geoengineering against any side-effect harms that may be generated. Similarly, because geoengineering involves primarily intentional decisions and actions that can result in significant harm, the distinction between acts and omissions may become an important ethical consideration. For example, it could be far worse to create significant harm than it would be to allow significant harm to happen.

Justice Considerations

As with many global solutions, there is more to consider beyond simply the existence of harms and benefits. It is also important to examine how these outcomes may be distributed, who is allowed to participate in the regulation and implementation of geoengineering, and how these impacts affect not only the present-day world, but the world of future generations as well. Deferring in part to the broader climate justice literature (cite from volume), we can break justice concerns over geoengineering into at least four broad issue areas:

Distributional Issues (e.g. Global Poor)

Inasmuch as distributional justice is concerned with how benefits and burdens are distributed across the global population, those concerned with the ethics of geoengineering must be concerned as well about additional harms generated, especially global ones, which are likely to disproportionately affect already disadvantaged groups (Gardiner 2011; McLaren 2018). Any given geoengineering scheme will affect geographically distinct parts of the globe differently. One scenario may alleviate desertification pressures in the global south, for instance, but exacerbate ocean acidification pressures that afflict coastal populations. In such cases, one would need to take special care to ensure that the global poor are not forced to bear a greater burden than they already are (Moellendorf 2014). For example, Matthews and Caldeira (2005) predict that sulfate aerosol injection could lead to decreased precipitation in over much of the world's landmass, but the worst effects are predicted to be located in South America, Africa, and southeast Asia. A decrease in precipitation in regions that already are hot and/or arid could have serious repercussions on agriculture and clean water availability. Matthews and Caldeira speculate that the decrease in precipitation could be more significant than if climate change were left unabated over the same time period (2007).

Recognition Issues (e.g. Tribal Claims)

Where distribution concerns relate primarily to disparate benefits and burdens, recognition concerns integrate the histories and voices of disadvantaged groups into decision-making processes and strive to ensure that policy decisions are respectful of and responsive to the cultural needs of disadvantaged people. With regard to geoengineering, the primary concern may be over whether histories and cultures are adequately considered, either substantively or procedurally.

Marion Hourdequin (2018), for instance, argues that geoengineering technologies, apart from simply maintaining the status quo, could easily affect cultures, places, and practices that groups have hitherto regarded as home. Consider an ocean iron fertilization effort that results in widespread ocean hypoxia. Damaged ocean ecosystems could force people who consider islands or coasts to be home to move. A focus on recognition would require greater deference toward the places disadvantaged groups call home, recognizing that in many cases their homes may previously have been threatened or not respected.

Likewise, changes in the earth's processes could affect the cultural traditions and norms of disadvantaged groups (Hourdequin 2018). In the above example, people who have traditionally relied on fishing for sustenance or economic survival could be forced to abandon those traditions. Cultural celebrations or events that involve the ocean, as a consequence, may be changed or abandoned. There is, of course, some precedent for this already with respect to climate change, but the problem is amplified when geoengineering technologies are added to the mix.

For example, some Native American tribes that were forced onto reservations in one part of the country previously lived in regions with more temperate climates. These same tribes have been forced to adapt cultural practices and rituals to a warmer, more arid climate (Maldonado et al 2013). While many argue that they have a grievance against those who are responsible (broadly speaking) for climate change, attribution of wrongdoing is a challenge (Whyte 2013). As Hourdequin argues, with the more focused agential relationship between a geoengineering technology and its deployment, claims of wrongdoing against the acting agent will be much easier to justify (Hourdequin 2018). Concern for recognition would place value on cultural traditions and seek to avoid damaging the cultural practices of disadvantaged groups. Recognition does not mean that all cultural change is forbidden, but it does mean that the homes and cultures of disadvantaged groups play a role in the decision-making process.

Participation Issues (e.g. Informed, Democratic Participation in Decision-Making)

There are additional procedural or participatory concerns to worry about with geoengineering. Participatory justice endeavors to involve all or as many people who might be affected by a policy as informed and equal stakeholders in a robust democratic process. Since geoengineering technologies would likely require substantial energy, resources, and infrastructure, there is a considerable risk that the wealthy and powerful would have outsized power over creating geoengineering policy and controlling deployment. Mechanisms to ensure the full and equal participation of disadvantaged groups and the individuals that compose them would need to be in place to arrive at just policy decisions.

There are a number of important participatory considerations with respect to geoengineering in particular. The overarching question is broad in scope: that if all human beings living on the planet, and perhaps non-humans as well, may be affected by geoengineering, the burden of creating a fully inclusive global decision-making process – considerate of the rights, needs, and interests of all affected parties – would be a challenge unlike any to come before it. In scenarios where some populations are enfranchised where others are not, it is more likely that the deployment would be conducted in a manner that confers greater benefits on the enfranchised (McLaren 2018). For example, if the global north were to deploy geoengineering without sufficient input and assent of the global south, the distributional issues mentioned above may be a consequence of disparate participatory involvement.

Complicating matters even further, given that geoengineering often involves complex changes to complex, interconnected earth systems, barriers to participation for "informed" stakeholders may be too great. Deep disagreements among stakeholders of the sort already reflected in debates about climate science promise only to grow more heated as the community of participants grows larger. Because of the seriousness and scope of potential geoengineering proposals, if justice were to require such a robust process to legitimize a geoengineering decision, it is easy to imagine that the participatory process alone would be overly demanding or unachievable.

Intergenerational Issues (e.g. Future Generations)

The potential impacts of geoengineering are not only large and asymmetrical in geographic scope, but they could extend for decades, even centuries, into the future. It is likely, for instance, that for many geoengineering methods to be effective, deployment systems may have to be installed permanently and continue into perpetuity (Gardiner 2010). Furthermore, responsibilities to maintain these systems may amplify over successive generations, leading to deeper concerns over distributional burdens.

But there are other concerns over future generations as well. Much like the moral hazard arguments discussed above, implementation of a global climate engineering technology in perpetuity may not only encourage present people to abide by the status quo, but may also generate entirely new consumption and extraction pathways that themselves generate additional future problems. These technologies may create institutional inertia that encourages further geoengineering use (Gardiner 2010). Because geoengineering has the potential to require perpetual implementation, and because geoengineering would likely do little to change the underlying consumption and pollution that lead to climate change, society could become "locked-in" to geoengineering to fight climate change. It could even lead to more expansive geoengineering or multiple kinds of geoengineering being deployed.

On top of these concerns about disenfranchised populations, there remains the backward-looking question of whether individuals and governments have responsibilities or obligations to redress past wrongs.

Governance

More practically, some have suggested that geoengineering might serve as an expedient method for circumventing legislative gridlock. Unfortunately, geoengineering opens countless new questions of jurisdiction and legitimacy (Gardiner 2011). Some of these concerns were discussed above with respect to participatory justice, but there are further, serious concerns related to respecting national sovereignty that regard the process of making international decisions. As a beginning question, does respecting national sovereignty require that there must be unanimity among nations in order to deploy geoengineering? (Jamieson 1996, 2013).

Under current international law systems, given the extensive effect that geoengineering could have on the sovereignty of all nations, unanimity would seem to be necessary for a geoengineering decision to be legitimate (Jamieson 1996, 2013). Absent an international governing body to regulate and monitor the geoengineering effort, it is not difficult to imagine that sovereign nations would have grievances against deploying nations. It is thus important, before moving down the path of deploying a geoengineering technology, to establish governance systems to ensure that all affected parties are engaged. If private companies were to develop geoengineering methods, it is easy to imagine the raft of intellectual property concerns that could arise as various parties jockey to deploy the technology (Dilling and Hauser 2013).

More worryingly, absent such a governing body, but in the face of potentially catastrophic climate shifts, nations or maybe even individuals may have incentives to deploy geoengineering without agreement from stakeholders. For example, a wealthy super-philanthropist might sponsor a geoengineering effort without buy-in from affected parties, or a disgruntled nation could unilaterally deploy a geoengineering technology to circumvent lengthy and complex decision-making processes (Gardiner 2010; Hale 2012). "Rogue Actor" problems such as these manifest first as problems of pure coordination. Consider a case in which we have two or more rogue actors, each of whom acts independently to deploy a full-scale geoengineering technology without consulting others. Two geoengineering technologies deployed simultaneously by independent but well-meaning parties, absent coordination, could result in twice or more the intended effect. In such a scenario, even well-meaning rogue actors could set off a catastrophic chain of events that dwarfs any anticipated negative impacts of climate change (Gardiner 2018).

A more extreme kind of Rogue Actor problem may manifest as a form of eco-terrorism, in which rogue actors use geoengineering to try to hold the rest of the world hostage. This could take the form of unilateral deployment of geoengineering by nations or other large groups with the express intention of intimidating others, damaging their interests, or even killing them (Keith 2000; Shepherd 2009; Bodansky 2013; Preston 2013). The more accessible and inexpensive these technologies become – and, in particular, if small groups could access geoengineering technologies – the greater the potential for eco-terrorism (Gardiner 2010; Hale 2012). Unlike other forms of terrorism, however, geoengineering-based eco-terrorism would require extensive planning, resources, and coordination in order to carry out on anything beyond the smallest of scales.

Fortunately, there are several international treaties and laws that may serve as mechanisms to discourage rogue actors from developing or deploying geoengineering without the participation of other parties. The Environmental Modification (ENMOD) Convention of 1976 prohibits the "military or any other hostile use of environmental modification techniques" (United Nations General Assembly 1976). Environmental modification techniques are defined as "the deliberate manipulation of natural processes -- the dynamics, composition or structure of the Earth, including its biota, lithosphere, hydrosphere and atmosphere, or of outer space" (United Nations General Assembly 1976; Banerjee 2011; Bracmort and Lattanzio 2013). ENMOD permits the modification of weather for "peaceful purposes." But these governing documents raise just as many questions as they resolve. If there are catastrophic risks associated with these technologies, can geoengineering be considered a peaceful purpose? Is there a reason to worry that the presence of ENMOD might stimulate, rather than quell, an international arms race (Gordin and Ten Have 2011; Gardiner 2018)? How might signatories to ENMOD seek to modify the agreement to serve their own interests?

To date, no nation has been found to be in violation of ENMOD (Banerjee 2011). Still, depending on how the "hostile use" provision is interpreted (for example, if a hostile use were determined to be the non-unanimous decision to implement geoengineering), ENMOD could represent a stringent set of limitations against executing geoengineering projects (Banerjee 2011).

Conclusion

While some geoengineering proposals hold the promise of addressing the symptoms of climate change, it is clear that geoengineering is no silver bullet. There are countless concerns that must first be addressed before geoengineering can be responsibly utilized, including concerns about costs, uncertainty, distribution, and governance. For starters, the costs of geoengineering are likely to be high, and future generations could be burdened by the continuous expense of perpetual deployment. Additionally, geoengineering easily could result in unintended, irreversible environmental harms far worse than those it purports to fix. There is no guarantee that the benefits and burdens of geoengineering would be justly distributed, and it could instead lead to deepening global inequality. Taken together, these concerns only scratch the surface of the complex governance challenges that do not neatly mesh with existing international decision-making structures. These challenges may prove too daunting to overcome.

Nevertheless, the promise of geoengineering remains tantalizing to some, and the pressure to take more drastic action will only increase if we cannot effectively mitigate and adapt to the escalating effects of climate change. Scientific research and innovation efforts are ongoing, endeavoring to improve geoengineering methods and our understanding of their effects. This scientific innovation must be coupled with governance and policy innovation – which is becoming an active topic of academic interest. These efforts, however, do not guarantee that geoengineering will be successful. Furthermore, the very need for geoengineering could be circumvented with a more committed world effort toward mitigation and adaptation. Should we ultimately choose to deploy geoengineering, our world could look quite different. Space mirrors could become fixtures in our sunsets. Artificial, CO_2-capturing trees might become more common in our cities than real trees. Oceans may appear more green than blue due to the proliferation of iron-fed algae. If such a world is to ever exist, we must take care to adequately address the potential risks and harms of making such significant changes to our environment.

References

Akbari, Hashem et al. "Global Cooling: Increasing World-Wide Urban Albedos to Offset CO2." *Climatic Change*, vol. 94, no. 3, June 2009, pp. 275–86, doi:10.1007/s10584-008-9515-9.

Banerjee, Bidisha. "The Limitations of Geoengineering Governance in a World of Uncertainty." *Stanford Journal of Law, Science, and Policy*. vol. IV, 2011, pp.15–36.

Barrett, Scott. "The Incredible Economics of Geoengineering." *Environmental and Resource Economics*, vol. 39, no. 1, 2008, pp. 45–54.

Beck, Ulrich, Anthony Giddens, and Scott Lash. *Reflexive Modernization: Politics, Tradition and Aesthetics in the Modern Social Order.* Stanford: Stanford UP, 1994.

Bodansky, Daniel. "The Who, What, and Wherefore of Geoengineering Governance." *Climatic Change*, vol. 121, no. 3, Dec. 2013, pp. 539–51, doi:10.1007/s10584-013-0759-7.

Bracmort, Kelsi and Richard K. Lattanzio. "Geoengineering: Governance and Technology Policy" *Library of Congress. Congressional Research Service. Resources, Science, and Industry Division*, R41371, 2013, pp. 336–350.

Caldeira, Ken, Govindasamy Bala, and Long Cao. "The Science of Geoengineering." *Annual Review of Earth and Planetary Sciences*, vol. 41, no. 1, 2013, pp. 231–231.

Dilling, Lisa and Rachel Hauser. "Governing Geoengineering Research: Why, When and How?" *Climatic Change*, vol. 121, no. 3, Dec. 2013, pp. 553–65, doi:10.1007/s10584-013-0835-z.

Fleming, James Rodger. *Fixing the Sky: The Checkered History of Weather and Climate Control*. New York: Columbia University Press, 2010.

Gardiner, Stephen. "Is' arming the future' with Geoengineering Really the Lesser Evil? Some Doubts about the Ethics of Intentionally Manipulating the Climate System." In: Gardiner, S.M., Caney, S., Jamieson, D., Shue, H. (eds) *Climate Ethics: Essential Readings*. Oxford: Oxford University Press, 2010, pp. 284–313.

Gardiner, Stephen M. "Some Early Ethics of Geoengineering the Climate: A Commentary on the Values of the Royal Society Report." *Environmental Values*, vol. 20, no. 2, 2011, pp. 163–188.

Gardiner, Stephen M. and Augustin Fragnière. "The Tollgate Principles for the Governance of Geoengineering: Moving Beyond the Oxford Principles to an Ethically More Robust Approach." *Ethics, Policy & Environment*, vol. 21, no. 2, 2018, pp. 143–174, DOI: 10.1080/21550085.2018.1509472

Gordijn, B. and H. Ten Have. "Ethics of Mitigation, Adaptation and Geoengineering." *Medicine, Health Care, and Philosophy*, vol. 15, no. 1, 2012, pp. 1–2. doi:http://dx.doi.org.colorado.idm.oclc.org/10.1007/s11019-011-9374-4.

Hale, Benjamin. "Remediation vs. Steering: An Act—Description Approach to Approving and Funding Geoengineering Research." In: Basl, J., Sandler, R. (eds) *Designer Biology: The Ethics of Intensively Engineering Biological and Ecological Systems*. Lanham, MD: Lexington Books, 2013, pp. 195–216.

Hale, Benjamin. "The World that Would Have Been: Moral Hazard Arguments against Geoengineering." *Engineering the Climate: The Ethics of Solar Radiation Management*, vol. 113, 2012.

Hale, Benjamin and Lisa Dilling. "Geoengineering, Ocean Fertilization, and the Problem of Permissible Pollution." *Science, Technology, & Human Values*, vol. 36, no. 2, 2011, pp. 190–212.

Hardin, Russell. *Indeterminacy and Society*. Princeton: Princeton UP, 2003.

Harris, Elisa D., ed., *Governance of Dual-Use Technologies: Theory and Practice*. Cambridge, MA: American Academy of Arts & Sciences, 2016.

Hourdequin, Marion. "Geoengineering Justice: The Role of Recognition." *Science, Technology, & Human Values*, 2018, pp. 16224391880289.

Irvine, Peter J., Ben Kravitz, Mark G. Lawrence, Dieter Gerten, Cyril Caminade, Simon N. Gosling, Erica J. Hendy, Belay T. Kassie, W. Daniel Kissling, Helene Muri, Andreas Oschlies, and Steven J. Smith. *Towards a Comprehensive Climate Impacts Assessment of Solar Geoengineering*. United States: N. Web, 2016. doi:10.1002/2016EF000389.

Jamieson, Dale. "Ethics and Intentional Climate Change." *Climatic Change*, vol. 33, no. 3, 1996, pp. 323–336.

Jamieson, Dale. "Some Whats, Whys and Worries of Geoengineering." *Climatic Change*, vol. 121, no. 3, 2013, pp. 527–537.

Keith, David W. "Geoengineering the Climate: History and Prospect." *Annual Review of Energy and the Environment*, vol. 25, no. 1, 2000, pp. 245–284.

Keutsch Research Group. "SCoPEx." 2019, https://projects.iq.harvard.edu/keutschgroup/scopex

Ladyman, James and James Lambert. "What Is a Complex System." *European Journal for Philosophy of Science*, vol. 3, no. 1, 2012, pp. 33–67. http://philsci-archive.pitt.edu/9044/4/LLWultimate.pdf.

Lazarus, Richard. "Super Wicked Problems and Climate Change: Restraining the Present to Liberate the Future." *Cornell Law Review*, vol. 94, no. 5, 2009, pp. 1153–1233.

Leonhardt, David. "A Battle over the Costs of Global Warming." *New York Times*, Feb 21, 2007. https://www.nytimes.com/2007/02/21/business/21leonhardt.html

Maldonado, J.K., C. Shearer, R. Bronen, K. Peterson, and H. Lazrus. "The Impact of Climate Change on Tribal Communities in the US: Displacement, Relocation, and Human Rights." In: Maldonado, J.K., Colombi, B., Pandya, R. (eds) *Climate Change and Indigenous Peoples in the United States*. Springer, Cham, 2013. https://doi.org/10.1007/978-3-319-05266-3_8

Matthews, H. Damon and Ken Caldeira. "Environews Spheres of Influence." *Environmental Health Perspectives*, vol. 113, no. 7, 2005, https://www.ncbi.nlm.nih.gov/pmc/articles/PMC1257666/pdf/ehp0113-a00464.pdf.

McLaren, Duncan P. "Whose Climate and Whose Ethics? Conceptions of Justice in Solar Geoengineering Modelling." *Energy Research & Social Science*, vol. 44, 2018, pp. 209–221.

Moellendorf, Darrel. *The Moral Challenge of Dangerous Climate Change: Values, Poverty, and Policy*. New York: Cambridge University Press, 2014.

Moriyama, Ryo et al. "The Cost of Stratospheric Climate Engineering Revisited." *Mitigation and Adaptation Strategies for Global Change*, vol. 22, no. 8, Dec. 2017, pp. 1207–28, doi:10.1007/s11027-016-9723-y.

Morrow, David R. "Starting a Flood to Stop a Fire? Some Moral Constraints on Solar Radiation Management." *Ethics, Policy and Environment*, vol. 17, no. 2, 2014a, pp. 123–138.

Morrow, David R. "Why Geoengineering Is a Public Good, Even if it Is Bad." *Climatic Change*, vol. 123, 2014b, 95–100.

Nordhaus, William. "Projections and Uncertainties about Climate Change in an Era of Minimal Climate Policies." *American Economic Journal: Economic Policy*, vol. 10, no. 3, 2018, pp. 333–60.

Preston, C. "Carbon Emissions, Stratospheric Aerosol Injection, and Unintended Harms." *Ethics & International Affairs*, vol. 31, no. 4, 2017, pp. 479–493. doi:10.1017/S0892679417000466.

Preston, Christopher J. "Ethics and Geoengineering: Reviewing the Moral Issues Raised by Solar Radiation Management and Carbon Dioxide Removal: Ethics & Geoengineering." *Wiley Interdisciplinary Reviews: Climate Change*, vol. 4, no. 1, 2013, pp. 23–37.

Robock, Alan et al. "Benefits, Risks, and Costs of Stratospheric Geoengineering." *Geophysical Research Letters*, vol. 36, no. 19, 2009, doi:10.1029/2009GL039209.

Russel, Lynn et al. "Ecosystem Impacts of Geoengineering: A Review for Developing a Science Plan." *Ambio*, vol. 41, no. 4, 2012, pp. 350–369.

Shepherd, John, Ken Caldeira, Peter Cox, Joanna Haigh, David Keith, Brian Launder, Georgina Mace, Gordon, MacKerron, John Pyle, Steve Rayner, Catherine Redgwell, and Andrew Watson. *Geoengineering the Climate: Science, Governance and Uncertainty*. London: Royal Society, 2009.

Smith, Wake and Gernot Wagner. "Stratospheric Aerosol Injection Tactics and Costs in the First 15 Years of Deployment." *Environmental Research Letters*, vol. 13, 2018, http://iopscience.iop.org/article/10.1088/1748-9326/aae98d/meta.

Svoboda, Toby. "The Ethics of Geoengineering: Moral Considerability and the Convergence Hypothesis." *Journal of Applied Philosophy*, vol. 29, no. 3, 2012, pp. 243–256.

Svoboda, Toby. "Is Aerosol Geoengineering Ethically Preferable to Other Climate Change Strategies?" *Ethics and the Environment*, vol. 17, no. 2, 2012, pp. 111–135.

Svoboda, Toby and Peter Irvine. "Ethical and Technical Challenges in Compensating for Harm Due to Solar Radiation Management Geoengineering." *Ethics, Policy & Environment*, vol. 17, no. 2, 2014, 157–174, DOI: 10.1080/21550085.2014.927962

Svoboda, Toby, Klaus Keller, Marlos Goes, and Nancy Tuana. "Sulfate Aerosol Geoengineering: The Question of Justice." *Public Affairs Quarterly*, vol. 25, no. 3, 2011, pp. 157–180.

Tollefson, Jeff. "The Sun Dimmers." *Nature*, vol. 563, 2018, pp. 613–615.

Trenberth, Kevin E. "Atmospheric Moisture Residence Times and Cycling: Implications for Rainfall Rates and Climate Change." *Climatic Change*, vol. 39, no. 4, Aug. 1998, pp. 667–94, doi:10.1023/A:1005319109110.

United Nations Environment Programme, *Global Linkages: A Graphic Look at the Changing Arctic*, UN Environment and GRID-Arendal, 2019.

United Nations General Assembly, *Convention on the Prohibition of Military or Any Other Hostile Use of Environmental Modification Techniques*, 10 December 1976.

United States. Congress. House. Committee on Science, Space, and Technology. Subcommittee on Environment, and United States. Congress. House. Committee on Science, Space, and Technology. Subcommittee on Energy. *Geoengineering: Innovation, Research, and Technology: Joint Hearing before the Subcommittee on Environment & Subcommittee on Energy, Committee on Science, Space, and Technology, House of Representatives, One Hundred Fifteenth Congress, First Session, November 8, 2017*. U.S. Government Publishing Office, Washington, 2018.

Vaughan, Naomi E. and Timothy M. Lenton. "A Review of Climate Geoengineering Proposals." *Climatic Change*, vol. 109, no. 3, 2011, pp. 745–790.

Whyte, K.P. "Justice Forward: Tribes, Climate Adaptation and Responsibility." In: Maldonado, J.K., Colombi, B., Pandya, R. (eds) *Climate Change and Indigenous Peoples in the United States*. Springer, Cham, 2013. https://doi.org/10.1007/978-3-319-05266-3_2

Wong, Pak-Hang, Tom Douglas and Julian Savulescu. "Compensation for Geoengineering Harms and No-Fault Climate Change Compensation." *The Climate Geoengineering Governance Working Papers*, 2014.

25

SKEPTICISM AND DENIALISM

Jay Odenbaugh

Introduction

This essay is an overview of the many different scientific, psychological, and communicative issues concerning climate skepticism. First, we begin the essay exploring different types of climate skepticism and the evidence for anthropogenic climate change. Second, we consider the role of consensus and dissent in science and recent discussion of "Merchants of Doubt" and Climategate. Third, we turn to psychological issues concerning and American opinion on the topic, the relevance or irrelevance of scientific literacy to climate skepticism, the role of affect in environmental decision-making, and cognitive biases that inform views regarding climate change. Third, we consider climate communication and how we might most effectively motivate pro-environmental behavior and beliefs.

Skepticisms

Terminology

In scientific debates regarding anthropogenic climate change, we need clear terminology. Unfortunately, terminology can appear prejudicial when it has a moralistic tone (O'Neill and Boykoff, 2010). As philosophers would say, these can be *thick* concepts which both describe and evaluate (Williams, 2011; Vayrynen, 2013). We will regiment our vocabulary as neutrally as we can. I will do so as follows. First, I will focus our discussion on the following claim:

(C) Average global temperatures are increasing in part because of human greenhouse gas emissions.

Anthropogenic climate change includes much more than C, but it is crucial for understanding climate skepticism. Second, how we understand the term "skeptic" is extremely important in debates over C. One way of thinking about skeptics is that they form their beliefs in proportion to the evidence. Another way of thinking about skeptics is that they form their beliefs without regard for evidence. The reason we have disparate accounts of skepticism is because it can be a term of praise or derision. Those who accept C often think of skeptics as ignoring evidence, and those who deny C think of them as closely following the evidence. Given an individual, a belief, and a

DOI: 10.4324/9781315768090-30

body of evidence, we can ask, "Is that belief is proportional to the evidence?" For our purposes, we will disambiguate these two uses as follows:

A *skeptic* believes (or disbelieves) a claim in proportion to the evidence for it.[1]

In this sense, scientists are customarily thought of as skeptics.

A *contrarian* believes (or disbelieves) a claim regardless of whatever evidence there is for the claim.[2]

Finally,

A *doubter* is either a skeptic or contrarian.

For example, if one denies that cigarette smoking is a positive causal factor in developing cancer regardless of the evidence, then one is a contrarian regarding cigarette smoking and cancer. They simply disregard the ascertainable evidence. Given the distinctions drawn, we should not assume that those who challenge *C* are contrarians. It is an empirical matter as to whether they are skeptics or contrarians – it depends on the relevant beliefs and body of evidence. In this way, our terminology is neutral.

There is a rich debate in philosophy between William James and W. K. Clifford as to whether we should be skeptics (Clifford, 1886; James, 2000). On James' view, sometimes we should not be skeptics. James rejects *evidentialism*. Simply put, evidentialism is the claim that one should believe a claim if, and only if, there is sufficient evidence for it (Conee and Feldman, 2004). Consider a "hypothesis" that God exists and suppose there is no evidence for or against the hypothesis. Insofar as the hypothesis is live versus dead, forced versus avoidable, and momentous versus trivial, James thought that one is permitted to believe it.[3] If there is no evidence for or against the existence of God and one would be happier believing that there is a God, then it is permissible for them to do so. W. K. Clifford disagreed with James' pragmatism. He argued that "It is wrong always, everywhere, and for anyone to believe anything on insufficient evidence." His concern was that believing irrespective of the evidence was morally and epistemically irresponsible. For example, the consequences of people forming their beliefs independent of evidence would be disastrous because it would undermine our trust in what each other asserts.

It is important to note that James' voluntarism is circumscribed. If a belief is neither confirmable nor disconfirmable by evidence, then one may believe what one likes regarding it. But James didn't think one could believe whatever one likes with regard to a claim when it is confirmable or disconfirmable. In the case of *C*, we are dealing with claims that are confirmable or disconfirmable. However, there is a Jamesian wrinkle we will discuss below.

The Evidence

Skepticism regarding *C* is associated with the following statements.

1   Average global temperature is increasing.
2   Average global temperature is increasing in part due to human fossil fuel emissions.
3   Average global temperature is increasing in part due to human fossil fuel emissions and the impacts of this increase will be mainly negative.
4   Average global temperature is increasing in part due to human fossil fuel emissions, the impacts of this increase will be mainly negative, and mitigation of those emissions will not lower average global temperature.

Trend doubters reject (1). Since (3) and (4) individually imply (2), and (2) implies (1), one who rejects (1) rejects the others. This is the most radical rejection. *Attribution doubters* accept (1) but deny that it is due to human fossil fuel emissions; e.g. they say that it is due to an increase in solar irradiance. *Impact doubters* accept (1) and (2), but suggest that the impacts are mostly positive. As one example, one might argue that southern Canada would be capable of increased agriculture, which is beneficial and generalize considerations like these. *Regulation doubters* agree with (1)–(3) but deny that mitigation could do anything about it. One reason offered is that it is such a serious collective action problem that international agreements reducing our emissions are practically unachievable. It is also worth noting that (1)–(4) are *factual* claims concerning our climate system. One could accept (1)–(4), but yet we think should not do anything about it. That is, they could agree with the facts but deny certain *normative* claims. It might be argued that nation states have minimal moral obligations to other national states and hence mitigation is unethical. For example, President George W. Bush said in debate with former Vice President Al Gore Jr.:

> I'll tell you one thing I'm not going to do, I'm not going to let the United States carry the burden for cleaning up the world's air, like the Kyoto treaty would have done.

It is important to note that the epistemic profile of doubters can look very similar. Suppose Richard and Fred deny (1)–(4). Further suppose Richard is a skeptic and Fred a contrarian. How would we tell the difference? Suppose we provided both of them with evidence for (1). Richard would change his mind if he were presented with such evidence of increasing global temperatures, whereas Fred would not. Thus, the chief difference between skeptics and contrarians concerns counterfactuals like this:

> All things being equal, if one were presented with evidence in favor of a claim which one does not already believe, then they would change their belief in proportion to the evidence.

We can empirically evaluate whether one is a skeptic or a contrarian. If they would change their beliefs in proportion to the evidence, then they are skeptics; if they would not, then they are contrarians.

With regard to *C*, we have three basic empirical questions:

- Is the planet warming up?
- Are we causing the planet to warm up?
- What will happen if we continue to warm the planet up?

Let's consider how climate scientists have answered them.

The first question is answered in the affirmative by climate scientists. First, global average surface temperature shows a warming of 0.85°C in the period 1880–2012 based on several independent sets of data. Nine of the ten warmest years recorded have occurred since 2000, and 2014 was the warmest recorded (Stocker et al., 2014, §2.4). Second, the glacial record since 1800 shows that mean glacier length has been declining at an accelerating rate (Stocker et al., 2014, §4.3). Third, sea level rise has increased; tide gauges show that over the twentieth century, sea level rose by about 1.7mm between 1901 and 2010 (Stocker et al., 2014, §3.7). Fourth, there has been sea ice decrease; e.g. annual mean Arctic Sea ice extent has decreased from 1979 to 2012 with a rate very likely between 3.5 and 4.1% per decade (Stocker et al., 2014, §4.2). Thus, we should not be trend doubters.

The second question is also answered in the affirmative by climate scientists.[4] The first component of the science concerning global climate change is the greenhouse effect which was first discovered by John Tyndall (Tyndall, 1861), though Jean-Baptiste Joseph Fourier had documented the atmosphere's effect of the Earth's temperature (Fourier, 1827). The greenhouse effect occurs as incoming solar radiation is reflected by the surface of the Earth and that radiation is trapped by greenhouse gases. The greenhouse gases include carbon dioxide (CO_2), methane (CH_4), nitrous oxide (N_2O), and water vapor. This effect has been known for quite some time and has been demonstrated experimentally. Atmospheric concentrations of CO_2, CH_4, and N_2O have all increased since 1750 due to human activity as was clearly demonstrated in the case of CO_2 by Charles Keeling (Keeling, 1960). In 2011, they were 391 ppm, 1,802 ppm, and 324 ppb, respectively, exceeding pre-industrial levels by approximately 40%, 150%, and 20%. Amazingly, concentrations of CO_2, CH_4, and N_2O are greater than the highest concentrations found in ice cores for the past 800,000 years (Stocker et al., 2014, §5.2, 6.1, 6.2).

Suppose one does accept that the Earth is warming but denies that the primary explanation is human fossil fuel emissions. What would the alternatives be? One alternative is orbital variations, i.e. Milankovitch cycles. The Milankovitch cycles are caused by changes in the shape of the Earth's orbit around the sun, the tilt of the Earth's rotational axis, and its axial tilt. However, climate scientists deny that these cycles can explain the temperature increase since there is a mismatch of time scale. The orbital variations happen on the order of 100,000, 41,000, and 26,000 years, respectively, and the warming we have observed is over the last 150 years (Dessler, 2011, §7.3). Another alternative is a volcanic activity. It certainly is true that volcanoes can change the Earth's temperature; however, they would have to occur with the right frequency and magnitude and observed volcanic eruptions do not fit the pattern we find. Additionally, as originally shown by Hans Seuss (Suess, 1955), plants prefer ^{12}C to ^{13}C where ^{12}C makes up 99% of the carbon in the atmosphere and ^{13}C makes up 1%.[5] Fossil fuels are derived from ancient plants and thus if the burning of fossil fuels was contributing to increases in CO_2 emissions, we should see the ratio $^{13}C/^{12}C$ increase which is exactly what we find (Dessler, 2011, §5.5). Last, some hypothesize that solar variation is the driver of temperature increase. However, this is not so for several reasons. First, there has been a 11-year solar cycle measured over past decades but the ocean's thermal inertia is insensitive to it. Second, if increased solar output were warming the atmosphere ,it would do so uniformly, but in fact the stratosphere (atmosphere above 10 km) is cooling (Dessler, 2011, §7.2). Thus, we should not be attribution doubters.

The observed and projected impacts of anthropogenic climate change are on the whole negative though a few are thought to be positive (Field et al., 2014, 1.3.2). These impacts include changing precipitation; melting snow and ice; extinction, migration, distribution, and interactions of terrestrial, freshwater, and marine species; decreasing crop yields; climate-related extreme weather, including heat waves, droughts, floods, tornadoes, and wildfires. Thus, we should not be impact doubters.

The Climate Wager

Let's now turn to the Jamesian wrinkle. *C* is an empirical claim which as we have seen has strong evidence in its favor. However, it concerns both belief and action.[6] When deciding what to do, we must include what we believe but also what we value. Decision theory (including game and social choice theory) is that framework in which rational decision-making is often formulated. Traditionally, decisions are a function of the probabilities of various ways the world might be and the utilities associated with our preferences or values. It is thought that an action is rational just in case it maximizes expected utility. Consider an extremely simplified decision matrix for *C*. For ease of exposition, we will use a touch of formalism. Let *p* be the probability that *C* occurs and $(1 - p)$ be the probability that *C* does not occur; let u_{AC} be the utility of acting *A* as to stop *C*, given *C* occurs; $u_{A\neg C}$ be the utility of acting *A* as to stop *C*, given *C* does not occur; $u_{AC\neg}$ be the utility of not acting $\neg A$ as to stop *C*, given *C* occurs; and $u_{\neg A\neg C}$ be the utility of not acting as to stop *C*, given *C* does not occur. We thus have the following matrix (Table 25.1).

Table 25.1 The climate decision

	C	$\neg C$
A	pu_{AC}	$(1-p)u_{A\neg C}$
$\neg A$	$pu_{\neg AC}$	$(1-p)u_{\neg A\neg C}$

Thus, acting as to stop C maximizes expected utility if, and only if,[7]

$$pu_{AC}+(1-p)u_{A\neg C} > pu_{\neg AC}+(1-p)u_{\neg A\neg C}$$

Of course, this matrix is oversimplified since C occurring or not does not represent all of the relevant states of the world for us, and there are a variety of other actions available to us not represented. However, this simple example makes an extremely important point. Suppose contrary to the evidence above, p is extremely low as trend and attribution doubters allege. Nevertheless, the rational course of action may be to act if the utilities associated with not acting are sufficiently bad compared to the utilities associated with acting.

Consider a famous example from Blaise Pascal. Suppose the probability that God exists is close but not equal to zero, the utility of believing God exists, given God does exist, is $+\infty$ (i.e. heaven), the utility of believing that God exists when God does not exist is a negative finite value, the utility of not believing God exists when God does is $-\infty$ (i.e. hell), and the utility of not believing God exists when God does not exist is a positive finite value. It follows that the expected utility of believing is $+\infty$ and the expected utility of not believing is $-\infty$. Pascal argued that even if the probability of God existing is extremely low, the stakes are so great that it is obvious what the rational choice is; one should believe.[8]

To use a more flatfooted example, rational homeowners purchase insurance in the case of a fire. The probability of one's home burning to the ground is extremely low, but the utility of it occurring is extremely bad. Hence, if we take it to be rational to purchase home insurance to prevent our homes burning to the ground, would it not also be rational to act as if C is true since it is far more likely? Incidentally, it is no accident that the insurance industry is very much concerned about C since there has been a tenfold increase in economic losses between the 1950s and the 1990s due to extreme weather (Houghton, 2009, 4). Thus, when deciding how we should act, we must be Jamesians and include not just evidence that beliefs are correct, but also the utilities associated with the consequences of being correct and incorrect. *Even if* doubters are right that the evidence for C is weak, it does not follow that we should not act as if C is occurring, given the stakes involved (Haller, 2002).

In this section, first we distinguished between skeptics and contrarians along with different types of doubt regarding anthropogenic climate change. Second, we considered the evidence for C and other claims associated with it and found that there is good evidence on their behalf. Third, given that C concerns action and not just belief, we considered a pragmatic argument for action even when our confidence in C is low. In the next section, we consider issues related to the consensus regarding C, the public, and detractors.

Consensus and All That

The Scientific Consensus

The American public is divided over C, but climate scientists are not. Several studies have demonstrated this consensus, but the most famous of these is Oreskes (2004). Naomi Oreskes argued that there is a consensus regarding this claim among climate scientists:

Average surface temperatures are increasing in part because of human greenhouse gas emissions.

She and her assistants surveyed over 928 abstracts of articles in peer-reviewed journals with the search term "global climate change" through the Web of Science. Each essay was placed in one of the following categories.

1   Those explicitly endorsing the consensus position,
2   Those explicitly refuting the consensus position,
3   Those discussing methods and techniques for measuring, monitoring, or predicting climate change,
4   Those discussing potential or documenting actual impacts of climate change, those dealing with paleoclimatic change, and
5   Those proposing mitigation strategies.

Oreskes found that there were no papers in category (2). That is, she found no essay which disagreed with the claim that "Global climate change is occurring, and human activities are at least part of the reason why."

Some have criticized the study (Figure 25.1). Roger Pielke Jr. argued that the study does not represent the variation of belief consistent with the consensus (Pielke Jr, 2005). Climate scientists could agree with regard to Oreskes' focal statement but disagree over much else regarding climate change. Of course, that is true, but irrelevant to the study's question. Additionally, one might argue that peer-reviewed research paper's consistency with the consensus does not mean the authors accept that consensus. Oreskes provides a persuasive response to this point.

> If a conclusion is widely accepted, then it is not necessary to reiterate it within the context of expert discussion. Scientists generally focus their discussions on questions that are still disputed or unanswered rather than on matters about which everyone agrees.
>
> *(Oreskes, 2007, 72)*

One does not see common descent argued for in the journal *Evolution* since it is taken for granted by evolutionary biologists. If we only included papers that explicitly accept common descent, one

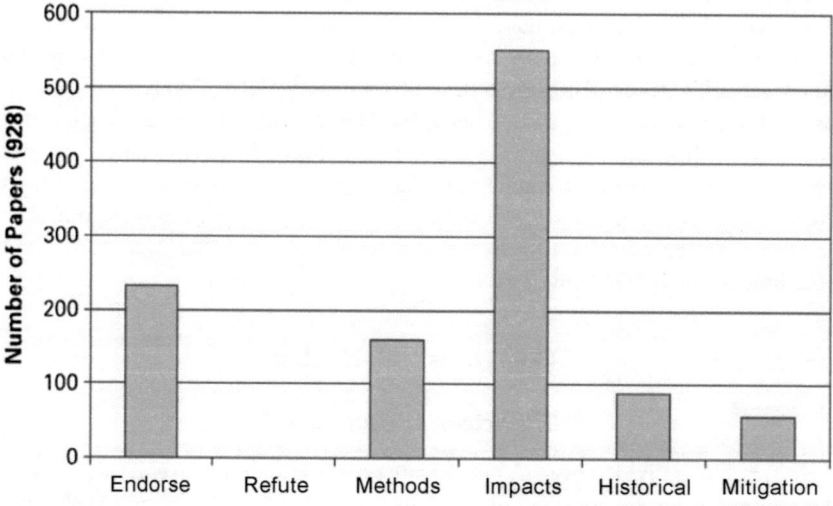

Figure 25.1   Oreskes' 2004 Consensus Study

would underestimate the acceptance of evolution. The same point holds with regard to anthropogenic climate change.

Peter Doran and Maggie Zimmerman did another study of the consensus among climate scientists (Doran and Zimmerman, 2009). They sent a survey to 10,257 earth scientists who were geosciences faculty, researchers at state geological facilities associated with local universities, researchers at the U.S. research facilities, and the U.S. Department of Energy national laboratories (Doran and Zimmerman, 2009, 21). They asked two questions:

- When compared with pre-1800s levels, do you think that mean global temperatures have generally risen, fallen, or remained relatively constant?
- Do you think human activity is a significant contributing factor in changing mean global temperatures?

Doran and Zimmerman determined that 90% answered "risen" to the first question and 82% "yes" to the second. When considered only the answers of those who listed "climate science" as their area of expertise and published more than 50% of their recent peer-reviewed papers in this area, then 96.2% answered "risen" to the first question and 97.4% "yes" to the second question (Figure 25.2).

Another study by William Anderegg et al. selected 908 from 1,372 climate scientists who had published at least 20 climate science papers and who had signed petitions opposing or supporting the positions taken by the IPCC, or had co-authored one or more reports associated with the IPCC (Anderegg et al., 2010). They determined that 97–98% agreed that:

> [I]t is "very likely" that anthropogenic greenhouse gases have been responsible for "most" of the "unequivocal" warming of the Earth's average global temperature in the second half of the 20th century.
>
> *(Anderegg et al., 2010, 12109)*

As one last example, John Cook et al. tried to add quantitative detail to our understanding of the consensus position (Cook et al., 2013). Through a Web of Science search, they considered a variety of claims, including the following "explicit endorsement with quantification."

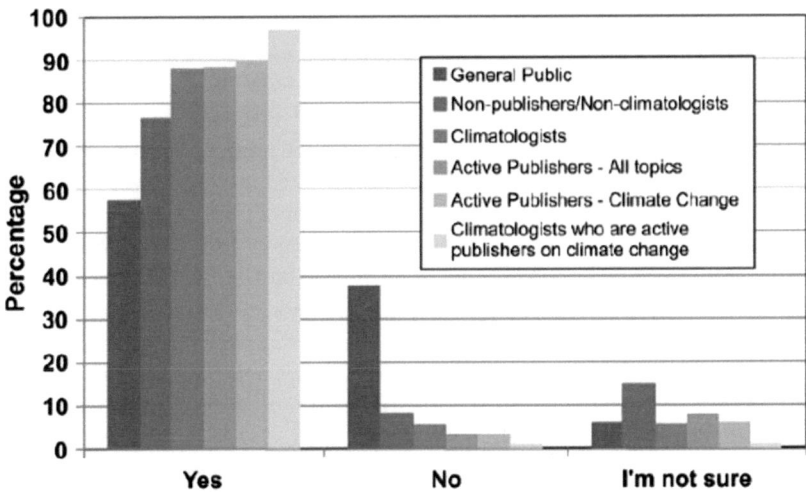

Figure 25.2 Doran and Zimmerman's 2009 Study

The global warming during the 20th century is caused mainly by increasing greenhouse gas concentration especially since the late 1980s.

(Cook et al., 2013, 3)

Of the 11,944 peer-reviewed climate science, only 75 in total either explicitly endorsed quantitatively or explicitly rejected quantitatively anthropogenic climate change with 87% endorsing the consensus view with 75 in total falling in these two categories. With regard to qualitative endorsement, 96% endorsed the consensus positions. Recently, David Legates has contested this finding (Legates et al., 2013). They write:

> But by taking into account that more than one-third of the 64 abstracts do not, in fact, endorse the quantitative hypothesis in Cook et al. (2013), the true percentages endorsing that hypothesis are 0.3% and 1.0%, respectively. Accordingly, their stated conclusion is incorrect.
>
> *(Legates et al., 2013, 309)*

But this is a fallacious argument. Cook et al. asked of those peer-reviewed climate science papers that either explicitly endorse or reject the quantitative statement of the consensus, what percentage endorsed it? Legates et al. changed the question to this: of those papers that either implicitly or explicitly endorse or reject the statement of the consensus, what percentage explicitly endorse the quantitative statement of the consensus? This is an irrelevant comparison.

We can safely conclude from the work of Oreskes, Doran and Zimmerman, Anderegg et al., and Cook et al. that there is a consensus regarding *C* among the experts.

Dissent in Science

Doubters of *C* correctly note that scientific consensus about a claim does not make or guarantee that it is true. As examples, the majority of scientists believed that the Earth was at the center of the solar system, continents do not drift, and species are not related by common descent. That is, scientists *en masse* believed claims later demonstrated to be false. As climate scientist Richard Lindzen writes:

> With respect to science, the assumption behind consensus is that science is a source of authority and that authority increases with the number of scientists. Of course, science is not primarily a source of authority. Rather, it is a particularly effective approach to inquiry and analysis. Skepticism is essential to science; consensus is foreign.[9]

Philosophers of science have recognized that dissent is fundamental to scientific progress. Different research groups are motivated out of self-interest to defend their hypotheses against incompatible ones of other research groups. Moreover, research groups whose work depends on those hypotheses will be motivated to do the same. The point is that objective inquiry can emerge out of base motivations, and dissent is critical to objectivity in science (Kitcher, 1995).

As we have seen, there is a consensus among the relevant scientists.[10] One might argue that just because there is a scientific consensus concerning *C* doesn't make *C* true. This is true but also irrelevant. The consensus matters especially with regard to policymakers rather than climate scientists. Suppose you are a policymaker working on climate-related issues and you ask, should I believe *C*? You of course have an extremely difficult time evaluating the evidence for *C* with its tree rings, glacial retreat, ice cores, satellite measurements, and so forth. The most reasonable way to form your beliefs would be to find expert opinion and proportion in your beliefs in accordance.

One way of making this clear is through Condorcet's Jury Theorem. Suppose that truth of an expert's or research group's collective belief regarding *Q* is probabilistically independent of other such experts. Further suppose that the probability that an expert or group i is correct about Q is p_i. Finally, assume for i, $\frac{1}{2} < p_i < 1$. We can prove that as $i \to \infty$, then $\Pr(Q) \to 1$. That is, as the number of independent experts that agree with Q, the greater the probability that Q is true. Of course, experts and research groups are not strictly speaking independent of one another, but analogues of the theorem can be proved where statistical dependence is included (Ladha, 1992; Estlund, 1994; Hawthorne, 2001). Hence, if the assumptions of the Condorcet Jury Theorem apply to expert's opinion regarding C, we have an argument that scientific consensus about C among climate scientists would be evidence for policymakers. It is also worth noting that this argument applies to non-expert scientists too – if you are an economist or evolutionary biologist, you should do the same.

If we suppose then that consensus regarding C is truth-indicative, then policymakers should proportion their degree of belief in accordance to the expert's degree of belief in C. The IPCC is an organization that represents the expert's degree of belief in this consensus position.[11] It does not conduct research on its own, but collates, evaluates, and synthesizes climate science done around the world. A panel of delegates from the represented countries elect the IPCC Chair and Bureau, and governments and organizations nominate experts that are ultimately chosen by the Bureau. There are three Working Groups, including Working Group I that considers the physical science behind climate change, Working Group II that considers the impacts, adaption, and vulnerability due to climate change, and Working Group III that considers mitigating the climate change that occurs. The Working Groups have Coordinating Lead Authors, one from a developing and another from a developed country, who coordinate their respective chapters. The Lead Authors work as a team to produce the content in their respective chapters and are supported by Contributing Authors whose research is published in peer-reviewed journals. After the authors prepare a first-draft of their chapters, experts are nominated by governments or organizations to evaluate the draft for its completeness and accuracy. The generated comments by these experts are then brought together by the Technical Support Unit, and a second-draft is written, and additionally a draft of the Summary for Policymakers is written. The final draft is submitted for acceptance by the Working Group responsible for it and the Summary for Policymakers is then edited line-by-line for approval. Lastly, the Panel approves the Summary and notes any remaining disagreement.

Each IPCC involves enormous numbers of expert scientists. IPCC Assessment Report Five (IPCC AR5) had over 830 Lead Authors from 85 countries with 301 of which were from developing nations. There were 179 female authors and 63% of the authors did not participate in IPCC AR4. As one example, Working Group I had 21,400 comments given on its first-draft by 659 experts and had 31,422 comments given on its second-draft by 800 experts and 26 governments. Similar results occurred with regard to the other Working Groups. This assessment done by the IPCC is remarkable. But we should recognize that other scientific bodies have done assessments and have come to the same conclusions, including the American Academy of Science, the American Meteorological Association, the American Geophysical Union, and the American Association for the Advancement of Science.

Merchants of Doubt

There have been several arguments offered for dismissing doubters and I will consider two. One such argument is given by Ross Gelbspan (1998). He provides documentation that climate doubters are contrarians since they have been funded by the oil and gas industry and many in this industry deny C. As one example, consider Patrick Michaels who is the Director for the Center for the Study of Science at the Cato Institute (Gelbspan, 1998, 40–44). Gelbspan claim that Michaels has received more than $115,000 from coal and energy interests. A quarterly publication *World*

Climate Review was founded by Michaels and was funded by the contrarian group Western Fuels. Finally, he was paid $100,000 by the electric utility Intermountain Rural Electric Association which also denies *C*. Though this argument is suggestive, one can argue that it is a circumstantial ad hominem. Surely, Michaels might have been convinced that *C* was false *before* he was paid to offer his views in public fora. If the Gelbspan argument is to be convincing, one must show that Michaels denies *C because* he received money to do so.

More recently, Naomi Oreskes and Erik Conway have chronicled the history of climate denial (Oreskes and Conway, 2010). Their story centers around the George C. Marshall Institute founded in 1984, and scientists whose careers were built around the Cold War nuclear weapons. Robert Jastrow was an astrophysicist who headed the Goddard Institute of Space Studies, William Nierenberg who as a nuclear physicist who worked for the Manhattan Project and directed the Scripps Institute of Oceanography, and Fred Seitz a physicist who was the president of the U.S. National Academy of the Sciences and worked on the atomic bomb. In the 1980s, they worked together on U.S. President Ronald Reagan's Strategic Defense Initiative. Since many believed that the "Star Wars" program would destabilize the Cold War, 6,500 American scientists and engineers signed a petition boycotting funding for the program. Jastrow, Nierenberg, and Seitz began writing articles, opinion pieces, and white papers defending the program – all three had strong anti-communist and pro-free market views. In 1979, Seitz began to work for the R. J. Reynolds Tobacco Corporation as a consultant in which he distributed more than $45 million to scientists for research that could cast doubt on the negative effects of cigarettes. After 1989, a new target was found; they had environmentalists in their sights. The "watermelons" (green on the outside and red on the inside) would then be subject to the same strategies. The Marshall Institute would drum up skepticism regarding acid rain, DDT, the ozone hole, and anthropogenic climate change. The techniques involved cherry picking data, stressing "balanced" opinions, and highlighting scientific uncertainty. Oreskes and Conway do not claim that there were financial incentives that motivated them directly. Rather, they claim that the deception arose due to a "free market fundamentalism" and its commitment to deregulation.

We can see this specifically in the work of Fred Singer who was a nuclear physicist employed by the Reagan administration to cast doubt on acid rain. In the 1990s, he also worked with Nirenberg and Seitz to challenge scientific opinion about the ozone hole and climate change. Additionally, he began to work for the Philip Morris tobacco company arguing that second-hand smoke was not bad for our health. The Environmental Protection Agency authored a report claiming that passive smoke was carcinogenic, and an independent scientific panel reviewed the evidence and concurred. Singer and lawyer Kent Jeffreys challenged the EPA's report writing:

> If we do not carefully delineate the government's role in regulating dangers, there is essentially no limit to how much government can ultimately control our lives.
>
> (Oreskes and Conway, 2010, 249)

Jeffreys himself was affiliated with the Cato Institute and the Competitive Enterprise Institute. Both are think tanks with a mission of defending freedom, free markets, and deregulation. With the U.N. Framework Convention on Climate Change and the Kyoto Protocol, we saw an ever-expanding group of think tanks who spread doubt regarding anthropogenic climate change.

Oreskes and Conway's claim then is that because these scientists doubted consensus positions across distinct scientific issues, and yet they held the same deregulation, pro-free market positions, the driver of their opinions was almost entirely political and not scientific. Put differently, it is extremely improbable for them to take the same doubting and deregulation positions across independent scientific issues if the political ideologies were not doing most of the work.

Climategate

Climate doubters have argued that climate scientists themselves have been guilty of deceit and distortion. The most discussed instance is "Climategate." During November 2009, the Climate Research Unit (CRU) of the University of East Anglia was hacked with 1,073 emails made public. CRU is known for its global record of instrumental temperature measurements and for publishing reconstructions of pre-1850 temperatures based on tree rings. Doubters claimed that CRU scientists had engaged in deceitful and faulty science. In order to appreciate what occurred, let's consider three examples of emails that were leaked.[12] Climate scientist Phil Jones writes:

> I've just completed Mike's Nature trick of adding in the real temps to each series for the last 20 years (ie from 1981 onwards) and from 1961 for Keith's to hide the decline.

Some believed that this was an admission by Jones of attempting to hide the recent decline of temperatures. However, the phrase "Mike's Nature trick" refers to a technique for integrating recent instrumental data and reconstructed data. Moreover, "decline" refers to decreasing reliability of inferring temperatures from tree ring data after 1960, which is known as the "divergence problem" in the scientific literature. Thus, an attentive reading of the email and scientific discussion removes doubt regarding Jones' email.

> A second email is from climate scientist Kevin Trenberth. He writes,
> The fact is that we can't account for the lack of warming at the moment and it is a travesty that we can't.

Some take this email to reference the claim that global warming has ceased. However, again it is clear from context that he is discussing a paper he had published regarding the difficulty of tracking energy flows in the climate system that lead to short-term cooling trends.

As one last example, in 2003, aerospace engineer Willie Soon and astronomer Sallie Baliunas published a paper in the journal *Climate Research* that argued after reviewing over 200 climate proxy studies that the twentieth century is not the warmest of the last millennium. Rather, the Medieval Warming Period between 800 and 1300 and the Little Ice Age of 1300–1900 were. Phil Jones wrote an email to Michael Mann saying:

> I think the skeptics will use this paper to their own ends and it will set paleo back a number of years if it goes unchallenged. I will be emailing the journal to tell them I'm having nothing more to do with it until they rid themselves of this troublesome editor, a wellknown skeptic in NZ. A CRU person is on the board but papers get dealt with by the editor assigned by Hans von Storch.

Climate scientist Michael Mann wrote back:

> This was the danger of always criticising the skeptics for not publishing in the "peer-reviewed literature." Obviously, they found a solution to that – take over a journal! So what do we do about this? I think we have to stop considering "Climate Research" as a legitimate peer-reviewed journal. Perhaps we should encourage our colleagues in the climate research community to no longer submit to, or cite papers in, this journal. We would also need to consider what we tell or request of our more reasonable colleagues who currently sit on the editorial board...

Some doubters argued that Jones and Mann were attempting to stifle the peer review process and prevent papers challenging *C* from being published. However, Soon and Baliunas' paper has been sharply criticized since they did not quantitatively analyze the data and their definition of "climate anomaly" had nothing to do with temperature, which was flawed, given the topic of the paper. Subsequently, five editors of *Climate Research* resigned from the board (which comprised one half of the editorial board). The editor-in-chief Hans Von Torch ultimately claimed that the review process had failed to uncover serious methodological flaws in the paper and that criticisms of it were correct.[13]

Given these emails along with a few others and the controversy that swirled around them, various institutions independently investigated them and Pennsylvania State University, the United Kingdom's House of Commons Science and Technology Committee, and the National Science Foundation all found that there was no wrongdoing. Most importantly however, suppose that these emails revealed deception and fabrication on the part of these scientists. This would not impugn the arguments we mentioned above in favor of *C* nor the consensus that was discussed above.

In this section, we first considered the scientific consensus regarding *C*. Second, we looked into the "merchants of doubt." Finally, we explored the controversy surrounding Climategate. We can know consider the psychological issues surrounding *C*.

Psychology

American Public Opinion

Let's now consider what Americans actually believe with regard to *C*. In the most recent Gallup poll, Americans clustered into three groups, *concerned believers*, *mixed middle*, and *cool skeptics*.[14] Concerned believers are those who attribute climate change to human actions and are worried about it. Cool skeptics are those who are not worried about climate change and believe that it is due to natural changes. The mixed middle is that group who are mixed with regard to concern, cause, and effects. Pollsters found that as of 2019, 51% of Americans are concerned believers, 30% are in the mixed middle, and 20% are cool skeptics (Figure 25.3).

Remarkably, this is the first time since 2001 that the majority of Americans are concerned believers. The cook skeptics have remained the same for the last four years, and the mixed middle has shrunk from 37% in 2015. Interestingly, the pollsters found that the opinion groups differed by gender, age, and politics. Concerned believers are more likely to be women than men; cool skeptics are even more likely to be male. The majority of concerned believers are younger than 50, whereas the largest percentage of cool skeptics are 50–64. Finally, 77% of concerned believers are Democrats or lean toward the Democratic Party, and 52% of cool skeptics are Republicans or lean toward the Republican Party.[15] The last interesting feature of this poll is that there are educational differences between concerned believers and cool skeptics. For example, 60% of concerned believers, 20% of the mixed middle, and 20% of cool skeptics are college graduates. We see that 58% of concerned believers, 24% of the mixed middle, and 18% of cool skeptics have post–graduate education. As we shall see, these data and others like them will be important for our discussion below.

Scientific Literacy and Cultural Cognition

One might assume that *if* the public understood climate science better, then they would accept *C*. Such an assumption would be incorrect according to the work of Dan Kahan et al. (2012). They attempted to test two hypotheses regarding public opinion regarding *C*, the *science comprehension thesis* (*SCT*) and the *cultural cognition thesis* (*CCT*).

(*SCT*) People fail to appreciate climate change because they lack scientific comprehension.

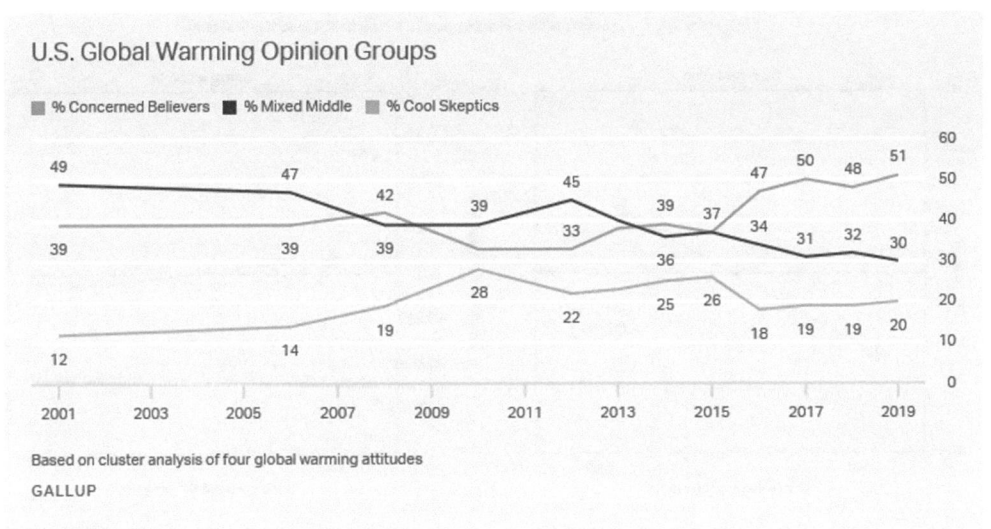

Figure 25.3   Gallup Global Warming Opinion Groups

(*CCT*) People fail to appreciate climate change because they identify risks through their social groups (i.e. conforming of beliefs to those that predominate within one's group).

With 1,540 subjects, they were asked to rank climate change risk on a scale of 0–10. Additionally, they distinguished between two social groups, *hierarchical individualists* (*HI*) and *egalitarian communitarians* (*EC*).

(*HI*) They think authority is tied to those with conspicuous social rankings and denies collective interference with such authorities.
(*EC*) They favor less regimented forms of social organizations and greater collective attention to individual need.

Kahan et al. reasoned that if *SCT* is correct, then increasing scientific literacy and numeracy will increase concern regarding climate change. Likewise, if *SCT* is correct, then cultural cognition should decrease with the most literate and numerate even among hierarchical individualists. Finally, if *CCT* is correct, then hierarchical individualists will show less concern for climate change than egalitarian communitarians. However, as respondent' science-literacy scores increased, concern with climate change decreased. There was also a negative correlation between numeracy and climate change risk (Figure 25.4). Among *EC*, literacy and numeracy were positively correlated with concern. Among *HI*, they are negatively correlated with concern (Figure 25.5).
Kahan et al. write:

> For the ordinary individual, the most consequential effect of his beliefs about climate change is likely to be on his relations with his peers. A hierarchical individualist who expresses anxiety about climate change might well be shunned by his co-workers at an oil refinery in Oklahoma City. A similar fate will probably befall the egalitarian communitarian English professor who reveals to colleagues in Boston that she thinks the scientific consensus on climate change is a hoax.

(2012, 3)

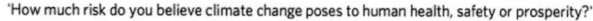

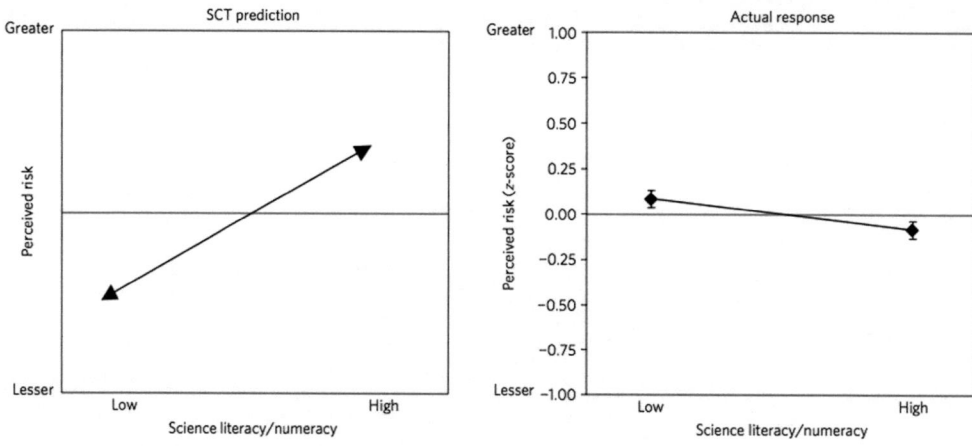

Figure 25.4 Kahan et al. Scientific Literacy Study

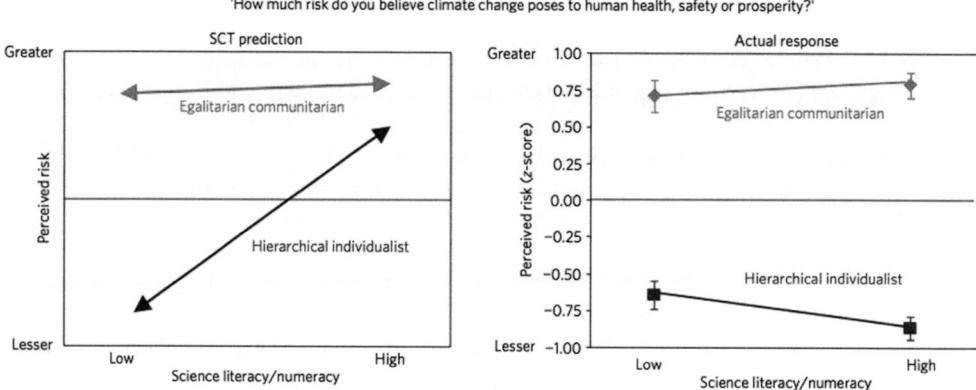

Figure 25.5 Kahan et al. Scientific Literacy Study

Given that we materially depend on others and we have no impact on climate change, then it is morally reasonable for people to form their beliefs about climate change in accordance with their community.[16]

One implication of this study is that often scientific literacy can provide one with skills for arguing against claims one already denies for moral or political reasons. As we shall see, we have good reason to think that moral judgments are based at least in part on our emotions, and are subject to a variety of cognitive biases that are difficult to combat.

Emotion and Moral Psychology

What is the nature of moral judgment? Sentimentalists claim that moral judgments necessarily involve emotions at least in part. Rationalists deny this. If sentimentalism is correct, then it has important implications for how we view disagreement among the American public over climate change since it is a moral issue. We will first consider the evidence for a sentimentalist moral psychology.

First, neuroimaging studies show that brain areas correlated with emotion are active during moral judgment (Greene and Haidt, 2002; Heekeren et al., 2003; Moll et al., 2003; Sanfey et al., 2003; Singer et al., 2006). When subjects were asked to consider moral sentences versus neutral sentences, were offered inequitable versus equitable payoffs in the ultimatum games, or violations of social rules like spitting food at dinner as opposed to spitting due to choking, the parts of subject's brains associated with emotion were far more active than in the non-moral cases. Likewise, Greene et al. (2001) showed using fMRI that emotions were involved in "trolley cases." Consider the following two cases.

Switch problem: A runaway trolley is headed for five people who will be killed if it continues on it present course. However, you can hit a switch, which turn to trolley to a different track killing one bystander. Should you hit the switch?

Footbridge problem: A trolley threatens to kill 5 people. You are standing next to very large stranger on a footbridge above the track. However, the only way to save the five is to push the one onto the track. Should you push them?

Most people say "Yes" and "No," respectively. Why? Customarily, the Footbridge case involved *killing a bystander*, whereas the Switch case involves *letting a bystander die*, and the former is thought always to be worse than the latter. Greene et al.'s hypothesis is that pushing someone is emotionally more salient than merely flipping a switch (2001, 2106). In one of the experiments conducted, they constructed 60 dilemmas some of which were moral and others non-moral. They then sorted those moral cases into "moral personal" and "moral-impersonal." Nine participants responded to each of the 60 dilemmas while undergoing fMRI scans. Those areas of the brain associated with emotion were significantly more active in "moral-personal" than in "moral-impersonal" cases (Greene et al., 2001, 2107).

Second, if a subject's disgust is aroused, they judge actions to be morally worse. The arousal of disgust can even be aroused unconsciously (Schnall et al., 2008). Schnall et al. asked subjects to morally evaluate stories while sitting on a desk which was clean or very dirty (e.g. has an old pizza box, chewed pencils, and a dirty cup). Subject's moral judgments were much harsher at the dirty desk in comparison to the tidy one. As another example, Wheatley and Haidt (2005) hypnotized subjects to be disgust whenever the words "take" or "often" were heard. When those words were heard, their moral judgments associated increased in severity.

Third, psychopaths have difficulty in distinguishing between moral and conventional transgressions due to emotional deficits (Blair, 1995). Moral norms are authority-independent, general, and bad to violate, and conventional norms are authority-dependent, specific, and are not particularly bad to violate (Turiel, 1983). Blair found that psychopaths have extreme difficulty discerning the difference between moral and conventional norms. Psychopaths also suffer deficits in affect. Thus, Blair and others have hypothesized that psychopath's failure to pass the moral/conventional task is that they lack certain emotions or affective capacities. Hence, they lack the ability to offer moral judgments.[17]

Fourth, when we are morally dumbfounded, we do not change our opinions (Haidt, 2003). Consider Haidt's cannibal story:

Jennifer works in a medical school pathology lab as a research assistant. The lab prepares human cadavers that are used to teach medical students about anatomy. The cadavers come from people who had donated their body to science for research. One night Jennifer is leaving the lab when she sees a body that is going to be discarded the next day. Jennifer was a vegetarian, for moral reasons. She thought it was wrong to kill animals for food. But then, when she saw a body about to be cremated, she thought it was irrational to waste perfectly edible meat. So she cut off a piece of flesh, and took it home and cooked it. The person had died recently of a heart attack, and she cooked the meat thoroughly, so there was no risk of disease. Is there anything wrong with what she did?

Subjects when asked claimed that eating human flesh is morally wrong. However, subjects could not give reasons that were not defeated by features of the case. Still, they would not change their minds. They were morally dumbfounded.[18] Moral judgments are a product of emotion and not just reason (for a critical discussion, see May 2018).

If emotions are associated with moral judgment, then this includes judgments regarding environmental issues (Ferguson and Branscombe, 2010; Mallett, 2012; Harrison and Mallett, 2013; Harth et al. 2013; Mallett et al. 2013; however, see Täuber et al. 2015). As one example, consider the study by Rees et al. (2014). Guilt, like shame, is self-directed emotion which occurs when one culpably violates a norm that generates reparative behavior. With guilt, the object of the emotion is one's action; with shame, the object of the emotion is one's self (Tangney and Dearing, 2003). Rees et al. randomly assigned participants to either the experimental human-caused or the controlled naturally occurring condition. First, guilt was more highly pronounced with regard to the human-caused manipulation than the naturally occurring manipulation. Second, willingness to sign a petition regarding environmental pollution was higher among the human-caused condition (88.1%) than the naturally occurring condition (70.9%). Moral emotions like guilt are clearly relevant to motivating pro-environmental behavior.[19]

Cognitive Biases

Starting with Herbert Simon and extending through the work of Daniel Kahneman and Amos Tversky, psychologists have argued that we make decisions using heuristics, and those heuristics are subject to cognitive biases. Biases of the sort they described are very relevant to thinking about climate change.

The *availability heuristic* concerns our tendency to only consider options that easily come to mind (Tversky and Kahneman, 1974; Greenberg et al., 1989; Gardner and Stern, 2002). For example, it is much easier to consider the extinction of polar bears as opposed to changes in albedo. Hence, the options we conceive of can be unrepresentative. The *anchor and adjustment heuristic* concerns how we anchor options with an example ranking the others in light of it (Tversky and Kahneman, 1974). For example, if we think of climate mitigation as requiring an extreme carbon tax, then mitigation will also seem extreme. So, the severity of an option depends on our anchoring. The *loss aversion heuristic* concerns the fact that for some good or service, we are more strongly averse to losing it as opposed to gaining it (Kahneman and Tversky, 1979). This heuristic is related to the *framing effect heuristic* in which the same information is associated with aversion or preference depending on framing (Tversky and Kahneman, 1981). The *temporal discounting heuristic* concerns our tendency to prefer rewards that occur sooner rather than later (Hendrickx and Nicolaij, 2004).

Different cognitive heuristics and biases are relevant to how we think and act regarding climate change. Our pro-environmental behaviors depend on the options we consider, their anchoring, their being represented as losses or gains, and how close or remote their impacts are in time. As we saw, how we view climate change depends not simply on how much information we possess, and our educational backgrounds. The cognitive heuristics and biases go deep, and we do not make poor decisions simply because of the absence of information (Nisbet and Mooney, 2009).

Our best account of moral psychology suggests several things. First, motivating pro-environmental behavior requires that we present considerations people care about; otherwise, they will lack motivation for behavioral change. Second, we must be on guard for cognitive biases. Arguments for action regarding climate change should relate to what really moves us in the short run concerning losses anchored in non-extreme ways with appropriate framing.

In this section, we considered recent work on American's opinions regarding *C*, and the modest relevance of scientific literacy on public opinion regarding climate change. Cultural cognition is especially relevant. Second, we examined the evidence for a moral psychology in which emotion plays an important role in moral judgment. Third, we considered various cognitive biases that affect our opinions on myriad topics but most certainly climate change. If we are to address doubts regarding *C*, it will involve a lot more than just climate science narrowly construed.

Communicative Strategies

We have come to several conclusions. First, the American public has polarized views of *C*. Second, the polarization is not due to scientific incompetence and hence is not to be remedied by increasing scientific literacy. Third, moral judgments are dependent in part on emotions and are affected by cognitive biases. Thus, how can one change public opinion to be responsive to the scientific consensus? How do we motivate the public morally regarding *C* (Gardiner, 2011; Jamieson, 2014)?

Climate change is an interdisciplinary subject of course. One area of inquiry which is directly relevant to our topic is rhetoric and communications.[20] We will first start by considering mistakes in climate communication that should be avoided. Then, we will consider how we can respond to climate contrarians. Last, we will consider positive strategies to try to communicate about *C* in more effective ways.

The American Psychological Association suggests that there are several impediments to motivating pro-environmental behavior regarding climate change. First, uncertainty, and the public's lack of understanding of it, gets in the way of "green behavior." Second, many people do not believe the message of risk from scientists and the government. Third, a minority of people deny anthropogenic climate change. Fourth, after study of more than 3,000 people in 18 countries, studies have demonstrated that many people believe our environmental conditions will get worse in the next 25 years. However, this framing may incline people to think that the relevant changes can be made later. Fifth, people often believe that their actions are inefficacious with regard to anthropogenic climate change. Sixth, we are creatures of habit, and ingrained behaviors can only be changed slowly and are extremely resistant to change.

With regard to contrarians, there are several strategies that are recommended by Susanne C. Moser and Lisa Dilling. First, we must be familiar with the arguments and strategies of contrarians. For example, contrarians routinely confuse weather and climate. Weather concerns the conditions of atmosphere variables over a short period of time in relatively small spatial scales, and climate is how the atmospheric variables change over relatively long periods of time over relatively large spatial scales. Second, debates over *C* largely concern value conflicts and not science per se. Hence, it is best to remove science from the discussion and focus on the relevant normative issues. Third, contrarians, because of their indifference to evidence, are best engaged in limited and focused ways. Skepticism as we have seen is healthy but contrarianism is not. We should focus on the mixed middle. Lastly, debates over *C* and whether they can be "won" depends on how the debate is framed. Hence, it is crucial to frame the debate in terms that do not prejudice the audience against *C* at the outset. Moreover, there are many different ways to frame climate change, and so we have to attend to our audiences.

I now want to mention several strategies for trying to change public opinion. These are strategies recommended by those who study effective communications, but the degree to which they will succeed is empirically unclear. First, effective communication regarding *C* requires that one knows their audience. For example, in order to be trusted by a group one speaks to, it is best to be a member of that very group. If one is speaking to evangelical Christians about *C*, then one will be trusted more so if one is an evangelical Christian (e.g. Sir John Houghton or Katharine Hayhoe). Second, it is important to get the attention of one's audience by choosing a relevant frame. Some individuals respond to risk and thus framing C in terms of risks will more attention grabbing than alternative frames. However, some are more influenced by focusing on benefits and thus one might focus on profits of alternative energy instead. Third, we should translate scientific claims into concrete experiences specifically making the scientific claims place-based. If we are in Oregon, for example, one can talk about the ever-lessening snowpack in the Cascades and the effects that will have. Third, we must be wary of the overuse of excessive emotional appeals. If we instill excessive fear in an audience, this can paralyze listeners. Additionally, if C concerns changes that are extremely difficult to achieve for individuals as opposed to collectives, they will feel impotent. If we couple extreme fear with impotence, we are bound to fail only adding to "climate fatigue." Fourth, we need to address scientific uncertainties since the public often has a naive view of science. Specifically, they often assume that scientific theories make precise predictions that are easily confirmed or disconfirmed. This means that model projections regarding 2050 or 2100, which are increasingly more complex at smaller scales, is simply very difficult for them to grasp. Thus, it is important to address uncertainties upfront since this can undercut their confidence and instead focus on what we know. Fifth, it is crucial to engage the social identifies of the groups one engages. As an example, among some, "environmentalism" is a negatively valanced word. Thus, we can simply avoid the term and have replaced it with other appropriate notions. Sixth, insofar as individuals need to change their behavior, those changes need to be as easy as possible to make. Consider a college student who can contribute some of their fees to carbon offsets. We might have an "opt-in" policy by which they have to select this option or we might have an "opt-out" policy which uses those fees unless the student selects for this not to occur. The latter is far more effective because it asks less of the student.

This is not a complete list of strategies for communicating and achieving change regarding people's beliefs and behavior regarding C. We live in a time in which we find ourselves desperate for ways to motivate change. The elements of this list are also not mutually exclusive. Motivating change regarding C will be more effective when as many of these strategies are employed as possible.

Conclusions

In this essay, we have covered many different topics. First, we began with different types of skepticism, regimented our terminology, examined the evidence for anthropogenic climate change, and pragmatic arguments for action even when we are uncertain. Second, we considered issues concerning the scientific consensus and role of dissent, the "merchants of doubt," and Climategate. Third, we turned to psychological issues regarding cultural cognition and the relative unimportance of scientific literacy, moral psychology, cognitive biases, and climate communication. Climate change is truly an interdisciplinary topic and thinking about how to think about climate skepticism is not easy. On the contrary, contrarians require us to bring all of our tools to the discussion.

Notes

1 Instead of a threshold for justified or warranted belief, one could follow Bayesianism and suppose there are degrees of belief. Assuming we can represent degrees of belief as probabilities, one's degree of belief in a claim simply is the probability of that claim, given the ascertained evidence. The Intergovernmental Panel on Climate Change (IPCC) appears to use a Bayesian approach to probability (http://www.ipcc. ch/ipccreports/tar/wg2/index.php?idp=106).
2 I will use the term "contrarian" for a person who believes (or disbelieves) a claim regardless of the evidence. The term "denier" might be used instead, but it unfortunately only concerns disbelieving some claim as opposed to what one might positively believe.
3 For James, the reason why the choice is forced is because he contrasts theism with a mutually exclusive alternative, including both atheism and agnosticism.

4 For an excellent collection of the original papers on which current climate science is based, see Archer and Pierrehumbert, 2011 and for a history of the same science, see Weart, 2008. Interestingly, doubters of evolution have a "common core" of texts to reference, including books by Michael Behe, William Dembski, Michael Denton, Duane Gish, Philip Johnson, and Henry Morris. There is no such "common core" with climate doubters. Their works are spread through multifarious publications in both print and online. However, this collection presents the varieties of climate doubt by the main figures (Moran, 2015).

5 ^{12}C and ^{13}C are different isotopes of carbon C. The former has six protons and six neutrons, whereas the latter has six protons and seven neutrons.

6 A pragmatist like James seems to think of belief as a type of action (Gale, 1999). If James is correct, then the argument below is much easier to make.

7 To determine the expected utility of an action, we multiply the probability and utility of each cell along the row, and then sum across the columns. The action which has the greatest expected utility is the rational one. Of course, it is *really, really hard* to determine these probabilities and utilities, but this is how would use decision theory to make such a decision if practically possible.

8 Actually, provided that the probability of God's existence is greater than zero, no matter how small it is, it is rational to believe, given the utilities. Pascal's wager is generally considered unpersuasive since the utilities reflect only the concept of a Christian God. But, if we can have next to no evidence for such a theology which he himself grants, then why assume those utilities and not others?

9 http://marshall.wpengine.com/wp-content/uploads/2013/08/Lindzen-Climate-Alarm-Where-Does-It-Come-From.pdf

10 Some doubters argue that there are many skeptics with regard to C. One purported example is and the "Oregon Petition" created by Arthur Robinson, which has over 30,000 signatories. But it is worth noting that only approximately 9,000 of them have PhDs. with the vast majority of those being engineers. Thus, they lack the relevant scientific background to evaluate C. It is also notable that Frederick Seitz wrote a cover letter endorsing the content of the petition.

11 The information that is discussed in this subsection can be found here http: //www.ipcc.ch/activities/activities.shtml

12 One can find the most discussed emails here: http: //www.telegraph.co.uk/news/earth/environment/globalwarming/6636563/University-of-East-Anglia-emails-the-most-contentious-quotes.html

13 For more information on this debate, see http://chronicle.com/article/Storm-Brews-Over-Global/27779/ and http://www.gpo.gov/fdsys/pkg/CHRG -108shrg92381/html/CHRG-108shrg92381.htm

14 http://www.gallup.com/poll/168620/one-four-solidly-skeptical-global -warming.aspx

15 It is also worth noting that less than 0.5% of cool skeptics are Democrats.

16 Druckman and McGrath (2019) have argued that observed results that have been taken to support "directional motivated reasoning" are consistent with models where individuals aim for accurate beliefs. However, those individuals differ over what are credible sources of evidence. For example, conservatives do not take the opinions of the climate scientists as reliable.

17 For critical response to this argument, see Aharoni et al. 2012; Roskies 2003.

18 For an interesting critical response to Haidt's work, see Jacobson 2012 and Railton 2014.

19 There are two provisos worth mentioning about the studies. First, the environmental issues included air pollution, fossil fuel consumption, and waste dumps which might not pertain to the effects of human-caused climate change alone. Second, the target population was one from Germany which might not export to the United States.

20 There is quite a bit written on climate communication. Some recent books that are worth examining are Hulme 2009; Norgaard 2011; Markowitz and Shariff 2012; Marshall 2014; Moser and Dilling 2006. There are also many very worthwhile online resources, including the following: http://guide.cred. columbia.edu/pdfs/CREDguide _full-res.pdf,http://www.apa.org/science/about/publications/climate -change.aspx,http://papers.ssrn.com/sol3/papers.cfm?abstract_id=1407958, andhttp://www.isse.ucar. edu/communication/docs/Environ_32-46a.pdf,http: //www.climateaccess.org/sites/default/files/Climate%20Crossroads% 20Guide.pdf.

References

Aharoni, E., W. Sinnott-Armstrong, and K. A. Kiehl (2012). Can psychopathic offenders discern moral wrongs? A new look at the moral/conventional distinction. *Journal of Abnormal Psychology* 121(2), 484.

Anderegg, W. R., J. W. Prall, J. Harold, and S. H. Schneider (2010). Expert credibility in climate change. *Proceedings of the National Academy of Sciences* 107(27), 12107–12109.

Archer, D. and R. Pierrehumbert (2011). *The Warming Papers*. John Wiley & Sons.

Blair, R. J. R. (1995). A cognitive developmental approach to morality: Investigating the psychopath. *Cognition 57*(1), 1–29.

Clifford, W. K. (1886). *Lectures and Essays*. Macmillan and Co.

Conee, E. and R. Feldman (2004). *Evidentialism: Essays in Epistemology: Essays in Epistemology*. Oxford University Press.

Cook, J., D. Nuccitelli, S. A. Green, M. Richardson, B. Winkler, R. Painting, R. Way, P. Jacobs, and A. Skuce (2013). Quantifying the consensus on anthropogenic global warming in the scientific literature. *Environmental Research Letters 8*(2), 024024.

Dessler, A. (2011). *Introduction to Modern Climate Change*. Cambridge University Press.

Doran, P. T. and M. K. Zimmerman (2009). Examining the scientific consensus on climate change. *Eos, Transactions American Geophysical Union 90*(3), 22–23.

Druckman, J. N. and M. C. McGrath (2019). The evidence for motivated reasoning in climate change preference formation. *Nature Climate Change, 9*(2), 111–119.

Estlund, D. M. (1994). Opinion leaders, independence, and Condorcet's jury theorem. *Theory and Decision 36*(2), 131–162.

Ferguson, M. A. and N. R. Branscombe (2010). Collective guilt mediates the effect of beliefs about global warming on willingness to engage in mitigation behavior. *Journal of Environmental Psychology 30*(2), 135–142.

Fourier, J. (1827). On the temperature of the terrestrial sphere and interplanetary space. *Mémoires de l?Académie Royale des Sciences 7*, 569–604.

Gale, R. M. (1999). *The Divided Self of William James*. Cambridge University Press.

Gardiner, S. M. (2011). *A Perfect Moral Storm: The Ethical Tragedy of Climate Change*. Oxford University Press.

Gardner, G. T. and P. C. Stern (2002). Human reactions to environmental hazards: Perceptual and cognitive processes. In *Environmental Problems and Human Behavior*, pp. 205–252. Allyn & Bacon.

Gelbspan, R. (1998). *Theheatison: The Climate Crisis, the Cover-up, the Prescription*. Da Capo Press.

Greenberg, M. R., D. B. Sachsman, P. M. Sandman, and K. L. Salomone (1989). Network evening news coverage of environmental risk. *Risk Analysis 9*(1), 119–126.

Greene, J. and J. Haidt (2002). How (and where) does moral judgment work? *Trends in Cognitive Sciences 6*(12), 517–523.

Greene, J. D., R. B. Sommerville, L. E. Nystrom, J. M. Darley, and J. D. Cohen (2001). An FMRI investigation of emotional engagement in moral judgment. *Science 293*(5537), 2105–2108.

Haidt, J. (2003). The emotional dog does learn new tricks: A reply to Pizarro and bloom. *Psychological Review 110*(1), 197–198.

Haller, S. F. (2002). *Apocalypse Soon?: Wagering on Warnings of Global Catastrophe*. McGill-Queen's Press-MQUP.

Harrison, P. R. and R. K. Mallett (2013). Mortality salience motivates the defense of environmental values and increases collective ecoguilt. *Ecopsychology 5*(1), 36–43.

Harth, N., C. Leach, and T. Kessler (2013). Are we responsible? Guilt, anger, and pride about environmental damage and protection. *Journal of Environmental Psychology 34*, 18–26.

Hawthorne, J. (2001). Voting in search of the public good: The probabilistic logic of majority judgements.

Heekeren, H. R., I. Wartenburger, H. Schmidt, H.-P. Schwintowski, and A. Villringer (2003). An FMRI study of simple ethical decision-making. *Neuroreport 14*(9), 1215–1219.

Hendrickx, L. and S. Nicolaij (2004). Temporal discounting and environmental risks: The role of ethical and loss-related concerns. *Journal of Environmental Psychology 24*(4), 409–422.

Houghton, J. (2009). *Global Warming: The Complete Briefing*. Cambridge University Press.

Hulme, M. (2009). *Why We Disagree about Climate Change: Understanding Controversy, Inaction and Opportunity*. Cambridge University Press.

Field, C. B., Barros, V. R., Dokken, D. J., Mach, K. J., Mastrandrea, M. D., Bilir, T. E., ... & Genova, R. C. (2014). *AR5 Climate Change 2014: Impacts, Adaptation, and Vulnerability, Global and Sectoral Aspects*, Working Group II Contribution to the Fifth Assessment Report of the Intergovernmental Panel on Climate Change.

Jacobson, D. (2012). Moral dumbfounding and moral stupefaction. *Oxford Studies in Normative* Ethics, Volume 2. Oxford University Press, pp. 289–316.

James, W. (2000). *Pragmatism and Other Writings*. Penguin.

Jamieson, D. (2014). *Reason in a Dark Time: Why the Struggle against Climate Change Failed–and What it Means for Our Future*. Oxford University Press.

Kahan, D. M., E. Peters, M. Wittlin, P. Slovic, L. L. Ouellette, D. Braman, and G. Mandel (2012). The polarizing impact of science literacy and numeracy on perceived climate change risks. *Nature Climate Change* 2(10), 732–735.

Kahneman, D. and A. Tversky (1979). Prospect theory: An analysis of decision under risk. *Econometrica: Journal of the Econometric Society*, 47(2), 263–291.

Keeling, C. D. (1960). The concentration and isotopic abundances of carbon dioxide in the atmosphere. *Tellus 12*(2), 200–203.

Kitcher, P. (1995). *The Advancement of Science-Science without Legend, Objectivity without Illusions*. Oxford University Press.

Ladha, K. K. (1992). The condorcet jury theorem, free speech, and correlated votes. *American Journal of Political Science*, 36(3), 617–634.

Legates, D. R., W. Soon, W. M. Briggs, et al. (2013). Climate consensus and "misinformation": A rejoinder to agnotology, scientific consensus, and the teaching and learning of climate change. *Science & Education*, 24(3), 1–20.

Mallett, R. K. (2012). Eco-guilt motivates eco-friendly behavior. *Ecopsychology* 4(3), 223–231.

Mallett, R. K., K. J. Melchiori, and T. Strickroth (2013). Self-confrontation via a carbon footprint calculator increases guilt and support for a proenvironmental group. *Ecopsychology* 5(1), 9–16.

Markowitz, E. M. and A. F. Shariff (2012). Climate change and moral judgement. *Nature Climate Change* 2(4), 243–247.

Marshall, G. (2014). *Don't Even Think About It: Why Our Brains Are Wired to Ignore Climate Change*. Bloomsbury USA.

May, J. (2018). *Regard for Reason in the Moral Mind*. Oxford University Press.

Moll, J., R. de Oliveira-Souza, and P. J. Eslinger (2003). Morals and the human brain: A working model. *Neuroreport 14*(3), 299–305.

Moran, A. (2015). *Climate Change: The Facts*. Stockade Books.

Moser, S. C. and L. Dilling (2006). *Creating a Climate for Change*. Cambridge University Press.

Nisbet, M. C. and C. Mooney (2009). Framing science. *Understanding and Communicating Science: New Agendas in Communication*, 316(5821), 56–56.

Norgaard, K. M. (2011). *Living in Denial: Climate Change, Emotions, and Everyday Life*. MIT Press.

O'Neill, S. J. and M. Boykoff (2010). Climate denier, skeptic, or contrarian? *Proceedings of the National Academy of Sciences of the United States of America 107*(39), E151.

Oreskes, N. (2004). The scientific consensus on climate change. *Science 306*(5702), 1686–1686.

Oreskes, N. (2007). The scientific consensus on climate change: How do we know were not wrong? *Climate Change: What it Means for Us, Our Children, and Our Grandchildren*, eds. Joseph F. C. DiMento and Pamela Doughman, 65–100, MIT Press.

Oreskes, N. and E. M. Conway (2010). *Merchants of Doubt: How a Handful of Scientists Obscured the Truth on Issues from Tobacco Smoke to Global Warming*. Bloomsbury Publishing USA.

Pielke Jr, R. A. (2005). Consensus about climate change? *Science (New York, NY) 308*(5724), 952.

Railton, P. (2014). The affective dog and its rational tale: Intuition and attunement. *Ethics 124*(4), 813–859.

Rees, J. H., S. Klug, and S. Bamberg (2014). Guilty conscience: Motivating pro-environmental behavior by inducing negative moral emotions. *Climatic Change 130*(3), 439–452.

Roskies, A. (2003). Are ethical judgments intrinsically motivational? Lessons from "acquired sociopathy". *Philosophical Psychology 16*(1), 51–66.

Sanfey, A. G., J. K. Rilling, J. A. Aronson, L. E. Nystrom, and J. D. Cohen (2003). The neural basis of economic decision-making in the ultimatum game. *Science 300*(5626), 1755–1758.

Schnall, S., J. Haidt, G. L. Clore, and A. H. Jordan (2008). Disgust as embodied moral judgment. *Personality and Social Psychology Bulletin*.

Singer, T., B. Seymour, J. P. O'Doherty, K. E. Stephan, R. J. Dolan, and C. D. Frith (2006). Empathic neural responses are modulated by the perceived fairness of others. *Nature 439*(7075), 466–469.

Stocker, T., D. Qin, G.-K. Plattner, M. Tignor, S. K. Allen, J. Boschung, A. Nauels, Y. Xia, V. Bex, and P. M. Midgley (2014). *Climate Change 2013: The Physical Science Basis*. Cambridge University Press.

Suess, H. E. (1955). Radiocarbon concentration in modern wood. *Science 122*(3166), 415–417.

Tangney, J. P. and R. L. Dearing (2003). *Shame and Guilt*. Guilford Press.

Täuber, S., M. van Zomeren, and M. Kutlaca (2015). Should the moral core of climate issues be emphasized or downplayed in public discourse? Three ways to successfully manage the double-edged sword of moral communication. *Climatic Change 130*(3), 453–464.

Turiel, E. (1983). *The Development of Social Knowledge: Morality and Convention*. Cambridge University Press.

Tversky, A. and D. Kahneman (1974). Judgment under uncertainty: Heuristics and biases. *Science 185*(4157), 1124–1131.

Tversky, A. and D. Kahneman (1981). The framing of decisions and the psychology of choice. *Science 211*(4481), 453–458.

Tyndall, J. (1861). Xxiii. On the absorption and radiation of heat by gases and vapours, and on the physical connexion of radiation, absorption, and conduction? The Bakerian lecture. *The London, Edinburgh, and Dublin Philosophical Magazine and Journal of Science 22*(146), 169–194.

Vayrynen, P. (2013). *The Lewd, the Rude and the Nasty: A Study of Thick Concepts in Ethics.* Oxford University Press.

Weart, S. R. (2008). *The Discovery of Global Warming: Revised and Expanded Edition.* Harvard University Press.

Wheatley, T. and J. Haidt (2005). Hypnotic disgust makes moral judgments more severe. *Psychological Science, 16*(10), 780–784.

Williams, B. (2011). *Ethics and the Limits of Philosophy.* Taylor & Francis.

PART V

Energy and Extraction

26

FOSSIL FUELS

Kian Mintz-Woo

Introduction

Fossil fuels are the accretions of unimaginably vast generations of algae and plankton (mostly not dinosaurs, despite what some might believe) that slowly settled on ocean beds, compacting and sedimenting over geological timescales. Under intense heat and pressure, the solar energy stored in those microscopic beings has been condensed into fossil fuels. They are fossil in at least two senses: they are extracted from the earth and they belong to a previous era. When we burn oil today, we are burning this long-contained accumulation of solar energy. One estimate is that it took 23 metric tons of ancient plant matter to produce one liter of gasoline (Dukes 2003). Think how quickly one can burn through that liter.

Not only has the volume of energy contained in fossil fuels been released in a short period of time, the number of firms that have facilitated the bulk of this extraction is also surprisingly small. Richard Heede (2014) traced the emissions catalogued in the historical record of 90 large fossil fuel producers, both public and private and reflecting both mergers and acquisitions. His conclusion was that, up until 2010, 63% of *all* estimated anthropogenic, or human-caused, historical emissions could be traced to these 90 producers (also cf. Frumhoff et al. 2015; Ekwurzel et al. 2017; Shue 2017). He called these producers "carbon majors", and they span well-known brands all the way to petrostates (Hormio 2017; Collins 2020; Grasso and Vladimirova 2019; Moss 2020). In short, thinking about fossil fuels reveals two kinds of compression: temporal, from the development of these fuels to the speed of their use; and distribution, from the variety of ways we all use fossil fuels to the relatively few entities that bring the majority of them out of the ground.

Our relationships with fossil fuels likewise span a variety of scales. To give a sense of these, this chapter discusses a range of relationships with fossil fuels. The purpose of this chapter is to show a range of debates that environmental ethicists have engaged with regarding fossil fuels.

First, with respect to our personal relationship to fossil fuels, it introduces arguments about whether we should or even can address our own usage of fossil fuels. This involves determining whether offsetting emissions is morally required and practically possible. Second, with respect to our relationship with fossil fuels at the national level, it discusses forms of local resistance, especially divestment and pipeline protesting. Finally, with respect to our relationship with fossil fuels at the international level, it considers two types of policy. On the one hand, some have argued that we should stop most trade in oil, on the basis that most oil that is traded is not subject to the control of citizens. On the other hand, some have argued that we should price the costs of fossil fuels so that there are market incentives to avoid digging them out of the ground. The chapter ends with a short conclusion.

On Our Personal Relationship to Fossil Fuels

Each one of our lives is entwined with fossil fuel emissions. Besides direct fuel use, necessary for activities like driving or flying, our choices of energy have significant consequences for our emissions and we also consume vast number of oil-produced products every day. If you are at the computer, most likely your keyboard and monitor are made of plastic produced by oil. Your food containers and cell phone are made from oil. It's likely in your lip balm and your luggage. You walked home on sidewalks made from it. And the list goes on. Most of our lives, especially in the wealthy, industrialized world, depend on fossil fuel emissions. [Average people in lower-income countries use considerably fewer fossil fuels, meaning that they could scale up fossil fuel use significantly and still generate fewer emissions than average people in high-income countries (Chakravarty et al. 2009).] These emissions together with those of many others will lead to major harms to current and future people. Should we reduce, or *mitigate*, our personal emissions by reducing the amount of fossil fuels we, directly and indirectly, consume?

On the one hand, it is very unlikely that any given person's emissions will make a recognizable or morally important contribution to climate change (Sinnott-Armstrong 2005; Nefsky 2011; Cripps 2013; Maltais 2013; Budolfson 2019; Kingston and Sinnott-Armstrong 2018). After all, even if one's emissions are significant for an individual, they are part of massive systems, including the emissions of many others. If one individual changed their own emissions, it might seem to make no difference with respect to climate outcomes. One important point to keep in mind is that we have to carefully distinguish claims that individual emissions make no difference *at all* and that they make negligible or *very small* differences. In order for this to be a special problem, it has to be the case that each individual's emissions make no difference at all.

On the other hand, if all the emissions make a difference, within those emissions, *some* of them had to be the ones that led to the difference. If so, there must be some likelihood that any given person will make a difference by producing those relevant emissions. For any given individual, that difference may be unlikely to occur at her hand, or hard to attribute to her, but if the value of the difference is great enough, a relatively low likelihood of making a difference might still make a significant difference in expectation (Hiller 2011; Kagan 2011; Nolt 2011; Broome 2019). This is because even a small likelihood of making a difference could be highly relevant if the size of that difference is large enough. For instance, it might be very unlikely that the emissions from my plane ride cause the strengthening of a heatwave from climate change, but if it did, the harm from that slightly more intense heatwave could be very large so the product of the likelihood and the disvalue of the outcome could be significant. It could also be that one's difference could be magnified by demonstrating integrity between one's political ideals and personal actions, as Marion Hourdequin (2010) argues.

But even if that were right, how could one person reduce their emissions? We might think that it is inevitable that the fossil fuels will be dug up and used (Hale 2011). After all, they are there in the ground, we know where many of them lie, and they are valuable in the marketplace. Also, it might seem we cannot live our lives without being responsible for some emissions, so why does it matter if it is somewhat more or somewhat less?

In response to this challenge, John Broome (2012) makes a surprising argument which relies on a distinction between *net* emissions and emissions *simpliciter*. To begin, he says that it does make a difference if you make more or less of a contribution of emissions; his claim is that it makes a difference in terms of what he calls your *duties of justice*. If your emissions will lead to some harms and you owe it—as a matter of justice—not to harm others, then you owe it to others not to emit at all!

But this leads to a problem. Let us grant that you cannot live at all while reducing your emissions to zero. Broome's claim is that, while you cannot reduce your emissions to zero *simpliciter*, you can reduce your *net* emissions to zero.

How could you reduce your net emissions to zero? If you have positive emissions, then to get to net zero you have to *offset* these emissions, usually by paying others to emit less or to pursue projects that involve fewer emissions than there would have been otherwise. This is very difficult to verify; in particular, it is difficult to determine whether those emissions reductions really change the outcome relative to some baseline (this issue is called *additionality*).

If these challenges can be met and there are reliable offsetting schemes, Broome claims that you can make zero net emissions by using such schemes. Given that greenhouse gases are to a first approximation global, meaning that they quickly mix well throughout the atmosphere, it does not matter whether you personally were the source (in whatever region you are in) as opposed to the offsetting project (in whatever region it is in). Economists might say that emissions appear to be *fungible*, meaning that there is no change associated with replacing one set of emissions with a different set of emissions of the same quantity.

Will this defense work? Can we reduce our emissions to net zero via offsetting, even if we cannot reduce them to zero simpliciter? Dale Jamieson (2014) denies that this line of reasoning succeeds; he thinks that replacing some emissions with others *does* make a difference. If we think again about the formation of the fossil fuels from which those emissions come, we recall that this involves a long slow physical and chemical process. Clearly, the offsetting schemes do not *reverse* that process. One of the original suggestions for offsetting, for instance, was planting trees. First of all, while this would sequester some carbon in the wood, it would take a very long time so there is a temporal mismatch between your emissions today and those sequestered emissions over the coming decades. Second, while it might sequester carbon, it is very different from the carbon that was stored in fossil fuels. One important way that these differ is that the carbon stored in fossil fuels underground is physically stable and unlikely to re-enter the atmosphere. However, carbon stored in trees is only retained there for the life of the tree. So there is a second sense in which there is a temporal mismatch. These objections might not apply to other forms of offsetting, but they demonstrate how difficult it is to truly offset any emissions.

There are good reasons to be suspicious of offsetting. However, there are companies that are trying to provide offsets which attempt to satisfy additionality (e.g. atmosfair, myclimate, and Cool Effect) and their offsets are certified via various standards.

In this section, we considered arguments about whether you should offset and whether your emissions matter. Some authors claim that individual emissions do not matter; some claim that they do. If they do, it could be because you owe it as a duty to reduce your net emissions to zero. But it's not clear whether this is possible. If you try to do so, make sure to read widely and choose carefully. But everyone can be mindful of their emissions and try to reduce their personal unnecessary uses of fossil fuels. For some, that might be enough. For others, more significant actions are called for; we turn to these in the next section.

On Our National Relationship to Fossil Fuels

While the harms of burning fossil fuels are, to a first approximation, global in that they have far more global effects than domestic, there are several important ways that we interact with the fossil fuels that are in our own countries or local regions. In this section, I will discuss two. First, we have the ability to politically affect fossil fuels by acting locally. Here, we can consider, for instance, protest and divestment movements that have arisen in North America. These can complement more familiar (but not unimportant) voting and letter-writing contributions to political discourse. Second, burning fossil fuels is not only relevant in terms of climate change; over closer spatial scales, we also face co-harms involving local pollutants and their effects on, inter alia, human health. This is important for many reasons, but one is political: governments can claim credit for improving local health much more easily than for preventing harms either internationally or in the future. Let us consider these, in turn.

Popularizing work done by the group Carbon Tracker (Gunningham 2017), the journalist and environmentalist Bill McKibben (2012) drew a large amount of attention to the impact of proven fossil fuel reserves, or reserves which were expected to be burned and catalogued as assets for the companies that control them. In short, the carbon contained in fossil fuels that firms intended to extract and burn was far more than was consistent with keeping climate change to two degrees Celsius, an internationally recognized limitation. That means that our best science tells us we cannot extract and burn the fossil fuels that companies and countries expect to.

In 2012, McKibben reported that the carbon contained within proven fossil fuel reserves, 2,795GtC (gigatons of carbon), was more than five times the amount estimated in the carbon budget. The carbon budget to maintain two degrees Celsius was estimated at 565GtC. This implies that, if we do manage to keep to this budget, we need to keep the vast majority of already proven fossil fuels in the ground. (A more recent scientific study found more tolerance than McKibben estimated; consistent with the two degree warming guideline, the study found that a third of oil reserves could be burned (McGlade and Ekins 2015).)

This point helped galvanize significant political opposition, especially in North America. Using slogans like "Keep it in the ground", many activists became concerned about the potential for all of this stored carbon to blow past the two degree target. Although other chapters (64 and 65) address lawbreaking and civil disobedience (also cf. Garcia-Gibson 2022), two particular forms of opposition are worth noting because they have to do with local control (Kyllönen 2014). While these can complement more familiar ways of registered political positions, such as voting, they deserve extra attention here precisely because they are less familiar and have been influential with respect to fossil fuels.

One innovation of the group that McKibben founded, 350.org, was in specifying the importance of fighting fossil fuel companies—and drawing attention to them via the tactic of *divestment* (Gunningham 2017). The idea here is that individuals can challenge fossil fuel companies by getting large local institutions to divest (un-invest) from these energy companies, regardless of where those companies or fossil fuels are physically located. In this way, individuals can send a signal about their engagement with fossil fuels from within their local organizations, such as universities and religious institutions. While divestment has limited direct financial impact on fossil fuel companies, it can be valuable in signaling moral disapproval and drawing attention to the potential for climate damages (Ansar et al. 2013). This tactic should be contrasted with activist shareholding, where one retains investment in companies and generates coalitions to shift firm behavior. Using terms popularized by Hirschman, the former is a form of *exit*; the latter is a form of *voice*.

The other type of opposition that has received attention in North America is protesting pipelines. This is a classic case where local action has global effects; the theory of change is that, without ways to get energy products to global markets, the fossil fuels will not be extracted. Many of these protests have been led or supported by Indigenous or First Nations people (for instance, see Whyte 2017; Tallbear 2019). Once again, this is possible only when the extraction or transportation takes place within one's state or country.

Another important local aspect of fossil fuel extraction—an aspect which has not received as much attention as it deserves by the philosophical community—is the role of local co-harms that accompany fossil fuel extraction and emissions. Globally, and over time, the major impact of fossil fuel use is climate change. Stephen Gardiner calls the *dispersion of causes and effects* the phenomenon that the effects of climate change are remote in time and space (Gardiner 2006). This makes climate effects more difficult to address.

However, climate change is not the only effect of extracting and burning fossil fuels. Furthermore, if we can prevent these additional effects, which are more local, this can be politically beneficial as well as important in changing the narratives we might expect about fossil fuels (Bain et al. 2016).

These other effects are called *co-harms* (and *co-benefits*), since they accompany the effects of emissions that are usually considered (i.e. climate change). However, co-harms and co-benefits of fossil fuel reductions are important because they can greatly strengthen the case for limiting fossil fuel use. For instance, when they are incorporated into models measuring the costs of climate change, emissions reductions have net benefits immediately (as opposed to when delayed climate benefits occur) and these immediate benefits can be very significant (Ürge-Vorsatz et al. 2014; Scovronick et al. 2019; Karlsson et al. 2020).

This is a powerful narrative; the health co-harms from fossil fuel use are more temporally and regionally immediate than climate harms. Furthermore, these benefits are politically useful because they mostly occur within the country which is emitting, instead of being felt internationally. Governments that encourage reduction of fossil fuel use can claim immediate health benefits (assuming we count avoided health costs). Since we do not have direct control over the actions of governments in different countries, this is an important connection to fossil fuel use in one's own country. As for our connections to fossil fuels internationally, there are many other policy options, some of which are considered in the next section.

On Our International Relationship to Fossil Fuels

Fossil fuels are extracted and sold across global borders—indeed, oil is perhaps first among equals as the most geopolitically important commodity sold. Blood and treasure are spent on control and access to it. Given this background, how should those of us in the wealthy, industrialized world relate to fossil fuels and the governments and regimes that sell it to us? There is a wide array of policy options in response to this question, but they are anchored on both sides by the question of whether such buying and selling are impermissible or permissible. If it is impermissible, then the question is on what grounds it would be impermissible and what practically could be done to put an end to it. This could involve bans and injunctions. If it is permissible, then the question is under which conditions it would be permissible and how to impose those conditions. This could involve carbon pricing and compensation. Consider these, in turn.

The first answer, as argued by Leif Wenar (2016), is that, when faced with the prospect of buying oil from certain kinds of repressive states, we should simply refrain from buying it. In particular, his claim is that countries that support rule of law should not purchase natural resources like oil from regimes that prevent citizens from discovering, controlling, and authorizing the sales of these resources. This *popular resource sovereignty* holds that the resources belong to the citizens. It requires that there is press freedom and other access to information about resource management, that there is the ability for citizens to intervene and object to various uses of resources, and a situation where the government does not manipulate public opinion by, for instance, making objecting dangerous or unduly costly. Given that many resources are extracted and sold in ways that are inconsistent with the citizens' endorsement or even knowledge, Wenar concludes that these resources are, morally speaking, stolen. He allows that they may not be *legally* stolen, since the legal regime that governs natural resources is governed by "effectiveness" or the idea that whoever controls the resources owns them. Wenar summarizes this doctrine as "might makes right". Just as we would not traffic in stolen goods or in any other way legitimize the ownership of thieves, Wenar argues that we should not engage in trade for oil and other natural resources under these conditions.

How could this look in policy terms? Wenar proposes a Clean Trade Act which makes purchase of such resources illegal and prevents regimes that trade in such goods from benefitting from commerce within the borders of the Act's state. He points out that, while a strong response, it need not involve any interference with the regimes' sovereignty and so involves far less intrusion than options such as military action. Which states are excluded from trade could be governed by non-governmental organization indices.

Recall that the question is how to respond to the international usage and sale of fossil fuels. While Wenar thinks that the appropriate response to trade is to stop it, one might think that the problem is not that fossil fuels are being traded *simpliciter*, but that they are being traded *without reflecting the real costs to society*. This *externalization* of the social costs (not being borne by those making the trade) could ground a new approach to climate ethics (Mintz-Woo and Leroux 2021). The issue is that, when a transaction can affect a third party, and those effects are not priced into the transaction, we expect the number of transactions to be non-optimal from the social point of view. In this case, extraction and purchase of fossil fuels involve harms from climate change as well as co-harms from local air pollution. Perhaps the appropriate policy would be to price carbon, with the appropriate price leading to the socially optimal level of use. When things are more expensive, on the margin, people will consume less of them.

Three important things are worth noting up front. First, there are many methods for pricing carbon: the most common are a *carbon tax* that imposes a set cost per unit of emission and a *cap-and-trade system* that divides a target amount of total emissions into *emissions allowances* to be traded among emitters (Mintz-Woo 2022). In principle, with sufficient information, these options are in most cases equivalent even though there may be significant differences with respect to political or social acceptance of the respective methods (Klenert et al. 2018).

Second, almost all fossil fuel policies will engender a price on carbon; the issue is how transparent those prices are to consumers. For instance, a portfolio standard policy which requires a certain percentage of state energy to be generated by renewables will demand some renewables when they are not the least cost energy option. That will lead to a greater cost for energy, some of which will fall on the consumers. The point is not that this is necessarily unjustified; it is that energy policies, even those without obvious price introductions, ultimately make energy more expensive. So the question is not whether to adopt policies that increase the price of carbon-intensive activities or those which do not—it is whether to adopt ones that price carbon explicitly or those that price it implicitly (Mintz-Woo 2022).

Finally, people do respond to price signals with respect to fossil fuels (Mintz-Woo 2021c). For instance, Murray and Rivers (2015) analyzed the impact of a carbon tax that was introduced in British Columbia in 2008. The carbon tax went to $30CAD/tCO$_2$. They found that the tax reduced carbon emissions relative to expectations (by roughly 5–15%), was politically popular, and had not disadvantaged British Columbia economically relative to other provinces in general. More generally, while many people find it counterintuitive, carbon taxes do shift behavior (for a similar case, raising taxes on tobacco robustly decreases the number of cigarettes purchased on the margin) (Klenert et al. 2018).

Having made those initial points, we can turn to the question of how to set the price. Economists have the concept of the *social cost of carbon*, or the cost to society of a marginal increase of a ton of CO$_2$. If the carbon price that policy sets matches the social cost of carbon, then the appropriate amount of fossil fuels will be burned under ideal circumstances. Note that the appropriate amount might not be zero—for instance, the emissions needed to rush a person to hospital might outweigh the harms that the fossil fuel extraction is associated with. Note also that many of our targets are in terms of *net*-zero, not zero anthropocentric emissions simpliciter, since there are very substantial carbon sinks in the system—on the order of a third of anthropocentric emissions (Keenan and Williams 2018). However, estimating the social cost of carbon requires taking a stand on both difficult empirical questions and a number of morally contested questions (Mintz-Woo 2021b). While the empirical questions include how the overall warming is driven by additional emissions and how society would respond to new warmer conditions, the moral questions include how to compare harms over time, what scope of harms is covered, which population ethics assumptions to adopt, and whether to include non-human impacts. One overall conclusion about these considerations is that the standard tools for estimating the social cost of carbon are more flexible than is often assumed (Fleurbaey et al. 2019). This section will focus on these moral considerations.

The consideration that has received the most philosophical attention is how to compare costs and benefits over time, which we call the *discounting problem* (Broome 1994; Davidson 2006; Kelleher 2017a, b; Mintz-Woo 2019; Rendall 2019). Greaves (2017) and Mintz-Woo (2021a) provide introductions and overviews of this problem. The issue is how to compare both pure goods (like utility or welfare) and impure goods (like money and material things) across times. This difference matters because the benefits of selling fossil fuels are mostly immediate, whereas the costs can stretch far into the future. In short, the question is how to compare future people and future things with their current present counterparts.

Another consideration concerns which *present* people we are including. Do we include all the global impacts or just our local or national impacts? There is a case to be made that it is more consistent with other types of policies to consider only national impacts. In response, Mintz-Woo (2018) argues that climate change is special in a couple of ways which justify considering global effects: the climate problem is predominantly global and the mechanisms which govern it are well-understood.

Other considerations have to do with *population ethics*, the moral positions that we make about the evaluative importance of comparing outcomes with different people with different levels of welfare. So, for instance, a standard way of aggregating pure values like utility or welfare is simply to sum them. This type of *total utilitarianism* is morally objectionable to some because it could be that an exceedingly large population with very low average level of welfare could be valuable (if we had a choice between creating a populous world with very low levels of welfare, this would be better than a less populous world with high levels of welfare). These worries have led some to an alternative, the *average utilitarian* view, where the value of a world is given by the average pure value. This abstract population ethics assumption can have major effects on the appropriate carbon price, as significant as which answer one gives to the discounting problem (Scovronick et al. 2017). Another population ethics alternative would be to weight the utility of those with lower utility more heavily, yielding some form of prioritarianism with respect to risk (Adler et al. 2017; Fleurbaey et al. 2019). These abstract moral positions could make significant differences with respect to the ideal carbon price (although Arrhenius, Budolfson, and Spears (2021) argue that there may be more agreement than initially appears).

Finally, with respect to a consideration of especial interest to environmental ethicists, we have the question of what to value: just humans, which is the default in these analyses (*anthropocentrism*); all beings with the capacity to suffer (*sentientism*); all living beings (*biocentrism*); or ecosystems themselves (*ecocentrism*) (Jamieson 2008). As incorporating non-humans into such social costs of carbon is an emerging topic, the space is ripe for possible methods of comparing impacts across species (Fleurbaey et al. 2019).

While we have focused on the moral aspects involved in estimating the level of the carbon price, it is also worth noting that there are distributive justice considerations involved in dealing with the *revenue*. Most forms of carbon pricing involve tax revenue, which can be used for one of several purposes, or even a combination of purposes (Gajevic Sayegh 2019; Singer and Mintz-Woo 2020; Mintz-Woo et al. 2021). For instance, if revenue is distributed as equal per capita lump-sum payments, it will actually be net progressive in most places (meaning that more than the poorest half of society will end up overall paying less taxes and some of the wealthiest will end up overall paying more taxes) (Mintz-Woo 2022). An alternative purpose is to remove burdensome distortionary taxes like labor taxes. Yet, another purpose is to spend revenue on public and green priorities. Economists call the two effects (incentivizing less pollution and reducing harmful taxes) the *double dividend hypothesis*. It shows that there can be multiple advantages to pricing carbon.

Conclusion

While we engage with fossil fuels on a daily basis in myriad ways, recognizing those relationships can reveal the opportunity for more conscious actions. These can involve trying to restrict or offset our own fossil fuel use; engaging in voting, protest, or other political activities to keep fossil fuels in the ground; and considering global policies that would limit our impact on the environment. This is by no means an exhaustive list, but it provides some idea of the discussions that environmental ethicists have been and could be engaging with on the topic of this buried fossil energy.

Related Topics

13. Property
20. Moral Bases of Responses to Climate Change
22. Climate Change Mitigation
23. Climate Justice and Equity
25. Skepticism and Denialism
30. Renewables Energy
32. Energy Poverty
49. Pollution and Polluter Pays
63. Environmental Justice
64. Environmental Civil Disobedience
65. Lawbreaking and Ecoterrorism

References

Adler, M.D., Anthoff, D., Bosetti, V., Garner, G., Keller, K. and Treich, N. (2017) "Priority for the Worse-Off and the Social Cost of Carbon," *Nature Climate Change*, 7(6), pp. 443–449. Available at: https://doi.org/10.1038/nclimate3298.

Ansar, A., Caldecott, B. and Tilbury, J. (2013) "Stranded Assets and the Fossil Fuel Divestment Campaign: What does Divestment Mean for the Valuation of Fossil Fuel Assets?" Technical Report, Oxford: Smith School. Available at: https://www.smithschool.ox.ac.uk/publications/reports/SAP-divestment-report-final.pdf.

Arrhenius, G., Budolfson, M. and Spears, D. (2021) "Does Climate Change Policy Depend Importantly on Population Ethics?" in T. McPherson, M. Budolfson and D. Plunkett (eds.) *Philosophy and Climate Change*, Oxford: Oxford University Press.

Bain, P.G., Milfont, T.L., Kashima, Y., Bilewicz, M., Doron, G., Garðarsdóttir, R.B., et al. (2016) "Co-benefits of Addressing Climate Change Can Motivate Action Around the World," *Nature Climate Change*, 6(2), pp. 154–157. Available at: http://doi.org/10.1038/nclimate2814.

Broome, J. (1994) "Discounting the Future," *Philosophy and Public Affairs*, 23(2), pp. 128–156.

——— (2012) *Climate Matters: Ethics in a Warming World*, New York: W. W. Norton.

——— (2019) "Against Denialism," *The Monist*, 102(1), pp. 110–129. Available at: https://doi.org/10.1093/monist/ony024.

Budolfson, M.B. (2019) "The Inefficacy Objection to Consequentialism and the Problem with the Expected Consequences Response," *Philosophical Studies*, 176(7), pp. 1711–1724. Available at: https://doi.org/10.1007/s11098-018-1087-6

Chakravarty, S., Chikkatur, A., de Coninck, H., Pacala, S., Socolow, R. and Tavoni, M. (2009) "Sharing Global CO_2 Emission Reductions Among One Billion High Emitters," *Proceedings of the National Academy of Sciences of the United States of America*, 106(29), pp. 11884–11888. Available at: https://doi.org/10.1073/pnas.0905232106.

Collins, Stephanie (2020) "Corporations' Duties in a Changing Climate," in J. Moss and L. Umbers (eds.) *Climate Justice and Non-State Actors: Corporations, Regions, Cities, and Individuals*, Oxford: Routledge.

Cripps, E. (2013) *Climate Change and the Individual Agent*, Oxford: Oxford University Press.

Davidson, M.D. (2006) "A Social Discount Rate for Climate Damage to Future Generations Based on Regulatory Law," *Climatic Change*, 76(1), pp. 55–72. Available at: https://doi.org/10.1007/s10584-005-9018-x.

Dukes, J.S. (2003) "Burning Buried Sunshine: Human Consumption of Ancient Solar Energy," *Climatic Change*, 61(1–2), pp. 31–44. Available at: https://doi.org/10.1023/A:1026391317686.

Ekwurzel, B., Boneham, J., Dalton, M.W., Heede, R., Mera, R.J., Allen, M.R. and Frumhoff, P.C. (2017) "The Rise in Global Atmospheric CO_2, Surface Temperature, and sea Level from Emissions Traced to Major Carbon Producers," *Climatic Change*, 144(4), pp. 579–590. Available at: http://doi.org/10.1007/s10584-017-1978-0.Fleurbaey, M., Ferranna, M., Budolfson, M.B., Dennig, F., Mintz-Woo, K., Socolow, R., Spears, D. and Zuber, S. (2019) "The Social Cost of Carbon: Valuing Inequality, Risk, and Population for Climate Policy," *The Monist*, 102(1), pp. 84–109. Available at: https://academic.oup.com/monist/article/102/1/84/5255707.

Frumhoff, P.C., Heede, R. and Oreskes, N. (2015) "The Climate Responsibilities of Industrial Carbon Producers," *Climatic Change*, 132(2), pp. 157–171. Available at: https://link.springer.com/article/10.1007/s10584-015-1472-5.

Gajevic Sayegh, A. (2019) "Pricing Carbon for Climate Justice," *Ethics, Policy & Environment*, 22(2), pp. 109–130. Available at: https://www.tandfonline.com/doi/full/10.1080/21550085.2019.1625532.

Garcia-Gibson, F. (2022) "Undemocratic Climate Protests," *Journal of Applied Philosophy*, 39(1), pp. 162–179. Available at: https://doi.org/10.1111/japp.12548.

Gardiner, S.M. (2006) "A Perfect Moral Storm: Climate Change, Intergenerational Ethics and the Problem of Moral Corruption," *Environmental Values*, 15(3), pp. 397–413. Available at: http://www.jstor.org/stable/30302196.

Grasso, M. and Vladimirova, K. (2020). "A Moral Analysis of Carbon Majors' Role in Climate Change," *Environmental Values*, 29(2), pp. 175–195. Available at: https://doi.org/10.3197/0963271 19X15579936382626.

Greaves, H. (2017) "Discounting for Public Policy: A Survey," *Economics and Philosophy*, 33(3), pp. 391–339. Available at: https://doi.org/10.1017/S0266267117000062.

Gunningham, N. (2017) "Building Norms from the Grassroots Up: Divestment, Expressive Politics, and Climate Change," *Law & Policy*, 39(4), pp. 372–392.

Hale, B. (2011) "Nonrenewable Resources and the Inevitability of Outcomes," *The Monist*, 94(3), pp. 369–390. Available at: https://doi.org/10.5840/monist201194319.

Heede, R. (2014) "Tracing Anthropogenic Carbon Dioxide and Methane Emissions to Fossil Fuel and Cement Producers, 1854–2010," *Climatic Change*, 122(1–2), pp. 229–241. Available at: http://link.springer.com/10.1007/s10584-013-0986-y.

Hiller, A. (2011) "Climate Change and Individual Responsibility," *The Monist*, 94(3), pp. 349–368.

Hormio, S. (2017) "Can Corporations Have (Moral) Responsibility Regarding Climate Change Mitigation?" *Ethics, Policy & Environment*, 20(3), pp. 314–332. Available at: http://doi.org/10.1080/21550085.2017.1374015.

Hourdequin, M. (2010) "Climate, Collective Action and Individual Ethical Obligations," *Environmental Values*, 19(4), pp. 443–464.

Jamieson, D. (2008) *Ethics and the Environment: An Introduction*, Cambridge: Cambridge University Press.

——— (2014) *"Climate Matters: Ethics in a Warming World* Book Review," *Ethics & International Affairs*, 28(2), pp. 263–265. Available at: https://www.ethicsandinternationalaffairs.org/2014/climate-matters-ethics-in-a-warming-world-by-john-broome/.

Kagan, S. (2011) "Do I Make a Difference?" *Philosophy and Public Affairs*, 39(2), pp. 105–141.

Karlsson, M., Alfredsson, E. and Westling, N. (2020) "Climate Policy Co-Benefits: A Review," *Climate Policy*, 20(3), pp. 292–316. Available at: http://doi.org/10.1080/14693062.2020.1724070

Keenan, T.F. and Williams, C.A. (2018) "The Terrestrial Carbon Sink," *Annual Review of Environment and Resources*, 43, pp. 219–243. Available at: http://doi.org/10.1146/annurev-environ-102017-030204.

Kelleher, J.P. (2017a) "Descriptive Versus Prescriptive Discounting in Climate Change Policy Analysis," *Georgetown Journal of Law & Public Policy*, 15, pp. 957–977.

——— (2017b) "Pure Time Preference in Intertemporal Welfare Economics," *Economics and Philosophy*, 33(3), pp. 441–473. Available at: https://doi.org/10.1017/S0266267117000074.

Kingston, E. and Sinnott-Armstrong, W. (2018) "What's Wrong with Joyguzzling?" *Ethical Theory and Moral Practice*, 21(1), pp. 169–186.

Klenert, D., Mattauch, L., Combet, E., Edenhofer, O., Hepburn, C., Rafaty, R. and Stern, N. (2018) "Making Carbon Pricing Work for Citizens," *Nature Climate Change*, 8(8), pp. 669–677. Available at: https://doi.org/10.1038/s41558-018-0201-2.

Kyllönen, S. (2014) "Civil Disobedience, Climate Protests and a Rawlsian Argument for 'Atmospheric' Fairness," *Environmental Values*, 23(5), pp. 593–613. Available at: https://doi.org/10.3197/0963271 14X13947900181671.

Maltais, A. (2013) "Radically Non-Ideal Climate Politics and the Obligation to at Least Vote Green," *Environmental Values*, 22(5), pp. 589–608. Available at: https://doi.org/10.3197/096327113X13745164553798.

McGlade, C. and Ekins, P. (2015) "The Geographical Distribution of Fossil Fuels Unused When Limiting Global Warming to 2°C," *Nature*, 517(7533), pp. 187–190.

McKibben, B. (2012) "The Reckoning," *Rolling Stone*, 1162, pp. 52, 54–58, 60. Available at: https://www.rollingstone.com/politics/politics-news/global-warmings-terrifying-new-math-188550/.

Mintz-Woo, K. (2018) "Two Moral Arguments for a Global Social Cost of Carbon," *Ethics, Policy & Environment*, 21(1), pp. 60–63. Available at: https://doi.org/10.1080/21550085.2018.1448038.

——— (2019) "Principled Utility Discounting Under Risk," *Moral Philosophy and Politics*, 6(1), pp. 89–112. Available at: https://doi.org/10.1515/mopp-2018-0060.

——— (2021a) "A Philosopher's Guide to Discounting," in T. McPherson, M. Budolfson and D. Plunkett (eds.) *Philosophy and Climate Change*, Oxford: Oxford University Press.

——— (2021b) "The Ethics of Measuring Climate Change Impacts," in T. Letcher (ed.) *The Impacts of Climate Change*, Ch. 23, Oxford: Elsevier. Available at: https://doi.org/10.1016/B978-0-12-822373-4.00023-9

——— (2021c) "Will Carbon Taxes Help Address Climate Change?" *Les ateliers de l'éthique/The Ethics Forum*, 16(1), pp. 57–67. Available at: https://doi.org/10.7202/1083645ar.

——— (2022) "Carbon Pricing Ethics," *Philosophy Compass*, 17(1), e12803. Available at: https://doi.org/10.1111/phc3.12803.

Mintz-Woo, K., Dennig, F., Liu, H. and Schinko, T. (2021) "Carbon Pricing and COVID-19," *Climate Policy* 21(10), pp. 1272–1280. Available at: https://doi.org/10.1080/14693062.2020.1831432.

Mintz-Woo, K. and Leroux, J. (2021) "What do Climate Change Winners Owe, and to Whom?" *Economics and Philosophy* 37(3), pp. 462–483. Available at: http://doi.org/10.1017/S0266267120000449.

Moss, J. (2020) "Carbon Majors and Corporate Responsibility for Climate Change," in J. Moss and L. Umbers (eds.) *Climate Justice and Non-State Actors: Corporations, Regions, Cities, and Individuals*, Oxford: Routledge.

Murray, B. and Rivers, N. (2015) "British Columbia's Revenue-Neutral Carbon Tax: A Review of the Latest 'Grand Experiment' in Environmental Policy," *Energy Policy*, 86(C), pp. 674–683.

Nefsky, J. (2011) "Consequentialism and the Problem of Collective Harm: A Reply to Kagan," *Philosophy and Public Affairs*, 39(4), pp. 364–395.

Nolt, J. (2011) "How Harmful Are the Average American's Greenhouse Gas Emissions?" *Ethics, Policy & Environment*, 14(1), pp. 3–10.

Rendall, M. (2019) "Discounting, Climate Change, and the Ecological Fallacy," *Ethics*, 129(3), pp. 441–463.

Scovronick, N., Budolfson, M.B., Dennig, F., Errickson, F., Fleurbaey, M., Peng, W., Socolow, R., Spears, D. and Wagner, F. (2019) "The Impact of Human Health Co-Benefits on Evaluations of Global Climate Policy," *Nature Communications*, 10(1), p. e126. Available at: https://doi.org/10.1038/s41467-019-09499-x.

Scovronick, N., Budolfson, M.B., Dennig, F., Fleurbaey, M., Siebert, A., Socolow, R., Spears, D. and Wagner, F. (2017) "Impact of Population Growth and Population Ethics on Climate Change Mitigation Policy," *Proceedings of the National Academy of Sciences of the United States of America*, 114(46), pp. 12338–12343.

Shue, H. (2017) "Responsible for What? Carbon Producer CO_2 Contributions and the Energy Transition," *Climatic Change*, 144(4), pp. 591–596. Available at: https://link.springer.com/article/10.1007/s10584-017-2042-9.

Singer, P. and Mintz-Woo, K. (2020) "Put a Price on Carbon Now!" *Project Syndicate*. Available at: https://www.project-syndicate.org/commentary/low-oil-prices-ideal-time-for-carbon-tax-by-peter-singer-and-kian-mintz-woo-2020-05 [Accessed May 25, 2020].

Sinnott-Armstrong, W. (2005) "It's Not My Fault: Global Warming and Individual Obligations," in W. Sinnott-Armstrong and R. Howarth (eds.) *Perspectives on Climate Change* (pp. 285–307), Oxford: Elsevier.

Tallbear, K. (2019) "Badass Indigenous Women Caretake Relations," in N. Estes and J. Dhillon (eds.) *Standing with Standing Rock* (pp. 13–18), Minneapolis: University of Minnesota Press.

Ürge-Vorsatz, D., Herrero, S.T., Dubash, N.K. and Lecocq, F. (2014) "Measuring the Co-Benefits of Climate Change Mitigation," *Annual Review of Environment and Resources*, 39(1), pp. 549–582. Available at: http://doi.org/10.1146/annurev-environ-031312-125456.

Wenar, L. (2016) *Blood Oil: Tyrants, Violence, and the Rules that Run the World*, Oxford: Oxford University Press.

Whyte, K.P. (2017) "The Dakota Access Pipeline, Environmental Injustice, and U.S. Colonialism," *Red Ink*, 19(1), pp. 154–169.

27

MINING

Jessica M. Smith

Contemporary large-scale mining is emblematic of the Anthropocene, an epoch distinguished by massive human alterations of the Earth's natural resources and climate. "Mining moves more earth than any other human endeavor" (Kirsch 2014: 3), raising fundamental questions of how the industry's existing and future impacts on ecosystems and human populations can and should be managed. Mining will continue to play a role in even the greenest new economies, as smartphones, laptops, wind turbines, electric cars, and LED lights all require the expanded production of rare earth minerals.

The ethical dilemmas posed by mining are increasingly framed in terms of the contested concept of corporate social responsibility (CSR), in which businesses seek to "do well" by "doing good" (Smith 2021). The extractive industries of mining, oil, and gas are major players in the global CSR movement since—depending on one's position in relation to these industries—the concept provides either seductive language for smoothing over public critique of operations or strategies for reconciling business interests with the well-being of both the environment and people living where industries operate. Assessments of CSR, its ethical frameworks, and its potential to address the challenges of industry practice are highly contested both within the business world and within the academy. This chapter provides an overview of these debates as related to the mining industry by examining the moral dilemmas CSR seeks to solve and the ethical implications of using CSR-based solutions for attempting to ameliorate them.

The chapter primarily draws on social science and humanities research about the mining industry. Though not all mines or mining companies are direct producers of energy, insights about community relations for hard rock mines, for example, shed light on the developments they share in common with their counterparts in coal and uranium. When appropriate, I also selectively engage research on CSR in oil and gas industries, since these companies' CSR discourses, policies, and practices resonate closely with those found in mining, keeping in mind when the distinct material qualities of the resources matter in the specific environmental and social challenges they engender.

Moral Dilemmas of Mining Coal

Mining comes under fire when the communities closest to production bear the burden of social, environmental, and bodily harms for economic development that ends up benefiting national elites, governments, and corporations. Coal and uranium are the two mining sectors directly involved in energy production. Coal's share of the U.S. electricity market has steadily decreased

DOI: 10.4324/9781315768090-33

since 2008, hovering at about 21% in 2021, due to the expansion of increasingly cheap natural gas coupled with public pressure for the electricity sector to reduce carbon emissions. On a worldwide scale, coal production has remained steady, buoyed by increased consumption in India and China.

The continued reliance on coal-fired electricity around the world is contentious, since it contributes to both local and global environmental problems. Coal-fired power plants are one of the largest sources of human-created carbon dioxide emissions, making coal a prime contributor to anthropogenic climate change. Climate change directly impacts the sustainability of ecosystems and the health of humans who depend on them, especially as place-based indigenous populations grapple with the sociocultural and economic disruptions engendered by major environmental shifts (Crate and Nuttall 2009).

The coal industry is also infamous for damaging local environments and inspiring passionate environmental movements. The most recent and high-profile controversies in the United States surround Appalachian mountaintop removal mining, in which coal is accessed by large heavy equipment removing the overburden (a mixture of rocks and dirt covering the coal) comprising the summit of a mountain and then placing it into adjacent valleys. The overburden should be replaced on the ridge, and then compacted and contoured to reflect the approximate original shape of the mountain, but many companies receive waivers from the state that allow the site to stay flat, if a case can be made for economic development . Peer-reviewed research and Environmental Protection Agency studies link the mining process and its attendant burial of headwater streams with a loss of biodiversity and toxification of watersheds (Palmer et al. 2010). Human contact with toxins in streams and the air is correlated with higher rates of chronic heart, lung and kidney disease, hypertension, lung cancer, and birth defects (Palmer et al. 2010). Social scientists demonstrate that these environmental harms simultaneously injure the social fabric of communities (McNeil 2011; Scott 2010).

Coal proponents have tried to rescue the industry by hailing improved technology to reduce greenhouse gas emissions (such as carbon capture and storage), industry best practices to mitigate local pollution, and the economic value of coal production, specifically in providing well-paying blue-collar jobs and relatively "cheap" energy (when not accounting for negative externalities). The actual number of employed miners has plummeted from its historic high of over 700,000 workers in 1923, to under 500,000 in the 1930s and 1940s, to just over 100,000 in the 1990s, to its current level of 50,000 (Jones 2020). Job losses have been particularly severe in Appalachia (McNeil 2011; Scott 2010). In the 1980s, changes in technology and environmental policy shifted the center of production to open pit mines in the American West, where vast amounts of lower sulfur coal in Wyoming's Powder River Basin could be quickly mined by comparatively fewer miners operating giant-heavy equipment. President Donald Trump took advantage of this increasing structural precarity in his election campaign, promising to bring back coal jobs as a strategy to perform his support for blue-collar whites and the middle classes who feared their own economic demise (Smith 2019).

The health conditions of workers in the industry have been critiqued by generations of scholars. The oppressive working and living conditions of miners in 19th century England, where coal powered the industrial revolution, prompted Friedrich Engels to pen the seminal "The Condition of the Working-Class in England in 1884." The iconic image of a headlamp-wearing, pickax toting male miner, whose eyes peer out from behind a face blackened by coal dust, continues to signify the bodily stresses endured by many rank-and-file workers (Duke 2002), though it does not correspond with contemporary labor conditions in the United States, where most miners operate heavy industrial equipment such as haul trucks and shovels. In the United States, Appalachia stands out for the exploitation endured by workers, who formed militant unions to protect their safety and guarantee a living wage. Surface mining in the American West is generally safer and does not expose workers to the all-too-familiar lung diseases that ravage underground miners. Yet, higher than average rates of injury, arthritis, and hearing impairment are found among Western coal miners in comparison with non-miners in those same areas (Madsen et al. 1998). In the Powder River Basin specifically, the mines have achieved industry-leading safety records, though operating heavy equipment engenders chronic bodily harms (Rolston 2013) and the requisite rotating shiftwork stresses family relationships among the rank-and-file (Rolston 2014). Scholars trace how technology and company policies shape workers' well-being, as well as how deindustrialization threatens to fragment individual and community identities (Campbell 1984; Charlesworth 2000).

Moral Dilemmas of Mining Uranium

Nuclear power plants provide around 10% of the world's electricity, including 75% of national electricity in France and about 20% in the United States, the United Kingdom, and Russia. The world's largest uranium mines are located in Canada, Australia, Niger, and Kazakhstan. Mining uranium poses uniquely concerning problems of workers' exposure to radiation and radon gas. Rather than coming through the direct handling of solid concentrates, this exposure is most likely through inhalation or ingestion of contaminated water, air, and dust associated with work areas, ore stockpiles, waste rocks, tailings repositories, and processing plants (Lottermoser 2010: 299). Up until the 1980s, however, dosimeter film badges and pens worn by workers to detect external exposure could not detect the radiation emitted by inhaled radon daughters, and the larger ambient instruments positioned throughout the mine could not account for "hot spots" of exposure (Hecht 2009: 900). In the United States, Canada, and Australia, government standards currently regulate employee exposure, and risks are mitigated through the installation of covers over tailings, dust suppression and ventilation systems, dose monitoring, and protective personal equipment and clothing (Lottermoser 2010: 300).

The history of uranium mining reveals that even minimal safety precautions were not integrated into the mining process, especially when the workforce was drawn from marginalized populations. For example, the French leadership in nuclear energy production was built upon uranium mining in Madagascar. During the colonial period, the application of the French nuclear authority's prescriptions for monitoring radiation in Madagascar was "uneven at best," and personal dosimeters were not used in mines operated by private contractors (Hecht 2009: 905). The Malagasy miners who did wear personal dosimeters were not informed of their significance or the results of their measurements, and therefore experienced personal dosimeters as disciplinary devices rather than as instruments for ensuring their own safety (Hecht 2002). Occupational illness, injury, and fatality rates in the Madagascar mines were not included in the French nuclear authority's reports on exposure monitoring programs (Hecht 2009). Even in the metropole, French uranium miners contested celebratory government reports of zero instances of over-exposure, pointing to faulty ventilation and troubling hot spots as evidence that monitoring practices were unevenly implemented (Hecht 2009: 903).

Occupational safety for uranium miners was also a concern in the United States, though uranium production has all but halted. The vast majority of these mines were located in the American West, where Native American men joined the workforce along with Anglos. Approximately 10,000 Navajo men alone worked in the mines on their land in Arizona, New Mexico, and Utah between the 1950s and 1970s. The miners and their families initially welcomed the jobs, but came to experience severe health problems, including lung cancer, as a result of working in poorly ventilated mines with rudimentary technology. For example, 133 of the 150 Navajo uranium miners who worked at a mine in Shiprock, New Mexico, until 1970 died of lung cancer or other forms of fibrosis by 1980 (Ali 2009). Though the federal government and mining companies were aware of potential health hazards, the Navajo miners were neither informed of the risks nor provided protective equipment (Brugge et al. 2006; Johnston et al. 2007).

Native American communities also disproportionately experience the negative health and environmental effects of nuclear waste disposal. Native American resistance to the now stymied Yucca Mountain waste repository in Nevada shows the tenacious cultural significance of land sacred to Indian tribes but deemed to be a barren wasteland by Anglo government and industry officials, as well as how current concerns about the negative environmental and health impacts of waste storage are embedded within longer histories of those communities suffering from weapons testing fallout, for which they were neither informed nor prepared (Kuletz 1998). The more recent targeting of the Skull Valley Goshutes in Utah to accept high-level nuclear waste from commercial power reactors emerges from the confluence of this history of environmental racism and exercises of Native American sovereignty, creating an uneasy form of "new capitalism" that makes environmental sacrifice zones perversely profitable (Clarke 2010; Davies 2012). When coupled with the literature on the bodily harms experienced by Native American uranium miners (Brugge et al. 2006), these cases illustrate the concept of nuclear colonialism (Kuletz 1998), a type of environmental racism perpetuated by federal energy and nuclear agencies against Native American lands and people. Memories of the 1979 Church Rock tailing dam failure in New Mexico, which resulted in the largest unintentional release of radioactive material in the U.S. history and is almost entirely traceable to government negligence, coupled with the more than 1,000 abandoned and unreclaimed mines that continue to contaminate the landscape, serve as a potent reminder of this colonialism (Ali 2009: 81; Brugge et al. 2006). These tragedies illustrate the systemic harms experienced by indigenous communities as a result of mining activity (Kirsch 2014).

Controversies in Corporate Social Responsibility

There is no single definition of CSR that encompasses all of the policies and programs implemented in its name, inspiring two key scholars in the field to argue that the term has "become so broad as to allow people to interpret and adopt it for many different purposes" (Blowfield and Frynas 2005: 503). The interpretation of CSR varies by country and industry, but is best understood as:

> an umbrella term for a variety of theories and practices all of which recognize the following: (a) that companies have a responsibility for their impact on society and the natural environment, sometimes beyond legal compliance and the liability of individuals; (b) that companies have a responsibility for the behavior of others with whom they do business (e.g. within supply chains); and (c) that business needs to manage its relationship with wider society, whether for reasons of commercial viability or to add value to society.
>
> *(Blowfield and Frynas 2005: 503)*

The diversity of programs captured under this umbrella definition makes it useful to distinguish among forms of CSR. "New CSR" reforms a firm's core business practices by internalizing negative social and environmental externalities, while "old CSR" diverts attention from those concerns by the strategic use of philanthropy (Auld et al. 2008). These approaches rest on different modes of community participation and result in differing degrees of effectiveness for firms in comprehending and addressing the concerns of local communities.

Though "there is no agreement among observers on why the concept of CSR has risen to prominence in recent history" (Blowfield and Frynas 2005: 500), the majority of social scientists agree that the intense public criticism leveled against extractive industries has prompted mining, oil, and gas companies to play a leading role in the development of global CSR discourses, institutions, and practices. CSR rose to dominance in the industry during the 1990s and 2000s as it expanded into territories in the developing world that were home to indigenous and agrarian communities without established extractive economies. Some of these communities aligned themselves with transnational environmental advocacy groups, resulting in increased power to protest against and sometime shut down operations (Ballard and Banks 2003; Kirsch 2014; Szablowski 2007). Yet, it is prudent to avoid romanticizing this resistance as antidevelopment, as some individuals and communities seek to benefit economically from the industry, even as they critique its practices (Gardner et al. 2012; Kirsch 2007; Welker 2014). For example, the activists behind the indigenous social movement that fought against the Ok Tedi Mine, infamously known for dumping more than 2 billion tons of mine waste into the river system rather than building the required tailings dam, were called into question when they sought economic benefits from mining rather than expressing a more simplistic antidevelopment stance toward the industry (Kirsch 2007).

The mining industry and its proponents position CSR as a strategy to reconcile the pursuit of profit with the well-being of employees, communities, and environments. The key concepts associated with CSR are intuitively appealing: sustainability, partnership, empowerment, capacity building, transparency, etc. Yet, CSR itself is fundamentally a "moral mechanism through which [corporate] authority is extended over the social order" (Rajak 2011: 13). Critical appraisals of CSR and its potential to ameliorate the harms associated with mining require analyzing the specific kind of authority it engenders is at the expense of other forms of governance.

Expansion of Neoliberal Governance

One key critique of CSR is that it inappropriately takes over the responsibilities of states to ensure the well-being of citizens. These responsibilities increasingly include building and maintaining infrastructure, providing social services, and setting and enforcing rules regulating industry, especially though not exclusively in the developing world, where most new mining projects are located (Campbell 2012). This reorganization can be traced to two interrelated developments. First, the expansion of mining in the developing world meant that companies began operating in places with political institutions weakened by years of colonial and neocolonial rule, making it necessary for them to "fill the gaps" in state governance to function (Blowfield and Frynas 2005: 508; Cheshire 2010; Szablowski 2007: 59). Second, increasing resistance to mining projects created a need for them to garner a "social license to operate," a term used (often uncritically) by industry to designate social acceptance of mining projects (Cheshire 2010; Owen and Kemp 2013; Smith 2021). Building roads, schools, hospitals, and community centers and implementing literacy and health programs served as a strategy to shore up community support and thereby reduce the risk of conflict-induced delays in start-up and production.

The assumption of state functions by corporations raises a series of ethical dilemmas. Though CSR is "presented as a universal good that can be embraced by different sections of the political spectrum" (Blowfield and Frynas 2005: 505), specific programs are steeped in a universalizing vision of development hinged upon the attempted creation of neoliberal capitalist subjects (Rajak 2011; Welker 2014). In contexts where this economic ethos collides with non-market economies, even well-intentioned CSR programs can transform cultural and social structures that provided economic security, replacing them with heightened economic inequality and community divisions. In Papua New Guinea, for example, the compensation and "community development" programs of mining and oil companies eroded longstanding cooperative and generally egalitarian sociopolitical exchange networks by encouraging entrepreneurial activity and the individual pursuit of wealth within more vulnerable cash economies (Gilbertthorpe and Banks 2012).

The neoliberal vision of development stressing self-reliance and entrepreneurship to "help people help themselves" also forecloses community desires for different kinds of connections with industry. Underlining CSR's definition of sustainable development is a "process of disconnection: the donor withdraws and the good works continue" (Gardner et al. 2012: 173). Though ensuring that the economic and social health of communities continues beyond the life of a mine is a laudable goal, communities may understand the obligations of corporations from a different cultural perspective. The corporate "ethic of detachment" documented in Chevron's Bangladesh operations resonates with similar developments in mining (Welker 2014). Chevron did not fulfill their promises of jobs or development to the local community, which subscribed to a model of assistance based on patronage by landowners and wealthy villagers visiting from abroad. What local people desired from Chevron was not sustainability predicated on distance, but connection: "not just to officials who might act as patrons, but to the long term benefits of global capitalism and the modernity it is supposed to bring" (Gardner et al. 2012: 174). Very little research documents the possibilities for communities to actively shape CSR agendas to match their own development desires.

Responsibility for Regulation

The assumption by corporations of responsibilities previously held by the state has weakened state structures and sovereignty (Campbell 2012: 139), which is particularly concerning for the issue of enforcing mining regulations designed to protect employees, communities, and workers. The CSR premise that corporations voluntarily regulate their own practices and go above and beyond the requirements of law represents raises challenges for corporations, governments, and civil society. At a basic level, while there are multiple voluntary international standards for best business practices, there is little standardization in mining corporations' reporting on CSR metrics in their publications, especially related to outcomes rather than simply expenditures on programs (Frynas 2009; Yakovleva 2005).

Perhaps even more problematically, asking corporations to regulate themselves and contribute to community development represents a change in focus for institutions otherwise dedicated to the pursuit of profit. The industry's harshest critics argue that CSR implies that the "very corporations which, for years, freely polluted and contributed minimally to the communities where they operated, have suddenly morphed into philanthropic organizations" (Hilson 2012: 132). In contrast, proponents of CSR point to the "business case" for responsible practice, arguing that environmental stewardship and positive community relationships actually enhance profitability by minimizing risk (Smith 2021). Proving the business case is difficult, and more research is needed to tease out if firms that invest heavily in CSR are more profitable or if profitable firms are those who are able to invest heavily in CSR. Either way, even the 2002 *Mining, Minerals, and Sustainable Development (MMSD)* report commissioned by the International Council on Mining and Metals warns: "Win-win solutions are not always possible; voluntary approaches alone are insufficient where there is a compelling priority but little or no business case to justify the additional expenditures needed to meet it" (IIED 2002). Better understanding these cases is crucial for wrestling with the opportunities and challenges of CSR, as they pose the greatest challenges to reconciling profitability against social and environmental well-being (Auld et al. 2008; Hilson 2012: 132).

Models of Participation

The voluntary nature of corporate community development, self-reporting, and self-regulation raises fundamental questions about the participatory models underlining CSR and state-based alternatives. CSR programs and publications rarely use the language of rights, obligations, or citizenship (Doane 2005; Gilberthorpe and Banks 2012). The absence of rights-based language and programs potentially circumscribes the ability of communities to shape programs and policies stemming from CSR models, leaving industry the power to define sustainable development and decide how to foster it (Gilberthorpe and Banks 2012; Owen and Kemp 2013; Szablowski 2007:

86). The frequent framing of CSR programs as gifts—and corporations and their personnel as beneficent givers—is one mechanism that entrenches corporate power despite discourses of community empowerment and participation: individuals and communities must appeal and be deferent to corporations rather than utilizing democratic processes and making rights-based claims of citizenship on them (Rajak 2011).

The growing circulation of the term "social license to operate" within industry represents a positive development of companies recognizing the importance of community acceptance, but its intangible and informal nature reveals the limitations of CSR-style governance in comparison to democratic processes. In the term's current use, it is unclear who can grant or revoke a social license—elected officials, local communities, groups within communities, or society at large—or if a license can be said to be in place when those groups have differing opinions of industry. Even industry proponents recognize that social licenses and their absences are notoriously difficult to define and measure (Owen and Kemp 2013). For communities, social licenses are more difficult to enforce and revoke, especially given the vast inequalities in power between major multinationals and the places where they operate in the developing world. For example, BHP, Rio Tinto, and Newmont have higher market values than Papua New Guinea, Madagascar, and Ghana, the countries in which they have major operations, respectively (Hilson 2012: 133). Perhaps most concerning, the social license's grounding in *consulting* communities falls short of the free, prior, and informed *consent* established in international law for indigenous populations (Campbell 2012; Coumans 2011; Kirsch 2014; Slack 2012). Consultation does not guarantee a community's right to say no to a proposed project, only to be informed about its potential impacts.

These limitations lead to the frequent argument that for CSR to be effective, it must be complemented by strong state regulation and democratic processes (Blowfield and Frynas 2005; Frynas 2009; Hilson 2012; Slack 2012). The primary challenge is that those regulations and democratic processes are not simply under pressure from broad neoliberal political and economic reforms that weaken state power, but are sometimes actively undermined by regulation-weary corporations and industry trade groups themselves through CSR-style discourses and practices.

Impact of CSR on Environmental and Labor Performance

CSR raises the question of what roles states and corporations ought to play in the regulation of extractive activities and the promotion of community development. Studies from mining as well as oil and gas suggest that CSR can result in improved environmental performance, perhaps because it draws on the technical strengths of company personnel (specifically engineers) and because it can enhance profitability (Frynas 2009; Gilberthorpe and Banks 2012, but compare Ottinger 2013). In this arena, CSR has the potential to address the widely recognized shortcomings of follow-up environmental monitoring, which is required by Environmental Impact Assessments but rarely rigorously implemented in practice (Fidler and Hitch 2007; Noble and Birk 2011). To realize its potential, CSR programs need to go beyond the voluntary framing of follow-up monitoring to make reporting more standardized, rigorous, and accountable to government regulations.

Assessing social performance is considerably more difficult, given the lack of established metrics and limitations of quantifying qualitative information such as community acceptance of a project. Though labor features prominently in many mining firms' CSR reports, very little scholarship examines the impact of CSR on labor. The most prominent study, which focuses on the platinum giant Anglo-American, documents the limitations of grounding CSR programs related to employee nutrition, housing, and HIV/AIDS treatment in paternalistic labor regimes of previous eras (Rajak 2011). Other work in Peru demonstrates that companies cordon off the "productive" side of mining operations from the CSR side of the house, with the effect of channeling community demands away from jobs to neoliberal community development (Smith and Helfgott 2009).

More research is needed to determine possible associations between particular types of CSR and improved safety performance, one of the key concerns for labor. A quick comparison of the incidence rate (including fatalities, nonfatal days lost cases, and medical treatment cases) between coal mines in Wyoming's Powder River Basin, recognized for strong CSR programs, and mines east of the Mississippi, long criticized for paternalistic and negligent labor practices, is stark, even when controlling for the smaller number of employees and production hours in Wyoming. The federal Mine Safety and Health Administration tracks incidence rates, or the number of injuries times 200,000 divided by the number of employee-hours worked. Since 1983, the highest incidence rate per mine in Wyoming was 4.4 in 1987, with the lowest being 0.73 in 2008. For all mines east of the Mississippi, in contrast, the highest rate was 11.74 in 1988 with lowest being 4.0 in 2011. From 1983 to 2011, the average rate was 2.14 in Wyoming and 7.9 in the East (Rolston 2014). While CSR has not prevented every fatality or injury, it would be useful to identify correlations between safety performance, specific program features, and regulatory agency involvement.

The lack of attention to labor in relation to CSR is partially due to a more general shift in social science research about mining from labor to new social movements, communities, and NGOs in the 1990s (Rolston 2013). Social scientists critique the "social thinning" of the industry in the developing world, in which increased capitalization and the need for highly educated workers to operate the equipment obviate the need for large masses of local labor (Ferguson 2006). Companies therefore turn to Fly-In, Fly-Out work organizations, more commonly associated with the oil and gas industry, in which expatriate or urban workers are transported to a minesite for an extended work period and then transported home for a period of rest (Cheshire 2010; Smith and Helfgott 2010). These shifts make large numbers of local employment unlikely. In fact, aboriginal employment in Australia's robust mining sector has remained virtually flat since the late 1960s despite mines being located adjacent to those communities (Barker 2006), and in the former incubator of militant indigenous labor movements in the Andes, "the new mining operations being developed since the boom of the 1990s essentially do not require local labor" (Szablowski 2007: 41). While transitions to market economies pose their own challenges and threats, a dearth of local employment is also a lost opportunity for both communities seeking economic development and companies seeking local support. These trends lead one observer to urge mining companies to stop raising community expectations by promising jobs in the period in which they are seeking to establish support for their operations (Hilson 2012: 134).

Conclusion

There is great potential for CSR to partially reconcile some of the environmental impacts of large-scale mining, since responsible environmental management can sometimes benefit both corporations and communities. Efforts to reduce spills, materials use, and greenhouse gas emissions can generate efficiencies and therefore profit for firms, providing clear examples of the business case for CSR. These kinds of initiatives are also welcomed by firms since they build on the technical skill sets of extractive industry employees, especially scientists and engineers.

Yet, it is prudent to remember that ecological and environmental objectives do not always overlap, and that businesses may privilege commercial interests over environmental protection (Frynas 2009: 186; Kirsch 2014). This observation points to a key problem in justifying CSR-style programs in a business case for their ultimate profitability: what happens when the most "responsible" environmental decision for communities is not the most financially lucrative one for industry? It is here where the differential power of corporations and communities, especially indigenous groups in the developing world who are currently bearing the brunt of new mining development, to define environmental responsibility is particularly worrisome. On a local level, mining companies can define environmental risks in ways that allow them to position themselves as capable of avoiding, managing, or mitigating them (Li 2009). On a broader scale, critics argue that the mining industry's use of the term sustainability has shifted its original ecological sense to almost an entirely economic one, with companies promising financial returns in exchange for resource depletion (Kirsch 2010). This redefinition is partially accomplished through firms enrolling industry-friendly environmental advocacy groups to lend legitimacy to their operations, choosing to work with what the business community calls "light greens"—environmentalists who view the market as the solution to environmental problems—rather than "dark greens"—those who view the market as part of the problem and remain critical of corporate claims to responsibility (Kirsch 2010; Rajak 2011; Smith 2021; Welker 2014). By positioning corporations and their allies to both define and solve environmental problems associated with the industry's practice, CSR raises the specter of allowing "business to appropriate the meaning of ethics" (Blowfield and Frynas 2005: 512).

Thus, for CSR to contribute to improved environmental conditions, the people who live close to mines must play an active role in defining what "good" environmental performance looks like and how it is best achieved. This requires companies to listen to their (often internally diverse) concerns and integrate them into technical decision-making on the minesite. For example, the location of a tailings pit can have major implications for an agricultural community's ability to continue providing food for itself, but mining engineers may not take those concerns into account when designing the most economically efficient mine plan since they are viewed as "social" concerns that are the proper domain of community relations rather than engineering teams (Kemp and Owen 2013). Listening to local residents from the very beginning of a project opens up an opportunity to find mutually agreeable and therefore ultimately more sustainable solutions, but it also raises the possibility of them rejecting mining activities outright, such as when the small Peruvian town of Tambogrande held a referendum that halted a controversial proposed open pit copper and gold mine. Yet if industry is committed to "creating shared value" with the communities where they operate, there must be acknowledgment that local values might conflict with the ones promoted by mining.

Related Topics

9. Mountains: Rethinking Thinking Like a Mountain
26. Fossil Fuels
30. Renewable Energy
31. Natural Gas and Fracking
32. Energy Poverty

References

Ali, S. H. (2009). *Mining, the Environment, and the Indigenous Development Conflicts.* Tucson: University of Arizona Press.

Auld, G., Bernstein, S., and Cashore, B. (2008). The New Corporate Social Responsibility. *Annual Review of Environment and Resources, 33,* 413–435.

Ballard, C., and Banks, G. (2003). Resource Wars: The Anthropology of Mining. *Annual Review of Anthropology, 32,* 287–313.

Barker, T. (2006) Employment Outcomes for Aboriginal People: An Exploration of Experiences and Challenges in the Australian Minerals Industry. Centre for Social Responsibility in Mining Research Paper No. 6. Sustainable Minerals Institute, University of Queensland, Australia.

Blowfield, M., and Frynas, J. G. (2005). Setting New Agendas: Critical Perspectives on Corporate Social Responsibility in the Developing World. *International Affairs, 81*(3), 499–513.

Brugge, D., Benally, T., and Yazzie-Lewis, E. (2006). *The Navajo People and Uranium Mining.* Albuquerque: University of New Mexico Press.

Campbell, Bea. (1984). *Wigan Pier Revisited: Poverty and Politics in the Eighties.* London: Verso.

Campbell, Bonnie. (2012). Corporate Social Responsibility and Development in Africa: Redefining the Roles and Responsibilities of Public and Private Actors in the Mining Sector. *Resources Policy, 37*(2), 138–143.

Charlesworth, S. J. (2000). *A Phenomenology of Working Class Experience*. Cambridge: Cambridge University Press.

Cheshire, L. (2010). A Corporate Responsibility? The Constitution of Fly-In, Fly-Out Mining Companies as Governance Partners in Remote, Mine-Affected Localities. *Journal of Rural Studies, 26*, 12–20.

Clarke, T. (2010). Goshute Native American Tribe and Nuclear Waste: Complexities and Contradictions of a Bounded-Constitutive Relationship. *Environmental Communication: A Journal of Nature and Culture, 4*(4), 387–405.

Coumans, C. (2011). Occupying Spaces Created by Conflict: Anthropologists, Development NGOs, Responsible Investment, and Mining: with CA Comment by Stuart Kirsch. *Current Anthropology, 52*(S3), S29–S43.

Crate, S. A., and Nuttall, M. (2009). *Anthropology and Climate Change: From Encounters to Actions*. Walnut Creek, CA: Left Coast Press.

Davies, L. (2012). Skull Valley Crossroads: Reconciling Native Sovereignty and the Federal Trust. *Maryland Law Review, 68*(2), 290.

Doane, D. (2005). Beyond Corporate Social Responsibility: Minnows, Mammoths and Markets. *Futures, 37*(2–3), 215–229.

Duke, D. C. (2002). *Writers and Miners: Activism and Imagery in America*. Lexington: The University Press of Kentucky.

Engels, F. (1845/2009). *The Condition of the Working Class in England*. London: Penguin Books.

Ferguson, J. (2006). *Global Shadows: Africa in the Neoliberal World Order*. Durham: Duke University Press.

Fidler, C. R., and Hitch, M. (2007). Impact and Benefit Agreements: A Contentious Issue for Environmental and Aboriginal Justice. *Environments Journal, 35*(2), 49–69.

Frynas, J. G. (2009). Corporate Social Responsibility in the Oil and Gas Sector. *The Journal of World Energy Law & Business, 2*(3), 178–195.

Gardner, K., Ahmed, Z., Bashir, F., and Rana, M. (2012). Elusive Partnerships: Gas Extraction and CSR in Bangladesh. *Resources Policy, 37*(2), 168–174.

Gilberthorpe, E., and Banks, G. (2012). Development on Whose Terms?: CSR Discourse and Social Realities in Papua New Guinea's Extractive Industries Sector. *Resources Policy, 37*(2), 185–193.

Hecht, G. (2002). Rupture-Talk in the Nuclear Age: Conjugating Colonial Power in Africa. *Social Studies of Science, 32*(5–6), 691–727.

Hecht, G. (2009). Africa and the Nuclear World: Labor, Occupational Health, and the Transnational Production of Uranium. *Comparative Studies in Society and History, 51*(4), 896–926.

Hilson, G. (2012). Corporate Social Responsibility in the Extractive Industries: Experiences from Developing Countries. *Resources Policy, 37*(2), 131–137.

IIED. (2002). *Breaking New Ground: The Report of the Mining, Minerals and Sustainable Development Project* (pp. 16–30). London: International Institute for Environment and Development and World Business Council for Sustainable Development. Retrieved from http://www.iied.org/mmsd/mmsd_pdfs/finalreport_01.pdf

Johnston, B. R. (2007). Half-Lives, Half-Truths, and Other Radioactive Legacies of the Cold War. In B. R. Johnston (Ed.), *Half-Lives, Half-Truths, and Other Radioactive Legacies of the Cold War* (pp. 1–24). Santa Fe, NM: SAR Press.

Jones, C. (2020). The Coal Industry Has Lost Almost One Thousand Jobs Since Trump Became President. *Forbes*. Retrieved July 3, 2020, from https://www.forbes.com/sites/chuckjones/2020/03/07/the-coal-industry-has-lost-almost-one-thousand-jobs-since-trump-became-president/

Kemp, D. and Owen, J. R. (2013). Community Relations and Mining: Core to Business but Not "Core Business". *Resources Policy, 38*(4), 523–531.

Kirsch, S. (2007). Indigenous Movements and the Risks of Counterglobalization: Tracking the Campaign against Papua New Guinea's Ok Tedi Mine. *American Ethnologist, 32*(4), 303–21.

Kirsch, S. (2010). Sustainable Mining. *Dialectical Anthropology, 34*, 87–93.

Kirsch, S. (2014). *Mining Capitalism: Dialectical Relations Between Corporations & Their Critics*. Berkeley: University of California Press.

Kuletz, V. (1998). *The Tainted Desert: Environmental and Social Ruin in the American West*. New York: Routledge.

Li, F. (2009). Documenting Accountability: Environmental Impact Assessment in a Peruvian Mining Project. *PoLAR: Political and Legal Anthropology Review, 32*(2), 218–236.

Lottermoser, B. G. (2010). *Mine Wastes: Characterization, Treatment and Environmental Impacts*. Berlin: Springer.

Madsen, G., James, D., Dawson, S., and Hunt, W. (1998). Injuries, Arthritis, and Hearing Impairment: A Case Study of Chronic Health Problems among Western Coal Miners. *Society and Natural Resources*, *11*(8), 775–694.

McNeil, B. T. (2011). *Combating Mountaintop Removal: New Directions in the Fight against Big Coal*. Urbana: University of Illinois Press.

Noble, B., and Birk, J. (2011). Comfort Monitoring? Environmental Assessment Follow-Up Under Community–Industry Negotiated Environmental Agreements. *Environmental Impact Assessment Review*, *31*(1), 17–24.

Ottinger, G. (2013). *Refining Expertise How Responsible Engineers Subvert Environmental Justice Challenges*. New York: New York University Press.

Owen, J. R., and Kemp, D. (2013). Social Licence and Mining: A Critical Perspective. *Resources Policy*, *38*(1), 29–35.

Palmer, M. A., Bernhardt, E. S., Schlesinger, W. H., Eshleman, K. N., Foufoula-Georgiou, E., Hendryx, M. S., and Wilcock, P. R. (2010). Mountaintop Mining Consequences. *Science*, *327*(5962), 148–149.

Rajak, D. (2011). *In Good Company: An Anatomy of Corporate Social Responsibility*. Palo Alto: Stanford University Press.

Rolston, J. S. (2013). The Politics of Pits and the Materiality of Mine Labor: Making Natural Resources in the American West. *American Anthropologist*, *115*(4), 582–594.

Rolston, J. S. (2014). *Mining Coal and Undermining Gender: Rhythms of Work and Family in the American West*. New Brunswick, NJ: Rutgers University Press.

Scott, R. (2010). *Removing Mountains: Extracting Nature and Identity in the Appalachian Coalfields*. Minneapolis: University of Minnesota Press.

Slack, K. (2012). Mission Impossible?: Adopting a CSR-Based Business Model for Extractive Industries in Developing Countries. *Resources Policy*, *37*(2), 179–184.

Smith, J., and Helfgott, F. (2010). Flexibility or Exploitation? Corporate Social Responsibility and the Perils of Universalization. *Anthropology Today*, *26*(3), 20–23.

Smith, J. M. (2021). *Extracting Accountability: Engineers and Corporate Social Responsibility*. Cambridge, MA: The MIT Press.

Smith, J. M. (2019). Boom to Bust, Ashes to (Coal) Dust: The Contested Ethics of Energy Exchanges in a Declining US Coal Market. *Journal of the Royal Anthropological Institute*, *25*(S1), 91–107. https://doi.org/10.1111/1467-9655.13016

Szablowski, D. (2007). *Transnational Law and Local Struggles: Mining Communities and the World Bank*. Oxford: Hart Publishing.

Welker, M. (2014). *Enacting the Corporation: An American Mining Firm in Post-Authoritarian Indonesia*. Berkeley: University of California Press.

Yakovleva, N. (2005). *Corporate Social Responsibility in the Mining Industries*. Burlington, VT: Ashgate.

28

NUCLEAR POWER

John Nolt

The Debate So Far

Paul B. Thompson, writing in the early 1980s, elegantly summarized the debate over nuclear power: proponents argue that it is both necessary and safe; opponents argue that it is neither, and each group rejects virtually every substantive premise of the other (1984: 57–58). Three decades later, that summary remains largely accurate.

Nuclear power is necessary, say its proponents, because demand for electricity is increasing, especially in developing nations. Ian Hore-Lacy, writing for the nuclear industry, contends that "One-third of the world's population does not have access to electricity supply, and a further third does not enjoy reliable supply. There is a huge need to address these shortcomings and expectations" (2012: 8). Fossil fuels cannot meet this need, proponents say, because their reserves are dwindling, their cost is increasing, and they are the primary sources of climate change. Renewable energy sources, such as wind and sunlight, are available only intermittently—and, say proponents, unlike nuclear power, they cannot be scaled up fast enough to meet the challenge of climate change. Energy efficiency is not a solution, since it decreases energy costs and hence encourages more consumption. It is, moreover, unrealistic to expect consumption to decrease, for that would require

> ...an attitude to energy use and lifestyle which is increasingly conservation-oriented, so that the rate of increase in overall energy consumption remains depressed after the initial easy fixes have been achieved. Despite popular acceptance of environmental ideas, there is little evidence of such an attitude taking precedence over comfort and amenity anywhere in the world.
>
> *(Hore-Lacy 2012: 9)*

There is thus, according to nuclear proponents, no prospect of meeting growing global demand, addressing climate change, and reducing world poverty without nuclear power. This argument is forceful. If foregoing nuclear power would indeed condemn large swaths of humanity to poverty, long-term climate catastrophe, or both, then necessity trumps safety, and proponents prevail on grounds of necessity alone.

Yet, proponents also argue that nuclear power is safe. According to their risk assessments, the probability of a serious accident is extremely low. They contend that nuclear power causes, on average, fewer fatalities annually than coal, natural gas, or hydropower (Hore-Lacy 2012: 93).

DOI: 10.4324/9781315768090-34

They maintain that past accidents, such as those at Three Mile Island, Chernobyl, and Fukushima, have resulted from design flaws and/or procedural inadequacies that can be or have been corrected (Murray 2009: 306–312; Hore-Lacy 2012: 98–102). They argue that nuclear waste can be treated and disposed of safely (Ferguson 2011: ch. 7). They cite studies indicating that nuclear plants are robustly protected from terrorist attacks (Chapin et al. 2002; Hore-Lacy 2012: 102–103). And they argue that, with adequate international safeguards, increased nuclear power production need not foster nuclear weapons proliferation (Ferguson 2011: ch. 4; Hore-Lacy 2012: 104–113).

Opponents of nuclear power dispute nearly all of these safety claims, arguing that they are based on analyses from researchers whose ties to the nuclear industry constitute conflicts of interest. They contend that the actual frequency of accidents belies past risk assessments, and that (primarily because of the long latency period for radiation-induced cancers) fatalities from those accidents are orders of magnitude higher than industry estimates. They reject industry assurances that reactors are adequately protected against attack, that they need not contribute to nuclear weapons proliferation, and that nuclear waste can be processed, transported, and stored safely. And, in support of their own safety concerns, they note that banks are reluctant to loan money for nuclear reactors, that credit rating agencies downgrade utilities that operate them, and that utilities build them only when given generous legal protection from liability for accidents (Shrader-Frechette 2011a: chs. 2–4, 2011b).

Opponents of nuclear power rebut the necessity argument by maintaining that energy needs can be met quickly enough to address climate change through conservation, efficiency, and renewable energy sources alone. They contend, moreover, that nuclear power plants take longer to build and are more expensive than renewable power sources—and, furthermore, that they are much more carbon-intensive over their life cycle (Shrader-Frechette 2011a: ch. 6).

In sum, Thompson's characterization of the debate remains accurate: proponents still argue that nuclear power is both necessary and safe, opponents still argue that it is neither, and each group still rejects the other's premises. Their arguments—complex, technical, and dependent on disputed empirical claims from the start—have not grown less so. To attempt a comprehensive assessment here would therefore be futile.

This much, however, is evident: in the three decades since Thompson wrote, escalating climate change, population growth, and energy demand have strengthened the case for necessity, and events at Chernobyl and Fukushima have weakened the case for safety. Hence, a third, more disquieting position—one that Thompson (1984: 68) also, presciently, considered—has gained credibility: nuclear power is necessary but not safe. This third position, unlike the first two, has no enthusiastic proponents.

Academic environmental ethicists have contributed little to this debate. Few have written authoritatively on nuclear power (Shrader-Frechette is a notable exception), and those who have typically echo the arguments of the opposition. How might we contribute more?

The next three sections attempt to answer that question for environmental ethics generally. Seeking inclusiveness, they reason from three aims that I assume to be shared by virtually all environmental ethicists. The first of these sections, "What Environmental Ethics Could Contribute," articulates criteria for the achievement of these aims, which the second, "Potential Harms of Power Sources," uses to assess nuclear power and its alternatives. The third, "Policy Implications," deduces implications for nuclear energy policy. Finally, "Non-anthropocentric Nuclear Power Ethics" jettisons the inclusiveness and advocates a richer, more partisan non-anthropocentric view.

What Environmental Ethics Could Contribute

What *could* environmental ethics contribute? In a word: perspective. Its practitioners share at least three laudable aims: (1) to protect nature, (2) to protect people from the harms of environmental degradation, and (3) to do both sustainably. When coupled with well-defined criteria for their accomplishment, these aims force us to take a deep temporal perspective and yield substantive implications for energy policy. Such criteria are, of course, always works-in-progress; yet without them, thought and action flounder. This section proposes criteria for accomplishment of each of the three aims.

Criteria for Protecting Nature

The first aim is to protect nature. "Nature"—insofar as that term may still be used in the anthropocene era (Crutzen and Stoermer 2000)—can be understood as lands or waters that have been relatively undamaged by human activities, or where surviving species exist in ecologically suitable habitat. Success in this first aim is therefore indicated in part by the geographic extent and degree of protection of more or less "natural" lands and waters, and by low extinction rates. This is nothing new; such criteria are standard among conservationists.

Criteria for Protecting People

The second aim is to protect people from the harms of environmental degradation. Low fatality, injury, and illness rates—that is, low casualty rates—from environmental disruption are rough but practical indicators of success in this aim. A morally significant consideration for any form of power generation, then, is the number of casualties or fatalities it produces—or that it could produce under various scenarios, weighted, of course, by the likelihood of those scenarios.

The risks of power production must be balanced against the risks of failure to produce adequate power. Draconian energy conservation measures might lower casualties from environmental disruption but raise them overall by increasing poverty. Policymakers must avoid sins of omission, as well as sins of commission.

One way *not* to assess bodily harms to people is to translate them into monetary terms, as economists and risk assessors often do, for this involves a host of distortions. Monetary values reflect willingness to pay and so give greater weight to the preferences of the rich. Moreover, the preferences that influence them may be ill-informed, irrational, manipulated, short-term, and/or poorly reflective of people's fundamental values (Sagoff 1988; O'Neill et al. 2008; Hausman and McPherson 2009; Nolt 2015a, sec. 3.2). For bodily harms, objective measures, such as casualty or fatality rates, provide much greater accuracy and moral relevance (Nolt 2014).

Criteria for Sustainability

Third aim of environmental ethics—sustainability—extends the temporal reach of the other two aims. To sustain a particular value, such as human safety or biodiversity, is to achieve its adequacy now without impairing posterity's ability to do the same later. There is no single ethically defensible time limit on what counts as posterity. If we knowingly do something with a certain probability of killing people or causing extinctions, then regardless of whether those deaths or extinctions occur today or in a decade, a century, or a millennium, the moral transgression is the same (Nolt 2010, 2015a: sec. 4.1).

Hence, any ethical effort toward sustainability must reject another common economic practice: discounting, that is, the devaluation of all future events, both harmful and beneficial, toward insignificance at a standard annual rate, as a function of their temporal distance from us. This practice, which unjustifiably discriminates against future people, has been widely condemned by ethicists (see, for example, Goodin 1980: 429–430; Parfit 1984: Appendix F; Broome 2004: 70–71, 92–94, 230–231; O'Neill et al. 2008: 57–58; Nolt 2015a: sec. 4.1.3).

While moral responsibility is not limited by temporal separation *per se*, it is limited. We are not morally responsible for harms over which we have no predictable influence (Partridge 1981; Nolt 2015a: sec 4.8). But, as the next section explains, our escalating knowledge and technological abilities have extended the predictable consequences of our power generation—and with it our moral responsibility—far into the future.

Potential Harms of Power Sources

The previous section's criteria for achievement of environmental ethical aims differ from the values most frequently invoked in the nuclear power debate primarily in their temporal reach. In the foreseeable future, many of the economic, political, and technological controversies that now occupy center stage in the nuclear power debate will largely be forgotten. People who live during the next few millennia—and who will likely be far more numerous than we are—will not be concerned with what consumption options were available to us, how much we paid per kilowatt hour, which researchers were subject to conflicts of interest, why financial institutions were wary of nuclear power, or even how we dealt with poverty. In the long run, what will matter most about our energy policies is how they affect overall possibilities for life on Earth. Such a long-term, wide-scope perspective is what environmental ethics can bring to the debate.

Over the next few decades, humanity will generate large amounts of power somehow. Much of it will be distributed through electrical grids, and the benefits of its use will be similar regardless of its source. What most significantly distinguishes the various forms of power generation, then, is the harm they may do. This section assesses that harm, using the criteria outlined above.

Potential Harms of Nuclear Power

All nuclear power operations, from mining though power generation to waste disposal, utilize radioactive materials. These materials pose health hazards via emission of ionizing radiation—subatomic particles energetic enough to strip electrons from atoms upon impact. Such particles normally do not travel far from their sources, but dangerous quantities of radioactive materials can be dispersed into the environment from power plants or nuclear waste repositories as a result of accident (as at Chernobyl or Fukushima) or deliberate attack.

Though all life has evolved with and adapted to low-level "background" radiation from cosmic rays and a variety of natural terrestrial sources, very high radiation doses are quickly lethal, and no amount is absolutely safe. Ionizing radiation harms organisms by damaging the machinery of cells, most notably DNA and RNA molecules. Cells can repair some of this damage, but what is not correctly repaired may cause mutations or cancers. The probability of many types of cancer increases with an individual's cumulative lifetime exposure to ionizing radiation (National Research Council 2006: 11–12).

The radioactivity of nuclear materials decreases over time. How quickly it decreases depends on the radioisotopes they contain. The rate of decrease for a given isotope is measured by its half-life—the average time required for half of its atoms to decay into a different isotope (which itself may or may not be radioactive). The most harmful contaminants from nuclear weapons explosions or reactor melt-downs are cesium 137 and strontium 90, each with a half-life of about 30 years, and iodine 131, whose half-life is about eight days. Radiation from iodine 131 decreases quickly; 99% decays away within two months. But to eliminate 99% of the radiation from cesium 137 or strontium 90 takes a couple of centuries. Soils, flora, and fauna in parts of Europe and the former Soviet Union are still dangerously contaminated with cesium 137 from the Chernobyl melt-down in 1986, forcing the inhabitants of these regions to take special precautions with food sources (Yablokov et al. 2009: chs. I and IV). Melt-downs and nuclear explosions also produce longer-lived radioisotopes—including plutonium 239, with a half-life of 24,100 years—though in much smaller quantities (Yablokov et al. 2009: 19).

High-level nuclear waste (e.g. spent fuel rods from nuclear power plants) also contains long-lived radioisotopes—including several isotopes each of uranium and plutonium (Murray 2009: 366–367). Unless the waste is reprocessed, it takes about 10,000 years for its radioactivity to subside to the level of the ore from which the fuel originated (NEA 1989).

High-level waste, most of which is currently stored onsite at power plants, must constantly be protected against such natural dangers as earthquakes, tsunamis, or floods, and against military attack. It cannot be used to make nuclear explosives, although it might be incorporated into "dirty bombs" that use conventional explosives to disperse radioactive contaminants. The likelihood that, over thousands of years, high-level waste somehow will produce casualties or environmental damage is, no doubt, high, but the geographical extent of such damage would probably not be large.

Power plants have in operation belied their proponents' assurances of safety. Chernobyl and Fukushima were the worst accidents, but there have been a number of others (Shrader-Frechette 2011a: 117–122). Reactor melt-downs have caused many deaths—hundreds of thousands, mostly from radiation-induced cancers, according to high-end estimates for Chernobyl (Yablokov et al. 2009: 192–216; Shrader-Frechette 2011a: 64–65)—though industry numbers are much smaller. Nuclear accidents have also done considerable biological damage. In a survey of 521 biological studies of the Chernobyl event, von Wehrden et al. (2012) document widespread deaths of individual organisms and radiation damage to habitat and natural systems. They do not, however, report any species extinctions.

Some elements of nuclear power infrastructure (e.g. uranium enrichment facilities or fast breeder reactors) can be used to manufacture weapons-grade fissionable material. Nuclear power development may therefore spur nuclear weapons proliferation and hence increase the risk of nuclear war. Nuclear conflicts can, of course, vary widely in severity. A thermonuclear exchange among superpowers would kill billions of people and deplete biodiversity worldwide. Atmospheric disturbances, including perhaps nuclear winter, would continue for years, and radioactive contaminants, mostly cesium 137 and strontium 90, would, though decaying, create hazards peppering much of the globe for more than a century. Many lesser forms of nuclear conflict are also possible (Pittock et al. 1986: 14–16 and 208–210; Robock et al. 2007; Toon et al. 2007; Robock 2010; Robock and Toon 2012).

But nuclear weapons can be—and, indeed, originally were—produced and used militarily in the absence of nuclear power. Consequently, what matters for the nuclear *energy* debate is not the probability of nuclear conflict *per se*, but how much that probability is increased by the dissemination of nuclear energy. Clearly, the widespread development of nuclear energy does not make nuclear conflict inevitable, but it does, no doubt, make it more likely.

Potential Harms of Fossil Fuels

Though nuclear power is dangerous, fossil power is more dangerous, and its dangers are more immediate, more prolonged, and more certain. Worldwide, emissions from fossil plants now cause hundreds of thousands of deaths annually. A substantial fraction of the world's outdoor air pollution, which kills well over a million people annually (Silva et al. 2013)—some 500,000 in China alone (DARA 2012: 256–258)—is produced by coal-fired power plants.

The energy supply sector produces about a quarter of the world's greenhouse gas emissions (IPCC 2007: Figure 2.1: 36), chiefly in the form of emissions from fossil-fueled power plants, and the various effects of climate change have been estimated to kill about 400,000 people annually (DARA 2012: 17). Hence as many as 100,000 of these deaths, too, are attributable to fossil plants.

Annual fatalities from nuclear power are not of the same order of magnitude, even assuming the highest of nuclear opponents' death toll estimates for nuclear accidents. Of course, fossil plants, being more numerous, produce about five times as much electricity as nuclear plants do. But, even when that is taken into account, fossil plants still cause many more deaths each year per unit of electricity. Indeed, it has been argued that nuclear power is already saving millions of lives by preventing air pollution that would otherwise occur from the burning of fossil fuels (Kharecha and Hansen 2013).

On the positive side, fossil-fueled climate change increases crop production in some regions, in the short term. This may save lives, but, globally and in the long term, these agricultural gains will be swamped by agricultural losses, so that they cannot be accounted an overall advantage of fossil fuels (Nelson et al. 2009; Liu et al. 2013).

On the positive side, too, are the substantial benefits of the generated electricity, especially for developing nations, but, as was noted above, these are the same regardless of the source, and hence do not differentiate nuclear from fossil power or renewables.

The further into the future we look, the more the picture darkens. Climate change mortality is forecast to reach 700,000 annually by 2030 (DARA 2012: 17). Hence, even ignoring air pollution, the cumulative death toll by the end of the century from fossil plants (which produce about a quarter of the greenhouse gases) is likely to be in the tens of millions. Climate disruption will, however, last far longer than that, even if we stop burning fossil fuels within decades. Assuming the likely climate sensitivity of 3°C for doubling of CO_2, elevated temperatures could persist, according to one recent estimate (Zeebe 2013), for 23,000 to 165,000 years. Burning all fossil fuel reserves could render most of the planet uninhabitable by humans (Hansen et al. 2013).

The total number of casualties resulting over the coming millennia from emissions during the relatively brief fossil fuel era is in any case likely to be astronomical (Nolt 2011 2013a, 2014). Among the dangers of nuclear power, only a large-scale nuclear war could produce comparable numbers. Yet, the expansion of nuclear power generation does not inevitably entail nuclear war, and the harmful physical effects of nuclear war would dissipate within centuries—as opposed to tens of millennia for climate change. (It is noteworthy in this regard that Hiroshima and Nagasaki, the only places on Earth ever to suffer nuclear attack, are thriving cities today.)

Climate change is, moreover, making it increasingly difficult to protect natural areas, at least in their historic form. As the Earth heats up, species ranges are shifting to higher elevations or toward the poles (SCBD 2010: 10). This causes ecosystems to disintegrate as their current inhabitants move out at different rates and new inhabitants move in. Species that do not move or adapt quickly enough will be lost. Extinctions will, predictably, accelerate.

Continued burning of fossil fuels, along with continued habitat destruction, could ultimately precipitate a mass extinction—loss of three-quarters or more of Earth's species. Biodiversity loss from mass extinctions, five of which are known from the fossil record, is extremely long-lasting. Recovery generally takes millions of years (Barnosky et al. 2011: 51; Bellard et al. 2012; Zeebe and Zachos 2013: 13).

A large-scale nuclear war would, no doubt, produce extinctions, but no responsible treatment that I know of suggests the possibility of mass extinction (Harwell and Hutchinson 1985: 252–253; ch. 7; Robock et al. 2007; Robock 2010; Robock and Toon 2012; Westing 2013).

In sum, the harms of fossil power greatly exceed in probability, severity, and spatiotemporal scale than those of nuclear power.

Potential Harms of Renewables

The main alternatives to nuclear and fossil power are solar, wind, hydroelectric, tidal, plant-based, and geothermal energy sources—collectively known as "renewables." None of these are as harmful by any of the environmental criteria outlined above as fossil-fueled energy sources. None poses risks as great as those of nuclear power.

It may seem, then, that we should generate all of our power with renewables. But here the necessity argument intrudes. About 68% of the world's electricity and 81.3% of its energy are now supplied by fossil fuels (IEA 2013: 7, 24). The finitude of fossil reserves and the catastrophic consequences of burning them all make deep reductions in fossil fuel use both urgent and inevitable. Yet, energy demand and use are still escalating, primarily in developing nations, and must continue to increase if we are to eradicate poverty in a global population that is likely to continue growing through the end of this century (Gerland et al. 2104). Nuclear advocates argue that we cannot generate nearly enough electricity with renewables alone, since wind and sunlight vary geographically and produce power only intermittently. Only nuclear reactors, they say, can generate enough reliable base-load to replace fossil fuels (Fox 2014: 73–117).

Opponents of nuclear power question whether nuclear reactors can be built quickly enough and whether, if they could, they would be reliable. They maintain, moreover, that energy demand can be reduced by conservation and efficiency, and that renewables alone can be scaled up quickly enough to address the problem of climate change (Shrader-Frechette 2011: ch. 6).

Once again, the premises of both sides turn on empirical matters that cannot be adjudicated here. Yet, as the next section shows, even without answers to these empirical questions, assessment of the competing energy sources in accord with long-term environmental ethical criteria yields substantive policy implications.

Policy Implications

To summarize: judged by the criteria of preserving habitat, protecting species, and sustaining the possibility of low extinction and casualty rates, fossil fuels are, among the available energy options, likely to produce by far the worst long-term consequences. Hence, fossil fuel use must be drastically reduced and ultimately eliminated well before reserves run out. Since the how hot the Earth gets will be roughly proportional to humanity's cumulative carbon emissions (Stocker 2013), since harm varies continuously with temperature, and since temperature elevation will last for millennia, the more carbon we emit, the more casualties, extinctions, and habitat disruption we will produce (Nolt 2014). Fossil fuels, once extracted, will almost inevitably be burned. We must, therefore, stop using fossil fuels relatively quickly and leave the remainder in the ground (Nolt 2015a, sec. 7.1.1).

These considerations reveal the importance of the kind of long-term perspective that environmental ethics could bring to the nuclear energy debate. Relative to narrower and shorter-term concerns, the needs of today's poor might, for example, seem to justify the construction of new fossil plants in developing regions. But given the moral equivalence of present and future people and the vast and prolonged damages of climate change, it makes little sense to reduce poverty now by measures that will worsen the human condition over millennia.

But if we eliminate fossil fuels without replacing them with other energy sources, then we will not only fail to alleviate global poverty, but radically deepen it. Whether *nuclear* power should be part of the replacement mix depends upon whether we can deploy conservation and efficiency measures and renewables quickly enough to take the place of fossil energy. If we can't, then nuclear power is necessary not only to alleviate poverty, but to protect both nature and humanity, since without it we will almost certainly persist in burning fossil fuels, with catastrophic consequences. Therefore, rather than fighting nuclear power, environmentalists ought to oppose fossil fuels and support conservation, efficiency, and renewables, with the aim of minimizing the need for nuclear power.

These conclusions depend on several assumptions that could turn out to be wrong. Among the most important are:

- Nuclear fission, fossil, and renewable sources are the only large-scale energy options available, at least for the next few decades.
- Despite efforts at carbon capture and storage, most large-scale fossil power systems will continue to produce large carbon emissions.
- The effects of carbon emissions will not be significantly reduced by geo-engineering.
- Human population will not decrease over the remainder of the fossil fuel era.

These assumptions are all probable, but if any of them proves false, then the conclusions of this section would have to be modified accordingly.

Non-anthropocentric Nuclear Power Ethics

Thus far, I have sought inclusivity, starting from aims that I assume nearly all environmental ethicists share. This final section takes a more partisan view. Environmental ethicists are often divided into anthropocentrists and non-anthropocentrists. Non-anthropocentrists hold that:

1   There is objective value (i.e. goodness independent of human valuing) in nature, and
2   We ought sometimes to protect this value even if protecting it does not serve human interests.

Ethical anthropocentrists may or may not accept 1, but they deny 2. Just as an egoist thinks that he ought to act on behalf of others only if doing so is in his interest, so the anthropocentrist holds that humans ought to act on behalf of non-human beings only if doing so serves human interests (Nolt 2013b). And just as the egoist need not waver from his conviction even if he concedes that other people are sources of value independent of him, so the anthropocentrist denies 2 even if she concedes that non-humans are sources of value independent of humans.

My reasoning so far has been consistent with anthropocentrism, but my own ethic is non-anthropocentric. Non-anthropocentrists hold that harms to non-humans may be moral wrongs in addition to, and independently of, their effects on humans. To the extent that human and non-human interests coincide—and to a great extent in such matters as climate change or nuclear war they do—non-anthropocentrism merely reinforces the conclusions reached in the previous section. But if that coincidence were disrupted, non-anthropocentric ethics could have distinctive implications for energy policy.

One thing that might disrupt it is the commercial implementation of power generation by nuclear fusion. Fusion reactors, already under development for decades, will probably not be viable energy sources for decades more. Still, in the long run, they promise substantial advantages. Unlike fission reactors, which generate power by splitting atoms of heavy radioactive elements such as uranium or plutonium, and leaving as a byproduct high-level radioactive waste, fusion reactors combine isotopes of hydrogen into harmless helium. (The reaction does, however, irradiate the reactor itself, causing it to become radioactive.) Fusion reactions are hard to maintain and hence easily stopped, and there are no fuel rods to melt down. For these reasons and others, proponents see fusion as a safe and practically limitless power source.

Suppose they are right. Still, that may be cold comfort to the non-anthropocentrist; for with such power, humans might ultimately transcend ecological limitations and redesign the entire planet at great cost to non-human life. Today's conservationists, of course, would resist any such tendency—many of them on the basis of their own anthropocentric preferences. But even now as cities grow, humanity spends less time outdoors, and familiarity with non-human life fades, such preferences seem to be dwindling. If, eventually, we no longer need much non-human life, conservation may become passé.

An analogy may help to clarify this non-anthropocentric worry. Imagine a technologically supreme ethical egoist who denies the moral importance of other people and who therefore sees

345

them as expendable, so long as he can survive, warm and satisfied, in an automated pleasure machine. Ethical anthropocentrists would dissent, of course, because they regard all people as objective and morally significant sources of value. But future anthropocentrists, fusion-fueled and supreme, might likewise regard most non-human life as expendable—so long as they could survive, warm and happy, on a re-engineered Spaceship Earth. The non-anthropocentrist, however, would still dissent, for she sees anthropocentrism as morally insidious in just the way that egoism is (Nolt 2013b).

So what nuclear energy policy would non-anthropocentrism recommend? Not, I think, wholesale rejection of fusion. If fusion reactors become viable quickly enough, then for the sake of all living things, both human and not, they should be deployed in place of the more dangerous fossil or fission plants wherever better alternatives are unavailable—at least insofar as the power is needed to prevent morally intolerable poverty. Beyond that, non-anthropocentrism would set some limits to energy production. To generate enough power to re-engineer the planet and cleanse it of much of its non-human life would clearly violate those limits, even if that could be done without harm to humanity, and even if it were what future people wanted (Nolt 2015b).

Non-anthropocentrism might thus bring not just perspective but a dose of humility and temperance to the nuclear power debate.

Related Topics

20. Moral Bases of Responses to Climate Change
26. Fossil Fuels
30. Renewable Energy
32. Energy Poverty

References

Barnosky, A. D. et al. (2011) "Has the Earth's Sixth Mass Extinction Already Arrived?" *Nature* 471: 51–57.
Bellard, C., C. Bertelsmeier, P. Leadley, W. Thuiller, and F. Courchamp (2012) "Impacts of Climate Change on the Future of Biodiversity," *Ecology Letters* 15(4): 365–377.
Broome, J. (2004) *Weighing Lives*, Oxford: Oxford University Press.
Chapin, D. M. et al. (2002) "Nuclear Power Plants and Their Fuel as Terrorist Targets," *Science* 297: 1997–1999.
Crutzen, P. J., and E. F. Stoermer (2000) "The 'Anthropocene'," *Global Change Newsletter* 41: 17–18.
Development Assistance Research Associates (DARA) (2012) "Climate Vulnerability Monitor," 2nd ed., http://daraint.org/climate-vulnerability-monitor/climate-vulnerability-monitor-2012/report/, accessed November 1, 2013.
Ferguson, C. D. (2011) *Nuclear Energy: What Everyone Needs to Know*, Oxford: Oxford University Press.
Fox, M. H. (2014) *Why We Need Nuclear Power: The Environmental Case*, Oxford: Oxford University Press.
Gerland, P. et al. (2104) "World Population Stabilization Unlikely this Century," *Science* 10, 346 (6206): 234–237.
Goodin, R. (1980) "No Moral Nukes," *Ethics* 90: 417–449
Hansen, J., M. Sato, G. Russell and P. Kharecha (2013) "Climate Sensitivity, Sea Level and Atmospheric Carbon Dioxide," *Philosophical Transactions of the Royal Society A* 371: 20120294.
Harwell, M. A. and T. C. Hutchinson (1985) *Environmental Consequences of Nuclear War, Volume II: Ecological and Agricultural Effects*, Chichester: John Wiley & Sons.
Hausman, D. S. and M. S. McPherson (2009) "Preference Satisfaction and Welfare Economics," *Economics and Philosophy* 25: 1–25.
Hore-Lacy, I. (2012) *Nuclear Energy in the 21st Century: The World Nuclear University Primer*, 3rd ed., London: World Nuclear University Press.
Intergovernmental Panel on Climate Change (IPCC) (2007) *Climate Change 2007: Synthesis Report*, Cambridge; Cambridge University Press.

International Energy Agency (IEA), *Key World Energy Statistics 2013*, http://www.iea.org/publications/free-publications/publication/name, 31287,en.html, accessed January 19, 2014.

Kharecha, P. A. and J. E. Hansen (2013) "Prevented Mortality and Greenhouse Gas Emissions from Historical and Projected Nuclear Power," *Environmental Science & Technology* 47: 4889–4895; doi 10.1021/es305119.

Liu, J., C. Folberth, H. Yang, J. Röckström, K. Abbaspour, and A. J. B. Zehnder (2013) "A Global and Spatially Explicit Assessment of Climate Change Impacts on Crop Production and Consumptive Water Use," *PLoS ONE* 8(2): e57750. doi:10.1371.

Murray, R. L. (2009) *Nuclear Energy*, 6th ed., Amsterdam: Elsevier.

National Research Council (U.S.) Committee to Assess Health Risks from Exposure to Low Level of Ionizing Radiation (2006) *Health Risks from Exposure to Low Levels of Ionizing Radiation: BEIR VII, Phase 2*, Washington, DC: National Academies Press.

Nelson, G. C. et al. (2009) *Climate Change: Impact on Agriculture and Costs of Adaptation*, Washington, DC: International Food Policy Research Institute.

Nolt, John (2010) "Sustainability and Hope," in *Sustainability Ethics: 5 Questions*, ed. Evan Selinger, Ryan Raffelle and Wade Robison. Copenhagen: Automatic/VIP Press, 143–170.

—— (2011) "How Harmful Are the Average American's Greenhouse Gas Emissions?" *Ethics, Policy and Environment* 14(1): 3–10.

—— (2013a) Replies to Critics of "How Harmful Are the Average American's Greenhouse Gas Emissions?" *Ethics, Policy and Environment* 16(1): 111–119.

—— (2013b) "Anthropocentrism and Egoism," *Environmental Values* 22(4): 441–459.

—— (2014) "Casualties as a Moral Measure of Climate Change," *Climatic Change*, DOI 10.1007/s10584-014-1131–1132, published online April 26, 2014.

—— (2015a) *Environmental Ethics for the Long Term*, London: Routledge.

—— (2015b) "Non-Anthropocentric Nuclear Power Ethics," in *Ethics of Nuclear Power: Risk, Justice and Democracy in the post-Fukushima Era*, ed. Behnam Taebi and Sabine Roeser. Cambridge: Cambridge University Press, 157–175.

Nuclear Energy Agency of the Organisation for Economic Co-operation and Development, (NEA) (1989) 'The Disposal of High-Level Radioactive Waste,' *NEA Issue Brief* 3, http://www.oecd-nea.org/brief/brief-03.html, accessed January 19, 2014.

O'Neill, J., A. Holland and A. Light (2008) *Environmental Values*, London: Routledge.

Parfit, D. (1984) *Reasons and Persons*, Oxford: Clarendon Press.

Partridge, E. (1981) "Why Care about the Future?" in *Responsibilities to Future Generations*, ed. Ernest Partridge. Buffalo: Prometheus Books.

Pittock, A. B., T. P. Ackerman, P. J. Crutzen, M. C. MacCracken, C. S. Shapiro, and R. P. Turco (1986) *Environmental Consequences of Nuclear War, Volume I: Physical and Atmospheric Effects*, Chichester: John Wiley & Sons.

Robock, A. (2010) "Nuclear Winter," *WIREs Climate Change* 1: 418–427.

Robock, A. and O. B. Toon (2102) "Self-Assured Destruction: The Climate Impacts of Nuclear War," *Bulletin of the Atomic Scientists* 68: 66–74.

Robock, A., L. Oman, and G. L. Stenchikov (2007) "Nuclear Winter Revisited with a Modern Climate Model and Current Nuclear Arsenals: Still Catastrophic Consequences," *Journal of Geophysical Research* 112: 1-14.

Sagoff, M. (1988) *The Economy of the Earth: Philosophy, Law and the Environment*, Cambridge: Cambridge University Press.

Secretariat of the Convention on Biological Diversity (SCBD) (2010) *Global Biodiversity Outlook 3*. Montréal, http://www.cbd.int/doc/publications/gbo/gbo3-final-en.pdf, accessed January 19, 2014.

Shrader-Frechette, K. (2011a) *What Will Work: Fighting Climate Change with Renewable Energy, Not Nuclear Power*, Oxford: Oxford University Press.

—— (2011b) "Fukushima, Flawed Epistemology, and Black-Swan Events" *Ethics, Policy and Environment* 14(3): 267–272.

Silva, R. A. et al. (2013) "Global Premature Mortality Due to Anthropogenic Outdoor Air Pollution and the Contribution of Past Climate Change," *Environmental Research Letters* 8(3): 34005.

Stocker, T. F. (2013) "The Closing Door of Climate Targets," *Science* 339: 280–282.

Thompson, P. B. (1984) "Need and Safety: The Nuclear Power Debate," *Environmental Ethics* 6: 57–69.

Toon, O. B., A. Robock, R. P. Turco, C. Bardeen, L. Oman, and G. L. Stenchikov (2007) "Consequences of Regional-Scale Nuclear Conflicts," *Science* 315: 1224–1225.

Von Wehrden, H., J. Fischer, P. Brandt, V. Wagner, K. Kummerer, T. Kuemmerle, A. Nagel, O. Olsson and P. Hostert (2012) "Consequences of Nuclear Accidents for Biodiversity and Ecosystem Services," *Conservation Letters* 5: 81–89.

Westing, A. H. (2013) *Arthur H. Westing: Pioneer on the Environmental Impact of War*, Dordrecht: Springer.

Yablokov, A. V., V. B. Nesterenko and A. V. Nesterenko (2009) *Chernobyl: Consequences of the Catastrophe for People and the Environment*, New York: New York Academy of Sciences.

Zeebe, R. E. (2013) "Time-Dependent Climate Sensitivity and the Legacy of Anthropogenic Greenhouse Gas Emissions," *Proceedings of the National Academy of Sciences* 110(34): 13739–13744.

Zeebe, R. E. and J. C. Zachos (2013) "Long-Term Legacy of Massive Carbon Input to the Earth System: Anthropocene versus Eocene," *Philosophical Transactions of the Royal Society* A 371: 1–17.

29

HYDROPOWER

Jonas Anshelm and Simon Haikola

Sweden is one of the world's major hydropower producers, relatively speaking, with hydro making up around 40% of its electricity mix. However, most forecasts made midway through the previous century would have counted on hydro making up a much higher proportion of the electricity mix today. Thus, the interesting aspect of Swedish hydropower history from a social science point of view is not the existing hydropower infrastructure, but rather that which was never built. Swedish hydropower infrastructure has not expanded to any substantial degree since the middle of the 1960s, and several pieces of legislation have since been put in place to safeguard the remaining rivers from exploitation. These legislative decisions were all taken with parliamentary majorities at times when rational economic calculations pointed to further hydropower expansion being the obvious way forward, and a united front of trade organizations, workers' unions and the politically dominant Social Democratic Party supported and promoted the expansion plans. The crucial question which must be asked about this development is: How could an opposition built largely on an environmental-ethical argumentation outmanoeuvre such a formidable political alliance and reach into parliament and the national legislation?

That is the question that makes the Swedish case worthy of study from an environmental ethics point of view, and the question that will be addressed in this essay. It is not an attempt merely to apply concepts from environmental-ethical theory to a historical narrative, nor to analyse Swedish hydropower history as a case of a universal hydropower logic. Hydropower development frequently gives rise to controversy and resistance that is both socially and environmentally motivated (Goldsmith & Hildyard 1986; Khagram 2000, 2004; Wood 2007; Fink & Cramer 2008; Diduck et al. 2013; Sharma & Awal 2013), and there are indeed certain grievances and dualisms that seem to be common for most hydropower-related conflicts. One prevalent feature is the perception by populations in the periphery of being exploited by the central power (Andersen & Midttun 1985). Such grievances are usually driven and fuelled by mistrust towards the groups that are promoting hydropower development, and in some regions of the world often further strengthened by demands on peripheral populations to resettle. Another common feature is the polarization of the conflict into an issue of 'environmentalism' versus 'developmentalism', in which proponents of hydropower expansion claim to be promoting modernization and industrial progress, whereas opponents raise arguments based on more intangible values such as an unspoiled nature, biodiversity and cultural heritage (Bellette Lee 2013). However, our purpose with this essay is not to analyse such parallels. Nor will we stake a claim to universal conclusions regarding hydropower developments. Instead, our aim is to tell the story of Swedish hydropower resistance as a story about the success of an ethically motivated environmental movement. In two thematically ordered sections,

DOI: 10.4324/9781315768090-35

we explain the historical development of the resistance movement first in terms of progressiveness and conservatism, and then in terms of elite participation and the social constitution of the movement. We then conclude with some final remarks.

From Tragedy to Triumph – Conservative and Progressive Ethics

As Harvey (1996) notes, every interaction between humans and the environment implies the creation of value, in the sense that every human's relation to another thing involves a valuation of that thing. Since values may be incommensurable, all human–environment interaction therefore also entails the possibility of conflict. The history of hydropower development in Sweden, and resistance to it, might well be understood as the conflict between different views of what value and meaning should be attached to the nation's streaming waters. In this process of value creation and conflict, the ideas of progressiveness and modernity have been of special importance.

Up until the Second World War, hydropower was largely an uncontroversial issue in Sweden in regard to environmental protection. The inherent conflict in energy-related political decisions between environmental concerns and industrial growth soon became apparent to everyone involved in the industrialization of Sweden, but champions of environmental values resigned themselves to the dominant view that industrial interests must take precedence. Following Zev Trachtenberg, the situation might be described as *tragic,* in that environmental conservationists were forced to 'sacrifice some values for the sake of one that is held to be preeminent' (Trachtenberg 1997, p.67). The chroniclers of Swedish hydropower history, Evert Vedung and Magnus Brandel, do however find 'a lonely voice calling in the desert' against the 1919 parliamentary decision to allow hydropower development within the bounds of a national park; that voice belonged to Thor Högdahl, the secretary of the Swedish Society for Nature Conservation (SSNC). Högdahl argued explicitly for a policy where hydropower expansion, and the industrial interests behind it, was not automatically valued higher than unspoiled waters (Vedung & Brandel 2001, p.39). His critique is worthy of mention because its singularity is evidence of the futility of taking a conservative stance on the issue of hydropower expansion. The tragedy for the environmental conservationists is only more apparent in this case, where the supremacy of the opposed value – industrial growth – is not recognized. Thus, Högdahl stands as a uniquely early representative of a tragic environmental ethic (see further Jakobsson 1996). However, this tragedy would eventually be turned into triumph, through an important recasting of the critique.

In the 1950s, the foundation was laid by the SSNC and other organizations engaged in nature conservation for what would eventually be a successful resistance to hydropower expansion. The line of reasoning underpinning the expansion plans at the time from the Swedish State Power Board (SSPB) was firmly rooted in a fatalistic view of industrial growth, as an ever-growing industrial sector was taken for granted. The further expansion of hydropower infrastructure was promoted as an absolute necessity by the directors at SSPB, whose projections showed an increase in energy usage by 6–7% during the 1950s and 1960s. This fatalism was not new but a direct continuation of the arguments from the first decades of hydropower development. What was new, however, was that those favouring hydropower expansion deemed it necessary to actually express these arguments in the public debate. This signalled a shift towards a new framing of the issue, whereby the legitimacy of hydropower expansion was no longer uncontroversial.

The critique against further hydropower expansion that was growing in the 1950s was initially more or less a continuation of Högdahl's lonely call for protection of the environment against the rampaging Mammon. Industrial and economic growth was posited as a threat against certain experiences of crucial significance for modern man; such experiences could only be found in unspoiled nature, which must therefore be safeguarded. An important part of the critique also took the form of appeals to the value of local history, tradition and the cultural landscape. Thus, this early form of organized resistance to hydropower had a significant moral, aesthetic and emotional tint, claiming the human right to the protection of local culture and the right to sublime experiences of the beauty of untamed nature. There was a significant shift in the debate on Swedish hydropower, however, as the opposition changed tactics in the 1960s and adopted the rationalism of their opponents, that is, an argumentative logic founded on economic, scientific and technical premises. Three arguments were central to the new line of reasoning: (1) the rivers held unique values as reference material for future research; (2) the rivers were needed as crucial resources for recreation and existential enrichment in a society characterized by rapid technological and industrial change; (3) nuclear power would soon render hydropower superfluous as an energy source. It seems likely that the underlying pathos for the local culture and environment of the 1950s had not disappeared as motivating ideas, but instead, regardless of their continued existence in the ideological world of the hydropower opposition, such arguments were being tactically downplayed.

The tactical shift can be seen in the subtle change in how the rivers were constituted as concrete, geographical places: while in the 1950s, they were construed as *sources* of abstract human experience, the 1960s saw a move into a discourse where they were rather promoted as *resources* to be instrumentally used by modern man for the purposes of recreation (Anshelm 1993).

Nuclear promised a future of electricity abundance, and the scientific and the recreational values of the rivers provided critics with an arsenal of arguments which representatives of the expansion interests could not easily trump. Of crucial importance in the debate throughout the 20th century seems to have been the ability of either side to portray itself as progressive and, conversely, the other side as conservative. Up to the Second World War, there was simply no real debate about the justifications for expanding hydropower infrastructure, since the industrial need for more electricity was taken for granted, and more hydropower seemed the only way forward for a rapidly industrializing and modernizing country. When Högdahl called for an environmental ethic in opposition to the industrial fatalism of the expansion interests, even his colleagues in the SSNC went out of their way to distance themselves from his comments (Vedung & Brandel 2001, p.40). As the opposition took form in the 1950s, the expansion interests still had the prerogative of speaking for modern society, as the moral, aesthetic and emotional arguments of the critics could easily be brushed aside as irrelevant in a discussion about Sweden's future as an advanced industrial nation. However, as the opposition adopted its new argumentative logic in the 1960s, the momentum gradually swung over to the other side. The first real sign of a power shift in the debate came with the so-called Sarek Peace in 1961, which was an agreement between the expansion interests and the opposition under which a couple of streams and rivers were protected from future exploitation and the first ever nature conservation precautions were written into the planning process. While far from an earth-shattering event at the time, it stands out as a watershed moment when the history of Swedish hydropower is read from the vantage point of the present. It was the first time the expansion interests had acknowledged the arguments of the opposition and the legitimacy in the claim that certain streams should be protected against exploitation. Those favouring expansion took it to be an agreement that would guide Swedish energy politics for decades to come, a strategic concession that would calm the high-pitched debate and allow for security in the planning of future hydropower installations. It would soon be clear, however, that the opposition regarded it merely as first blood drawn (Anshelm 1993).

While the Sarek Peace could have been regarded as a minor concession made by the expansion interests at the time, the conclusion to the events that has come to be called 'the battle of the Vindel River' was an undeniable triumph for the hydropower opposition. Initial plans to exploit the river were made by Vattenfall, the state power company, at the beginning of the 1960s, and they were vigorously opposed almost from the start. As the decade and the planning progressed, the future of the Vindel River in northern Sweden grew into an issue of national concern. Crucially, the power constellations in parliament shifted from a majority supporting expansion plans into a majority opposed, which led to the abandonment of the plans by the Social Democratic government, which had consistently promoted an expansion, in 1970. Unprecedentedly, the state power company Vattenfall had to abort a project that was already well under way in its initial phases and had received significant investment.

The defeat for the industry, the government and the trade unions in the battle for the Vindel River must be understood in the context of the changing debate. When the hydropower critics could point to nuclear power as the energy source of the future, which could be implemented without undue destruction of environmental, scientific and recreational values, it was suddenly the expansion interests that seemed conservative, clinging to a form of electricity production that now had an air of the past about it. There is a sense of the expansion interests being hoisted on their own petards, as the industrial fatalism became adopted by their critics and turned against them (Anshelm 1993). As the 1960s turned into the 1970s, the power companies completely abandoned their position as spokespersons of industrialization and modernity. Instead, they tried to turn hydropower expansion into an issue of regional development and job creation. However, the opposition countered their arguments with labour political arguments of their own, claiming that hydropower projects created only very temporary jobs. In 1972, the parliament protected the four remaining untouched main rivers against further exploitation in the national planning of the use of land and water. This was the second major triumph for the hydropower opposition, and it meant that its ethical arguments were written into the national legislation. Thus, in the early 1970s, the voices of dissent had garnered broad support in parliament and environmental ethics had been transformed into regulatory politics. History would reveal it to be a point of no return, as there was no significant hydropower expansion thereafter.

Yet, the latter half of the 1970s brought several factors into the energy political debate that made the position of the hydropower critics precarious. The oil crisis of 1973 and unforeseen cost overruns in nuclear power projects reflected favourably on the comparably consistent and safe hydropower. Energy reports revised their previously very optimistic forecasts about nuclear technology, and there was a growing consensus that hydropower would serve as a complement to nuclear power for the foreseeable future, rather than being replaced by it. However, the image of hydropower as a technology of yesterday, associated with the destruction of unique national values, had become fixed in the national consciousness and in parliament, and further regulation of the main rivers was deemed unacceptable. Instead of a return to hydropower, the critique from the 1970s onwards increasingly focused on visions of social transformation, first in the form of a low-energy society where neither hydropower nor nuclear power would have to increase on a consistent basis, and later in the form of the sustainable society built on renewables. This has led to the paradoxical situation of the present day. Further expansion of hydropower is forbidden by law on the grounds that it is environmentally unsustainable, whereas the already existing hydropower is hailed as a renewable and therefore environmentally sound source of energy.

From the 1980s onwards, the opposition also gained further arguments, centred on biodiversity and endangered species, in an argumentative change that could be described as a shift from a firmly anthropocentric ethic to a biocentric ethic, i.e. from an ethical view of the world in which humans are perceived as the central species and deemed more valuable than all other organisms, to a worldview that entails a more symmetrical valuation of all living beings (see Dobson 1995). Sweden had signed the Bern Convention in 1979 and was accordingly obliged to preserve endangered species that inhabited the unexploited rivers. These species had a right to exist, it was argued, and Sweden would appear hypocritical if it should decide to spoil the last wild rivers of Northern Europe while at the same time calling for the protection of rainforests and their habitats.

Thus, the history of opposition to hydropower in Sweden can be read as a narrative of tragedy turned into triumph. The conservative ethics that were tragic when promoted by Högdahl in the

1910s eventually became triumphant as they could be reformulated as progressive ethics. When the remaining unexploited rivers were written into the new Natural Resources Law in 1987, it was regarded as an act of progressiveness rather than conservatism. However, it should also be said that a foundational theme in the critique of hydropower has remained unchanged throughout. Even while the opposition changed tactics and adopted a more instrumental approach in the 1960s, the underlying narrative of the need to safeguard the invaluable, sublime experiences offered by an unspoiled nature, and of local culture and tradition being threatened by the brutal expansion of industrial modernity, has been a constant. While the adoption of a set of ethics of progress was crucial for the success of the hydropower opposition, it also seems likely that this underlying, largely unspoken and genuinely conservative set of ethics was of significance for the critique gaining a wider audience, and certainly as a driving force for the actors deeply engaged in this lifelong struggle.

Whose Ethics? – The Role of Authoritative Voices

While the change in argumentative tactics by those opposed to further hydropower expansion was crucial for their eventual success, their argumentation would never have reached parliament without the support of authoritative groups in Swedish society. A significant fact when one looks at the history of Swedish hydropower opposition in the postwar era is the prevalence of voices from the upper social echelons – cultural, scientific, political and juridical – arguing for the need to safeguard the unique rivers against undue industrial exploitation. All the while, the critics enjoyed broad support among the general public.

By the 1950s, the fatalistic debate climate of the 1920s had changed, and calls for river protection that had been doomed to the extreme margins 30 years earlier were starting to receive attention as serious claims for action rather than the negligible opinions of reactionaries. The discursive shift can be detected, for instance, in the magazine for employees of Vattenfall, in which hydropower expansion and nature conservation became recurrent themes in the second half of the decade. It became common for managers at Vattenfall and representatives of the industry to go out of their way to argue for the need for further hydropower expansion on the basis of forecasts showing rapidly growing electricity demand. While the rhetoric employed by the expansion interests was familiar, the fact that they deemed it pertinent to justify their point of view was not. Apparently, they had identified their opponent as a force to be reckoned with, which indeed it was, when analysed from a sociological perspective. The groups that became organized as the first coherent opposition to hydropower expansion, and alerted politicians and the general public to the destructive effects of dam building and river diversion, were far from the margins of society. Quite to the contrary, they might be described as belonging to a societal elite. Several of the most prominent lawyers and professors in the country took the lead and the ranks were filled by engineers, scientists, journalists, writers and doctors. The 1950s was largely a period of opinion formation and organizing, but in the 1960s, the civic and professional capital of the opposition members started to become converted into political victories. The struggle was conducted through respected organizations such as the SSNC, the Society for Landscape Conservation, the Swedish Tourism Organization and the Organization for the Preservation of Rivers, and the opposition made good use of journals and public calls for action, as well as opportunities to comment on official political reports. It also employed extensive lobbying to good effect, for instance, by directly courting politicians and asking them to sign political motions. Each group within the organized opposition provided expertise within its respective domain: engineers provided knowledge about the technical aspects of nuclear power and hydropower, economists about profitability, botanists and zoologists about the consequences for the flora and fauna, lawyers about legal permutations and writers about the emotional dimensions.

Then, during the early 1970s, a new oppositional phenomenon arose: the local action group. As the nationally organized opposition focused on the main rivers in northern Sweden, activist protesters in areas with minor streams that were harder to defend from exploitation on the basis of unique preservation value took matters into their own hands. From that point onwards, these local action groups flanked the already established, elite-based opposition, providing a separate source of leverage. In the first instance, they exerted pressure on municipal and regional politicians through protest marches, name lists and civil disobedience, but such pressure would reach into parliament as the issue became a deciding factor in local elections.

Importantly, as well, the detrimental effects on the culture and ways of life of the indigenous Sami population were also incorporated into the hydropower resistance, as they were acknowledged as an important argument against further hydropower expansion for the first time in a governmental report in 1976 (Anshelm 1992). While the Sami struggle against encroachments by the central government had been a constant feature of hydropower expansion throughout the century (Össbo & Lantto 2011), indigenous rights had never been a part of the opinion formed by the SSNC.

A striking aspect of the opposition to hydropower expansion is its consistent, strategic focus on the parliamentary venue. Judging from its actions, it seems that the opposition had identified the enrolment of parliamentary politicians as crucial at an early stage. Without progress in Stockholm, and without conversion into political capital of the capital – civic, professional and cultural – they possessed in abundance, the hydropower critics would have fought for nothing. When the development of the issue in parliament is analysed, it becomes evident how the opposition, after having won a political victory, immediately upped the stakes and strove for further gains. The developments following the Sarek Peace in 1961 are telling, as the hydropower opposition quickly moved against the expansion plans for the Vindel River, which had been explicitly approved as an object of exploitation for power purposes in the Sarek agreement; it was a move that left the expansion interests feeling betrayed. Also, it must be said, once the issue had become a factor in the election tactics of the political parties, it took on a momentum of its own. The first parliamentary breakthrough came through the enrolment of the parliamentary oppositional parties during the battle for the Vindel River, as the Left-Communist Party, the liberals and the conservative parties adopted the environmental ethics of the hydropower opposition. As the debate over the expansion plans raged on, the Social Democratic government also suffered several notable withdrawals from the official party line, weakening its position.

The establishment of its line of argument in parliament proved crucial for the hydropower opposition in the later part of the 1970s, as the intense criticism of nuclear power eroded the very basis of its case. The Social Democrats, allied with the national organization for workers' unions, had never dropped its positive stance towards further expansion, and the 1980 referendum on nuclear phase-out gave fresh impetus to their calls for more hydropower. The three state reports during the 1970s had, however, largely adopted the argumentation of the hydropower opposition, and set a rather modest target for future expansion. As the Social Democratic Party and the other expansion interests worked throughout the 1980s to raise this target in the light of the nuclear power referendum, they consistently failed to breach the resistance from the Left-Communist Party and the conservatives. This resistance, coupled with significant vote losses in municipal elections because of their stance on the hydropower issue, eventually led the Social Democrats to revise their policy. In the latter half of the decade, the Social Democratic government's proposition for a Natural Resources Law contained far-reaching protection for rivers and streams, in line with the demands of the hydropower opposition. The law was instituted in 1987. At the beginning of the 1990s, all parliamentary parties declared that no more rivers would be targeted for exploitation. In 2005, the final nail was hammered into the coffin of the expansion interests, as the remaining, unexploited major rivers were elevated to the status of National Rivers, thereby gaining constitutional protection.

Finally, having stressed the crucial importance of social elites for the political success of the hydropower opposition, it should also be said that the opposition had already had strong public support from the beginning. All across Sweden, there are streams that have either been exploited or the object of expansion plans. Therefore, each locally anchored opposition could be elevated into national awareness, making a large part of the general public responsive to the critique that had been formulated by the organized opposition. Also, the irreversible and visually striking characteristics of hydropower infrastructure facilitated its creation as an issue of national concern, as it easily lent itself to mass media representation, in, for example, the TV-news. Furthermore, as the environmental movement arose during the 1960s, hydropower took on an added meaning as a unifying symbol for a widespread, abstract unease about the damaging effects of industrial modernity on biodiversity and ecosystems. While it is true that the expansion interests were indeed a formidable political power, being made up of an alliance between the dominant political party, the workers' unions and industry, it is also true that the opposition had the backing of expert knowledge, prominent societal elites and a significant part of the general public. Thus, rather than the David-and-Goliath narrative that is suggested by a superficial glance at the history, the hydropower issue would be better described as a struggle in which a well-organized and well-anchored opposition, armed with highly persuasive and ethically committing arguments, managed to win several important political victories by employing a consistent strategy and a pragmatic, tactical approach.

Concluding Discussion

We have attempted to explain how a social-environmental movement managed to successfully bring an argumentation based on environmental ethics into parliament and into the constitutional legislation, against the efforts of a formidable alliance of industry, workers' unions and a series of Social Democratic governments. Our explanations – the shift from an aesthetical to a science-based argumentation, the enrolment of social elites and a broad anchoring among local resistance groups, the strategic focus on parliamentary politicians, the ability to refer to the coming expansion of nuclear power, the strikingly visual character of the environmental changes associated with hydropower expansion – are well aligned to standard sociological theories on social movements. These tend to focus on either 'resource mobilization' (McCarthy & Zald 1977) – that is, the enrolment and involvement of influential people (or money) for the movement's cause; 'political opportunity structures' (Meyer 2004) – which in our case would be the responsiveness of the parliamentary opposition to the hydropower opposition's arguments (for tactical reasons or otherwise), and the promise of nuclear power; or 'framing' (Benford & Snow 2000), which explains the fortunes of a social movement by how successful it is in representing its core issue in a way that has resonance among a wider constituency. Without venturing into a cause-effect explanation, we could conclude that, certainly, 'reframing' in the 1960s was crucial for the Swedish hydropower opposition's success, just as elite participation, the existence of nuclear power as a plausible alternative for large-scale electricity supply and the political opening in parliament were prerequisites for the movement's ethical arguments being written into the national legislation.

What lessons may be drawn from the Swedish case and applied to hydropower development in other national contexts? Superficially at least, there are some clear parallels between the Swedish history of hydropower resistance and international developments. For example, in the USA, just as in Sweden, a widespread critique of hydropower in the latter half of the 20th century led to a casting of hydro as an environmentally unsustainable power source and to a lack of further expansion. In both countries, the primacy of the climate threat on the environmental agenda in the recent two decades has rendered an ambivalent image of hydropower as a power source that is simultaneously sustainable and unsustainable (O'Connor 2013).

This ambivalence has been a recurring theme globally, with sustained resistance in developing nations contributing to the World Bank curtailing funding for hydropower developments, and to the World Commission of Dams' landmark declaration in 2000 that large-scale dams were environmentally and socially unsustainable (World Commission on Dams 2000; also McCully 2001; Dwivedi 2002; Rest 2012). The Commission's report has since been used by local resistance groups around the world as a potent weapon in their struggles against hydropower developments, giving further evidence of the importance for local environmental resistance groups of being able to translate their opposition into scientific and universal arguments (Finley-Brook & Thomas 2011, see also Anshelm & Haikola 2016).

Subsequently, however, hydropower has come to be seen as a sustainable source of power for many developing nations facing the challenges of electrification and industrialization in the context of global warming, thus renewing the impetus for large-scale hydropower developments while triggering further resistance (Baruah 2012; Matthews 2012; Cole et al. 2014; Ahlers et al. 2015; Huber & Joshi 2015).

In many developing countries, hydropower resistance is now inextricably tied to struggles for indigenous and human rights in a way that makes it different from the Swedish case, where Sami rights were largely kept separate from the environmental opinion formed around, and by the SSNC in the 1960s (Simpson 2013; Sikor et al. 2019).

Looking into specific cases, it's possible to find other parallels. To name just a few examples, resistance groups formed in opposition to the Nu River hydropower project in China (Han 2013), as well as those used by local action groups organizing against the Belo Monte and Rio Madeira projects in the Brazilian Amazon (McCormick 2010) made use of enrolment and argumentative tactics similar to those deployed by the Swedish resistance from the 1960s onwards. In Myanmar, as in Sweden, much resistance against hydropower has drawn force from being cast as struggles of the local community against centralized government (Simpson 2013).

Regardless of important similarities – and key differences – however, this is not the place to analyse in detail reasons for why environmental resistance based in ethics may succeed or fail to have impact. Geographical, cultural and politico-institutional specifics for any case of ethics-based environmental resistance will always make comparisons precarious, and superficial similarities can fail to explain why a certain opposition was successful at a specific point in time, while hiding distinctive features.

Thus, being cautious about staking universalizing claims about the effectiveness of environmental resistance, we shall conclude with a lesson about historical change. It is indeed – somewhat ironically, given our previous reservations about universal claims – a highly general lesson, as we believe it to be relevant for any kind of energy-related infrastructure development, in the present as well as in the future. The story about hydropower resistance in Sweden gives evidence that entrenched political and industrial powers might actually be swayed by ethical arguments that conflict with strictly rational, economic considerations. As our historical analysis shows, the development of Swedish energy politics between 1950 and 1990 was never a foregone conclusion. Indeed, the outcome of the struggle would have seemed highly improbable, if not unthinkable, if suggested by anyone before 1972. A possible, deterministic counter argument, that the expansion interests surrendered their plans simply because there were no rivers left worth exploiting or because they found other forms of power production to focus on, is untenable. Throughout the period covered in this essay, there was a manifest will to further expand the hydropower infrastructure in Sweden, and when the national infrastructure plans were laid in the middle of the 1970s, the Department of Industry, Vattenfall and other power companies all agreed that at least 30% of exploitable hydro resources remained.

Regardless of whether the hydropower opposition was 'right' or not, the fact that the course of environmental politics is not predetermined by the established interests of incumbents or economic 'common-sense'-concerns seems heartening in the face of coming transition challenges.

Related Topics

20. Moral Bases of Responses to Climate Change
23. Climate Justice and Equity
30. Renewable Energy
31. Natural Gas and Fracking
32. Energy Poverty

References

Ahlers, R., Budds, J., Joshi, D., Merme, V. & Zwartaveen, M. (2015) 'Framing hydropower as green energy: Assessing drivers, risks and tensions in the Eastern Himalayas', *Earth System Dynamics*, 6, 195–204.

Andersen, S. & Midttun, A. (1985) 'Conflict and local mobilization: The Alta hydropower project', *Acta Sociologica*, 28, 317–335.

Anshelm, J. (1992) *Vattenkraft of naturskydd: En analys av opinionen mot vattenkraftsutbyggnaden i Sverige 1950–1990*. Linköpings universitet, Institutionen för Tema - Teknik och social förändring.

Anshelm, J. (1993, ed) *Modernisering och kulturarv: Essäer och uppsatser*. Stockholm: Bruno Östlings bokförlag, Symposion.

Anshelm, J. & Haikola, S. (2016) 'Power production and environmental opinion: Environmentally motivated resistance to wind power in Sweden', *Renewable and Sustainable Energy Reviews*, 57, 1545–1555.

Baruah, S. (2012) 'Whose river is it anyway? Political economy of hydropower in the Eastern Himalayas', *Economic and Political Weekly*, 47, 41–52.

Bellette Lee, Y. (2013) 'Global capital, national development and transnational environmental activism: Conflict and the Three Gorges Dam', *Journal of Contemporary Asia*, 43, 102–126.

Benford, R. & Snow, D. (2000) 'Framing processes and social movements: An overview and assessment', *Annual Review of Sociology*, 26, 611–639.

Cole, M., Elliott, R. & Strohl, E. (2014) 'Climate change, hydrodependency, and the African dam boom', *World Development*, 60, 84–98.

Diduck, A., Pratap, D., Sinclair, J. & Deane, S. (2013) 'Perceptions of impacts, public participation and learning in the planning, assessment and mitigation of two hydroelectric projects in Uttarakhand, India', *Land Use Policy*, 33, 170–182.

Dobson, A. (1995) *Green political thought*. 2nd edn. London: Routledge.

Dwivedi, R. (2002) 'Displacement, risks and resistance: Local perceptions and actions in the Sardar Sarowar', *Development and Change*, 30, 43–78.

Fink, M. & Cramer, A. (2008) 'Towards implementation of the World Commissions on Dams recommendations', in Scheumann, W., Neubert, S. & Kipping, M. (eds.) *Water politics and development cooperation: Local power plays and global governance*. Berlin: Springer.

Finley-Brook, M. & Thomas, C. (2011) 'Renewable energy and human rights violations: Illustrative cases from indigenous territories in Panama', *Annals of the Association of American Geographers*, 101, 863–872.

Goldsmith, E. & Hildyard, N. (1986) *The social and environmental effects of large dams*. San Francisco, CA: Sierra Club Books.

Han, H. (2013) 'China's policymaking in transition: A hydropower development case', *Journal of Environment and Development*, 22, 313–336.

Harvey, D. (1996) *Justice, nature and the geography of difference*. Oxford: Blackwell.

Huber, A. & Joshi, D. (2015) 'Hydropower, anti-politics, and the opening of new political spaces in the Eastern Himalayas', *World Development*, 76, 13–25.

Jakobsson, E. (1996) *Industrialisering av älvar: Studier kring svensk vattenkraftutbyggnad 1900–1918*. Göteborgs universitet, Historiska Institutionen.

Khagram, S. (2000) 'Towards democratic governance for sustainable development: Transnational civil society organizing around big dams', in Florini, A. (ed.) *The third force: The rise of transnational civil society*. Washington, DC: Carnegie Endowment for International Peace.

Khagram, S. (2004) *Dams and development: Transnational struggles for water and power*. Oxford: University Press.

Matthews, N. (2012) 'Water grabbing in the Mekong basin: An analysis of the winners and loser of Thailand's hydropower development in Lao PDR', *Water Alternatives*, 5, 392–411.

McCarthy, J. & Zald, M. (1977) 'Resource mobilization and social movements: A partial theory', *American Journal of Sociology*, 82, 1212–1241.

McCormick, S. (2010) 'Damming the Amazon: Local movements and transnational struggles over water', *Society and Natural Resources*, 24, 34–48.

McCully, P. (2001) *Silenced rivers: The ecology and politics of large dams*. 2nd edn. London: Zed Books.

Meyer, D. (2004) 'Protest and political opportunities', *Annual Review of Sociology*, 30, 125-145.

O'Connor, G. (2013) 'America's quest for clean energy: Is hydropower relevant?', *The Electricity Journal*, 26, 15–24.

Össbo, Å. & Lantto, P. (2011) 'Colonial tutelage and industrial colonialism: Reindeer husbandry and early 20th-century hydroelectric development in Sweden', *Scandinavian Journal of History*, 36, 324–348.

Rest, M. (2012) 'Generating power: Debates on development around the Nepalese Arun-3 hydropower project', *Contemporary South Asia*, 20, 105–117.

Sharma, R. & Awal, R. (2013) 'Hydropower development in Nepal', *Renewable and Sustainable Energy Reviews*, 21, 684–693.

Sikor, T., Satyal, P., Dhungana, H. & Maskey, G. (2019) 'Brokering justice: Global indigenous rights and struggles over hydropower in Nepal', *Canadian Journal of Development Studies*, 40, 311–329.

Simpson, A. (2013) 'Challenging hydropower development in Myanmar (Burma): Cross-border activism under a regime in transition', *The Pacific Review*, 26, 129–152.

Trachtenberg, Z. (1997) 'The takings clause and the meanings of land', in Light, A. & Smith, J. (eds.) *Space, place, and environmental ethics*. Lanham: Rowman & Littlefield.

Vedung, E. & Brandel, M. (2001) *Vattenkraften, staten och de politiska partierna*. Nora: Nya Doxa.

Wood, J. (2007) *The politics of water resource development in India: The Narmada dams controversy*. New Delhi: Sage.

World Commission on Dams (2000) *Dams and development: A new framework for decision-making*. London: Earthscan.

30

RENEWABLE ENERGY

Anne Schwenkenbecher and Martin Brueckner

How Shifting to Renewable Energies Is Morally Mandatory

Anthropogenic greenhouse gas emissions are the major contributor to climate change, which is generating severe disruptions to human and non-human lives on this planet. Risks to humans include those associated with an increasing number of extreme weather events, heat exposure, water and food security, sea level rise and subsequent disappearing islands and coastal regions, and the spread of infectious diseases; many of these climate-induced effects are already being felt and are likely to worsen over the course of this century even if warming could be kept to 1.5°C (IPCC 2018a).

The largest proportion of our greenhouse gas emissions results from the way we generate and use energy, in particular from the burning of fossil fuels (IPCC 2014). Fossil-fuel-based energy generation and consumption are thus the main drivers of climatic change. In order to mitigate climate change, we must reduce our greenhouse gas emissions substantially. This reduction will only be possible if we manage to profoundly change the way we use and generate energy (IEA 2020).

There are a number of ways to reduce emissions in order to mitigate climate change. According to Mark Diesendorf (2011), greenhouse gas emissions have three drivers:

1   Consumption per person
2   Population
3   Technology choice

Total emissions (C) are a product of the number of people on the planet (P), the energy use per person (E/P), and the emissions related to each unit of energy (C/E):

$$C = P \times E/P \times C/E \text{ (ibid.)}$$

Diesendorf argues that in order to reduce emissions substantially, we must address each factor (2011: 562). While the first factor – population size – is politically highly problematic, many people are already addressing the second and third factors to reduce their individual carbon footprint. Individual emissions matter (Nolt 2011), and each of us can reduce their emissions by using more efficient appliances (C/E), but also by using public transport instead of driving a car, refraining from air travel, or conserving electricity domestically. But while it is necessary that individuals, especially those living in the world's ten richest countries (Oxfam 2015), consume less energy and adopt environmentally friendly behaviors, there is only so much energy each of us can save – the emissions problem cannot be resolved through individual carbon footprint reductions alone

DOI: 10.4324/9781315768090-36

(Schwenkenbecher 2014). What is needed, ultimately, is a structural response (IRENA 2019). In order to reduce emissions on a significant scale, society as a whole must shift to lower-emission energy technologies (Climate Council of Australia 2018; Diesendorf 2010, 2011; IEA 2020).

There are a number of technology options for lowering GHG emissions while still satisfying the global demand for energy services (IPCC 2011). These include:

- Renewable energy (RE) technologies: wind, solar, biomass, tidal and wave, geothermal;
- "Clean" fossil-fuel-based energy technologies which produce few or no greenhouse gas emissions: nuclear, carbon capture and storage (CCS)

However, not all of these options are equally effective in mitigating climate change to the required extent, and some come with serious problems attached. The risks of nuclear energy are well known (Shrader-Frechette 2011), notwithstanding continued political appetite for nuclear energy solutions (Neumann et al. 2020; Pampel 2011). But what about so-called "clean-coal" technologies such as CCS when juxtaposed with renewable forms of energy generation?

Renewables versus "Clean" Fossil-Fuel-based Technologies

Despite its potential for being a viable future technology and various CCS technologies reaching maturity (Bui et al. 2018), the roll-out of CCS globally has been hindered by a slow technological change and a lack of complementary and targeted policy measures (IPCC 2018b; IEA 2020). As continuing global fossil-fuel dependence seems likely over the medium term (Bui 2019), however, CCS is deemed a critical technology for most of the IPCC's mitigation pathways, stressing the need for further development of the technology in the near term (IPCC 2018b; Budinis 2018).

Irrespective, various problems remain with CCS. First, the technology harbors the risk of prolonging our reliance on fossil fuels, locking in the suite of environmental, social, and political problems resulting from the extraction of fossil fuels. In addition, there are risks associated with the required carbon dioxide (CO_2) storage and potential leakage of CO_2 (Leung et al. 2014). Large quantities escaping would severely impact on greenhouse gas concentrations in the atmosphere (diminishing the technology's impact on mitigation), the local environment, and health of local populations. Diesendorf (2003) points to detrimental impacts on living organisms in waterways and ground water, given that CO_2 can dissolve to form a weak acid in water. Similarly, soil microbes can be negatively affected by carbon dioxide release into the ground. This may have impacts on local ecosystems as a whole (2003: 10). Zoback and Gorelick (2012: 10164) have argued that "there is a high probability that earthquakes will be triggered by injection of large volumes of CO_2 into the brittle rocks commonly found in continental interiors", concluding that large-scale CCS is too risky a strategy to be adopted in the reduction of GHG. While some experts judge the changes of leakage to be vanishingly small (Alcade et al. 2018), many admit that there is still a significant amount of uncertainty surrounding the risks associated with CCS projects (Anderson 2017; van der Zwaan & Smekens 2009; Vinca, Emmerling & Tavoni 2018).

Finally, if CCS were employed on a large scale, in contrast to renewable energies (REs), the environmental costs of harvesting coal, including methane release, the loss of carbon sequestration capacity through vegetation removal, the emissions from extraction machinery and transport (see Keith et al. 2012: 23), would, in fact, increase. Capturing carbon dioxide is energy-intensive and consumes roughly 20–25% of the energy a CCS plant produces. Therefore, CCS plants require more coal per energy unit produced than current coal-fired power plants, because some of the produced energy will be used for the capture and storage process. This means that a large-scale employment of CCS is only possible with increased coal production (Biegler 2009: 66f).

Furthermore, unlike wind and solar energy, coal as a resource is finite: "clean-coal" technologies postpone the problem of long-term energy security, instead of solving it. According to the Intergovernmental Panel on Climate Change, successful mitigation is more likely to be achieved if a complete substitution of fossil-fuel-based energy generation takes place: "Individual studies indicate that if RE deployment is limited, mitigation costs increase and low GHG concentration stabilizations may not be achieved" (IPCC 2011: 24). In light of the current carbon lock-in globally "clean coal" technologies are considered vital for effective mitigation (IPCC 2018b). Nonetheless, there are good reasons to think that in order to achieve effective and timely GHG mitigation, while also securing long-term energy security, we must shift to RE technologies, away from fossil fuels (for a third option – geo-engineering – see chapter on *Geoengineering*).

Wider Benefits of Renewable Technologies

Apart from being instrumental in reducing emissions and stabilizing greenhouse gases in the atmosphere, there are other morally compelling reasons for shifting from conventional, fossil-fuel-based technologies to REs. The IPCC (2011: 7) asserts that RE "can provide wider benefits. RE may, if implemented properly, contribute to social and economic development, energy access, a secure energy supply, and reducing negative impacts on the environment and health". For these reasons, RE investments are currently being promoted for the economic recovery from the Covid-19 pandemic as part of a Green New Deal in countries such as Germany and Korea (Jung 2020).

Compelling reasons for endorsing renewables arise from concerns about the negative environmental impacts of conventional energy generation. The environmental impact of energy technologies is usually measured by way of producing the so-called "lifecycle assessments" (LCAs). LCAs attempt to quantify the environmental impact of technologies over an entire lifecycle, from resource extraction, via manufacturing to operation and disposal (for a discussion of LCA, see IPCC 2011: 730). As a general tendency, RE technologies are clearly favored in LCA (IPCC 2011). Even for solar photovoltaic (PV) systems, which traditionally had poorer LCAs (Varun et al. 2009), technological changes in recent years have greatly improved the environmental performance of PV (Muteri et al. 2020).

Environmental impacts have a direct influence on human health. Conventional power plants (coal-, gas-, and oil-fired) emit "thousands of tons of emissions of sulphur dioxide, nitrogen oxides, carbon monoxide, particulate matter, hydrocarbons, mercury, and other pollutants" (Rosenberg 2008: 523) every year, while REs such as wind and solar produce no such emissions when in operation (see also Kaswan 2009: 1146). Moving away from fossil-fuel-based energy technologies will reduce these so-called co-pollutant emissions, which, in turn, will have a positive impact on human health (WHO 2012).

Commonly, the health burdens of conventional energies are very unequally distributed. Given that pollution is always limited to certain areas, the distribution of health impacts often exacerbates existing inequalities. For decades, environmental justice advocates have been raising awareness of the problem of unequal distributions of exposure to environmental hazards within populations. According to Kaswan (2009: 1146), emissions and air pollution "are disproportionately concentrated in disadvantaged areas, since many of the most significant emission sources, like refineries, power plants, transportation corridors, and other industrial land uses, are located in poor and minority neighborhoods". High-income communities or neighborhoods, in contrast, are less affected by air, water, or soil pollution. Technologies that do not cause such pollution, like RE technologies, are inherently fairer. In this sense, a shift from conventional energy generation to RE will be especially beneficial to those who are disproportionately affected by these adverse impacts. Kaswan argues that "[b]y reducing fossil fuel combustion, greening the grid could serve a critical environmental justice function" (ibid.; see also Rabinowitz 2012 and McCauley et al. 2019).

Apart from promoting environmental justice, reducing pollution can entail substantial economic benefits for all. According to Kaswan, "The consequences of these public health threats fall not only on those directly exposed, but on society as a whole through higher medical costs, lost school and work days, and lower productivity" (2009: 1147, see also Biegler 2009; Gielen et al. 2019; Groosman et al. 2011: 600).

RE can – subject to policy settings (Monyei et al. 2019; Samarakoon 2019) – also play a crucial role in the eradication of poverty and for reaching the United Nations' *Sustainable Development Goals* (Swain & Karimu 2020). "Historically, economic development has been strongly correlated with increasing energy use and growth of GHG emissions, and RE can help decouple that correlation" (IPCC 2011: 18, see also Cherni & Hill 2009). The availability of affordable energy is critical to the economic development of a community. According to the IPCC, under favorable conditions, RE will in some locations be cheaper than non-RE technologies, for instance, through avoiding expensive energy imports. This often makes them the only feasible option in remote and poor rural areas (ibid.). This means that "RE can help accelerate access to energy, particularly for the 1.4 billion people without access to electricity and the additional 1.3 billion using traditional biomass" (ibid., see also Florini 2012: 297). RE is seen by many as a key factor in ensuring that economic development is sustainable and equitable (Stephens 2019). The IPCC *Special Report on Renewable Energy* mentions other aspects in which RE can aid development: "RE deployment might reduce vulnerability to supply disruption and market volatility if competition is increased and energy sources are diversified" (IPCC 2011).

In fact, long-term energy security is not only desirable for developing economies, but for all economies. Just how much of an improvement independence from fossil fuels would be becomes clear when considering the economic and political costs of securing continuing supply, for instance, through military presence and political engagement in oil-producing countries (Babson 2019). Furthermore, even large fossil-fuel deposits that will last for several centuries from now will eventually be exhausted or too expensive to extract; fossil-fuel energy-return-on-investment ratios are already beginning to fall (Brockway 2019). In the meantime, those who control the resources can control those who depend on it, economically and politically (Lehmann 2019). REs, such as wind and solar, provide long-term energy security and independence. They are potentially unlimited, available at no (extraction) cost, and harvesting them does not diminish their availability to anyone else, neither present nor future people.

In sum, there are morally compelling reasons for shifting to RE other than mitigation of dangerous global warming (Florini 2012; IPCC 2011; Jamieson 2011; O'Neill 1993; Shue 2005). A broad scale adoption of RE can generate considerable public health benefits and broader environmental benefits, bear the potential for sustainable and just economic development and equitable energy access, and, finally, provide long-term energy security.

The Feasibility of a Zero-Carbon Economy

Even though overwhelming moral and prudential reasons seem to speak in favor of shifting to REs and away from conventional fossil-fuel-based technologies, some may doubt that this shift is politically, economically, or technologically feasible. In the following, we will briefly address each aspect, in turn.

Is the shift *technologically* feasible? There are an increasing number of low-emission and zero-emission technologies available for large-scale deployment, including hydro, solar PV, and concentrated solar thermal (CST), wind, biomass, and geothermal power. Expert opinions on the viability and capacity of the currently available low-emission and zero-emission technologies differ, but these differences concern merely the timeframe within which conventional energies can be completely substituted by low-emission and zero-emission technologies; they do not usually doubt that this substitution is technologically feasible (see Diesendorf 2011, 2019; Lund 2007; Rissman et al. 2020).

RE pioneers such as Denmark, Finland, Iceland, Norway, and Sweden are already covering large portions of their energy needs with renewables and

> [e]ach has a series of longstanding policy goals; each has binding climate targets; each are attempting to become entirely or mostly "fossil fuel free" or "carbon neutral," with Denmark, Sweden, and Norway committed to 100% renewable energy penetration, Finland 80%, and Iceland 50–75%.
>
> *(Sovacool 2017)*

But some experts, in fact, think that even for a country with the energy needs of Australia, a complete shift to RE is technologically feasible within 10 years to 20 years (The Australian Greens 2019; Diesendorf 2019; ZCA 2010; for an overview of zero-carbon blueprints, see Wiseman & Edwards 2012).

Across the world, cities and municipalities have adopted ambitious emission reduction plans: the Copenhagen Climate Plan (2009) aims at carbon neutrality for the city of Copenhagen by 2025. The Australian Capital Territory, home to Australia's capital Canberra, envisages it for 2060 (ACT Government 2012); indeed, various jurisdictions in Australia have made pledges to carbon neutrality (ClimateWorks Australia 2020).

But on a large scale, globally, is such a shift possible in a short time while maintaining current levels of energy consumption? A common argument against a radical shift is that we critically depend on GHG emissions – we cannot currently have decent lives without emitting. Importantly, transition plans to RE will usually see a major role for increasing energy efficiency. However, the German Advisory Council (WBGU) finds that "even in a world characterized by rapidly growing energy consumption, it is possible to transform global energy systems such that they become sustainable" (WBGU 2009: 129). Since WBGU operate with very high growth rates of energy use and economic development, they arrive at a more conservative model which achieves 90% coverage of energy needs by renewables by 2100. Further, as the 2011 IPCC Special Report points out: "The theoretical potential for RE greatly exceeds all the energy that is used by all economies on Earth" (p.165).

What about the *economic* feasibility of a shift to renewables? According to the IPCC, in principle, the employment of REs does not hinder economic growth or limit projected future energy demand:

> The global technical potential of RE sources will also not limit continued market growth. A wide range of estimates are provided in the literature but studies have consistently found that the total global technical potential for RE is substantially higher than both current and projected future global energy demand.
>
> *(IPCC 2011: 165)*

It should be noted, though, that some scholars argue that in order to successfully mitigate climate change and create sustainable future societies, fundamental changes to our economic system are imperative (Diesendorf 2010, 2019).

Another way of thinking about economic feasibility of transitioning to RE is the comparative economic cost of "business as usual". The *Stern Review on the Economics of Climate Change* focuses on the cost of climate change adaptation and mitigation on a global scale and concludes that timely mitigation is less costly than later adaptation:

> The Review estimates that if we don't act, the overall costs and risks will be equivalent to losing at least 5% of global GDP each year, now and forever. If a wider range of risks and impacts is taken into account, the estimates of damage could rise to 20% of GDP or more. In

contrast, the costs of action – reducing greenhouse gas emissions to avoid the worst impacts of climate change – can be limited to around 1% of global GDP each year.

(*Stern 2006*)

These figures, as confirmed by other studies (Garnaut 2008), indicate that, from an economic perspective, substantial climate change mitigation is preferable to business-as-usual.

Finally, let us very briefly look into the question of *political* feasibility. It is obvious that for some countries, the shift is (or has been) possible, with some countries, including the Nordic Five mentioned above, meeting a large proportion of the energy demands with RE. Yet, it may be argued that for most countries and on a global scale, while the shift is not *impossible*, it is politically *unfeasible*. But what does that mean?

According to Holly Lawford-Smith (2012: 14), "[f]easibility is a concept that treads a fine line between possibility, on the one hand, and likelihood, on the other". She suggests a scalar understanding of feasibility: "the probability of the outcome given the best (or best equal) action" (ibid. 13). Outcomes that are possible are made less probable and therewith less feasible through the so-called soft constraints, the most common of which are economic, institutional, and cultural constraints (ibid.). However, it is possible to influence these constraints – for instance, by attempting to change the culture or values of a given society: "In some instances, if we want the reforms badly enough, we will have to be prepared to really manipulate people's incentives in order to secure success" (ibid.).

On this analysis, the shift *is* politically feasible in that it has a positive probability, given the best available action. We can also see that the best available actions are not currently taken by many governments. Why this may be so will be discussed in the following. Second, it also reveals that in order to increase the feasibility of the shift (and its probability), given the best action, one must address those soft constraints: facts about the current political and economic system, facts about the current system of energy governance and ownership, but also facts about people's values regarding the environment and climate change.

But why is so little happening? Why do politicians mostly fail to take the "best" action for steering our societies toward RE? Why do some countries embrace these new technologies more than others? The reasons for the lack of appropriate policy responses in so many countries are diverse and cannot be discussed in any detail here, but we will indicate some possible explanations. Richard Norgaard (2011) argues that part of the problem lies in the economic analysis of the problem of climate change itself and its theoretical foundations. Cost-benefit analysis – a major decision-making tool for public policy – is best used in small-scale contexts, but is not suitable as a decision-making tool when it comes to addressing global problems such as climate change where costs and benefits are dispersed over time and space (p. 191). According to Norgaard, because not all relevant outcomes can be adequately captured, decisions made on the basis of traditional cost-benefit analyses are flawed (ibid.).

Ann Florini considers defects in institutional design and the lack of power one of the greatest problems of energy governance: "almost no country has a coherent and sensible energy policy implemented by a well-designed set of institutions" and existing institutions do not have the "necessary institutional clout" (2012: 299). In addition, there exist strong vested interests in the oil and coal industries. Florini remarks that "Transparency International's 'Bribe Payer's Index' has ranked the oil and gas sector as the fourth worst sector (out of 19) for bribing public officials" (ibid., 298). On the global level, there is currently no comprehensive energy organization and no coherent global energy governance: "the current system of global energy governance is a mess, with many actors, many priorities, little coherence, and limited effectiveness" (ibid.). Furthermore, governments have been passing the buck on climate leadership for some time in order to avoid possible national disadvantages from making unilateral efforts (Schwenkenbecher 2013; Shue 2011).

Why some countries have managed to overcome these problems cannot be discussed here, unfortunately. However, it does not seem far-fetched to assume that at least some of the political determination to execute the shift to renewables resulted from necessity. For instance, Denmark's economy suffered enormously during the 1970s oil crises as it was extremely dependent on oil imports. Together with a strong anti-nuclear movement, this led to a complete shift in domestic energy policy and cleared the way for renewables. Hence, it appears to have been partly a need for greater levels of energy security which prompted the Danes to adopt those changes (Rüdiger 2019). The political feasibility of shifting to renewables depends on countries' geographic and socioeconomic contexts, including the ability of political actors to intervene in the economy in a range of interdependent ways (Jewell & Cherp 2020).

Even though we have merely hinted at some of the obstacles to national and global sustainable energy regimes, it is safe to say that the transition to RE is economically, technologically, and politically feasible (at least in many of the OECD countries) at this point in time. However, in order to significantly increase the likelihood of such a transition, what is needed is a political shift, or rather many political shifts on the national, regional, and local levels. We can and must work toward such a shift in our own societies and globally.

The Ethical Problems of (Some) RE

Yet, even if there is agreement on the necessity and feasibility of a transition to an RE regime, we will still be required to make choices between different possible pathways toward that goal. Each of these choices will involve some undesirable consequences that we will need to balance. Some of these undesirable consequences will result from technology choices and others from policy and economic measures. Some decisions will concern different ways of living: Do we have to give up our current living standard for a sustainable (energy) future? If a trade-off is needed, how does the value of high living standards compare against that of a sustainable energy regime? Fundamentally, changing the way we generate and use energy will entail tough choices, sometimes between equally desirable aims, and sometimes between options that seem impossible to compare.

In the following, we will not conduct a comprehensive analysis of all available RE technologies, but rather provide a number of examples of tough ethical choices entailed by solar, wind, and hydro power. We will also attempt to rebut some of the common misconceptions regarding RE technologies.

Hydropower – Green Electricity in Exchange for Environmental Destruction?

Hydropower is a paradigmatic example of a controversial RE technology (Diesendorf 2011; Jamieson 2011). On the positive side, it generates energy at extremely low cost and offers "significant potential for carbon emissions reductions" (IPCC 2011: 442). If replacing conventional fossil-fuel-based energy technology, hydropower helps secure environmental benefits, because it does not cause air pollution and soil contamination. Furthermore, in contrast to open pit mines, dams often provide recreational benefits for humans. On the negative side, dams severely interfere with landscapes and ecosystems, altering existing waterways and disrupting a river system's ecology with possibly devastating effects on fauna and flora (see Kahn, Freitas & Petrere 2014). Depending on local conditions, hydropower may even potentially result in higher emissions than the fossil-fuel-based power generation it seeks to replace (Giles 2006). The decision to build and use a hydropower plant necessarily entails a decision as to which of these conflicting aspects are more important: significant emission reductions at low cost or (possibly) the preservation of an existing ecosystem; further, there are often significant impacts on local communities and their livelihoods to consider (Moran et al. 2018). Whether or not building any particular hydro power plant is overall the ethically best choice will depend on the specific circumstances. In some cases, the potential to gain access to cheap energy may override environmental concerns and the interests of the local community (where these conflict with the project). It should also be noted that not all hydropower systems are equally problematic from an environmental point of view. For example, run-of-river hydropower systems do not alter a river's flow regime in the way systems with reservoirs do and have therefore more benign ecological impacts (IPCC 2011: 463).

Solar Photovoltaic – A Sham Package?

Solar PV is a versatile RE technology and often used for small-scale, self-supporting electricity generation. It enables households and communities to have greater independence from the electricity grid. While governments all over the world have provided economic incentives for homeowners to install solar PV systems, these measures have been criticized for reinforcing inequality as people in low-income groups are less likely to benefit from them (see Hitzeroth, Jehling & Brueckner 2017; Macintosh & Wilkinson 2011). However, while this may be true for relatively well-off urban and suburban residents, small-scale solar PV can provide enormous benefits to remote communities subject to country-context and policy setting (Baurzhan & Jenkins 2016; Okoye & Oranekwu-Okoye 2018). It can provide those who have not formerly had access to electricity with affordable energy, therewith improving their living standard and benefiting those who are usually disadvantaged in society (IPCC 2011: 66).

One downside of solar PV is that, in contrast to wind and hydro technologies, it has been a comparatively costly way to generate electricity (IPCC 2011: 188, Figure 1.19), though this is changing rapidly. However, the relative cost would vary depending on circumstances: "In some applications, PV systems are already competitive with other local alternatives (e.g. for electricity supply in certain rural areas in developing countries)" (ibid. p. 68). In terms of the environmental record of PV, lifecycle emissions of first-generation PV systems have been considerable and were, on average, greater than those from hydro and wind (Varoun et al. 2009). Improvements in manufacturing processes and PV performance, however, have seen dramatic reductions in PV systems' environmental footprint and overall energy amortization (Muteri et al. 2020).

On a large scale, solar PV has less clear economic and ethical benefits than CST and wind power. It is both more expensive and less effective in terms of climate change mitigation (Desideri & Campana 2014). Depending on the subsidy structure, it may often benefit primarily those who are already well-off. However, it can play a very important role as a small-scale technology in providing remote communities with electricity and resulting benefits, therewith in some cases benefiting those who are often comparatively worse-off.

In contrast, CST has overwhelming ethical benefits and few downsides. Employed on a large scale, it can contribute substantially to emission reductions and climate change mitigation (Sonawane & Bupesh Raja 2018). It is furthermore inexpensive, safe, and effective. However, it is not suited for small-scale employment, but rather as a substitute for conventional power plants.

Wind Power – Saving the Climate While Destroying the Landscape?

Wind energy, after hydropower, is the world's most widely used RE resource (Ritchie & Roser 2020), and is a desirable way of generating electricity in many respects. Wind turbines have low lifecycle emissions (AER 2009: 52–53, Biegler 2009) and wind power "has significant potential to reduce (and is already reducing) GHG emissions" (IPCC 2011: 99). Wind energy is currently the cheapest of the REs available. Notwithstanding some noted negative economic impacts (Dorrell & Lee 2020), implementing wind energy on a large scale in many places has been shown to positively influence the job market and manufacturing sector, creating new jobs and establishing a (or expand the existing) renewable technology industry. (See e.g. IPCC 2011: 719; Ortega-Izquierdo & Río 2020; for the U.S., this is confirmed by Wei et al. 2010. See also Patterson 2012 and *Zero Carbon Australia Stationary Energy Plan* 2020.) Also, wind farms can constitute a source of income for the rural population, safeguarding them from the impacts of droughts or other unforeseeable events (IPCC 2011:195, see also Rosenberg 2008: 525f). It has been suggested that the value of properties can be negatively affected by views of wind turbines (Sunak & Madlener 2016), but also that this impact is correlated with acceptance of (or opposition to) wind power (Vyn 2018).

While generating electricity from wind instead of from fossil fuels avoids a number of health hazards and environmental damage resulting from conventional energy generation (e.g. GHG emissions, air, water, and soil pollution and degradation), wind turbines have been suspected to sometimes have adverse health effects on the people living in the immediate vicinity of the turbines. Problems have been said to result from infrasound noise, electromagnetic interference, shadow flicker, and blade glint. Yet, according to several studies surveyed by the National Health and Medical Research Council (NHMRC 2010), there is no evidence for a positive link between wind turbines and adverse health effects: "There are no direct pathological effects from wind farms and … any potential impact on humans can be minimized by following existing planning guidelines" (ibid). The German Advisory Council comes to a similar conclusion: "[p]rovided adequate distances to settlements are maintained, noise emissions from modern wind power plants are therefore no longer a problem" (WBGU 2004: 64). This suggests that rather than being the result of the actual impact of wind turbines, health problems that people experience in the vicinity of turbines seem to be resulting from anxiety surrounding the turbines and be correlated with people's attitudes toward them (see CSIRO 2012; NHMRC 2010). Anxiety issues and negative attitudes may successfully be addressed by involving local populations early in the decision-making process and providing them with the relevant information to prevent such problems (see Schwenkenbecher 2017).

Wind farms do have some adverse environmental impacts, too (for an overview, see Nazir et al. 2020). For example, there is proven interference of rotating blades with local fauna, especially birds and bats (Baisner et al. 2010; Biegler 2009; Drewitt & Langston 2008; Mindermann et al. 2012). However, studies suggest that careful planning can avoid or minimize these effects (IPCC 2011: 100, see also Australian Greenhouse Office and Australian Wind Energy Association 2004: 3). Overall, the environmental impacts appear to be negligible compared to those of conventional energies: "attempts to measure the relative impacts of various electricity supply technologies suggest that wind energy generally has a comparatively small environmental footprint" (IPCC 2011: 99).

The only impact of wind farms that cannot be mitigated is their visual impact on the landscape. Wind farms usually feature prominently in the landscape and alter the visual composition of their surroundings (and this applies to hydro power plants, and to a lesser extent to CST, too). Concerns about landscape alterations seem to be at the heart of a lot of opposition to wind farms (Wolsink 2007). Bell et al. (2005: 470) have argued that "there is no 'technical fix' for the problem of landscape impact. Instead, the only way of accommodating people's landscape concerns is to site wind farms in places that people find more acceptable". People's concerns regarding landscape impacts of wind turbines can be addressed in a number of ways. The Australian Greenhouse Office and Australian Wind Energy Association (2004) suggest that communities should always be consulted on turbine placement and important viewpoints should be agreed upon early in the process (see also Auswind & ACNT 2007 and Gross 2007). Another possibility is to compensate communities affected by RE developments, including wind farms, for loss of "environmental qualities that people might otherwise have expected to keep" (Cowell et al. 2012: 12; see also Cowell et al. 2011). Compensation should hence be combined with consultation to ensure that legitimate complaints are heard and that compromises are found if possible. It appears then that many of the seemingly negative aspects of wind power are based on attitudes and prejudices, which can be moderated or overcome if stakeholders are involved in the decision-making process (Bell and Rowe 2012; see also Toke 2002 & Zoellner et al. 2008 and Schwenkenbecher 2017).

Finally, another downside of renewables, in particular wind and solar power, is the intermittent energy supply and resulting issues of energy security. We cannot go into any detail, but it should be noted that the technological challenges of intermittency appear to be perfectly resolvable (Jacobson et al. 2015) and that from a socio-political perspective, RE can be seen as affording greater energy security than fossil fuels (Valentine 2011; Viviescas et al. 2019).

In sum, there are downsides and limitations to each RE technology which must not be ignored and – where possible – must be mitigated. Addressing them will ensure that the shift is as equitable, socially just, and environmentally sustainable as possible. Those orchestrating the transition must be sensitive to these goals – the process should be fair and ecologically sound so as not to replicate many of the energy justice problems of conventional modes of energy generation.

Concluding Remarks

The moral, economic, and prudential reasons for a transition to REs globally and domestically not only, but most urgently, to mitigate climate change are overwhelming. Some countries are pioneering this shift in an exemplary way. Other countries, especially large and historically large GHG emitters must follow in due course and help developing nations in their efforts to shift to RE. The major obstacles to doing what needs to be done are not technological, but seem to result from lacking political will and power as well as vested economic interests in perpetuating the status quo.

Acknowledgments

Research for this chapter was undertaken as part of the "Zero Carbon Australia and Social Justice" project with the Social Justice Initiative at the University of Melbourne (2012–2013) led by Jeremy Moss.

Related Topics

20. Moral Bases of Responses to Climate Change
29. Hydropower
32. Energy Poverty
33. Urban Sustainability

References

Alcalde, J., Flude, S., Wilkinson, M., Johnson, G., Edlmann, K., Bond, C. E., Scott, V., Gilfillan, S. M., Ogaya, X., & Haszeldine, R. S. (2018) Estimating Geological CO2 Storage Security to Deliver on Climate Mitigation. *Nature Communications* 9(1): 2201.

Anderson, S. T. (2017) Risk, Liability, and Economic Issues with Long-Term CO2 Storage—A Review. *Natural Resources Research* 26(1): 89–112.

Australian Capital Territory (ACT) Government (2012) *AP2- A New Climate Change Strategy and Action Plan for the Australian Capital Territory.* Retrieved from: https://www.environment.act.gov.au/__data/assets/pdf_file/0006/581136/AP2_Sept12_PRINT_NO_CROPS_SML.pdf.

Australian Energy Regulator (AER) (2009). *State of the energy market.* Retrieved from: https://www.aer.gov.au/system/files/State%20of%20the%20energy%20market%202009%C3%A2%E2%82%AC%E2%80%9D-complete%20report.pdf.

Australian Greenhouse Office & Australian Wind Energy Association (AGO & AusWEA) (2004) *Wind Farm Fact Sheets.*

Australian Wind Energy Association and Australian Council of National Trusts (AusWEA & ACNT) (2007) Wind Farms and Landscape Values; *National Assessment Framework; Foundation Report.* Melbourne: Australian Wind Energy Association.

Babson, E. (2019) *US Oil Dependence. More Drilling Does Not Make the US More Energy Secure.* American Security Project. Retrieved from: https://www.americansecurityproject.org/wp-content/uploads/2019/05/US-Oil-Dependence.pdf.

Baisner, A. J., Andersen, J. L., Findsen, A., Yde Granath, S. W., Madsen, K. O., & Desholm, M. (2010) Minimizing Collision Risk Between Migrating Raptors and Marine Wind Farms: Development of a Spatial Planning Tool. *Environmental Management* 46(5): 801–808.

Baurzhan, S., & Jenkins, G. P. (2016) Off-Grid Solar PV: Is it an Affordable or Appropriate Solution for Rural Electrification in Sub-Saharan African Countries? *Renewable and Sustainable Energy Reviews* 60: 1405–1418.

Bell, D., Gray, T., & Haggett, C. (2005) The 'Social Gap' in Wind Farm Siting Decisions: Explanations and Policy Responses. *Environmental Politics* 14(4): 460–477.

Bell, D., & Rowe, F. (2012) *Are Climate Policies Fairly Made.* Joseph Rowntree Foundation. Retrieved from: https://www.jrf.org.uk/file/39251/download?token=zywefI1O&filetype=viewpoint.

Biegler, T. (2009) *The hidden costs of electricity: Externalities of power generation in Australia*, Report for the Australian Academy of Technological Sciences and Engineering (ATSE). Retrieved from: https://apo.org.au/sites/default/files/resource-files/2009-02/apo-nid4196.pdf.

Brockway, P. E., Owen, A., Brand-Correa, L. I., & Hardt, L. (2019) Estimation of Global Final-Stage Energy-Return-On-Investment for Fossil Fuels with Comparison to Renewable Energy Sources. *Nature Energy* 4(7): 612–621.

Budinis, S., Krevor, S., Dowell, N. M., Brandon, N., & Hawkes, A. (2018) An Assessment of CCS Costs, Barriers and Potential. *Energy Strategy Reviews* 22: 61–81.

Bui, M., Adjiman, C. S., Bardow, A., Anthony, E. J., Boston, A., Brown, S., ... Hallett, J. P. (2018) Carbon Capture and Storage (CCS): The Way Forward. *Energy & Environmental Science* 11(5): 1062–1176.

Bui, M., & Mac Dowell, N. (Eds.) (2019) Introduction. In: *Carbon Capture and Storage.* Croydon: The Royal Society of Chemistry, pp. 1–7.

Cherni, J. A., & Hill, Y. (2009) Energy and Policy Providing for Sustainable Rural Livelihoods in Remote Locations: The Case of Cuba. *Geoforum* 40(4): 645–654.

City of Copenhagen (2009) *Copenhagen Climate Plan.* Retrieved from: http://www.energycommunity.org/documents/copenhagen.pdf.

Climate Council of Australia (2018) Clean & Reliable Power: Roadmap to a Renewable Future. Retrieved from: https://www.climatecouncil.org.au/wp-content/uploads/2018/03/Clean-and-Reliable.pdf

ClimateWorks Australia (2020) Net Zero Momentum Tracker. Local Government Sector. Retrieved from: https://www.climateworksaustralia.org/wp-content/uploads/2020/01/ClimateWorks_NZMT_Local-Government-Report_Jan20.pdf

Cowell, R., Bristow, G., & Munday, M. (2011) Acceptance, Acceptability and Environmental Justice: The Role of Community Benefits in Wind Energy Development. *Journal of Environmental Planning & Management* 54(4): 539–557.

Cowell, R., Bristow, G., & Munday, M. (2012) *Wind Energy and Justice for Disadvantaged Communities.* Joseph Rowntree Foundation. Retrieved from: https://www.jrf.org.uk/file/41945/download?token=00JwwUnc&filetype=viewpoint.

CSIRO (Commonwealth Scientific and Industrial Research Organisation) (2012) *Exploring Community Acceptance of Rural Wind Farms in Australia: A Snapshot.* https://doi.org/10.4225/08/584ee83bb6373.

Dennekamp, M., & Carey, M. (2010) Air Quality and Chronic Disease: Why Action on Climate Change Is also Good for Health. *New South Wales Public Health Bulletin* 21(6): 115–121.

Desideri, U., & Campana, P. E. (2014) Analysis and Comparison Between a Concentrating Solar and a Photovoltaic Power Plant. *Applied Energy* 113: 422–433.

Diesendorf, M. (2003) *Australia's Polluting Power: Coal-Fired Electricity and its Impacts on Global Warming.* Sydney: WWF Australia.

——— (2010) Strategies for Radical Climate Mitigation. *Journal of Australian Political Economy* (66): 98–117.

——— (2011) Redesigning Energy Systems. In D. Schlosberg, R. B. Norgaard & J. S. Dryzek (Eds.), *The Oxford Handbook of Climate Change and Society.* Oxford: Oxford University Press, pp. 561–580.

——— (2019) Energy Futures for Australia. In *Decarbonising the Built Environment.* Singapore: Palgrave Macmillan, pp. 35–51.

Dorrell, J., & Lee, K. (2020) The Cost of Wind: Negative Economic Effects of Global Wind Energy Development. *Energies* 13: 3667.

Drewitt, A. L., & Langston, R. H. W. (2008) Collision Effects of Wind-Power Generators and Other Obstacles on Birds. *Annals of the New York Academy of Sciences* 1134: 233–266.

Figueroa, R. M. (2011) Indigenous Peoples and Cultural Losses. In D. Schlosberg, R. B. Norgaard & J. S. Dryzek (Eds.), *The Oxford Handbook of Climate Change and Society*. Oxford: Oxford University Press, pp. 232–247.

Florini, A. (2012) The Peculiar Politics of Energy. *Ethics & International Affairs* 26.3. Retrieved from: https://www.ethicsandinternationalaffairs.org/2012/the-peculiar-politics-of-energy-full-text/.

Gardiner, S. M. (2011) Cost-Benefit Paralysis. In Gardiner, S. M. (ed.) *A Perfect Moral Storm: The Ethical Tragedy of Climate Change*. New York; Oxford: Oxford University Press, pp. 247–298.

Garnaut, R. (2008) *Garnaut Climate Change Review*. Final Report. Canberra.

German Advisory Council on Global Change (WBGU) (2004) *World in Transition: Towards Sustainable Energy Systems*. Retrieved from: https://www.wbgu.de/fileadmin/user_upload/wbgu/publikationen/hauptgutachten/hg2003/pdf/wbgu_jg2003_engl.pdf.

Gielen, D., Boshell, F., Saygin, D., Bazilian, M. D., Wagner, N., & Gorini, R. (2019) The Role of Renewable Energy in the Global Energy Transformation. *Energy Strategy Reviews* 24: 38–50.

Giles, J. (2006) Methane Quashes Green Credentials of Hydropower. *Nature* 444: 524.

Gross, C. (2007) Community Perspectives of Wind Energy in Australia: The Application of a Justice and Community Fairness Framework to Increase Social Acceptance. *Energy Policy* 35(5): 2727–2736.

Groosmann, B., Muller, N. Z., & O'Neill-Toy, E. (2011) The Ancillary Benefits from Climate Policy in the United States. *Environmental and Resource Economics* 50(4): 585–603.

Hanna, E. G. (2011) Health Hazards. In D. Schlosberg, R. B. Norgaard & J. S. Dryzek (Eds.), *The Oxford Handbook of Climate Change and Society*. Oxford: Oxford University Press, pp. 217–231.

Hitzeroth, M., Jehling, M., & Brueckner, M. (2017) Apples and Oranges? A Multi-Level Approach Explaining Social Acceptance of Renewable Energy in Germany and Australia. *International Journal of Global Energy Issues* 40(3–4): 141–165.

Holland, B. (2008) Justice and the Environment in Nussbaum's "Capabilities Approach": Why Sustainable Ecological Capacity Is a Meta-Capability. *Political Research Quarterly* 61(2): 319–332.

IEA. (2009) *Technology Roadmap: Carbon Capture and Storage*. Retrieved from: https://www.iea.org/reports/technology-roadmap-carbon-capture-and-storage-2009

——— (2020a) *Clean Energy Innovation*. Retrieved from: https://www.iea.org/reports/clean-energy-innovation.

——— (2020b) *CCUS in Power*. Retrieved from: https://www.iea.org/reports/ccus-in-power.

Intergovernmental Panel on Climate Change (IPCC) (2011) *Special Report on Renewable Energy Sources and Climate Change Mitigation*. Cambridge: Cambridge University Press.

——— (2013) Summary for Policymakers. In: T. F. Stocker, D. Qin, G.-K. Plattner, M. Tignor, S. K. Allen, J. Boschung, A. Nauels, Y. Xia, V. Bex & P. M. Midgley (Eds.), *Climate Change 2013: The Physical Science Basis. Contribution of Working Group I to the Fifth Assessment Report of the Intergovernmental Panel on Climate Change*. Cambridge and New York: Cambridge University Press.

——— (2014) Energy Systems. In: *Climate Change 2014: Mitigation of Climate Change. Contribution of Working Group III to the Fifth Assessment Report of the Intergovernmental Panel on Climate*. Cambridge and New York: Cambridge University Press.

——— (2018a) *Global Warming of 1.5°C. An IPCC Special Report on the Impacts of Global Warming of 1.5°C Above Pre-Industrial Levels and Related Global Greenhouse Gas Emission Pathways, in the Context of Strengthening the Global Response to the Threat of Climate Change, Sustainable Development, and Efforts to Eradicate Poverty*. Geneva, Switzerland: IPCC.

——— (2018b) Mitigation Pathways Compatible with 1.5°C in the Context of Sustainable Development. In: *Global Warming of 1.5°C. An IPCC Special Report on the Impacts of Global Warming of 1.5°C above Pre-Industrial Levels and Related Global Greenhouse Gas Emission Pathways, in the Context of Strengthening the Global Response to the Threat of Climate Change, Sustainable Development, and Efforts to Eradicate Poverty*. Geneva, Switzerland: IPCC.

IRENA (2019) *Global Energy Transformation: A Roadmap to 2050*. Abu Dhabi: International Renewable Energy Agency.

Jacobson, M. Z., Delucchi, M. A., Cameron, & Frew, B. A. (2015) Stabilizing Grid with 100% Renewables 2050. *Proceedings of the National Academy of Sciences* 112(49): 15060–15065.

Jamieson, D. (1996) Ethics and Intentional Climate Change. *Climatic Change* 33(3): 323–336.

——— (2011) Energy, Ethics, and the Transformation of Nature. In D. G. Arnold (Ed.), *The Ethics of Global Climate Change*. Cambridge; New York: Cambridge University Press, pp. 16–37.

Jewell, J., & Cherp, A. (2020) On the Political Feasibility of Climate Change Mitigation Pathways: Is it too Late to Keep Warming Below 1.5°C? *WIREs Climate Change*; 11: e621.

Jung, Ys (2020) Renewable Energy: EU Green Deal and Korea Green New Deal. BusinessKorea. (25 June). Retrieved from: http://www.businesskorea.co.kr/news/articleView.html?idxno=48066

Kahn, J. R., Freitas, C. E., & Petrere, M. (2014) False Shades of Green: The Case of Brazilian Amazonian Hydropower. *Energies* 7: 6063–6082.

Kaswan, A. (2009) Greening the Grid and Climate Justice. *Environmental Law* 39(4): 1143-1160.

Keith, G. et al. (2012) *The Hidden Costs of Electricity: Comparing the Hidden Costs of Power Generation Fuels.* Prepared for the Civil Society Institute.

Lawford-Smith, H. (2012) Understanding Political Feasibility. *Journal of Political Philosophy* 21(3): 243–259.

Lehmann, T. C. (2019) Honourable Spoils? The Iraq War and the American Hegemonic System's Eternal and Perpetual Interest in Oil. *The Extractive Industries and Society* 6(2): 428–442.

Leung, D. Y. C., Caramanna, G., & Maroto-Valer, M. M. (2014) An Overview of Current Status of Carbon Dioxide Capture and Storage Technologies. *Renewable and Sustainable Energy Reviews* 39: 426–443.

Lund, H. (2007) Renewable Energy Strategies for Sustainable Development. *Energy* 32(6): 912–919.

Macintosh, A. and D. Wilkinson (2011) Searching for public benefits in solar subsidies: A case study on the Australian government's residential photovoltaic rebate program. *Energy Policy* 39(6): 3199-3209.

McCauley, D., Ramasar, V., Heffron, R. J., Sovacool, B. K., Mebratu, D., & Mundaca, L. (2019) Energy Justice in the Transition to Low Carbon Energy Systems: Exploring Key Themes in Interdisciplinary Research. *Applied Energy* 233–234: 916–921.

Melbourne Sustainable Society Institute. (2012) *2020 Vision for a Sustainable Society.* The Melbourne Sustainable Society Institute. Parkville, Victoria.

Minderman, J., Pendlebury, C. J., Pearce-Higgins, J. W., & Park, K. J. (2012) Experimental Evidence for the Effect of Small Wind Turbine Proximity and Operation on Bird and Bat Activity. PloS One 7(7): e41177.

Monyei, C. G., Sovacool, B. K., Brown, M. A., Jenkins, K. E., Viriri, S., & Li, Y. (2019) Justice, Poverty, and Electricity Decarbonization. *The Electricity Journal* 32(1): 47–51.

Moran, E. F., Lopez, M. C. Moore, N. Müller, N., & Hyndman, D. W. (2018) Sustainable Hydropower in the 21st Century. *Proceedings of the National Academy of Sciences* 115(47): 11891–11898.

Muteri, V., Cellura, M., Curto, D., Franzitta, V., Longo, S., Mistretta, M., & Parisi, M. L. (2020) Review on Life Cycle Assessment of Solar Photovoltaic Panels. *Energies* 13: 252.

National Health and Medical Research Council (NHMRC) (2010) *Wind Turbines and Health: A Rapid Review of the Evidence.* Retrieved from: https://puc.sd.gov/commission/dockets/electric/2018/EL18-003/testimony/dakotarange/rExhibit2.pdf (accessed 24 March 2022).

Nazir, M. S., Ali, N., Bilal, M., & Iqbal, H. M. N. (2020) Potential Environmental Impacts of Wind Energy Development: A Global Perspective. *Current Opinion in Environmental Science & Health* 13: 85–90.

Neumann, A., Sorge, L., von Hirschhausen, C., & Wealer, B. (2020) Democratic Quality and Nuclear Power: Reviewing the Global Determinants for the Introduction of Nuclear Energy in 166 Countries. *Energy Research & Social Science* 63: 101389

Nolt, J. (2011) How Harmful Are the Average American's Greenhouse Gas Emissions? *Ethics, Policy & Environment* 14(1): 3–10. doi:10.1080/21550085.2011.561584

Norgaard, R. B. (2011) Weighing Climate Futures: A Critical Review of the Application of Economic Valuation. In D. Schlosberg, R. B. Norgaard & J. S. Dryzek (Eds.), *The Oxford Handbook of Climate Change and Society*. Oxford: Oxford University Press, pp. 190–204.

Nugent, Daniel, and Benjamin K. Sovacool. (2014) Assessing the Lifecycle Greenhouse Gas Emissions from Solar Pv and Wind Energy: A Critical Meta-Survey. *Energy Policy* 65(0): 229–244.

Okoye, C. O., & Oranekwu-Okoye, B. C. (2018) Economic Feasibility of Solar PV System for Rural Electrification in Sub-Sahara Africa. *Renewable and Sustainable Energy Reviews* 82: 2537–2547.

O'Neill, J. (1993) *Ecology, Policy, and Politics: Human Well-Being and the Natural World*. Psychology Press.

Ortega-Izquierdo, M., & Río, P. D. (2020) An Analysis of the Socioeconomic and Environmental Benefits of Wind Energy Deployment in Europe. *Renewable Energy* 160: 1067-1080

Oxfam (2015) *Extreme Carbon Inequality. Why the Paris Climate Deal Must Put the Poorest, Lowest Emitting and Most Vulnerable People First.* Retrieved from: https://oi-files-d8-prod.s3.eu-west-2.amazonaws.com/s3fs-public/file_attachments/mb-extreme-carbon-inequality-021215-en.pdf

Pampel, F. C. (2011) Support for Nuclear Energy in the Context of Climate Change: Evidence From the European Union. *Organization & Environment* 24(3): 249–268.

Patterson, J. (2012) Jobs vs. Public Health: A False Dilemma. *Environmental Forum* 29(2), p. 53.

Rabinowitz, D. (2012) Climate Injustice: CO2 from Domestic Electricity Consumption and Private Car Use by Income Decile. *Environmental Justice (19394071)* 5(1): 38–46.

Rissman, J., Bataille, C., Masanet, E., Aden, N., Morrow, W. R., Zhou, N., … Helseth, J. (2020) Technologies and Policies to Decarbonize Global Industry: Review and Assessment of Mitigation Drivers through 2070. *Applied Energy* 266: 114848

Ritchie, H., & Roser, M. (2020) *Renewable Energy*. Retrieved from: https://ourworldindata.org/renewable-energy

Rosenberg, R. H. (2008) Diversifying America's Energy Future: The Future of Renewable Wind Power. *Virginia Environmental Law Journal* 26: 505–544.

Rüdiger, M. (2019) From Coal to Wind. *Scandinavian Journal of History* 44(4): 510–530.

Samarakoon, S. (2019) A Justice and Wellbeing Centered Framework for Analysing Energy Poverty in the Global South. *Ecological Economics* 165: 106385.

Schwenkenbecher, A. (2013) Bridging the Emissions Gap: A Plea for Taking up the Slack. *Philosophy and Public Issues* 3(1): 273–301.

——— (2014) Is there an Obligation to Reduce One's Individual Carbon Footprint? *Critical Review of International Social and Political Philosophy* 17(2): 168–188.

——— (2017) What Is Wrong with Nimbys? Renewable Energy, Landscape Impacts and Incommensurable Values. *Environmental Values* 26(6): 711–731.

Seligman, P. (2012) Energy. In Melbourne Sustainable Society Institute (2012), pp. 37–46.

Shrader-Frechette, K. (2011) *What Will Work: Fighting Climate Change with Renewable Energy, Not Nuclear Power*. New York: Oxford University Press.

Shue, H. (1995) Avoidable Necessity: Global Warming, International Fairness, and Alternative Energy. In Ian Shapiro & Judith Wagner DeCew (Eds.), *Theory and Practice, NOMOS XXXVII*. New York: NYU Press, pp. 239–64.

——— (2005) Responsibility to Future Generations and the Technological Transition. In W. Sinnott-Armstrong & Richard B. Howarth (Eds.), *Perspectives on Climate Change, Vol. 5*. Oxford: Elsevier, pp. 265–283.

——— (2011) Face Reality? After You!—A Call for Leadership on Climate Change. *Ethics and International Affairs* 25(1): 17–26.

Sonawane, P. D., & Bupesh Raja, V. K. (2018) An Overview of Concentrated Solar Energy and its Applications. *International Journal of Ambient Energy* 39(8): 898–903.

Sovacool, B. K. (2017) Contestation, Contingency, and Justice in the Nordic Low-Carbon Energy Transition. *Energy Policy* 102: 569–582

Stephens, J. C. (2019) Energy Democracy: Redistributing Power to the People through Renewable Transformation. *Environment: Science and Policy for Sustainable Development* 61(2): 4–13.

Stern, N. (2006) The economics of climate change: the Stern review. Cambridge, UK: Cambridge University Press.

Sunak, Y., & Madlener, R. (2016) The Impact of Wind Farm Visibility on Property Values: A Spatial Difference-In-Differences Analysis. *Energy Economics* 55: 79–91.

Swain, R. B., & Karimu, A. (2020) Renewable Electricity and Sustainable Development Goals in the EU. *World Development* 125: 104693

The Australian Greens (2019) *Renew Australia 2030 Powering Past Coal to a Clean Future for all of us*. Retrieved from: https://greens.org.au/sites/default/files/2019-03/Greens%202019%20Policy%20Platform%20-%20Renew%20Australia.pdf.

The Royal Society (2009) *Geoengineering the Climate: Science, Governance and Uncertainty*. London. Retrieved from https://royalsociety.org/-/media/Royal_Society_Content/policy/publications/2009/8693.pdf.

Toke, D. (2002) Wind Power in UK and Denmark: Can Rational Choice Help Explain Different Outcomes? *Environmental Politics* 11(4): 83–100.

United Nations Environment Programme (UNEP) (2007) *Report on Indigenous and Local Communities Highly Vulnerable to Climate Change Inter Alia of the Arctic, Small Island States and High Altitudes, With a Focus on Causes and Solutions*. Retrieved from: https://www.cbd.int/doc/meetings/tk/acpow8j-02/official/acpow8j-02-03-en.pdf.

van der Zwaan, B., and Smekens, K. (2009) CO2 Capture and Storage with Leakage in an Energy-Climate Model. *Environmental Modelling & Assessment* 14: 135–148.

Varun, I. K. B. and R. Prakash (2009). LCA of renewable energy for electricity generation systems—A review. *Renewable and Sustainable Energy Reviews* 13(5): 1067–1073.

Voigt, T., et al. (1998) Air Pollution in the Latrobe Valley and Its Impact Upon Respiratory Morbidity. *Australian and New Zealand Journal of Public Health* 22(5): 556–561.

Vinca, A., Emmerling, J., & Tavoni, M. (2018) Bearing the Cost of Stored Carbon Leakage. *Frontiers in Energy Research* 6(40). doi:10.3389/fenrg.2018.00040

Viviescas, C., Lima, L., Diuana, F. A., Vasquez, E., Ludovique, C., Silva, G. N., … Paredes, J. R. (2019) Contribution of Variable Renewable Energy to Increase Energy Security in Latin America: Complementarity and Climate Change Impacts on Wind and Solar Resources. *Renewable and Sustainable Energy Reviews* 113: 109232.

Vyn, R. J. (2018) Property Value Impacts of Wind Turbines and the Influence of Attitudes toward Wind Energy. *Land Economics* 94(4): 496–516.

Wei, M. (2010) Putting Renewables and Energy Efficiency to Work: How Many Jobs Can the Clean Energy Industry Generate in the US? *Energy Policy* 38(2): 919–931.

Wiseman, J., & Edwards, T. (2012) *Post Carbon Pathways. Reviewing Post Carbon Economy Transition Strategies.* Melbourne Sustainable Society Institute & Centre for Policy Development.

Wolsink, M. (2007) Planning of Renewables Schemes: Deliberative and Fair Decision-Making on Land-scape Issues Instead of Reproachful Accusations of Non-Cooperation. *Energy Policy* 35(5): 2692–2704.

World Health Organization (WHO) (2005) *Ecosystems and Human Well-Being.* Health Synthesis. Retrieved from: https://apps.who.int/iris/bitstream/handle/10665/43354/9241563095.pdf?sequence=1&isAllowed=y.

World Health Organization (WHO) (2011) *Health Co-benefits of Climate Change Mitigation – Transport.* Executive Summary. Retrieved from: https://climateandhealthalliance.org/wp-content/uploads/2018/02/Occupational-Health-Health-in-the-Green-Econmy.pdf (accessed 24 March 2022).

———— (2012) *Health Indicators of Sustainable Energy.* Retrieved from: http://www.who.int/hia/green_economy/indicators_energy2.pdf (accessed 4 February 2014).

Zero Carbon Australia (ZCA) (2010) *Zero Carbon Australia Stationary Energy Plan.* The University of Melbourne Research Institute & Beyond Zero Emissions. Retrieved from: https://bze.org.au/wp-content/uploads/2021/02/stationary-energy-plan-bze-report-2010.pdf.

Zoback, Mark D., and Gorelick, Steven M. (2012) Earthquake Triggering and Large-Scale Geologic Storage of Carbon Dioxide. *Proceedings of the National Academy of Sciences* 109(26): 10164–10168.

Zoellner, J., Schweizer-Ries, P., & Wemheuer, C. (2008) Public Acceptance of Renewable Energies: Results from Case Studies in Germany. *Energy Policy* 36(11): 4136–4141.

Further Reading on the Transition to Renewable Energy

Centre for Alternative Technology (2019) *Zero Carbon Britain: Rising to the Climate Emergency.* Retrieved from: https://www.cat.org.uk/download/35541/ (accessed 3 August 2020).

Committee on America's Energy, F., National Academy of, S., National Academy of, E., & National Research, C. (2009) *America's Energy Future: Technology and Transformation.* Washington, DC: The National Academies Press. Retrieved from: https://www.nap.edu/login.php?record_id=12091&page=https%3A%2F%2Fwww.nap.edu%2Fdownload%2F12091(accessed 3 August 2020).

United Nations Development Programme (UNDP) (2012) *Transforming On-Grid Renewable Energy Markets.* Retrieved from: https://www.undp.org/content/dam/undp/library/Environment%20and%20Energy/Climate%20Strategies/UNDP_FIT_Port_TransformingREMarkets_15Nov2012%20(B).PDF (accessed 3 August 2020).

World Wildlife Fund (WWF) Report (2017) *Planning to Succeed. How to Build Strong 2050 Climate and Energy Development Strategies.* Retrieved from: http://d2ouvy59p0dg6k.cloudfront.net/downloads/wwf_maximiser_planning_to_succeed_guidance_report_1.pdf (accessed 3 August 2020). See Annex 1 for more literature.

31

NATURAL GAS AND FRACKING

Adam Briggle

Definition, History, and Overview

The term *fracking* is used more or less broadly to refer to various aspects of oil and gas extraction. Most narrowly, it is shorthand for *hydraulic fracturing*, a well stimulation technique that uses pressurized liquids (mixed with various chemicals and sand) to induce and prop open cracks in rock formations in order to maximize recovery of hydrocarbons. The US Environmental Protection Agency (EPA) has defined this process more broadly to include the "acquisition of source water, well construction…and waste disposal," in addition to well stimulation (EPA 2010, p. 1).

In order to address a more comprehensive range of ethics and policy issues, I will use an even broader definition of fracking: the intensification of hydraulic fracturing and its convergence with other exploration and production innovations, such as 3-D seismic imaging and horizontal drilling. Beginning roughly at the turn of the twenty-first century, this intensification and convergence has made oil and gas production from unconventional reserves (shale gas, tight gas, tight oil, and coal seam gas) economically viable, which has led to an increase in oil and gas production, especially in the US, where the vast majority of oil and gas wells are now fracked (see King 2012).

This chapter will mainly focus on the US, because it has the most experience with fracking. It is important to note, though, that the energy boom unleashed by fracking is spreading to other parts of the world, bringing with it a number of policy- and ethics-related controversies (see Howarth, Ingraffea, and Engelder 2011; Resources for the Future 2012; Briggle 2015). Some of the most important disputes pertain to fracking's impact on the economy, national security, jobs, surface and ground water, air and climate, seismic activity, and surrounding property values (for example, see Colborn et al. 2011; HIS Global Insight 2011; Howarth, Santoro, and Ingraffea 2011; Schmidt 2011; Urbina 2011; Clark, et al. 2012; Lomborg 2012; McKenzie et al. 2012; Freyman and Salmon 2013). In response to such controversies, many nations have adopted policies with regard to fracking that range from a total ban (France) through cautious embrace (UK) to ardent support (Poland).

Large-scale development of oil and gas began in the mid-nineteenth century. Many wells in this era tapped into conventional reservoirs that flowed simply from the change of pressure caused by drilling. But beginning as early as the 1860s, experiments in fracking began on hard rock formations in the eastern part of the US (see Montgomery and Smith 2010; Zuckerman 2013). Initial technology used explosive nitroglycerin to increase flow and ultimate recovery. This practice was successful but hazardous.

DOI: 10.4324/9781315768090-37

In the 1930s, petroleum engineers began experimenting with non-explosive means of increasing hydrocarbon extraction. Initially, operators used acids because "acid etching" creates channels in the rocks through which oil and gas can flow. (Acidizing remains a common well stimulation technique.) In 1947, Stanolind Oil and Gas Company conducted the first experimental treatment of a nonacid hydraulic fracturing (called "Hydrafrac") in Kansas, using 1,000 gallons of thickened gasoline. That experiment was largely a failure. But by 1950, Halliburton Oil Well Cementing Company had hydraulically fractured 332 wells with an average production increase of 75%. Halliburton used a mixture of crude oil and gasoline (averaging 750 gallons of fluid per well) to fracture the rock along with sand (averaging 400 pounds per well) to prop open the fractures.

In 1967, the US Atomic Energy Commission's Project Gasbuggy detonated a nuclear bomb underground to loosen shale formations. This stimulated flow, but the gas was too radioactive to be usable. In the 1970s and 1980s, as domestic oil and gas production declined in the US, the federal government launched a major R&D effort (the US Unconventional Gas Research Programs) in partnership with several private companies to improve recovery from unconventional formations (Burwen and Flegal 2013; Wang and Krupnick 2013). Around the turn of the twenty-first century, this work paid off as the production of shale gas and oil became commercially viable.

The development of contemporary fracking entailed several ways of intensifying the basic process of breaking up rock formations. First, more chemicals have been added to the fracking fluids to serve a variety of purposes (e.g. manage viscosity and pH, kill bacteria, and prevent corrosion). In the 1990s, water emerged as the fracking fluid of choice, because it yielded higher recovery rates than gels and foams. As a result, much contemporary fracking is more precisely labeled *slick-water* hydraulic fracturing. Second, greater quantities of water are now used, which has led some to refer to contemporary practices as *high-volume* hydraulic fracturing. On average, a fracked well now requires roughly 4 million gallons of water. Third, greater quantities of sand (with some wells requiring over 4 million pounds) are used. Fourth, much larger pumping engines and more pump trucks are used.

In the 1990s, engineers merged this intensified version of hydraulic fracturing with other technologies – most notably, horizontal drilling. Turning the wellbore horizontally allows it to run through the production zone, thereby maximizing the surface area for collecting oil and gas. The industry continues to increase the length of the "laterals" (the horizontal portion of the wellbore). It is now common to see laterals approaching and even exceeding one-mile long. This increases contact with production zones while minimizing surface disturbances, because more minerals can be extracted from fewer wells. Taken together, this convergence and intensification of fracking has precipitated a rapid boom in the US production. This is reshaping the geopolitics of energy, particularly insofar as the US is now able to rival oil production figures from OPEC countries and seeks to export liquefied natural gas to Asian and European markets. Indeed, in 2018, the US became the world's largest global crude oil producer (EIA 2018). Fracking can be seen as a key part of what some have called "extreme energy," signifying ever more technically and resource-intense and risky efforts to exploit hard-to-access energy reserves (usually fossil fuels) in harsh environments under hazardous working conditions (see Klare 2009).

There are so many ethics and policy questions related to fracking that the challenge is to order them into a coherent structure. In the popular literature about fracking, the usual approach is to list the benefits and costs. There seems to be an assumption that somehow these could be compared and tallied to yield a conclusion about which policies would maximize utility or happiness. This utilitarian approach may be a useful first blush, but it suffers from a "God's eye view" delusion in supposing that we could establish some neutral identification and weighting of the costs and benefits.

Here, I use the notion of the "technological wager" as my ordering principle, in part because it avoids this problem of weighting. In the following section, I re-tell the history of fracking in terms of the technological wager, argue that fracking is essentially a real-world experiment, and sketch three conditions that should be in place to make real-world experiments ethical. The remaining three sections focus on one of these three conditions and the related ethics and policy issues at stake in the ongoing development and regulation of fracking.

Fracking and the Technological Wager

Most unconventional oil and gas deposits are at least 5,000 feet below the surface. Accessing them required improvements not just to hydraulic fracturing and horizontal drilling, but also to Measurement While Drilling (MWD). Like instruments in a cockpit, MWD allows engineers to "pilot" the wellbore to a target. This is crucial, because the extreme heat and pressure a mile below the surface of the earth turn the pipe into a string of spaghetti that can flail about in unexpected ways.

In the 1970s and 1980s, responding to fears about a looming energy crisis, engineers in the US attempted to combine these techniques to make production from unconventional shale gas formations economically viable. Notes from a field trial in West Virginia show the frustrations involved:

> drilling operations proceeded as planned, until a depth of 3459 feet was reached and the drill string stuck while reaming the hole. We...tried to work the drill string loose for 12 hours before deciding to sidetrack the hole. We pulled back up to 3200 feet and cemented the old hole and kicked off again. On the second try, we built angle too quickly and were going to come in about 20 feet higher than the target zone. We elected to sidetrack again to be able to hit the target zone when we had reached 3600 feet... drilling proceeded on the planned trajectory, but at an inclination of 74 degrees we were having a severe problem with lifting the cuttings out of the hole...
>
> *(Yost 1989, p. 41)*

One project in West Virginia had the objective of getting a directional well to reach an inclination of 60°. At first, they only achieved 42°. Four years later and a few counties north, another team tried again and got 52°. Six years later, in 1982, another attempt across the border in Ohio only got 25°. Similar kinds of field trials happened with hydraulic fracturing. George P. Mitchell's company on the Barnett Shale in Texas, for example, sunk millions of dollars over 18 years into one failed experiment after another (see Steward 2007).

Eventually, these experiments paid off insofar as they accomplished the rather narrow technical goal of making unconventional oil and gas production economically viable. Eventually, greater understanding brought about a suite of technologies that were cost-effective (although market viability was also aided by a weak dollar, which had inflated the price of natural gas). But there were, of course, several unanswered and unintended social, environmental, and ethical implications of this technical success. Engineers exercise power over reality by ignoring most of its complexity. Their goal is functional adequacy, not mirroring the world. This means that important things are almost always overlooked (Martin and Schinzinger 1989). Automobiles can be made to transport people without modeling effects on the character of cities (think sprawl) and the geopolitics of oil. Industrial fertilizers can grow corn absent any thought about algal blooms in the Gulf of Mexico. In the case of fracking, the goal is to get the fossil fuels out of the ground at a profit. In order to accomplish that, there is no need to account for the long-term fate of chemicals, water and air quality, or community and climate impacts.

In this way, fracking exemplifies the technological wager, by which I mean a gamble or even a faith that we can transform the world in the pursuit of narrowly defined goals and successfully manage the broader unintended consequences that result (see Ellul 1963). We can manage the unintended consequences of technology through further technological development. And this raises the central ethical question: What should we do about the fact that freedom to pursue intended desired outcomes also creates unintentional harms?

One response is to ask engineers and their corporate employers to take more factors into account *prior* to enrolling their technologies into society. This approach is usually couched in terms of the precautionary principle, which sees the technology as guilty until proven innocent (see Stirling 2007; Von Schomberg 2012). In New York, where there was a moratorium on fracking for six years, proponents were forced to establish that fracking is safe prior to taking action. There was no fracking while the debate continued (and eventually New York prohibited the use of hydraulic fracturing, claiming that the risks were "too great" and not "fully known" (Kaplan 2014)).

But there is no way to *prove* that a technology is innocent or to "fully know" the risks in advance. In the early stages of development, it is impossible to predict how it will play out. If we want the intended benefits of a technology, we must also risk getting unintended harms. At some point, we must roll the dice and move the experiment out of the lab into the world. Innovation becomes a real-world experiment.

That's why it is important to ask not just about how much knowledge we should have prior to committing to a technology but also about the structure of real-world experiments – that is, about how knowledge relates to ongoing action (Holbrook and Briggle 2014). There are three conditions that must be in place to make them reasonable and fair.

- First, those most vulnerable to the unintended harms should give their consent to the risk or, at the very least, they should be compensated for any harms incurred.
- Second, there should be a robust system for monitoring and learning from the real-world experiment.
- Third, the experiment should be modifiable when problems are identified through the learning process, that is, the original innovation should be readily renovated.

These ethical criteria for guiding and evaluating policies about fracking are derived from the *proactionary principle*, which was originally conceived by the transhumanist Max More as an alternative to precautionary politics. More summarized the essence of proaction as follows:

> Let a thousand flowers bloom! By all means, inspect the flowers for signs of infestation and weed as necessary. But don't cut off the hands of those who spread the seeds of the future.
>
> *(More 2005)*

According to the proactionary principle, rather than avoiding error, we should take risks in the pursuit of profound truths and great rewards. In contrast to precaution, which situates scientific study prior to policy action, proaction frames policy as itself an extension of scientific study. Like the engineers in West Virginia and Texas, policymakers must learn from and modify the ongoing technological experiment. Unlike those engineers, however, the task for public policymaking is to find ways to account for the unintended consequences and not just the narrow technical goal of profitable production.

Proaction is a call to embrace risk, not recklessness. It has a set of ethical norms that, although different from precautionary norms, establish standards for evaluating policies related to fracking. We must "inspect" (monitor) and "weed" (renovate) and, as More acknowledges, must compensate those who are harmed. Without those conditions, any "principled" alternative to precautionary politics is no more than rapacious corporate capitalism and its unbridled pursuit of profits. The proactionary principle, at least in theory, is an alternative both to strict precaution and to ruthless market utilitarianism.

Informed Consent

Every new technology will create both winners and losers – those who gain and those who are harmed. While the precautionary principle seeks to minimize harms prior to action (see Von Schomberg 2012), the proactionary principle suggests two other (perhaps complementary) ways of handling this reality. First, those who are most vulnerable to potential harms should give their consent. Second, those who are harmed should be compensated.

In the case of fracking, those most vulnerable to harms are those who live near the industrial operations at oil and gas pad sites. Over 15 million Americans now live within one mile of an oil or gas well (Gold and McGinty 2013). Pad sites emit unknown quantities of toxins into the air, as operators are not required to submit emissions inventories to state regulatory agencies. There are several sources of onsite air emissions (see Alvarez and Paranhos 2012). The fracking fluids, with trade secret ingredients, pose risks to ground and surface water (Lustgarten 2008). Other impositions are the trucks hauling sand, chemicals, and water, and the constant churning of diesel engines during drilling and hydraulic fracturing, which can last for months at a time.

With any proposed fracking operation, there is sure to be some harm involved. But will it just be a temporary nuisance, a minor health problem, a major health problem, or even an explosion? Like so much of policymaking, the actors involved have to muddle through on incomplete information. Things are uncertain. How much risk is there and how much is acceptable?

There are more or less reasonable answers to these questions. But there is no right answer to them in the way there are right answers to math questions. There is no expert with the "one best solution," as is presupposed with the utilitarian approach of tallying costs and benefits. The right balance depends on what you value, your tolerance for risks, and how you are situated in relation to the costs and benefits. What counts as the *right* decision depends on your point of view. Thus, the question that matters most is not "*what* is the right decision?" but "*who* should make the decision?"

Consider this question in light of the proactionary view of policy as a real-world experiment. Decisions about fracking pad sites are analogous to experiments on pharmaceuticals or other research trials involving human subjects. In both cases, the expected gains can only come about by subjecting people to potential harms. In the former case, it is those near the industrial sites. In the latter case, it is the research subjects who take the experimental drug.

There is a distinction in medicine between "therapy" and "research." The term *therapy* typically applies to practices that are intended to promote the health and well-being of the patient. *Research*, by contrast, is an activity designed to test a hypothesis and contribute to generalizable knowledge. When you are a patient of a therapy, the goal is to benefit you. When you are a participant in a research trial, the goal is to use you to benefit others. Medical researchers have concluded that someone can only be used in this way if they first give their informed consent to the experiment. Since this constitutes good science and we are thinking about policy as an extension of science, the same condition of informed consent should hold for anyone exposed to potential harms from fracking.

When it comes to fracking policies, there are two kinds of decisions, private and public. The private decisions pertain mostly to the leasing of mineral rights. The public decisions pertain to federal, state, and local regulations. We can consider each briefly to see what informed consent does, or should, look like.

When it comes to private, or market, decisions, those who own the minerals hold the power to dictate whether or not to conduct a real-world experiment at a given site. Those who do not own the minerals are disenfranchised. One way to approach the condition of informed consent, then, would be to ensure that those living on the surface near the site (and exposed to potential harm) own the mineral rights so that they are empowered to decide whether or not to accept the risk.

Unfortunately in places that sever the mineral and surface estates, this situation often does not exist. Further, even if someone owns minerals, their consent is not always necessary for a project to proceed. Thirty-nine states have some form of "compulsory integration" or "forced pooling" law that compels holdout mineral owners into joining lease agreements (Baca 2011). Even some staunch drilling advocates object to forced pooling as an illegitimate version of "private eminent domain." Yet, there are also practices that increase informed consent, such as the bargaining collectives that mineral owners sometimes form in order to garner more leverage in negotiating leasing terms with operators (Wilber 2012).

There are many other thorny issues at play with informed consent. For example, what counts as consent in situations where someone is in dire economic straits? In some depressed rural areas of Pennsylvania or North Dakota, people may "consent" to fracking even though they would never do so if they could afford to say no (see McGraw 2011). This is the same fuzzy border between consent and exploitation that occurs with questions of payments to entice people to participate in drug trials or even to sell their organs.

Another dilemma with fracking has to do with what it means to be informed. Especially in the early part of the boom, many mineral owners were signing leases with very little understanding of what they were getting themselves into. Landmen (employees of oil and gas companies responsible for securing mineral leases) are able to take up huge amounts of money that could be made while downplaying all the fine print in the leases, even though that fine print may amount to a near total takeover of one's property (Urbina and McGinty 2011). And across the country, major land developers are retaining the mineral rights and not informing prospective home buyers about existing wells and the probability of future fracking activity (Conlin and Grow 2013).

When it comes to public decisions, fracking is primarily regulated at the state level. As the forced pooling laws suggest, state regulatory agencies are mostly concerned about fostering and promoting the development of mineral resources. From the state's perspective of running a massive and complex technological system, the informed consent of mere amateurs who happen to live in proximity to it is irrelevant. It won't improve the functioning of the system any more than getting the consent of research subjects which will improve the validity of a study's conclusions. Indeed, in both cases, the requirement of informed consent can throw a major wrench in the works. Some of the Nazi experiments (for example, those on hypothermia) were sound science but they involved such pain that they would never have been run if people had the choice to opt out of them. If we really respect autonomy, then some technoscientific projects just won't happen.

When it comes to technology policy, achieving the informed consent of those made vulnerable to harms requires a different kind of public sphere with broader participation and representation of values. It needs a public sphere oriented not toward system functionality (getting the minerals out of the ground at a profit) but toward protection of goods like health, safety, beauty, and community integrity that might be sacrificed in the name of functionality. In the case of fracking, that public sphere is local town and municipal governments and its primary instruments are home rule authority, zoning powers, and drilling and production ordinances.

Local government is the voice of those living on the surface and made vulnerable to the harms caused by real-world experiments. This is why municipalities have become the most important flash point in the politics of fracking: they represent a different moral order, one that is rooted in place and community rather than the engineering and economic logics of commodity production (Negro 2012; Briggle 2013a; Briggle 2013b).

Monitoring

"Informed consent" does not appear at all in More's version of the proactionary principle. Rather, he lowers the bar and writes that "the Proactionary Principle allows for handling mixed effects through compensation and remediation" (More 2005; see also Briggle 2014a). In order for compensation to be an ethical response to harms caused by technologies, there needs to be a monitoring system in place capable of producing independent and reliable evidence of those harms. In the absence of such a system, corporations can appeal to uncertainties about harms in order to ward off financial liability and further regulations.

Those claiming that they ought to be compensated can always be charged with the questionable-cause fallacy. Fracking companies can argue that rash, nosebleed, chemicals in the water, drop in home value, or earthquake merely correlate with their activities but are not *caused by* their activities. To establish causation, and thus justify compensation and/or increased regulation, not only would evidence need to be collected but it would have to meet or exceed established data quality standards (see Mitka 2012). Corporations with vested interests in status quo policies can seek to raise the standards such that any evidence collected would be deemed insufficient or too biased to justify action (see Sargent 2008). (Of course, this tactic can also be used in precautionary policy settings where opponents of a proposed technology can delay action indefinitely.) Corporations can also perpetuate doubt about any accusations of harm by discrediting science and hiding information (see Oreskes and Conway 2012).

In his short film "The Sky is Pink," anti-fracking filmmaker Josh Fox argues that the oil and gas industry has sustained a campaign of obstruction and misinformation to conceal facts about the risks fracking poses to ground water (Fox 2012). This campaign is made more effective by the large sums of corporate money infused into politics (see CREW 2013). Some have argued that pressures to stifle evidence of harms from fracking have grown to the point where federal studies about water contamination are being abandoned (Lustgarten 2013).

But determining whether fracking has caused harms is not as simple as identifying the color of the sky. The complex systems under study can generate such large and diverse bodies of knowledge that any interest group can find a set of scientifically legitimated facts to support their position. Some disagreements may result from one side distorting or silencing the truth. But other disagreements may be the result of a lack of coherence among competing, and equally valid, scientific understandings (see Sarewitz 2000; Sarewitz 2004). In such cases, everyone can claim to have the facts on their side, and the more facts accumulated, the greater the ensuing gridlock. The result is the scientization of policy as the adjudication of values disputes is warped into a debate supposedly just about facts.

Even with the proviso in mind that there can be such a thing as too much knowledge, monitoring is important to any real-world experiment. Indeed, it hardly counts as an experiment, in the scientific sense, if it just entails a new activity without any data collection. But how much monitoring is needed? Here too, reasonable people can disagree and such disagreements are reflected in various policies. For example, there are dozens of municipal ordinances on the Barnett Shale in Texas regulating gas drilling and production. Some cities, like Flower Mound and Southlake, require air and water monitoring. Other cities do not have any monitoring requirements and rely on state regulatory agencies to conduct this work. State-level monitoring tends to happen on an ad hoc basis in response to citizen odor complaints. But some states do also operate a limited number of permanent air monitors (Rawlins 2014). The federal government also plays some role in monitoring, especially methane. As part of its wider deregulatory agenda, the administration of US President Donald Trump has sought to relax monitoring requirements for methane leaks and oil and gas facilities (see Popovich et al. 2020).

Other important policy questions pertain to the kinds of monitoring technologies that can be employed. Different technologies can detect different amounts, types, and concentrations of chemicals. Air monitoring technologies, for example, range from fairly standard summa canisters (that employ vacuums to grab a sample of air as it passes by) to real-time mobile laboratories with a set of equipment capable of detecting more chemicals across a wider spatial range and coupling this information with air dispersion models. Even the more basic monitoring equipment can be expensive (a 30-second summa canister air grab, for example, can cost nearly $400), which raises the question of who should pay. It is not uncommon for concerned neighbors near fracking sites to pool their own money to purchase monitoring equipment from non-governmental monitoring organizations such as ShaleTest. Even if they find data to suggest a violation, it is difficult to overcome charges that data collected by "amateur scientists" are flawed.

In the case of water monitoring, data collection is often made difficult by the industry's use of trade secret chemicals (see Elgin et al. 2012). Disclosure requirements vary, but most states allow companies to claim that certain proprietary ingredients used in their fracking fluids do not need to be disclosed to regulators or the public. Further, although several parts of the federal Superfund, Emergency Planning, and Community Right to Know Programs apply to the oil and gas industry, they are not subject to the requirements of the Toxic Release Inventory (sec. 313 see also sec. 372). The industry can also make a claim of trade secrecy and replace the specific chemical identity with "the generic class or category of the chemical claimed as trade secret" on the Material Safety Data Sheets used by state emergency planners and local fire departments (sec. 350.5). A nation-wide chemical disclosure registry named "Frac Focus" has been established, which provides well-specific information about the chemicals used. However, the use of this site by operators remains largely voluntary and even when a well is listed, some of the chemical constituents still remain undisclosed and labeled simply "proprietary."

This issue is related to the so-called "Halliburton Loophole" in the 2005 Energy Policy Act, which states that the underground injection of fluids and propping agents associated with hydraulic fracturing are not subject to the Safe Drinking Water Act. Drilling waste has been labeled "non-hazardous" and is thus permitted to be pumped down Class 2 wells, which are not subject to the same standards and oversight as Class 1 wells that handle hazardous waste (Lustgarten 2012).

Another obstacle in the collection of evidence of harms is the frequent use of legal non-disclosure agreements. Landowners who claim that fracking operators polluted their water, for example, often reach settlements that involve money or property buyouts for the landowners in exchange for their silence about the case. This strategy keeps data from policymakers, scientists, and the public, which helps to perpetuate doubt about any claims regarding water pollution from fracking.

Renovation

The third condition for an ethical real-world experiment requires that the ongoing innovation be amenable to renovation. In other words, as monitoring and experience produce evidence of problems, the system must be flexible or adaptive enough to accommodate the necessary changes. Of course, the same basic challenge exists here insofar as those profiting from the system are empowered to maintain status quo and ward off renovations. In this section, I first provide some concrete examples of fracking renovations and then discuss three kinds of barriers to renovation: scientific, legal, and material.

We could consider the historical development of fracking as itself one long sequence of renovations. But it is more accurate to reserve this term for changes to technological systems designed to account (at least in part) for unintended problems rather than (solely) to make improvements to intended functionality. Two examples of renovation through regulation are the addition of sound barriers around fracking sites near homes to mitigate noise and increased setback distances between fracking sites and homes or parks to mitigate nuisances more generally.

Many other renovations take advantage of "best available technologies." The EPA Natural Gas STAR Program is a voluntary partnership that encourages companies to adopt technologies that reduce air emissions, especially of methane. Examples include reduced emissions (or "green") completions, low-bleed pneumatic valves, composite wraps for pipelines, vapor recovery units for tanks, and a variety of improvements to operational and maintenance procedures. Other renovations pertain to water, including closed-loop systems that can reduce the need for open pits and water recycling technologies that can reduce overall water consumption.

Of course, the ideal renovation is a win–win in terms of both improving system functionality and mitigating unintended harms. This is the case, at least in theory, with many methane emission reduction technologies and similar efficiency measures that can save money and reduce impacts. In such "tech-fix" cases, scientific controversies may be less of a roadblock to action, because both sides will favor the action and thus are less likely to engage in battles of dueling experts or the politics of uncertainty. In the case of setback distances, however, there will be no such détente, because every extra foot of distance means less ground for oil and gas companies to exploit. That's why defining a safe distance from a fracking site is one of the most controversial issues for cities and towns (Fry 2013).

This highlights a more general question about whether enough of these kinds of renovations can make fracking safe or acceptable. Here too, the answer will depend on different perspectives. For some, the practice is intrinsically dangerous or wrong, such that "safe fracking" amounts to an oxymoron (Steingraber 2012), whereas others see ways to "frack responsibly" (IEA 2012; Pickens 2013).

The first barrier to renovation continues the conversation from the preceding section, because it pertains to the challenges of establishing sufficient evidence to justify a need for renovation. For example, in the Dallas-Fort Worth metroplex on the Barnett Shale, there has been an ongoing controversy about the impacts of fracking on health, air pollution (especially ozone), and surrounding property values. For every study, there seems to be a counter-study, which undermines the case for renovation and prolongs status quo policies (see Briggle 2014b).

The second barrier to renovation is anchored in the legal doctrine of vested rights. In Texas, state law says that a regulatory agency shall consider the approval of a permit application on the basis of the rules in effect at the time the original application for a permit is filed or when a plan for development or plat application was filed. The rationale is that it would be unfair for regulators to change the rules on a business in the middle of their project (unless the rules were to fix immediate threats to safety, such as when it becomes apparent that the existing electrical or fire code is flawed).

In the case of fracking, the industry uses this doctrine to ward off renovations. For example, the City of Denton, Texas passed a new ordinance that increased its original setback distance of 500 feet to 1,200 feet from a fracking pad site to a neighboring-protected use, like a home, school, or park. But EagleRidge Energy was allowed to frack two new gas wells only 720 feet from a park, because they claimed that those wells were part of an existing project that is vested under the older setback distance requirements. This was a major hurdle in Denton's ordinance revision process as it came to learn that its new rules – based on experience and information gleaned from hosting 280 gas wells – would not apply to any of the existing pad sites, including "new" wells drilled on those sites (Briggle 2014c). The vested rights issue was a major reason why the citizens of Denton voted to ban hydraulic fracturing in November 2014, a ban that was subsequently overturned by the Texas state legislature.

The third significant barrier to renovation stems from the material infrastructure itself and the economic dependencies that build up around it. Natural gas is often touted as a "bridge fuel" (Moniz et al. 2011) to an alternative energy future, but each additional gas well and pipeline represents further entrenchment and path dependency along the trajectory defined by the existing system. More infrastructure creates more rigidity and more challenges for changing course. This, in turn, marginalizes renovations to the status of changes *within* the system, closing off possibilities for changes *to* the system.

By way of summary, the proactionary principle offers an account of the ethics of innovation that casts the controversies over fracking in new light. It pivots a discussion about costs and benefits into one about who has the power to decide and which knowledge is legitimate. In opposition to the precautionary principle, it invites the kind of risk-seeking mentality that predominates our high-tech society, but it tempers this attitude with requirements to consider the vulnerable, monitor activity, and adjust course when problems arise. In this way, the proactionary principle is an alternative to both of the main voices in the fracking debate, namely, a naysaying precautionary stance on the one hand and a ruthless market utilitarianism on the other.

Unfortunately in the US, the latter extreme has come to dominate the politics of fracking. A principled proactionary approach has been eclipsed under the Trump administration's "Energy Dominance" agenda, which selectively focuses on just the positive impacts of fracking and discounts climate change almost entirely (see Briggle and Sherrod 2019). In the future – indeed if we are to have a future – the US and all nations will need to make a more sober and responsible assessment of the gamble that is fracking.

Important Federal Studies Related to Fracking

US Department of Energy (DOE). 2007. "DOE's Unconventional Gas Research Programs 1976– 1995: An Archive of Important Results." http://www.netl.doe.gov/kmd/cds/disk7/disk2/ Final%20Report.pdf.

US Department of Energy (DOE). 2009. "Modern Shale Gas Development in the United States: A Primer." http://www.netl.doe.gov/technologies/oil-gas/publications/EPreports/Shale_ Gas_Primer_2009.pdf.

US Energy Information Administration (EIA). 1993. "Drilling Sideways: A Review of Horizontal Well Technology and Its Domestic Application." http://www.eia.gov/pub/oil_gas/natural_ gas/analysis_publications/drilling_sideways_well_technology/pdf/tr0565.pdf.

US Energy Information Administration (EIA). 2012. *Annual Energy Outlook 2012: With Projections to 2035.* Washington, DC: Author. http://www.eia.gov/forecasts/aeo/pdf/0383(2012).pdf.

US Energy Information Administration (EIA). 2013. *Annual Energy Outlook 2013: With Projections to 2040.* Washington, DC: Author. http://www.eia.gov/forecasts/aeo/pdf/0383(2013).pdf.

US Energy Information Administration (EIA). 2018. *Short-Term Energy Outlook.* Washington, DC: Author. https://www.eia.gov/todayinenergy/detail.php?id=37053.

US Environmental Protection Agency (EPA). 2004. "Evaluation of Impacts to Underground Sources of Drinking Water by Hydraulic Fracturing of Coalbed Methane Reservoirs: Final Report." http://water.epa.gov/type/groundwater/uic/class2/hydraulicfracturing/wells_ coalbedmethanestudy.cfm.

US Environmental Protection Agency (EPA). 2010. "Hydraulic Fracturing Research Study." http://www.epa.gov/safewater/uic/pdfs/hfresearchstudyfs.pdf.

US Environmental Protection Agency (EPA). 2011. "Investigation of Ground Water Contamination Near Pavillion, Wyoming." http://www2.epa.gov/sites/production/files/documents/ EPA_ReportOnPavillion_Dec-8-2011.pdf.

US Environmental Protection Agency (EPA). 2012. "The Potential Impacts of Hydraulic Fracturing on Drinking Water Resources: Progress Report." http://www2.epa.gov/sites/production/files/documents/hf-report20121214.pdf.

US Environmental Protection Agency (EPA). 2013. "Inventory of U.S. Greenhouse Gas Emissions and Sinks: 1990–2011." http://www.epa.gov/climatechange/Downloads/ghgemissions/US-GHG-Inventory-2013-Main-Text.pdf.

Related Topics

13. Property
14. Water Quality and Availability

27. Mining
30. Renewable Energy
49. Pollution and Polluter Pays
56. Cost–Benefit Analysis

References

Alvarez, R. A. and Paranhos, E. (2012) 'Air pollution issues associated with natural gas and oil operations', *EM (Air and Waste Management Association Magazine)*, June, pp. 22–25.

Baca, M. (2011) 'Forced pooling: When landowners can't say no to drilling', *ProPublica* [Online]. Available at: http://www.propublica.org/article/forced-pooling-when-landowners-cant-say-no-to-drilling

Briggle, A. (2013a) 'Tilting at gas wells', *Truthout* [Online], 27 November. Available at: http://truth-out.org/opinion/item/20183-tilting-at-gas-wells-whats-the-best-way-to-defend-your-community-from-fracking.

Briggle, A. (2013b) 'Fracking? not in my back yard (or yours)', *The Conversation* [Online], 23 April. Available at: http://theconversation.com/fracking-not-in-my-back-yard-or-yours-13185.

Briggle, A. (2014a) 'Proaction and the politics of fracking', *The Guardian* [Online], 28 January. Available at: http://www.theguardian.com/science/political-science/2014/jan/28/proaction-policy-fracking.

Briggle, A. (2014b) 'Nature or neoliberalism? Two views on science and the persistence of environmental controversies', *Interdisciplinary Environmental Review*. Forthcoming.

Briggle, A. (2014c) *How Denton got fracked: The story of a city on the shale* [Online]. Available at: http://www.youtube.com/watch?v=tPcjIto7c_o.

Briggle, A. (2015) *A field philosopher's guide to fracking*. New York: Liveright.

Briggle, Adam and Sherrod, Callie. (2019) 'Postmodern prometheus: A discourse analysis of energy dominance', *Sustainable Communities Review*, 12(1), pp. 50–69. Available at: http://scrjournal.org/SCR%20Spring%202019/Briggle_Sherrod.pdf.

Burwen, J. and Flegal, J. (2013) 'Unconventional gas exploration and production', *American Energy Innovation Council* [Online]. Available at: http://americanenergyinnovation.org/wp-content/uploads/2013/03/Case-Unconventional-Gas.pdf.

Citizens for Responsibility and Ethics in Washington (CREW) (2013) *Natural cash: Fracking industry contributions to Congress*. Available at: http://www.citizensforethics.org/pages/natural-cash-fracking-industry-contributions-to-congress.

Clark, C. E., Burnham, A. J., Harto, C. B., and Horner, R. M. (2012) 'The technology and policy of hydraulic fracturing and the potential environmental impacts of shale gas development', *Environmental Practice*, 14(4), pp. 249–261.

Colborn, T., Kwiatkowski, C., Shultz, K., and Bachran, M. (2011) 'Natural gas operations from a public health perspective', *Human and Ecological Risk Assessment*, 17(5), pp. 1039–1056.

Conlin, M. and Grow, B. (2013) 'Special report: U.S. Builders hoard mineral rights under new homes', *Reuters* [Online], 9 October. Available at: http://www.reuters.com/article/2013/10/09/us-usa-fracking-rights-specialreport-idUSBRE9980AZ20131009.

Elgin, B., Haas, B., and Kuntz, P. (2012) 'Fracking secrets by the thousands keep U.S. clueless on wells', *Bloomberg* [Online], 30 November. Available at: http://www.bloomberg.com/news/2012-11-30/frack-secrets-by-thousands-keep-u-s-clueless-on-wells.html.

Ellul, J. (1963) 'The technological order', excerpted in Mitcham, C. and Mackey, R. (eds.) (1983). *Philosophy and Technology*. New York: The Free Press.

Environmental Protection Agency. (2010). "Hydraulic Fracturing Research Study," *Science in Action*, Office of Research and Development, June.

Freyman, M. and Salmon, R. (2013) 'Hydraulic fracturing and water stress: Growing competitive pressures for water', *Ceres* [Online]. Available at: http://www.ceres.org/resources/reports/hydraulic-fracturing-water-stress-growing-competitive-pressures-for-water.

Fox, J. (2012) *The sky is pink* [Online]. Available at: http://vimeo.com/44367635.

Fry, M. (2013) 'Urban gas drilling and distance ordinances in the Texas Barnett Shale', *Energy Policy*, 62, pp. 79–89

Gold, R. and McGinty, T. (2013) 'Energy boom puts wells in America's backyards', *Wall Street Journal*, 25 October [Online]. Available at: http://stream.wsj.com/story/latest-headlines/SS-2-63399/SS-2-365197/.

Holbrook, J. B. and Briggle, A. (2014) 'Knowledge kills action: Why principles should play a limited role in policy formation', *Journal of Responsible Innovation*, 1(1). Forthcoming.

Howarth, R. W., Ingraffea, A., and Engelder, T. (2011) 'Should fracking stop?', *Nature*, 477(7364), pp. 271–275.

Howarth, R. W., Santoro, R., and Ingraffea, A. (2011) 'Methane and the greenhouse-gas footprint of natural gas from shale formations', *Climatic Change*, 106(4), pp. 679–690.

IHS Global Insight. (2011) *The economic and employment contributions of shale gas in the United States* [Online]. Available at: http://www.ihs.com/info/ecc/a/shale-gas-jobs-report.aspx.

International Energy Agency (IEA). (2012) *Golden rules for a golden age of gas - World Energy Outlook special report* [Online]. Available at: http://www.worldenergyoutlook.org/goldenrules/.

Kaplan, T. (2014) 'Citing health risks, Cuomo bans fracking in New York State', *New York Times*, 17 December [Online]. Available at: http://www.nytimes.com/2014/12/18/nyregion/cuomo-to-ban-fracking-in-new-york-state-citing-health-risks.html?_r=0.

King, G. E. (2012) *Hydraulic fracturing 101: What every representative, environmentalist, regulator, reporter, investor, university researcher, neighbor, and engineer should know about estimating frac risk and improving frac performance in unconventional oil and gas wells*, paper presentation, Society of Petroleum Engineers' Hydraulic Fracturing Technology Conference, The Woodlands, Texas, delivered 6–8 February 2012. Available at: http://fracfocus.org/sites/default/files/publications/hydraulic_fracturing_101.pdf.

Klare, M. (2009) 'The era of xtreme energy: Life after the age of oil', *Huffington Post*, 22 September [Online]. Available at: http://www.huffingtonpost.com/michael-t-klare/the-era-of-xtreme-energy_b_295304.html.

Lomborg, B. (2012) 'A fracking good story', *Project Syndicate*, 13 September [Online]. Available at: http://www.project-syndicate.org/commentary/a-fracking-good-story-by-bj-rn-lomborg.

Lustgarten, A. (2008) 'Buried secrets: Is natural gas drilling endangering U.S. water supplies?' *ProPublica*, 13 November [Online]. Available at: http://www.propublica.org/article/buried-secrets-is-natural-gas-drilling-endangering-us-water-supplies-1113.

Lustgarten, A. (2012) 'The trillion-gallon loophole: Lax rules for drillers that inject pollutants into the earth', *ProPublica*, 20 September [Online]. Available at: http://www.propublica.org/article/trillion-gallon-loophole-lax-rules-for-drillers-that-inject-pollutants.

Lustgarten, A. (2013) 'EPA's abandoned Wyoming study one retreat of many', *ProPublica*, 3 July [Online]. Available at: http://www.propublica.org/article/epas-abandoned-wyoming-fracking-study-one-retreat-of-many.

Martin, M. and Schinzinger, R. (1989) *Ethics in engineering*, 2nd ed. New York: McGraw-Hill.

McGraw, S. (2011) *The end of country: Dispatches from the frack zone.* New York: Random House.

McKenzie, L. M., Witter, R. Z., Newman, L. S., and Adgate, J. L. (2012) 'Human health risk assessment of air emissions from development of unconventional natural gas resources', *Science of the Total Environment*, 424, pp. 79–87.

Mitka, M. (2012) 'Rigorous evidence slim from determining health risks from natural gas fracking', *Journal of the American Medical Association* [Online], 307(20), pp. 2135–2136. Available at: http://jama.jamanetwork.com/article.aspx?articleid=1167312.

Moniz, E. et al. (2011) *The future of natural gas: An interdisciplinary MIT study.* Available at: http://mitei.mit.edu/publications/reports-studies/future-natural-gas.

Montgomery, C. T. and Smith, M. B. (2010) 'Hydraulic fracturing: History of an enduring technology', *Journal of Petroleum Technology*, 62(12), pp. 26–32.

More, M. (2005) 'The proactionary principle', *Extropy Institute* [Online]. Available at: http://www.extropy.org/proactionaryprinciple.htm.

Negro, S. E. (2012) 'Fracking wars: Federal, state, and local conflicts over the regulation of natural gas activities', *Zoning and Planning Law Report*, 35(2), pp. 1–14.

Oreskes, N. and Conway, E. (2012) *Merchants of doubt.* New York: Bloomsbury Press.

Pickens, T. B. (2013) 'A free-market approach that works for energy', *Politico*, 13 March [Online]. Available at: http://www.politico.com/story/2013/03/a-free-market-approach-that-works-for-energy-88825.html.

Popovich, Nadja, Albeck-Ripka, Liva and Pierre-Louis, Kendra. (2020) 'The Trump administration is reversing nearly 100 environmental rules. Here's the full list', *New York Times*, 6 May. Available at: https://www.nytimes.com/interactive/2020/climate/trump-environment-rollbacks.html.

Rawlins, R. (2014) 'Planning for fracking on the Barnett Shale', *Virginia Environmental Law Journal*, 31, pp. 226–306.

Resources for the Future. Center for Energy Economics and Policy. (2012) *Risk matrix for shale gas development* [Online]. Available at: http://www.rff.org/centers/energy_economics_and_policy/Pages/Shale-Matrices.aspx.

Sarewitz, D. (2000) 'Science and environmental policy: An excess of objectivity', in Froderman, R. (Ed.), *Earth matters: The earth sciences, philosophy, and the claims of community.* Upper Saddle River, NJ: Prentice Hall, pp. 79–98.

Sarewitz, D. (2004) 'How science makes environmental controversies worse', *Environmental Science & Policy*, 7, pp. 385–403.

Sargent, R. (2008) 'Philosophy of science in the public interest: Useful knowledge and the common good', *Philosophy of Science Association 21st Biennial Meeting (Pittsburgh, PA) Contributed Papers* [Online]. Available at: http://philsci-archive.pitt.edu/4300/

Schmidt, C. W. (2011) 'Blind rush? Shale gas boom proceeds amid human health questions', *Environmental Health Perspectives*, 119(8), pp. a348–a353. doi:10.1289/ehp.119-a348.

Steingraber, S. (2012) 'Safe hydrofracking is the new jumbo shrimp', *Huffington Post*, 4 June [Online]. Available at: http://www.huffingtonpost.com/sandra-steingraber/safe-hydrofracking_b_1520574.html.

Steward, D. (2007) *The Barnett Shale play: Phoenix of the Fort Worth Basin: A history.* Fort Worth, TX: Fort Worth Geological Society.

Stirling, A. (2007) 'Risk, precaution and science: Towards a more constructive policy debate', *EMBO Reports*, 8(4), pp. 309–15.

Urbina, I. (2011) 'Insiders sound an alarm amid a natural gas rush', *The New York Times*, 25 June [Online]. Available at: http://www.nytimes.com/2011/06/26/us/26gas.html

Von Schomberg, R. (2012) 'The precautionary principle: Its use within hard and soft law', *European Journal of Risk Regulation*, 2, pp. 147–56.

Wang, Z. and Krupnick, A. (2013) 'A retrospective review of shale gas development in the United States', *Resources for the Future* [Online]. Available at: http://www.rff.org/RFF/documents/RFF-DP-13-12.pdf.

Wilber, T. (2012) *Under the surface: Fracking, fortunes, and the fate of the Marcellus Shale.* Ithaca, NY: Cornell University Press.

Yost, A. (1989) 'Session 2A: Eastern gas shales research', *DOE Conference Proceedings*. Available at: http://www.fischer-tropsch.org/DOE/_conf_proc/MISC/Conf-89_6103/doe_metc-89_6103-2A.pdf.

Zuckerman, G. (2013) *The frackers: The outrageous inside story of the new billionaire wildcatters.* New York: Portfolio.

32

ENERGY POVERTY

Robin Attfield

Introduction

Energy poverty affects multitudes of people in developing countries whose consumption of energy depends on polluting fuels or who have no access to a regular electricity supply. The phrase 'energy poverty' is also used in a different sense in parts of Europe to refer to fuel poverty, the situation of people whose income is too small for them to be able to afford to heat their homes adequately in cold weather; in the UK, 'fuel poverty' means that a household has to spend more than 10% of its income to heat their home. This is a significant and increasing problem of social justice, affecting millions of people, despite the ample resources of most of the countries where it is prevalent. But energy poverty in the distinctive sense employed in this chapter strikes at both the health and the life-prospects of billions of people (see below), and, if it is not remedied, threatens to affect yet greater numbers in parallel ways in coming generations.

The nature and scale of the problem of energy poverty begin to emerge when it is realised that out of a global population of between 7 and 8 billion people, 1.3 billion (over one sixth of the global total) lack access to electricity, and 2.7 billion (over one third of that total) rely on traditional biomass (wood or dung) for cooking fuel, sometimes at the expense of forests, and more often at the expense of their own health (Hildyard et al. 2012b: 21; see also International Energy Agency 2011: 7). One foreseeable response will be to regard this as a problem of rights and of inter-generational justice, and to consider ways in which access to electricity can be increased, without contributing to other global problems such as climate change. Putting increased national resources into renewable energy generation is likely to figure among proposed solutions, as is increased international aid to facilitate such changes. Indeed, there can be little doubt that policies such as these must be included among those needing to be adopted.

However, such problems should not be considered in isolation. Thus, problems of energy poverty should not be addressed in complete separation from ecological problems, such as the erosion of forests around human settlements, and from other problems of poverty such as food poverty, scarce supplies of fresh water, and inadequate resources for health-care and for education. The goal of halving poverty by 2015 was included among the internationally agreed Millennium Development Goals (United Nations 2000), and was partially attained, insofar as the threshold level of income was at issue, but was not attained with respect to the number of those suffering from hunger. (For the problems of attaining this goal, see United Nations 2012; Gallatsidas and Sheehy 2015.) The various global problems just specified often reinforce one another, and should ideally be tackled together. This is one of the themes of development ethicists such as David Crocker, who emphasise the importance of integrated policies of development, including enhanced local

DOI: 10.4324/9781315768090-38

participation and self-help (Crocker 1996, 2010). This integrated and participatory approach not only makes us step back from viewing forms of poverty such as energy poverty in isolation, but also bids us cease to regard the people of poor countries as passive recipients of aid and recognise them as potential contributors to the development of their own communities and countries.

Further, reflection on development brings to light the importance of whatever new systems are put into effect being sustainable (or capable of being maintained into the indefinite future). For unsustainable systems are likely to fail both the current generation and coming ones, and to generate problems at least as broad and deep as those of the present for our successors to address as if from scratch. The crucial nature of sustainability has been recognised at least since the Rio Summit of 1992 (Granberg-Michaelson 1992), although the process of agreeing policies to put it into practice widely lags behind this recognition, not least in the field of climate change mitigation. Thus, policies of energy generation that may be introduced to address energy poverty need to be sustainable, and so do the policies of development that would have to be introduced to address both this and other forms of poverty.

Justice, Biocentrism, and Energy Poverty

It is sometimes maintained that such policy proposals should be shown to be supported by a general theory of justice and a related theory of normative ethics. In the coming section, more is said about which areas of ethics are or are not centrally relevant to the problems of energy poverty. Beyond that, this is not the place to present, let alone defend, a general theory of justice, although I have done so elsewhere (Attfield 1995, 2012: 88–90, 126–129). This leaves readers free to relate the policies suggested here to their own preferred theory of justice, or alternatively to reflect anew on the shape of such a theory in the light of energy poverty and related problems.

As for a general theory of normative ethics, I have argued elsewhere for a form of practice-consequentialism called 'biocentric consequentialism' (Attfield 1999: 38–42, 157–162, 2003, 2012: 49–52, 81–89, 2014: 43–52, 2018: 44–60), and have sought to ensure that what is argued here is fully consistent with that theory. (Biocentrists hold that all living creatures warrant moral consideration, and that their flourishing has intrinsic value, while practice-consequentialists hold that it is the difference made to the world by social practices that make the actions that comply with them right or wrong. Biocentrists regard the impacts of our actions and practices on nonhumans as relevant to that difference, as well as the impacts on human well-being.) To some, this may be a surprise, given that biocentrists maintain that nonhuman interests need to be taken into account in ethical decision-making, although not all biocentrists treat these interests as having equal weight either with each other or with human interests (and I am one of those who assign different weight to the interests of creatures according to their different capacities). What may possibly occasion surprise is the centrality given in this essay to certain human interests (albeit by a biocentrist), and in particular the interest in not suffering from poverty.

Yet, there should be no surprise when biocentrists focus on certain human interests, such as the interest not to be subjected to poverty, in view of the range of capacities characteristic of human beings, and the range and degree of deprivation liable to arise when circumstances prevent those capacities being realised. Biocentrists can consistently focus on human deprivation just as readily as on the neglect or maltreatment of nonhuman creatures, and can also consistently give priority to human interests over those of (say) plants when greater interests are at stake (as arguably is the case when the human interest in cooking food and in heating one's home conflicts with the interest of trees in being allowed to flourish without being cut down for fuelwood). The adherence of biocentrists to a broader-than-usual theory of moral standing and of intrinsic value is no barrier to an undiminished concern to avert avoidable human suffering, both in the present and in the future. There are of course limits from a biocentrist perspective to the destruction of trees, habitats, and forests. But most people's ethical intuitions would incline them to adopt a closely similar view. Being a biocentrist, then, need neither involves misanthropic policies, nor distracts anyone from concern about energy poverty.

Poverty Present and Future

All this said, however, it should also be recognised that energy poverty is not only a problem of rights and of intra-generational justice, for, short of effective policies to counter it, energy poverty is likely to affect future generations as well as people currently alive. Those who regard ethical reflection on the future as a luxury or (as Aristotle once contended) an undue complication should take into account the likelihood that in the course of the next 200 years, the number of people prone to be subject to energy poverty is more than three times the number of people currently alive, unless global catastrophes extinguish all or most of humanity before that period has elapsed, while at least a corresponding number of people would, subject to the same proviso, be at risk in the 200 years following. Nor is the problem of future energy poverty confined to the numbers at risk, for the environment and the quality of life of whoever lives are also at stake, and are significantly dependent on current actions and policies as well.

Energy poverty is thus an issue not only of intra-generational justice but also one of justice between generations; if present-day energy poverty is not addressed, it will continue across the generations until action is taken at last to confront this cluster of problems. But that approach may itself not be the best way of understanding the issue of current responsibilities with regard to future people. For, as Derek Parfit has argued, the identities of most future people are not yet known, and are largely dependent on current actions and policies. We should not think of future individuals as bearers of rights (at least in the present) and capable of faring better or worse accordingly. Rather, the individuals who will live if one set of policies are adopted in the present would not live at all if another set of policies are adopted, and different individuals would live instead (Parfit 1983: 166–179, 1984).

Some philosophers respond to this realisation with the claim that accordingly we have no responsibilities with regard to future people at all, since there are no individuals to whom we could owe them. But Parfit persists in defending the widespread intuition that our responsibilities with regard to the future remain in place. They are not responsibilities owed to identifiable future others, but responsibilities with regard to whoever there will be, to prevent their quality of life being avoidably lower than it could have been. And since future energy poverty is one form that an avoidably low future quality of life could take, this is directly relevant to the responsibilities of many current agents. However, since such responsibilities differ considerably from ones concerned with the satisfaction of rights or related forms of justice, it is probably best to think of them as responsibilities grounded not so much in justice as in future well-being. To think of them as concerned with inter-generational justice can be misleading, implying as it appears to do that we know which people future generations will consist in, or at least how numerous they will be, and what they are owed accordingly. One reason why this is misleading is that the size of coming generations is in part dependent on current actions and policies, just as their quality of life is, and the environment in which they will live, as well as the identity of their constituent individual members. Rather than being an issue of rights or of inter-generational justice, the issue of future energy poverty should best be understood as an issue of the future level of well-being or the future scope of opportunities and quality of life that the present generation makes possible for our successors of the tracts of the future that we can affect. Even if future rights are not at issue, the well-being of countless future people whom we will never meet but can profoundly affect is at stake. These are people who will be aware of what we, their predecessors, either did or left undone.

There may be some whose response is that we, the current generation, have little or no reason to care about future generations, who, in Joseph Addison's famous words, 'have never done anything for us' (Addison 1714). But as Thomas Thompson, who apparently inclines to this view himself, comments, the issue of why we should care about future people is the same issue as that of why we should care about others, or be moral at all. In any case, as Douglas McLean adds, the value of future people's lives is one of the factors that makes life in the present worth living. Because future people are capable of continuing our pursuits, practices, and values, caring about them is constitutive of lives that we ourselves in the present can regard as well lived (McLean 1983: 180–197). (I have discussed this topic in greater detail elsewhere (Attfield 1999: 62–74, 2015: 65–77).) So rather than adopting the stance of the moral sceptic, we should take seriously the issues of future opportunities and quality of life, together with that of the quality of life of our contemporaries, and thus the impact on quality of life (present and future) of phenomena such as energy poverty.

Discounting Future Costs?

With this in mind, can we justify discounting future costs and benefits, as conventional economics encourages us to do? Standard defences of discounting, such as most people's preference for earlier rather than later benefits, clearly fail to justify such behaviour. (For example, using up all known resources of rare metals over the next two decades would not remotely be justified by our preference to enjoy the benefits in the present, thus depriving our successors of access to them.) There are, certainly, better defences, such as the uncertainty of future benefits and costs, defences that probably succeed in defending practices of selective discounting. There again, it is sometimes suggested that future people will find substitutes for resources, and that for this reason we cannot be expected to conserve existing resources for their sake. This defence too could justify selective discounting, in cases where there is evidence that substitutes are likely to be found (for example, through the progress of ongoing research into finding them).

But as the philosophers who contribute to McLean and Brown's collection show, these defences are no defence of across-the-board discounting at all. For some future costs are far from uncertain, just as some present ones are far from certain, and human beings are unlikely to find substitutes for resources of every kind; to discount future costs and benefits by a uniform discount-rate is thus unreasonable (Parfit 1983: 31–37). If, in accordance with much current practice, future costs are discounted annually at (say) 5%, then the costs of 30 years hence will be given negligible weight, and those of a century hence effectively none. But this would involve total disregard for (say) the foreseeable energy poverty of people of the twenty-second century, even though it probably could be significantly mitigated by present decisions and resulting policies.

Besides, while some probabilities are apparently so low that this is conclusive against taking the related costs seriously, such as the probability that any nameable individual will suffer from energy poverty in 2120, others are both more relevant and of a different order. Thus, the probability of energy poverty affecting the people of Asia, Africa, and Latin America in the twenty-second century (that is, whoever lives in those continents in that century) if policies are not put in place during this century to prevent it is almost certainly high enough to justify preventive action. So blanket, across-the-board discounting (which treats such costs as negligible on grounds such as uncertainty) should be rejected, and future costs treated on a par with present costs to the extent that their probability is comparable. (This, as it happens, is Parfit's own conclusion in his essay on discounting in the same collection (Parfit 1983: 33).)

Sustainable Solutions

The ethical issues thus give rise to questions such as how current agents can make a difference to the probable quality of life of the people of the coming centuries as well as that of contemporary people. These are not the only ethical issues, since the prospects of nonhuman creatures are also at stake, and related responsibilities need to be factored in when policies are being decided. But the interests of nonhuman creatures (other than animals farmed indoors) concern not an electricity supply, but the preservation of those ecosystems on which they depend. Energy policies, then, should be compatible with preserving existing forests, grasslands, wetlands, seas, and coral reefs, and preventing pollution of the land, the air, and the oceans. That, however, is compatible with a range of technologies devised to make provision for current and future human energy needs, particularly renewable technologies.

Indeed, it will be some form or forms of sustainable and renewable energy generation that is/are capable of reducing energy poverty both in the present and across the coming centuries, without generating additional problems. (Likely examples include hydro-electric schemes, and wider use of wave-power and of solar-energy technology.) Humanity must not continue with large-scale carbon-based energy generation, since it has already used up more than half of its carbon budget of a trillion tonnes if it is to avoid an anthropogenic increase of over 2° (Celsius) to average global temperatures (Meinhausen et al. 2009: 1158–1163; Shue 2011). Nuclear energy generation, although in some respects sustainable, generates new problems because of the absence of safe ways to dispose of waste products and decommission power stations. The successful attachment of processes of Carbon Capture and Storage to conventional carbon-based power stations could in theory make this form of energy generation sustainable, but this technology is nowhere in the world operative as yet, and carries large and numerous risks of geological damage and of leakage, if conducted on a sufficient scale (Hildyard et al. 2012b: 27). Accordingly, it is non-nuclear renewable energy to which resort will be needed, that is, solar, hydro-electric, wind, wave, tidal, or geothermal forms of energy generation.

Current Energy Poverty

When the two forms taken by energy poverty, lack of access to electricity and reliance for energy on polluting fuels, are considered, they are often found to go together. Energy consumption in the US and Canada turns out to be 20 times as high as in India and about 50 times as high as in the poorest countries of sub-Saharan Africa (Hildyard et al. 2012b: 24–25). But it is in sub-Saharan Africa and in India that dependence on fuelwood is most concentrated.

The gathering of fuelwood is often held to be a major cause of deforestation (Hildyard et al. 2012b: 24). Yet, most fuelwood comes from areas other than forests, such as scrub, bush fallow, and common lands, that easily regenerate, and damage to forests is not as widespread as is frequently assumed (Hildyard et al. 2012b: 24; Pacala and Socolow 2004: 968–972). While the damage to the lands and species of urban fringes should not be disregarded, the central problems arising from reliance on fuelwood probably lie elsewhere.

Fuelwood consumption is probably most harmful as a threat to human health through respiratory illness. Improved stoves are nowadays available, and their use has been taken up in towns in countries such as Ethiopia and Kenya. But many rural households have not adopted the improved stoves, despite their time-saving and energy-efficient properties, because of the cost of buying them ((Hildyard et al. 2012b: 24). It has also been found that women in India regard reducing respiratory illness as not as high a priority as other issues such as enhancing water supply and tackling sanitation problems, and this epitomises the importance of addressing all aspects of poverty, and not energy poverty alone (Hildyard et al. 2012b: 24). Yet since fuelwood consumption seems unlikely to be displaced for many years to come, this suggests that a higher priority should be given both by developing country governments and by aid agencies to publicising the advantages of improved stoves and to making them more affordable. In the longer term, energy policies should aim at the universal availability of cheap electricity generated from renewable sources. But more difference can be made to many people's quality of life for the coming decades through wider use of improved stoves.

At the same time, forest policy needs to be modified (e.g. by deliberately growing fuelwood as a crop) to facilitate access to fuelwood for those who continue to rely on it, without these people being obliged to damage forests. This is particularly important in Africa, where land privatisation is disrupting access to fuelwood and other forms of biomass. Improved access to fuelwood (together with access to improved stoves) can alleviate the problems of those dependent on fuelwood collection and use (Hildyard et al. 2012b: 24; Smil 2010: 108), and thus of energy poverty, as long as such policies do not distract attention from the long-term goal of improved electricity supplies available to all, which could avert the problems mentioned above, including the erosion of forests and the health problems attendant on use of biomass for heating and cooking.

It is difficult for the beneficiaries of a reliable electricity supply to grasp what the absence of such a supply means for the affected communities. Among other problems is the lack of adequate lighting after nightfall. This, in turn, spells danger for vulnerable people, including women, and also deprivation of the opportunity to read during nighttime, let alone to study. (I was once taken to see a large rural Nigerian town after dark, where the only sources of light were kerosene lamps and candles.) This deprivation means a large-scale loss of human creativity and potential, through the narrowing of options, together often with the other forms of poverty that often accompany energy poverty, such as unemployment, hunger, and high rates of morbidity. Accordingly, policies need to be formed to make the supply of electricity both comprehensive and reliable, using renewable sources that will not at the same time exacerbate climate change.

One of the forms of renewable energy generation is hydro-electric energy, readily available in Himalayan areas of India and in Nepal, a country where there are '6,000 or so rivers cascading down the Himalayas' (Hildyard et al. 2012b: 25). But there has been a tendency to concentrate on the construction of large dams there, partly because this suits the interests of the large dam industry and of international development agencies such as the World Bank. Outcomes include short spells of excess capacity, followed by several years of 'brownouts' due to shortages as demand for electricity increases. Opening up the energy sector in 1990 to small producers (often at the level of village co-operatives) and mini-hydro schemes led both to an increase in electricity supply and to large reductions in its cost, and thus to a major redistribution in access to the grid (Hildyard 2012b, 25–26). The previous scarcities turned out to have been artificial products of unwise energy policies. Nepal has experienced many other problems, but with demand for electricity likely to increase, energy policies will need to remain sensitive to local needs as well as the interests of government, companies, and international agencies.

In India, according to a BBC Radio 4 report (anonymous 2013a), large parts of the rural population, including some half of India's villages, remain without an electricity supply, and it is unlikely that the grid can be upgraded sufficiently to connect them to it within decades without unprecedented efforts being taken. In these circumstances, local electricity generation has in places been able to supply electricity, from sources such as solar and hydro-electric energy, independently of the national grid. In view of the difference this can make to human well-being, it is clear that India cannot wait for the grid to be made comprehensive, although that should remain a key national goal. That is what future generations are likely to need, but local, decentralised production is the best hope of many for this and several decades to come.

Local activism clearly has a place in making good the gaps in India's electricity supply. The same is likely to hold good of many other developing countries; for example, micro-generation on the part of a co-operative in the Mount Halimun area of West Java (Indonesia) is supplying electricity more cheaply than the national grid, and thus making it available to residents who would otherwise struggle to afford it (anonymous 2013b: 9). Such initiatives are going to need expertise, some of it from voluntary agencies, and credit from schemes such as micro-credit. Long-term solutions will need to be more comprehensive, but significant opportunities remain for local initiatives to show the way.

Energy Security and Energy for All

Policies for structural, long-term change are needed, through which renewable energy can be massively expanded, energy can as far as possible be made available to all by 2030, and much greater proportion of remaining fossil fuels can be kept in the ground. Nothing less that this would comprise a genuine global policy of energy security that deserved this name. These policies contrast markedly with many current policies of so-called 'energy security', designed to conserve existing structures and secure them for the world's richer consumers against all challenges and threats. (One of the first philosophers to recognise that developing countries need to increase the amount of electricity they generate, so as to satisfy the unsatisfied basic needs of their peoples, was Henry Shue. (See Shue 1995: 385–392.))

Making energy available to all will be expensive but remains affordable. As the International Energy Authority wrote in 2011: 'We estimate that, in 2009, around $9 billion was invested globally to provide first access to modern energy [sc. energy from sustainable forms of electricity rather than from biomass], but more than five times this amount, $48 billion, needs to be invested each year if universal access is to be achieved by 2030' (International Energy Agency 2011: 7). If, however, humanity fails to rise to this challenge, then the responsibilities of our successors with regard to making energy supply universal will be all the greater.

Increases to renewable energy generation will not be without their environmental impacts, which will on occasion outweigh the value of particular projects. Not even deserts, which are sometimes proposed as the sites of vast arrays of solar panels, are unwanted lands, and even desert ecosystems should where possible be preserved. Similarly, enhancements to energy efficiency (important as it would be to curtail overall electricity consumption) could involve the retrofitting of old buildings on a global scale, and even the redesign of whole cities, involving some widespread adverse impacts on urban landscapes and also on the ecosystems they support (Hildyard et al. 2012a: 29).

Likewise ecological problems associated with planting fuelwood as a crop, problems, such as the erosion of forests and threats to biodiversity, must not be forgotten or neglected. The same applies to the widespread forms of poverty that frequently accompany energy poverty such as food poverty, scarcity of fresh water, and insufficient resources for health-care and for education. Yet, tackling energy poverty remains crucial, and must remain a high priority for the national and international policy-makers of the present and the future.

Related Topics

14. Water Quality and Availability
23. Climate Justice and Equity
41. Sustainable Agriculture
63. Environmental Justice

References

Addison, Joseph. (1714). *The Spectator*, no. 583, 20 August
Anonymous. (2013a). BBC Radio 4 broadcast, Spring. (Source irretrievable)
Anonymous. (2013b). 'Micro-Hydro in West Java', *Action* (Magazine of the World Development Movement), Autumn
Attfield, Robin. (1995). *Value, Obligation and Meta-Ethics*, Amsterdam and Atlanta, GA: Rodopi
Attfield, Robin. (1999, 2015). 'The Ethics of Extinction', in *The Ethics of the Global Environment*, Edinburgh: Edinburgh University Press and West Lafayette, IN: Purdue University Press, pp. 62–74 (first edition); pp. 65–77 (second edition)

Attfield, Robin. (2003, 2014). *Environmental Ethics*, Cambridge: Polity and Malden, MA: Blackwell

Attfield, Robin. (2012). *Ethics: An Overview*, London and New York: Continuum/Bloomsbury

Attfield, Robin (2018). *Environmental Ethics: A Very Short Introduction*, Oxford: Oxford University Press

Crocker, David. (1996). 'Hunger, Capability and Development', in W. Aiken and H. LaFollette (eds), *World Hunger and Morality*, 2nd edn., Upper Saddle River, NJ: Prentice-Hall, pp. 211–230

Crocker, David. (2010). 'Hunger, Capability and Development', in Des Gasper and Asuncion Lera St. Clair (eds), *Development Ethics*, Farnham, UK and Burlington, VT: Ashgate, pp. 383–402

Gallatsidas, Achilleas and Finnbarr Sheehy (2015), 'What have the Millennium Goals Achieved? *The Guardian*, 6 July. https://www.theguardian.com/global-development/datablog/2015/jul/06/what-millennium-development-goals-achieved-mdgs

Granberg-Michaelson, Wesley. (1992). *Redeeming the Creation: The Rio Earth Summit: Challenges for the Churches*, Geneva: WCC Publications

Hildyard, N., Lohmann, L., and Sexton, S. (2012). *Energy Security: For Whom? For What?*, Sturminster Newton, UK: The Corner House

Hildyard, N., Lohmann, L., and Sexton, S. (2013). *Energy Alternatives: Surveying the Territory*, Sturminster Newton, UK: The Corner House

International Energy Agency. (2011).*World Energy Outlook 2011 (Executive Summary)*, Paris: IEA, http://www.iea.org/Textbase/npsum/weo2011sum.pdf.

McLean, Douglas. (1983). 'A Moral Requirement for Energy Policies', in Douglas McLean and Peter G. Brown (eds), *Energy and The Future*, Totowa, NJ: Rowman & Littlefield, pp. 180–197

Meinhausen, M., Meinhausen, N., Hare, W. et al. (2009). 'Greenhouse Gas Emission Targets for Limiting Global Warming to 2° C,' *Nature*, Vol. 458 (30 April), pp. 1158–1163

Pacala, S., and Socolow, R. (2004). 'Stabilization Wedges: Solving the Climate Problem for the Next 50 Years with Current Technologies', *Science*, Vol. 305, No. 5685, 13 August, pp. 968–972

Parfit, Derek. (1983). 'Energy Policy and the Further Future: The Identity Problem', in Douglas McLean and Peter G. Brown (eds), *Energy and the Future*, Totowa, NJ: Rowman & Littlefield, pp. 166–179

Parfit, Derek. (1984). *Reasons and Persons*, Oxford: Clarendon Press

Shue, Henry. (1995). 'Equity in an International Agreement on Climate Change', *Proceedings of IPCC* (sc. Inter-governmental Panel on Climate Change) *Workshop, Nairobi, 1994.* Nairobi: ICIPE Science Press, 1995, pp. 385–392

Shue, Henry. (2011). 'Climate Hope: Fair Shares in the Exit Strategy' (unpublished paper presented at 'Justice and Climate Change' Conference, University of Oxford, September 2011)

Smil, Vaclac. (2010). *Energy Transitions: History, Requirements, Prospects*, Santa Barbara, CA: Praeger

United Nations. (2000). *United Nations Millennium Development Goals*, http://www.un.org/millenniumgoals/

United Nations. (2012). *UN-Habitat and the Millennium Development Goals*, http://www.unhabitat.org/categories.asp?catid=312

PART VI

Cities

33

URBAN SUSTAINABILITY

Steven A. Moore and Meghan Kleon

Because more than half of us now live in "places" already classified as cities (Steiner 2008), and the United Nations predicts that the bulk of future population growth will be in cities (Dugger 2007), achieving environmental sustainability will require sustainable urbanism. Yet, discussion of *urban sustainability* in North America inevitably reveals the contested definition of the concept. For the purposes of this chapter, we define the term "sustainable" to mean *the ability of a system to reproduce itself without unjustly consuming external resources.* (The dominant definition of "sustainable development," authored by the World Council on Economic Development in 1987, is: "…development that meets the needs of the present without compromising the ability of future generations to meet their own needs." We accept that definition, but use the term "sustainable" in this chapter in the context of systems theory. We also recognize that the concept of "justice" is articulated differently and contested in all cultures. Without entering that debate, we will simply associate our use of the term with the utilitarian tradition most commonly used in the United States. In that context, "just" and "fair" are synonymous terms. And by the term "external" to the system, we mean any space or resource that is simultaneously claimed by another system or social group. However, this definition holds that cities need not be contiguous physical space with a firm boundary and an "ecological footprint" of 1:1 (Wackernagel and Rees 1996), but can be city-regions with extended, porous, and just trading relationships.)

Even on the basis of this abstract definition, many people will still perceive "sustainable urbanism" to be a contradiction in terms. For example, Rees and Wackernegel (1997: 307) hold that "no city or urban region can be sustainable on its own. 'Sustainable city' – at least as we presently define cities – is an oxymoron." They question, quite reasonably, how the freeways, skyscrapers, transmission lines, and sewers that dominate urban landscapes could possibly be part of sustainable nature. The assumption behind this common question is that the concept of sustainability describes a state of sustained *nature* that excludes, or at least forcefully restrains, human impact, and thus cities, on it. Yet, other scholars argue that cities present the greatest opportunity for both sustainable development and a sustainable nature. Answering the question of whether urban sustainability is possible, and whether cities can be a functional part of nature, is at the heart of this chapter. We begin in Part 1 by considering changing historical perspectives on the relationship between cities and nature; in Part 2, we consider alternative perspectives about ecology and human inhabitation; and in Part 3, we consider if a balance is possible between humans and nature before we conclude with a few strategies to move us forward.

DOI: 10.4324/9781315768090-40

Part 1: The Modern Nature/City Dualism

Divine Nature gave the fields, human art built the cities.

Marcus Terentius Varro, De Re Rustica (116 BC–27 BC)

Cities are the abyss of the human species.

Jean-Jacques Rousseau, Émile (1762)

All cities are mad: but the madness is gallant. All cities are beautiful: but the beauty is grim.

Christopher Morley, Where the Blue Begins (1922)

Cities: Anti-Nature, Unsustainable

Politicians, philosophers, and poets have given voice to a categorical division and dichotomy between the city and nature. For Terentius, the city is the domain of "art"; for Rousseau it is an "abyss"; and for Morley, the city embodies a "grim beauty." Although these three characterizations of cities are far from the same, they share at least one agreement—that cities are other than nature and they operate under the unnatural rules of human civilization.

Much of historic city planning and design theory was founded on the assumption that cities are unhealthy because they are set apart from, and even anti-nature. Van Den Berg et al. (2007) note that "pro-rural and anti-urban ideology gained additional influence during the 1800s when the devastating living conditions in cities in England during the industrial revolution provided the fuel for a mass social reform movement (Figure 33.1). This movement inspired a number of utopian urban visions... [that] sought some way to join the best of the city with the best of the countryside, and to address the worst of the city (the roots of the anti-urban bias) and the worst of the countryside (which rendered some romantic views of nature)" (Van Den Berg et al. 2007: 82). Among these were Ebenezer Howard's 1898 proposal for the Garden City, which sought to limit the growth of dense metropolises, transforming them into a "decentralized but closely interrelated network of garden cities" (Ward 2005: 2) all surrounded by green space. Radial garden cities, according to Howard, should be limited to 30,000 residents, with separate residential, industrial, and agricultural zones that would allow occupants a connection to nature. Another example is Daniel Burnham's City Beautiful movement, which originated with the Columbian Exposition of 1893 in Chicago. City Beautiful meant to reform the "unnatural" city through beautification, improving the living conditions and thus the civic virtue of residents.

A century later, in 1997, Charles Waldheim "coined the term 'landscape urbanism' to describe the practices of many designers for whom landscape had replaced architectural form as the primary medium of citymaking," advocating for the use of landscape urbanism as an "interstitial design discipline, operating in the spaces between buildings, infrastructural systems, and natural ecologies" (Waldheim 2006: 59). Although landscape urbanism is focused on the development of sustainable and integrated landscapes, the field still assumes a fundamental separation between city and nature.

Characterization of the city in opposition to nature has also dominated the history of the modern environmental movement in North America, in what David Owens calls the "anti-city ethos" of the environmental movement. Owens writes that "the hostility of many environmentalists toward densely populated cities is a manifestation of a much broader phenomenon, a deep antipathy to urban life which has been close to the heart of American environmentalism since the beginning" (Owens 2009: 18). In his discussion of "The Moral Journey of Environmentalism" (2010), Andrew Light summarizes the transformation of North American environmentalism

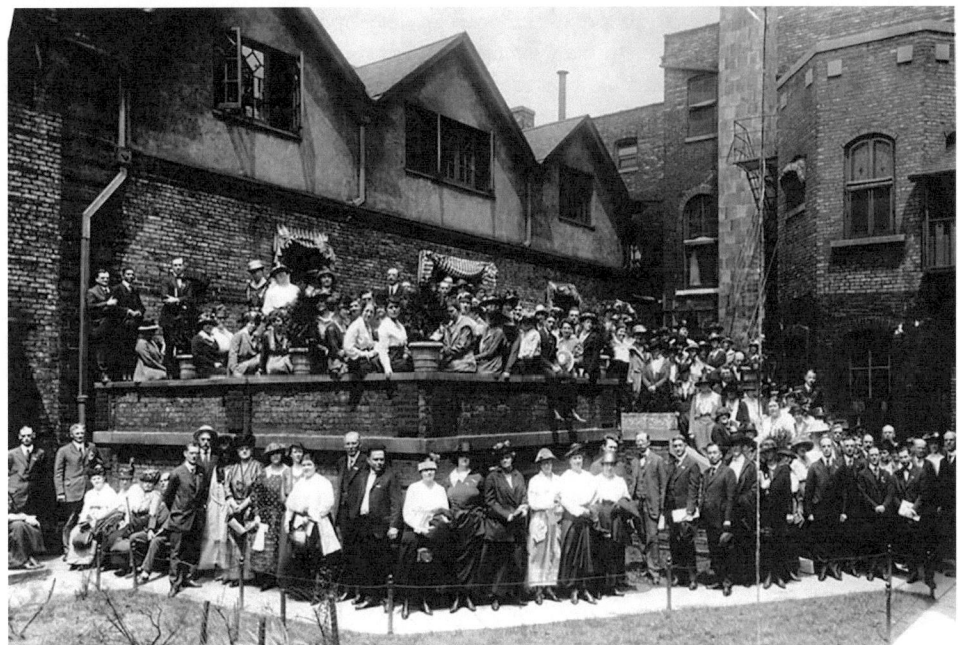

Figure 33.1 Changing Neighborhoods: Photographs of Social Reform from 7 Chicago Settlement Houses
Source: Courtesy, The University of Illinois at Chicago, JAMC 0000-0038-588.

through three stages of development, all of which have been dominated by a focus on preserving "wilderness, now understood as those vast untrammeled swaths of what is presumed to be original nature" (Light 2010: 140) while neglecting social and cultural concerns. Light argues that by idealizing wilderness and conceiving of it as external to human values and relationships, we have unintentionally emptied the landscape of information that makes nature relevant to our everyday lives—we have made nature distant and abstract and thus more, not less, vulnerable to the abstraction of modern economics.

Cities: Preserving Nature, Paragon of Sustainability

Instead of categorizing cities as "anti-nature" many scholars now understand the ability of cities to enhance social and cultural vibrancy as well as ecological functions. Tim Beatley articulates this view when he writes that: "in contrast to the historic opposition of things *urban* and things *natural*, cities are fundamentally embedded in a natural environment. They can, moreover, be reenvisioned to operate and function in natural ways – they can be restorative, renourishing, and replenishing of nature, and in short like natural ecosystems" (Beatley 2000: 197).

Some scholars take this reasoning further, arguing that cities are not only a part of the natural environment, but are the most inherently sustainable form of settlement. Douglas Farr writes in *Sustainable Urbanism: Design with Nature*, "the conventional view in America is to think of cities as the source of the pollution that is causing climate change. Indeed, per unit of land area, cities generate a great deal of pollution. However, on a per capita basis, city dwellers generate the least CO_2"; furthermore, "per capita environmental impacts, across the board, decrease with increasing density" (Farr 2008: 25–26). Thus, these scholars argue, even the most "sustainable" individual buildings are not truly sustainable if they are located outside of established urban centers. This position, which runs counter to proposals for exurban development, is also supported by activist organizations like the National Resource Defense Council (NRDC 2016). In their view, exurban development of so-called "sustainable houses" disrupts agriculture and natural ecosystems while also catalyzing the spread of infrastructures though which humans consume more energy and water than they do in cities.

Those in the field of historic preservation, or heritage conservation, make a similar argument, proposing that historic towns and city centers are "inherently sustainable." Building on Carl Elefante's assertion that "The Greenest Building Is... One that is Already Built" (2007), many historic preservation scholars and practitioners note that historic buildings and infrastructure have significant embodied energy, are durable and repairable, use indigenous materials, and were designed for passive survivability and to take advantage of local climatic conditions. Good examples of this claim would be New York's building stock of 19th-century warehouses that have now been transformed to new uses several times, or that city's stone aqueducts that still transport water. In addition, many historic urban areas and districts are walkable, dense, transit-oriented, and mixed-use—all desirable attributes of sustainable cities.

However, there is a danger in taking the stance that existing cities and buildings are "inherently" sustainable. Such an argument can create the impression that since urban centers are already sustainable, nothing further needs to be done to enhance human and environmental health within them, beyond incentivizing development in established urban centers. This position remains stuck in the city/nature dualism.

Part 2: Emergent Perspectives on the Relationship between Nature and City

There are at least two emergent views about the relationship of cities to nature that deserve attention: (1) cities as dynamically balanced ecosociotechnical systems, and (2) cities as inhabited infrastructure. We will consider each, in turn.

Cities as Dynamically Balanced Ecosociotechnical Systems

Andrew Light holds that the false dichotomy of nature and city might be countered by reframing "place" not on a scale between wilderness and urban, but as a "conception of location imbued with a storied relationship between people and the things around them" (Light 2010: 143). If, as Light suggests, the value of place is to be found in its importance to "valuers" like us, then sustainability efforts might be best focused on creating intentional communities of people dedicated to the places they inhabit, rather than on the "preservation of bits of nature on the periphery of civilization" (Light 2010: 146). Light's proposal takes us from modern assumptions about nature, in which humans are separate from nature, to the postmodern notion that humans, and the cities we create, are now and always have been a part of nature. If the city/nature dualism is, in fact socially constructed, hybrid options become possible.

To move beyond the city/nature dichotomy, rather than focusing on what cities *are*, as the authors cited at the beginning of this chapter did, we might more productively focus on what cities *do* (Leatherbarrow 2010). Instead of categorizing the city as inherently natural or unnatural, divine or grim, sustainable or unsustainable, it is more helpful to reframe the conversation and consider cities in an alternate way—as *dynamically balanced ecosociotechnical processes, or systems*, made up of integrated ecological, social, and technical forces in a state of constant change.

This possibility was considered early on by Sir Patrick Geddes in 1915 and further developed a generation later by Lewis Mumford in *Technics and Civilization* (1934) and elsewhere. Mumford considered not only the state of nature, but also how humans act on nature through technology. He classifies four overlapping eras, or "complexes" of technological change: the *eotechnic, the paleotechnic, the neotechnic,* and the *biotechnic.* In simple terms, these eras can be sequentially characterized as "a water and wood complex; … a coal and iron complex, … an electricity-and-alloy complex" and a hard-to-imagine biological complex (1934: 110). But Mumford (1934: 267) makes it clear that what distinguishes these eras from each other is not just the technologies with which we work on nature, but also the "institutional forms" that control and direct them. If the eotechnic era was "politically checkered" and a time "of deepening degradation of the industrial worker" it was also "brilliant" in the achievements made by skilled technical artisans (1934: 111). By contrast, the paleotechnic era was dominated by the institutions of "private capitalists" constituted by "the bankers, the business men, and the politicians" who generally served their own interests rather than those of the community. What Mumford envisioned in the possibility of a *neotechnic* era was a society in which technologies were "our servants, not our tyrant" (1934: 427). And lastly, the *biotechnic* is an idea that Mumford himself could only vaguely grasp, but one that we can now describe as a *dynamically balanced ecosociotechnical system,* in which the built environment is itself alive and productive.

The first *eotechnic* era, or "water and wood" complex, is nicely associated by Mumford with the "Goose-quill pen, sharpened by the user" (1934: 110). As opposed to the mass produced steel pen manufactured in the following *paleotechnic* era, the goose-quill is made by an individual craftsman, adapts to the *penmanship* of the user and is closely associated with agriculture. As satisfying as that intimate relationship of goose-to-maker-to-writer-to-ink-to-paper may sound, however, the quill pen also linked the writer to class-based politics and land-use practices that we now reject as not yet democratic or civilized, which takes us to the second option.

Cities as Inhabited Infrastructure

Mumford's characterization of the *paleotechnic* city would be familiar to anyone who experienced, say, Pittsburgh in the 1940s and 1950s. It was a dark and grimly beautiful place where drivers routinely turned on their headlights at midday to navigate through the smoke and ash disgorging from the mills. The three great rivers of that city were understood and used as the infrastructure required to move vast quantities of iron ore, coal, coke, and silica into the mills, as well as the steel and glass that came out from the other end. Likewise, the streets, electric lines, trains, and trolleys were understood and used as the infrastructure required to fuel the engines of capital, which, in turn, fueled private consumption not only in the mansions of Shady Side but also in the tenements of Homestead. Modern cities like Pittsburgh at that time were experienced as the public means of heavy infrastructure that would, in the end, enable private wealth accumulation.

Pittsburgh has, of course, since been transformed. And, like Pittsburgh, the infrastructural means of contemporary North American or European cities no longer disgorge smoke and ash because our infrastructure has become far more *efficient,* meaning that even with less input there is more output. The concept of "infrastructure" itself has, however, not changed very much—we still commonly understand infrastructure to be technological systems that enable the ends we desire. And it is the presence of infrastructure that distinguishes the city from nature. The term, "infrastructure" initially referred only to the military context of supplying material to the war front. But, by the early decades of the 20th century, the term also referred to "the roads, bridges, rail lines, and similar public works that are required for an *industrial economy,* or a portion of it, to function" [emphasis ours]. And by mid-century, the term was expanded beyond physical infrastructure to include "basic social services such as schools and hospitals" (AHD 1981). Most recently the term has expanded once again to include organizational capacity such as corporate branch offices or the franchises of fast food restaurants and the internet.

Although its definition has expanded to include social as well as physical systems, when we speak of infrastructure in the present we still refer to what is required for an *industrial economy,* or a portion of it, to function. The societal assumption implicit in this definition is that life- and environmental-diminishing means, like coal-fired electrical generators or ten-lane freeways, are required to enable the distant ends we imagine to be the "good life." We accept that *paleotechnic* infrastructures may have to be unpleasant, and far from ideal, but we also accept it as seemingly necessary means. We have been slow to recognize the fundamental means/ends conflict embedded in modernity.

The public infrastructure we build, however, is not just invisible or instrumental means to an end—infrastructure *itself* constitutes the ecosociotechnical systems in which we live. Being "stuck in traffic," for example, may not be the life to which we aspire, but it is a *kind of life*. Technology is not a disengaged mechanistic means, but is literally the places we in*habit*—the habits and tools we live, in the process of place-making. The width of residential streets in the United States, for example, is typically determined by technological codes specifying width equal the turning radius of a fire truck. The purpose of the code is to optimize homeowner protection. This instrumental logic may be effective in rare circumstance, but it also suppresses consideration of the everyday experience of children playing or pedestrians seeking shade or shelter from the wind, as well as the negative impact of increased stormwater volume on adjacent waterways.

Given this technocratic reality, how might we (re)conceive of the city as *inhabited infrastructure* rather than as a place fouled by means that might enable a better life lived elsewhere? Such a city would necessarily be (re)conceived as a living system in which interrelated ecological, social, and technological complexes are collectively (re)designed to positively, rather than negatively influence each other. If cities and their infrastructure are made up of arrangements of ecological, social, and technical forces, then it follows that a truly sustainable urban environment must account for not only the natural environment, or deal with technology and physical infrastructure, but must also be concerned with social sustainability (Oden 2016). Since all three factors, working in unison, constitute the infrastructure of the urban environment, all three must be addressed holistically to create an ecosociotechnical system that can function as a life-enhancing urban ecosystem. An existing example that almost meets this standard is the Boston *Fens*, sometimes referred to as *The Emerald Necklace,* design by Frederick Law Olmstead (our nation's first landscape architect) and completed in 1879 (Figure 33.2). The Fens is simultaneously a waterway system that manages storm drainage, a public transit system, a public park system, a wildlife corridor, and a sanitation system. In short, it is an interrelated *system of systems* that is also beautiful and much loved by the citizens of Boston. It is our best model for inhabited infrastructure.

Consider the ordinary house as a less successful example. We typically understand the modern house as a unit of consumption existing at the end of a long delivery chain of infrastructure. Houses are bought, sold, and taxed on the basis of size, or square footage. Larger floor area equates with more space to put, not only family members, but the things we buy—many of which consume energy in their manufacture and transport but also in their operation. Most residential utility bills itemize the units of water, electricity, and other fuels (e.g. gas) consumed on a monthly basis. Those rates of consumption drive the costs of municipal solid waste collection as well as associated costs assessed for wastewater and stormwater flows for each home. After the mid-1990s, increasing home sizes, and increasing costs and rates of energy and water consumption began to influence the market value of the home because discerning buyers preferred an "efficient" home (Porteous 1997). Rates of consumption can be both positively and negatively correlated with home value, but consumption remains the measure.

The modern infrastructure that supports the household as an individual unit of consumption is also itself at odds with sustainability. Electricity is delivered to our homes from distant power plants that consume large quantities of drinkable water and spew carbon into the atmosphere via

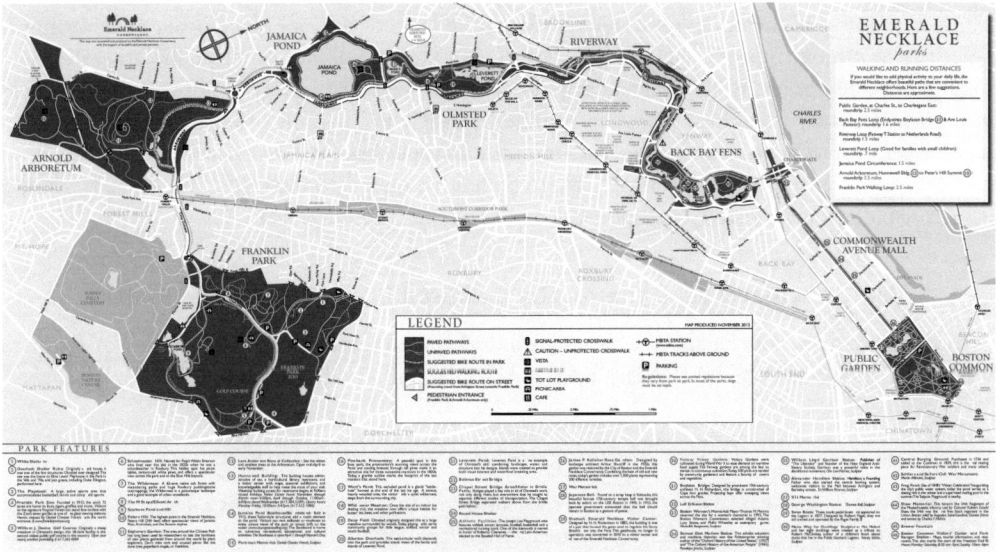

Figure 33.2   The Fenway and Simmons College, Boston, Mass. Designed by America's first landscape architect, Frederick Law Olmstead, "the Fens" is a historic example of an ecosociotechnical system

Source: Photo from Detroit Publishing Co/US Library of Congress, LC-D4-73371.

power lines that lose as much as 65% of the potential energy along the way. Water is delivered via pumps and pipes that require significant energy consumption to move it up and down hills. After home use, that water is then flushed to mechanical treatment sites that consume yet more energy to clean it up. And the unwanted packaging materials we bring home from shopping trips, or receive via Amazon, are subsequently trucked away to solid waste landfills that consume significant land area and expend even more energy to dispose of mass that might have been harvested. Although such unsustainable linear systems of production and consumption may have made economic sense in the era of cheap power when they were conceived, they no longer do today— nor do they make environmental or social sense. Environmentally, these infrastructural systems threaten air and water quality by externalizing costs to other sectors of the economy (e.g. public health). Socially, space that is ecologically degraded by infrastructures also becomes economically degraded, meaning cheap, and thus accessible to our most vulnerable citizens—the elderly and the poor—as places to live. Finally, living adjacent to the municipal land fill, a major highway, or the airport further degrades citizen health (Turnock 2009). This modern vision of single-use infrastructure requires us to question our acceptance of enabling means as well as the unsustainable end-products they produce and support.

"Sustainable urbanism" requires that we reconceive not only the ordinary house, but our office buildings, schools, hospitals, museums, and shops as inhabited infrastructure of both consumption *and* production. Each building or landscape could contribute, for example, by producing energy, sequestering rain water, filtering storm water before returning it to the aquifer below, or processing solids as biofuels. This is not the anarchist dream of the autonomous house, but an interconnected and sustainable urban ecology that is already being implemented in communities in Denmark and elsewhere in Northern Europe (Jensena 2008). A related concept, that of "eco-districts" within existing cities such as Malmo, Sweden (Fitzgerald and Lenhart 2015) has stimulated planners to integrate heating and water systems at the scale of the city block using novel technologies that not only improve the efficiency of those systems, but add other public amenities like small parks and "green walls" (Alexandri and Jones 2008) that insulate buildings from the cold in the winter, cool urban heat islands in summer, catalyze pollination and delight pedestrians. In such places, infrastructure is no longer demonized as unpleasant but necessary "anti-nature," and seen as a single-purpose conveyance system to be hidden underground or behind walls. Rather, it is sustainable infrastructure—multipurpose and life-enhancing in novel ways.

Part 3: Achieving a Dynamically Balanced Ecosociotechnical System

Achieving a dynamically balanced ecosociotechnical system requires that we consider (a) design as well as scientific thinking, (b) regulation, (c) public talk, and (d) the concept of "building culture."

Design Thinking and the Sustainable City

Mumford wrote nearly a century ago that the transition to a *biotechnic* city is a *design* problem. But by the term "design," he did not refer to the iconic works of great architects, or to the technological hubris of "high modernism" (Scott 1999), but to the highly social and political process of, as Herbert Simon (1969) put it, "transforming existing conditions into preferred ones."

Dwight Eisenhower is generally credited with saying that "the very idea of planning anything as dynamic and complex as a city is ridiculous. Yet, we should always be planning cities." Implicit in his insight is a humble, incremental, and pragmatic attitude toward the design of cities, rather than the overly rational and frequently hubristic attitude displayed by modern architects and planners. For example, le Corbusier in his 1930 proposal (Figure 33.3) for *Ville Radieuse* (the

Figure 33.3   Le Corbusier's *Plan Voisin* for Paris, 1922
Source: Photo by SiefkinDR, CC BY-SA 4.0. Also see: https://www.flickr.com/photos/27922834@N05/4936960279.

Radiant City), presents a utopian design for a city that would "completely supplant its predecessor," replacing any existing cityscape with standardized buildings and districts that adhered to a strict, abstract formal order and geometry that disregards human scale (Scott 1998: 104). That abstract order was roughly based on the human body and the assumption that rational spatial order could produce ration human order. This kind of architectural determinism was roundly rejected by Mumford, but did influence the planning of other cities such as Brasilia—the Brazilian capital designed by Lúcio Costa and Oscar Niemeyer in 1956.

Although the modern discipline of urban planning is crowded with theories, it is globally dominated by the concept of "rational planning," or the idea that "our political and social system had the capacity to plan comprehensively and rationally for the attainment of commonly held goals" (Rothblatt 1971: 26). As implied in this definition, application of the rational, or scientific method to solving urban problems requires: first, the collection of data which accurately describes existing conditions; and second, the use of deductive and inductive reasoning to construct valid hypotheses for action, based on the relationships reconstructed from these data. Traffic engineers, for example, use the principles of fluid dynamics to calculate the required number of traffic lanes from the most recent traffic count, plus an estimated amount to satisfy the current rate of growth.

The primary problem with this so-called "rational method" is that, as Eisenhower intuitively recognized, it depends on the *ceteris paribus* clause of the scientific method—that "all other things being equal or held constant," the calculation of this or that variable will be correct (Feenberg 2010). The problem is that modern urban conditions change so fast, any hypothesis constructed from data mined from the past will become instantly outdated. Another way to make Eisenhower's point is to argue, after landscape architect James Corner, that engineers, architects, and planners need a new paradigm of thought. Corner holds that it is "only through the imaginative reordering of the design disciplines and their objects of study might we have some potential traction on the formation of the contemporary city," and that integration between the city and nature will only be possible through a "transformation of the design disciplines themselves" (Waldheim 2006: 16). So, rather than giving up rational planning altogether as political conservatives are prone to do, an alternative is for highly transdisciplinary professional *and* citizens to conduct rapid, incremental interventions in the city that can be quickly assessed and modified as they unfold (Lerner 1996). This kind of design thinking is not rationally deductive and inductive as in the scientific method used by engineers and scientists, but *abductive*—a term coined by Charles Sanders Peirce (1839–1914).

By abduction, Peirce (1967: L75:286–287) meant that "familiar knowledge"—or, knowledge long-tested in the course of our lives in specific contexts—can anticipate surprising possibilities, or opportunities to achieve preferred conditions (Haraway 1995; Magnani 2001). In other words, rapidly unfolding social and environmental conditions may open up opportunities for action that lead to "preferred conditions," if we act humbly, collectively, quickly, and prudently, rather than allowing the opportunity presented by changing conditions to pass us by in pursuit of the perfectly rational five-year plan. Under this dynamic design approach, in which effective city design relies on abductive, as well as inductive and deductive reasoning, design must be a highly social process in which many people have "a controlling disposition to behave as if it were true" (Peirce 1967; 625:15). In this sense, a design proposal for collective action is a hypothesis that makes a tentative causal claim— "if we build in this particular way, in this particular place, we expect x, y, and z to occur." The effective social construction of a design hypothesis is, then, one that can catalyze collective action, or "behavior," and produce preferred social and environmental conditions in a "complex adaptive system" (Lanham et al. 2016).

For example, to return to Mumford's terms, a common *paleotechnic* response to periodic urban flooding is the building of ever-larger underground storm water sewers to divert ever-larger volumes of run-off from upstream development. A *neotechnic* or *biotechnic* approach is almost the

Figure 33.4 Bioswale Detail for Elmer Avenue Retrofit

Source: Courtesy, Landscapeperformance.org. Retrieved at: https://www.google.com/search?q=bioswale&sxsrf=ALe-Kk02fqTYQdnhKitdkHwbmmjIsHSneoQ:1594133244857&tbm=isch&source=iu&ictx=1&fir=Ed8W9hWPR8BUZ-M%252Cu25mNhpqAlfEZM%252C_&vet=1&usg=AI4_kSs1P7RMKHdQ_szrsA1UJlu9ykfw&sa=X&ved=2a-hUKEwjL36KOsbvqAhUlZjUKHRhgCdsQ_h0wAnoECAQQCA&biw=719&bih=438&dpr=3.16#imgrc=nySB-mAmmXkxIVM. 07 July 2020.

opposite. Rather than speed up flows of water to remove it more quickly by calculating the correct pipe size through the principles of fluid dynamics, *neotechnic* designers slow down water flows by constructing "bio-swales" and "rain-gardens" (Figure 33.4). These living infrastructure technologies allow water to be filtered and cooled as it returns to the ground. And when planted with native vegetation, they are not only beautiful, but provide other services (e.g. plant pollination) upon which a thriving urban ecology depends. In the context of traditional civil engineering, such living technologies are "surprising" and "foreign to the data"—they are the result of abductive reasoning by designers who gained the support of local communities to get *many* things, not a *single* thing done.

Eisenhower's proposal for continuous and incremental urban planning by interdisciplinary groups is consistent with Peirce's proposal for abductive reasoning (paired with deduction and induction) in city design. Both proposals lead us to a significant shift in how we think about buildings, cities, and places—not as anti-nature, but subsystems *of* nature. In order to put such logic into practice, we must establish how best to assess whether design interventions are achieving preferred social and environmental conditions. In the context of sustainable urbanism, this means considering how we can measure whether ecosystems, society, and technology are collectively moving toward an urbanism that can "sustain" itself—which we defined above as the ability to reproduce a system without unjustly appropriating resources external to it.

Assessing and Regulating Urban Sustainability

Moore and Wilson (2013) have documented the development of the standards, codes, and regulations that attempt to order urban development in the United States. Although it is now politically popular to decry how government has attempted to regulate every facet of American life, from the height of handrails to the wearing of seatbelts, Moore and Wilson's findings did not support such populist complaint. Although government does regulate some aspects of the built environment, they found that it is trade organizations, which comprised primarily corporate members, who establish the standards and codes that regulate most industrial production. In this unique arrangement forged by our Founding Fathers and their successors, it is corporate stockholders, rather than government or citizens, who get to vote on our technological choices and our urban infrastructure choices—it is their influence that decides, for instance, whether we should ride trolleys or buy a driverless car.

In the pursuit of sustainable urbanism, the United States government has only minimally established standards, such as those for "clean air" (through the 1970 Clean Air Act) and "clean water" (through the 1972 Clean Water Act). And because industry has generally demonstrated itself to be unreliable with regard to self-policing, or creating effective environmental standards for its members, non-governmental organizations (NGOs), or civil-society groups, have emerged to define what "sustainable urbanism" might look like. A plethora of organizations, including the US Green Building Council (USGBC), Green Globes, BREEAM, the Living Building Challenge, the Sustainable Sites Initiative (SSI), and many others have attempted to fill the vacuum in technological governance, each from its own perspective and using a variety of measurement tools. To date, however, most of the rating or sustainability certification tools in use can be described as encyclopedias of "best practices" determined inductively by expert groups, and their reliability is questionable (Murphy 2009). For example, the National Building Institute (NBI) conducted an audit of projects rated various shades of "green" by the US Green Building Council's LEED standard in 2008, on the basis of design-phase documents. They found post-occupancy measurement of actual conditions to vary significantly from design-phase computer models (NBI 2008). Only a few sustainability certification tools require post-occupancy evaluation, or POE (Preiser 2005), which require rigorous empirical third-party evaluation methods to determine if outcomes actually match design intentions. The Living Building Challenge, for example, will not issue its certification of any project until pre-construction energy models are validated by post-construction empirical data. In general, it is fair to say that the field of sustainable urban design is evolving rapidly as civil-society groups vie to dominate the field of regulation left nearly vacant by government. In other words, the concept of "sustainable urbanism" is now being *socially constructed*.

Public Talk and Building Culture

The biggest problem we find in the current situation is not the variety of sustainability rating or certification tools, or the pluralism (Guy and Moore 2005) of values embedded in the various sustainability measurement systems, but the *opacity* of those values. How do we know which technological standard is best, or most consistent with our own values and desires for a sustainable city? An even more basic question is against what *benchmark*, or what scientifically "preferred condition," is any proposal for place-making measured? One method to resolve this problem may be to give up the idea that some single, "context-independent standard" (Moore and Karvonen 2007) of sustainability can be determined scientifically and expressed as an abstract quantity of energy per square foot or a particular human density per square mile.

In lieu of seeking long-term scientific certainty, we might instead measure the gap, or the degree to which there is one, between the explicitly stated intentions of the "design team"—meaning the client and anyone who exerted agency in design decision making—and the "reception" (Holub 1984) of the community served by what is built. By making *context-dependent* design intentions explicit by plainly stating "we hope to accomplish x, y and z," and using those stated goals as a benchmark for assessment, both the intended and unintended consequences of design and technologies become *visible* and *measurable*. We can then use those pre- and post-occupancy measurements as empirical data to inform future design proposals, and to better achieve preferred social and environmental conditions. Such an approach to assessing urban sustainability could contribute significantly to the democratization and efficacy of standard- and technology-making.

We contend that urban sustainability will not emerge fully formed from the singular works of great architects, although those works may indeed open new possibilities for citizens to consider. Rather, as Davis (2006: 5) argues, urban sustainability can only be self-consciously constructed by those who constitute the "building culture"—"the coordinated system of knowledge, rules, procedures, and habits that surround the building process in any given place and time." That mobilization of agency will require a great deal of *public talk* and *social learning*. By the term "public talk" (Barber 1984), we refer to rational public debate over matters of common concern that determine our technological choices. Through this process, urban infrastructures will stem not from the seemingly scientific choices of experts, but the informed choices of citizens. And by "social learning" (Minteer 2002: 45), we refer to the process through which social groups transcend the limits of their individual knowledge by producing new and often unexpected proposals for action. Or, as Holden (2008: 3) would have it, social "learning [is] achieved through the practice of collective enquiry and public deliberation' over public goals"—in this case, goals concerning sustainable urbanism. We contend that urban sustainability must be, then, the product of a highly social process, not the dictate of experts in abstract and scientific terms.

The twin concepts of *public talk* and *social learning* requires that cities create for themselves spaces of "experimental thinking" (Dewey 1916: 334) where new technological norms, codes, designs, and regulations can be invented, tested, and proliferated on the basis of local empirical knowledge, conditions, and preferences. In the United States, a few cities have self-consciously created experimental municipal spaces that are working toward sustainable urbanism. One very good example is *The Oberlin Project* in Oberlin, Ohio, led by activist and scholar David Orr. The goal of that project is to "revitalize the local economy, eliminate carbon emissions, restore local

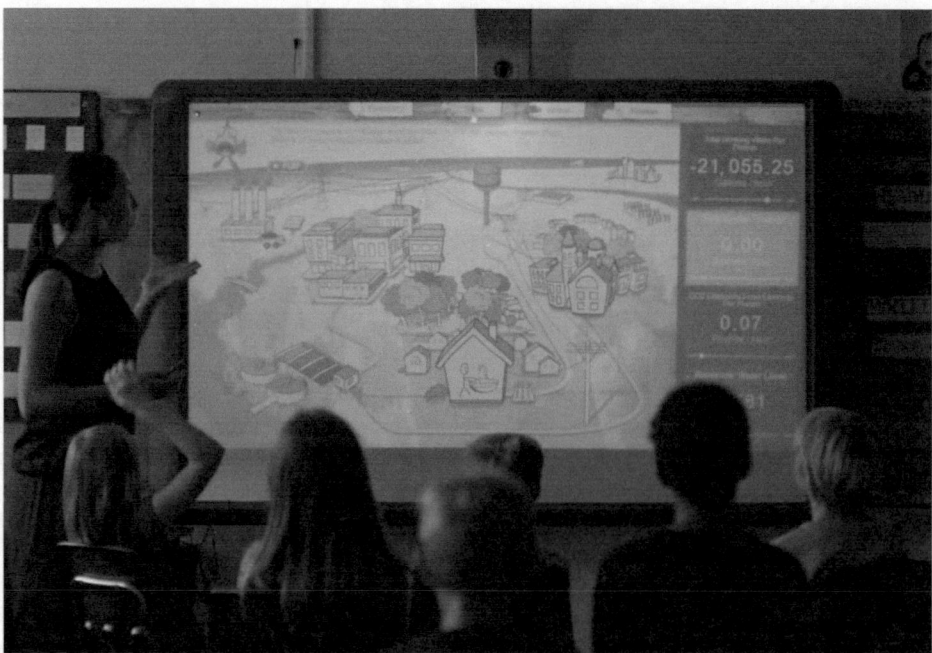

Figure 33.5 Oberlin, Ohio Primary School – Urban Sustainability Education via "Environmental Dashboard"

Source: Retrieved at: Oberlin Dashboard, March 23, 2022. https://environmentaldashboard.org/press-page.

agriculture, food supply and forestry, and create a new, sustainable base for economic and community development" (Orr 2014). The Oberlin Project (Figure 33.5) is slowly achieving its stated intentions through the work of a diverse community of learners, including the City of Oberlin, Oberlin College, and a consortium of business, educational, community, and institutional leaders that regularly engage in *public talk* in designated spaces, on and off the Oberlin College Campus, and disseminate *public learning* in an accessible on-line blog format (TOP 2016) to which citizens and experts of all stations contribute.

Conclusion

Boston's Fenway, the examples cited from Northern Europe, Scandinavia, and The Oberlin Project, are a few examples demonstrating that cities need not be experienced as antithetical to nature. In these cases citizens have moved beyond the modern city/nature dualism and reconceived of their cities as *dynamically balanced systems-of-systems, constituted of integrated ecological, social, and technological systems in a state of constant change, adaptation, and planning.* This characteristic may prove to be the definition of Mumford's *biotechnic* city in which highly democratic "institutional forms" both invent and respond to novel natures and technologies.

We have not, however, demonstrated that these exemplars are capable of "justly reproducing themselves" as we proposed in the introduction. That test must be left to others when there is sufficient evidence and knowledge to do so. For the moment, we can only suggest that (1) sustainable urbanism is our best opportunity to produce and reproduce ecosociotechnical systems justly, and (2) sustainable urbanism will emerge through design thinking, as applied to the hard work of public talk, social learning, technological experimentation, and the transformation of building cultures.

Related Topics

2. Eating
10. Wilderness
13. Property
23. Climate Justice and Equity
32. Energy Poverty
35. Suburbs and Exurbs
36. Transportation
41. Sustainable Agriculture
42. Community Gardens

References

Alexandri, E. and Jones, Phil. (2008). 'Temperature decreases in an urban canyon due to green walls and green roofs in diverse climates', *Building and Environment,* 43(4), pp. 480–493.
American Heritage Dictionary of the English Language. (1981). Edited by W. Morris. Boston, MA: Houghton Mifflin Company.
Barber, B. (1984). *Strong democracy: participatory politics for a new age.* Berkeley: University of California Press.
Carroon, J. (2010). *Sustainable preservation: greening existing buildings.* Hoboken: John Wiley & Sons.
Davis, H. (2006). *The culture of building.* New York: Oxford University Press.
Dewey, J. (1916). *Essays in experimental logic.* Chicago: University of Chicago.
Dugger, C. W. (2007). 'U.N. predicts urban population explosion', *The New York Times,* 28 June.
Elefante, C. (2012). 'The greenest building is ... one that is already built', *Forum Journal,* 27(1), pp. 62–72.
Farr, D. (2008). *Sustainable urbanism: urban design with nature.* Hoboken, NJ: Wiley.

Feenberg, A. (2010). 'Incommensurable paradigms: values and the environment', in S. A. Moore (ed.) *Pragmatic sustainability: theoretical and practical tools*. London: Routledge.

Fitzgerald, J. and Lenhard, J. (2015). 'Eco-districts: can they accelerate urban climate planning?" *Environment and Planning C: Politics and Space*, 34(2), pp. 364–380.

Guy, S. and Moore, S. A. (eds.) (2005). *Sustainable architectures: cultures and natures in Europe and North America*. New York and London: Routledge.

Haraway, D. (1995). 'Situated Knowledges: The Science Question in Feminism and the Privilege of Partial Perspective', in Feenberg, A. and Hannay, A. (ed.) *Technology and the Politics of Knowledge*, Bloomington, Indiana: Indiana Univeristy Press.

Holden, M. (2008). 'Social learning in planning: Seattle's sustainable development codebooks', *Progress in Planning*, 69, pp. 1–40.

Holub, R. C. (1984). *Reception theory: a critical introduction*. London: Methuen.

Jensena, J. O. K. G.-H. (2008). 'Ecological modernization of sustainable buildings: a Danish perspective', *Building Research & Information*, 36(2), pp. 146–158.

Lanham, H., Jordan, M., and McDaniel Jr., R. R. (2016). 'Sustainable development: complexity, balance and a critique of rational planning', in S. Moore (ed.) *Pragmatic sustainability: dispositions for critical adaptation*, New York and London: Routledge, pp. 48–66.

Leatherbarrow, D. (2005 [2010]). 'Architecture's unscripted performance', in B. and A. M. M. Kolarevic (eds.) *Performative architecture: beyond instrumentality*. New York and London: Spon Press.

Lerner, J. (1996). 'Change comes from the cities', *Human Settlements Our Planet* [online]. Available at: http://www.ourplanet.com/imgversn/81/lerner.html (Accessed: 29 March 2014).

Light, A. (2010). 'The moral journey of environmentalism: from witness to place', in S. A. Moore (ed.) *Pragmatic sustainability: theoretical and practical tools*. New York and London: Routledge.

Magnani, L. (2001). *Abduction, reason, and science: processes of discovery and explanation*. New York: Kluwer Academic/Plenum Publishers.

Minteer, B. A. (2002). *Deweyan democracy and environmental ethics: democracy and the claims of nature: critical perspectives for a new century*. B. and B. P. T. Minteer. Lanham, MD: Rowman & Littlefield, pp. 33–48.

Moore, S. A. (ed.) (2016). *Pragmatic sustainability: dispositions for critical adaptation*. 2nd ed. New York and London: Routledge.

Moore, S. A. and Wilson, B. B. (2013). *Questioning architectural judgment: the problem with codes in the United States*. London: Routledge.

Mumford, L. (1963 [1934]). *Technics and civilization*. New York: Harvest/HBJ.

Murphy, P. (2009). 'Special report: Part 1 LEEDing from behind: the rise and fall of green building', *New Solutions*. Yellow Springs, OH: The Arthur Morgan Institute for Community Solutions.

NBI. (2008). *Energy performance of LEED for new construction buildings—final report*. White Salmon, WA: New Buildings Institute.

NRDC. (2016). *Communities* [Online]. National Resource Defense Council. Available at: https://www.nrdc.org/issues/communities (Accessed: 14 December 2016).

Oden, M. (2016). 'Equity: the awkward E in sustainable development', in S. A. Moore (ed.) *Pragmatic sustainability: dispositions for critical adaptation*. 2nd ed. London: Routledge, pp. 30–49.

Orr, David. (2014). *The Oberlin Project* [online]. Available at: http://www.oberlinproject.org/. (Accessed: 15 February 2014).

Owens, D. (2009). *Green metropolis: why living smaller, living closer, and driving less are the keys to sustainability*. New York: Riverhead Books.

Peirce, C. S. (1967). *Houghton Library Peirce manuscripts numbered in accordance with the R. Robin's Annotated Catalogue of the Papers of Charles S. Peirce. Ketner*. Cambridge, MA: Harvard University.

Porteous, Colin. (1997). 'Energy efficiency in housing', *Urban Studies*, 34(9), p. 1519.

Preiser, Wolfgang, F. E., and Vischer, J. C. (2005). *Assessing building performance*. Oxford: Elsevier.

Rothblatt, D. N. (1971). 'Rational planning reexamined', *Journal of the American Institute of Planners*, 37(1), pp. 26–37.

Scott, J. C. (1999). *Seeing like a state*. New Haven, CT: Yale University Press.

Simon, Herbert. (1969). *The sciences of the artificial*. 1st ed. Cambridge, MA: MIT Press.

Steiner, F. (2008). *The living landscape: an ecological approach to landscape planning*. 2nd ed. Washington, DC: Island Press.

TOP. (2016). *The Oberlin Project* [online]. Available at: http://www.oberlinproject.org/current-news/recent-headlines (Accessed: 15 December 2016).

Turnock, B. J. (2009). *Public health: what it is and how it works*. 4th ed. Sudbury, MA: Jones & Bartlett.

Van Den Berg, A. E., Hartig, T., and Staats, H. (2007). 'Preference for nature in urbanized societies: stress, restoration, and the pursuit of sustainability', *Journal of Social Issues*, 63, pp. 79–96.

Waldheim, C. (2006). *The landscape uranism reader*. New York, NY: Princeton Architectural Press.

Ward, S. (2005). *The Garden City: Past, present and future*. New York and London: Routledge.

Further Reading

Achterhuis, H., (ed.) (1999 [2001]). *American philosophy of technology*. The Indiana Series in the Philosophy of Technology. Bloomington and Indianapolis: Indiana University Press.

Beatley, T. (2000). *Green urbanism learning from European cities*. Washington, DC: Island Press.

Ben-Joseph, E. (2005). *The code of the city: standards and the hidden language of place making*. Cambridge, MA and London: MIT Press.

Farr, D., (2008). *Sustainable urbanism: urban design with nature*. Hoboken, NJ: Wiley.

Geddes, P. (1915). *Cities in evolution: an introduction to the town planning movement and to the study of civics*. London: Benn.

Jackson, K. T. (1987). *Crabgrass frontier: the suburbanization of the United States*. New York: Oxford: Oxford University Press.

Jacobs, J. (1992). *The death and life of great American cities*. New York: Vintage Books.

Jenks, M. and Dempsey, N. (2006). *Future forms and design for sustainable cities*. New York and London: Routledge.

McHarg, I. L. (1994). *Design with nature*. New York: J. Wiley.

Moore, S. A. (2007). *Alternative routes to the sustainable city: Austin, Curitiba, and Frankfurt*. Lanham, MD: Lexington Books.

Newman, P. and Jennings, I. (2008). *Cities as sustainable ecosystems principles and practices*. Washington, DC: Island Press.

Owen, D. (2009). *Green metropolis: why living smaller, living closer, and driving less are keys to sustainability*. New York: Riverhead Books.

Rees, W. E. (1997). 'Is 'sustainable city' an oxymoron?', *Local Environment*, 2, pp. 303–310.

Register, R. (2006). *EcoCities: Rebuilding cities in balance with nature, revised edition*. Gabriola Island, BC: New Society Publishers.

Ritchie, A. and Thomas, R. (2013). *Sustainable urban design: an environmental approach*, 2nd ed. Oxfordshire, UK: Taylor & Francis.

Wackernagel, M. and Rees, William. (1996). *Our ecological footprint: reducing human impact on earth*. Gabriola Island, BC: New Society Publishers.

WCED. (1987). *Our common future*. New York: United Nations World Council on Economic Development.

34

URBAN PARKS AND OPEN SPACE

Roger Paden

This essay examines the value of urban parks from a philosophical perspective. It looks at several justifications that have been given for parks to uncover the values implicit in them. Special attention is paid to the views of Frederick Law Olmsted. These values, I will argue, should be directly connected to the idea of a city. Thus, the exploration of urban parks and open spaces casts some light on the nature and value of cities and urban life. Uncovering these values will not only help parks in the annual municipal budget battles, but it should also help guide the design of parks.

Our house sits next to a park that is adjacent to a lower-income, historically black community in Montgomery County, Maryland. My wife was working in our small garden one Saturday afternoon, when a neighbor stopped by with news that the staff of the county school system had decided that a new middle school should be built in our local park, thereby effectively destroying it. The School Board was to vote on their proposal the following Tuesday. Luckily, a controversy over cutting funds for high school bands delayed the decision and gave the neighborhood a chance to organize. A great deal of organizational work was done in the park itself. In hearings a month later, members of the Board of Education's staff made it clear that one of the attractions of using park land for schools was that this land was "free," as it was already owned by the county and would, thus, cost the school system nothing to acquire. In addition, park land was "vacant," and indeed virtually the only vacant land in the "down county" area close to Washington DC. Responding to a question as to how many of the county's larger down county parks would be needed for schools, given projected population growth, one member of the staff replied, "All of them."

With crucial help from the county's Parks Department and the regional Planning Board, we won this battle. Moreover, as a result of our work, the School Board realized that, in the future, they would not be able to seize parks easily and that, consequently, they would need to change their practices regarding both land acquisition and school design. Specifically, they would now have to acquire privately owned land and build more compact urban schools rather than sprawling suburban schools. In addition, a new community organization was formed that turned its attention to other issues, such as zoning and transit.

During the battle, we were accused by the school staff of NIMBYism. These charges were mistaken, but often neighborhood groups arguing for the protection of nearby parks can seem reactionary rather than forward-looking, and self-interested rather than public-spirited. To avoid these charges, arguments appealing to the public good of parkland are needed. Arguments for the preservation of urban parks and open spaces need to be based on a set of civic values, and these arguments require a clearly articulated set of values that are realized by or embodied in urban parks and open spaces.

DOI: 10.4324/9781315768090-41

But what are these values? What might justify the existence – and the expansion – of urban parks and open spaces, given that these spaces can always be used for alternative purposes? These alternatives include more than just more McMansions and strip malls; they could include schools, affordable housing, and public transportation, alternatives which themselves realize important civic values. But do urban parks and open spaces realize important and substantive civic values? This is an important question because, in addition to helping to preserve parks, a theory of the value of urban parks and open spaces might give us some insight into how best to design them.

Traditional Park Values

The question of the value of urban parks is at least as old as the oldest grand urban parks in the United States. In 1880, Frederick Law Olmsted, the father of American Landscape Architecture and designer of many of the best urban parks in the United States (including Central and Prospect Parks), published an article entitled "The Justifying Value of a Public Park" in which he mentioned a number of values that could be used to justify the existence of parks and guide their design (Olmsted 1880). These values included the protection of nature, the facilitation of leisure, and the creation of beauty.

Arguments justifying urban parks on the basis of the fact that they help to preserve nature are similar to justifications for national parks, which are often defended as wilderness habitats that preserve nature in a pristine state. These arguments, when aimed at the preservation of national parks, may be sound, but, when used to defend urban parks, they are flawed. It is true that some urban parks play a role in preserving nature by providing shelter for migrating birds and butterflies or year-round habitats for various types of flora and fauna. But, except for some particularly large or well-placed parks, most urban parks are too small to afford much protection. In addition, many parks – for example, recreational parks – make no pretense at protecting nature. Indeed, most urban parks, not to mention other urban open spaces, do not preserve pristine nature at all, as, contrary to common belief, they are almost entirely unnatural and artificial – the product of their designer's art (Gobster 2001). Moreover, all things being equal, urban parks take up space in the city that might be used for other structures, and if those structures are simply built elsewhere, then urban parks merely relocate "nature" from rural areas at the edge of cities into the city itself. Parks simply push development further out, thereby aggravating urban sprawl, increasing pollution, and worsening congestion. Finally, because smaller urban parks are often isolated bits of nature (shards of ecosystems that are too small to be self-sustaining), if they are created at the expense of more rural and complete natural habitats, the overall effect would be to degrade the environment.

Olmsted often defended parks by arguing that they provide a quite refuge away from the hustle and bustle of city life. Parks, he claimed, play a role in reducing the "vital exhaustion" and "nervous irritation" caused by city life (Olmsted 1880: 156). Today, the term "leisure" appears in many arguments of this sort. Parks are said to provide "leisure services" to the inhabitants of a city. They are examined by departments of "Leisure Studies." The term, "leisure," however, has a number of unfortunate connotations readily apparent in standard dictionary definitions, where "leisure" is defined in terms of freedom from duties and responsibilities or is connected to unoccupied hours that can be filled with hobbies and other relatively aimless and unimportant activities. As such, the term reeks of the insignificant, the unnecessary, and the amoral. All things being equal, the value of leisure activities seems low relative to many important civic values, such as economic expansion or efficient transportation, and, by itself, is unlikely to help preserve parks.

Justifying urban parks based on their aesthetic value also has a long tradition that can be traced back to Olmsted. Typically, these arguments are based on the idea that cities are ugly, noisy, unattractive places, while nature is quiet and beautiful. Nature, in this view, can play a role as an antidote to the ugly artifice of the city and, while large expanses of nature are best, it is good to have more local and immediately available (and therefore small) parks located in the city. There are problems with this argument. Among the most important is that policy analysts typically treat aesthetic experience as less important than economic growth and development. In part, this is because aesthetic values are often said to be purely personal, and beauty itself to be in the eye of the beholder and, thus, difficult to measure (Sagoff 1988). Setting aside the radical relativism of these claims, it is not clear if designers of parks should aim at creating parks that are beautiful or whether instead, they should aim at creating parks that are picturesque. In any case, it is often claimed that aesthetic values can only be fully appreciated by those whose more basic needs have been satisfied: beauty, like leisure, is truly available only to those who are not pursuing more important things. Archibald Alison, for example, claimed that beauty could only be experienced by those whose minds are "vacant and unemployed" (Alison 1815: 10). If so, then it can be argued that public monies should be expended to fight hunger or to advance education, rather than on expensive urban ornaments that will be enjoyed mainly by the wealthy (Rosenzweig and Blackmar 1992: 211–237). Finally, it should be noted that not all parks (for example, recreational parks) are intended to be aesthetically pleasing.

Given the difficulties inherent in arguments based on these somewhat abstract and/or subjective values, John Crompton has argued that, if parks are to be defended in a period of tight budgets, it is necessary to "reposition" parks relative to other, more easily quantifiable, values (Crompton 2008). Similarly, Chris Walker, in "The Public Value of Urban Parks," urges policy makers to move "beyond recreation" to take "a broader view of urban parks," one that emphasizes a new set of essentially public values (Walker 1999: 1). Crompton and Walker discuss several types of values that might be used to justify parks. These include promoting health, increasing safety, and boosting economic activity.

It was common in the past to refer to parks as "the lungs of the city," as it was thought that parks acted to clean city air of "miasma," the "bad air" that produces cholera and other diseases. Parks, on this view, can detoxify city air and should be built to prevent the epidemics that were common in cities (Frank *et al.* 2003: 14). Today, this "miasmic theory" has been rejected, but parks are still held to be useful in promoting "healthy lifestyles" by providing sites for recreation and exercise (Chadwick 1966). As Crompton points out, there is a great deal of evidence that parks do have this effect. For example, studies show that one of the factors that lead city dwellers to exercise is the availability of nearby parks. And the health benefits provided by parks are not just physical; instead, recreation also provides psychological benefits that are correlated with better health (Crompton 2008: 48–50).

There is also a great deal of evidence that parks help to decrease crime. It is often thought that parks "attract crime" by providing places where criminals can linger while searching for victims (Groff and McCord 2012). Studies have shown, however, that the presence of parks in communities is associated with lower overall crime rates, and, indeed, crimes committed in well-maintained parks are relatively rare. In the past, this effect was attributed to the fact that parks provide adult men with alternatives to taverns and the alcoholism and crime they produce. Today, however, attention has shifted to juveniles, and studies have shown that parks and athletic facilities can provide young people with opportunities to engage in active recreation (an alternative not otherwise widely available in cities). And this provides them with a chance to socialize with peers and to engage in athletic competition, thereby reducing juvenile delinquency (Trust for Public Land 1995; Crompton 2008: 52–54).

Finally, parks are said to be valuable because they promote economic development. They can provide job opportunities by employing people to build and maintain them and to provide essential services in them, e.g. policing, groundskeeping, and coaching. As Crompton points out, "Park and conservation work is relatively labor intensive" (Crompton 2008: 55). Parks have also been shown to increase property values for nearby residences. Walker cites studies that show that property values near Pennypack Park in Philadelphia "increased [by] from about $1000 per acre at 2,500 feet from the park to $11,500 per acre at 40 feet from the park" (Walker 1999: 1). Olmsted often argued (and his parks demonstrated) that new parks pay for themselves through increased property tax revenue generated by these increased property values.

Arguments based on these more quantifiable values are sound. They are generally well-grounded in scientific research. And they connect parks with important public values. Perhaps more important, they are often thought persuasive by local government officials and are, therefore, particularly effective in budget debates. But it is not clear if parks are the best way to realize these values. Increasing the number of health clubs and giving more time to physical education classes in schools might better bring about improved health. Crime might be more efficiently reduced through more attractive recreation centers and increased policing. And new commercial districts or increases in education and/or research and development budgets might better stimulate economic development. These arguments, that is to say, do not show that there is an *essential* connection between these values and urban parks.

Ruralism

These arguments also posit no essential connection between urban park values and cities. Instead, they imply that there is a conflict between these values and the city as a normative ideal, as they reflect an underlying set of assumptions that are associated with a position that Avner de-Shalit has called "ruralism" (de-Shalit 1996). According to de-Shalit, ruralism involves "the glorification of country life, and a dissatisfaction with urbanism [in part]… because [the city] is said to represent an inferior moral condition, or even a state of degeneration" (de-Shalit 1996: 50). Cities, that is, are foul places. Physically, they are dirty, and the air and water found in them are unhealthy. Socially, they are overcrowded, alienating centers of commerce and industry. Most importantly, psychologically, they produce people – "the rabble" – who lack virtue, and, consequently, cities have relatively high crime rates and high rates of delinquency. Moreover, most people in cities are economically dependent on their employers and, as a result, do not possess the autonomy needed to be good citizens of a democracy.

Dale Jamieson has traced the history of this idea in the United States, finding evidence of it in the writings of Jefferson, who not only celebrated the "yeoman farmer," the ideal citizen of the new democracy, but who wrote that "I view great cities as pestilential to the morals, the health, and the liberties of man" (Jamieson 1984: 40). It can also be found in the writings of the American Transcendentalists who celebrated nature, in Josiah Royce's praise of New England town meetings as the quintessence of democracy, and in Frank Lloyd Wright's comparison of cities to tumors (Jamieson 1984: 38–43). Traces of it can also be found in contemporary writings on environmental ethics (Light 2003).

Ruralism incorporates many of the ideas first developed by Romantic thinkers. Like Romanticism, ruralism tends to understand nature as a unified, balanced, and harmonious whole. These ideas are also applied to virtuous people. Thus, ruralists celebrate individuals who possess such characteristics ("virtues") as integrity, unity, harmony, and grace, along with traditional societies that, by embodying those values, produce virtuous people. Conversely, ruralists condemn modern society. They reject industrial capitalism not only for destroying traditional society, but also for destroying the virtues associated with those societies. On this view, the modern commercial city is the epitome of a diseased society. It undermines the diverse organic forms of earlier communities and replaces them with a "cosmopolitan," mechanical, and alienating society peopled by isolated, self-interested individuals possessing impoverished conceptions of the good (de-Shalit 2003). The economic poverty and dependency characteristic of urban society destroy the true freedom of its inhabitants, replacing it with the unregulated license that Enlightenment philosophers mistook for liberty. Finally, this society imposes a scientific mode of understanding on its inhabitants that undermines their judgment and blinds them to their true situation.

Ruralism underpins the traditional park values discussed above. Ruralism justifies urban parks as antidotes to the problems of city life. The city is unnatural, but parks provide a bit of pristine nature at its center. The commercial life of the city drives people to exhaustion and isolation, but parks provide an opportunity for leisure and communal recreation. The city is ugly, but parks are beautiful. The city is unhealthy, but parks provide a place to breath clean air. The city is rife with crime, but parks reduce it. Thus, ruralism understands the true value of urban parks to be remedial in nature. They provide an opportunity for urbanites to experience the good life characteristic of the countryside and to be healed by that experience. On this view, the value of parks is that they realize values that cannot otherwise be realized in the city. They help cure or minimize the problems inherent in city life. This is an anti-urban justification of urban parks.

Urban Values and Urban Parks

It is possible to conceive of an alternative "urbanist" justification for urban parks and open spaces. Unlike the ruralist justification, such an argument would attempt to justify parks as integral parts of cities that contribute to the realization of essentially urban values. This justification is based on the simple idea that urban parks and open spaces are *in* cities, and, therefore, if they are to be justified, they should be justified by reference to the values realized by cities: parks should add to the urban experience and reinforce the values associated with the city, rather than simply repairing the damage cities supposedly inflict on their inhabitants. But this raises the question of the nature of the "urban values" upon which such a justification would rest. And this, in turn, raises the question of the true nature of a city, of the city as a normative ideal.

In the first book of the *Politics*, Aristotle contrasts the city with the village. A village (which Plato, in the second book of the *Republic*, referred to as a "city of pigs") is, in Aristotle's view, a community composed of people who are united by natural and hierarchical biological ties. Paradigmatically, a village is a large family, naturally ruled by a king, who, as its "father," guides and commands its inhabitants. A city, however, is composed of men who are not united by such natural ties, but who instead stand in a new relationship based on equality. This relationship – "*isonomia*" – is to be found in the fact that all *citizens* are both authors and subjects of the law. Understood in this way, Aristotle stands at the beginning of a republican tradition in political theory, a tradition that is so expansive as to include diverse figures such as Niccolò Machiavelli, Hannah Arendt, and Iris Marion Young (Pettit 1997). What is central to these various thinkers is the idea that a city should be thought of as a collection of strangers whose pursuit of the good life is bound by rules which they devise to make this possible; and that citizens of such a city are "free" because they are bound only by rules that they have deliberatively determined. Philosophical republicanism of this sort demands that people play an active role in the political life of their cities and in this it is distinguished from philosophical liberalism. Republican societies require the existence of institutions that reward people for their participation in public life and that equip them with the capacities and dispositions – the "civic virtues" – to do so. These virtues include respect for others, civility, curiosity, and honesty (a willingness to search for and base one's views on publicly tested ideas), deliberative rationality, and patriotism (a willingness to be involved in the political process and work for the good of the society) (Dagger 1997). Richard Dagger has argued that some urban designs – namely, those that create time-consuming sprawl – undermine republican society by making civic engagement difficult (Dagger 2003). Conversely, some urban designs can enhance republicanism, by providing sites for political activities, by providing topics for political debate, and by helping citizens acquire and practice civic virtues. I would argue that, because they can help in this process, urban parks and open spaces can play a role in fostering republican governance.

Historically, parks and squares have played a central role in politics as *sites* for political activity. Many political movements are identified with public squares where people come to protest, to organize, and to share ideas. The "Arab Spring" was identified with the protests held in Tahrir Square, while "Occupy Wall Street" was identified with Zuccotti Park. Squares and parks have also been the *objects* of political activity. There has been an ongoing political debate concerning the design of Central Park; in the sixties, students in Berkeley created and fought to protect a "People's Park"; and the recent political crisis in Turkey began with protests over plans to destroy Istanbul's Gezi Park. On a much smaller level, my local park served as a site in which to organize protests aimed at its preservation.

But urban parks and open places have played an even more important role in creating the pre-requisites for republican political life. There is a very long history of urban planning in which cities (together with their parks and open spaces) were designed with this end in mind. For example, Hippodamus, mentioned by Aristotle as the father of both urban planning and political philosophy, designed cities with central squares to support *isonomia* (Paden 2001a). Similarly, during the Renaissance, Leon Battista Alberti, the author of the second oldest existing book on architecture and urban planning in the Western canon, based his plans on the idea that central plazas play an important role in shaping civic values (Paden 2001b). In their plans, open spaces, particularly city squares and agoras, were designed to play an essential role in the political process. Although these open places often have other purposes – for example, an agora was a public market and squares demonstrate the wealth of a city – that serve to attract people to them, they also serve more important republican purposes.

Parks and open spaces can also play a role in establishing a political community. Well-designed parks are places where people from different families and members of different occupational classes can meet. In parks, they can see and be seen, and interact in unstructured ways. In doing this, visitors can observe others on promenade. They can come to appreciate the value of new activities and other ways of life (by, for example, watching amateur athletes passionately pursue an unknown sport). They can learn to make reasonable accommodations for others' pursuits (as different activities are arranged across a meadow). They can share activities (joining with strangers in a pick-up game). They can practice negotiating *Modus vivendi* agreements (as they take turns posing for pictures in front of a particularly beautiful tree). In doing so, they can come to value the public goods that make this possible. In this way, parks provide a model of urban life and allow citizens to develop and practice civic virtues on a small scale. Parks and open spaces can be a microcosm of the city, and the actions that take place in them can reflect the essential nature of urban life.

Republicanism and the History of Public Parks: Olmsted's Legacy

I have argued that urban parks and open spaces can play an important role in nurturing a republican society and that some parks can be justified on these grounds. While this may seem an odd claim – arguments based on "traditional park values" are much more familiar – it can be traced back to the origins of the public park movement in the United States and specifically to Olmsted's writings. Most large parks in America consciously or unconsciously follow a design philosophy developed and popularized by Olmsted and his associates, and this design philosophy, I argue below, was based, in part, on the idea that parks could play a role in strengthening republican society and its virtues. Unfortunately, the values implicit in this design philosophy have largely been forgotten. What follows is an essay in retrieval.

Olmsted began his working life as a ruralist. His first career was farming, but he began to transition away from that occupation and its ruralist ideology when he toured England and visited many of its private gardens and public parks. He found both interesting: the private gardens because they reminded him of Uvedale Price's book, *An Essay on the Picturesque*, which he had read as a youth; and the public gardens because they called attention to the lack of such gardens in the United States. This tour led to his first book, published in 1852, *Walks and Talks of an American Farmer in England*. This book persuaded Henry Raymond, the editor of the *New York Daily Times*, to offer Olmsted a job traveling the American South and reporting on its economy, society, and culture, and on the effect that slavery had on all three. Olmsted signed his reports, which appeared in the *Times* from 1852 to 1857, "Yeoman," in what must have been both a reference to his own past and, increasingly, an ironic reference to Jefferson's ruralism. Olmsted was hired in part because Raymond believed that since Olmsted favored, not the immediate abolition of slavery, but a more gradual and ordered transition toward freedom, he would produce a neutral account of the South. As it turned out, however, Olmsted was appalled both by the savagery of slavery and by the lack of culture exhibited by the Southern gentry. Olmsted blamed both these "barbarisms" partly on slavery itself and partly on the South's rural character. The fact that many Southerners were ruralist who believed their culture superior to that of the more urbanized North convinced Olmsted of the folly of this view. Nevertheless, when one of his hosts, John Allison, argued in defense of slavery and the South that members of the Northern urban working classes were no more cultured than Southern slaves, Olmsted was unable to dispute him. Eventually, Allison's argument led him to adopt the view that members of the urban working classes needed to be "civilized," and that this type of education was necessary for the political health of the republic. Olmsted wrote that, in order to improve the "democratic condition of *society* as well as *government*," it was necessary to provide the poor with an education that would furnish them with the "taste and the mental and moral capital of gentlemen" (Rybczynski 1999: 105–127). Further, Olmsted came to believe that parks should play an important role in this project. Olmsted's abhorrence of slavery led him to reject ruralism, but this only made him more conscious of the problems facing Northern cities, and this led him to meditate on urban values and how parks might play a role in their realization.

In 1857, Olmsted got a job supervising the workmen then beginning construction of Central Park. Early in his tenure, a competition was announced for a new design for the park to replace the flawed plans developed by the Park's Chief Engineer. This competition was won by Olmsted and the architect, Calvert Vaux, an expatriate Englishman. Olmsted spent much of the next six years as the "Architect in Chief" supervising the construction of the park. Forced to resign in 1863, Olmsted took a job directing operations at the Mariposa Mining Estates in California. While in California, he not only served as the president of the commission overseeing the creation of Yosemite Park, centered in a nearby valley, but he worked on a never-finished book, to be titled "Civilization," in which he was to set out his urbanist philosophy.

In notes for that book, Olmsted drew a key contrast between the rural "frontier" and urban "civilization." On Olmsted's view, the frontier forces people to look after their own interests, leading to "a dangerous form of selfish individualism," which he thought was common in both the South and the West. In contrast, in a civilized state, which Olmsted associated with true republicanism, "the ruling spirit is benevolent and cooperative" (Kalfus 1990: 260). Olmsted believed that the future of America lay in its cities, which were destined to grow at a great pace at the expense of more rural areas. In general, he regarded this as a good thing, as cities are a great source of creative energy and the locus of true democracy. But often their citizens' pursuit of wealth undermines the essential characteristic of civilized life, which Olmsted called "communitiveness." This was the ability to deliberate about public values while being governed by the civic virtues prized by the republican tradition. He argued that as this characteristic is essential to republican self-rule, cities must be designed in such a way as to nurture it. To this end, it is necessary that cities establish such institutions as parks, gardens, museums, and schools to help integrate their inhabitants into a cosmopolitan and republican life (Kalfus 1990: 164).

Olmsted was not alone in his beliefs concerning the civilizing nature of parks and, in fact, he derived the idea from other sources. The most immediate influence on him was Andrew Jackson Downing, who may have been the real father of American Landscape Architecture. Downing, the editor of *The Horticulturist* magazine, championed the idea that republican cities need large parks and it was only natural that he was asked to produce a plan for National Mall in Washington. He almost certainly would have been hired to design Central Park, if not for his tragic drowning in 1852. Olmsted was a friend of Downing, who not only read his magazine regularly, but who published several articles in it.

In several articles published in *The Horticulturist*, Downing contrasted the politics and culture of Europe and the United States, arguing that while "the French and the Germans – difficult as they find it to be republican in a political sense – are practically [i.e., socially and culturally] far more so..., than Americans." Troubled by this gap between our politics and culture, he argued, "We owe it to ourselves and our republican professions to set about establishing a larger and more fraternal spirit in our social life." This can be done, he claimed, "mainly by establishing refined places of resort – parks and gardens, galleries, libraries, museums, etc." Thus, he invited his readers to contemplate "the *social* influence of a great park in New York," claiming that "among the topics discussed by the advocates and opponents of [a central park], none seems so poorly understood as the social aspect of the thing." If we understand the effects of parks correctly, we will see that "out of this enjoyment of public grounds grows also a *social freedom*, and an easy and agreeable in-tercourse of all classes" (Downing 2012: 213–248). Such a result would be possible, if parks were properly designed to take advantage of the "elevating influences of the beautiful in nature and art" and if they are "enjoyed in common by... all classes without distinction" (Downing 2012: 241).

When Olmsted wrote of the social value of parks, he discussed them in terms of "democracy." Vaux, too, wrote that his goal in designing Central Park was to "translate Democratic ideas into Trees and Dirt" (Rosenzweig and Blackmar 1992: 136). Downing also championed Central Park by referring to its "republican" effects. More recently, social scientists have studied the role that parks can play in establishing "community" (Coley, Sullivan and Kuo 1997). Critically, all three landscape architects were influenced on this point – as well as on many matters of design – by Uvedale Price who, with other British theorists of the picturesque, made a similar argument by contrasting authoritarian French gardens and politics with the more natural, complex, egalitarian, and picturesque English gardens (Paden 2013). Thus, the idea that urban parks and open spaces should play an important role in nurturing a republican society, and that parks can be justified by the fact that they can do just this is an old one.

Olmsted and his associates not only established the profession of Landscape Architecture in the United States, but they played a central role in shaping Americans' ideas on park design. Central Park was used as a template in the design for most important American urban parks, from Buffalo to San Francisco. The design philosophy that shaped these parks was based on the notion of "the Picturesque" which, itself, was informed by a broadly republican political outlook. Thus, from the beginning, our parks have been shaped by a specific political philosophy and justified by appeals to specific political virtues. They still should be.

From Values to Design

If Price and his heirs in America are right, the possible republican effects of parks should play some role in their design. It is important, however, to distinguish between immediate benefits and long-term effects. Even if parks and open spaces are ultimately justified by their social effects, that is not why people visit them; instead, they go to parks to commune with nature, to experience natural beauty, to engage in recreation, or merely to while away the hours watching other people. Parks must satisfy these immediate needs. In part, they are valuable because they provide these benefits, but while they depend on these immediate benefits, the more important social value of parks is found elsewhere. Yet, if people are not drawn to parks, then parks will not be able to have the proper effect on people. It is important, therefore, that parks be designed so that people, leisurely pursuing their own desires and activities, are brought together where they can see and be seen and, thereby, be brought into each other's lives.

Price argued that an aesthetically valuable, "picturesque" garden would be characterized by mixed and contrasting patterns, in which various elements would serve both as a beautiful fore-grounded object and as a part of the background for other such objects (Price 1796). A picturesque garden would be composed of mixed elements, of contrasting patterns, and of sometimes-abrupt changes. A picturesque garden would thus contain many discrete spaces with their own unique character which, at the same time, are parts of other larger spaces. As Olmstead realized, a park designed on picturesque principles would be made up of many adjacent and overlapping areas that could be used to support a variety of different activities, from strolling, to bird watching, to athletics, to quiet contemplation. Such a park could then play a valuable role in supporting repub-lican politics. Central Park was designed to be both aesthetically and socially picturesque in just this way, and, as a result, it is both aesthetically and socially valuable, not just for its immediate uses, but for the republican values it cultivates. Central Park then served as a paradigm for many American parks, all of which can play a complex role in our society supporting republican values. It is, in part, in terms of the values championed by Olmstead, that urban parks can be justified.

Urban parks – like my park – can be both sites and objects of political activity, and they can, if designed well, play a role in fostering civic virtues. The central kind of value that these parks help re-alize is civic value. Urban parks, though related to more rural national parks, are different and distinct. They are built in cities by city dwellers, and they have their own unique value. City parks should be aesthetically pleasing, should provide a place to exercise and rest, but while, as Olmsted realized, these characteristics are valuable in themselves, they also provide a foundation for other, more important, quintessentially urban values, or at least they could if designed according to picturesque principles.

Related Topics

10. Wilderness
11. National Parks
12. Landscape
35. Suburbs and Exurbs
42. Community Gardens
44. Restoration
47. Rewilding

References

Alison, A. (1815) *Essays on the Nature and Principles of Taste*, Edinburgh: Constable.

Chadwick, G. F. (1966) *The Park and the Town: Public Landscape in the 19th and 20th Centuries*, New York: F. A. Praeger.

Coley, R., Sullivan, W., and Kuo. (1997) "Where Does Community Grow? The Social Context Created by Nature in Urban Public Housing," *Environment and Behavior* 29: 486–494.

Crompton, J. (2008) "Empirical Evidence of the Contributions of Leisure Services to Alleviating Social Problems: A Key to Repositioning the Leisure Services Field," *World Leisure* 4: 243–258.

Dagger, R. (1997) *Civic Virtues: Rights, Citizenship, and Republican Liberalism*, Oxford: Oxford University Press.

——— (2003) "Stopping Sprawl for the Good of All," *Journal of Social Philosophy* 34: 28–43.

de-Shalit, A. (1996) "Ruralism or Environmentalism?" *Environmental Values* 5: 47–58.

——— (2003) "Philosophy Gone Urban: Reflections on Urban Restoration," *Journal of Social Philosophy* 34: 6–27.

Downing, A. J. (2012) *Andrew Jackson Downing: Essential Texts*, edited by Robert Twombly, New York: W. W. Norton.

Frank, L., Engelke, P., and Schmid, T. (2003) *Health and Community Design: The Impact of the Built Environment on Physical Activity*, Washington, DC: Island Press.

Gobster, Paul. (2001) "Visions of Nature: Conflict and Compatibility in Urban Park Restoration," *Landscape and Urban Planning* 56: 35–51.

Groff, E. and McCord, E. (2012) "The Role of Neighborhood Parks as Crime Generators," *Security Journal* 25(1): 1–24.

Jamieson, D. (1984) "The City Around Us," in T. Regan (ed.) *Earthbound: Introductory Essays in Environmental Ethics*, Prospect Heights, IL: Waveland Press.

Kalfus, M. (1990) *Frederick Law Olmsted: The Passion of a Public Artist*, New York: New York University Press.

Light, A. (2003) "Urban Ecological Citizenship," *Journal of Social Philosophy* 34: 44–63.

Olmsted, F. L. (1880) "The Justifying Value of a Public Park," *Journal of Social Science* 12: 150–171.

Paden, R. (2001a) "The Two Professions of Hippodamus of Miletus: On the Relationship Between Philosophy and Urban Planning," *Philosophy and Geography* 4: 25–48.

——— (2001b) "Values and Planning: The Argument from Renaissance Utopianism," *Ethics, Place, and the Environment* 4: 5–30.

——— (2013) "A Defense of the Picturesque," *Environmental Philosophy* 10: 1–21.

Pettit, P. (1997) *Republicanism: A Theory of Freedom and Government*, Oxford: Oxford University Press.

Price, U. (1796) *An Essay on the Picturesque as Compared with the Sublime and the Beautiful: And on the Use of Studying Pictures for the Purpose of Improving Real Landscape*, London: J. Robson.

Rosenzweig, R. and Blackmar, E. (1992) *The Park and the People: A History of Central Park*, Ithaca, NY: Cornell University Press.

Rybczynski, W. (1999) *A Clearing in the Distance: Frederick Law Olmsted and America in the Nineteenth Century*, New York: Scribner.

Sagoff, M. (1988) *The Economy of the Earth*, Cambridge: Cambridge University Press.

Sagoff, M. (1988) "Some Problems with Environmental Economics," *Environmental Ethics* (10) 55–74.

Trust for Public Land. (1995) "Healing America's Cities: How Urban Parks Make Cities Safe and Healthy," *Children's Environments* 12(1): 65–70.

Walker, C. (1999) *The Public Value of Urban Parks*, Washington, DC: Urban Institute.

35

SUBURBS AND EXURBS

Robert Kirkman

When I talk to other people in the Atlanta region about where I live, I have to be cautious. If I tell them I live "inside the Perimeter" – ITP, for short – they are likely to jump to all sorts of conclusions about me, my political leanings, my intelligence, my education, my vocation and hobbies, and perhaps even the state of my soul. To be fair, if someone replies that he or she lives "outside the Perimeter" – OTP – I am likely to leap to one or two conclusions, myself.

The Perimeter is the major ring-road around the core of the Atlanta region, Interstate 285. Up close, it is impressive enough: on some stretches, there are five or six lanes of traffic in each direction, and it features dramatically tangled interchanges with other highways. The Perimeter is not just asphalt and concrete and steel, though: it is a cultural barrier, a mental divide, between the Inside and the Outside. Sometimes, it seems to me almost to be a membrane around the core of the city, one that often can be traversed only with some difficulty and sometimes with a sense of peril.

Those who live Inside and Outside, respectively, tend to harbor all sorts of stereotypes of those who live on the other side of the divide – stereotypes rooted in the old cultural divide between the city and the suburbs. Viewed from Outside, the city, ITP, is the realm of crime and dirt and graffiti and poor people and minorities and yuppies and pretentious hipsters with unusual piercings. Viewed from Inside, the suburbs, OTP, are the domain of characterless subdivisions, long commutes, ugly commercial sprawl, segregation, social conformity and shallow, wasteful consumerism. So run the stereotypes.

I must admit, though, that since my daily life is lived entirely ITP, oriented toward Midtown and dependent on Atlanta's meager system of public transit, it sometimes seems to me that anything OTP might just as well be on the far side of the moon.

The problem with these stereotypes is that, in a sense, there is no such thing as "the suburbs," at least not in any form that can really justify the ITP/OTP dualism or the mutual contempt of those on either side of the highway itself. There may be patterns of development and habitation that correspond roughly to the traditional vision of the suburbs – single-use zoning, single-family houses surrounded by lawns, commercial strips, office parks and congested arterial roads – but instances of those patterns are as easy to find Inside as they are Outside. Conversely, patterns more often associated with urbanism can be found Outside as well as Inside, in older town centers and in the newer "edge cities" as identified by Joel Garreau (1991), to one of which he gave the mind-bending name of Perimeter Center.

The distribution of populations and cultures across the region further blurs the significance of the Perimeter, as poor, minority and immigrant communities have increasingly found places for themselves OTP and as stereotypically urban problems of drug use and crime find their way

DOI: 10.4324/9781315768090-42

Outside, while at the same time many areas ITP are gentrifying and becoming more affluent – and perhaps introducing stereotypically suburban attitudes of social conformity and conspicuous consumerism, as well.

In short, there seems little point in talking about "the suburbs" as places or landscapes wholly distinct from "the city," except in the almost mythical terms of ideal types. In practice, on the ground, suburbs and also exurbs – the lower-density patterns of development even further out from the core – are really just the city by other means, the city broken up, its functions separated and dispersed across metropolitan regions (Fishman 1987: 5).

As they have developed through the twentieth century and into the twenty-first, metropolitan regions in the United States do have a particular character and tendency that make them of real concern for environmentalists and environmental ethicists, and to anyone who takes an interest in the sustainability of the project of human civilization. Put simply, metropolitan regions expand.

What drives the expansion of metropolitan regions is a complex system with social, technological and natural components (Kirkman 2009: 237–238, 2010b: 60–71). Growth and movement of human populations is a factor, but so are the practices and institutions of markets and political decision-making. The norms and habits embodied in cultures play a role: the suburbs may not exist, as such, but the *myth* of the suburbs still has compelling force for many people, who keep moving further and further out into the countryside to catch up to the retreating ideal (Stern 1986: 125). All of this is intertwined with the techniques and practices of transportation, communication, energy production and distribution, and construction – everything from machine nails and balloon-frame construction (Jackson 1985: 124–128) to fiber-optic telecommunications networks.

From the point of view of many environmentalists and perhaps most conventional environmental ethicists, metropolitan growth is a calamity. Not only do metropolitan regions sprawl out across the countryside, tearing up and paving over rural and wild lands alike, they depend on systems of resource extraction and processing that disrupt ecological systems far and wide and generate untold waste and pollution. Indeed, from this perspective, the demands of expanding metropolitan regions must be implicated in ills from groundwater pollution to global warming.

It would be difficult to argue for the wisdom of continuing to allow or to pursue metropolitan growth as it has developed in the United States, and it is not my intention to make such an argument. In fact, I would join in the call for a sober assessment of the current state of the metropolis and for a serious reconsideration of how we humans organize our lives in the landscape. I take the urgent need for such an assessment and reconsideration to be a given. What is not given, just yet, are the terms on which the assessment is to be carried out.

My aim here is to point out two ways in which environmentalists and, perhaps especially, environmental ethicists of a conventional sort might go astray in criticizing how we currently organize our lives in the landscape: both the myth of the suburb and the reality of metropolitan growth pose a serious challenge to conventional environmentalist discourse about metropolitan regions and their tendency to expand.

First Challenge: The Myth of the Suburbs and the Myth of the Wild

As it first arose in the nineteenth century, the suburban vision was of a pastoral middle landscape between the city on the one side and the country and the wild on the other, bringing the best of each together in a new model of human habitation, available to people who had attained a certain social standing (Stern 1986: 12, 126–127; Stilgoe 1988: 53–55). Llewellyn Park, New Jersey and Riverside, Illinois stand as fine examples of the type, with tasteful houses set in what amounts to a park – the latter designed by Olmsted and Vaux, no less! – not all of which is carefully manicured. That is to say, in addition to lawns and other park-like features, there are inclusions of the wild in the idea suburban landscape, as in the Ramble at Llewellyn Park (Fishman 1987: 125).

Of course, one might argue that this was merely an aesthetic vision, an expression of Romanticism in landscape design, which does not pay due attention to the interconnections by which truly natural landscapes function. Riverside was developed decades before the first stirrings of ecology as a science, after all, so it can hardly be surprising that the designers of the suburban ideal so easily substituted merely apparent pastoral harmony for the reality of ecological fitness.

It could also be said that such a vision, however appealing, could only have a chance of coming to be if a select few could enjoy it, perhaps at the expense of those less fortunate who remain stuck in the dirty and corrupt city. In the twentieth century, beginning after World War I and accelerating after World War II, more and more Americans found themselves financially and technically capable of pursuing the suburban ideal, with the automobile leading the charge out into the countryside. The landscapes they helped to create could be seen as poor, mass-produced imitations of the original vision – there is a world of difference between Riverside and Levittown, whatever the common threads between them – but as more people pursued the vision of the middle landscape between the city and the country, the economic and environmental consequences of the suburban myth magnified accordingly. That other functions of urban life, from retail stores to industrial plants, followed the mass migration to arcadia further magnified those consequences while at the same time subverting and destroying the vision itself. The seemingly endless commercial strips that have become symbolic of suburban life could as easily be taken as a betrayal of the suburban vision: there's nothing pastoral about a strip mall or a big-box "power center."

Nevertheless, there is a clear and deep connection between the first impulses of the suburban ideal and the first impulses of conservationism and environmentalism in the United States (Kirkman 2004: 53–57). In some of its variants, environmentalism is predicated on a strict geographic dualism (Light 2001: 17) that draws a hard distinction between the domestic and the wild, between that which is shaped by humans for our own ends and that which is pristine and untouched, and hence natural, wild and free. That which is wild and natural is said to have value in itself, regardless of any use humans might make of it, a value that is necessarily destroyed as soon as humans interfere or alter it in any way. Conventional environmental ethicists have filled books and journals with all manner of contorted arguments for the intrinsic value of natural entities and systems, sometimes with the explicit aim of mounting a defense of the wild against further incursions of the domestic.

This would seem to place humans who care for natural value in something of a bind: How are we to live in the world when anything we do to make our own way necessarily destroys natural value? At worst, this seems to result in a "static, halt-and-withdraw strategy" (King 2000: 115–116; see also Gunn 1998: 348), according to which humans must confine most of their activity to some subset of landscapes we have already developed, leaving the rest of the world to be wild and free.

A somewhat more generous approach might be that we should seek ways of organizing our lives in the landscape that are more attentive and harmonious, ways of balancing the domestic and the wild, preserving natural value while realizing human values. Such an approach would seem to be in keeping with the land ethic of Aldo Leopold (1949: 226), who noted that we will hardly relinquish the shovel, metaphorically speaking, for all that we need to be more thoughtful in using it.

Read this way, there would seem to be a common thread between the current discourse of wildness-oriented environmentalism and the early suburban ideal: the possibility of reconciling the domestic and the wild under certain conditions.

One of those conditions may be the exclusion of most human beings from full participation in the desired landscape. Consider, for example, the problem of wilderness access. Light (2001: 15–16) points out that Holmes Rolston valorizes direct experience of wilderness as a necessary condition for a full and rich human life, but that universal human access to what wilderness can be said to remain would quickly obliterate whatever natural value it is supposed to retain. All those people would simply trample wilderness to death, just as so many Americans pursuing the dream overdeveloped the suburb to death. The direct experience of wilderness, like the early ideal suburbs, only work if they are reserved for a tiny elite of the affluent or the deserving or the wise – as they might style themselves – who understand the vision. It is a commonplace among planners and others who study metropolitan landscapes that everyone wants to be the last to move to the suburbs.

An alternative, one that should trouble the thought of many environmentalist and conventional environmental ethicists, is to recognize that the wilderness idea and, with it, the geographic dualism of the domestic and the wild, is just as much a myth – a social construct, as Cronon (1995: 69) would have it – as the ideal suburb; as a consequence, the geographic dualism of city and suburb, of ITP and OTP, is likewise a myth.

Taking this step would be difficult for those who have invested themselves in the discourse of wildness, but the alternative to geographic dualism need not be a collapse into shallow, human-centered, short-term economic thinking they might associate with the domestic realm. A third way would be to adopt a richer and more nuanced approach to value and choice within the landscapes we actually inhabit, including the vast, messy and expanding metropolitan regions even environmentalists call home (Kirkman 2010b: 72–74).

The last point draws out another deep connection between environmentalist discourse and the suburban myth: the modern environmental movement is itself very much a product of the suburbs (Sellers 1999: 31–34; see also Rome 2001). One of the seminal works in the formation of the movement, *Silent Spring* (Carson 1962), was written in response to concerns about the practice of spraying DDT along residential streets in suburban Long Island, New York. It was Love Canal, a suburban subdivision near Buffalo, New York, that brought the issue of toxic waste disposal into the consciousness of the American public.

According to the myth, suburbs are supposed to be clean and safe, good places to raise children. Obvious departures from that ideal, especially threats to the healthy development of children, helped to stir outrage and activism.

Likewise, the suburbs are supposed to be green and leafy, so the appearance of yellow bulldozers in the (second-growth) woodlots adjacent to subdivisions would likely be met with dismay and outrage at the "destruction" of "nature" – and the rise of another subdivision for people trying to get out of the city and closer to nature. Still, the image of the bulldozer in the trees became, for many, the symbol of environmental destruction – as fraught with irony as that symbol must be when invoked by people living in subdivisions first cleared and graded by bulldozers (Rome 2001: 13, 150; Kirkman 2010b: xi–xiii).

The trouble with the symbolism of the bulldozer in its suburban context is that it places the focus of environmentalism on the concerns of predominantly white and affluent suburbanites. When the environmental justice movement arose in the 1990s to call out the prevalence of environmental racism and the unequal protection afforded to poor and minority communities, its activists were quick to point out the degree to which the modern environmental movement was created by and existed for the benefit of middle-class white suburbanites trying to make sure nothing undesirable happened in their own backyards (Newton 1996: 17).

I should emphasize that I am not trying to establish guilt by association, the thesis that environmental ethics is somehow necessarily implicated in whatever harms and wrongs have arisen from pursuit of the suburban ideal. The point is rather that the cultural heritages of environmentalism and of the suburban ideal are complicated and intertwined, and that the simple dualism of domestic and wild, which runs parallel to the simple dualism of suburb and city, ought very seriously to be reconsidered.

Against a blanket condemnation of suburbs and exurbs, I have argued elsewhere for a much more fine-grained consideration of built environments that is broadly pluralistic and rooted in lived moral experience: as people pursue their various projects in particular places, all sorts of very particular values and disvalues may be in play (Kirkman 2010b: 8–14). In this, I have built upon the work of others who have departed from the conventional discourse of environmental ethics, not least Bryan Norton's (2007: 29–31) appeal to adaptive management as a model for environmental policy making.

In this way, it is possible to acknowledge whatever is good and right and appealing about the suburban ideal and the real metropolitan landscapes it has helped to produce, as well as what is bad and wrong and ugly about them. Where people disagree they might, with a richer array of values and finer tools for inquiry, be able to move past disagreement toward the establishment of landscapes that are better, by one measure and another.

Or so I would hope.

Second Challenge: The Metropolitan System

Environmental ethics is a hopeful enterprise, after all.

Confronted by an apparent ecological crisis, one that may well pose an existential threat to human civilization and to life on Earth as we know it, a number of philosophers seized up on a solution. The crisis, they argued, is rooted in ways of thinking and ways of valuing, so all we need to do is to change the way we think and the way we value. If we can convince enough people of the need for such a change, preferably by a reasoned argument, we can hope to dispel the crisis and restore some balance and harmony to our lives in the world.

Surely, there is something right about this premise: how we humans think and how we value, how we make sense of our lives in the world, is bound to play a role in how we actually structure our lives in the landscape, how we see those landscapes in relation to our values, and how we alter landscapes to get what we are after. Conventional environmental ethicists may go astray in focusing so much on arcane questions of meta-ethics, but it is hard to fault the intention behind their work.

Or, at least, it would be hard to fault such intentions, but the hope of a purely or mainly intellectual and cultural solution to the environmental crisis may simply be naïve.

Thinking back to the suburban ideal, it would not be enough to show that the ideal is a mirage resulting from a distant and hazy understanding of "nature" and even of "the city." I have already indicated how difficult it would be for those engaged in conventional environmentalist discourse to point out and to critique this mirage, given their own commitments regarding "nature" and "the city." Even if they could, though, and even if there could be a widespread public debate over the meaning of the suburban myth leading to a shift in the way people think about and value landscapes, very little might happen as a result.

The myth of the suburbs is one component in a large, complex and surprisingly resilient sociotechnical ensemble: the metropolitan region (Kirkman 2009: 237–239). That is to say, the metropolitan region is a *system* which comprised natural, social and technical components. Like all systems, the metropolis has emergent properties that make it more than the sum of its parts. Like all systems, the metropolis includes feedbacks that tend either to stabilize or to destabilize it when something changes. To say the system is resilient is to say that it tends to return to a relatively stable state after a disruption.

From the point of view of someone living in a metropolitan region, the system itself may be out of sight and out of mind. What enters most directly into my experience are my own projects and the opportunities and constraints I find in my environment in relation to those projects (Kirkman 2005: 43–47, 2010b: 35–39). What I may miss in this is that my own way of making sense of my environment and the values and projects I pursue may themselves incorporate opportunities and constraints from the system of which I am, however unwittingly, a part.

Put simply, I can cope quite well with the constraints imposed on me by the metropolis if my imagination is likewise constrained, if it never occurs to me to want or value or pursue something the system will not afford me in any case (Kirkman 2010a: 135–136). In that circumstance, I might have the impression of perfect freedom: I am free to choose a house in a subdivision with a golf course or a house in a subdivision with tennis courts.

The automobile is an especially powerful artifact within the metropolitan system, taking on the status of a "boundary object" (Bijker 1995: 282–287; see also Hommels 2005: 331–332). A boundary object presents an all-or-nothing choice; once I accept the object, I am drawn inside its "technological frame," which structures the meaning of my environment accordingly. So, by choosing to use the automobile, my environment takes on automobile-oriented meaning, and I begin to confront problems that can only arise for someone inside the frame: how to avoid a particular traffic jam, where to buy gas, where to park. To someone outside the frame, those problems are supposed simply not to arise.

Once inside the automotive frame, it can be extremely difficult to get back out of it again, perhaps especially if the landscape itself – the hard infrastructure and established institutions and cultural tropes of the metropolis – has been reconfigured over decades by people who are in the frame for people who are in the frame. In the United States, which has been dominated by the culture – not to say the cult – of the automobile for so long, it could be argued that there is no practicable way for someone to avoid being in the frame. Even for an individual who chooses not to drive, gas prices and traffic jams still matter because the whole, deeply interconnected system is configured for those who drive.

All of this is leading to the insight that sociotechnical ensembles can come to exhibit obduracy (Pinch and Bijker 1987: 40–46; Bijker 1995: 282; Hommels 2000: 323–324). Even though they are, in some sense, the products of human choice and human activity, they are constituted by a convergence of many choices and many activities that, in their interaction, give rise to emergent properties. Because the system is not the product of any one choice or activity, it is unlikely to be altered by any one choice or activity, however deep and widespread might be the commitment to that choice.

To use Thomas Hughes' term for it, as a large sociotechnical system becomes better established, it can take on a kind of momentum (1994: 15). Even if we all were to change our minds, we would be hard pressed to change the course and speed of the system except, perhaps, for a minor deflection one way or another – and maybe not the way we intended.

To those who coined the term, the roots of obduracy are to be found in culture: the meanings of artifacts become fixed, the technological frames rigid; we invest ourselves and our values into them as well as our resources and time, and so we have trouble thinking our way out of them. Consider that the standard response to the problems of resource depletion and climate change tied to petroleum-powered automobiles is typically to find some new way to power automobiles so we can keep on driving, rather than to reconfigure our lives in the landscape so that we are no longer so dependent on automobiles. It is not just that Americans love our cars, but that we cannot even imagine the alternatives nor, really, the depth of our dependence. As with the choice of subdivisions, the choice of gasoline or diesel or electric or hybrid feels like freedom.

I am convinced that cultural factors alone are not sufficient to account for the obduracy of the metropolis (Kirkman 2009: 242–245). Following Hughes, I would draw attention to the brute physicality of the system, all the asphalt and concrete, all the tracks and pipes and wires already in place. The physical infrastructure does represent a massive investment of time and energy and materials made over decades. The difficulty of changing or abandoning the configuration of that infrastructure is not only due to our cognitive or emotional reluctance to lose our investments, but also to the simple fact that elements of the infrastructure are hard and heavy. Tearing up the infrastructure and reestablishing it in a new and better configuration would require an even larger investment of time and energy and material – none of which are as plentiful now as they once seemed to be.

Not to put too fine a point on it, but we are, to an alarming degree, stuck with the metropolitan system, including the persistent myth of the suburb. At the very least, environmental ethicists must face up to the fact that the landscapes we inhabit are not transparent to our will, and that even our own minds may not be entirely in our control. This puts serious limits on the efficacy of environmental ethics as a force for change.

This state of affairs is alarming because the core motivation of modern environmentalism and of environmental ethics is at least partly correct: human civilization is confronted by a crisis, one that may well pose an existential threat to all our human projects and all that we value. An understanding of the obduracy of sociotechnical ensembles can help to bring that crisis more clearly into focus.

Everyday human activity can usually be framed within a timescale of hours and days and weeks, and at a spatial scale of meters and kilometers. Long-term planning may encompass months and years and involve connections at wider scales, but day-to-day life is generally close to home. These activities can have spill-over effects on systems that work at larger spatio-temporal scales, ecological and climatic systems that span decades and centuries, from regions to continents to the biosphere as a whole (Norton 1996: 67). An accumulation of spill-over effects, magnified by the various sociotechnical ensembles we have set up for ourselves, can begin to shift these larger systems in significant ways, but ways that may not be immediately evident down at the level of daily human concern (Kirkman 2007: 25–32).

Global warming, ecological disruption and displacement, extinction, invasive species and other biogeographic shifts are changing the landscapes on which the metropolitan system depends for its continuation. That landscape is, metaphorically and sometimes literally, slipping away, shifting out from under the metropolis – and yet the metropolitan system is slow to change in response to it, because of its momentum.

The image of an obdurate metropolis in a transitory landscape suggests a way of framing the problem of sustainability: in order to respond to shifts in the larger systems on which we depend, we humans need an agile and adaptable way of organizing our lives in the landscape, so that the most important human projects can continue – including the project of human civilization – even in the face of environmental change. What we have instead seems to be a solid and sluggish system of organization and a stubborn commitment to its forms and to its various short-term advantages.

We may be in for a devastating crash (regarding which see Kunstler 2005: 20–21).

To bring this back around to the main point, the problem of sluggishness would persist even if environmental ethicists could, in the time we have left before serious disruption of all human projects, bring about the cultural transformation they hope for.

Closing Reflection: Hope

I do not mean to counsel despair or defeat. I would instead suggest that, in thinking seriously about metropolitan regions as the native habitat of many humans, and about the myth of the suburbs, it may be possible to establish a more practical approach to environmental ethics and, with it, fresh grounds for modest hope.

The first condition is to drop geographic dualism in both of the variants considered here: city versus suburb/exurb and domestic versus wild. In its place, take up our human environment in all its variety, across metropolitan regions and beyond; take it as a continuum across which human projects may intertwine and come into conflict. Take up also the intertwining of the systems that constitute and support these landscapes – natural, social and technological – in all their richness and complexity.

428

The second condition is to adopt a finer-grained, more concrete approach to the values at stake in human projects, including the natural values that catch the imaginations of conservationists and environmentalists – and not only them. Spell out these natural values in specific detail and range them alongside all the other values, also in their specificity: human thriving, justice and the legitimacy of decision-making processes.

The third condition – as suggested by others before me, but perhaps especially by Norton (2007), as already noted – is to adopt an experimental, adaptive and inclusive approach to the problem of reconciling and integrating the many values that are at stake in decisions about our shared environment. Such an approach is at least more likely than a more abstract and perhaps dogmatic approach to yield the kind of agility we may need to respond to the problem of obdurate metropolitan regions in transitory places.

What modest hope we can find in such an approach lies in the fact that obduracy is not monolithic. Some components of or interactions within a sociotechnical system may be more responsive or pliable than others, more responsive to change in the face of shifting human values and habits, new market trends or an emerging political will.

Efforts to retrofit low-density and car-dependent patterns of development are one small but promising trend (Dunham-Jones and Williamson 2009). Even as I write this, a gap along the main street of Decatur, Georgia is being closed. The site has for decades been occupied by a suburban-style office block, set well back from the street and turning a blind side to pedestrians going by. Construction is now underway on a new building in front of the tower, with retail space fronting directly on the sidewalk and four stories of condominiums above: more dwellings for people who will have access to transit and to a small walking city, and more destinations for existing foot traffic.

The effort may well turn out poorly, the good intentions behind it may be turned aside, one way or another, by the obduracy of the metropolitan system. It is notable, for example, that retail in the center of Decatur now consists mainly of restaurants and of boutiques that can only be described as "twee." Even so, the change in the feel and the function of that stretch of Ponce de Leon Avenue is at least a demonstration that small changes are possible, changes that might accumulate and begin to develop some counter-momentum of their own.

The idea of retrofitting the metropolis, one project at a time, does strike me as the basis for a modest and, it must be said, fragile kind of hope. It is entirely possible that, even if we face up to the myth of the suburbs and the myth of the wild and begin to see our predicament more clearly, our timing may simply be off. Shifts in the natural systems on which our systems of habitation depend may already have accelerated to the point that, even with a focused, concerted effort on our part, we could not shift the metropolis quickly enough to keep up. We may still be headed for a devastating crash.

But that does not strike me as an excuse not to make the attempt.

Related Topics

33. Urban Sustainability
36. Transportation

References

Bijker, W. E. (1995) *Of Bicycles, Bakelites, and Bulbs: Toward a Theory of Sociotechnical Change*, Cambridge, MA: MIT Press.

Carson, R. (1962) *Silent Spring*, New York: Houghton-Mifflin Company.

Cronon, W. (1995) The Trouble with Wilderness; or, Getting Back to the Wrong Nature. *In:* Cronon, W. (ed.) *Uncommon Ground: Rethinking the Human Place in Nature.* New York: W.W. Norton & Co.

Dunham-Jones, E. & Williamson, J. (2009) *Retrofitting Suburbia: Urban Design Solutions for Redesigning Suburbs*, Hoboken, NJ: Wiley.

Fishman, R. (1987) *Bourgeois Utopias: The Rise and Fall of Suburbia*, New York: Basic Books.

Garreau, J. (1991) *Edge City: Life on the New Frontier*, New York: Anchor Books.

Gunn, A. S. (1998) Rethinking Communities: Environmental Ethics in an Urbanized World. *Environmental Ethics*, 20, 341–360.

Hommels, A. (2000) Obduracy and Urban Sociotechnical Change: Changing Plan Hoog Catharijne. *Urban Affairs Review*, 35, 649–676.

—— (2005) Studying Obduracy in the City: Toward a Productive Fusion between Technology Studies and Urban Studies. *Science, Technology & Human Values*, 30, 323–351.

Hughes, T. P. (1994) Technological Momentum. *In:* Smith, M. R. & Marx, L. (eds.) *Does Technology Drive History?: The Dilemma of Technological Determinism*. Cambridge: MIT Press.

Jackson, K. T. (1985) *Crabgrass Frontier: The Suburbanization of the United States*, Oxford: Oxford University Press.

King, R. J. H. (2000) Environmental Ethics and the Built Environment. *Environmental Ethics*, 22, 115–131.

Kirkman, R. (2004) Rousseau in the Suburbs: Geography, Environment, and the Philosophical Tradition. *In:* Backhaus, G. & Murungi, J. (eds.) *Earth Ways: Framing Geographical Meanings*. Lanham, MD: Lexington Books.

—— (2005) Ethics and Scale in the Built Environment. *Environmental Philosophy*, 2, 38–52.

—— (2007) A Little Knowledge of Dangerous Things: Human Vulnerability in a Changing Climate. *In:* Hamrick, W. & Cataldi, S. (eds.) *Merleau-Ponty and Environmental Philosophy: Dwelling on the Landscapes of Thought*. Albany: SUNY Press.

—— (2009) At Home in the Seamless Web: Agency, Obduracy and the Ethics of Metropolitan Growth. *Science Technology & Human Values*, 34, 234–258.

—— (2010a) Did Americans Choose Sprawl? *Ethics and the Environment*, 15, 123–142.

—— (2010b) *The Ethics of Metropolitan Growth: The Future of Our Built Environment*, London: Continuum.

Kunstler, J. H. (2005) *The Long Emergency: Surviving the End of Oil, Climate Change, and Other Converging Catastrophes of the Twenty-First Century*, New York: Grove Press.

Leopold, A. (1949) *A Sand County Almanac and Sketches Here and There*, New York: Oxford University Press.

Light, A. (2001) The Urban Blind Spot in Environmental Ethics. *Environmental Politics*, 10, 7–35.

Newton, D. E. (1996) *Environmental Justice: A Reference Handbook*, Santa Barbara, CA: ABC-CLIO.

Norton, B. G. (1996) Integration or Reduction: Two Approaches to Environmental Ethics. *In:* Light, A. & Katz, E. (eds.) *Environmental Pragmatism*. London: Routledge.

—— (2007) Ethics and Sustainable Development: An Adaptive Approach to Environmental Choice. *In:* Atkinson, G., Dietz, S. & Neumayer, E. (eds.) *Handbook of Sustainable Development*. Cheltenham, UK: Edward Elgar.

Pinch, T. J. & Bijker, W. E. (1987) The Social Construction of Facts and Artifacts: Or How the Sociology of Science and the Sociology of Technology Might Benefit One Another. *In:* Bijker, W. E., Hughes, T. P. & Pinch, T. J. (eds.) *The Social Construction of Technological Systems: New Directions in the Sociology and History of Technology*. Cambridge, MA: MIT Press.

Rome, A. (2001) *The Bulldozer in the Countryside: Suburban Sprawl and the Rise of American Environmentalism*, Cambridge: Cambridge University Press.

Sellers, C. (1999) Body, Place and the State: The Makings of an 'Environmentalist' Imaginary in the Post-World War II U.S. *Radical History Review*, 74, 31–64.

Stern, R. A. M. (1986) *Pride of Place: Building the American Dream*, Boston and New York: Houghton Mifflin and American Heritage.

Stilgoe, J. R. (1988) *Borderland: Origins of the American Suburb, 1820–1939*, New Haven: Yale University Press.

36

TRANSPORTATION

Lisa Schweitzer

The transportation sector poses major problems for ecology, particularly motorized transport such as cars, trucks, and airplanes. These activities consume about 61.5% of all the oil used each year on the planet, the largest share of that goes to passenger mobility, or cars (Rodrigue et al. 2013). Motorized transport accounts for roughly a quarter of global carbon monoxide (related to climate change emissions) (OECD IEA 2009). The human health effects of road transport (both freight and cars) account for nearly $1 trillion in human health costs globally in 2010 (OECD 2014). Vehicular crashes cause 1.24 million deaths worldwide, while an additional 20–50 million are injured. Road crashes are the leading cause of death among young people aged 15–29, and the second leading cause of death worldwide among young people aged 5–14. In sum, every year, the planet loses 1,000 young people (under the age of 25 die on the world's roads) *each day*.

Granted these problems, it is tempting to conclude that personal cars have little role to play in a truly good society. But cars have also greatly enhanced economic and social lives. Cars still dominate cities and transport in the US despite four decades of public policy and urban planning that have attempted to get people to choose alternatives means of mobility. Auto consumption continues to grow around the globe (Sperling & Gordon 2009). Cars are convenient, comfortable, and provide both flexibility and speed for the people who have them. Car users also enjoy considerable advantage in labor markets in many cities. Cars may kill people and make us sick, but they have proved too useful for many people to give up easily.

Thus, the quest in sustainable transport has been to formulate strategies whereby individuals and economies can enjoy the car's mobility benefits while eliminating (or at least lessening) the environmental problems. This chapter discusses the key environmental ethics in transportation planning and policy. I will begin with an overview of key transportation terms, move into a discussion of the specific environmental problems, and, finally, describe the ethical tensions in play in transportation policy and planning.

Before I begin, I should clarify my argument and approach. Most normative research on transportation and the environment adopts a consequentialist and distributive justice approach to evaluating what is good. Consequentialists judge right and wrong in moral choice based on the outcomes of the decision, whereas distributional justice examines how these outcomes—costs and benefits, mostly—are allocated among different social groups. These approaches have dominated transportation because the field itself has been populated mostly by engineers and economists far more than any other field, and these disciplines tend to view right and wrong in utilitarian calculations about what, on balance, benefits or costs the most. In addition, when access to

DOI: 10.4324/9781315768090-43

opportunities determines economic distributions, the most important goods that can be distributed are access and mobility.

Nonetheless, activists and urban planners have begun to argue for a "right" to mobility or access that transcends cost-benefit analysis and prioritizes access to opportunity, particularly among those who are impoverished (Lucas 2004). These approaches often advocate for public transit, but questions persist about whether transit, granted the labor market advantages of the car in many urban contexts, simply keeps impoverished workers at a disadvantage, whereas having access to a car would not. Those who advocate for car access among impoverished workers note that fewer cars would benefit the environment; it would undermine auto use among impoverished workers, contributing to economic injustice (Blumenberg & Smart 2010). Still others argue that the environmental problems from car use fall hardest on low-income and minority populations in cities, so that failing to constrain the car, and its attendant environmental problems, itself poses an environmental justice problem far worse than policies that constrain auto use (Schweitzer & Valenzuela 2004). That tension—between wanting to secure access to opportunity as a basic right while also protecting the environment—leads to the basic questions in this essay: if cars harm environments, but they are also important to job access and economic security, what are the right public policies to pursue, and what is the right choice to make as an individual traveler?

A Basic Vocabulary

"Transportation" covers many different concepts, but most policy and planning centers on passenger and freight travel. Passenger mobility involves transporting people to places they desire to go, including work and educational opportunities. Freight concerns shipping parcels and commodities. Throughout, I will focus primarily on cars, but virtually all of the environmental problems associated with cars also occur with aviation and freight, though the types of emissions and magnitude of the problems differ. Passenger travel also includes modes like walking, skateboarding, and biking. These modes are generally considered to be positive for the environment, although these modes, too, can affect wildlife and ecologies in natural areas if overused or poorly designed. But that impact is small and more readily addressed with design compared with walking and biking as means to help alleviate the environmental problems of the car in urban settings.

Within passenger travel, scholars further distinguish between mobility and access. Mobility involves moving things around physically. Access means the ability to be somewhere or participate in things that offer the social and economic opportunities that one both wants and needs (Cervero 1997). In general, mobility is the means, access is the end: one uses mobility in order to have access, although there are those who enjoy walks, train rides, and driving for fun. The distinction between mobility and access divorces being able to do things people want to do—like get and keep a job, attend a concert, or go shopping—from having to be mobile.

Familiarity with a few more terms also helps clarify the issues. "Mode" describes travel by different means, such as driving alone, riding the bus or walking. Mode share describes the breakdown of all travel according to mode, usually given in percentages, and varies between cities. In the US, for example, the national personal car mode share is close to 83% for commutes in the US, though cities, like Chicago and New York, have much greater shares undertaken by foot than the US aggregate (Schweitzer 2017). Further, a "mode shift" means that people move from taking one mode to another. One key public policy goal has been to get a mode shift from cars to other, less environmentally damaging modes like walking, biking, and transit.

Whether people can use cars, bikes, transit, or walking safety hinges on the social environment. Racist policing in the US means that Black Americans face state violence during routine traffic stops, on transit, or on the sidewalk (Butler 2020). The same is true of LGBTQIA travelers who face violence when they enter public spaces (Lubitow et al. 2017). Police, policy, and design

432

actively harm people with disabilities, and especially when a traveler faces multiple, intersecting oppressions of race and disability. Race, gender, and disability structures (a) the experiences and consequences for individuals using different modes and (b) how various solutions might work, or not, in specific neighborhoods.

The Environmental and Urban Issues of Car Use

A "lifetime" environmental perspective on cars helps enable a discussion for why cars, regardless of their engine technology, pose serious environmental problems. Most automobiles on the road around the world are internal combustion engines running either gasoline or diesel fuel. Refining technologies could make fuel much cleaner. Similarly, engine technologies can use energy more efficiently and create fewer local problems for air quality. Moreover, most automobile parts can be scraped, reused, and recycled readily. But battery production and disposal for electric vehicles poses thorny environmental problems, and their environmental impact hinges on whether the electricity they will use is generated with coal or other feedstocks.

Currently, global transportation activities consume about 61.5% of all the oil used each year on the planet; the largest share of that goes to passenger travel like cars. Rail transit, for instance, is four times more energy-efficient than automobiles. The life-cycle costs of oil include the many environmental problems associated with extraction. Of those, perhaps the most serious and well-documented concern the ecological and political strife that accompanies oil extraction throughout the world (O'Rourke & Connolly 2003). Economists have documented the "curse of oil" that describes how this commodity splits people within oil-rich countries into the resource-controlling, super-rich elite, and the poverty-stricken, environmentally devastated local residents. At best, Ross (2012) argues that oil-rich countries have governments that are less accountable to their citizenry, and they have worse outcomes for child mortality and nutrition, lower literacy and school-enrollment rates, and poor marks on the United Nations' Human Development Index. At worst, the quest for oil wealth increases inter-ethnic tensions and violence, as in Nigeria and Sudan, due to inter-ethnic violence over the unjust distribution justice of oil wealth and the environmental problems of extraction (Kaldor et al. 2007). Local communities and ecosystems near oil wells, pipelines, and other oil transport face risk from potentially devastating spills and accidents (Bushell & Jones 2009). Major oil spills, however infrequent, are disastrous to marine ecosystems, commercial and subsistence fisherman, tourism, and local community residents (Palinkas et al. 1993; Garza-Gil et al. 2006). The chronic, small spills associated with oil production and pipelines also pollute groundwater, agricultural land, and nearby animal populations that local populations may need (Amadi et al. 1996).

Refining crude oil into gasoline and diesel fuel for use in transportation, too, has serious environmental consequences, particularly for communities near refineries (Berman and Bui 2001). Due to industry consolidation, most European and North American refineries handle large volumes of industrial chemicals. Air pollution produced during the refining process includes toxic and hazardous pollutants, such as carcinogens (benzene, naphthalene, 1,3-butadiene), and polycyclicaromatic hydrocarbons (PAH). In data released in 2011, the oil and natural gas industry was the second largest industrial polluter, with refining at third (Fahim et al. 2010). Accidents, fires, and leaks at refineries can cause toxic chemicals held or combusted onsite to enter local air and water supplies.

As with refining, carbon-based fuels pollute through driving. There are many possible health effects, and virtually all major regulated pollutants affect the heart, lungs, and airways, but the geographic area of risk exposure varies to some degree. Freeways, bus barns, and truck corridors are the places where large numbers of vehicles assemble and operate, and where they are likely to run in more polluting engine modes, like idling (Kinsey et al. 2007; Karner et al. 2010).

Beyond local "hotspot" effects from traffics, pollutants actively move throughout the atmosphere. Carbon monoxide tends to remain near roadways and ground level; it reduces oxygen delivery through the blood to the heart, brain, and other tissues, increasing the risks of heart attack and stroke (Phalen & Phalen 2013). Other pollutants can move significant distances, and some react with other atmospheric chemicals, so that their potential health effects can occur both right by transportation facilities and farther away (Seinfeld & Pandis 2012).

Sulfur dioxide prompts airway inflammation in healthy people. Oxides of nitrogen contribute to ozone formation, as well as act as respiratory irritants themselves. Particulate matter from combustion comes in many forms and sizes, and their effects range from respiratory irritation to, at the scale of ultra-fine particles, carcinogenic effects. These pollutants impair respiration, which affects those with cardiovascular and respiratory ailments. About 13% of carbon dioxide, associated with global warming, comes from the transport sector (IPCC 2007).

Cars also affect urbanization scale and scope. Cars take up a lot of space, both when they are in motion and when they are stored. On-street parking, parking structures, and parking lots consume urban land in hardscape that, unless carefully designed, contributes to runoff issues and urban heat island effects (Shoup 2011). Urban heat islands occur when the hard surfaces in cities dominate over greenspace and plant life in such a way that causes cities to have higher ambient temperatures than otherwise. Along with housing polices and suburbanization, cars are also tied to urban sprawl because they help spread urban populations out, such that human settlement consumes more land than higher density, transit-oriented development (TOD) (Brueckner 2000). Sprawl reinforces auto dependence, and all the ills that I just enumerated, along with additional ecological problems like habitat destruction and fragmentation (Ewing et al. 2005).

Sprawl and suburbanization, and its supporting technology, the auto, are also associated with residential segregation by race, class, and ethnicity and social exclusion from jobs and other means of economic mobility among those who do not have cars. Residential segregation is associated with social and economic isolation for those in low-income and minority communities. John Kain (1968) observed that suburbanization isolated African American men in inner cities from suburban job opportunities. As metropolitan regions continued to spread and decentralize from 1970 until today, subsequent generations of researchers have built on Kain's observations, noting the effect that sprawl and auto-oriented development have on women, single-parent wage earners, welfare recipients, and immigrants (Blumenberg & Manville 2004; Blumenberg & Smart 2010). Beyond exclusion from economic opportunities, sprawl and segregation contribute to exclusion from urban amenities, like parks and school funding (Powell 2007).

Some advocate to overcome exclusion by investing more in transit, to connect impoverished workers with job opportunities located throughout metropolitan regions. But job accessibility via transit has proved difficult to deliver. Even after four decades of public investment in transit in the US, those who cannot afford cars or who cannot drive currently face disadvantage in labor markets in the US regions, even in regions where there is fairly good transit, walking, and biking. Even in New York, by far best transit supplier in the US, the jobs accessible by car still outnumber those accessible by transit 10 to 1 for a 20-minute commute, and 6 to 1 for a 40-minute commute. These data were calculated by the author using data from the University of Minnesota's Accessibility Observatory.

In other regions, the job access advantages of commuting by car are even greater, and the commutes by transit much, much more time-consuming. The downside, of course, is that owning a car is very expensive for impoverished families, even if cars are helpful in getting and keeping a job (Smart & Klein 2015).

The labor market advantages of car ownership mean that public policy and planning cannot simply assume that people will "get out of their cars" to benefit the environment. People need employment, and they cannot be expected to choose transit if that means considerably more time away from their families and other priorities. The labor market advantages of the car also mean that policies designed to change how much people use the car, to the degree that those policies might "sanction" or "cost" car drivers, can dampen employment and mobility for impoverished people who need cars. In turn, public policy and planning have two strong ethical imperatives, each in tension with each other: (1) the environmental costs of automobile dominance are major, and these can fall disproportionately on low-income communities of color (Schweitzer & Valenzuela 2004) and (2) the costs of discouraging auto use can also fall disproportionately on impoverished workers and job seekers (Schweitzer 2009). I shall explore these issues more in the next section.

Proposed Policy and Planning Solutions

For smaller, mid-sized, or less motorized cities internationally, American auto dominance has had a clear lesson: invest in transit and other modes, like walking and biking, as cities grow. Most major Asian and Latin American cities have actively invested in transit in the past three decades as their urban populations have swelled. Nonetheless, growing wealth globally has prompted greater auto ownership in places like metropolitan India and China (Sperling & Gordon 2009). In rapidly growing metropolitan regions that also see growth in wealth among at least some classes of residents, all modes tend to get more use.

Policy and planning solutions in regions that *already* have auto dominance, however, have proven much more difficult. Four major policy and planning approaches have received the most attention in research and practice: (1) boost transit, walking, and biking infrastructure to supply alternatives; (2) joint development models of transit and land use (TOD); (3) price auto use higher through parking charges, tolls, fuel taxes, or user fees; and (4) technological fixes to make cars cleaner.

Investing in Infrastructure Alternatives to the Auto

The US policy has, for the last five decades, relied almost exclusively on increasing supply of transit, walking, and biking infrastructure. Nonetheless, across the US, mass transit (like subways and buses) serve only 5% of all trips (National Household Travel Survey 2010). This number includes rural areas where mass transit is not particularly viable. The low number is itself a problem, but more worrying is just how stable that mode share has been at low levels for decades (United States Census Bureau 2009), even as population growth in the US urban areas far outpaced national growth rates. If anything, urbanization should alter mode shares between cars and transit because with urbanization, population growth occurs more in cities, where transit is supplied, than in rural areas, where transit is less viable. But in the US case, mass urbanization in the last 50 years has not greatly altered the overall share that transit serves. Certainly, more people are riding transit, and that is a positive thing, but transit's *share* of total travel has moved little, even with substantial investments.

One reason why the mode share numbers have been so difficult to shift concerns the job accessibility advantage of cars. Perhaps the US has not yet invested sufficiently in public transit for it to be truly competitive with autos. Proponents of this approach note that the US transit investment has been so much lower than that of its investment in automobiles that judging mass transit's performance against automobiles damns transit unfairly (Litman 2015). That said, Los Angeles has spent roughly US$20 billion on building its regional rail system, and its mode shift has been small to nonexistent even as LA transit systems serve millions of trips a day. Other large US cities have also invested in mass transit with relatively small changes in overall mode shares even as they improve their service.

Skeptics of the "more investment" approach note that automobile access is already so much better than transit access in all but a few places, so that investing more on mass transit does not really help either the environment or impoverished people. Instead, expecting low-income commuters to depend on mass transit puts them in a permanent economic disadvantage. It may be that the car's geographic and mobility advantages are so difficult to counter that transit accessibility will never mirror that of cars. Proponents hope, however, that greater investment may reduce the differences in job access between having a car and not having one. For proponents, transit's comparative disadvantage, however, is evidence of urban injustice because low levels of transit investment have reinforced "second class" citizenship for low-income transit riders (Chen 2007). Given that African Americans have been among the US mass transit most loyal customers, lower public investment in transit, thus, represents both racism and classism in urban services, and therefore, that low investment is itself an injustice that cities and regions should rectify.

Transit-Oriented Development

For some transit experts like Robert Cervero (1997), one reason why transit investment over the last three decades has failed to change mode shares has been that the US land development patterns do not support the population densities required for transit service. Instead, land development in the US has suburbanized, with an emphasis on single-family homes rather than more urban-style, multi-family buildings with more residents.

TOD relates to transit investment by focusing on real estate development practices around transit service in cities. By intensifying land development near stations with a mix of different types of activities, TOD seeks incorporate high-density housing (multi-family units, apartments, condos) with retail and workspace so that people have ready access via transit, walking, and biking to many different things they need, from employment to grocery stores. Nearly all contemporary sustainable development models advocate for higher residential densities, mixed land uses, and transit access as opposed to single-use zoning where residential housing is geographically separated from workplaces and shopping areas by municipal regulation.

Taken as such, these models promise quite a bit for both the environment and social justice. Increasing development by transit stops and stations means more access for transit users. De-emphasizing the automobile during development should, in theory, reduce environmental ills. In practice, however, the results are far more mixed. One first concern is just how much transit use these new developments foster. New businesses, new housing opportunities, new shops—new activities in general—can provide wonderful amenities to transit users, but they can also attract additional car trips right along with new transit trips.

These developments have done well in real estate markets, so well in some regions that contemporary urban political conflicts center on the role TOD may have in sparking gentrification, a process where more affluent residents displace current residents from a neighborhood (Song & Knaap 2003). Because most new development tends to serve comparatively more affluent renters and shoppers, renters who need access to transit the most can get priced out of neighborhoods where TODs get built. Displaced transit dependents can wind up with longer transit trips than before in these situations, and relatively affluent TOD residents may do little to increase transit's mode share. Rather, high rises of luxury condos near subway stations can wind up with floor after floor of underground parking for the BMWs and Land Rovers. Careful thought has to go into changing urban land with gentrification pressures at the ready.

Finally, higher residential densities near existing communities have proven to be politically controversial, and it is not clear how much democratic support exists for higher population densities and neighborhood change outside urban planners, architects, and people with a taste for urban living. While European cities often receive praise for their transit and street life, few North American advocates for European-style cities go so far as to suggest that Americans, in particular, should also pursue the social policies that have supported European patterns of urbanization around public transit, walking, and biking. For example, while European cities restrict automobiles and invest in transit, many also support subsidized and public housing that can allow low-income residents to remain competitive in expensive, highly urbanized housing markets proximate to transit. By contrast, public housing or housing support vouchers in the US have sought to disperse impoverished residents throughout metropolitan regions, rather than keep them near transit.

Pricing Approaches

Unlike transit investment and TOD strategies, which may take years to implement, pricing policies could, in theory, be enacted quickly. Pricing strategies render automobile use much costlier through different mechanisms so that people are more likely to choose another, cleaner mode instead. Mechanisms include price floors on gasoline (Sperling & Gordon 2009); higher gasoline taxes (Parry & Small 2005); carbon taxes (Hammer & Jagers 2007); usage tolls or fees (Schweitzer & Taylor 2008); and parking charges (Shoup 2011). Given pervasive auto use, and that the auto enables accessibility, many worry that additional fees for car usage simply punishes low-income, working poor who require cars to obtain or hold onto work. For some locations and at some income levels, good-quality transit exists and provides a ready substitute for cars. In other locations, however, higher auto use costs could conceivably push those on the margin of employment out of the workforce because transit is an unreliable substitute, or make an owning a car in order to keep a job additionally financially burdensome. Transition policies can help ameliorate these potential problems.

Environmental economists suggest two possible approaches to deal with these concerns. First, the revenue from pricing cars could be simply returned to people in a lump-sum credit at tax time as a rebate. That approach would raise the cost of the car relative to other modes, but households would not be worse off, in sum, over the long term. Rebate policies are, however, rarely implemented.

Another possible means to help out low-income drivers might be to invest the revenues from environmental pricing strategies into enhancing transit service quickly so that people priced out of the cars might have a viable alternative quickly. When Transport London implemented a high cordon toll to dampen traffic in central London, they also put 300 extra buses on the road to help those priced out of their cars get where they needed to go (BBC News 2003). In Los Angeles, new toll facilities on freeways dedicate at least part of the revenue to boosting up transit service in the corridor. These kinds of approaches assume that transit is really a substitute for driving, and for some families, that may not be true. In that case, some may wind up just paying the additional costs of their mobility and cutting back other necessaries.

Technology Solutions

Changing technologies perhaps do not get the respect they deserve from environmentalists or urbanists, who tend to want to get people to stop driving and start walking, biking, and taking transit. Environmentalists note that "technology won't save us." Rather, cars with better fuels and improved engine technologies, which would help with local air quality and climate emissions, still kill people and animals, contribute to urban sprawl, and cause congestion. For many detractors, cars are aesthetically displeasing and reinforce undemocratic, segregated urban environments, unlike transit, which can be both attractive and socially unifying (Powell 2007).

Less abstract critiques focus on the environmental limitations to new technologies. Fleetwide engine changes, however effective they may be, appear to be decades away from implementation.

Furthermore, life-cycle assessment of batteries, electricity, and other potential energy sources for different technologies shift environmental problems from one place to another, rather than solve them outright (Gallagher 2006). Plug-in battery electronic vehicles (BEVs) promise to be quiet and virtually nonpolluting in *local* environments where they operate but, if powered by coal-fired power-plants elsewhere, environmental and resource depletion problems persist where those plants are located. Finally, low-income groups tend to be the last group to adopt new technologies because these are costly, and thus they may be the last group to benefit from new engine technologies.

Nonetheless, existing vehicle and refining technologies have reduced emissions and crash deaths significantly, with spillover benefits for low-income neighborhoods that face air quality problems. Because the global auto fleet is already so large, changing engines to improve fuel economy or moving to clean energy sources for vehicles could have immense benefits in fuel conservation and pollution control (Sperling and Gordon 2009). The same plug-in battery electric vehicles above could be much cleaner if their electricity comes from solar or other renewables.

New technologies also appear to be changing how people use cars in urban environments (Levinson and Krizek 2015). Digital technologies employed by companies like Uber and Lyft have made it much easier to share vehicles, which may mean fewer cars on the road. People seem to use these services in tandem with transit, and thus, they might serve to reduce auto ownership and use overall, particularly among young people. Nonetheless, rideshare labor conditions are a persistent problem; rideshare cars cruising while they look for passengers can increase driving and fossil fuel consumption; and many worry that rideshare is creaming passengers away from mass transit.

Automatic routing and vehicle control technologies could further lessen morbidity and mortality from crashes. These technologies may well be a boon to transit as well cars, by reducing the operating costs associated with running bus fleets and allowing local agencies to supply additional transit. These technologies may also enable car subscription services that reduce car ownership through periodic renting. These services have long been considered environmentally and socially beneficial: they allow people to use cars as necessary without owning one. In the latter case, car use increases tremendously when compared with subscription services that require planning and per-trip use.

Conclusions for Environmental Ethics

What to take away from the discussion, in terms of environmental ethics? Policy design requires effort and care to protect both the environment and access to opportunity, but the last 40 years of research and debate about the public ethics of owning and using cars has resulted in feasible and multi-objective policy options.

Transit advocacy has been the most successful in marketing transit as a solution, but perhaps the least successful in really understanding justice concerns and questions about rights to access. Advocates for pricing policies have for decades faced criticisms that pricing is unfair, and thus they have had to figure out equity issues by developing strategies to send price signals that discourage car use and, at the same time, protect low-income drivers. By contrast, transit advocates could assume until very recently that their arguments for additional transit spending, particularly for constructing new projects, served everybody in cities well, and that land uses were a local or secondary problem. But civil rights lawsuits, chronic financial instability of transit agencies, poor-quality service, skyrocketing fares, and station-area gentrification have demonstrated that building more transit is not, necessarily, good for everybody, despite the relative environment benefits (Schweitzer 2017). If anything, transit and TOD strategies require more thinking and new innovations on how to make transit-, walking-, and bicycle-accessible housing and employment more available for those facing both affordability and discrimination problems in urban housing and labor markets.

Even though I spend most of this essay discussing public ethics, I do not wish to entirely ignore individual choice. Transportation professionals tend to discuss an individual's decision to take a car over transit, bike, or walking as a simple economic choice that weighs time and money. That framing emphasizes consumer sovereignty, which is the idea that individuals should be able to pick what they prefer. To the degree that cars possess advantages that people value, such speed and comfort, alternative modes can have difficulty competing for many trips. That notion that individuals have no obligation beyond their own transportation needs places the responsibility on cities and regions to create environments and transit supply roughly equal or better to that of what might be provided in private cars. That is a very difficult endeavor.

Instead, individuals who care about the environment and cities should feel more obligation than a simple consumer choice suggests. Here, the idea of an ethical consumer model for transportation might be quite useful in helping people to understand that every time they get out of their car, they are making a moral choice in addition to a market choice. In turn, every time people make a choice about where to live, whether they select a location where alternatives to cars are a viable option or not, they also choose social and environmental consequences for their global and urban peers.

Individual moral choices, like public policy choices, are not so straightforward. Just as it is likely an undue burden, if not impossible, to expect cities or regions to supply transit supply everywhere and anywhere to suit individual preferences, so, too, individuals can face some real constraints in choosing what is morally better in transport. Perfection is less important than the realization that transit, walking, and biking work much better for some people in the city than others, and that among those of us who can afford to live near transit, and who can perhaps spend more time on our commutes on transit, really do have the obligation to do so even if those are not yet ideal. Getting out of a car is a virtue, not just a consumer choice; those of us who can demonstrate that virtue should do so, and that virtue is conditioned on differences in what people are really able to do, given cities, housing markets, and transportation systems as they are, not as we hope they will be in future.

Related Topics

33. Urban Sustainability
35. Suburbs and Exurbs
49. Pollution and Polluter Pays

References

Amadi, A., Abbey, S. & Nma, A. (1996) 'Chronic effects of oil spill on soil properties and microflora of a rainforest ecosystem in Nigeria,' *Water, Air, and Soil Pollution*, 86(1–4), pp. 1–11.

BBC News (2003) 'First congestion fines to go out,' *World Edition*. Retrieved from http://news.bbc.co.uk/2/hi/uk_news/england/2774271.stm

Berman, E. & Bui, L. (2001) 'Environmental regulation and productivity: Evidence from oil refineries,' *Review of Economics and Statistics*, 83(3), pp. 498–510.

Blumenberg, E. & Manville, M. (2004) 'Beyond the spatial mismatch: Welfare recipients and transportation policy,' *Journal of Planning Literature*, 19(2), pp. 182–205.

Blumenberg, E. & Smart, M. (2010) 'Getting by with a little help from my friends… and family: Immigrants and carpooling,' *Transportation*, 37(3), pp. 429–446.

Brueckner, J. (2000) 'Urban sprawl: Diagnosis and remedies,' *International Regional Science Review*, 23(2), pp. 160–171.

Bushell, S. & Jones, S. (2009) *The spill: Personal stories from the Exxon Valdez disaster*. Kenmore, WA: Epicenter Press

Butler, T. (2020) 'Why we must talk about race when we talk about bicycling,' *Bicycling Magazine*, June 9, 2020. Retrieved from https://www.bicycling.com/culture/a32783551/cycling-talk-fight-racism/

Cervero, R. (1997) 'Tracking accessibility,' *Access*, 11, pp. 27–31

Chen, D. (2007) 'Linking transportation equity and environmental justice with smart growth,' in Bullard, R. (ed.) *Growing smarter*. Cambridge, MA: MIT Press, pp. 299–322.

Ewing, R., Kostyack, J., Chen, D., Stein, B. & Ernst, M. (2005) *Endangered by sprawl. How runaway development threatens America's wildlife*. Washington, DC: National Wildlife Federation.

Fahim, M. A., Al-Sahhaf, T. & Elkilani, A. (2010) *Fundamentals of petroleum refining*. Amsterdam; London: Elsevier Science.

Gallagher, K. 2006. 'Limits to leapfrogging in energy technologies? Evidence from the Chinese automobile industry,' *Energy Policy*, 34(4), pp. 383–394.

Garza-Gil, M., Prada-Blanco, A. & Vázquez-Rodríguez, M. (2006) 'Estimating the short-term economic damages from the Prestige oil spill in the Galician fisheries and tourism,' *Ecological Economics*, 58(4), pp. 842–849.

Hammer, H. & Jagers, S. (2007) 'What is a fair CO2 tax increase? On fair emission reductions in the transport sector,' *Ecological Economics*, 61(203), pp. 377–387.

International Panel on Climate Change (IPCC) (2007) *Contribution of working group I to the fourth assessment report of the intergovernmental panel on climate change*. Retrieved from http://www.ipcc.ch/publications_and_data/ar4/wg3/en/contents.html (Accessed 18 October 2014),

Kain, J. (1968) 'Housing segregation, Negro employment, and metropolitan decentralization,' *The Quarterly Journal of Economics*, 82(2) pp. 175–197

Kaldor, M., Karl, T. & Said, Y. (2007) 'Introduction,' in Kaldor, M., Karl, T. & Said, Y. (eds.) *Oil wars*. London: Pluto, pp. 1–40.

Karner, A., Eisinger, D. & Niemeier, D. (2010) 'Near-roadway air quality: Synthesizing the findings from real-world data,' *Environmental Science & Technology*, 44(14), pp. 5334–5344.

Kinsey, J., Williams, D., Dong, Y. & Logan, R. (2007) 'Characterization of fine particle and gaseous emissions during school bus idling,' *Environmental Science & Technology*, 41(14), pp. 4972–4979.

Levinson, D. & Krizek, K. (2015) *The end of traffic*. Minneapolis: Network Design Lab. Amazon Digital Services, Inc.

Litman, T. (2015) 'Analysis of public policies that unintentionally encourage and subsidize urban sprawl,' *New Climate Economy*. Retrieved from http://static.newclimateeconomy.report/wp-content/uploads/2015/03/public-policies-encourage-sprawl-nce-report.pdf

Lubitow, A., Carathers, J., Kelly, M. & Abelson, M. (2017) 'Transmobilities: Mobility, harassment, and violence experienced by transgender and gender nonconforming public transit riders in Portland, Oregon,' *Gender, Place & Culture*, 24(10), pp. 1398–1418.

Lucas, K. (2004) 'Locating transport as a social policy problem,' in Lucas, K. (ed.) *Running on empty: Transport, social exclusion, and environmental justice*. Chicago: University of Chicago Press.

National Household Travel Survey (2010) 'National Household Travel Survey,' *U.S. Department of Transportation - Federal Highway Administration*. Retrieved from http://nhts.ornl.gov/

OECD (2014) *The cost of air pollution: Health impacts of road transport*. Paris: OECD Publishing. Retrieved from http://dx.doi.org/10.1787/9789264210448-en

OECD IEA (2009) *Transport, energy, and CO_2*. Paris: OECD Publishing. Retrieved from https://www.iea.org/publications/freepublications/publication/transport2009.pdf

O'Rourke, D. & Connolly, S. (2003) 'Just oil? The distribution of environmental and social impacts of oil production and consumption,' *Annual Review of Environment and Resources*, 28(1), pp. 587–617

Palinkas, L., Petterson, J., Russell, J. & Downs, M. (1993) 'Community patterns of psychiatric disorders after the Exxon Valdez oil spill,' *American Journal of Psychiatry*, 150(10), pp. 1517–1523;

Parry, I. & Small, K. (2005) 'Does Britain or the United States have the right gasoline tax?' *The American Economic Review*, 95(4), pp. 1276–1289.

Phalen, R. & Phalen, R. (2013) *Introduction to air pollution science: A public health perspective*. Burlington, MA: Jones & Bartlett Learning.

Powell, J. (2007) 'Race, poverty, and urban sprawl: Access to opportunities through regional strategies,' in Bullard, R. (ed.) *Growing smarter*. Cambridge, MA: MIT Press, pp. 51–72.

Rodrigue, J. & Comtois, C. (2013) *The geography of transport systems*. New York: Routledge.

Ross, M. (2012) *The oil curse: How petroleum wealth shapes the development of nations*. Princeton, NJ: Princeton University Press.

Schweitzer, L. (2009) *The empirical research on the social equity of gas taxes, emissions fees, and congestion charges. Special Report 303*. Washington, DC: Transportation Research Board.

Schweitzer, L. (2017) 'Mass transit,' in Giuliano, G. and Hanson (eds.) *The geography of urban transport.* 5th edition. New York: Guilford Press, pp. 187–216.

Schweitzer, L. & Taylor, B. (2008) 'Just pricing: Equity and the future of road finance,' *Transportation*, 6(85), pp. 797–812

Schweitzer, L. & Valenzuela, Jr., A. (2004) 'Environmental justice and transportation: The claims and the evidence,' *Journal of Planning Literature*, 18(4), pp. 383–398.

Seinfeld, J. & Pandis, S. (2012) *Atmospheric chemistry and physics: From air pollution to climate change.* New York: John Wiley & Sons.

Shoup, D. (2011) *The high cost of free parking.* Chicago, IL: Planners Press, American Planning Association.

Smart, M. & Klein, N. (2015) *A longitudinal analysis of cars, transit, and employment outcomes. Mineta Transportation Institute Report # 12–49.* San Jose, CA: Mineta Transportation Institute.

Song, Y. & Knaap, G.-J. (2003) 'New urbanism and housing values: A disaggregate assessment,' *Journal of Urban Economics*, 54(2), pp. 218–238

Sperling, D. & Gordon, D. (2009) *Two billion cars: Driving toward sustainability.* Oxford: Oxford University Press.

United States Census Bureau (2009) American community housing survey, community in America. Retrieved from http://www.census.gov/prod/2011pubs/acs-15.pdf

37

WASTE AND CONSUMPTION

Jen Everett and Rich Cameron

Consider our stuff: our clothes, cell phones, computers, toys, furniture, toothbrushes, books, rugs, diapers, cleaning products, kitchen gadgets, yard equipment, picture frames, appliances, cat litter, bags to carry it all home, shipping boxes and packing materials to bring it undamaged to our door, and bins to haul it to the curb when we throw it "away," making room for *newer and better* clothes, cell phones, computers, toys, furniture, etc. This movement of consumer products from stores, through our lives, and toward their fated disposal is both normal and normative in consumer societies. Indeed, this way of life is increasingly the norm in the majority of the world's cities, not just those in rich countries. According to mainstream political and economic thinking, this is a good thing – a sign of progress, if not the very definition of development.

As our trajectory toward climate catastrophe becomes ever clearer, however, environmental scholars increasingly finger this same economic system as a (if not *the*) primary driver of rising greenhouse gas emissions (Cafaro 2011a: 192–193, citing IPCC 2007). Back in 2006, leading UK economist Sir Nicholas Stern noted that emissions "have been, and continue to be, driven by economic growth," although on a hopeful note, he claimed that if the global community were to take "strong, early action," then stabilizing atmospheric concentrations of greenhouse gases would be "feasible and consistent with continued growth" (2006: xi). Unfortunately, since international action on climate change can only be described as weak and late at best, the window for economically palatable pathways to such stabilization is closing fast. UK climate researchers Kevin Anderson and Alice Bows conclude unflinchingly that "avoiding dangerous (and even extremely dangerous) climate change is no longer compatible with economic prosperity" (2011: 40; see also Klein 2014).

Climate change has given greater urgency to such assessments, but environmentalists have challenged consumption-driven economics for a very long time. Aldo Leopold's much revered land ethic explicitly repudiated commodity-centered conceptions of value, and he held no admiration for a life oriented around consumer goods. Indeed, many of his writings challenged orthodox economic thinking, including the logic of perpetual growth (Lin 2013). Likewise, Kenneth Boulding's classic 1966 essay, "The Economics of the Coming Spaceship Earth," argued that the "cowboy" view of the economy – as having "infinite reservoirs from which material can be obtained and into which effluvia can be deposited" – must give way to a view of the economy as traveling on a spaceship "without unlimited reservoirs of anything, either for extraction or for pollution." As he explains:

> The difference between the two types of economy becomes most apparent in the attitudes toward consumption. In the cowboy economy, consumption is regarded as a good thing and production likewise; and the success of the economy is measured by the amount of the

DOI: 10.4324/9781315768090-44

throughput .... The gross national product is a rough measure of this total throughput.... By contrast, in the spaceman economy, throughput is by no means a desideratum, and is indeed to be regarded as something to be minimized rather than maximized. The essential measure of the success of the economy is not production and consumption at all, but the nature, extent, quality, and complexity of the total capital stock, including in this the state of the human bodies and minds included in the system. In the spaceman economy ... less production and consumption ... is clearly a gain.

(1966: 9–10)

So the idea that an economic system based on consumerism poses a profound threat to ecological systems is neither new nor controversial, at least among environmentalists. One might therefore expect searching moral critiques of consumptive, wasteful practices to figure prominently among the key works in environmental ethics. Yet, with but few exceptions (e.g., Cafaro 2001, 2011a), this hasn't been the case. Because we believe a prime directive of environmental ethics should be to contribute moral and political understanding to practical efforts to pull civilization back from the suicidal/ecocidal abyss, we hope in this chapter to help refocus the energies of environmental ethicists on consumer society as a central and pressing moral problem.

We begin in Part I by reviewing the environmental and social impacts of prevailing waste and consumption practices, and go on to argue, using Annie Leonard's *Story of Stuff* (2010) as a scaffold, that these practices must be understood in the context of the materials economy as a whole. The environmental ethics of consumption and waste, then, are but pieces of what we call the *environmental ethics of stuff* – an examination of the right and wrong we do, the good and bad we bring about, the justice and injustice we perpetuate, and the virtue and vice we cultivate as we remove materials from natural systems as "resources," transform and exchange them through consumer economies as "goods," and discard them from our lives as "waste."

The traditional philosophical aims of such a project might be to root out fundamental errors in moral reasoning, judgment, and perception that perpetuate consumerism and the throw-away society. Our contribution to an environmental ethics of stuff, however, is pitched more directly to the discourse of environmental activism. We argue that environmental activism around waste and consumption must take into account the complex, systemic nature of the materials economy, rather than focusing on piecemeal metrics such boosted recycling rates. Calling for transformation of a complex and well-entrenched economic system sounds – and is – daunting. But as Leonard points out, this system is broken in so many ways that there are endless opportunities to go to work fixing it. In Part II, we survey a broad array of political, economic, and cultural strategies for shifting the global materials economy toward a just and sustainable "zero waste" ideal.

Finally, in Part III we address the ethical quandary that stuff provokes for us on a personal, individual level. Many activists appear to think that ethics requires us to "be the change we wish to see in the world" by minimizing the environmental impacts of our lifestyles and consumption choices wherever possible. Critics of such strategies, by contrast, argue that such efforts make very little empirical difference and may even be counterproductive from a pragmatic point of view. Rather than policing our individual choices, they urge us to become more politically and civically active on a collective level. Both views have their merits: lifestyle activism reflects an admirable effort to grapple with our personal complicity in the wrongs perpetrated by the consumer economy in which we are enmeshed. At the same time, the critics are right to cast doubt on incrementalist assumptions about social change that are often assumed by lifestyle activists. Ultimately, we hope to complicate the distinction between acting as consumers and acting as citizens in ways that enrich pathways for individuals to act both ethically and effectively.

Waste, Consumption, and the Stuff Upstream

It's easy to get the impression that one of our most important moral failings in the environmental domain is that we throw away too much stuff. Recycling is often the first environmental lesson taught to kids in public schools, and among adults, claiming to "recycle everything" is a familiar way of conveying one's environmental concern. For reasons discussed more fully below, we believe it's possible to become over-fixated on household waste as an environmental problem and over-enamored with recycling as a solution (MacBride 2012). But this is not because garbage doesn't matter. To the contrary, waste both reflects and exacerbates interconnected environmental and social problems of grave moral importance (Preston and Corey 2005). In philosophical terms, we can characterize these problems in at least three overlapping ways – as *harms* to persons, other animals, and nature; as *injustices* to those who are disproportionately burdened by waste and/or unfairly underrepresented in decisions regarding how to manage it; and/or as failures of moral *character*.

The most familiar problem, of course, is the sheer *quantity* of waste we produce: the world's cities generate more than 3.8 million tons of solid waste per day, more than half a pound per human being. By the end of the century, daily global waste production is expected to triple; per capita waste production will more than double (Hoornweg et al. 2013: 616). According to official estimates, the U.S. is responsible for nearly 18% of global waste production, contributing 250 million tons per year or about 4.4 pounds per day per person (EPA 2011a), for an average personal lifetime legacy of more than 60 tons of trash. Indeed, it may be even worse. Some waste analysts argue that these official estimates are grossly understated, with the real figures nearer to 7 pounds per person per day in the U.S., for a lifetime legacy of around *100 tons* per U.S. citizen (van Haren et al. 2010; Humes 2013).

That's a lot of trash. And it is never really thrown "away"; there is no such place. It all goes *somewhere* on earth – to land where we bury it, to water as streams carry litter out to sea, and to air as particulate and gaseous emissions when we burn it. The trick is to minimize the harm done along the way. In the U.S. municipal waste system's ideal model, households separate recyclables and compostables from irredeemable rubbish; all three streams are collected at curbside; loads are consolidated at transfer stations and transported to appropriate and well-regulated facilities and disposal sites. If this system worked perfectly, compostables and recyclables would return to beneficial uses, and the remaining trash would end up in a safe and well-regulated "sanitary landfill" or waste-to-energy incinerator.

Of course, as every avid recycler knows, one way in which reality falls short of this ideal is that landfills and incinerators receive much more than just "irredeemable rubbish." Rates of diversion of reusable, recyclable, and compostable materials range from pretty good (e.g., on the range of 60–80% in San Francisco, depending on accounting methods) to pathetic (e.g., somewhere between 4% and 10% in Indianapolis) (Quraishi 2009; MacBride 2013). Enormous quantities of materials that might with care and mindfulness be rendered useful are instead treated as disposable – either burned or buried. This is one reason why our waste stream appears to evidence a throw-away mindset – the vice of profligacy, a failure to value things properly. And this seems to suggest that, along with ensuring adequate environmental regulation of landfills and incinerators, the most important thing we can do about waste is to improve recycling rates.

But notice this: even if *all* materials that could be reused, recycled, or composted were removed from the waste stream, even if *no* discards escaped the collection system, and even if landfills and incinerators were *perfectly* benign, collecting and hauling millions of tons of materials from place to place would remain a massive physical and economic undertaking with its own tremendous energy demands, carbon emissions, and diesel exhaust pollution (Weitz et al. 2002). Those of us who live in middle- and upper-class neighborhoods can easily fail to notice the scale of this constant flow of refuse through our cities or register it as a distinct environmental problem; a truck comes through our neighborhoods once a week, and the waste is gone. People who live near or work in recycling centers, waste transfer stations, landfills, and incinerators, however – typically the poor and people of color – see dozens or hundreds of diesel trucks per day, suffering elevated rates of afflictions caused by chronic exposure to diesel exhaust and particulate matter – ranging from headaches, dizziness, and eye/nose/throat irritation to asthma, cardiopulmonary disease, and lung cancer (EPA 2002). The hauling requirements of a throw-away society with inequitably sited transfer stations would thus be inherently more harmful for these already economically disadvantaged populations even in a world of perfect recycling, perfect landfills, and perfect incinerators – which is to say nothing of the additional health risks associated with imperfect, under-monitored, and/or under-regulated facilities. Moreover, as David Pellow (2002) has argued, the recycling industry, for all its noble associations, cannot boast that its laborers enjoy noble working conditions. Recycling jobs in many cities – held disproportionately by people of color – are dirty, repetitive, dangerous, and poorly paid, with high turnover rates and often no union representation. In short, recycling as a practice is neither harm-free nor immune to injustices that intersect with other segments of the waste management industry.

So we produce too much waste, irrespective of how much of it we recycle. But what makes the *size* of the waste stream especially harmful and unjust has more to do with its *composition*. In the most general terms, the reason we have to worry about toxic emissions from incinerators and about toxic landfill leachate is that there are toxins in the things that end up in the trash – a problem rooted in product design to which we will return below. But we can also identify discrete tributaries in the waste stream that raise distinct ethical concerns. Electronic waste is the fastest growing category and one of the worst, since it contains rare and precious metals that are destructive and/or oppressive to mine, hard to extract for recycling, and profoundly poisonous when released – toxic burdens that rich countries have historically exported to poor countries (Basel Action Network 2002; Byster & Smith 2006; Pucket 2006; EPA 2011b). Another troubling stream, plastic, is also notoriously difficult to recycle and produces dioxin when burned. Because it is "made to last forever, yet designed to be thrown away," plastic pollution is accumulating throughout the biosphere, particularly in the oceans (UNEP 2005, 2009a, 2011; Allsop et al. 2006; Arthur & Baker 2012). To take just one more example, food waste has begun to receive tremendous attention (Smil 2004; Gustavsson 2011; Gunders 2012; Buzby et al. 2014). That nearly 40% of the food grown for human consumption in the U.S. is not eaten is a moral travesty in the face of food insecurity. It represents unnecessary application of fertilizer, pesticides, and herbicides; unnecessarily incurred soil erosion and stream eutrophication; and unnecessary expenditures of energy and water in a time of droughts and climatic uncertainty. To make matters worse, as food waste and other organic materials decompose in landfills, they produce methane, a potent greenhouse gas.

So, yes, waste does matter a great deal, both morally and environmentally. But because waste is but the output of our patterns of *consumption*, we can't do much about the former unless we tackle the latter. And consumption patterns typical of our globalized consumer society are morally problematic for reasons over and above their role in creating mountains of trash. Once again, these can be characterized in three ways – in terms of harm, in terms of injustice, or in terms of flawed conceptions of the good life.

One such reason – consumption's role in fueling climate change – has already been mentioned. The IPCC identifies economic growth as a primary driver of anthropogenic climate change, but politicians and economists depend on the globalization of consumer society – with ever-increasing consumption at its heart – to drive that growth. Consumption practices are centrally implicated, therefore, in the "perfect moral storm" that is climate change (Gardner 2013).

A second, and in some ways deeper ethical concern is that the consumption practices of the world's rich appropriate a grossly *unfair* share of the world's limited ecological space. To put this in ecological footprint terms, there are about 1.8 global hectares (gha) of bioproductive land available for each person on earth. The level of consumption of an average resident of the U.S., by contrast, requires about 8 gha. If everyone on earth consumed at those levels, we would need about four and a half earths (Ewing et al. 2010) to sustainably support the earth's *current* population. Since (to state the obvious) we have only one, since human population will continue to grow for a few more decades, and since much of the world's population currently *under-consumes* the necessities for a decent life, basic fairness clearly dictates that the world's rich must consume much less (Worldwatch 2004, 2008; Jackson 2009; Assadourian 2012).

Finally, a growing body of multi-disciplinary research poses fundamental challenges to old orthodoxies about both the relationships between macroeconomic growth and social well-being, and between individual consumption and happiness. At the macroeconomic level, rising Gross Domestic Product (GDP) – for all intents and purposes the end-all, be-all policy goal of governments everywhere – is itself a measure of all the spending that takes place in a given economy. It is supposed to provide a rough proxy for utility. And it is true that comparisons across countries suggest that, for poor countries, very small increases in GDP do indeed result in very large increases in well-being. But beyond a fairly low level (around $15,000 per capita), there ceases to be any relationship between a country's GDP and the life satisfaction of its citizens. Indeed, citizens of many countries with lower GDP per capital than the U.S. enjoy greater life satisfaction. Moreover, within rich countries, economic growth does not appear to have improved well-being over time. Again, in some cases just the opposite has occurred – reported happiness has declined while incomes have grown (Diener and Seligman 2004: 3; Jackson 2009: 41–42). On reflection, this isn't all that surprising, since GDP does not measure well-being (or even utility) directly. Instead, whatever increases market activity adds to GDP, regardless of whether its impacts on life prospects are good or bad: increasing cancer and incarceration rates, wars, natural disasters, and industrial accidents all contribute positively to GDP growth. Further, GDP is blind to any kind of value that is not traded on the market, regardless of how essential to our well-being – e.g., intact ecosystems, wildlife, parental care, health, happiness, friendship, and justice. At the microeconomic level, the disconnect between consumption and well-being is also quite clear. There is, of course, more nuance in the empirical research than we can summarize here, but as Tim Kasser and his colleagues in psychology put it, the upshot seems to be that "The more people value materialistic aspirations and goals, the less they are happy with their personal lives and the more they act in ways that are socially and ecologically damaging" (Kasser 2006: 201).

If these lines of reasoning are on the right track, then consumer society as we know it appears indefensible. It produces billions of tons of trash, with the pollution, waste, and inequities that go along therewith. It exceeds the sustainable biocapacity of the planet, fueling climate change and destroying the planet's capacity to sustain life for all species. The world's rich are consuming far more than their fair share of this capacity. And our pursuit of growth for its own sake poorly serves human happiness and well-being. We thus have both moral and prudential reasons to reduce the ecological footprints of consumer societies dramatically and to support policies that will bring the world's poorest people up to levels of consumption compatible with a decent standard of living.

But the life cycle of material goods goes beyond waste and consumption. As Leonard explains in *The Story of Stuff*, the products we consume and throw away actually go through a five-stage life cycle: extraction of raw materials, production of consumer goods, distribution of those goods to retail outlets, consumption at the point of purchase, and disposal from households to landfills and incinerators (see also EPA 2009). So if we are to think seriously about consumption and waste, we need to keep this whole system (and the various actors at each node) in mind. Such systems thinking reveals essential insights into the ethics of stuff.

For example, one of the lessons from *The Story of Stuff* is that the social and environmental impacts of household consumption and waste disposal – great as they are – are *miniscule* in comparison to the impacts caused "upstream" during extraction, production, and distribution. There does not appear to be reliable data on the quantity of industrial waste produced each year – itself an environmental policy failure discussed in depth in Samantha MacBride's highly acclaimed book *Recycling Reconsidered* (2012). But even without official reporting, it is clear that the vast majority of the material that goes to landfills and incinerators, perhaps as much as 98%, is industrial waste or construction and demolition waste. Leonard (2010) cites one study claiming that, for each pound of household waste we produce, at least 40 pounds of industrial waste is produced upstream, and another claiming that the ratio is 1:71. "Either way," she writes,

> the point is that there's a whole lot more waste being made upstream beyond the Stuff we haul to our curbs each week, so if we really want to make a dent in our waste production, we need to be looking upstream to where the bulk of it is generated.
>
> *(2010: 186–187; see also MacBride 2012:88–123)*

Another lesson from *The Story of Stuff* is that our power to alleviate the impacts of our consumption and waste practices at the household level is significantly constrained by decisions made upstream during the processes of extraction, production, and distribution and by the public policies that regulate (or fail to regulate) those processes. We don't have a municipal solid waste stream full of toxic material because consumers prefer products containing toxic ingredients. Rather, the products we buy and throw away are (often unknown to us) toxic because of decisions made at the design and production stages of the story. We don't seek out products that will wear out quickly and be expensive or impossible to repair. Rather, in order to keep consumer demand high, producers opt for disposability over durability; products are designed with short lifespans so that consumers will continue to upgrade. Besides designed obsolescence, producers work to foster perceived obsolescence – dissatisfaction with the things we already have because of the allure of the next-generation product or the fear of falling conspicuously out of fashion. Certainly our individual choices and desires as consumers play a contributing role; we are not passive puppets. But as individuals we have extremely limited control over the range of options from which we are able to choose and the wider cultural context in which our choices are made.

The scope of an environmental ethics of stuff, then, must include not only waste and consumption but also upstream processes and subprocesses and the policy contexts in which they occur. Decisions about product *design*, for example, are enormously important from a social and environmental point of view – determining how long products will last, whether they are designed for repair or replacement, to be thrown away or recycled, what resources must be extracted in order to make them, whether they will contain dangerous toxins, etc. Moreover, it is clear that profession of *marketing* has a great deal to answer for in fueling destructive consumption habits all over the world. Less frequently noted is the role that *retailers* play in the process. In the e-waste documentary *Terra Blight* (2012), scenes of boys using manifestly unsafe methods for recovering scrap metal from toxic electronic waste in Ghana are juxtaposed with a sales pep talk at an electronics store. When the store manager is asked what happens to the goods they're selling when they become obsolete, his honest answer is that he has no idea. If, as activists often insist, the individuals who purchase electronics have a responsibility to know how these products are constructed and disposed, then what reasons can there be for thinking that actors at other nodes in the supply and disposal chain do not have similar responsibilities? To us, this teachable moment suggests that an environmental ethics of stuff should include an environmental ethics of product design, supply chain management, manufacturing, marketing, distribution, and retail – indeed, of each identifiable subprocess or profession that helps perpetuate the current unsustainable and unjust system of stuff.

Seeking Systemic, Transformative Change

So much for the *scope* of an environmental ethics of stuff. We've argued so far that the harms, injustices, and failings of virtue and value associated with waste and consumption are enormous in themselves yet only pieces of a much larger complex system whose total impacts often dwarf those apparent when looking at household consumption and waste alone. It makes little sense to try to develop an environmental ethics focusing on household waste and consumption in isolation from a critique of the larger system in which these practices take place, and a vision for a just and sustainable alternative.

Increasingly, the term for that vision is "zero waste" (Connett 2013). Rather than a take-make-waste linear materials economy, we can imagine a world in which humanity minimizes extractive pressures on the natural world because the economy is designed to meet central human needs, not to satisfy manufactured human wants, and because producers design durable products that contain no toxic inputs, that are made from feedstocks of recycled materials, and that are readily repairable, recyclable, or biodegradable. So how do we get there? As many have noted, there is no single solution or silver bullet; we must experiment with a wide variety of different approaches.

Still, not every well-intended effort is well-advised. For example, consider New Jersey-based TerraCycle, a "highly-awarded, international upcycling and recycling company that collects difficult-to-recycle packaging and products and repurposes the material into affordable, innovative products." Their goal of finding a way to recycle *anything*, no matter how difficult it may be, has required them to elicit corporate support to fund the collection of hard to recycle waste streams such as food pouches, cigarette butts, and single-use coffee capsules. Moreover, since TerraCycle often works with producers to ensure that its "upcycled" products prominently feature brand logos (TerraCycle 2010), their efforts to turn difficult-to-recycle Capri Sun drink pouches into sellable products seem likely to exacerbate rather than solve the underlying problem of excessive drink pouch packaging waste. Unfortunately, cases of well-intentioned but counterproductive efforts are easy to multiply. Annie Leonard, for example, cites strategies to incentivize recycling (e.g., via contests) that instead reward increased consumption (2010: 229–231). Samantha Mac-Bride (2012) argues that recycling enthusiasts' insistence on curbside collection of glass ("a low-value, relatively environmentally benign material") while ignoring the challenge of reclaiming waste textiles ("a high-value, environmentally significant material of equal presence") has been both economically and ecologically irrational.

What, then, do more promising systemic strategies look like? We suggest three broad arenas for activism below, noting of course that their borders are fuzzy and the categorizations are not mutually exclusive. Broadly, we can look to reforms in the *political* sphere, to shifts in *economic* theory and practice, and to efforts to shift the *culture* of consumerism.

First, opportunities for effective and achievable political change abound. At a seemingly modest scale, for example, much post-consumer food waste appears to be driven by confusion about the meaning of expiration dates. People throw away food that is perfectly edible because it is past a "best by," "use by," or "sell by" date, although these dates have no consistent meaning or discernable connection with food safety. Establishing consistent, unambiguous date-labeling regulations is one piece of "low-hanging fruit" that could significantly reduce the amount of still-edible food we throw away (Lyndhurst 2011; Gunders 2012; Leib 2013). Policies for reducing plastic pollution (such as taxes, fees, and bans on single-use plastic bags and plastic microbeads) are also developing quickly (ten Brink et al 2009; UNEP 2011; Gold et al. 2013; Herzog et al. 2013). Appropriately scaled carbon taxes are an efficient means of giving all producers, buyers, and sellers an incentive to reduce a key component of current waste streams (Spross 2014).

A more expansive policy approach is "extended producer responsibility" or EPR. The core idea behind EPR is to develop policy solutions that make designers and producers responsible for the entire life cycle of the product that they sell, cradle-to-grave (or better, cradle-to-cradle), rather than having the onus of source separation, material collection, and recycling fall on individuals and municipalities who have no effective say over the nature of the materials that enter the waste stream. EPR policies thus give producers incentives to design for durability, reparability, and recyclability, because at the end of their product's life the producers themselves will be responsible for its fate. EPR laws have been passed in 34 states governing a variety of product categories, including auto switches, batteries, carpets, cell phones, electronics, fluorescent lighting, mercury thermostats, paint, pesticide containers, and mattresses. Some municipalities have even passed pharmaceutical EPR ordinances. Increasingly, activists are targeting producers of unrecyclable packaging – those Capri Sun pouches again – with "make it, take it" campaigns urging them to either phase out such materials or take responsibility for their collection and recycling or disposal (Prindiville 2014, 2015).

Other large-scale policy solutions worth pursuing appear less "environmental" at first glance, but may be essential given the current state of dysfunction, corporate capture, and gridlock crippling the political system. For example, the success of intelligent environmental policy reforms may depend on campaign finance reform, revocation of corporate personhood, redrawing gerrymandered voting districts, and/or other efforts to restore the democratic process itself. Expanding on a point made by David Orr (2004) and Frances Moore Lappé (2005), we must come to see threats to democracy as environmental threats.

Finally, prominent scientists such as James Hansen and mainstream environmentalists such as Bill McKibben and his 350.org are increasingly urging citizens to take to the streets (Hansen 2012). Many environmental ethics texts engage with questions about the justifiability of "Earth First"-type resistance, but broadening the scope of these discussions to develop better understandings of successful movement building and social change related to our current economic, environmental, and social predicament will clearly be central to resolving problems of consumption and waste (Stephenson 2013).

In addition to working in the political sphere, we need profound changes in our *economic* system. Some ecological economists have concluded that problems surrounding consumption and waste are so severe that we must slow down, stabilize, or even reverse the growth of our economy (Assadourian 2012). As Tim Jackson (2009) argues, we need a macroeconomics of sustainability, models for steady-state or even de-growth economies that don't lead to crippling unemployment. These are no doubt heretical ideas (Cafaro 2011a). But as the dangers of growth-driven economics have become increasingly obvious, we can see signs that this orthodoxy may be weakening. Given the shortcomings of GDP for measuring what we care about with regard to economic progress, policymakers need new metrics for evaluating what works in an ecologically endangered world characterized by injustice. Fortunately, interesting work is being done on many different indicators, ranging from the "Happy Planet Index" to the "Genuine Progress Indicator," among others (Costanza et al. 2014).

Short of such grand revolutions in economic thought, numerous practical and local strategies can be considered. Rob Hopkins (2008) and Shaun Chamberlin (2009), leading voices in the Transition Town movement, outline concrete steps we can take to create resilient local economies that are not dependent on far-flung sources of fuel, food, goods, and services, and that are as prepared as can be for climate shocks. The focus on buying locally and the growth of farmers' markets are examples of such strategies. A more recent and potentially revolutionary idea is that of a *sharing economy* (Agyeman 2013). Sharing economy enthusiasts call for a decline in the centrality of private ownership. For example, as Rachel Botsman and Roo Rogers (2013) point out, the average power drill is in use for less than 20 minutes in all the years it is owned. "Why buy a drill," they ask, "if what you want is a hole?" In a sharing economy, citizens can get what they want out of product ownership (i.e., holes) through tool libraries and the like without the wastefulness of individual ownership. In short, in addition to pursuing changes in laws and public policies that govern the materials economy (by direct action if necessary), there are also myriad ways we can change our local economies on the ground.

Finally, in addition to changes in public policies and economic structures, Eric Assadourian of Worldwatch points out that the consumer economy rests not just on faulty political and economic structures but on a *culture* of consumerism (Gardner & Assadourian 2004; Assadourian 2010a, 2010b, 2012, 2013. See also Fairbanks 2010). Cultures are ways of life, patterns of practices (eating, speaking, celebrating, worshiping, etc.) that express and embody peoples' beliefs and values, especially through the domains of *civil society* – religion, the arts, education, professions, avocational affiliations, and so on. To say we have a *consumeristic* culture, then, is to say that consumption is profoundly normative for us and permeates many of these domains. Shopping is a leisure activity unto itself; life success is commonly measured in terms of material possessions; the creative arts are largely put to the service of increasing consumption; the educational system is expected to prepare students to succeed (in conventional, consumeristic terms) in the market, and so on. Assadourian is suggesting that we learn from anthropologists and sociologists, from marketing and media experts, from missionary movements – from whomever we can – about the mechanics of shifting values. We need better, more empirically informed creative thinking about how to cultivate the cultural change we so urgently need. As J. Baird Callicott (1995) has argued, even academic philosophy can contribute to this effort – particularly, we would add, when it brought into public discourse (Light 2002).

The Ethics of "Being the Change"

Because we've argued that the ethics of consumption and waste must ultimately be seen as part of a more systemic ethics of stuff, we have focused so far on arenas of activism aimed at shifting the materials economy as a whole. But what (if anything) ought we to do about the stuff in our own lives? We can certainly do our own part to keep recyclable and compostable materials out of the trash. But, as Leonard puts it, that is really a "no-brainer"; our five-year-old can manage this much. *Waste reduction* should be equally obvious: the goal should not be to get our 35 weekly pounds of waste into the recycling and compost bins instead of the trash, but to minimize the amount that needs to go into any of these bins in the first place – using refillable water bottles instead of buying bottled beverages, avoiding products like drink boxes that come in unrecyclable packaging, and taking care to buy only as much produce as we will eat before it spoils, for instance.

Another possibility is ethical consumption – making purchases to some extent on the basis of moral commitments rather than solely on the basis of taste or cost. (*Ethical* consumption, in other words, is contrasted with *non*ethical consumption, not with *un*ethical consumption.) Ethical consumption is sometimes called "voting with our dollars" – whether by boycotting those products, producers, and retailers that conflict with our values, or by "buycotting" – that is, favoring – those that are working to express or embody them. Avoiding carbon-intensive meat, shampoos tested on animals, soaps containing plastic microbeads, and so on, all constitute practices of ethical consumption, as does favoring fair trade chocolate, organic and/or locally grown produce, and products made with post-consumer recycled materials. Product certifications, rating systems, and screening tools are increasingly available to assist consumers in such efforts (see, e.g., EPA 2013; Good Guide 2014).

More challengingly, people may aspire to *ethical* non*consumption*. In "The Art of Buying Nothing," Barbara Kingsolver (2007) defends her aspiration to live by Wendell Berry's criteria for purchasing any new thing: that it should cost less than the old one, be smaller, be repairable locally, be purchasable locally from a small shop, do better work, use less energy, run off renewable energy, and do no harm to good things that already exist. Kingsolver and Berry encourage us to reconceive consumption as work we do, work that can be done well or badly, haphazardly or with discipline and according to principles (see also Farrell 2014). Colin Beavan, of the famed documentary and book *No Impact Man* (2009), and Bea Johnson, author of the *Zero Waste Home* (2013), are well-known exemplars of ethical nonconsumption. Beavan undertook a year-long experiment in which he and his wife attempted to reduce their family's negative environmental impact to zero – zero trash, zero unnecessary consumption, zero greenhouse gas emissions. Johnson's family manages to produce just a single mason jar's worth of garbage each year. Many other sustainable lifestyle experiments (e.g., plastic-free lives, tiny houses, 100-mile diets, 100-thing challenges, dumpster diving, etc.) have been featured in documentaries, print, and social media in recent years. What unifies the efforts described above is that they ask us to embody an environmental ethics of stuff in our personal lifestyle choices.

What should we make of such exercises in lifestyle activism? Are they obligatory? Or, to put it in Aristotelian terms, is this what environmental virtue looks like? Committed environmentalists disagree on such matters. Some clearly regard this as a matter of integrity. Phil Cafaro (2011b) exhorts environmental ethicists to "commit ourselves more fully to *living* our environmental philosophies" (16) on grounds that doing so will help us raise children more capable of sustainable living themselves as well as making us better teachers. Kevin Anderson of the Tyndall Center argues that climate change academics and activists must lead by example, foregoing air travel whether for personal or professional purposes (2013). Flying, he maintains, "sends a very clear market signal: expand your airport," whereas slow travel (e.g., by train) "forces us to travel much less, to be much more selective in what events we attend, and to endeavor to get more out of those trips we do take" (3; see also Dwyer 2013).

Other, equally serious environmental thinkers are dubious about calls for individuals to make such significant lifestyle changes. In *Recycling Reconsidered* (2012), Samantha MacBride decries lifestyle activism as resting on an implicit, incrementalist theory of social change – "every-little-bit-helps-ism," as she calls it – that is politically naïve, factually false, and pragmatically counterproductive (2012: 226). As Annie Leonard puts it:

> Framing environmental deterioration as the result of poor individual choices – littering, leaving the lights on when we leave a room, failing to carpool – not only distracts us from identifying and demanding change from the real drivers of environmental decline. It also removes these issues from the political realm to the personal, implying that the solution is in our personal choices rather than in better policies, business practices, and structural context.
>
> *(2013: 247)*

Similarly, environmental politics scholar Michael Maniates worries that efforts by well-meaning environmentalists to help us green our homes and individual lifestyles frame problems in a way that limits our "environmental imaginations" and thus subverts our capacities to develop real solutions.

> Confronting the consumption problem demands, after all, the sort of institutional thinking that the individualization of responsibility patently undermines. It calls too for individuals to understand themselves as citizens in a participatory democracy first, working together to change broader policy and larger social institutions, and as consumers second. By contrast, the individualization of responsibility, because it characterizes environmental problems as the consequence of destructive consumer choice, asks that individuals imagine themselves as consumers first and citizens second.
>
> *(2001: 46–47)*

Thus, instead of (or, to the extent feasible, in addition to) using our power as *consumers* to address environmental problems, Maniates and Leonard urge us to use our power as *citizens* to create institutional rather than merely personal solutions to the crises we face. Ethically, we need to make much more vivid and real to ourselves, in our thinking, our acting, and our planning, the extent to which our individual actions are shaped, constrained, and enabled by social forces. We must learn to see structural problems and solutions that go far beyond the individual actions that come so readily to mind. (See also Sagoff 1981; Anderson 1995; Klein 1999 on the consumer/citizen distinction.)

In our view, the critics of lifestyle activism make a legitimate point. A sustainable society cannot be created on an individual-by-individual basis. Real solutions require more than individual changes in buying and disposal habits, yet much environmental discourse encourages us to think that far and no further. Focusing on cleaning up our individual act is appealing because this locus of responsibility is familiar and within our control. Collective action and political engagement, by contrast, are for many of us obscure and mystifying (Maniates 2003, 2013). Indeed, this is no accident. As Leonard and MacBride remind us, the packaging industry funded the infamous *Keep America Beautiful* "crying Indian" campaign in order to shift public perception of responsibility for litter firmly onto individuals just at the political moment that citizens were demanding bottle bills and drawing broader attention to the packaging industry's responsibility for creating so much trash (2010: 196, 2013: 244). Such industry-funded public relations campaigns have been stunningly effective in shaping public perception of waste as a matter of individuals' ethical failures rather than those of producers. The same story could be told in many arenas of environmental politics: for example, tailpipe emissions have non-accidentally come to be seen as the responsibility of drivers and their preferences for certain cars/SUVs rather than as the responsibility of those who drill for and profit from the production and use of oil, the auto industry itself, poor community planning, weak regulation, and the like.

Insofar as each of us has obligations arising out our dire environmental predicament, we must count participation in collective efforts to achieve broader change as among our most pressing responsibilities. And we recognize that discussions of the ethics of lifestyle choices easily risk distracting us from the onerous project of working toward systemic change. But as philosophers, we must push back a bit on behalf of the Socratic concern for living an examined life. Challenging ourselves to consume and waste as little as possible need not constitute navel-gazing avoidance of messy political thinking, nor a delusion about how to go about creating a more sustainable world. Such efforts may instead represent serious engagement with the moral and existential question of how best to live one's own life, a project worth defending in even in times of crisis. After all, as David Schwartz (2010) and Christopher Kutz (2007) argue, it is not just the causal effects of our consumption decisions that determine their morality, but also the degree to which our consumption decisions make us complicit with systems to which we have strong moral objections. Further, lifestyle activism provides experiential learning in the travails we all face in transitioning to a truly sustainable society, and can even generate unexpected pleasures, as Colin Beavan found in the course of his "no impact" experiment. Our moral imaginations are improved when we see and reflect on a wide variety of individual and social experiments for living more lightly on the planet.

Finally, just as we need to think more systemically about the causal structure of the materials economy, we must also learn to think more systemically about its moral structure. It is true, and important, that individual consumers do not bear the lion's share of the responsibility for our environmental predicament. Corporations, industries, and governments exert much greater influence over the materials economy than we do as consumers. It is also true, and important, that changing these institutions and structures will require us to act collectively as citizens, not just as consumers. But we want to emphasize that our roles *as individuals* in the system of stuff are not exhausted by the categories of (political) citizen and (private) consumer. We occupy many roles in civil society – in businesses, schools, religious organizations, professional circles, clubs, teams, social networks, and so on – each of which offers possibilities for exercising environmental responsibility. When complex systems generate significant social and environmental problems, peoples' individual roles in producing those outcomes are typically banal rather than intentional or driven by malice. We are just "doing our part" within larger structures, and too often fail to ask whether our environmental predicament has any bearing on what "our part" within these larger entities should be. It is therefore essential, as Hannah Arendt (2003) noted, for individuals to interrogate their own involvement in such structures. Our suggestion, in other words, is that the environmental ethics of stuff permeates realms where we usually expect and are expected to subordinate our "personal" moral convictions to other organizational priorities. Thinking critically about what complicity in destructive institutions says about oneself, about what lines one will or will not cross, and about what kind of a person one is willing to be constitutes a crucial moral resource for resisting such complicity (Glover 2001; see also Luban 2003). Even as we agree that individualized action in the sphere of consumption is an inadequate remedy for a broken materials economy and that we need to grow our citizen muscles by acting collectively in the public sphere, we would insist that cultivating this internal moral resource is not only compatible with but essential to our growth as citizens and human beings.

Related Topics

2. Eating
14. Water Quality and Availability
20. Moral Bases of Responses to Climate Change
33. Urban Sustainability
41. Sustainable Agriculture
42. Community Gardens
49. Pollution and Polluter Pays
61. Everyday Aesthetics
62. Community Participation

References

Agyeman, J., McLaren, D., & Schaefer-Borrego, A. (2013) *Sharing Cities*, Briefing for Friends of the Earth. http://www.foe.co.uk/sites/default/files/downloads/agyeman_sharing_cities.pdf.

Allsop, M., Walters, A., Santillo, D., & Johnston, P. (2006) *Plastic Debris in the World's Oceans*, Greenpeace International. http://www.greenpeace.org/international/Global/international/planet-2/report/2007/8/plastic_ocean_report.pdf.

Anderson, C. (2011) "Zero Waste Theory and Practice," *Kinesis* 38:2, 59–70.

Anderson, E. (1995) *Value in Ethics and Economics*. Cambridge, MA: Harvard University Press.

Anderson, K. (2013) "Hypocrites in the Air: Should Climate Change Academics Lead by Example?" *kevinanderson.info*. http://kevinanderson.info/blog/hypocrites-in-the-air-should-climate-change-academics-lead-by-example/.

Anderson, K., & Bows, A. (2011) "Beyond 'Dangerous' Climate Change: Emission Scenarios for a New World," *Philosophical Transactions of the Royal Society A* 369, 20–44.

Arendt, H. (2003) *Responsibility and Judgment*. New York: Schocken Books.

Arthur, C., & Baker, J. (2012) *Proceedings of the Second Workshop on Microplastic Marine Debris*. Silver Springs, MD: NOAA Marine Debris Program.

Assadourian, E. (2010a) "The Rise and Fall of Consumer Cultures," in Worldwatch Institute, *State of the World 2010: Transforming Cultures: From Consumerism to Sustainability*. New York: W. W. Norton & Company, 3–20.

Assadourian, E. (2010b) "Cultural Change for a Bearable Climate," *Sustainability: Science, Practice, & Policy* 6:2, 1–5.

Assadourian, E. (2012) "The Path to Degrowth in Overdeveloped Countries," in Worldwatch Institute, *State of the World 2012: Moving Toward Sustainable Prosperity*. Washington, DC: Island Press, 22–37.

Assadourian, E. (2013) "Re-engineering Cultures to Create a Sustainable Civilization," in Worldwatch Institute, *State of the World 2013: Is Sustainability Still Possible?* Washington, DC: Island Press, 113–125.

Basel Action Network (2002) *Exporting Harm: The High-Tech Trashing of Asia*. http://www.ban.org/E-waste/technotrashfinalcomp.pdf.

Beavan, C. (2009) *No Impact Man: The Adventures of a Guilty Liberal Who Attempts to Save the Planet, and the Discoveries He Makes About Himself and Our Way of Life in the Process*. New York: Farrar, Straus, and Giroux.

Botsman, R., & Rogers, R. (2010) *What's Mine Is Yours: The Rise of Collaborative Consumption*. New York: HarperCollins.

Boulding, K. (1966) "The Economics of the Coming Spaceship Earth," in H. Jarrett (ed.), *Environmental Quality in a Growing Economy*. Baltimore, MD: Resources for the Future/Johns Hopkins University Press, pp. 3–14.

Buzby, J., Wells, H., & Hyman, J. (2014) *The Estimated Amount, Value, and Calories of Postharvest Food Losses at the Retail and Consumer Levels in the United States EIB-121*. U.S. Department of Agriculture, Economic Research Service, February 2014.

Byster, L., & Smith, T. (2006) "The Electronics Production Lifecycle. From Toxics to Sustainability: Getting Off the Toxic Treadmill," in T. Smith, D. Sonnenfeld, & D. Pellow (eds.), *Challenging the Chip: Labor Rights and Environmental Justice in the Global Electronics Industry*. Philadelphia: Temple University Press, 205–214.

Cafaro, P. (2001) "Economic Consumption, Pleasure, and the Good Life," *Journal of Social Philosophy* 32:4, 471–486.

Cafaro, P. (2011a) "Beyond Business as Usual: Alternative Wedges to Avoid Catastrophic Climate Change and Create Sustainable Societies," in D. Arnold (ed.), *The Ethics of Global Climate Change*. Cambridge University Press, 192–215.

Cafaro, P. (2011b) "Taming Growth and Articulating a Sustainable Future: The Way Forward for Environmental Ethics," *Ethics & the Environment* 16:1, 1–23.

Callicott, J.B. (1995) "Environmental Philosophy is Activism: The Most Radical and Effective Kind," in D. Marietta & L. Embree (eds.), *Environmental Philosophy and Environmental Activism*. Lanham, MD: Rowman & Littlefield, 19–35.

Chamberlin, S. (2009) *The Transition Timeline for a Local, Resilient Future*. White River Junction, VT: Chelsea Green Publishing.

Connett, P. (2013) *The Zero Waste Solution*. White River Junction, VT: Chelsea Green Publishing.

Costanza, R., Kubiszewski, I., Giovannini, E., Lovins, H., McGlade, J., Pickett, K., Vala Ragnarsdóttir, K., Roberts, D., De Vogli, R., & Wilkinson, R. (2014) "Time to Leave GDP Behind," comment in *Nature* 505, 283–285.

Diener, E., & Seligman, M. (2004) "Beyond Money: Toward an Economy of Well-Being," *Psychological Science in the Public Interest* 5, 1–31.

Dwyer, James. (2013) "On Flying to Ethics Conferences: Climate Change and Moral Responsiveness," *IJFAB: International Journal of Feminist Approaches to Bioethics* 6:1, 1–18.

EPA (2002) *Health Assessment Document for Diesel Engine Exhaust*. Washington, DC: United States Environmental Protection Agency. http://cfpub.epa.gov/ncea/cfm/recordisplay.cfm?deid=29060.

EPA (2009) *Opportunities to Reduce Greenhouse Gas Emissions through Materials and Land Management Practices*. Washington, DC: United States Environmental Protection Agency.

EPA (2011a) *Municipal Solid Waste Generation, Recycling, and Disposal in the United States: Facts and Figures for 2011*. Washington, DC: United States Environmental Protection Agency.

EPA (2011b) *Electronics Waste Management in the United States through 2009*. Washington, DC: United States Environmental Protection Agency.

EPA (2013) "Introduction to Ecolabels and Standards," http://epa.gov/greenerproducts/standards/.

Ewing, B., Moore, D., Goldfinger, S., Oursler, A., Reed, A. & Wackernagel, M. (2010) *The Ecological Footprint Atlas*. Oakland: Global Footprint Network.

Fairbanks, Sandra Jane. (2010) "Environmental Goodness and the Challenge of American Culture," *Ethics & the Environment* 15: 79–102.

Farrell, James. (2014) "Good Work and the Good Life: Vocation as What We Do," in K. Schwehn & L. Lagerquist (eds.), *Claiming Our Callings: Toward a New Understanding of Vocation in the Liberal Arts*. London & New York: Oxford University Press. 29–47.

Gardner, G., & Assadourian, E. "Rethinking the Good Life," in Worldwatch Institute, *State of the World 2004: Special Focus: The Consumer Society*. New York: W. W. Norton & Company, 164–179.

Gardner, S. (2013) *A Perfect Moral Storm: The Ethical Tragedy of Climate Change*. Oxford: Oxford University Press.

Glover, J. (2001) *Humanity: A Moral History of the Twentieth Century*. New Haven, CT: Yale University Press.

Gold, M., Mika, K., Horowitz, C., Herzog, M., & Leitner, L. (2013) "Stemming the Tide of Plastic Marine Litter: A Global Action Agenda," *Pritzker Environmental Law and Policy Brief No. 5*. Emmett Center on Climate Change the Environment, UCLA School of Law.

GoodGuide, Inc. (2014) "Methodology," http://www.goodguide.com/about/methodologies.

Gunders, D. (2012) *Wasted: How America Is Losing Up to 40 Percent of Its Food from Farm to Fork to Landfill*, NRDC Issue Paper 12-06-B. National Resources Defence Council.

Gustavsson, J., Cederberg, C., & Sonesson, U. (2011) *Global Food Losses and Food Waste: Extent, Causes, & Prevention*. Rome, Italy: Food and Agriculture Organization of the United Nations.

Hansen, James. (2012) "Why I Must Speak out about Climate Change. TED Talks," http://www.ted.com/talks/james_hansen_why_i_must_speak_out_about_climate_change.

Herzog, M., Howe, A., Oh, T., & Parekh, J. (2013) *Federal Actions to Address Plastic Marine Pollution: Preventing Marine Plastic Pollution through Upstream Controls and Lifecycle Management*, briefing for U.S. Senate Ocean Caucus, presented at Natural Resources Defense Council by Surfrider Foundation and UCLA's Frank G. Wells Environmental Law Clinic. http://www.surfrider.org/coastal-blog/entry/marine-plastic-pollution-legal-policy-solution-briefings-for-u.s.-senate-oc.

Hoornweg, D., Bhada-Tata, P., & Kennedy, C. (2013) "Peak Waste Must End This Century," *Nature* 502: 615–617.

Hopkins, R. (2008) *The Transition Handbook: From Oil Dependency to Local Resilience*. White River Junction, VT: Chelsea Green Publishing.

Humes, E. (2013) *Garbology: Our Dirty Love Affair with Trash*. New York: Avery Trade.

IPCC (2007) "Technical Summary," in B. Metz, O. R. Davidson, P. R. Bosch, R. Dave, & L. A. Meyer (eds.), *Climate Change 2007: Mitigation. Contribution of Working Group III to the Fourth Assessment Report of the Intergovernmental Panel on Climate Change*. Cambridge and New York: Cambridge University Press.

Jackson, T. (2009) *Prosperity without Growth: Economics for a Finite Planet*. United Kingdom: Earthscan.

Johnson, B. (2013) *Zero Waste Home: The Ultimate Guide to Simplifying Your Life by Reducing Your Waste*. New York: Scribner.

Kasser, T. (2006) "Materialism and Its Alternatives," in M. Csikszentmihalyi & I. S. Csikszentmihalyi (eds.), *A Life Worth Living: Contributions to Positive Psychology*. London, New York: Oxford University Press, 200–214.

Kingsolver, B. (2007) "The Art of Buying Nothing," in J. Peters (ed.), *Wendell Berry: Life and Work*. Lexington, KY: University Press of Kentucky: 287–295.

Klein, N. (1999) *No Logo: Taking Aim at the Brand Bullies*. New York: Picador. http://www.newstatesman.com/2013/10/science-says-revolt.

Klein, N. (2013) "How Science is Telling Us All to Revolt," *New Statesman* Oct. 29. http://www.newstatesman.com/2013/10/science-says-revolt.

Klein, Naomi. (2014) *This Changes Everything: Capitalism vs. The Climate*. New York: Simon & Schuster.

Kutz, C. (2007) *Complicity: Ethics and Law for a Collective Age*. Cambridge: Cambridge University Press.

Lappé, F. (2005) *Democracy's Edge: Choosing to Save Our Country by Bringing Democracy to Life*. San Francisco: Jossey Bass.

Leib, E. (2013) *The Dating Game: How Confusing Food Date Labels Lead to Food Waste in America*. NRDC Report 13-09-A, Harvard Food Law and Policy Clinic and National Resources Defence Council.

Leonard, A. (2010) *The Story of Stuff: How Our Obsession with Stuff Is Trashing the Planet, Our Communities, and Our Health - and a Vision for Change*. New York: Free Press.

Leonard, A. (2013) "Moving from Individual Change to Societal Change," in Worldwatch Institute, *State of the World 2013: Is Sustainability Still Possible?* Washington, DC: Island Press, 244–252.

Light, A. (2002) "Taking Environmental Ethics Public," in D. Schmidtz & E. Willott (eds.), *Environmental Ethics: What Really Matters? What Really Works?* Oxford: Oxford University Press, 556–566.

Lin, Q. (2013) "Aldo Leopold: Reconciling Ecology and Economics," *Minding Nature* 6:1, 23–34.

Luban, D. (2003) "Integrity: Its Causes and Cures," *Fordham Law Review* 72, 279–310.

Lyndhurst, B. (2011) *Consumer Insight: Date Labels and Storage Guidance*, WRAP, http://www.wrap.org.uk/sites/files/wrap/Technical_report_dates.pdf.

MacBride, S. (2012) *Recycling Reconsidered: The Present Failure and Future Promise of Environmental Action in the United States*. Cambridge, MA: The MIT Press.

MacBride, S. (2013) "San Francisco's Famous 80% Waste Diversion Rate: Anatomy of an Exemplar," *Discard Studies*. http://discardstudies.com/2013/12/06/san-franciscos-famous-80-waste-diversion-rate-anatomy-of-an-exemplar/.

Maniates, M. (2001) "Individualization: Plant a Tree, Buy a Bike, Save the World?" in T. Princen, M. Maniates, & K. Conca (eds.), *Confronting Consumption*. Boston, MA: The MIT Press, 41–66.

Maniates, M. (2003) "Civic Virtue and Classroom Toil in a Greenhouse World," in M. Maniates (ed.), *Encountering Global Environmental Politics: Teaching, Learning, and Empowering Knowledge*. Lanham, MD: Rowman & Littlefield, 129–147.

Maniates, M. (2013) "Teaching for Turbulence," in Worldwatch Institute, *State of the World 2013: Is Sustainability Still Possible?* Washington, DC: Island Press, 255–268.

McKibben, Bill. (2007) "Why Having More No Longer Makes Us Happy," *Mother Jones*.

Orr, David W. (2004) "The Problem of Education," in *Earth in Mind: On Education, Environment, and the Human Prospect*, 2nd ed. Washington, DC: Island Press, 26–34.

Pellow, D. (2002) *Garbage Wars: The Struggle for Environmental Justice in Chicago*. Boston, MA: The MIT Press.

Preston, C., & Corey, S. (2005) "Public Health and Environmentalism: Adding Garbage to the History of Environmental Ethics," *Environmental Ethics* 27, 3–21.

Prindiville, M. (2014) "The Politics of Extended Producer Responsibility: Moving Forward in Hard Times," *UPSTREAM*. http://upstreampolicy.org/download/3931/.

Prindiville, M. (2015) "Flexible Packaging and Other Conundrums – 'What are we gonna do with all this stuff?" *UPSTREAM*. http://upstreampolicy.org/flexible-packaging-and-other-conundrums-what-are-we-gonna-do-with-all-this-stuff/.

Puckett, J. (2006) "High Tech's Dirty Little Secret: The Economics and Ethics of the Electronic Waste Trade," in T. Smith, D. Sonnenfeld, & D. Pellow (eds.), *Challenging the Chip: Labor Rights and Environmental Justice in the Global Electronics Industry*. Philadelphia: Temple University Press, 225–233.

Quraishi, J. (2009) "Recycling: Curb Your Enthusiasm," *Mother Jones*. http://www.motherjones.com/environment/2009/05/recycling-curb-your-enthusiasm.

Sagoff, M. (1981) "At the Shrine of Our Lady of Fatima: Or, Why Political Questions are Not All Economic," *Arizona Law Review* 23, 1283–1298.

Schwartz, D. (2010) *Consuming Choices: Ethics in a Global Consumer Age*. Lanham, MD: Rowman & Littlefield.

Smil, V. (2004) "Improving Efficiency and Reducing Waste in Our Food System," *Environmental Sciences* 1:1, 17–26.

Spross, J. (2014) "Is The Solution To Climate Change In Vancouver?" *Climate Progress*. http://thinkprogress.org/climate/2014/02/27/3198521/vancouver-carbon-price/.

Stephenson, W. (2013) "The New Abolitionists: Global Warming Is the Great Moral Crisis of Our Time," *The Phoenix*, Feb. 20. http://thephoenix.com/Boston/news/151670–new–abolitionists–global–warming–is–the–great/.

Stern, N. (2006) *Stern Review: The Economics of Climate Change* (pre-publication edition). "Executive Summary," HM Treasury, London. http://webarchive.nationalarchives.gov.uk/20130129110402/http://www.hm-treasury.gov.uk/d/Executive_Summary.pdf.

ten Brink, P., Lutchman, I., Bassi, S., Speck, S., Sheavly, S., Register, K., & Woolaway, C. (2009) *Guidelines on the Use of Market-based Instruments to Address the Problem of Marine Litter*. Institute for European Environmental Policy (IEEP), Brussels, Belgium, and Sheavly Consultants, Virginia.

TerraCycle (n.d.) "Products Index," https://www.terracycle.com/en-US/products.html.

TerraCycle (2010) "Brigade Overview 9.1," http://www.maala.org.il/warehouse/userUploadFiles/File/events/pdf/TerraCycle%20-%20Brigades%209.1.pdf.

UNEP (2005) *Marine Litter: An Analytical Overview*. Nairobi, Kenya: United Nations Environment Programme.

UNEP (2009) *Marine Litter: A Global Challenge*. Nairobi, Kenya: United Nations Environment Programme.

UNEP (2011) "Plastic Debris in the Ocean," *UNEP Yearbook 2011: Emerging Issues in Our Global Environment*. Nairobi, Kenya: United Nations Environment Programme, 21–34.

van Haren, R., Themelis, N., & Goldstein, N. "The State of Garbage in America," *BioCycle*, October.

Weitz, K., Nishtala, S., Yarkosky, S., and Zannes, M., & Thoreloe, S., 2002 "The Impact of Municipal Solid Waste Management on Greenhouse Gas Emissions in the United States," *Journal of the Air & Waste Management Association* 52, 1000–1011. http://gcsusa.com/pdf%20files/US%20EPA%20-%20The%20Impact%20of%20Municipal%20Solid%20Waste%20Management%20on.pdf.

Worldwatch Institute (2004) *State of the World 2004: Special Focus: The Consumer Society*. New York: W. W. Norton & Company.

Worldwatch Institute (2008) *State of the World 2008: Innovations for a Sustainable Economy*. New York: W. W. Norton & Company.

Further Reading

Dauvergne, P. (2010) *The Shadows of Consumption: Consequences for the Global Environment*. Cambridge, MA: The MIT Press.

Melosi, M. (2005) *Garbage in the Cities: Refuse, Reform, and the Environment*. Revised Edition. Pittsburgh: University of Pittsburgh Press.

Scanlan, J. (2005) *On Garbage*. London: Reaktion Books.

Strasser, S. (2000) *Waste and Want: A Social History of Trash*. New York: Holt Paperbacks.

PART VII

Agriculture

38

FOOD

David M. Kaplan

This chapter will examine the role of narratives in our understanding of the relationship between food, agriculture, and the environment. Narratives are the most comprehensive way of representing things that have a historical dimension. They focus on the central actors, select the key events, and create meaningful accounts of what happened and what could happen. They are crucial for putting events into context, portraying characters, and depicting scenarios. I will argue that environmental ethics needs to embrace the "narrative turn" in order to account for the diversity of ethical issues surrounding food, agriculture, and the environment, as well as to connect the countless actors, animals, and circumstances that make up our food systems. I will then identity several narratives commonly seen in the US that frame the way we understand food and environmental issues. A shift from theories to narratives might lead us away lead away from complex theoretical issues in environmental ethics toward more accessible, non-academic approaches that might help wider publics understand how our food relates to and impacts the environment.

Narrative Knowing

Narratives put things into context by organizing parts into meaningful wholes. Anything that takes place in time can be recounted in a story that connects characters, events, and things. I take story and narrative to be synonymous. Stories create meaningful accounts of what happened by configuring sequences into episodes that answer basic journalistic questions: who, what, where, why, and how. Narratives combine descriptions and explanations to take a longer, broader, more encompassing view than any theory can. They can also depict actors and their complex relationships to situations in greater detail than any shorter unit of meaning, for example a word or a proposition. The case for the epistemic and normative function of narratives is not new. In the last 40 years, we've seen the importance of narratives for epistemology (MacIntyre 1981; Ricoeur 1984; Danto 1985; Swirski 2007), ethical theory (Blum 1980; Nussbaum 1992; Walker 1998), bioethics (Frank 1995; Nelson 1997; Charon 2002), personal identity (Ricoeur 1992), and philosophy of mind (Fireman 2003; Hutto 2008). Narratives have also become a part of the methodological diet of the social sciences, education, and psychotherapy. Their role is well established.

There are five things non-fictional stories do particularly well that are relevant to understanding how food, agriculture, and the environment are related.

1   **Stories Interpret Events.** They are always told from a perspective, which means that any account of what happened is interpretive rather than objective. This is not to say that all

DOI: 10.4324/9781315768090-46

interpretations are equally valid but rather to note that alternate interpretations of events are always possible. Unlike simple propositions about states of affairs or logical necessity, which can be understood without stories, human affairs and historical events are essentially interpretive. As such, they have a narrative form: unfolding in time, occurring in episodes, and always understood in a partial, limited, and prejudiced manner. Stories depict events, give meaning to series of occurrences, and create interpretive frameworks that influence how we understand and evaluate things.

2   **Stories Portray Characters.** A story can flesh out a character and detail one's inner life and outer manifestations. We come to see whom someone is by how he or she thinks, relates to others, and responds to situations. The identity of a person unfolds over time – even a lifetime – with all the beginnings and endings, successes and failures, digressions and sub-plots, and everything else that makes up our stories. The answer to the question "who?" is not merely a set of identifying references but a whole person whose life is like a character in a story. A narrative builds on episodes to create a portrait of who someone is. We understand the life of person in a story as we would any historical occurrence.

3   **Stories Depict Scenarios.** Stories not only recount the past but also forecast the future. They are not only interpretive but also inventive. Scenario analysis – or predictions that try to take as many social variables into account as possible – has become a standard practice in fields that require strategic planning. Stories are the glue that stick the present to projected events and possible outcomes. We use these stories to test possible response by imagining how events might play out. Often moral deliberation involves telling stories in order to flesh out the ethical dimensions of characters, situations, and decisions. The temporal dimension of narration reaches back in time to recount what happened, through the present to relate what is happening now, into the future to imagine what might happen.

4   **Stories Make Arguments.** Stories can enhance and even stand in for an argument. The two work together in order to make a potentially compelling case. A narrative framework delimits a context of relevance, depicts situations, portrays characters, and imagines consequences in defense of truth claims and normative claims. In turn, arguments often rely on stories for illustration. For example, President Reagan used to tell stories of "real people" in his speeches in support of his policy agendas, such as the alleged welfare queens who leached off the system at taxpayer expense, or the earnest small businessman hampered by government regulations. The stories justify policy proposals, and the policy proposals provide alternative endings to the stories. It is often not enough to make claims and marshal evidence. Often stories are the best way to persuade people. They never replace arguments but rather act as vehicles for them by helping to establish facts and make judgments about things that unfold over time (Kaplan 2003; Griffin 2013).

5   **Stories Humanize Characters and Make Events Relatable.** Perhaps the greatest virtue of stories is that they can engage us in ways that theories cannot. They bring us into their world and let us see things from the perspectives of others. Stories can draw us in and make us care about persons and things we otherwise might never have considered. And they can do so in ways that are related and accessible, especially to those who have no specialized expertise. Stories help us to identify with others, comprehend other people's motives and choices, and challenge conventional wisdom and values (Nussbaum 1998).

Food and Environment Narratives

Although environmental ethics has to some extent recognized the importance of stories (Cheney 1989; King 1999; Liszka 2003), very little of this work has been applied to food and agriculture (Frasier 2001; Sanford 2011). This is unfortunate. The relationship between food, agriculture, and the environment is well suited to be framed in terms of narratives because of the countless particulars and changing circumstances that can never be fully captured in detail by theories, descriptions, or explanations. There is a temporal and historical element in food production, distribution, and consumption that is best captured in narrative form. Although there are no formulas for telling good stories and giving good readings, there are patterns: recurring narratives about food and the environment that are common in academic literature, the media, and advocacy outreach. With practice, we can become attuned to these underlying narratives and learn to recognize how they interpret actors and events, and what they presuppose about how the world works and what we should do.

Here are 11 of the most common narratives found in the US that relate food and agriculture to the environment.

The Scientific Narrative. The dominant approach to questions concerning food and environment is a scientific discourse of explanations and predictions governing causal interactions of physical matter. The scientific narrative is based on the presumption that methodological precision and empirical evidence are the only ways to secure not only objectivity but also reasoned consensus about human actions on the environment. The main actors are the scientists themselves; the narrative arc begins with questions and ends with certainty; the main conflict is with either nature itself, which seemingly hides its truths, or the forces of ignorance that stand in the way of reason. The scientist and his or her methodological precision rescues us; hard facts redeem us. This narrative is found wherever data-driven analyses and scientific expertise are privileged, including environmental impact assessments and lifecycle analysis. The science narrative underlies the dominant utilitarianism and risk assessment discourses, both of which rely on research and experimentation to establish better or worse outcomes for humans, animals, and environments. Science alone, on this reckoning, has to power to understand the workings of nature and to solve our practical problems.

We find this story in, for example, scientific attempts to resolve the controversies surrounding genetically modified foods (Freedman 2013). What many see as a political issue of labeling, choice, and food sovereignty is translated into a technical issue of food safety and environmental risks (*Eurobarometer* 2014). Although facts matter, the genetically modified food debates are legitimately political and, therefore, reasonably contested on the grounds of moral and political convictions as well as established facts. The scientific narrative, however, treats it as an exclusively scientific matter, where the main actors are clear-minded scientists educating a naïve public.

Advantages: This story has the virtues of precision, clarity, authority, rhetorical force, and evidence. It is indispensable for establishing environmental hazards and harms, health and safety risks, economic costs and benefits, and other quantifiable indicators. It also forms a solid basis for potential consensus over contested claims over states of affairs.

Disadvantages: It is difficult for non-experts to understand, impossible for non-experts to contribute to, and omits elements not suitable for scientific explanation, such as qualitative properties, human experience, and ethics. It is incapable of accounting aspects of our experience that do not lend themselves to causal explanations (e.g., voluntary actions, motives, interpretations). It presents itself as seemingly authoritative and objective when in fact based on often hidden presuppositions, choices, and interests. The science story can also silence as much as support public deliberation: data is no more or less likely to resolve an environmental issue than involving the public in meaningful decision-making (Depoe, Delicath, Esenbeer 2004).

The Techno-Utopian Narrative. The techno-utopian narrative underlies optimistic assessments of our technical prowess to make food production as environmentally sound as possible. The promise of technology on this reading is that human-made things make our lives better. Whatever practical or political obstacles we face we believe that there will eventually be a device, an app, or material to save the day. We find the techno-utopian narrative in the discourses on genetically modified foods, renewable energy, and many accounts of sustainable agriculture. The techno–utopian narrative has faith in progress and endless optimism of the ability of technologies to solve our problems and improve our lives: we will eventually engineer whatever we need to fix (Dyson 2000). Private industry and pro-development interests typically invoke techno-utopianism. Monsanto's website, for example www.monsanto.com, is filled with optimistic assessments of what new technologies can do to feed the hungry and clean the environment. The General Electric website (www.ge.com) describes its site as a "global destination for GE storytelling and content discover. Its technology-first, social approach creates a fluid home for GE content of various origin and format." We are invited to explore "stories of innovation from around the world."

Advantages: The utopian story is often true since technologies sometimes solve not only technical problems but also social problems (e.g., energy efficient machinery, pharmaceuticals, safety technologies). It fits into the familiar narrative of Enlightenment progress and therefore squares with other progressive sentiments. It is plausible and backed by evidence of past successes, and therefore encouraging, motivating, and affirmative; it gives hope that alternatives are possible and that the future can actually be better than the past.

Disadvantages: It ignores non-technological factors in social or environmental problems. It treats everything as capable of technological modification, correction, or improvement, and falsely takes technology to be the driving (even determining) feature of a society. The narrative disempowers non-experts and treats technology designers and engineers as the important actors while most of us are positioned as helpless (or at least not helpful). It also focuses attention away from underlying (often longstanding and systemic) causes in favor of technological solutions.

The Techno-Dystopian Narrative. Fear of technology run amok often underlies concerns about the specter of new technologies or criticism about old ones. According to the techno-phobic narrative, technology is out of control; humans no longer control it; instead, it controls us. Technological development is an independent force that follows its own rules and imperatives; we have no choice but to adapt to it. For example, unpopular production practices involving transgenic agriculture, factory farming, and food irradiation are seen as inevitable, as are undesirable consequences such as water pollution, topsoil erosion, and food contamination (W.K. Kellogg Foundation 2012). This dystopian narrative sometimes animates the discourses that criticize new technologies, such as genetically modified food, aquaculture, and other proposed technical solutions to social and environmental problems. On this reckoning, the problems we face are due to increasingly complex technical systems that are geared toward efficiency, higher yields, and profit. The alarmism that surrounds the genetically modified food debates often relies on a techno-phobic narrative – cautionary tales of the horrors of modernization (Chiles 2002). The Center for Food Safety website (www.centerforfoodsafety.org) features several stories that have a predictably grim take on agricultural biotechnologies. For example, in their story, "Tell Pop Secret to Stop Killing our Bees!" we are told that the popcorn industry uses bee-killing chemicals in their seeds. Another story titled "Unregulated Tricks in Your Nano Treats" warns that "there's a threat to your child so small, you can't even see it coming." Even if there is truth to their warnings, the website tends to commerce in techno-dystopian fears.

Advantages: This story is often true since technological innovations typically appear without prior public debate about their desirability even in the face of public disapproval. People have come to accept health and environmental harms caused by food production as inevitable consequences of industrialization. The techno-dystopian story of the forces of modernization that are beyond our control is the common framework for understanding how food systems work. The story explains the resignation and helplessness people feel in the face of large-scale enterprises over which they have no control.

Disadvantages: Like the techno-utopian narrative, the dystopian narrative falsely treats technology development as inevitable. It is a tired and stale story steeped in 1940s critiques of scientific reductivism as the root cause of technological excesses – better suited for science fiction than serious public deliberation. Dystopian narratives can also work against themselves and inadvertently make the case for more technological development rather than condemning techno-fixes as such. For example, the moral of a dystopian story can be a utopian one: given the harms caused by untrammeled development (e.g., farming relies too much on pesticides), the solution is new and improved development (e.g., safe pesticides have not yet been invented). The techno-dystopian story also cannot account for environmentally harmless agricultural practices such as hydroponics or wastewater recycling.

The Romantic Narrative. The techno-phobic narrative often evokes Romantic themes of human versus nature, life out of balance, and the desire to restore lost harmony with nature. According to the Romantic narrative, science and technology assume a detached, objectifying perspective that corrupts our food and pollutes the environment. The more we tamper and tinker with traditional food the more we make nature, society, and the quality of food worse. On this reckoning, connection with nature is good and alienation from it is bad. The Romantic narrative, however, is about more than just the folly of science and technology. It is a holistic worldview about the virtues of an emotional or spiritual connection with nature and the vices of disconnection. We find this narrative in the local food movement and a lot of organic food literatures, as well as other calls to eschew industrial agriculture and processed food in favor of sustainable living and traditional foodways. Michael Pollan's food writing relies on the Romantic narrative when he tells us to eat the kinds of things our grandparents might have grown up eating rather than processed foods. He criticizes prepared foods and exhorts us to cook for ourselves, even to hunt and forage for our food. On Pollan's Romantic narrative, a broad, holistic view of the relationship of food to nature and culture is good, and a narrow reductivist view is bad (Pollan 2009).

Advantages: This narrative serves as a counterweight to the juggernaut of advanced industrialized agriculture, and helps organize disparate convictions about farming, animals, nature, diet, and community into a coherent worldview. It counters the helplessness associated with techno-pessimism by offering a viable alternative that includes practical guidance (e.g., buy local, promote community supported agriculture, act as an ethical consumer). It helps people reflect on an otherwise hidden relationship between the food we eat and its environmental footprint. It also taps into and reinforces religious convictions about the sacredness and connectedness of life.

Disadvantages: The story is overly theoretical, if not metaphysical, and potentially off-putting to secularists and atheists. It can fetishize nature and demonize technology. The concepts of connection and alienation only make sense within a holistic worldview, which is presupposed rather than defended. It also focuses attention away from moral-political causes of environmental harms in favor of deep, conceptual explanations. It can, therefore, be difficult to translate the concepts of connection and alienation into the liberal vocabulary of politics and law. The Romantic story is not as practical as it seems.

The Agrarian Narrative. Agrarianism stresses the role of farming and ranching in the formation of moral character and in preserving culture and traditions. By living a rural lifestyle connected to the climate and soil, we acquire a sense of identity and place that can only come about by direct contact with the land. That is to say, communities and the environment thrive only when farming is done properly using traditional methods; both suffer when farming is over-mechanized. The agrarian narrative celebrates the virtues of a farming lifestyle, such as care for the land, animals, and our neighbors. We need to recover these virtues to mitigate the social and economic blight, pollution, and destruction of traditions brought about by large-scale industrial agriculture (Freyfogle 2001; Wirzba 2004). We are all involved in agriculture because our daily food choices affect how the land is treated. Communities, food, and the environment are all connected in the act of eating – and when we eat the right foods in the right way, we strengthen these connections. When Wendell Berry says that eating is an agricultural act, he relates food consumption with the realities of food production and farming life. He invites us to reflect on where our food comes from and how our choices can make life better or worse for rural farming communities (Berry 1991).

Advantages: The agrarian narrative often underlies the discourses in support of local food, farmers' markets, community-supported agriculture, and sustainable agriculture. Like Romanticism, it provides a unify thread to a range of convictions – but with a decidedly American flavor. It taps into longstanding populist sentiments, it links a critique of capitalist consumerism with a critique of agricultural practices, it illuminates how food and the environment are related, and it presents preferable alternatives.

Disadvantages: Agrarian narratives invariably flirt with Romanticism and nostalgia for simpler ways of living. They celebrate the virtues of rural life but at the expense of city life, where 81% of the US population resides (US Census Bureau 2015). The story legitimately defends the moral significance or rural life, neighborhoods, and communities but neglects the moral significance of state and federal powers. It valorizes regionalism, traditionalism, and provincialism and is, therefore, difficult to apply to urban or to international contexts. The agrarian narrative has the same advantages and disadvantages of the virtue ethics theories it depends on.

The Political-Economy Narrative. The political narrative treats food as a commodity within a global economic system governed by various regulatory bodies. This narrative frames food and the environment in relation to things like profit motives, industrial enterprises, and public policies. It highlights the producers and consumers, regulators and lobbyists, advertisers and marketers, and other political actors. On a liberal version of the narrative, food production and consumption are influenced by market forces, cultural practices, and laws; on a different telling, by under-staffed regulatory agencies, food industry lobbyists, and co-opted nutritionists (Nestle 2003). On a Marxist version, governing bodies serve the interests of capital accumulation rather than the health and welfare of workers and consumers. When agriculture and food production are subject to market forces wages will be cut, the environment polluted, food cheapened, animals made to suffer, costs externalized while benefits are privatized (Foster 2000; Magdoff 2000). Food scares are retellings of the familiar story of the indifference of capitalist industrial power; so are stories of industrial pollutants in air and water caused by lack of regulations or oversight. We find this narrative more in journalism and advocacy literature than in the philosophical literature (Kirby 2010; Hauter 2012; Robin 2012). We find it in food documentaries as well, such as *Cowspiracy* (2014), *Hungry for Change* (2012), and *That Sugar Film* (2014).

Advantages: This story is a potent narrative animated equally by facts and norms – *realpolitik* and appeals to justice. It explains how commercial systems function in relation to markets and governments and reveals often hidden agents and patients; exposes naked power. It is amenable to scientific discourses yet not entirely dependent upon them. It frames food and agricultural issues in practical, secular terms and hence does not depend on additional epistemic or metaphysical commitments. Finally, it identifies real causes and often supplies real solutions. Although often daunting, the political-economy narrative can be empowering.

Disadvantages: It can also be disempowering when the relationships among food, agriculture, and the environment are framed in relation to such immoveable forces as capitalism and government. It can breed cynicism and despair, even apathy and indifference. Its use of facts opens the story to refutation (even when its basic plot remains valid), often crossing over with the activist literature, which may or may not be reliable. And it overlooks non-political or economic elements, such as the metaphysical, religious, or aesthetic dimensions to food and the environment.

The Developing World Narrative. The developing world narrative highlights the effects of the global economy, the fate of people in poor countries, and the stark contrasts to life in Western nations. It is ultimately a Marxist narrative of capital-driven exploitation of poor countries by Western industrialized nations and corporations. The developing world narrative focuses on the lives of roughly 2.5 billion or 40% of the world's population who are farmers in developing countries, and who produce most of the world's food, including 90% of the food produced in Africa. In addition, most of these farmers are women (FAO 2013). The life expectancy, literacy, and standard of living of these farmers are low; the technologies, access to markets, and available infrastructure are poor (FAO 2013a). The environmental problems in developing countries include overuse of natural resources, inability to invest in sustainable technologies, and poor environmental protections. The developing world narrative is rarely heard in the US except for the advocacy and outreach literatures. The narrative also is under-represented in food and agriculture documentaries, where the focus is usually on the problems of industrialized agriculture and what we in the West might do about it. However, the website of the Food and Agriculture Organization of the United Nations (www.fao.org) is replete with information from global perspective. So is the website for the non-profit organization Heifer International (www.heifer.org).

Advantages: This narrative expands our understanding of food and the environment by telling the stories of usually overlooked peoples and places and broadens the horizons for those in the developed world. It puts the experiences of Westerners into context by showing how the same political economy that benefits the privileged also worsens the lives of the poor. Like agrarianism, the developing world narrative connects eating to agriculture but to international agriculture and to the institutions and technological systems that enable it. It helps people in the US recognize the huge footprint we leave on the rest of the world.

Disadvantages: This story can, however, reduce complex phenomenon to a simple story of exploitation and under-development at the expense of cultural, regional, and other differences. It can also reinforce stereotypy of the developing world by emphasizing the worst cases of malnutrition and environmental degradation. Storytellers should exercise caution whenever telling the stories of others, especially when the stories fit into familiar frameworks that can be used to defend the ostensible superiority of Western societies.

The Travelogue. The story of food begins on the farm and ends at the table. Along the way there is change, development, and transformation at every stage: planting, harvesting, rearing, slaughtering, processing, transporting, preparing, serving, spoiling, and so on. This kind of narrative follows the development of a food item in relation to the environment in order to highlight the key events or actors along the way. Often it serves a critical function by revealing what is hidden from consumers: we learn what really happens on farms, ranches, processing plants, and other places most of us are ignorant of. The narrative works in two directions: as we witness the path that food takes from farm to table, we are invited to other worlds where we might empathize with other humans and animals, and to see for ourselves how our seemingly private acts of shopping and eating are connected to broader social and environmental issues. Examples include *Omnivore's Dilemma* (Pollan 2006), *Fast Food Nation* (Schlosser 2001), and *The Jungle* (Sinclair 1906). These travelogues bring us to places we've never been so that we might be amazed and horrified by the origins and consequences of our food.

Advantages: Typically told by journalists, these stories are not philosophical; rather they are accessible and informative. They make their arguments usually by describing events without complex theories or intricate explanations. Simply telling the history of a food item can be an effective way to educate and motivate people.

Disadvantages: These stories are not philosophical. Travelogues can reduce food to agriculture and, therefore, reduce a complex matter into a simple question of origin. They are necessarily partial and incomplete, often slanted and reticent to tell more than a few stories at a time. There is often a legitimate industry counter-story that never gets told.

The Ethical Consumerism Narrative. Like the travelogue, the ethical consumerism narrative tells the story of the origins of a food item in order to highlight the exploitation of humans, animals, and the environment in its production or distribution. But unlike the travelogue, ethical consumerism places the buyer of a product squarely in the story. We are urged to do the right thing by buying goods or boycotting products based on moral considerations. The story follows two paths, with the reader positioned as key actor. Readers either make the wrong choice to buy goods from companies that are only self-interested thereby contributing to a litany of harms, or they make the right choice to boycott the bad companies and/or support the good companies that also benefit the greater good. The appeal is to vote with our dollars, something we can do every day without waiting for elections to happen. The documentary *Food Inc.* (2008), for example, ends by suggesting we vote with our dollars to change the nature of our food systems to produce healthier, more humane, more environmentally responsible food.

Advantages: It reminds us that food is a moral issue with consequences for others, animals, and the environment yet the story is empowering and practical: it tells people what and what not to buy. It calls attention to the global reach of our food systems, which often affects peoples and environments thousands of miles away. The story is a good prompt to reflect on related food issues such as labeling, marketing, and child labor.

Disadvantages: This story places moral burden squarely on consumers, not on producers. It makes it seem as we can shop our way to a better future. Yet if the health and environmental problems we face stem from over-consumption, the solution cannot be to consume more of the right products but rather to reverse the effects of consumerism itself. The story of ethical consumerism is a best narrow, limited solution – the very least we can do to lessen harms.

Religious Narratives. Each of the major religions has an account of the relationship of food and agriculture to the environment framed in terms of an overarching narrative that explains the natural order and prescribes appropriate conduct. Religious narratives typically tell us about origins, our relationships, and our fate. The Christian tradition emphasizes the sanctity of creation, environmental stewardship, our unique responsibilities to protect the land, to protect animals, and to refrain from over-consumption (Fick 2008, Davis 2004). The Jewish tradition also stresses the sacredness of the Earth, the giftedness of food, the dignity of farmers and shepherds, and the commandment not to destroy or to waste (Zamore 2011). The Islamic tradition emphasizes planting trees and conservation, avoiding over-consumption, tending to animals, recycling and preserving water, and eating wholesome foods (Abdul-Matin 2010). Hinduism stresses the sanctity of all life, the sacredness of nature, the unity of humanity with the natural world, and the special role of food plays in spiritual life (Sanford 2011). Buddhism stresses the connectedness of beings, respect for life, moderation and compassion, and harmonious living (Tucker and Williams 1998). Other religions such as Shintoism, Sikhism, and Jainism also have explanations of the food-environment relationship, and, of course, there are divisions within each major religion.

Advantages: Religious narratives are persuasive, motivating, unifying, and authoritative. They are time tested, entrenched in social practices, and accessible to people without education or means. They often square with secular reasons to pursue or avoid actions albeit for different reasons. Religious narratives have the power to resonate among believers more than ordinary appeals to reasons.

Disadvantages: Religious narratives are not binding on non-believers. They require serious metaphysical commitments.

Meat Narratives. Eating meat figures into each of the other major stories of food in one way or another but it also has its own narratives that justify or criticize killing animals for their taste. Both marshal facts recount the past and trace out a range of consequences for humans, animals, and environments. Meat-eating narratives recount who eaters are, what a good life for an animal is, and what the fate of the planet might be. No other food (except for maybe alcohol) has such elaborate justifications or endures such on-going criticism.

One common meat-eating defense is a naturalistic story about human evolution (with special attention to our teeth). The story follows an inevitable arc of life: all creatures are born, eat, reproduce, and die. It's futile and unnatural to think that we are any different from any other animal, so we should eat our natural, dentally appropriate, omnivorous diet. The naturalistic defense is basically the circle of life song from the *Lion King*. It combines religious conceptions of life after death, Romantic longing for unity with nature, and Darwinian evolution theory. In short, it is natural for us to eat meat because everything fits together in a food chain, and it would be unnatural to disrupt it (Keith 2009).

Critics of meat-eating connect the dots between industrial food animal production and consumers to show the bad consequences usually for our health or the environment. Like a travelogue, these stories follow the lifecycle of food animal production, distribution, and consumption to make the case that it is shot through with injustice and harms for animals, slaughterhouse workers, ecosystems, and our health. When framed as more than a dietary choice about the kind of cheap meat most of us eat, the meat narrative is a generally depressing story (Singer and Mason 2007).

Advantages: For those who believe it is wrong to eat animals for pleasure, meat narratives make a compelling case. They explain how the seemingly innocuous act of eat is, in fact, morally fraught from beginning to end – from farm to table. These stories answer the basic question critics pose: do you know where that meat came from? Typically, recounting the events that led up to the meal is enough to turn anyone off. Or the story might justify meat eating when the story paints a different picture of happy animals on happy farms; or when different facts are marshaled to reveal surprising connections, for example, between livestock grazing and grassland biodiversity. Both the case for and against eating meat is bound up with narratives.

Disadvantages: It depends which meat narrative you tell. One story might show how food animal production is bad for the environment, while another shows it is good for the environment. One story relates animal agriculture to soil erosion, deforestation, and loss of biodiversity, while another story relates grazing to proper land management (FAO 2006; Schwartz 2013). The stories themselves are indeterminate. It depends on *which* story is told about *what* meat.

The narratives mentioned above are by no means the only ones that shape our understanding of food, agriculture, and the environment. Nor are they mutually exclusive; it is possible for events to be framed by more than one narrative at a time. Nor are narratives incompatible with traditional philosophical theories, moral judgments, or scientific explanations. The challenge of a narrative analysis of the relationship between food, agriculture, and the environment is to examine how they influence philosophical debates, public discourse, policy making, or anywhere else stories might influence thought and action. A valuable research project would be to uncover implicit food narratives to assess their truth and moral claims, their advantages and disadvantages, and their prejudices and presuppositions, that is to say, to interpret them. Such a research project will inevitably lead one outside of philosophy to other disciplines in the humanities, social and natural sciences but so be it: the relationship between food and the environment is, after all, more than a philosophical issue.

So, the next time you read an article about the relationship between food and the environment, look for the story behind facts. Try to identify which basic storyline is framing the debate, who the key actors are, and what plotline is at play. Once you figure out where the author is coming from, you will probably be able to figure out where the story is going. You will be able to see the big picture better once you see how a story is slanted. Before long, you start to see patterns: yet another science narrative, yet another dystopian narrative, yet another agrarian narrative. Or another meat story that may make sense for one kind of meat but not for another. The fact is, there are only so many stories out there. Once we recognize their advantages and disadvantages, we'll be in a better position to decide what public policies and which food products deserve our support.

Related Topics

2. Eating
39. Industrial Agriculture
41. Sustainable Agriculture
62. Community Participation

References

Abdul-Matin, I. (2010). *Green Deen: What Islam Teaches about Protecting the Planet*, San Francisco, CA: Berrett-Koehler.

Berry, W. (1991). "The Pleasures of Eating," in *What Are People For?* New York: Counterpoint, p. 145.

Blum, L. (1980). *Friendship, Altruism and Morality*, London: Routledge and Kegan Paul.

Center for Food Safety. www.foodsafety.com. [Accessed 15 September 2015].

Charon, R. and Montello, M. (2002). *Stories Matter: The Role of Narratives in Medical Ethics*, New York: Routledge.

Cheney, J. (1989). "Postmodern Environmental Ethics: Ethics as Bioregional Narrative." *Environmental Ethics* 11: 117–134.

Chiles, J. R. (2002). *Inviting Disaster: Lessons from the Edge of Technology*, New York: Harper Business.

Danto, A. (1985). *Narration and Knowledge*, New York: Columbia University Press.

Davis, E. F. (2004). *Scripture, Culture, and Agriculture: An Agrarian Reading of the Bible*, Albany: SUNY Press.

Depoe, S., Delicath, J. W. and Aepli Esenbeer, M.-F. (eds.) (2004). *Communication and Public Participation in Environmental Decision Making*, Albany: State University of New York Press

Dyson, F. (2000). *The Sun, the Internet, and the Genome*, New York: Oxford Press.

"Europeans, Agriculture, and the Common Agricultural Policy," (2014). *Eurobarometer* 80(2).

Fick, G. W. (2008). *Food, Farming, and Faith*, Albany: SUNY Press.

Fireman, G. D., McVay, T. E., and Flanagan, O. J. (eds.) (2003). *Narrative and Consciousness: Literature, Psychology, and the Brain*, New York: Oxford University Press.

Food and Agriculture Organization of the United Nations (Rome, 2006). "Livestock's Long Shadow: Environmental Issues and Options."

Food and Agriculture Organization of the United Nations (Rome, 2013). "FAO Policy on Gender Equality: Attaining Food Security Goals in Agriculture and Rural Development."

Food and Agriculture Organization of the United Nations (Rome, 2013a). "FAO Statistical Yearbook 2013: World Food and Agriculture."

Foster, J. B. (2000). *Marx's Ecology: Materialism and Nature*, New York: Monthly Review Press.

Frank, A. (1995). *The Wounded Storyteller: Body, Illness, and Ethics*, Chicago: The University of Chicago Press.

Frasier, V. (2001). "What's The Moral of the GM Food Story?" *Journal of Agricultural and Environmental Ethics* 14(2): 147–159.

Freedman, D. H. (2013). "The Truth about Genetically Modified Food." *Scientific American* 309(3): 80–85.

Freyfogle, E. T. (ed.) (2001). *The New Agrarianism: Land, Culture, and the Community of Life*, Washington, DC: Island Press.

Griffin, L. K. (2013). "Narrative, Truth, and Trial." *Georgetown Law Journal* 101(281): 1–51.

Hauter, W. (2012). *Foodopoly: The Battle over the Future of Food and Farming in America*, New York: New Press.

Hutto, D. B. (2008). *Folk Psychological Narratives: The Sociocultural Basis for Understanding Reasons*, Cambridge: MIT Press.

Kaplan, D. M. (2003). *Ricoeur's Critical Theory*, Albany: SUNY Press.

Keith, L. (2009). *The Vegetarian Myth: Food Justice and Sustainability*, Oakland: PM Press.

King, R. (1999). "Narrative, Imagination, and the Search for Intelligibility in Environmental Ethics." *Ethics and the Environment* 4(1): 23–38.

Kirby, D. (2010). *Animal Factory: The Looming Threat of Industrial Pig, Dairy, and Poultry Farms to Humans and the Environment*, New York: St. Martin's Press.

Liszka, J. (2003). "The Narrative Ethics of Leopold's Sand County Almanac." *Ethics and the Environment* 8(2): 42–70.

MacIntyre, A. (1981). *After Virtue: A Study in Moral Theory*, Notre Dame: University of Notre Dame Press.

Magdoff, F., Foster, J., and Buttel, Frederick H. (eds.) (2000). *Hungry for Profit: The Agribusiness Threat to Farmers, Food, and the Environment*, New York: Monthly Review Press.

Nelson, H. (ed.) (1997). *Stories and Their Limits: Narrative Approaches to Bioethics*, New York: Routledge.

Nestle, M. (2003). *Food Politics*, Berkeley: University of California Press.

Nussbaum, M. (1992). *Loves Knowledge: Essays on Philosophy and Literature*, New York: Oxford.

—————— (1998). *Cultivating Humanity: A Classical Defense of Reform of Liberal Education*, Cambridge: Harvard University Press.

"Perceptions of the U.S. Food System: What and How Americans Think about their Food." *W.K. Kellogg Foundation*, 2012, pp. 48–53.

Pollan, M. (2006). *Omnivore's Dilemma: A Natural History of Four Meals*, New York: Penguin Press.

—————— (2009). *In Defense of Food: An Eater's Manifesto*, New York: Penguin.

Ricoeur, Paul (1984). *Time and Narrative, Vol. 1*, trans. Kathleen McLaughlin and David Pellauer, Chicago: University of Chicago Press.

———— (1992). *Oneself as Another*, trans. David Pellauer Chicago: University of Chicago Press, p. 143.

Robin, Maria-M. (2012). *The World According to Monsanto*, New York: New Press.

Sanford, A. W. (2011). *Growing Stories from India: Religion and the Fate of Agriculture*, Lexington: University of Kentucky Press.

Sanford, W. (2011). "Ethics, Narrative, and Agriculture: Transforming Agricultural Practice through Ecological Imagination." *Journal of Agricultural and Environmental Ethics* 24(3): 283–303.

Schlosser, E. (2001). *Fast Food Nation: The Dark Side of the All-American Meal*, New York: Houghton Mifflin.

Schwartz, J. (2013). *Cows Save the Planet and Other Improbably Ways of Restoring Soil to Heal the Earth*, White River Junction: Chelsea Green Publishing.

Sinclair, U. (1906). *The Jungle*, New York: Doubleday.

Singer, P. and Mason, G. (2007). *The Ethics of What We Eat: Why Our Food Choices Matter*, Emmaus: Rodale Press.

Swirski, P. (2007). *Of Literature and Knowledge*, New York: Routledge.

The United States Census Bureau, https://ask.census.gov. Accessed September 15, 2015.

Tucker, M. E. and Williams, D. R. (1998). *Buddhism and Ecology: The Interconnection of Dharma and Deeds*, Cambridge: Harvard University Press.

Walker, M. U. (1998). *Moral Understandings: A Feminist Study in Ethics*, New York: Routledge.

Wirzba, N. (ed.) (2004). *The Essential Agrarian Reader: The Future of Culture, Community, and the Land*, New York: Counterpoint.

Zamore, M. L. (ed.) (2011). *The Sacred Table: Creating a Jewish Food Ethic*, New York: CCAR Press.

39

INDUSTRIAL AGRICULTURE

Paul B. Thompson

Part dream, part nightmare, industrial agriculture is a cultural imaginary residing just beyond the boundaries of "official" environmental philosophy. Like U.S. Supreme Court Justice Potter Stewart noted with respect to obscenity, people seem to know industrial agriculture "when they see it," even if they would be hard pressed to define it. And environmentalists are wont to associate industrial agricultural with a whiff of obscenity, as well. Nevertheless, industrial agriculture has not really made it into the list of topics routinely covered in environmental ethics.

Environmental ethics has emphasized generalized debates over valuation of species, ecosystems and environmental goods. Battles among advocates of biocentric, ecocentric and anthropocentric axiology have filled out a literature sprinkled with occasional forays into key problems such as energy or restoration ecology, as well as essays on politics and policy. The practical side of environmental ethics has been waxing in the era of climate change, but forays into agriculture have never become mainstream. Yet the book often credited with sparking the contemporary environmental movement was an exposé of chemical use in post-war agriculture. Developed with defense funding during World War II, DDT was cheap to produce and singularly effective at killing a broad spectrum of insects. In 1962, Rachel Carson's *Silent Spring* brought the specter of bioaccumulation into public consciousness and political discourse. Songbirds were declining as their consumption of pesticide-contaminated grain or insects softened their eggshells. Agricultural science responded with umbrage, shock and disbelief at Carson's accusations. Indeed, evaluating the trade-offs between agricultural technology's ability to "feed the world," on the one hand, and its impact on biodiversity, ecosystem services and public health, on the other hand, is the dominant paradigm for discussions of industrial agriculture in the present day.

Despite Carson's clarion call to reform pesticide use in agriculture in the 1960s, few environmental philosophers have conducted thematic inquiries into the nature and function of agriculture. As a rule, their understanding of food and fiber production is that of urbanites whose relationship to agriculture is mediated by restaurants and grocery stores, moderated by an occasional encounter with gardening. Simply understanding industrial agriculture and situating it within a more comprehensive environmental philosophy is a compelling philosophical inquiry, so that is where this chapter begins. It is useful to start by reminding ourselves of some basic premises before moving on to consider the philosophical orientations implicit in contemporary agricultural science, policy and practice. Only with that in mind can a review of industrial agriculture's faults be placed in an appropriate perspective. The chapter concludes with a concise overview of recent arguments made against industrial agriculture, as well as in its defense.

DOI: 10.4324/9781315768090-47

Some Basics

Agriculture itself is old. It is thought to have evolved from hunting and gathering at several independent junctures in prehistory. Archeologists seem to revise the story of agriculture's origins with increasing frequency. Drawing a line between some pre-agricultural era for *Homo sapiens* and the introduction of crop production or animal husbandry is becoming less and less plausible. This has implications for environmental philosophy. Philosophers should not presume that domestication of other species is anything more than an adaptive response to a particular environment. They should not even presume that it is uniquely human (Cossins, 2015).

Agriculture may be thought of as a perturbation of ecosystem processes or as a particularly spectacular example of mutualism in which human populations exist in a symbiotic and reciprocally beneficial relationship with populations of other species. Swidden (or slash and burn) systems are among the oldest forms of traditional agriculture, but involve the temporary creation of fertile plots for cultivation by destroying a forest or savannah ecosystem through fire. Swidden agriculture is sustainable, however, so long as human population density is low and cultivated areas have sufficient periods of "rest." Relatively more settled approaches to cultivation were especially crucial for the emergence of European-style civilization, but these, too, relied on fallow periods when cleared fields might be occupied by domesticated livestock. Achieving sustainability in settled agriculture proved to be much more challenging, though history provides a few examples of relatively successful approaches (Mazoyer and Roudart, 2006). Though diverse in their specifics, such regionally and climatically adapted methods represent the *traditional agriculture* to which industrial agriculture is a contrast case. Traditional agriculture, however, should not be presumed environmentally benign. If previously existing ecosystem interactions are valorized, *all* forms of agriculture are inherently disruptive.

The early industrial revolution (1760–1840) initially had relatively little direct effect on traditional farming methods, but it did increase the demand for agricultural commodities. The gradually but steadily growing population of industrial workers needed food. In addition, fiber commodities, such as cotton and hemp, were the raw materials for the textile factories that were the first bloom of mass production. Prior to the 19th century, improvements in agricultural methods were labor intensive, but horse-drawn mechanical implements, and then steam- and combustion-powered tractors began to appear. Improved shipping offered farmers the opportunity to purchase fertility in bags of guano. World War I created an infrastructure for synthetic nitrogen production intended first for bombs, but then used for fertilizer. Luther Burbank (1849–1926) achieved fame for his genetic improvements to numerous agricultural plants. Large-scale wheat and corn monocultures were common throughout North America by the first decades of the 20th century. As noted already, the capacity to produce synthetic pesticides grew during World War II. While there does not appear to be any conceptually or environmentally decisive way to distinguish traditional farming from industrial agriculture, the mid-20th century marks a point when the accumulation of technical change seemed to take on greater significance.

Stephen Stoll's *Larding the Lean Earth* makes it clear that alarms were sounded about the cascade of environmental consequences issuing from environmentally exploitative farming practices long before *Silent Spring*. Eighteenth century American writers bemoaned the practice of "skimming" fertility, then abandoning farms and moving ever westward in search of virgin soil (Stoll, 2002). Yet the idea that industrial agriculture embodied a unified *philosophy* of agriculture is a more recent phenomenon. Critics such as Wendell Berry (1977) and Wes Jackson (1980) have defined it in terms of its deficiencies; specifically, it neglects the quality and value of farm work, seeking ease and efficiency. This bias has led industrializing farmers to choose motorized methods, ignoring both the skill and pleasure of working teams of draft animals, as well as the ecological benefits of animal traction. Attention to the peculiarities of place and ecology is given no intrinsic value, while specialization in pursuit of technical and economic efficiencies becomes the measure of success. This transition has caused an expansion of monocultures well beyond pre-war practice, so much so that Everyday aesthetics farmers have abandoned the production of fruits and vegetables even for home consumption and local markets.

Early 20th century reformers touted efficiency in the kitchen, as well as in the field, and argued that diets should be based on scientific principles. Ideals of safe and efficient eating were exploited first by food processors and then by retailers who were steadily concentrating control over the entire food system in the hands of an ever decreasing number of incorporated firms. First additives and preservatives were used to facilitate these commercial activities, though justified in terms of benefits to consumers that took the form of cheaper and more uniform products. Foods became more and more synthetic throughout the 20th century as industrial technology for processing, packaging, distributing and retailing food products progressed. From cured meats or canned fruits and vegetables to frozen meals and finally wholly synthetic products (margarine and trade names like Splenda® or Olestra®), food has increasingly become a manufactured good (Vilesis, 2008). The food industry views agriculture as but one link in a complex and globalized supply chain. This chain begins not on the farm but with input companies that produce the machinery, chemicals, seeds and animals that are simply "grown out" on farms. Processors and retailers dictate farming methods in accordance with the need for inputs into a wholly industrialized manufacturing process. Given this, many contemporary food scholars would insist that a focus on "industrial agriculture" is misplaced. The topic is *industrial food systems* (see McMichael, 1994).

What is the philosophical significance of these transitions and transformations? Criticism and critique of industrial food systems is an obvious launching point, and it is certainly the topic with the largest literature (duly reviewed below). However, my main thesis is that critiques generally neglect three interrelated themes that are of deeper significance. First, there is the epistemic and perceptual change in how people see the world and conceptualize their problems associated with modernism itself. The general perspective on reality that is associated with modern scientific worldviews gives rise to the agricultural sciences and defines their mission in the specific terms that lead to an industrial food system. Second, there is the emergence of a broad orientation to governance of the industrial economy. This social philosophy sees agriculture as a sector of the economy and creates a way to specify general norms that apply equally to *all* sectors of an industrialized society. In fact, most debate over industrial agriculture fails to challenge these norms and is situated *within* the very philosophical framework that has given rise to an industrially organized food system. Finally, there is philosophical significance to whatever it was about food and farming that was threatened or displaced by the industrialization of agriculture. (A discussion of this last topic exceeds the remit for an overview of industrial agriculture and is largely absent from the present chapter.) In short, "industrial agriculture" is an inherently vague term used to indicate tools and techniques associated with contemporary food systems, many of which justifiably merit critique. There are, however, more subtle issues for environmental philosophy that are concealed by such critiques, mainly because they generally adopt the main assumptions of an *industrial philosophy of agriculture*.

The Industrial Philosophy of Agriculture: Part I

I associate modernism with a mentality or outlook that is broadly characteristic of the first two centuries of growth in the natural sciences and the period of philosophy that extends roughly from Descartes to Mill. It gave rise to the technological innovations and accompanying social transformations we associate with the industrial revolution through to the end of the 19th century. Interest in mechanisms that might lead to improvements in food systems was present at the very birth of the modern period. Francis Bacon (1561–1626) died from pneumonia contracted while conducting experiments on the preservation of meat. A series of innovations in husbandry, crop rotation and agricultural technology accompanied the enclosures defended by John Locke (1632–1704), arguing that one who "...has a greater plenty of the conveniences of life from ten acres, than he could have from an hundred left to nature, may truly be said to give ninety acres to mankind" (Locke, 1689). One can discern a pattern of reasoning that would blossom into full flower with the arrival of utilitarianism: actions that make productive use of resources (including and perhaps especially science-based innovations) are ethically justified when the net value of benefits reaches an optimum. Much of the philosophical debate occurs in unpacking what is meant by benefit and by net value, but that is to get ahead of our story.

There was a social, political and environmental history of agriculture well before there was much critical philosophical reflection on it. The existence of precursors notwithstanding, the industrial *philosophy* of agriculture emerged with Thomas Malthus (1766–1834) and the publication of his *Essay on the Principle of Population* in successive revised editions between 1798 and 1830. The *Essay* itself made rather broad assumptions about agriculture that displayed relatively little cognizance of underlying biological processes. Malthus famously assumed that if humans are left to their own devices, growth in population would inevitably outstrip growth in food production. His theoretical interests were more aligned with explaining why starvation was *not* the actual limiting factor on human population growth. His findings anticipated current views on a demographic transition that follows industrial growth, as incentives for unconstrained birthrates and large families shift. Nevertheless, it is the simple claim that population grows exponentially while agricultural production grows arithmetically that articulates the basis for an industrial philosophy of agriculture. Though internalized among many agriculture specialists much earlier, the implications of this model were not fully articulated until the 1960s when population ecologists such as Paul and Anne Ehrlich or Garrett Hardin wrote of impending famines and inevitable environmental decline in the decades to come (see Ehrlich and Ehrlich, 1990; Hardin, 1995 for definitive treatments of this work). Hardin is known for arguing that it is morally wrong to offer food aid to the victims of famine, because doing so only encourages them to persist in unsustainable rates of reproduction. He deploys utilitarian principles to argue for the lesser of two evils: a fixed number of people dying today is better than a multiple of that number dying tomorrow (Hardin, 1974).

The population ecology arguments were formative for environmental philosophy and will not be re-rehearsed here. What has been less well appreciated is that the basic Malthusian picture, amplified and given increased urgency by population ecologists, also had significant implications for the other side of Malthus' original equation: the rate of growth in food production. On the empirical side, Malthus' original assumptions about the divergence between rates of population growth and growth in agricultural production were simply wrong. Although population had in fact started to grow at something like the rate that Malthus speculated it would absent the controlling factors of "misery and vice," agricultural production has, for 200 years since the *Essay* was published, been able to keep up with it. The numbers convincingly demonstrate that the mechanization of the 19th century, the motorization of the early 20th century and the chemical technologies that begin with synthetic fertilizers and eventually extend to pesticides had, with genetic improvements made possible by scientific breeding, grown at least as fast as population (see Mazoyer and Roudart, 2006).

Although debates on the empirical accuracy of Malthus' food/population equation have elements of philosophical interest, it is the normative implications that concern us here. Specifically, if advances in agricultural science and technology *have* allowed food production to outpace population growth (forestalling the misery predicted by Malthus), and if (as the Erlichs and Hardin predicted) population growth can be expected to continue unchecked, then continuing to find new ways to increase global food production becomes a moral imperative. It would be difficult to overstate how deeply agricultural scientists internalized this vision of agriculture, or how profoundly it has shaped the conduct of agricultural input firms (such as seed and chemical companies), and even farmers in industrialized economies, themselves.

The Industrial Philosophy of Agriculture: Part II

Our story so far suggests how someone might come to see agriculture in terms of its ability to generate a stock of food (measured generally on a national or global basis). The Malthusian vision implies that it is morally important for society as a whole to be attentive to this stock, and to increase the flow of future commodities into the stock at a rate commensurate with the growth in human population. The picture thus functions as a script in which science and industrial technology are well suited to be the actors who will play the heroic role of saving future generations by delivering the tools to increase the total global harvest, yet again. Taken from one vantage point, it is a Sisyphusian task, but to any given individual who can see the problem from the perspective of societal needs and who thinks of themselves as having the power to make a meaningful contribution to increasing yields, it is a serious and sacred responsibility. Yet the industrial era also ushers in a substantially different way of understanding the role that any actor in an industrial economy is expected to play.

In *The Agrarian Vision*, I argue that both utilitarian and rights-based theories of ethics and governance have converged on very general way of characterizing the moral responsibility of firms in an industrial economy. For-profit actors are expected to do two things. First, they are expected to conduct their activity in a manner that does not harm third parties, and second, they are expected to engage in efficiency-increasing competition with other for-profit firms. The emphasis on not harming third parties implies that for-profit actors engage with one another as well as with individuals in voluntary contracts or exchanges. Well-known political philosophies contest the details and power dimensions of exchange, but "side effects," externalities and forms of harm to persons *not* party to the exchange are widely agreed to justify ethically based constraints on the competitive process. My interest here resides in the generality with which these presuppositions apply to firms in varying sectors of the industrial economy. The second norm has been of less direct interest to those who take a rights view, but it is the element of economic thought that is, perhaps, least contentious. If we can be assured that the rights of third parties are being respected, surely the innovations that lower prices and multiply the quantity of goods available to us are a good thing.

My objective here is to neither presume nor defend this normative framework. My point is that it accommodates much of the criticism leveled against industrial food systems. To take the case of agriculture itself, the issue with chemical pesticides is clearly a case of limiting impact on third parties. Chemical companies strike their bargain with farmers, but then the actual costs are born by people whose bodies bear the burdens of exposure, not to mention the songbirds that were Rachel Carson's chief concern. Philosophical questions arise as to whether non-humans can count as third parties, to be sure, but these questions do not challenge the presumption that farmers and chemical companies alike would be within their moral rights to transact business however they like so long as no harm ensues. Furthermore, these debates about environmental harms and moral culpability are characterized by a classical distinction between rights-based approaches in moral theory and consequentialist approaches that are more amenable to decision-making that focuses on making appropriate trade-offs.

To wit, a "rights-based" or deontological approach will place most of its chips on limiting harm. Whether articulated in terms of human, animal or ecosystem rights (or more typically in terms of intrinsic or sacred values that must not be transgressed), much of environmental philosophy identifies constraints that effectively define harm, then stipulate norms prohibiting acts that impose a harm so defined. For agriculture, this kind of position has been most evident in critiques of animal production, with some authors arguing that individual animals have rights that are violated by the practices of concentrated animal feeding operations (CAFOs). This feature makes a livestock production system in which animals are confined indoors with limited opportunity to engage in species-typical behavior morally unacceptable, without regard to the benefits that these systems may have (Rollin, 1990; Dieterle, 2008). Of course, for some advocates of animal rights, *no* livestock production system will be morally acceptable simply because raising animals for food involves morally unjustifiable confinement, at a minimum, and, more generally, their death (Regan, 1980). While a more expansive "animal rights" interpretation of the "do not harm" claim militates against any form of livestock production, the focus here is limited to critiques of industrialized methods.

In contrast to deontological theories, many observers are more willing to countenance the possibility that improvements in the efficiency of industrial food systems might offset some of the harms to third parties. This point of view is argued most pointedly by economists such as Luther Tweeten, who points out that a decrease in the proportion of a person's budget spent on food is doubly important from an ethical perspective. First, *any* such decrease is a good thing for anyone, because it frees up resources for expenditure on other goods that will, presumably, increase the quality of life. Spending less on food means more for clothing, more for education, more for the increasingly necessary accouterments of modern life that run from cell phones to tablet computers. Second, and perhaps more importantly, a decrease in the cost of food is especially good for the poor, since the poor spend a proportionally larger share of their income on food. Food is not, for the most part, a discretionary expense, and economies in the food sector that translate into lower food costs thus serve the Rawlsian/egalitarian goal of delivering benefits to the worst-off group. For this reason, many economists follow Tweeten to argue that these increased efficiencies balance out harms to third parties (Tweeten, 1992). This type of argument makes a moral appeal to the potential Pareto improvement and exemplifies a classic instance of utilitarian cost/benefit trade-offs.

Critics of the Industrial Food System

The chapter now concludes with a synoptic overview of the way that harm to third parties has been characterized in critiques of the industrial food system. Harm implies an impact or outcome that is contrary to the interest of the affect party. It thus blends a normative judgment (What makes this outcome bad?) with empirical claims to the effect that (a) such outcomes occur and (b) they are caused by one or more aspects of the industrial food system. The key harms associated with the industrial food system can be classified under four types:

A Outcomes that degrade human health or cause disease;
B Outcomes that affect non-human animals
C Outcomes that are adverse to the environment
D Social impacts that are viewed as adverse or unwanted

This schema is developed at some length in my book *Food Biotechnology in Ethical Perspective* with respect to the use of gene transfer technology—in recent years a signature tool of the industrial food system. In this context, the goal is to note just a few exemplary cases.

A *Human Health.* Agricultural chemicals are unquestionably associated with adverse impact on human health, both through environmental exposure and through residues in food. Environmental exposure is particularly significant for agricultural workers, who tend to have low socio-economic status and are thus especially vulnerable to exploitation (Krimsky, 2000; Wright, 2005). Furthermore, the industrial food system has led to a shift in the types and quality of foods that people eat (Scrinis, 2013), linking industrial technologies to a global increase in the incidence of obesity and to increasing obesity among school-age children (Cawley and Kirwan, 2011). Key philosophical issues are beginning to emerge in the literature on obesity, as social scientists have begun to question both assumptions about nutritional science and the methodologies that have been used to study obesity (Lupton, 2012). However, there is as yet relatively little work on this class of harms by environmental philosophers, perhaps because the normative dimensions are so uncontroversial.

B   *Non-human Animals.* As noted briefly above there is a significant literature on the ethics of industrial livestock production that is ancillary to an even larger number of publications on human obligations to non-human animals. Recent philosophical interest in this topic was stimulated by the U.K. publication of Ruth Harrison's book *Animal Machines: The New Factory Farming Industry* in 1964. The first edition of Peter Singer's *Animal Liberation* included a chapter focused exclusively on the suffering of animals being raised in CAFOs. This chapter (which is included with revisions in subsequent editions) figures pointedly in Singer's argument for vegetarianism as a form of moral protest (Singer, 1975). Similar descriptions now appear routinely in works by philosophers. Lori Gruen's *Ethics and Animals* has a similar chapter with headings such as "The evolution of industrial agriculture" and "Living and dying on factory farms" (Gruen, 2011). One obvious philosophical issue here is simply the moral standing of non-human animals. Singer's first edition of *Animal Liberation* made the false claim that animal producers denied the fact of animal pain on Cartesean grounds, but this has been withdrawn in subsequent editions. The more reasonable accusation against producers is that they uncritically assumed that the parameters crucial to them (growth, morbidity and mortality) were also the keys to animal welfare. This assumption is systematically reviewed and rebutted with respect to most industrial production systems by Bernard Rollin (1995).

C   *Environmental Impact.* Accounting for the value of impacts on non-humans, ecosystems and biodiversity has been a key issue for environmental philosophy since the 1970s. As made clear in other contributions to this volume (See Chapter 1: Animal Cognition and Moral Status; Chapter 5: Species and Wildlife; Chapter 10: Wilderness), the philosophical issues turn upon whether and how the relevant entities can be characterized as subjects capable of sustaining harm, on the one hand, and alternative ways of tying these impacts to human interests (including aesthetic and cultural interests), on the other. Relatively little of this work by philosophers singles out industrial agriculture, though expansion of agriculture into fragile ecosystems is occasionally cited as a case in point (Callicott, 1990). However, having drawn the attention of philosophers for its impact on farm animal welfare, industrial livestock production has also been critiqued for its environmental impact (for a review see Rossi and Garner, 2014) and the environmental impact has in turn been linked to obesity (Cafaro et al., 2006). Vandana Shiva, who, though trained as a philosopher, often represents herself as a physicist, has made repeated allegations that industrial farming methods are destructive to biodiversity in developing countries (Shiva, 1993, 1999). Given the extensive environmental footprint of the industrial food system, ecologists, geographers and other environmental scientists have shown a high level of cognizance that agriculture in general is an important source of the impacts that have been at the center of debate in environmental philosophy.

D   *Social Impact.* Though perhaps less central to the concerns of mainstream environmental philosophers, the social impact of industrial food systems has been a central theme of rural social science for many decades. The work of anthropologist Walter Goldschmidt is a centerpiece. Goldschmidt's study of two California farming communities linked large-scale monocultures with declines in schools, local business and other community services (Goldschmidt and Nelson, 1978). Although a few scholars have attempted to integrate the social impact theme into environmental philosophy (see Ebenreck, 1983; Comstock, 1988), this theme has only recently begun to achieve resonance. As Wendell Berry's writings have begun to be cited by environmental philosophers, there has been a growth of sensitivity to the way that industrial food systems are connected to alterations in the pattern of social life that can be regarded as forms of harm (Thompson, 2017; Cheney and Weston, 1999). Shiva has tied industrial farming to the bankruptcy and suicide of smallholders, many of whom are women (Shiva and Jalees, 2006). The industrial food system is also criticized for the poverty-level wages of many field laborers and food service workers, but given the unsavory history of slavery, sharecropping and bondage in traditional agriculture this social justice critique may not be unique to the industrial era (see Thompson, 2015).

The response to these harms has been to argue for local, sustainable production systems that return to more traditional production methods. To the limited extent that philosophers have spoken on this issue, the strategy has been to emphasize "food ethics," or a consumer-driven approach where reform is effected by making dietary choices that do not implicate individuals in such forms of harm. Writing with Jim Mason, Peter Singer has again been at the forefront of sketching a dietary reform approach to the problems of industrial food systems. Singer and Mason discuss the environmental and animal welfare impacts of a number of different farming systems, urging their readers to applying their own values in deciding what is ethical to eat, and what not (Singer and Mason, 2006).

Defending Industrial Agriculture

To be sure, there is a literature (largely outside philosophical journals) of response to these criticisms. The blending of utilitarian and egalitarian claims demonstrated by Tweeten exemplifies the general pattern of argument that defenders of industrial food systems mount. It is vital to recognize that this argument accepts the normative vision of agriculture that arose in response to Malthusian predictions: increasing the amount of food that is available on a global basis is an overriding moral necessity. It is instructive to consider a few examples that defend industrial farming against the claim that it has unacceptable impacts on the environment. According to Ron Herring, Shiva's claims are viewed as, at best, restating critiques of early "green revolution" development that have been made since the 1970s. Current efforts to adapt industrial methods in developing countries are significantly more nuanced and less destructive, while those damages that do occur are more than offset by improvements in the livelihoods of very poor people. Herring claims that given the progress being made in addressing the ethical and environmental challenges of agricultural development, Shiva's critiques should be viewed as politically motivated overstatements that are careless with the truth (Herring, 2008).

Broader defenses of industrial food systems do not respond directly to its critics. The Food and Agricultural Organization (FAO) of the United Nations has published a series of reports, with *Livestock's Long Shadow* among the most impactful and widely cited. The report focused on the global contribution of food animal production to processes of climate change. Although referenced by philosophers who oppose CAFOs (see Ilea, 2009), the report is actually an argument for further utilization of CAFOs because they have significantly lower greenhouse gas emissions per volume of product than do traditional livestock production systems (Steinfeld and coauthors, 2006). A series of detailed studies on the environmental impact of industrial farming methods have been performed by Judith Capper. Like the FAO report, Capper finds that when the entire life-cycle of food products is taken into consideration, industrial farming methods have lower per pound impact than did methods in use over 50 years ago. Although total impact may have increased, industrial methods have reduced what that impact would have been had industrial technologies been rejected (Capper and coauthors, 2008; Capper, 2011, 2013).

These rebuttals to the critics of industrial agriculture are not philosophically sophisticated, but they should be interpreted within the context of the more comprehensive philosophical framework that has been described above. *Given* an increase in global population, the Malthusian minimum of food availability is rising. Hence, increasing the efficiency of food production systems actually serves environmental values. Environmental benefits from industrialization are quantifiable in the form of water conservation and reduction in the overall chemical burden and greenhouse gasses per unit of food produced. Even if the total environmental impact of world agriculture remains high, it is much lower than it would have been if traditional methods had been used to produce the same total quantity of food. The most plausible reading is that these defenders of industrial methods are making a utilitarian argument to the effect that continued ability to "feed the world" offsets whatever harms are being noted by critics. They do not deny the normative assumptions that critics make when analyzing the impact of industrial food systems as a forms of harm. My point here is not to declare which side is winning this battle of harms and trade-offs, but simply to note how both sides seem to be content with mounting the debate within the framing assumptions of an industrial philosophy of agriculture.

The key point to recognize at this juncture is that there is nothing particularly significant about the fact that it is food or agricultural practices that generates harm to third parties in these critiques. Very similar debates surround energy, manufacturing, transportation or sanitation and waste disposal. Although there are philosophical points to debate concerning whether non-humans can *be* ethical subjects, the substance of these debates has been articulated in philosophical studies that do not take agriculture or food systems to be more than a case study for more general aspects of human attitudes and conduct. The upshot is that there is not really anything like a "philosophy of food systems" at work in the critique of industrial agriculture. There is a very broad philosophical approach to ethics and policy that might be equally well applied to environmental impact and to a host of long-standing social problems such as housing, education or health care. The defenders, however, *are* claiming a special status for the agricultural sector. While we might get along with substantial reductions in the output of our entertainment, communication, education, manufacturing, financial services and perhaps even energy and healthcare sectors, we are still going to need food. This creates an implied "trumping" factor that is presumed to override the force of critiques.

Conclusion: The Agrarian Alternative?

In considering agriculture to be just one sector of the economy among many, contemporary critics and defenders alike are adopting an attitude quite unlike that of philosophers writing before the modern era. In contrast, Greek, Roman and Medieval philosophers would have presumed that the food system is deeply implicated in patterns of culture and personal character. Entire societies would have been thought to develop habits of virtue and vice in light of the way that they organized their subsistence activities. Indeed, this way of thinking about food and farming survived much of the modern era as it was described above, penetrating into works of the Scottish Enlightenment and German Romanticism, alike. Its most famous advocate was, of course, Thomas Jefferson, though traces of it surface in the writings of Ralph Waldo Emerson or Henry David Thoreau. The idea that the environment itself could play a role in forming moral character and establishing the habits of ordinary people would have been quite familiar to any of the thinkers working in these traditions. A chapter on industrial food systems must note the existence of alternatives to the industrial philosophy of agriculture, but it cannot discuss them to any degree of sophistication. As I have argued at more length elsewhere, the great tragedy of industrial food systems may well be that even critics of industrial farming seem deaf to the alternative arguments today (Thompson, 2010, 2018).

The take-home message for policymakers is that agriculture has the potential to deliver much more than food. The aesthetic amenities of a pleasing landscape are only the first dimension. Cuisine is widely recognized to be an important carrier of cultural identity, and European societies have long recognized how the flavors, aromas and the accompanying emotional valence of food can be connected to factors such as climate, season and soil type. These components of the food system *can* produce widely shared cultural norms that support a variety of environmental initiatives, not just reducing the chemical or carbon pollution and resource consumption of industrial agriculture. The take-home message for scholars is that this broader multi-functional approach to food production is above and beyond the citizenship virtues and political solidarities that have long been associated with agrarian philosophy. Experiments with farmers' markets, community-supported agriculture and farm-to-table food systems should be evaluated from a perspective that reflects an integrative and relational philosophy of food systems, rather than one which focuses only on food availability and the mitigation of adverse environmental impact.

Related Topics

2. Eating
20. Moral Bases of Responses to Climate Change
33. Urban Sustainability
37. Waste and Consumption
41. Sustainable Agriculture
42. Community Gardens

References

Berry, W. (1977) *The unsettling of America: Culture and agriculture*. San Francisco, CA: Sierra Club Books.

Cafaro, P.J., Primack, R.B., and Zimdahl, R.L. (2006) 'The fat of the land: Linking American food over-consumption, obesity, and biodiversity loss', *Journal of Agricultural and Environmental Ethics*, 19(6), pp. 541–561.

Callicott, J.B. (1990) 'The metaphysical transition in farming: From the Newtonian-mechanical to the Eltonian ecological', *Journal of Agricultural Ethics*, 3, pp. 36–49.

Capper, J.L. (2011) 'The environmental impact of beef production in the United States: 1977 compared with 2007', *Journal of Animal Science*, 89, pp. 4249–4261.

Capper, J.L. (2012) 'Food security and the environment: Animal protein production impacts and trends', in *A sustainability challenge: Food security for all: Report of two workshop*. National Academies Press, pp. 52–55.

Capper, J.L., Cady, R.A. and Bauman, D.E. (2009) 'The environmental impact of dairy production: 1944 compared with 2007', *Journal of Animal Science*, 87, pp. 2160–2167.

Carson, R. (1962) *Silent spring*. Boston: Houghton Mifflin.

Cawley, J. and Barrett, K. (2011) 'Agricultural policy and childhood obesity', in J. Cawley (ed.) *The Oxford handbook of the social science of obesity*. New York: Oxford University Press, pp. 480–491.

Cheney, J. and Weston, A. (1999) 'Environmental ethics as environmental etiquette', *Environmental Ethics*, 21, pp. 115–134.

Comstock, G. (ed.) (1988) *Is there a moral obligation to save the family farm?* Ames: Iowa State University Press.

Comstock, G. (1994) 'Some virtues and vices of agricultural technology', in P. Hartel, K. Paxton George, and J. Vorst (eds.) *Agricultural ethics: Issues for the 21st century: Proceedings of a symposium sponsored by the Soil Science Society of America, American Society of Agronomy, and the Crop Science Society of America in Minneapolis, MN*, Oct. 31–Nov. 5, 1992. Madison, WI: SSSA.

Cossins, D. (2015) 'Amazing animal farmers that grow their own food', *BBC-Earth*, Accessed May 11, 2020 at http://www.bbc.com/earth/story/20150105-animals-that-grow-their-own-food

Dieterle, J.M. (2008). 'Unnecessary suffering', *Environmental Ethics*, 30: 51–67.

Ebenreck, S. (1983) 'A partnership farmland ethic', *Environmental Ethics*, 5: 33–45.

Ehrlich, P.R., and Ehrlich, A.H. (1990) *The population explosion*. New York: Simon and Schuster.

Goldschmidt, W. and Nelson, G. (1978) *As you sow: Three studies in the social consequences of agribusiness*. Montclair, NJ: Allanheld, Osmun.

Gruen, L. (2011) *Ethics and animals: An introduction*. Cambridge: Cambridge University Press.

Hardin, G. (1974) 'Living on a lifeboat', *Bioscience*, 24, pp. 561–568.

Hardin, G. (1995) *Living within limits: Ecology, economics, and population taboos*. New York: Oxford University Press.

Herring, R.J. (2008) 'Opposition to transgenic technologies: Ideology, interests and collective action frames', *Nature Reviews Genetics*, 9, pp. 458–463.

Ilea, R.C. (2009) 'Intensive livestock farming: Global trends, increased environmental concerns, and ethical solutions', *Journal of Agricultural and Environmental Ethics*, 22(2), pp. 153–167.

Jackson, W. (1980). *New roots for agriculture*. Lincoln: University of Nebraska Press.

Krimsky, S. (2000) *Hormonal chaos: The scientific and social origins of the environmental endocrine hypothesis*. Baltimore, MD: Johns Hopkins University Press.

Lupton, D. (2012) *Fat*. London: Routledge.

Mazoyer, M. and Roudart, L. (2006) *A history of world agriculture: From the neolithic age to the current crisis*. London: Earthscan.

McMichael, P. (ed.) (1994). *The global restructuring of agro-food systems*. Ithaca, NY: Cornell University Press.

Regan, T. (1980) Utilitarianism, vegetarianism and animal rights, *Philosophy and Public Affairs*, 9, pp. 305–324.

Rollin, B.E. (1990) 'Animal welfare, animal rights and agriculture', *Journal of Animal Science*, 68(10), pp. 3456–3461.

Rollin, B.E. (1995). *Farm animal welfare: Social, bioethical, and research issues*. Ames, IA: Iowa State University Press.

Rossi, J. and Garner, S.A. (2014) 'Industrial farm animal production: A comprehensive moral critique', *Journal of Agricultural and Environmental Ethics*, 27, pp. 479–522.

Scrinis, G. (2013) *Nutritionism: The science and politics of dietary advice*. New York: Columbia University Press.

Shiva, V. (1993) *Monocultures of the mind: Biodiversity, biotechnology and scientific agriculture*. London: Zed Books.

Shiva, V. (1999) 'Monocultures, monopolies, myths and the masculinization of agriculture', *Development*, 42(2), pp. 35–38.

Shiva, V. and Jalees, K. (2006) *Seeds of suicide: The ecological and human costs of seed monopolies and globalisation of agriculture,* 4th ed. Rev. New Delhi: Navdanya.

Singer, P. (1975) *Animal liberation*. New York: HarperCollins.

Singer, P. and Mason, J. (2006) *The way we eat: Why our food choices matter*. Emmaus, PA: Rodale Press.

Steinfeld, H., Gerber, P., Wassenaar, T., Castel, V., Rosales, M., and De Haan, C. (2006) *Livestock's long shadow*. Rome: FAO.

Stoll, S. (2002) *Larding the lean earth: soil and society in nineteenth-century America*. New York: Hill and Wang.

Thompson, P.B. (2010) *The Agrarian vision: Sustainability and environmental ethics*. Lexington: University Press of Kentucky.

Thompson, P.B. (2015) *From field to fork: Food ethics for everyone*. New York: Oxford University Press.

Thompson, P.B. (2017). *The spirit of the soil: agriculture and environmental ethics,* 2nd Ed. New York: Routledge.

Thompson, P.B. (2018) 'Farming, the virtues and agrarian philosophy', in A. Barnhill, M. Budolfson and T. Doggett (eds.) *The Oxford handbook of food ethics*, New York: Oxford University Press, pp. 53–66.

Tweeten, L. (1991) 'The costs and benefits of bGH will be distributed fairly', *Journal of Agricultural and Environmental Ethics*, 4, pp. 108–120.

Vilesis, Ann. (2008) *Kitchen literacy: How we lost knowledge of where food comes from and why we need to get it back*. Washington, DC: Island Press.

Wright, A. (2005) *The death of Ramon González: The modern agricultural dilemma*. Revised Edition. Austin: University of Texas Press.

40

BIOTECHNOLOGY

Dane Scott

The first commercial genetically modified (GM) crops were approved for use in the United States in 1995. Four years later half of the corn and soybean crops grown in the United States were GM and by 2012 the number had jumped to 90%. Despite the rapid and widespread adoption of GM crops in the United States and in many other countries around the world, agricultural biotechnology has been met with strong and consistent opposition: activists have burned experimental GM crops, there is a large and vocal international anti–GM movement, and many countries in Europe, for example, have all but rejected GM crops and food. Two decades after its initial commercialization, agricultural biotechnology remains highly contentious, especially among environmentalists, and there are no signs that this controversy is coming to an end any time soon.

Many environmental groups have focused on this technology as an area of special concern. The Sierra Club, for instance, has called for a moratorium on GM crops on the basis of the precautionary principle; their position in genetic engineering should not be given the benefit of scientific doubt in regard to safety.

> The Genetic Engineering Committee believes that genetically engineered farm crops are wrongly given the benefit of the doubt in the regulatory process, and that, under the precautionary principle, they should not be released into the environment or allowed to be part of the food supply.
>
> *(Sierra Club, n.d.)*

The Sierra Club's website states: "The long-term impacts of GMOs are unknown, and once released into the environment they cannot be recalled" (Sierra Club). Greenpeace International has mounted an aggressive campaign against the biotech industry. Like the Sierra Club, Greenpeace advocates for a precautionary approach and the organization asserts on its website: "GMOs should not be released into the environment since there is not an adequate scientific understanding of their impact on the environment and human health" (Greenpeace). Many environmental organizations see this powerful new technology as a threat to the earth's biodiversity and the integrity of the world's food supply.

While some environmental groups and scientists worry about the unknown environmental dangers of GM crops, ironically, other scientists, especially those in the field of agricultural biotechnology, see this technology as having tremendous potential for reducing the environmental impacts of industrial agriculture. In his history of the biotech industry, Daniel Charles remarks, "[the young genetic engineers] saw themselves as 'green' revolutionaries fighting against the

DOI: 10.4324/9781315768090-48

entrenched power of chemists.... Chemicals represented a dirty and regrettable past, and biology was a savior" (Charles 2001: 24–25). The genetic engineers working for Monsanto in the 1980s believed the scientific breakthroughs they were making would be good for the environment by reducing the need for harmful chemical inputs. To some, biotechnology is a much-needed tool to lessen the environmental impacts of agriculture. To others, it is but the latest example of a tech-no-fix that threatens the earth's biodiversity. There are sincere people (and no doubt cynical ones) on both sides of the biotechnology debate who believe their position is grounded in ethical values.

The GM debate raises important ethical issues that expand the sphere of environmental ethics to overlap with agricultural ethics. However, with a few notable exceptions, environmental phi-losophers have not written much about agriculture, despite its enormous environmental impacts. Agriculture arguably has more impacts than any other human activity. For example, growing food occupies 50% of the earth's habitable land, comprises 69% of water use, and it is the largest user of industrial chemicals (World Wildlife Fund). With the pressure of a rapidly growing world popu-lation combined with an increasing environmental footprint of current agricultural technologies, few believe that existing production practices are sustainable. Many have championed biotech-nology as an essential tool for solving the challenges of 21st century agriculture. The goal of this chapter is to outline three perspectives for considering the relationship between crop biotech-nology and environmental ethics: anthropocentric agricultural ethics, ecocentric environmental ethics, and pragmatic environmental ethics.

In the field of environmental ethics there is a much-discussed distinction between anthropo-centric, or human-centered ethics, and ecocentric, or biocentric ethics. Until the advent of the fields of animal ethics and environmental ethics in the second half of the 20th century, academic ethics was almost exclusively focused on human values, and only indirectly con-sidered the values of the non-human world. For example, the influential moral philosopher John Rawls writes in *A Theory of Justice* about the limits of his theory: "We should recall here the limits of a theory of justice. Not only are many aspects of morality left aside, *but no account is given* of right conduct in regards to animals and the rest of nature" (Rawls 1971: 512, emphasis added). One of the founders of the field of academic environmental ethics, Holmes Rolston, published several influential papers in the 1980s and 1990s that aimed to give an account of right conduct toward the rest of nature. For instance, in a 1985 article, "Duties to Endangered Species," Rolston explored the idea of direct obligations to the non-human world (Rolston 1985). Rolston's goal was to explore new frontiers in ethics beyond the exclu-sive, human-centered perspective. He did this by investigating connections between biology, particularly ecology, evolution and natural history, and ethics to discover how we might find direct duties to species, ecosystems, and the earth. He writes: "We can, if we insist on being anthropocentrists, say that it is all valueless except as our human resource. But we will not be valuing Earth objectively until we appreciate this marvelous natural history. (Rolston 1994: 27). To some extent, from its beginnings the field of academic environmental ethics has been preoccupied with the distinction between anthropocentric ethics and ecocentric ethics. This distinction will serve to organize the following discussion on environmental ethics and ag-ricultural biotechnology.

Anthropocentric Agricultural Ethics

Numerous scientists, economists, policy experts, and philosophers have defended GM crops against environmental critics from the perspective of an anthropocentric ethics. For example, luminaries such as the billionaire, philanthropist, Bill Gates and the Nobel Prize-winning agrono-mist, the late Norman Borlaug, have campaigned for biotechnology as a much needed tool to feed the world's poor. In their books, philosophers Gregory Pence (2002) and R. P. Thompson (2011) advocate for biotechnology from a strictly human-centered agricultural ethics. Both philosophers argue that scientific technological agriculture is key for promoting social justice. Further, they believe that the current largely profit-oriented, corporate-driven model for research and development, which has produced most GM crops, is the way forward toward solving the problem of world hunger. Neither philosopher has sympathy for an ecocentric environmental ethics. Gregory Pence devotes a chapter of his book attempting to undermine the field of environmental ethics. For example, Pence believes Holmes Rolston's efforts to discover a non-anthropocentric ethics are logically flawed and starts one down a slippery slope leading to "ecofascism" (Pence 2002: 125–129). For his part, R. P. Thompson simply ignores environmental ethics in his list of the relevant ethical theories for considering biotechnology; there is no mention of environmental ethics in his book on philosophy and agro-technology. However, while attacking or disregarding environmental ethics, both philosophers argue that GM crops will have positive environmental benefits. Their basic position in biotechnology will help solve the problem of hunger while reducing the environmental footprint of intensive agriculture, which is currently threatening our life-support system. Further, weighing the economic, health, and environmental benefits of a GM crop versus the economic costs and scientifically determined environmental and health risks provides the framework for decision-making.

The humanitarian successes of the Green Revolution have greatly influenced advocates like Gates, Borlaug, Pence, and R. P. Thompson for biotechnology. The Green Revolution roughly denotes the period from the 1960s to the 1980s when technological inputs and high-yield varieties of wheat, rice, and corn greatly increased food production in South Asia and South America. Norman Borlaug won the Nobel Peace Prize for his essential contributions to the Green Revolution. The Bill and Melinda Gates Foundation website on agricultural development begins with a reference to the Green Revolution, stating: "[The Green Revolution]…helped to double food production and saved hundreds of millions of lives" (Bill and Melinda Gates Foundation). Pence and R. P. Thompson devote sections of their books to framing their arguments for biotechnology in terms of extending the successes, and correcting the defects, of the Green Revolution with a Gene Revolution. In general, these supporters of genetic engineering in agriculture have a sanguine interpretation of the modern history of technologically driven industrial agriculture. For instance, R. P. Thompson begins his book by quoting Jeffry Sachs, who writes: "I believe that the single most important reason why prosperity spread, and why it continues to spread, is the transmission of technologies and the ideas underlying them" (Sachs 2005: 41–42). From the perspective of anthropocentric agricultural ethics, scientific and technological progress aimed at agricultural efficiency is *the* way to promote social justice by making food more affordable, available, and abundant.

There are, of course, other, less sanguine, interpretations of the history of modern technological agriculture. For example, as is well known, the high yields generated by the Green Revolution with its industrial monocultures have come with equally high environmental costs: habitats have been consumed at an alarming rate leading to a loss of biodiversity, soils have been degraded, and ground and surface waters have been polluted. In an article in *Nature* assessing the legacy of intensive production practices, Tillman et al. highlight "the need for more sustainable agricultural methods" (Tillman et al. 2002: 672). The narrow focus on increasing yields and lowering the costs of inputs is unsustainable in its present form. Proponents of technologically intensive agriculture and biotechnology are, of course, well aware of this. For example, the prominent biotechnologist, Anthony Trewavas, admits that "technological progress driven by the forces of technological change, economic growth, and trade is a prime cause of the problems facing biodiversity" (Trewavas 2001: 175). Yet later in the same essay he writes: "Technological progress to solve the above problems is now necessary to help ensure that a growing human population does not squeeze out the rest of nature in the process" (Trewavas 2001: 175). The defenders of biotechnology from the perspective of an anthropocentric agricultural ethics are convinced it would be a grave mistake to abandon the current trajectory of scientifically driven technological agriculture. Further progress in technology, notably biotechnology, will allow agriculture to solve the environmental problems previously created by agricultural technologies: in other words, there is nothing wrong with the current system that cannot be fixed by the system.

Most of the GM crops planted in the world today are dominant by two traits, resistance to the herbicide, glyphosate (Roundup®) and *Bt* crops, crops that contain genetic material from the bacteria (*bacillus thuringiensis*), which conveys insecticidal properties. R. P. Thompson argues that these first-generation GM crops provide evidence for the idea that biotechnology can help solve the environmental problems of technological agriculture. For example, glyphosate is a broad-spectrum herbicide that readily binds to soils and is degraded by microbes; it does not tend to accumulate in the environment and aquatic life (Thompson 2011: 139). Herbicide-resistant GM crops (e.g., Monsanto's Roundup Ready® products, typically cotton, corn, and soybean) when sprayed with glyphosate have less environmental impacts than conventional crops when sprayed with other, less benign herbicides (in a series of papers since 2013, MIT scientist, Stephanie Seneoff, has speculated on a connection between glyphosate and the increasing number of children with autism). Moreover, the use of glyphosate-resistant GM crops can limit the need for tilling as a means of weed control. No, or low till agriculture has many environmental benefits, such as limiting the loss of topsoil due to erosion (Thompson 2011: 139). As mentioned, the other major GM corps contain bacteria gene that allows crops to have insecticidal properties. *Bt* powder is commonly used in the organic industry as an insecticide and poses few health or environmental risks. The use of this GM crop can greatly limit the need for spraying with environmentally harmful pesticides and its use reduces the fossil fuel required by tractors in repeated sprayings (Thompson 2011: 139). While critics of biotechnology have raised many questions about the long-term benefits of both GM crops, however, at least in the short term, a case can be made that these GM crops have had environmental benefits when compared to conventional herbicides and pesticides.

Proponents of biotechnology, like those mentioned above, are very optimistic that the second, and future generations, of GM crops will have traits that will increase production, lower the costs of inputs, lessen the environmental footprint of agriculture, and help feed the world's poor. Some of traits are crops that have increased nutritional value (e.g., the famous "Golden Rice" that contain beta carotene, the precursor vitamin D. Vitamin D deficiency is a major cause of childhood disease and blindness in the developing world), crops that can be grown in saline soils (e.g., a salt tolerant tomato has recently been developed), and staple crops (e.g., maize, wheat, peas, rice) that have greater tolerance to water stress, longer and more severe droughts, which will be more prominent due to climate change.

Again, the many supporters, like philosophers Pence and R. P. Thompson, of crop biotechnology are convinced that corporate-centered, technological progress is a necessary tool for civilization to successfully meet the challenges of 21st century agriculture. However, in the next sections it will be seen that some environmental critics of biotechnology believe that faith in corporations and scientific-technology's ability to fix the harmful environmental legacy of industrial agriculture is fundamentally misguided. Further, these critics believe that an exclusively human-centered perspective is deeply flawed and does not provide an adequate vision for an ecologically sustainable culture.

Ecocentric Environmental Ethics

In contrast to the anthropocentric agricultural ethics just described, which values non-nature for the benefits it provides to humans, in an ecocentric ethics humans have direct duties and obligations to non-human nature. Modern thinkers and activists influenced by ecocentric ethics focus on preserving wilderness and eliminating human interference with evolutionary and ecological processes. This preoccupation with preservation creates an inherent tension with, or antipathy toward, horticulture and agriculture, whose purpose is to manipulate nature for human ends. While agriculture is not a central topic for philosophers in the preservationist tradition, at least one philosophical program within this tradition, Deep Ecology, gestures toward agricultural ethics.

Deep Ecology, whose founder was the late Norwegian philosopher Arne Naess, is an ambitious philosophical program that aims to completely transform humanity's relationship with the natural environment. Deep Ecology champions a non-anthropometric worldview where the well-being of "non-human life on Earth have value in themselves…that are independent of the usefulness… for human purposes" (Naess 2001: 189). It follows that such a complete transformation of industrial society must include agriculture. Deep Ecology was designed with the hope that it would become a global social and political movement. To support that goal, Arne Naess and George Session authored the Eight-Points Deep Ecology Platform to provide adherents with tenants for political activism. It should come as no surprise that for Deep Ecologists the environmental problems created by corporate-driven, intensive agriculture practices cannot be corrected with crop biotechnology, as envisioned in the previous section. Anthropocentric agricultural ethics must be replaced by an "ecosophy" of agriculture. To the best of my knowledge, no philosopher has developed such an ecosophy.

It is, however, clear what Deep Ecology of agriculture would *not* look like. Deep Ecology is founded on a sharp distinction between "Deep Ecology" and "shallow ecology." The anthropocentric arguments for biotechnology discussed in Section "Anthropocentric Agricultural Ethics" would be a clear examples of a "shallow ecology." Shallow solutions like biotechnology are merely techno-fixes that address symptoms without getting to the root of the problem. Deep solutions address the underlying causes: the anthropocentric worldview undergirding, for instance, corporate-driven technological agriculture. An ecosophy of agriculture rejects the narrow focus on maximizing human welfare and promoting social justice. Deep Ecology focuses on eliminating human impacts on natural systems to the greatest extent possible. From this view, solutions to the enormous ecological footprint of industrial agriculture would include reducing the pressures created by overconsumption and population growth. The Deep Ecology platform states: "The flourishing of nonhuman life *requires* a smaller human population" (Naess 2001: 189). Again, for Deep Ecologists, the root of the problem is that the shallow anthropocentric ethics of technological civilization is blind to the intrinsic value of non-human nature.

The writer and activist, Jerry Mander, cofounder of the Foundation for Deep Ecology, has commented on industrial agriculture and biotechnology. Mander has devoted his career to combating cooperate-driven, industrial civilization in general, which leads him to oppose industrial agriculture in particular. He writes: "The immediate need is to abandon the industrial model as quickly as possible and seek to apply such principles and practices as express a reciprocal relationship with nature" (Mander 2002: 91). For Deep Ecology the industrial civilization and agriculture has been an ecological catastrophe. In regard to biotechnology, Mander writes, "Some corporations say the solutions lies with biotechnology. They call it the 'ecologically sound solution.' But biotechnology operates by the same monoculture principles" (Mander 2002: 89). Insofar as biotechnology is a product of an industrial philosophy of agriculture, a Deep Ecology of agriculture would strongly oppose it. It seems that from this view, that if the whole is corrupt, the part is also corrupt.

From the perspective of Deep Ecology and ecocentric ethics, there is a consistent critique of the human domination of nature with scientific technology. A key distinction used by those who argue against the technological domination of nature is between natural objects and artifacts. For example, the environmental philosopher, Eric Katz, employs this distinction to argue against using ecological restoration as policy to justify the development of natural resources

(Katz 1992). Katz rejects the idea that humans can restore nature using science and technology. Once we use technology to manipulate nature, regardless of our motives, we imbue it with human purposes and transform it into an artifact. Characteristic of ecocentric preservationist thought, Katz believes that value exists in nature to the extent that it avoids modification by human technology (Ibid.).

The philosopher Keekok Lee makes a similar move to Katz when she employs the natural/artifactual distinction to oppose biotechnology. An artifact would be any object that has been manipulated for human purposes (e.g., a piece of obsidian rock that is shaped by human hands for the purpose of hunting) while a natural object would be formed by unhindered natural forces (e.g., a piece of obsidian that is shaped by geological forces only). In her argument, Lee makes a distinction between "nature-polluting technologies" and "nature-replacing technologies" (Preston 2008: 27). In exact opposition to advocates of GM crops mentioned in Section "Anthropocentric Agricultural Ethics," who see biotechnology as more environmentally benign than chemical pesticides and herbicides, Lee sees biotechnology as a deeper, ontological threat to nature. Chemical pesticides are "nature polluting"; they are harmful to the environment but their effects can be mitigated and reversed. Lee believes that biotechnology "manipulates nature at such a fundamental level [that it poses] an altogether different kind of problem" (Preston 2008: 28). By manipulating nature at the level of the DNA molecule, humans are "systematically transforming naturally occurring beings…to become artificial ones" (Lee 1999: 1). Christopher Preston remarks, Lee seems to think that "*Bt* corn is…the deepest kind of biotic artifact. Its very genome is the product of human intention" (Preston 2008: 28). Nature-replacing technologies represent a threat to the ontological category of natural. Biotechnology systematically and irreversibly replaces morally valuable objects (natural objects, e.g., wild strawberries) with ones deemed less valuable (artifacts, e.g., the Flavr Savr tomato, a genetically engineered tomato with a fish gene to preserve freshness during transport).

While it is clear what a Deep Ecology of agriculture would not look like, it is less clear what a positive philosophy of agriculture would look like. Perhaps one reason for this lack of clarity is that a fundamental commitment preservation of unaltered nature seems to be commit ecocentric ethics to the myth of a lost Arcadian past that occurred prior to the development of agriculture-based civilization and the human domination of nature. For example, in a book that influenced Deep Ecology thinking, *Coming Home to the Pleistocene*, Paul Shepard asserts, "If there is a single complex of events responsible for the deterioration of human health and ecology, agricultural civilization is it" (Shepard 1998: 103). Dave Foreman, the founder of the Deep Ecology inspired organization, Earth First!, looks back to a time when humans were a part of nature. The environmental historian, William Cronon, quotes Foreman as observing that with the advent of agriculture came an ever-widening rift between the "wilderness that created us and the civilization created by us" (Cronon 1998: 488). There is a sense in an ecocentric preservationist ethic, like Deep Ecology, that the domestication of plants and animals necessarily leads to the loss of biodiversity. Farms are seen as the battlefields for a war against nature (Taylor 2000: 273).

The Arcadian myth makes developing a positive Deep Ecology philosophy of agriculture a difficult task when faced with a growing world population that is expected to be over 10 billion by 2100. Further, the rejection of industrial monocultures does not *necessarily* require one to reject biotechnology outright. It is true that the current context for most research and development of GM crops supports corporate-driven industrial agriculture. However, as will be seen in the next section, there is no logical reason why biotechnology could not be used to support a more ecological model for agriculture. Without a positive philosophy of agriculture, it seems hard for ecocentric thinkers to develop a nuanced view on crop biotechnology in relations to the difficult and complex realities of 21st century agriculture.

Pragmatic Environmental Ethic

Arguably, the most important thinker in the short history of environmental ethics is Aldo Leopold, the originator of the land ethic. Aldo Leopold's land ethic can more easily include activities like agriculture than Deep Ecology, as described above. Leopold practiced farming and discussed agriculture at length, for example, in his essay "The Farmer as Conservationist" (Leopold 1992). In what follows I will use Leopold's land ethics as a starting point for discussing a pragmatic environmental ethics, which will then be used as a third, and final, perspective for discussing crop biotechnology.

It must be cautioned that philosophers disagree as to whether or not Leopold's land ethic is best interpreted as being ecocentric or pragmatic. This is not merely an intramural debate among Leopold scholars. These contrasting interpretations of the land ethic would likely lead to different judgments on biotechnology. Baird Callicott takes Leopold's famous dictum that human actions are right when they uphold the "integrity, stability, and beauty" of the biotic community as a universal moral principle (Leopold 1949: 224–225). Callicott sees this principle as being derived from a moral theory that is committed to the intrinsic value of non-human nature and ecocentrism, similar to Deep Ecology. Bryn Norton argues that the land ethic falls within the tradition of pragmatic ethics. Pragmatic ethics understands moral principles as instruments to guide decisions and actions in morally perplexing situations. In pragmatic ethics, moral principles, like scientific laws, are not necessarily unchangeable and absolute. They can be adjusted or replaced to accommodate new experiences and insights. Because of this, pragmatic ethics is pluralistic; it is open to using multiple abstract moral principles to resolve concrete problems. Again, for the purpose of this discussion, Leopold's land ethic will be interpreted as pragmatic. The main reason being that a pluralistic and pragmatic interpretation of the land ethic seems to better-fit Leopold's remarks on conservation and farming.

In very broad terms, Leopold argues that insights from ecology will allow humans to take the next step in the evolution of ethics. This does not necessarily imply an ecocentric worldview. Rather, it means that along with utilitarian, economic goods, farmers, for instance, should list semi-economic goods and non-economic goods in their practical and moral accounting. Leopold writes: "Can a farmer afford to devote land to woods, marsh, pond, windbreaks? These are semi-economic land uses—that is, they have utility but they also yield non-economic benefits" (Leopold 1939: 258). When Leopold looks at the results of modern scientific agriculture, he does not just see overuse, but awkward, anti-ecological use. The blinders of production and efficiency have created research and farming practices that fail to see wider issues that affect land health. Leopold comments: "Conservation, then, is keeping the resource in working order, as well as preventing over-use. Conservation, therefore, is a positive exercise of skill and insight not merely a negative exercise of abstinence or caution" (Leopold 1939: 257). Unlike an ecocentric ethics that is focused on preservation, a pluralistic and pragmatic land ethic can more seamlessly integrate farming into an environmental ethics.

Leopold's land ethic has left a profound legacy on agricultural ethics. Wes Jackson, an influential leader in the sustainable agricultural movement, remarks, "No person in the twentieth century was more responsible for the intellectual underpinnings of sustainable agriculture than Aldo Leopold" (Jackson 1999: 89). He further notes that Leopold provided the "intellectual framework leading toward the eventual marriage of ecology and agriculture" (Jackson 1999: 89). The philosopher Paul B. Thompson (to be distinguished from R. P. Thompson discussed in Section "Anthropocentric Agricultural Ethics" of this essay) has written insightfully and extensively on agricultural biotechnology. He relies heavily on Leopold's land ethic for his work on sustainable agriculture and environmental ethics. In his most recent book, *The Agrarian Vision, Sustainability and Environmental Ethics*, Paul B. Thompson emphasizes his debt to Leopold. He writes:

> Leopold's land ethic stresses the importance of our expanding notion of community so that it is inclusive of land, of place, of our environment. My attempt to frame the social goals of agriculture, to examine food and community, and to explore the various meaning of sustainability have all been undertake in service to Leopold's hoped-for expansion.
>
> *(Thompson 2012: 291)*

Leopold's land ethic provides the context for Thompson's case for biotechnology from environmental ethics. Thompson's argument has two parts: (1) he argues that environmental ethicists *do not* have good reasons for blanket opposition to biotechnology; and (2) he provides evidence that biotechnology *can* support ecological agriculture. However, before discussing Paul B. Thompson's case for biotechnology from environmental ethics, it will be helpful to briefly note his critique of the narrow anthropocentric ethics of industrial agriculture, mentioned in the first section.

One major theme in Thompson's many writings on agriculture and environmental ethics is the need to examine the social context of scientific research and technological development (Thompson 2003: 201). When critiquing agricultural technology, environmental philosophers would do well to step back and analyze the institutional culture and research paradigm that produces particular innovations, like biotechnology. In his 1995 book, *Sprit of the Soil*, Thompson provides a detailed philosophical critique of the institutional culture of technological agriculture as it is currently practiced in research institutions in the United States, and elsewhere. He labels this culture the "productionist paradigm," a "world view that governs research and development that is characterized by a total confidence in production enhancing agricultural technologies" (Thompson 1995: 60). While Thompson notes that the productionist paradigm cannot hold up to philosophical scrutiny, it continues to barrel ahead because of tremendous institutional inertia that resists fundamental change to this culture. It has, Thompson writes, become the "the headlong and unreflective application of industrial technology for increasing production" (Thompson 1995: 70). He concludes, "if agriculture is to become truly sensitive to environmental quality, it will require conceptual resources not present in the productionist ethic" (Thompson 1995: 71). Thompson's consistent message over the last 30 years is that environmental philosophers should participate in dialogues with scientists, economists, decision makers, and others, aimed at reshaping the narrow anthropocentric world view of the productionist paradigm that currently dominates agricultural research.

While Thompson provides a careful critical analysis of the narrow anthropocentrism of industrial agriculture, discussed in Section "Anthropocentric Agricultural Ethics," he also critiques a narrow ecocentric ethics that rejects GM crops out of hand, discussed in Section "Ecocentric Environmental Ethics." His arguments aim to show that GM crops cannot be rejected simply because of their association with industrial agriculture or because they the pose an ontological threat to the intrinsic value of nature. Thompson identifies two implications of an ecocentric ethics that sees "agriculture as *inimical* to nature preservation" (Thompson 2003: 191). Ecocentric ethics, as described in Section "Ecocentric Environmental Ethics," seems to be "committed to the twin goals of (1) minimizing the extent of land given over to agriculture, and (2) isolating agriculture from the rest of nature" (Thompson 2003: 191–192). Thompson argues that when followed to its logical conclusion, "this is a recipe for the most thoroughly industrialized agriculture we can possibly imagine" (Thompson 2003: 192–193). Further, Thompson quickly dismisses the argument against agricultural biotechnology using the natural/artifactual distinction as simply being confused. It is unclear how someone could accept conventional plant breeding, without which agriculture and civilization would not exist, and reject GM technology using this distinction. All agricultural crops and domesticated animals, not just GM crops, bear the mark of human purpose through thousands of years of selective breeding. This line of argument makes any agricultural crop or domesticated animal an ontological threat against nature. Deep Ecology seems to be logically, if not practically, committed to an Arcadian myth, a vision of humanity living in pre-agricultural arrangements, something resembling human life in the Pleistocene. It seems that Deep Ecology's objections to GM crops logically extend to agriculture (and therefore civilization) in general.

In the second part of his essay, Thompson provides several examples of GM crops that could potentially support ecological agriculture. Thompson notes the farmers in the tropics frequently save insect carcasses; they then pulverize the carcasses, which they spread on their crops. To some extent this works as an insecticide by spreading a virus that killed one generation of pest to the next generation (Thompson 2003: 203). The virus is naturally occurring and specific to the insects causing problems for the farmers. If scientists could identify the DNA sequence of the virus and insert it into the crops, the new GM crops would prevent damages to the corps from insects, while mimicking the practice of spreading pulverized insect carcasses, but more effectively and with less work. Furthermore, the GM crop containing DNA from the virus would be a more sustainable and benign alternative to chemical pesticides, which tropical farmers are rapidly adopting.

The other example is virus-resistant GM papaya (Davidson 2008). Papaya ringspot virus devastated the papaya industry in Hawaii in the 1990s. In response, a virus-resistant GM papaya was widely adopted by growers and has effectively addressed this problem. More to the point, many poor people in developing countries grow papaya for household consumption. Papaya is highly nutritious and plays a key role in a healthy diet. It does not cost more to grow than non-GM papaya and it does not require any changes to traditional ways of growing the fruit. The argument here is that the virus-resistant GM papaya supports ecologically sound, small-scale farming practices. It is an example of a GM crop that is not necessarily tied to industrial monocultures. It must be clear that Thompson does not think these examples are knockdown arguments for the pragmatic environmental ethics case for crop biotechnology. Rather, they demonstrate that environmental ethicists should not sweepingly dismiss biotechnology.

Thompson's overall argument is that environmental philosophers should be engaged in with "agricultural researchers who want to pursue environmental goals but who may lack either the right conceptual underpinnings or the right sociopolitical philosophy to do so effectively" (Thompson 2003: 213). It is through dialogue with scientists and policymakers that a pragmatic environmental ethics might be used to align research priorities with an ecological agriculture. Thompson's pragmatic-land ethics sees the goal for environmental ethics as an aid to help "build networks of communities that work to building ecologically sound and sustainable agriculture" (Thompson 2003: 213). There are, unfortunately, immense obstacles to this goal being realized. On one side, there is the institutional inertia (educational, financial, cultural) built into research institutions that resists fundamental change. On the other side, there is inertia in environmental thought that has difficulty including agriculture in an ethical system at the scale needed to feed over 10 billion people that will live on the earth by 2100.

Conclusion

The point of this chapter is to raise awareness about the basic connections between the field of environmental ethics and the debate over GM crops. In very broad terms, this chapter introduced three ethical perspectives, there are others that were not mentioned for the sake of space, for examining this powerful new technology. The first two perspectives take unambiguous, pro and con, stances on agricultural biotechnology. From the perspective of an anthropocentric agricultural ethics, biotechnology means progress toward promoting human welfare and social justice. This vision does not seek to change the trajectory of industrial agriculture; it seeks to extent it with better technologies. However, this ethical perspective fails to directly incorporate the non-human world into its ethical vision. From the perspective of ecocentric environmental ethics, biotechnology is the latest, and a very dangerous, example of the ruthless attitude toward the non-human world of corporate-driven industrial agriculture. This vision highly criticizes industrial agriculture, but it does not seem capable of imagine agricultural systems capable of feeding 10 billion people. It is difficult to imagine these strong pro and con ideological positions ever being brought into fruitful dialogue; this fact has no doubt contributed to the long life of the controversy over biotechnology. The third perspective, a pragmatic environmental ethics attempts to bring agricultural ethics and environmental ethics into fruitful dialogue by examining the worldview (ethical, philosophical, political, economic social structures) that guides researchers and research institutions in hopes of effecting positive change. This pragmatic and pluralistic ethics seeks to use Leopold's land ethics to help shape the future directions of research in biotechnology toward a more ecological and humanistic agriculture.

Related Topics

38. Food
39. Industrial Agriculture
41. Sustainable Agriculture

References

Bill and Melinda Gates Foundation. (n.d.) *What we do* [Online]. Available at: http://www.gatesfoundation.org/What-We-Do/Global-Development/Agricultural-Development (Accessed: 17 December 2014).

Charles, D. (2001) *Lords of the harvest, biotech, big money, and the future of food*. Cambridge, MA: Perseus Publishing.

Cronon, W. (1998) 'The trouble with wilderness; or, getting back to the wrong nature' in Callicott, J.B. and Nelson, M.P. (eds.) *The great new wilderness debate*. Atlanta: The University of Georgia Press, pp. 471–499.

Davidson, S.N. (2008) 'Forbidden fruit: transgenic papaya in Thailand', *Plant Physiology* (147) 487–493.

Greenpeace. (2015) *Genetic engineering* [Online]. Available at: http://www.greenpeace.org/international/en/campaigns/agriculture/problem/genetic-engineering (Accessed: 17 December 2014).

Jackson, W. (1999) 'Preparing for sustainable' in Meine, C.D. and Knight, R.L (eds.) *The essential Aldo Leopold: quotations and commentaries*. Madison: The University of Wisconsin Press, pp. 87–98.

Katz, E. (1992) 'The call of the wild', *Environmental Ethics* (14) 267–273.

Lee, K. (1999) *The natural and the artifactual: the implications of deep science and deep technology for environmental philosophy*. New York: Lexington Books.

Leopold, A. (1949) *A Sand County almanac, and sketches here and there*. New York: Oxford University Press.

Leopold, A. (1991) 'The farmer as conservationist' in Flader, S.L. and Callicott, J.B. (eds.) *The river of the Mother of God, and other essays*. Madison: The University of Wisconsin Press, pp. 255–265.

Leopold, A. (1992) 'The farmer as conservationist,' in Flader, S. L. and Callicott, J. B. (eds.) *The river of the mother the god and other essays by Aldo Leopold*. Madison: University of Wisconsin Press, pp. 255–265.

Mander, J. (2002) 'Machine logic: industrializing nature and agriculture' in Kimbrel, A. (ed.) *The fatal harves reader: the tragedy of industrial agriculture*. Foundation for Deep Ecology, pp. 87–91.

Naess, A. (2001) 'The deep ecological movement: some philosophical aspects' in Zimmerman, M.E., Callicott, J.B., Sessions, G., Warren, K.J., and Clark, J. (eds.) *Environmental philosophy: from animal rights to radical ecology*. Upper Saddle River, NJ: Prentice Hall, pp. 185–203.

Pence, G. (2002) *Designer foods, mutant harvest or breadbasket to the world?* Landham, MD: Rowman and Littlefield Publishers.

Preston, C. (2008) 'Synthetic biology: drawing a line in Darwin's sand', *Environmental Values* (17) 23–39.

Rawls, J. (1971) *A theory of justice*. Cambridge, MA: Harvard University Press.

Rolston, H. (1985) 'Duties to endangered species', *BioScience* 35(11) 718–726.

Rolston, H. (1994) 'Value in nature and the value of nature' in Attfield, R. and Belsey, A. (eds.) *Philosophy and the natural environment, Royal Institute of Philosophy Supplement: 36*. Cambridge: Cambridge University Press, pp. 13–30.

Sachs, J. (2005) *The end of poverty, economic possibilities for our times*. London: Penguin Books.

Shepard, P. (1998) *Coming home to the Pleistocene*. Washington, DC: Island Press/Shearwater Books.

Sierra Club. (n.d.) *Biotech* [Online]. Available at: http://vault.sierraclub.org/biotech/ (Accessed: 7 December 2014).

Taylor, B. (2000) 'Deep ecology and its social philosophy: a critique' in Katz, E., Light, A., and Rothenberg, D. (eds.) *Beneath the surface, critical essays in the philosophy of deep ecology.* Cambridge, MA: The MIT Press, pp. 269–289.

Thompson, P.B. (1995) *The spirit of the soil, agriculture and environmental ethics.* London and New York: Routledge.

Thompson, P.B. (2003) 'The environmental ethics case for crop biotechnology, putting science back into practice' in Light, A. and DeShalit, A. (eds.) *Moral and political reasoning in environmental practice.* Cambridge, MA: The MIT Press, pp. 187–218.

Thompson, P.B. (2010) *The agrarian vision: sustainability and environmental ethics.* Lexington: The University of Kentucky Press.

Thompson, R.P. (2011) *Agro-technology, a philosophical introduction.* Cambridge: Cambridge University Press.

Tillman, D., Cassman, K.G., Matson, P.A., Naylor, R., and Polasky, S. (2002) 'Agricultural sustainability and intensive production practices', *Nature* (418) 671–676.

Trewavas, A. (2001) 'The population/biodiversity paradox, agricultural efficiency to save wilderness', *Plant Physiology* (125) 174–179.

World Wildlife Fund (n.d.) *What we do* [Online]. Available at: http://wwf.panda.org/what_we_do/footprint/agriculture/impacts/ (Accessed: 17 December 2014).

Suggested Further Reading

Ruse, M. and Castle, D. (eds.) (2002) *Genetically modified foods: debating biotechnology.* Amherst, NY: Prometheus.

Thompson, P.B. (2007) *Food biotechnology in ethical perspective*, 2nd ed. Dordrecht, NL: Springer.

41

SUSTAINABLE AGRICULTURE

Alastair Iles

Introduction

Experiments with sustainable agriculture span the planet – from smallholders practicing agroecology in El Salvador, to farmers preserving their hedgerows in Europe, to urban farmers in Detroit seeking to restore their communities. These efforts promise to create more socially and ecologically nourishing alternatives to industrial agriculture. Nonetheless, the production of this food may not be as ethically sound as many people would presume. Farmers confront difficult 'choices' over what methods they should adopt: for example, whether they should use crops genetically modified to grow more efficiently under drought conditions, or draw on locally developed agroecological techniques instead. Workers are exploited in many attempts at alternative agriculture. Localizing food production appears desirable, yet may conceal pervasive community politics and environmental damage, while undercutting fair trade with farmers in other countries. Fair trade also seems appealing, yet certification schemes rarely guarantee workers a livable wage.

Elsewhere in this book, Paul Thompson has appraised industrial agriculture which causes a legion of environmental and social problems. Industrial agriculture has progressed to the point where it threatens biodiversity globally, emits 30–35% of global greenhouse gases, and consumes a sizable share of freshwater. Industrial agriculture has also reinforced a human diet that contains high concentrations of sugar, salt, and fat in the form of processed foods. Partly in response, over the past 40 years, many strands of alternative agriculture activities have emerged. Often driven by social movements and community organizations, they include organic farming, fair trade coffee, agroecology, urban gardens, community-supported agriculture, and farmer markets.

I consider the ways in which attempts to achieve 'sustainable' agriculture pose significant ethical concerns, in both developing and industrial country contexts. Ethical concerns are created and connected across places and communities by the diffusion of technologies, farming methods, economic structures, political and market ideologies, and global trade patterns. These concerns can be evaluated by applying any of several theories of ethical reasoning, but people can hold differing opinions of whether agriculture is ethical or not, based on which theory they favor. To be sustainable, I argue, agriculture must incorporate social and economic justice, not only ecological viability. I also suggest that we adopt an approach that assesses the politics and processes of making ethical choices (who chooses, for whom, why, and where). Taking a critical lens to alternative agriculture can help expose such hidden dimensions more publicly.

First I review some broad ethics issues pertaining to agriculture, such as the definition of sustainability, and the seeming dichotomy between humans and environment. I then work through

DOI: 10.4324/9781315768090-49

three examples of ethical conundrums (farming methods, labor, and localization of food). Finally, I conclude with suggested guidelines that people can use in deciding whether specific forms of agriculture are acceptable.

Integrating Ethics into Sustainable Agriculture

Typically, definitions of sustainability emphasize alleviation of environmental harm. This includes reducing waste from intensive operations, habitat destruction leading to biodiversity loss, and use of external inputs like pesticides and fertilizers (e.g., Pretty 2008). Retaining family farms and smallholders in the era of growing industrial consolidation preserves rural landscapes and traditions. More recently, greenhouse gas emissions have surged to prominence as agriculture is recognized both as a sizable source and sink for carbon. A strong temporal element is often present: agriculture should be sustainable for future, not just present, generations of humans.

Philosophical arguments can be made on this broad basis for pursuing sustainable agriculture. In comparison to industrialized agriculture, sustainable agriculture in itself is assumed to be substantially more ethical in character because it strives to diminish ecological harm and improve life for exploited animals (Palmer 2008; Thompson 2010). Simply pursuing sustainability is an ethical act in itself. Surprisingly few scholars have considered the ethics of sustainable agriculture beyond this level of argument.

Nonetheless, such environmentally focused definitions are limited. In 1991, agrarian sociologists Patricia Allen and Carolyn Sachs asked, "Who and what do we want to sustain?" (Allen and Sachs 1991). They pointed to widespread inattention to racial, ethnic, socio-economic, and gender equity, as well as power disparities, across food systems. Farmers and workers, for example, may diverge in their interests and power. The scholars also highlighted failures by scholars and practitioners to meet head-on the problems of hunger and poverty. The larger social, economic, and political structures within which agricultural activities occur – not simply farms – are part of what needs to be made sustainable. In the past decade, substantial new work has focused on the notion of 'just sustainability'. Julian Agyeman and others have argued on deontological grounds that sustainability must include social justice (e.g., Dobson 1998; Agyeman 2003; Jaffee 2014). Sustainability cannot be authentic without equity. This stance has emerged from the environmental justice movement. In other words, organic agriculture cannot be dubbed sustainable if farmers continue to abuse their workers or to inflict health damage on rural communities. Integrating justice into sustainability means that not all forms of environmentally improved agriculture may be acceptable.

To some extent, this critical appraisal is being heard. For example, in 2010, the US National Academy of Sciences Committee on 21st Century Agriculture suggested that sustainable agriculture entails a process of making progress toward four goals. These were satisfying human food, feed, and fiber needs; enhancing environmental quality and the resource base; sustaining the economic viability of agriculture; and improving the quality of life for farmers, farm workers, and society as a whole (NRC 2010). That is, sustainability is not merely ecology or resource use but encompasses human welfare. But the committee's detailed explanation omits much of the complex socio-economic, racial, and cultural dimensions that can make sustainability unfair for some groups.

By contrast, as Allen and Sachs (1991) say, the idea of 'sustainability' is always contested. Sustainability includes the politics and processes of deciding what will be sustainable. Many alternative agriculture activities exemplify what has been called 'food democracy' – in which diverse peoples express their desire and power to choose their own food systems (Lang and Heasman 2015), instead of acquiesce to what they consider a dominant, corporate-run industrial food regime. Yet other alternative agricultures can simply be an attempt to make industrial agriculture more environmentally benign. The underlying technologies, productivist impulses, and industrial consolidation may hardly budge. Producers can still exploit workers and animals in immoral ways. The biosphere may be subordinated to human desires. Allen and Sachs (1991) suggest that sustainability is frequently mistaken as economic viability: to thrive, farmers must be profitable.

What is of interest, then, is whether and how policy-makers, farmers, eaters, scientists, companies, and many other actors evaluate specific forms of agriculture as ethical. Their reasoning will affect the sustainability of agricultural systems, now and in the future. As the NAS explains, sustainable agriculture is not an endpoint but an open-ended process of numerous actors making decisions and taking actions to use farming methods, invest in technologies, choose particular crops, develop marketing systems, and manage food waste. Food systems are complex, multi-layered, many-sited things. A single choice is unlikely to lead to wide-ranging changes. Considering sustainable agriculture in this way, as processual and systemic, illustrates the nature of how food is now produced. Simply focusing on farm activity is no longer sufficient; many other scales must be included, such as the connections between farms and supermarket supply chains, the effects of national government policies on the local level, and scientific research in laboratories distant from farming regions. Conversely, from the consumers' perspective, many decisions and actions take place far away from dinner tables and supermarket shelves, largely invisible to the unassuming eater.

Who are making the ethical judgments, if any? With whose experiences and criteria? Whose actions and values have greater power and influence? Whose approaches to sustainability will prevail? Ethical agency – namely, viewing oneself as an actor capable of learning, acting, and changing the food system – is central to sustainable agriculture.

Several themes pertaining to ethics are noteworthy. First, what theory should we use to evaluate 'sustainable' agriculture? A commonly invoked philosophical framework is utilitarianism (FAO 2004; Thompson 2008). The consequences of undertaking an action are decisive: does it provide the largest possible benefit to the most people, and do the benefits clearly outweigh the costs? That is, sustainable agriculture should be used only when it is obviously feeding people in a comparable way to industrial farming, and also creating environmental gains (FAO 2004). Yet this moral calculus is invoked to justify countless industrial farming practices – a highly dissonant tension that can discredit the use of utilitarianism. In contrast, a rights-based framework suggests that participants in food systems have their own rights and obligations (FAO 2004; Thompson 2008). Many proponents assert that sustainable agriculture will protect human health and welfare, implicitly invoking the idea that humans have rights to health, food, and livelihoods. Farmers and eaters are entitled not to be exposed to harmful pesticides, while governments and companies are obliged to avoid using such pesticides. Nonetheless, it may not be easy to reconcile rights that appear to conflict, such as the right that companies often assert to profit and make decisions, vis-à-vis the right that food sovereignty movements claim to control their own food systems. In practice, these theories are often blended in pragmatic application. They need to be supplemented by appreciation of the conditions under which unethical behaviors occur.

In turn, science studies has much insight to offer to ethical analysis. Science and technology can embody veiled values that influence what sorts of sustainability are sought in farming. The reliance on technological advances and scientific progress to achieve sustainability can be a form of technological determinism (Wyatt 2008). That is, policy-makers, researchers, and industry may presume that technical innovations will resolve the ecological and social problems of farming, when social, political, and cultural changes are also required and may be more critical. They may overlook the place-specific knowledge and experiences of farmers, workers, and eaters in favor of using a-historical and objective technical expertise to make decisions (Glenna, Jussaume and Dawson 2011). Science studies also suggest that particular agricultural systems can become 'standards' in their own right – models that should be emulated in many other places. The sociologist

Laurence Busch (2005) argues that productivism itself is a standard. A productivist system combines agri-chemicals, machinery, monoculture fields, high fossil fuel use, and other features to produce more and more food; all other farming systems are then compared to its seemingly efficient production within a confined optics and are, unsurprisingly, judged to be obsolete and inadequate.

Finally, illusory dichotomies pervade sustainable agriculture. One important split occurs between humans and environment. The work of making agriculture sustainable invariably takes place within an anthropocentric frame. For example, farmers deserve a fair return on the time, knowledge, and labor that they dedicate to their land, rather than being subjugated to impersonal market forces demanding the lowest possible cost. Protecting ecosystem services such as pollination and water cycling will maintain production of food, fiber, and fuel primarily for human benefit. In contrast, a biocentric perspective emphasizes the rights and existence of animals and ecosystems within and beyond the agricultural system. Animals are exploited on an enormous scale in industrial agriculture, while they are deprived of habitat through land clearing. Ecosystem functions are weakened, resulting in greater species vulnerability, and in biosphere decline. Complicating this seeming divide is the tendency of both agriculture and ecology to blot 'culture' out altogether. This denies the dynamic social-ecological interactions that have shaped agriculture from the very beginnings of crop domestication. There is no agriculture without culture; there is no ecological landscape that farmers have not reworked. Both ecological and human life should be integrated together through recognizing this biocultural heritage.

The following selective examples give a sense of the breadth of ethical conundrums that can arise in sustainable agriculture, and the themes that need to be considered.

Farming Methods: Sustainable Intensification and Agroecology

A key battleground features the types of agriculture used across the world. One example of how scientists, governments, industry, and farmers contest farming methods is through the rivaling philosophies of sustainable intensification and agroecology. Over the past decade, 'sustainable intensification' has appeared as a popular idiom in national and international policy circles. In 2009, the Royal Society in Britain published a prominent report that defined sustainable intensification as "agriculture in which yields are increased without adverse environmental impact and without the cultivation of more land" (Royal Society 2009: ix). The Food and Agricultural Organization (FAO) then created a program to implement sustainable intensification among small farmers. Sustainable intensification is pointedly catholic in scope: its advocates contend that nothing is ruled out and what counts is whether something succeeds locally or regionally. Thus genetically modified crops and computer-controlled irrigation will be considered alongside agroecological methods and rainwater harvesting.

According to many policy-makers and scientists, agriculture faces the herculean challenge of expanding food production by 50–70% by 2050 to feed a population of 9 billion humans (Tomlinson 2013). Agriculture, in their reckoning, must become more productive to satiate hunger and maintain human life – or be immoral. Even now, hundreds of millions of people go hungry or lack functional nutrients. Intensification can also allow 'land sparing': if more output can be generated from the same or less land, then other land can be reserved for conservation purposes (Phalan et al. 2011). Instead of agriculture continuing to engulf ecologically valuable land, it can be confined or shrunk. Moreover, intensification can help overcome 'yield gaps': the difference in output between best-performing cropping systems and the system in question. Commonly, farms using intensive inputs of pesticides, fertilizers, and irrigation are said to 'out-perform' farms without such inputs. Responding to these concerns, many scientists propose that intensification itself should be sustainable. The Royal Society notes: "If a technology improves production without adverse ecological consequences, then it is likely to contribute to the system's sustainability" (Royal Society 2009: 7). Intensification should foresee how climate change might disrupt the rhythms of growing crops: declining yields demand greater productivity to compensate. 'Climate-smart' and 'eco-efficient' agricultures are thusly other synonymous monikers seen in World Bank and Gates Foundation reports.

Such approaches are ostensibly beneficial: they promise benign forms of agriculture that can feed the world while protecting ecosystems. Yet technological fixes and growing yields 'will' meet the indefinitely continuing trends of population growth and dietary demands such as meat. All of these assumptions are contentious. In particular, fallacious, unsubstantiated analysis underlies the projection of 50–70% production growth, which has circulated widely among credulous policy-makers and knowing companies (Tomlinson 2013). Facing severe criticism, advocates of sustainable intensification now admit that capacious room already exists to improve food system performance prior to any attempt to intensify farming (e.g., Garnett et al. 2013). Food waste can be diminished; meat consumption can be moderated. They also agree that more attention should be dedicated to overcoming problems in distribution, by making food more affordable and accessible.

Nonetheless, considerable evidence suggests agricultural R&D and policy remains tilted toward maintaining the productivist system (Friends of the Earth 2012; Loos et al. 2014). The Royal Society, in their 2009 report, had called for greater investment in GM crops, genomic assays of plants and animals, enhanced irrigation technologies, and technical extension – even while endorsing locally developed agroecological methods. The Friends of the Earth NGO later found the British and US governments, FAO, and the Royal Society favored technology solutions in their research funding. These technologies aim in particular to close yield gaps among farmers in Africa, supposedly neglected by the Green Revolution. Yet the focus on overcoming yield gaps ignores the externalization of the adverse costs and the lack of government investment in developing alternative agricultural science. The ethical debates regarding GM crops now follow ritualized lines of opposition and support. Almost all engineered crops currently used only feature resistance to herbicides or pests, reinforcing farmer dependence on external inputs. While some evidence suggests pesticide usage has declined, overall agri-chemical applications have soared, raising concerns over weed resistance, pollinator mortality, and pollution of soils, waterways, and workers. Many farmers in developing countries are entrapped in cycles of debt through being required by industrial supply chains, biotechnology companies, and financial lenders to use costly GM crops, such as Bt cotton in India (Stone 2010). How does a moral calculus weigh short-term economic gains through increased yield against the longer-term ability to renew seeds?

The coming generations of GM crops are infinitely more genetically complex; crops edited with CRISPR-Cas9 technology promise adaptation to drought, salinity, disease, submergence in water, and other environmental conditions. They could overcome nutritional deficiencies by incorporating the capacity to make substances such as Vitamin D. This research, proponents say, could apply biotechnology more ethically, by aiding farmers to adjust to future conditions. For instance, Cornell University researchers are already developing an 'improved' version of RuBisCo, an enzyme essential to photosynthesis (Le Page 2014). Due to their evolutionary history, most crop plants depend on a 'slow and wasteful' form of RuBisCo. Free-ranging cyanobacteria are more efficient. By finding ways to integrate their genes into plants, the scientists could boost photosynthetic capacity by 25%. Such plants might result in substantially greater crop growth while limiting water use. To their credit, the scientists consider the arguments for and against introducing this technology. Yet they emphasize land sparing, without acknowledging the potential for further simplification of the agricultural landscape, or the possibility that food supplies dependent on a few super-crops may fail unexpectedly. They suggest that using the renovated enzyme will enable plants to grow more lushly around farm fields, thus feeding wildlife, without addressing the problem of why habitat is vanishing. There is little consideration of how such technologies might advance the broader mission of sustainability. This technological determinism, as Loos et al. (2014) argue, is characteristic of sustainable intensification advocates. They rarely address whether a technological innovation actually enhances human well-being through supplying food to those who need it.

In contrast, farmers have historically used many sustainable farming methods. Agroecology seeks to promote diversified farms, based on applying ecological principles to the creation and management of sustainable agriculture and food systems (Méndez et al. 2013; Gliessman 2014; Altieri et al. 2015). Through recycling biomass, nutrients, and energy, and promoting beneficial interactions of organisms in the farming system, agroecological practices reject input/output orthodoxy. Instead, enhanced biological functions can be internally regenerated – sustained, that is, within the agroecosystem. Farmers using agroecology limit their use of external fossil fuel and chemical inputs – instead relying on beneficial insects, for example, to reduce pest pressures, cover crops to enhance soil fertility and moisture, and habitat features such as hedgerows to provide nesting grounds for pollinators (Gliessman 2014; Perfecto and Vandermeer 2015).

Agroecological science has learned much from the longstanding practices of traditional and indigenous farmers. One example is the 'three sisters' method used in *milpa* agriculture in Mexico, where indigenous farmers grow winter squash, maize, and climbing beans (Altieri 2002). While the bean plant roots fix nitrogen from the air for the other plants, the ground-loving squash competes with 'weeds' and provide mulch. All three crops complement each other in terms of human nutrition: maize provides starch, beans give protein, and squash seeds are concentrated source of oil. In contrast to the technical extension that sustainable intensification programs envisage, agroecology is often learned through participatory, experiential processes of peer to peer sharing. The *campesino-to-campesino* movement has emerged across much of Latin America as an effective way for farmers to teach each other and experiment with agroecological techniques (Holt-Giménez 2006).

The emphasis is on farmer-led regeneration. One example of such renewed relationships is glimpsed through the changing perceptions and experiences of the ensemble of Montana legume growers that Liz Carlisle describes in her book, *Lentil Underground* (Carlisle 2015). Many of these growers report that they view farming as a process of 'working with' their soils, plants, water, seeds, and beneficial insects. Farming is less an extractive activity than a collaborative endeavor. Sustainable agriculture, then, can be understood as an ethics of nurturing and revering a farmer's relationships with the various parts of the farm. This view resonates with both Aldo Leopold's land ethics (Leopold 1949) and Carolyn Merchant's feminist-inspired relational ethos (Merchant 2014). It also connects to the long histories of indigenous farmers working with their landscapes to create complex 'biocultural' realms where livelihoods go hand-in-hand with ecologically rich habitats.

Nonetheless, diversified and agroecological farming remains on the margins. Within the dominant productivist regime, this farming is depicted as inefficient and obsolete, even though it may maintain ecological health and be more productive overall than monoculture farming. Those who desire sustainable agriculture must ask: what sorts of farming methods will be used, and according to whose criteria? Is it moral to depend uncritically on technological fixes by the very system that is creating sustainability problems?

The Place of Labor

Another critical ethical conundrum is the treatment of workers in ostensibly sustainable agriculture operations. While vast imaginative energy is devoted to making farming more ecologically beneficial, worker welfare is typically omitted in both policy and practice (Gottlieb and Joshi 2010). For example, organic farming has emerged as a popular alternative to industrial agriculture (Obach 2015). Its methods are associated with bucolic scenes of pesticide-free fields, healthy soils, thriving pollinators, and nutritious foods. Organic agriculture is painted as offering consumers the ability to quarantine themselves and their families from health risks (Szasz 2007). Many mainstream supermarkets now stock organic foods in an attempt to benefit from a lucrative new market, in which some eaters are willing to pay substantially more for products that are certified as 'organic'. Several retailers, such as Whole Foods, have based their business models on offering primarily organic foods.

Nonetheless, it was organic agriculture that provoked the first queries about the 'sustainability' of alternative farming systems, at least in the US Agrarian sociologist Julie Guthman drew attention to a materializing 'conventionalization' trend for many organic producers to emulate industrial practices: they were expanding their scales of operation, growing crops on monoculture fields, and eliminating biodiversity (Guthman 2002). Indeed, under US government organic standards, farmers can use a few pesticides. Equally important, organic operations at all scales frequently exploit their farm workers, because they must reduce costs to gain access to industrial supply chains, or to survive at all in the larger economic milieu. Workers can be subjected to poor pay, substandard housing, food insecurity, and lack of health care (Minkoff-Zern 2014). Sociologist Christy Getz has discovered that organic growers may demand that workers use tools and practices that harm their bodies (Shreck, Getz and Feenstra 2006; Getz, Brown and Shreck 2008). In 'stoop weeding', for instance, laborers use hoes to strip out weeds instead of spraying pesticides, leading to severe muscleskeletal problems. It is commonplace for organic farm workers to face food security concerns: they may find the crops they harvest to be unaffordable. Many growers also benefit from the powerlessness of undocumented workers in the broken US immigration regime (Minkoff-Zern 2014).

Similarly, retailers and restaurants purporting to sell sustainable foods can provide deplorable conditions for their own workers. In the US, Walmart has increased its organic offerings yet pays its workers so poorly that they depend on government food aid. Whole Foods has a record of being exploitative toward its labor, rejecting union organization. Indeed, 'ethical' retailers tend to force their farmer suppliers to cut their prices or pay for the privilege of stocking products. Union organizers recently found that restaurant workers in the San Francisco Bay Area "who served organic or ingredients were 22% more likely to be food insecure compared to other Bay Area restaurant workers" (Food Chain Workers Alliance et al. 2014: 19). Several farm worker movements are demanding economic justice: the Immokalee tomato worker campaign in Florida has successfully increased pay by a penny per pound of picked tomatoes (Estabrook 2012). Beginning in 2012, fast food workers have organized an increasingly visible nationwide movement to urge that the minimum wage be boosted. Such protests have not yet spread to the sustainable food industry

Many farmers face a larger challenge of how to make a living. While sustainable agriculture is described as psychologically and spiritually satisfying, this may not translate into tangible livelihood or communal bond. Jaclyn Moyer wrote about her befuddlement at how her neighboring growers near Sacramento, California were apparently flourishing even though she and her husband were financially stressed (Moyer 2015). But through gossip, the local farmers realized that they were all almost destitute and competing for the same eaters. Sustainable farming – indeed, farming more generally – consumes vast time and energy that is not adequately compensated by the market. Some sustainable producers make highly lucrative livings but this can result from using productivist practices (as seen above) and from securing access either to very affluent consumers or to high-end supermarkets. The significant socio-economic inequalities that pervade sustainable farming are rarely recognized.

In the US and Europe, community-supported agriculture (CSA) has appeared as a novel mechanism for connecting eaters and producers while strengthening ties between local producers. In essence, depending on the scale of operation, farmers pool their crops and sometimes livestock in a cooperative that consigns boxes of food to eaters in return for a monthly subscription. This can yield more reliable income for farmers. Geographer Ryan Galt finds that CSA often relies on the moral economy that farmers are willing to build and contribute to (Galt 2013). The prices that CSAs charge may not be adequate, particularly if eaters are unwilling to pay more than a certain threshold. To cope, farmers may ask their families, friends, and community members to donate their labor and money for reasons of ethical commitment and social solidarity. They may also communally share harvesting equipment to mitigate some costs. Without this moral economy, much sustainable agriculture probably would not exist. Is it fair to rely so much on the volunteer spirit?

Tying these seemingly different aspects together is the cost structure of food systems. Prices must be as low as possible for the majority, or else be high and inaccessible to most. Are eaters and retailers prepared to pay more for worker welfare and farmer survival? Can the cost structure be transformed so that agriculture rewards sustainability and communal contributions?

Making Food Local *and* Fair?

A popular approach to making food systems more sustainable is to 'localize', rather than 'globalize', the production of food (Halweil 2002; Hinrichs and Lyson 2007). Many cities, schools, hospitals, and restaurants now call for their offerings to be sourced from within 100 kilometers. The idea is that producing foods locally can diminish food miles significantly, generate new livelihoods, build vibrant local economies, and reduce food safety risks. Sourcing food locally makes it easier for farmers to connect with consumers through direct marketing mechanisms, thus bridging the vast distances between producers and consumers that pervade industrial food chains. Producers and eaters can be made more accountable to one another. Importantly, local foods can mean that communities are better able to exercise their food sovereignty, or the right of nations and peoples to control their own food systems (Wittman 2011). They can establish governance institutions at the local or regional level that give eaters a democratic voice in how agriculture should evolve next. Local production can also be less ecologically destructive, depending on what methods are adopted. In short, localizing agriculture is replete with ethical advantages.

Nonetheless, localizing food production raises several moral issues that must be resolved for just sustainability. One argument for localization is grounded in the concept of food miles, which measures the distance that foods travel from farm to dining table (Pirog and Benjamin 2003; Iles 2005). In Britain, for example, vegetables can be transported 12,000 kilometers from New Zealand, while fish can be imported from Africa. In Britain, some supermarket chains have introduced labels indicating a product's food miles to encourage eaters to buy locally. In recent years, critics have pointed out that transportation may provide only a fraction of agricultural GHG emissions (Coley, Howard and Winter 2011). Farming locally may contribute a far larger share: energy-intensive glasshouses in colder climates, use of fossil fuels to make chemicals and fertilizers, and soil depletion all emit carbon. Such problems can be tackled with the use of appropriate farming practices and constraints on people's diets (namely, people cannot expect to eat some foods year-round). Importantly, local does not inevitably mean sustainable.

In turn, localization could mean that the benefits of agriculture accrue only to certain regions where sizable populations of ethical or affluent eaters live, and where there is demand for change (Ballingall and Winchester 2010). If foods are only sourced from within a local foodshed, then this means that producers who grow foods sustainably in more distant locations will be excluded from the foodshed (Carlisle 2015). These remote producers may scrap to find consumers in their vicinity and rely on nurturing markets in large metropolitan areas. It is morally important to resurrect local economies everywhere, not simply in more privileged regions. Similarly, localizing foods can undercut the concept of fair trade across national borders. Since the 1990s, fair trade certification has emerged as a new alternative agri-food network that allows consumers predominantly in industrial countries to pay more for foods from producers in developing countries that are verified as treating their workers responsibly (Jaffee 2014). Fair trade coffee, cocoa, chocolate, and cotton have become increasingly popular in Europe and the US. While there are now substantial doubts whether fair trade schemes are empowering smallholder farmers (Bacon 2005), the concept of local foods can conflict with desirable international trade.

Moreover, localization can be associated with exploitation, power inequalities, and discrimination. According to sociologists Clare Hinrichs (2003), Julie Guthman (2008), and others, local foodshed regimes may be prone to exclusionary practices that sideline some less powerful and visible social groups. These practices are grounded in dominant assumptions about a locality's homogeneity and cohesive social identity. Who defines what 'local foods' are and whose criteria are used? This phenomenon can be seen in the rise of farmer markets and regional food hubs across Europe and the US that reflect the preferences of an unrepresentative wealthier and higher-class clientele (Alkon and Agyeman 2011). Local sourcing rules may not incorporate support of racial, ethnic, class, and cultural diversity. Namely, businesses run by racially diverse people may be rejected as not falling within the foodshed radius. How 'local foods' are defined may also disenfranchise minority groups from having their culturally diverse cuisines recognized. Immigrant communities frequently grow their own indigenous crops at home – a practice that can be invisible (Minkoff-Zern 2012). In the US, there is a growing African-American urban farming movement, with gardens in cities from Detroit to New York City (White 2011). Its members can struggle to secure markets because they are effectively segregated from their larger communities.

In short, localization offers many advantages over industrial foods but its practice must be carried out with an eye to the ethical issues. Reflexive localization is possible, DuPuis and Goodman (2005) suggest, in which instead of presuming localization is inherently ideal, participants in local food systems continually acknowledge the politics in their amidst, and work toward assuring a democratic voice for all.

Conclusion

Greater mindfulness is needed in appraising whether or not sustainable agriculture is indeed ethical. As we have seen, 'sustainable' agriculture may not be fair, and may be conducted in ways that simply reinforce the dominant productivist regime. An array of powerful industry, scientific, financial, and government actors have vested stakes in trying to temper this regime; their proposals for making farming more sustainable must be evaluated carefully. Likewise, many alternative agriculture movements can overlook the importance of economic and racial/ethnic justice, in advocating for the localization of food and demanding greater availability of alternative foods.

Practitioners can apply broad ethical principles when they are pursuing business opportunities, making policy, developing technology, or practicing farming. Examples include:

- Seek 'just sustainability' in both alternative and conventional agriculture: not merely biological or ecological sustainability but social justice too.
- Avoid reinforcing the dominant productivist system even under the guise of 'improving' its sustainability and create authentic alternatives.
- Treat workers and farming communities fairly.
- Avoid unequal distribution of the benefits and costs of sustainable agriculture across populations and geographical areas.
- Avoid assuming that localizing food systems is automatically beneficial.
- Give voice to a diversity of farmers and consumers, not only elite researchers and industry managers.

Achieving truly sustainable agriculture calls for our nuanced, multi-dimensional, and practical ethical agency.

Related Topics

2. Eating
38. Food
39. Industrial Agriculture
42. Community Gardens
62. Community Participation

References

Agyeman, J. (2003). *Just Sustainabilities: Development in an Unequal World*. Cambridge, MA: MIT Press.

Alkon, A. H., & J. Agyeman. (2011). *Cultivating Food Justice: Race, Class, and Sustainability*. Cambridge, MA: MIT Press

Allen, P. & C. Sachs. (1991). *What Do We Want to Sustain? Developing a Comprehensive Vision of Sustainable Agriculture*. Issue Paper 2. The Center for Agroecology and Sustainable Food Systems, University of California, Santa Cruz.

Altieri, M. A. (2002). Agroecology: The science of natural resource management for poor farmers in marginal environments. *Agriculture, Ecosystems & Environment, 93*(1), 1–24.

Altieri, M. A., C. I. Nicholls, A. Henao, & M. A. Lana. (2015). Agroecology and the design of climate change–resilient farming systems. *Agronomy for Sustainable Development, 35*(3), 869–890.

Bacon, C. (2005). Confronting the coffee crisis: Can fair trade, organic, and specialty coffees reduce small-scale farmer vulnerability in northern Nicaragua? *World Development, 33*(3), 497–511.

Ballingall, J., & N. Winchester. (2010). Food miles: Starving the poor? *The World Economy, 33*(10), 1201–1217.

Busch, L. (2005). Commentary on "Ever since Hightower: The politics of agricultural research activism in the molecular age". *Agriculture and Human Values, 22*(3), 285–288.

Carlisle, L. (2015). *Lentil Underground—Renegade Farmers and the Future of Food in America*. New York: Gotham Books.

Coley, D., M. Howard, & M. Winter. (2011). Food miles: Time for a re-think? *British Food Journal, 113*(7), 919–934.

Dobson, A. (1998). *Justice and the Environment: Conceptions of Environmental Sustainability and Dimensions of Social Justice: Conceptions of Environmental Sustainability and Dimensions of Social Justice*. New York: Oxford University Press.

DuPuis, E. M., & D. Goodman. (2005). Should we go "home" to eat?: Toward a reflexive politics of localism. *Journal of Rural Studies, 21*(3), 359–371.

Estabrook, B. (2012). *Tomatoland: How Modern Industrial Agriculture Destroyed Our Most Alluring Fruit*. New Jersey: Andrews McMeel Publishing.

Food and Agriculture Organization. (2004). *The Ethics of Sustainable Agricultural Intensification*. Rome: FAO.

Food Chain Worker Alliance, Food First, and Restaurant Opportunities Centers. (2014). *Food Insecurity of Restaurant Workers*. Oakland, CA.

Friends of the Earth. (2012). *Wolf in Sheep's Clothing? An Analysis of the 'Sustainable Intensification' of Agriculture*. Available at: www.foei.org/en/wolf-in-sheeps-clothing.

Galt, R. E. (2013). The moral economy is a double-edged sword: Explaining farmers' earnings and self-exploitation in community-supported agriculture. *Economic Geography, 89*(4), 341–365.

Garnett, T., M. C. Appleby., A. Balmford, I. J. Bateman, T. G. Benton, P. Bloomer, & M. Herrero. (2013). Sustainable intensification in agriculture: Premises and policies. *Science, 341*(6141), 33–34.

Getz, C., S. Brown, & A. Shreck. (2008). Class politics and agricultural exceptionalism in California's organic agriculture movement. *Politics & Society, 36*(4), 478–507.

Glenna, L. L., R. A. Jussaume Jr, & J. C. Dawson. (2011). How farmers matter in shaping agricultural technologies: Social and structural characteristics of wheat growers and wheat varieties. *Agriculture and Human Values, 28*(2), 213–224.

Gliessman, S. R. (2014). *Agroecology: The Ecology of Sustainable Food Systems*. Baton Rouge: CRC Press.

Gottlieb, R., & A. Joshi. (2010). *Food Justice*. Cambridge, MA: MIT Press.

Guthman, J. (2002). *Agrarian Dreams: The Paradox of Organic Farming in California*. Berkeley: University of California Press.

——— (2008). Neoliberalism and the making of food politics in California. *Geoforum, 39*(3), 1171–1183.

Halweil, B. (2002). *Home Grown: The Case for Local Food in a Global Market* (Vol. 163). Worldwatch Institute.

Hinrichs, C. C. (2003). The practice and politics of food system localization. *Journal of Rural Studies* 19(1), 33–45.

Hinrichs, C. C., & T. A. Lyson. (Eds.). (2007). *Remaking the North American food system: Strategies for Sustainability*. Omaha: University of Nebraska Press.

Holt-Giménez, E. (2006). *Campesino a Campesino: Voices from Latin America's Farmer to Farmer Movement*, Oakland, CA: Food First.

Iles, A. (2005). Learning in sustainable agriculture: Food miles and missing objects. *Environmental Values*, 14(2), 163–183.

Jaffee, D. (2014). *Brewing Justice: Fair Trade Coffee, Sustainability, and Survival*. Berkeley: University of California Press.

Lang, T., & M. Heasman. (2015). *Food Wars: The Global Battle for Mouths, Minds and Markets*. London: Routledge.

Leopold, A. (1949). *A Sand County Almanac*. New York: Oxford University Press.

Le Page, M. (2014). Turbocharge our plants. *New Scientist*, 224(2989), 26–27.

Loos, J., D. J. Abson, M. J. Chappell, J. Hanspach, F. Mikulcak, M. Tichit, & J. Fischer, (2014). Putting meaning back into "sustainable intensification". *Frontiers in Ecology and the Environment*, 12(6), 356–361.

Méndez, V. E., C. M. Bacon, & R. Cohen, (2013). Agroecology as a transdisciplinary, participatory, and action-oriented approach. *Agroecology and Sustainable Food Systems*, 37(1), 3–18.

Merchant, C. (2014). *Earthcare: Women and the Environment*. New York: Routledge.

Minkoff-Zern, L. A. (2012). Pushing the boundaries of indigeneity and agricultural knowledge: Oaxacan immigrant gardening in California. *Agriculture and Human Values*, 29(3), 381–392.

——— (2014). Subsidizing farmworker hunger: Food assistance programs and the social reproduction of California farm labor. *Geoforum*, 57, 91–98.

Moyer, J. (2015). What nobody told me about small farming. Available at: http://www.salon.com/2015/02/10/what_nobody_told_me_about_small_farming_i_cant_make_a_living

National Research Council (NRC). (2010). *Toward Sustainable Agricultural Systems in the 21st Century*. National Research Council Report. Washington, DC: The National Academies.

Obach, B. K. (2015). *Organic Struggle: The Movement for Sustainable Agriculture in the United States*. Cambridge, MA: MIT Press.

Palmer, C. (Ed.). (2008). *Animal Rights*. London: Ashgate Publishing Company.

Perfecto, I., & J. Vandermeer. (2015). *Coffee Agroecology: A New Approach to Understanding Agricultural Biodiversity, Ecosystem Services and Sustainable Development*. New York: Routledge.

Phalan, B., M. Onial, A. Balmford, & R. E. Green. (2011). Reconciling food production and biodiversity conservation: Land sharing and land sparing compared. *Science*, 333(6047), 1289–1291.

Pirog, R. & A. Benjamin. (2003). *Checking the Food Odometer: Comparing Food Miles for Local Versus Conventional Produce Sales to Iowa Institutions*. Ames, Iowa: Leopold Center for Sustainable Agriculture.

Pretty, J. (2008). Agricultural sustainability: Concepts, principles and evidence. *Philosophical Transactions of the Royal Society of London B: Biological Sciences*, 363(1491), 447–465.

Royal Society. (2009). *Reaping the Benefits: Science and the Sustainable Intensification of Global Agriculture*. London: Royal Society.

Shreck, A., C. Getz, & G. Feenstra. (2006). Social sustainability, farm labor, and organic agriculture: Findings from an exploratory analysis. *Agriculture and Human Values* 23(4), 439–449.

Stone, G. D. (2010). The anthropology of genetically modified crops. *Annual Review of Anthropology*, 39, 381–400.

Szasz, A. (2007). *Shopping Our Way to Safety: How We Changed from Protecting the Environment to Protecting Ourselves*. Minneapolis: University of Minnesota Press.

Thompson, P. B. (2008). *The Ethics of Intensification: Agricultural Development and Cultural Change*. Dordecht: Springer.

——— (2010). *The Agrarian Vision: Sustainability and Environmental Ethics*. Lexington: University Press of Kentucky.

——— (2015). *From Field to Fork: Food Ethics for Everyone*. New York: Oxford University Press.

Tomlinson, I. (2013). Doubling food production to feed the 9 billion: A critical perspective on a key discourse of food security in the UK. *Journal of Rural Studies*, 29, 81–90.

White, M. M. (2011). D-Town farm: African American resistance to food insecurity and the transformation of detroit. *Environmental Practice*, 13(04), 406–417.

Wittman, H. (2011). Food sovereignty: A new rights framework for food and nature? *Environment and Society: Advances in Research*, 2(1), 87–105.

Wyatt, S. (2008). Technological determinism is dead: Long live technological determinism. E. Hackett, O. Amsterdamska, M. Lynch, and J. Wajcman (eds.), *The Handbook of Science and Technology Studies*, 165–180. Cambridge, MA: MIT Press.

42

COMMUNITY GARDENS

Stephanie Ross

Introduction

Imagine walking by an urban plot, fenced, gated, and rife with plants. There are some flowers within, but vegetables predominate. No overarching design unites the regular beds into which the ground is subdivided. Paths penetrate the area, and a few picnic benches are clustered by the entrance. In all likelihood, you are in the presence of a community garden. In 2013 the National Gardening Association determined that 42 million American households grew food at home or in community gardens, with 2 million more households taking part in community gardens compared to 2008 (http://garden.org/learn/articles/view/3819/). The numbers have surely continued to rise since then. And in fact, community gardens are so common nowadays that they merit a Wikipedia entry, which opens with the following definition: "A community garden is a single piece of land gardened collectively by a group of people."

Community gardens are clearly worthy of philosophical reflection. They inhabit the intersection of two concepts – gardening and community – about which we feel singularly nostalgic. They also fuel idealistic visions of how we can live harmoniously with one another and live lightly on the land. Thus, the very idea of community gardens is alluring. Promoting, even institutionalizing, a practice of community gardening promises to answer a wide range of interests and needs. But can this promise be fulfilled? Or are we falling victim to that very nostalgia, buying into misleading notions of both component concepts? In what follows, I will briefly examine both core concepts in my title, gardening and community, and lay out the aspirations associated with each. Next I will consider a number of the needs community gardens seem poised to address – education, healthy diet, food security, social cohesion, individual growth, community activism – and assess the likelihood that they can succeed. I will close by noting some further liabilities that come with looking to community gardens to cure social and environmental ills but will conclude more hopefully by pointing to a near relative of the community garden that promises to reconfigure the relationship between inner-city residents and the food they consume.

Dissecting the Component Concepts: Gardens and Community

Gardens have been a nearly universal part of human culture; they are a feature of almost all non-nomadic societies. Gardens are traditionally places of respite and renewal. In addition to

DOI: 10.4324/9781315768090-50

providing beauty and utilitarian produce of various sorts, they perform social roles by offering shelter and places to gather (Ross 1998: 1–6). Gardens also provide contact with nature that is edifying, attuning us to the cycles of the seasons, the variety of plants and their differing needs, and the astounding complexity of the ecosystem constituted by even the smallest garden patch.

In the United States, the quintessential garden certainly contains flowers. But gardens can possess a wide range of additional features – vegetables, vines, shrubs, and trees; paths, terraces, walls, and benches; ponds, streams, canals, and fountains; temples, follies, grottoes, and green-houses; playing fields, mazes, orchards, and amphitheaters. Any garden with flowers provides visual beauty but also multi-sensory appeal. Gardens with water features provide special opportunities for contemplation and recreation. Walled gardens provide shelter and shade in inhospitable desert climes. Distinctive gardening styles are associated with different cultures – the Islamic Walled Garden, the French Formal Garden, the English Landscape Garden, the Japanese Zen Garden, and more. Each style adapts to the local/indigenous climate and terrain.

Perhaps because of the range of examples at hand, a succinct definition of "garden" is hard to come by. The philosopher Mara Miller offers this suggestion in her book *The Garden as an Art*:

> A garden is any purposeful arrangement of natural objects (such as sand, water, plants, rocks, etc.) with exposure to the sky or open air, in which the form is not fully accounted for by purely practical considerations such as convenience.
>
> *(Miller 1993: 15).*

In *Greater Perfections: The Practice of Garden Theory*, the garden historian John Dixon Hunt characterizes gardens as "concentrated or perfected forms of place-making" (Hunt 2000: 11). Here is a condensed version of the much more ambitious definition he goes on to offer:

> A garden will normally be out-of-doors, a relatively small space of ground … The specific area of the garden will be deliberately related through various means to the locality in which it is set … To the extent that gardens depend on natural materials, they are ever-changing…
>
> *(Hunt 2000: 14–15)*

Clearly both authors aspire to accommodate gardens we would consider works of art rather than mere utilitarian uses of land.

We should note that not all gardens' associations are happy. While the Garden of Eden was a paradise, its temptations occasioned humankind's Fall, and the Bible names the garden of Gethsemane as the site of Jesus's agony and betrayal. Mara Miller, in her book mentioned above, chronicles the gardener's never-ending struggle to maintain order against the entropic forces of nature. And radical environmentalists date the beginning of environmental degradation and decline to the advent of agriculture. At the very least, modern industrial-style monoculture can be charged with restructuring nature on a monumental scale. On a smaller scale, gardeners must constantly contend with pests, blight, inhospitable weather, and distressing die-offs. Still, the preponderance of our feelings toward gardens is surely positive.

Turn next to community. Social epistemology is a newly popular subfield, with philosophers attending to collective aspects of many traditional epistemological notions. Topics examined include our epistemic interactions with others – for example, testimony, expertise, disagreement – as well as aggregated versions of some standard epistemic categories – for example, group knowledge, group deliberation, group rationality. Community gardens fit this trend, as they can be sites for collective ownership, collective aims and projects, and the collective embrace of shared values. We ask community gardens to do a lot of work in this regard, as we look to them to both build and fix communities. Relatively affluent neighborhoods, where single-family homes sit on expansive lots, don't seem in need community gardens. Rather, it is neighborhoods experiencing hardships of one sort or another – inner city neighborhoods, neighborhoods with vacant lots or abandoned housing, industrialized neighborhoods with brownfield expanses – that seem candidates for the visual, psychological, and economic improvement that productive green spaces might bring. Will Allen and Ron Finley are two popular champions of this approach to gardening in the United States. Allen has won a McArthur Genius Grant for his work developing an urban farm, Growing Power, that serves low-income residents of Milwaukee, while Finley outlined his history as a Guerilla Gardener planting vegetables in vacant lots in South Central LA in a stunningly persuasive TED-talk video that soon went viral.

The aspect of community that is foregrounded by community gardens can vary with the charters governing each site. A community garden might just be a single plot of land worked in common by a group of people who share the resulting produce. More typically, community gardens are subdivided with participants – individuals, households, families, or collectivities of other sorts – presiding over their own allotments of land. Their freedom would vary with the specific set of rules in place. For example, there might be rules limiting what could be grown – vegetables only? vegetables and/or flowers? only "polite" varieties with limited height or sprawl? – as well as rules limiting certain approaches to gardening – no pesticides? only specified fertilizers? no GMO crops? Like any other collective enterprise, community gardens require a governance structure in order to flourish, a system to formulate, revise, and enforce the site's overarching rules. Many inner-city gardens have waiting lists. Presumably those who join are knowingly embracing a pre-existing community of which they would like to become a part, so they are likely to comply with the rules and practices in place. In addition to governance, members of community gardens are linked by shared social bonds. Many community gardens have some space set aside for social functions – resting and hanging out between chores, communal meals highlighting garden bounty, and more – and so they actively build and support a sense of community.

John Searle offers an accessible account of social organizations in his book *Constructing Social Reality*. He proposes three requisites for any socially constructed entity: assigned functions (he labels these observer-relative), collective intentions, and constitutive rules. Community gardens fit this model. A piece of land is a community garden if and only if there is a shared understanding within that community acknowledging its purpose. Thus, there is an assigned function for that acreage. The constitutive rules Searle requires are present either explicitly or implicitly, depending on how rigorous a governance structure is in place. The degree to which shared intentions unite those who garden in adjoining plots might depend on the charter of the larger garden. Some community gardens are explicitly political or ideological, devoted to the promotion of sustainability, the availability of non-GMO organic produce, the elimination of food insecurity, and similar goals. Others might simply allow individuals or groups with varying motivations to tend personal plots of land. Participants allotted plots in the more demanding gardens would be expected to accept and commit to the relevant goals; presumably those who complied would exhibit collective intentionality in pursuing them.

Philosophers disagree about how to characterize collective or "we" intentions. (Tuomela 2005) Searle insists that they are prior, primitive, and sui-generis: "Collective intentionality is a biologically primitive phenomenon that cannot be reduced to or eliminated in favor of something else" (Searle 1995: 24). This means that for Searle, we-intentions are a basic part of our psychological inventory, they don't have to be constructed from something else. Others define collective intentions in terms of ontologically prior individual intentions. For example, Michael Bratman's definition of intending some joint activity J begins as follows: "We intend to J if and only if: (a) I intend that we J and (b) you intend that we J," but immediately becomes wildly complex (Schweikard and Schmid 2013). (Readers interested in this topic are encouraged to browse in the rich and rewarding SEP article "Collective Intentionality.") The collective intentions uniting members of a community garden will vary with the originating history and governing rules of each site. But members of any community garden might well together intend that the plots would be cultivated and the produce used by those who signed up for each, that the planting and care would be in keeping with the guidelines agreed to by the group, and that the governance structure including required meetings and decision-making procedures would be observed. In what follows, I will discuss such intentions while staying neutral with regard to the question of how they might be realized. I trust that my claims will comport with the range of candidate theories taking on this analytical task.

Just as gardens have dual associations, we should note that the notion of community has a dark side as well. The so-called tragedy of the commons threatens to unfold in any communally owned greenspace that yields benefits but requires care. The label is owed to the biologist and population activist Garrett Hardin. In a 1968 article in the journal *Science*, Hardin argued that communal ownership of certain resources, a pasture, for example, could have bad consequences. Each herdsman would gain a positive unit of utility by adding an animal to the pasture but would only bear a fraction of the disutility resulting from overgrazing. Thus it would be in each herdsman's rational self-interest to contribute to the deterioration of the common space (Hardin 1968: 1244). Such imperatives would not take hold on privately owned land. The rules governing any communal garden could of course include provisions to avert negative consequences of this sort. But the classic free-rider problem remains live any time a communal activity accords benefits to participants. That is, individuals can always take advantage of the benefits that flow from the arrangement without performing their fair share of the supporting tasks.

A Variety of Functions

Community gardens have a long history in the United States. One account mentions a Moravian community garden dating back to the early 18th century in Bethabara, a settlement near present-day Winston Salem, N.C. (Wikipedia: Community gardening in the United States) while Laura J. Lawson notes in her book *City Bountiful* that "...people have organized to create places for people to garden in American cities since the 1890s" (Lawson 2005: 1). Lawson's overview indicates the wide variety of purposes such gardens have served: general education, ecological awareness, training in civics, economic opportunity, healing of various sorts. While Lawson initially proposes three themes to accommodate this range – nature, education, and self-help (Lawson 2005: 8) – the categories ramify as she leads us through the 20th- and 21st-century instantiations of this form. Community gardens contributed to morale building and ration supplementation in wartime and to job-training and neighborhood revitalization in periods of urban renewal. Today we continue to look to community gardens to perform a world of good – to help eliminate inner city food deserts, to combat the epidemic of obesity and diabetes among Americans, even to help in the fight against global warming (for example, transportation costs are lower when more individuals to "eat locally"). The job losses and consequent food insecurity triggered by the 2020 COVID-19 pandemic offer one last reminder of the crucial benefits community gardens can provide. In what follows, I will examine three functions community gardens are said to enhance – education, community activism, and individual growth – and comment on their suitability for each task.

(a) Education

In the early 1900s Fannie Griscom Parsons created a children's educational garden in New York City's DeWitt Clinton Park. It aimed to teach children civic virtues as well as inculcate the dignity of labor and love of nature (Wikipedia: Community gardening in the United States).

Nowadays community gardens most quintessentially dedicated to the goal of education are those that are on school grounds and integrated into the school curriculum. The most likely examples are associated with elementary schools, as older students would generally be engaged in more specialized sorts of environmental study – classes explicitly addressing biology, botany, ecology, and the like. A community garden associated with an elementary school promises many benefits. It gets children outside, teaches them about the life cycle of plants, shows them where at least some of their food comes from, and perhaps encourages them to embrace vegetables not featured in their family meals. Picky eating is a common trait in young children, so much so that it is addressed by countless websites such as WebMD; moreover, vegetables are the foods youngsters most often refuse. So growing their own vegetables in a school garden can help break these barriers. As Ron Finley declares in his TED talk, "If kids grow kale, kids eat kale" (Finley 2013). The bounty from a school garden can be shared, divided up to be taken home by the students. Alternatively, a "Farmer's Market" can be created in association with the garden, providing students with additional lessons about business and economic exchange. And finally, offering children hands on experience with a garden might be a way to proselytize for environmentalism and create a future Green generation.

In many ways, First Lady Michelle Obama's White House garden functioned virtually as an educational community garden for us all. Intended to support her *Let's Move* initiative encouraging American children to become healthier and more active, the garden featured 55 varieties of fruits and vegetables that found their way to the White House table. But the First Lady's garden was motivational and inspiring even for those who never received or sampled its bounty. The garden has its own book – *American Grown: The Story of the White House Kitchen Garden and Gardens Across America* (Obama 2012) – and a Wikipedia article situates it in a long tradition of White House garden plots including a first garden planted in 1800 by John and Abigail Adams and a WWII Victory Garden planted by Franklin and Eleanor Roosevelt. Since the first incarnation of Michelle Obama's White House garden was planted in 2009 with the aid of 23 students from Washington's Bancroft Elementary School, it fits within my first category under consideration, education. Moreover, those students had their own dedicated vegetable garden back at their school! The White House Garden was a didactic enterprise, a sort of preaching via plants. But in fact, any green space is health-promoting in and of itself. In addition to the benefits yielded when consumed, plants help ground complex ecosystems and do their part to clean the air, and research has shown that the mere view of green space, for example, from hospital windows, can promote healing.

As our nation ages and as the increasing prevalence of senility and dementia makes the typical life trajectory a Bell curve rather than a continuous ascent, much of what has just been claimed about the benefits of community gardens associated with schools can be applied with equal effectiveness to community gardens associated with nursing homes and retirement communities. There are, however, impediments to any of these proposed educational gardens. For a start, they need at least a small patch of land or an ambitious roof- or container-gardening set-up. The typical urban school with a chain-link fence surrounding an expanse of asphalt doesn't lend itself to bucolic transformation. Even when a plot of land is available, start-up costs are considerable. At the very least a garden requires seeds, tools, a source of water, fertilizer or soil enhancement, a system for pest control, and some sort of border or fence. When school budgets are being cut and seemingly inessential extra-curricular activities eliminated, it can be hard to lobby for educational gardening. We hear of teachers in beleaguered urban school districts reaching into their own pockets to support activities that would otherwise be eliminated. Some even buy their students such basics as books or paper. Thus school gardens can require supererogatory efforts on the part of sponsoring teachers who provide the money and, at times, the labor to sustain them. Job insecurity and burn-out threaten their efforts.

(b) Community Activism

One response to the problems just sketched is to move the educational community garden out of the schoolyard and into the neighborhood. This is the approach of community activists such as Ron Finley and Will Allen. Finley began gardening on the city-owned strip of land – in my area, it's called the "tree lawn" – in bleak and riot-ravaged South Central LA. The city objected, and Finley's initiative became increasingly politicized. He engaged members of the community, converting them to the benefits of locally grown healthful food.

This second sort of community garden aims to rehabilitate both neighborhoods and their residents. The dedication of vacant lots to growing healthful foods seems a way to avert blight and decay. It also benefits communities by alleviating "food desertification" – the phenomenon that befalls inner-city neighborhoods when major grocery store chains refuse to expand, leaving residents at the mercy of Quik Shops and similar establishments that tend to feature less healthy foods and snacks at higher prices. Working together to transform abandoned lots brings community members together. And the educational benefits that were claimed above for schoolground gardens are all the more effective when they flow not from a mandated curriculum with which possibly reluctant students must interact but from the voluntary choices of residents living near the garden.

While these promised benefits are heartening, elaborate rules and regulations must be put in place to create and support such gardens. Here is Lawson's description of the institutional structure supporting the Wattles Farm and Neighborhood Garden in downtown LA:

> To receive a plot, gardeners must sign an agreement and pay yearly dues. Rules and regulations cover cultivation practices as well as responsibilities that members bear as part of the group. The gardeners must agree to keep their plots and the surrounding paths free of overhanging plants, weeds, and debris. To avoid casting shade on other plots, gardeners are not allowed to plant tall trees, shrubs, or vines, or build obtrusive structures. Organic gardening is mandated and the use of chemical herbicides and pesticides is prohibited. Each gardener is expected to contribute at least one hour per month to community cleanup. The second weekend of the month is designated as cleanup time. There are also occasional general meetings to elect officers and discuss matters pertaining to the garden. A fourteen-member elected board of directors runs the garden. A subcommittee, called the Gardenmasters, oversees various responsibilities, including new member orientation and enforcement of rules and regulations. If a gardener is not maintaining his or her site, a Gardenmaster has the authority to first warn the gardening and then give a termination notice.
>
> *(Lawson 2005: 267–268)*

This is the bureaucratic apparatus needed to sustain a longstanding (in place since 1975) community garden on a 4.2-acre site. The institutional complexity is already dense, as the governance structure includes an elected board of directors but also a subcommittee charged with enforcement. Yet that order of complexity must ramify for an entity like Gateway Greening of St. Louis, an umbrella organization which oversees 200 community gardens within the greater metropolitan area while also maintaining an educational center/demonstration garden, an urban seed farm, an ongoing series of presentations and workshops, a lending library, a tool-sharing bank, assorted civic greening projects, and more (see http://www.gatewaygreening.org/).

Despite this complexity, many community gardens are also grassroots structures. That is, they aspire to be "bottom-up" entities guided by consensus building and participatory democracy. Consider the following declaration by Gardening Matters, Gateway Greening's counterpart in Minnesota's Twin Cities: "We embrace the richness of multiple perspectives, value all voices, encourage innovation, embody a culture of reflection and adaption, and lift up community wisdom" (http://www.gardeningmatters.org/about-us). This ringing statement describes co-learning, one of eight "Guiding Values" listed on the organization's website. The others are collaboration, community building, eco-responsibility, stewardship, equity, system change, and transformational. Overall then, some institutional apparatus is required for every community garden to permit effective self-governance and to ensure the perpetuation of the garden itself, but this structure must comport with the overall garden goals.

Unfortunately, there are further challenges to the idealistic vision just sketched. For a start, gardening can involve struggle, set-back, and heartbreak. While verdant gardens are wondrous, pests ravage, weather can be fickle, and plants can simply fail to thrive. In addition some community gardens are made possible in part by zoning changes, open space designations for example, that ease the use of land. Yet such support is not always available. And ironically the very community gardens that help anchor and revive a given neighborhood can fall victim to the subsequent gentrification they usher into place. When property values rise, owners are less willing to donate to community gardening lots that can be sold in a revived real estate market. Stories of thriving community gardens that lose their lease in this manner are all too common. For just one example, see the 12/28/13 *New York Times* article titled "Residents Outraged by Bulldozing of Brooklyn Community Garden." The garden, in place for 16 years, was leveled without warning to make room for a development project.

(c) Individual Growth and Enhancement

While community gardens are lauded for improving entire neighborhoods, they also have positive effects that emerge on an individual level. They benefit countless people who interact with them, from passers-by whose mood is enhanced by a glimpse of greenery, to workers, volunteer or paid, who take on the tasks of tilling, planting, weeding, and harvesting, to residents who gain the opportunity to consume fresh produce. These "distributed benefits," the good that a garden can do a community recalculated in terms of the effect on each individual member, range from the physical to the psychological to the moral.

A *New York Times* article on character education proposes that communal gardens inculcate virtue. Citing an educational studies professor who believes that schools should teach such virtues as self-control, curiosity, and "grit," the author presents his example of an effective means to this end, a WW1 era community garden:

> 'Planting, growing and distributing food taught many of the same traits that character education programs hope to instill,' he said, 'but it's all richly integrated into a task that has genuine purpose and that makes the students think beyond themselves'.
>
> *(North 2015)*

The mention at the close of the previous section of the many ways plants can fail to thrive underscores why gardening must be entered into in the spirit of perseverance and hope. The need to cooperate and work with others in accomplishing garden tasks is yet another reason that character education and moral growth take hold in this realm.

Gardens can support personal improvement in a more focused manner. A number of non-profit organizations operate community gardens in the greater Boston area. One such garden sponsored by the umbrella organization Victory Programs based in Dorchester was specifically intended to help the homeless and individuals struggling with substance abuse problems. In addition to the training and work opportunities afforded to those who tended the garden, the associated farmers' markets accepted SNAP vouchers and offered free produce to residents of specified shelters (Shemkus 2014: 22). Another nonprofit operating in the same locale, the Food Project (deemed "perhaps the oldest urban agriculture group in the city"), employed 150 young men and women from the community, assigning tasks that teach them "the importance of fresh local food along with leadership, teamwork, and civic engagement" (Shemkus 2014: 22). Thus gardens can be sites for vocational training. The subtitle of the article describing these ventures proclaims, "small-scale growing has the potential to create jobs, clean up blighted landscapes, and improve neighborhood residents' access to fresh, nutritious foods." This trend continues today. A recent article lauding a $100,000 grant awarded to support community gardening in Pontiac, MI listed various improvements the grant would fund including a 2,000 square foot greenhouse to allow gardening year-round, three full-time employees to perform the necessary work, and an apprenticeship program to introduce high school students to gardening tasks and possibilities for gardening careers (Broda: TheOaklandPress.com). The ambitious group organizing this venture also planned to open an associated Open Sprout Fresh Food Store to sell harvest from the garden, supplemented by wholesale purchase of non-seasonal produce such as oranges!

Gateway Greening in St. Louis, first mentioned in section (b) above, sponsors a variety of programs that benefit needy segments of the community. Its City Seeds Urban Farm spinoff runs both a Therapeutic Horticulture program and a job-training venture. The first is a 15-week workshop, run in conjunction with the St. Patrick Center, that aids residents struggling with homelessness, mental illness, and addiction. All clients, paired with caseworkers, "learn to grow food, to improve their nutrition literacy, and to build a strong connection to their recovery goals." This is accomplished "through classes on the '12 Steps of Gardening,' nature journaling, meditations and working collectively at City Seeds Urban Farm. Since 2006, over 245 clients have participated." Another venture was a Jobs Training workshop. This ten-week program provided "intensive training in landscaping and horticulture ... [Participants studied] plant biology, landscape design, pest management and more [and received] hands-on training in commercial mowing, ornamental plant maintenance, tree planting, [etc.]." This program has placed over 150 graduates in local jobs ranging from landscaping and recycling to garden retailing and green care (see http://www. gatewaygreening.org/grow/gardens/city-seeds-urban-farm/). These summary descriptions document astounding ways in which community gardens and their extended programs and services can benefit the neediest members of the surrounding urban community. And the list of Gateway Greenings' triumphs grows even longer when the organization's many food distribution activities are taken into account.

Unfortunately, we can continue the pattern of pointing out the negatives associated with each aspect of community gardens under discussion. On the individual level, assorted transgressions can occur. Jesse Hirsch's article "Thievery, Fraud, Fistfights and Weed: The Other Side of Community Gardens" chronicles the tendency of community garden participants to raid, thwart, and even destroy one another's plots. Under a teaser proclaiming: "This summer, a community garden in Queens, New York became a battlefield, as long-simmering tensions erupted into fistfights and death threats," Hirsch documents a variety of altercations (Hirsch 2013). Quarrels arose over procedure and protocol – times to water, permissible heights of vines, shade encroachments, and the like. But there were also more serious transgressions ranging from veggie pilfering to spite-motivated destruction of others' plants to the theft of expensive equipment – one Houston garden sustained a $4,000 loss! Criminality also took hold when gardeners used their plots to grow controlled substances. And finally, even the nostalgic aspects of gardening proved problematic when immigrants growing the distinctive foods and spices of their home cuisines generated cultural clashes within the gardening community.

(d) Some Related Practices

While I have organized my applied sections to indicate both pros and cons of community gardening, on balance, the benefits of the gardens and their associated activities are beyond doubt. I have not yet considered one further impediment that affects the St. Louis and Twin Cities ventures described above but not those in LA – the changing of the seasons. For all areas that experience a true winter, the growing season is limited, and the garden plots lie vacant and dormant for several months. Infrastructure enhancements like the greenhouse planned for Pontiac, MI may lie beyond the budget of most community garden associations. If the gardens in question successfully ameliorate problems like food desertification and poor nutrition in areas of the inner city, then it is all the more distressing that these good effects cannot be sustained year-round. Of course, garden groups can revive the old-time practices of pickling, canning and like methods for putting up garden produce. While these cannot provide a complete substitute for the perk of picking and consuming fresh produce, there are some related practices that might fill this gap. I have in mind the option of indoor hydroponic farming. Some visionaries see this as an especially apt remedy for the inner city, as blighted buildings provide ideal warehouse-style space for large-scale hydroponics. This technique involves growing plants not outdoors in soil but inside in water and under artificial lights that replicate the sun's natural wavelengths. Elaborate self-contained systems that include the element of fish farming can be designed. Here happy synergies allow the fish excrement to fertilize the plants whose roots inhabit the spaces/tanks where they swim. Such installations are admittedly high-tech; considerable energy is required to power the grow lights and water circulation systems. But vertical gardening systems located in urban areas eliminate the need to transport produce from distant farms, thus saving fuel costs. And solar panels can be effectively employed at some sites to provide the needed power.

Dickson Despommier, an emeritus professor of public health and microbiology at Columbia University, is a visionary who proposes vertical farming as an enlightened alternative to big agriculture as it is practiced today as well as a solution to food insecurity and poor nutrition in America's inner cities. Various YouTube videos show him presenting his brief for this proposal. Given the technology and hardware involved, the start-up cost for vertical farming is way higher than for standard community gardening space. Thus it is more likely that the initial ventures will be entrepreneurial and large scale rather than grass roots and single-family size. Even with this model, Despommier envisions considerable community benefit, as he proposes that area residents can obtain jobs within the hydroponic complex. He also imagines a cluster of related ventures – seed companies, produce vendors, flour mills, bakeries, and more – that will come to surround a flourishing vertical farm providing additional employment opportunities and overall economic stimulus for the neighborhood. It isn't beyond the pale to foresee a burgeoning movement with small-scale grassroots hydroponic ventures as well as individualized partitioning of larger vertical farming spaces once they are in place. Such changes would allow vertical farms to provide many of the benefits of present-day urban community gardens. Importantly, established vertical farms need not succumb to the limitations of the seasons. Moreover, they can be readily scaled up to support large and continuous harvests made available to city residents without excessive transportation costs and also without many of the environmental downsides of industrial agriculture as currently practiced. Admittedly this proposal lacks the romance of neighbors, side by side, scrabbling in soil. Nor does the hydroponic model seem equally suited to all crops. Corn, pumpkins, and watermelons all seem problematic inhabitants of high-rise hydroponic structures. But we said at the outset that considerable nostalgia has accumulated around the notion of community gardens, and perhaps some of that must be shed if the descendants of these plots are going to address and solve the many problems of 21st-century urban life. Perhaps some yet-to-be-realized hybrid of the humble and progressive community garden and the more hi-tech vertical garden represents the future of the current community garden movement.

Related Topics

34. Urban Parks and Open Space
35. Suburbs and Exurbs
39. Industrial Agriculture
41. Sustainable Agriculture

References

Broda, Natalie. "Pontiac Nonprofit Awarded $100,000 to Double Community Gardens, Grow Food Year Round," https://www.theoaklandpress.com/news/local/pontiac-nonprofit-awarded-100-000-to-double-community-gardens-grow-food-year-round/article_344af220-a9c9-11ea-bb65-873ce7d0e786.html

Finley, R. (2013) "A Guerilla Gardener in South Central LA," TED Talk, http://www.ted.com/talks/ron_finley_a_guerilla_gardener_in_south_central_la

Hardin, G. (1968) "The Tragedy of the Commons," *Science* 162:1243–1248.

Hirsch, J. (2013) "Thievery, Fraud, Fistfights and Weed: The Other Side of Community Gardens," *Modern Farmer*

Hunt, J. (2000) *Greater Perfections: The Practice of Garden Theory*, Philadelphia: University of Pennsylvania Press

Lawson, L. (2005) *City Bountiful: A Century of Community Gardening in America*, Berkeley: University of California Press.

Miller, M. (1993) *The Garden as an Art*, Albany: SUNY Press.

North, A. (2015) "Smarts vs. Personality in School," *New York Times*. http://op-talk.blogs.nytimes.com/2015/01/10/should-schools-teach-personality/?_r=0

Obama, M. (2012) *American Grown: The Story of the White House Kitchen Garden and Gardens Across America*, New York: Random House.

Ross, S. (1998) *What Gardens Mean*, Chicago: University of Chicago Press.

Schweikard, D. and H. Schmid (2013) "Collective Intentionality," *Stanford Encyclopedia of Philosophy*.

Searle, J. (1995) *The Construction of Social Reality*, New York: The Free Press.

Shemkus, S. (2014) "Urban Farming Takes Root," *Diversity Boston*, 20–22.

Tuomela, R. (2005) "We-Intentions Revisited," *Philosophical Studies*, 125:327–369.

Wikipedia, "Community Gardening in the United States," https://en.wikipedia.org/wiki/Community_gardening_in_the_United_States

PART VIII

Environmental Transformation

43

REMEDIATION

Marion Hourdequin

Today nearly one in six Americans lives within three miles of a major hazardous waste site, though few people could tell you where it is.

−*Voosen 2014*

[R]esidual contamination is a fact of life for the post-industrial world, in every American and European state.

−*Erdem and Nassauer 2013, 283.*

Introduction

The residues of industrial life surround us. Pollution, hazardous waste, and contamination are key features of many contemporary landscapes and environments. Environmental remediation, which focuses on cleaning up pollution and contamination, is thus a crucial and ongoing practice that helps to counteract the degradation of air, water, and land to protect human and ecological health.

From an ethical perspective, remediation is important, but overlooked. Remediation is important because decisions about environmental contamination and cleanup affect all of us, and because remediation raises numerous ethical questions: should all contaminated sites be remediated? What levels and kinds of remediation are acceptable? How can we minimize risks to human health and the environment from residual contamination? How can remediation practices avoid unfairly burdening certain groups and communities? Despite these concerns, environmental philosophers have paid little attention to remediation, in contrast to its related counterpart, ecological restoration. Whereas remediation focuses on reducing unwanted pollution or contamination (Clewell and Aronson 2013, 206), ecological restoration "is the process of assisting the recovery of an ecosystem that has been degraded, damaged, or destroyed" (SER 2004) and typically aims to return an ecosystem to a condition approaching its natural, undamaged state. Often, remediation is a precursor to restoration: environmental contamination is cleaned up or contained before ecological restoration begins. Yet despite the connections between restoration and remediation, it is restoration that has garnered the majority of attention from environmental philosophers.

This disparity in attention likely mirrors the priority environmental ethicists historically have placed on natural systems and wild landscapes, as opposed to human-occupied and built environments. Environmental ethics was born of the insight that ethical theory and practice too often center on humans, and humans alone. Hence, environmental philosophy began by shifting the focus from humans to non-human nature. Although the philosophy of ecological restoration sits

DOI: 10.4324/9781315768090-52

at the nexus of nature and culture because restoration requires human agency to return a site to something approaching its "natural state", it has traditionally focused on restoring *nature*, or systems as they existed prior to human influence (but see Elliot 1982 and Katz 1997 for a critique of the idea that restoration can bring back natural landscapes).

Remediation, in contrast, rarely aspires to such a purist ideal. Remediation is down and dirty: it emphasizes cleaning up polluted areas and mitigating health and environmental risks associated with lingering contamination. Because both ecologists and environmental philosophers historically have tended to prize undisturbed nature, and environmental remediation explicitly deals with disturbed and polluted systems, remediation doesn't neatly fit the goal of protecting nature at its most pure.

In an ideal world, one might think, remediation would be largely unnecessary because pollution would be avoided in the first place. Nevertheless, the time is ripe for philosophers to take more seriously the importance of remediation and to engage in the social, ethical, and policy issues it raises. With many arguing that we have entered "the Anthropocene," a new geological epoch in which the human footprint extends to all corners of the planet, environmental philosophers cannot afford to ignore urban environments, human-occupied landscapes, or the pervasive legacies of industrial activity.

From a practical perspective, the ethics of environmental remediation brings critical issues to the fore. There are important ties between remediation, environmental justice, and environmental health. There are questions about who bears the burdens of contamination, and who should pay for cleanup. There are issues involving disclosure, right to know, and right to participate in environmental decision-making. And there are the challenges of intergenerational justice, as many toxic legacies will remain for generations, centuries, or even millennia to come. Examining the ethics of environmental remediation also provides a practical context for developing richer theoretical perspectives and frameworks. However, linking the practical and theoretical is challenging, because it requires not only careful conceptual and philosophical thought, but an understanding of the policies, laws, and institutions that govern remediation, and the social, cultural, and economic issues at play.

Legal and Social Contexts for Remediation

The legal and social contexts for remediation are often complex, and remediation processes frequently involve public, political, and legal contention. In the United States, environmental remediation requires compliance with a number of environmental laws, including RCRA (the Resource Conservation and Recovery Act) and CERCLA (the Comprehensive Environmental Response, Compensation, and Liability Act), also known as the Superfund law. These laws, passed in 1976 and 1980 respectively, govern the management and control of hazardous waste (RCRA) and the cleanup process for unexpected toxic releases and abandoned hazardous waste sites (CERCLA). More recently, the Brownfields Law of 2002 amended CERCLA to incentivize redevelopment of "brownfields," lands where historical and residual contamination pose a barrier to development.

The Superfund law and its successor amendments provide the core federal legislative framework through which toxic sites in the United States are remediated. Whereas RCRA focuses on tracking and safe handling of hazardous materials and on prevention of environmental contamination, CERCLA was developed to address the thousands of abandoned hazardous waste sites, small and large, found throughout the United States. In this sense, RCRA is primarily a forward-looking law focused on avoidance of contamination, and CERCLA is a backward-looking law focused on remediation. CERCLA authorizes the Environmental Protection Agency to undertake cleanup at contaminated sites, then to recuperate costs from the responsible parties under a standard of strict liability. This means that corporations or other entities are legally responsible for pollution and contamination they generate, regardless of whether this pollution is the result of negligence. The "Superfund" that serves as the basis for CERCLA's nickname is a Congressionally authorized trust fund intended to provide monies for cleanup actions. Although federal actions to recover cleanup costs are intended to replenish the Superfund, at many sites, the parties responsible for the contamination are defunct or lack the resources to pay. Thus, Superfund originally required a tax on chemical producers to support cleanup at these "orphan sites." Although Congress reauthorized CERCLA in 1986 and increased the size of the federal financial commitment to the Superfund, the tax was repealed in 1995 (Voosen 2014), and congressional appropriations to the Superfund have steadily diminished since the mid-1990s (U.S. PIRG Education Fund 2005). Thus, the program is now hampered by a significant lack of resources (see U.S. PIRG Education Fund 2005), which seriously limits funding for remediation.

Although Superfund suffers from significant problems and limitations, the situation in other part of the world is often much worse. Like the United States, individual European countries and the European Union (EU) as a whole have regulatory mechanisms for environmental remediation. The EU's core overarching framework for regulating and responding to environmental damage is the Environmental Liability Directive, which is based on the polluter pays principle (see http://ec.europa.eu/environment/legal/liability/ and associated links). However, many less developed countries lack any systematic approach to cleanup of toxic contamination, and preventative legislation is minimal, lacking, or unenforced. This means that most hazardous and toxic pollution goes unremediated, and, in many cases, spreads beyond its initial source, contaminating food and water. In China, for example, widespread contamination of soils and crops has resulted from rapid industrialization with limited regulation, and there are few mechanisms in place to address this problem (He 2014). What's more, China, India, and many African countries are experiencing severe problems associated with electronic waste, due in large part to illegal transport of post-consumer electronics from developed to less developed countries (see Nuwer 2014, Rucevska et al. 2015). Children in these recipient nations often work to dismantle and recycle components, which exposes them to toxic chemicals (Carroll 2008). In addition, safe disposal practices are rare; thus, toxins associated with electronics recycling frequently leak into the surrounding environment.

Because successful remediation of toxic sites – whether former mines, industrial sites, or military production facilities – typically depends on careful assessment, planning, and monitoring, it is difficult to accomplish in the absence of sophisticated institutions that can facilitate cleanup. From an ethical perspective, the opportunities for engagement in remediation policy center on the design and reform of institutions that govern remediation processes and standards. For contexts lacking such institutions, such as the case of Chinese soil contamination discussed above, the core task is to develop them. In areas with more robust, existing standards and processes, such as the United States with its Superfund framework, the focus is often on enforcement and reform.

There are a number of key ethical concerns related to remediation. Some of the central issues involve (1) assignment of moral and legal responsibility, (2) establishment of remediation goals, standards, and outcomes, (3) remediation planning and decision-making, and (4) long-term management of remediated sites, especially where contamination poses risks to future generations. In addition, there are fundamental questions about why pollution is wrong and whether remediation constitutes an adequate response. Although pollution is commonly considered wrong because of the harms it causes to people, other living things, and the environment more broadly, philosopher Benjamin Hale (2013, see also Hale and Grundy 2009) argues that pollution is wrong because it *fails to respect others*. For Hale, polluters act unjustifiably because they trespass on the rights of others by polluting, even if the pollution is later cleaned up. Thus, the availability of remediation can't absolve polluters from the wrong of pollution, because the fundamental wrong is not creating bad consequences, it is about acting in ways that others would not be willing to accept.

For the purposes of this chapter, I focus primarily on the issues enumerated above, setting aside fundamental questions such as those that Hale raises. I will assume that at least one important moral reason for remediating polluted sites is that pollution causes harm to humans, animals, plants, and the environment. Remediation, therefore, should focus on reducing harm and risk. Issues of respect remain important, however, because respect for persons requires that remediation decisions and processes provide opportunities for affected parties to engage, and that remediation takes seriously the perspectives of all stakeholders, with particular attention to the vulnerable. In addition, as noted below, remediation intersects with concerns about historical memory, meaning, and interpretation, all of which play a role in shaping future land uses as well as people's attitudes toward and understanding of pollution and environmental damage.

The following section discusses the ethical dimensions of remediation in more detail, using illustrative examples to show how these issues have played out in the context of Superfund remediation at two former military sites near Denver, Colorado: Rocky Mountain Arsenal and Rocky Flats. Although the characteristics of these sites differ – both from one another and from other sites – many of the lessons they offer are relevant to remediation projects throughout the United States and beyond. As these cases show, ethics and institutions intersect in the context of remediation. Effective remediation requires well-designed institutions that take seriously questions of environmental justice, public engagement, moral and legal responsibility, transparency, and trust. Some ethical considerations are more easily institutionalized than others; thus, effective remediation will also depend on the establishment of constructive relationships between the agencies, groups, and individuals involved in remediation processes. For example, institutions can be designed to establish trust and transparency, on the one hand, or secrecy (and often, consequently, distrust), on the other. But no institutional design can guarantee the trustworthiness of those who act within it; thus, individual actors, relationships, and institutional cultures provide critical support to formal structures in establishing ethical remediation practices.

Rocky Flats and Rocky Mountain Arsenal: Superfund Remediation at Two Formerly Militarized Sites

Rocky Mountain Arsenal and Rocky Flats are sister sites along Colorado's Rocky Mountain Front, and they share parallel histories. Both were used to produce weapons for the U.S. military: chemical weapons at Rocky Mountain Arsenal, and plutonium triggers for nuclear weapons at Rocky Flats. Both sites experienced extensive contamination due to poor waste disposal practices and unintentional accidents and spills. Lastly, both sites have been remediated, transferred to the U.S. Fish and Wildlife Service, and converted to national wildlife refuges. Despite these similarities, the two sites differ in important ways. The Rocky Mountain Arsenal National Wildlife Refuge has been open to the public for more than two decades; Rocky Flats National Wildlife Refuge has been open for only four years. Public controversy over the Arsenal has mostly died down; controversy over Rocky Flats has not. For example, for many years, controversy has surrounded a proposed road on the east side of Rocky Flats, with opponents expressing concern about the potential for construction to mobilize radioactive dust and proponents dismissing those concerns as ill-founded. Additionally, former workers at the plant continue to seek compensation for toxic exposures and associated health effects, and concerns remain over safety and public access to the site.

The following discussion reviews the histories of contamination and remediation at Rocky Mountain Arsenal and Rocky Flats, and considers the ethical implications of their similar, yet divergent, trajectories.

Rocky Mountain Arsenal: "From Weapons to Wildlife"

Rocky Mountain Arsenal was established first, in 1942, to produce chemical weapons used in World War II. Production and stockpiling of nerve gas, incendiary chemicals, defoliants, and other chemical weapons continued throughout the Cold War, and portions of the site were leased to private companies for pesticide manufacturing from mid-century through the early 1980s. Over decades, liquid wastes were dumped in unlined basins and injected into deep wells, and solid wastes were disposed in trenches (CDPHE 2022). This caused significant and widespread contamination of the site's soil and water with solvents, pesticides, heavy metals, and various byproducts of the production processes (Tetratech n.d.). Evidence suggests that deep injection wells, used as one means to dispose of liquid wastes, triggered earthquakes in the Denver area (Evans 1966). In addition, by the 1950s, contaminated groundwater began to harm livestock on lands adjacent to the Arsenal (CDPHE 2022). The State of Colorado found in 1975 that contamination by multiple chemicals had migrated off-site, and began to push for cleanup (Peters, Perrault, and Smith 1993).

Serious efforts to address contamination at Rocky Mountain Arsenal began in the early 1980s, and in 1982, the U.S. Army, Shell Oil Company (the major commercial lessee of the site), the State of Colorado, and the U.S. Environmental Protection Agency signed a Memorandum of Agreement acknowledging that the requirements of RCRA and CERCLA applied to the site (Peters, Perrault, and Smith 1993). In 1984, the Army began to investigate contamination at the site under CERCLA, and the arsenal was added to National Priorities List – a list of sites slated for Superfund cleanup – in 1987 (GAO 1996). As the result of multiple lawsuits (the State of Colorado sued the U.S. Army, and the Army sued Shell Oil Company, which had used the site for pesticide production (GAO 1996)), the Army and Shell Oil jointly engaged in and financed a 2.1 billion dollar cleanup lasting until 2010 (Finley 2010). This effort involved demolition of numerous buildings, removal and containment of contaminated soil in lined hazardous waste landfills on site, and establishment of a groundwater treatment system that uses an underground barrier to prevent contaminated water from flowing off-site. Through this system, groundwater is pumped to the surface, treated, and then returned underground on the off-site side of the barrier (CDPHE 2022).

Remediation at Rocky Mountain Arsenal has been, from many perspectives, a success. The Arsenal has gone from being one of the country's most contaminated sites to a thriving national wildlife refuge. As part of its commitment, the Army paid for the initial stages of large-scale restoration of the site's short-grass and mixed-grass prairie (Bruce Hastings, pers. comm.). The Fish and Wildlife Service has reintroduced both bison and black-footed ferrets to the refuge, and the human-made ponds at the Arsenal provide habitat for a diverse array of fish, birds, and other wildlife. In addition, a state-of-the-art visitor center tells the story of the site and provides natural history information about the plants and animals that inhabit it.

From an ethical perspective, there are a number of things worth noting. First, with regard to legal and moral responsibility, the polluters did not provide the main impetus for remediation. Instead, cleanup was spurred by the discovery of off-site contamination affecting farms and communities nearby, and the State of Colorado and the public had to push for accountability in order to prompt remedial action. Second, the Army and Shell frequently disagreed with the State of Colorado and the EPA (and each other) during remediation planning, with the polluting parties pushing for less elaborate and cheaper remedies than the state and the EPA. A 1996 GAO report noted "extensive debate over cleanup remedies" and "the costliest study phase in DOD history" with approximately $354 million spent studying contamination at the site by the time the report was issued (GAO 1996). Debates over remediation objectives and standards were extensive, despite consensus that the site would be designated as a national wildlife refuge, which both acknowledged the site's value for wildlife and allowed cleanup to a less stringent standard than would be required for commercial or residential development. (Restricting land use in order to limit human exposure to residual pollutants – an approach known as "institutional control" – is a common strategy for managing sites where full cleanup is impossible or prohibitively expensive.)

Third, not only was there controversy over specific remedies, the State also had to fight in court to retain its jurisdiction to enforce state hazardous waste laws at the Arsenal once the site was listed for federal cleanup (Peters, Perrault, and Smith 1993). The legal system thus played a crucial role in spurring remediation and ensuring compliance with all of the relevant laws. Remediation planning at Rocky Mountain Arsenal also involved the public extensively throughout the decades. Based on interviews with local residents, the site's Community Involvement Plan found a high level of satisfaction regarding community engagement, communication, and cleanup and restoration at the site (Remediation Venture Public Relations Office 2008). In general, because the interests of polluting corporations or agencies (such as the Department of Defense) are rarely aligned with those of the public or primarily concerned with human and environmental health, oversight and enforcement mechanisms are crucial to ethically adequate remediation. Remediation processes that provide genuine opportunities for oversight and engagement by the public – or by public agencies lacking vested interests – play an important role in ensuring that cleanup protects human health and the environment and considers the interests of all affected parties.

A fourth ethical issue for the Arsenal, and for almost all remediation sites, involves long-term monitoring and management of residual contamination. As part of the remediation effort, monitoring was established both on-site and off-site to address ongoing concerns. The State of Colorado's Medical Monitoring Program tracked birth defects before and following remediation, and compared these to rates of birth defects in the state of Colorado as a whole (CDPHE 2010). This study found no significant increase in birth defects during the remediation period. The Tri-County Health Department (n.d.) monitors groundwater adjacent to the Arsenal. And on site, the Fish and Wildlife Service established monitoring stations to periodically sample the blood and tissues of resident birds (for discussion, see NIEHS 2010). Results from this sampling have been used to pinpoint residual contamination hot spots, and to investigate and address them. Despite these efforts, new issues will undoubtedly arise as time passes – further contamination will be uncovered, or the hazardous waste landfills will require maintenance and updating. For this reason, it is important that remediation sites not only establish responsibility for the period of the cleanup, but that they identify long-term stewardship plans.

At some decommissioned military sites, minimal cleanup occurs before transfer of ownership to the Fish and Wildlife Service or other agencies, and after the Army walks away, the new owners may have few resources and little recourse for addressing ongoing environmental contamination. This is of particular concern for military-to-wildlife refuge conversions, as these sites move from the jurisdiction of a large agency with deep pockets (the U.S. Army) to a significantly smaller one with minimal funds (the U.S. Fish and Wildlife Service) (Hourdequin and Havlick 2011).

Though the site will never be fully "clean," remediation at Rocky Mountain Arsenal has been effective in many respects, even in the face of widespread contamination. There are a number of particular features of this case that contributed to this success:

- Public and political pressure: The site is in close proximity to a large urban center, with a large interested public constituency and significant opportunities for media attention. This generates public pressure and greater accountability than at less visible sites. Rocky Mountain Arsenal was located adjacent to the historically low-income neighborhoods of Commerce City, Colorado, and dedicated advocacy within Commerce City, in conjunction with leverage from more advantaged populations in Denver and its suburbs, proved a powerful combination in securing remedial action.

- Legal enforcement: State and national environmental laws provided effective leverage to hold the U.S. Army and Shell Chemical responsible, though persistence in pursuing legal remedies was essential to the ultimate settlement and remediation of the site.
- Financial resources: Due to the extent of the contamination, the lawsuits surrounding the case, and the significant financial resources of the Army and Shell Oil, over $2 billion was allocated for cleanup.
- Institutional control: As a federally owned site, Rocky Mountain Arsenal was available for transfer to another federal agency following its closure as a military base. Conversion to a national wildlife refuge provided a form of institutional control (discussed above) that allowed for cleanup to a less stringent standard than would have been required for other forms of ownership and development. This, of course, is a double-edged sword: while designation as a wildlife refuge makes it easier to limit human access to contaminated areas, it allows for a greater degree of residual contamination at the site.
- Effective public engagement processes during and following remediation: Although there are surely ways in which public involvement could have been improved, the Army, Environmental Protection Agency, Colorado Department of Public Health and Environment, and Fish and Wildlife Service made significant efforts to inform and involve the public throughout the remediation process.

Many of these features are lacking at other remediation sites. For example, many contaminated sites occur in residential areas where exposure to contamination cannot be significantly limited by institutional controls. Additionally, contaminated sites are more common in low-income neighborhoods and in areas with higher proportion of racial and ethnic minorities (see, e.g., Bullard et al. 2008). These communities are often politically disempowered due to structural inequalities that affect access to the media, public officials, and the courts, three institutions that play an important role in accountability for and remediation of environmental damage.

Even at the Rocky Mountain Arsenal, where tremendous resources were spent studying and remediating environmental damage, the site remains contaminated. Portions have been cleaned, but the overarching goal of the cleanup was to *contain* toxic wastes. Here and elsewhere, remediation compensates for but does not fully undo the effects of pollution. Remediation rarely, if ever, returns a site to its pre-contaminated state. The focus at the former Rocky Mountain Arsenal has shifted from weapons to wildlife – a narrative that media stories and government publications embrace – but the residues of chemical weapons production remain, encapsulated in landfills, permeating the groundwater, and trapped in sediments of the site's ponds. More than 2.5 million yards of contaminated or hazardous materials remain in two landfill sat the center of the site (U.S. Department of the Army 2021).

A core lingering concern of local communities near Rocky Mountain Arsenal thus centers around the site's history and its presentation to the public and to wildlife refuge visitors. As the cleanup has come to a close, the visible remnants of the chemical weapons plant have gradually disappeared. The buildings are gone, the site's iconic water tower was felled and removed, the refuge entrance – previously at the old military guard station – has been moved, and the hazardous waste landfills appear as large, gentle mounds, covered with prairie grasses. A visitor who takes a driving tour of the site might never know – but for the name – the purposes this site once served. Is this knowledge important? Does it matter whether people know what came before? Remediation cannot turn back the clock, and one of the key questions for remediated sites is how to acknowledge and take account of the past while making decisions for the future (Havlick et al. 2014, Hourdequin 2013, Hourdequin and Havlick 2013).

The Nuclear Legacy of Rocky Flats

At Rocky Flats, this question – how to take account of the past while planning for the future – remains both significant and controversial. Rocky Flats was established in 1951 as a nuclear production facility, manufacturing the plutonium "pits," or triggers, at the core of U.S. nuclear bombs. For reasons of national security, many aspects of production at the site were classified, shrouded in a veil of secrecy. Workers typically focused on just one element of production, and had little knowledge of activities in other part of the site. Over the years, many workers were exposed to significant doses of radiation – through both accidents and routine work. Two major fires, in 1957 and 1969, released plumes of radioactive smoke, and surrounding communities were not warned or informed of the hazard (Iversen 2012). Poor waste disposal practices led to soil and water contamination, both within and beyond the plant site (for an overview of off-site contamination, see Rocky Flats Stewardship Council 2014). The site was polluted by radioactive and chemical waste, and including Pu-239, which has a half-life of 24,000 years (EPA 2021).

Reports of inadequate safety and waste management procedures led the FBI to raid Rocky Flats in 1989 to investigate environmental violations there. This led to a subsequent grand jury investigation, whose results were deemed non-actionable. There were no prosecutions, the grand jury files were sealed, and those who served on the grand jury were sworn to secrecy. Many on the grand jury were dissatisfied with the outcome, though they had little power to do anything about it (Iversen 2012).

Despite the secrecy surrounding the site, operations at Rocky Flats generated significant political controversy and became a political flashpoint in the later decades of the Cold War. Activists from Boulder, Denver, and around the country gathered at Rocky Flats to protest the U.S. nuclear program and to block shipments of materials to and from the site. Many protesters were arrested, and longtime activist LeRoy Moore conducted a 24-day hunger strike in 1989 in an effort to garner support from the governor of Colorado to close the site (Ackland 1999, 218). Two popular books – Len Ackland's *Making a Real Killing* (1999) and Kristen Iverson's *Full Body Burden* (2012) – chronicle the history of lax safety, irresponsible waste disposal, secrecy, and cover-ups at Rocky Flats.

The site has now been cleaned – faster than expected and significantly under budget, which some view as a success and others eye with skepticism – yet the site remains the subject of ongoing controversy. Like Rocky Mountain Arsenal, Rocky Flats has been converted to a national wildlife refuge and now falls under the jurisdiction of the U.S. Fish and Wildlife Service. However, a number of local opponents objected to the decision to open the site to the public in 2018, citing health concerns over residual contamination and the potential for plutonium exposure. These lingering concerns are intertwined with disagreements about how the site's history is portrayed. The triumphal narrative of successful environmental cleanup, for example, clashes with a counter-narrative of mistrust and skepticism, which reflects the history of secrecy and non-disclosure at the site. A third narrative focuses not so much on the success of remediation (though it perhaps presupposes that) as on the conversion to a wildlife refuge as marking a new phase in the site's history – from this perspective, the move from "weapons to wildlife" represents the closure of an earlier chapter in the site's history and the opening of a new one.

The tensions between these views are exemplified by a 2006 controversy over the proposed text for signs at visitor entry points to the refuge, as well as in disagreements about whether the site should be opened to the public at all. (Due to U.S. Fish and Wildlife Service website updates, the Rocky Flats sign plan and comment letters are now difficult to locate on the internet. The author retains pdf copies of the Rocky Flats National Wildlife Refuge Step-Down Plan for Site History/Safety Signs and comment letters on file.) The sign controversy is particularly striking because while public land managers are accustomed to disagreements over issues such as whether and how to control invasive species, fire management, recreational access, significant public engagement surrounding informational and interpretive signs is much less common. Although the number of comment letters received on the proposed entrance sign text was relatively small (14), respondents included the Jefferson County Commissioners, the Rocky Flats Stewardship Council, the Boulder Area Trail Commission, the City of Arvada, the City of Westminster, the Boulder County Commissioners, the City of Golden, and a number of individuals. The pivotal issue focused on the proposed signs' characterization of the site's history. Some argued against what they viewed as the proposed signs' triumphalist rhetoric and the euphemistic description of the site's work to develop "deterrent weapons" (a.k.a. nuclear bombs). Others felt that the signs stressed mundane and ubiquitous risks of tripping or falling while failing to give sufficient attention to the site's unique risks – those related to lingering chemical and nuclear contamination. Still another writer supported the Fish and Wildlife Service's proposed text, noting that any action at Rocky Flats "elicits hysterical responses from the people who have made careers criticizing the process."

Although one might think that entrance signs are relatively unimportant compared to the quality of the cleanup process or the ongoing management of the site, the disagreement here exemplifies the tensions that continue to surround Rocky Flats. Not only do opinions diverge regarding wildlife refuge's opening to the public; development at the edge of the site has sparked controversy as well. For example, the homepage for the Candelas development along the southern edge of Rocky Flats greets visitors like this:

> Welcome to Candelas, the next great place on the Front Range. With vast open space, a wide array of community amenities and beautiful homes being crafted by premier homebuilders, Candelas presents a life full of the very things people love most about Colorado. We invite you to come live life wide open.
>
> *(Candelas: Life Wide Open 2020)*

This depiction stands in stark contrast to that offered by the counter-site "Candelas Glows: Development Next to a Nuclear Waste Superfund Site. Come for the view, stay for the exposure." The authors of this website argue:

> Candelas…[is] part of an alarming trend of forgetting about its neighbor, Rocky Flats– a former Nuclear Weapons Plant & now an active Department of Energy (DOE) managed Superfund site. When the Rocky Flats weapons plant was in operation, it built plutonium triggers or 'pits' for over 70,000 nuclear bombs. The area in and around the former Nuclear Weapons Plant site has been contaminated with plutonium, uranium, americium, beryllium, and, according to the Department of Labor, over 1,000 other carcinogenic chemicals.
>
> …We believe Rocky Flats needs to be remembered for what it is with plant workers recognized as the veterans they are. The "Wildlife Refuge" designation needs to be immediately stripped and NOT opened to the public for recreation. We believe the site should be memorialized, calling on artists to help us build permanent structures that speak to the site's past much the way other historical tragedies are memorialized. A memorial could also commemorate the workers and neighbors who have been deeply impacted by the legacy of the site.
>
> *(Candelas Glows, "About Us," 2020)*

The Candelas Glows group expresses two core concerns: first, they believe that the site is not safe for public use or nearby development, and second, they worry that the site's complex history and its many workers will be forgotten.

This is a case where every argument seems to spawn a counterargument, and due to questions about Rocky Flats' safety, the Candelas developers host their own distinct website focused on refuge cleanup and environmental quality (Candelas 2015). In 2015, this site described the cleanup process and its outcome, "an environmentally sound open space":

> The Rocky Flats cleanup effort created an open space that exceeds every standard for being environmentally sound. Water streaming from the area, for example, is reported by the U.S. government to be 100 times cleaner than federal drinking water standards. And the DOE reports that the ground on and around the old site carries no increased risk of exposure to harmful contaminants for humans or animals.
>
> *(Candelas 2015)*

In recent updates, the website addresses ongoing safety concerns, including those associated with the opening of the refuge to the public in 2018, insisting that those who oppose public access "ignore the science of safe" (Candelas 2020).

These divergent perspectives reflect different interpretations of Rocky Flats' history, its remediation, and the site's (and its environs') future use and management. At the root of the controversy is significant distrust between groups such as Candelas Glows and the Rocky Flats Nuclear Guardianship project, on the one hand, and government agencies and private developers, on the other. Another group, the Rocky Flats Stewardship Council, was formed in 2006 by federal Congressional mandate, "to provide ongoing local government and community oversight of the post-closure management of Rocky Flats, the former nuclear weapons plant northwest of Denver" (Rocky Flats Stewardship Council 2015). This group offers fact sheets on Rocky Flats, disseminates information to the public, and solicits public engagement on management issues such as a recent proposal to control weeds on the refuge through the use of prescribed fire.

Despite these efforts, polarization remains surrounding the site, and those who oppose keeping the refuge open to the public remain skeptical about the adequacy of the cleanup. Due to this distrust, additional studies "proving" that the site is now safe are unlikely to satisfy skeptics. Whyte and Crease (2010, 418) describe such situations as *poisoned-well cases*, "where trust and distrust in experts is an explicit and irreducible element in multilateral negotiations over scientific and technical issues." In such cases, they argue, "there is no hope for a technical argument to succeed" (Whyte and Crease 2010, 418).

There is no easy way to resolve the challenges associated with distrust and the polarization it can generate. Transparency on the part of government agencies and experts may help, but doesn't cut to the core of the problem, because even when studies and data on environmental risk are publicly shared, an atmosphere of distrust leads some to discount the relevance or reliability of this information. In cases where different groups embrace significantly different narratives, they may view issues surrounding remediation from entirely different perspectives and worldviews.

Differences at the level of these synoptic lenses can cause opponents to talk past one another and fail to genuinely grasp one another's concerns. Danielle Endres (2012) describes a clear case of such failure in the case of the proposed Yucca Mountain nuclear waste disposal site in Nevada. Endres argues that the U.S. government (or more specifically, the Department of Energy) viewed the site as a national sacrifice zone, whereas the Paiute and Shoshone tribes understood it as sacred land. The public participation process provided information about health and environmental risks associated with waste disposal at the site, and the government was prepared to consider comments related to these risks. However, they were not well positioned to take into account arguments based on the cultural and spiritual value of the land to local peoples. The Department of Energy's framework rested on quantitative risk assessment, and lacked the means to adequately register Paiute and Shoshone concerns (Endres 2012).

Something similar seems to be at stake in the Rocky Flats case. The government and private developers insist that the site is "safe" and cleanup meets all applicable federal standards.

Opponents doubt the credibility of the federal standards. They further question the rationale for opening the site to the public and rebranding it as a national wildlife refuge. From the perspective of the Rocky Flats Nuclear Guardianship project, for example, the Rocky Flats site is not just a place that happens to have been contaminated by nuclear waste and other chemicals, and that is now cleaned up (Rocky Flats Nuclear Guardianship 2015). Instead, it symbolizes the irrationality of the Cold War and nuclear proliferation, as well as the willingness to sacrifice human and environmental health for geopolitical advantage. The Guardianship project emphasizes the need for ecological stewardship and intergenerational justice, encouraging people to take responsibility for the dangers nuclear production has unleashed. Although remediation, scientific monitoring, and other technical approaches to controlling nuclear contamination at the site are crucial, the idea of nuclear guardianship is tied to a larger cultural shift that opposes militarization and embraces an ethic of responsibility for future generations (Rocky Flats Nuclear Guardianship 2011). This requires not only remediation, but remembrance and recognition of past damage. Thus, the whole idea of cleaning up and moving on – opening a new chapter in the history of Rocky Flats in which the old chapter is largely left behind – is anathema to the core ideas of nuclear guardianship. The guardianship perspective also emphasizes the need to steward radioactive waste sites "for the millennia," raising questions about institutional constancy in societies characterized by short-term thinking and rapid change (LaPorte and Keller 1996). Effective long-term stewardship requires that institutions maintain commitments across multiple generations, along with effective actions to fulfill those commitments. However this is difficult, as it requires institutions to sustain these commitments through changes in personnel, fluctuations in funding, institutional reorganizations, new exigencies, and other factors that disrupt continuity (LaPorte and Keller 1996).

The aim of this chapter is not to resolve disputes between those who seek to focus on the future of Rocky Flats and those who remained deeply concerned about its past, but to highlight the conflict, and in doing so, to illuminate the contested nature of Rocky Flats, and of many remediated sites more generally. All remediation is a response to environmental damage, so remediation typically takes place in layered landscapes with complex socio-environmental histories (see Hourdequin and Havlick 2015). At these sites, understanding and engaging the diverse perspectives of stakeholders may be just as important to the development of effective remediation processes, post-remediation plans, and ongoing site management as risk assessments and scientific monitoring.

Looking Ahead: Environmental Ethics and Environmental Remediation

Although this chapter has focused in depth on two cases, its fundamental aim has been to reveal the ethical complexity of environmental remediation, and the importance of institutionally informed and case-based normative inquiry on this topic. As these cases show, risk and harm clearly represent critical ethical concerns in environmental remediation; however, the management of risk and harm has complex ethical, political, legal, social, and historical dimensions. Remediation is not only about outcomes; it involves questions about processes as well: who is included? Who is excluded? Who decides what level of cleanup is "enough"? Who decides whether lingering health, safety, and ecological concerns are justified or "irrational"? The design of remediation institutions and decision-making processes thus plays a crucial role in determining the substantive outcomes of remediation, as well as the degree to which decision-making processes recognize the diverse perspectives at stake. In highlighting the contested narratives of Rocky Flats, I also aim to show that at certain sites, more is at issue than the degree to which the cleanup removes contamination and reduces the risk of harm to humans, animals, and the environment. Remediation also can involve constructing and reconstructing the meaning and significance of a place. At sites with local, national, and international significance, such as Rocky Flats, these meanings are contested:

is it sufficient to "re-brand" the site as a refuge, leaving its nuclear history to fade into the background? Or should the remediated site serve in part to recognize and critically question the U.S. role in the nuclear arms race, and the sacrifices that required? Remediation invites transition in the appearance, uses, and meanings of land. In some cases, what is needed is a reprieve from environmental damage and the physical, psychological, and aesthetic burdens it carries (cf. Ingram 2015). In other cases, what is called for is an ongoing effort to understand and grapple with the sources that generated that damage in the first place.

Environmental ethicists and other scholars have a vital role to play in illuminating the complexities of the issues involved in environmental remediation, and in drawing out important considerations in particular cases. However, because remediation goals and outcomes depend so crucially on historical, social, environmental, and economic contexts, no simple algorithm can guide all aspects of remediation. Human health, environmental protection, and consideration of long-term and multi-generational risks are key, as are laws, policies, and institutions that create accountability for environmental damage and establish baseline expectations for cleanup. However, specific approaches to remediation need to be developed in context, at particular sites, through open, inclusive processes for making remediation decisions – processes that take seriously the perspectives of all affected parties, with particular attention to those who are most vulnerable. Just, inclusive remediation processes, together with regulatory structures that ensure protection of human and ecological health, are thus critical to developing ethical remediation strategies for the future.

Related Topics

20. Moral Bases of Responses to Climate Change
27. Mining
30. Renewable Energy
31. Natural Gas and Fracking
44. Restoration
47. Rewilding
48. Novel Ecosystems

References

Ackland, L. (1999) *Making a Real Killing: Rocky Flats and the Nuclear West*. Albuquerque: University of New Mexico Press.

Bullard, R.D., Mohai, P., Saha, R. and Wright, B. (2008) 'Toxic wastes and race at twenty: why race still matters after all of these years', *Environmental Law*, 38(2), pp. 371–411.

Candelas. (2015, 2020) *Rocky Flats National Wildlife Refuge* [Online]. Available at: http://www.candelas-rockyflats.com/ (Accessed 2 September 2015; 12 June 2020)

Candelas Glows. (2020) *Candelas Glows* [Online]. Available at: http://candelasglows.com/ (Accessed 2 September 2015; 12 June 2020)

Candelas: Life Wide Open. (2015, 2020) *Candelas - Life Wide Open* [Online]. Available at: http://www.candelaslife.com/ (Accessed 2 September 2015; 12 June 2020)

Carroll, C. (2008) 'High-tech trash', *National Geographic*, January.

CDPHE (Colorado Department of Public Health and Environment). (2010) *Rocky Mountain Arsenal Medical Monitoring Program Surveillance for Birth Defects Compendium* [Online]. Denver, CO: Colorado Department of Public Health and Environment. Available at: https://www.colorado.gov/pacific/sites/default/files/HM_RMA-Rocky-Mountain-Arsenal-Medical-Monitoring-Program-Surveillance-for-Birth-Defects-Compendium.pdf (Accessed 12 June 2020)

CDPHE (Colorado Department of Public Health and Environment). (2022) *Rocky Mountain Arsenal* [Online]. Available at: https://www.colorado.gov/pacific/cdphe/rocky-mountain-arsenal (Accessed 24 March 2022)

Clewell, A.F. and Aronson, J. (2013) *Ecological Restoration: Principles, Values, and Structure of an Emerging Profession*. Washington, DC: Island Press.

Elliot, R. (1982) 'Faking nature', *Inquiry*, 25(1), pp. 81–93.

Endres, D. (2012) 'Sacred land or national sacrifice zone: the role of values in the Yucca Mountain participation process', *Environmental Communication: A Journal of Nature and Culture*, 6(3), pp. 328–345.

EPA (Environmental Protection Agency). (2021) Radionuclide Basics: Plutonium [Online]. Available at: https://www.epa.gov/radiation/radionuclide-basics-plutonium (Accessed 24 March 2022)

Erdem, M. and Nassauer, J.I. (2013) 'Design of brownfield landscapes under different contaminant remediation policies in Europe and the United States', *Landscape Journal*, 32(2), pp. 277–292. DOI: 10.3368/lj.32.2.277.

Evans, D.M. (1966) 'The Denver area earthquakes and the Rocky Mountain arsenal disposal well', *The Mountain Geologist*, 3(1), pp. 23–26.

Finley, Bruce. (2010) 'Rocky mountain arsenal ready for its post-superfund life', *Denver Post*, 10 September [Online]. Available at: https://www.denverpost.com/2010/09/17/rocky-mountain-arsenal-ready-for-its-post-superfund-life/ (Accessed 12 June 2020)

GAO. (1996) *Environmental Cleanup: Progress in Resolving Long Standing Issues at Rocky Mountain Arsenal*. GAO/NSIAD-96-32. Washington, DC: U.S. Government Accounting Office.

Hale, B. (2013) 'Can we remediate wrongs?' In Hiller, A., Ilea, R. and Khan, L. (eds.) *Consequentialism and Environmental Ethics*. New York: Routledge, pp. 147–163.

Hale, B., and Grundy, W.P. (2009) 'Remediation and respect: do remediation technologies alter our responsibility?', *Environmental Values*, 18(4), pp. 397–415.

Havlick, D.G., Hourdequin, M. and John, D. (2014) 'Examining restoration goals at a former military site: Rocky Mountain Arsenal, Colorado (USA)', *Nature + Culture*, 9(30), pp. 288–315.

He, G. (2014) 'China's dirty pollution secret: the boom poisoned its soil and crops', *Yale Environment*, 360 [Online]. Available at: http://e360.yale.edu/feature/chinas_dirty_pollution_secret_the_boom_poisoned_its_soil_and_crops/2782/ (Accessed 12 June 2020)

Hourdequin, M. (2013) 'Restoration and history in a changing world: a case study in Ethics for the Anthropocene', *Ethics and The Environment*, 18(2), pp. 115–134.

Hourdequin, M. and Havlick, D.G. (2011) 'Ecological restoration in context: ethics and the naturalization of former military lands', *Ethics, Policy, and Environment*, 14(1), pp. 69–89

Hourdequin, M. and Havlick, D.G. (2013) 'Restoration and authenticity revisited', *Environmental Ethics*, 35(1), pp. 79–93.

Hourdequin, M. and Havlick, D.G. (eds.) (2015) *Restoring Layered Landscapes: History, Ecology, and Culture*. New York: Oxford University Press.

Ingram, M. (2015) 'Material transformations: urban art and environmental justice', In Hourdequin, M. and Havlick, D.G. (eds.) *Restoring Layered Landscapes: History, Ecology, and Culture*. New York: Oxford University Press, pp. 222–238.

Iversen, K. (2012) *Full Body Burden: Growing up in the Nuclear Shadow of Rocky Flats*. New York: Crown Publishers.

Katz, E. (1997) 'The big lie: human restoration of nature', In *Nature as Subject: Human Obligation and Natural Community*. New York: Rowman and Littlefield, pp. 93–108

LaPorte, T.R. and Keller, A. (1996) 'Assuring institutional constancy: requisite for managing long-lived hazards', *Public Administration Review*, 56(6), pp. 535–544.

NIEHS (National Institute of Environmental Health Sciences). (2010) *Wildlife Biomonitoring at Hazardous Waste Sites: Superfund Research Program*. Available at: https://www.niehs.nih.gov/research/supported/dert/programs/srp/phi/archives/ecology/wildlife/index.cfm. (Accessed 12 June 2020)

Nuwer, R. (2014) 'Eight million tons of illegal e-waste is smuggled into China each year', *Smithsonian SmartNews*, 28 February [Online]. Available at: http://www.smithsonianmag.com/smart-news/eight-million-tons-illegal-e-waste-smuggled-china-each-year-180949930/?no-ist (Accessed 12 June 2020)

Peters, V.L., Perrault, L.E. and Smith, S.M. (1993) 'Can states enforce RCRA at superfel sites? The Rocky Mountain Arsenal Decision', *Environmental Law Reporter*, 23 ELR 10419 [Online]. Available at: http://elr.info/sites/default/files/articles/23.10419.htm (Accessed 12 June 2020)

Remediation Venture Public Relations Office. (2008) *Community Involvement Plan for Rocky Mountain Arsenal Contamination Cleanup*. Available at: http://www2.epa.gov/sites/production/files/documents/rma_cip2008.pdf (Accessed 12 June 2020)

Rocky Flats Nuclear Guardianship. (2011) 'Nuclear guardianship ethic'. Available at: https://www.rocky-flatsnuclearguardianship.org/about (Accessed 12 June 2020)

Rocky Flats Nuclear Guardianship. (2015) *Rocky Flats Nuclear Guardianship: A Project of the Rocky Mountain Peace and Justice Center*. Available at: http://www.rockyflatsnuclearguardianship.org (Accessed 12 June 2020)

Rocky Flats Stewardship Council. (2014) *Off-site Lands: Contamination and Risk*. Available at: http://www.rockyflatssc.org/fact_sheets.html (Accessed 12 June 2020)

Rocky Flats Stewardship Council. (2015) *Homepage*. Available at: http:www.rockyflatssc.org (Accessed 12 June 2020)

Rucevska, I., Nellemann, C., Isarin, N., Yang, W., Liu, N., Yu, K., Sandnæs, S., Olley, K., McCann, H., Devia, L., Bisschop, L., Soesilo, D., Schoolmeester, T., Henriksen, R. and Nilsen, R. (2015) *Waste Crime – Waste Risks: Gaps in Meeting the Global Waste Challenge. A UNEP Rapid Response Assessment*. United Nations Environment Programme and GRID-Arendal, Nairobi and Arendal. Available at: https://www.grida.no/publications/166 (Accessed 12 June 2020)

Society for Ecological Restoration International, Science & Policy Working Group (SER). (2004) *The SER International Primer on Ecological Restoration*. Available at: https://cdn.ymaws.com/www.ser.org/resource/resmgr/custompages/publications/ser_publications/ser_primer.pdf (Accessed 12 June 2020)

Tetratech. (n.d.) *Rocky Mountain Arsenal Remediation*. Available at: https://www.tetratech.com/en/projects/rocky-mountain-arsenal-remediation (Accessed 24 March 2022)

Tri-County Health Department (Adams, Arapahoe, and Douglas Counties, CO). (n.d.) *Rocky Mountain Arsenal Oversight Program*. Available at: http://www.tchd.org/268/Rocky-Mountain-Arsenal (Accessed 12 June 2020)

U.S. Department of the Army. (2021) Rocky Mountain Arsenal: Draft Fifth Five-Year Review Report for the Rocky Mountain Arsenal, Commerce City, Adams County, Colorado, revision E. Available at: https://www.rma.army.mil/Portals/32/May%202021%20updates/Draft%20Fifth%20FYRR%20Rev.%20E%20(Part%201%20-%20Text).pdf?ver=2D7DRXN7FqZdRnmRVNFSiQ%3D%3D (Accessed 24 March 2022)

U.S. PIRG Education Fund. (2005) *Empty Pockets: Facing Hurricane Katrina's Cleanup with a Bankrupt Superfund*. Available at: http://www.uspirg.org/sites/pirg/files/reports/Empty_Pockets_USPIRG.pdf (Accessed 12 June 2020)

Voosen, P. (2014) 'Superfund sites', *National Geographic*, December [Online]. Available at: https://www.nationalgeographic.com/magazine/2014/12/superfund/ (Accessed 12 June 2020).

Whyte, K.P. and Crease, R.P. (2010) 'Trust, expertise, and the philosophy of science', *Synthese*, 177(3), pp. 411–425.

44

RESTORATION

Mark Woods

Introduction

In 2000, the US State of Florida mandated a massive restoration effort known as the Comprehensive Everglades Restoration Plan (CERP). CERP's goal is to capture unused water that flows to the Gulf of Mexico and the Atlantic Ocean and redirect the majority of this water to the native ecosystems of the Everglades region of south Florida, ecosystems starved for water that has been diverted elsewhere for cities, farms, and other human developments. The centerpiece of CERP is the construction of 300 aquifer storage and recovery wells in clusters around Lake Okeechobee, the Hillsboro Canal, and the Caloosahatchee River Basin. These wells collectively will inject up to 1.7 billion gallons of surface water into the underground Floridan Aquifer, where this water can be stored for years, retrieved, and channeled to help restore historic water flows. Over 200 more miles of canals and levees in the Everglades are to be removed. To clean up polluted waters from agriculture, wetland treatment/holding areas are being built. Supplementing these water restoration efforts, the state of Florida and the US federal government have financed the acquisition of numerous lands in the Everglades region in order to increase the spatial extent of protected areas.

There are a number of reasons to doubt that these efforts will be successful. The aquifer storage and recovery wells are largely untested technology, and the strength of the limestone walls of the Floridan Aquifer is not uniform. The clusters of wells could create cumulative pressures, limestone walls could crack, and the stored underground water could get out of control, thus undermining the entire process of aquifer storage and recovery. The phosphorus content of the waters of Lake Okeechobee is 130 parts per billion, and further south it is 200–500 parts per billion. It's not clear how water treatment areas south of the lake are supposed to reduce phosphorus to a recommended 10 parts per billion, and sugarcane growers in south Florida have successfully opposed water cleanup efforts. One important restoration problem remains largely unaddressed. Recreating some semblance of historic water flows in the Everglades from the Kissimmee River Basin to Florida Bay requires a flow of water from Lake Okeechobee south to Everglades National Park. In order for this to occur, some significant part of the 700,000 acre Everglades Agricultural Area north of the park must be reclaimed as wetlands. CERP calls for 50,000–60,000 acres to be reclaimed, but something like 150,000–200,000 acres might be required to begin to recreate a historic flow of water. These massive restoration efforts will cost billions of dollars.

Ecological restoration raises fundamental issues of what nature is, what environmental values most warrant our attention, how environmental values should be protected, and how humans are related to nature. Philosophical issues of ecological restoration are more than mere applications

DOI: 10.4324/9781315768090-53

of environmental philosophy because these issues help define the very subject matter of environmental philosophy itself. In its broadest form, ecological restoration is a process whereby humans assist the recovery of some impaired tract of nature. The Society for Ecological Restoration defines ecological restoration as "the process of assisting the recovery of an ecosystem that has been degraded, damaged or destroyed" (SER 2004: 3). There are important issues about what it means to say that an ecosystem or tract of nature has been somehow impaired, the goals of recovery, the process of recovery, and how precisely humans assist in this process.

Describing Acts of Ecological Restoration

What is an act of ecological restoration (ER)? The term "restoration" and the verb "restore" are themselves ambiguous. The *Oxford English Dictionary* lists nine meanings for the verb "restore" that include giving back, making amends or compensation, building up again, renewing, recovering, and bringing back to a previous or normal condition (OED 1971: 2517). Acts of ER are sometimes described as, and sometimes contrasted with, acts of assisted colonization or migration, creation, dialogue with nature, domination, ecological engineering, ecological tinkering, ecosystem management, extitution, fabrication, gardening, manipulation, participation, reclamation, reconstruction, recovery, rehabilitation, reintroduction, remediation, repair, restitution, revegetation, rewilding, and ritual. I discuss all of these in this chapter.

Recovery is central to acts of ER. Recall that the Society for Ecological Restoration defines ER as assisting in the recovery of a degraded, damaged, or destroyed ecosystem. Degradation occurs when there is prolonged or gradual, but not severe, human disturbance such as deforestation, grazing, and killing specific organisms. Damage occurs when there is acute or severe human impact or disturbance that transforms an area of nature into an altered state. Destruction occurs when most or all organic matter is removed, sometimes accompanying human-caused topographical changes (Clewell and Aronson 2013: 38–39). Mountain-top removal mining, open-pit mining, and strip or surface mining are examples of this. These three kinds of disturbances can be placed on a continuum such that degradation can lead to damage which can lead to destruction.

Acts of revegetation, reintroduction, and assisted colonization/migration are more properly subtypes of activities that might occur when an area of nature is being restored. Revegetation involves replanting native species and might involve controlled burning and/or removing exotic plant species. Reintroduction typically involves rebuilding populations of locally extinct animal species. This approach can be controversial as seen in the reintroduction of North American gray wolves in Yellowstone National Park in 1995 and Mexican gray wolves in Arizona and New Mexico in 1998 when some stakeholders, including some ranchers, opposed such reintroductions. Assisted colonization/migration, sometimes called translocation, involves moving a species into a habitat where it doesn't currently exist or has not been known to exist in recent history, moving a species beyond its natural range to compensate for lost habitat due to something such as climate change, or using different species as functional analogs to replace locally extinct species. There is considerable debate about this (cf. Sandler 2009; Parker et al. 2010; Seddon 2010; Jørgensen 2011; Dalrymple and Moehrenschlager 2013).

Revegetation and reintroduction raise what Eric Katz originally called the substitution problem (Katz 1985): can nature be restored by swapping out parts (specific organisms), or can value be restored by substituting component parts and replacing them with functional (but different) component parts that restore health or overall function to a particular ecosystem? Assisted colonization involves what some call the new substitution problem: we can't know if an ecosystem is working after restoration or if value has been restored because we can't know what role substitute components play when new environments are formed—because of new species combinations—or whether substitute components maintain value in a new system that otherwise would not exist

(Lee et al. 2014). The new substitution problem is raised especially by the specter of climate change.

Some of the different definitions of the verb "restore" stipulate making compensation or amends for past wrongs. In the case of ER this means to repair an area impaired by human disturbance. To compensate and right the ecological wrongs, ER can be characterized as a reparation in terms of reparative or restorative justice, or as environmental moral repair (Oksanen 2008; Almassi 2017). ER also can be characterized as a form of restitution (Basl 2010). Richard Sylvan coins the term "extitution" which involves making re*stitution* for human *ex*tractive activities such as clear-cutting where ER activities might include recontouring the ground in addition to replanting trees (Sylvan 1994: 50). Restoration mandated by laws such as the US Surface Mining Control and Reclamation Act of 1977, which regulates the environmental effects of coal mining, is called reclamation—literally reclaiming the land. Left alone, many mining lands might never develop natural vegetal cover, and some people claim that reclamation of mining sites might more properly be labeled creation rather than reclamation or restoration (Baldwin 1994). The US Clean Water Act requires compensatory mitigation for projects that destroy wetlands; it might be problematic to label such mitigation as reclamation because wetlands are often not reclaimed in an impacted area but can be created anew somewhere else. Finally, remediation involves reducing or eliminating contaminants. This can be an act that is done to help prepare a site for restoration or simply to clean up a polluted site. Phytoremediation consists of removing toxic metals or other substances from soil or substrate by using plants that accumulate the unwanted substances; bioremediation consists of using bacteria to metabolize petroleum or using other microorganisms to remove contaminants from soils and water.

ER is sometimes called rehabilitation (Attfield 1994; Ladkin 2005). The Society for Ecological Restoration defines rehabilitation as the "reparation of ecosystem processes, productivity, and services rendered without regard to achieving the fullest possible reestablishment of preexisting biota in terms of its species composition and community structure" (Clewell and Aronson 2013: 203). The goal is to improve productivity, and there is an assumption that ecosystem services and former functionality can be recovered by substituting other species for past species—putting aside the substitution problems. Rehabilitation might be more apt in working landscapes where protection of nonhuman nature is mixed with continuing human activities. In Spanish-speaking and Portuguese-speaking countries, rehabilitation is sometimes called *recuperación*.

Rewilding is a type of massive landscape restoration that goes far beyond rehabilitation, reintroduction, and revegetation. Rewilding was first proposed in the 1990s as a way to preserve functional ecosystems and biodiversity on a continental scale across North America. The Wildlands Project (now the Wildlands Network) proposed restoring something along the order of 25% of North America in wilderness-style core protected areas surrounded by buffers and connected by corridors for wildlife movement (Soulé and Noss 1998). Variously known as continental conservation (Soulé and Terborgh 1999), rewilding as a conservation strategy (Foreman 2004), or Pleistocene rewilding (Donlan et al. 2006), smaller versions of rewilding have been proposed for parts of every continent except Antarctica.

Goals and Values of Ecological Restoration

Controversy exists over the goals of ER, and this controversy largely is driven by the values we seek to restore. End-state values include biodiversity (or some components of biodiversity such as particular species combinations), services such as ecosystem services or productivity, welfare (particularly human welfare), and aesthetics. Historical values include authenticity, fidelity, naturalness, and wildness. End-state and historical values will help determine the reference state or goal of a particular restoration project. Ecological integrity can be either an end-state or a historical value. Although it is often listed as a stand-alone value or goal of ER, ecological integrity might be better understood as a reference dynamic that sets process parameters for how to get to a reference state. If the restoration is successful, the reference state and reference dynamic will set a reference trajectory that will continue indefinitely into the future.

Determining the reference state is one of the central philosophical problems for ER. While it is impossible to literally restore a tract of nature to the condition it was in prior to human disturbance, there supposedly must be *some* connection between a future reference trajectory and a reference state prior to human disturbance in order for a series of actions to count as ecological *restoration*. This is clearly true for those concerned with historical environmental values, but it is also true for those concerned with end-state values because without some pre-disturbance reference state, we might be mired in a form of relativism in which almost anything goes.

There are at least five baseline problems associated with determining a reference state. First, ecological restoration is typically steered by natural historical values to determine a baseline (Lee et al. 2014). But what pre-disturbance state should serve as the reference and why? The forests of northern Minnesota in the Boundary Waters Canoe Area Wilderness (BWCAW) consisted of glacial ice replaced with tundra after the last ice age, followed by spruce forests, followed by jack pine and red pine forests, followed by paper birch and alder forests, followed by white pine forests, and finally followed by spruce, jack pine, and white pine forests. Daniel Botkin raises the problem of which of these states constitutes the natural state of the BWCAW for preservation, but, following deforestation, which of these states might best serve as the pre-disturbance baseline reference state for ER and why (Botkin 1990: 58–59)? To pick one particular state might seem arbitrary (Choi 2007).

The second baseline problem is the problem of mixed cultural/natural landscapes. Given problems of alternating over-populations and under-populations of elk in Yellowstone National Park, the US National Park Service commissioned five wildlife biologists to write a report about how best to manage national parks. Their 1963 report "Wildlife Management in National Parks" (better known as "The Leopold Report") recommended that national parks be maintained or recreated "as nearly as possible in the condition that prevailed when the area was first visited by the white man" so that national parks could represent vignettes of primitive America (Leopold et al. 1963: 32). National parks in the US were previously settled and/or utilized by Native American Indians and their ancestors. Virtually all landscapes on Earth outside of Antarctica were previously settled and/or utilized by people in the past. How should previous human impacts factor into "pre-disturbance" baseline reference states (Callicott 2002)? Attending to the issue of ER in mixed cultural and natural landscapes is the subject of a number of philosophical discussions (Brook 2018; Drenthen 2018), and these discussions undoubtedly will continue.

The third baseline problem concerns the issue of disturbances themselves. As per the disturbance paradigm that now prevails in ecology, nature itself is fundamentally characterized by nonhuman disturbances (Pickett et al. 1994). Even if we could sort out human from nonhuman disturbances, it is possible that in the time between impairment and restoration, a tract of nature could have been on a different nonhuman trajectory that is different from the new restoration-induced trajectory.

The fourth baseline problem concerns the problem of knowledge. Ecological knowledge of nonhuman and cultural pre-disturbance conditions of a particular area to be restored is rarely complete. Restoration ecologists might have to rely upon extrapolating from ecosystem remnants of a smaller or different area and/or secondary information ranging from previous natural history accounts to old photographs and journals. It can be difficult to generalize ecological conditions from smaller and/or different areas, and the accuracy of relying upon a picture or journal entry to determine a scientifically sound account of what an area of nature used to be is tricky at best. It's important to note that reference states or models gleaned from historical accounts provide more than just scientific-technical guidelines and that ER expresses both ecological and social values (Prior and Smith 2019).

The fifth baseline problem is associated with climate change. Many restoration strategies aim to restore native species to an area, or make the habitat of an area more favorable to native species. As such, ER tends to rely upon native species prioritization. Such prioritization becomes increasingly problematic as climate-driven habitat conditions change, forcing populations to move and ultimately shift their species habitat. Populations unable to move such as marmots living at or near the tops of mountains in the Rocky Mountains of North America or unable to move quickly enough such as quiver trees in the arid west of Namibia and South Africa face local extinction. Because of climate change, ecosystems are changing so fast that it's not clear that trying to return an ecosystem to its past baseline is desirable or even possible. Because there are so many possible moving targets of biota to restore, Andrew Light argues for rethinking of baselines in terms of functions, such that ER should attempt to recreate some aspect of the prior function of an ecological reference system (Light 2009). Stephen Jackson and Richard Hobbs suggest that when historical baselines cannot be met, ER might need to aim toward engineered and novel ecosystems (Jackson and Hobbs 2009). The relationship between novel ecosystems and ER needs to be explored further (Lennon 2017).

In light of these five baseline problems, how important is it to rely upon past history to determine the reference state for a particular act of ER? Eric Higgs argues that historical fidelity (or historicity) is one of the four keystone concepts of ER (Higgs 2003: 4–5). In addition to ecological integrity, wild design (that helps natural processes kick back in), and social and cultural values (ER as a focal practice), historical fidelity is important for three reasons. First, we're drawn to old baseline reference ecosystems for reasons of nostalgia; second, narrative continuity is important; and third, older baseline reference ecosystems give us time depth because they tend to be rarer, and we value rarity (pp. 143–158). Today's climate change-driven world, however, can move historical fidelity from being less of a fixed reference point of the past to more of an anchor or link to process-oriented configurations of the future. Accordingly, Higgs claims that there are seven types of historical fidelity for ER today: (1) ER that redresses specific disturbances, (2) ER that mimics art restorations, such as preserving a particular battlefield of the past, (3) ER that encompasses a historical range of variability, (4) ER that resets ecosystems to a condition of greater ecological integrity or health, (5) ER that serves as a reference point for how ecosystems have functioned in the past and might function under new conditions, (6) ER that serves to embody historical human practices and beliefs, and (7) ER that serves as a governor on our exuberant ambitions (Higgs 2012: 91–93).

The accelerated rate and magnitude of climate change-driven ecological changes are leading a number of people to rethink the wisdom of historical fidelity for ER. Young D. Choi argues that ER should be re-envisioned as futuristic restoration (Choi 2004). Choi outlines four goals for futuristic restoration: (1) ecosystems should be established that are able to be sustained in future environments, (2) there should be multiple alternative goals and trajectories for endpoints that are unpredictable, (3) there should be a focus on the rehabilitation of ecosystem functions instead of recomposing particular species or landscape surface cosmetics, and (4) it should be acknowledged that futuristic restoration is a value-laden applied science set within economically and socially acceptable frameworks (Choi 2007). Emma Marris argues that ER should jettison the idea of a historic baseline and replace this with a new goal of designer ecosystems (Marris 2011: 126–127). Before we go this far, however, Marion Hourdequin argues that there is a middle ground between ER grounded in historical fidelity of the past and something entirely different grounded in the future (Hourdequin 2013). This middle ground articulates a story between what has happened in the past and what ought to happen in the future:

> What we ought to do, then, depends on what came before. History has value as part of the narrative we are working to construct: it is not just what *informs* what we ought to do by

helping us see the best means to our current ends; it constrains what we do by being an integral part of the whole story which itself is assessable from an ethical point of view.

(Hourdequin 2013: 124, italics in the original)

Drawing from narrative ethics, Hourdequin develops a narrative account of ER that retains historical authenticity as part of the story for any particular place where ER plays a role in the continuation of the storied place across time and into the future.

If there is still a role for historical fidelity in ER, this raises the philosophical issue of ontological authenticity: what is the fundamental, true state of nature? People such as Robert Elliot claim that it is naturalness (Elliot 1997). However, the idea of naturalness has fallen upon hard times (Aplet and Cole 2010). The term "natural" is ambiguous, and it can imply a state of nature that existed before humans, a state that is increasingly difficult to find as we discover ubiquitous human impacts for millennia on every continent except Antarctica. Stephen Woodley argues that ecological integrity should replace naturalness as a restoration goal, defining ecological integrity in terms of ecosystems that are "whole, complete, intact, sound, and unimpaired" (Woodley 2010: 121). It's not clear, however, that ecological integrity as an end-state environmental value completely jettisons naturalness as a historical value when Woodley defines integrity in terms such as ecosystems having complex, intact trophic levels and having a full complement of native species. Erika Zavaleta and Stuart Chapin argue that resilience should replace naturalness, but resilience is defined in part by native biodiversity—another possible nod to naturalness (Zavaleta and Chapin 2010). Nigel Dudley argues that restoration increasingly will become more important in today's humanizing world, but naturalness still has an important role to play:

> [N]aturalness has its own unique values and by cutting it out of the picture we are left planning and implementing ecosystem conservation in the absence of any yardstick against which to measure progress. If we want to avoid sliding inexorably towards ecosystems dominated by weeds, brown rats and feral pigeons, and if we want to continue drawing on ecosystem services from natural ecosystems, we need to be able to understand concepts of naturalness in a rapidly changing world, which means that we also need a definition that fits into the 21st century.
>
> *(Dudley 2011: 154)*

Andre Clewell underscores Dudley's plea for naturalness by arguing that natural authenticity should be a restoration goal (Clewell 2000). Understanding naturalness to denote what has not been intentionally planned or cultured, Clewell argues that the primary task of restoration should be to initiate or accelerate natural autogenic processes that allow for self-renewal. This is not historical authenticity for Clewell because such authenticity remains tethered too tightly to a past that cannot always be restored. Rather, natural authenticity is future-oriented such that naturalness washes back into an ecosystem after restoration and persists as much as possible on its nonhuman own. For example, controlled burns might be appropriate in a fire-suppressed ecosystem, but the controlled burns should be considered temporary and should end as natural lightning-caused fire regimes wash back into the ecosystem.

Moving on to the reference dynamic, the Society for Ecological Restoration lists nine attributes of restored ecosystems. While the baseline reference state is the goal of a particular restoration, these nine attributes are supposed to characterize any given restored ecosystem.

1 The restored ecosystem contains a characteristic assemblage of the species that occur in the reference ecosystem that provides appropriate community structure.
2 The restored ecosystem consists of indigenous species to the greatest practicable extent.

3   All functional groups necessary for the continued development and/or stability of the restored ecosystem are represented, or if they are not, the missing groups have the potential to colonize by natural means.

4   The physical environment of the restored ecosystem is capable of sustaining reproducing populations of the species necessary for its continued stability or development along the desired trajectory.

5   The restored ecosystem apparently functions normally for its ecological stage of development, and signs of dysfunction are absent.

6   The restored ecosystem is suitably integrated into a larger ecological matrix or landscape, with which it interacts with abiotic and biotic flows and exchanges.

7   Potential threats to the health and integrity of the restored ecosystem from the surrounding landscape have been eliminated or reduced as much as possible.

8   The restored ecosystem is sufficiently resilient to endure the normal possible stress events in the local environment that serves to maintain the integrity of the ecosystem.

9   The restored ecosystem is self-sustaining to the same degree as its reference ecosystem, and has the potential to persist indefinitely under existing environmental conditions (SER 2004: 3–4).

There are at least four things to note about this list. First, the second item refers to indigenous (or native) species. There are significant philosophical problems concerned with defining exotic species and their values that remain to be resolved (Woods and Moriarty 2001; Simberloff 2012). And, as discussed above, native species prioritization becomes increasingly problematic in light of climate change. Second, several items refer to normal functions and normal stress events. "Normal" is typically interpreted in terms of pre-human or nonhuman disturbance regimes, and to make sense of this, ER seems to rely upon a definition of what is natural in a nonhuman sense. Third, the seventh item refers to ecosystem health and integrity. Ecosystem health is further defined as dynamic attributes expressed in normal ranges which bring us once again to naturalness (SER 2004: 7). Ecosystem integrity refers to species composition, community structure, and, once again, "normal" functioning (SER 2004: 7). Fourth, there are several references to resilient, self-sustaining, and sustainable ecosystems which reflect the ecological trajectory. Are there time scales associated with this trajectory, and how much continual human action to maintain this trajectory is acceptable? Insofar as philosophers actually engage with what restoration ecologists say and do, there are undoubtedly further issues to discuss in regard to the values and goals of ER.

The Elliot-Katz Challenge for Ecological Restoration

Philosophers Robert Elliot and Eric Katz raised a number of important challenges to ER in the 1980s and 1990s. In 1982 Elliot wrote the article "Faking Nature" in which he argued that natural values could not be restored through an act of ER because what is significant about naturalness is its nonhuman origin and causal continuity. An act of ER abrogated such origin and continuity. He likened tracts of restored nature to art forgeries—hence the concept of "faking nature." In his 1997 book of the same title, he more fully developed his argument that the value-adding property of naturalness cannot be restored because of a human-caused break with a natural area's causal continuity with the past. He did not condemn all acts of ER but claimed that if such acts aim at the restoration of naturalness, they fail because naturalness cannot be restored or reconstructed by people.

Katz took Elliot's argument further. Similar to Elliot's claim that restored nature was akin to an art forgery, Katz called ER "the big lie" (Katz 1992a) and argued that acts of ER necessarily asserted the human domination of nature (Katz 1992b). Somewhat confusingly, Katz equivocated between the properties of naturalness and wildness and claimed that what is significant about the former is that natural entities pursue independent, nonhuman, and unplanned courses of development; what is significant about the latter is nonhuman authenticity and autonomy. Humans could restore neither naturalness nor wildness, and tracts of nature that were "restored" were nothing more than human artifacts. Katz has developed his position in more detail and, more recently, arguing that the key issue is the meaning of ER: it fundamentally connotes an artifactual system that dominates and usurps the nonhuman autonomy of nature (Katz 2012, 2015). After ER, human artifactuality forever remains because nature has been set by a different, human trajectory. There is no such thing as a positive act of ER for Katz.

Katz and Elliot lead us into a paradox. We begin by picking an area of nature that has been degraded, damaged, or destroyed. ER seems to be preferable to further degradation, damage, or destruction, or to merely leaving the area on its own following human impacts. Although some people disagree (Mathews 1999), it also seems preferable to autogenic recovery—often nothing more than secondary succession—that takes place without direct human intervention because such recovery might be too slow and/or inadequate. For many people, ER seems to be aimed at restoring naturalness, wildness, or simply the way the area existed before human impacts. But if Elliot and Katz are right, then naturalness, wildness, or the pre-disturbance state already has been lost because of previous human impacts, and none of these can be restored by further human actions. While ER seems to be preferable or required to right human-caused wrongs to nature, ER ultimately fails to restore what is significant about nature—its naturalness, wildness, or nonhuman pre-disturbance existence.

Philosophers have proposed a variety of different ways to address the Elliot-Katz paradox. One thing to note up front is that this paradox and some responses to it might equivocate between descriptions and prescriptions—descriptions of what ER is and assessments of whether acts of ER are good or bad in a social or moral sense (Light 2009).

Alastair Gunn takes issue with Elliot's claim that restored nature is like an art forgery (Gunn 1991). Unlike restored tracts of nature, art forgeries can exist simultaneously with the original artwork. Gunn argues that ER is an obligatory response to environmental destruction and trumps the problem that nature cannot be restored exactly to the way it was before; he focuses on alleviating previous human impacts in a forward-looking sense rather than an Elliot-Katz backward-looking sense. Robin Attfield argues that we should bite the Elliot-Katz bullet by admitting that what is historically distinct about nature cannot be restored, and we should redefine what we are doing as rehabilitation—instead of restoration—to increase biodiversity and improve habitat following previous human disturbance (Attfield 1994). Sylvan offers another way of saying this: rather than trying to restore nature to the way it was when it was pristine before human impacts, the correct baseline for ER should be the point of environmental destruction, such that ER makes restitution to destroyed nature by helping nature move past this point (Sylvan 1994). For example, instead of trying to restore a degraded or damaged grassland to the baseline condition it was in prior to degradation or damage, we should simply make the grassland more habitable for more organisms, populations, and species in the future. William Throop argues that human actions themselves don't forever destroy wildness and that wildness can reemerge over time (Throop 1997). Throop proposes a pattern criterion as a goal for ER: rather than trying to restore wildness itself, the goal of ER should be to restore similar patterns of species that are in characteristic relations to each other and to the abiotic elements of an ecosystem. Throop also describes ER in terms of a healing metaphor in which humans assist an ecosystem itself to heal, in contrast to using a gardening or an engineering metaphor that puts misguided emphasis on what is done *to* an ecosystem rather than *with* an ecosystem (Throop 2012).

I make distinctions between naturalness, wildness, and freedom and argue that human-assisted ER—to the degree that it moves from active to passive assistance—can remove barriers to freedom for animals and plants, allow wildness to reassert itself over various time scales, and allow naturalness to wash back in over much longer time scales (Woods 2005, 2017: Chapter 9). For example, a restoration project began in 2002 to restore at least three miles of the original stream channel of Karnowsky Creek and adjacent wetlands in the US state of Oregon. After excavating the creek's meandering channel, plugging old drainage ditches, and transplanting logs of large trees to the channel and floodplain, shrubs, trees, and wetland vegetation were planted. By 2008, replanted alders and willows had reached what restoration ecologists called a "free to grow" stage. Without further human assistance, beaver returned to the creek, meander cutoffs in the creek formed, coho smolt salmon returned, and a seasonal flooding regime returned (Clewell and Aronson 2013: 187–189). The Karnowsky Creek restoration began as an active process in 2002 using bulldozers and helicopters, but by 2008 the restoration was proceeding by itself without human assistance. Freedom and wildness were reasserted by the flora, fauna, and creek hydrology, and over time naturalness will hopefully wash back into the riparian area.

Andrew Light argues that a distinction between malicious and benevolent restoration can help us move past the Elliot-Katz paradox (Light 2000). The context for Elliot's original "Faking Nature" discussion was a mining company's plan to restore beach sands after rutile mining, and Elliot was concerned that ER itself was being used to justify the mining operation as long as the beach was "restored." Light claims that this is a malicious case of ER, and the ER itself should not justify the mining. In other cases where the promise of ER itself was not used to justify environmentally destructive acts, ER can have some positive value insofar as it repairs destruction. Katz's claim that all acts of ER necessarily dominate nature and are negative is problematic for Light because acts of ER do not necessarily create negative values and can, in contrast, create positive values. For example, in the case of the rutile mining, if the mining company did not try to justify rutile mining by claiming that it would restore any damage caused by the mining and simply mined rutile and left, any subsequent acts of ER to repair the damage caused by the mining would add positive values. Donna Ladkin agrees that there can be benevolent acts of ER and develops another account of ER as rehabilitation (Ladkin 2005). Acts of ER can consist of non-dominating, collaborative human-land relationships that assist with the rehabilitation of ecosystem health; there can be a return to an ecological alignment, and naturalness might wash back into an ecosystem.

Michael Scoville makes a useful distinction between historical and end-state environmental values that helps us classify the above responses to the Elliot-Katz ER paradox (Scoville 2013). Both Elliot and Katz champion historical environmental values insofar as what is valuable about nature is its nonhuman origin and causal continuity with the past. Gunn, Attfield, Throop, Light, and Ladkin champion end-state environmental values when ER helps alleviate previous human impacts to get to a desired future state (Gunn), makes an area more biodiverse and habitable for plants and animals (Attfield), restores similar patterns of species relations (Throop), benevolently repairs destruction (Light), and rehabilitates ecosystem health (Ladkin). Throop and I champion modified historical environmental values such that these values of naturalness and/or wildness (and freedom for me) can wash back into restored tracts of nature. Sylvan offers a historical value alternative by redefining the historical trajectory to the point of destruction and offering ER as restitution. The Elliot-Katz paradox and responses to it help show different accounts of the values of nature and the goals of ER.

Ecological Restoration, Wilderness Preservation, and Environmental Paradigms

A debate began in the 1990s over the relationship between wilderness preservation and ER (Baldwin et al. 1994). This debate was related to a larger critique of the concept of wilderness and the practice of wilderness preservation (Callicott and Nelson 1998). The concept of wilderness that connoted quintessential nonhuman nature was criticized from opposite directions for being inconsistent with naturalism and for being inconsistent with social constructivism. The concept of wilderness that denoted natural areas empty of people was criticized as naïve in light of past human impacts virtually everywhere on Earth, and it was problematically racist and ethnocentric given its origins in colonizing European, Euro-American, and Euro-Australian cultures. The practice of wilderness preservation was problematized as an ecologically naïve way of freeze-framing nature, and morally condemned for relying upon imperiums of power that displaced, dispossessed, removed, and killed native peoples to create empty places for wilderness designation.

Proponents of ER such as William Jordan and Frederick Turner offer ER as a replacement paradigm for wilderness preservation (Jordan 1994; Turner 1994). The latter is criticized for being passive, allowing no place for human interaction, freeze-framing nature, and being inadequate in light of persistent human impacts that cannot be held at bay by merely designating some nonhuman areas as wilderness. ER is championed for being active, offering positive human-nature interactions, embodying a garden metaphor to help nature grow, and being the most realistic way to take care of nature that is under persistent human attack. Insofar as wilderness preservation might have been the ruling motif of the traditional environmental movement, ER might help inform a new paradigm for environmentalism for the 21st century (Minteer and Pyne 2012). Jordan goes so far as to reduce preservation to restoration: "As *the* means by which preservation is actually achieved, it [ER] will determine both the existence and the quality or character of the classic landscapes of the future" (Jordan 2003: 108).

Some wilderness proponents counter these ER critiques of wilderness preservation by arguing that ER might be based upon or embody a domination model of nature. Examples of this are not hard to find. Turner argues that we should not leave nature alone because we are the future, purpose, and lords of nature; we should understand restored landscapes as invented landscapes that we actively create and maintain much as a gardener maintains a garden (Turner 1994). Carl Pletsch argues that we should abandon the idea that nature is independent and instead exert human sovereignty over nature through ER (Pletsch 1994). John Harper argues that ER works by taking nature apart, much as a watchmaker might take a watch apart, in order to restore and control nature (Harper 1987). G. Stanley Kane comments:

> By holding that humans are the lords of creation, restoration metaphysics tolerates no enclaves anywhere kept free of human domination and control; by maintaining that maker's knowledge gives us the best understanding of things, restoration epistemology tolerates no mystery. But mystery and the preservation of places kept free of human domination are at the heart of the preservationist program and philosophy.
>
> *(Kane 1994: 83)*

For Kane, wilderness preservation serves as a foil to the domination of nature. Sheila Lintott defends the so-called passivity of preservation by criticizing restorationists that reduce nature to a passive presence to be manipulated and argues that passive preservation—and not active ER—can be a strength and a more appropriate way to engage with an independent, nonhuman nature (Lintott 2011). Colette Palamar offers an ecofeminist critique of ER. She agrees that the practice of ER can be based on tacit acceptance of oppression and domination insofar as it forces nature into a specifically human vision (Palamar 2006). She offers what she calls "restora*shyn*" as a more benign, ecofeminist version of ER that acknowledges nature as a distinct and active player with its own destiny.

In some respects, wilderness preservation and ER can be compatible and complementary. Wilderness can provide baseline reference states for ER (Willeke 1994), the wilderness management concept of limits of acceptable change can help guide ER efforts (Brunson 2000), preservation can work together with ER for regional conservation (Noss 1994), and ER can help alter trajectories of decline in impaired wilderness areas to help reset self-willed, autochthonous landscapes (Oelschlaeger 2002). The larger issue today, however, is whether ER and wilderness preservation are still meaningful in today's world of the Anthropocene (Marris 2011). I have argued that wilderness preservation is needed more now than ever (Woods 2017), and others have argued that ER is also needed now more than ever (Covington and Vosick 2015).

Ecological Restoration and Restoration Ecology

ER is the practice of restoring impaired areas of nature, while restoration ecology is the science upon which ER is based. Restoration ecologists are the community of practitioners who provide concepts, methodologies, models, and tools that are used for ER. Especially with the advent of the Society for Ecological Restoration, restoration ecologists have become a professional society of practitioners. Many restoration ecologists in the US trace their roots to the Curtis Prairie at the University of Wisconsin Arboretum in Madison and the restoration project begun there in the mid-1930s. The fact that the famous ecologist and nature writer Aldo Leopold was involved with this project and with his own ecological restoration project at his family farm near Baraboo, Wisconsin in the 1930s almost makes Leopold into a sort of American patron saint of restoration ecology. Australian restoration ecologists can trace their roots to restoration projects in New South Wales at Lumley Park in Alstonville and at Broken Hill, restoration projects that also began in the 1930s (Jordan and Lubick 2011: 72).

Philosophers of science have paid scant attention to restoration ecology, but the philosophy of science can be used pedagogically to ask questions about the science of restoration ecology. First, what exactly do restoration ecologists restore, or what natural entities do they claim to use? Preliminary answers to this question have been offered above. William Jordan and George Lubick claim that ecological restoration should be regarded first and foremost as ecocentric restoration that is focused on restoring land for nonanthropocentric reasons in contrast to what they call meliorative land management which makes landscapes better for people (Jordan and Lubick 2011). This distinction is not without controversy.

Second, what constitutes restoration ecology knowledge, and how is restoration ecology related to ecology as a whole? Restoration ecologists give various answers to this question in an anthology edited by William Jordan et al. (1987b). While many regard restoration ecology as a practical science that focuses on techniques for restoring nature, Jordan et al. (1987a) claim that restoration ecology also offers techniques for doing basic ecological research. That is, empirical ER experiments can test both techniques for restoring nature and basic ideas about ecological concepts themselves. A.D. Bradshaw argues that restoration ecology tests basic theories and models about how nature is organized and constitutes an acid test for ecology (Bradshaw 1987). Jared Diamond argues that the science of ecology puts constraints on restoration ecology because the kinds of experiments that can give far-reaching insights into how nature works would be illegal, immoral, or impractical because of the need to test perturbations on large temporal and spatial scales (Diamond 1987).

Third, is there a distinctive restoration ecology method, or what distinctively do restoration ecologists do? Restoration ecologists Clewell and Aronson argue that ER is distinct from ecosystem management because the latter consists of continual attempts to maintain various levels of ecological integrity, while the former is a shorter-term attempt to help set a trajectory that ecosystem managers might use (Clewell and Aronson 2013: 202). Some restorations consist of fabrication or creation in which a previously occurring ecosystem is entirely replaced following radical changes in physical conditions, such as when dune ecosystems and tidal marshes are created on islands by depositing dredge spoils along riverbanks (p. 207). Some restorations can be classified as ecological engineering in which living organisms, natural materials, and the physical-chemical environment are used to solve technical problems and achieve specific human goals; examples include bioremediation, compost engineering, and phytoremediation (pp. 209–210). Eric Higgs argues that good ER consists of focal practices directed toward restoring ecological integrity and historical fidelity, steered by intention or a wild design to allow naturalness to wash back into an ecosystem.

> [F]ocal restoration [consists of] practices that create a stronger relationship between people and natural processes…. A focal restoration is one that centers the world of the restorationist, expresses the commanding presence of nature, and demonstrates continuity between that particular act of restoration and other activities on the landscape.
>
> *(Higgs 2003: 242)*

Carolina Murcia and James Aronson propose a new classification system for types of restorations: (1) careless tinkering: using haphazard trial and error to enact short-term repair, (2) amateur intelligent tinkering: using basic inquiry and hypothesis testing to address a site-specific problem, (3) professional intelligent tinkering: using scientific methodology to address a site-specific problem, and (4) a scientific approach: using theory testing to extrapolate and advance generalizable knowledge (Murcia and Aronson 2014). They argue that ER is best practiced as professional intelligent tinkering to institutionalize it (as professional) and to distinguish it from scientific research. ER for Murcia and Aronson should consist of specific case practices based on logical, careful, well-documented, ecological intuition where there is no intention to extrapolate information to other sites or systems but simply to get a restoration job done. This sounds similar to the approach of conservation biologists who claim that because of the immediacy of extinctions and other human harms to nature, they are focused with on-the-ground approaches to getting the job of conserving biodiversity done, sometimes with incomplete scientific knowledge. Restoration ecology provides fertile ground for future work in the philosophy of science.

Ecological Restoration and People

What is the human relationship to nature? Are humans a part of nature, apart from nature, or both? How does ER embody or not embody relationships between people and nature? While not explicit, these questions have been lying beneath the surface of much of the above discussion.

Many people argue that ER is a way for people to participate with nature. Insofar as nature is not merely a passive recipient awaiting restoration actions, ER might be construed as a type of dialogue with nature (Tanasescu 2017). Beyond the restoration practitioners themselves, many restoration projects involve public participation with various levels of voluntariness ranging from prison inmates and students to individual volunteers. Unlike wilderness preservation, there is a positive place for people in nature, and public participation can create positive human-nature values. Light and Higgs argue that the positive values of public participation in restoration projects have an inherent potential for participatory, democratic citizenship; restoration projects conducted solely by professionals with no opportunity for public participation are instances of bad restoration (Light and Higgs 1996). William Throop and Rebecca Purdom provide a counterargument that public participation in restoration projects in wilderness settings is usually inappropriate because it adds more human influence than necessary and is contrary to letting wilderness ecosystems follow their own inherent, natural, and wild trajectories (Throop and Purdom 2006).

For Higgs (2005), public participation in restoration also raises what is sometimes called the "two cultures problem" that originally stems from Charles Snow's (1960) division of people into a scientific elite versus a political and cultural elite educated in the humanities, ignorant and even contemptuous of science. Higgs notes that the two cultures of restoration consist of restoration ecology thoroughly grounded in science and a broad constituency of non-science ER that includes cultural, economic, and political dimensions. Higgs is worried that the practice of ER will become narrowed to restoration ecology with a focus on efficient, technological restoration projects that lock out everyone except restoration ecologists; he argues that restoration needs more of a moral center that bridges the two cultures of restoration ecology and non-science ER. The two cultures problem can be seen in the Chicago restoration controversy in the 1990s when the Natural Areas Management Program was working to restore 7,000 acres of mostly forested land holdings into oak savanna and tallgrass prairie (Gobster 2000). There was public outcry over the cutting of trees and accusations that the restoration project was being directed by a small private elite. More opportunities for public dialogue might have prevented the controversy.

In addition to arguing for more public participation in restoration projects, Higgs argues that good restoration is a focal practice that allows us to be mindful and fully engaged with restoration projects and that is more than mere technological mastery (Higgs 2003). In a related vein Jordan argues that good restoration can allow for ritual participation with nature that helps us heal ourselves and the damage we cause to nature (Jordan 2003). He also claims more controversially that ER is a way for us to confront the existential shame we have brought upon ourselves for harming nature. Ned Hettinger argues that while ER is sometimes necessary, we should not forget that it can be grandiose, hubristic, and ultimately regrettable (Hettinger 2021). Other philosophers characterize ER in more positive terms of environmental virtues such as humility, openness, patience, and self-restraint (Sandler 2012; Throop 2012).

Finally, Light and Higgs draw attention to the politics *of* restoration and politics *in* restoration (Light and Higgs 1996). The latter concerns the political use to which ER is put. The former concerns the political context in which ER is done. To serve as a foil against the increasing corporatization of professional ER projects, ER needs more democratic participation. ER might help define the fundamental human-nature relationship.

Conclusion

I have discussed how to describe acts of ER, the goals and values of ER, the Elliot-Katz challenge for ER, how ER is related to wilderness preservation, how ER is related to restoration ecology, and the relationship between people and ER. Consider again the CERP Everglades restoration project I introduced at the beginning of this chapter. As of 2015, the CERP is behind schedule. Meaningful water redistribution seems unlikely until at least 2035 with completion of the project sometime after that. What is the best way to describe a project such as this? Are we trying to recover lost water for the Everglades, repair the ecosystems of south Florida, reconstruct historic water flows, ecologically reengineer water flows, rehabilitate the Everglades, or simply provide a reliable source of potable water for the human residents of south Florida (Light 2012: 112)? Is CERP an act of restitution, or is it another way for humans to dominate nature? What immediate goals steer CERP, and what is CERP's ultimate goal? Are we trying to restore end-state or historical environmental values? Contra Elliot and Katz, can we restore natural water patterns or native ecosystems? Does the attempt to restore the Everglades ecosystems and historic water flows help or undermine wilderness preservation in the federally designated Marjory Stoneman Douglas Wilderness of Everglades National Park? What are restoration ecologists doing in south Florida? Are they attempting to set a water flow trajectory that ecosystem managers in the Everglades can use (Clewell and Aronson 2013), practicing professional intelligent tinkering (Murcia and Aronson 2014), or engaging in focal practices directed toward restoring ecological integrity and historical fidelity, steered by intention or wild design (Higgs 2003)? Finally, what is the proper role for the public in CERP—should more active participation of people be encouraged, or should people stand by and watch restoration ecologists do their thing? Philosophers by themselves cannot provide fully adequate answers, but they can at least help identify the questions.

Related Topics

10. Wilderness
43. Remediation
45. Assisted Migration and Reintroduction
47. Rewilding
48. Novel Ecosystems
59. Adaptive Management
62. Community Participation

References

Almassi, B. (2017) "Ecological Restorations as Practices of Moral Repair," *Ethics & the Environment* 22: 19–40.

Aplet, G.H. and Cole, D.N. (2010) "The Trouble with Naturalness: Rethinking Park and Wilderness Goals," in Cole, D.N. and Yung, L. (eds) *Beyond Naturalness: Rethinking Park and Wilderness Stewardship in an Era of Rapid Change*, Washington, DC: Island Press, pp. 12–29.

Attfield, R. (1994) "Rehabilitating Nature and Making Nature Habitable," in Attfield, R. and Belsey, A. (eds) *Philosophy and the Natural Environment*, Royal Institute of Philosophy Supplement: 36, New York: Cambridge University Press, pp. 45–57.

Baldwin, A.D. Jr. (1994) "Rehabilitation of Land Stripped for Coal in Ohio—Reclamation, Restoration, or Creation?" in Baldwin, A.D. Jr., de Luce, J., and Pletsch, C. (eds) *Beyond Preservation: Restoring and Inventing Landscapes*, Minneapolis: University of Minnesota Press, pp. 181–191.

Baldwin, A.D. Jr., de Luce, J., and Pletsch, C. (eds) (1994) *Beyond Preservation: Restoring and Inventing Landscapes*, Minneapolis: University of Minnesota Press.

Basl, J. (2010) "Restitutive Restoration: New Motivations for Ecological Restoration," *Environmental Ethics* 32: 135–147.

Botkin, D.B. (1990) *Discordant Harmonies: A New Ecology for the Twenty-first Century*, New York: Oxford University Press.

Bradshaw, A.D. (1987) "Restoration: an Acid Test for Ecology," in Jordan, W.R. III, Gilpin, M.E., and Aber, J.D. (eds) *Restoration Ecology: A Synthetic Approach to Ecological Research*, New York: Cambridge University Press, pp. 23–29.

Brook, I. (2018) "Restoring or Re-storying the Lake District: Applying Responsive Cohesion to a Current Problem Situation," *Environmental Values* 27: 427–445.

Brunson, M.W. (2000) "Managing Naturalness as a Continuum: Setting Limits of Acceptable Change," in Gobster, P.H. and Hull, R.B. (eds) *Restoring Nature: Perspectives from the Social Sciences and Humanities*, Washington, DC: Island Press, pp. 229–244.

Callicott, J.B. (2002) "Choosing Appropriate Temporal and Spatial Scales for Ecological Restoration," *Journal of Bioscience* 27: 409–420.

Callicott, J.B. and Nelson, M.P. (eds) (1998) *The Great New Wilderness Debate: An Expansive Collection of Writings Defining Wilderness from John Muir to Gary Snyder*, Albany: University of Georgia Press.

Choi, Y.D. (2004) "Theories for Ecological Restoration in Changing Environment: Toward 'futuristic' Restoration," *Ecological Research* 19: 75–81.

Choi, Y.D. (2007) "Restoration Ecology to the Future: A Call for New Paradigm," *Restoration Ecology* 15: 351–353.

Clewell, A.F. (2000) "Restoring for Natural Authenticity," *Ecological Restoration* 18: 216–217.

Clewell, A.F. and Aronson, J. (2013) *Ecological Restoration: Principles, Values, and Structure of an Emerging Profession* (2nd ed), Washington, DC: Island Press.

Covington, W.W. and Vosick, D.J. (2015) "Restoration, Preservation, and Conservation: An Example for Dry Forests of the West," in Minteer, B.A and Pyne, S.J. (eds) *After Preservation: Saving American Nature in the Age of Humans*, Chicago: University of Chicago Press, pp. 133–145.

Dalrymple, S.E. and Moehrenschlager, A. (2013) "'Words matter.' A Response to Jørgensen's Treatment of Historic Range and Definitions of Reintroduction," *Restoration Ecology* 21:156–158.

Diamond, J. (1987) "Reflections on Goals and on the Relationship between Theory and Practice," in Jordan, W.R. III, Gilpin, M.E., and Aber, J.D. (eds) *Restoration Ecology: A Synthetic Approach to Ecological Research*, New York: Cambridge University Press, pp. 329–336.

Donlan, C.J., Berger, J., Bock, C.E., Bock, J.H., Burney, D.A., Estes, J.A., Foreman, D., Martin, P.S., Ro-emer, G.W., Smith, F.A., Soulé, M.E., and Greene, H.W. (2006) "Pleistocene Rewilding: An Optimistic Agenda for Twenty-First Century Conservation," *American Naturalist* 168: 660–681.

Drenthen, M. (2018) "Rewilding in Layered Landscapes as a Challenge to Place Identity," *Environmental Values* 27: 405–425.

Dudley, N. (2011) *Authenticity in Nature: Making Choices about the Naturalness of Ecosystems*, London: New York: Earthscan.

Elliot, R. (1982) "Faking Nature," *Inquiry* 25: 81–93.

Elliot, R. (1997) *Faking Nature: The Ethics of Environmental Restoration*, New York: Routledge.

Foreman, D. (2004) *Rewilding North America: A Vision of Conservation for the 21st Century*, Washington, DC: Island Press.

Gobster, P.H. (2000) "Introduction: Restoring Nature: Human Actions, Interactions, and Reactions," in Gobster, P.H. and Hull, R.B. (eds) *Restoring Nature: Perspectives from the Social Sciences and Humanities*, Washington, DC: Island Press, pp. 1–19.

Gunn, A.S. (1991) "The Restoration of Species and Natural Environments," *Environmental Ethics* 13: 291–310.

Harper, J.L. (1987) "The Heuristic Value of Ecological Restoration," in Jordan, W.R. III, Gilpin, M.E., and Aber, J.D. (eds) *Restoration Ecology: A Synthetic Approach to Ecological Research*, New York: Cambridge University Press, pp. 35–45.

Hettinger, N. (2021) "Nature Restoration as a Paradigm for the Human Relationship with Nature," in Thompson, A. and Bendik-Keymer, J. (eds) *Ethical Adaptation to Climate Change: Human Virtues of the Future*, Cambridge, MA: The MIT Press, pp. 27–46.

Higgs, E. (2003) *Nature by Design: People, Natural Process, and Ecological Restoration*, Cambridge, MA: The MIT Press.

Higgs, E. (2005) "The Two-Culture Problem: Ecological Restoration and the Integration of Knowledge," *Restoration Ecology* 13: 159–164.

Higgs, E. (2012) "History, Novelty, and Virtue in Ecological Restoration," in Thompson, A. and Bendik-Keymer, J. (eds) *Ethical Adaptation to Climate Change: Human Virtues of the Future*, Cambridge, MA: The MIT Press, pp. 81–101.

Hourdequin, M. (2013) "Restoration and History in a Changing World: A Case Study in Ethics for the Anthropocene," *Ethics & the Environment* 18: 115–134.

Jackson, S.T., and Hobbs, R.J. (2009) "Ecological Restoration in the Light of Ecological History," *Science* 325: 567–569.

Jordan, W.R. III (1994) "'Sunflower Forest': Ecological Restoration as the Basis for a New Environmental Paradigm," in Baldwin, A.D. Jr., de Luce, J., and Pletsch, C. (eds) *Beyond Preservation: Restoring and Inventing Landscapes*, Minneapolis: University of Minnesota Press, pp. 17–34.

Jordan, W.R. III (2003) *The Sunflower Forest: Ecological Restoration and the New Communion with Nature*, Berkeley: University of California Press.

Jordan, W.R. III, Gilpin, M.E., and Aber, J.D. (1987a) "Restoration Ecology: Ecological Restoration as a Technique for Basic Research," in Jordan, W.R. III, Gilpin, M.E., and Aber, J.D. (eds) *Restoration Ecology: A Synthetic Approach to Ecological Research*, New York: Cambridge University Press, pp. 3–21.

Jordan, W.R. III, Gilpin, M.E., and Aber, J.D. (eds) (1987b) *Restoration Ecology: A Synthetic Approach to Ecological Research*, New York: Cambridge University Press.

Jordan, W.R. III and Lubick, G.M. (2011) *Making Nature Whole: A History of Ecological Restoration*, Washington DC: Island Press.

Jørgensen, D. (2011) "What's History Got to Do with It? A Response to Seddon's Definition of Reintroduction," *Restoration Ecology* 19: 705–708.

Kane, G.S. (1994) "Restoration or Preservation? Reflections on a Clash of Environmental Philosophies," in Baldwin, A.D. Jr., de Luce, J., and Pletsch, C. (eds) *Beyond Preservation: Restoring and Inventing Landscapes*, Minneapolis: University of Minnesota Press, pp. 69–84.

Katz, E. (1985) "Organism, Community, and the 'Substitution Problem'," *Environmental Ethics* 7: 241–256.

Katz, E. (1992a) "The Big Lie: Human Restoration of Nature," *Research in Philosophy and Technology* 12: 231–241.

Katz, E. (1992b) "The Call of the Wild: The Struggle against Domination and the Technological Fix of Nature," *Environmental Ethics* 14: 265–273.

Katz, E. (2012) "Further Adventures in the Case against Ecological Restoration," *Environmental Ethics* 34: 67–97.

Katz, E. (2015) *Anne Frank's Tree: Nature's Confrontation with Technology, Domination, and the Holocaust*, Cambridge: White Horse Press.

Ladkin, D. (2005) "Does 'Restoration' Necessarily Imply the Domination of Nature?" *Environmental Values* 14: 203–219.

Lee, A., Hermans, A.P., and Hale, B. (2014) "Restoration, Obligation, and the Baseline Problem," *Environmental Ethics* 36: 171–186.

Lennon, M. (2017) "Moral-Material Ontologies of Nature Conservation: Exploring the Discord Between Ecological Restoration and Novel Ecosystems," *Environmental Values* 26: 5–29.

Leopold, A.S., Cain, S.A., Cottam, C.M., Gabrielson, I.N., and Kimball, T.L. (1963) "Wildlife Management in the National Parks" ["The Leopold Report"], in Trefethen, J.B. (ed) *Transactions of the Twenty-Eighth North American Wildlife and Natural Resources Conference*, Washington, DC: Wildlife Management Institute, pp. 28–45.

Light, A. (2000) "Ecological Restoration and the Culture of Nature: A Pragmatic Perspective," in Gobster, P.H. and Hull, R.B. (eds) *Restoring Nature: Perspectives from the Social Sciences and Humanities*, Washington, DC: Island Press, pp. 49–70.

Light, A. (2009) "Ecological Restoration: From Functional Descriptions to Normative Prescriptions," in Krohs, U. and Kroes, P. (eds) *Functions in Biological and Artificial Worlds: Comparative Philosophical Perspectives*, Cambridge, MA: The MIT Press, pp. 147–161.

Light, A. (2012) "The Death of Restoration?" in Thompson, A. and Bendik-Keymer, J. (eds) *Ethical Adaptation to Climate Change: Human Virtues of the Future*, Cambridge, MA: The MIT Press, pp. 105–121.

Light, A. and Higgs, E.S. (1996) "The Politics of Ecological Restoration," *Environmental Ethics* 18: 227–247.

Lintott, S. (2011) "Preservation, Passivity, and Pessimism," *Ethics & the Environment* 16: 95–114.

Marris, E. (2011) *Rambunctious Garden: Saving Nature in a Post-Wild World*, New York: Bloomsbury.

Mathews, F. (1999) "Letting the World Grow Old: An Ethos of Countermodernity," *Worldviews: Global Religions, Culture, and Ecology* 3: 119–137.

Minteer, B.A. and Pyne, S.J. (2012) "Restoring the Narrative of American Environmentalism," *Restoration Ecology* 21: 6–11.

Murcia, C. and Aronson, J. (2014) "Intelligent Tinkering in Ecological Restoration," *Restoration Ecology* 22: 279–283.

Noss, R.F. (1994) "Wilderness Recovery: Thinking Big in Restoration Ecology," in Pilarski, M. (ed) *Restoration Forestry: An International Guide to Sustainable Forestry Practices*, Durango, CO: Kivakí Press, pp. 92–101.

OED (1971) *The Compact Edition of the Oxford English Dictionary*, Oxford: Oxford University Press.

Oelschlaeger, M. (2002) "The Politics of Wilderness Preservation and Ecological Restoration," *Natural Resources Journal* 42: 235–246.

Oksanen, M. (2008) "Ecological Restoration as Moral Reparation," *WCP 2008 Proceedings: Philosophy of Environment* 28: 99–105.

Palamar, C.R. (2006) "Restora*shyn*: Ecofeminist Restoration," *Environmental Ethics* 28: 285–301.

Parker, K.A., Seabrook-Davison, M., and Ewen, J.G. (2010) "Opportunities for Nonnative Ecological Replacements in Ecosystem Restoration," *Restoration Ecology* 18: 269–273.

Pickett, S.T.A., Kolasa, J., and Jones, C.G. (1994) *Ecological Understanding: The Nature of Theory and the Theory of Nature*, San Diego, CA: Academic Press.

Pletsch, C. (1994) "Humans Assert Sovereignty over Nature," in Baldwin, A.D. Jr., de Luce, J., and Pletsch, C. (eds) *Beyond Preservation: Restoring and Inventing Landscapes*, Minneapolis: University of Minnesota Press, pp. 85–89.

Prior, J. and Smith, L. (2019) "The Normativity of Ecological Restoration Reference Models: An Analysis of Carrifran Wildwood, Scotland, and Walden Woods, United States," *Ethics, Policy & Environment* 22: 214–233.

Sandler, R. (2009) "The Value of Species and the Ethical Foundations of Assisted Colonization," *Conservation Biology* 24: 424–431.

Sandler, R. (2012) "Global Warming and Virtues of Ecological Restoration," in Thompson, A. and Bendik-Keymer, J. (eds) *Ethical Adaptation to Climate Change: Human Virtues of the Future*, Cambridge, MA: The MIT Press, pp. 63–79.

Scoville, J.M. (2013) "Historical Environmental Values," *Environmental Ethics* 35: 7–25.

Seddon, P.J. (2010) "From Reintroduction to Assisted Colonization: Moving along the Conservation Translocation Spectrum," *Restoration Ecology* 18: 796–802.

Simberloff, D. (2012) "Nature, Nativism, and Management: Worldviews Underlying Controversies in Invasion Biology," *Environmental Ethics* 34: 2–25.

Snow, C.P. (1960) *The Two Cultures*, Cambridge: Cambridge University Press.

Society for Ecological Restoration (SER) (2004) *The SER International Primer on Ecological Restoration*, <http://www.ser.org/docs/default-document-library/english.pdf>.

Soulé, M.E. and Noss, R. (1998) "Rewilding and Biodiversity: Complementary Goals for Continental Conservation," *Wild Earth* 8: 19–28.

Soulé, M.E. and Terborgh, J. (eds) (1999) *Continental Conservation: Scientific Foundations of Regional Reserve Networks*, Washington, DC: Island Press.

Sylvan, R. (1994) "Mucking with Nature," in Sylvan, R. *Against the Main Stream: Critical Environmental Essays*, Discussion Papers in Environmental Philosophy, Number 20, Australian National University, pp. 48–78.

Tanasescu, M. (2017) "Responsibility and the Ethics of Ecological Restoration," *Environmental Philosophy* 14: 255–274.

Throop, W. (1997) "The Rationale for Ecological Restoration," in Gottlieb, R.S. (ed) *The Ecological Community: Environmental Challenges for Philosophy, Politics, and Morality*, New York: Routledge, pp. 39–55.

Throop, W.M. (2012) "Environmental Virtues and the Aims of Restoration," in Thompson, A. and Bendik-Keymer, J. (eds) *Ethical Adaptation to Climate Change: Human Virtues of the Future*, Cambridge, MA: The MIT Press, pp. 47–62.

Throop, W. and Purdom, R. (2006) "Wilderness Restoration: The Paradox of Public Participation," *Restoration Ecology* 14: 493–499.

Turner, F. (1994) "The Invented Landscape," in Baldwin, A.D. Jr., de Luce, J., and Pletsch, C. (eds) *Beyond Preservation: Restoring and Inventing Landscapes*, Minneapolis: University of Minnesota Press, pp. 35–66.

Willeke, G.E. (1994) "Landscape Restoration: More than Ritual and Gardening," in Baldwin, A.D. Jr., de Luce, J., and Pletsch, C. (eds) *Beyond Preservation: Restoring and Inventing Landscapes*, Minneapolis: University of Minnesota Press, pp. 90–96.

Woodley, S. (2010) "Ecological Integrity: A Framework for Ecosystem-Based Management," in Cole, D.N. and Yung, L. (eds) *Beyond Naturalness: Rethinking Park and Wilderness Stewardship in an Era of Rapid Change*, Washington DC: Island Press, pp. 106–124.

Woods, M. (2005) "Ecological Restoration and the Renewal of Wildness and Freedom," in Heyd, T. (ed) *Recognizing the Autonomy of Nature: Critical Essays in Environmental Philosophy*, New York: Columbia University Press, pp. 170–188.

Woods, M. (2017) *Rethinking Wilderness*, Peterborough: Broadview Press.

Woods, M. and Moriarty, P.V. (2001) "Strangers in a Strange Land: The Problem of Exotic Species," *Environmental Values* 10: 163–191.

Zavaleta, E.S. and Chapin, S. III (2010) "Resilience Frameworks: Enhancing the Capacity to Adapt to Change," in Cole, D.N. and Yung, L. (eds) *Beyond Naturalness: Rethinking Park and Wilderness Stewardship in an Era of Rapid Change*, Washington DC: Island Press, pp. 142–158.

Further Reading

Clewell and Aronson (2013) (Supplementing *The SER International Primer on Ecological Restoration*, this is the best single scientific account of ecological restoration written by restoration ecologists.)

Cowell, C.M. (1993) "Ecological Restoration and Environmental Ethics," *Environmental Ethics* 15: 19–32. (Cowell provides a wide-ranging philosophical critique of ecological restoration that addresses the Elliot-Katz paradox, the domination of nature, and human participation in restoration.)

Gobster, P.H. and Hull, B. (eds) (2000) *Restoring Nature: Perspectives from the Social Sciences and Humanities*, Washington, DC: Island Press. (Combining a good selection of articles from philosophers and social scientists, this anthology is motivated by the Chicago restoration controversy.)

Higgs, E. (2003) (Higgs offers a book-length discussion of ecological restoration written from his perspectives of restoration ecology and philosophy.)

Marris, E. (2011) *Rambunctious Garden: Saving Nature in a Post-Wild World*, New York: Bloomsbury. (In addition to ecological restoration, Marris muses on topics such as assisted migration, designer and novel ecosystems, exotic species, and rewilding, and she argues for a new hybrid of human management and wild nature.)

Throop, W. (ed) (2000) *Environmental Restoration: Ethics, Theory, and Practice*, Amherst, NY: Humanity Books. (A number of classic philosophical articles about ecological restoration can be found in this anthology.)

45

ASSISTED MIGRATION AND REINTRODUCTION

Ronald L. Sandler

This chapter considers the ethics of intentionally creating independent wild or *in situ* populations for species conservation purposes. After providing some background, I discuss the challenge that accelerated rates of climatic and ecological change pose for the core justifications for species reintroductions. I then evaluate the case for translocating species to new locations in order to forestall their extinction. Overall, I suggest that species reintroductions and assisted migrations can be justified under conditions of rapid ecological change in some cases, though much less frequently than is often supposed. Moreover, while particular introductions can be locally significant, the strategies do not represent scalable responses to the biodiversity crises that we face. The focus of this chapter is primarily animal reintroduction and translocation. However, much (but not all) of the discussion applies *mutatis mutandis* to plants.

Background

Species reintroductions and translocations are management strategies employed in response to population extirpation and endangerment. The background or historical rate of extinction is estimated to be less than 1 species/million/year (Baillie et al. 2004; De Vos et al. 2014). On most estimates, there are 10–20 million eukaryotic (plant and animal) species (Strain 2011; Vié et al. 2009). Thus, a "normal" number of extinctions would be less than 20 extinctions per year. However, many researchers believe that species extinction rates are currently thousands/year due to human-related activities, such as habitat destruction, extraction, pollution, and introduced species (Baillie et al. 2004; IUCN 2011). Anthropogenic climate change is expected to increase species extinction rates still further. One study found that 24–50% of bird species, 22–44% of amphibian species, and 15–32% of coral species have traits that make them "highly vulnerable" to climate change (Foden et al. 2013). Earlier studies projected that 15–37% of species will be committed to extinction by 2050 on mid-level climate change scenarios (Thomas et al. 2004), though there are significantly increased rates of extinction projected even on very optimistic scenarios (IPCC 2007). Overall, the Intergovernmental Panel on Climate Change (IPCC) concludes that "A large fraction of species face increased extinction risk due to climate change during and beyond the 21st century, especially as climate change interacts with other stressors (high confidence)" (IPCC 2014: 10). As more species become endangered and locally extirpated, there are more candidates for reintroduction and translocation.

Reintroductions involve intentionally releasing individuals of a species to an area within their recent historical range where there is not a currently existing population (IUCN 1995). Because

DOI: 10.4324/9781315768090-54

reintroductions are returns to historical ranges, they are by definition native species reintroductions. *Population reinforcement* or *augmentation* refers to releasing organisms – captive bred or wild captured from elsewhere – to an area where the population is depleted but not extirpated, as is commonly done with fish stocking and has been done with the Florida panther, for example. Species reintroductions are conducted with a wide variety of species – e.g. trees, flowing shrubs, epiphytes, mollusks, mammals, insects, reptiles, amphibians, birds, and fishes.

Native species reintroductions are frequently part of public and private ecological restoration and ecosystem management programs. They are sometimes successful. A review of 116 animal reintroductions found that "30 (26%) were classified as successful, 31 (27%) were classified as failures, while the outcome of 55 (47%) re-introductions were classified as unknown at the state of publication" (Fischer and Lindenmayer 2000: 7). An analysis of 249 plant species reintroductions found that "survival, flowering and fruiting rates of reintroduced plants are generally quite low (on average 52%, 19% and 16% respectively)" (Godefroid et al. 2011: 672). Examples of high-profile reintroductions include gray wolves in Yellowstone National Park (USA), European beavers in Wales and Scotland, black-footed ferrets in North America, California condors in the United States, European Lynx in Switzerland, peregrine falcons in the United States and Europe, golden tamarin in Brazil, and bull trout in Oregon (USA).

Intentional species *translocations* are also quite common. Plant and animal species are transported, raised, and cultivated for agriculture purposes, for ornamental and recreational purposes (e.g. hunting and gardening), and for ecosystem management purposes (e.g. biocontrol), for example. It is estimated that there are over 50,000 nonnative species in the United States (FWS 2009). In New Zealand, there are as many naturalized nonnative plant species as native plants (~2,000), and far more cultivated nonnative species (>22,000) (Sax and Gaines 2008). However, the term *assisted migration* (or *assisted colonization* or *managed relocation*) is reserved for the practice of intentionally moving individuals of a species to a location beyond their historical range in order to establish a viable independent population for the purpose of preventing the species from going extinct. Advocates of assisted migration believe that when a species is at risk of extinction due to habitat loss that cannot be arrested or prevented, the only (or best) way to maintain a wild population is to move individuals of the species to a new, ecologically suitable location. They argue that, given global climate change in particular, "the future for many species and ecosystems is so bleak that assisted colonization might be their best chance" (Hoegh–Guldberg et al. 2008: 346).

Support for translocation for conservation purposes has been growing in recent years (Camacho et al. 2010; Hoegh–Guldberg et al. 2008; Minteer and Collins 2010; NPS 2010; Richardson et al. 2009; Vitt et al. 2010). It now appears as a management option in the US National Park Service climate change response strategy. In the United Kingdom, two butterfly species have been translocated northward to sites that climate-species models suggest will be more conducive to their long-term survival than their prior ranges. In Canada, scientists have relocated dozens of tree species to locations beyond their recent historical range, and in the United States an environmental group called the Torreya Guardians has translocated specimens of a threatened conifer from its present range in Florida to more northerly locations. Assisted migration has been proposed for a wide variety of other species, from lobster to lynx.

Pleistocene rewilding is perhaps the most ambitious "reintroduction" program that has been suggested. Its advocates propose translocation of wild (or de-domestication of non-wild) tortoises, camels, cheetahs, horses, elephants, and lions from Asia and Africa, among other places, to expansive parks in the Great Plains and Western United States. (It has been proposed for Australia as well.) They believe that the species are appropriate ecological proxies for the large vertebrates that went extinct in North America 13,000 years ago, in part due to hunting by humans. Moreover, many of the proxy species are at risk in their current habitats, so it would contribute to species conservation by providing an intercontinental refuge (Donlan et al. 2005, 2006). There are other, more modest, rewilding projects. For example, in The Netherlands, Konic Ponies, Heck Cattle, and Galloway Cattle have been introduced as proxies for extinct herbivores in two comparatively small and lightly managed reserves. Several similar efforts are being planned throughout Europe.

In what follows, I discuss the justifications for species reintroductions and translocations. There is a body of scientific research on how best to conduct them – e.g. what types of species (and release stock) are good candidates, the processes and methods that are most effective, and what sorts of ecological, economic, legal, and social conditions are conducive to success (Cochran-Biederman et al. 2014; Fischer and Lindenmayer 2000; Godefroid et al. 2011; IUCN 1995). However, whether and when we ought to support reintroduction and translocation, not best practices for doing it, are the focus here.

Rapid Ecological Change and Species Reintroductions

Species reintroductions are an instance of *ecological restoration*, or actively intervening into an ecological system in order to increase its historicity – i.e. to make it in some respect more like it was or would have been prior to its anthropogenic degradation. As with ecological restoration generally, the justification for a species reintroduction can be forward-looking, backward-looking, or both.

Backward-looking justifications claim that we ought to engage in species reintroductions to make up for the harms or wrongs involved in anthropogenic species extirpations. There can be different views on who or what is harmed or wronged when a species is extirpated from part of its range – e.g. the system, the species, the population, the individual organisms, or the peoples that value or use them. But the core idea is that something of value is lost, the system's value as a whole is diminished, and/or a wrong has been committed. It is therefore the responsibility of those who caused or benefitted from the loss (or those who are in appropriate positions) to undo or make up for it. Reintroducing extirpated species is then considered an effective (or the only) way to discharge that responsibility. There are thus two key components to backward-looking (or restitutive) justifications: (1) a claim about the value (e.g. cultural, economic, natural, intrinsic, or aesthetic) that has been lost or the wrong that has been done with an anthropogenic extirpation; (2) a claim about why reintroduction is the appropriate response to that loss or wrong. (See Sandler [2013] for a discussion of an analogous form of argument regarding species extinctions and de-extinction.)

Forward-looking justifications claim that we ought to engage in species reintroductions because they are beneficial or value adding going forward. It is sometimes argued that reintroduction will help to forestall a possible future anthropogenic extinction, thereby preventing the loss of a unique naturally evolved form of life. It is often argued that reintroducing species to their former systems is ecologically beneficial to the system – e.g. increases its health or resilience – since reintroduction reestablishes co-evolved ecological relationships. In some cases, reintroduced species are expected to have economic value or provide ecosystem services. It has also been argued that engaging in restoration (including reintroductions) can benefit those who engage in it by improving their ecological awareness, knowledge, connection to place and character. Moreover, in many cases people have a preference, desire, or interest in having a species restored for cultural or aesthetic reasons, for example. Thus, a wide variety of values – intrinsic, instrumental, historical, aesthetic, cultural, and ecological – are appealed to in forward-looking reintroduction justifications.

Backward-looking and forward-looking justifications for reintroduction are often interconnected. The reason that reintroduction is thought to be an appropriate way to make up for past harms, wrongs, and losses associated with extirpation is in part that reintroduction recreates lost (or creates new) value by reestablishing past ecological relationships.

The case for species reintroductions also has a negative component. It must be that the social, ecological, economic, and animal welfare costs and risks involved with the reintroduction are not overly great and are justified by the potential value or restitution gains. Considerations that count against engaging in a reintroduction includes public opposition, that the species is likely to become ecologically disruptive, that the economic or opportunity costs are high, and that there will be considerable animal suffering and death involved. Thus, even once a reintroduction is deemed feasible and legal, there remain quite a lot of considerations relevant to determining whether it is well justified overall.

Given the foregoing, the most justified species reintroductions are highly valued or valuable populations that were lost from an ecological system that remains largely otherwise intact (so that its reintroduction is not disruptive, contributes to ecological integrity, and has historical continuity) and for which the causes of the extirpation no longer obtain (so that the species can be successfully reintroduced without large numbers of losses, intensive management, or overly high costs). Reintroduction of the gray wolf to Yellowstone fits these criteria, for example. Yellowstone is a large and intact protected ecological system in which the cause of the wolf extirpation (a large-scale government-supported elimination campaign) no longer exists. There was strong support for the reintroduction federally, from many environmental organizations, and from some Native American tribes. The wolf is an apex predator and a keystone species within the system. The reintroduction is widely considered a success, both for the wolves and for the system, despite there having been some associated human-caused mortality.

However, in many cases, each of the elements of the justification for reintroduction – making up for past wrongs, creating value, and not having overly high costs and risks – is weakened by rapid and high-magnitude ecological change. The reason for this is that they largely depend upon the background ecological conditions of the reintroduction being similar to what they were when the species thrived in that location in the past. But increasingly the cause of extirpation is that the background ecological conditions are changing faster than populations are able to adapt, with respect to such things as temperature ranges, precipitation patterns, and the composition of air, water, and soil. When these changes are driven by macroscale factors, such as global climate change, which cannot be addressed through local management plans, reintroduction will not reestablish the lost relationships. In such cases, reintroduction will not be an effective approach to making up for past wrongs or for generating value, and it will be more risky, costly, and less likely to succeed.

The justification for prioritizing native species in ecosystem management rests primarily on two considerations. The first is that native species are conducive to maintaining ecological integrity and historical continuity (natural and cultural). The second is that most of the types of value that wild species possess are tied to their ecological and evolutionary situatedness. Both of these justifications for native species reintroductions are undermined by rapid and high-magnitude change, since they both depend upon the relative stability of background ecological and climatic conditions (see Sandler 2012, ch. 2 for elaboration on these points). Climate change in combination with other large-scale and intensive anthropogenic ecological impacts is likely to disrupt ecological systems in such a way that reintroduction will be less well justified. As the Yellowstone wolf reintroduction indicates, this will not be the case when extirpations have local and reversible causes, so case-by-case assessment is required. But, in general, the greater the magnitude and rate of climatic and ecological change, the more difficult and costly will be a reintroduction, the less likely it will contribute to ecological integrity, and the less restitution and value there is to be gained from it.

Rapid Ecological Change and Species Translocations

Macroscale anthropogenic change, and global climate change in particular, undermines the historically dominant place-based approaches to conserving species: creation of reserves and ecological restoration. These strategies depend upon ecological systems remaining sufficiently intact.

However, global climate change represents non-normal spatial and temporal change, in comparison to the recent past (~12,000 years). As a result, current habitats and species assemblages are coming apart at unusually high rates and they are doing so in ways and for reasons that cannot be addressed by ecosystem managers, since it is not the product of stressors local to the system that might be reduced or eliminated (e.g. pollution, habitat destruction, and extraction). Thus, to the extent that species are climate threatened, place-based reserves and restorations will be less effective for preserving species and maintaining species communities.

As discussed earlier, in response to this challenge, many conservationists have begun to advocate for translocating individuals of climate-threatened species beyond their historical range. Assisted migration is controversial because it involves intentional creation of independent nonnative species populations and is, therefore, in tension with maintaining historical continuity, non-intervention, and native species prioritization, which have been part of ecosystem management orthodoxy for decades and are part of the field's foundational normative commitments (Soule 1985). Nevertheless, many believe that unless such novel strategies are adopted conservation biology will become a field of "managing extinctions" (Donlan et al. 2005: 913–914).

However, the same sorts of difficulties that arose for reintroductions under conditions of rapid and open-ended ecological change apply to translocation to an even greater extent. Because the value of a population – e.g. ecological, instrumental, intrinsic, and cultural – is so tied to its ecological and cultural context, the value of the species will in most cases not be preserved through translocation (Sandler 2010). There can be exceptions to this – for example, species that are valued in ways or for reasons (e.g. economic) that do not depend upon their ecological situatedness or when the translocated species would be ecologically or instrumentally valuable in the recipient site (and this can be reliably predicted in advance). But, in general, very few assisted migrations would make up for harms that have been done or maintain important values, since they do not preserve ecological relationships.

Moreover, a background context of rapid and open-ended ecological change raises significant challenges to accomplishing successful assisted migrations. Effective assisted migration requires knowing where to relocate the target population in order for it to thrive. This requires identifying where suitable habitats are – e.g. where there is appropriate precipitation, temperature, soil chemistry, or other species – not just now, but extended into the future. However, among the distinctive features of global climate change are high levels of uncertainty and contingency regarding the ecological future. Because the ecological future depends so heavily on not-yet-made public policy decisions and not-yet-created and disseminated technological innovations, as well as on incompletely understood effects of greenhouse gas levels on climatic systems and ecological processes, it is not possible to predict the climatic and ecological future of particular places with high confidence. There are simply too many poorly understood and indeterminate (not just unknown, but unknowable) factors. Moreover, due to the abnormally high rates of climatic and ecological change and the elevated levels of uncertainty and indeterminacy, the range of possible ecological futures in any particular place is broader than in the recent ecological past. The difficulty of identifying an appropriate recipient site is particularly acute for populations of species that are less ecologically flexible, which are precisely those that would be most in need of assisted migration, since they will typically have less adaptive capacity. Thus, one reason successful assisted migrations are likely to be rare under conditions of global climate change is that it is not possible to predict with any confidence where suitable habitat for the target species will be 50 or 100 years from now. This difficulty is severely compounded by the many other anthropogenic drivers of rapid and macroscale ecological change, such as nitrogen eutrophication, unintentional species introductions, and shifting land-use patterns.

Furthermore, because the rate of climatic and ecological change associated with global climate change is elevated, even if a recipient site is appropriate now, it might become unsuitable for the species population in 50, 100, or 200 years. This difficulty, like the previous one, is particularly salient for species that would be the primary candidates for assisted migration, those whose populations have less adaptive capacity. As a result, under conditions of global climate change, even if an assisted migration is initially successful, it is likely to be unsustainable. The same features – global climate change and rapid ecological change – that are driving the need for translocation now will continue to obtain and, in the near future, will place many translocated species populations in much the same position as they are presently. Thus, the distinctive features of global climate change suggest not only that it will be difficult to successfully translocate in the short run, but that even temporarily successful translocations are often unlikely to be the basis for long-term population stability.

In addition, responsible translocations are likely to be difficult to accomplish, expensive to execute, and have high opportunity costs. A responsible assisted migration must attend to the relevant ecological, legal, social, and cultural considerations associated with a candidate species and recipient site (Richardson et al. 2009; Shirey and Lamberti 2009; Vilà and Hulme 2011). Executing a responsible translocation involves studying the potential effects on the recipient system – e.g. on other species and on ecosystem services. It involves assessing alternative conservation strategies with respect to feasibility and cost. It involves identifying and consulting stakeholders, with due process and diligence. And it involves addressing all relevant legal considerations, such as reviews, permits, and liability. This will typically require resource-intensive longitudinal study, on the part of both ecologists and social scientists, as well as extensive social, cultural, and legal engagement – and even then the translocation may not be supported.

If a candidate relocation is determined to be technically feasible, low in ecological risk, socially and culturally acceptable, legal, and preferable to alternative conservation strategies, it must still be executed. Depending upon the species, there could be considerable time and resource costs associated with capture, transport, and post introduction monitoring and management. Thus, overall evaluation, execution, and management costs, in terms of time, effort, and resources, will often be significant. They will differ by case – e.g. the risks and costs associated with a large mammal relocation are likely to be much different than those for an easily propagated plant species. Nevertheless, in all cases there will be non-trivial costs, and in many cases quite substantial ones (Hunter 2007).

Furthermore, successful assisted migrations will be detrimental to some individuals in the recipient system. A successful translocation results in the establishment of a population in the recipient system, the individuals of which consume resources – e.g. sunlight, water, and shelter – that individual organisms of other species would have used. To the extent that individuals of the translocated population out compete other individuals in the recipient system, they are detrimental to them. When a translocated species is predatory on individuals in the recipient system, the harm is more direct.

Translocation can also be detrimental to the individuals that are relocated. The process of capture, transport, release, and monitoring can be stressful, and injury is possible. This is particularly so for sentient animals that have psychological awareness of what is happening to them. In addition, the relocated individuals are taken from a familiar environment and placed in an unfamiliar environment, which may make meeting their basic needs more challenging, particularly if they were bred in captivity. Moreover, if predatory species are held in captivity for some time prior to their translocation – e.g. to increase their population or return members to good health – they will need to be fed individuals of other species (Bekoff 2010).

The foregoing considerations, taken together, suggest that assisted migrations, except in quite rare cases, are ill advised. With respect to only a small number of species is there even value to be preserved through a successful assisted relocation. But successful, responsible assisted migrations are themselves likely to be quite rare, given the distinctive features of global climate change and the characteristics of species that are likely to be most in need of relocation. Moreover, even in the rare cases of responsible, value preserving, and likely to be successful relocation, there may be significant disvalues, including opportunity costs and negative impacts on the welfare of individual organisms.

This does not imply that all assisted migrations are unjustified. It may be that there are candidate species: (1) that are high in value in ways that are not dependent upon their ecological and evolutionary situatedness; (2) that are likely to be ecologically successful in the recipient systems, though not invasive, now and into the future, even given high rates of ecological uncertainty and change; (3) whose relocations will not be overly costly, with respect to resources and welfare; and (4) whose relocations are widely supported. However, these are substantial criteria that taken in combination imply not only that such cases will be rare, but that the burden of establishing that a case is an exception is quite high.

Conclusion

I have argued that the ethical bases for species reintroduction and translocation are often undermined by rapid and open-ended ecological change. If this is correct, then the same processes that are expected to enormously increase the number of threatened species and local extirpations also undermine traditional and proposed strategies for conserving them in the wild. Moreover, reintroduction and translocation focus on only one or a few species at a time. They cannot scale to the magnitude of the population and species loss problem that we face. At most, they might enable us to forestall the extinction of some of the species that we care most about for some amount of time. Furthermore, the strategies do not address what is causing the species extinctions in the first place – e.g. climate change, habitat destruction, and over-extraction. The reality is that once species are imperiled at the rates and for the reasons predicted we do not have good conservation options. The implication of this is that the best, and perhaps only, way to adequately respond to our conservation challenges is to address their causes – i.e. by reducing our ecological impacts and the share of planetary resources that we use by changing our consumption patterns, decreasing our population, and innovating technologies.

Related Topics

6. Wild Animals
7. Hunting
9. Mountains: Rethinking Thinking Like a Mountain
10. Wilderness
20. Moral Bases of Responses to Climate Change
44. Restoration
46. Zoos and Conservation
47. Rewilding
48. Novel Ecosystems

References

Baillie, JEM et al. (2004) *A Global Species Assessment*, UK: IUCN [online]. Available at: http://data.iucn.org/dbtw-wpd/html/Red%20List%202004/completed/cover.html.

Bekoff, M (2010) 'Conservation and Compassion: First Do No Harm', *The New Scientist* [online]. Available at: http://www.newscientist.com/article/mg20727750.100-conservation-and-compassion-first-do-no-harm.html.

Camacho, AE et al. (2010) 'Reassessing Conservation Goals in a Changing Climate', *Issues in Science Technology* (26), pp. 21–26.

Cochran-Biederman, JL et al. (2014) 'Identifying Correlates of Success and Failure of Native Freshwater Fish Reintroductions', *Conservation Biology*, 29 (1) [online]. Available at: http://www.ncbi.nlm.nih.gov/pubmed/25115187.

De Vos, JM et al. (2014) 'Estimating the Normal Background Rate of Species Extinction', *Conservation Biology* [online]. Available at: http://onlinelibrary.wiley.com/doi/10.1111/cobi.12380/full.

Donlan, J et al. (2005) 'Re-Wilding North America', *Nature* (436), pp. 913–914.

Donlan, J et al. (2006) 'Pleistocene Rewilding: An Optimistic Agenda for Twenty-First Century Conservation', *The American Naturalist* (168), pp. 660–681.

Fischer, J & Lindenmayer, DB (2000) 'An Assessment of the Published Results of Animal Relocations', *Biological Conservation* [online]. Available at: http://www.cbsg.org/sites/cbsg.org/files/2013_AM/Fischer_Lindenmayer_Assessment_animal_relocations_2000.pdf.

Fish and Wildlife Service (FWS) (2009) *Frequently Asked Question About Invasive Species*, U.S. Fish and Wildlife Service [online]. Available at: http://www.fws.gov/invasives/faq.html#q7?.

Foden, WB et al. (2013) 'Identifying the World's Most Climate Change Vulnerable Species: A Systematic Trait-Based Assessment of all Birds, Amphibians and Corals', *Plos One* [online]. Available at: http://journals.plos.org/plosone/article?id=10.1371/journal.pone.0065427#s5.

Godefroid, S et al. (2011) 'How Successful are Plan Species Reintroductions?' *Biological Conservation*, 144, pp. 672–682.

Hoegh-Guldberg et al. (2008) 'Assisted Colonization and Rapid Climate Change', *Science*, 321 (5887), pp. 345–346.

Hunter, ML (2007) 'Climate Change and Moving Species: Furthering the Debate on Assisted Colonization', *Conservation Biology*, 21 (5), pp. 1356–1358.

Intergovernmental Panel on Climate Change (IPCC) (2007) *Climate Change 2007: Synthesis Report*. Geneva, Switzerland: IPCC.

Intergovernmental Panel on Climate Change (IPCC) (2014) *Climate Change 2014: Synthesis Report*. Geneva, Switzerland: IPCC.

International Union for the Conservation of Nature (IUCN) (1995) *IUCN/SSC Guidelines for Re-Introductions*, IUCN [online]. Available at: http://intranet.iucn.org/webfiles/doc/SSC/SSCwebsite/Policy_statements/Reintroduction_guidelines.pdf.

International Union for the Conservation of Nature (IUCN) (2011) *IUCN Biodiversity*, IUCN [online]. Available at: http://www.iucn.org/what/biodiversity/.

Minteer, B & Collins, J (2010) 'Move it or Lose it? The Ecological Ethics of Relocating Species Under Climate Change', *Ecological Applications*, 20, pp. 1801–1804.

National Park Service (NPS) (2010) *National Park Service Climate Change Response Strategy*. Fort Collins, CO: National Park Service Climate Change Response Program.

Richardson, DM et al. (2009) 'Multidimensional Evaluation of Managed Relocation', *Proceedings of the National Academy of Science*, 106, pp. 972–9724.

Sandler, R (2010) 'The Value of Species and the Ethical Foundations of Assisted Colonization', *Conservation Biology*, 24 (2), pp. 424–31.

Sandler, R (2012) *The Ethics of Species*. Cambridge: Cambridge University Press.

Sandler, R (2013) 'The Ethics of Reviving Long Extinct Species', *Conservation Biology*, 28 (2), pp. 354–360.

Sax, DF & Gaines, SD (2008), 'Species Invasion and Extinction: The Future of Native Biodiversity on Islands', *Proceedings of the National Academy of Sciences*, 105, pp. 11490–11497.

Shirey, PD & Lamberti, GA (2009) 'Assisted Colonization Under the U.S. Endangered Species Act', *Conservation Letters*, 3, pp. 45–52.

Soule, ME (1985) 'What Is Conservation Biology?', *BioScience*, 35 (11), pp. 727–742.

Strain, D (2011) '8.7 Million: A New Estimate for All the Complex Species on Earth', *Science*, 333 (6046), p. 1083.

Thomas, CD et al. (2004) 'Extinction Risk from Climate Change', *Nature*, 427, pp. 145–148.

Vié, J-C, Hilton-Taylor, C & Stuart, SN (eds.) (2009) *Wildlife in a Changing World: An analysis of the 2008 IUCN Red List of Threatened Species*. Gland, Switzerland: IUCN.

Vilà, M & Hulme, PE (2011) 'Jurassic Park? No Thanks', *Trends in Ecology and Evolution*, 26 (10), pp. 496–497.

Vitt, P et al. (2010) 'Assisted Migration of Plants: Changes in Latitudes, Changes in Attitudes', *Biological Conservation*, 143 (1), pp. 18–27.

46

ZOOS AND CONSERVATION

Ben A. Minteer, James P. Collins, and Aireona Bonnie Raschke

Introduction: Zoos and the Conservation Challenge

Zoos have long championed conservation as a core part of their institutional and public mission. The last decade, though, has seen a significant amplification of their commitment to save wild species from further decline and destruction. Indeed, in their role as "*ex situ* conservation centers" (and as partners in field conservation projects), zoos are expected to play an even more significant role in biodiversity conservation in this century (Conde et al. 2011; Barongi et al. 2015; Minteer, Maienschein, and Collins 2018). One high-profile example of zoos' intensifying dedication to species conservation is the "Saving Animals From Extinction" (SAFE) program launched by the Association of Zoos and Aquariums (AZA), the main accrediting body for zoological parks in the US. SAFE is a ten-year program targeting a designated set of threatened species around the world with the goal of organizing and promoting zoos' efforts to recover and conserve them over the long run (AZA 2015).

It's a well-timed development given the generally precarious status of global biodiversity, which by most estimates is in a state of accelerating decline. The 2020 Living Planet Index report authored by the World Wildlife Fund and the London Zoological Society, for example, paints a worrying picture: globally, on average, there has been a 68% fall in monitored populations of mammals, birds, amphibians, reptiles and fish between 1970 and 2016 (WWF 2020). These and similar other concurrent metrics of ecological loss feed into the conclusion that we are likely in the midst of a sixth mass extinction event on the planet, a time of "biological annihilation" (e.g., Kolbert 2014; Ceballos, Ehrlich, and Raven 2020).

The hard pivot toward a stronger conservation mission by zoos, however, raises a complex of historical, ethical, scientific, and institutional questions and challenges as these organizations seek a more significant role in safeguarding global biodiversity in the coming years. The modern American zoo that emerged in the late 19th century, for example, may have touted itself as a center of science, natural history, education, and conservation, but it was also (and remains) an entertainment enterprise. This part of their mission has led an array of zoo critics and skeptics to question both the presumption that keeping animals in captivity is morally justifiable (e.g., Jamieson 1985, 1995; Regan 1995) and the idea that zoos can play any sort of helpful conservation role moving forward (Tullis 2014).

With few exceptions (e.g., Norton et al. 1995; Minteer and Collins 2013; Keulartz 2015; Palmer, Kasperbauer, and Sandøe 2018) environmental ethicists have not paid much attention to the normative questions raised by the conservation activities and goals of zoological institutions,

DOI: 10.4324/9781315768090-55

especially when compared to the more traditional (and not coincidentally, comparatively more "natural") arenas of conservation, such as national forests, parks, and wilderness. Instead, most of the ethical scrutiny devoted to zoos traditionally focuses on the more salient and controversial animal welfare issues relating to animal captivity (e.g., Gruen 2014). As a result, we believe that the history and practice of zoo conservation is not very well known among students and scholars in the field.

In this chapter, we'll examine the evolving conservation agenda of zoological parks as they grew from relatively crude animal menageries into (in many cases) full-fledged conservation centers by the end of the 20th century. In doing so, we'll highlight the important role of captive (conservation) breeding and reintroduction in zoos' embrace of a more extensive and focused conservation effort. We'll conclude the chapter by exploring some of the philosophical and ethical issues raised by an assessment of zoos as *ex situ* conservation centers in an era of ecological change – and in a time of increasing human global influence on wildlife and ecosystems.

Science, Conservation, and the Growth of the Modern Zoo

The proto-modern version of the zoological garden that emerged in the mid-18th century is often depicted as reflecting (and reinforcing) an imperialist ethos and a profound human-nature dualism, with the display of exotic captive animals signifying human mastery and control by the colonial powers over both people and wildlife from their territories (Ritvo 1987). Yet, even in the early period, researchers at zoological gardens demonstrated a concern for scientific discovery and description of biological variety. Braverman (2015), for example, notes the intertwining histories of the early zoo and other collection-based institutions, such as natural history museums, which were united in an attempt to order and catalog nature (dead and alive). Indeed, by the 19th century, zoos in Europe and the US began to self-consciously position themselves as centers of science, especially as places of living natural history and education in ethology and zoology. This period saw the rise of what today we would consider to be the modern zoo model, a form exemplified by institutions such as the London Zoological Garden (aka, the London Zoo, est. 1828). Zoos such as the one in London and the later flagship American zoos were formed by scientific societies eager for these institutions to become hubs of scientific study and zoological education (Hochadel 2005).

The focus on science and learning was a way for many of the new, modern 19th century zoos to distinguish themselves from the ubiquitous animal menageries and traveling shows of the time, exhibits that offered exotic entertainment to a curious public but little more (Hanson 2002). At the same time, however, such menageries were inextricably linked to the rise of the modern zoological park, the roots of which can be traced back to these "unscientific" collections and exploitative displays of animals that predated even the ancient Greeks (Hoage and Deiss 1996). Furthermore, the deliberate association with science and education in early zoos was also an expression of a kind of social exclusivity, inasmuch as these places were originally intended for the privileged few, i.e., the cultural and scientific elite. The class character and nationalism permeating the early European zoological gardens would, however, soon give way to a more democratic model, which was at least partly due to the inability of zoological gardens to sustain themselves as institutions with a narrow appeal.

In the US, the zoological societies behind the first zoos, such as the Philadelphia Zoological Society (which established the Philadelphia Zoological Garden, aka the Philadelphia Zoo, in 1874) and the New York Zoological Society (which opened the New York Zoological Park, aka the Bronx Zoo, in 1899), strongly emphasized science and education in their plans for the new animal parks. For example, the New York Zoological Park was to contain "collections of North American and exotic animals" to encourage "interest in animal life, or zoology, amongst all classes of people, and the promotion of zoological science in general" (NYZS 1897: 13). The Philadelphia and New York zoos were also important sites for the early establishment of zoo-based and zoo-sponsored animal and wildlife research, including the first tropical field research program (in British Guyana) founded by the New York Zoological Society in 1916 (Kisling, Jr. 2001: 164).

In addition to zoology and natural history, wildlife conservation and protection has been a part of the mission of US zoological institutions from the very beginning. This was partly a product of timing. The emergence of American zoos in the late 19th century coincided with the growing national interest in the destruction of wildlife and the decline of the nation's American wilderness, a worry motivated by a range of scientific, economic, aesthetic, and moral concerns (Stott 1981; Barrow 2009; Nash 2014). Among other objectives, the new zoo boosters hoped that the emerging institutions, by allowing the American people to experience species that were fast disappearing from the landscape, would encourage them to support the preservation of wildlife and natural resources from further exploitation and destruction. This made for some intriguing (and often forgotten) associations between the new zoological institutions and some of the leading voices of the early American conservation movement.

The New York Zoological Society, for example, was formed by members of the Boone and Crockett Club, one of the first conservationist-sportsman organizations, which included among its members prominent conservationists such as Theodore Roosevelt and William T. Hornaday (Brinkley 2009). Hornaday, the first director of the Bronx Zoo and a legendary zealot for wildlife protection, had an ambitious vision for zoos, one that included a prominent place for wildlife preservation and scientific study as well as the exhibition and display of exotic animals for the public (Bridges 1974). His 1913 book, *Our Vanishing Wild Life*, is a classic early wildlife preservation jeremiad and a public plea to stop the "slaughter" of species by indiscriminate hunters and wildlife exploiters. The book, published by the New York Zoological Society, would prove influential in drawing public and professional attention to the growing challenge of protecting wildlife in the first half of the 20th century. Notably, it also had a profound impact on the great conservationist (and amateur environmental philosopher) Aldo Leopold, whose early interest in wildlife protection was partly shaped by his encounter with Hornaday's work (Meine 2010: 128–129).

In Hornaday's view, zoos could play an active and important role in saving wildlife from destruction by using methods such as captive breeding and reintroduction. These techniques, which would later become core conservation practices for zoological institutions, were famously employed by the Bronx Zoo to aid the recovery of American bison in 1907 (Kleiman 1989; Isenberg 2000). Flagship American zoos from the beginning were thus intended to be "surrogate Noah's arks," places where imperiled species would be protected and preserved as genetic resources for eventual repopulation of the landscape (Wirtz 1997: 70).

Not every attempt to save species from slipping into extinction, however, was as successful as the case of the American bison. The last passenger pigeon and Carolina parakeet, for example, both expired in the Cincinnati Zoo (in 1914 and 1918, respectively), with little systematic and coordinated attempt to save these species (Barrow 2009, 2018). Outside the US, a similar "lost opportunity" describes the plight of the thylacine (Tasmanian tiger), the last individual of which died in a zoo in Hobart, Tasmania in 1936 (Minteer 2018).

The original aspirational missions of zoological institutions in the US, then, included natural history, zoological research, science education, wildlife protection, and public entertainment. Although early zoos' scientific and conservation aims (i.e., focused on wildlife protection) mostly failed to translate into tangible wildlife outcomes in the early years beyond a few notable cases such as the bison, what would later be called the zoo conservation agenda gradually become more sophisticated, extensive, and effective. And zoos as a whole would also become more organized and professionalized, a trend marked by the creation of the AZA (est. 1924) and the World Association of Zoos and Aquariums (est. 1935). Among other impacts, these new societies spurred the development of improved standards of animal care and housing in zoos and aquariums (AZA 2014; WAZA 2014). Zoo animal husbandry and biology would soon also become more organized and institutionalized, supported by the growth of scientific journals, conferences, and new professional societies in the second half of the century (Kisling Jr. 2001).

During this same period, the design philosophy of the modern zoo was evolving rapidly to emphasize more natural and immersive exhibits. Most zoo historians trace these shifts back to the influence of German animal dealer and zoo entrepreneur Carl Hagenback, whose revolutionary design (unveiled at his zoo in Hamburg in 1907) dispensed with the familiar cages and bars in favor of more naturalistic enclosures, such as moated exhibits and the use of less patently artificial landscape barriers (Rothfels 2002). American zoos such as those in Denver and St. Louis were early adopters of this approach. The Bronx Zoo would also develop a number of naturalistic and immersive animal exhibits in the 1940s, attempting to recreate exotic landscapes (mostly modeled after African habitats) in part to inspire the public to care about habitat and wildlife protection (Mitman 1996: 118–119). As Hyson (2000) notes, however, zoo planners often made modifications to the new enclosures to ensure that the public could still have a clear view of the displayed animals, a tangible reminder that human aesthetic and recreational interests were still paramount.

In the late 1960s and 1970s the wildlife conservation goals of American zoological parks would be reinforced by the rise of the environmental movement and in particular by the flurry of federal and international environmental policy making bearing on wildlife conservation, especially threatened and endangered species. The Marine Mammal Protection Act (1972), the Endangered Species Act (1973), and the Convention on International Trade in Endangered Species of Wild Fauna and Flora, or CITES (1973) all carried implications for zoos and aquariums in the conservation policy arena (Hanson 2002; Kisling Jr. 2018). The founding of the Conservation Breeding Specialist Group in 1978 as part of the Species Survival Commission of the International Union for Conservation of Nature (IUCN) was another key scientific and institutional development, connecting zoos to the wider global network of conservation science and policy (Conway 2011).

This broad historical trend of zoos embracing a more explicit conservation agenda has had a number of interesting consequences, running from the rhetorical to the programmatic. The case of the Bronx Zoo is again instructive. The New York Zoological Society evolved into the Wildlife Conservation Society (WCS) in the 1930s (Goddard 1995). Today, the WCS is one of the best-known and most active global conservation organizations, with hundreds of field-based programs around the world. The conservation mission of the Bronx Zoo in the second half of the 20th century drove the institution's rebranding in the early 1990s, when its official name (i.e., the New York Zoological Park) was changed to the "International Wildlife Conservation Park" (even though the zoo still does business as the "Bronx Zoo"). It was a move that suggested the growing importance of a conservation mission and ethic within the zoo and aquarium community as well as the influence of the WCS vision for wildlife conservation at the Bronx Zoo (Mazur 2001). Another, more recent example of this rhetorical evolution is the Phoenix Zoo, which in 2014 renamed its governing body (formerly the Arizona Zoological Society) the "Arizona Center for Nature Conservation" to signal an intensified commitment to the conservation mission, both within and beyond zoo walls. It's a trend that indicates, perhaps, a growing dissatisfaction with the constraints of the "zoo" label, particularly that term's ability to describe adequately an emerging conservation identity in many zoological institutions.

But the commitment of zoos to conservation, supporters argue, is more than branding and rhetoric. Indeed, over the past several decades many zoological institutions have significantly enhanced their capacity to protect vulnerable species via a range of tactics and methods, including improving or developing species conservation programs. In the 1980s, for example, many zoos began to develop Species Survival Plans (SSPs), which coordinate breeding and population management programs for threatened and endangered animals among zoos worldwide. Such programs were responsible for assisting in the recovery of the black-footed ferret and California condor, among other species. The goal is to create healthy and genetically diverse animal populations of these species across the zoo community, an effort that can ultimately aid the conservation of the species in the wild. In addition, many zoos have developed extensive field and wildlife conservation programs (i.e., supporting *in situ* conservation), including strategic partnerships with field-based agencies and organizations such as the US Fish and Wildlife Service and the Nature Conservancy. The numbers are impressive. In its 2019 Annual Report on Conservation Science, the AZA documents $231.5 million spent on field conservation in 127 countries (AZA 2019).

Beyond the Ark: Conservation Breeding and Reintroduction

For decades one of the most visible conservation-centered activities in zoos and aquariums has been maintaining "captive assurance populations" via captive breeding, with the goal of reintroducing some individuals back into the wild to restore or expand lost or declining populations (Beck et al. 1994; Reid and Zippel 2008). This technique, which harkens back to programs such as the Bronx Zoo's bison reintroduction efforts mentioned above, has produced several notable conservation successes, including recovery of the Arabian oryx, the black-footed ferret, and the California condor. Given their knowledge of and experience in breeding and husbanding many exotic animals, including endangered species, zoos would seem to be in a unique position in the conservation community (Junhold and Oberwemmer 2011; Che-Castaldo, Grow, and Faust 2018).

In addition to providing biological insurance or augmenting populations in the wild, captive breeding programs in zoos and aquariums can also help lessen the direct impact that these institutions have on biodiversity. In the early years, there was little incentive to breed animals in zoos for anything other than conservation, as most collections could be replenished and maintained through the capture of new animals in the wild (Wayre 1969). But a combination of increased knowledge about exotic animal breeding requirements, more stringent federal and international regulations on the taking of wild species, and, in general, shifting ethical attitudes toward the wild sourcing of animals among both zoo professionals and the public has prompted accredited zoos to move away from this practice, and to build their collections via captive breeding programs. Still, many wild species are regularly collected for display in zoological settings. For example, collecting from the wild is still very common for reptiles, fish, and invertebrates, and it is becoming increasingly clear that this pressure is unsustainable for many species (Kawata 2011; Tlusty et al. 2013). Moreover, even species that do have breeding programs for captive populations may face collection pressures in the future, as many of them are not breeding to replacement in captivity (Lees and Wilcken 2009; Powell 2019).

For captive breeding to be true "conservation breeding" (and not just breeding to augment zoo populations), it should either result in the re-establishment of populations that have gone extinct in the wild, or the supplementation of extant *in situ* populations (Lees and Wilcken 2009). According to Hoffmann et al. (2010), reintroductions have played an important role in the recovery of 17 of the 68 vertebrate species whose level of endangerment has been reduced globally (see also Gilbert et al. 2017). Still, conservation breeding is not without its challenges (or its critics). To be effective, populations must be carefully maintained for success, with the right number of founders with high levels of genetic diversity, and populations must be expanded quickly to slow the rate of genetic diversity loss (Lees and Wilcken 2009). Then there are the practical constraints presented by the sheer numbers of species of conservation concern. Organizations such as Species 360 (formerly the International Species Information System) have been developed to help address these issues by aiding zoos, aquariums, and conservation organizations in collaborating with one another in attaining their biodiversity protection goals through maintaining detailed, shared datasets concerning captive animal populations (Conde et al. 2011).

Another concern, not surprisingly, is cost: breeding programs tend to be very expensive, and as long-term programs they're particularly vulnerable to changes in institutional management, budgeting, and/or shifts in the political climate of their home countries (Gippoliti and Carpaneto 1997; Monfort and Christen 2018). Physical space constraints, too, limit the number of species (and the percentage of global biodiversity) that can be bred to provide a "safety net" for species threatened in the wild (Martin et al. 2014). Finally, zoos do not always have the scientific capacity to deal with difficult reproductive challenges. Although zoos can claim extensive experience in breeding, animal husbandry, and veterinary science, as Monfort (2014) points out, the complexity of reproduction in captivity often overwhelms the scientific knowledge and capacity of zoo managers to produce successful outcomes. As a result, some conservation biologists, such as Balmford et al. (2011), have taken a much more pessimistic view toward conservation breeding, noting that captivity conditions can also introduce their own challenges to successful reintroduction by increasing exposure to novel diseases and reducing animals' fear of humans.

Although some regions and certain taxa have seen more success than others (e.g., projects involving bird species in New Zealand have up to an 80% success rate; see Conway (2011)), reintroduction is a complicated and often delicate enterprise. The causes for failure vary, but behavioral deficiencies, sociopolitical factors, cost, and technical difficulties (e.g., hybridization) are among the more common challenges to successful reintroductions (Conde et al. 2011).

To address these problems, zoological institutions involved in breeding and reintroduction programs recognize that they need to continue to work to improve *ex situ* conservation methods (including exchanging animals across zoos to avoid genetic bottlenecks), while learning from past failures and building from the lessons of successful reintroduction campaigns. Pilot reintroduction studies, for example, can help delineate challenges specific to different programs, investigate the long-term costs, and ascertain the readiness of the environment to support the introduced individuals (Griffiths and Pavajeau 2008). Greater use of demographic modeling to determine whether captive breeding and release of a candidate species back into the wild is likely to deliver significant conservation benefits has also been proposed as a way to improve the success rates (and prioritize conservation resources) for reintroduced populations (Dolman et al. 2015).

As mentioned above, one of the most critical constraints facing conservation breeding within zoos is space, especially for animals that mate more successfully in larger roaming herds. To address this challenge, a consortium of zoological institutions was formed in 2005 called "Conservation Centers for Species Survival," or C2S2 for short (Wildt et al. 2019). These facilities are AZA-accredited conservation breeding centers with the mission to cooperatively leverage their "unique resources for the survival of threatened species with special needs – large areas, natural group sizes, minimal public disturbance and research" (Sawyer 2012). The C2S2 coalition is diverse, running from the 1,700-acre Fossil Rim Wildlife Center in Glen Rose, TX to The Wilds, a sprawling (nearly 10,000 acre) safari park and breeding center located near Cumberland, OH. It's also an expanding network: a new 1,000-acre breeding center (to include herds of antelope okapi and Masai giraffe) is in the works south of New Orleans on the banks of the Mississippi, a partnership between the Audubon Nature Institute and San Diego Zoo Global (Robertson 2013). Interestingly, C2S2 facilities work together with the US Fish and Wildlife Service, a collaboration allowing for the sharing of expertise and experience with captive breeding and husbandry as well as endangered species conservation *in situ*.

Although some observers (e.g., Braverman 2015) have suggested that the C2SC centers are more interested in sustaining zoo collections than they are in enhancing conservation success *in situ*, the breeding centers in the network do have a track record of participation in successful wildlife reintroduction programs. To give just two examples, Scimitar horned oryx and Attwater's Prairie Chicken, two of the species bred at Fossil Rim, were reintroduced successfully in Tunisia and the Gulf Coast, respectively (see, e.g., Kimble 2014). Nevertheless, as Steven Monfort, former Director of the Smithsonian Conservation Biology Institute (a C2S2 unit at Front Royal, VA) has observed, the continued success of such programs will depend on maintaining a range of suitable social and environmental conditions, from establishing tight alliances among zoo and non-zoo conservationists, scientists, and animal managers, to securing additional space and facilities and conducting research needed to support conservation management over the long run (Monfort 2014).

The rise of the C2S2 model is an example of an intriguing trend: the convergence of *ex situ* and *in situ* conservation approaches as zoos become more involved in managing their animals not as closed systems, but as parts of larger metapopulations linking wildlife across a continuum of institutions and management domains (Redford et al. 2012; Conde et al. 2013). This shift in thinking and management (toward what might be called "pan *situ*" conservation; e.g., Minteer and Collins 2013) will likely be necessary to not only sustain the resiliency and health of zoo populations into the future, but improve conservation outcomes in the field by restoring the genetic diversity of small wild populations (Lacy 2012). It's a view that's reflected within the "One Plan" approach developed by the IUCN's Conservation Breeding Specialist Group, an effort to combine the resources and expertise of zoo professionals, wildlife biologists, and conservationists in the breeding, recovery, and conservation of vulnerable species across the *ex situ-in situ* spectrum (Traylor-Holzer, Leus, and Byers 2018). The more progressive and integrative models of zoo conservation, it seems, are increasingly occupying a biological, geographical, and normative space between the wild and the walled.

Zoological Ethics on a Changing Landscape

The trends discussed above suggest that many zoos are rapidly evolving as centers of biodiversity conservation, with a growing number of scientists and leaders in the zoological community calling for an even greater emphasis on conservation in the future (see, e.g., Hancocks 2001; Rabb and Saunders 2005; Fa et al. 2011; Mascarelli 2013; Minteer, Maienschein, and Collins 2018). Zoos' deepening commitment to the protection of nature and biodiversity is, however, also taking place at a time when conservation is becoming more challenging and more complicated. The mounting ecological and evolutionary pressures imposed by habitat loss, landscape fragmentation, urbanization, overexploitation, the spread of infectious diseases, and climate change threaten the viability of scores of wild species around the globe (Rands et al. 2010; Stokstad 2010; Dirzo et al. 2014; Ceballos, Ehrlich, and Raven 2020). The scale and speed of these threats, and the synergistic interactions among many of them have and will continue to present an exceedingly difficult challenge for conservation efforts in this century.

To the degree that zoos become more meaningfully engaged in conservation, however (both within their own walls and as partners or as participants in field-based projects), they'll be forced to grapple with the various constraints and legacies of their own scientific, cultural, and institutional histories – and to navigate the changing and often hard to predict landscape of public values, interests, and ethical expectations.

On this last point, and as mentioned above, zoos have attracted ethical scrutiny on a range of animal welfare and rights issues for some time. As far back as 1902, for example, Bronx Zoo director Hornaday derided the "cranks" and sentimentalists who objected to the capture and display of animals in zoos (Rothfels 2002: 67). Today, ethical concerns surrounding the well-being of zoo animals run the gamut from foundational questions regarding the moral acceptability of keeping wild animals in captivity (e.g., Jamieson 1985, 1995; Regan 1995; Gruen 2014), to more specific ethical arguments over captive breeding, manipulative zoo- and aquarium-based research, wild animal acquisition, population management, habitat enrichment, animal commercialization, and the use of emerging biotechnology (see, e.g., Norton et al. 1995; Davis 1997; Bekoff 2002; Kreger and Hutchins 2010; Friese 2018).

This perennial debate over zoo animal ethics was reignited in 2014 by the case of Marius the giraffe in the Copenhagen Zoo (Schwartz 2014; see also Minteer 2018). A young and healthy giraffe considered a "surplus animal" by zoo managers, Marius was shot and his body was dissected and fed to the Zoo's lions and polar bears. The Copenhagen Zoo officials emphasized that the decision was made primarily on the grounds of science, space, and efficiency; Marius's genes were already well represented in the zoo system and so he had little if no remaining conservation value.

Regardless of what one thinks about the Copenhagen case (and for the record, the AZA distanced itself from the decision), modern zoos have to make a complex set of decisions about animal population management, including planning for surplus animals. AZA-accredited zoos in the US generally avoid euthanasia for population management purposes, using contraception and other non-lethal forms of population management. Nevertheless, animal rights organizations such as the People for the Ethical Treatment of Animals (PETA) frequently criticize even accredited zoos for their population management policies, which can involve moving surplus animals outside of the AZA system into less stringently managed institutions lacking a professional ethic of care (PETA n.d.).

A fuller ethical consideration of zoos today, however, requires more than identifying and defending animal welfare concerns, as important as these are. It requires understanding and managing an emerging set of value-laden and ethical questions regarding our relationship with and responsibility to individual animals *alongside* our duties to conserve biodiversity, preserve nature, and promote wildness across a spectrum of rapidly changing *ex situ* and *in situ* contexts (Minteer and Collins 2013). As zoos become more invested in the ark model of biodiversity conservation, for instance, they face issues of conservation prioritization given limited space and resources. Which species should be selected to "ride" on the "ark" – and for how long? What is the value of a species that can only be viable in a zoo or an aquarium? These are far from academic questions.

Consider the "Amphibian Ark" (AArk), a global consortium of zoos, aquariums, universities, and conservation organizations created to address the crisis of global amphibian declines and extinctions (see, e.g., Collins and Crump 2009; Zippel et al. 2011; Stone 2013). Zoos and aquariums in the AArk network effectively serve as way stations for species threatened by the combined forces of habitat loss, infectious disease, and climate change. Yet, until the environmental threats facing the animals are mitigated (chiefly the spread of a deadly fungus) they will exist in a holding pattern, housed indefinitely in zoos and other *ex situ* facilities. The conservation mission of zoos is premised on the eventual recovery of populations *in situ* and so such scenarios – and the uncertainty about the endgame regarding the animals' return to their native habitat – raise difficult ethical questions about the acceptability of zoos' removal of animals from the wild without a clear and predictable plan for their reintroduction (Mendelson 2018).

But there is another interesting dynamic at work as zoos become more significant players in field-based research and *in situ* conservation practice. As many scientists and other observers have noted (e.g., Hobbs et al. 2011; Marris 2011; Biello 2016), the landscape will inevitably become further degraded by human activities in the coming decades. It's a conclusion that seems to suggest our nature preserves will therefore require more human control than they have in the past in order to deliver the same conservation benefits. As a result, the margin separating purportedly "wild" nature and the more conservation-themed zoo will get even thinner. As it does, our understanding of what zoos are and what we want them to be – entertainment destinations, science centers, conservation arks – will also change. So will our traditional views of the wild as those places in nature independent of significant human influence and control (Minteer 2018).

A number of intriguing ethical challenges confront zoological institutions working within this rapidly changing ecological and institutional environment – issues that relate directly to their role in this more intensive and manipulative form of conservation intervention in wild populations and systems. In this more activist and expanding conservation portfolio, for example, zoos will have to clarify and balance their responsibilities to: (1) preserve, recover, manipulate, and reintroduce animal populations (or to curtail such activities if they cross significant moral and scientific thresholds); (2) protect, change, and redesign natural habitats (or to refrain from doing so in certain cases due to preservationist commitments); and (3) create (or to avoid creating) novel environments suitable for species in a rapidly changing world (Minteer and Collins 2013; see also Thompson this volume). Perhaps ironically, then, as zoos push toward a less patently artificial model and toward a higher degree of naturalism and ecological design, they'll become implicated in the broader trend of more intensive and manipulative management of the wild, an evolution that raises deep philosophical and ethical questions about the future of nature preservation under tighter human control and influence (Minteer and Pyne 2015). What does it mean to "preserve nature" in a world increasingly shaped by human activities, especially one in which the distinction between the "artificial" realm of the zoo and the "natural" realm of the wild no longer seems as obvious or as clear as it once did?

The growing attention to "radical" interventionist conservation efforts such as "assisted colonization" and "Pleistocene rewilding" which involve the introduction of animal populations to new environments either to save them from predicted climate change impacts (assisted colonization) or to replace ecological and evolutionary processes driven by extinct megafauna (Pleistocene rewilding), also carry significant implications for *ex situ* conservation centers. Presumably, zoos would play a significant role in supporting such efforts in the future (Donlan et al. 2005; Pritchard et al. 2012; Greene 2015). These challenges, however, have and will continue to force conservationists – in zoos and aquariums, as well as in the field – to consider whether it is ethically acceptable to engage in such novel and highly manipulative conservation practices, efforts that in some cases will depart dramatically from traditional preservationist models focused on protecting native species *in situ* and minimizing human intervention in populations and ecosystems (Marris 2011; Minteer and Collins 2012; Minteer 2018; see also the chapters by Sandler and Turner in this volume).

Furthermore, it is clear that as zoos and aquariums become more serious partners in the study, recovery, management, and preservation of wildlife populations across the landscape (and across conservation institutions) they will likely have to make difficult decisions and consider uncomfortable trade-offs regarding the values and interests that have traditionally shaped their mission. For one thing, they will have to decide how to balance their longstanding entertainment interests with their growing commitment to scientific research and biodiversity conservation (Conway 2011). Many of the breeding centers managed under the C2S2 umbrella, for example, provide only limited public access given their primary scientific and conservation goals – a departure from the traditional entertainment and recreational values at the center of many zoo programs and policies. And then there is the fact that some of the most globally endangered animals are not the large-bodied, exotic, and highly charismatic species that reliably attract zoo visitors (Eveleth 2010; Martin et al. 2014). It's a tension, and a taxonomic bias, that might pit zoos' conservation values against their entertainment goals, and likely their financial bottom line.

As the historical survey above reminds us, however, this is far from a new challenge. Zoos have been struggling with the challenge of harmonizing their public, scientific, and conservation goals since at least the late 19th century. From the very beginning, zoos were forced to realize that they could not survive as purely scientific institutions for research and education; they had to build and sustain their popularity with an often fickle public, which was typically more interested in seeing new and exhilarating animal attractions than it was in learning scientific facts about natural history or about wildlife conservation (Hochadel 2005: 39–40). Old story or not, it's nevertheless a friction that leaves critics of zoo and aquarium conservation wondering whether the effort to synchronize entertainment, education, and conservation programs and values will in practice continue to stifle development of a strong and clear conservation philosophy in most zoological institutions (McKie 2020).

Conclusion

Zoos clearly face a significant set of institutional and practical constraints – namely space, capacity, resources, and in some cases, expertise – that will continue to complicate their ability to make headway on the global extinction crisis in the decades ahead. Still, as we've seen, the zoo community is also positioning itself to take on more of an activist role in biodiversity conservation in the coming years, a move consistent with their modern origins, even if their commitment to conservation has historically taken a backseat to other interests and values.

As those zoo landscapes devoted most seriously to conservation become more naturalistic and more park-like, however, and as natural areas and refugia become more intensively managed for conservation and other ends, we may witness the rise of a distinct, hybrid natural area-conservation facility – and a new multi-sector institutional form – devoted to the goals of wildlife science, conservation, and perhaps more limited display of animals for education and public entertainment. It will be interesting to consider how such places will impact traditional views toward the ethics of animal care, wildlife management, and nature preservation – as well as our familiar understanding of what a zoo is – as we continue to muddy the waters running between *ex situ* and *in situ* conservation.

Furthermore, given predictions about the high degree of "conservation reliance" of many vulnerable species today, i.e., their dependence on intensive intervention and management for the foreseeable future (Scott et al. 2010; Rohlf, Carroll, and Hartl 2014), it certainly stands to reason that the managerial presence at wildlife reserves and parks will only increase in the coming decades. One of the more daunting ethical challenges presented by this scenario is therefore whether we can, in this more actively managed environment, maintain our respect for the wildness and autonomy of nature that remains. Doing so might require restraining some of our more aggressive interventions that threaten to undermine these values even as we take the full measure of a world increasingly of our own making.

Related Topics

2. Eating
3. Experimentation
4. Companion Animals
6. Wild Animals
7. Hunting
44. Restoration
45. Assisted Migration and Reintroduction
47. Rewilding

References

Association of Zoos and Aquariums (AZA) (2014) *Association of Zoos and Aquariums*. Available at: https://www.aza.org/.

Association of Zoos and Aquariums (AZA) (2015) AZA SAFE: saving animals from extinction. *Association of Zoos and Aquariums*. Available at: https://www.aza.org/safe/.

Association of Zoos and Aquariums (AZA) (2019) 2109 Annual report on conservation and science. Available at: https://assets.speakcdn.com/assets/2332/aza_arcshighlights_2019_final_web.pdf.

Balmford, A. et al. (2011) 'Zoos and captive breeding', *Science*, 332, pp. 1149–1150.

Barongi, R., F. A. Fisken, M. Parker, and M. Gusset, eds. 2015. *Committing to conservation: the world zoo and aquarium conservation strategy*. Gland: World Association of Zoos and Aquaria Executive Office.

Barrow, M. V., Jr. (2009) *Nature's ghosts: confronting extinction from the age of Jefferson to the age of ecology*. Chicago: University of Chicago Press.

Barrow, M. (2018) 'Teetering on the brink of extinction: the passenger pigeon, the bison, and American zoo culture in the late nineteenth and early twentieth centuries' in Minteer, B. A., Maienschein, J., and Collins, J. P. (eds.) *The ark and beyond: the evolution of zoo and aquarium conservation*. Chicago: University of Chicago Press, pp. 51–64.

Beck, B. B. et al. (1994) 'Reintroduction of captive-born animals', in Olney, P. S., Mace, G. and Feistner, A. T. C. (eds.) *Creative conservation: interactive management of wild and captive animals*. London: Chapman & Hall, pp. 265–286.

Bekoff, M. (2002) 'Ethics and marine mammals', in Perri, W., Würsig, B., and Thewissen, H. (eds.) *Encyclopedia of marine mammals*. San Diego, CA: Academic Press, pp. 398–404.

Biello, D. (2016) *The unnatural world: the race to remake civilization in earth's newest age*. New York: Scribner.

Braverman, I. (2015) *Wild life: the institution of nature*. Stanford, CA: Stanford University Press.

Bridges, W. (1974) *Gathering of animals: an unconventional history of the New York Zoological Society*. New York: Harper & Row.

Brinkley, D. (2009) *The wilderness warrior: Theodore Roosevelt and the crusade for America*. New York: Harper Collins.

Ceballos, G., Ehrlich, P. R., and Raven, P. (2020) 'Vertebrates on the brink as indicators of biological annihilation and the sixth mass extinction', *Proceedings of the National Academy of Sciences* 201922686; DOI: 10.1073/pnas.1922686117.

Che-Castaldo, J. P., Grow, S. A., and Faust, L. J. (2018) 'Evaluating the contribution of North American zoos and aquariums to endangered species recovery', *Scientific Reports*, 8, https://doi.org/10.1038/s41598-018-27806-2.

Collins, J. P. and Crump, M. L. (2009) *Extinction in our times: global amphibian decline*. Oxford: Oxford University Press.

Conde, D. A. et al. (2011) 'An emerging role of zoos to conserve biodiversity', *Science*, 331, pp. 1390–1391.

Conde, D. A. et al. (2013) 'Zoos through the lens of the IUCN Red List: a global metapopulation approach to support conservation breeding programs', *PLoS ONE*, 8, e80311.

Conway, W. G. (2011) 'Buying time for wild animals with zoos', *Zoo Biology*, 30, pp. 1–8.

Davis, S. G. (1997) *Spectacular nature: corporate culture and the Sea World experience*. Berkeley: University of California Press.

Dirzo, R. et al. (2014) 'Defaunation in the Anthropocene', *Science*, 345, pp. 401–406.

Dolman, P. M. et al. (2015) 'Ark or park: the need to predict relative effectiveness of *Ex Situ* and *In Situ* conservation before attempting captive breeding', *Journal of Applied Ecology*, 52, pp. 841–850.

Donlan, C. J. et al. (2005) 'Re-wilding North America', *Nature*, 436, pp. 913–914.

Eveleth, R. (2010) 'Zoo illogical: ugly animals need protection from extinction, too', *Scientific American*. Available at: http://www.scientificamerican.com/article/zoo-illogical-ugly-animal/

Fa, J. E., Funk, S. M., and O'Connell, D. (2011) *Zoo conservation biology*. Cambridge: Cambridge University Press.

Friese, C. (2018) 'Cloning in the zoo: when zoos become parents', in Minteer, B. A., Maienschein, J., and Collins, J. P. (eds.) *The ark and beyond: the evolution of zoo and aquarium conservation*. Chicago: University of Chicago Press, pp. 267–278.

Gilbert, T. et al. (2017) 'Contributions of zoos and aquariums to reintroductions: historical reintroduction efforts in the context of changing conservation perspectives', *International Zoo Yearbook*, 51, pp. 15–31.

Gippoliti, S. and Carpaneto, G. M. (1997) 'Captive breeding, zoos, and good sense', *Conservation Biology*, 11, pp. 806–807.

Goddard, D. (1995) *Saving wildlife: a century of conservation*. New York: Harry N. Abrams.

Greene, H. (2015) 'Pleistocene rewilding and the future of biodiversity', in Minteer, B. A. and Pyne, S. J. (eds.) *After preservation: saving American nature in the age of humans*. Chicago: University of Chicago Press, pp. 105–113.

Griffiths, R. A. and Pavajeau, L. (2008) 'Captive breeding, reintroduction, and the conservation of amphibians', *Conservation Biology*, 22, pp. 852–861.

Gruen, L. (ed.) (2014) *The ethics of captivity*. Oxford: Oxford University Press.

Hancocks, D. (2001) *A different nature: the paradoxical world of zoos and their uncertain future*. Berkeley: University of California Press.

Hanson, E. (2002) *Animal attractions: nature on display in American zoos*. Princeton, NJ: Princeton University Press.

Hoage, R. J. and Deiss, W. A. (eds.) (1996) *New worlds, new animals: from menagerie to zoological park in the nineteenth century*. Baltimore, MD: The Johns Hopkins University Press.

Hobbs, R. J. et al. (2011) 'Intervention ecology: applying ecological science in the twenty-first century', *BioScience*, 61, pp. 442–450.

Hochadel, O. (2005) 'Science in the 19th-century zoo', *Endeavour*, 29, pp. 38–42.

Hoffmann, M. et al. (2010) 'The impact of conservation on the status of the world's vertebrates', *Science*, 330, pp. 1503–1509.

Hyson, J. (2000) 'Jungles of Eden: the design of American zoos', in Conan, M. (ed.) *Environmentalism in landscape architecture*. Washington, DC: Dumbarton Oaks Research Library and Collection, pp. 23–44.

Isenberg, A. C. (2000) *The destruction of the Bison*. Cambridge: Cambridge University Press.

Jamieson, D. (1985) 'Against zoos', in Singer, P. (ed.) *In defense of animals*. Oxford: Basil Blackwell, pp. 108–117.

Jamieson, D. (1995) 'Zoos revisited', in Norton, B. G., Hutchins, M., Stevens, E., and Maple, T. (eds.) *Ethics on the ark: zoos, animal welfare, and wildlife conservation*. Washington, DC: Smithsonian Institution Press, pp. 52–66.

Junhold, J. and Oberwemmer, F. (2011) 'How are animal keeping and conservation philosophy of zoos affected by climate change?' *International Zoo Yearbook*, 45, pp. 99–107.

Kawata, K. (2011) 'Of circus wagons and imagined nature: a review of American zoo exhibits, part II', *Der Zoologische Garten*, 80, pp. 352–365.

Keulartz, J. (2015) 'Captivity for conservation? Zoos at a crossroads', *Journal of Agricultural and Environmental Ethics*, 28, pp. 335–351.

Kimble, A. (2014) 'Fossil rim wildlife center celebrates 30 years', *The Glen Rose Reporter*, 17 May [online]. Available at: http://www.yourglenrosetx.com/visitors/article_dd315e87-69f2-5fea-b733-9a618a090d85.html).

Kisling, V. N., Jr. (ed.) (2001) *Zoo and aquarium history: ancient animal collections to zoological gardens*. Boca Raton, FL: CRC Press.

Kisling, V. N., Jr. (2018) 'Historic and cultural foundations of zoo conservation: a narrative timeline', in Minteer, B. A., Maienschein, J., and Collins, J. P. (eds.) *The ark and beyond: the evolution of zoo and aquarium conservation*. Chicago: University of Chicago Press, pp. 41–50.

Kleiman, D. G. (1989) 'Reintroduction of captive mammals for conservation', *Bioscience*, 39, pp. 152–161.

Kolbert, E. (2014) *The sixth extinction: an unnatural history*. New York: Henry Holt.

Kreger, M. D. and Hutchins, M. (2010) 'Ethics of keeping mammals in zoos and aquariums', in Kleiman, D. G., Thompson, K. V., and Baer, C. K. (eds.) *Wild mammals in captivity: principles and techniques for zoo management*. 2nd edition. Chicago: University of Chicago Press, pp. 3–10.

Lacy, R. C. (2012) 'Achieving true sustainability of zoo populations', *Zoo Biology*, 32, pp. 19–26.

Lees, C. M. and Wilcken, J. (2009) 'Sustaining the ark: the challenges faced by zoos in maintaining viable populations', *International Zoo Yearbook*, 43, pp. 6–18.

Marris, E. (2011) *Rambunctious garden: saving nature in a post-wild world*. New York: Bloomsbury.

Martin, T. E. et al. (2014) 'Mammal and bird species held in zoos are less endemic and less threatened than their close relatives not held in zoos', *Animal Conservation*, 17, pp. 89–96.

Mascarelli, A. (2013) 'Ecology: conservation in captivity', *Nature*, 498, pp. 261–263.

Mazur, N. (2001) *After the ark? Environmental policy-making and the zoo*. Victoria, VI: Melbourne University Press.

McKie, R. (2020) 'Is it time to shut down zoos?', *The Guardian*, February 2. Available at: https://www.theguardian.com/world/2020/feb/02/zoos-time-shut-down-conservation-education-wild-animals.

Meine, C. (2010, rev. ed.) *Aldo Leopold: his life and work*. Madison: University of Wisconsin Press.

Mendelson, J. (2018) 'Frogs in glass boxes: responses of zoos and aquariums to global amphibian extinctions', in Minteer, B. A., Maienschein, J., and Collins, J. P. (eds.) *The ark and beyond: the evolution of zoo and aquarium conservation*. Chicago: University of Chicago Press, pp. 298–310.

Minteer, B. A. (2018) *The fall of the wild: extinction, de-extinction, and the ethics of conservation*. New York: Columbia University Press.

Minteer, B. A. and Collins, J. P. (2012) 'Species conservation, rapid environmental change, and ecological ethics', *Nature Education Knowledge*, 3, p. 14. Available at: http://www.nature.com/scitable/knowledge/library/species-conservation-rapid-environmental-change-and-ecological-67648942.

Minteer, B. A. and Collins, J. P. (2013) 'Ecological ethics in captivity: balancing values and responsibilities in zoo and aquarium research under rapid global change', *ILAR Journal*, 54, pp. 41–51.

Minteer, B. A., Maienschein, J., and Collins, J. P. (eds.) (2018) *The ark and beyond: the evolution of zoo and aquarium conservation*. Chicago: University of Chicago Press.

Minteer, B. A. and Pyne, S. J. (eds.) (2015) *After preservation: saving American nature in the age of humans*. Chicago: University of Chicago Press.

Mitman G. (1996) 'When nature *is* the zoo: vision and power in the art and science of natural history', *Osiris*, 11, pp. 117–143.

Monfort, S. (2014) "'Mayday mayday mayday,' the millennium ark is sinking!" in Holt, W. V., Brown, J. L., and Comizzoli, P. (eds.) *Reproductive sciences in animal conservation: progress and prospects*. New York: Springer, pp. 15–31.

Monfort, S. and Christen, C. A. (2018) 'Sustaining wildlife populations in human care: an existential value proposition for zoos', in Minteer, B. A., Maienschein, J., and Collins, J. P. (eds.) *The ark and beyond: the evolution of zoo and aquarium conservation*. Chicago: University of Chicago Press, pp. 313–319.

Nash, R. F. (2014) *Wilderness and the American mind*. 5th edition. New Haven, CT: Yale University Press.

New York Zoological Society (NYZS) (1897) *First annual report*. New York: L. S. Foster, printer.

Norton, B. G., Hutchins, M. Stevens, E., and Maple, T. (eds.) (1995) *Ethics on the ark: zoos, animal welfare, and wildlife conservation*. Washington, DC: Smithsonian Institution Press.

Palmer, C. T., Kasperbauer, J., and Sandøe, P. (2018) 'Bears or butterflies: How should zoos make value driven decisions about the composition of their collections?', in Minteer, B. A., Maienschein, J., and Collins, J. P. (eds.) *The ark and beyond: the evolution of zoo and aquarium conservation*. Chicago: University of Chicago Press, pp. 179–191.

People for the Ethical Treatment of Animals (PETA) (n.d.) *Animal rights uncompromised: zoos*. Available at: http://www.peta.org/about-peta/why-peta/zoos/.

Powell, D. M. (2019) 'Collection planning for the next 100 years: what will we commit to save in zoos and aquariums?' *Zoo Biology*, 38, pp. 139–148.

Pritchard, D. J. et al. (2012) 'Bring the captive closer to the wild: redefining the role of ex situ conservation', *Oryx*, 46, pp. 18–23.

Rabb, G. B. and Saunders, C. D. (2005) 'The future of zoos and aquariums: conservation and caring', *International Zoo Yearbook*, 39, pp. 1–26.

Rands, M. R. et al. (2010) 'Biodiversity conservation: challenges beyond 2010', *Science*, 329, pp. 1298–1303.

Redford, K. H., Jensen, D. B., and Breheny, J. J. (2012) 'Integrating the captive and the wild', *Science*, 338, pp. 1157–1158

Regan, T. (1995) 'Are zoos morally defensible?' in Norton, B. G, Hutchins, M., Stevens, E., and Maple, T. (eds.) *Ethics on the ark: zoos, animal welfare, and wildlife conservation*. Washington, DC: Smithsonian Institution Press, pp. 38–51.

Reid, G. M. and Zippel, K. C. (2008) 'Can zoos and aquariums ensure the survival of amphibians in the 21st century?' *International Zoo Yearbook*, 42, pp. 1–6.

Ritvo, H. (1987) *The animal estate: the English and other creatures in the Victorian age*. Cambridge, MA: Harvard University Press.

Robertson, C. (2013) 'On Louisiana range, the giraffe and antelope will play', *The New York Times*, 14 January [online]. Available at: http://www.nytimes.com/2013/01/15/us/near-new-orleans-a-new-effort-to-breed-endangered-species.html?_r=0.

Rohlf, D. J., Carroll, C., and Hartl, B. (2014) 'Conservation-reliant species: toward a biology-based definition', *BioScience*, 64, pp. 601–611.

Rothfels, N. (2002) *Savages and beasts: the birth of the modern zoo*. Baltimore, MD: Johns Hopkins University Press.

Sawyer, R. (2012) 'Conservation centers for species survival aid species recovery', *U.S. Fish and Wildlife Service Endangered Species Program*. Available at: http://www.fws.gov/Endangered/news/episodes/bu-01-2012/c2s2/index.html.

Schwartz, N. D. (2014) 'Anger erupts after Danish zoo kills a 'surplus' giraffe', *The New York Times*, 9 February [online]. Available at: http://www.nytimes.com/2014/02/10/world/europe/anger-erupts-over-danish-zoos-decision-to-put-down-a-giraffe.html.

Scott, J. M. et al. (2010) 'Conservation-reliant species and the future of conservation', *Conservation Letters*, 3, pp. 91–97.

Stokstad, E. (2010) 'Despite progress, biodiversity declines', *Science*, 329, pp. 1272–1273.

Stone, R. (2013) 'A rescue mission for amphibians at the brink of extinction', *Science*, 339, p. 1371.

Stott, J. R. (1981) 'The historical origins of the zoological park in American thought', *Environmental Review*, 5, pp. 52–65.

Tlusty, M. F. et al. (2013) 'Opportunities for public aquariums to increase the sustainability of the aquatic animal trade', *Zoo Biology*, 32, pp. 1–12.

Traylor-Holzer, K., Leus, K., and Byers, O. (2018) 'Integrating ex situ management options as part of a one plan approach to species conservation', in Minteer, B. A., Maienschein, J., and Collins, J. P. (eds.) *The ark and beyond: the evolution of zoo and aquarium conservation*. Chicago: University of Chicago Press, pp. 130–141.

Tullis, P. (2014) 'When you walk into a zoo, are you helping animals or hurting them?' *Take Part*, 2 May [online]. Available at: http://www.takepart.com/feature/2014/05/02/do-zoos-matter.

Wayre, P. (1969) 'The role of zoos in breeding threatened species of mammals and birds in captivity', *Biological Conservation*, 2, pp. 47–49.

Wildt, D. et al. (2019) 'Breeding centers, private ranches, and genomics for creating sustainable wildlife populations', *BioScience*, 69, pp. 928–943.

Wirtz, P. H. (1997) 'Zoo city: bourgeois values and scientific culture in the industrial landscape', *Journal of Urban Design*, 2, pp. 61–82.

World Association of Zoos and Aquariums (WAZA) (2014) *About WAZA: history.* Available at: http://www.waza.org/en/site/about-waza/history.

WWF (2020) *Living planet report 2020 - bending the curve of biodiversity loss.* Almond, R.E.A., Grooten M., and Petersen, T. (eds.). Gland, Switzerland: WWF.

Zippel, K. et al. (2011) 'The amphibian ark: a global community for ex situ conservation of amphibians', *Herpetological Conservation and Biology*, 6, pp. 340–352.

47

REWILDING

Derek Turner

I would not be surprised to read someday that cheetahs are helping to control deer and that mesquite is being "overbrowsed" by rhinoceroses.

—*Michael Soulé (1990: 235)*

I take the stand as a character witness for wolf trees.

—*Charles Elliott (1945)*

Introduction: Rewilding as a Form of Environmental Nostalgia

Rewilding, very roughly, is any effort to return a landscape to what we imagine to be a wilder state. In recent years, popular writers have portrayed rewilding as an exciting new direction in Western environmental thought (Foreman 2004; Fraser 2009; Monbiot 2013a; Tree 2019). In this overview, I will focus less on the definitional questions—what exactly does "rewilding" mean?— and more on rewilding's relationship to different kinds of environmental nostalgia. (For further discussion of the meanings of "rewilding," see Jørgensen 2015; Prior and Ward 2016; Tananescu 2017; Gammon 2018; Keulartzt 2018.) Cultural theorist Svetlana Boym (2001) draws a distinction between *reflective* and *restorative* nostalgia:

> Restorative nostalgia stresses *nóstos* (home) and attempts a transhistorical reconstruction of the lost home. Reflective nostalgia thrives in *álgos*, the longing itself, and delays the homecoming—wistfully, ironically, desperately.
>
> *(Boym 2001: xviii)*

Restorative nostalgia is more active, seeking to recreate or re-establish an imagined past. By contrast, reflective nostalgia is more accepting and more contemplative, a kind of bittersweet dwelling on what has been lost. I will argue that Boym's distinction between these two modes of nostalgia tracks a distinction between two forms of rewilding: a more passive version that involves a commitment to letting non-human nature go its own way, as contrasted with more active, interventionist versions of rewilding. My goal here is a modest one: I will try to place these two forms of rewilding into productive tension with each other. Perhaps one way to temper enthusiasms for more ambitious interventionist forms of rewilding is to remind ourselves of the value of humbler, more reflective engagement with natural processes.

DOI: 10.4324/9781315768090-56

The most familiar experiences of nostalgia probably involve bittersweet reminiscences about earlier moments in our own lives. It is also possible, though, to be nostalgic for past times that we have never experienced, for an imagined past. Nostalgia often involves a tacit comparison of past and present, accompanied by the thought that something valuable has been lost, or that the past (at least, as we imagine it) was in some respects better (Howard 2012). One fascinating example of restorative nostalgia, in Boym's sense, is the popular paleo diet (for critical assessment, see Zuk 2014). Converts to the paleo diet try to eat, and in some cases exercise, like our ancestors who subsisted by hunting and gathering. Paleo dieters eschew grains and other agricultural products. As the paleodiet illustrates, sometimes restorative nostalgia can involve longing for a rather distant past that we ourselves never experienced. But if the paleo diet is an exercise in restorative nostalgia, then so, too, is much of the sustainable agriculture movement. As Michelle Neely argues, "sustainability's ideal future often involves the recovery of an idyll" (2020: 10). For many, this ideal future looks like a return to an imagined pre-industrial agricultural system. The difference here concerns the targets of restorative nostalgia: are you nostalgic for world before the rise of *industrial* agriculture? Or for a world before the rise of *agriculture*? How, then, do things look when what we're nostalgic for is a wilder, less human world?

In what follows, I survey some of the different things that "rewilding" has come to mean, while keeping Boym's distinction between reflective and restorative nostalgia in mind (Table 47.1). My survey begins with what I take to be the least controversial but perhaps also the least discussed sort of case—the unplanned, inadvertent rewilding of northeastern U.S. forests—and proceeds from there to consider more controversial proposals and cases, culminating with recent debates about the use of biotechnology to reverse extinctions. The restorative nostalgia gets more intense as we move from each of these proposals to the next. In a 2004 essay, "Letting the world do the doing," Freya Mathews argues that if we want to "return to nature" we should resist the impulse to remake landscapes according to our own designs and learn how to let things be (compare also Monbiot 2013b). This humbler, more passive attitude toward nature—a kind of environmental forbearance—is at odds with restorative nostalgia. But the more passive stance may be the best way to live in a relatively wilder world. Interestingly, critics of rewilding tend to focus more on active rewilding projects, and have had less to say about passive rewilding (Nogués-Bravo et al. 2016). Passive rewilding sometimes gets mentioned in the course of more general discussions of rewilding (Corlett 2016: 454), but proponents of rewilding often have more ambitious interventionist projects in mind. Where rewilding advocates do focus on cases of passive rewilding, they sometimes treat them as sources of data that can inform more active rewilding projects (see, e.g., Svenning et al. 2016: 901, on learning from "spontaneous wildlife comebacks").

Table 47.1 Varieties of rewilding

Reflective nostalgia	Restorative Nostalgia		
Passive rewilding	"Three C's" Rewilding	Pleistocene Rewilding	De-Extinction
Wolf trees in a northeastern woodland	Reintroduction of gray wolves to Yellowstone	Introducing ecological proxies for long extinct megafauna (e.g. elephants for woolly mammoths)	Using biotechnology to create ecological proxies that have the traits of extinct species (e.g. elephants with mammoth like traits)
Little or no human intervention	Intensification of intervention		

Unplanned Rewilding and "The Irony of Eastern Wilderness"

In some parts of the U.S., a certain amount of rewilding is happening naturally, without much deliberate planning or management (Klyza 2001; Davis 2015). John Elder writes that "The irony of eastern wilderness is that, while it may have seemed to receive that title as a courtesy, the vector of wildness may actually be more remarkable here than anywhere in the west" (2001: 257). During colonial times and up through the 19th century, most of the rural landscape in Connecticut, for example, was deforested. However, agriculture went into decline as people took up other economic activities and farmers packed up and moved west to places where farming was more promising. Over the 20th century, the forests began to come back. Today, the wooded areas protected in state forests and local land trust preserves are, to some degree, rewilded landscapes. Cockaponsett State Forest, in central Connecticut, is home to a number of "wolf trees"—huge, old sprawling oaks with branches spreading laterally, with younger maples growing around and sometimes straight up through them (Figure 47.1). When you see a wolf tree, sometimes also called a "pasture tree," you know that you are standing in someone's old pasture or farm field (Shaw 2015). If the oak had started its life in the middle of a forest, it would have grown more vertically. And of course, you cannot travel very far through the woods in Connecticut, even in areas that feel relatively wild, without encountering a stone wall. Occasionally you will also run across the remains of an old charcoal kiln, or the foundations of an old farmhouse.

In recent decades, wild turkeys, foxes, black bears, beavers, coyotes, fisher cats, and even a few moose have rebounded and/or returned to Connecticut. According to the Connecticut Department of Energy and Environmental Protection, there were over 5,000 black bear sightings in the state in 2021. The return of large animals is seen by many as a sign that Connecticut's woodlands are growing wilder.

Figure 47.1   A wolf tree in Cockaponsett State Forest, Connecticut, USA

As noted above, Boym (2011) contrasts restorative nostalgia with *reflective nostalgia*. This second variety of nostalgia does not seek to recreate a longed-for past. Instead, Boym writes that reflective nostalgia "savors details and memorial signs," and that it involves a "meditation on history and passage of time" (2011: 49). The contemplation of a wolf tree or a stone wall in a regrown forest is a good example of reflective nostalgia. There's no effort or even any desire to recreate the farmstead that the forest is slowly obliterating. But there is still something bittersweet in the experience of contemplating the wolf tree, as one is reminded that rewilding entails the gradual loss of traces of a more humanized landscape with which we can easily identify. Mathews (2004) captures this same feeling of reflective nostalgia when she writes that

> 'returning to nature' in an urbanized world means allowing this world to go its own way. It means letting the apartment blocks and warehouses and roads grow old … Gradually such a world, left to grow old, rather than erased for the sake of something entirely new, will be absorbed into the larger process of life on Earth. Concrete and bricks will become weathered and worn. Moss and ivy will take over the walls. Birds and insects may colonize overhangs and cavities within buildings … Left to itself, the living world reclaims its own.
>
> *(2004: 4)*

The difference between restorative and reflective nostalgia closely tracks the difference between wolf reintroduction and wolf trees. Whereas restorative nostalgia seeks to alter nature in accordance with human designs, reflective nostalgia lingers on the traces of a human world that nature has reclaimed. Of course, reflective nostalgia could in principle work the other way, too. One could also experience reflective nostalgia upon contemplating a remnant of wild nature in an area that is rapidly urbanizing. If anything, the contemplation of a wolf tree in the northeastern woods may involve a double reflection: the reflection on nature's reclaiming the landscape that Euro-American settler colonists had modified may also inspire reflection upon earlier periods of the landscape's history.

Connecticut, which is the fourth most densely populated state in the U.S., probably does not leap to mind as an example of a wild place. It has no national parks and no wilderness areas, although the Eightmile River, which feeds into the Connecticut River about eight miles above Long Island Sound, has received federal designation as a Wild and Scenic River. But the reforestation of the state over the last century has brought with it some unplanned, inadvertent rewilding, and that process is pretty far along in some places. These changes make the northeastern U.S. an excellent case study to think about when assessing more ambitious proposals for rewilding in other regions (Klyza 2001). Here we can get a good handle on some of the downsides of rewilding, which include increased potential for human-wildlife conflicts, as well as very serious habitat reduction for some species, such as birds that prefer meadows. But we can also see some of the advantages, which might include, in addition to some of the more obvious conservation benefits, the spiritual and aesthetic benefits of living in closer proximity to relatively wilder places.

Prior and Brady (2017) explore some of the connections between environmental aesthetics and rewilding. However, they focus more on active rewilding, defining it as "a process of (re)introducing or restoring wild organisms and/or ecological processes" (p. 34). This discussion of passive rewilding suggests a need for broadening the discussion of rewilding and aesthetic engagement. Reflective nostalgia is a distinctive way of engaging aesthetically with landscapes, and reflective nostalgia involves contemplating the history of the landscape. Understanding the history of a landscape can enhance our aesthetic engagement with it and help us to cultivate a sense of place (Turner 2019).

Rewilding as a Conservation Strategy: Cores, Corridors, and Carnivores

Unplanned rewilding in the northeastern U.S. is a far cry from what conservationists had in mind when they first introduced the term "rewilding" in the 1990s. In 1998, Michael Soulé and Reed Noss published a programmatic essay recommending rewilding as an approach to conservation biology (compare also Foreman 1998, 2004). Their original idea was not too radical, though as we'll see, others have in the meantime picked up on it and carried it much further. Soulé and Noss recommended an approach that emphasizes the three Cs: cores, carnivores, and corridors. The "cores" they envisioned were large, undeveloped wilderness areas, such as those established in the western U.S. under the Wilderness Act of 1964. Soulé and Noss also emphasized the reintroduction of keystone species, especially carnivores. This emphasis on reintroduction of animal species also helped distinguish rewilding from traditional restoration ecology, which tends to focus more on reconstructing plant communities. Soulé and Noss worried about trophic cascades that can happen when a top predator is removed from an ecosystem. Many biologists think that this happened, for instance, when wolves were removed from the greater Yellowstone ecosystem in the early part of the 20th century. The species that the wolves had preyed upon—especially elk and mule deer—increased in abundance with damaging effects on other populations. Finally, Soulé and Noss stressed the importance of connectivity among the core wilderness areas. Rewilding, they argued, should entail the creation of wildlife corridors so that populations do not get stranded in isolated pockets of protected wilderness. The corridors would facilitate the movement of populations between protected areas.

One important goal of rewilding, as Soulé and Noss conceived of it, was to build a bridge between biodiversity conservation and wilderness protection. One consequence is that rewilding inherits all the problems that theorists such as William Cronon have identified with the concept of wilderness (Cronon 1996; for more discussion, see the papers collected in Callicott and Nelson 1998). Protecting biodiversity and protecting wilderness are distinct environmental concerns. Sometimes, these values can pull in different directions. For example, protecting biodiversity might require human intervention in nature, ranging from efforts to eradicate invasive species to efforts to breed endangered animals in captivity. These sorts of interventions do not sit well with the idea that wild places have special value precisely because humans have not interfered with them. Rewilding, in the sense of Soulé and Noss, draws upon wilderness advocates' idea that we should be setting aside big roadless tracts of land—the "cores"—and placing them off limits for most human activities. They also argued that reintroducing big keystone species—the "carnivores"—would restore the "emotional essence" of wilderness (1998: 7). Finally, they claimed that rewilding would be an effective strategy for protecting biological diversity. For example, connecting wilderness cores is often crucial for the protection of migratory species. And the trophic cascades that result from the loss of top carnivores can also be bad for diversity. If fewer wolves kill fewer elk and mule deer, overbrowsing will have an impact on plant species.

Perhaps there is some internal tension in the idea that human beings can do anything at all to make a landscape wilder (Katz 1992). Some wilderness advocates might favor an immediate "hands off" approach: if we want wilder landscapes, the best approach is not to intervene at all, but rather to let nature go its own way. In reply to this concern, advocates for reintroducing keystone species might argue that their goal is, ultimately, to let nature go its own way, at least in protected, rewilded areas. The crucial issue is the starting point: should we take a "hands off" approach now and let a damaged and degraded landscape go its own way? Or should we actively undertake some restoration/reintroduction efforts first, with the aim of taking a more "hands off" approach later on when the system is in a better condition? Proponents of rewilding tend to favor the second approach of slowly managing our way back to a wilder landscape. Taking an immediate "hands off" approach might not be the best way to minimize the overall human impact on a landscape.

Although it may sound paradoxical, in a case where human activities have already significantly altered a landscape (say, by exterminating a keystone species), advocates of rewilding, in the sense of the three Cs, argue that some further human intervention might be needed to make the landscape wild again.

Notice how different Soulé and Noss's vision of big, interconnected wilderness areas is from the process of inadvertent rewilding that has occurred in the northeastern U.S. The rural landscape in Connecticut is arguably wilder than it used to be, but of course the state has no significant roadless wilderness areas. Some carnivores—especially foxes and coyotes—have rebounded or moved in. But there are (so far) no significant populations of wolves or mountain lions, though there are alleged sightings of big cats. At the very least, we need to distinguish between planned and unplanned rewilding, but intentionality (or lack thereof) is not the only difference between the two cases. Another difference is that in the northeast, the areas that are, in some sense, getting wilder are also fairly densely populated. The Soulé and Noss model inherits from the wilderness movement an emphasis on protecting places where nobody lives. This is too bad, because it ignores the fact that places where people live can also become relatively wilder. Soulé and Noss's proposal has inspired a variety of rewilding projects around the world (Fraser 2009; Manning 2009; Monbiot 2013a). And as we'll see, a few conservationists in Europe and North America have carried the idea to new extremes.

Pleistocene Rewilding

The hedge apple tree, or Osage orange (*Maclura pomifera*), is a lovely example of an evolutionary anachronism (Janzen and Martin 1982; Barlow 2002). It has marvelous bumpy green fruits the size of softballs. Plants that produce enticing fruit have typically co-evolved with animal species that serve as seed dispersers. The striking thing about hedge apples is that there are no wild animals in North America today that are big enough to eat them. The only animals that even come close are bison, but they are primarily grazers. Almost certainly, the original seed dispersers for the Osage orange were mammoths and mastodons, creatures that were extinct by the end of the Pleistocene, 11,000–12,000 years or so ago. It's tough to think about these evolutionary anachronisms—and there are many of them in the western hemisphere, from avocados to honey locust trees to pronghorn antelope—without getting the sense that something is missing from the ecosystem.

One challenge for restoration ecologists is to determine the historical reference conditions for a particular project (Callicott 2002; Higgs 2003). In general, and setting aside many philosophical complications, restoration ecologists start with ecosystems that human activities have damaged, and they seek to restore those ecosystems to an earlier, undamaged state. In some cases in North America, it may be tempting to try to restore environments to conditions prior to the arrival of European settlers and colonists. That approach, however, seems to assume that Indigenous North Americans did not impact their environments much, which is inaccurate. Certainly, when Europeans arrived, the North American landscape had already been used and modified by people for thousands of years. For example, at places like Cahokia, Illinois, there were large population centers with monumental architecture, supported by extensive agricultural systems and trading networks (Pauketat 2009). One might think that the more principled approach is to restore North American ecosystems to the condition they were in before any human beings showed up at all. This is the idea that animates Pleistocene rewilding. Just which people got to North America first, how they got there, and where they came from are all issues that archaeologists and anthropologists continue to investigate (Raff 2022). We know that people had occupied North America by 11,000–12,000 years ago, though there are controversial hints that they might have arrived considerably earlier. Although Pleistocene rewilding might seem more principled than other forms of restoration because the goal is to return the landscape to something like the condition it was in before human migration to the Americas, it is also vulnerable to the objection that it is just contributing to the erasure of Indigenous communities from North America.

One early advocate for Pleistocene rewilding, Paul Martin, also defended what is known as the Pleistocene overkill hypothesis (Martin 1966, 2005). According to this view, human hunting contributed to the end-Pleistocene extinction of the megafauna in North and South America, including everything from woolly mammoths to giant ground sloths. The hypothesis gains some plausibility from better understood cases where we know that humans rapidly exterminated large animals. For example, when the Maori arrived in New Zealand, it did not take long for hunting pressure to contribute to the extinction of the flightless moa (Perry et al. 2014). Nevertheless, the Pleistocene overkill story remains controversial as an explanation of continental extinctions (Koch and Barnosky 2006; Wolverton 2010; Cooper et al. 2015). There is evidence in the archaeological record that humans hunted mammoths, but many other species in North and South America went extinct at the end of the Pleistocene, and there's no clear evidence that humans hunted any of them. For present purposes, the important thing to see is that if the Pleistocene overkill hypothesis were to turn out to be correct, then it might appear to lend some support to the Pleistocene rewilding idea, at least when it's conjoined with certain normative assumptions. For starters, the Pleistocene overkill hypothesis would imply that human activities impacted North American ecosystems very early, perhaps not long after humans arrived. Some have the intuition that we have some special obligation to try to repair damage that our species has done. The Pleistocene overkill hypothesis gives rewilding the feel of atoning for past environmental sins.

What would Pleistocene rewilding actually look like in North America? Notwithstanding recent work on de-extinction, to which I will turn in the next section, the Pleistocene megafauna appear to be gone for good. But as Josh Donlan and colleagues (2005, 2006; Sandom et al. 2013) pointed out, the extinct megafauna of the western hemisphere have relatively close evolutionary cousins in Africa and Asia (compare also Martin and Burney 1999; Galetti 2004; Nicholls 2006). And some of those existing species are threatened by habitat loss and poaching. Occasionally, in ecological restoration, where one species is completely extinct one might introduce a closely related surrogate. Introducing large mammals from Africa and Asia into North American landscapes could be a way of hedging our bets against extinction. Even if the species disappear from their original ranges, they might hang on in "rewilded" North American landscapes. Lions and cheetahs, Bactrian camels, African elephants, and—who knows—perhaps even rhinoceroses could serve as surrogates for extinct evolutionary cousins that once roamed North America.

Pleistocene rewilding builds on the "three Cs" of Soulé and Noss (1998). The idea is to create a network of large, interconnected wilderness cores, while reintroducing keystone species, especially carnivores. Pleistocene rewilding just takes things further by seeking to introduce *proxies* for keystone species that have been extinct for 11,000–12,000 years. Critics, however, have raised a number of concerns (Smith 2005; Rubinstein et al. 2006; Caro 2007; Hintz 2007; Oliveira-Santos and Fernandez 2010; Sandler 2012: 91ff.; Minteer 2019). Some see this proposal as tantamount to an assisted biological invasion, and one that could do more ecological harm than good. It is not so easy to predict what sorts of ecological impacts elephants and rhinos might have in the American West. There is also great potential for human/wildlife conflict. (Re)introducing megafauna to North America could diminish people's appreciation of the biological diversity that's already there. It also sounds more than a little condescending to import wildlife from other parts of the world on the grounds that conservationists in the U.S. know better how to protect biodiversity. That aspect of Pleistocene rewilding has echoes of colonialism. And given the history of often violent displacement and erasure of Indigenous communities in North America by Euro-American settlers, Pleistocene rewilding can look like a problematic form of restorative nostalgia, like an effort to reverse the impacts that Indigenous people had on North American landscapes.

In addition to Donlan and colleagues' call for Pleistocene rewilding in western North America, Soulé and Noss's proposal has also been carried forward in fascinating ways by conservationists in Europe. One especially important test case for thinking about rewilding is the Oostvaardersplassen in the Netherlands (Marris 2009; Kolbert 2012). The site is on land reclaimed from the North Sea; in prehistoric times, it was under water. But conservationists working at the site have set out to challenge the view that prehistoric Europe was heavily forested. Their hypothesis is that big grazing animals—horses, aurochs (ancestral cattle that went extinct in the 1600s), deer, and wisents (European bison)—would have maintained grassland ecosystems. And they've set about recreating such an ecosystem at the Oostvaardersplassen, not too far from Amsterdam. They have brought in red deer as well as konik horses from Poland, which biologists think are closely related to the ancestral horses that roamed around prehistoric Europe. Most controversially, a breed of cattle known as Heck cattle are being used as proxies for the larger aurochs. The Heck cattle get their name from Heinz and Lutz Heck, two German scientists who in the 1930s set about to recreate the prehistoric aurochs by back breeding existing types of cattle (Lorimer and Driessen 2016). Part of the controversy has to do with the Heck brothers' close association with Nazism: the restorative nostalgia evident in their back-breeding program was affiliated with a more insidious political nostalgia for an imagined past characterized by racial purity. As it happened, the Heck brothers' efforts fell well short of recreating the aurochs anyway (Lorimer and Driessen 2013; Gremmen 2014).

Meanwhile, in Russia, a conservation biologist named Sergey Zimov has actually created a so-called "Pleistocene Park" in a remote part of Siberia (Zimov 2005; Lewis 2012). His stated goal is to recreate the ecosystem known as the "mammoth steppe." The mammoth steppe was a grassland that once stretched across much of Asia and North America, and was maintained by woolly mammoths and other big grazing animals. Today, in its place, we have tundra and taiga ecosystems. Pleistocene Park is currently home to a number of re(introduced) herbivores, such as Yakutian horses, musk oxen, reindeer, and even some bison imported from Canada. The thought is that over time, having healthy populations of big herbivores will convert some of the existing wet mossy tundra into a drier Pleistocene grassland. This could have the effect of reducing soil temperatures and slowing the loss of permafrost in an era of climate change. The Yakutian horses were the first species reintroduced, and a small population of them persists in the park today. According to the Park's website, six baby musk oxen were imported from Wrangel Island in 2010. All were males (Zimov 2014). It's not clear whether the project has achieved any major ecological results. Nor is it clear that the experiment in Pleistocene Park could be scaled up in a way that might make it relevant to climate change.

Part of the initial motivation for rewilding in the North American context is to restore the ecosystem to something resembling the condition it was in before humans arrived. In the European and Siberian contexts, however, humans and our close evolutionary ancestors (e.g. the Neanderthals and Denisovans) have impacted the ecosystems for tens of thousands of years. In those areas, humans were part of the Pleistocene landscape, and the motivation for rewilding may be somewhat different. Boym (2001: xiii) characterizes nostalgia as "a longing for a home that no longer exists." Our species has arguably spent much of its evolutionary history living in grassland environments—and those are the environments that people are trying to recreate in the Oostvaarderplassen and in Pleistocene Park.

De-Extinction

Pleistocene rewilding, as originally conceived, was an attempt to recreate Pleistocene environments using living animals as surrogates for the extinct megafauna. But if you think this is a good idea, you might also think that we should go a step further, and try to recreate the extinct animals. Although proponents of Pleistocene rewilding do not necessarily support de-extinction, any arguments in favor of recreating the mammoth steppe would also seem to lend at least some support to recreating the mammoths that once lived there. Indeed, Zimov's (2005) name for the rewilding project in Siberia—"Pleistocene Park"—hints at an ambition to bring back the extinct animals themselves, in the spirit of *Jurassic Park*. As it happens, the debate about Pleistocene rewilding unfolded in the 2000s at the same time that scientists began to develop techniques for paleogenomic sequencing. Ancient DNA research was gaining momentum (Jones 2022). In 2008, for example, scientists published a draft sequence of the genome of the extinct woolly mammoth (Miller et al. 2008). That immediately fueled speculations about using biotechnology to recreate the mammoths, and perhaps other extinct species (Nicholls 2008). Science writer Olivia Judson (2008) referred to this as "resurrection science," and it has since come to be known as de-extinction. The movement has gained considerable momentum, with a major TEDx conference on de-extinction sponsored by the National Geographic Society in the spring of 2013, as well as recent popular books (e.g. Shapiro 2016; Kornfeldt 2018; Wray 2019).

Except for small populations of holdouts on remote islands, woolly mammoths went extinct by 10,000 years ago. Carcasses preserved in permafrost contain genetic material that is in good enough condition for scientists to sequence. If we shift the focus to more recently extinct creatures, and to cases where humans' role in causing the extinction is less controversial, it turns out that natural history museums around the world contain a great many specimens whose fur and feathers could supply genetic material for sequencing (Wandeler et al. 2007). The Heck brothers' back breeding project, mentioned above, could be seen as an early and relatively unsophisticated attempt at de-extinction. In the last few years, scientists have begun to explore some other options that would take full advantage of the latest biotechnology.

The most straightforward approach would be to use genetic engineering (Shapiro 2016). If you could identify the woolly mammoth genes responsible for, say, long hair, those could theoretically be spliced into the genome of an Asian elephant (the mammoths' closest living relative). Remarkably, scientists have succeeded in using genetic engineering to create bacteria that produce mammoth blood protein (Campbell et al. 2010). Another more complicated approach that is probably much further over the technological horizon would be to recreate the full complement of chromosomes of some extinct species, enclose them in a cell nucleus, and then insert that nucleus into an enucleated egg cell taken from some living near relative of the extinct species. This approach uses the same technology that scientists used in the 1990s to clone Dolly the sheep, but with the added twist of using two different species. As it happens, scientists have already succeeded in cross-species cloning with two living species—African wildcats and housecats (Gomez et al. 2004; for another example, see Lanza et al. 2000).

De-extinction remains somewhat speculative, and it's entirely possible that in spite of the efforts of enthusiasts, the technology will never pan out. We're still a good way from being able to reintroduce extinct populations into the wild. We shouldn't assume because we know how to clone some species that cloning a woolly mammoth would be straightforward. Nicholls (2008) describes some of the technical challenges, starting with the difficulty of harvesting eggs from female elephants. However, it was not so long ago that cross-species cloning and genetic engineering using DNA from extinct species would have sounded to well-informed people like science fiction. So it behooves environmental philosophers to begin thinking seriously about de-extinction.

De-extinction raises some conceptual questions as well as ethical ones. For example, it puts some pressure on our intuitions about the meaning of "extinction" (Delord 2007, 2014; Siipi and Finkelman 2017). Is it an analytic truth that extinction is irreversible? De-extinction is also an interesting test case for the view that biological species are historical individuals. In his classic statement of that view, David Hull wrote that "if a species evolved which was identical to an extinct species of pterodactyl save origin, it would still be a new, distinct species" (1978: 349). This is a point about qualitative *vs.* numerical identity. The new species might be exactly like the extinct one, but it wouldn't be one and the same individual. The interesting issue here, from the perspective of philosophy of biology, is whether the historical connection between the mammoths of the Pleistocene and a mammoth-like animal created by biotechnologists would be the sort of connection that would ensure the continuity of the species *qua* historical individual (Siipi 2014 further explores these issues).

Setting aside these and other related issues in the philosophy of biology, philosophers have also begun to explore questions about the ethics of de-extinction (Salsberg 2000; Sherkow and Greely 2013; Gamborg 2014; Oksanen and Siipi 2014; Kasperbauer 2017; Sandler 2017; Minteer 2019). Two of the more serious problems with de-extinction have to do with animal welfare and resource allocation. Although the animal welfare concerns will depend upon the species under consideration as well as the techniques being used, the going proposals for de-extinction (e.g. cloning) would likely mean pain and distress for large numbers of animals (Turner 2017; Browning 2018). This issue is not unique to de-extinction, as other conservation methods, from captive breeding to the eradication of invasives, also raise animal welfare concerns. Second, de-extinction also represents "big science" requiring significant levels of funding that could be put to better use in the service of biodiversity conservation. A major concern is that the de-extinction drama is drawing private funding away from other conservation efforts where it would likely do more good (Bennett, Maloney, and Steeves et al. 2017).

The strongest argument in favor of de-extinction is one that I have elsewhere called "the restorationist argument" (Turner 2014). This argument is essentially the same as the one that proponents of rewilding have made for reintroducing locally extinct keystone species.

> P1. In general, it is a good thing to try to promote ecosystem health.
> P2. In some cases, the loss of some particular species is damaging to ecosystem health, and the reintroduction of that species, if successful, would help restore the system to health.
> C. Therefore, in those cases, it is a good thing to try to reintroduce species that have gone extinct.

Perhaps the most interesting feature of this argument is that it's neutral with respect to the difference between reintroducing a species that's extinct in some portion of its historic range and reintroducing a species that's extinct, full stop. This argument is vulnerable to objections. For example, it depends crucially on the notion of ecosystem health, which some have challenged (Jamieson 1995). But if we wanted to, we could replace "ecosystem health" in the above argument with some other set of ecological features that we might agree we wish to promote by reintroducing a species that has been absent from the system for some time. Note also that it's an empirical question whether reintroducing an extinct species would actually promote ecosystem health. It could well turn out that this is hardly ever the case. And even if it were the case, this line of argument could (and likely would) still be trumped by concerns about animal welfare and resource allocation. Another potential problem with the argument is that it might overshoot the mark and justify interventions that do not involve restoring lost biological diversity. If we could somehow show that adding a new species to an ecosystem would improve the health of that system—a big if, of course—then the argument might justify such an introduction. In presenting the argument here, my main goal is to make an observation about its structure: namely, that the case for de-extinction is structurally very similar to the original argument for reintroducing wolves into Yellowstone.

Many conservationists will, I'm sure, have a lot of sympathy for rewilding as Soulé and Noss originally conceived it. But many will want to jump ship once the conversation turns to Pleistocene rewilding or, even worse, de-extinction. A simple science fiction thought experiment will show how difficult it is to draw a principled line here. (See also Minteer 2019 for a rich and helpful discussion of some of these line-drawing problems.) Suppose that gray wolves have been totally extinct for decades. The reintroduction of wolves into Yellowstone never happened. But someone figures out how to use biotechnology to create a viable population of wolves, so that the reintroduction can go forward. Surely anyone who favors rewilding, in Soulé and Noss's sense, should also favor the reintroduction in this case (setting aside concerns about animal welfare and resource allocation). In both cases, there is significant human involvement in nature—trapping and relocating wild animals vs. recreating them—done for the sake of eventually letting nature go its own way.

De-extinction is a challenging topic that deserves more attention from philosophers. Here I have only introduced a few of the relevant ethical considerations. The main point I want to convey, however, is that de-extinction can be seen as an extension of earlier calls for rewilding. Earlier discussions of rewilding and especially Pleistocene rewilding are crucial context for understanding how the de-extinction debate has taken shape. The best available argument for de-extinction is just a version of the rewilders' argument for reintroducing keystone species. Although de-extinction advocates are now focusing on a variety of more recently extinct species, the sequencing of the mammoth genome was the big event that gave momentum to de-extinction. And the interest in woolly mammoths comes right out of the Pleistocene rewilding debate and Paul Martin's advocacy of the Pleistocene overkill hypothesis, not to mention Zimov's project of restoring the mammoth steppe. Remarkably, an article on de-extinction in *The New York Times Magazine* had a woolly mammoth on the cover, but the article scarcely mentioned mammoths at all. It was all about passenger pigeons (Rich 2014). The mammoth has become the emblem of de-extinction, even if it is not a terribly good de-extinction candidate. More recently extinct species whose habitats still exist would presumably be better candidates.

Finally, note that de-extinction research is an especially clear example of what Boym calls "restorative nostalgia." It's driven by a nostalgic impulse to bring back things from the past that have been lost, seemingly for good.

Conclusion: The Intensification of Restorative Nostalgia

In this overview, I've considered a spectrum of possible rewilding proposals, ranging from the less controversial to the more controversial. At the easy end of the spectrum we have cases of unplanned, inadvertent rewilding that few would find objectionable, although those do come with greater potential for human/wildlife conflict. I then considered Soulé and Noss's (1998) proposal, which in hindsight seems fairly mainstream, to create core wilderness areas, connected by corridors, with reintroduced keystone species. This has spawned a number of more ambitious proposals to recreate prehistoric ecological conditions, often using living animals as proxies for extinct keystone species that went extinct. Prehistoric rewilding leads inexorably to the idea of using biotechnology to recreate those extinct keystone species, and current work on de-extinction can be seen as an outgrowth of the rewilding movement.

At each stage of the narrative that I've constructed, the restorative nostalgia grows more intense. Indeed, at the first stage, the unplanned, *passive rewilding*, the urge to recreate past conditions is largely absent, and the rewilding has occurred as an accidental side effect of changes in land-use patterns. Unplanned rewilding might induce a more reflective nostalgia for the human-altered landscape that's disappearing, a reflective nostalgia that goes hand-in-hand with a willingness to "let the world do the doing" (Mathews 2004). The move toward purposeful, *active rewilding* comes with restorative nostalgia. Even in the original rewilding proposal that emphasized the three Cs—cores, corridors, and carnivores—the reintroduction of keystone species is tinged with restorative nostalgia, and a drive to put things back the way we imagine they were. The more radical proposals for Pleistocene rewilding represent nostalgic efforts to recreate Pleistocene environments. The movement to recreate extinct species indulges the restorative nostalgia even further.

The de-extinction movement is all about using the latest, fanciest, and most expensive biotechnology to fix perceived environmental problems. Indeed, if scientists could successfully cheat extinction, that would be a significant technological and scientific milestone. I myself have argued that the case for de-extinction might be a bit stronger than some critics realize, at least for some recently extinct species (Turner 2014; and see Seddon et al. 2014 for an attempt to prioritize de-extinction candidates). However, we should also bear in mind that de-extinction is completely antithetical to "letting nature itself decide much more and man decide much less." Whatever you think about the value of wildness, it's not even remotely plausible to say that a bioengineered herd of mammoth-like animals living in a heavily managed Pleistocene Park would be wild. The harder we try to make the world wilder, the further we get from the goal of living in a relatively wilder world.

Acknowledgments

Thanks to Lydia Dixon, Ben Hale, Norah Hannel, and Ned Hettinger for their helpful comments on an earlier version of this chapter. I have also benefitted from many conversations about rewilding with Chloe Mayhew.

Related Topics

41. Sustainable Agriculture
44. Restoration
48. Novel Ecosystems

References

Barlow, C. (2002) *The Ghosts of Evolution: Nonsensical Fruit, Missing Partners, and Other Ecological Anachronisms.* New York: Basic Books.

Bennett, J.R., Maloney, R.F., Steeves, T.E., et al. (2017) "Spending limited resources on de-extinction could lead to net biodiversity loss," *Nature: Ecology and Evolution* 1: Article number 0053.

Boym, S. (2001) *The Future of Nostalgia.* New York: Basic Books.

Browning, H. (2018) "Won't somebody please think of the mammoths? De-extinction and animal welfare," *Journal of Agricultural and Environmental Ethics* 31(6): 785–803.

Callicott, J.B. (2002) "Choosing the appropriate temporal and spatial scales for ecological restoration," *Journal of Biosciences* 27: 409–420.

Callicott, J.B., and Nelson, M.P. eds. (1998) *The Great New Wilderness Debate.* Athens: University of Georgia Press.

Campbell, K.L., et al. (2010) "Substitutions in woolly mammoth hemoglobin confer biochemical properties adaptive for cold tolerance," *Nature Genetics* 42(6): 536–539.

Caro, T. (2007) "The Pleistocene re-wilding gambit," *Trends in Ecology and Evolution* 22(6): 281–283.

Cooper, A., Turney, C., Hughen, K.A., et al. (2015) "Abrupt warming events drove late Pleistocene Holarctic megafaunal turnover," *Science* 349(6248): 602–606.

Corlett, R.T. (2016) "Restoration, reintroduction, and rewilding in a changing world," *Trends in Ecology and Evolution* 31(6): 453–462.

Cronon, W. (1996) "The trouble with wilderness: or, getting back to the wrong nature," *Environmental History* 1(1): 7–28.

Davis, J. (2015) "Letting it be on a continental scale: some thoughts on rewilding," in G. Wuerthner, E. Crist, and T. Butler (eds.) *Protecting the Wild: Parks and Wilderness, the Foundation for Conservation.* Washington, DC: Island Press, pp. 109–119.

Delord, J. (2007) "The nature of extinction," *Studies in History and Philosophy of Biology and Biomedical Sciences* 38: 656–667.

Delord, J. (2014) "Can we really recreate extinct species by cloning? A metaphysical analysis," in M. Oksanen and H. Siipi (eds.) *The Ethics of Animal Recreation and Modification: Reviving, Rewilding, Restoring.* New York: Palgrave MacMillan, pp. 22–39.

Donlan, J., et al. (2005) "Re-wilding North America," *Nature* 436: 913–914.

Donlan, J., et al. (2006) "Pleistocene rewilding: an optimistic agenda for twenty-first century conservation," *The American Naturalist* 168: 660–681.

Elder, J. (2001) "A conversation at the edge of wilderness," in C.M. Klyza (ed.) *Wilderness Comes Home: Rewilding the Northeast*. Middlebury, VT: Middlebury College Press, pp. 256–262.

Elliott, C. (1945) "Woodman, Spare that 'wolf' tree!" *American Forests* 51(10): 489–490.

Foreman, D. (1998) "The Wildlands Project and the Rewilding of North America," *University of Denver Law Review* 76(2): 535–553.

Foreman, D. (2004) *Rewilding North America: A Vision for Conservation in the 21st Century*. Washington, DC: Island Press.

Fraser, C. (2009) *Rewilding the World: Dispatches from the Conservation Revolution*. New York: Picador.

Galetti, M. (2004) "Parks of the Pleistocene: recreating the cerrado and the Pantanal with the megafauna," *Natureza & Conservação* 2: 93–100.

Gamborg, C. (2014) "What's so special about reconstructing a mammoth? Ethics of breeding and biotechnology in recreating extinct species," in M. Oksanen and H. Siipi (eds.) *The Ethics of Animal Recreation and Modification: Reviving, Rewilding, Restoring*. New York: Palgrave MacMillan, pp. 60–76.

Gammon, A. (2018) "The many meanings of 'rewilding': an introduction and the case for a broad conceptualisation," *Environmental Values* 27(4): 331–350

Gomez, M.C., et al. (2004) "Birth of African wildcat kittens born from domestic cats," *Cloning and Stem Cells* 6(3): 247–258.

Gremmen, B. (2014) "Just fake it! Public understanding of ecological restoration," in M. Oksanen and H. Siipi (eds.) *The Ethics of Animal Recreation and Modification: Reviving, Rewilding, Restoring*. New York: Palgrave MacMillan, pp. 134–149.

Higgs, E. (2003) *Nature by Design: People, Natural Processes, and Ecological Restoration*. Cambridge, MA: MIT Press.

Hintz, J. (2007) "Some political problems for rewilding nature," *Ethics, Place, and Environment* 10(2): 177–216.

Howard, S.A. (2012) "Nostalgia," *Analysis* 72(4): 641–650.

Hull, D.L. (1978) "A matter of individuality," *Philosophy of Science* 45(3): 335–360.

Jamieson, D. (1995) "Ecosystem health: some preventive medicine," *Environmental Values* 4(4): 333–344.

Janzen, D.H., and Martin, P. (1982) "Neotropical anachronisms: the fruits the gomphotheres ate," *Science* 215: 19–27.

Jones, E.D. (2022) *Ancient DNA: The Making of a Celebrity Science*. New Haven, CT: Yale University Press.

Jørgensen, D. (2015) "Rethinking rewilding," *Geoforum* 65: 482–488.

Judson, O. (2008) "Resurrection science," *The New York Times* 25 November 2008. Available online at http://opinionator.blogs.nytimes.com/2008/11/25/resurrection-science/, last accessed 29 March 2014.

Kasperbauer, T.J. (2017) "Should we bring back the passenger pigeon? The ethics of de-extinction," *Ethics, Policy, and Environment* 20(1): 1–14.

Katz, E. (1992) "The big lie: human restoration of nature," *Research in Philosophy and Technology* 12: 231–241.

Keulartzt, J. (2018) "Rewilding," *Oxford Research Encyclopedia, Environmental Science*. Oxford University Press. DOI: 10.1093/acrefore/9780199389414.013.545.

Klyza, C.M. (2001) "An eastern turn for wilderness," in C.M. Klyza (ed.) *Wilderness Comes Home: Rewilding the Northeast*. Middlebury, VT: Middlebury College Press, pp. 3–26.

Koch, P.L., and Barnosky, A.D. (2006) "Late quaternary extinctions: the state of the debate," *Annual Review of Ecology, Evolution, and Systematics* 37: 215–250.

Kolbert, E. (2012) "Recall of the wild" *The New Yorker* 24 December 2012, p. 50.

Kornfeldt, T. (2018) *The Re-Origin of Species: A Second Chance for Extinct Animals*. Chennai: Westland Publications.

Lanza, R.P., et al. (2000) "Cloning of an endangered species (*Bos gaurus*) using interspecies nuclear transfer," *Cloning* 2(2): 79–90.

Lewis, M.W. (2012) "Pleistocene park: the regeneration of the mammoth steppe," *Geocurrents* (12 April 2012). Available online at http://www.geocurrents.info/place/russia-ukraine-and-caucasus/siberia/pleistocene-park-the-regeneration-of-the-mammoth-steppe, last accessed 3 November 2014.

Lorimer, J., and Driessen, C. (2013) "Bovine biopolitics and the promise of monsters in the rewilding of Heck cattle," *Geoforum* 48: 249–259.

Lorimer, J., and Driessen, C. (2016) "From 'Nazi cows' to cosmopolitan 'ecological engineers': specifying rewilding through a history of Heck cattle," *Annals of the American Association of Geographers* 106(3): 631–652.

Manning, R. (2009) *Rewilding the West: Restoration in a Prairie Landscape*. Berkeley: University of California Press.

Marris, E. (2009) "Reflecting the past," *Nature* 462: 30–32.

Martin, P. (1966) "Africa and Pleistocene Overkill," *Nature* 5060(212): 339–32.

Martin, P. (2005) *Twilight of the Mammoths: Ice Age Extinctions and the Rewilding of America*. Berkeley: University of California Press.

Martin, P., and Burney, D. (1999) "Bring back the elephants," *Wild Nature* 9: 57–65.

Mathews, F. (2004) "Letting nature do the doing," *Australian Humanities Review* 33(August–October 2004). Available online at http://www.australianhumanitiesreview.org/archive/Issue-August-2004/mathews.html, last accessed 8 April 2014.

Miller, W., et al. (2008) "Sequencing the nuclear genome of the extinct woolly mammoth," *Nature* 456(20): 387–391.

Minteer, B. (2019) *The Fall of the Wild: Extinction, De-Extinction, and the Ethics of Conservation*. New York: Columbia University Press.

Monbiot, G. (2013a) *Feral: Searching for Enchantment on the Frontiers of Rewilding*. London: Allen Lane.

Monbiot, G. (2013b) "Accidental rewilding," *Aeon Magazine*, June 4, 2013.

Neely, M. (2020) *Against Sustainability: Reading Nineteenth-Century America in the Age of Climate Crisis*. New York: Fordham University Press.

Nicholls, H. (2006) "Restoring nature's backbone," *PLOS Biology*: DOI: 10.1371/journal.pbio.0040202

Nicholls, H. (2008) "Let's make a mammoth," *Nature* 456(20): 310–314.

Nogués-Bravo, D., Simberloff, D., Rahbek, C., and Sanders, N.J. (2016) "Rewilding is the new Pandora's box in conservation," *Current Biology* 26(3): R87–R91.

Oksanen, M., and Siipi, H. (2014) "Introduction: towards a philosophy of resurrection science," in M. Oksanen and H. Siipi (eds.) *The Ethics of Animal Recreation and Modification: Reviving, Rewilding, Restoring*. New York: Palgrave MacMillan, pp. 1–21.

Oliveira-Santos, L.G.R., and Fernandez, F.A.S. (2010) "Pleistocene rewilding, Frankenstein ecosystems, and an alternative conservation agenda," *Conservation Biology* 24(1): 4–6.

Pauketat, T. (2009) *Cahokia: Ancient America's Great City on the Mississippi*. New York: Penguin Books.

Perry, G.L.W., Wheeler, A.B., Wood, J.R., and Wilmshurst, J.M. (2014) "A high-precision chronology for the rapid extinction of New Zealand moa (Aves, Dinornithiformes)," *Quaternary Science Reviews* 105(1): 126–135.

Prior, J., and Brady, E. (2017) "Environmental aesthetics and rewilding," *Environmental Values* 26: 31–51.

Prior, J., and Ward, K.J. (2016) "Rethinking rewilding: a response to Jørgensen," *Geoforum* 69: 132–135.

Raff, J. (2022) *Origin: A Genetic History of the Americas*. New York: Twelve/Hachette Book Group.

Rich, N. (2014) "The mammoth cometh," *The New York Times Magazine*, 2 March 2014.

Rubinstein, D.R., et al. (2006) "Pleistocene park: does re-wilding North America represent sound conservation for the 21st century?" *Biological Conservation* 132: 232–238.

Salsberg, C.A. (2000) "Resurrecting the woolly mammoth: science, law, ethics, politics, and religion," *Stanford Technology Law Review* 1: 1–30.

Sandler, R. (2012) *The Ethics of Species*. Cambridge: Cambridge University Press.

Sandler, R. (2017) "De-extinction: costs, benefits, and ethics," *Nature: Ecology & Evolution* 1: Article number 0105.

Sandom, C., et al. (2013) "Rewilding," in D.W. Macdonald and K.J. Willis (eds.) *Key Topics in Conservation Biology 2*, Oxford: Wiley-Blackwell, pp. 430–451.

Seddon, P.J., et al. (2014) "Reintroducing resurrected species: selecting deextinction candidates," *Trends in Ecology and Evolution* 29(3): 140–147.

Shapiro, B. (2016) *How to Clone a Mammoth: The Science of De-Extinction*. Princeton, NJ: Princeton University Press.

Shaw, E. (2015) "The old in the forest: wolf trees in New England and farther afield," *Atlas Obscura*, February 25, 2015.

Sherkow, J.S., and Greely, H.T. (2013) "What if extinction is not forever?" *Science* 340(6128): 32–33.

Siipi, H. (2014) "The authenticity of animals," in M. Oksanen and H. Siipi (eds.) *The Ethics of Animal Recreation and Modification: Reviving, Rewilding, Restoring*. New York: Palgrave MacMillan, pp. 77–96.

Siipi, H., and Finkelman, L. (2017) "The extinction and de-extinction of species," *Philosophy and Technology* 30: 427–441.

Smith, C.I. (2005) "Re-wilding: introductions could reduce biodiversity," *Nature* 437: 318.

Soulé, M. (1990) "The onslaught of alien species, and other challenges in the coming decades," *Conservation Biology* 4: 233–239.

Soulé, M., and Noss, R. (1998) "Rewilding and biodiversity: complementary goals for continental conservation," *Wild Earth*, Fall 1998: 2–11.

Svenning, J.-C., et al. (2016) "Science for a wilder Anthropocene: synthesis and future directions for trophic rewilding research," *Proceedings of the National Academy of Sciences* 113(4): 898–906.

Tananescu, M. (2017) "Field notes on the meaning of rewilding," *Ethics, Policy, and Environment* 20(3): 333–349.

Tree, I. (2019) *Wilding: Returning Nature to Our Farm*. New York: New York Review Books.

Turner, D. (2014) "The restorationist argument for extinction reversal," in M. Oksanen and H. Siipi (eds.) *The Ethics of Animal Recreation and Modification: Reviving, Rewilding, Restoring*. New York: Palgrave Mac-Millan, pp. 40–59.

Turner, D. (2017) "Biases in the prioritization of candidates for de-extinction," *Ethics, Policy, and Environment* 20: 21–24.

Turner, D. (2019) *Paleoaesthetics and the Practice of Paleontology*. Cambridge: Cambridge University Press.

Wandeler, P., et al. (2007) "Back to the future: museum specimens in population genetics," *Trends in Ecology and Evolution* 22(12): 634–642.

Wolverton, S. (2010) "The North American Pleistocene overkill hypothesis and the re-wilding debate," *Diversity and Distributions* 16: 864–866.

Wray, B. (2019) *The Rise of the Necrofauna: The Science, Ethics, and Risks of De-Extinction*. Vancouver, Canada: Greystone Books.

Zimov, S.A. (2005) "Pleistocene park: return of the mammoth's ecosystem," *Science* 308: 796–798.

Zimov, S.A. (2014) "Musk ox situation," available online at http://www.pleistocenepark.ru/en/news/13/, last accessed 3 November 2014.

Zuk, M. (2014) *Paleofantasy: What Evolution Really Tells Us About Sex, Diet, and How We Live*. New York: W.W. Norton & Company.

48

NOVEL ECOSYSTEMS

Allen Thompson

Novel ecosystems are of growing interest in applied ecology and will be significant for restoration, conservation, and other land management practices through the 21st century and, I shall argue, well beyond. I explain what novel ecosystems are and one way they could be incorporated into an environmental ethic suited to the Anthropocene. "Today, humans are driving extensive and pervasive [environmental] change", according to Mascaro et al., "and novel ecosystems are the unambiguous response" (Mascaro et al. 2013: 55).

Orientation by Analogy

The International Union of Geological Sciences is considering a proposal that the present epoch, the Holocene, has ended and we are entering the Anthropocene. The idea of the Anthropocene is widely discussed across disciplines and embroiled in controversy. In this section, I will approach the subject of novel ecosystems by considering some of the controversy surrounding the Anthropocene and by appeal to two analogies.

First, there is debate about the name. Even if we are now in a new geological period, naming it after humans seems problematic. For one, it breaks with tradition. Of ten previous epochs that have a name, none are named after the *cause* of the changes marked. "Instead, all the names refer to the changed composition of species present in each epoch" (Suckling 2014). Second, this break with tradition seems to instantiate what many perceive as the objectionable self-centered orientation of anthropocentrism. Instead of representing the host of other species that characterize the period, "the Anthropocene" signifies human exceptionalism and transformative power. Third, the name suggests that humanity as a whole is responsible, which conveniently obscures the subset with real geopolitical power. It's the behaviors of a global elite, wealthy consumers, and corporations that drive greenhouse gas emissions, pervasive land-use changes, and accelerated extinction rates, not meeting subsistence needs of 1.2 billion people living in extreme poverty.

There is no significant controversy, however, that behaviors of at least some subset of humanity, mediated by contemporary technologies, have dramatically altered Earth systems on a global scale (Steffen et al. 2004). And aggregated, these alterations are widely viewed as undesirable; we are turning a lovely natural world into a monster. So consider an analogy, portrayed in Mary Shelley's 1818 novel, *Frankenstein, or, the Modern Prometheus* (Shelley 1818/1996).

Parallels start with naming. The monster is often mistakenly called Frankenstein, whereas this is really the name of its creator, Dr. Frankenstein. Likewise, "the Anthropocene" refers to the planet's bio- and geophysical condition, now seen as a degraded and unnatural monster arising

DOI: 10.4324/9781315768090-57

from the technological interventions of self-absorbed and conceited human agency; our name is wrongly projected onto the ugly beast of our creation. Further, Frankenstein's monster was not brought about by society *writ large* but only by the obsessive and maniacal Doctor. Likewise, perhaps the elites of global capitalism, even the modestly well-off in developed nations, are to blame for the monstrous condition of our planet, not the poor and powerless masses of humanity.

Or perhaps this whole comparison—creating a "Franken-earth"—seems hyperbolic. Rather, the Anthropocene is about unprecedented human influence on Earth systems, a descriptive fact that humans have become the central driver of change on the planet. It means the whole world has been hybridized to a greater or lesser extent, neither artifact nor wholly natural, but co-determined and so always everywhere exhibiting some effects of human intervention. If so, we would expect to find evidence of human activity across all scales, from the planetary and regional down to landscapes and specific sites. Consider, then, a decidedly more mundane illustration, the "coywolf".

Activities in North America by European settler colonists, specifically clearing forests for farmland and wolf eradication programs in southern Ontario, reduced reproductive opportunities for wolves, broadened the range of coyotes, and introduced domesticated dogs. Consequent interbreeding has led to a versatile type of beast now numbering in the millions, adept at hunting in forests (like wolves) and open terrain (like coyotes) and tolerant of people (like dogs), and occupying the continent's entire northeast territory, urban areas included.

Human activity, indirectly and unintentionally, has led to a novel animal, unknown in the historic record. The idea that this "naturally" recombinant DNA (0.25 wolf, 0.10 dog, and 0.65 coyote) constitutes a new species remains debated but the "mixing of genes that has created the coywolf has been more rapid, pervasive and transformational than many once thought" (The Economist 2015). Is the coywolf a monster, a hideous result of humans meddling with nature? It would seem not, no more than a Pug or Great Dane, for example, which are the products of intentional interbreeding and artificial selection, or maize (i.e., corn) for that matter, which is the result of human cultivation spanning about 10,000 years. Indeed, more clearly than these cases of intentionally altering nature to meet human purposes, the coywolf seems to represent nature's own autonomous response, responding to new selection pressures, to new constraints and opportunities brought about by human activity. If we retain the language of human/nature dualism, then the appearance of coywolves may be thought of as nature rebounding after some significant human interference; likely there would be no coywolves absent the activity of human beings. Coywolves represent nature adapting to new, anthropogenic environmental conditions.

The stage is now set to introduce novel ecosystems, which many in traditional, historically oriented nature conservation and restoration communities have viewed with contempt and disgust (Simberloff, Carolina, & Aronson 2015). First, consider a generic definition of an ecosystem: a community of living organisms (including plants, animals, insects, soil microbes, etc.) together with nonliving features of their environment (climate, hydrology, minerals, etc.) that may be described by reference to some particular set of compositional, structural, and functional properties. Different types of ecosystems, then, can be distinguished from one another by reference to differences in their compositional, structural, and functional profiles.[1]

Another property of ecosystems important for our discussion is *resilience*: the capacity of a system to return to normal functioning after a disturbance. A useful heuristic for thinking about the resilience is a marble resting at the bottom of a bowl. Disturbance to the bowl may displace the marble temporarily but the system (bowl, marble, gravity) will return itself to the normal, pre-disturbance condition. Of course, as will be discussed later, some disturbance may knock the marble out of the bowl, in which case the system cannot return itself to its previous state; the resilience of the system has been exceeded.

Formally defined, then, "a novel ecosystem is one of abiotic, biotic, and social components (and their interactions) that, by virtue of human influence, differ from those that prevailed historically, having a tendency to self-organize and manifest novel qualities without intensive human management" (Mascaro et al. 2013: 55). Let's anchor our imaginations with an example:

> On Santa Cruz Island in the Galapagos Islands, management of the humid highlands has taken a different path form what we might imagine....What began as a conventional restoration exercise of returning ecosystems to their historical conditions... became, over two decades, something quite different. Pervasive ecosystem change driven by invasion of non-native species such as the red quinine tree (*Cinchona pubescens*), black rats (*Rattus rattus*) and others, means that achieving the original goals is increasingly unrealistic. It is simply impossible, or at least practically impossible, to recover historical ecosystems. The focus of goals now rests on key species of conservation interests and involves a constantly adaptive approach to control invasive species without any realistic intention of eliminating them. Here is a novel ecosystem. It defies conventional management approaches, and demands a new way of thinking about our interventions in and responsibilities toward ecosystems.
>
> *(Hobbs, Higgs, and Hall 2013: 3)*

Novel ecosystems are examples of recombinant ecologies that, as a consequence of human intervention, exhibit at least a new compositional profile and may also exhibit new functional characteristics. As illustrated by the example of Santa Cruz Island, a common cause of ecosystem novelty (in the relevant sense) is a change in the ecosystem's composition brought about by non-native species, the presence of which is rightly attributable to some kind of human influence on the system. The ecosystems referred to exhibit compositional profiles that did not prevail historically. These sites in the Galapagos require *some* management, aimed at warding off additional non-natives and preserving a habitat for "key species of conservation interest", but they also exhibit resilience and do not require intensive, on-going management interventions to retain their current, but non-historic, compositional, and functional profiles. Management of these systems aims to protect them from disturbances that would exceed their capacity for resilience, but they are not sustained only in virtue of on-going human interventions as are, for example, productive agricultural sites.

Frankenstein's creature was an "unnatural" thing assembled from the disparate parts of other men, pieced together with purpose, intention, and ambition by another man. Yet it was alive and autonomous. The coywolf is a new canine, clearly robust and alive, composed of DNA from different species and arising from reproductive opportunities that were the unintentional byproduct of human activity, interventions in nature on landscapes and into animal populations. In both cases, something new appears as an assembly of other parts, parts that are brought together because of human behaviors, intentionally or not. And so with novel ecosystems—ecosystems with new parts and operations that are, by definition, the result of human activities but nonetheless are resilient and autonomous in the sense that they can persist without on-going human interventions.

Coywolves may be a new species, they number in the millions, and they appear well adapted to contemporary conditions, occupying a territory that is vast and uninhabitable to their purebred forbearers. Novel ecosystems are diverse and widespread. It's estimated that today novel ecosystems already occupy about 28–36% of the non-frozen terrestrial surface of the Earth (Perring and Ellis 2013: 78). Against a background of pervasive anthropogenic drivers of directional environmental change (including climate change, introduction of exotic species, nitrogen deposition and other forms of pollution, and land fragmentation), it's likely that this percentage will increase significantly. Novel ecosystems are distinct from historic, original, or "pristine" ecosystems, on one hand, and, on the other, landscapes that have been intensively modified and continuously managed, such as in agriculture. So, if by "nature" or "natural ecosystems" we mean the historic assemblages of (fundamentally) non-anthropogenic ecosystems and we contrast such natural places with built environments that require on-going and intensive human intervention (such as urban, suburban, and productive agricultural sites), then novel ecosystems represent a *new* kind of nature. They are autonomous and resilient, like the coywolf. The "new" nature of novel ecosystems will be one characteristic of the Anthropocene, when no ecosystems have a genesis wholly independent of human influence.

Theories of Ecological Change and Natural Ecosystem Novelty

Strange as it may sound, "ecological novelty is not new". Stephen T. Jackson imagines a group of contemporary ecologists who "touring the world 14 or 15 thousand years ago, while the ice sheets were in rapid retreat, would have seen many communities and ecosystems alien to their modern eyes". "Our ecologists", he continues, "would be flummoxed repeatedly as they encountered peculiar groupings of plants, insects, mullusks, birds, mammals and other groups from the tropics to the high latitudes" (Jackson 2013: 63) Quaternary paleoecologists, Jackson tells us, have been puzzling over these assemblages since the 1960s. "Variously called 'disharmonious' or 'intermingled' fauna and flora… and 'no-analog' assemblages…, they all represent ancient communities and ecosystems that lack precise modern counterparts (Jackson and Williams 2004)" (Jackson 2013: 63).

The characterization of an ecosystem as new or novel implies difference over time, or ecosystem *change*, which is exactly what Jackson reports having occurred between the late Pleistocene and the Holocene epoch. How should we understand ecosystem change? Arthur Tansley (1871–1955) introduced the term "ecosystem" (Tansley 1935) but among the most influential and early ecological theorists were Frederic Clements (1874–1945) and Henry Gleason (1882–1975) who each offered an influential theory. Clements's ideas about ecological succession are central to what is often referred to as the "balance-of-nature" paradigm in ecology while Gleason's view is central to the "flux-of-nature" paradigm (Callicott 2003: 249).

Clements provided an account of community-based succession, an "organismal" and teleological view of ecosystems according to which ecosystems move progressively toward a mature climax state (Clements 1916). While an ecosystem is subject to periodic disturbances, such as fires, flooding, disease or insect outbreaks, windstorms, and droughts—and significant disturbances will cause fundamental changes to the structure of an ecosystem—over time the ecosystem will recover, returning on a trajectory, in a predictable series of stages progressing toward its mature state.

This explanation of ecosystem change, as a progressive and orderly succession through determined stages, dominated early 20th century ecology and strongly influenced latter 20th century environmental ethics. Those who held a "balance-of-nature" view "referred to human modified ecosystems like farms as 'dis-climax'", a label which operates as a precursor of the more modern construction of calling a human-altered ecosystem "degraded", as both share the negative implication of diminishment in rank (Mascaro et al. 2013: 47). From the perspective of a climax-state theory of ecology, human influence can disrupt an ecosystem's natural progression, which results in dis-climax or degradation, judged by comparison to how the system *ought* to be, that is, how it *would* be without human interference.

To early environmental ethicists, especially those holding an eco-centric theory of value (a holist view attributing intrinsic value to environmental collectives, including species and ecosystems), the science of ecology appeared to be teaching a lesson: non-anthropogenic ecosystems exhibit a typology and have a teleological (i.e., goal-directed) structure, thus having a kind of interest in the conditions conducive to reaching or realizing their end state. This allowed a bridge between facts and values: if we are to avoid the bias of anthropocentrism, then we have to acknowledge the climax state of an ecosystem as its own autonomous good (that is, a value not connected with human ends or interests). Further, the thought continues, we should acknowledge that what is conducive to the realization of an ecosystem's good is morally relevant, thus generating prima facie reasons against harming the ecosystem. Harming an ecosystem would be whatever is contrary to the system reaching or maintaining its mature state. We ought not harm an ecosystem (unless we have overriding reasons for doing so), but instead we ought to let ecosystems develop into and then persist in their natural, mature state.

By comparison, Gleason offered an "individualistic" conception of ecosystem change according to which the ecological associations exhibited by component parts of a system were not progressive and holistic, as in Clement's view (Gleason 1926; McIntosh 1975). Instead, ecosystems are a consequence of the migration and resulting recombination of individual species, as different individual species of plants and animals respond differently to changing environmental conditions. This theory predicts, over the course of geologic time as conditions vary, that *all* species will migrate individually, not as a fixed community. Accordingly, disturbance events and subsequent recombination of species are the norm. Thus, particular token ecosystems are not an instance of some natural kind because types of ecosystems are not natural kinds. Ecosystem types are not showing us how the world is *really* structured, that is, they are not "nature carved at its joints". Rather, they reflect relatively transient or ephemeral associations brought about by contingent circumstances, changing environmental conditions, and the adaptive capacities of individual species.

The very possibility of a novel ecosystem, then, as it is composed of species in new association, depends on Gleason's theory of ecosystem change. If ecosystems could not be understood as recombinant collections of individual species, but rather must be explained by appeal to Clement's organismal view, then truly novel (new) ecosystems could not exist. No particular collection of biotic and abiotic components could be completely unprecedented; what would appear as strange to the time-traveling ecologists that Jackson imagined must be only hitherto unobserved stages of contemporary ecosystem types, or perhaps particular stages of ecosystem types that no longer exist. According to Mascaro et al., Gleason's view has since received "irrefutable proof", due to steady advances in ecological science (Mascaro et al. 2013: 46). The paradigm shift in ecology, from the balance-of-nature view to the flux-of-nature view, is now largely complete (Callicott 2003: 249).

Accordingly, we should expect recombinant ecologies to be the norm, if our temporal perspective is sufficiently long. Nature, we may say, has perpetually generated recombinant ecologies and a degree of biotic and abiotic difference (i.e., the novelty of the ecosystem) from present ecosystem assemblages (at any particular time) will be greater as one looks further into the past. Similarly, Gleason's theory predicts that the composition and function of ecosystems will be increasingly different the further one looks into the future. In this sense, then, novel ecosystems are not new; there always has been and always will be ecosystems that are novel.

However, the definition of novel ecosystems quoted above specifies that in order to count as a novel ecosystem in the relevant sense, ecosystem must differ from the properties that prevailed historically *in situ* by "virtue of human influence". And indeed, analogous to the rate of species extinctions, the rate at which novel ecosystems are emerging today is significantly higher than the normal, background rate (see Figure 7.1, Jackson 2013: 64). So, something new is happening in regard to recombinant ecologies. How can we better understand and respond to the emergence of novel ecosystems today, those new ecosystems attributable to a distinctively human influence?

Anthropogenic Novel Ecosystems: Degraded or Just Different?

In 1990 co-founder of the Society for Conservation Biology, Michael Soulé, discussed the challenges posed by new combinations of species that arose in the wake of human activities and called for a new ecological discipline to study them, what he called "mixoecology" (Standish et al. 2013:

296). Later, in 2002, a workshop devoted to changing ecologies produced the idea of an "emerging ecosystem", defined as "an ecosystem whose species composition and relative abundance have not previously occurred within a given biome" (Milton 2003; Mascaro et al. 2013: 48). Then, in a seminal paper, Hobbs et al. (2006) defined "novel ecosystems" as assemblages characterized by "(1) novelty: new species combinations with the potential for changes in ecosystem functioning; and (2) human agency: ecosystems that are the result of deliberate or inadvertent human action, but do not depend on continued human intervention for their maintenance" (Hobbs et al. 2006: 2).

In early modeling, Hobbs et al. (2006) conceived of novel ecosystems as located along a single axis, between historically intact ecosystems and intensively managed agriculture sites (see Figure 48.1). Here, novel ecosystems arise either by recent human agency disturbing historic ecosystems or by past human agency dominating sites, which have since been abandoned. In either case the result is an ecosystem with either novel compositional or functional attributes that are not under direct human control. Instead, the biotic and abiotic features of a novel ecosystem are a response of the biosphere to some form of human influence, a response that yields an ecosystem manifesting both resilience and self-organization.

Conceptual work continued and Hobbs et al. (2009) offered another model, locating novel ecosystems by reference to alterations away from historic conditions along two axes, biotic and abiotic (Hobbs et al. 2009: 601; see Figure 48.2). *Some* amount of variation to an ecosystem's biotic and abiotic features is consistent with the system's historic range of variation. But alterations beyond the historic range result in what Hobbs et al. refer to as a "hybrid ecosystem". Hybrid ecosystems exhibit non-historic composition but additional human interventions could push the system back into a state consistent with its historic range of variation. Novel ecosystems, then, are the product of even greater biotic and/or abiotic variation, thereby exhibiting new compositional features, and, further, may have new functional profiles as well. The crucial difference between hybrid and novel systems, on this model, concerns crossing some threshold, either social or ecological or both,

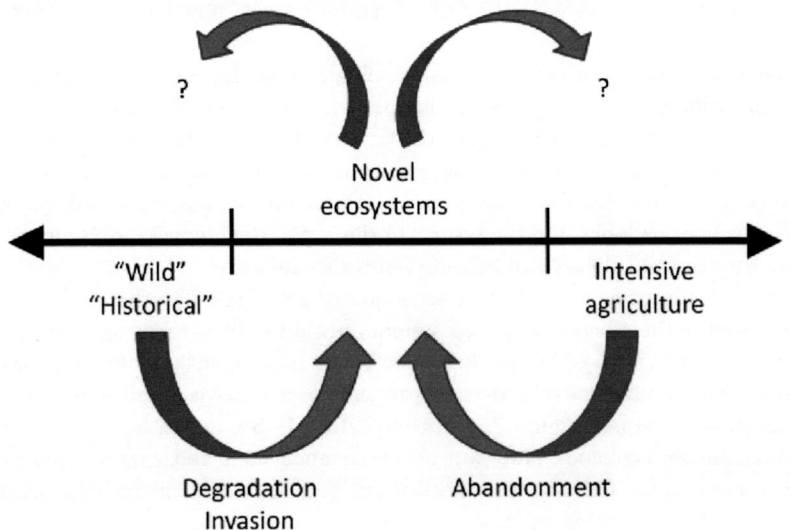

Figure 48.1   Novel ecosystems arising from either degraded historical ecosystems or abandoned agricultural sites. Arrows from novel ecosystems represent uncertainty about how to intervene or manage novel ecosystems or what the results would be

Source: From Hobbs et al. 2006. "Novel ecosystems: theoretical and management aspects of the new ecological world order". *Global Ecology & Biogeography* 15:1–7

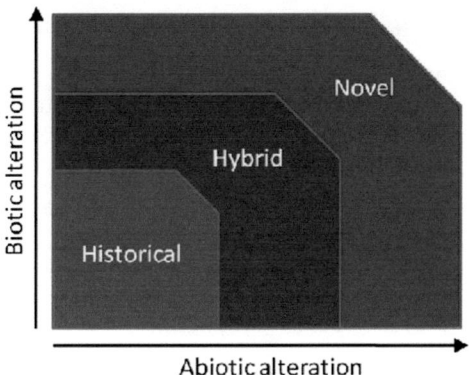

Figure 48.2   Historic, hybrid, and novel ecosystem types located along axes of biotic and abiotic alteration. Hybrid systems may be restored to historic conditions through intervention but novel systems are conceived to have passed over a (social and ecological) threshold making classic restoration impossible.

Source: From Hobbs, R. J., Higgs, E. and Harris, J. A. (2009). Novel ecosystems: implications for conservation and restoration. *Trends in Ecology & Evolution* 24: 599–605.

which effectively prevents the system from returning back to a historic condition *in situ*, as one that could then maintain the historic state without on-going human interference.

Recall the marble-in-a-bowl heuristic for imaging resilience. Since the pioneering work of C. S. Holling (1973) the concept of resilience has been used to describe the ability of a system to rebound after disturbances (also see Walker et al. 2004). Disturbances, or changes, that exceed the resilience of a system effectively knock the marble out of the basin, outstripping the system's ability to recover. The possibility of exceeding a system's resilience has long been a central concern of land management practices that aim at preserving "natural" ecosystems. This is so whether the goal is to preserve an ecosystem for its own sake or preserving it to provide specific habitats in the service of species conservation, thus protecting valued and endangered (or at least imperiled) biotic species by safeguarding the ecosystem habitats that they occupied historically. Extreme ecosystem disturbances of a non-anthropogenic origin, such as volcanic eruptions, could devastate a site, destroying critical species habitat. With reference to the system's role to provide habitat for the conservation of a valued species, then, we may evaluatively judge the post-disturbance system as degraded. But without reference to this instrumental value, it becomes more difficult to justify calling the altered system *degraded*, rather than simply *different*.

Now, instead of a marble in a solitary bowl, imagine a tray of adjacent, recessed cups. When a system, which has been occupying a particular historic range of variation vis-à-vis its biotic and abiotic properties, is subject to a disturbance that exceeds its capacity for resilience, the system "flips" into a new state—the marble gets displaced from one recession into another one. The result is a new ecosystem, a system with different compositional and functional features; it is a *different* ecosystem, not a *degraded* version of the previous ecosystem. "Degraded" is a comparative evaluation, a value judgment relative to some standard or norm. On the basis of what values could we defend such a judgment? To judge an ecosystem degraded simply because it's different seems unsatisfactory (Hobbs 2016). Novel ecosystems are different by definition. Are these anthropogenic novel ecosystems rightly judged as degraded simply because their genesis, again by definition, significantly involves some kind of human influence? Let's bracket this question for the time being. We will return to it after considering a dilemma that the rapid emergence of novel ecosystems poses for the way that conservation and restoration practitioners have traditionally conceived of the value of natural ecosystems.

Managing for Historic "Naturalness" in Conservation and Restoration

Whether justified on instrumental value to humans or by appeal to the intrinsic value of non-human collectives, such as species or ecosystems, nature conservation in the United States has sought to maintain the ecosystems and biodiversity that prevailed historically *in situ*. Colloquially, we may say the aim of conservation has been to maintain significant portions of nature *as it had been received*. Indeed, the document establishing America's National Park Service identifies the agency's mandate:

> …to conserve the scenery and the natural and historic objects and the wild life therein and to provide for the enjoyment of the same in such manner and by such means as will leave them unimpaired for the enjoyment of future generations.
>
> *(NPSOA 1916)*

Nearly half a century later, in 1963, a report titled Wildlife Management in the National Parks (i.e., the "Starker Leopold Report") famously advised

> …that the biotic associations within each park be maintained, or where necessary recreated, as nearly as possible in the condition that prevailed when the area was first visited by the white man. A national park should represent a vignette of primitive America.
>
> *(Leopold et al. 1963)*

In the wake of environmental historian William Cronon's influential work on the active land management undertaken by pre-Columbian Native Americans, there need be no pretense that these federal mandates direct us to imagine and preserve "pristine" nature, perfectly unaltered by the activity of human hands (Cronon 1995). But the historic orientation remains overriding; the mandate is to preserve historic conditions.

Likewise, diverse practices of ecological restoration have long held the notion of *historic fidelity* in high esteem. But attention to novel ecosystem began with restoration ecologists who recognized that for some sites, where human intervention had brought about significant biotic or abiotic change, there was little hope achieving the traditional goal of ecological restoration: resetting the system to some state within its historic range of variation and allowing it, without ongoing and intensive human intervention, to continue along its pre-disturbance trajectory. Ecosystem recovery, as may be ideally hoped for by environmentalists, once the anthropogenic disturbance is abated, frequently requires the assistance of additional and ongoing restorative interventions. Yet against a background of pervasive anthropogenic drivers of directional environmental change, the fully *autonomous* recovery of the ecosystem's historic state is frequently impossible. Additionally, while traditional restoration interventions may reestablish historical composition and function, given pervasive background anthropogenic drivers of environmental change, there is little hope that properties of the altered ecosystem will remain within the system's historic range of variation without on-going and increasingly intensive human interventions. In other words, in apparently increasing number of cases, the system restored to a historical condition will not be resilient. Historic ecosystems, whether original or restored, will repeatedly be subject to anthropogenic conditions which drive the alteration of their biotic and abiotic properties outside their range of historic variation.

Thus, management and restoration interventions aimed at historic "natural" conditions face a dilemma. In a world with pervasive anthropogenic drivers of environmental change, we will increasingly be unable to maintain ecosystems in historical states without increasing levels of human intervention and control. If under the guise of promoting naturalness we value *historic* ecosystem states, then we will have to tolerate increasing levels of human *control* over nature. However, if we value nature as "wild and free", a realm largely independent of intentional human intervention, then increasingly we will have to tolerate non-historic, novel ecosystems. Land management practices that have traditionally valued both the historic condition and the autonomy of "natural" ecosystems are under increasing pressure because, in an increasing number of cases, you simply can't have both. Are there alternative values or visions that could provide a moral ground for nature conservation and restoration? What management goals could replace historic fidelity? Where return to past conditions is not a reasonable, what *should* replace historical fidelity as a management goal?

Modeling a Space of Ecosystem States and Various Pathways of Change

In this section I address two unresolved strands of the discussion so far and how they are connected. First, there are questions about the role of human influence in the recent and rapid emergence of novel ecosystems. Ecosystem novelty over time is implied by a Gleasonian theory, so why does the recently developed definition of novel ecosystems, cited above, specify human influence? What is new about *anthropogenic* ecosystem novelty and why does it matter? Second, there are questions about the value of "natural" ecosystems; are they valuable because they are autonomous from human control and purposeful intervention or because they exhibit historic compositional and functional properties? What alternative values, if any, should guide land management? How should we value anthropogenic novel ecosystems, if at all? To address these questions, I review how progressive modeling of anthropogenic novel ecosystems has portrayed dimensions of *human influence* and *ecosystem alternation* before presenting the earlier models as synthesized into a single model that represents a three-dimensional map of possible ecosystem states and vectors of anthropogenic change.

Let's begin with a quick review. Confronted with challenging sites, conservation managers and restoration ecologists needed to refine the idea of the "new" kind of ecosystem that they increasingly face. Hobbs et al. (2006) began with two factors, novelty and human activity, and located novel ecosystems along a single axis, between "wild" or "historic" systems, and "intensively managed" sites. Novelty was characterized by reference to new species composition and the potential for new functional characteristics; human activity was characterized by deliberate activity on-site ("degradation") or the cessation of intensive human control ("abandonment"). Refer again to Figure 48.1.

Later, Hobbs et al. (2009) further refined the idea of novel ecosystems, now represented along two axes, of biotic alteration and abiotic alternation. Points near zero along either axis represent the system's historical range of variation; points midway represent hybrid systems, characterized by new species composition but remaining amenable to human interventions aimed at restoring historic conditions. Points further out along either axis capture novel ecosystems, now exhibiting new functional relations and postulated to have crossed an ecological or social threshold that effectively prevents the restoration of historic composition. Presumably the cause of biotic and abiotic alterations is anthropogenic, although the precise nature of human influence is not specified nor given further conceptual clarification. Refer again to Figure 48.2.

Now consider a third graphic representation (see Figure 48.3). Along the horizontal, x-axis we can plot the degree of divergence of an ecosystem's biotic and abiotic properties from historical states and along the vertical, y-axis we can plot the amount of human design and control. Traditional conceptions of "natural" ecosystems will occupy points approaching zero along both axes, historical sites with no intentional human design. Traditional ecological restoration remains near zero on the x-axis but only in virtue of human design and purposeful, controlling interventions, thus ranging higher along the y-axis. Non-historic ecosystems brought about by human design occupy space plotted by high values along both axes, while novel ecosystems may also emerge with little to no intentional human design, the result of background anthropogenic drivers of change.

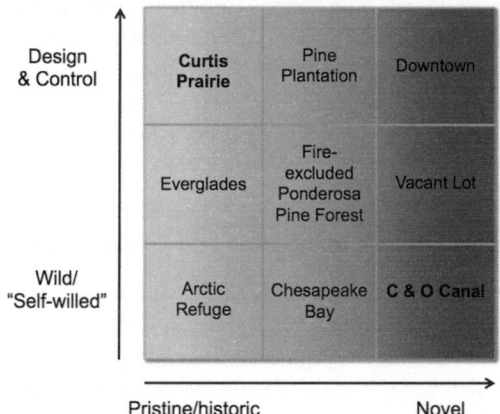

Figure 48.3 Landscape types, with examples, along axes of increased ecological novelty and increased human design and control.

Source: Modified from Aplet and Cole, "The Trouble with Wilderness: Rethinking Park & Wilderness Goals" in Cole and Yung, eds. *Beyond Naturalness* (Island Press, 2010).

Synthesizing these three figures, Mascaro et al. (2013: 52) represent a three-dimensional "*continuous* space in which to ordinate novel ecosystems versus other ecosystems by quantifying the essential components of *novelty* and *human agency*".

A close and careful study of Figure 48.4 will be essential in preparing to discuss environmental values, land management policy, and the implications of novel ecosystems for the development of a conservation ethos that is calibrated to our times.

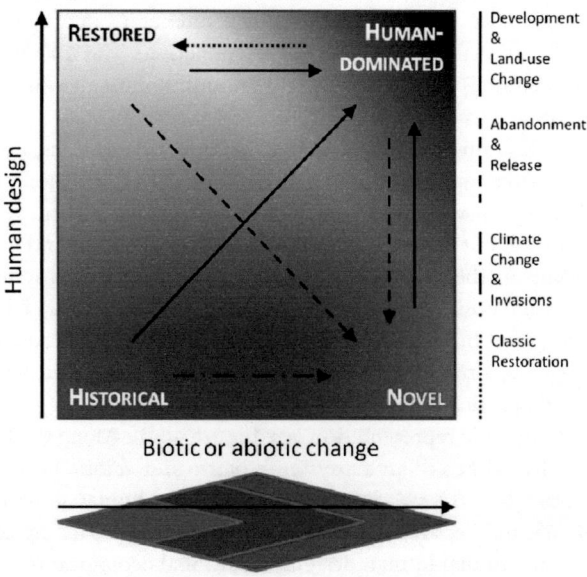

Figure 48.4 A three-dimensional representation of possible ecosystems states with regard to novelty and human agency. The "floor" of the space is provided in Figure 48.2, across which the arrow represents the left-hand section of Figure 48.1. Also shown are four "pathways" of state change due to distinct anthropogenic drivers.

Source: From Mascaro et al. 2013: 52.

Remember where we began, in Figure 48.1, by locating novel ecosystems with regard to only two ecosystem types: historical and intensively used sites. The attribution of value was binary as well: intrinsic or instrumental. Ecosystems that are natural, wild, and historic are valued for their own sake (or their instrumental role in species conservation) or we have ecosystems degraded by human activities, which may or may not result in productive sites of instrumental value. That transformation is represented in Figure 48.4, from the lower-left toward the upper-right, from "Historical" to "Human-Dominated". Along this trajectory, we move via human activities of "development and land-use change", through a progressive loss of historical conditions and subsequent gain of human intentional design and control. Intensively used systems of human design, such as urban, suburban, or agricultural sites, typically lack resilience and thus require on-going human intervention.

Once we have arrived at either a fully "Human-Dominated" or simply some hybrid site (along the way), it may be possible to return the site to historical conditions through the interventions of "classical restoration". But notice that the only possible vectors returning a site to historic conditions require increasing human design and intervention. All "Historical" or "Restored" sites in a historical condition (i.e., all those along the vertical, y-axis) will be affected by background anthropogenic drivers of biotic and abiotic change, represented in Figure 48.4 by "climate change & invasions", the causal origin of which lies primarily off-site and the ecological consequences of which have been either unintentional or unforeseen, or both. The "unambiguous" consequence of these background drivers of environmental change is an increase of novel ecosystems (Mascaro et al. 2013: 55).

Let's consider the other pathways of change represented in Figure 48.4. The vector of "abandonment and release", beginning from restored sites and moving away from historic conditions, is driven by pervasive background anthropogenic phenomena (i.e., "climate change and invasions", nitrogen depositions, etc.). Alternatively, novelty may spontaneously arise following the "abandonment and release" of designed systems, whether "Restored" or "Human-Dominated". Notice, again, that all four pathways of change except for "classic restoration" move toward non-historic, novel states.

Overall, the *location* of an ecosystem within this three-dimensional state-space is plotted by reference to three factors: (i) biotic and (ii) abiotic anthropogenic alteration from historic conditions and (iii) to what extent the extant assemblage is a consequent of intentional human design and intervention or, alternatively, occurs independently of human intention. The *processes of change* represented here, from one compositional/functional state into another, are all anthropogenic. The space is continuous and the four corners represent conditions that are measurable and knowable "even though they may be rarely (if ever) observed". Thresholds would be represented along the x-axis but would be unique for each system, preventing human interventions from driving an ecosystem's "biotic or abiotic characteristics in the direction of a historical ecosystem" (Mascaro et al. 2013: 52).

Far from a simple continuum running between a historical state of "pristine" nature and a site degraded by human use or abuse, the sheer volume of possible ecosystem space-states represented in Figure 48.4, and the socio-ecological narrative about how each particular system arrived at the space it occupies, provides a far more realistic and thus useful model for thinking about environmental values and ecosystem management in the Anthropocene.

Environmental Values for "New" Nature

The ecological world is changing and humans are a primary driver of these changes. Environmental ethics is the sub-discipline of philosophy dedicated to thinking clearly about the ethical dimensions of human activity vis-à-vis the non-human world. Phenomena of the so-called

Great Acceleration impressed upon a first generation of environmental ethicists that much of the non-human world, specifically the natural environment as it had been received in the early 20th century, was changing rapidly and often irreversibly due to human activity. The central, driving factors include human population growth, technological innovations, and a global, capitalist, consumer economy. Humanity was wrecking the world and Clement's theory of ecological change supported the claim that (in many cases) this was morally bad, independent of its consequences for human beings. The view that environmental collectives, including ecosystems or species, were valuable for their own sake as they were naturally—intrinsically valuable—was an ethos that provided grounds for nature conservation and ecological restoration. The explanation of why certain activities with significant environmental consequences were morally wrong was that they contribute to a loss of value, thought of as "natural value", the intrinsic value of a historic and wild nature. Nature conservation was justified by preserving natural value and restoration could be justified by returning, or attempting to replace, lost natural value (see Lee, Hermans, and Hale 2014). In short, natural environments bear an intrinsic value and "unnatural" human interventions thereby degrade nature.

Two developments have undercut the viability of this ethical framework: pervasive anthropogenic drivers of directional environmental change and Gleason's theory of ecological change. The moral terrain has changed with ecological realities—it has gotten significantly more complicated. The paradigm shift away from Clements's climax-state ecology undercut the teleological ontology supporting the attribution of intrinsic goodness to particular, mature states of an ecosystem while background anthropogenic drivers of ecological change increase the difficulty of sustaining historic ecosystems without on-going and increasing human intervention. If the (intrinsic) value of natural ecosystems arises from their wildness, that is, their freedom from human control, then novel ecosystems could be just as valuable as a historic ecosystem. However, if the intrinsic value of natural ecosystems arises from either retaining their historical condition and/or having a genesis free from human intervention, then novel ecosystems cannot be valuable in this way.

Figure 48.4, however, shows that the possible state of an ecosystem, in regard to human influence and departure from historic conditions, demands that we abandon a polarized and binary valuation of ecosystems as either intrinsic or instrumental. If we wish to retain a moral accounting of ecosystems and human interventions in terms of the value of ecosystems, then it seems we must adopt a more pluralistic axiology. I argue that we should adopt the perspective that the value profile of an ecosystem could be as diverse as any of the possible locations within the three-dimensional space represented in Figure 48.4 *and* colored by the socio-ecological narrative history of how the system arrived at its particular state. A brief catalog of alternative kinds of values will provide some examples. But such a listing will not be exhaustive; a central point here is that we no longer have a good basis for restricting environmental values to any closed and finite system.

Let's "stake out" the four corners of Figure 48.4: (A) historical, (B) restored, (C) human-dominated, and (D) novel. First, consider (A). We see that assignment of intrinsic value to a historic and wild ecosystem will vary as the system exists at some distance or another from the absolute "zero" of both historic, biotic and abiotic composition and freedom from human design. This point is already accepted by traditional environmental philosophers who acknowledge there is no longer any "pristine", non-humanized nature. Instead these philosophers claim that "natural value" or "wildness" exists along a continuum; there is more or less natural value in a site. But, of course, on a view of value in nature that only recognizes intrinsic or instrumental value, wherever natural value is less—where natural, intrinsic value has been lost—it has been replaced by the disvalue of a degraded nature, or perhaps in the best case exhibiting instrumental value for human ends. Either restoration serves to return some intrinsic natural value—of historic conditions—or there is degradation. This position recognizes only the single axis in Figure 48.1, from "wild and historic" to "intensively used". But instead of grappling with novel ecosystems in the middle, any interventions toward historic conditions are restorative and thus good, and any anthropogenic drivers of change moving away from historic conditions are degrading and thus bad.

Second, consider (B), classic restoration. In addition to restorative interventions returning an intrinsic value, alternatives have been defended by others. Andrew Light and Eric Higgs have described cultural or human values associated with participation in the practices of ecological restoration (Light and Higgs 1996; Light 2000; Higgs 2003). John Basal and William Jordan have approached restoration practice via virtue ethics and the betterment of human character (Basal 2010; Jordan 2012), and Martin Drenthen understands the norms of restoration practice and the value of restored sites through a hermenutical method, attending to an ability to read the landscape and proposing that we guide interventions with an intention to continue an appropriate narrative (Drenthen 2013). Additionally, restored ecosystem may provide a variety of valuable ecosystem services to humans and revive various biological functions that benefit species of high conservation value.

Third, consider (C). Heavily "human-dominated" landscapes, both non-historic and designed, may be instrumentally valuable in a wide variety of ways. Urban and suburban environments provide residential, commercial, and social space for human beings, of course. People care for the storied *places* they live in a wide variety of ways. Industrial and agricultural sites provide employment, productive capacity, and food. Much-discussed "ecosystem services" are diverse, initially framed to harness utilitarian sensibilities in the pursuit of protecting ecosystems from development and conserving biodiversity (Norton 2015: 174), and include several categories of benefit to humans: provisioning, regulating, supporting, and cultural services. Of course, such ecosystem services are not limited to nor even predominately associated with highly designed and novel ecosystems. But the point is that "use" value is diverse and can be associated with a wide variety of historic and human influenced ecosystem types.

Finally, consider (D). Novel ecosystems could exhibit a wide range of variation, from the highly designed (if a "designer" ecosystem becomes resilient and self-organizing) to nature's spontaneous organization in response to a wide variety of possible human influences. Far from simply "degraded", if one already values the traditional goals of ecosystem management "such as preserving biodiversity, maintaining stability, or enhancing resilience", then novel ecosystems should be valued in a variety of ways (Light, Thompson, and Higgs 2013: 258). First, they may provide the continuation of, or new and improved, benefits for people. Novel ecosystems may exhibit a higher biological diversity than historical systems or, when compared to states with greater historic fidelity, they may be seen as more suitable for controlled and uncontrolled experiments designed to increase our ecological understanding and develop innovative management techniques. Further, allowing or even encouraging the emergence of novel ecosystems will increase resilience to further anticipated ecological changes. So novel ecosystems could provide additional conservation capacity, providing non-traditional habitat to endangered species, or perhaps they could function in association with the exercise of more controversial conservation practices, such as assisted migration or de-extinction. Further, novel ecosystems may be judged to have aesthetic or even "natural" heritage value, culturally assimilated into people's lives and seen as places that are important to people. Many of the landscapes in European countries, for example, are beloved but novel. Finally, as nature rebounding from human influence but yet unguided by human design, novel ecosystems can be valued as a new "wild" nature (Light, Thompson, and Higgs 2013)

Additionally, we can begin to imagine possible combinations of all these kinds of values, since all various *kinds* of ecosystems have, and each *particular* ecosystem has, an individual history of diverse anthropogenic influence and natural processes and a wide variety of resulting compositional states and functional profiles. Figure 48.5 represents possible path trajectories through the state-space of ecosystems discussed previously and represented in Figure 48.4. We can now understand this three-dimensional space also as a diverse space of pluralistic values, values tied to various positions and specific socio-ecological histories and narratives of ecosystem change.

595

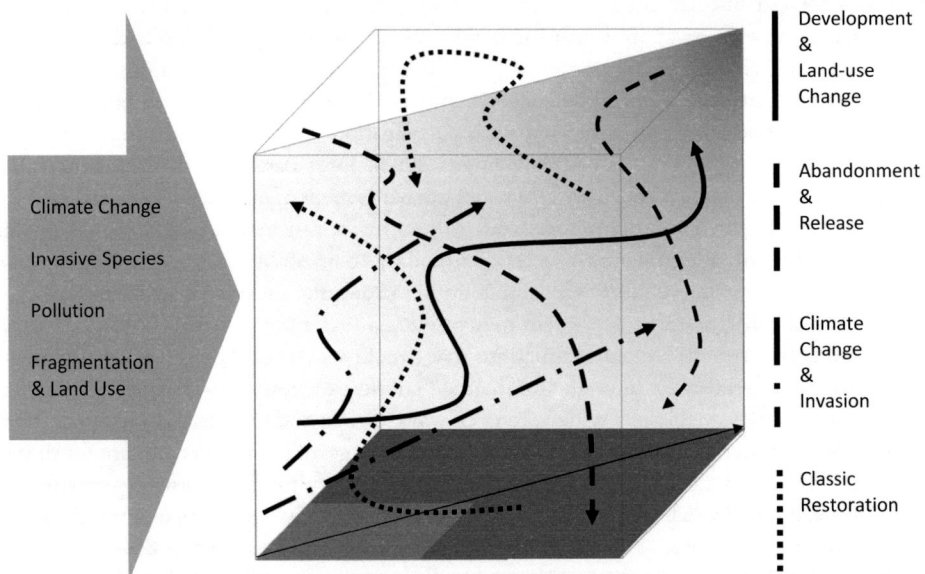

Figure 48.5   Possible value-space of different ecosystems, in which value dimensions are sensitive to kinds of environmental value (discussed in text) relative to (i) location in a state-space (referenced to degree of biotic and abiotic alternation and degree of human intention, see Figure 48.4) and (ii) the particular narrative history of how the ecosystem *in situ* came to occupy that state-space via a specific combination of various drivers of change.

Adopting this perspective, we can see the implausibility of maintaining that environments with such diverse conditions and historical narratives possess only one or both of two types of value (i.e., either intrinsic value, the idea highlighted in traditional environmental ethics, or instrumental value, a use-value which has been the focus of environmental economics). On these grounds I conclude that value pluralism is a more appropriate axiology for deliberations concerning ecosystem management into the Anthropocene.

Other arguments for value pluralism about the environment can be found in the literature (Light, O'Neill, and Holland 2008; Light 2010). Recently, for example, Kia Chan and colleagues have argued against monistic theories of value in nature and instead defend the notion of relational value, which is pluralistic (Chan et al. 2016). First, they characterize the two monistic views

1   Instrumental value = being in/seeing nature brings people pleasure or satisfaction
2   Intrinsic value = nature has value independent of people
     Then they identify eight types of value that reflect the relations between people and elements of "nature" (the non-human environment), which are divided into two categories:
3   Relational values (involving the human collective)

   a   *Cultural Identity* = place is important to my people, who we are as a people
   b   *Social Cohesion* = being in nature is a vehicle of me to connect with people
   c   *Social Responsibility* = caring for ecosystems is crucial to caring for my fellow humans, present and future
   d   *Moral Responsibility to Non-Humans* = caring for all life forms and physical forms is a moral necessity

4   Relational values (primarily individual)

  a   *Individual Identity* = this place is important to me, to who I am as a person
  b   *Stewardship Eudaimonic* = my care for this land fulfills me, helps me lead a good life
  c   *Stewardship Principle/Virtue* = keeping the land healthy is the right thing to do

Clearly, the possible space for novel ecosystems as "new" nature is significant and will further complicate this scheme of relational values. This has consequences for the traditional methods, policies, and tools of conservation practice and ecosystem management, which will have to be supplemented with new approaches. This work, too, is underway. Space prohibits anything but a brief mention of this work, which is summarized nicely in the abstract to Hobbs et al. (2014), which I quote at length:

> The reality confronting ecosystem managers today is one of heterogeneous, rapidly trans-forming landscapes.... A landscape management framework that incorporates all systems, across the spectrum of degrees of alteration, provides a fuller set of options of how and when to intervene, uses limited resources more effectively, and increases the chances of achieving management goals. That many ecosystems have departed so substantially from their historical trajectory that they defy conventional restoration is not in dispute. Acknowledging novel ecosystems need not constitute a threat to existing policy and management approaches. Rather, the development of an integrated approach to management interventions can provide options that are in tune with the current reality of rapid ecosystem change.

To conclude, it's important to remember that novel ecosystems are not ontologically deep; the type does not represent some natural kind, as ecosystems were mistakenly portrayed under Clements's "organismal" theory. Instead, the category of novel ecosystems, like restored, historical, intensively used, "pristine" or wild, is just a tool for a nominal taxonomy, used only to ordinate a particular ecosystem in relation to human influence on its composition and function across time. How or why we intervene is up to us, guided by values that are constituted through various relations, to each other and the places we live, and embedded in a rich diversity of narratives. Perhaps the most important point is that interventions aimed at restoring historic conditions based on intrinsic value will not always be appropriate, nor are such interventions the only way to promote significant environmental values.

Finally, recall that ecosystem novelty is not new but follows directly from Gleason's "species-individualist" theory. A novel ecosystem cannot be a new, unnatural, and grossly anthropogenic monster, as many think of the creation that Dr. Frankenstein failed to love (Latour 2012). What *is* new, however, is a matter of scale, specifically the rate at which rapidly emerging, anthropogenic novel ecosystems are introducing ecosystem novelty into the global biosphere. And this *is* scary, but this is the Anthropocene. So, unless we can somehow prevent the Anthropocene from unfolding, we better begin further incorporating the concept of novel ecosystems into environmental science, planning, policy, and ethics.

Note

1 "*Composition* describes the parts of each biodiversity component in a given area (e.g., habitat types, species present, genetic populations within species). *Structure* refers to the physical characteristics supporting that composition (e.g., size of habitats, forest canopy structure, etc.). *Function* means the ecological and evolutionary processes affecting life within that structure (e.g., natural disturbances, predator-prey relationships, species adaptation over time)". (http://www.biodiversitybc.org/EN/main/where/131.html. Italics added). In the reminder of the chapter, I will simply refer to composition and function.

References

Basal, J. (2010). Restitutive restoration: new motivations for ecological restoration. *Environmental Ethics*, 32/2, 135–147.

Callicott, J. B. (2003). The Implication of the "Shifting Paradigm" in Ecology for Paradigm Shifts in the Philosophy of Conservation. In B. A. Minteer and R. E. Manning (eds.) *Reconstructing Conservation: Finding Common Ground*, Washington, DC: Island Press, 239–261.

Chan, K. M. A., Balvanera, P., Benessaiah, K., Chapman, M., Diaz, S., Gomez-Baggethun, E., Gould, R., Hannaha, N., Jax, K., Klain, S., Luck, G.W., Martin-Lopez, B., Muraca, B., Norton, B., Ott, K., Pascual, U., Satterfield, T., Tadaki, M., Taggart, J., and Turner, N. (2016). Why protect nature? Rethinking values and the environment. *Proceedings of the National Academy of Sciences of the United States of America*, 113/6, 1462–1465.

Clements, F. E. (1916). Plant succession: an analysis of the development of vegetation. *Carnegie Institution of Washington Publication*, 242 (1916).

Cronon, W. (ed.) (1995). *Uncommon Ground: Rethinking the Human Place in Nature*, New York: W. W. Norton & Co.

Drenthen, M. (2013). New Nature Narratives: Landscape Hermeneutics and Environmental Ethics. In F. Clingerman, M. Drenthen, B. Treanor, and D. Utsler (eds.) *Interpreting Nature, the Emerging Field of Environmental Hermeneutics*. New York: Fordham University Press.

Gleason, H. A. (1926). The individualistic concept of the plant association. *Bulletin of the Torrey Botanical Club*, 53/1.

Higgs, E. (2003). *Nature by Design: People, Natural Process, and Ecological Restoration*. Cambridge, MA: MIT Press.

Hobbs, R. J. (2016). Degraded or just different? Perceptions and value judgments in restoration decisions. *Restoration Ecology*, 24/2, 153–158.

Hobbs, R. J., Arico, S., Aronson, J., Brown, J. S., Bridgewater, P., Cramer, V. A., Epstein, P. R., Ewel, J. J., Klink, C. A., Lugo, A. E., Norton, D., Ojima, D. Richardson, D. M., Sanderson, E. W., Valladares, F., Vila, M., Zamora, R., and Zobel, M. (2006). Novel ecosystems: theoretical and management aspects of the new ecological world order. *Global Ecology & Biogeography*, 15, 1–7.

Hobbs, R. J., Higgs, E., and Hall, C. (eds.) (2013). *Novel Ecosystems: Intervening in the New Ecological World Order*. Hoboken, NJ: Wiley-Blackwell.

Hobbs, R. J., Higgs, E., Hall, C. M., Bridgewater, P., Chapin III, F. S., Ellis, E. C., Ewel, J. J., Hallett, L. M., Harris, J., Hulvey, K. B., Jackson, S. T., Kennedy, P. L. Kueffer, C., Lach, L., Lantz, T. C., Lugo, A. E., Mascaro, J., Murphy, S. D., Nelson, C. R., Perring, M. P., Richardson, D. M., Seastedt, T. R., Standish, R. J., Starzomski, B. M., Suding, K. N., Tognetti, P. M., Yakob, L., and Yung, L. (2014). Managing the whole landscape: historical, hybrid, and novel ecosystems. *Frontiers in Ecology and the Environment*, 12, 557–564. Retrieved from http://dx.doi.org/10.1890/130300

Hobbs, R. J., Higgs, E., and Harris, J. A. (2009) Novel ecosystems: implications for conservation and restoration. *Trends in Ecology & Evolution*, 24, 599–605.

Holling, C. S. (1973). Resilience and stability in ecological systems. *Annual Review of Ecology and Systematics*, 4, 1–23.

Jackson, S. T. (2013). Perspective: Ecological Novelty Is Not New. In R. Hobbs, E. Higgs, and C. Hall (eds.) *Novel Ecosystems: Intervening in the New Ecological World Order*, Hoboken, NJ: Wiley-Blackwell.

Jordan, W. R. (2012). *The Sunflower Forest: Ecological Restoration and the New Communion with Nature*. Berkeley, CA: University of California Press.

Latour, B. (2012). Love Your Monsters: Why We Must Care for Our Technologies As We Do Our Children. In T. Nordhuas and M. Shellenberger (eds.) *Love Your Monsters: Postenvironmentalism and the Anthropocene*, Oakland, CA: Breakthrough Institute

Lee, A., Hermans, A., and Hale, B. (2014). Restoration, obligation, and the baseline problem. *Environmental Ethics*, 36/2, 171–186.

Leopold, A. S. et al. (1963). The Goal of Park Management in the United States. *Wildlife Management in the National Parks*. National Park Service.

Light, A. (2000). Ecological Restoration and the Culture of Nature: A Pragmatic Perspective. In P. Gobster and B. Hull (eds.) *Restoring Nature: Perspectives from the Social Sciences and Humanities*. Washington, DC: Island Press.

Light, A. (2010). Methodological Pragmatism, Pluralism, and Environmental Ethics. In D. Keller (ed.) *Environmental Ethics: The Big Questions*. Oxford: Blackwell Publishers.

Light, A., and Higgs, E. (1996). The politics of ecological restoration. *Environmental Ethics*, 18/3, 227–247.

Light, A., O'Neill, J., and Holland, A. (2008). *Environmental Values*. New York: Routledge.

Light, A., Thompson, A., and Higgs, E. (2013). Valuing Novel Ecosystems. In R. Hobbs, E. Higgs, and C. Hall. (eds.) *Novel Ecosystems: Intervening in the New Ecological World Order*. Hoboken, NJ: Wiley-Blackwell, 257–268.

Mascaro, J., Harris, J., Lach, L., Thompson, A., Perring, M., Richardson, D., and Ellis, E. C. (2013). Origins of the Novel Ecosystem Concept. In R. Hobbs, E. Higgs, and C. Hall. (eds.) *Novel Ecosystems: Intervening in the New Ecological World Order*. Hoboken, NJ: Wiley-Blackwell. 45–57.

McIntosh, R. P. (1975). H. A. Gleason: 'Individual ecologist' 1882–1975: his contributions to ecological theory. *Bulletin of Torrey Botanical Club*, 102, 253–273.

Milton, S. J. (2003). 'Emerging ecosystems': a washing-stone for ecologists, economists and sociologists? *South African Journal of Science*, 99, 404–406.

National Park Service Organic Act (NPSOA) (1916). Retrieved from http://web.archive.org/web/20080413094946/ http://www.nps.gov/legacy/organic-act.htm

Norton, B. (2015). *Sustainable Values, Sustainable Change: A Guide to Environmental Decision Making*. Chicago: University of Chicago Press.

Perring, M. P., and Ellis, E. C. (2013). The Extent of Novel Ecosystems: Long in Time and Broad in Space. In R. Hobbs, E. Higgs, and C. Hall (eds.) *Novel Ecosystems: Intervening in the New Ecological World Order*. Hoboken, NJ: Wiley-Blackwell.

Shelley, M. (1818/1996). *Frankenstein: A Norton Critical Edition*. Ed. J. Paul Hunter. New York: W.W. Norton & Company.

Simberloff, D., Carolina, M., and Aronson, J. (2015). 'Novel ecosystems' are a Trojan horse for conservation. Ensia. Retrieved from http://ensia.com/voices/novel-ecosystems-are-a-trojan-horse-for-conservation/

Standish, R., Thompson, A., Higgs, E., and Murphy, S. (2013). Concerns about Novel Ecosystems. In R. Hobbs, E. Higgs, and C. Hall (eds.) *Novel Ecosystems: Intervening in the New Ecological World Order*. Hoboken, NJ: Wiley-Blackwell, 296–309.

Steffen, W., Sanderson, A., Tyson, P. D., Jäger, J., Matson, P. A., Moore III, B., Oldfield, F., Richardson, K., Schellnhuber, H. J., Turner, B. L., and Wasson, R. J. (2004). *Global Change and the Earth System: A Planet Under Pressure*. New York: Springer-Verlag.

Suckling, K. (2014). Against the Anthropocene. Retrieved from http://blog.uvm.edu/aivakhiv/2014/07/07/against-the-anthropocene/

Tansley, A. G. (1935). The use and abuse of vegetational concepts and terms. *Ecology*, 16, 284–307.

The Economist (Oct. 31st, 2015). Greater than the sum of its parts. Retrieved from http://www.economist.com/news/science-and-technology/21677188-it-rare-new-animal-species-emerge-front-scientists-eyes

Walker, B., Holling, C. S., Carpenter, S. R., and Kinzig, A. (2004). Resilience, adaptability and transformability in social–ecological systems. *Ecology and Society*, 9/2, 5. Retrieved from http://www.ecologyandsociety.org/vol9/iss2/art5

PART IX

Policy Frameworks and Response Measures

49

POLLUTION AND POLLUTER PAYS

Aaron Lercher

Introduction

The slogan "Polluter pays" is here taken to mean that, in certain laws and regulations, polluters are required to pay damages regardless of their degree of fault for pollution. This liability standard is part of two US environmental laws. The 1980 Comprehensive Environmental Response, Compensation, and Liability Act (CERCLA) creates a federal government power to respond to threats posed by toxic materials. The 1990 Oil Pollution Act (OPA) creates a federal power to respond to oil spills. CERCLA is popularly known as Superfund, but since the emphasis here is on the liability provisions not the trust fund, the law will be referred to as CERCLA. For a history of US environmental law, written by an authority on CERCLA, see Lazarus (2007).

The slogan "Polluter pays" can also be used more generally to mean that external costs of pollution should be internalized by polluters in some way. These economic strategies are discussed in Chapter 55 on Economic Instruments in this volume. This chapter focuses narrowly on explaining the liability provisions in CERCLA and OPA.

An agent is at fault for wrongdoing when the agent either intentionally does wrong, or is reckless or negligent. But these laws hold those who handle toxic materials and oil to the standard of "strict" liability, which does not require fault. In order to explain the liability provisions of CERCLA and OPA, we need to address two ethical questions. These questions are posed in an anthropocentric setting, since one organism's toxin is another's food or pleasant living environment. First, when can an act or activity be wrong, even when the actor is not at fault? Second, when can an activity be wrong, even when it is not harmful and only poses a risk? In addition, there are analogous legal questions. Should an act or activity be held illegal, even when there is no intentional, reckless, or negligent wrongdoing? Should an activity be illegal because it poses risks of some kind, although it is not (yet) harmful? We also want our moral demands to be at least consistent with a plausible way of running society. Although the latter requirement is weak, it plays a role in the following argument.

I shall argue that acts and activities should be held illegal for posing risks of certain kinds, even when the agent is not at fault, and even though there might not be any moral prohibition against such acts or activities. In some cases polluters should be legally required to pay, and indeed they morally ought to pay, even though they may have done nothing wrong. This argument is deontological, and perhaps counterintuitive, so I shall address an objection from a consequentialist perspective that the laws framed on this basis may be irrational.

DOI: 10.4324/9781315768090-59

Fault and Wrong-Doing in Causally Complex Cases

The kind of situation we are interested in occurs in legal cases. Here is a hypothetical case quoted from a standard legal reference work:

> **Malloy**: The Malloy Corporation produces components for computers that are essential to the modern economy. Its manufacturing plant is located in a community almost all of which is residential. Its manufacturing process generates a toxic chemical as a byproduct. Malloy stores this chemical in storage bins pending shipment of the chemical to an off-site disposal facility. This storage arrangement complies with the requirements of reasonable care and likewise with applicable public regulations. Even during normal and proper operation, it is often necessary to open the lids on these bins for periods of time. Wind conditions may then arise that can disperse the chemical from the storage bins to the property of Malloy's neighbors; over time, such dispersion is quite likely but not certain. When and if dispersion occurs, the toxic fumes emanating from the chemicals can easily induce serious illness in those living on the property.
>
> *(American Law Institute 2010: 240)*

In this case, the Malloy Corporation is not at fault. It exercises reasonable care. For simplicity, let us stipulate, more strongly, that Malloy is exercising all reasonable care that is currently possible. The Malloy Corporation's activities have not, as far as this case tells us, caused harm. These activities pose a serious risk of harm, however. By the (weak) law of large numbers, a predictable rate of harmful results will emerge over time.

Malloy's activities are also probably not what the neighbors expect or understand well. Malloy's chemical processes are parts of complex chains of events, including random events such as wind conditions, as well as the complex physiological events that would happen if someone were harmed.

The Malloy example, although it represents a common type of example, nevertheless goes against any intuitive belief that wrongdoing implies fault. As long as the reader agrees that Malloy is doing something wrong, this example counts as evidence against the otherwise plausible claim that fault is necessary for wrongdoing.

In this case, however, it is unclear what exactly is wrong here. Malloy is not at fault as long as it is exercising reasonable care and we have more strongly stipulated that it is exercising all possible and reasonable care. But perhaps Malloy should not have built its factory at that location in the first place, and perhaps it was at fault then. Being at fault in the distant past does not fit well in most legal processes, except perhaps the most monstrous crimes. Also, the surrounding community may be held equally at fault for negligence in its zoning law. So we want to focus on the immediate case.

Why then should we disregard fault in the Malloy case? It seems that the complexity of a causal chain can, in many cases, be a reason for disregarding fault.

> **Day's End**: When *B* returns home from work in the evening, he turns on the light switch in the front room of his apartment. Due to circumstances *B* could not foresee or control, one night this ignites a fire in the apartment next door, which quickly kills his neighbor.
>
> *(Thomson 1986: 229)*

Again, it is methodologically helpful to note that this hypothetical case does not simply narrate my assumptions. Instead, it provides the reader an opportunity to confirm a moral judgment. Another assumption that I suggest we make here, following Thomson, is that if *B* ought not to turn on the light switch, then "ought" is meant objectively, rather than describing any moral obligation *B* may have perceived or not. Thus if Day's End were in a movie, and the audience had been shown the circumstances connecting the light switch to the fire, someone in the audience might call out, "Don't!" when *B* arrives home and then reaches for the light switch. So I suggest that *B* objectively ought not to turn on the light, and in this sense is doing wrong when he turns on the light.

In the Malloy case, chemical processes constitute a paradigmatically complex causal chain. The length of the causal chain could be extended further, if the chemicals were buried and then uncovered after many years, as in the toxic waste sites covered by CERCLA. The complex processes of mining and transporting oil, covered by OPA, also count as complex causal chains. The long causal chains in making consumer goods have led in the US to strict liability in product liability (Moss 2002: 216–252). Analogously, since most employees lack control over workplace conditions, this fact has led to strict liability for worker's compensation in the US (Moss 2002: 152–169).

Risk Imposition and the Right to Self-Defense

Malloy has not caused any harm yet, as described in the hypothetical case. Cases in which harm is clearly caused by an agent tend not to be regarded as environmental pollution cases. These are cases of poisoning, asphyxiation, burning, etc. Environmental pollution, in contrast, involves risks of harm, and resulting harms are not usually clear. Malloy, then, has imposed risks on its neighbors. Is this wrong?

In the Malloy case, the fumes really are toxic, not merely perceived as toxic. But in this case, no harm has yet occurred. The combination of lack of harm or at least lack of *proof* of harm, together with risk of harm, is common in cases of toxic pollution (National Research Council 1991; Cranor 1993; Tesh 2000). There is no uniform method for determining which risks are real, and how to compare them (National Research Council 2007: 105–111). There are statistical, toxicological, and epidemiological reasons why it is difficult to prove that even highly toxic materials have caused harm in people. The exposed population is normally too small to allow a statistical inference using standard methods. It is difficult to make inferences from animal experiments to human effects. It is difficult to tell how much exposure a population has received, and by what pathways.

In cases in which there is no proof of harm, it may also be argued that there is no real risk (Wildavsky 1995; Sunstein 2002). It is difficult to know which risks are real. But it would be a serious mistake to infer from this difficulty that risks are not real unless people are harmed.

Consider then only real risks. Which of these risks are wrong to impose? This is the ethical problem of risk imposition. This problem is difficult to state clearly and has persistently eluded an adequate solution (Lewens 2007; Hayenhjelm and Wolff 2012). For this problem, we need to clarify what counts as a relevant harm or loss, as well as the situations in which increasing the probability of this harm or loss is wrong. In the Malloy case there are immediate risks of illness for inhabitants of the neighboring property. But beyond that there are other risks, such as loss of property values, inhabitants' abilities to fulfill work, and family obligations. For the narrow topic in this chapter, we shall see how we can bypass the general ethical problem of risk imposition.

Nevertheless, despite the difficulty of the ethical problem of risk imposition, sometimes it is wrong to impose risks. The following kind of example is often used in ethics literature for discussing problems of risk imposition:

> **Russian roulette**: *B* plays Russian roulette on *A* who does not agree to this and does not know about it. (*B* has various revolvers, with number of cylinders ranging up to very large numbers, which *B* uses in different cases.) In this case, there is a single bullet in *B*'s six-chamber revolver, but the bullet is not under the firing the pin when *B* pulls the trigger.
>
> *(Nozick 1974: 79)*

The reader is likely to agree that *B*'s activity is clearly wrong. But even if it is, it is unclear whether *B* has violated *A*'s rights. *B* has not harmed *A*, and has not even frightened *A*. It may be wrong to impose risks in some cases, but it is difficult to explain which risks and for what reason. So instead of attempting to answer the question of which risks are wrong to impose, I shall argue that *A* has a right of self-defense. The advantage of this approach requires some analysis to explain.

In a Hohfeldian analysis of rights, the most basic form of right is a claim-right: *A* has a claim-right against *B* that *B* carries out action *P*. Other rights are built from these units. For every claim-right that *A* has against *B* that *B* carries out *P*, there is a corresponding duty that *B* has toward *A* of carrying out *P*. Presumably, for example, *A* has a right against everyone that none of us should poke *A* in the eye, and everyone has the duty not to poke *A* in the eye. In Thomson's (1990) account of Hohfeldian rights, claims against bodily incursion are fundamental rights. Many such claim-rights would have to be specified in an account of a right against risk impositions, and we would need to specify many acts.

Self-defense in a Hohfeldian analysis, however, is a privilege (Thomson 1990; Thomson 1991; Doggett 2011). A privilege is a lack of claims, and does not imply the existence of any claim. This considerably lightens the burden of moral argument. If *A* has a privilege of doing *P* with respect to *B*, this is analyzed: *B* does not have any claim against *A* that *A* should not carry out action *P*.

Assume that *A* has a right against being physically harmed. Then in the Russian roulette case, there are some actions for which *B* has no claims that *A* should not carry out such actions. In particular *B* has no claim against *A* that *A* should not carry out actions of the kind that would prevent *B* from playing Russian roulette on *A*. In such a case, Thomson remarks that the relevant thesis about self-defense is: "No one has a right that we *let* him infringe our rights. No one has a right that we shall not prevent him from infringing on our rights" (1986: 161). In the Russian roulette case, *A* has a privilege of self-defense. I suggest that Malloy's neighbors also have a privilege of self-defense.

The Right of Recourse to Government Action

Suppose then that by imposing risks on its neighbors, Malloy has not (yet) infringed the rights of any of its neighbors. Nevertheless, the privilege of self-defense allows the residents of the neighborhood to take some action. What action? We should worry whether the response is proportionate with the threat. Again, there is a problem of evidence. We need to worry a lot about what kind of evidence one needs to have about the risk in order to act (Cranor 1993).

But again, like the questions about the extent of rights, if any, against risk imposition, we can bypass these difficult ethical questions here. Perhaps the residents are entitled to vigilante action as soon as they learn what Malloy is doing, or perhaps not. Instead we can assume that there is a government, or something like one, which is capable of acting their place.

This assumption about the existence of a government can be made in many ways. We may assume the neighbor residents are capable of convening themselves as self-governing, and as a group they form a government. Or we might assume there is some other authority for the neighborhood residents to call on. We need not assume anything about the nature of the government, as long as the neighborhood residents end up connected with an agency of some sort that is capable of acting in their place. Then the neighborhood residents' right of self-defense does not disappear, but the residents do not act on it. Instead, their right of self-defense is converted into what I call a right of recourse.

We do not need to analyze this right of recourse in order to understand it well enough for our purposes here. There are only a few possible risk management policies that can be employed by a government, whatever government this may be. There are three basic kinds of risk management policy: risk reduction, risk spreading, and risk shifting. Risk management policies then are composed of combinations of these three basic policies (Calabresi 1970; Moss 2002). (It is said that the government of an industrial state is an "insurance company with an army." To the extent this is true, Moss (2002) is an indispensable overview of the modern state.)

Risk Reduction

Risk reduction is not relevant to the Malloy case, if Malloy is already exercising all possible and reasonable care, as was stipulated. But if its degree of care is merely reasonable and non-negligent, there may be room for improvement. Then a negotiation, if feasible, would attempt to find an agreement on whether the neighborhood residents should pay Malloy for an increased level of safety, or whether Malloy should compensate the residents for the risk they are running.

> **Bribing Malloy**: If Malloy installs vapor-capturing mechanisms on its chemical storage bins, the problem of toxic fumes will be eliminated. Suppose the vapor-capture mechanism costs less than the costs of illnesses caused by Malloy's toxic fumes. Then it is efficient for the neighborhood residents to pay Malloy's cost of installing the vapor-capturing mechanisms. None of the neighborhood residents complain. Every neighborhood resident willingly contributes to the bribery fund.

If such negotiations fail, and if Malloy avoids the costs, then action by some other authority, acting as a government, would be required for an efficient outcome. This authority would compel Malloy to install the vapor-capturing mechanisms. It might impose the cost on Malloy or else collect it from the neighborhood residents. Either way, the agreement would reduce risks. Also, both choices are equally efficient. That is, no further rearrangement of goods would improve things for either Malloy or the residents, without also being to someone's detriment. The classic economic analysis of this situation is provided by Coase (1960).

The reader will probably agree that it is unlikely that the neighborhood residents and Malloy would be able to come to an agreement to reduce the risk due to the toxic fumes, without having to appeal to another authority capable of compelling Malloy to install vapor-capture mechanisms. Instead it is likely that obstacles of various kinds will intervene, including lack of knowledge, delays, suspicion, uncooperativeness, and the expense of lawyers and courts. These are collectively labeled "transaction costs" by economists and economics-minded lawyers (Calabresi 1970). But suppose that by an agreement or by compulsion, an authority is successful in reducing risks. Then success in this sense is measured by an outcome, installing vapor-capturing mechanisms. Possibly, there could also be measurable health benefits. But that is not assumed here.

Risk Spreading

Risk spreading is done by insurance or else something that functions like insurance. Given a probability p of incurring a cost C each year, and given a group of people with the same probability and potential cost, everyone in the group pays pC, plus administrative overhead, each year to an administrator, who pays C to each member when she incurs the cost. The mathematics of the weak law of large numbers implies the existence of a predictable annual cost for a group of individuals with similar risks.

Spreading risks of non-compensable harms, such as illness or injury, would amount to a threat of a harm to everyone in a group. Such threats may be common, but they violate due process rights. So risk spreading spreads costs.

Spreading risks is often rational even if individuals are not risk adverse. Suppose that a hypothetical illness strikes by chance on average once every four years. Then there is only 1/16 chance

(0.0625) of getting it in two years in a row. But over three years, there is a bit less than 1/5 chance (0.1875) of getting it two years in a row, this chance rises to over 1/5 (0.2109) over four years, and it keeps rising. For individuals who are not risk adverse, insurance is a rational way of avoiding the excess savings needed to cover short runs of "bad luck."

> **Insuring Against Malloy**: Malloy's neighbors agree to spread the risk of exposure to toxic fumes. Exposure depends on wind direction when Malloy opens the bins of toxic material. Not all the neighbors are exposed at once. The neighborhood forms a mutual insurance organization, spreading risk among themselves.

If every neighborhood resident buys insurance, this spreads the risk. If a toxic emission occurs, then the affected residents will be awarded a payment. Again, the reader will probably agree that, even if this policy were acceptable and rational, it is unlikely that every neighborhood resident will enter such an agreement, without some push from other residents or the government.

Risk Shifting

Neither the policies of risk reduction nor those of risk spreading provide everything that one would want from risk management. Risk reduction is an important goal. But achieving it is indifferent to whom bears the cost, and the cost may be borne by neighborhood residents. Risk spreading pays victims after the fact. But potential victims bear the cost of risk spreading. Neither of these policies is a substitute for the residents' privilege of self-defense, which gives them the right to act prior to being harmed in order to prevent their being harmed.

Risk shifting is the only other policy that a government can provide. It is not generally accepted that Malloy's owners should be exposed to the risks of exposure to toxic fumes in place of the neighborhood residents. So shifting risk means shifting costs. The usual rationale is efficiency. Malloy is better able to anticipate risks and has more control over them than the neighborhood residents. So Malloy should bear the risks. To the extent the risks are managed by insuring or investing in safer technology, the risk becomes a cost of doing business. Fault may be irrelevant if risk shifting merely allocates this cost either on the neighbors or on Malloy. Another reason for disregarding fault is the complexity of the causal chain leading to harm. Yet another reason is that the burden of proving Malloy is at fault may be too difficult for neighborhood residents, even if Malloy in fact is at fault.

Risk shifting, finally, is a plausible substitute for neighborhood residents' right of self-defense. Risk shifting gives the residents something, at least, in exchange for giving up their right of self-defense. In exchange, Malloy is made to give up one of its defenses. Malloy is made to give up the defense that it was not at fault. Without that recourse, residents' options are limited. They can just put up with the risks, bribe the risk imposer, move away, or undertake vigilante action. Gibbs (1998) and Bullard (2000) give influential primary accounts of citizen action in this situation. Szasz (1994) analyzes the political context of these events.

Risk shifting assures residents they will not pay the costs of Malloy's risk imposition. Risk shifting can be an efficient means for inducing risk reduction. Malloy is in a better position to reduce risks than the neighborhood residents. But if Malloy is able to spread its risk of compensation for harm to residents by buying liability insurance, this lowers its incentive to reduce risks.

Strict Liability in Environmental Law

Now we can explain the liability provisions of CERCLA and OPA. Both these laws entitle the government to act in response to threats. In CERCLA, government recourse is provided for toxic threats to people. In OPA, government recourse is provided for threats to natural resources. In CERCLA and OPA, risk shifting is combined with risk reduction. In these laws, liability shifts risk, and risk is reduced when the federal or state government orders a liable party to clean up a hazard it has caused. Risk shifting is explained by the need for self-defense, which in practice emerges in a demand for government recourse.

Risk shifting need not be efficient, although efficiency is one argument in favor of shifting risks with complex causes onto those who cause the risks, and presumably are better able to control them. But the goal of risk reduction does not explain the risk-shifting provisions in CERCLA and OPA. Indeed, Congress abandoned one important risk-shifting provision of CECLA in 1995 by failing to reauthorize a tax on the chemical and oil industries, which allows some of the risks of these industries to shift back onto the taxpaying public, where they have since remained. A policy of risk reduction alone does not require polluters to pay. But the liability provisions of CERCLA remain in effect.

Liability in CERCLA is for releases or threatened releases of hazardous materials. Liability is strict, joint and several, and retrospective. Strict liability, as has been explained, is liability that applies regardless of any degree of fault. Liability that is "joint and several" means that every person who contributes to a hazard can be found liable for the whole problem. Then presumably secondary suits will sort out the degree of liability after an initial civil action by the United States or a State. The purposes of this harsh "joint and several" liability are to obtain efficient recovery of costs and to force action.

The terms "strict," "joint and several," and "retrospective" do not occur in the text of CER-CLA or OPA. Instead, the laws set out liability in the form of lists of those covered, and the defenses they are entitled to. The text of CERCLA (42 USC §9607(a)) gives an expansive list of persons covered by the law, including any "owner" or "operator" of any entity in charge of hazardous materials. The text of CERCLA, at the same place, gives a very narrow range of defenses, none of which depends on degree of fault. The defenses are based on (1) "an act of God," (2) "an act of war," and (3) "an act or omission of a third party." That is, the only defense is that someone or something other than the defendant caused the problem. The definition of liability in the 1980 CERCLA is nearly the same as 1970 Water Quality Improvement Act covering oil spills, which later was incorporated in the 1972 Clean Water Act (33 USC §1321), to which the 1990 OPA liability rules (33 USC §2703) made no changes (Murchison 2011).

Retrospective liability may seem unfair. But, like strict liability, it makes sense for assigning liability in complex causal chains. For example, at the Love Canal toxic waste site that initially prompted the 1980 CERCLA law, the current owner of the site was the Niagara Falls School Board, which had built an elementary school on top of a thinly covered toxic waste site in the middle of a residential neighborhood. The purchase of the site from the Hooker Corporation (later part of Occidental Chemical) proves, if anything, that the School Board was not capable of understanding and managing the risks, not that it should be held responsible. In this case, remedial action need not depend on who, if anyone, is at fault, whether the School Board or Hooker (U.S. vs. Hooker Chemical and Plastics Corporation 1994). Hooker (Occidental) was required to pay for a cleanup simply because it caused the hazard. In order to achieve this outcome at Love Canal, the neighborhood residents sought the protection of the State and federal governments and did not try to negotiate with Hooker or get compensation from the company or School Board (Gibbs 1998). This strategy is explained by the model of self-defense and government recourse.

Consequences of Strict Liability

The consequences of CERCLA and OPA are not guaranteed to be efficient. The success of CER-CLA might be measured by its effectiveness in cleaning up the sites on the National Priorities List of sites requiring Environmental Protection Agency attention. As of May, 2020, there were 1,335 sites on the NPL, and 424 had been deleted from the NPL as not needing further EPA work. At 1,215 sites, remedial construction had been completed, after which sites sometimes remain on the NPL and sometimes are deleted (EPA 2020).

The NPL and the costs of CERCLA are a good way to measure the success of CERCLA. Also there are sites that are currently unknown but will be put on the NPL in the future, as well as sites that actually need cleanups but will remain unknown. There is a continuing threat due to toxic chemical waste against which self-defense and government recourse are needed, even if every site currently on NPL is eventually cleaned up so well as to be deleted from the list.

Success is more difficult to measure for OPA, since oil spills are infrequent. The 2010 Deepwater Horizon drilling rig explosion offers an unwelcome opportunity to check on the outcomes of OPA. Environmental, social, and legal consequences of the massive spill that followed the 2010 explosion persist ten years afterward. The explosion initially killed 11 drilling rig workers, resulting in a criminal manslaughter conviction for BP. In September 2014, BP was found to have been grossly negligent, which removed limits on BP's civil liability (Robertson and Krauss 2014). On October 5, 2015, a \$5.5 billion OPA civil penalty was imposed by a US federal judge, together with \$15.3 billion damages (EPA 2015; US Department of Justice 2016).

Risk Shifting in Context

Risk shifting is unavoidably disruptive to polluters. In the hypothetical case, Malloy might not be able to afford the vapor-capturing mechanisms that we have imagined would reduce risks of exposure to toxic fumes. Then Malloy would either have to go out of business, or else sell its business to a larger company. Political conflict is inevitable, since polluters will seek to avoid such outcomes. This kind of political conflict explains the OPA liability limit. OPA limits liability for offshore releases to removal costs plus \$75 million (33 US §2704) to cover other damages, including economic damages.

This penalty seems far too low in comparison with the cost of the 2010 Deepwater Horizon oil spill. In addition to OPA liability and damages imposed in 2015, BP created a \$20 billion fund for settling economic damage claims, among other costs stemming from the spill. The low liability limit protects smaller oil drillers who are politically powerful in oil producing states. Shifting risks onto these drillers would be disruptive to them, although it seems likely that drillers can efficiently reduce risks. Aldy (2011) and Murchison (2011) discuss risk spreading by drillers.

The disruptive character of risk shifting may make it unattractive to anyone who hopes for a decision procedure that avoids political conflict and which instead aims directly and consciously toward efficient decisions (Sunstein 2002). Risk assessment and cost-benefit analysis, in particular, are ways of planning for efficient outcomes. A consequentialist objection to the deontology of rights and self-defense is that self-defense and government recourse, in comparison, seem haphazard and irrational, and may lead to worse outcomes than good planning. A response to the objection can, I think, be framed more directly and straightforwardly when based on the account of self-defense and government recourse used in this chapter than when given in terms of rights, understood more generally or in a more idealized way.

The response is that self-defense and formation of social authorities for carrying it out is a basic feature of human life. At the beginning of this chapter, I said that the setting was anthropocentric. Other kinds of organisms might not be driven in the same way to defend themselves, and do not form societies in the same ways. This response to the irrationality objection is more naturalistic than a response based on more idealized moral theories would be. It should be emphasized that the self-defense justification for government action is limited to cases, such as Molloy, in which citizens' self-defense would be justified.

The argument here has shown how the choice of strict liability policies for managing risks of toxic pollution and oil spills is nearly forced on us. A basic right to self-defense, in a context of government management of risks of industrial activity, drives this choice of policy, even if some lingering ethical questions about fault and risk imposition remain unresolved. The reason for strict liability policies does not depend on how we resolve such lingering questions.

Related Topics

20. Moral Bases of Responses to Climate Change
32. Energy Poverty
33. Urban Sustainability
37. Waste and Consumption
50. Constitutional Rights
58. Precautionary Principles
63. Environmental Justice

References

Aldy, J. (2011) 'Real-time economic analysis and policy development during the BP Deepwater Horizon oil spill', *Vanderbilt Law Review*, 64(6), pp. 1795–1817.

American Law Institute. (2010) *Restatement of the law third, torts, liability for physical and emotional harm*. St. Paul, MN: American Law Institute Publishers.

Bullard, R. D. (2000) *Dumping in Dixie: race, class, and environmental quality*. Boulder, CO: Westview Press. (First edition 1990).

Calabresi, G. (1970) *The costs of accidents: a legal and economic analysis*. New Haven, CT: Yale University Press.

Coase, R. H. (1960) 'The problem of social cost', *The Journal of Law and Economics*, 3, pp. 1–44.

Cranor, C. F. (1993) *Regulating toxic substances: a philosophy of science and the law*. New York: Oxford University Press.

Doggett, T. (2011) 'Recent work on the ethics of self-defense', *Philosophy Compass*, 6(4), pp. 220–233.

Environmental Protection Agency (EPA) (2015) Consent Decree for Deepwater Horizon – BP Gulf of Mexico Oil Spill [online]. Available at: https://www.epa.gov/enforcement/consent-decree-deepwater-horizon-bp-gulf-mexico-oil-spill

Environmental Protection Agency (EPA) (2020) *National Priorities List* [online]. Available at: https://www.epa.gov/superfund/superfund-national-priorities-list-npl

Gibbs, L. M. (1998) *Love Canal: the story continues*. Gabriola Island, BC: New Society Press (First edition 1982).

Hayenhjelm, M. and Wolff, J. (2012) 'The moral problem of risk impositions: a survey of the literature', *European Journal of Philosophy*, 20, E26–E51.

Lazarus, R. J. (2007) *The making of environmental law*. Chicago: University of Chicago Press.

Lewens, T. (ed.) (2007) *Risk: philosophical perspectives*. London: Routledge.

Moss, D. A. (2002) *When all else fails: government as the ultimate risk manager*. Cambridge, MA: Harvard University Press.

Murchison, K. M. (2011) 'Liability under the oil pollution Act: current law and needed revisions', *Louisiana Law Review*, 71(3), pp. 917–956.

National Research Council (1991) *Environmental epidemiology: vol. 1: public health and hazardous*. Washington, DC: National Academies Press.

National Research Council (2007) *Scientific review of the proposed risk assessment bulletin from the Office of Management and Budget*. Washington, DC: National Academies Press.

Nozick, R. (1974) *Anarchy, state, and utopia*. New York: Basic Books.

Robertson, C. and Krauss, C. (2014) 'BP may be fined up to $18 billion for spill in Gulf', *New York Times*, 4 September.

Sunstein, C. R. (2002) *Risk and reason: safety, law, and the environment*. Cambridge: Cambridge University Press.

Szasz, A. (1994) *Ecopopulism: toxic waste and the movement for environmental justice*. Minneapolis: University of Minnesota Press.

Tesh, S. N. (2000) *Uncertain hazards: environmental activists and scientific proof.* Ithaca, NY: Cornell University Press.

Thomson, J. J. (1986) *Rights, restitution, and risk.* Cambridge, MA: Harvard University Press.

Thomson, J. J. (1990) *The realm of rights.* Cambridge, MA: Harvard University Press.

Thomson, J. J. (1991) 'Self-defense', *Philosophy & Public Affairs*, 20(4), pp. 283–310.

United States Department of Justice (2016) *Deepwater Horizon* [online, with links to Summary and Fact Sheet on Consent Decree]. Available at: https://www.justice.gov/enrd/deepwater-horizon

U.S. vs. Hooker Chemical and Plastics Corporation (1994) 850 F. Supp. 993.

Wildavsky, A. B. (1995) *But is it true?: A citizen's guide to environmental health and safety issues.* Cambridge, MA: Harvard University Press.

50

CONSTITUTIONAL RIGHTS

Kristian Skagen Ekeli

Introduction[1]

Constitutional environmental rights can be regarded as rights that aim to protect the environment for the sake of present and/or future generations. During the last three decades, a large number of states from around the globe have incorporated procedural and substantive rights into their constitutions. *Procedural rights* regulate decision-making procedures or the way decisions should be made. They are typically rights that govern legal, political or administrative decision-making processes. Examples are the right to information concerning the environment and the right of individuals or organizations (e.g. environmental organizations) to initiate legal proceedings in environmental cases. The aim of *substantive rights* is usually to protect and promote a certain level of environmental quality, such as clean water and air or an environment adequate for the health and well-being of human beings. The right to a healthy or adequate environment is the most common substantive right found in existing constitutions. Here is an example of such a substantive right, and this is combined with a procedural right to initiate legal proceedings.

> *The state constitution of Illinois, article XI, section 2*
> Each person has the right to a healthful environment. Each person may enforce this right against any party, governmental or private, through appropriate legal proceedings.[2]

The purpose of this chapter is to consider the question of how constitutional democracies can and should protect present and future interests by means of constitutional environmental rights. I will argue that constitutional democracies should constitutionalize judicially enforceable environmental rights.[3] (A right is judicially enforceable if courts have the power (or a procedural power-right) to enforce the right in cases brought before the court, and if certain agents have the procedural power-right (or legal standing) to initiate legal proceedings on the basis of the right at issue.[4]) However, I will argue that certain procedural rights that govern legislative procedures are more desirable from a normative point of view than substantive constitutional environmental rights. One important problem with substantive rights – such as a right to a healthy environment – is that they tend to be very vague. This vagueness or indeterminacy makes the judicial enforcement of such rights problematic with regard to central democratic values and rule of law values.

This chapter proceeds as follows. In Section 2, I will explain what is characteristic of constitutional rights compared with ordinary laws. This section will also give a brief account of the role constitutional rights play with regard to political and legal processes, because this is important

DOI: 10.4324/9781315768090-60

in order to understand the rationale for constitutionalizing environmental rights. The next two sections will consider how present and future interests can and should be protected by means of different types of constitutional environmental rights. Section 3 will focus on substantive rights, while the subject of Section 4 is procedural rights.

Constitutional Rights

Constitutional rights are rights conferred and protected by constitutional laws. In contemporary constitutional democracies, constitutional laws typically have a special legal and political status compared with ordinary laws. From a *legal* point of view, constitutional laws are *lex superior* (i.e. higher law) in relation to ordinary laws. Thus, in cases of conflict, constitutional laws take precedence over ordinary laws. Most contemporary constitutional democracies have a system of judicial review, and such a system gives courts an important political role in their enforcement of constitutional rights. A system of judicial review grants courts the right to review the constitutionality of laws, regulations and administrative actions, and the power to set aside or invalidate ordinary laws and regulations deemed unconstitutional. Such a system empowers courts to block political decisions that they regard as unconstitutional. There are two main types of systems of judicial review of the constitutionality of legislation. In a *centralized system*, judicial review is the exclusive competence of a special constitutional court. Many European states, such as Austria and Germany, have this system. In a *decentralized system*, every court in the hierarchy of courts has the power to exercise judicial review, but the supreme court is the final court of appeal on constitutional matters. The best-known example of the latter system is the United States. Both of these systems provide important institutional devices through which courts can serve as guardians of constitutional rights.

From a *political* point of view, it is more difficult to change the constitution than to enact ordinary legislation. While simple majorities in the legislature can pass ordinary laws, constitutional amendment must be made on the basis of special and more demanding decision-making procedures. In current jurisdictions, one can find a number of different amendment procedures, and these are often combined. Among the most important mechanisms of constitutional amendment are (1) supermajority rules (e.g. a two-thirds or a three-fourths majority in the legislature), (2) delays (e.g. amendments must be passed by two successive parliaments or they must be proposed during one parliament and adopted during another), (3) referendums, (4) state ratification (in federal systems), and (5) absolute entrenchment of parts of the constitution. Constitutional rights can be important instruments to protect the fundamental interests of citizens and steer the direction of future politics, precisely because such rights are more difficult to change than ordinary laws. Since constitutional rights can only be amended by special and more demanding procedures that favour the *status quo*, such rights will place constraints on the exercise of legislative and executive power. First, this means that present legislators and other state organs are bounded by the decisions of earlier electorates. Second, such restraints imply that decisions that are in conflict with constitutionally entrenched rights (such as a constitutional environmental right) are withdrawn from the immediate control of present majorities and governmental agencies.

Substantive Constitutional Rights

Substantive constitutional environmental rights typically aim to protect or promote a certain level of environmental quality – such as clean water and air, an environment adequate for human health and well-being or a sustainable development. Some substantive constitutional environmental rights mention future generations explicitly, whereas others do not. Nevertheless, I will assume that in most cases the aim of such rights is to protect and promote the interests of *both* present and future people. Here are two examples of substantive rights that explicitly refer to future generations:

The Constitution of Norway, article 112, subsection 1

Every person has the right to an environment that is conducive to health and to a natural environment whose productivity and diversity are maintained. Natural resources shall be managed on the basis of comprehensive long-term considerations which will safeguard this right for future generations as well.

The Constitution of South Africa, section 24

Everyone has the right a. to an environment that is not harmful to their health or well-being; and b. to have the environment protected, for the benefit of present and future generations, through reasonable legislative and other measures that i. prevent pollution and ecological degradation; ii. promote conservation; and iii. secure ecologically sustainable development and use of natural resources while promoting justifiable economic and social development.

In existing constitutions, the most common version of substantive rights is what can be called "the right to a healthy or adequate environment", and this right can be formulated like this: all human beings have a right to an environment adequate for their health and well-being. About 60 nations have included a substantive constitutional right of this type (i.e. similarly worded substantive rights),[5] and Tim Hayward has recently set out a very interesting case for such a constitutional right (see Hayward 2005).

Judicially enforceable substantive rights can, as Hayward argues, be useful and reasonably effective means to protect the environment for the sake of present and future generations. For example, such rights place substantive limits on majority decisions, in the sense that courts can block majority rule decisions in the legislature on the basis of such rights. However, judicial enforcement of substantive environmental rights, such as those mentioned above, faces some important problems that must be considered more closely in order to assess the desirability of such rights from a normative point of view.

Judicial Review and the Legal Indeterminacy of Substantive Environmental Rights

One important problem with substantive constitutional environmental rights is that they tend to be very vague or unclear. Examples are terms or formulations such as "an environment adequate for human health and well-being" and "secure ecologically sustainable development". Since the meaning and general aims of such substantive rights are so vague, it is difficult to see how they can provide clear guidelines for judges in cases brought before the court. The vague language in the formulations of these substantive rights is an important source of legal indeterminacy, because the requirements of the law in particular cases will often be unclear. In legal philosophy and theory, it is a widely held assumption that "the law is indeterminate when there is no single right answer to a question of law, or to a question of the application of the law to the facts of the case" (Endicott 2000: 2). In view of this vagueness or indeterminacy, two important objections can be made against judicial enforcement of substantive rights. The first is the objection from democratic values, and the second is the objection from rule of law values (see Ekeli 2007).

(1) *The democratic problem*: The representative legitimacy of elected legislators is different from that of appointed judges who are not directly elected by the people. In contrast to judges, voters can directly elect and replace legislators by means of periodic elections. This means that citizens can authorize their elected representatives to act on their behalf and can hold them politically accountable by punishing or rewarding them during elections. Thus, periodic elections constitute

an important institutional mechanism for popular control of legislators that is absent when it comes to judges. Against this background, it is widely acknowledged both in political philosophy and democratic theory that elected and accountable legislators should have the right to enact laws, while courts should be empowered to enforce (i.e. interpret and apply) the laws passed by the legislative assembly. This assumption is also a central aspect of the principle of the separation of powers in democratic states.

Some degree of law-making and policy-making discretion and power is, however, inherent in judicial enforcement (i.e. interpretation and application) of laws. This means that if the people and their elected representatives employ judges to enforce laws, some power to govern will be transferred to the judges. One important reason for this is that judges will always have some leeway of interpretation, and this will give them some degree of law-making discretion. Martin Shapiro has called this "the interpretation trap" – that is, "whoever is assigned to interpret text to some degree makes the text" (Shapiro 2002: 178). This is unavoidable, and I believe it is unproblematic from a democratic point of view (see also Kelsen 1931: 586; and Shapiro 2002: 178–179).

The situation is different when it comes to judicial enforcement of very vague laws. If the laws that judges are empowered to interpret and apply are very vague or unclear, they will have a high degree of law-making discretion and power. Since substantive constitutional environmental rights tend to be very vague or unclear, the judicial enforcement of such laws is problematic from a democratic point of view. The reason for this is that judicial enforcement of substantive rights implies that a high degree of law-making discretion and power will be transferred from elected and accountable legislators to the courts.

(2) *The rule of law problem*: The rule of law is often contrasted with despotism or arbitrary use of state power. In a state regulated by the rule of law, everyone – both those who govern and those who are governed – is subject to the law, and all are regarded as equals before the law. This means that government officials must operate within a limiting framework of law. Furthermore, it means that citizens are only subject to the law, and not the arbitrary will or judgement of those who wield government power. An important dimension of the rule of law is also that the exercise of state power should be regulated by laws that meet certain conditions. Laws should, for example, be general, publicly promulgated, not retroactive, clear and understandable, and relatively stable over time. One important value of a legal system that respects the rule of law is that it enables individual autonomy or freedom. Under the rule of law, it is possible for people to plan their lives and to predict the consequences of their actions in light of a pre-existing framework of law without fear of arbitrary state intervention. In this way, the rule of law reduces uncertainty that restricts people's ability to plan for their future.

The rule of law is an ideal that can be approximated to a greater or lesser degree. Complete realisation of the ideal is impossible, in part because some vagueness in law is inescapable. Moreover, some controlled (or restricted) judicial discretion (whatever its source) can be desirable (see Raz 1979: 222). It is, however, widely recognized that too much judicial discretion (e.g. resulting from vague laws) is a deficit in the rule of law. Therefore, judicial review of laws, regulations and administrative decisions based on *very* vague or unclear substantive constitutional environmental rights can be problematic in view of the central rule of law ideals outlined above. The reason for this is that the rights in question do not provide clear constraints on judicial decisions. This will make judicial review on the basis of such rights highly discretionary, and it can lead to unpredictable judicial decisions that will undermine the value of the rule of law. To the extent that this happens, citizens are not subject to clear and understandable laws, but to the judgement of judges.

The outlined problems are serious, but they do not necessarily imply that constitutional democracies should not constitutionalize judicially enforceable substantive rights. There are at least two ways to meet these problems, and these approaches can be combined.[6] The first approach is to try and come up with new substantive environmental rights that are sufficiently clear to serve as an adequate basis for judicial enforcement. The second approach is to find an alternative way to involve courts in the enforcement of substantive environmental rights. Instead of granting courts the power-right to block laws passed by a majority in elected legislatures, constitutional courts (i.e. a special constitutional court or some similar body such as a state's supreme court) can be given certain procedural rights, such as the following: (1) the right to require that state authorities undertake environmental impact assessments – e.g. require the assessment of the environmental impact of a law proposal before the legislature makes a final decision; (2) the right to delay legislation until a new election has been held; and (3) the right to require a referendum on the issue under consideration. These procedural rights empower courts to slow down the political decision-making process in cases where the courts assume that a majority decision is in conflict with a substantive right. One important advantage of such procedural constitutional rights is that they can avoid the democratic problem. The aim of these procedural rights is primarily to give courts certain tools to create more public awareness and improve the democratic process of deliberation and decision-making about environmental issues. Similar procedural rights can be ascribed to legislators (e.g. minorities of legislators or a second chamber in a bicameral legislature), and they will be discussed in more detail in the next section.

Procedural Constitutional Rights

Procedural constitutional environmental rights are typically rights that govern legal, political (e.g. legislative) or administrative procedures – that is, they regulate decision-making procedures or the way decisions should be made. A number of states (about 35) have incorporated procedural environmental rights into their constitutions. The most common types of procedural rights found in existing constitutions are the following: (1) rights to information concerning the environment; (2) rights to participation (e.g. a right to participation in planning and decision-making activities that might have an impact on the environment); (3) rights to access to justice (e.g. a right to initiate legal proceedings in environmental cases or a right to compensation for damage caused to health or property by ecological violations).

One can, however, imagine other procedural constitutional rights that might have more interesting and far-reaching effects with regard to important political decision-making processes than these procedural rights. In this section, I will present and consider two such proposals of constitutional reform that set out alternative systems of procedural rights that govern legislative decision-making processes. The first has been proposed by Tine Stein, and it can be called the ecological council model (Stein 1998a, 1998b). The second, which I have recently proposed, can be termed the submajority rule model (Ekeli 2009, 2016). These two models can, in different ways, serve as useful means to promote more thorough and future-oriented political deliberations and decisions about environmental issues in representative democracies.

The Ecological Council Model

Stein's proposal is to establish an ecological council ("ein ökologisher Rat") that should function as a consultative chamber of the legislature. The role and procedural power-rights of the ecological council can in certain respects be compared with that of the British House of Lords. First, it should have the power to review the impact of law proposals and regulations on the environment and recommend revisions or amendments. Second, the council should be granted suspensive veto power – that is, the right to delay legislation. If the ecological council reaches the conclusion that a given bill passed by the legislature can cause serious environmental harms, it can use its suspensive veto power to slow down the process of deliberation and decision-making. The hope is that this can induce more thorough public debate about the issues in question and lead to better and more enlightened decision outcomes. By withholding their approval or assent, the ecological council might over time convince a sufficient number of legislators to change their minds and introduce alternative laws that might have better and more desirable environmental effects.

Stein suggests that the members of the ecological council ought to be elected by the legislature. This means that the voters cannot hold the members of the council directly accountable by means of periodic elections. This move can be regarded as problematic from a democratic point of view, but Stein views this as an advantage because the purpose is to establish a deliberative chamber of the legislature, where the members should not have to worry about their popularity and re-election. Her aim is to design a system that can help secure the independence of the members of the council from political parties and the political branches of government. In this connection, she suggests two devices. First, the members of the council should not have the opportunity to be re-elected. Second, they should be elected for longer terms than ordinary legislators who are usually elected for four- or five-year terms. More precisely, she recommends that the members of the council can stay in office for nine-year terms.

The delay device that Stein proposes can have some of the same positive effects as the two submajority rules that I will consider in the next section. Having said that, Stein's ecological council model is open to two important objections. Since these are relevant for an assessment of the ecological council model compared with the submajority rule model, it is worth considering them before proceeding. First, Stein does not describe the function and procedural rights of the council in detail. For example, she does not specify how long the council should have the right to delay legislation. Neither does she address the issue of whether the council should have the right to initiate legislation. The second objection is related to the selection of the members of the council and whether her model is adequate to secure independence. If the goal is to establish an independent deliberative body that should be shielded from the struggles between political parties and partisanship, Stein's proposed election procedure seems to face at least one serious problem. Since the members of the council are elected by the legislature, it is not unlikely that the party alignments and blocs in the legislature will be reproduced in the council. The more power the council is granted (e.g. the longer the council has the power to delay the final enactment of bills), the stronger incentives political parties will have to select people who are likely to support their party's views and policies. If party alignments are reproduced in the council, it will lose any plausible claim to independence. This can be an obstacle to the establishment of an independent deliberative forum that is not marked by party politics. Furthermore, if the ecological council becomes a clone of a legislative assembly dominated by partisan struggles, this would undermine the new system, since much of the point of introducing this reform seems to disappear.

Even if Stein's proposal is open to the outlined objections, the idea of establishing a second "ecological" or "future-oriented" chamber in legislatures is interesting and worth exploring in more detail. There are alternative ways to set up or design the ecological council model with regard to both the selection of its members and the specification of its powers. In this area, there is, however, a need for more research on issues of institutional design.

The Submajority Rule Model

The aim of the submajority rule model is to give minorities of legislators, who are selected and accountable through ordinary periodic elections, certain political tools to represent and protect the interests of future generations. The common denominator of the constitutional reforms that I have proposed is that they represent examples of submajority rules that grant predefined minorities of legislators two procedural rights.

(1) The Right of Minorities to Require Delays

A minority of at least one-third of the legislators should be granted the right to require that the final enactment of a law proposal should be delayed until a new election has been held, if they believe that the law in question can inflict serious harm upon posterity. Thus, a minority is empowered to demand that the bill can only be enacted after an intervening election.

(2) The Right of Minorities to Require Referendums

A minority of at least one-third of the legislators should be granted the right to require a referendum on a bill that can have a serious adverse impact on the life conditions of future generations. Since it is important that the electorate has sufficient time to gather relevant information as well as to consider and discuss the bill, there should be a time interval of at least one year from the minority calls for the referendum until it is held. However, in order to avoid a too time-consuming process, there should be a limit to the interval allowed – e.g. two years.

At this point, it is important to make three clarifications with regard to the proposed submajority rules and their application. First, the submajority rules do not privilege any particular minority of legislators. Any predefined numerical minority of legislators will be given these tools, and the hope is that future-oriented legislators will use them in order to protect future interests. Second, a minority of legislators should only be allowed to demand a delay or a referendum if they present a *prima facie* case for the assumption that the law proposal in question can inflict serious harm or risk upon posterity. Thereafter, the burden of proof should shift to those legislators who reject the minority's harm scenario. Third, conflicts about the reliability of competing harm scenarios should be resolved by a constitutional court. In cases where controversies arise, the legislators who want to prevent a delay or a referendum (for instance 10% of the legislators) should be allowed to initiate legal proceedings. But, as pointed out above, in such cases the onus of proof should rest with those who reject the minority's harm scenario after the minority has presented their *prima facie* case. It should, however, be underlined that when controversial cases are brought before a court, the court has the power to reject a delay or a referendum in cases where it does not find the *prima facie* case offered by the minority convincing – that is, if the court assumes that the law in question does not expose posterity to risks that can seriously harm their living conditions, or if the court suspects that a minority abuses the submajority rules for strategic reasons.

In what follows, I will argue that the submajority rules can have important effects with regard to processes of agenda-setting, intrapersonal and interpersonal deliberation, [7] exchange of information (i.e. distribution and dissemination of information) and citizen involvement. The proposed procedural rights also provide a future-oriented system of checks and balances that can guard future generations against myopic majority decisions that neglect their vital interests.

(1) *Agenda-setting*: The distribution of power to place issues on the formal political voting agenda, and the distribution of procedural rights to determine how those issues are to be decided play an important role in political decision-making. The proposed submajority rules will affect the distribution of agenda-setting power, because they will, to some extent, have the effect of distributing the power to control the agenda away from majorities to minorities. First, minorities are empowered to influence how long an issue should be on the agenda, and to decide how the issue should be placed on the formal voting agenda. In this way, minorities have the power to force the majority to pay more attention to certain issues affecting posterity. Second, the suggested procedural rights can give minorities of legislators the opportunity to increase the political visibility and the public awareness of the issues in question. These points will be elaborated below.

(2) *Processes of deliberation*: An important aim of the proposed procedural rights is to improve the process of deliberation and decision-making about important intergenerational issues. The

purpose is to improve the basis of information and enhance the level of reflection among legislators and voters. Since both submajority rules empower minorities to slow down the decision-making process, they can promote a more thorough and well-informed process of intrapersonal and interpersonal deliberation about certain issues or law proposals. First, to the extent that minorities use their right to demand delays or referendums, this will ensure that both the electorate and politicians have the opportunity to consider the proposals in question more closely before a decision is made through majority rule in the legislature or a referendum. Second, in this process they will have time to gather and distribute new and relevant information which can in turn affect the subsequent process of deliberation, agenda-setting and decision-making.

Third, if a minority requires a delay or a referendum, decision-makers will have more time to come up with, discuss and consider alternative courses of action, which might have more desirable consequences with regard to future generations than the bills that were initially introduced. The hope is that this can initiate a dynamic educative process of public deliberation where relevant decision-makers and publics are exposed to a diversity of ideas, proposals and problems. Moreover, compared with decision procedures which make it possible to make more hasty and less visible decisions in the legislature (i.e. the status quo in most democratic states), the proposed decision rules can also improve the quality of collective decisions, since decision-makers (both citizens and legislators) are given a better opportunity to pool their knowledge, insight and experience prior to voting. In these ways, the proposed procedural rights can lead to a process of intrapersonal and interpersonal deliberation that might lessen the problem of bounded rationality – the problem that our knowledge, imaginations and reasoning abilities are limited and fallible.

(3) *Citizen involvement*: The procedural rights to demand delays and referendums can, as indicated above, induce more public awareness and engage citizens more directly when it comes to political issues affecting future generations. First, if a minority requires a delay, the people have the opportunity to consider the law proposal more closely during election campaigns. Citizens will also be given the chance to determine the composition of the legislative assembly before the law proposal can be enacted through majority vote.

Second, a direct democratic device such as a referendum can provide a useful institutional mechanism for engaging citizens more directly in public deliberations about important intergenerational issues. One noteworthy reason for this is that referendums can change the demand for political information and the supply of it (see Benz and Stutzer 2004: 33–34). On the demand side, referendums can increase the incentives of voters to gather information, partly as a result of more intense public discussions before the popular vote. On the supply side, referendums increase the incentives of politicians and the media to provide information. If politicians and interest groups want to win a referendum, they are forced to inform the public about the reasons why they are for or against the policy in question. During referendum campaigns, these political actors have to provide information on the issue at stake, and they must publicly discuss and critically scrutinize the arguments and information offered by their opponents.

(4) *A future-oriented system of checks and balances*: Systems of checks and balances are usually introduced in order to guard against abuse of state power and despotism (or arbitrary use of state power). One central aim of checks and balances is to provide a guard against the danger that the rulers (e.g. majorities of legislators) use the power that is conferred on them against the ruled. The submajority rule model can be regarded as a future-oriented system of checks and balances, the purpose of which is to guard future generations against myopic majority decisions that neglect their vital interests and needs. The main aim of the proposed procedural rights is to empower minorities of future-oriented legislators to function as watchdogs for posterity in present political debates and struggles. They grant a predefined numerical minority of future-oriented legislators, who are elected and accountable through periodic

elections, the power to police and control myopic majorities in the legislature. Thus, the submajority rule model has an important power-checking function, in the sense that it can prevent the process of decision-making about issues affecting posterity from being subject to the immediate and unlimited control of myopic majorities of legislators. Since future generations cannot themselves influence present political decisions that can have a serious impact on their living conditions, minorities of legislators should be granted the proposed procedural rights to represent and protect posterity.

Perhaps the most important problem facing the submajority rule model is that the proposed procedural rights can be abused by minorities for strategic or egoistic reasons. It might be tempting for a minority in the legislature – who are not concerned for the well-being of future generations – to require for instance a delay in the hope that a bill they dislike or oppose would never be passed. In order to analyse how the problem of strategic abuses affects the desirability of the submajority rule model, it is worth keeping the following considerations in mind. First, the requirement that the minority has to present a *prima facie* case for demanding a delay or a referendum will presumably reduce the danger of strategic abuse, because it would make it difficult for a strategically motivated minority to come up with cogent public justifications for demanding delays or referendums for the sake of future generations. Second, constitutional courts have, as mentioned above, the power to reject a delay or a referendum in cases where they believe that a minority abuses the submajority rules for strategic reasons. Third, minorities who more or less openly abuse the suggested devices will expose themselves to the risk of being punished during elections. Fourth, even if certain minorities abuse the procedural rights, the submajority rules can also in such cases create a more thorough and future-oriented process of deliberation about important intergenerational issues.

Conclusion

In this chapter, I have considered the question of how constitutional democracies can and should protect present and future interests by means of constitutional environmental rights. I have argued that constitutional democracies should constitutionalize judicially enforceable environmental rights. More precisely, I have argued that certain procedural constitutional rights that govern legislative procedures have some advantages over substantive rights, such as a right to a healthy or adequate environment. Substantive rights can serve as useful means to protect the environment for the sake of present and future generations. However, they face a serious problem of vagueness – that is, substantive rights tend to be *very* vague or unclear, and this makes the judicial enforcement of such rights problematic with regard to both democratic values and rule of law values. The ecological council model and the submajority rule model set out alternative systems of procedural rights that can avoid the problem of vagueness. Moreover, these two models can, in different ways, serve as useful political tools to promote more thorough and future-oriented political deliberations and decisions about environmental issues in representative democracies. However, I have argued that the submajority rule model has more far-reaching and desirable effects than Stein's ecological council model.

Related Topics

51. Libertarianism
54. Command and Control
60. Education
64. Environmental Civil Disobedience
65. Lawbreaking and Ecoterrorism

Notes

1 I would like to thank Benjamin Hale, Andrew Light and an anonymous referee for valuable comments.
2 The federal U.S. constitution does not include environmental rights, but several state constitutions do. For an interesting discussion of environmental rights found in state constitutions throughout the United States, see May and Romanowicz (2011).
3 Roughly speaking, a right can be constitutionalized (a) through formal amendment procedures prescribed by the constitution (see also section 2) or (b) by judicial interpretation – i.e. courts can create or invent new rights when they interpret the constitution or the constitutional document (see also Marmor 2005: ch. 9; Shapiro 2002; and May and Daly 2011). In this chapter, I will assume that in constitutional democracies environmental rights are typically constitutionalized through formal amendment procedures.
4 The issue of legal standing in environmental cases is discussed more closely in Stone (1974), Hayward (2005: 98–101 and 208–210) and Ekeli (2006).
5 See May and Daly (2011: 332).
6 For a more thorough discussion of these approaches, see Ekeli (2007).
7 *Interpersonal deliberation* refers to the process of discussion with others or interpersonal communications – e.g. debates in legislatures. *Intrapersonal deliberation* refers to an individual's internal reflections (or considerations), for instance, on political issues – e.g. when we read a newspaper or watch a political discussion on TV and deliberate about the pros and cons of alternative policies.

References

Benz, M. and Stutzer, A. (2004) "Are Voters Better Informed When They Have a Larger Say in Politics?" *Public Choice* 119(1–2): 31–59.

Ekeli, K.S. (2006) "The Principle of Liberty and Legal Representation of Posterity," *Res Publica* 12(4): 385–409.

——— (2007) "Green Constitutionalism: The Constitutional Protection of Future Generations," *Ratio Juris* 20(3): 378–401.

——— (2009) "Constitutional Experiments: Representing Future Generations Through Submajority Rules," *Journal of Political Philosophy* 17(4): 440–461.

——— (2016) "Electoral Design, Sub-Majority Rules, and Representation for Future Generations," in I. González-Ricoy and A. Gosseries (eds.), *Institutions for Future Generations*, Oxford: Oxford University Press, pp. 214–227.

Endicott, T. (2000) *Vagueness in Law*, Oxford: Oxford University Press.

Hayward, T. (2005) *Constitutional Environmental Rights*, Oxford: Oxford University Press.

Kelsen, Hans (1931) "Wer soll der Hüter der Verfassung sein?" *Die Justiz* 6: 576–628.

Marmor, A. (2005) *Interpretation and Legal Theory*, Portland OR: Hart Publishing.

May, J.R. and Daly, E. (2011) "Constitutional Environmental Rights Worldwide," in J.R. May (ed.), *Principles of Constitutional Environmental Law*, Chicago: American Bar Association, pp. 329–357.

May, J.R. and Romanowicz, W. (2011) "Environmental Rights in State Constitutions," in J.R. May (ed.), *Principles of Constitutional Environmental Law*, American Bar Association, pp. 305–327.

Raz, J. (1979) "The Rule of Law and Its Virtue," in *The Authority of Law*, Oxford: Oxford University Press, pp. 210–229.

Shapiro, M. (2002) "The Success of Judicial Review and Democracy," in M. Shapiro and A. Stone Sweet , *On Law, Politics and Judicialization*, Oxford: Oxford University Press, pp. 149–183.

Stein, T. (1998a) "Does the Constitutional and Democratic System Work? The Ecological Crisis as a Challenge to the Political Order of Constitutional Democracy," *Constellations* 4(3): 420–449.

——— (1998b) *Demokratie und Verfassung an den Grenzen des Wachstums*, Opladen: Westdeutscher Verlag.

Stone, C.D. (1974) *Should Trees Have Standing?*, Los Altos, CA : William Kaufman.

51

LIBERTARIANISM

Matt Zwolinski

Introduction

Libertarianism is a species of political philosophy distinguished by its commitment to strong rights of private property, free markets, and strictly limited government (Mack and Gaus, 2004, Zwolinski, 2008, Zwolinski and Tomasi, forthcoming). Some libertarians defend these institutions by appeal to strong natural rights of self-ownership and property in external resources (Rand, 1961, Nozick, 1974, Rothbard, 1982a). Others defend them by appeal to the supposedly beneficial consequences they produce (Friedman, 1962, Friedman, 1989, Epstein, 1998). Most libertarians, however, believe them to be justified on *both* pragmatic and moral grounds. So, they think, free markets are to be praised not merely because they produce efficient outcomes, but because they embody respect for individual rights. Libertarians believe that individual rights to private property and freedom of contract are relatively inviolable, and place strict limits on the scope of permissible regulation by government. My right to my justly acquired property does not give way merely because you happen to need it more, or even because the welfare of society as a whole could be greatly improved by overriding it. If seizing my land through the exercise of eminent domain is the only effective way of building a public highway then, libertarianism must conclude, so much the worse for the public. Or at least for its highway.

Because of its support for strong rights of private property and relatively unregulated capitalism, libertarianism is often perceived as being fundamentally incompatible with the kinds of policy goals demanded by a thoroughgoing commitment to environmentalism. And many libertarians themselves, by their often hostile rhetoric toward environmentalism, seem to have gone out of their way to confirm this perception (Rand, 1999).

Despite this common perception, however, a libertarian regime of strong private property rights might actually better serve environmental values than commonly proposed policies that undercut or infringe upon such rights. It is true that respect for libertarian property rights would in some cases serve as a block against certain forms of environmental regulation. Libertarians believe, after all, that the government may not legitimately restrict what a company does with the land or natural resources in which it has a valid property right. It is vitally important to remember, however, that it is not merely the property rights of *business* that libertarians are committed to protecting. Other people have property rights too – both in whatever land and other resources they might own and, by virtue of their self-ownership, a kind of property right in their *person* that makes it impermissible to send unwanted pollution into their bodies without their consent. Businesses have a right to do what they want with their *own* property, but they have no such right when it comes to the property of *others*.

DOI: 10.4324/9781315768090-61

This simple fact has profound implications for the libertarian position on environmental issues, especially when it comes to the issue of pollution. Indeed, the libertarian commitment to property rights is so absolute, and so far-reaching in its implications, that it actually flips our initial worry about libertarianism on its head. Once we consider the full implications of respect for libertarian property rights, it appears that the real problem with libertarianism isn't that it's not sensitive enough to environmental considerations, but that it is *too sensitive by far.*

This chapter will examine the implications of libertarian political philosophy for the problem of pollution. Its focus will be on the kind of rights-based libertarianism embodied in the work of Robert Nozick, Murray Rothbard, and Eric Mack, though the concluding section will also briefly discuss the more consequentialist libertarianism derived from the work of Friedrich Hayek and Ronald Coase. Section II will begin with a discussion of the libertarian emphasis on property rights, and the implications of those rights for environmental policy. Section III will go on to explain why those implications are more radical, and more implausible, than they might at first appear. Section IV examines various libertarian attempts to avoid those implausible conclusions. And Section V concludes with a discussion of consequentialist libertarian approaches to the problem of pollution, and how those considerations can and must be a part of, rather than a substitute for, a coherent libertarian theory of environmental justice.

Taking Property Rights Seriously

Among academic philosophers, the best-known version of libertarianism is that articulated by Robert Nozick in his 1974 book, *Anarchy, State, and Utopia*. In that book, Nozick defends a form of minimal-state libertarianism based on a roughly Lockean conception of individual natural rights. For our purposes, two features of that conception are particularly relevant: Locke's commitment to each individual's right of self-ownership, and his commitment to their right to acquire private property in external resources through a process of homesteading and voluntary exchange.

For Locke, the most fundamental sort of property right is each person's ownership of his or her own person. "[E]very man has a property in his own person: this no body has any right to but himself. The labour of his body, and the work of his hands, we may say, are properly his" (Locke, 1952, Book 5, sect. 27). Because each person has property in her own self, and because all are governed by a law of nature that "teaches all mankind...that...no one ought to harm another in his life, health, liberty, or possessions," it follows that each person has a right against others not to have harms imposed on her body without her consent (Locke, 1952, Book 2, sect. 6).

Each person's right of self-ownership also, for Locke, grounds their moral right to acquire legitimate property rights in external resources such as land, minerals, and crops. Because each individual owns the labor of her body, she can come to own these external resources by "mixing her labor" with them. "The labour that was mine, removing them out of that common state they were in, hath fixed my property in them" (Locke, 1952, sect. 28). But because God has given the earth to humankind for their *common* use, no individual may appropriate so much to herself that she fails to leave "enough, and as good...for others." Subject to this proviso, however, private property in external resources is not only permissible, but *necessary* in order for individuals to preserve themselves and flourish.

For Robert Nozick, Lockean rights to self-ownership and private property take the more specific form of "side constraints" against aggression (Nozick, 1974). These side constraints strictly limit the actions others may permissibly take in pursuit of either *their* own good, the good of the *right-holder* his or her self, or any conception of a *greater social good*. Predation, paternalism, and utilitarian trade-offs are thus prohibited in almost all circumstances, save perhaps for those (presumably quite rare) cases in which respect for rights would lead to what Nozick described as "catastrophic moral horror" (Nozick, 1974).

Nozick invites us to imagine individual rights as a kind of "line (or hyper-plane) [that] circum-scribes an area in moral space around an individual" (Nozick, 1974). Actions that "transgress the boundary or encroach upon the circumscribed area" are to count as infringements of those rights. And, in general, such infringements may be justly prohibited. Murder, theft, and assault all involve the crossing of an individual's moral boundaries without his or her consent, and are all therefore properly criminalized by the libertarian minimal state.

The Nozickian theory of rights as side constraints is a controversial doctrine. Many theorists, including but certainly not limited to consequentialists, will worry that the absolutism of these rights renders them too incapable of responding to the kind of trade-offs that seem necessary in a morally complicated world. But at least a certain level of immunity seems to be an important part of our ordinary understanding of both rights in general and property rights in particular. The common law tort of battery, for instance, prohibits others from touching your body without your consent, regardless of whether that touching causes you "harm" or not. And common law trespass prohibits anyone from entering your land, or causing some object or person to enter your land, without your consent (Glannon, 2010, Chapter 2). Neither of these rights is conditional on the cost-benefit analysis turning out the right way. The benefit you receive from touching me without my consent might be greater than the harm I suffer. And society as a whole might be better off if you could trespass across my land. But neither of these facts, even if uncontestably true, would serve as a valid defense against the tort.

On its face, the libertarian commitment to strong rights of private property would seem to provide a strong basis for opposing environmental pollution. For many of the most worrisome forms of pollution can be understand as a violation of property rights. In the words of libertarian and former Reagan White House official Martin Anderson,

> Just as one does not have the right to drop of a bag of garbage on his neighbor's lawn, so does one not have the right to place any garbage in the air or the water or the earth, if it in any way violates the property rights of others.
>
> *(Anderson, 1989)*

A pro-business conservative might be willing to tolerate some pollution of this sort. After all, pollution is often a byproduct of productive economic activity, and productive economic activity is the key to long-term economic growth. But a principled libertarian should have no truck with this argument. Just as property rights may not be sacrificed for government regulations that purportedly serve the public good, neither may they be sacrificed for business interests that do so. Taking property rights seriously means taking pollution seriously. Anderson thus has little patience for the policy of Tradable Pollution Permits:

> Now some even seriously propose that we should have economic incentives, to charge pol-luters a fee for polluting – and the more they pollute the more they pay. But that is just like taxing burglars as an economic incentive to deter people from stealing your property, and just as unconscionable... What we need are tougher clearer environmental laws that are enforced – not with economic incentives but with jail terms.
>
> *(Anderson, 1989)*

In his rejection of economic cost-benefit analysis in favor of a principled opposition to pollu-tion, Anderson mirrors the doctrine of Murray Rothbard. Rothbard, one of the most influential libertarian theorists of the 20th century, frequently defined libertarianism in terms of its com-mitment to the "non-aggression axiom," a principle which holds that "no man or group of men may aggress against the person or property of anyone else," where aggression is understood as "the initiation of the use or threat of physical violence against the person or property of anyone else" (Rothbard, 1973).

Like Anderson, Rothbard held that pollution was to be condemned as a violation of property rights. In his 1973 book, *For a New Liberty*, Rothbard wrote that the "vital fact" with respect to air pollution is that

> the polluter sends unwanted and unbidden pollutants—from smoke to nuclear fallout to sulfur oxides—*through* the air and into the lungs of innocent victims, as well as onto their material property. All such emanations which injure person or property constitute aggression against the private property of the victims. Air pollution, after all, is just as much aggression as committing arson against another's property or injuring him physically. Air pollution that injures others is aggression pure and simple.
>
> *(Rothbard, 1973)*

Like Anderson, Rothbard had little patience for the Chicago School argument that pollution could be justified by a kind of cost-benefit analysis, writing that such arguments are "as reprehensible as the pre-Civil War argument that the abolition of slavery would add to the costs of growing cotton" (Rothbard, 1973). For Rothbard, matters of basic moral principle trump merely pragmatic considerations. If stopping aggression sets back economic progress, then so be it. Such is the price of respect for human rights.

In many ways, then, the libertarian commitment to strong rights of private property actually seems capable of providing much *stronger* grounds for opposing pollution than alternative political ideologies, such as Rawlsian liberalism, are capable of providing (Taylor, 1992). Insofar as industrial pollution violates the property rights of landowners downriver, nearby foresters whose trees are damaged by acid rain, or individual persons whose lungs are intruded upon by airborne pollutants, that pollution constitutes aggression and should be prohibited. Indeed, as long as we make the not unreasonable assumption that some of the harms associated with global climate change constitute a violation of property rights, there is no reason why the libertarian framework could not be extended to address even this most difficult and challenging of environmental problems (Singer, 2004, chapter 2, Dolan, 2006, Shahar, 2009).

Ending Pollution: An Impossible and Implausible Demand

Libertarianism's radical critique of environmental pollution is exhilarating, and an important reminder of the way in which its principled support for free markets and private property stands sharply at odds with a conservative apology for the corporatist status quo (Chartier and Johnson, 2012). Indeed, upon reflection it may appear that libertarianism's real problem is not that it is not radical enough when it comes to the protection of the environment, but that it is *too* radical.

David Friedman, himself a libertarian but a critic of the kind of natural rights position espoused by Locke, Nozick, and Rothbard, was the first to point out the problem:

> [C]arbon dioxide is a pollutant. It is also an end product of human metabolism. If I have no right to impose a single molecule of pollution on anyone else's property, then I must get the permission of all my neighbors to breathe. Unless I promise not to exhale.
>
> *(Friedman, 1989)*

Light pollution, too, constitutes a physical, nonconsensual invasion of one's neighbor's property. This is obvious enough if we consider, as Friedman goes on to ask us to do, a case in which you fire a thousand megawatt laser beam at your neighbor's front door. But is shining a flashlight at your house any less a violation of your right against unconsented-to boundary crossings? Is lighting a match within eyesight of your property?

There is a sense in which questions like Friedman's are puzzles for *any* view that takes individual rights seriously. Libertarians are not the only theorists who believe in property rights, and any view that takes property rights seriously will have to draw a line between permissible and impermissible infringements on those rights.

Nevertheless, the problem is especially pressing for libertarians, for reasons having to do the distinctive and relatively *absolutist* character of libertarian property rights. On the libertarian view, property rights can be justifiably overridden only in very unusual situations and only by the very weightiest of competing moral concerns, if at all. Libertarians of the Rothbardian/Nozickian type hold, for instance, that a person's property right in his justly acquired wealth is so strong that it would be impermissible to steal (or tax) even a nickel from him, no matter how great a social benefit we could produce by doing so (Nozick, 1974). But this seems to commit libertarians to the extremely demanding view that it is equally wrong to impose a nickel's worth of damage on one's neighbor through pollution. In the same way that the libertarian's property absolutism leads him unable to recognize any morally significant difference between excessive and acceptable levels of taxation, so too is he unable to distinguish between relatively serious and relatively trivial forms of pollution.

It is hard to see how libertarians could back away from the (unwanted) radical implications of respect for property rights on the issue of pollution while maintaining the (wanted) radical implications with respect to theft, taxation, and regulation. But it is not at all hard to understand why libertarians would be very strongly motivated to do just this. After all, to describe the environmental implications of a strict libertarian respect for property rights as "radical" is a severe understatement. The demands that libertarianism seems to place on us go far beyond asking us to recycle more of our waste products, bicycle more to work, or pass a few more laws for the protection of wilderness areas. Indeed, not even the complete abolition of the automobile or electrical power would seem to go far enough. The only way to ensure that the actions of one human being do not cross the moral boundaries surrounding any other human being in even the slightest of ways is to abolish human society altogether (Friedman, 1992). Libertarianism's apparent demand that we eliminate pollution altogether thus appears to be an impossible and grossly implausible goal.

Backing Away from Environmental Radicalism

There is still relatively little literature addressing the problem of pollution from the perspective of natural rights libertarianism. However, a few prominent libertarians have recognized the scope and importance of the problem, and have made arguments that purport to show that libertarianism is not, in fact, committed to the radical and deeply counterintuitive conclusions described in the last section. In this section, I examine some of those arguments.

a. Robert Nozick and the Attenuation of Rights

Given Nozick's insistence that individuals are protected by strict side constraints against aggression, one might have expected him to take a hard line against pollution. Instead, however, Nozick embraces a position that looks less like what one would expect from a principled advocate of a natural and inviolable right of private property, and much more like the kind of cost-benefit analysis one would expect from a straightforward utilitarian – a moral theory that Nozick famously and decisively rejected just one chapter prior to his discussion of the problem of pollution (Nozick, 1974). Nozick's position is that pollution should be permitted whenever the "benefits are greater than the costs," and prohibited when they are not (Nozick, 1974). Permissible pollution ought to be able to "pay its way" in the sense that those who benefit from the activity could in principle compensate those who are harmed by it.

It is true that, unlike a utilitarian, Nozick believes that such compensation ought actually to be paid, not merely that such payment be possible in principle. If the pollution you create crosses my moral boundaries and sets back my interests, then you ought to pay me a sufficient amount of money to fully compensate me for that harm. But even with the requirement of compensation, Nozick's principle of "cross and compensate" still represents a significant weakening of the conception of rights as moral side constraints (Sobel, 2012). Nozick's position seems to abandon the intuitive idea with which he began that individuals are "inviolable," and replace it with the idea that violating individual rights is perfectly fine, so long as you're willing to pay the price. Such a principle appears to be the moral equivalent of a policy of eminent domain – a policy that libertarians have vocally and consistently opposed.

Nozick's position on pollution is somewhat less surprising when one considers the context in which it occurs. The discussion of pollution occurs in the first part of *Anarchy, State, and Utopia*, in which Nozick is attempting to refute the individualist anarchist and show that a minimal state could evolve out of a state of nature by morally permissible means. One of the most important moments in that argument occurs when Nozick asks what steps the "Dominant Protective Association" (DPA) could permissibly take to deal with the risks posed by competing "independent" protective associations enforcing their clients' rights in possibly unreliable ways. What right does such a firm have to protect itself from others attempting to enforce their *own* rights in potentially risky ways? Nozick's surprising answer is that the DPA may justly prohibit the risky activity, but *only* if it compensates those who are disadvantaged by that prohibition (Nozick, 1974, chapter 5, especially pp. 101–113).

This move is absolutely essential for Nozick's "invisible hand" argument for the minimal state. But its success depends on an important switch in the way in which rights are understood, from a conception of rights as

> claims protected by property rules (that forbid boundary-crossings) to a conception of rights as being at their core claims that are protected by liability rules (that allow crossings as long as liability for due compensation is paid).
>
> *(Mack, 2011)*

As Eric Mack notes, this "attenuation of rights" seems deeply inconsistent with the moral core of Lockean/Nozickian libertarianism. The imposition of slavery is wrong (at least in part) because it violates a person's ownership rights over his or her own person. It would not cease to be wrong "if only we could be sure that the slave is receiving at least as much compensation as would have induced him to agree to enslavement" (Mack, 2011).

Nozick's move thus succeeds at avoiding an implausibly radical view regarding the impermissibility of pollution, but arguably does so only at the cost of abandoning the moral core of libertarianism. Once the status of libertarian rights has been softened from property clams to mere liability claims, it is unclear what grounds the libertarian has for maintaining his distinctively steadfast opposition to *other* kinds of violation – particularly those involved in paternalistic interferences with an individual's person or property for his or her own good (Sobel, 2012, 2013).

b. Murray Rothbard and the Reformulation of Torts

In an ambitious 1982 paper, Murray Rothbard tries to address the problem of pollution by re-conceptualizing the law of torts from a libertarian perspective (Rothbard, 1982b). Starting with the Non-Aggression Principle, Rothbard proceeds to consider issues of liability, causation, and evidence. Three ideas are particularly relevant to the problem of pollution.

1 **Trespass/Nuisance distinction** – Rothbard builds on common law doctrine to distinguish between trespass and nuisance. He defines trespass as involving "visible and tangible or 'sensible' invasion, which interferes with possession and use of the property." Nuisance, however, involves "invisible, 'insensible' boundary crossings that do not" interfere in this way (Rothbard, 1982b). Trespass, Rothbard notes, is and ought to be regarded as illegal *per se*, even if it causes no harm to the owner beyond the mere act of invasion itself, while nuisance ought to be illegal only if harm can be demonstrated. Thus, "[a]ir pollution, consisting of noxious odors, smoke, or other visible matter, definitely constitutes an invasive interference. These particles can be seen, smelled, or touched, and should therefore constitute invasion *per se*" (Rothbard, 1982b). However,

> [a]ir pollution...of gasses or particles that are invisible or undetectable by the senses should not constitute aggression per se, because being insensible they do not interfere with the owner's possession or use. They take on the status of invisible radio waves or radiation, *unless* they are proven to be harmful.

> (Rothbard, 1982b)

2 **Strict causal connection** – In demonstrating that A harmed B, and is thus the proper object of judicial coercion, statistical correlation is not enough. It is not enough, in other words, to demonstrate that A emitted pollutants, and that B and others exposed to A's pollution suffered a certain kind of harm much more frequently than the general population. Instead, plaintiffs must prove beyond a reasonable doubt that "a *strict causal connection* between the defendant and his aggression" (Rothbard, 1982b).

3 **Homestead easements** – Finally, A will not be liable for harm caused to B if A has *homesteaded* the right to emit that pollution. Rothbard believes that pollution rights, like property rights in land, can be obtained by possession and use. If A builds a factory on a piece of land and starts emitting pollution on to unoccupied neighboring land, he acquires what Rothbard describes as a "pollution easement" (Rothbard, 1982b). Should that neighboring land eventually come to be put to residential use, homeowners would have no right to enjoin the pollution since, in effect, the pollution was there first.

Taken together, Rothbard's conditions make effective action against mass torts such as air pollution caused by automobile emissions virtually impossible. We might know, as a general matter, that automobiles emit pollutants, and that these pollutants can be harmful to human beings in various ways. But Rothbard's conditions entail that no individual B can bring legal action against any particular driver A unless he can demonstrate *beyond a reasonable doubt* that the pollution emitted by *A himself* caused demonstrable harm to B. Nor can B bring action against automobile manufacturers, since those manufacturers are not themselves individually responsible for B's harm. Thus, in the vast majority of cases, even those in which B has suffered real physical harm as a result of automobile emissions, she will lack effective recourse to enjoin the pollution or to demand compensation for the harm she has suffered. In these cases – which represent a *wide* range of environmental harms – the victim must, in Rothbard's words, "assent uncomplainingly" to the harm he or she suffered (Rothbard, 1982b).

Quite apart from this implausible implication, however, Rothbard's analysis suffers from several serious weaknesses. First, the insistence that plaintiffs establish a "strict causal connection" between the action of the defendant and the harm suffered by the victim seems both intuitively much too strong and ill-suited for coping with the kinds of complexly interconnected network of causes and effects that characterize environmental phenomena. Consider an analogy. Suppose a group of 100 individuals celebrate the New Year by firing their rifles into the air in the middle of a residential neighborhood. Several minutes later, a person standing in his backyard a few blocks away is struck and killed by a bullet falling from the air. Let us stipulate that ballistics tests are insufficient to determine from which particular gun the bullet was fired. It's reasonably certain that the bullet came from *one* of the hundred guns fired by the crowd. But we have no way of knowing whose. Must we conclude, then, that *nobody* can be prosecuted for the murder? Or even – since endangerment without harm is apparently no crime on Rothbard's view – for reckless endangerment?

Rothbard's distinction between trespass and nuisance is even more problematic. Rothbard relies heavily on this distinction to block the deeply counterintuitive conclusion that *all* pollution, even harmless radio waves, must be prohibited out of respect for the absolute rights of property owners to control their property. But the basis on which he draws this distinction appears to be entirely arbitrary. Rothbard *might* have plausibly argued that what matters is that individuals maintain the right to *exclusive control* over their property, and that any "invasion" that undermines that control is impermissible, whether harmless or not. Or he might have plausibly argued that actual harm is what really matters, such that invasions that do not result in harm really don't count as invasions at all. But what Rothbard actually argues is that harm to the victim determines the legality of *invisible* invasions, but not for visible ones. And this is difficult to make sense of.

If harm is relevant to legality, then it ought to be relevant across the board, and Rothbard ought to conclude that harmless invasions by detectable masses are just as permissible as harmless invasions by undetectable ones. Likewise, if harm is *irrelevant* to legality, then it ought to be irrelevant across the board, with harmless undetectable invasions being just as impermissible as harmless detectable ones. Drawing the line at detectability helps Rothbard avoid the extremely counterintuitive conclusion that radio waves (and CO_2 emissions, and thermal radiation) are the proper subjects of prohibition. But it does so in a way that seems entirely *ad hoc* and unprincipled.

c. Eric Mack and "Live and Let Live"

A more recent, and more promising, attempt to resolve the problem has been presented by Eric Mack (Mack, 2015). For Mack, the libertarian commitment to rights of private property is itself grounded in a more basic moral "ur-claim" of each individual to be allowed to pursue her good in her own way (Mack, 2010). Since, Mack argues, the ability to acquire, transform, and use external resources is essential to an individual's ability to pursue her good, and since this ability is greatly extended by a regime of private property, a right to private property (or, more precisely, a right not to be excluded from living under a justifiable rule-constituted practice of private property) flows out from the moral ur-claim.

But what is the precise nature of the property rights that flow out of that ur-claim? To some degree, Mack argues, this is a matter to be settled by convention. The ur-claim doesn't specify the precise shape, for instance, that riparian rights ought to take. There are many ways in which conventions might specify the contours of such rights compatible with each individual's ur-claim to pursue her own good in her own way. But the ur-claim does rule *some* conventions out: namely, those that would *not* be compatible with this basic claim.

Turn now to the problem of pollution. Suppose we understood property rights to be so absolute as to rule out even the most minor kinds of infringement. Understood in this way, property rights would render impermissible even the most ordinary activities of human life. Breathing, radiating heat, and lighting matches visible from your neighbor's property would all be violations of the property rights of others, and hence impermissible. The result, Mack notes, would be a kind of moral paralysis, or "hog-tying." Property rights, which were supposed to *facilitate* individuals' ability to pursue their own good, on this understanding instead wind up making that pursuit *impossible.*

For Mack, this is sufficient reason to reject this understanding of libertarian property rights. Any specification of rights, according to Mack, must be consistent with what he calls the "elbow room postulate," which holds that "a reasonable delineation of basic moral rights must be such that the claim-rights that are ascribed to individuals do not systematically preclude people from exercising the liberty-rights that the claim-rights are supposed to protect" (Mack, 2015). When applied to the problem of externalities, Mack argues, this postulate supports a principle much like the principle invoked by the majority in the case of *Bamford v. Turnley*, an 1862 case in which the plaintiff sued his neighbor for introducing noxious fumes into his home by his operation of a brick kiln. The court ruled that Turnley was liable for the damages caused by his brick kiln, but took pains to distinguish Turnley's actions from other more reasonable uses of one's property that might nevertheless "annoy" one's neighbors. In his separate opinion in support of the majority, Baron Bramwell wrote that

> those acts necessary for the common and ordinary use and occupation of land and houses may be done, if conveniently done, without submitting those who do them to an action... It is as much for the advantage of one owner as of another; for the very nuisance the one complains of, as the result of the ordinary use of his neighbor's land, he himself will create in the or-dinary use of his own, and the reciprocal nuisances are of a comparatively trifling character. The convenience of such a rule may be indicated by calling it a rule of give and take, live and let live...
>
> (*Bamford v. Turnley, 122 Eng. Rep. (1862), pp. 32–33*)

Mack's analysis appears to cope nicely with the problem posed by the *absolutism* of libertarian property rights. Mack avoids the implausibly radical implications of this position by saying, in effect, that minor everyday crossings of other people's (physical) boundaries do not really count as infringements of their (moral) rights at all. Property rights remain absolute, but those rights are *specified* in such a way that they give people no claim to be immune from minor, unintentional intrusions (Shafer-Landau, 1995).

Moreover, the elbow room postulate appears to provide a sensible rationale for distinguishing between the unintentional minor harms caused by pollution and the intentional minor harms caused by stealing a nickel. After all, a system of rights that prohibited all minor, unintentional boundary crossings would result in moral paralysis. But no such paralysis results from the prohibi-tion of minor but "wanton and malicious" intrusions. Indeed, that kind of prohibition would seem to *advance* the goal of allowing each individual to pursue her own good in her own chosen way.

But while Mack's theory copes well with the problem of absolutism, it has nothing to say about a second major problem posed for libertarianism by the problem of pollution – the problem of *interconnectedness*. Mack's analysis, like Rothbard's, is focused entirely on the harm caused by indi-vidual acts. So long as the harm caused by an action falls below a certain threshold, it is deemed permissible. But this exclusive focus on the outcome of individual actions leaves many of the most serious problems posed by environmental pollution entirely unaddressed. Air pollution in the Los Angeles basin is not the result of any single individual's actions. Nor are the problems of global cli-mate change, species extinction, river pollution, and so on. Any particular action by an individual, considered in itself, makes only a miniscule contribution to the overall problem. Either no one is harmed at all by such actions (the harm resulting only once the cumulative amount of pollution crosses a certain threshold), or the harm produced is minimal (becoming significant only when it is added up with all the other harms resulting from other individuals' actions).

Intuitively, a morality of individual rights ought to have something to say about actions of this sort. If I violate your rights by poisoning you, then I also violate your rights when I, along with nine of my friends, each add 1/10 of a lethal dose of poison to your drink. Does the rights violation disappear if the poisoning is performed by a million people, instead of ten? Or if their "coopera-tion" in poisoning you is the unintended (but foreseeable?) byproduct of their actions, rather than their intended aim? (Friedman, 1992)

Contemporary environmental problems arise largely as a result of the complexly interre-lated activities of large numbers of dispersed individuals. Libertarian intuitions about indi-vidual rights and coercion, in contrast, seem most at home at the "micro" level of discrete interactions between one person and another. Whether those intuitions can be developed and extended in a way that is adequate to deal with the kind of non-linearities and threshold effects that characterize the problem of environmental pollution remains to be seen. But the skepticism of environmentally sensitive philosophers like Peter Railton on this matter is not entirely without warrant.

> [I]f we take seriously the fact that we find ourselves situated in, and connected through, an environment, we are soon impressed with the inaptness of a conception of morality that pic-tures individuals as set apart by propertylike boundaries, having their effect upon one another largely through intentional action, limiting their intercourse by choice, and free to act as they please within their boundaries, although absolutely constrained by them. The result of this conception is gross restrictiveness here, gross latitude there, and, in general, an inadequate vocabulary for debating, or even expressing, a number of pressing moral issues concerning the environment.
>
> *(Railton, 1985)*

Free-Market Environmentalism

So far, this chapter has focused on those libertarian approaches to pollution that emphasize issues of rights and justice rather than economic efficiency. But not all libertarians approach questions of institutional design from that particular moral perspective. And even those libertarians who do employ a rights-based perspective on morality often concede that consequentialist considerations can play both a justifying and constraining role in shaping those rights (Nozick, 1974). Before this chapter comes to a close, then, it is worth devoting some attention to those considerations and their relation to the libertarian approach to pollution. What follows is a very brief survey.

First, libertarians have long stressed the vital role of private property in providing individuals with an incentive to care for natural resources. Pollution is a problem mainly when individuals are able to externalize the costs of the harmful byproducts of their activities onto *other* people's property or bodies. When the benefits of use are internalized, but the costs are externalized, trag-edies of the commons ensue (Hardin, 1968, Schmidtz, 1994). But when resources have an owner, those owners have a strong incentive to protect their property from harmful invasion by others, and to ensure that their own use is sustainable (Anderson and Leal, 1991, chapter 9). Libertarians thus recommend extending property rights over as wide a range of objects as possible – not just land and trees but lakes, highways, and even oceans (Block, 1990b, 1998). Such property rights would not only provide owners with incentives to use their own resources efficiently, they also serve as a necessary prerequisite for the formation of a market and of a market price system, which most libertarians regard as essential mechanisms for coordinating the activities of large numbers of dispersed individuals in their use of scarce resources (Hayek, 1945, Anderson and Leal, 1991, chapter 2). In cases where technological constraints make the establishment of full property rights impractical, as with rights to air or atmospheric space, some libertarians favor the kind of "quasi-" property rights embodied in a policy of tradable pollution credits (Dolan, 1990, 2006, Penning-ton, 2005). But neither the desirability nor the feasibility of such policies is uncontroversial among libertarians (Cordato, 1997, Block and McGee, 2011).

Second, libertarians argue that much existing pollution is due not to the excesses of the un-trammeled free market, but to various forms of "government failure." In crafting both particular regulations and the more fundamental definition and enforcement of property rights, the government is prone to being unduly influenced by powerful economic interests (Tullock, 1967, Krueger, 1974). Those interests influence the government's policy for its own benefit, often at the expense of the rights or interests of the broader citizenry. In terms of environmental policy, this means that businesses are often given a license to pollute by the state, at least when that pollution is perceived to serve some important interest of key decision-makers of the state (Coase, 1960, Block, 1990a). Moreover, due to its extraordinary legal status, the state itself is able to externalize costs with relative impunity (Pennington, 2005). There is thus relatively little direct incentive for governments to take effective action to curb their own polluting activity.

Third and finally, libertarians note that a variety of private, voluntary solutions exist to the problem of pollution. Following the ideas of Ronald Coase, many libertarians argue that so long as property rights are clearly defined and transaction costs are low, resources will be put to efficient use in the sense that the potential externality will be internalized by whichever party is able to do so in the most cost-effective way (Coase, 1960, Friedman, 1997). And following the work of Elinor Ostrom and others, many libertarians argue that even in the absence of well-defined property rights, communities can develop social norms to cope effectively with the problems posed by costly externalities (Ostrom, 1990, Ellickson, 1991, 1993).

Taken together, these three considerations make a compelling case for expanding the role of market and other non-governmental responses to the problem of pollution. At best, however, these considerations can serve as a *supplement* to a theory of rights, and never as a *substitute* for one. Considerations of efficiency such as those embodied in the Coase theorem, or in cost-benefit analyses of mitigation vs. adaptation to global climate change, are important to take into account in formulating public policy. But they are far from sufficient. Even if adaptation is *cheaper* than mitigation as an approach to dealing with the problems of climate change, it might nevertheless be deeply *unjust*. After all, we have good reason to care not merely about the total costs and benefits of alternative policies, but about how those costs and benefits are distributed among separate parties. And libertarians have *especially* good reason to care if some of those costs are imposed through the violation of property rights, such as would seem to occur when the activities of power plants in the Midwestern United States contribute to the flooding of a farmer's property in Bangladesh (Dolan, 2006, Adler, 2009, Shahar, 2009).

In this respect, at least, the Rothbardian approach to thinking about pollution in terms of homesteading has an intuitive advantage over the pure Coasian law and economics approach (Rothbard, 1982a). In deciding which party ought to bear the responsibility for dealing with pollution, what matters is not just who can do so more cheaply, but who got there first. From the perspective of justice, if not efficiency, there is a world of difference between a person who buys a home near an already existing airport and complains about the noise, and a one who has an airport built next to the home in which he'd lived for the last 20 years (Coase, 1960). Priority of use might not be the *only* moral principle that matters in determining rights and liabilities, but it is certainly one such principle, and sufficient to demonstrate the inadequacy of efficiency considerations by themselves.

Conclusion

To say that rights matter, of course, does not yet tell us anything in particular about *which* rights matter. And, as the survey in the first part of this chapter revealed, there is considerable debate even among libertarians about how precisely the various rights and responsibilities relevant to the problem of pollution should be understood. Still, it is worth noting the considerable overlap among otherwise diverse libertarians on the principle of "Live and Let Live" as expressed by Justice Bramwell in *Bamford v. Turnley*. Not only is this principle endorsed by the Nozickian Eric Mack (2015); something like it is also endorsed by the Rothbardian Walter Block (Block and McGee, 2011), and by the consequentialist Richard Epstein (Epstein, 2009). In this respect, there appears to at least be considerable consensus among libertarians about where they want to end up. Libertarians, by and large, want a system in which property rights are taken seriously, but in which the minor but ubiquitous "dust" that is kicked up as part of people going about the everyday business of life is not regarded as a violation of those rights. Still, both the theoretical underpinnings of this principle and its practical application to difficult problems such as those posed by global climate change remain a fruitful subject for future research.

References

ADLER, J. H. 2009. Taking Property Rights Seriously: The Case of Climate Change. *Social Philosophy and Policy*, 26, 296.

ANDERSON, M. 1989. George Bush Environmentalist. *The Christian Science Monitor*, January 4.

ANDERSON, T. L. & LEAL, D. R. 1991. *Free Market Environmentalism*, San Francisco, CA, Pacific Research Institute for Public Policy.

BLOCK, W. 1990a. *Economics and the Environment: A Reconciliation* Vancouver, Canada, The Fraser Institute.

BLOCK, W. 1990b. Environmental Problems, Private Property Rights Solutions. *In:* BLOCK, W. (ed.) *Economics and the Environment: A Reconciliation*. Vancouver, Canada, The Fraser Institute.

BLOCK, W. 1998. Environmentalism and Economic Freedom: The Case for Private Property Rights. *Journal of Business Ethics*, 17, 1887–1899.

BLOCK, W. & MCGEE, R. W. 2011. Pollution Trading Permits as a Form of Market Socialism and the Search for a Real Market Solution to Environmental Pollution. *Fordham Environmental Law Review*, 6, 51–77.

CHARTIER, G. & JOHNSON, C. W. 2012. *Markets Not Capitalism: Individualist Anarchism Against Bosses, Inequality, Corporate Power, and Structural Poverty*, Brooklyn, NY, Autonomedia.

COASE, R. H. 1960. The Problem of Social Cost. *Journal of Law and Economics*, 3, 1.

CORDATO, R. E. 1997. Market-Based Environmentalism and the Free Market: They're not the Same. *The Independent Review*, 1, 371–386.

DOLAN, E. G. 1990. Controlling Acid Rain. *In:* BLOCK, W. E. (ed.) *Economics and the Environment: A Reconciliation*. Vancouver, Canada, The Fraser Institute.

DOLAN, E. G. 2006. Science, Public Policy, and Global Warming: Rethinking the Market-Liberal Position. *Cato Journal*, 26, 445.

ELLICKSON, R. 1991. *Order Without Law: How Neighbors Settle Disputes*, Cambridge, Harvard University Press.

ELLICKSON, R. 1993. Property in Land. *Yale Law Journal*, 102, 1315–1400.

EPSTEIN, R. A. 1998. *Principles for a Free Society: Reconciling Individual Liberty with the Common Good*, New York, Basic Books.

EPSTEIN, R. A. 2009. Property Rights, State of Nature Theory, and Environmental Protection. *New York University Journal of Law and Liberty*, 4, 1–35.

FRIEDMAN, D. 1989. *The Machinery of Freedom: Guide to Radical Capitalism*, La Salle, IL, Open Court.

FRIEDMAN, D. 1997. *The Swedes Get it Right* [Online]. Available: http://www.daviddfriedman.com/Academic/Coase_World.html [Accessed 5/15 2014].

FRIEDMAN, J. 1992. Politics or Scholarship? *Critical Review*, 6, 429–445.

FRIEDMAN, M. 1962. *Capitalism and Freedom*, Chicago, University of Chicago Press.

GLANNON, J. W. 2010. *The Law of Torts: Examples and Explanations*, Austin, TX, Wolters Kluwer.

HARDIN, G. 1968. The Tragedy of the Commons. *Science*, 162, 1243–1248.

HAYEK, F. A. 1945. The Use of Knowledge in Society. *American Economic Review*, 35, 519–530.

KRUEGER, A. O. 1974. The Political Economy of the Rent-Seeking Society. *The American Economic Review*, 64(3), 291–303.

LOCKE, J. 1952. *The Second Treatise of Government*, New York, MacMillan.

MACK, E. 2010. The Natural Right of Property. *Social Philosophy and Policy*, 27, 53.

MACK, E. 2011. Nozickian Arguments for the More-Than-Minimal State. *In:* BADER, R. M. & MEADOWCROFT, J. (eds.) *The Cambridge Companion to Nozick's Anarchy, State, and Utopia*. New York, Cambridge University Press.

MACK, E. 2015. Elbow Room for Rights. *Oxford Studies in Political Philosophy*, 1, 194–221.

MACK, E. & GAUS, G. 2004. Classical Liberalism and Libertarianism: The Liberty Tradition. *In:* GAUS, G. & KUKATHAS, C. (eds.) *Handbook of Political Theory.* London, Sage.

NOZICK, R. 1974. *Anarchy, State, and Utopia*, New York, Basic Books.

OSTROM, E. 1990. *Governing the Commons: The Evolution of Institutions for Collective Action*, Cambridge University Press.

PENNINGTON, M. 2005. Liberty, Markets, and Environmental Values. *Independent Review*, 10, 39–57.

RAILTON, P. 1985. Locke, Stock, and Peril: Natural Property Rights, Pollution, and Risk. *In:* GIBSON, M. (ed.) *To Breathe Freely.* Trenton, NJ, Rowman and Littlefield.

RAND, A. 1961. Man's Rights. *The Virtue of Selfishness.* New York, Signet.

RAND, A. 1999. The Anti-Industrial Revolution. *In:* SCHWARTZ, P. (ed.) *Return of the Primitive: The Anti-Industrial Revolution.* New York, Meridian.

ROTHBARD, M. N. 1973. *For a New Liberty*, New York, Collier.

ROTHBARD, M. N. 1982a. *The Ethics of Liberty*, Atlantic Highlands, NJ, Humanities Press.

ROTHBARD, M. N. 1982b. Law, Property Rights, and Air Pollution. *Cato Journal*, 2, 55–99.

SCHMIDTZ, D. 1994. The Institution of Property. *Social Philosophy and Policy*, 11, 42–62.

SHAFER-LANDAU, R. 1995. Specifying Absolute Rights. *Arizona Law Review*, 37, 209–224.

SHAHAR, D. C. 2009. Justice and Climate Change: Toward a Libertarian Analysis. *The Independent Review*, 14, 219–237.

SINGER, P. 2004. *One World: The Ethics of Globalization*, New Haven, CT, Yale University Press.

SOBEL, D. 2012. Backing Away from Libertarian Self-Ownership. *Ethics*, 123, 32–60.

SOBEL, D. 2013. Self-Ownership and the Conflation Problem. *In:* TIMMONS, M. (ed.) *Oxford Studies in Normative Ethics.* New York, Oxford University Press.

TAYLOR, R. 1992. The Environmental Implications of Liberalism. *Critical Review*, 6, 265–282.

TULLOCK, G. 1967. The Welfare Costs of Tariffs, Monopolies, and Theft. *Economic Inquiry*, 5, 224–232.

ZWOLINSKI, M. 2008. *Libertarianism* [Online]. The Internet Encyclopedia of Philosophy. Available: http://www.iep.utm.edu/libertar/.

ZWOLINSKI, M. & TOMASI, J. forthcoming. *The Individualists: Radicals, Reactionaries, and the Struggle for the Soul of Libertarianism*, Princeton, NJ, Princeton University Press.

52

PREDICTION AND FORECASTING

Arielle Tozier de la Poterie and Meaghan Daly

Introduction

Climate variability has long had an impact on human activities and well-being. Climate and weather influence decisions as important as what kinds of crops to grow and as mundane as what clothing to wear or whether to put snow tires on your car. Many of us have gotten used to the convenience of being able to look up the daily weather, but longer-range climate information, including inter-annual predictions (months in advance) and climate projections (several decades into the future) accompanied by advisories or suggested actions, also has the potential to influence decisions of much greater consequence. Such products are broadly referred to as climate services.

Decision-makers ranging from farmers, to business owners, to government policy-makers, and even to average citizens have an interest in knowing the likelihood of heavy rains and flooding, thunderstorms, hurricanes, drought, wildfires, heat waves, and other hydro-meteorological events. For example, insurance companies want to understand the likelihood of floods and hurricanes that may damage the properties they insure; water managers want to know how big to build dams to ensure adequate water supplies into the future, particularly in areas potentially affected by drought; development and climate change professionals want to know how climate change will affect emerging economies and the livelihoods of the world's poor; and farmers want to know the likelihood of rains in their area so that they can decide what to plant and when. Seasonal forecasts and other climate services present the possibility of providing answers to these questions and are therefore commonly cited as a means to better, more informed decision-making (Bailey 2013; Braman et al. 2010; Cane et al. 1994; Coughlan de Perez et al. 2014; Government Office for Science 2012; Haile 2005; Hansen 2002; Roncoli 2002; Suarez and Tall 2010; Tall et al. 2012; Visbeck 2007; Ziervogel 2004). For example, in response to climate services humanitarian organizations could procure and preposition supplies before floods or droughts, leading to better response times; farmers could change their agricultural practices according to anticipated precipitation. This potential, coupled with increasing forecasting skill, better technology, and intensifying national and international policy debates about the impacts of and solutions for climate change, has led to increasing demands from a range of actors for more accessible scientific climate information. This includes the desire for scientists to provide predictions of both short-term variability and long-term climate trends and, perhaps more importantly, to couple these with sectoral advice or warnings in the form of climate services.

DOI: 10.4324/9781315768090-62

The humanitarian, development, and climate change communities in particular have been quick to tout the potential benefits of climate services for the world's poor. The Global Framework for Climate Services (GFCS) is perhaps the most prominent effort to promote the use of climate services to improve decision-making at a variety of levels in order to help the world's most vulnerable populations respond and adapt to climate variability and change. The initiative establishes a guiding framework for coordinating the development of climate services in order to "empower the most vulnerable" (WMO 2011). It also seeks to shape the trajectory of investment in and capacity development for climate services, particularly in developing countries that may lack the capacity to develop such services without external support. As with other climate service initiatives, the GFCS emphasizes the need to engage stakeholders, who may include scientists, humanitarian organizations, and government representatives, in the development of climate services to ensure that such services eventually meet user needs. Tailoring climate services to user needs is intended to make the climate services more "usable" (Buizer et al. 2010) so that they can inform the decisions of actors ranging from government ministries and policy-makers, to international aid organizations or individual farmers. As part of capacity development, the GFCS seeks to facilitate dialogue between scientists and technical experts (commonly referred to as the "producers" of climate services) and decision-makers within governments, humanitarian and development agencies, and individual households (the potential "users" of climate services). By bringing together stakeholders at the global, regional, national, and local levels, initiatives under the GFCS aim to encourage two-way dialogue about the kinds of services that are needed to address climate variability and change at a variety of scales.

Despite enthusiasm for developing climate services across various disciplines and sectors, the benefits of climate services to the world's most vulnerable are not guaranteed (Bruno Soares, Daly, and Dessai 2018; Gerlak et al. 2020), and efforts to develop such services raise a number of ethical questions. Who actually benefits from climate services? How should climate data be shared between nations? Who gets to decide what kinds of climate information are made available? Who should participate in processes of climate services production?

In recognition of the ethical challenges facing promoters of climate services, Adams et al. (2015) advanced principles that they believe should guide the ethical development and application of climate services. This chapter builds upon this effort by examining three real, ethical challenges that shape the trajectory of climate service production, dissemination, and use in developing countries: climate service financing, data sharing and capacity development, and stakeholder participation. Although not often discussed in explicitly ethical terms, each of these current debates has clear ethical implications. We examine these three issues using insights from rights-based and deliberative justice in order to demonstrate where current discourses may be falling short and how more explicit consideration of the ethical nature of these problems may help to align service development with stated goals.

We argue that unless international actors pay more attention to how climate services are financed, which countries or forecasting centers have the capacity to develop such services, and which stakeholders are able to contribute to climate service development, efforts to use climate services to address climate change and "empower the most vulnerable" (a goal espoused by the GFCS's foundational documents) are likely to fall short.

Background

In order to provide context for our examination of the three ethical debates outlined above, the following sections provide background on the production of climate services and a brief overview of several principles of rights-based and deliberative justice.

What Are Climate Services?

While the term "climate services" was first employed at least several decades ago (e.g., Changnon et al. 1990), broad interest in formally developing climate services has been a relatively recent phenomenon, spurred by national and international efforts to address emerging issues associated with climate change. In recent years, researchers and practitioners have applied multiple definitions of climate services (Brasseur and Gallardo 2016), reflecting attention directed toward climate services development at multiple levels and involving different sets of actors. For example, while the US National Research Council defines climate services under national initiatives as "the timely production and delivery of useful climate data and knowledge to decision makers" (2007, as cited in Brasseur and Gallardo 2016, p. 80), the European Commission has embraced a more comprehensive definition, which includes the "transformation of climate-related data…into customized products such as projections, forecasts, information, trends, economic analysis, assessments, counseling on best practices, development and evaluation of solution, and other services in relation to climate that may be of use for society at large" (European Commission 2015, p. 10). At the global level, the Global Framework for Climate Services (GFCS)—an international framework that aims to guide and coordinate the development of climate services in all 192 WMO member countries, with a particular emphasis on supporting climate services in developing countries—emphasizes that climate services involve "…the provision of climate information in such a way as to assist decision-making" (Hewitt et al. 2012, p. 831). According to the GFCS, climate services should be based on scientifically credible information and expertise, developed with engagement from users and providers, and result in effective access of potential users to information and services.

These multiple definitions have important practical implications that, as we will illustrate, are also relevant to ethical considerations involved in developing "just" climate services. The timescales at which climate services should be provided is, for example, an implicit issue. Further, differences in how climate services are defined present challenges in making broad statements about the benefits of climate services and also require different human resources, data, observation systems, technical and institutional capacities, and other factors needed to support effective climate service development. For example, in many developed countries the observation networks, institutional arrangements, and human capacities needed to effectively produce climate services are well established. However, there are many challenges in developing countries, such as insufficient monitoring networks, computing capacities, and institutional and human resources, all of which make the successful development of climate services a much greater challenge. Unless these disparities are explicitly and effectively addressed, it is unlikely that the most vulnerable populations in developing countries will benefit from climate services.

A commonality of nearly all definitions of climate services is the desire to shift from providing data or information about weather and climate to an explicit emphasis on directly informing *decision-making and actions* that reduce climate-related risks. This shift includes providing advice about how information should be interpreted, as well as specific options for acting in response to this information. Each of these aspects of climate services presents important ethical stumbling blocks and justice-considerations that have, thus far, been relatively unaddressed within the practical implementation and use of climate services (see Adams et al. 2015 for an exception).

Climate Change and the Development of "Just" Climate Services

The rise of climate services has been, in part, a response to the anticipated impacts of climate change and the perceived need for better information in order to help people cope with a changing climate. Rich nations have contributed to greenhouse gas (GHG) emissions disproportionately relative to the size of their populations. GHG emissions are already causing rising temperatures, and this will lead to changing climate patterns around the world, with serious implications for government decision-making and for the well-being of individual citizens. In the pursuit of their own economic benefits, developed nations have emitted greenhouse gases that are likely to contribute to harm in developing countries (Agarwal and Narain 1990). Because those who are most vulnerable to the impacts of climate change are also those who have contributed least to the problem, issues of fairness and justice are central to debates about adaptation frameworks, policies, and funding (Caney 2010). Many argue that because it is wrong to harm others (Lyons 1994) and because these harms were inflicted unilaterally by developed nations upon developing nations, developing nations are owed some sort of compensation. For example, the United Nations Framework Convention on Climate Change (UNFCCC) includes mechanisms to deal with "loss and damage" related to climate change; however, there is still no consensus about how developing countries should be compensated for the adverse impacts of climate change. Recognizing that climate change is likely to have serious implications for the livelihoods of people in developing nations (IPCC 2014) as well as economic, social, and environmental policy in all nations, scientists and other advocates argue that climate services are a fundamental tool to minimize, or in some cases mitigate, the impacts of climate change in both developing and developed countries.

In this context, several conceptualizations of justice might be considered relevant to decisions regarding climate change adaptation and climate service development (Baer 2006; Caney 2010; Ikeme 2003; Jamieson 2010; Ringius et al. 2002). In this chapter, we choose to focus on principles from rights-based, capabilities, and deliberative approaches to justice because they shed light on the contradictions present in current climate service policy debates. While there are many ways in which these conceptualizations of justice could be brought to bear on climate service development, we apply them in the context of three ethical problems to introduce the potential ethical dilemmas that may arise. We apply a rights-based lens to the exploration of who funds service development and who benefits. We use concepts from the capabilities approach to explore issues of data sharing and capacity building; and we examine calls for inclusive, stakeholder-driven service development using principles of deliberative justice. We summarize each approach briefly below. These approaches do not necessarily agree on what is just or ethical. One focuses on process, while the others are oriented toward outcomes. However, each highlights interesting ethical issues related to climate services that are worthy of consideration going forward.

Rights-Based Justice

Various approaches to rights-based justice assert that people should be guaranteed minimum standards of liberty, opportunity, and capabilities necessary for them to secure their well-being (Nussbaum 2011; Rawls 1971). Because developing nations are already at an economic disadvantage, the extra burden imposed due to climate change may prevent those who live below these minimum standards from advancing to a reasonable standard of living. Furthermore, the impacts of climate change may push those who would otherwise have been able to achieve a reasonable standard of living below this minimum threshold. As such, climate change is fundamentally an issue of human rights: the right to a minimum level of well-being that is worthy of human dignity (Nussbaum 2011).

Within the tradition of rights-based justice, Rawls (1971) believes, justice entails the protection of liberties and the distribution of wealth and income that enable *opportunities*. Rawls is primarily concerned with determining how social and economic inequalities can be addressed through distributive justice, which seeks the best allocation of burdens and benefits. He argues that people should not be prevented from acquiring skills that would afford them the opportunity to pursue their interests purely because of arbitrary factors such as the family or society one is born into. Such principles imply that if climate change makes some countries worse-off, to the extent that the negative effect limits people's opportunities, others have a duty to take redistributive action to level the playing field so that everyone will have equal opportunity. Developed nations have the obligation to assist developing nations to ensure that conditions they have created through climate change do not result in additional inequality of liberties, opportunities, and capabilities.

Through this lens, developing and improving climate services in developing nations represents a potential avenue for helping developing nations adapt to changing climates caused by economic activities in developed nations, thereby restoring equality of opportunity. Climate services offer one potential avenue for developed nations to right the wrongs imposed on developing countries through the provision of information and knowledge that can inform appropriate adaptation measures. Such information can be a means of ensuring that individuals or groups that currently live below thresholds of minimum well-being will, using such information, be able to lift themselves above the minimum threshold of justice in the future. Such a rationale for the development of climate services also implies an imperative to ensure that the provision or use of climate services does not lead to additional harms.

A Capabilities Approach to Rights

Nussbaum's capabilities approach to rights-based justice asserts that people ought to have equal access to the *opportunities* to pursue their interests, and that society has an obligation to provide an *enabling environment* in which people can actually realize opportunities (Nussbaum 2011). The capabilities approach to rights-based justice is concerned with what people are "effectively able to do and to be," whether as individuals or as members of a group or of a nation (Robeyns 2005). From this perspective, the extent to which climate change jeopardizes people's ability to realize opportunities—perhaps by creating greater insecurity within or between nations or by diverting government resources from social programs—those with more resources have an obligation to help establish sufficient capabilities around the world.

Although at their most basic level, capabilities may be the elements people need to escape poverty, the capabilities framework can also be applied to analyze social institutions and public goods and measure inequalities within or across more affluent societies (Robeyns 2005, p. 101). Therefore, the need to protect "conditions or states of human enablement" (Nussbaum 2011) has implications for the development of climate services, as it demands that such services must also build societal capacity to help people and nations realize their own paths. It also introduces the possibility that nations have an obligation to help each other build these necessary capacities. The development of climate services might, then, be seen as an essential capacity to support effective climate mitigation and adaptation. Further, a rights-based approach that is focused on human capabilities also calls attention to (sometimes stark) disparities in the abilities of different individuals or groups to equally use and benefit from climate information.

Deliberative Justice

In contrast to rights-based approaches to justice, deliberative justice focuses on process rather than outcomes. In this tradition of ethics, the process by which allocations of benefits and burdens are determined is more important than the final distribution. Deliberative justice differs from distributive justice because it stipulates that only fair negotiations, whether between nations or between groups and individuals within nations, can ensure good outcomes. Deliberative principles stress public participation and the absence of coercion between parties. All parties affected by a decision ought to be included in deliberations prior to the decision. When this condition is met, fair, deliberative, democratic processes will arrive at just outcomes and ensure that governments address priorities articulated by their citizens (Dryzek and List 2003).

Deliberative justice is applicable to the development of climate services because it means that the development of such services should engage all relevant stakeholders to ensure that the final product meets a variety of needs and does not disadvantage any groups. Because climate change will affect global-, regional-, and national-level policy, as well as individuals, deliberative approaches to justice would require policies for development and dissemination of climate services that promote equality between nations and meet a variety of different needs in order to ensure that they contribute to well-being across society.

Examining Current Climate Services Discourses in Light of Principles of Justice

In light of climate change and using principles from the approaches to justice discussed above, we now examine three real debates surrounding the development of climate services. In each, we highlight how stakeholders currently frame and represent these ideas and areas in which principles of justice might productively contribute going forward. The objective is not to argue for any single perspective or approach but to demonstrate how more explicit consideration of the ethical implications of various policies can help to align climate services with their stated goals, potentially providing relevant insights for climate services development going forward.

Ethical Issue 1: Climate Services Funding—Who Pays? Who Benefits?

One of the central ethical questions surrounding the development and use of climate services concerns who should invest in the development of these services and to what ends. This issue is particularly salient when climate services are framed as a means of supporting climate change adaptation in developing countries. Many advocates of climate services justify calls for increased investment by claiming that such services will improve the plight of the world's most vulnerable (WMO 2011). At the same time, other proponents call for increasing private investment in climate services, developing markets, and making the business case for climate service investments (Brasseur and Gallardo 2016; Lourenço et al. 2015; WMO 2015).

For example, the European Commission's (EC) *Roadmap for Climate Services* is an international effort to promote the development of climate services as a means to addressing the threats posed by climate change. It evolved in response to global efforts, led by the World Meteorological Organization (WMO), to implement the GFCS. The Roadmap proposes nine activities meant to guide discussion and development of climate services in the European Union toward "a market for climate services that provides benefits to society" (p. 7). These activities include stimulating private demand for climate services, establishing networks to help connect producers and users of climate information, and fostering cooperation in product development and quality control.

While on the surface efforts to increase private investment in climate services and to address the needs of the world's most vulnerable seem complementary, the assumptions behind these common narratives deserve greater investigation. A market approach assumes that climate services should be developed and sold to consumers instead of being a public good that is provided by governments or for the benefit of the most vulnerable.

Economic Arguments for Climate Services

Market-based approaches and justifications for climate services development based upon demand and cost-benefit analysis (CBA—also called benefit-cost analysis, BCA, or socioeconomic benefit studies, SEB) reflect an economic rationale that is commonly used in policy-making. Economic approaches assign monetary value or use willingness to pay (WTP) as metrics for measuring value to society. In these approaches, price or hypothetical WTP is considered a reflection of value. Everything from human life to natural systems can be assigned either quantitative or qualitative value, and hence entered into an equation to determine if costs outweigh benefits or vice versa (Sunstein 2005).

A recent WMO report attempts to bolster the economic justification for climate services by advocating socioeconomic benefit studies as a means of demonstrating that "the benefits of [weather and climate services] are significantly larger than the costs to produce and deliver them" (WMO 2015, p. xv). According to this report, by cataloguing and comparing the costs and benefits of climate services, national meteorological and hydro-meteorological services (NMHSs) around the world can make the case for increased funding for their programs.

Limitations of Economic Approaches

A focus on developing private, market-oriented demand for climate services has the potential to conflict with justifications of investment in climate services based on public benefit or societal vulnerabilities. Calls for market-based climate services neglect important policy-relevant ethical considerations, including the distribution of benefits and how such services affect people's abilities to meet their basic needs. As critics of economic arguments have pointed out, WTP is not necessarily reflective of the benefits parties derive (Sagoff 2004). When speaking of people's assets and livelihoods, such methods raise difficult questions about the value of life and the resources required to sustain it in different places (Sunstein 2005). For example, property and incomes in the United States may have higher dollar value than homes and livelihoods in some areas of Africa or Asia. While from an economic perspective it might make more sense to invest in preventing disasters where people earn more money and have more assets, from an ethical perspective such behavior is questionable at best. It would also be contrary to the goal of using climate services to help the world's most vulnerable.

Market-based development of climate services based on CBA or WTP has the potential to neglect important distributive concerns. As in other areas of science policy, a market-driven focus has the potential to "shift the discourse away from the political question of 'why?' and 'to what end?' to economic questions of 'how much?'" (Bozeman and Sarewitz 2005, p. 120). Likewise, as with any technology, when the broad societal benefits of climate services are taken for granted, without regard to how those benefits are distributed, the focus may shift from "how can we best achieve our goal of serving the most vulnerable?" to "how can we ensure the development of climate services?" The assumption that climate services will inevitably contribute toward vulnerability reduction, or that economic calculations will automatically lead to the greatest societal benefits—regardless of how exactly markets are configured and investments are made—is never tested. Under such circumstances, questions of who wins and who loses, though essential from both the policy and ethical perspectives, become obscured.

There is ample evidence from other areas of science policy demonstrating that leaving decisions about science funding to markets alone can push development toward what is more profitable rather than what will result in the greatest societal benefit or human well-being (Bozeman and Sarewitz 2005; Morss and Hooke 2005; Sarewitz and Pielke 2007). AIDS funding provides an excellent example; while from the perspective of drug companies, development of AIDS drugs represented significant market success, public benefit lagged because those in need of drugs cannot afford them (Bozeman and Sarewitz 2005; Reich 2000).

In the case of climate services, markets may succeed by economic measures while still failing to produce public goods or to meet important goals, such as helping the most vulnerable adapt to climate change. Just as increasing the market for weather services threatens to undermine the public-good status of weather services (Morss and Hooke 2005), investments by private actors could shift development toward those who can pay rather than those most in need. Webber and Don-ner (2016) have argued that there is a need to shift away from commercialized models of climate services development in order to enable actors in developing countries to equally use and benefit from climate services. Privately funded climate services have the potential to produce valuable information that is neither accessible nor usable by the vast majority of people affected by climate variability and change. Even GFCS stakeholders acknowledge that "it is unlikely that [a user-pays] approach could fully meet user needs, including responding to high levels of climate sensitivity and achieving sustained development" (WMO 2011, p. 175). Unless market-based approaches are adjusted to account for differences in financial and political power, free market principles run the risk of skewing climate service development toward those who are already comparatively well-off. This would be contrary to the principles of distributive justice. Therefore, a more justice-oriented approach to climate services development would require some form of oversight to ensure that development of climate services proceeds in such a way that, at a minimum, they are not used in ways that privilege those who are already better off thereby increasing existing inequalities.

How information is produced and disseminated influences who is able to receive and act on information and, hence, benefit (Broad et al. 2002; Carr and Thompson 2014; Lemos and Dilling 2007; Lemos et al. 2002; Miller 2004; Vogel and O'Brien 2006). Climate information exacerbates power imbalances and can be used to advance specific agendas rather than address the greatest needs. There is a disconnect between claims that climate services will benefit the poor and calls for stimulating private markets: those who are most vulnerable are likely to be those *least* able to pay. Who pays for such services is therefore important, as it is likely to dictate the kinds of services that are developed, who receives them, and, by extension, the stakeholders who are most likely to benefit.

The lack of empirical evidence that climate services can benefit the world's poor (Bruno Soares, Daly, and Dessai 2018; Vaughan and Dessai 2014) compounds distributive concerns. Studies have repeatedly demonstrated that because of dissemination channels, differential access to informa-tion, differential ability to act on such information, and other factors (Broad et al. 2002; Carr and Thompson 2014; Lemos and Dilling 2007; Lemos et al. 2002; Miller 2004; Vogel and O'Brien 2006), seasonal forecasts have the potential to exacerbate existing inequalities rather than bestow-ing broad-based benefits. Ensuring equality of opportunity and capabilities among individuals (Nussbaum 2011; Rawls 1971) requires that climate services efforts should be designed and funded so that climate services do not compound existing inequalities. Application of theories of rights-based or distributive justice also implies an obligation to guarantee that climate services will allow individuals greater access to their basic needs and reduce existing inequalities. Finally, attention to distributive justice obliges developed nations to invest in adaptation efforts that will enable people to meet their basic needs and contribute toward equality.

Ethical Issue 2: Data Sharing and Capacity Building within and between Nations

While the previous discussion focuses on issues of individual justice and ability to pay, the devel-opment of climate services also has important implications for economic and technical equality between nations. These issues manifest most prominently in ongoing discussions of how national climate data should be shared, as well as how technical capacity should be built and distributed among nations.

After World War II the International Meteorological Organization, now the World Meteo-rological Organization (WMO), created a system of international data sharing based upon the idea that meteorological services were public goods and that the data should be international public property (Landis 2001; Zillman 1997). Although increasing financial pressures and the commercialization of weather services began to erode commitments to data sharing in Europe beginning in the 1980s (Landis 2001; Saarikivi et al. 2000; Zillman 1997), the WMO reaffirmed principles of data sharing through the adoption of Resolution 40 in 1995. The resolution requires "essential" data to be shared but allows nations to decide what data are essential and what data are "additional," and hence subject to fees (Resolution 40 in Saarikivi et al. 2000).

Despite this resolution, the debate about how freely to share data continues both within developed nations and between developed and developing nations. A decade ago, this debate surfaced when representatives of developing nations expressed concern over language about data sharing included in the founding documents of the GFCS (Permanent Mission of India to the UN 2010).

Those in favor of public data argue that the free exchange of data helps national met services provide the best possible services to their citizens, that the WMO stands as a model of international cooperation, and that open exchange is the key to sustained competition and research progress (Saarikivi et al. 2000; Venema n.d.; Zillman 1997). Indeed, the EC Roadmap mentioned above envisions a world in which open access to data and climate models serves as the foundation for "public and private climate service operators [to] develop a variety of customized high added-value services with and for users" (European Commission 2015, p. 7). Because "weather is global and all nations need data from many countries" (Saarikivi et al. 2000, p. 832), many scientists argue that all nations should contribute to a global reservoir of information for mutual benefit. From this perspective, there is a need for continued data sharing and coordination among and between scientists and governments so that the science of climate change can be thoroughly studied and, thereby, contribute to "human safety and welfare and to the formulation of sound public policy at the local, national, and global level" (Zillman 1997, p. 1086).

What seems on the face to be an argument about the well-being of global society overlooks the distributive and deliberative implications of global agreements for open data sharing in the context of historical financial and technical inequality between nations. Indeed, the fact that "open and unrestricted exchange of observational data produced and collected around the world" has been "one of the guiding principles of the international meteorological community from its origins" (Saarikivi et al. 2000, p. 831) has not prevented great discrepancies in the technical capacity among developed and developing nations. Furthermore, at the time that data-sharing norms were established, it was largely developed nations (Japan, the USA, the USSR, European countries) who were producing and sharing data (Landis 2001). In the context of global frameworks incorporating actors with different levels of technical and economic capacity, agreements about data sharing have different implications.

Although the data-sharing argument is framed in terms of global well-being, individual nations do, and will continue to, prioritize their own interests. Those countries with more resources are better positioned to use global data to advance their own national priorities. As discussed in the first example above, those with more resources are able to make investments in services that benefit them while neglecting the needs of those less able to pay. While least developed countries often benefit from the capacity-building initiatives implemented by the WMO, this comes along with expectations about the obligation to share data and metadata with other WMO member states (WMO 2016). Yet, the capacities of all countries to utilize data are not equal. Most global prediction centers are located in the Global North. This puts developing countries in the Global South at a disadvantage within international climate negotiations (Mahony 2013). In this way, developing countries are contributing data for the "global good," while they continue to be disadvantaged in their ability to use the information themselves. Therefore, requiring developing nations to freely provide their data to other governments, absent significant efforts to boost capacities to apply these data at the same level as wealthier countries in the global North, is unjust.

Capacity Building and a Capabilities Approach

If climate services are to be used as a means of reducing vulnerability and rectifying the harms caused by climate change, the capabilities approach argues that developed nations have an obligation to ensure equal capacity to develop climate services that meet national needs and priorities and allow for the realization of individual capacities and well-being. To achieve this goal, policies and programs need to ensure that investments in climate services address people's basic needs and do not result in greater inequality between nations. Rather than focusing on the responsibility of each nation to contribute to global well-being, as though each nation has equal capacities and opportunities, a capabilities approach to well-being might instead focus on the degree to which individual nations, and people within those nations, have the capabilities to develop and use climate services to inform their decisions and pursue fulfilling lives.

A capabilities approach is reflected in arguments advanced by the Like-minded Group of Developing Countries (LMDC). In discussions surrounding the foundations of the GFCS, the LMDC argued that the GFCS should advance the principle that "...all people have the right to access climate services that will assist them to know how climate will affect them and to know how to defend themselves from global climate impacts" (Third Meeting of HLT 2010 in Permanent Mission of India to the UN 2010 p 3). They called for the GFCS to focus on enhancing national-level prediction capacities in developing countries to enable them to develop their own, "appropriate" climate services (Permanent Mission of India to the UN 2010 p 1). The LMDC also emphasized that GFCS policies should not override national sovereignty and control over the data they produce, but instead "conform" to national policies (Permanent Mission of India to the UN 2010 p 3). Such arguments emphasize, as does the capabilities approach, that justice is a matter of individual nations developing "conditions or states of human enablement" to pursue their own priorities (Robeyns 2005) rather than relying on the benefits of technical capacities in other countries to trickle down. From such a perspective, the development of equal capabilities among nations ought to take precedence over efforts to convince disadvantaged countries to share data with nations that have higher financial and technical capabilities.

Ethical Issue 3: Who Participates? Stakeholder Deliberation, Justice, and Co-production of Climate Services

Another challenge facing proponents of climate services concerns how they are produced and who is involved in such processes. The structure of such processes is important because it dictates for and by whom the information is developed, how useful the information is likely to be for different groups, and what kinds of knowledge and information are deemed legitimate. There is growing consensus that in order for scientific knowledge to be useable for societal decision-making, it should be co-produced jointly with "users" (Cash et al. 2006), a shift that has taken hold particularly in the realm of climate information and services (e.g., Dilling and Lemos 2011; Lemos and Morehouse 2005; Lemos et al. 2012; Meadow et al. 2015). Indeed, regional and global efforts to produce climate services emphasize the need for structured, sustained interaction between "producers" and "users" of climate information. The process through which various stakeholders interact and collaboratively produce information is commonly referred to as *co-production*. The concept of co-production reflects a broader deliberative turn in environmental governance (Hegger et al. 2012) which recognizes the importance of public involvement in the processes of creating knowledge related to climate risks and adaptation. While it is possible for current perspectives regarding the imperative for co-production (i.e., the need for increased participation *and* the enhanced usability of climate services) to be complementary to achieving "just" climate services, this is not always the case. Despite the proliferation of the term co-production, facilitating equitable processes and outcomes that respect divergent stakeholder values and knowledge in practice is challenging. Indeed, co-production has been recognized as an inherently political activity (Daly and Dilling 2019; Turnhout et al. 2020).

Part of the challenge is that co-production is understood and applied differently across literature and practice (Miller and Wyborn 2018). There are several key areas in which the concept of co-production has developed: public services (e.g., Ostrom 1996; Parks et al. 1981), natural resource management (e.g., Armitage et al. 2011; Puente-Rodriguez et al. 2015), science policy (e.g., Cash et al. 2006), and science and technology studies (e.g., Jasanoff 2004). However, the theoretical foundations of these different perspectives are very different. Some perspectives of co-production are primarily concerned with the fundamental relationships between scientific knowledge and society and the embedded normative implications. In other cases, co-production is concerned with epistemic debates involved within efforts to integrate multiple knowledges. Still other approaches to co-production are more concerned with achieving particular (often pre-determined) objectives, thus serving an instrumental purpose. It is generally the latter two approaches to co-production that have dominated in the case of climate services development, and there is growing concern about the potential for co-production processes to be co-opted for purely instrumental ends (Bremer et al. 2019; Goldman, Turner, and Daly 2018).

In keeping with the tenets of deliberative justice, the goal of co-production of climate services is to better ensure that those who have a stake, or who could be potentially impacted by climate services, are able to participate in their production for the purposes of making it more usable for decision-making (Lemos et al. 2012). This is in contrast to dominant approaches to climate science in which the scope of the problems is defined primarily by scientists, often without any accountability to the various publics that may seek to use such information. In theory, co-production is intended to set up the ideal space in which deliberative processes can take place. For example, it has been argued that a key component of co-production of knowledge involves jointly defining the scope of the problem to be solved (Armitage et al. 2011; Brugnach et al. 2014).

A deliberative justice lens also draws attention to the power inequalities that manifest themselves within all knowledge production. Within the tradition of deliberative justice, legitimacy is derived from the ability of groups to come together, debate issues of societal import, and to reach decisions based on the exchange of values and ideas. It is important that all interested parties have access to such deliberations and have a willingness (and ability) to voice concerns and consider alternative perspectives. The absence of coercion is also essential to such processes, as power differentials have the potential to silence minority voices and distort outcomes (Dryzek and List 2003). While deliberative models aim to create the space in which all participants are able to participate equally, this is often not the reality within efforts to co-produce climate services (Daly and Dilling 2019). In general, climate scientists are often the ones convening the co-production process, and within formal policy-making processes they are also considered the sole authority on weather and climate. In these situations, it can be difficult for non-scientists to participate on equal footing (Steynor et al. 2016). While it has been well-recognized that there is a wealth of indigenous knowledge on the topics of weather and climate (Green and Raygorodetsky 2010), such knowledge is often discounted on the grounds that it is not systematic or scientific. This creates fundamental challenge to models of co-production that adhere to the basic tenets of deliberative approaches (Daly and Dilling 2019). The inability of non-scientists to voice their concerns on equal footing with scientists is at odds with the principles of deliberative justice, and how to remedy such inequalities of power and prestige remains a significant challenge to the co-production of climate services.

Another challenge with the deliberative approach is that, given that climate services ostensibly have the potential to benefit nearly all individuals or groups, it becomes difficult to determine who exactly should be included within efforts to facilitate co-production of climate services (Vincent et al. 2018). Under strict deliberative models, co-production of climate services would require opening these processes to all potential users. This is clearly an impractical suggestion, since it is not possible to include every single potential user. A deliberative model therefore requires some system of representation in which individuals or groups are asked to speak for others. How such individuals or groups should be selected is an inherently ethical question, one that practitioners are still grappling with and that demands transparency.

Building on the deliberative challenges of co-production, a rights-based lens can help to elucidate further problems with co-production processes. In the first sense, not all potential participants in co-production have the same opportunities or capacities to participate in these processes. Efforts to engage with small-holder farmers or livestock keepers mean that participants must sacrifice a day's work. For many, this is not a viable option since the opportunity cost is much too high. Additionally, individuals or groups who have better access to resources are more likely to be able to take advantage of or co-opt co-production processes (Parks et al. 1981). In rural areas, this may mean that elites can position themselves to benefit most from co-production processes. As another example, large, well-funded, international humanitarian organizations may be better positioned to request and participate in processes of co-production of climate services compared to governmental actors in developing countries who may already be over-burdened with heavy workloads and minimal funding.

Thus, while co-production has been treated as a cure-all to the challenges of producing just climate services, the issues raised here illustrate that it is not a silver bullet. There are many ethical dangers to using co-production instrumentally, including continued marginalization of vulnerable populations, increased inequality, and the devaluation of non-scientific ways of knowing (Goldman, Turner, and Daly 2018). Scholars have argued that co-production is a fundamentally normative activity because it speaks not only to how human societies are organized but also to how they *should* be organized (Jasanoff 2004). Yet, normative considerations are often left by the wayside in deference to prescribed policy, program, or project objectives that do not reflect the goals and desires of intended beneficiaries. Thus, the challenges of achieving truly equitable co-production remain.

Conclusion

The discussion above suggests that, given prevailing economic, social, and technical inequalities that exist at multiple levels, current approaches are insufficient to produce climate services that will improve the lives of the "most vulnerable." A focus on a climate services market may place problematic emphasis on those who have the ability to "get their money's worth" over the needs of these vulnerable populations. Data sharing without provisions for capacity building may disadvantage nations with less technical and financial capabilities or perpetuate existing inequalities of opportunity and capacity. Finally, efforts at stakeholder involvement through co-production of climate services do not automatically alleviate imbalances in power or opportunity, and in some instances, they can serve to reinforce the perceived primacy of science.

The justice-based arguments explored above highlight the need to more explicitly answer the following questions in order to steer climate services toward desired, equitable outcomes: what is the *real* goal in developing climate services? Who do we want to benefit from climate services? What other (possibly more efficient or effective) ways might there be to achieve these goals? If climate services are appropriate, what is the best way to develop such services? Who should participate in the development of these services? How should data be shared and technical capacity be built in order to achieve these goals? If proponents of climate services wish to justify investments in climate services as a means of addressing the wrongs of climate change or of building the resilience of those most vulnerable to climate change, they will need more consistent answers to these questions and more explicit attention to ensuring that efforts to develop climate services do indeed contribute to these goals.

Related Topics

21. Climate Modeling
53. Disaster Response
56. Cost-Benefit Analysis
57. Risk Assessment

Works Cited

Adams, P., Eitland, E., Hewistson, B., Vaughan, C., Wilby, R., & Zebiak, S. (2015). Toward an ethical framework for climate services: A white paper of the climate services partnership working group on climate services ethics. Climate Services Partnership. https://hdl.handle.net/10568/68833

Agarwal, A., & Narain, S. (1990). *Global Warming in an Unequal World: A Case for Environmental Colonialism.* New Delhi: Centre of Science and the Environment.

Armitage, D., Berkes, F., Dale, A., Kocho-Schellenberg, E., & Patton, E. (2011). Co-management and the co-production of knowledge: Learning to adapt in Canada's Arctic. *Global Environmental Change*, 21(3), 995–1004. http://doi.org/10.1016/j.gloenvcha.2011.04.006

Baer, P. (2006). *Adaptation: Who Pays Whom? Fairness in Adaptation to Climate Change.* Ed. W. Adger, J. Paavola, S. Huq, and M. Mace. Cambridge: MIT Press, 131–153.

Bailey, R. (2013). *Available at: Managing Famine Risk Linking Early Warning to Early Action.* A Chatham House Report. https://www.chathamhouse.org/sites/files/chathamhouse/public/Research/Energy%2C%20 Environment%20and%20Development/0413r_earlywarnings.pdf

Braman, L. M., Suarez, P., & van Aalst, M. K. (2010). Climate change adaptation: Integrating climate science into humanitarian work. *International Review of the Red Cross*, 92(879), 693–712. doi:10.1017/ S1816383110000561

Brasseur, G. P., & Gallardo, L. (2016). Climate services: Lessons learned and future prospects. *Earth's Future*, 4, 79–89. doi:10.1002/2015EF000338

Bozeman, B., & Sarewitz, D. (2005). Public values and public failure in US science policy. *Science and Public Policy*, 32(2), 119–136.

Bremer, S., Wardekker, A., Dessai, S., Sobolowski, S., Slaattelid, R., & van der Sluijs, J. (2019). Toward a multi-faceted conception of co-production of climate services. *Climate Services*, 13, 42–50.

Broad, K., Pfaff, A. S. P., & Glantz, M. H. (2002). Effective and equitable dissemination of seasonal-to-inter-annual climate forecasts: Policy implications from the Peruvian fishery during El Niño 1997–98. *Climatic Change*, 54(4), 415–438. http://doi.org/10.1023/A:1016164706290

Brugnach, M., Craps, M., & Dewulf, A. (2014). Including indigenous peoples in climate change miti-gation: Addressing issues of scale, knowledge and power. *Climatic Change*. http://doi.org/10.1007/ s10584-014-1280-3

Buizer, J., Jacobs, K., & Cash, D. (2010). Making short-term climate forecasts useful: Linking science and action. *Proceedings of the National Academy of Sciences of the United States of America.* www.pnas.org/cgi/ doi/10.1073/pna s.0900518107.

Cane, M. A., Eshel, G., & Buckland, R. W. (1994). Forecasting Zimbabwean maize yield using eastern equatorial Pacific sea surface temperature. *Nature*, 370(6486), 204–205. http://doi.org/10.1038/370204a0

Caney, S. (2010). Climate change, human rights, and moral thresholds. In S. M. Gardiner, D. Jamieson and H. Shue (Eds.), *Climate Ethics Essential Readings.* New York: Oxford University Press, 163–177.

Carr, E. R., & Thompson, M. C. (2014). Gender and climate change adaptation in Agrarian settings: Cur-rent thinking, new directions, and research frontiers. *Geography Compass*, 8(3), 182–197.

Cash, D. W. (2006). Countering the loading-dock approach to linking science and decision making: Com-parative analysis of El Nino/Southern Oscillation (ENSO) forecasting systems. *Science, Technology & Human Values*, 31(4), 465–494. http://doi.org/10.1177/0162243906287547

Changnon, S., Lamb, P., & Hubbard, K. (1990). Regional climate centers: New institutions for climate services and climate-impact research. *Bulletin of the American Meteorological Society*, 71, 527–537, doi: 10.1175/1520-0477(1990)071<0527:RCCNIF>2.0.CO;2.

Coughlan de Perez, E., van den Hurk, B., van Aalst, M. K., Jongman, B., Klose, T., & Suarez, P. (2014). Forecast-based financing: An approach for catalyzing humanitarian action based on extreme weather and climate forecasts. *Natural Hazards and Earth System Sciences Discussions*, 2(5), 3193–3218. doi:10.5194/ nhessd-2-3193-2014

Daly, M. (2016). *Co-production and the Politics of Usable Knowledge for Climate Adaptation in Tanzania* (Doctoral Dissertation). University of Colorado, Boulder, USA.

Daly, M., & Dilling, L. (2019). The politics of "usable" knowledge: Examining the development of climate services in Tanzania. *Climatic Change*, 157(1), 61–80.

Dilling, L., & Lemos, M. C. (2011). Creating usable science: Opportunities and constraints for climate knowledge use and their implications for science policy. *Global Environmental Change*, 21(2), 680–689. doi:10.1016/j.gloenvcha.2010.11.006

Douglas, H. (2009). *Science, Policy, and the Value-Free Ideal.* Pittsburgh: Pittsburgh University Press.

Dryzek, J. S., & List, C. (2003). Social choice theory and deliberative democracy: A reconciliation. *British Journal of Political Science*, 33, 1–28.

European Commission (2015). A European research and innovation roadmap for climate services. Luxembourg. http://ec.europa.eu/research/index.cfm?pg=events&eventcode=552E851C-E1C6-AFE7-C9A99A92D4104F7

Gerlak, A. K., Mason, S. J., Daly, M., Liverman, D., Guido, Z., Soares, M. B., Vaughan, C., Knudson, C., Greene, C., Buizer, J., & Jacobs, K. (2020). The Gnat and the bull do climate outlook forums make a difference? *Bulletin of the American Meteorological Society*, 101(6), E771–E784.

Goldman, M. J., Turner, M. D., & Daly, M. (2018). A critical political ecology of human dimensions of climate change: Epistemology, ontology, and ethics. *Wiley Interdisciplinary Reviews: Climate Change*, 9(4), e526.

Government Office for Science. (2012). *Reducing Risks of Future Disasters, Priorities for Decision Makers, Final Project Report. The Government Office for Science.* London, pp. 1–139.

Green, D., & Raygorodetsky, G. (2010). Indigenous knowledge of a changing climate. *Climatic Change*, 100(2010), 239–242.

Haile, M. (2005). Weather patterns, food security and humanitarian response in sub-Saharan Africa. *Philosophical Transactions of the Royal Society B: Biological Sciences*, 360(1463), 2169–2182. http://doi.org/10.1098/rstb.2005.1746

Hansen, J. W. (2002). Realizing the potential benefits of climate prediction to agriculture: Issues, approaches, challenges. *Agricultural Systems*, 74(3), 309–330. doi:10.1016/S0308–521X(02)00043-4

Hegger, D., Lamers, M., Van Zeijl-Rozema, A., & Dieperink, C. (2012). Conceptualising joint knowledge production in regional climate change adaptation projects: Success conditions and levers for action. *Environmental Science and Policy*, 18, 52–65. http://doi.org/10.1016/j.envsci.2012.01.002

Hewitt, C., Mason, S., & Walland, D. (2012). The global framework for climate services. *Nature Publishing Group*, 2(12), 831–832. http://doi.org/10.1038/nclimate1745

IPCC (2014). *Climate Change 2014: Impacts, Adaptation, and Vulnerability. Part A: Global and Sectoral Aspects. Contribution of Working Group II to the Fifth Assessment Report of the Intergovernmental Panel on Climate Change.* Ed. C. B. Field, V. R. Barros, D. J. Dokken, K. J. Mach, M. D. Mastrandrea, T. E. Bilir, M. Chatterjee, K. L. Ebi, Y. O. Estrada, R. C. Genova, B. Girma, E. S. Kissel, A. N. Levy, S. MacCracken, P. R. Mastrandrea, and L. L. White. Cambridge, UK and New York: Cambridge University Press, 1132 pp.

Jasanoff, S. (2004). *States of Knowledge: Science, Power, and Political Culture.* Routledge.

Landis, R. (2001). The road to Resolution 40 and beyond: Evolution of international atmospheric data exchange. *Case Study Presented to the 2001 AMS Summer Policy Colloquium.* www.ametsoc.org/atmospolicy/documents/PartI-DataExchange.pdf.

Lemos, M. C., & Dilling, L. (2007). Equity in forecasting climate: Can science save the world's poor? *Science and Public Policy*, 34(2), 109–116. http://doi.org/10.3152/030234207X190964

Lemos, M. C., Finan, T. J., Fox, R. W., Nelson, D. R., & Tucker, J. (2002). The use of seasonal climate forecasting in policymaking: Lessons from Northeast Brazil. *Climatic Change*, 55(4), 479–507.

Lemos, M. C., Kirchhoff, C. J., & Ramprasad, V. (2012). Narrowing the climate information usability gap. *Nature Climate Change*, 2(11), 789–794. http://doi.org/10.1038/nclimate1614

Lemos, M. C., & Morehouse, B. J. (2005). Co-production of science and policy in integrated climate assessments. *Global Environmental Change*, 15, 57–68.

Lourenço, T. C., Swart, R., Goosen, H., & Street, R. (2015). The rise of demand-driven climate services. *Nature Climate Change*, 6(1), 13–14. http://doi.org/10.1038/nclimate2836

Lyons, D. (1994). *Rights, Welfare, and Mill's Moral Theory.* New York: Oxford University Press.

Mahony, M. (2013). The predictive state: Science, territory and the future of the Indian climate. *Social Studies of Science*, 0(0), 1–25.

Meadow, A. M., Ferguson, D. B., Guido, Z., Horangic, A., Owen, G., & Wall, T. (2015). Moving toward the deliberate coproduction of climate science knowledge. *Weather, Climate, and Society*, 7(2), 179–191. http://doi.org/10.1175/WCAS-D-14-00050.1

Miller, C. A. (2004). Resisting empire: Globalism, relocalization, and the politics of knowledge. In S. Jasanoff and M. Long-Martello (Eds.), *Earthly Politics: Local and Global in Environmental Governance*. Cambridge: MIT Press, 81–102.

Miller, C. A., & Wyborn, C. (2018). Co-production in global sustainability: Histories and theories. *Environmental Science & Policy*. https://doi.org/10.1016/j.envsci.2018.01.016

Morss, R., & Hooke, W. (2005). The outlook for U.S. Meteorological Research in a commercializing world: Fair early, but clouds moving in? *Bulletin of the American Meteorological Society*, 86(7), 921–936.

Nussbaum, M. (2011). *The Central Capabilities. Creating Capabilities: The Human Development Approach*. Cambridge: Belknap Press of Harvard University.

Ostrom, E. (1996). Crossing the great divide: Coproduction, synergy, and development. *World Development* 23(6), 1073–1087.

Parks, R. B., Baker, P. C., Kiser, L., & Oakerson, R. (1981). Consumers as coproducers of public services: Some economic and institutional considerations. *Policy Studies Journal*, 9(7), 1001–1011.

Permanent Mission of India to the UN. (2010). No. GEN/PMI/GFCS/2010.

Puente-Rodríguez, D., van Slobbe, E., Al, I. A. C., & Lindenbergh, D. E. D. (2016). Knowledge co-production in practice: enabling environmental management systems for ports through participatory research in the Dutch Wadden Sea. *Environmental Science and Policy*, 55(3), 456–466.

Rawls, J. A. (1971). *A Theory of Justice*. Cambridge, MA: Harvard University Press.

Reich, M. R. (2000). The global drug gap. *Science*, 287(5460), 1979–1981. http://science.sciencemag.org/content/287/5460/1979.short

Robeyns, I. (2005). The capability approach: A theoretical survey. *Journal of Human Development*, 6(1), 93–117.

Roncoli, M. C., Ingram, K. T., & Kirshen, P. H. (2002). Opportunities and constraints for farmers of west Africa to use seasonal precipitation forecasts with Burkina Faso as a case study. *Agricultural Systems*, 74(3), 331–349.

Saarikivi, P., Soderman, D., & Newman, H. (2000). Free information exchange and the future of European Meteorology: A private sector perspective. *Bulletin of the American Meteorological Society*, 81(4), 831–836.

Sagoff, Mark (2004). *Price, Principle, and the Environment*. United Kingdom: Cambridge University Press.

Sarewitz, D., & Pielke, R. A., Jr. (2007). The neglected heart of science policy: Reconciling supply of and demand for science. *Environmental Science & Policy*. http://doi.org/10.1016/, DanaInfo=dx.doi.org+j.envsci.2006.10.001

Suarez, P., & Tall, A. (2010). Towards forecast-based humanitarian decisions: Climate science to get from early warning to early action. *Paper Commissioned by the Humanitarian Futures Programme*.

Sunstein, C. R. (2005). Cost-benefit analysis and the environment. *Ethics*, 115(2), 351–385.

Steynor, A., Padgham, J., Jack, C., Hewitson, B., & Lennard, C. (2016). Co-exploratory climate risk workshops: Experiences from urban Africa. *Climate Risk Management*, 13(2016), 95–102.

Tall, A., Mason, S. J., van Aalst, M. K., Suarez, P., Ait-Chellouche, Y., Diallo, A. A., & Braman, L. (2012). Using seasonal climate forecasts to guide disaster management: The Red Cross experience during the 2008 West Africa floods. *International Journal of Geophysics*, 2012(2), 1–12. doi:10.1016/S0308–521X(02)00044-6

Turnhout, E., Metze, T., Wyborn, C., Klenk, N., & Louder, E. (2020). The politics of co-production: Participation, power, and transformation. *Current Opinion in Environmental Sustainability*, 42, 15–21.

Vaughan, C., & Dessai, S. (2014). Climate services for society: Origins, institutional arrangements, and design elements for an evaluation framework. *Wiley Interdisciplinary Reviews: Climate Change*, 5(5), 587–603. doi:10.1002/wcc.29

Venema, V. (n.d.). Free our climate data: From Geneva to Paris. http://variable-variability.blogspot.com/2015/06/Sharing-climate-data-global-framework-climate-services.html

Vincent, K., Daly, M., Scannell, C., & Leathes, B. (2018). What can climate services learn from theory and practice of co-production? *Climate Services*, 12, 48–58.

Visbeck, M. (2007). From climate assessment to climate services. *Nature Geoscience*, 1(1), 2–3. http://doi.org/10.1038/ngeo.2007.55

Webber, S., & Donner, S. (2016). Climate service warnings: Cautions about commercializing climate science for adaptation in the developing world. *WIREs Climate Change*. doi: 10.1002/wcc.424.

Wilby, R. (2016). Climate service sector needs robust standards. *SciDevNet*. http://www.scidev.net/global/climate-change/opinion/climate-service-sector-standards.html

World Meteorological Organization (2011). *Climate knowledge for action: A global framework for climate services – empowering the most vulnerable*. http://library.wmo.int/pmb_ged/wmo_1065_en.pdf.

WMO. (2015). *Valuing weather and climate: Economic assessment of meteorological and hydrological services*. Geneva, Switzerland. https://www.gfdrr.org/sites/gfdrr/files/publication/Valuing-Weather-and-Climate-Economic-Assessment-of-Meteorological-and-Hydrological-Services.pdf

WMO (2016). WMO Data Policy (Resolutions 40, 60) and update on CBS discussions. Geneva. http://www.wmo.int/pages/prog/sat/meetings/documents/CM-13_Doc_03-01_WMO-Data-Policies.pdf

Ziervogel, G. (2004). Targeting seasonal climate forecasts for integration into household level decisions: the case of smallholder farmers in Lesotho. *The Geographical Journal*, 170(1), 6–21. http://doi.org/10.1111/j.0016-7398.2004.05002.x

Zillman, J. W. (1997). Atmospheric science and public policy. *Science*, 276, 1084–1086.

53

DISASTER RESPONSE

Bruce Jennings

Introduction

My own work on public health emergency preparedness and response began against the backdrop of the flooding of New Orleans in the aftermath of Hurricane Katrina in 2005. Years later, the research and writing of this chapter has taken place against the backdrop of widespread infectious disease threats involving influenza, Ebola, Zika, and most recently the COVID-19 pandemic. COVID-19 is reminding us that achieving and sustaining public health protection, economic prosperity, and social justice require public trust and concerted public action. These have been the stakes in the COVID-19 pandemic since 2020, and they will be the stakes in the health, economic, and moral threats that loom with climate change in the coming decades.

In this chapter, I turn attention from the immediate pain and drama of coping with disaster and focus instead on the organizational challenges and ethical reflection required by planning and preparing for disaster. The world was not well prepared for the extraordinarily infectious and lethal virus that has transformed the world since 2020. Will we be better prepared for future pandemics to come? Will our response be Keystone Kops or well-informed and judicious leadership inspiring trust?

To weather future pandemics and to balance medical risks against economic disruptions and loss, we need to sign a new social contract with public health. That involves disaster preparedness planning as an ongoing—not merely a periodic—activity. How well we plan ahead of time affects how well we can respond later. And if we do not plan together it is likely that we will not be able to sustain justice and mutual aid when we must respond in the face of disaster. I suggest that we re-think disaster planning so that it becomes a civic practice. If we do so, then a disaster preparedness planning process will come to be seen as an expression of the entire community about the value of the lives and health of its members. A disaster plan can be an agreement to be entered into by all that establishes commitments of responsibility for each, if it arises from a process of inclusive deliberation, solidarity, and mutual respect.

The Spectrum of Disaster

Disasters happen in many ways. We might think of this along a spectrum from what I shall call the "eruptive" to the "emergent." Some disasters erupt suddenly, with warning signs coming over a few days and local or regional devastation wrought in a matter of hours. Some emerge slowly, with warning signs, often unrecognized or misinterpreted, coming over years or decades

DOI: 10.4324/9781315768090-63

and cumulative disruption that festers into acute devastation in various places from time to time. Examples of eruptive disasters would be avalanches, floods, violent windstorms, earthquakes, and volcanic eruptions. Emergent disasters have a wider range of temporal duration and biophysical scale. They include pandemic disease, long-lasting drought, topsoil erosion, freshwater depletion, deforestation, biodiversity loss, and global climate change affecting both atmospheric temperature and oceanic acidity, temperature gradients, and currents (Gardiner 2011; Kolbert 2014).

Societies and individuals respond to the prospect and actual occurrence of disaster in many ways. In this chapter when I refer to "disaster response" and "disaster ethics" I intend these terms to encompass three distinct moments: preparedness and planning for disaster before it happens, coping with the immediate circumstances of disaster in progress, and recovery in the extended aftermath of disaster. In addition to fixing terms, three general points are important to bear in mind in any discussion of disaster.

First, disasters tend to be associated with environmental degradation and poor ecosystemic functioning. Sudden disruptive events harm the natural environment as well as the social and the built environments within which human beings live. And the state of environmental systems at the time of such disasters plays a significant role in determining the extent of the disruption suffered by societies and the options open for effective recovery from a disaster. The still widespread view that sees human beings as separate from and in control of nature—and that sees nature as simply raw material for human use—is problematic when thinking intelligently about virtually any topic, but it is particularly misleading for the subject of disaster response (Jennings 2016).

Second, it is important to recognize that disasters can take place at various time scales. A slowly developing and progressing situation that poses incremental and cumulative challenges to social and natural systems may be more devastating and significant in its ultimate effects than a sudden emergency. The prolonged drought in the American Great Plains during the 1930s called the Dust Bowl was a decade-long disaster made possible by an earlier period of inadvisable agricultural practices (Worster 2004). The recognition of emergent, relatively slow-motion disasters—disasters as processes and not simply as delimited events—should be considered in any environmental ethics of disaster response (Paton and Johnston 2006).

Third, both eruptive and emergent disasters involve the interaction of natural and cultural factors (Hoffman and Oliver-Smith 2002). It is tempting to call an earthquake a "natural disaster," and something like warfare a "cultural disaster" (Kapur and Smith 2011). Yet even in the case of warfare, the extent of damage, loss of life, and the requirements of recovery after the disruptive event(s) are all largely determined by the prior ecological resilience of the affected areas. The magnitude seven earthquake that struck Haiti in 2010 was made more devastating by a century of systematic neocolonial economic and ecological exploitation. This provides one particularly striking example of the nature-culture interaction at work in disasters (Macintyre 2010). The response to flooding in New Orleans after Hurricane Katrina in 2005 provides another (Hartman and Squires 2006).

In sum, disasters represent a particular pattern of interaction between nature and culture—that is, between biophysical tolerances and ecosystem functions, on the one hand, and human social institutions, behavioral patterns, and purposive actions and choices, on the other (Honig 2009; Scarry 2011).

That which human beings cannot cause, foresee, or prevent is usually thought to be outside the boundaries of ethical responsibility. In the contemporary world, the number of disaster events that fall outside the purview of ethics in this way—that are genuinely "natural disasters"—is vanishingly small. Growing scientific evidence concerning the anthropogenic impact on the functioning of planetary biophysical systems suggests that ethical responsibility for severe destabilizing and disruptive environmental events can be assigned to human activity (Hamilton, Bonneuil, and Germenne 2015). Increasingly, what I have called eruptive disasters are in fact related to emergent disasters working in the background, such as climate change. As greenhouse gas (GHG) emissions lead to global temperature increase, thereby altering earth systems and cycles, climate change will generate disasters and instigate severe challenges to public health—violent storms, flooding in some areas, drought in others, fatal heat waves, the contamination and depletion of fresh water supplies, the spread of zoonotic disease, aggravation of chronic conditions such as allergies and asthma, and large-scale human migration and its epidemic side effects (Melillo, Richmond, and Yohe 2014; National Institute of Environmental Health Sciences 2015; WHO 2016).

The severity of the social and human impact of disasters is due to multiple factors, including poverty, inadequate urban design construction standards, and poor transportation, sanitation, and public health infrastructure (Cooper and Block 2006; Daniels, Kettl, and Kunreuther 2006). Disasters strike especially hard against children and displaced persons, and others who have special needs or vulnerabilities like the elderly, the mentally ill, and those with physical or cognitive impairments (United Nations 2011).

Similarly, disasters often disrupt natural ecosystems and nonhuman beings in many of the same ways that they affect social systems and human beings. The same disruptions that trouble human communities may trigger significant change in the behavior of many species. For entire ecosystems, new patterns of organization often emerge during and after a disaster, such as significant changes in predator-prey balances and the influx of non-native plant species better suited to thrive under the post-disaster conditions. If these changes are not permanent or overwhelmingly widespread, ecosystems may reestablish themselves in sustainable and healthy ways, a capability of ecosystems known as resilience (Walker and Salt 2006). The response of forest ecosystems to fire is one instructive example of ecosystemic resilience that shows its multifaceted and complex pattern. Alternatively, they may be drastically degraded and simplified, as is the case with massive deforestation, topsoil erosion, and transformation of a biodiverse ecosystem into a desert due to prolonged drought. As with their counterparts in human social systems, the response and recovery of ecosystems in the face of significant stress and functional disruption are dynamic and complex processes.

The Ethical Situation of Disaster Response

Ethical dilemmas abound in disaster response activities and decisions. Complex and acute dilemmas arise, as one would expect, in the immediate aftermath of a disaster event, when resources are scarce, and responders are under stress. Consider the following situation:

> Eight weeks after the earthquake in Haiti, humanitarian health care practitioners were working at a field hospital about two hours outside of the capital Port-au-Prince. One late afternoon, an acutely sick baby was brought in with respiratory distress. ... The child needed to be transferred to another facility in Port-au-Prince. The drive would take about two or two-and-a-half hours if the roads were clear and there was not too much traffic, so the people who accompanied the baby would not be able to return until the next day. The stretch of road to Port-au-Prince was also a dangerous kidnapping area. Because of the time of day, using a vehicle to transport the baby meant leaving the team with only one vehicle overnight. This was against unit protocol because one vehicle was not enough to evacuate the team if that was needed due to an aftershock or for security reasons.
>
> *(Schwartz et al. 2014: 45)*

Emergency care of an infant, difficult but manageable in normal times, becomes an ethical imperative challenged and blocked at multiple points by a surrounding context of natural and human-made dysfunctions. Responders are faced with choosing between an existential response to compelling need and a disciplined, coordinated form of behavior that may serve the greatest good for the greatest number, but at the sacrifice of individuals. How important is an individual life in the context of disaster? Advance disaster planning is ethically important because it should serve to minimize tragic choices such as this one during the emergency response phase. Post-disaster reconstruction should aim to heal community, support grieving, promote reconciliation, and lessen the probability that similar ethical tragedies will happen again when the next disaster strikes.

Although disasters are frequent and significant, they are difficult to define. There is no uniform system for classifying them, and there is no standard language for describing them. The Centre for Research on the Epidemiology of Disasters in Brussels (CRED 2011) categorizes disasters into one of three groups: "natural disasters" (e.g., floods, earthquakes, mudslides), "technological disasters" (e.g., industrial accidents, transport accidents), and "complex emergencies" (e.g., combinations of natural and technological disasters). Disasters can also be conflict-related (e.g., war or terrorism).

From the point of view of the spectrum of disasters discussed above, one can question the conceptual underpinnings of this type of classification scheme. Virtually all disasters are actually "complex emergencies" in the CRED terminology. Disasters upset and unsettle assumptions about behavior and society that ordinarily do not have to be questioned. They startle us into new ways of seeing: we observe things previously unnoticed, such as the fragility of available social capital and indigenous support systems in certain communities, and we interpret the meaning and significance of experience in a new way. As disasters change our understanding of present reality—both social and environmental reality—around us, they can also alter our sense of the past and especially our expectations concerning the future. Disasters are therefore not simply events or states of affairs to be observed from the outside, from a third-person perspective. Both the structure of their lived experience from a first-person perspective and their effect on the patterns and meanings of interpersonal relationships in a society, the second-person perspective, are essential to an understanding of disaster ethics (Darwall 2006).

Disasters affect people outwardly in terms of their bodies, health, and choices; yet living through disaster also provides—or more accurately, imposes—new terms for self-understanding. Powerful disaster novels, like Albert Camus' *The Plague* and José Saramago's *Blindness* (which describes the collapse of social order following an epidemic of a mysterious infectious disease that causes sudden and total loss of sight), explore this. The incorporation of this way of seeing, this viewing of the world and ourselves through the lens of disaster, may be ethically dangerous, but it also has a potential for a growth of moral awareness. In *The Plague*, we find the following exchange between Raymond Rambert, who secretly plans to leave the quarantined city and is struggling to justify this to himself, and Dr. Bernard Rieux, a physician trying to convince Rambert to stay and help by appealing to the notion of a good larger than the needs of any one person:

> "Ah, I see now!" Rambert exclaimed. "You'll soon be talking about the interests of the general public. But public welfare is merely the sum total of the private welfares of each of us."
> The doctor seemed abruptly to come out of a dream.
> "Oh, come!" he said. "There's that, but there's much more to it than that..."
>
> *(Camus 1991: 88)*

Later in the novel, Camus adds: "What's true of all the evils in the world is true of the plague as well. It helps men to rise above themselves" (Camus 1991: 125).

Norms may either break down or reassert themselves in surprising ways—both fear and selfishness may increase during an emergency, but so may altruism, compassion, and solidarity (Solnit 2009). In an emergency, new patterns of organization often emerge, and then give way to the reconstruction of the preexisting social order— "a return to normal." But in some cases, disaster response and recovery can lead to significant social transformations—for good or ill. Communities rebuild in improved, more resilient and more just ways, or communities become demoralized, depopulated, and stagnant. Disaster response is history accelerated, so to speak.

Ways of Understanding the Ethics of Disaster

There is a large and growing literature on ethical aspects of disaster response—preparing, coping, and recovering. (See Further Reading below.) Many of these studies concentrate on ideal ethical principles or rules and then apply those general rules to specific policies or decisions in disaster planning and response. Ethical standards and guidelines of this kind are important and can be useful to planners and to emergency responders faced with difficult allocation decisions, such as the triage and rationing of mechanical ventilators or the use of operating rooms. These studies help us understand what might be called the *ethics of coping with disaster.* Less commonly discussed is what might be called the *ethics of preparing for disaster.* Studying the ethics of preparing for disaster involves paying attention, not so much to the rules that should govern coping with emergencies, but rather to the values revealed by the manner in which a society faces its own vulnerability and confronts the prospect of disaster existentially. Disasters remind us forcefully about living with constraints and limits; they might be thought of as the shock therapy of an ecological awareness. How are ethical values such as rights and obligations, justice, respect, and dignity built into the institutions of disaster preparedness and into the self-understandings of planners and the vulnerable people they strive to protect?

These ways of confronting disaster can be called "frames," by which I mean ways of seeing, feeling, and knowing. Frames emerge from interpretation and meaning making; they converge in a cultural ethos that has been called the "social imaginary" (Taylor 2003). People partake of this imaginary and use frames sometimes explicitly, often tacitly, to make sense of what they are doing and being asked to do. In their study of ethical reasoning and decision making, Schwartz and Sharpe call attention to the importance of such framing: "Frames tell us what is important and help us establish what should be compared with what.… Framing is pervasive, inevitable, and often automatic. There is no "neutral," frame-free way to evaluate anything" (Schwartz and Sharpe 2010: 63). Such interpretive frameworks are especially important when the activity in question is intellectual and socially complex, requiring assessment of many sources of information and the consideration of many diverse interests, viewpoints, and values (Fischer 2003; Schön and Rein 1994).

The study of ethics itself involves the process of cognitive framing. Ethical analysis conceptually organizes the modes of thought and action that comprise disaster response by defining and identifying the key values at stake, by clarifying the array of choices and options available to disaster planners and responders in a given situation, and then by evaluating and prioritizing those choices. In disaster response, as in most of life, simply because something can be done, it does not follow that ethically it should be done. Moreover, even if something will most likely be effective and save lives or prevent harm—such as mandatory HIV screening and quarantining all those found to be HIV positive in restricted facilities in order to control the transmission of AIDS— doesn't mean that it should be done, either (Bayer and Healton 1989). Judgments of rightness or wrongness of given actions depend on the specific circumstances and the meaning of the situation as a whole.

The role of ethics in the planning phase before a crisis, as in the recovery phase afterward, is to define reasonably just, humane, and responsible parameters for action and decision making (Upshur 2002). In general, I believe that ethically sound and practically effective disaster response planning must define human health and safety as environmentally contextualized and located health and well-being. It should bring health into contact with other key values such as respect, dignity, liberty, and equality, and with principles of right relationship and right recognition among human beings and between human beings and the natural world. Ethical disaster response is "place-based" (attentive to local landscape and built environment) and requires broad, inclusive community participation and deliberative planning. Disasters are always destructive and disruptive, but their harm can be lessened by prior planning and by having a robust public health infrastructure already in place.

For its part, the ethics of coping with disaster is about proscribing some inherently wrong or causally harmful types of action, no matter how expedient those actions may seem or whose powerful interests they may serve. Further, the ethics of disaster response is also about positive moral learning even in dark times: it is about creating the proper kind of sensibility, motivation, and moral commitment in people.

A study by the Nuffield Council on Bioethics (2007) addressed several issues pertinent to disaster response. It argued that responders should employ the least restrictive alternative to achieve emergency health and safety goals as a general rule. Nonetheless, the use of coercion or deliberately withholding information from the public cannot be ruled out categorically.

Let's consider this point in more detail. The ethical assessment of extraordinary measures under extraordinary circumstances is one of the most difficult facets of ethics in disasters, particularly in the foggy and time-pressured situation of response activities in the immediate aftermath of a disaster event. Ethical examples from philosophy textbooks spring to life in the collapsed building, the flooded neighborhood, and the hospital emergency room. Rescue of some may have to be intentionally forgone in order to rescue many others. Individual civil rights and freedoms may have to be curtailed in order to preserve public order, prevent panic, or tamp down the spread of infectious disease. There may not be time to collect all the relevant information before making a decision or to consider a wide range of options so as to choose the least restrictive, coercive, or disrespectful one. Mistakes will be made. Atonement, restitution, reparations, and reconciliation may be needed in the aftermath of a disaster.

If rules cannot specify exactly the right thing to do under disaster circumstances, then rules can provide for fair and orderly procedures, and accountability need not be sacrificed, although in truth unfair actions often accompany chaotic conditions and must be rectified by compensation after the fact. Here are some examples. Mandatory evacuation measures or quarantine may be unavoidable and ethically justified under extreme circumstances (Fairchild, Colgrove, and Jones 2006; Gostin 2006). Withholding information from the public may be necessary in order to prevent panic and counterproductive behavior on a large scale. Normal standards of medical practice may have to be altered in response to mass casualty events.

Three ethical frameworks that represent different approaches to disaster ethics—both the ethics of preparing for disaster and the ethics of coping with disaster—will be discussed below. The first is what I call the "ethical engineering" frame. It casts disaster ethics as a matter of the application of general moral principles to specific action situations. The second is an "ecological being" frame. It focuses on various ways in which the material and living world affected by a disaster situation can be represented in normative terms, in particular the human body during an outbreak of infectious disease. The third is what I call the "civic practice" frame. While the first two of these frames tend to focus on the ethics of coping with disaster, the civic practice frame is primarily concerned with the ethics of preparedness and planning for disaster.

Ethical Engineering—Applying Principles to Practice

Disaster ethics studies norms and values that are pertinent to how disaster response should be conducted in a generic sense and in regard to specific types of hazards or emergencies that might arise. In the current literature on disaster ethics, it is common to find a kind of moral engineering grounded in positive science and in a logic of deductive reasoning (moving from general principles to specific decisions or actions). Actions and decisions are assessed in light of general ethical norms, and recommendations are made concerning educational training and procedural or institutional reforms that may lead to improved compliance with these general norms or principles in the future.

Applied to areas other than disaster ethics, this engineering frame has served practical ethics well for several decades as it has examined diverse arenas in biomedicine, health policy, and biomedical research and technology (Beauchamp and Childress 2013). However, the engineering frame has been criticized on several fronts (Arras 2017). In its utilitarian variants it is subject to the objection that it ignores individual rights and the dignity of the human person. Applying principles in a way that most effectively promotes population health and safety during an emergency may—and very often does—require the ethical subordination or disregard of particular individuals or groups. It is also resolutely anthropocentric, although it could easily accommodate principles of a more ecocentric kind. Environmental interests or values can be factored into this frame indirectly through the impact that environmental pollution or ecosystem dysfunction can have on the health and interests of human beings.

Ecological Being—Ethical Dilemmas of Embodiment

The ethical engineering frame has been the most influential; however, other frames of moral discourse are beginning to appear. Generally speaking, the more the issue under discussion involves emergency acute clinical medical care, or the logistics of social control, the more likely it is that the ethical engineering frame will be used.

As one moves away from the medicalized aspects of disaster response toward the processes of communication, cultural meaning, and group dynamics, one finds patterns of activity and practices that are themselves meaningful and value-laden for the participants involved in them. Specific decisions or choices to which general ethical principles can be applied become less pertinent than the lived experiences or "lifeworld" of human relationships and the interaction with the surrounding environment. Even in the scenario of acute medical emergency following disaster in Haiti presented earlier, these relationships are the essence of the ethical problem: physician relationships with administrators; on-site personnel interacting with policy rules and off-site authorities; the need to venture on roads washed out by flooding or vulnerable to criminal or terrorist attack. These relationships are constituted by their own internal norms and value narratives. Ethical analysis and evaluation from an external standpoint give way to ethical interpretation and critique rooted in the internal purposes and ideals of a lifeworld under stress. In short, some aspects of disaster ethics involve the application of the ethical rules of everyday life to an unusual, and unusually demanding, set of circumstances. Other aspects of disaster ethics, however, require an approach that is focused on the struggle human beings face when their moral lifeworld, settled expectations, and identities are confronted by a large-scale anomaly to which they must respond.

In their important work on disaster ethics, *The Patient as Victim and Vector*, Battin, Francis, Jacobson, and Clark (2009) question the value of the engineering frame and present what may be called an ecological ontology frame in their study of emergency situations involving infectious disease control. They argue that disaster ethics must be embedded in a complex web of historical, cultural, and emotional meanings given to the forces of nature and the biological fragility of the human body. Human beings are caught up in a biological web that contains both beneficial symbiotic and pathogenic microorganisms. Our bodies and activities are integral parts of the mobility and life cycle of these microorganisms, and hence we are what Battin and colleagues refer to as "way station" selves.

This ecological being frame identifies moral ambiguity inherent in the biology of infectious disease—namely, that individuals may be both victims to be cared for and carriers of the pathogens, or "vectors," who must be controlled. (The same may be said with nonhuman animals and even states of an ecosystem.) Care and control, concern and coercion, are two sides of the same coin in an infectious disease emergency. Ethical engineering involving the deductive application of general principles will provide little guidance in these contexts. What is needed is an interpretive framework through which to articulate what the ethical questions really are when emergency preparedness and response measures are looked at from an ecological perspective, and when human beings are seen as ecologically embodied as well as socially and relationally embedded selves.

Civic Practice—Ethical Membership and Solidarity

Another way of framing the ethics of disaster is illustrated by work I have done with John Arras (Jennings and Arras 2016). Like the ecological being frame, this approach to disaster response emphasizes an ecological perspective, but in a different way. Rather than see the issue in terms of a dialectic between culture, the biology of infectious disease, and social control of human bodies, this frame takes on board an ecological perspective through a stress on relationality and interdependence. This ethical perspective argues that the construction, support, and maintenance of resilient social and natural communities is an ethical imperative in a world increasingly prone to incidents of disaster.

Although focusing largely on including all human beings in communities within the moral ambit of disaster, this perspective is not merely instrumentally anthropocentric in its implications for environmental ethics more broadly defined. On the contrary, the resilience of the nonhuman living world is ethically valuable and important in its own right and not merely because it extrinsically affects human interests, and not merely because (some) human beings may happen to value natural resilience for its own sake. Thus insofar as the ethics of disaster response is concerned, persons not only have ethical responsibilities to support and maintain certain properties and values in social systems, they also have cognate responsibilities concerning the care and maintenance of natural systems before, during, and after an emergency.

In sum, this civic practice framing of disaster ethics permits us to see preparing for disaster as an integral part of community building—communities of inclusive social justice and communities of environmental conservation. This viewpoint stands in stark contrast to the ethical engineering view in which protection from death or harm in times of emergency is a service that individual members of a society purchase with their taxes, their cooperation, and their own personal and family efforts to achieve safety when disaster occurs. Rather than community building, then, disaster preparedness is perceived as a rational solution to a collective action problem, in which the value of communal planning and preparation is understood ultimately in terms of individual self-preservation and self-interest. The ethical engineering frame ultimately depends on the viewpoint of an impartial spectator to adjudicate this scrutiny (Zack 2009). The civic practice frame is more first-person plural, a "We" rather than a "They" orientation. It is concerned with how an effective participatory and deliberative form of democracy could be made practical in our societies and how this could be done under the sign of vulnerability to disaster. What type of motivation would lead to genuine deliberation and not simply special interest advocacy? What kinds of institutional forms would facilitate enhanced grassroots participation and discussion of key public health (and other social) issues that disaster preparedness planning brings to the fore? Disaster ethics framed as civic practice emphasizes key, but often neglected, topics, such as community capacity and resilience, community participation, civic responsibility, and public trust.

The following example—which has been at the center of the struggle against COVID-19— illustrates this point. Public health measures during pandemics or epidemics designed to control the transmission of infectious disease include so-called "social distancing" plans that call for people to remain in their own homes, closing schools, and the prohibition of mass gatherings. Such measures require individuals to forgo or temporarily suspend some ordinary civil liberties and freedoms for the sake of the public good and the health of others. The exercise of political and legal authority is involved, of course. But so is the requisite conditions of public understanding, support, and voluntary compliance since coercive enforcement of these measures on large numbers of people is both not feasible for long and socially disruptive in its own right. Therefore, in the planning phase prior to the onset of an emergency, proposed restrictions must be fully explained and justified. Indeed, if the planning and its directives are deliberative, transparent, and publicly justified, emergency preparedness can turn into a kind of cultural growth providing a sense of trust and community while also enhancing health and safety. However that may be, this pattern of disaster response to infectious disease events suggests an important theme applicable in other types of situation as well—namely, that the ethical acceptability of an emergency plan is a function both of its substantive content (what it tells people to do and what the consequences of that are) and of the process through which that content is discussed, formulated, argued about, and ultimately agreed to.

Rather like the ecological being frame, the civic practice frame is based on the notion that ethics involves an interpretation of what disaster preparation or adaptation require of institutions and individuals and an interpretation of the meaning of this activity for those engaged in it and affected by it. As a civic practice, disaster ethics is not a series of discrete decisions, but is an ongoing narrative or drama unfolding, often surprisingly, over time.

Another key element of the civic practice frame is its emphasis on democratic participation and pragmatic social learning in the disaster preparedness process. This is one way in which preparing for disaster can help not only to sustain community resilience, but also to strengthen it (Benjamin 2006; Schafer et al. 2008). Disasters have few silver linings. But one is the surprising way they can bring altruism and solidarity to the fore, even in societies whose official ideology purports to find such traits unlikely, especially at such times (Solnit 2009). Another is the possibility of some lasting ethical growth and transformation as a result of living through and enduring a disaster. If such growth in communal and moral awareness is often ephemeral and fades as normal life returns, perhaps it is because we make few efforts to institutionalize this sense of community and thereby squander its collective possibilities.

Let me develop this point by returning to the contrast suggested earlier between disaster response as a collective action problem (the engineering frame) and as a community-building one (civic practice). Now consider the parallel contrast between consumerism and citizenship—the consumer is oriented toward private interest and the citizen is oriented toward the common good. From a civic practice perspective disaster response is a mutual endeavor aiming at the common good of all members of the threatened moral community, which may include not only human individuals and social systems but also nonhuman organisms and ecosystems (Goodpaster 1978). As a consumerist activity, by contrast, disaster planning and response is primarily consumption by individuals of goods and services—police and fire protection, emergency rescue services, medical and nursing expertise, the expertise of trained planners and managers, and political leadership. These goods and services are more dramatic and urgent than commodities procured during normal times, to be sure, but they are still fundamentally a consumer purchase, nonetheless.

Disaster planning as civic practice is activity that ordinary citizens ought to engage in and support out of a sense of membership and solidarity, or mutual support, care, and concern, rather than solely out of individualistic self-interest. The sense of membership allows us to see that everyone is a part of a community of common need and vulnerability. Solidarity perceives that we have a responsibility for others and for the health of our shared community as a whole. Disaster preparation recognizes this moral community and becomes a vital part of it in our risky world. Disaster planning as consumerism constructs social insurance more than it grows out of moral community.

Individuals do—and should—purchase it (as taxpayers) for their own protection. This is fine as far as it goes and is far superior ethically to a laissez-faire or every-man-for-himself approach. But it still falls short of the ethical potential that disaster response and planning activities hold. As Camus' Dr. Rieux said, "There's that, but there's much more to it than that."

In my view, it is overly confining to see disaster response merely as a commodity to be exchanged between a consumer with an interest and a provider with the expertise to fulfill that interest. A disaster plan is not the property of those who create it or those who purchase it. It is not simply used by the people who benefit from it. It is an expression of the entire community about the value of the lives and health of its members. It establishes reciprocal obligations of respect, care, and mutual aid for each and all (Vawter et al. 2010; 2011; Zack 2009). Hence from the civic point of view, it is not only appropriate but vital to emphasize broad, inclusive participation and community engagement because disaster response inherently is an exercise of power and social control. Disasters—and certain types of response to them—can threaten democratic values, to be sure, but more openly conducted in keeping with the values central to the civic practice frame, disaster response planning both depends on and helps to strengthen strong democratic communities (Garrett et al. 2009; 2011).

Implementing Disaster Preparedness Planning as a Civic Practice

Disaster planning is a practical task, and people will not take advantage of disaster planning engagement opportunities in their community, assuming that planning elites provide such opportunities, unless they find the activities and issues meaningful in their own lives, and unless they believe that their involvement will actually make a difference. Disaster situations from both cultural and natural causes are now at the forefront of public awareness, but attention to these matters ebbs and flows, and it is unclear if fear and concern can be transformed into constructive community response informed by membership and solidarity. Over the last few years, during the extraordinary global refugee crisis and the growing political power of authoritarian, ethnically nationalistic parties in many countries, it is difficult to be optimistic on this score. Nonetheless, disaster planning continues apace around the world and will gain momentum as the effects of climate change become more salient. So it remains important to pay critical attention to the frames of ethical discourse that surround those activities.

Regarding the civic practice frame, I close by identifying some benchmarks for implementing its insights and tenets in ongoing planning practice in local communities. Certainly, I cannot develop a full specification of how disaster civic practice should be conducted ethically in this chapter (see Jennings and Arras 2016). Instead, I offer some points that are ethically salient and practically achievable. These benchmarks are not currently the standard of practice in disaster preparedness planning, which tends to reflect a public safety and law enforcement orientation more than a public health orientation. But they are not utopian. Ethically, if not politically, they are low hanging fruit.

- Emergency plans and mitigation activities should have clearly defined, widely understood, and realistic goals that are informed by a process of community consensus building. (By "mitigation activities" I mean the policies, protocols, and practices specified in a disaster preparedness plan.)
- These goals should be pursued and implemented as effectively as possible, given existing resources and information.
- Officials and planners should attempt to identify in advance the known or potential burdens of the mitigation activity and identify the segments of the population upon whom those burdens are likely to fall.

- Planners and policymakers should attempt to minimize the burdens of the mitigation activity. They should avoid imposing undue burden on particular groups that may be marginalized or stigmatized. Instead, approaches that reconcile equity with efficiency and effectiveness should be used.
- Public trust is key to the success of any emergency planning, and deliberative public participation is one important key to securing and sustaining public trust. If the reasons behind proposed mitigation activities are explained clearly and are defensible both scientifically and ethically, the likelihood of public acceptance and voluntary compliance with these provisions will increase.
- Planning processes should be transparent and accessible. Multiple venues for deliberative citizen participation should be provided for.
- Meaningful two-way communication, bottom-up communication as well as top-down communication, is essential.

Conclusion

We should by now understand that extractive and exploitative social and ecological relationships do not enrich but endanger us. Making common cause with ecosystemic health and resiliency is the key to protecting and strengthening our own health and resilience, to say nothing of our own flourishing and justice. Etymologically the word "disaster" means being under a bad star or an evil disarray in the heavens. It also means being amidst an evil disarray on earth. Possibly, however, disaster is also a good sign, a sign that can teach us to recognize the common vulnerability and fate of human life and all life. Such commonality is the ethical basis for human rights, and it can be the basis for a new kind of ecological solidarity, too.

Related Topics

23. Climate Justice and Equity
33. Urban Sustainability
52. Predication and Forecasting
54. Command and Control
58. Precautionary Principles
59. Adaptive Management
62. Community Participation
63. Environmental Justice

References

Arras, J. D. (2017). *Methods in Bioethics: The Way We Reason Now*. New York: Oxford University Press, 45–74.

Battin, M. P., L. P. Francis, J. A. Jacobson, and C. B. Smith. (2009). *The Patient as Victim and Vector: Ethics and Infectious Disease*. New York: Oxford University Press.

Bayer, R., and C. Healton. (1989). "Controlling AIDS in Cuba. The Logic of Quarantine," *New England Journal of Medicine* 320, 15 (April 13): 1022–1024.

Beauchamp, T., and J. Childress. (2013). *Principles of Biomedical Ethics*. 7th ed. New York: Oxford University Press.

Benjamin, C. G. (2006). "Putting the Public in Public Health: New Approaches," *Health Affairs* 25: 1040–1043.

Camus, A. (1991). *The Plague*, translated by Stuart Gilbert. New York: Vintage Books.

Cooper, C., and R. Block. (2006). *Disaster: Hurricane Katrina and the Failure of Homeland Security*. New York: Times Books.

CRED. (2011). Killer Year Caps Deadly Decade—Reducing Disaster Impact Is "critical" Says Top UN Disaster Official. Available at http://www.unisdr.org/archive/17613

Daniels, R. J., D. F. Kettl, and H. Kunreuther, eds. (2006). *On Risk and Disaster: Lessons from Hurricane Katrina*. Philadelphia: University of Pennsylvania Press.

Darwall, S. (2006). *The Second Person Standpoint: Morality, Respect, and Accountability*. Cambridge, MA: Harvard University Press.

Fairchild, A. L., J. Colgrove, and M. M. Jones. (2006). "The Challenge of Mandatory Evacuation: Providing for and Deciding for Others." *Health Affairs* 25(4): 958–967.

Fischer, F. (2003). *Reframing Public Policy: Discursive Politics and Deliberative Practices*. New York: Oxford University Press.

Gardiner, S. M. (2011). *The Perfect Moral Storm: The Ethical Tragedy of Climate Change*. New York: Oxford University Press.

Garrett J. E., D. E. Vawter, A.W. Prehn, D. A. DeBruin, and K. G. Gervais. (2009). "Listen! the Value of Public Engagement in Pandemic Ethics," *American Journal of Bioethics* 9(11): 17–19.

Garrett J. E., D. E. Vawter, K.G. Gervais, A.W. Prehn, D. A. DeBruin, F. Livingston, A. M. Morley, J. Liaschenko, and , R. Lynfield. (2011). "The Minnesota Pandemic Ethics Project: Sequenced, Robust Public Engagement Process," *Journal of Participatory Medicine* 3(6). https://participatorymedicine.org/journal/evidence/research/2011/01/19/the-minnesota-pandemic-ethics-project-sequenced-robust-public-engagement-processes/. Accessed March 17, 2022.

Goodpaster, K. (1978). "On Being Morally Considerable," *Journal of Philosophy* 75(6): 308–325.

Gostin, L. O. (2006). "Federal Executive Power and Communicable Disease Control: CDC Quarantine Regulations," *Hastings Center Report* 36(2): 10–11.

Hamilton, C., C. Bonneuil, and F. Germenne, eds. (2015). *The Anthropocene and the Global Environmental Crisis: Rethinking Modernity in a New Epoch*. London: Routledge.

Hartman, C., and G. D. Squires, eds. (2006). *There Is No Such Thing as a Natural Disaster*. New York: Taylor and Francis.

Hoffman, S. M., and A. Oliver-Smith, eds. (2002). *Catastrophe and Culture: The Anthropology of Disaster*. Santa Fe, NM: School of American Research Press.

Honig, B. (2009). *Emergency Politics: Paradox, Law, Democracy*. Princeton: Princeton University Press.

Jennings, B. (2016). *Ecological Governance: Toward a New Social Contract with the Earth*. Morgantown: West Virginia University Press.

Jennings, B., and J. D. Arras. (2016). "Ethical Aspects of Emergency Preparedness and Response," in B. Jennings, J. D. Arras, D. H. Barrett, and B. A. Ellis, eds. *Emergency Ethics: Public Health Preparedness and Response*. New York: Oxford University Press, 1–103.

Kapur, G. B., and J. P. Smith, eds. (2011). *Emergency Public Health: Preparedness and Response*. Sudbury, MA: Jones and Bartlett Learning.

Kolbert, E. (2014). *The Sixth Extinction: An Unnatural History*. New York: Henry Holt.

Macintyre, B. (2010). "The Fault Line in Haiti Runs Straight to France," *The Times*, January 21. Available at: http://www.timesonline.co.uk/tol/comment/columnists/ben_macintyre/article6995750.ece

Melillo, J. M., T. Richmond, and G. W. Yohe, eds. (2014). *Climate Change Impacts in the United States: The Third National Climate Assessment*. U.S. Global Change Research Program, 841 pp. doi:10.7930/J0Z31WJ2. Available at: http://nca2014.globalchange.gov/downloads

National Institute of Environmental Health Sciences. (2015). "Health Impacts of Climate Change" Available at: http://www.niehs.nih.gov/research/programs/geh/climatechange/health_impacts/

Nuffield Council on Bioethics. (2007). *Public Health: Ethical Issues*. London: Author.

Paton, D., and D. Johnston, eds. (2006). *Disaster Resilience*. Springfield, IL: Charles C. Thomas Publisher.

Scarry, E. (2011). *Thinking in an Emergency*. New York: W. W. Norton.

Schafer W. A., J. M. Carroll, S. R. Haynes, and S. Abrams. (2008). "Emergency Management Planning as Collaborative Community Work," *Journal of Homeland Security and Emergency Management*; 5: Article 10.

Schön, D., and M. Rein. (1994). *Frame Reflection: Toward the Resolution of Intractable Policy Controversies*. New York: Basic Books.

Schwartz, B., and K. Sharpe. (2010). *Practical Wisdom: The Right Way to Do the Right Thing*. New York: Riverhead Books.

Schwartz, L., M. Hunt, L. Redwood-Campbell, and S. de Laat. (2014). "Ethics and Emergency Disaster Response. Normative Approaches and Training Needs for Humanitarian Health Care Providers," in O'Mathúna, D. P., B. Gordijn, and M. Clarke, eds. *Disaster Bioethics: Normative Issues When Nothing Is Normal*. Dordrecht: Springer, 33–48.

Solnit, R. (2009). *A Paradise Built in Hell: The Extraordinary Communities That Arise in Disaster*. New York: Viking.

Taylor, C. (2003). *Modern Social Imaginaries*. Durham, NC: Duke University Press.

United Nations. (2011). *Global Assessment Report on Disaster Risk Reduction*. Available at: http://www.unisdr.org/we/inform/publications/19846

Upshur, R. E. (2002). "Principles for the Justification of Public Health Intervention." *Canadian Journal of Public Health* 93(2):101–103.

Vawter, D. E., J. E. Garrett, K. G. Gervais, A. W. Prehn, and D. A. DeBruin. (2010). "Dueling Ethical Frameworks for Allocating Health Resources," *American Journal of Bioethics* 10(4):54–56.

Vawter, D. E., J. E. Garrett, K. G. Gervais, A. W. Prehn, and D. A. DeBruin. (2011). "Attending to Social Vulnerability when Rationing Pandemic Resources," *Journal of Clinical Ethics* 22(1):42–53.

Walker, B., and D. Salt. (2006). *Resilience Thinking: Sustaining Ecosystems and People in a Changing World*. Washington, DC: Island Press.

World Health Organization (WHO). (2016). "Climate Change and Health: Fact Sheet" Available at: http://www.who.int/mediacentre/factsheets/fs266/en/

Worster, D. (2004). *Dust Bowl: The Southern Plains in the 1930s*. New York: Oxford University Press.

Zack, N. (2009). *Ethics for Disaster*. Lanham, MD: Roman and Littlefield Publishers.

Further Reading

The ethics of disaster response covers a broad spectrum, from the triage of scarce resources to the long-term planning and justice issues that pertain to building an infrastructure of disaster mitigating capabilities and resilience. Several important works of this kind are now available. N. Zack, *Ethics for Disaster* (Lanham, MD: Roman and Littlefield Publishers, 2009) considers the approaches of utilitarianism, virtue ethics, and social contract theory as a basis for an ethical approach to disaster-related problems. She develops a contractarian approach and stresses the ethical importance of equitable treatment of all in disaster response. M. P. Battin, L.P. Francis; J. A. Jacobson, and C. B. Smith, *The Patient as Victim and Vector* (New York: Oxford University Press, 2009) focuses on outbreaks of infectious disease to challenge the emphasis on autonomy and the application of general principles that is prominent in bioethics. This work exemplifies the frame of ecological ontology discussed in this chapter. R. Solnit, *A Paradise Built in Hell: The Extraordinary Communities That Arise in Disaster* (New York: Viking, 2009) is an important study of disasters from different areas and historical periods, showing the complexity of human response and underscoring the capacity for compassion and solidarity. E. Scarry, *Thinking in an Emergency* (New York: W.W. Norton, 2011) demonstrates through many emergency examples that clear thinking by ordinary people and quick action required in a disaster situation are not mutually exclusive. She uncovers the threat to democratic values contained in many authoritarian perspectives. D. P. O'Mathúna, B. Gordijn, and M. Clarke, eds. *Disaster Bioethics: Normative Issues When Nothing Is Normal* (Dordrecht: Springer. 2014) is an important collection of essays on health ethics during disasters and on the ethics of conducting medical and behavioral research during disaster situations. B. Jennings, J. D. Arras, D. M. Barrett, and B. A. Ellis, eds. *Emergency Ethics: Public Health Preparedness and Response* (New York: Oxford University Press, 2016) provides an example of emergency response understood as a form of civic practice and contains essays on protecting the vulnerable, research ethics in emergencies, community engagement, the obligations of responders and medical professionals, and approaches to justice and priority setting in disaster situations.

PART X

Regulatory Tools

54

COMMAND AND CONTROL

Joshua Preiss

This chapter considers the ethics and economics of command-and-control environmental policy. Command and control refers to environmental policy that relies on direct regulation (permission, prohibition, standard setting) as opposed to policy that relies on economic or market-based incentives. Such policies are so named because they "command" a particular environmental outcome, and control how firms or individuals must meet that outcome. This approach dominated American environmental policy for many decades, and remains a central tool which legislatures utilize to attempt to reach desired environmental outcomes. Nonetheless, the past three decades of legal and economic discussion of command-and-control policy center on the intrinsic limitations and inefficiencies of this approach, with some even going so far as to equate it to "Soviet-style" regulation and "socialist central planning" (Ackerman and Stewart 1985, Stewart 1992, 1993, Sunstein 1997). The purpose of this chapter is not to defend command-and-control regulation from such criticisms. Instead, this chapter highlights several normative and practical concerns with market-based alternatives. These concerns, in many cases, provide reason for policy makers to continue to favor command-and-control regulation. In addition, this analysis calls for greater scrutiny and public discussion of the notions of efficiency theorists and policy makers utilize to determine whether command-and-control or market-based environmental policy is more efficient.

Environmental regulation of the 1960s, 1970s, and 1980s, including the landmark Clean Air and Clean Water Acts (and several subsequent amendments), relied heavily on command-and-control regulations (Ackerman and Stewart 1985, Baldwin, Cave, and Lodge 2011, Rose 2005). Such policy attempted to meet quality standards by placing explicit emissions limitations on specific sources of pollution, including large industrial factories. New factories and automobiles, then, were subject to "New Source Performance Standards" (NSPs) that, in turn, required firms to utilize government-mandated and industry uniform "Best Available Technology" (BAT). At times, divergence between the treatment of new and old technology, as well the practice of tying levels of regulation to firm solvency, raised concerns of fairness, as more profitable firms were subject to more stringent regulatory standards. Most criticism, however, particularly in legal and economic discourse, focused on the apparent inefficiency of the BAT approach. In their widely cited indictment, Ackerman and Stewart argue that "Uniform BAT requirements waste many billions of dollars annually by ignoring variations among plants and industries in the cost of reducing pollution and by ignoring geographic variations in pollution effects" (Ackerman and Stewart 1985: 1335). In short, the rigidity of these standards denied firms the flexibility and space for innovation that markets promise. Creating a market in tradable permits, by contrast, would provide firms with economic incentives to deliver goods and services with the least possible pollution. Other firms,

DOI: 10.4324/9781315768090-65

which because of technological, geographical, capital, or other limitations could not fully use and utilize the least polluting technologies, could buy permits from lesser polluting firms. The market-based approach, moreover, would capitalize on the informational advantages the individual business managers possess, relative to the "inevitably ill-informed bureaucrats," leading to more "transparent," "democratically accountable," and "bureaucratically effective" version of environmental law (Ackerman and Stewart 1985).

Before considering the merits of market alternatives, it is important to emphasize that market-based approaches do not entail "non-governmental approaches." Market-based environmental policies also rely on a governmental command of a desired outcome, which government agents control by mandating and enforcing procedures for monitoring and regulation. Often, they require inventing and legally enforcing a particular property rights regime where one had heretofore never existed. Indeed, to be effective in combatting the most pressing environmental problems of our age, including global climate change, market-based strategies will rely upon what are in many ways unprecedented government involvement in the institutions that structure global markets. In many cases (such as tradable permits in emissions, wetlands, forest, fisheries, and other natural resources) the implementation of a market-based approach entails further regulation than in the status quo, rather than the removal of the shackles of government regulation (Cole 2012). For those who lament the inherent inefficiency of government, regulatory capture and the perverse incentives that government regulation can provide for legislators and market actors alike, then, market-based approaches may provide relatively little comfort.

This central and substantial role for government regulatory instruments distances market or property rights approaches from the claim that, rather than commanding and controlling regulatory policy, societies should rely on corporate social responsibility and accountability in the marketplace. Robert Keyes, President of the Canadian Council for International Business, articulates this orientation with great clarity. He states, in an interview for the award-winning documentary *The Corporation,*

> Does there need to be some measure of accountability? Yes. And, I think the business community recognizes that. But that accountability is in the marketplace, it's with their shareholders. It's with the public perception and the public image that they are projecting. If companies don't do what they should be doing, they're going to be punished in the marketplace, and that's not what any company wants.

Though widespread in both business and management literature and the public discourse of environmental regulation, there are clear, fundamental problems with the accountability in the marketplace strategy, and the tendency to equate freedom with the absence of government intervention. To see why, we need to look no further than the writings of the 20th century's most celebrated defender of free markets: Nobel Prize-winning economist Milton Friedman. When discussing what he calls "neighborhood effects," Friedman writes,

> An obvious example is the pollution of a stream. The man who pollutes a stream is forcing others to exchange good water for bad. These others might be willing to make the exchange at a price. But it is not feasible for them, acting individually, to avoid the exchange or to enforce appropriate compensation.
>
> *(Friedman 1962: 33)*

Imagine that the man in Friedman's example is the representative of a mid-sized, publicly traded company in Pennsylvania that supplies paper to publishers in cities around the U.S. This company disposes of the effluent from paper production in a nearby stream, which is not controlled by regulation. How would the marketplace hold such a company accountable? We can imagine a number of possibilities. First, the managers of various publishers might punish the supplier. Assuming that polluting the stream is clearly the most cost-effective (in terms of profit for the firm rather than total social cost) means of disposing of the waste, this form of accountability presumes that managers of the various publishers are more interested in doing business with environmentally friendly suppliers than the cost of the paper they acquire – more interested in the health of a stream in Pennsylvania than profits for shareholders. For the shareholders of the publishing companies to hold the paper company accountable, they would have to (1) know a good deal about the operations of the company that they own stock in and (2) be more interested in the conduct of that companies suppliers than the cost and quality of their product, and the returns on their investment. Indeed, with ownership increasing bundled together in mutual funds and other financial instruments, shareholders are less likely to know not only the conduct of but even the identity of the companies they have stock in.

Finally, prospective book buyers might hold the paper company accountable. It is this sort of accountability that Keyes and other advocates of accountability in the marketplace celebrate – business ethics through consumer power. In order for consumers to play this role, however, they would need to be expected both to inquire into the day-to-day operations of the companies that supply the paper to the publisher of the book they are considering purchasing, and to care that such a company is legally polluting a stream as part of the production process more than the cost and content of the book itself. As a generalization, Keyes and countless others are surely right shareholders do not want the appearance of impropriety to lead consumers to punish their firm in the marketplace. In the Friedman-inspired case, however, I think it is safe to say that such accountability is unlikely. The same reasoning that disciples of public choice theory (Buchanan and Tullock 1962, Posner 1974) utilize to critique command-and-control regulation reveals the Panglossian farce of relying mostly or exclusively on corporate social responsibility and consumer accountability: consumers typically lack the kind of knowledge and incentives necessary to make such accountability a reality. It also highlights the problem with understanding freedom and coercion as the absence of government interference. My example is hardly far-fetched, and the circumstances that structure the possibility of marketplace accountability are far from uncommon.

Friedman's example also helps to explain the normative appeal of command-and-control regulation. Such regulation typically relies on what Rose calls "behavior-based" strategies, focusing on regulating the "*activities* or *performance* of the various environmental players" (Rose 2006). These policies target the most obvious forms of pollution by placing limitations on the specific sources of the pollution itself. In addition, they establish minimum requirements for consumer products, including the use of catalytic converters for automobile production. One might characterize the moral motivation of such legislation in Kantian or deontological (as opposed to consequentialist) terms. Large industrial polluters, as in Friedman's example of a man who pollutes a stream, cause clear (if diffuse) harm to numerous humans and other living creatures. They also threaten the freedom of many people by, in effect, forcing them to exchange clear water, air, and so on for polluted. These forced trades can have disastrous effects on the lives and livelihood of coerced parties to this transaction.

From a deontological perspective, then, the economic incentives approach, and the kinds of cost-benefit analysis characteristic of that approach, appears to miss the normative core of environmental law. In his discussion of the non-aggression principle Matt Zwolinski (this volume) quotes libertarian and Reagan White House official Martin Anderson on Tradable Pollution Permits,

> Now some even seriously propose that we should have economic incentives, to charge polluters a fee for polluting – and the more they pollute the more they pay. But that is just like taxing burglars as an economic incentive to deter people from stealing your property, and just as unconscionable… What we need are tougher, clearer environmental laws that are enforced – not with economic incentives but with jail terms.
>
> *(Anderson 1989)*

To understand this repulsion to markets in pollution, consider the workings of Ankh-Morpork, the main city in novelist Sir Terry Pratchett's celebrated *Discworld* series. When he takes over the city widely known for both commerce and crime, the patrician of Ankh-Morpork, Havelock Vetinari, implements a number of sweeping reforms. Instead of policing and punishing such crimes as theft and murder, as was the tradition, he legalizes a number of guilds, such as the Guild of Thieves and the Guild of Assassins. These non-governmental organizations, then, agree to only commit a certain number of thefts or murders per year (they always leave a receipt) and seek out and punish the unlicenced practice of their trade. This arrangement is a constant source of frustration to the Commander of the City Watch, Samuel Vimes, even if he must grudgingly acknowledge that "it works."

If, like Commander Vimes, we believe that just law holds those who harm others responsible for their actions, rather than merely generating the most efficient outcomes according to a given standard, market-based approaches to environmental law may appear fundamentally misguided (Goodin 1994). Such concerns apply to any understanding of personal or public morality that rejects the consequentialist foundations of the economic incentives approach, including those that give primary normative status to property rights, non-aggression, freedom, or individual self-ownership. These principles remain central to most liberal and libertarian understandings of justice and freedom. From a civic republican perspective, Michael Sandel worries that "turning pollution into a commodity to be bought and sold removes the moral stigma that is properly associated with it... [and] may undermine the sense of shared responsibility that increased global cooperation requires" (Sandel 2005: 94–95). Though a use right need not be the same as a property right (Caney 2010, Ott and Sachs 2002) others argue that natural resources, including living creatures and the habitats essential for their survival, ought to count among the kinds of goods that should not be owned or traded (Andre 1995). Many worry that market-based environmental policy encourages an increasingly instrumental understanding of our relationship to the natural world and other living creatures. It may also impact our relationship to our fellow humans, undermining even the moral sentiments – honesty, fairness, reciprocity – that enable markets function effectively in the first place (Gintis et al. 2005). Command-and-control legislation, for all of these reasons, may preserve the normative core of environmental law in way that market-based approaches cannot, whatever their claims to efficiency.

The clear rejoinder to these concerns is that the normative core of environmental law is conservation. Market-based or economic incentives approaches to environmental policy, then, are normatively preferable in any and all cases that they are the most effective means of achieving this goal. Even if we take this normative claim as given, however, the example from *Capitalism and Freedom* anticipates fundamental challenges for the market-based or property rights approach to environmental policy, and in turn the continued efficacy of command-and-control regulation. The reason why polluting the river is a clear example of a neighborhood effect is, in Friedman's words, that "strictly voluntary exchange is impossible" when "the actions of individuals have effects on other individuals for which it is impossible to recompense them" (Friedman 1962: 33). In such cases, the firm is able to externalize the costs of production by forcing non-parties to the exchange to pay part of the cost of production. Much of the persistent appeal of command-and-control legislation, then, reflects cases where it is difficult or impossible to successfully force firms to internalize the costs of production – in short, where it is difficult to establish and enforce the rules of a property rights regime.

While much literature, following Demsetz and Hardin, suggests that property rights are a natural and organic response to externalities and the so-called "tragedy of the commons" (Demsetz 1967, Hardin 1968), the reality of property rights construction is one of highly varied property regimes, which depend substantially upon local context and values, and are fraught with conflict and domination (Cole and Ostrom 2012, Eggertson 1990, 2012, Pistor 2019).

Attempts to establish and maintain a property rights regime for the protection of natural resources face a number of challenges. First, these efforts require a great investment of time and resources, and usually take place in contexts with a persistent risk of free-riding on these efforts (Rose 2009: 6–7). Second, the natural resources, and the goods that we aim to protect through the protection of these resources, themselves often involve highly complex systems of interaction and interdependencies. Consider a dominant folk and philosophical paradigm for the just acquisition and transfer of property rights: Lockean acquisition (Locke 1967). Locke posits an egalitarian justification of property rights, whereby all should be able to acquire property by mixing their labor with the world (Becker 1976). A right to property, then, is derivative of a right to yourself and the fruits of your labor (Christman 1986, Cohen 1995). Whether conceived as natural or God-given, this right is pre-political, and therefore serves as a normative standard with which to evaluate a given property rights regime. According to what is now referred to as the Lockean Proviso, one has no reason to complain about such acquisition, which essentially claims a kind of private dominion over what was formerly in common, insofar as there is enough, and as good, left for others to acquire.

I am not interested in this chapter in defending or rejecting the Lockean account, or different conceptions of the Lockean proviso. What is crucial to recognize is that the Lockean ethics of property, like most such accounts, entails a clear bias toward production (rather than conservation). Mixing your labor with nature, after all, means changing nature – making use of it by chopping down trees and building a house, plowing a field, capturing a fish, and so on. Such use, moreover, signals clear (if controversial) markers of ownership. This signal is crucial because in order to be effective, property rights must command the respect of non-owners (Smith 2003). This signal is comparatively easy in the case of land – put a fence around it – but water, air, wildlife stocks, ecosystem sustainability, and so on are much harder to mark. Even in the case of land, large, unexploited sections of prairie or old-growth forest will appear unclaimed to those on the ground, and non-physical methods of marking ownership (accounting and registration) are likely to be least effective precisely in the remote areas where environmental resources may be most valuable (Rose 2009: 14). The less clear (or more complex) a particular property right marker is, the harder it will be to create a market for those resources.

If the resources are non-fungible, moreover, it is difficult to establish a market in way that preserves much of what makes them valuable in the first place. Rose describes the example of a market-based approach to wetland preservation. She writes, "For example, if a wetland is to provide such ecosystem services as flood control or fish spawning, its location matters, as does the consistency its plant and animal life." By contrast to tradable environmental property regimes in sulfur dioxide emissions,

> Trades have been far more problematic for wetlands or habitat. These more complex resources cannot be traded against one another without significant alterations in the very features that make them valuable: location makes a difference, for example, to a wetland's ability to tame floodwaters or provide fish for spawning grounds.
>
> *(Rose 2009: 15)*

The specifications required to assure that any given trade doesn't result in the loss of these distinctive ecosystem values both greatly limit the pool of potential trading possibilities (Salzman and Ruhl 2000) and substantially increase the implementation and regulation costs of the market-based approach. In some cases, existing technology does not permit effective market-based approaches to environmental policy (Rose 2005). For all of these reasons, command and control measures often remain the best option, even though they appear clearly inefficient relative to simulation-based analysis of market alternatives (Cole and Grossman 1999).

Perhaps most challengingly, the creation of property rights raises obvious distributional concerns. From an economic efficiency perspective, who gets what in terms of initial endowment is of comparatively little concern (Coase 1960). For actual human beings, of course, these decisions matter a great deal. In the U.S., for example, broadly Lockean concepts of *discovery* and *improvement* defined the 18th and 19th century invention of individual property rights holders in land. In this way, through a series of legal rulings Native Americans became squatters on the land they had long inhabited, providing legal title to past capture and conquest (Pistor 2019: 34–35). In the former Soviet Union, controversy over the privatization of Soviet Union's vast state-owned resources, steeped in normative claims of unfairness, continues to infect attempts to establish liberal and democratic institutions (McFaul 1998a, 1998b, Rose 2006). In order to be effective, markets in natural resources will often need to upset pre-existing, socially and institutionally recognized understandings of property rights. It is not surprising that people respond strongly, and at times with hostility and even violence, to environmental regulation that upsets psychologically engrained norms of property – whether these changes involve taking a commons and issuing use rights to individual actors, or placing restrictions on how individuals or firms may make use of what they have long considered their own property (Hale 2008). Distributional issues take center stage in countless disputes over market-based approaches to the environment. In the case of global climate change, they arguably serve as the chief obstacle to addressing this central, civilizational, and biosphere-altering problem (Helm 2010). In addition to the demonstrable normative issues involved, these disputes greatly impact the feasibility and efficiency of market-based approaches to environmental policy (Libecap 1989). The uniformity of command and control, by contrast, makes it less likely to raise objections of distributional fairness (Hsu 2004).

Of course, command-and-control legislation also has distributive implications. Some firms will be better able to reform production and implement best available technology than others, some goods and services can be more easily substituted than others, and so on. There is no distribution-neutral form of regulation. Moreover, as Friedman's example demonstrates, doing nothing in the case of environmental regulation can make coercion, rather than free exchange, the norm. The point, instead, is that as political and economic institutions, environmental polices structure the freedom and opportunities of individuals. While in theory it is often possible to use taxing and spending to enable a more equal sharing of the costs and benefits of economic institutions, in political practice such redistribution is difficult to achieve. After all, we live in a time when many decry and reject out of hand as "socialism" the kind of redistribution implicit in both Kalder-Hicks improvements and the second welfare theorem of economics. For this reason distributional concerns ought to comprise a central component of debates between command-and-control and market-based approaches to environmental policy.

Finally, policy makers, theorists, and activists ought to be as clear and precise as possible about what they mean when they claim that given form of environmental regulation is more or less efficient. In popular discourse and philosophical, legal, and economic theory, many criticize egalitarian economic policy on the basis of efficiency. This assertion is terminologically muddled, as "efficiency" is simply a measure of different attempts to reach particular ends. If equality is one of these ends, then efficiency is the measure of our effectiveness in promoting equality, rather than something we trade-off with equality (Dietz 2010, LeGrand 1990). How do economists understand efficiency? Mainstream welfare economics relies upon a welfarist, consequentialist conception of the good. For reasons of normative parsimony, many economists understand this good in terms of individually and collectively revealed preference. Leaving aside widely articulated problems with this understanding (Beckerman 2011, Broome 1999, Hausman, McPherson, and Satz 2016, Preiss 2014, Sen 1977), this approach provides very limited guidance for environmental policy. Unlike consumer choices, it is extremely difficult to infer revealed preferences in many issues of environmental policy, in particular global climate change (Dietz, Hepburn, and

Stern 2008). Moreover, mainstream economic analysis remains overwhelmingly anthropocentric, in contrast to substantial, foundational work in environmental ethics that trumpets the intrinsic value of nature and non-human animals (Goodpaster 1978, Naess 1973, Rolston III 1975, Routley 1973, Singer 1977). Recognizing the intrinsic value of animals and the natural world does not undermine the importance for cost-benefit analysis. Instead, it reveals the profound limitations of any analysis that relies upon the revealed preferences of humans in real or imagined markets. Debates over environmental regulation, whether that regulation entails the uniform adoption of best available technology, the establishment of tradable permits, or some other mechanism, cannot avoid addressing difficult normative questions of how to understand and balance human welfare with animal welfare, habitat preservation, freedom, responsibility, rights, and other values. Increased reliance on economic incentives and market-based environmental policy in no way eliminates the need for such essentially public and (in the case of climate change and other fundamental environmental concerns) essentially global discussion.

References

Ackerman, B. A. and R. B. Stewart (1985) "Reforming Environmental Law," *Stanford Law Review* 37: pp. 1332–1365.

Anderson, M. (1989) "George Bush Environmentalist," *The Christian Science Monitor*, January 4.

Andre, J. (1995) "Blocked Exchanges' A Taxonomy," In Miller, D. and M. Walzer (eds.) *Pluralism, Justice, and Equality*. Oxford: Oxford University Press, pp. 171–196.

Baldwin, R., M. Cave, and M. Lodge (2011) *Understanding Regulation: Theory, Strategy, and Practice*. 2nd Edition. Oxford: Oxford University Press.

Becker, L. (1976) "The Labor Theory of Property Acquisition," *Journal of Philosophy* 73(18): pp. 653–664.

Beckerman, W. (2011) *Economics as Applied Ethics*. New York: Palgrave MacMillan.

Broome, J. (1999) *Ethics Out of Economics*. Cambridge: Cambridge University Press.

Buchanan, J. M. and G. Tullock (1962) *The Calculus of Consent: Logical Foundations of Constitutional Democracy*. Ann Arbor: University of Michigan Press.

Caney, S. (2010) "Markets, Morality, and Climate Change: What, if Anything, Is Wrong with Emissions Trading?" *New Political Economy* 15(2): pp. 204–205.

Christman, J. (1986) "Can Ownership Be Justified by Natural Rights?" *Philosophy and Public Affairs* 15(2): pp. 156–177.

Coase, R. H. (1960) "The Problem of Social Cost," *Journal of Law and Economics* 3(1): pp. 1–44.

Cohen, G. A. (1995) "Marx and Locke on Land and Labor," in *Self-Ownership, Freedom, and Equality*. Cambridge: Cambridge University Press.

Cole, D. (2012) "Property Creation by Regulation: Rights to Clean Air and Rights to Pollute," in Cole, D. C. and E. Ostrom (eds.) *Property in Land and Other Resources*. Cambridge, MA: Lincoln Institute of Land Policy Press.

Cole, D. and E. Ostrom (2012) "The Variety of Property Systems and Rights in Natural Resources," in Cole, D. C. and E. Ostrom (eds.) *Property in Land and Other Resource*. Cambridge, MA: Lincoln Institute of Land Policy Press.

Cole, D. and P. Grossman (1999) "When Is Command-and-Control Efficient? Institutions, Technology, and the Comparative Efficiency of Alternative Regulatory Regimes for Environmental Protection," *Wisconsin Law Review* 887, pp. 88–938.

Demsetz, H. (1967) "Toward a Theory of Property Rights," *American Economic Review* 57: pp. 347–359.

Dietz, S. (2010) "The Equity-Efficiency Trade-Off in Environmental Policy: Evidence from Stated Preferences," *Land Economics* 86(3): pp. 423–443.

Dietz, S., C. Hepburn, and N. Stern (2008) "Economics, Ethics, and Climate Change," In Basu, K. and R. Kanbur (eds.) *Arguments for a Better World: Essays in Honour of Amartya Sen (Volume 2: Society, Institutions and Development)*. Oxford: Oxford University Press.

Eggertson, T. (1990) *Economic Behavior and Institutions*. Cambridge: Cambridge University Press.

Eggertson, T. (2012) "Opportunities and Limits for the Evolution of Property Rights Institutions," In Cole, D. C. and E. Ostrom (eds.) *Property in Land and Other Resources*. Cambridge, MA: Lincoln Institute of Land Policy Press.

Friedman, M. (1962) *Capitalism and Freedom*. Chicago: University of Chicago Press.

Gintis, H., S. Bowles, R. Boyd, and E. Fehr (2005) "Moral Sentiments and Material Interests: Origins, Evidence, and Consequences," In Gintis, H., S. Bowles, R. Boyd, and E. Fehr (eds.) *Moral Sentiments and Material Interests.* Cambridge, MA: MIT Press.

Goodin, R. (1994) "Selling Environmental Indulgences," *Kyklos* 47(4): pp. 573–596.

Goodpaster, K. E. (1978) "On Being Morally Considerable," *Journal of Philosophy* 75: pp. 308–325.

Hale, B. (2008) "Private Property and Environmental Ethics: Some New Directions," *Metaphilosophy* 39(3): pp. 302–421.

Hardin, G. (1968) "The Tragedy of the Commons," *Science* 162: pp. 1243–1248.

Hausman, D. M., M. McPherson, and D. Satz. (2016) *Economic Analysis, Moral Philosophy, and Public Policy,* 3rd Edition. Cambridge: Cambridge University Press.

Helm, D. (2010) "Climate-Change Policy: Why Has So Little Been Achieved?" In Helm, D. and C. Hepburn (eds.) *The Economics and Politics of Climate Change,* Oxford: Oxford University Press.

Hsu, S. (2004) "Fairness vs. Efficiency in Environmental Law," *Ecology Law Quarterly* 31: pp. 303, 375–276.

LeGrand, J. (1990) "Equity vs. Efficiency: The Elusive Tradeoff," *Ethics* 100(3): pp. 554–568.

Libecap, G. (1989) *Contracting for Property Rights.* Cambridge: Cambridge University Press.

Locke, J. (1967) *Two Treatises of Government,* Book II, P. Laslett (ed.) Cambridge: Cambridge University Press.

McFaul, M. A. (1998a) "Russia's 'Privatized State' as an Impediment to Democratic Consolidation: Part I," *Security Dialogue* 29(2): pp. 191–199.

McFaul, M. A. (1998b) "Russia's 'Privatized State' as an Impediment to Democratic Consolidation: Part II," *Security Dialogue* 29(3): pp. 315–332.

Naess, A. (1973) "The Shallow and the Deep, Long-Rand Ecology Movements: A Summary," *Inquiry* 16: pp. 95–100.

Ott, H. E. and W. Sachs (2002) "The Ethics of International Emissions Trading," In Pinguelli-Rosa, Luiz and Mohan Munasinghe (eds) *Ethics, Equity, and International Negotiations on Climate Change.* Cheltenham: Edward Elgar.

Pistor, K. (2019) *The Code of Capital: How the Law Creates Wealth and Inequality.* Princeton: Princeton University Press.

Posner, R. A. (1974) "Theories of Economic Regulation," *RAND Journal of Economics* 5(2): pp. 335–358.

Preiss, J. (2014) "Global Labor Justice and the Limits of Economic Analysis," *Business Ethics Quarterly* 24(1): pp. 55–83.

Rolston III, H. (1975) "Is There an Ecological Ethics?" *Ethics* 85(2): pp. 93–109.

Rose, C. M. (2005) "Environmental Law Grows Up (More or Less), and What Science Can Do to Help," *Lewis & Clark Law Review* 9(2): pp. 273–294.

Rose, C. M. (2006) "Privatization- The Road to Democracy?" *Saint Louis University Law Journal* 50(3): pp. 691–720.

Rose, C. (2009) "Liberty, Property, and Environmentalism," *Social Philosophy and Policy* 10: pp. 1–25.

Routley, R. (1973) "Is there a Need for a New, an Environmental Ethic?" *Proceedings of the 15th World Congress of Philosophy* 1: pp. 205–210.

Salzman, J. and J. B. Ruhl (2000) "Currencies and the Commodification of Environmental Law," *Stanford Law Review* 53: pp. 607–637.

Sandel, M. (2005) "Should We Buy the Right to Pollute?" in *Public Philosophy: Essays on Morality in Politics,* Cambridge, MA: Harvard University Press.

Sen, A. (1977) "Rational Fools: A Critique of the Behavioral Foundations of Economic Theory," *Philosophy & Public Affairs* 6(4): pp. 317–344.

Singer, P. (1977) *Animal Liberation: A New Ethics for Our Treatment of Animals.* New York: Avon.

Smith, H. (2003) "The Language of Property: Form, Content, and Audience," *Stanford Law Review* 55: pp. 1105–1191.

Stewart, R. B. (1992) "Models for Environmental Regulation: Central Planning Versus Market-Based Approaches," *Boston College Environmental Affairs Law Review* 19: p. 547

Stewart. R. B. (1993) "Environmental Regulation and International Competitiveness," *The Yale Law Journal* 102(8): pp. 2039–2106.

Sunstein, C. (1997) *Free Markets and Social Justice,* Oxford: Oxford University Press.

55

ECONOMIC INSTRUMENTS

Joseph E. Aldy

Introduction

Economic instruments are a class of policy tools a policy maker may employ as the means to attain an environmental goal. Such instruments provide pollution sources with flexibility and discretion – thereby tapping incentives for innovating and cost cutting – in reducing pollution. They stand in contrast to traditional forms of regulation that may prescribe specific technological fixes or proscribe specific behavior. As a result, economic instruments are frequently referred to as market-based instruments.

Before addressing the ethical implications of economic instruments, let me start by illustrating the concept of market-based instruments through a description of cap-and-trade programs and pollution taxes. In these illustrations and throughout this chapter, I will use power plants and their associated air pollution as an example. These tools are generally applicable to a wide array of environmental problems, from regulating the total allowable catch in commercial fisheries to per unit fees on garbage collection to taxing pollution in wastewater discharges.

A cap-and-trade program establishes an aggregate emission cap – i.e., the total amount of emissions among all sources covered by the program. This cap is divided into emission allowances, which grant the holder of the allowance the right to emit a specific amount of pollution. For example, a cap of 1 billion tons could be divided into 1 billion emission allowances that each grants the right to emit one ton of pollution. The government then allocates these emission allowances to the economy through a public auction or by giving them away for free. In this latter case, pollution sources may receive allowances from the government as a function of their historic emissions. Firms may buy and sell pollution allowances among each other, and subsequently a price emerges in the market for allowances that reflects the supply and demand of allowances (just as prices represent supply and demand for goods in markets more generally). The government requires firms to surrender allowances to cover their emissions over the past year. Thus, a power plant that emits annually 1 million tons of a pollutant must submit 1 million allowances to the government. If the market price for allowances is $25, then the value of the surrendered allowances is $25 million.

While cap-and-trade programs set an aggregate quantity (the "cap") that makes the right to emit scarce, which is revealed in the price of allowances, a pollution tax sets the price to emit another unit of pollution. A pollution tax imposes a per unit fee – e.g., dollars per ton – on the emission of a pollutant from a given source. Thus, a power plant that emits annually 1 million tons of a pollutant taxed at $25 per ton must submit annual tax payments of $25 million. While a pollution tax imposes a price on pollution and cap-and-trade restricts the quantity of pollution,

DOI: 10.4324/9781315768090-66

this example of a power plant illustrates how they can result in the same outcome. Under each instrument, noncompliance can trigger penalties, just as the failure to comply with conventional regulations or conventional tax requirements does.

Cap-and-trade and pollution taxes have the appealing characteristic of delivering cost-effective pollution reductions. These instruments create the incentive to minimize the total cost of attaining a given environmental goal. Under cap-and-trade, a firm with high pollution abatement costs may buy allowances from a firm with low abatement costs. In effect, the high-cost firm finances pollution abatement at the low-cost firm. In the market for emission allowances, the profit motive of firms drives them to abate pollution until the marginal cost of pollution abatement is equal to the price of allowances traded in the allowance market. Likewise, under a pollution tax, the profit motive drives firms to abate pollution until the marginal cost of pollution abatement is equal to the per ton tax. As a result, each firm pays the same amount for the last unit of emission reduction, a necessary condition for minimizing total emission abatement costs (see Aldy 2020, and Aldy and Stavins 2012a, 2012b for more details).

These approaches differ from conventional command-and-control regulation, such as mandating specific pollution-control technologies. Requiring that all power plants install a sulfur scrubber, for example, may result in some plants bearing high costs – because their equipment is not well designed for the addition of a scrubber – and in other plants bearing low costs – because they can easily install a new scrubber. The standard economic argument for market-based instruments is that they deliver a given level of environmental quality at the lowest possible cost and typically at lower cost than command-and-control regulation.

Why can't the government design a command-and-control regulation to minimize costs, i.e., to effectively replicate the outcome of the market-based approaches to regulation? In most environmental policy contexts, the regulator suffers from an information disadvantage. The sources of emissions likely understand their opportunities and costs for reducing pollution better than the government. The regulator would have to undertake quite time- and resource-intensive research and analysis on a source-by-source basis to eliminate this information disparity. Then, it would impose effectively different standards on each source in order to minimize total societal costs of attaining the environmental goal. The politics of such an invasive information collection program and the tailoring of regulations on a source-by-source basis would likely be quite challenging. Instead, a market-based approach creates the incentive for firms to seek out information on the lowest-cost ways of reducing emissions and to exploit them. It also serves as the basis for another appealing advantage of market-based instruments – they provide a stronger incentive to invest in technological innovation. This investment may enable even more ambitious environmental objectives in the future as new innovations become commercially viable.

Economists have traditionally supported market-based instruments for environmental protection because of the outcome or consequence of such policy implementation: minimizing costs of attaining a given environmental objective. Moreover, another consequence of market-based instruments – that all sources face the same marginal abatement cost – is a necessary condition for maximizing social welfare as defined through benefit-cost analysis. An active scholarly debate has addressed the question of whether benefit-cost analysis accurately reflects social welfare (e.g., Adler and Posner 2006) and whether benefit-cost analysis can appropriately represent individuals' values of the environment (e.g., Sagoff 2000). Therefore, this chapter focuses on economic instruments as the means for policy implementation and not the larger question of their role in normative decision rules for environmental policy.

Of course, in focusing on consequences, such an approach avoids consideration of potentially intrinsic values that may affect how individuals and society view and would prefer to address environmental problems. Indeed, much of the debate – in academic as well as policy circles – over the potential use of economic instruments reflects these two, very different ethical perspectives on environmental policy. For those interested in consequences of policy actions and holding an instrumental view of the environment, market-based instruments represent an appealing means to implement environmental objectives. For those who believe in the intrinsic value of the natural environment, the social opprobrium associated with bans and regulatory mandates represents alternatives preferred to economic instruments.

This chapter proceeds with a discussion of how the use of economic instruments in environmental policy reflects the polluter pays principle, which has been a key principle in domestic and international environmental policy for decades. Then I address the objection that market-based policies provide polluters with a license to pollute. I return to the consequentialist perspective associated with three dimensions of the design and use of economic instruments: environmental justice and "hot spots," regulatory capture by "Baptists" and "Bootleggers," and the enabling of more ambitious environmental goals.

Polluter Pays Principle

> National authorities should endeavor to promote the internalization of environmental costs and the use of economic instruments, taking into account the approach that the polluter should, in principle, bear the cost of pollution....
> *– Principle 16, United Nations Rio Declaration on Environment and Development 1992*

A.C. Pigou (1920) first wrote about the "external" impacts of economic activity that may necessitate government intervention. Such externalities – such as pollution emitted from a power plant smokestack – are not accounted for in the private decision making of firms because the costs are external to the firm. An electric utility maximizes its profits by generating power with the lowest possible costs for labor, fuel, equipment, and so on, but does not consider the adverse impacts that its air pollution has on human and ecosystem health downwind. Requiring the utility to bear – or internalize – the cost of the external damages would encourage it to find ways to reduce pollution, just as high wages and high fuel costs create incentives for firms to invest in labor-saving technology and more fuel-efficient combustion equipment.

Such an approach treats pollution as a cost of production, and an extensive literature has evolved to quantify and monetize the damages associated with pollution (e.g., Freeman 2003; Viscusi and Aldy 2003). Even if policy does not focus on estimating the economic value of pollution externalities, there is a widely held view that environmental policy should be implemented such that the polluter pays for the effort to prevent or clean up the pollution. At the 1992 Earth Summit in Rio de Janeiro, Brazil, the international community incorporated the polluter pays principle into the conference declaration's principle 16. This principle reflects a preference for the sources of pollution to pay for pollution control instead of the general public either through government subsidies or by bearing the harm from unabated pollution. Nash notes that this "principle underlies much of modern environmental law" (2000: 466).

Of course, an environmental policy that imposes technology mandates on all power plants – thereby requiring utilities to purchase new scrubber equipment, for example – would make the polluter pay for emission abatement just as a pollution tax would. The international community defined the polluter pays principle to include the use of economic instruments. As such, the principle acknowledges the need to minimize the costs of attaining a pollution reduction goal in light of competing claims on scarce resources. In this sense, it reflects the dual objectives of the Earth Summit to advance environmental protection and economic development.

The use of economic instruments in implementing the polluter pays principle may raise two potential ethical objections. First, the design of some economic instruments may result in some polluters paying more than the damages they impose on the environment and others paying less than the damages they impose on the environment (Nash 2000). If the damage associated with pollution varies geographically, then a common price – either through cap-and-trade or through a pollution tax – could deliver this result. In line with this critique, Muller and Mendelsohn (2009) show that the acid rain cap-and-trade program that regulates the emissions of sulfur dioxide from power plants yields quite heterogeneous environmental benefits across the eastern half of the United States. Such heterogeneity could result in so-called pollution "hot spots" (discussed below). In addition, the free allocation of emission allowances may convey an economic windfall to the sources of pollution (see discussion of "Baptists" and "Bootleggers" below). The emission allowances may carry quite significant economic value. An analysis of the European Union's Emission Trading Scheme for carbon dioxide emissions shows that free allocation yielded windfall profits among the most pollution-intensive sources (Bushnell et al. 2013). While these sources implemented costly emission abatement strategies, they passed on most of these costs to their consumers in the form of higher product prices. For those sources that had emissions below their free allocation of allowances could sell their unused allowances to other sources for a profit. Careful design of economic instruments can ensure that cost of reducing emissions is in line with the damages the pollution imposes on society.

Second, there is a question about who actually pays when a polluter is regulated by environmental policy. Even though the polluter pays principle specifies that the polluter should pay, in practice the polluter can pass along these costs to its consumers, just as it may pass along higher labor costs to consumers through the price of its products. In response to a regulation or a market-based instrument, pollution sources may pay for pollution abatement equipment, emission allowances, or emission taxes. Thus, on its face, the polluter is paying. As a result of market forces, the consumers may bear the cost of the pollution through higher product prices. Thus, the economic incidence – who ultimately bears the cost of the policy – of an economic instrument (or, for that matter, a command-and-control regulatory mandate) may fall on individual consumers. Distributive justice concerns may arise for environmental policies that concentrate on reducing pollution from power plants (and other sources of energy). Since energy expenditures represent a disproportionately larger share of low-income households' budgets, an increase in energy prices as utilities pass along the costs of environmental policy would impose the greatest harm on these low-income households (Burtraw et al. 2009). Again, the design of an economic instrument can mitigate potential adverse impacts associated with the economic incidence of the policy (Farber 2012). For example, the Government of British Columbia implemented a carbon tax in 2008 and coupled it with a "low income climate action tax credit" to offset the higher costs for energy and goods and services for low-income households in the province.

Selling the Right to Pollute

> [T]he fee is a license to pollute.
>
> – *Senator Edmund Muskie 1971*

Market-based instruments, by their design, impose a cost on a firm's pollution with the intent of driving emission reductions by the firm. It takes pollution and converts it into a commodity – explicitly in the creation of an emission allowance and implicitly under a pollution tax. For analysts who view pollution as morally wrong, this effectively removes the "social stigma" that should be associated with pollution (Drury et al. 1999; Sandel 2012). One scholar has described market-based instruments as the "selling of environmental indulgences," akin to the sale of "God's grace" through indulgences in the Catholic Church (Goodin 1994). The environmental indulgence reflects the view that allowing the polluter to pay for the right to pollute absolves it of its "environmental sins." Market-based instruments effectively convert behavior, perceived by some, as subject to a moral standard into that which can be resolved through an economic transaction.

In the debate over water quality legislation in the early 1970s, Senator Muskie noted that a proposal to impose fees, in lieu of regulatory mandates, on water pollution effectively granted the right to pollute. This occurred at a time when pollution was viewed by many as morally wrong and Congress established ambitious (arguably infeasible) goals, such as eliminating all pollution discharges into navigable waters by 1985. But why is pollution considered wrong? Is pollution intrinsically immoral, or is it considered wrong because of the adverse impacts associated with it? Why should pollution be stigmatized? Sandel (2012) argues that implementing market-based policies advance "market norms," which may crowd out non-market norms. Thus, he argues that society should consider more than just efficiency and distributive justice in determining whether creating an environmental market to address pollution should move forward.

Such non-market norms may reflect a variety of perspectives. First, some may believe in the rights of nature. The natural environment may have an intrinsic value that is divorced from the instrumental value humans may have for it (Goodin 1994). Paehlke (2013) notes, however, that debates about the rights of nature have not meaningfully informed public policy on the environment. Second, some may believe that individuals should have the right to clean air, clean water, etc. as a basic human right. Just as one may object to buying and selling the right to free speech, one may object to buying and selling the right to clean air. Third, there may be cultural norms that identify a clear incongruence between market-based approaches and the sanctity of the natural environment (e.g., Drury et al. [1999] quote Chief Seattle on this point).

The objection to economic instruments on the grounds that they grant a license to pollute could apply just as easily, however, to conventional command-and-control regulatory policies. Consider that the vast majority of environmental policy limits, but does not ban, pollution. Complying with a regulatory mandate – e.g., to install a sulfur scrubber – could be simply viewed as the cost of doing business. In addition, power plants in the United States must acquire operating permits from state regulatory authorities that specify their allowable emissions, which typically reflect existing Federal regulations such as technology and performance standards. It's not clear how this differs ethically or practically from an emission allowance granting a power plant the right to emit a ton of pollution.

In his critique of "environmental indulgences," Goodin concludes by noting that "[h]ow attractive we find green taxes and the 'polluter pays' principle more generally depends, in large part, upon what we see as their alternative" (1994: 594). If the alternative is that pollution is eliminated, then economic instruments would appear as inferior options. But if the alternative is letting polluters off without bearing the cost of their pollution, then they may look "relatively attractive." The environmental consequences of the use of such instruments are important, which I address in the final section.

Hot Spots and Environmental Justice

> Pollution trading in Los Angeles has led to concentrated toxic air emission hot-spots that have shackled low-income and minority communities with the region's air pollution.
> — *Drury et al. 1999*

By design, market-based instruments provide regulated firms with discretion on how to comply with the policy. Some may invest in pollution abatement equipment, some may reduce production, and some may purchase more allowances or pay a greater pollution tax. Since the government does not prescribe specific actions to individual pollution sources, the emissions will vary from source to source. If a number of firms with pollution sources located near each other decide to purchase more allowances (under a cap-and-trade program) or pay more in taxes (under a pollution tax), then pollution in this location may be higher than it would have been under a command-and-control regulatory approach. As a result, economic instruments may give rise to these pollution "hot spots."

Concerns over hot spots emerging in low-income and minority neighborhoods have long colored debates over cap-and-trade approaches to air pollution in California. Drury et al. (1999) describe how major purchasers of emission allowances under southern California's "RECLAIM" air pollution cap-and-trade program were disproportionately located in low-income and minority neighborhoods. Fowlie et al. (2012), however, found that RECLAIM-covered facilities reduced their emissions 20% relative to similar non-RECLAIM facilities covered by command-and-control regulations and that these pollution reductions occurred across all neighborhoods, regardless of socio-economic characteristics. In the early years of California's carbon dioxide cap-and-trade program, greenhouse gas emissions increased for about half of the covered emission sources with similar increases in associated particulate matter and air toxic pollution. The communities near these sources with higher greenhouse gas emissions were more likely to have higher proportions of residents of color and higher rates of poverty (Cushing et al. 2018).

These examples pose important questions about distributional fairness that are often associated with the issue of environmental justice. What constitutes a "fair" distribution of pollution? Should the market or the government determine the distribution of pollution? Recognizing these concerns, in 1994 President Clinton issued Executive Order 12898, "Federal Actions to Address Environmental Justice in Minority Populations and Low-Income Populations." This action required all federal agencies to identify and address potential adverse environmental impacts of their policies on disadvantaged communities.

"Hot spots" in low-income or minority communities are not an intrinsic characteristic of cap-and-trade programs and pollution taxes (Farber 2012). First, some pollutants mix on large spatial scales (e.g., carbon dioxide and ozone-depleting substances mix globally, in contrast to many hazardous air pollutants that mix at city or even smaller scales) so a local concentration of emissions does not necessarily result in higher local ecosystem or human health damages. Second, the profit motive at play with market-based instruments creates a strong incentive to reduce emissions where it is cheapest to do so. High pollution abatement costs are a function of a source's existing technology and its opportunities for replacing or upgrading the technology. These characteristics are not intrinsically correlated with a source's presence in a low-income or minority community. Third, economic instruments can be designed in a way to mitigate the potential for hot spots. For example, a cap-and-trade program could limit trading between regions or create trading ratios that result in greater emission reductions with higher volume trading (Berck and Helfand 2005). The need for such restrictions, however, may limit the cost savings of market-based instruments relative to command-and-control regulations. Finally, frequent emissions monitoring and public transparency can serve to identify the emergence of any potential hot spots and inform potential policy reforms. For example, the U.S. Environmental Protection Agency launched the Toxics Release Inventory program in 1989. This information disclosure program required sources of toxic pollutants to release publicly data on their emissions; there were no regulatory requirements. The transparency on air toxic emissions adversely impacted publicly traded firms' stock prices and induced substantial reductions in pollutant emissions (Khanna et al. 1998; Konar and Cohen 1997).

Concerns about hot spots illustrate a fundamental challenge for the application of economic instruments to some kinds of environmental problems. If a population's health is directly affected by a single source – for example, a hazardous waste incinerator adjacent to a residential neighborhood – then it may be inappropriate to implement a market-based instrument to address this problem. Under a cap-and-trade program, this incinerator may have high abatement costs and thus purchase emission allowances, thereby imposing health risks on the local population. Under a pollution tax, this incinerator may respond to its high abatement costs by continuing to pollute at current levels, pay a high tax bill, thereby imposing health risks on the local population. In this context, direct, prescriptive regulation may be necessary to deliver public health benefits in lieu of a market-based approach.

"Baptists and Bootleggers"

'Baptists' point to the moral high ground and give vital and vocal endorsement of laudable public benefits promised by a desired regulation.... 'Bootleggers,' who expect to profit from the very regulatory restrictions desired by Baptists, grease the political machinery with some of their expected proceeds.

– Bruce Yandle 1999

Some firms do not view environmental regulations as the cost of doing business but rather as an opportunity for their business. Bruce Yandle (1999) describes the "Baptists and Bootleggers" phenomenon, in which an advocacy group stakes out a social objective on moral grounds (e.g., the Baptists who traditionally opposed Sunday liquor sales in the United States) and businesses support the cause out of self-interest (e.g., the bootleggers who would face less competition with a ban on liquor sales). In many cases, the supporters for the moral cause stake out an ambitious regulatory goal and the business interests focus on the details of implementation to maximize their profits.

For example, in the late 1970s, owners of high-sulfur coal supported tighter sulfur pollution regulations at power plants – in line with many environmental advocacy groups – so long as these regulations required the installation of scrubbers on power plant smokestacks. Such an approach ensured that the regulation would not disadvantage high-sulfur coal relative to low-sulfur coal. Such concerns were well placed – under the sulfur dioxide cap-and-trade programs launched in the 1990s, many utilities found it cheaper to purchase low-sulfur coal from the Western United States than to retrofit a sulfur scrubber and burn high-sulfur coal.

Economic instruments are not immune to such Baptist and Bootlegger unions. For example, in the lead up to the 2009 Congressional debate over climate change legislation, a half-dozen leading environmental groups joined more than 20 major corporations to create the U.S. Climate Action Partnership that proposed a legislative framework for a greenhouse gas emissions cap-and-trade program. The 2009 Waxman-Markey Bill, passed in the U.S. House of Representatives, reflected many elements of the proposal made by this environmental-business coalition. A number of analysts remarked that the bill was a "Rube Goldberg contraption" and its complexity masked profit-making opportunities for some businesses (e.g., Friedman 2009).

Rent seekers seek rents. New environmental regulations – whether command-and-control or market-based instruments – create rents, and those who receive the rents are the winners, while others who may experience a decline in profits are the losers. These rents provide strong incentives for businesses to engage in the policy process – from the development of and passage of legislative authority to the design and implementation of regulations. This extensive engagement, especially with agencies responsible for promulgating regulations, yields a framing more in line with utilitarianism and a divorce from other lines of environmental ethics (Purdy 2013).

Saving the Planet

The approach that seems most promising... is one of harnessing market forces to spur both technological advance and sustainable management of national and global natural resources. Indeed, a more economically efficient approach will allow more ambitious policies.

– Senators Timothy Wirth and John Heinz 1988

Implementing environmental policy through market-based instruments may facilitate the setting and attainment of more ambitious environmental goals than would be possible under command-and-control regulatory instruments. From an economic (utilitarian) perspective, if economic instruments lower the total cost of environmental pollution abatement, then net social benefits could be maximized with a more stringent environmental goal. Game-theoretic research on multilateral climate change policy has shown that moral objections to international emission trading could lead to less ambitious domestic emission targets and higher global greenhouse gas emissions (Eyckmans and Kverndokk 2010).

From a political economy perspective, lower pollution control costs may weaken the opposition by pollution sources to new or more ambitious environmental policy. The prospect of lower pollution control costs could also mobilize advocates for environmental policy to push for tougher goals that impose greater costs on polluters. Moreover, the design of economic instruments could be oriented toward eliciting political support, such as through targeted free allocation of emission allowances under cap-and-trade (Stavins 2007).

The prospect of lowering the cost of reducing pollution through the use of economic instruments and securing more ambitious environmental goals may be especially appealing in the context of global climate change. The United Nations Environment Programme (2013, 2019) has repeatedly noted the substantial "emissions gap" between current actions and what may be necessary to avoid serious, abrupt, and potentially catastrophic climate change. Many governments at the national, state, and provincial level – including those in Australia, Canada, China, Europe, New Zealand, the United States – have implemented or are designing greenhouse gas emission cap-and-trade or carbon tax programs (World Bank 2019).

Some of those who have held strong moral views on the use of economic instruments have changed their mind in light of the challenge of combating climate change. In 1997, Jonathan Wiener noted that

> Europe, in particular Germany, may be guided more by a Kantian perspective in which the solution to pollution is moral conduct (cease polluting) rather than a Benthamite perspective in which pollution is seen as a market failure to be corrected by market pragmatism (34).

Despite European opposition to allowance trading at the time, the 1997 Kyoto Protocol included several provisions to permit the use of international market-based instruments. In 2005, the European Union implemented what has become the world's largest greenhouse gas emissions cap-and-trade program. This transition may have reflected an evolution in European ethics and politics. It highlights in part how environmental policy has played out in weighing alternative ethical bases to address the most daunting environmental challenge of the 21st century.

There are fundamental tensions between a focus on consequences, which lends strong support to economic instruments in environmental policy, and a perspective that the environment has intrinsic value and thus pollution is morally wrong. Yet, in the design of public policy, if such economic instruments enable greater preservation of the environment, then it may elicit begrudging support even by those who hold the latter view.

Related Topics

22. Climate Change Mitigation
49. Pollution and Polluter Pays
54. Command and Control
56. Cost-Benefit Analysis
63. Environmental Justice

References

Adler, M.D. and E.A. Posner (2006) *New Foundations of Cost-Benefit Analysis*. Cambridge, MA: Harvard University Press.

Aldy, J.E. (2020) "Pricing Pollution Through Market-Based Instruments," in D. Konisky (ed) *Handbook on U.S. Environmental Policy*, Northampton, MA: Edward Elgar, 202–216.

Aldy, J.E. and R.N. Stavins (2012a) "The Promise and Problems of Pricing Carbon: Theory and Experience," *Journal of Environment and Development* 21(2): 152–180.

Aldy, J.E. and R.N. Stavins (2012b) "Using the Market to Address Climate Change: Insights from Theory and Experience," *Daedalus* 141(2): 45–60.

Berck, P. and G.E. Helfand (2005) "The Case of Markets versus Standards for Pollution Policy," *Natural Resources Journal* 45: 345–368.

Burtraw, D., R. Sweeney, and M. Walls (2009) "The Incidence of U.S. Climate Policy: Alternative Uses of Revenues from a Cap-and-Trade Auction," *National Tax Journal* 62: 497–518.

Bushnell, J.B., H. Chong, and E.T. Mansur (2013) "Profiting from Regulation: An Event Study of the EU Carbon Market," *American Economic Journal: Economic Policy* 5(4): 78–106.

Cushing, L., D. Blaustein-Rejto, M. Wander, M. Pastor, J. Sadd, A. Zhu, and R. Morello-Frosch (2018) "Carbon Trading, Co-Pollutants, and Environmental Equity: Evidence from California's Cap-and-Trade Program (2011–2015)," *PLoS Medicine* 15(7): e1002604.

Drury, R.T., M.E. Belliveau, J.S. Kuhn, and S. Bansal (1999) "Pollution Trading and Environmental Injustice: Los Angeles' Failed Experiment in Air Quality Policy," *Duke Environmental Law and Policy Forum* 9: 231–289.

Eyckmans, J. and S. Kverndokk (2010) "Moral Concerns on Tradable Pollution Permits in International Environmental Agreements," *Ecological Economics* 69: 1814–1823.

Farber, D.A. (2012) "Pollution Markets and Social Equity: Analyzing the Fairness of Cap and Trade," *Ecology Law Quarterly* 39: 1–56.

Fowlie, M., S.P. Holland, and E.T. Mansur (2012) "What Do Emissions Markets Deliver and to Whom? Evidence from Southern California's NO_X Trading Program," *American Economic Review* 102(2): 965–993.

Freeman, A.M. (2003) *The Measurement of Environmental and Resource Values: Theory and Methods*. Washington, DC: RFF Press.

Friedman, T.L. (2009) "Just Do It," *The New York Times*, July 1, p. A33.

Goodin, R.E. (1994) "Selling Environmental Indulgences," *Kyklos* 47: 573–596.

Khanna, M., W.R.H. Quimio, and D. Bojilova (1998) "Toxics Release Information: A Policy Tool for Environmental Protection," *Journal of Environmental Economics and Management* 36: 243–266.

Konar, S. and M.A. Cohen (1997) "Information as Regulation: The Effect of Community Right to Know Laws on Toxic Emissions," *Journal of Environmental Economics and Management* 32: 109–124.

Muller, N.Z. and R. Mendelsohn (2009) "Efficient Pollution Regulation: Getting the Prices Right," *American Economic Review* 99(5): 1714–1739.

Nash, J.R. (2000) "Too Much Market? Conflict Between Tradable Pollution Allowances and the 'Polluter Pays' Principle," *Harvard Environmental Law Review* 24: 465–535.

Paehlke, R.C. (2013) "Ethical Challenges in Environmental Policy," in S. Kamienicki and M.E. Kraft (eds) *The Oxford Handbook of U.S. Environmental Policy*, Oxford: Oxford University Press.

Pigou, A.C. (1920) *The Economics of Welfare*. London: Macmillan and Company.

Purdy, J. (2013) "Our Place in the World: A New Relationship for Environmental Ethics and Law," *Duke Law Journal* 62(4): 857–932.

Sagoff, M. (2000) "Environmental Economics and the Conflation of Value and Benefit," *Environmental Science and Technology* 34: 1426–1432.

Sandel, M.J. (2012) *What Money Can't Buy: The Moral Limits of Markets*. New York: Farrar, Straus and Giroux.

Stavins, R.N. (2007) "A U.S. Cap-and-Trade System to Address Global Climate Change," The Hamilton Project Discussion Paper 2007–13. Washington, DC: The Brookings Institution.

United Nations Environment Programme (2013) *The Emissions Gap Report 2013*. November. Nairobi.

United Nations Environment Programme (2019) *Emissions Gap Report 2019*. November. Nairobi.

Viscusi, W.K. and J.E. Aldy (2003) "The Value of a Statistical Life: A Critical Review of Market Estimates Throughout the World," *Journal of Risk and Uncertainty* 27(1): 5–76.

Wiener, J.B. (1997) "Designing Global Climate Policy: Efficient Markets versus Political Markets," *Policy Study Number 143*. Center for the Study of American Business. St. Louis, MO: Washington University.

Wirth, T.E., J. Heinz, and R.N. Stavins (1988) *Project 88: Harnessing Market Forces to Protect Our Environment: Initiatives for the New President: A Public Policy Study* (Vol. 52). Washington, DC: Environmental Policy Institute.

World Bank (2019) *State and Trends of Carbon Pricing 2019.* Washington, DC: The World Bank Group.

Yandle, B. (1999) "Bootleggers and Baptists in Retrospect," *Regulation* 22(3): 5–7.

56

COST-BENEFIT ANALYSIS

David Schmidtz

The rule of law is a complex product of negotiation and compromise, continuously evolving in ways only partly foreseen and only partly intended. One source of law is a society's court system. Another source of law, what we unreflectively imagine to be *the* source, is the government's legislative branch. In ideal theory, we imagine that the law of the land is in some way perfect. In truth, no one has power to perfect, not even legislators. A legislature can only hope to influence an ongoing process by which an accumulating body of law drifts, in hope of turning that process in a healthy direction. Experienced legislators are aware that unintended consequences are the rule, not the exception. So, they craft bills as best as they can, compromising continuously with other legislators who have diverse perspectives and are honor-bound to represent different constituencies. They vote. If different legislative branches pass different versions of a bill, reconciliation follows. Something gets treated as having passed, and legislators get back to the campaign trail, praying their bill of legislation turns into something to be proud of.

That sounds complex, but it is the tip of an iceberg. We may find that, once the bill's final shape is resolved, legislators have only a day to prepare to vote on a 12,000-page bill that no one in the world has read, not even the hundreds of legislative staffers and lobbyists who each wrote a few pages of it.

To this, we add yet another layer of mind-boggling complexity. After passage, it falls not to the legislature but to regulatory agencies of the executive branch to sort through the rubble and make sense of what just became law. Regulators puzzle out how to enforce the law. Subject to judicial challenge, law-as-enforced will be law-as-interpreted by regulatory agencies of the executive branch.

So, regulatory agencies need to decide. When unsure, regulators may weigh pros and cons. Occasionally, regulators make the weighing explicit, listing pros and cons and assigning numerical weights. What could go wrong? In fact, things can go terribly wrong.

For example, consider the case of *Peeveyhouse vs. Garland Coal* (Morriss 2000, 144). Having completed a strip-mining operation on Peeveyhouse property, Garland Coal refused to honor its contractual promise to Peeveyhouse to restore the land to its original condition. Garland estimated that it would cost $29,000 to restore the land to its previous condition and that the restoration would add only $300 to the value of the land. Therefore, abiding by the original contract would not be cost-effective. Nevertheless, Peeveyhouse wanted the land restored and Garland Coal had promised to do it. Garland Coal's counter-argument was to admit that Garland had damaged the Peeveyhouse family by refusing to honor the contract, and that Garland was willing to pay for the damages, but the damages amounted to $300, not $29,000.

DOI: 10.4324/9781315768090-67

Incredibly, the Oklahoma court awarded Peeveyhouse only the $300, judging that Garland Coal could not be held liable for a restoration when such restoration would cost more than it was worth. The Court's verdict generally is regarded as utterly mistaken. One way of understanding the mistake is to see it as failing to understand the limits of CBA's legitimate scope.

Somehow, the fact that a promise was made seems relevant. Moreover, it seems relevant in a way that does not depend on the fact that the parties entered into the contract knowing that keeping the promise would cost more than it was worth. Where does that leave us? Do we want to acknowledge that costs and benefits do matter, yet still insist that not everything that matters can be categorized as a cost or a benefit? Does that even make sense?

This chapter considers what cost-benefit analysis can do, and what it cannot.

What Is CBA, and What Is It For?

Many critics of cost-benefit analysis (henceforth CBA) seem driven by a gut feeling that CBA is heartless. They think, in denouncing CBA, they are taking a stand against heartlessness. This is unfortunate. In truth, weighing a proposal's costs and benefits does not make you a bad person. What makes you a bad person is *ignoring* costs—costs you impose on others.

Problems arise not when decision makers take costs and benefits into account, but rather when they *neglect* to do so. The problem in general terms is a problem of *external* cost. External costs are those that decision makers ignore, leaving them to be paid by someone else.

We naturally are prone to ignoring external costs. Every time you drive a car, you are risking other people's lives, and you probably never feel guilty about it. (And just like you, industrial polluters defend themselves by saying, "But everybody does it!") It is only human.

No ethicist would defend forms of CBA that explicitly set aside external cost. What is controversial is whether there exists any defensible other form of CBA. Those with expertise in accounting draw fine-grained distinctions among variations on the basic theme of CBA. Full Cost Accounting, for example, refers to an attempt to carry out CBA in such a way as to take *all* known costs, external as well as internal, into account. From here on, except where otherwise noted, when I speak of CBA, I will be referring to cost-benefit analysis with Full Cost Accounting. (The technical terms here are not quite standardized. What I call Full Cost Accounting is sometimes called Multiple Accounts Analysis or Life Cycle Analysis. Whatever term we use, I have in mind CBA that does not deliberately ignore any cost whatever.)

Under what circumstances, then, should we want policy makers to employ CBA? Two answers come to mind: first, when one group pays the cost of a regulation while another group gets the benefit; second, and more generally, whenever regulators have an incentive not to take full costs into account. When benefits of political decisions are concentrated while costs are dispersed, we see special interest groups pushing through policies even when costs to the larger population outweigh benefits. To contain the proliferation of unconscionable policies, we may require that policies be justified by the lights of a proper CBA. Making a CBA available for public scrutiny is one way of trying to teach regulators to take environmental costs into account. We want social, cultural, and legal arrangements that encourage people to be aware of the full cost of what they do.

If a business pollutes, would it be wrong to insist that the business should pay the true full cost of its operation? No. The fundamental argument for CBA is that a proper CBA is not merely an accounting method. It is a commitment to being accountable for the consequences of one's actions. As a mechanism for holding regulators publicly accountable for external costs, CBA can constrain projects that are not worthwhile when external costs are taken into account, thereby make CBA potentially a friend of the environment.

Current environmentalist opinion, however, generally is anti-CBA. From a philosopher's perspective, CBA seems like the most crudely numerical way of implementing a moral philosophy, namely utilitarianism, that is itself controversial at best. This is a point of real contention, and we will come back to it. The first thing to say, however, is that CBA *per se* is merely an analytical tool, no more, no less. Consider this argument. Everything we care about is either a cost or a benefit. So to say we should do CBA is simply to say we should take what we care about into account. There is a grain of truth to this argument, but also two problems.

First, we can say everything we care about is a cost or a benefit, but saying it does not make it so. When we add up costs and benefits, we add up weights. Yet, not everything is a weight. This is one of the most tragically misunderstood points in moral philosophy. The truth is that treating rights as weights fails to respect the fact that rights are a status, not merely a weight. Your saying no does not imply that I should put great *weight* on your refusal! Rather, the implication of you being the rightholder is that how much weight to put on your interests is your call, not mine. No means no. End of story. For example, we can agree that it is better for a doctor to save five patients than to save one, while also thinking that this is not even the beginning of a proof that doctors have a right to *kill* one patient to save five, even though the weights arguably are the same in both cases: five lives versus one.

A second problem with being too cavalier about costs and benefits is this: we are only human, and humans are wired to seek evidence confirming what they want to believe. When we deal with weights, we aren't built to be impartial. We exaggerate weights in favor of what we want to believe, and discount anything weighing against what we want to believe. Critics and defenders of CBA alike do this, just like anyone else. With that caveat, the following sections consider reasons (some cogent, some not) for distrusting CBA.

Is CBA Anthropocentric?

Is it only human interests that CBA can take into account? Is CBA essentially anthropocentric? No. CBA as construed here is partly an accounting procedure, and partly a way of organizing public debate. In no way is it a substitute for philosophical debate. Animal liberationists who think full costs must include pain suffered by animals must argue for that point in philosophical debate with those who think otherwise. If CBA presupposed one or the other position, thereby falsely implying that there is no room for philosophical debate, that would be a real flaw.

Does CBA Presuppose Utilitarian Moral Theory?

Current utilitarian moral theory holds that X is right if and only if X maximizes utility, where maximizing utility is a matter of producing the best balance of benefits over costs. Does CBA presuppose utilitarianism? No. CBA is a way of organizing a public forum expressing respect for persons: not only persons present at the meeting but other persons as well, on whose behalf those present can speak (citizens of faraway countries, future generations, etc.). For that matter, those present at the forum will speak not only on behalf of other people but on behalf of whatever they care about: animals, trees, canyons, historic sites, and so on. The forum therefore is defensible on utilitarian grounds, but does not presuppose utilitarianism. A CBA ignoring external cost would be endorsed by neither deontologists nor utilitarians, but CBA with Full Cost Accounting could be endorsed by either.

Note that utilitarian (including rule-utilitarian) moral theories are theories about what to do and what to choose. CBA is not a moral theory in that sense; neither does it presuppose a moral theory. CBA is analysis of how things work. (David Hume might have called it a theory about why people see things as useful and agreeable. See Schmidtz 2022.) It is a process of checking apparent costs and benefits, but CBA per se does not tell us to sum them in the way that a

20th-century utilitarian would. Neither does it tell us to use the sum in deciding what to do in the way that a 20th-century utilitarian would.

Does CBA Tell Us to Sacrifice the One for the Sake of the Many?

We can imagine advocates of CBA assuming that actions are justified *whenever* benefits exceed costs. That would be a mistake.

Suppose a doctor, contemplating killing one patient to save five, performs CBA and concludes that, well, five is more than one. Does that mean killing one patient is permitted? Required? No. It would be only human if we tried to get the answer that we *want* by fiddling with the calculation, but if we did that, we would miss what is fundamental. Namely, CBA offers relevant guidance when our mandate is to promote the best balance of costs and benefits, but not all situations are mandates to maximize value. Maximizing value is not always the best way of respecting it. There are times when morality calls on us not to maximize value but simply to respect it.

I argued that CBA does not presume the truth of utilitarianism. Now I may appear to be arguing that CBA presumes utilitarianism is false! Not so. Here is an example of how it sometimes is right to respect value rather than promote it even from a broadly utilitarian perspective. Consider what a terrible idea it would be to give doctors or anyone else a license to kill whenever they think they can do a lot of good by killing. Some institutions have utility precisely by taking utilitarian calculation out of the hands of citizens. Hospitals, for example, cannot serve their purpose unless people can trust hospitals to treat people as rights-bearers. Respecting people's rights is part of what helps make it safe to visit hospitals. And making it safe to visit hospitals is a prerequisite of hospitals functioning properly. Therefore, even a utilitarian should not try to justify killing one patient to save five simply by saying five is more than one. Sometimes, even a utilitarian should see that numbers do not count. Sometimes, refusing to count—acknowledging that one has no right to count—is good policy.

Think again about the case of *Peeveyhouse vs. Garland Coal*. We live in a society where companies like Garland Coal are reasonably expected to honor their contracts, and normally that is exactly what they do. Thus, we know where we stand.

The central and crucial point is this: we do not want to force our fellow citizens to be perpetually preparing to prove before a tribunal that strip-mining their land or killing them to save other patients is *inefficient*. That is not the kind of world we want to live in, if for no other reason than that a society where people think about efficiency in that way would be grossly inefficient. Instead, what makes our society work is that we have a simple right to say no. That simple right to say no empowers us to live with dignity in a community. In giving us moral space that we govern by right, our laws limit how much energy we need to waste trying to influence regulators, fighting to keep what belongs to us, fighting to *take* what belongs to others.

Crucially, our ability to say no teaches regulators and other people to search for ways to get what they want in such a way as to benefit everyone. CBA in its crudest form allows sacrificing the few for the sake of the many. However, the proper purpose of CBA is not to give us a license to sacrifice the few. If we see CBA as indicating when takings are permissible, we will have a problem: breaking contracts, or taking things from people (especially their lives!) whenever benefit exceeds cost is not a way of respecting people. However, if we instead treat CBA as a *constraint* on takings, ruling out inefficient takings without licensing efficient takings, then it is not disrespectful. So, when one ascertains that winners are gaining more than losers are losing, the properly circumspect conclusion is not that we thereby have a license to take but only that the taking has passed one crucial test.

CBA so construed becomes a way of preventing people from treating each other as mere means. Requiring people to offer an accounting of the true costs and benefits of their operations is a way of holding them publicly accountable for failing to treat fellow citizens as ends in themselves. CBA will *not* filter out every proposal that ought to be filtered out, but it will help to filter out many of the most flagrantly disrespectful proposals, and that is its proper purpose.

Must CBA Treat All Values as Mere Commodities?

As Mark Sagoff notes, there are people who believe that "neither worker safety nor environmental quality should be treated as a commodity to be traded at the margin for other commodities, but rather each should be valued for its own sake" (Sagoff 1981, 1288–1289). Sagoff may be correct, but note that CBA is perfectly compatible with the idea that worker safety and environmental quality should be valued for their own sake. Suppose a recycling process improves environmental quality, but inevitably poses some risk to workers. Workers risk getting their hands caught in the machines, and so on. Notice: if we treat both environmental quality and worker safety as ends in themselves, we still have to weigh the operation's costs and benefits. Is the increment of environmental quality worth the risk? We would be ignoring the question if we said "environmental quality is valued for its own sake."

CBA would be needed even if environmental quality were the *sole* value at stake. Suppose recycling saves paper (and therefore trees), but only at the cost of all the water and electricity used in the process. Trucks use gasoline collecting paper from recycling bins. Therefore, recycling has environmental costs as well as environmental benefits. I suggest that we need to know these costs and benefits just in case environmental quality *matters*. "Recycling" is a politically correct word, but does that mean we support any operation using the word in its title, even if the operation is environmentally catastrophic? Or should we instead stop to think about costs and benefits? Those concerned only to maintain politically correct environmentalist appearances do not worry about such things, but stopping to think can be a way of showing respect.

In a nutshell, we sometimes find ourselves in situations of conflicting values. Critics of CBA sometimes seem to say: when values are important, that is when we should *not* think hard about costs and benefits of resolving conflicts in one way rather than another (Sagoff 1981, for example. See also Kelman 1981 or Anderson 1995). That seems backward.

Sagoff asserts that CBA "treats all value judgments other than those made on its behalf as nothing but statements of preference, attitude, or emotion" (Sagoff 1981, 1290–1291). Many things are going on in this passage; I will mention only two. First, Sagoff considers it a mistake to see all values as reducible to costs and benefits. I agree. On one hand, an economist's job is to go as far as possible in treating values as preferences, and within economics narrowly construed, this reductionist bias serves a purpose. On the other hand, when we switch to philosophical analysis, we cannot treat all values as mere preferences, as if valuing honesty were on a par with valuing chocolate. We cannot safely jump from economic to philosophical discussion without acknowledging that what is taken for granted in one discussion cannot be taken for granted in the other.

Second, Sagoff is saying that CBA treats all values as mere preferences. Perhaps, if Sagoff means to say CBA *typically* does so. But saying CBA *necessarily* does so would be incorrect. CBA is about weighing costs and benefits. It does not presume everything is either a cost or a benefit. We have to decide what to treat as mere preferences, costs, or benefits, and what to treat separately, as falling outside the scope of CBA. CBA itself does not decide for us.

To be clear, it is true by definition that to care about X is to have a preference regarding X. However, we can care about X without thinking X *itself* is merely a preference. CBA assumes nothing about the nature of values, other than that values sometimes conflict. Further, CBA does not assume choice is unproblematic; the only thing it assumes is that we sometimes do, after all, need to choose.

Can CBA Handle Qualitative Values?

Steven Kelman says CBA presupposes the desirability of being able to express all values as dollar values. As Kelman correctly notes, converting values to dollars can distort the nature of the values at stake. On the other hand, it would be incorrect to think CBA *requires* us to represent all value as dollar values. If we care about elephants and do a CBA of alternative ways of protecting them, nothing in that process even suggests we have reduced the value of elephants to dollars.

More generally, we sometimes put a dollar value on X despite knowing perfectly well that X's value to us is essentially different from the value of dollars. Philosophers distinguish between instrumental and intrinsic value, and the distinction makes a difference here. An object's instrumental value is the value that it has as a tool. So, a paintbrush's instrumental value is a matter of what I can use it for. I can paint with it. If I produce a painting, the painting may have an instrumental value as well. For example, if I can sell the painting, and thereby use it as a tool for raising the money to buy a car, then the painting has instrumental value in that way. A painting, however, can have a second kind of value as well, namely an intrinsic value. The value that the painting has to me in and of itself, simply because it is a beautiful painting, is its intrinsic value to me. (See Schmidtz 2015 for an extended discussion.)

Both values are real. Neither is necessarily huge. In particular, having intrinsic value does not entail being priceless. Suppose I sell a painting. Kelman says, "selling a thing for money demonstrates that it was valued only instrumentally" (Kelman 1981, 39). Not so fast. The money I receive from the sale is the painting's instrumental value to me, but does my deciding to sell imply that the painting had no intrinsic value? No. Suppose I love the painting, but I desperately need to quickly raise a large sum of money, so I sell. The implication is not that the painting lacks intrinsic value but rather that the instrumental value of selling it can outweigh the intrinsic value of keeping it, at least in desperate circumstances.

Incommensurability of values is not an insurmountable obstacle to CBA. We can make up numbers when assessing the value of a public library we could build on land that otherwise will remain a public park. Maybe the numbers will mean something, maybe not. More often, even when we can accurately predict a policy's true costs and benefits, it does not entail that there will be any bottom line from which we simply read off what to do. When competing values are incommensurable—meaning that they cannot be reduced to a number or any other common measure without distortion—that makes it harder to know the bottom line. It may even mean there is no unitary or quantitative bottom line to be known. Sometimes the only honest way to represent the bottom line is simply to say that one precious and irreplaceable thing is gained while another precious and irreplaceable thing is lost. Even so, that does not mean there is a problem with the very idea of taking costs and benefits into account. It just means we should not assume too much about what kind of bottom line we can expect to see.

Crucially, there often is no point in trying to convert a qualitative balancing into something that *looks* like a precise quantitative calculation and thus *looks* scientific but in fact remains the same qualitative balancing, only now its qualitative nature is disguised by attaching made-up numbers to it.

Some Things Are Priceless. So What?

Critics of CBA think they capture the moral high ground when they say some things are beyond price. They miss the point. For example, choosing to classify elephants as priceless does not settle what is to be done about them. We still need to look at costs and benefits of protecting them in one way rather than another. First, we want our protection to be an effective way of spending whatever dollars we have available to spend on protection. Second, we need to know whether we would be saving them at a cost of sacrificing something equally priceless.

If baby Jessica has fallen into an abandoned well in Midland, Texas and it will cost 9 million dollars to rescue her, is it worth the cost? It seems wrong even to ask; after all, it is only money. But it is not wrong. If it would cost 9 million to save Jessica's life, what would the 9 million otherwise have purchased? Could it have been used to support peace-keeping forces in South Sudan where it might have saved 9,000 lives? Consider an even more expensive case. If a public utility company in Pennsylvania (in the wake of a frivolous lawsuit blaming high-voltage power lines for a child's leukemia) calculates that burying its power lines will cost 2 billion dollars, in the process maybe preventing one or two deaths from leukemia, is it only money? Is it obvious a decent person would not even *think* about it?

Critics like to say not all values are economic values. Yes. In fact, no values are purely economic values. Even money itself is never only money. In a small town in Texas in 1987, a lot of money was spent to save a baby's life—money that took several lifetimes to produce. It was not only money. It did after all save a baby's life. It also gave a community a chance to show the world what it stands for. These are not trivial things. Neither are many other things that on which 9 million could have been spent.

There are things so valuable to us that we view them as beyond price. What does this plain fact imply? Not much. When we must make tradeoffs, should we ignore items we consider priceless, or take them into account? The hard fact is, priceless values sometimes conflict. When that happens, and we try rationally to weigh our options, we in effect put a price on what is priceless. In that case, CBA is not the problem. It is a response to the problem. The world has handed us a painful choice, and trying rationally to weigh our options is our way of trying to cope.

Although critics often speak of incommensurable values, incommensurability is not strictly the issue. Consider the central dilemma of the novel, *Sophie's Choice* (1979). Sophie's two children are about to be executed by a concentration camp commander. The commander says he will kill both unless Sophie picks one to be killed, in which case the commander will spare the other. Now, to Sophie, both children are beyond price. She does not value one more than the other. In some way, she values each more than anything. Nevertheless, she does in the end pick one for execution, thereby saving the other. The point: although her values were incommensurate, she was still able to pick in a situation where the cost of failing to pick would have meant losing both. The values were incommensu*rate*, but not incommensu*rable*. To Sophie, both children were beyond price, but when forced to put a price on them, she could.

Of course, as the sadistic commander foresaw, the process of ranking Sophie's previously incommensurate values was devastating. At some level, commensuration is *always* possible, but there are times when something (our innocence, perhaps) is lost in the process of commensuration. Perhaps that explains why some critics want to reject CBA; they deem it a mechanism for ranking values that should not be ranked.

Elizabeth Anderson voices this concern when she says life and environmental quality are not "mere commodities. By regarding them only as commodity values, cost-benefit analysis fails to consider the proper roles they occupy in public life" (Anderson 1995, 190). But if we *blame Sophie* for treating her children as commodities subject to tradeoffs, we blame the victim. Sophie's treating her children this way is unquestionably a catastrophic failure to honor their value. But when Sophie does CBA, her calculation is not *causing* the catastrophe so much as acknowledging and coping with it. Reducing values to mere commodities can indeed be a terrible thing, but sometimes the world forces tradeoffs on us that in a better world we would have been lucky enough to avoid. Although we can hope people like Sophie will never need to rank their children and can instead go on thinking of each child as having infinite value, and although we can wish we never had to choose between worker safety and environmental quality, or between different aspects of environmental quality, the fact remains that we live in a world that sometimes requires tradeoffs.

Does CBA Work?

When agencies engage in CBA, they typically ask how much they should be willing to pay. That is an obvious and legitimate question because they are, after all, constrained by their budget. Sometimes, though, legislators ask how much they are willing to make *other* people pay. That is a problem. Situations where we are not fully accountable—where we have the option of not paying for our decisions—tend not to bring out the best in us. CBA is potentially a smokescreen for the real action that takes place before numbers get added.

Can anything guarantee that CBA will not be subject to the same political piracy that CBA was supposed to limit? Probably not. I mentioned that the verdict in *Peeveyhouse* generally is regarded as mistaken. What I did not mention is that, as Andrew Morriss notes, "Shortly after the *Peeveyhouse* decision, a corruption investigation uncovered more than 30 years of routine bribery of several of the court's members" (Morriss 2000, 144). CBA per se does not correct for corrupted inputs. Neither does CBA stop people from applying CBA to cases where CBA has no legitimate role. However, if the process is public, with affected parties having a chance to protest when their interests are ignored, public scrutiny will have some tendency to correct for biased inputs. It will also encourage planners to supply inputs that can survive scrutiny in the first place. If the process is public, people can step forward to scrutinize not only valuations but also the list of options, suggesting possibilities that planners have concealed or overlooked.

Even if we know the costs and benefits of any particular factor, it does not guarantee that we have considered everything. In the real world, we must acknowledge that for any actual calculation we perform, it very often turns out that there is some cost or benefit or risk we have overlooked. What can we do to avoid overlooking what in retrospect will become painfully obvious? Although it is no guarantee, the best thing I can think of is to open the process to public scrutiny.

Kelman (1981) says CBA presumes we should spare no cost in enabling policy makers to make decisions in accordance with CBA. Kelman is right to mock such a presumption, for CBA is itself an activity with costs and benefits. Analyzing costs and benefits can be a waste of time. It is not always warranted on cost-benefit grounds, and therefore proper CBA on its own grounds recognizes limits to CBA's legitimate scope.

Must CBA Measure Valuations in Terms of Willingness to Pay?

CBA often is depicted as requiring us to measure a good's value by asking how much people would pay for it. One problem: willingness to pay is a function not only of perceived values but also of resources available for bidding on those values. Poorer people show up as less willing to pay even if, in some other sense, they value the good as much.

Another problem is that surveys designed to measure willingness to pay often fail to take willingness to pay seriously. What they ask subjects to declare is not willingness to pay but *hypothetical* willingness to pay. Such surveys spuriously justify building waste treatment plants in poorer neighborhoods because we *judge* that poorer people would not pay as much as richer people would pay to have the plant built elsewhere. Critics call this environmental racism (because minorities tend to live in poorer neighborhoods). Whatever we call it, we can concede that it does indeed look preposterous.

Is there an alternative that would be more respectful of neighborhoods that provide the most likely building sites? Suppose we initially choose sites by random lottery. Suppose that by random luck, Beverly Hills is selected as the site of a new waste treatment plant. We then ask Beverly Hills's rich residents what they are willing to pay to site the plant elsewhere. Suppose they say they jointly would pay 10 billion dollars to locate the plant elsewhere. Suppose we then announce that the people of Beverly Hills are actually, not just hypothetically, offering 10 billion to any neighborhood willing to serve as the alternate host of that waste treatment facility. Suppose a poor neighborhood accepts the bid. Would that be respectful?

Or instead, suppose no one accepts the Beverly Hills offer, so the plant is built in Beverly Hills. Is anything wrong with richer residents leaving, selling their houses to poorer people happy to live near a waste treatment plant if that means being able to own far nicer houses than they otherwise could afford? If siting a waste treatment plant drives down property values so that poorer people can afford to live in Beverly Hills, while rich people take their money elsewhere, is that bad?

Note that even a random lottery inevitably will produce nonrandom results. Regardless of matter where waste treatment facilities are built, home buyers who opt to move in, accepting the nuisance in order to have a nicer house at a lower price, will tend to be poorer than buyers who opt to pay higher prices to live farther from the nuisance. One thing will never change: waste treatment facilities will be found in poorer neighborhoods. *Not even putting them all in Beverly Hills* could ever change that.

Critics presume the process of siting waste treatment facilities will *not* be conducted in a respectful manner. They presume politicians will site waste treatment facilities in response to calculations about what will minimize adverse effects on campaign contributions and ultimately on reelection bids. I concede that the critics (Bullard 1990, for example) may be right. Under those circumstances, the point of subjecting a CBA to public scrutiny is to lead (quite possibly racist) politicians not to recalculate answers so much as to start asking the right questions.

Must Future Generations Be Discounted?

In financial markets, a dollar acquired today is worth more than a dollar we acquire in a year. Dollars acquired today can be put to work immediately. At worst, dollars put in the bank can collect interest of a few percent. Therefore, if you ask me how much I would pay today to be given a dollar a year from now, I certainly would not pay as much as a dollar. I might pay a few percent less. Properly valued, then, future dollars sell at a discount. Therefore, borrowing to get a profitable project off the ground can be rational, even when the cost of borrowing a thousand dollars now will be more than a thousand later.

Here is the catch. There is nothing wrong with taking out a loan, so long as we *pay it back*. But there is something wrong with taking out a loan we have no intention of repaying. Discounting is one thing when the cost of raising capital is internalized, and something else when we borrow against *someone else's* future rather than our own. Let me stress: the problem here is not discounting; it is the same problem of external cost with which this chapter began. We have no right to discount the price that *others* will have to pay for our projects.

In any case, if we undertake a CBA to evaluate the merits of borrowing against our futures, we decide how or whether to discount. CBA will not decide for us.

Conclusions

I talked about CBA with Full Cost Accounting, but no mechanical procedure can be guaranteed to take all costs into account. For any mechanical procedure we devise, there will be situations where that procedure overlooks something important. This is reason not to reject the very idea of CBA, though, but rather to be wary of the desire to make decisions in a mechanical way. We cannot wait for someone to devise a perfect procedure, guaranteed to give everything its proper weight. Whatever procedures we devise for making decisions as individuals or as a community, we need to exercise judgment. At some point we draw the line, make a decision, and get on with our lives, realizing that any real-world decision procedure inevitably will be of limited value. It will not be perfect. It never will be beyond question.

CBA is an important response to a real problem. However, it is not magic. There is a limit to what it can do. CBA is a way of organizing information. It can be a forum for eliciting further information. It can be a forum for correcting biased information. It can be a forum for giving affected parties a voice in community decision making, thereby leading to better understanding of, and greater acceptance of, the tradeoffs involved in running a community. CBA can be all of these good things, but it is not necessarily so. CBA can constrain a system's tendency to invite abuse, but CBA is prone to the same abuse that infects the system as a whole. It is no panacea. It is an antidote to abuse that is itself subject to abuse.

CBA is not inherently biased, but if inputs are biased, then so will be the outputs, generally speaking. However, although the method does not inherently correct for biased inputs, if the process is conducted publicly, so that people can publicly challenge suppliers of biased inputs, there will be some tendency for the process to correct for biased inputs as well. We can hope there will be adequate opportunity for those with minority viewpoints to challenge mainstream biases, but we cannot guarantee it.

The most we can say is this: a CBA that weighs in favor of a policy or piece of legislation is to that extent morally defensible. That is, the benefits exceed the costs. But we need to understand that this defense will not be morally conclusive. It is a moral defense offered en route to a political compromise. Moral compromise is a bad thing, but political compromise is not. We are moving toward a political compromise when we realize that what we want is not what our fellow citizens want, and that we do not want our fellow citizens to be enemies. Rather, we want our fellow citizens to be our partners in building communities that do not require all of us to want the same thing. We want to send them away with a compromise that may not have given them everything they wanted, but that did give them something, and that leaves them with a sense that they are better off with us than without us. If you have to resort to a CBA, you are probably already in a situation where no one is going to get everything they want, but if you can resort to CBA, then at least you have a peace-maker. Namely, if CBA is the basis for your political compromise, then the fact that none of you got everything you wanted does not imply that any of you are second-class citizens. It means that what you wanted was on the table, but so was the price tag of what you wanted, and that a CBA done in public view gave your democracy a fighting chance to operate as democracies are supposed to operate.

Acknowledgments

This chapter radically condenses, substantially revises, and generally supersedes "A Place for Cost-Benefit Analysis," *Philosophical Issues* (*Noûs* annual supplement) 11(2001) 148–171. My work on this chapter was supported by a grant from the John Templeton Foundation. The opinions expressed here are those of the author and do not necessarily reflect the views of the John Templeton Foundation. I'm also grateful to the Georgetown Institute for the Study of Markets and Ethics for their hospitality when I was in residence as a Visiting Scholar in the fall of 2016, and to the Property and Environmental Research Center for generously supporting my work in the summer of 2014. Thanks also to Ben Hale and Lydia Lawhon for helpful suggestions on a penultimate draft.

References

Anderson, E. (1995) *Value in Ethics and Economics*. Cambridge: Harvard University Press.

Bullard, R. (1990) *Dumping in Dixie: Race, Class, and Environmental Quality*. Boulder, CO: Westview Press.

Kelman, S. (1981) 'Cost-Benefit Analysis: An Ethical Critique', *Regulation* 5(1), pp. 33–40

Morriss, A. (2000) "Lessons for Environmental Law from the American Codification Debate," in Meiners, R. and Morriss, A. (eds.) *The Common Law and the Environment*. Lanham, MD: Rowman & Littlefield, pp. 130–157.

Sagoff, M. (1981) 'At the Shrine of Our Lady of Fatima, or Why Political Questions are not all Economic', *Arizona Law Review* 23, pp. 1283–1298.

Schmidtz, D. (2015) 'Value in Nature,' in Olson, J. and Hirose, E. (eds.) *Oxford Handbook of Value Theory*. New York: Oxford University Press, pp. 381–398.

Schmidtz, D. (2022) *Living Together: Reinventing Moral Science*. New York: Oxford University Press.

57

RISK ASSESSMENT

Sven Ove Hansson

- High levels of air pollutants have been measured in a city center. They can be reduced by restricting motor traffic in the city. However, that would be costly and inconvenient for many of those who work and live in the city. Are the risks from air pollution serious enough to warrant such measures?
- Some plastic toys have been shown to leak a toxic substance in quantities so low that they have only recently become measurable. Can such a small exposure pose a risk to children playing with the toys?
- Modern forestry has drastically reduced the habitats suitable for several species of woodpeckers and other birds living on xylophagous (wood-eating) insects and their larvae. How large is the risk of extinction for these bird species?
- An underground depository for nuclear waste has been proposed. Anti-nuclear activists point to risks that future leakages of radiotoxic substances can harm humans and the environment at some future time. How large is that risk, and how should we assess risks that will only materialize thousands of years into the future?

In these and many other cases, environmental policies need to be informed by careful studies of potential risks. It has often been assumed that risk assessments can be performed independently of values and ethical considerations. However, experience shows that risk assessment is closely connected with ethical issues. The ethical issues of risk should be dealt with openly and systematically.

The Dichotomous Model of Risk Assessment and Management

In the late 1960s, increased public attention to environmental risks gave rise to a wave of academic activities related to risk. Scientists and scholars from a wide range of disciplines, often in new interdisciplinary combinations, started to investigate risks and assess their potential impact. It was soon realized that decision-making on risk-taking and countermeasures is a complex process that involves both scientific judgments and value-based policy positions. There can be scientific answers to questions about what types of dangerous events can happen, and how plausible or probable they are, but science alone cannot tell us what risks we should accept.

Based on the lessons learned from various encounters between science, policy making, and public opinion, risk analysts proposed what can be called the *dichotomous model* for decision-making about risks. Its most famous expression was a 1983 report by the American National Academy of Sciences (NAS). The basic idea in this model is that the decision procedure for major risk

DOI: 10.4324/9781315768090-68

management decisions should be divided into two distinct parts to be performed consecutively. The two parts correspond roughly to the distinction between facts and values. The first part, commonly called *risk assessment*, is described as a fact-based, scientific undertaking. It consists of collecting and assessing the relevant information about the risks and, based on that information, characterizing their nature and magnitude. The second part is called *risk management*. It is based on the outcome of risk assessment, to which it adds economic, political, and social information. Based on all the available information, a decision is made on what measures – if any – should be taken to reduce the risk. An essential difference between risk assessment and risk management, according to this view, is that social and ethical values only appear in risk management. Risk assessment is perceived as (at least ideally) a value-free process.

This dichotomous model has served important social purposes, such as protecting science from undue pressure from decision-makers and making the values on which decisions are based more explicit. However, the model also has serious problems. Most fundamentally, the separation of values from risk assessment is usually only partially attainable. This is largely because judgments of risks tend to be complexly interwoven combinations of value statements and statements of fact (Hansson 2010). Risk has the double nature of being both fact-laden and value-laden. The statement that you risk losing your leg if you tread on a landmine has both a factual component (landmines tend to dismember people who tread on them) and a value component (it is undesirable that you lose your leg). In principle, once we have made the value-based (but in this case rather uncontroversial) decision that it is undesirable for people to be hurt by landmines, an assessment of how often and under what circumstances this will happen should be performable in a strictly fact-based way, with no additional references to values. However, in practice, the separation of value issues from fact issues is often difficult, and not performable as a single, preparatory measure before risk assessment begins (Hansson and Aven 2014). In controversial areas, participating parties tend to present their value-based standpoints in the guise of purely fact-based statements. It may then be quite difficult to disentangle the two components (Wagner 1995).

The consecutive principle in the dichotomous model can also be problematic. According to the model, scientists are expected to finish their risk assessment before the policy makers take over. This approach is often inadequate since policy makers tend to repeatedly ask for additional information. Therefore, an iterative and dialogical process is usually more suitable. Another problem for the model is that decision-makers often prefer to have clean hands, and therefore delegate "impossible" decisions to experts. In consequence, expert committees are often required not only to determine what the risks are, but also to draw the line between acceptable and non-acceptable exposures, which is of course by definition a risk management task (Clausen Mork and Hansson 2007).

The Standard Methodology

Risk assessments are performed on many types of risk, and they are therefore based on many types of expertise: medical expertise for health risks, biological expertise for environmental risks, climatological expertise for climate-related risks, economic expertise for economic risks, etc. However, in spite of this multifariousness we can identify a common core of methodologies that can be summarized in terms of two major components.

The first of these is *scenario development*, which consists in determining what undesirable events can possibly happen. Any risk assessment worth the name must include a serious attempt to find out what can go wrong and what the consequences can then be. In technological risk assessment, this usually takes the form of a *fault tree analysis*, i.e. a careful investigation of various chains of events that may lead to failures. A fault tree analysis is often an efficient way to identify weaknesses in a complex technological system. For instance, if failures in a particular component turn out to have an important role in several chains of events that lead to an accident, then that is a good reason to improve the component in question and/or make the system less dependent on its functioning. But it must be recognized that in a complex system an exhaustive list of dangerous event chains cannot be obtained. Therefore, accident prevention should not be based on the assumption that only the identified accident scenarios can take place. Many serious accidents have resulted from chains of events that were not foreseen in pre-accident fault tree analyses. One example of this is the Fukushima Daiichi nuclear disaster in 2011. It was caused by a tsunami that was larger than those included in the fault tree analyses that were used for the design of the plant.

The second major component of risk assessment is *probabilistic analysis*. For some risks we have fairly good probability estimates that are based on empirically observed frequencies. For instance, we know the overall probabilities of fatal accidents in different transportation systems. But in many other cases we do not have sufficient basis for such, experience-based estimates of probabilities. This applies in particular to rare but large and potentially disastrous accidents. The common procedure is then to ask experts to assign probabilities to the events in question. Unfortunately, such probability estimates are more uncertain and error-prone than what some risk assessors seem to believe. Although it is reasonable to assume that experts make better estimates than laypeople, there is no lack of examples of inaccurate expert estimates. For instance, early estimates by nuclear energy experts of the probability of a core damage in a nuclear reactor were too low. (The highly influential WASH-1400 report in 1975 predicted that the frequency of core damages would be 1 in 20,000 reactor years. In the first 15,000 reactor years there were 10 accidents with core damages, i.e. about 1 in 1,500 reactor years. See Cochran 2011; Escobar Rangel and Lévêque 2014; Ha-Duong and Journé 2014.)

Risk analysts combine empirical information with probability estimates to calculate the statistical *expectation value* of the unwanted events under analysis. The synonymous terms *probabilistic risk assessment (PRA)* and *probabilistic safety assessment (PSA)* are used to denote such calculations. The expectation value (often but confusingly called "the risk") is the number obtained by multiplying the probability of an unwanted event by a measure of its disvalue (negative value). If only death risks are considered (which is a surprisingly common approach), then this means that risk is identified with the statistically expected number of deaths caused by a possible event or class of possible events. Hence, if 200 deep-sea divers perform an operation in which the individual risk of death is 0.1% for each individual, then the expected number of fatalities from this activity is $0.001 \times 200 = 0.2$. Expectation values have the important property that they can be added up in a meaningful way. Suppose that a certain activity is associated with a 1% probability of an accident that will kill five persons, and also with a 2% probability of another type of accident that will kill one person. Then the total expectation value is $0.01 \times 5 + 0.02 \times 1 = 0.07$ deaths. In similar fashion, the expected number of deaths from a nuclear power plant is equal to the sum of the expectation values for each of the various types of accidents that can occur in the plant.

Some technological risks that have given rise to much public opposition, such as those associated with nuclear reactors and nuclear waste disposal, have much lower expectation values of lethal outcomes than some other risks, such as those associated with road traffic, which have encountered much less public opposition. Some risk analysts see this as a sign of "irrationality" among the public, and call the expectation values "objective risks" and the public's risk appraisals "subjective risks." However, many of the "objective risks" are to a large extent based on subjective estimates (by experts, but expertise is no guarantee against bias or cognitive failures).

Probabilistic risk assessments of technological systems have often been based on fault tree analyses. In the 1970s and 1980s, leading risk analysts assumed that by adding up the expectation values from all identified event trees leading to an accident, they could obtain a reasonable approximation of the total expected number of deaths from accidents in a complex system such as a nuclear reactor. However, there are several reasons why this does not work. As already mentioned, it is not possible to identify all accident scenarios in a complex technological system. Accidents can happen in more ways than we can foresee. Furthermore, many of the probabilities involved are not known empirically, and therefore uncertain expert judgments have to be used.

Yet another problem with such calculations is that the probabilities of different events depend on each other in highly intricate ways. For example, suppose that an accident will occur if two safety valves fail. We have reasons to believe that the probability is 1 in 500 that a valve of this construction will fail during a period of one year. This may lead us to believe that the probability that both will fail in that period is 1/500×1/500, i.e. 1/250,000. However, this may be a gross underestimate of the probability since failures in the two valves are not independent events. For instance, if the valves are of the same type, then an unknown manufacturing defect is likely to affect both of them. Failure to follow the maintenance plan can lead to malfunction in both, and so can a fire that damages both of them. Therefore, the probability of failures in both of them will have to be assigned a higher estimate than what we obtained by just multiplying the two probabilities. Such dependencies make probabilistic risk assessment a much more complex and uncertain enterprise than what was originally thought.

This does not mean that probability calculations of accidents are meaningless. To the contrary, when competently performed, such calculations can be very useful tools to identify potential failures in a technological system that risk managers should attend to. In modern probabilistic risk assessment, the focus is on the contributions of various types of failures rather than on the total sum (Michal 2000).

It should be emphasized that these considerations refer to the risk assessment of rare events in complex technological systems. This is an application in which it is extremely difficult to make reliable probability estimates. In many other areas of risk assessment, reasonably reliable estimates of the relevant probabilities are obtainable. This applies not least to many health risks. We know the sizes of the fatal risks associated with many of the environmental and occupational exposures that threaten our health. For instance, there is abundant evidence that half of the smokers die prematurely due to tobacco-related disease (Boyle 1997; cf. Doll et al. 2004).

The Ethical Deficit in Risk Assessment

Traditional risk assessment is entirely focused on probabilities and on the size of losses (negative values). Probabilistic risk assessment, currently the dominant methodology in the field, treats risks as impersonal entities and pays little attention to how they are distributed or how their distribution is related to that of the benefits they are supposedly associated with. This impersonal approach is almost diametrically different from how risks are discussed in most other contexts than professional risk assessment. When policy makers or the public discuss risks, persons and their relations to each other tend to be at the center of interest. In such discussions, risks are related to the persons who take them, are exposed to them, or expose others to them. This "personalized" approach to risk is closely associated with attention to the moral issues of risk. Outside of professional risk assessment, the question is not only what the risks are and how large they are. Much attention is also paid to issues such as who exposes whom to the risks, with what right they do so, and for what purpose. In short, when professional risk assessors discuss risks in terms of numbers, the surrounding society largely discusses them in moral terms.

Expert assessments that are based on quantitative risk analysis or risk-benefit analysis have repeatedly been mistrusted by the public. This mistrust has often been attributed to inefficient communication or lack of public understanding. In some cases it may instead have resulted from the inability of an established analytical framework to deal adequately with ethical and normative issues to which members of the public attach great importance. Standard (probabilistic) risk assessment has an *ethical deficit* that has much to do with its exclusive focus on probabilities and impersonal consequences. The following is a brief list of considerations for which standard risk assessment is badly equipped:

- Individuals may have *rights* not to be exposed to risks. According to some ethical views, rights and duties may be more important than consequences in determining what risks you are permitted to expose another person to. For instance, suppose that your neighbor applies pesticides in his garden in such a careless way that your tomatoes are also sprayed. You may very well claim that he had no right to do this, even if he can prove that the potential health risk from eating your sprayed tomatoes is minuscule.
- We usually regard *involuntary* risk-taking as more problematic than voluntary risk-taking. It makes a big difference if someone is hurt when jumping into the water from a high bridge for his own pleasure, or when being forced by others to do. However, the distinction between voluntary and involuntary risk exposure is unclear and seems to depend largely on social conventions that are taken for granted without much reflection.
- To *intentionally* expose someone else to a risk is usually seen as more problematic than doing so unintentionally. Compare, for instance, the act of throwing down a brick on a person from a high building to the act of throwing down a brick from a high building without first making sure that there is nobody beneath who can be hit by the brick. Most people would regard the former act as morally more blameworthy even if the probability of someone being harmed is the same in both cases.
- The treatment of *sensitive individuals* is often ethically problematic. Risk assessments tend to focus on individuals with average sensitivity to the exposure in question. In many cases there are identifiable groups that run larger risks than others. For instance, the cancer risk from ionizing radiation is around 40% higher for women than for men at any given level of exposure (ICRP 2007: 210). Exposure limits are based on a population average, rather than subpopulations. The ethical implications of this practice are seldom discussed (Hansson 2009; Hansson and Schenk 2016).
- The *distribution* of risks is a critical issue in many policy contexts. For instance, measures against the tobacco epidemic have been much less successful in disadvantaged subpopulations than in the more privileged ones (Graham 2012). In many countries, underprivileged ethnic groups have a disproportionate exposure to multiple health risks as well as to social and economic risks (Adeola 2000; Bullard 2001; Cureton 2011; Mohai et al. 2009).

One way to deal with the ethical deficit is to introduce ethical components into standard risk analysis. This can often be done adequately with distributional analysis. Total risk estimates can be supplemented with estimates for various subpopulations, and policy makers can then choose whether to focus on the total sum or on disadvantaged populations. For instance, decisions on urban air pollution can be based not only on the health risks for average city dwellers but also on the (much higher) risks for people with respiratory diseases such as asthma (Guarnieri and Balmes 2014).

Other ethical aspects, such as those concerning rights and intentions, cannot be so easily dealt with by such additions to standard risk analysis. Its framework is thoroughly consequentialist in a way that makes it difficult to introduce these types of considerations. Therefore, the alternative approach of performing an entirely separate ethical risk assessment has considerable advantages.

A Method for Ethical Risk Assessment

The method of ethical risk assessment to be presented here has its focus on inter-individual relations (Hansson 2018; for earlier treatments, see Hansson 2016, 2017; Hermansson and Hansson 2007; Wolff 2010). In every risk management problem there are people involved in different ways.

The vague term "stakeholder" is often used to cover all such groups. For the purposes of ethical risk assessment we can identify three major groups of stakeholders. When there is risk exposure, there are people exposed to the risk, the *risk-exposed*. There are also people who make decisions affecting the risk, the *decision-makers*. Since non-trivial risks are rarely taken unless they are associated with some benefit, there are almost invariably also people who gain from the risk being taken; we can call them *beneficiaries*.

The relationships among those who have these three roles are determinative for the ethical aspects of a risk management problem. Two types of such relationships are particularly important. First, how are the roles combined? Are (some of) the risk-exposed also beneficiaries? Do the risk-exposed take part in the decision-making, and if so to what extent? Are the decision-makers also beneficiaries, i.e. do they gain from the risk exposure? Second, does one of the parties depend on another party in some way or other? For instance, are the risk-exposed economically dependent on the decision-makers, or do they depend on them for information about the risk?

The combinations of roles are indicated in Figure 57.1. Let us consider each of the seven possible combinations.

Risk role 1, only beneficiary: In this position we find individuals who gain from others being exposed to a risk, without being themselves either risk-exposed or decision-makers. This would typically be a favorable position for the individual. However, it is also an indication of potential distributional problems in the society, in particular if the risk is serious. One example of people in this position is consumers in Western countries who buy cheap goods produced under dangerous working conditions in Third World countries. We may well ask whether it is acceptable to gain in this way from others being exposed to a risk.

Risk role 2, beneficiary and decision-maker: Persons in this position do not themselves take the risk, but they make decisions that lead to others being exposed to the risk. This group includes most of the people in privileged positions whose decisions lead to risks for workers, consumers, and the environment. They gain from risk-taking, which means that they have incentives to act in ways that lead to more risks being taken. This is of course a problematic situation from the viewpoint of risk management. In many cases, various types of institutional arrangements have been made either to avoid this combination of roles, or to prevent or alleviate its negative effects. When the risks are economic in nature, these measures have often been quite strict. One example of this is regulations against insider trading. Many countries prohibit persons in certain decision-making positions from taking actions that would benefit themselves economically while putting other investors at risks.

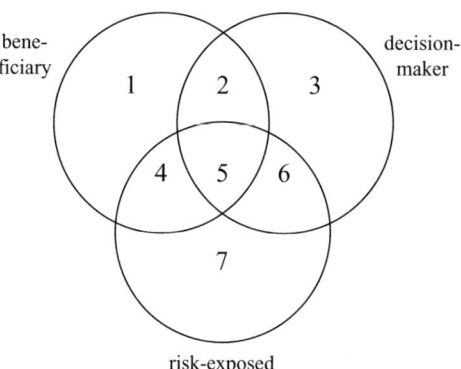

Figure 57.1   Combinations of roles

Countermeasures are also taken when the risks are environmental or health-related, but in these cases the measures are usually milder. Businesspeople are not prohibited from making decisions in which they themselves gain economically from exposing others to risk. Instead, their decisions are hedged by restrictions on how large risks they may impose on others, for instance in the form or regulations on workplace health and safety, consumer protection, and environmental protection. These restrictions are enforced by other decision-makers, such as government officials (risk role 3) and workers' safety representatives (risk role 6).

Risk role 2 is a privileged place to be in, but the privilege comes with a price. People in this position often find themselves distrusted and criticized. They often perceive themselves as having communication problems. However, the reasons for public distrust usually go deeper than to communication problems. The incentives structure – that some people gain from exposing others to risk – is in itself problematic.

Risk role 3, only decision-maker: This is the position of a "pure" decision-maker, someone who makes decisions without being exposed to the risks, and also without being one of those who gain from the risk being taken. A typical example would be (at least in the ideal case) a government official whose job is to make impartial decisions that restrict risk exposure, and/or balance the interests of the risk-exposed against those of the beneficiaries. This is the type of decision-making that we would expect of environmental protection agencies.

Another example is a physician or other proxy who decides on the treatment of a patient who is unable to make that decision herself. All treatment decisions involve a weighing of the expected positive effects of treatments against the risks of negative side effects. This should be done entirely in the interest of the patient. As patients we are anxious that the physician has a risk role of the third rather than the second type in our classification. Commercial healthcare organizations can indeed be found to have risk role 2, but medical professionalism, ethics, and various forms of oversight are expected to make them behave as if they had risk role 3.

Risk role 4, risk-exposed and beneficiary: In this position we find persons who are exposed to the risk and also receive benefits from it, but do not take part in the decisions. This can be a favorable position if the risk is outweighed by the benefit, but if the opposite is the case, then it is of course an unfavorable position. An unconscious person who arrives at the hospital in need of acute care is in this position; whether it is favorable depends on the proficiency of the clinical decision-makers (whom we find in risk role 3). However, from a procedural point of view, risk role 4 is always disadvantageous in the sense that the person is precluded from making or at least participating in decisions concerning herself.

Risk role 5, risk-exposed, beneficiary and decision-maker: This would normally be seen as the ethically ideal position people can be in with respect to risk. If a person carries the risk herself, receives the advantages associated with the risk, and makes the decisions herself, can there be any ethical problems to worry about?

Well, in practice there usually can. This is because in addition to the person in risk role 5, there are usually other affected persons in other risk roles. There may be others at risk as well, and perhaps some of them do not have a say, or do not have a share in the benefits. (They can then have risk roles 4, 6, or 7.) Off-piste skiing is a clear enough example of this. You may have risk role 5 with respect to your decision to take a dangerous route on the mountain, but if you trigger an avalanche, then others are at danger, and they certainly did not have risk role 5 with respect to your decision. If you have a family or other dependents, then they may be affected by your decision to take a risk. John Stuart Mill, whose *On Liberty* is rightly lauded as a milestone in the defense of human liberty, did not hesitate to assert that we have a duty to preserve our own health since "by squandering our health we disable ourselves from rendering to our fellow-creatures the services to which they are entitled" (Mill [1865] 1969: 340).

Equally importantly, even if an exposed person is a decision-maker, there are usually also other persons who have made decisions that contributed to the risk. (They can have risk roles

2 or 3.) In most of the situations when a person is exposed to a risk, several decisions have been made that contribute, directly or indirectly, to the risk exposure. Typically, some, but not all, of these decisions have been made by the risk-exposed person her- or himself. For instance, when a smoker dies from smoking, this is the result of a complex combination of causes, including the smoker's decision (usually as a minor many years ago) to smoke her first cigarette, but also many decisions by cigarette manufacturers and other "background decision-makers" who contribute to the continued sale of a deadly and highly addictive product. Unsurprisingly, those who contribute to the risk exposure of others tend to deemphasize the causal role of their own decisions, and try to create the impression that the decisions by the risk-exposed persons themselves are all that matters. Tobacco companies have sponsored anti-paternalist campaigns that focus on the right of smokers to smoke (Fox 2005). The (implicit) inference seems to be that if the smoker has a moral right to harm herself, or risk harming herself, then the tobacco company has a moral right to aggressively market products with which she can do so. This, however, is a *non sequitur*. A person's moral right to expose herself to a risk or a harm does not necessarily imply a right for others to facilitate or contribute to her doing so (Hansson 2005). In an ethical risk analysis it is important to include all important decision-makers that contribute to a risk, not least those that try to absolve themselves of responsibility.

Risk role 6, risk-exposed and decision-maker: In this position we find persons who are both risk-exposed and decision-makers, but do not gain anything from the risk they take. This seems to be a rather uncommon combination of risk roles. Perhaps the best example is that of a bystander who takes a risk to save someone from a danger. This does not seem to be a particularly common role combination. Joe Wolff introduced the term "maternalism" for this type of self-sacrificing risk-taking, since it is presumably a particularly common behavior for parents (Wolff 2010: 156).

Risk role 7, only risk-exposed: In the position at the bottom of the diagram, we find risk-exposed persons who neither take part in the decisions nor have any share in the benefits that result from their risk-taking. In other words, this is the position of uncompensated and powerless risk-exposed persons. This is of course a highly undesirable position to be in. Unfortunately, a large share of the risks that are taken in modern societies are borne by people in this position. This applies not least to workplace risks, air pollution, and other environmental exposures.

According to standard risk analysis (and utilitarian ethics), it is not necessarily a problem if people are in this position. To the contrary, such an arrangement is satisfactory and even laudable to the extent that the total benefits add up to a greater sum than the total risks. Indeed, this type of argument has often been used in attempts to make underprivileged people accept their risk exposure. So-called NIMBY (not in my backyard) cases are highly illustrative of this. These are cases when people oppose a planned facility, such as a polluting incineration plant, in the vicinity of their homes. They are then often accused of just wanting it to be placed in someone else's backyard instead of their own. In typical cases, society at large will (at least allegedly) gain from building the plant, but its neighbors will lose. However, this argumentation is based on the highly dubious premise that a disadvantage to one person is automatically outweighed, from a moral point of view, by a larger advantage received by others. This is a morality for exploitation, not for social justice. Most of the grievances concerning health and the environment that have been dismissed as "NIMBY" have in fact referred to injustices that could and should have been avoided (Hermansson 2007; Luloff et al. 1998; Wolsink 2006).

In major environmental issues, such as those concerning biodiversity and the climate, the majority of the risk-exposed persons, namely the future persons that will inhabit this planet after us, are all in position 7. They do not have a say, and whatever gains we make from burning fossil fuel or chopping down the rain forest will not be enjoyed by them. It is up to us to take their interests into account.

Adding up all this, the two most problematic positions in this diagram are those in risk roles 2 and 7. The identification of groups with these two risk roles is a crucial task in ethical risk analysis. This is because unethical risk exposures tend to be common when there are decision-makers who reap the benefits of risks taken by others, and/or risk-takers who have no share in the decision-making or in the benefits associated with the risk. Another important task in ethical risk analysis, mentioned above in relation to risk role 5, is the identification of background decisions-makers who try to tone done their own roles and responsibilities in relation to the risk.

When Is a Risk Acceptable?

Unfortunately, we cannot live risk-free lives. What we can do, however, is to arrange our societies so that risk-taking is ethically acceptable, both in terms of the distribution of risks and benefits and in terms of how decisions on risk-taking and its distribution are made. The following criterion is an attempt to summarize what it takes to make it acceptable to expose others to a risk:

> Exposure of a person to a risk is acceptable if (i) this exposure is part of a persistently justice-seeking social practice of risk-taking that works to her advantage and which she de facto accepts by making use of its advantages, and (ii) she has as much influence over her risk-exposure as every similarly risk-exposed person can have without loss of the social benefits that justify the risk exposure.
>
> *(Hansson 2013: 108)*

The first clause makes it clear that societal risk exposure cannot be justified by advantages to others than those who are exposed to risk. However, it allows for *fair risk exchanges*. In modern society, with its division of labor and its complex production of commodities, we engage in different work tasks and other activities, which are associated with different types of risks and benefits. We may then be said to exchange risks and benefits with each other. According to the proposed principle, for this to be acceptable, the total effects of all these exchanges should be to the benefit of all concerned, and society should strive to make the distribution as fair as possible.

The second clause is concerned with the decision-making processes. At least in theory, the first clause could be realized by a well-meaning dictator distributing risk impositions among the population. However, this would be disrespectful to the risk-exposed persons. People should be allowed to make their own decisions in their private lives, and they should have a say and a vote in issues affecting society at large. The principle of *maximal equal influence* that is laid down in this clause is intended to ensure this combination of private freedom and an equal share of influence in societal decision-making.

Conclusion

Philosophers did not have a big role in the early development of risk analysis. Most of the philosophical contributions to the area were in fact outsiders' criticisms of risk analysis. In the early phases of academic risk analysis there was a tendency to overemphasize the objective component of risk assessment, and downplay the value issues. Risk assessments were often presented as objective scientific statements, even when they took a stand on value-laden issues such as risk acceptability. Most of the early philosophical work on risk had as its main purpose to expose the value-dependence of allegedly value-free risk assessments (Cranor 1997; Hansson 1998; MacLean 1985; Shrader-Frechette 1991; Thomson 1985). This was an important task, and it was undertaken with some success. Although hidden value assumptions are still common in risk assessments, risk analysts and their audiences are now much more aware of their presence. Today it may be time to extend the moral analysis of risks to more constructive tasks (Hansson 2013). Using the tools introduced above we can go beyond the simple aggregations of conventional risk analysis and provide a richer basis for risk management decisions that takes the rights and the interests of everyone affected into account.

References

Adeola, Francis O. (2000) "Cross-national environmental injustice and human rights issues a review of evidence in the developing world," *American Behavioral Scientist* 43:686–706.

Boyle, P. (1997) "Cancer, cigarette smoking and premature death in Europe: a review including the recommendations of European cancer experts consensus meeting, Helsinki, October 1996," *Lung Cancer* 17(1):1–60.

Bullard, Robert D. (2001) "Environmental justice in the 21st century: race still matters," *Phylon* 49:151–171.

Clausen Mork, Jonas and Sven Ove Hansson (2007) "Eurocodes and REACH – differences and similarities," *Risk Management* 9:19–35.

Cochran, Thomas B. (2011) "Statement on the Fukushima nuclear disaster and its implications for U.S. nuclear power reactors", Joint Hearings of the Subcommittee on Clean Air and Nuclear Safety and the Committee on Environment and Public Works, United States Senate, April 12, 2011. Downloaded March 22, 2015. Available at: http://www.nrdc.org/nuclear/files/tcochran_110412.pdf.

Cranor, C.F. (1997) "The normative nature of risk assessment: features and possibilities," *Risk: Health, Safety & Environment* 8:123–136.

Cureton, Shava (2011) "Environmental victims: environmental injustice issues that threaten the health of children living in poverty," *Reviews on Environmental Health* 26:141–147.

Doll, Richard, Richard Peto, Jillian Boreham och Isabelle Sutherland (2004) "Mortality in relation to smoking: 50 years' observations on male British doctors," *BMJ* 328(7455):1519–1533.

Escobar Rangel, Linda, and François Lévêque (2014) "How Fukushima Dai-ichi core meltdown changed the probability of nuclear accidents?," *Safety Science* 64:90–98.

Fox, Brion J. (2005) "Framing tobacco control efforts within an ethical context," *Tobacco Control* 14(Suppl 2):38–44.

Graham, Hilary (2012) "Smoking, stigma and social class," *Journal of Social Policy* 41:83–99.

Guarnieri, Michael and John R. Balmes (2014) "Outdoor air pollution and asthma," *Lancet* 383(9928): 1581–1592.

Ha-Duong, Minh, and Venance Journé (2014) "Calculating nuclear accident probabilities from empirical frequencies," *Environment Systems and Decisions* 34:249–258.

Hansson, Sven Ove (1998) *Setting the Limit. Occupational Health Standards and the Limits of Science.* New York: Oxford University Press.

Hansson, Sven Ove (2005) "Extended antipaternalism," *Journal of Medical Ethics* 31:97–100.

Hansson, Sven Ove (2009) "Should we protect the most sensitive people?," *Journal of Radiological Protection* 29:211–218.

Hansson, Sven Ove (2010) "Risk – objective or subjective, facts or values?," *Journal of Risk Research* 13:231–238.

Hansson, Sven Ove (2013) *The Ethics of Risk. Ethical Analysis in an Uncertain World.* New York: Palgrave Macmillan

Hansson, Sven Ove (2016) "Managing risks of the unknown," pp. 155–172 in Paolo Gardoni, Colleen Murphy and Arden Rowell (eds.) *Risk Analysis of Natural Hazards.* Cham: Springer.

Hansson, Sven Ove (2017) "Ethical risk analysis," in press in Sven Ove Hansson (ed.) *The Ethics of Technology. Methods and Approaches.* London: Rowman and Littlefield.

Hansson, Sven Ove (2018) "How to perform an ethical risk analysis (eRA)," *Risk Analysis* 38(9):1820–1829.

Hansson, Sven Ove and Terje Aven (2014) "Is risk analysis scientific?," *Risk Analysis* 34:1173–1183.

Hansson, Sven Ove and Linda Schenk (2016) "Protection without discrimination: pregnancy and occupational health regulations," *European Journal of Risk Regulation* 7:404–412.

Hermansson, Hélène (2007) "The ethics of NIMBY conflicts," *Ethical Theory and Moral Practice* 10:23–34.

Hermansson, Hélène and Sven Ove Hansson (2007) "A three party model tool for ethical risk analysis," *Risk Management* 9:129–144.

ICRP (2007) "The 2007 Recommendations of the international commission on radiological protection publication," *Annals of the ICRP* 103.

Luloff, A.E., S.L. Albrecht, and L. Bourke (1998) "NIMBY and the hazardous and toxic waste siting dilemma: the need for concept clarification," *Society and Natural Resources* 11:81–89.

MacLean, Douglas, ed. (1985) *Values at Risk*. Totowa, NJ: Rowman and Allanheld.

Michal, R. (2000) "The nuclear news interview. Apostolakis: On PRA," *Nuclear News* 43(3):27–31.

Mill, John Stuart ([1865] 1969) *Auguste Comte and Positivism* [1865], pp. 261–368 in John Mill, *Collected Works*, volume 10. J.M. Robson, editor. Toronto: University of Toronto Press.

Mohai, Paul, David Pellow, and J. Timmons Roberts (2009) "Environmental justice," *Annual Review of Environment and Resources* 34:405–430.

National Research Council (1983) *Risk Assessment in the Federal Government: Managing the Process*. Washington, DC: National Academy Press.

Shrader-Frechette, K. (1991) *Risk and Rationality. Philosophical Foundations for Populist Reforms*. Berkeley: University of California Press.

Thomson, P.B. (1985) "Risking or being willing: Hamlet and the DC-10," *Journal of Value Inquiry* 19:301–310.

Wagner, Wendy E. (1995) "The science charade in toxic risk regulation," *Columbia Law Review* 95(7):1613–1723.

Wolff, Jonathan (2010) "Five types of risky situation," *Law, Innovation and Technology* 2(2):151–163.

Wolsink, Maarten (2006) "Invalid theory impedes our understanding: a critique on the persistence of the language of NIMBY," *Transactions of the Institute of British Geographers* 31(1):85–91.

Further Reading

Ahteensuu, Marko, and Per Sandin (2012) "The precautionary principle." In *Handbook of Risk Theory: Epistemology, Decision Theory, Ethics, and Social Implications of Risk*. Edited by Sabine Roeser, Rafaela Hillerbrand, Per Sandin, and Martin Peterson, 962–978. Dordrecht, The Netherlands: Springer. (An overview of the history, definition(s), normative underpinnings, and applications of the precautionary principle.)

Cranor, Carl F. (2017) *At Risk: How and Why We Are Exposed to Toxic Chemicals*. New York: Oxford University Press. (An in-depth study of the problems arising when scientific information is used in risk assessments for legal purposes.)

Hansson, Sven Ove (2013) *The Ethics of Risk: Ethical Analysis in an Uncertain World*. New York: Palgrave Macmillan. (Argues that traditional moral theories lack resources to deal with problems involving risk and uncertainty, and proposes new principles for that purpose.)

Heinzerling, Lisa (2000) "The rights of statistical people," *Harvard Environmental Law Review* 24:189–207. (A critique of the practice in cost-benefit analysis to assign monetary values to human lives.)

Shrader-Frechette, Kristin (2005) *Environmental Justice: Creating Equality, Reclaiming Democracy*. New York: Oxford University Press. (Argues against current practices that make minority groups and the poor the worst-affected victims of environmental risks.)

58

PRECAUTIONARY PRINCIPLES

Kevin C. Elliott

Introduction

Most contemporary environmental debates involve conflicts over whether there is enough information available to justify potentially expensive actions to prevent environmental threats. Climate change provides the most obvious example, insofar as skeptics have insisted for decades that we do not have enough evidence to justify taking preventive actions that could harm the global economy. A host of other debates follow a similar pattern. Proponents of genetically modified crops argue that they have been shown to be safe, while critics insist that the evidence is inadequate. Beekeepers call for eliminating neonicotinoid pesticides, while the manufacturers insist that there is not enough information to justify doing so. Nanoparticles are being employed in sunscreens and athletic equipment and fabrics, while critics insist that we should not be using them until we have more information about their safety. The US Fish and Wildlife Service (FWS) argues that wolves no longer need to be protected as endangered species, while others argue that the FWS has appealed to flawed and inadequate evidence.

The precautionary principle (PP) was developed to provide guidance in cases like these. It states that precautionary actions should be taken to prevent significant threats to humans or the environment, even if the available scientific information about the threats is limited. In recent decades, various formulations of this principle have received a great deal of attention. The PP has been established as a guiding principle in national and international documents such as the Montreal Protocol of 1987, the Earth Summit's Rio Declaration of 1992, the European Union's Maastricht Treaty of 1992, and the Cartagena Protocol on Biosafety of 2003 (Fisher et al. 2006). Nevertheless, it has also received widespread criticism from some academics and policy makers. One of the most common complaints is that it is hopelessly vague and that it falls prey to a dilemma when it is specified. On one hand, if it is interpreted in a strong fashion, such that it calls for banning any potentially harmful activities, then it is unrealistic and paralyzing. For example, critic Cass Sunstein (2005) argues that one could appeal to strong interpretations of the PP as a basis for banning genetically modified (GM) crops and also as a basis for banning efforts to avoid GM crops, because both activities could have harmful consequences. On the other hand, if the PP is interpreted in a weak fashion, such that it merely calls for taking cost-effective steps to lessen the severity of potential threats, then it is obvious and trivial. For example, almost everyone can agree that it makes sense to promote energy efficiency in order to alleviate climate change because people will save money in the process; the difficult question is whether it makes sense to take more expensive steps to address climate change.

DOI: 10.4324/9781315768090-69

In this chapter, I suggest that one can make sense of both the popularity of the PP and the widespread criticisms of its vagueness by recognizing that various formulations of the PP are unified more by what they are *against* rather than what they are *for*. Those who call for application of the PP are unified in their frustration with current approaches for handling uncertainty and ignorance in environmental research and policy making. They are much less unified in the solutions that they propose for handling these problems in a better fashion. Thus, I propose that one can develop a much better understanding of the debates surrounding the PP by regarding it as a concept that rallies people around the recognition of significant problems that need to be addressed. It is crucial to note that this interpretation of the PP does not preclude other, more precise formulations of the principle as well. Rather, it shows that in addition to the more substantive and precise formulations of the PP that other authors have provided, it is enlightening to recognize the more symbolic role of the PP as a rallying point for criticizing the current state of environmental science and policy making.

The next section of the chapter provides an overview of previous attempts to characterize the PP, along with the common criticisms that are leveled against it. In response to these difficulties, I propose that the PP can be fruitfully interpreted as a rallying point for criticizing current approaches for handling ignorance and uncertainty in environmental policy. Section 3 then highlights four of the most significant problems that have generated concerns with current approaches and that have stimulated interest in the PP. Section 4 shows how the range of proposals that have been developed under the umbrella of the PP can be understood more coherently when they are seen as varied efforts to respond to the problems identified in Section 3.

Precautionary Principles

The origin of the PP is often traced to Germany in the 1970s (see, e.g., Jordan and O'Riordan 1999). In the German context, the PP was often used to justify implementing new technologies to prevent pollution at its source. From Germany, it began to spread into other European Union countries and even into international statements such as the Rio Declaration of 1992. But even though it has been cited as a guiding principle in numerous treaties and agreements, both proponents and opponents of the PP frequently acknowledge that it is vague and difficult to specify (Elliott 2010; Jordan and O'Riordan 1999; Manson 2002; Sunstein 2005; Tickner 1999). Part of the problem is that it is not clear whether the PP actually refers to a single principle, and even if it does refer to a particular principle it is not clear what that principle states.

Many proponents of the PP include a wide variety of concepts and ideas under its umbrella. For example, Joel Tickner's (1999) attempt to operationalize the PP includes (1) a general duty to take precautionary action; (2) the setting of goals for reducing hazardous substances; (3) shifting the burden of proof to those who engage in hazardous activities; (4) a structure for making decisions under uncertainty; (5) prevention-based tools for precautionary action; (6) the polluter pays principle; (7) assessment of alternatives; (8) ongoing monitoring, investigation, and information dissemination about hazards; (9) methods for participative and democratic decision making; and (10) strong enforcement. Similarly, Andrew Jordan and Timothy O'Riordan (1999) claim that the PP incorporates a number of themes, including willingness to take action in advance of proof, recognition of the interests of nonhuman entities, shifting the burden of proof onto those who initiate changes, and greater concern for the impacts of policies on future generations. Others have argued that the PP also calls for significant changes to scientific practice so that it is more likely to uncover environmental and public-health hazards (Tickner 2003).

Given all this complexity, it is no wonder that many figures try to narrow down the PP to a more specific statement or principle. But even when it is narrowed in this way, the content of the principle remains elusive. Marko Ahteensuu and Per Sandin (2012) point out that various formulations of the PP fall into at least three categories: (1) decision rules for making choices among environmental policies, (2) procedural requirements for constraining how to make decisions, and (3) epistemic rules that specify how to make scientific inferences under uncertainty. Even when it is interpreted primarily as a decision rule, various commentators have pointed out that different formulations of the PP vary in terms of the types of threats that are taken to justify precautionary action, the level of knowledge needed to justify action, and the specific actions that are called for (Manson 2002; Sandin 1999). In order to alleviate some of this confusion, Daniel Steel (2013, 2015) recently proposed that various "meta" versions of the PP should be distinguished from the PP proper. The meta versions require that environmental decision rules not be paralyzed by scientific uncertainty, whereas the PP itself requires that precautionary actions be taken in a manner that is appropriately proportional both to the level of harm and to the knowledge about potential threats.

While there is undoubtedly value to these attempts to interpret the PP in a more coherent and precise manner, my contention in this chapter is that there is also much to be gained by considering the full diversity of ways in which the PP is used in practice. Why do various formulations of the PP refer to so many different sorts of activities and strategies, and why is it so difficult to regiment the use of this concept? My contention is that the PP is used in so many ways because it was developed primarily as a way to criticize previous approaches to making environmental policy decisions. Calling for decision making "in accordance with the precautionary principle" is akin to calling for decisions that are more public-health friendly, that can be taken even when scientific evidence is ambiguous, and that do not marginalize the concerns of affected stakeholders. But there are all sorts of strategies for improving environmental decisions in this way, which explains why the PP is so difficult to analyze precisely.

In this respect, the PP is much like the concept of sustainability (see Jordan and O'Riordan 1999). Both concepts are notoriously vague, but they have become common buzzwords in environmental discourse (Jamieson 1998; Vucetich and Nelson 2010). In both cases, a number of commentators have attempted to develop more precise formulations of these concepts (see, e.g., Hartzell-Nichols 2012; Norton 2005; Steel 2015; Thompson 2007). While these clearer interpretations can help prevent confusion, it is also worth keeping the broader interpretation of these concepts in mind. Efforts to regiment their use can eliminate some of their important connotations and rhetorical value, whereas the broader uses of these concepts provide ways of rallying people around the desire to change the status quo. It may not always be clear exactly what people are demanding when they use these concepts in a broader sense, but their rhetorical power and their ability to generate cohesion among a variety of stakeholders should not be underestimated. And even if these concepts are ultimately set aside because they become too vague, the major problems that they underscore should not be forgotten. The remainder of this chapter elucidates a number of problems that proponents of the PP are concerned to address and some of the major strategies for addressing these problems.

Problems with Addressing Ignorance and Uncertainty

Questions about how best to respond to ignorance and uncertainty are central to environmental science and policy making. Some decision theorists carefully distinguish between ignorance and uncertainty, such that decisions under uncertainty occur when the possible outcomes of an action are known but it is not possible to assign numerical probabilities to those outcomes, whereas decisions under ignorance occur when the possible outcomes of an action are not even known (Elliott and Dickson 2011). In this chapter I will use the terms more loosely, so that "ignorance" refers to lack of knowledge in general and "uncertainty" refers to limitations in the ability to predict or characterize particular phenomena. Thus, uncertainty in this sense is a type of ignorance and can be generated by various forms of ignorance. The significance of ignorance and uncertainty in the environmental context is that we often know just enough about various environmental challenges to recognize that we face significant potential threats. Nevertheless, we often do not know enough to make confident claims about the likelihood of the threats or the precise factors that cause them or the best ways of mitigating them.

As noted in Section 2, many academics, environmental organizations, and government bodies have called for the PP to be employed as a way of alleviating problems with our current approaches to handling ignorance and uncertainty. These concerns fall into four broad categories: (1) selective ignorance; (2) asymmetries of knowledge, power, and policy; (3) overreliance on quantitative analyses; and (4) lack of respect for local forms of knowledge. Each of these categories encompasses a variety of more specific problems. By examining these four general categories of concern, we can better understand what motivates proponents of the PP. We can also better understand why the principle has been elaborated in so many different ways and what solutions may need to be developed, whether those solutions are ultimately classified under the PP or not.

The first category is selective ignorance. A number of historians, philosophers, and sociologists have argued that we need to pay more attention to the social and political forces that cause us to remain ignorant about socially important topics, such as the life experiences of minoritized groups or the public-health effects of toxic chemicals (see, e.g., Frickel et al. 2010; Gross and McGoey 2015; Kourany and Carrier 2020; Proctor and Schiebinger 2008; Tuana 2006). The concept of *selective* ignorance refers to the notion that scientists often have to make methodological choices that lead them to collect extensive information about some topics and questions while they remain ignorant about other important topics and questions (Barrett and Raffensperger 1999; Elliott 2013; Tickner 2003). The partial and selective understandings that result from these methodological choices can have significant ramifications for society.

Consider four common forms of selective ignorance that proponents of the PP have frequently emphasized. First, those who develop new products are typically very concerned to identify and study their potential beneficial effects and uses but often much less interested in identifying their potential harmful effects (Cranor 1999). Second, when the producers of potentially toxic substances do study the harmful effects of their products, they often do so in ways that are either intentionally or unintentionally designed to minimize the chance of identifying any problems (e.g., Michaels 2008; Myers et al. 2009; vom Saal and Hughes 2005). Third, proponents of the PP argue that scientific research on environmental threats often revolves around performing extensive risk assessments of worrisome substances and practices while largely failing to develop safer alternative practices (Elliott 2013; O'Brien 1999; Sarewitz 2009). Fourth, risk assessors often focus on well-known pathways of exposure to hazardous substances while failing to consider pathways and practices that are particularly important to low-income or marginalized members of society. For example, Maria Powell and Jim Powell (2011) argue that, despite their good intentions, state regulatory agencies sometimes fail to recognize the unique pollution hazards faced by low-income subsistence anglers, who often eat more fish and different parts of fish than regulators anticipate.

A second category of problems for our current approaches to resolving ignorance and uncertainty stem from significant asymmetries associated with knowledge, power, and policy making. Carl Cranor (1999) has highlighted a number of these asymmetries. Some of them, such as the incentives for manufacturers to study the beneficial effects of new products rather than their harmful effects, are also examples of selective ignorance. But many of the other asymmetries identified by Cranor are distinct. For example, he notes that the beneficial uses of a toxic chemical are often very obvious, but the harmful effects are often difficult to identify, in part because the diseases caused by chemicals frequently have long latency periods and are often difficult to pinpoint as being related to particular chemical exposures. This asymmetry is exacerbated by further asymmetries in the norms of science. The scientific community generally prefers to make false-negative errors about the existence of effects or phenomena rather than false-positive errors. In the case of environmental threats, this means that scientific norms are designed to prevent scientists from falsely pronouncing the existence of threats, even if this means failing to identify some threats that really exist.

Besides these asymmetries associated with our knowledge, Cranor (1999) identifies others that have to do with power and policy making. For example, the manufacturers of potentially harmful products often have a great deal of money and a very strong interest in continuing to sell them, whereas those who will be harmed by the products are often much more diffuse, less organized, less wealthy, and perhaps even unaware of the threats that they face. This power asymmetry is exacerbated by the fact that many of the statutes that regulate toxic substances employ a post-market regulatory framework (Cranor 2011). In other words, the manufacturers of these substances do not bear the burden of proof to show that their products are safe before marketing them; instead, citizens or government agencies bear the burden of proof to show that the products are harmful in order to pull them from the market. These asymmetries of power and policy making exacerbate the asymmetries of knowledge and science mentioned in the previous paragraph, insofar as they provide further incentives to avoid collecting detailed knowledge about environmental health threats and to keep engaging in potentially harmful activities until more knowledge is forthcoming.

A third set of problems emphasized by proponents of the PP concerns the potential for scientists or policy makers to be over-reliant on or overconfident in quantitative analyses of threats. For example, Brian Wynne (2005) has argued that since the latter half of the 20th century, debates over new technologies have almost always been framed as questions about risk, as if the only relevant question is whether new technologies are safe enough. But this tendency to focus on risk can marginalize or hide legitimate questions about justice and equity and the broader social ramifications of new technologies and the available alternatives to them. Even when risk analyses are appropriate and helpful, an additional problem is that they frequently incorporate quantitative information that is based not solely on empirical data but rather on experts' subjective estimates. It is often very difficult to determine how much credence to place in these estimates (Elliott and Dickson 2011). Scientific experts are sometimes seriously overconfident in the accuracy of their subjective estimates, and psychologists have found that experts are prone to a variety of heuristics, biases, and cultural influences (Elliott and Resnik 2015; Solomon 2007). As a result, it is often difficult to determine who to trust when various experts disagree or when members of the public challenge experts.

A fourth worry about decision making under ignorance and uncertainty is related to this difficulty of deciding whether to trust publics or experts when their perspectives conflict. Namely, many proponents of the PP argue that policy makers frequently do not pay adequate attention to the local knowledge and expertise of citizens. Scholars such as Brian Wynne (1989) and Alan Irwin (1995) have drawn attention to cases in which particular groups of citizens had unique knowledge of hazards that scientific experts failed to appreciate. For example, Wynne (1989) has famously documented how, after the Chernobyl nuclear accident of 1986, expert predictions of radioactive contamination in sheep failed to take account of important details of sheep behavior and soil type that were well known to farmers. Similarly, Irwin (1995) has discussed how risk analyses of pesticides have been challenged by agricultural workers because they allegedly did not reflect the on-the-ground conditions under which the pesticides are actually used. This phenomenon is related to the problem of selective ignorance, insofar as experts may not recognize the conditions faced by people who are marginalized or who face unique health threats (Corburn 2005; Ottinger and Cohen 2011).

Precaution as a Response to Problems with Ignorance and Uncertainty

Once one recognizes the diverse range of problems that various proponents of the PP are concerned about, it should be no wonder that the PP is not easily expressed as a single, straightforward solution to these problems. Perhaps the most common interpretation of the PP is as a guideline for making decisions under ignorance. As Section 2 noted, this guideline typically specifies particular sorts of threats that should be mitigated or eliminated when the available information about their likelihood reaches a particular level (even if that information is still significantly limited). But Section 2 also highlighted a number of other approaches that are often included in discussions of the PP. For example, in his efforts to operationalize the principle, we saw that Joel Tickner (1999) provides a framework that includes a wide variety of activities, including setting goals for the reduction of hazardous substances, shifting the burden of proof to those who want to maintain these activities, assessing alternatives, using the "polluter pays" approach, engaging in ongoing monitoring and information dissemination, and promoting participative and democratic decision making. This section shows how these activities, although not otherwise very coherent or unified, are all geared toward addressing the problems discussed in Section 3.

Consider first the narrow interpretation of the PP (as a guideline for justifying actions to mitigate or eliminate threats under uncertainty). It is helpful for addressing at least three of the problems mentioned in Section 3: overconfidence in quantitative analyses, lack of respect for local knowledge, and asymmetries in knowledge and power. If decision makers can justifiably take various sorts of precautionary actions even when they do not have precise quantitative information about a threat's likelihood, it eliminates much of the pressure for them to create numerical estimates of risk that go beyond the available data. As Section 2 clarified, interpreting the PP as a decision guideline still leaves room for a great deal of ambiguity because groups who employ various formulations of the PP can disagree about what sorts of threats justify action, how much knowledge is needed, and what actions to take in response. Nevertheless, many formulations of the PP as a decision guideline would allow decision makers to take action to mitigate threats in response to qualitative information (as long as it is of adequate quality). These formulations of the PP also frequently open the door for citizens with local knowledge to provide information that can justify preliminary steps to start addressing threats.

The interpretation of the PP as a decision guideline also helps to address asymmetries of knowledge and power, because it typically lessens the standards of evidence required for taking protective action. If, as Section 3 noted, the norms of the scientific community and the nature of our social context are such that it is very difficult to obtain compelling information that particular substances or activities are harmful, then it is unwise to require this sort of information before taking action. To do so could have dire effects on society. The PP allows decision makers to take action in response to more realistic kinds of information.

But while the narrow interpretation of the PP as a decision guideline helps to promote clarity about the nature of the principle, it does not fully address the range of problems that proponents of the PP have been keen to address. In particular, it is primarily reactive, in the sense that it provides decision makers with justification for addressing threats despite limited knowledge, but it does not provide clear mechanisms or incentives for alleviating this lack of information. Even Daniel Steel's (2013, 2015) meta-PP, which he includes in addition to the narrow interpretation of the PP, focuses only on rejecting decision-making procedures that are paralyzed by scientific uncertainty. With this in mind, we can see why it is valuable to consider a richer and more diverse range of solutions to the problems considered in Section 3. The remainder of this section briefly considers four of the major solutions that are often discussed by proponents of the PP: (1) deliberative, participatory approaches to scientific research and policy making; (2) alternatives assessment and goal setting; (3) shifting the burden of proof in science policy making; and (4) policies for aggressively monitoring, investigating, and disseminating information about hazards.

First, deliberative and participatory approaches to scientific research and policy making have the potential to address all four problems discussed in Section 3. In recent years, there have been increasing efforts to create innovative mechanisms and forums for this purpose. These can include citizens' juries and panels, public hearings, surveys, focus groups, interactive technology-based approaches, consensus conferences, science shops, alternative dispute resolution, citizens' advisory committees and task forces, activist movements, and various efforts at promoting citizen science (Elliott 2011, 2017; Kleinman 2000). Some of these efforts are more focused on influencing the course of scientific research on environmental issues, while others are geared primarily toward influencing the development of public policy. Proponents of these strategies argue that they have at least three virtues: (1) making environmental policies and decisions more democratically responsive; (2) producing better-informed and higher-quality decisions; and (3) promoting greater social acceptance and adherence to environmental policies (Fiorino 1990).

Unlike the narrow interpretation of the PP as a decision guideline, deliberative and participatory approaches to science and policy are well placed to alleviate selective ignorance in environmental research. For example, in an analysis of selective ignorance in agricultural research, I have previously argued that one of the best mechanisms for uncovering important bodies of information that are being neglected in current research is for NGOs and activist groups to draw attention to these lacunae (Elliott 2013). Efforts at promoting science shops and citizen science are also designed to make academic research more responsive to the needs and concerns of local communities (see, e.g., Brown and Mikkelsen 1990; Corburn 2005; Elliott and Rosenberg 2019; Irwin 1995; Ottinger and Cohen 2011). Deliberative and participatory approaches are also helpful for alleviating overconfidence in quantitative information and lack of respect for local knowledge. When scientific findings are scrutinized in deliberative bodies, it provides opportunities for the participants to share their unique perspectives and to highlight questionable assumptions embedded in scientific results. Finally, even though there are no easy ways to eliminate the asymmetries in knowledge and power that plague public policy making, the creation of deliberative and participatory forums provides at least some opportunities for citizen groups to influence regulatory policies so that they are less biased in favor of those who generate potentially hazardous substances or activities.

Another strategy that is often discussed under the umbrella of the PP is alternatives assessment (O'Brien 1999; Tickner 1999). The idea behind this strategy is to aggressively look for alternatives to potentially hazardous activities and to set goals for shifting toward these alternatives. The Massachusetts Toxic Use Reduction Act (TURA) of 1989 is a frequently discussed example of a regulatory approach that emphasizes alternatives assessment. Under the TURA, companies that use large quantities of worrisome chemicals are required to publicly report their use of those chemicals as well as to create a plan that documents why they need to use them and whether there are alternatives. By requiring companies to explore alternatives, this legislation has cut the use of potentially toxic substances and even saved companies money in the process (Tickner 1999). By encouraging those who engage in hazardous activities to consider their options and to look for alternatives, this strategy can help to alleviate the problem of selective ignorance discussed in Section 3. It also addresses some of the asymmetries discussed in Section 3, insofar as it increases the incentives for manufacturers to move away from worrisome products and to look for safer ways to achieve their goals.

Another strategy associated with the PP that is crucial for alleviating asymmetries of knowledge and power is to shift the burden of proof for identifying and regulating hazards (Cranor 1999). One significant way to create this shift is to move from post-market regulatory regimes to pre-market regimes. For example, whereas the Toxic Substances Control Act (TSCA) in the

US has historically allowed the manufacturers of industrial chemicals to place them on the market without providing evidence that they are safe, the Registration, Evaluation, Authorisation, and Restriction of Chemicals (REACH) legislation in the European Union requires evidence of safety before products can go on the market (Cranor 2011). The polluter pays principle provides another influential strategy for shifting burdens of proof and thereby altering current asymmetries of knowledge and power. According to this principle, those who create pollution must pay for the social costs that it generates. This makes it much less appealing for polluters to continue producing harmful products, and it provides incentives for them to collect information about their hazardous qualities and the available alternatives.

Finally, proponents of the PP often argue that it calls for aggressively monitoring, investigating, and communicating information about potential hazards (Tickner 1999, 2003). For example, in an issue of *Human and Ecological Risk Assessment* (Vol. 1, No. 1, 2005) that focused on the implications of the precautionary principle for environmental health research, Philippe Grandjean argued that

> The PP has … been misunderstood as anti-science. However, what has been called for [by advocates of the PP] is not an embargo of science, but rather the initiation of 'new science.'… The ways in which science can support PP-based decisions is … likely to differ from the science that has supported traditional risk assessment.
>
> *(Grandjean 2005, 14)*

Articles throughout the issue argued that the PP calls for altering traditional scientific practices that favor making false-negative errors over false positives. They also argued that scientists should put more effort into pursuing studies that are of importance to public policy makers (even when those studies cross traditional disciplinary boundaries and yield results that are somewhat difficult to interpret) and finding ways to communicate those results effectively. These strategies are obviously designed to alleviate selective ignorance and asymmetries of knowledge and power.

Conclusion

This chapter has argued that it is fruitful to understand the PP as a very general concept around which people can rally to call for more environmentally protective approaches to handling ignorance and uncertainty in environmental research and policy making. Proponents of the PP have highlighted at least four major problems with current approaches: (1) selective ignorance; (2) asymmetries of knowledge, power, and policy making; (3) overreliance on quantitative analyses; and (4) lack of respect for local forms of knowledge. In response to these problems, they call for a variety of solutions. These include (1) using a decision guideline that allows for regulatory action in the face of uncertainty (the narrow PP); (2) developing deliberative and participatory approaches to scientific research and policy making; (3) implementing alternatives assessment and goal setting; (4) shifting the burden of proof in policy making; and (5) aggressively monitoring, investigating, and disseminating information about potential hazards.

For the sake of clarity, one might argue that the term "precautionary principle" should refer only to decision guidelines that call for action under uncertainty. Proponents of the PP could still call for other strategies, such as alternatives assessment and participatory approaches to research and policy, but these would not be part of the PP itself. But as I suggested in Section 2, the problem with this approach is that it buys clarity about the PP at the expense of failing to do justice to the full rhetorical power and the variety of uses for the concept. Frequently, when commentators call for policy making in accordance with the PP, they are rallying people around new ways of doing science, broader public involvement in policy making, and less dependence on quantitative analyses of hazards.

There may come a point when the PP becomes sufficiently unpopular or polarizing or confusing that it is better to jettison the broader interpretation in favor of more precise interpretations or new principles altogether. But the fruitfulness of the broader interpretation should not be dismissed too quickly. Like the concept of sustainability, it has significant rhetorical power for bringing a variety of stakeholders together to change current practices for the better. And even if the PP is ultimately set aside or made more precise, the problems that it highlights and the solutions that it champions should continue to receive sustained attention.

References

Ahteensuu, M. and P. Sandin (2012) "The Precautionary Principle," in S. Roeser, R. Hillerbrand, P. Sandin, and M. Peterson (eds) *Handbook of Risk Theory*, New York: Springer.

Barrett, K. and C. Raffensperger (1999) "Precautionary Science," in C. Raffensperger and J. Tickner (eds) *Protecting Public Health and the Environment: Implementing the Precautionary Principle*, Washington, DC: Island Press, pp. 106–122.

Brown, P. and E. Mikkelsen (1990) *No Safe Place: Toxic Waste, Leukemia, and Community Action*, Berkeley: University of California Press.

Corburn, J. (2005) *Street Science: Community Knowledge and Environmental Health Justice*, Cambridge, MA: MIT Press.

Cranor, C. (1999) "Asymmetric Information, the Precautionary Principle, and Burdens of Proof in Environmental Health Protections," in C. Raffensperger and J. Tickner (eds) *Protecting Public Health and the Environment: Implementing the Precautionary Principle*, Washington, DC: Island Press, pp. 74–99.

—— (2011) *Legally Poisoned: How the Law Puts Us at Risk from Toxicants*, Cambridge, MA: Harvard University Press.

Elliott, K. (2010) "Geoengineering and the Precautionary Principle," *International Journal of Applied Philosophy* 24: 237–253.

—— (2011) *Is a Little Pollution Good for You? Incorporating Societal Values in Environmental Research*, New York: Oxford University Press.

—— (2013) "Selective Ignorance and Agricultural Research," *Science, Technology, and Human Values* 38: 328–350.

—— (2017) *A Tapestry of Values: An Introduction to Values in Science*, New York: Oxford University Press.

Elliott, K. and M. Dickson (2011) "Distinguishing Risk and Uncertainty in Risk Assessments of Emerging Technologies," in Torben Zülsdorf, Christopher Coenen, Arianna Ferrari, Ulrich Fiedeler, Colin Milburn, and Matthias Wienroth (eds) *Quantum Engagements: Social Reflections of Nanoscience and Emerging Technologies*, Heidelberg: AKA Verlag, pp. 165–176.

Elliott, K. and D. Resnik (2015) "Scientific Reproducibility, Human Error, and Public Policy," *BioScience* 65: 5–6.

Elliott, K. and J. Rosenberg (2019) "Philosophical Foundations for Citizen Science," *Citizen Science: Theory and Practice* 4(1): 1-9.

Fiorino, D. (1990) "Citizen Participation and Environmental Risk: A Survey of Institutional Mechanisms," *Science, Technology, and Human Values* 15: 226–243.

Fisher, E., J. Jones, and R. von Schomberg (2006) "Implementing the Precautionary Principle: Perspectives and Prospects," in E. Fisher, J. Jones, and R. von Schomberg (eds) *Implementing the Precautionary Principle: Perspectives and Prospects*, Northampton, MA: Edward Elgar.

Frickel, S., S. Gibbon, J. Howard, J. Kempner, G. Ottinger, and D. Hess (2010) "Undone Science: Charting Social Movement and Civil Society Challenges to Research Agenda Setting," *Science, Technology, and Human Values* 35: 444–473.

Grandjean, P. (2005) "Implications of the Precautionary Principle for Public Health Practice and Research," *Human and Ecological Risk Assessment* 11: 13–15.

Gross, M. and L. McGoey (2015) *The Routledge International Handbook of Ignorance Studies*, London: Routledge.

Hartzell-Nichols, L. (2012) "Precaution and Solar Radiation Management," *Ethics, Policy & Environment* 15: 158–171.

Irwin, A. (1995) *Citizen Science: A Study of People, Expertise, and Sustainable Development*, New York: Routledge.

Jamieson, D. (1998) "Sustainability and Beyond," *Ecological Economics* 24: 183–192.

Jordan, A. and T. O'Riordan (1999) "The Precautionary Principle in Contemporary Environmental Policy and Politics," in C. Raffensperger and J. Tickner (eds) *Protecting Public Health and the Environment: Implementing the Precautionary Principle*, Washington, DC: Island Press, pp. 15–35.

Kleinman, D. (2000) *Science, Technology, and Democracy*, Albany: State University of New York Press.

Kourany, J. and M. Carrier (eds) (2020) *Science and the Production of Ignorance: When the Quest for Knowledge Is Thwarted*, Cambridge, MA: MIT Press.

Manson, N. (2002) "Formulating the Precautionary Principle," *Environmental Ethics* 24: 263–274.

Michaels, D. (2008) *Doubt Is Their Product: How Industry's Assault on Science Threatens Your Health*, New York: Oxford University Press.

Myers, J., F. vom Saal, B. Akingbemi, K. Arizono, S. Belcher, T. Colborn, I. Chahoud, et al. (2009) "Why Public Health Agencies Cannot Depend on Good Laboratory Practices as a Criterion for Selecting Data: The Case of Bisphenol A," *Environmental Health Perspectives* 117: 309–315.

Norton, B. (2005) *Sustainability: A Philosophy of Adaptive Ecosystem Management*, Chicago: University of Chicago Press.

O'Brien, M. (1999) "Alternatives Assessment: Part of Operationalizing and Institutionalizing the Precautionary Principle," in C. Raffensperger and J. Tickner (eds) *Protecting Public Health and the Environment: Implementing the Precautionary Principle*, Washington, DC: Island Press, pp. 207–219.

Ottinger, G. and B. Cohen (2011) *Technoscience and Environmental Justice: Expert Cultures in a Grassroots Movement*, Cambridge, MA: MIT Press.

Powell, M. and J. Powell (2011) "Invisible People, Invisible Risks: How Scientific Assessments of Environmental Health Risks Overlook Minorities—and How Community Participation Can Make Them Visible," in G. Ottinger and B. Cohen (eds) *Technoscience and Environmental Justice: Expert Cultures in a Grassroots Movement*, Cambridge, MA: MIT Press, pp. 149–178.

Proctor, R. and L. Schiebinger (eds) (2008) *Agnotology: The Making and Unmaking of Ignorance*, Stanford: Stanford University Press.

Sandin, P. (1999) "Dimensions of the Precautionary Principle," *Human and Ecological Risk Assessment* 5: 889–907.

Sarewitz, D. (2009) "A Tale of Two Sciences," *Nature* 462: 566.

Solomon, M. (2007) *Social Empiricism*, Cambridge, MA: MIT Press.

Steel, D. (2013) "The Precautionary Principle and the Dilemma Objection," *Ethics, Policy, and Environment* 16: 321–340.

——— (2015) *Philosophy and the Precautionary Principle: Science, Evidence, and Environmental Policy*, New York: Cambridge University Press.

Sunstein, C. (2005) *Laws of Fear: Beyond the Precautionary Principle*, New York: Cambridge University Press.

Thompson, P. (2007) "Agricultural Sustainability: What It Is and What It Is Not," *International Journal of Agricultural Sustainability* 5: 5–16.

Tickner, J. (1999) "A Map Toward Precautionary Decision Making," in C. Raffensperger and J. Tickner (eds) *Protecting Public Health and the Environment: Implementing the Precautionary Principle*, Washington, DC: Island Press, pp. 162–186.

Tickner, J. (ed) (2003) *Precaution, Environmental Science, and Preventive Public Policy*, Washington, DC: Island Press.

Tuana, N. (2006) "The Speculum of Ignorance: The Women's Health Movement and Epistemologies of Ignorance," *Hypatia* 21: 1–19.

vom Saal, F. and C. Hughes (2005) "An Extensive New Literature Concerning Low-Dose Effects of Bisphenol A Shows the Need for a New Risk Assessment," *Environmental Health Perspectives* 113: 926–933.

Vucetich, J. and M. Nelson (2010) "Sustainability: Virtuous or Vulgar?" *BioScience* 60: 539–544.

Wynne, B. (1989) "Sheep Farming After Chernobyl: A Case Study in Communicating Scientific Information," *Environment* 31: 10–39.

——— (2005) "Risk as Globalizing 'Democratic' Discourse? Framing Subjects and Citizens," in M. Leach, I. Scoones, and B. Wynne (eds) *Science and Citizens: Globalization and the Challenge of Engagement*, London: Zed Books, pp. 66–82.

59

ADAPTIVE MANAGEMENT

R. Bruce Hull

Introduction

The biosphere—humanity's habitat—is changing, rapidly. Humanity, over the next few decades, is on course to add 2 billion people, double the global middle class purchasing power, migrate to cities at a rate of 1 million a week, increase agriculture production by 70% to feed a larger and wealthier population, warm the climate, deplete aquifers, over-fish the oceans, and drive species to extinction (Steffen et al. 2015). As change accelerates and uncertainty grows, human civilization will depend upon our ability to adapt. Theory and practices of Adaptive Management will be keys to our success.

The term Adaptive Management (AM) refers to an evolving and contested approach/theory for planning and managing natural resources, natural areas, and related environmental challenges. The National Academy of Sciences offers a state-of-the-art review of it (National Research Council 2004). At its essence, AM is a method of intentional trial-and-error. It is a process where stakeholders collaborate to define goals, propose and implement a strategy for achieving those goals, measure the results and compare them to what was expected, revise and adapt both the goals and the strategies based on what was learned, and try, try again. Similar approaches/theories of learning by doing exist in other domains of planning and management including business management (e.g., Walton 1986), urban planning (e.g., Friedmann 1973), and social entrepreneurship (e.g., Kania and Kramer 2011). Limited cross-referencing among these literatures exists (Lane 2001). An excellent literature about AM explains the rationale for doing it and the theory behind it (e.g., Lee 1993; Norton 2005; Walters and Holling 1990).

History of Adaptive Management

Command-and-control was the dominant approach of natural resource managers during the first half of the 20th century. This approach—administrative rationalism is its more technical name—assumed rational, scientific, expert-led processes would produce sufficient understanding of environmental systems to guide management interventions to optimize healthy conditions and desired resource flows (Dryzek 2005; Hays 1959). For an example, consider fire. The US Forest Service sought to control all forest fires so as to eliminate the waste and damage they caused. Decades of implementing that policy produced unhealthy forests, an alienated public, a bankrupted and dysfunctional agency, and larger and wilder fires (Pyne 2004). It turns out fire cannot be *controlled* and should not be *eliminated* from the forest. Instead, humans most find ways to co-exist with fire's wildness—we must manage fire adaptively.

DOI: 10.4324/9781315768090-70

The hubris of command-and-control management became increasingly evident by the 1970s as ecologists and managers realized that ecological systems were open, dynamic, unpredictable, and inextricably interdependent with cultural and economic systems that were also open, dynamic, and unpredictable (Norton 2005; Worster 1994). The bio-cultural systems that needed managing were too complex to understand, too uncertain to predict, and too dynamic to manage with techniques that worked in the past. Managers and scientists teamed up to practice an adaptive style of management where management interventions became testable hypotheses about how systems would respond. The learn-by-doing philosophy breathed life and hope into a field beset by controversy, uncertainty, and failure. The failures and frustrations that demoralized and paralyzed command-and-control management became empowering learning opportunities to perfect management craft, which in turn helped to legitimize the manager's role that had been increasingly drawn into question (Lee 1993; Ludwig 2001; Walters and Hollings 1990).

However, by the 1980s it also became clear that expert-led processes, even adaptive ones, were insufficient. The natural resource management community began to accept that they were confronted with "wicked" problems (Rittel and Webber 1973). The desired conditions managers aimed to create could not be identified without engaging stakeholders, who typically could not agree on desired outcomes because of inherently conflicting needs and values. Wicked problems are political as well as technical, so science, rationality, and expertise are insufficient (Schon 1983). *Collaborative* AM efforts emerged in response. These efforts sought participation and inclusivity. Scientists, managers, and other stakeholders shared decision-making power and co-produced the problem and solutions (Weber 2003).

Leadership Practices for Adaptive Management

It is not enough to understand why environmental problems are wicked and what AM seeks to accomplish. AM must be implemented. This chapter builds on the theory and rational of AM but emphasizes leadership practices that help implement AM. It builds on the recent book by Hull and colleagues about leadership practices for wicked situations (2020). Successful practice of AM requires actors that are humble and courageous, collaborative and disruptive, and reflective and action-oriented.

Humble and Courageous

Managers, scientists, and other professionals derive their salaries and identities from offering expert advice and solving problems. Their legitimacy is threatened by uncertainty about how systems will respond to interventions and by failures to produce desired results (Schon 1983). Yet, uncertainty and failure are conditions of complex adaptive systems. Actors involved in AM must accept that their expertise is partial and fleeting, that problems have outgrown their science and technology, that techniques and tools that worked in the past may not work in the future, and that the management interventions they recommend and implement will inevitably fail. Actors involved in AM efforts therefore must be humble and courageous: humility is needed to deal with the uncertainty and courageousness is needed to deal with the failure (Westley et al. 2006).

The uncertainty that leads to failure emerges from at least three sources: systems, stakeholders, and strategies.

System Uncertainty: Natural ecological systems are dynamic. The myth of a stable or balanced nature is busted (Worster 1994). Species and ecosystems evolve. Disturbance is normal. New system conditions emerge. The future is unpredictable. Additional complexity, openness, instability, and unpredictability are added through the increasing integration of natural ecological systems with cultural and economic systems that are also open, dynamic, unstable, and unpredictable

(Norton 2005). Aldo Leopold, the early 20th century ecologist-manager-philosopher, called for humility in his pleas to listen to the mountain, save all the pieces, and preserve the stability, beauty, and integrity of the biological systems (Meine 1988). By the early 1970s, theories about general systems, chaos, complexity, evolution, and quantum physics confirmed that human and natural systems were beyond management control, that cause–effect science was insufficient, and that outcomes would be uncertain. By the end of the 20th century, even senior members of the American Association for the Advancement of Science acknowledged that system uncertainty was too much for traditional science to overcome and that society's most pressing management challenges had "radically outgrown" science's ability to solve them (Jasanoff et al. 1997).

Experts of all types therefore face a dilemma of relevance. Their scientifically proven techniques and knowledge apply to a smaller and smaller set of relevant problems. Professionals can stay on the high, firm ground where positivist, science-proven solutions are known to work or wade into the swampy, messy lowlands where uncertainty, uniqueness, and unpredictability prevail. The high ground is where professionals can solve problems with certainty and confidence; however, those problems are often of limited social relevance. The mounting and urgent problems of sustainable development are down in the swamp where uncertainty reigns. AM professionals need humility and courage to wade there (Schon 1983: 42).

Stakeholder Uncertainty: A defining characteristic of AM is that stakeholders don't agree on the problem. Stakeholders desire different and sometimes conflicting outcomes. They also possess competing values and are grounded in different understandings about how a system operates. These differences produce a class of problems that have been called *wicked* because credible, scientific understanding is helpful but insufficient for decision-making—no amount of science, no matter how valid and reliable, will resolve the different values, understandings, and expectations (Norton 2005; Rittel and Weber 1973; Schon 1983). Instead, the decisions are political, enmeshed in values, history, and power. Moreover, many of the decisions require making tragic choices: someone, some place, or some plant, animal, or ecosystem will lose. Win–win solutions don't exist. People lose jobs or a species goes extinct. One community gets a park and another gets a waste dump. Intractable, heated conflicts persist when core values are threatened—religion, patriotism, heritage—because these values support identities that people don't compromise (Lewicki, Gray, and Elliott 2003). Values derived from identities resist compromise and cannot be swapped with monetary compensation or traded for alternatives. For example, if I value species as God's creation, then no amount of money will compensate for the sin of me allowing their extinction.

Stakeholders learn from the experience of planning their future (Throgmorton 2003). This experience adds uncertainty to AM. The goals stakeholders state at the beginning of the planning process are hypotheses about their values that are just as refutable as hypotheses about ecological functions and management techniques. For example, stakeholders in a community may agree to maximize wealth and freedom through real estate development. After years of pursuing these values, residents may become frustrated by traffic congestion, fossil fuel dependence, and the loss of local foods, open space, and biodiversity. That is, our values change as we experience the consequence of the landscape and lifestyle those values create. Uncertainty exists because it is not until after we create the future that we can decide if we want to live there. AM facilitates this type of social learning. It is forward-looking, assuming that truth lay in the future and is found through learn-by-doing (Norton 2005).

Strategy Uncertainty: Even if stakeholders agree on goals, the strategies to achieve those goals may not yet exist. Another defining characteristic of AM is that the scale and scope of the challenges exceeds the capacity of known strategies. Not only must stakeholders construct agreement on what the problem is, they must also negotiate, develop, test, and refine strategies to solve it. Conflict emerges when stakeholders differ in the strategies they favor. Professions, organizations, and individuals can get so embedded in one strategy that they miss opportunities to collaborate with stakeholders with whom they align on ends but not means. Dryzek (2005) reviews types of strategies in his classic book: *The Politics of Earth*. A simple listing of strategies illustrates the potential for conflict: government-led law and regulation, civil society-led education and behavior change, market-led innovation and efficiency, and technological advances that intensify production and extend ecological limits. Vastly different assumptions and values are embedded in each of those strategies. Different professions advocate different strategies. Thus, additional uncertainty arises from not being able to predict the intervention strategies a group of diverse stakeholders agree to apply.

Clearly, AM requires humility to accept the limits of expertise, to make decisions with imperfect information, and to navigate conflicting values. It also requires courage. Actors need the courage to implement their plan with the expectation that it will fail in two important ways. First, it will fail to sustain the intended outcomes because the bio-cultural system is unpredictable, dynamic, and unstable—any intervention will change the system and likely require additional interventions. Second, the plan will fail because stakeholders will learn and change and re-evaluate their goals for a desired future (this last point is developed more fully below in a discussion of double-loop learning).

Collaborative and Disruptive

The need for collaboration should be obvious. No person, profession, organization, or institution possesses the power and insight to solve the major challenges of sustainable development. Government agencies, businesses, and civil society organizations must collaborate. Engineers must collaborate with ecologists and accountants. Young must collaborate with old, rich with poor, patriots with globalists, conservatives with liberals, managers with stakeholders. Collaborating across such diversity is daunting. Three attributes of successful collaboration are leading from where you are, influencing without authority, and overcoming differences. Each will be introduced below before discussing what might be less obvious: the need to be disruptive.

Leading from where you are: AM efforts do not lend themselves to hierarchical, command-and-control organization or leadership. Power, responsibility, and roles are dispersed among stakeholders that differ in their agendas and obligations. No one stakeholder has the authority to tell others what to do. Everyone must collaborate to achieve the goals, process, and implementation of the AM effort. Successful collaboration in these situations boils down to three tasks—direction, alignment, and commitment (Drath et al. 2008; Hull et al. 2020). Rather than defer to a leader from above to force collaboration, AM actors lead from where they are by facilitating direction, alignment, and commitment among other actors.

Direction, Alignment, and Commitment (DAC) can make the difference between individuals, organizations, and civilizations that survive and those that fail (Antonovsky 1987; Diamond 2005; Heifetz 1994). Antonovsky focuses on individuals who overcome profound personal challenges, everything from divorce and unemployment to imprisonment in a concentration camp. Heifetz focuses on organizations and institutions struggling to remain relevant to rapidly evolving markets and cultures. Diamond focuses on entire civilizations that have collapsed (Easter Island and Anasazi Native Americans) or face collapsing (China and Australia). Three reasons permeate the many powerful lessons learned in these classic studies: failure to adapt is caused by lack of direction, lack of alignment, and lack of commitment—a lack of DAC. Those that fail to adapt do so because they ignore or misperceive the challenges they face. When they do finally see them, they can't agree on what to do—they lack *direction*. If they do agree on what to do, they can still fail to adapt if they cannot coordinate resources. That is, they lack *alignment* of time, talent, and resources needed to mount a successful response to the opportunity or challenge. Most insidiously, people, organizations, and civilizations fail to adapt when they lack *commitment*. Actors must commit to developing a shared vision and coordinating their resources to address the agreed upon problems. Commitment gets stifled by self-filling feedback loops such as helplessness, pessimism, fear of failure, and analysis paralysis.

Importantly—and this cannot be overemphasized—everyone can and must work to facilitate DAC (Hull et al. 2020). Successful AM requires people to lead from where they are in their organizational hierarchy to promote and achieve DAC. This lead-from-where-you-are philosophy is empowering and makes everyone responsible. The success of an AM effort is the responsibility of all actors, not just those with authority, a title, or a corner office.

Influence without Authority: Achieving DAC often requires influencing collaborators without having authority over them because collaborators are employed by different organizations and/or are accountable to different constituents. Several strategies of influence work in these situations, but none are foolproof and all require nuance and practice: rationalizing, asserting, negotiation, inspiring, and bridging (Musselwhite and Plouffe 2012).

- *Rationalizing* uses logic, facts, and reasoning to present ideas and persuade others. This influence strategy is the default for many natural resource managers and scientists who appeal to the authority of science to convince others. But, it can be ineffective when it overwhelms others with detail or makes others feel the speaker values data more than principles and ethics. It can alienate others if it makes them feel that their perspectives will not be respected because they lack a scientific background.
- *Asserting* uses laws, authority, and personal confidence to influence others and challenge ideas. Actors accustomed to being in positions of authority or enforcing laws and regulations may default to this style. It doesn't work when people feel pressured and the actor is aggressive or heavy-handed. It may be especially ineffective in AM because most actors have veto power and can sabotage the process just by withdrawing their participation.
- *Negotiating* emphasizes compromises, concessions, and tradeoffs to mitigate conflict and keep the process moving. Facilitators and planners may default to this style. However, it may be ineffective when others in the process become confused about an actor's agenda and values. Furthermore, too much compromise may raise concerns that the process will degenerate to the lowest common denominator people can agree to rather than a decision of consequence.
- *Inspiring* encourages others to share a position by communicating a sense of shared mission and exciting possibility with inspirational appeals, stories, and metaphors. Mission-oriented NGO stakeholders may default to this style. It doesn't work when there is a lack of transparency on the part of the inspirer or insufficient basis for trust.
- *Bridging* employs peers, personal relationships, and other connections in the actor's network to bring influence. It is ineffective when stakeholders feel manipulated or forced to rely on these indirect indicators of legitimacy rather than have time to study and be persuaded by the issues.

For most of us, our default is to use on others the influence style that works best on us, which means our ability to influence others will be limited until we learn to use influence styles that are effective on others. No influence style is inherently better than another. Selecting an effective influence style requires assessing the actors involved and the norms and culture defining the context. That is, a style's effectiveness or ineffectiveness is completely situational and the same argument may be "heard" differently by different people and in different situations. It is no wonder that the collaboration required of AM is so difficult to achieve.

Overcoming Differences: Real and perceived differences among stakeholders present barriers to collaboration. At least two types of differences must be overcome: interests and practices. Different interests exist because stakeholders in an AM process are accountable to different constituents who hold different values and want different outcomes. Different practices exist because stakeholders come from different professions, disciplines, and organizations that have different norms and ways of communication—jargon is the classic example. Strategies to overcome the barriers these differences create include Interest-Based Negotiation (Fisher and Ury 1991), Reaching Higher Ground (Dukes, Piscolish, and Stephens 2000), and Boundary Spanning Leadership (Ernst and Chrobot-Mason 2011). These classic and tested strategies for overcoming differences are described in the literature cited and taught in countless workshops and classrooms. Although they differ importantly in purpose and approach, they share many attributes, including focusing on the interests and values that stakeholders share, creating the trust and space to inspire finding alternative solutions, and respecting the capacities and constraints each stakeholder brings to the table. When these conditions exist, stakeholders are more likely to collaborate, be creative, discover acceptable solutions they did not know existed, and forge the commitment needed to realize those solutions.

Disruptive Double-Loop Learning: While actors engaged in AM need to be collaborative for all the reasons mentioned above, they also need to be disruptive. The challenges of sustainable development exist in part because the status quo is creating and sustaining problems. The people, organizations, professions, and even governments might be seeking the wrong outcomes—core assumptions and values might need to change. Challenges to foundational assumptions and core values make all of us uncomfortable, and the challengers get characterized as quacks and renegades to be dismissed or threats to be managed. Most people prefer stability to change, clarity to uncertainty, and orderliness to conflict. But actors in AM must sometimes be disruptive agents, provocateurs, and opponents of the status quo. AM, at least in theory, allows us to ask these difficult questions through the process of double-loop learning. AM actors must facilitate and engage in this type of learning (Goldstein 2009; Throgmorton 2003).

To understand double-loop learning, let's first examine single-loop learning. AM promotes both. Single-loop learning occurs through the familiar, science-based, adaptive-management process of monitoring a system's response to management interventions, evaluating the response against expectations to improve conceptual and empirical models of how the system works, and then fine-tuning the management intervention so that it more effectively delivers desired outcomes. While single-loop learning focuses on outcomes, double-loop learning focuses on goals. Double-loop learning occurs when we re-evaluate the goals we thought we wanted. By participating in a collaborative process with diverse stakeholders, articulating and explaining one's values, and accepting the uncertainty and functions of bio-cultural systems, participants can develop new and different hopes, dreams, values, and expectations of what they expect of their system (Goldstein 2009).

Double-loop learning provides a powerful opportunity for stakeholders of AM efforts to be disruptive. The planning, implementation, and decision-making processes of AM involve stakeholders in the unfolding story of a region's development trajectory, giving people roles, defining their goals, and giving them direction (Throgmorton 2003). It thereby motivates and engages people in creating their future. By articulating desired future conditions, people identify and refine their values. As the plans become realized, people learn about whether their values and hopes were appropriate. With a view toward the future and achieving the good life, people can understand the changes and sacrifices being asked of them today. They can become disruptive or at least understand the need for disruption.

Reflective and Action-Oriented

Schon (1983) studies the practices of a wide range of professionals. He argues that successful practitioners, regardless of whether they are doctors, lawyers, accountants, architects, engineers, or natural resource managers, are able to combine reflection and action. The best practitioners reflect on and learn from their actions. They "learn by doing," "think on their feet," "keep their wits about them," and "get a feel for" a situation by practicing an endless cycle of action-reflection-adjustment-reflection when embedded in a decision-making process. Lessons learned from this sort of reflection differ from the technical, rational, rules and principles derived from science taught in school, textbooks, and journal articles. The lessons learned from this sort of reflection also differ from the social learning about values and questions that occur through double-loop learning discussed above. Instead, the key lessons from reflective practice are about *skills of practice*—lessons about how to act during the process of AM—skills that are difficult to describe but skills we recognize and appreciate when we encounter them in the professionals we engage in the everyday aspects of our lives.

Practitioners learn these skills by being reflective about their practice. They question their actions and assumptions regarding systems, stakeholders, and strategies. Each situation is unique, so questions should not be limited by accepted theory or science or even by what worked elsewhere. Reflective practitioners must simultaneously juggle questions, analysis, and actions. They must attend to norms and differences of stakeholders, question their own assumptions and expectations, and adopt, test, and continuously change their actions in response to responses their actions generate. Although this process of reflection-in-action seems extraordinary, it is not rare. Its practice and application, and the lessons it generates, form the essence of professional practice and what distinguishes successful from unsuccessful professionals (Schon 1983: 62–69).

Analysis is also required. In situations requiring AM, there certainly is plenty to analyze, including the complexity of bio-cultural systems; the nuance, assumptions, values, and hidden agendas of stakeholders, including your own; and the uncertainty around and lack of testing of intervention strategies. Too much analysis, however, leads to analysis paralysis. Federal land management agencies such as the US Forest Service found their staff and budgets nearly consumed by a rational planning approach that assumed collecting the "right" information would lead to indisputable, objective answers. No such answers were found and the agency became impotent (Behan 1981; Fukuyama 2014). Many natural resource professionals, scientists, and technical experts have a bias for analysis (Kennedy, Dombeck, and Koch 1998). They want to understand the system, feedback loops, causes and effects. They want to quantify, model, and test predictions. They privilege understanding over action; thus, they risk analysis paralysis because they will never act because they will never fully understand the system.

While analysis paralysis can be a problem, so is action without reflection. The remainder of this section presents an analytic approach that builds on the discussion above: it recommends continuous attention to and analysis of stakeholders, strategies, systems, and outcomes. In just about every case of AM, stakeholders will use strategies to influence systems to produce desired outcomes. This suggestion builds on the work of Heifetz's (1994) theory of adaptive leadership and Senge, Hamilton, and Kania's (2015) theory of systems leadership. Complex adaptive problems require co-defining both the problem and the solution, an interactive process that creates a moving target that requires analytical, reflective, and, ultimately, adaptive. Take as an example the following system: a city with development stifled by long commutes full of traffic jams. An infinitude of problem definitions exist. AM actors must collaboratively frame, narrow, bound, and define the problem that will inform the strategy for solving the problem and achieving the desired outcomes. The problem could get framed and defined as road construction, mass transit development, fuels subsidies, car taxes, home mortgage tax deductions, or development impact fees (to name just a few commonly competing frames for defining the problem of sprawl and smart growth); depending upon how the problem is framed, very different strategies and solution sets emerge. Change occurs while these issues are being sorted out: immigrants arrive, development occurs, and commuting preferences change. As an example of a different system consider a declining fishery that threatens biodiversity, livelihood, and food security. The problem could get framed or defined as preserving biodiversity, finding alternative employment for out of work commercial fishers, or changing diets and finding alternative sources of protein. Again, different system framing will lead to different solution sets. Practitioners of AM can exercise considerable influence by helping frame the problem; doing so requires analysis and reflection.

Stakeholder analysis is another useful analytic to apply early and often. As described above, value conflicts must be managed and navigated. For each stakeholder, it is helpful to know what they stand to win or lose, how much influence they can exert, what outcomes they desire, what values and principles they hold in highest esteem, where their loyalties lay, and who their allies are. Some if not all of these qualities may change as stakeholders learn from and about one another.

Analysis of strategies also can be useful. As described above, numerous solution strategies exist, system interventions can occur at many places and times, and stakeholders often differ in which strategies for intervention they think are appropriate. Conflict among stakeholders can limit the choice of strategies. For example, some stakeholders may oppose any government-led solution and favor a market-based strategy instead. AM practitioners should be aware of and navigate these differences, which are also likely to change and stakeholders learn from and about one another.

Analysis of outcomes also is important. A key part of any AM process is defining outcomes that are the goals of management, selecting indicators used to measure those goals, and monitoring those indicators to evaluate the success of the management actions. Many sets of indicators exist, each with their own embedded assumptions and values. Some indicators emphasize social capital, others natural capital; some are generalizable to many locations, others are specific to one bio-cultural location; and some will speak to the interests of one stakeholder group while ignoring the interests of another. AM practitioners will have greater influence if they have access to an extensive menu of possible indicators and can articulate criteria about what makes an indicator useful to stakeholder groups.

Conclusion

Adaptive Management is a process for addressing complex, conflict-laden challenges that lack clarity and certainty. It accepts uncertainty, questions assumptions, reveals values, and creates new understandings and trust needed to sustain development. Similar problem solving approaches exist in business, planning, and other professions and disciplines that also face challenges where technical, rational knowledge is insufficient because systems are open, unique, dynamic, unstable, and uncertain, and because diverse stakeholders must collaborate to achieve direction, alignment, and commitment. Ultimately, AM and approaches like it depend on the actions and practices by actors—professionals who participate in and practice AM. The quality of their practices determines the success or failure of AM efforts. It has been argued here that qualities leading to success include being able to juggle and walk the fine line between humility and courageousness, collaboration and disruption, reflection and action.

Related Topics

22. Climate Change Mitigation
54. Command and Control
55. Economic Instruments
56. Cost-Benefit Analysis
58. Precautionary Principles

References

Antonovsky, A. (1987) *Unraveling the mystery of health: how people manage to stress and stay well*. San Francisco, CA: Jossey Bass.

Behan, R. W. (1981) 'RPA/NFMA--time to punt', *Journal of Forestry*, 79, pp. 802–805.

Diamond, J. (2005) *Collapse: how societies choose to fail or succeed*. New York: Viking.

Drath, W. H., McCauley, C. D., Palus, C. J, Van Velsor, E., O'Connor, P. M. and McGuire, J. B. (2008) 'Direction, alignment, commitment: toward a more integrative ontology of leadership', *The Leadership Quarterly*, 19, pp. 635–653.

Dryzek, J. S. (2005) *The politics of the earth: environmental discourses*. 2nd edition. New York: Oxford.

Dukes, F. E., Piscolish, M. A. and Stephen, J. B. (2000). *Reaching for higher ground in conflict resolution*. San Francisco, CA: Jossey Bass.

Ernst, C. and Chrobot-Mason, D. (2011) *Boundary spanning leadership*. New York: McGraw Hill.

Fisher, R. and Ury, W. (1991). *Getting to yes*. 2nd edition. Penguin Group London.

Friedmann, J. (1973) *Retracking America: a theory of transactive planning*. New York: Doubleday Anchor.

Goldstein, B. (2009) 'Resilience to surprises through communicative planning', *Ecology and Society*, 14, p. 33.

Hays, S. P. (1959) *Conservation and the gospel of efficiency*. Pittsburgh: University of Pittsburgh.

Heifetz, R. A. (1994) *Leadership without easy answers*. Cambridge: Harvard University Press.

Hull, R. B., Robertons, D. P. and Mortimer, M. (2020) *Leadership for sustainability: strategies for tackling wicked problems*. Washington, DC: Island Press.

Kania, J. and Kramer, M. (2011) 'Collective impact', *Stanford Social Innovation Review*, 9, pp. 36–41.

Kennedy, J. J., Dombeck, M. P. and Koch, N. E. (1998) 'Values, beliefs and management of public forests in the western world at the close of the twentieth century', *Uasylva*, pp. 16–24.

Fukuyama, F. (2014) 'America in decay: the sources of political dysfunction', *Foreign Affairs*, Sept/Oct. Available at: http://www.foreignaffairs.com/articles/141729/francis-fukuyama/america-in-decay

Jasanoff, J. et al. (1997) 'Conversations with the community: AAAS at the millennium', *Science*, 278, pp. 2066–2067.

Lane, M. B. (2001) 'Affirming new directions in planning theory: comanagement of protected areas', *Society & Natural Resources*, 14(8), pp. 657–671.

Lee, K. N. (1993) *Compass and gyroscope: integrating science and politics for the environment*. Washington, DC: Island Press.

Lewicki, R., Gray, B. and Elliot, M. (2003). *Making sense of intractable environmental conflicts*. Washington, DC: Island Press.

Ludwig, D. (2001) 'The era of management is over', *Ecosystems*, 4, pp. 758–764.

Meine, C. (1988) *Aldo Leopold: his life and works*. Madison, WI: University of Wisconsin Press.

Musselwhite, C. and Plouffe, T. (2012) 'When your influence is ineffective', *Harvard Business Review*, 28 March. Available at: http://blogs.hbr.org/2012/03/when-your-influence-is-ineffective/

National Research Council (2004) *Endangered and threatened fishes in the Klamath River Basin: causes of decline and strategies for recovery*. Washington, DC: National Academies Press.

Norton, B. (2005) *Sustainability*. Chicago: University of Chicago Press.

Pyne, S. J. (2004) *Tending fire: coping with America's wildland fires*. Washington, DC: Island.

Rittel, H. and Webber, M. (1973) 'Dilemmas in a general theory of planning', *Policy Sciences*, 4, pp. 155–169.

Schon, D. (1983) *The reflective practitioner: how professionals think in action*. New York: Basic Books.

Senge, P., Hamilton, H. and Kania, J. (2015) 'The dawn of system leadership', *Stanford Social Innovation Review*, Winter. Available at: http://www.ssireview.org/articles/entry/the_dawn_of_system_leadership

Steffen, W., Broadgate, W., Deutsch, L., Gaffney, O. and Ludwig, C. (2015). 'The trajectory of the Anthropocene: the great acceleration', *The Anthropocene Review*, 2, pp. 81–98.

Throgmorton, J. A. (2003) 'Planning as persuasive storytelling in a global-scale web of relationships', *Planning Theory and Practice*, 2, pp. 125–151.

Walters, C. and Holling, C. S. (1990) 'Large-scale management experiments and learning by doing', *Ecology*, 71, pp. 2060–2068.

Walton, M. (1986). *The deming management method*. The Putnam Publishing Group.

Weber, E. P. (2003) *Bringing society back in: grassroots ecosystem management, accountability, and sustainable communities.* Cambridge, MA: The MIT Press.

Westley, F., Zimmerman, B. and Patton, M. Q. (2006). *Getting to maybe.* Toronto, ON: Vintage Canada.

Worster, D. (1994) *Nature's economy: a history of ecological ideas.* Cambridge: Cambridge University Press.

PART XI

Advocacy and Activism

60

EDUCATION

Matt Ferkany

Introduction

Environmental education is a term of art ordinarily intended to refer to any education that is in, about, or for the environment in some way (Lucas 1980). Outdoor or place-based education programs are examples of environmental education in the first intended sense, ecology or environmental studies examples of the second, and environmental advocacy or public campaigning of the third. The distinction between these is soft. Any instance of environmental education can be in, about, or for the environment all at once and any education intended to be merely in or about the environment can have incidental pro-environmental advocacy effects (i.e. an agriculture student can become more pro-environmental just by taking a field ecology course). Because environmental education in any of the intended senses can take place informally or as part of formal schooling, environmental education does not refer exclusively to institutionalized primary, secondary, or post-secondary schooling. A large proportion of environmental education in the United States takes place at nature centers, parks, and the like. Messages about the environment that individuals (especially developing youth) receive in the public space—in the media, from friends and family, or from government or businesses—can also educate them.

Most ethical controversy about environmental education concerns education for the environment in formal settings, public schools especially. Two big issues are perennial favorites:

- *The aims problem*: What ideally is the purpose of environmental education? What should environmental educators aim to teach? Should they aim to impart knowledge and understanding in natural or environmental sciences, to strengthen students' pro-environmental values, attitudes, or behaviors, or to foster environmental decision-making capacity?
- *The legitimacy problem*: How, if at all, can environmental education be legitimate insofar as it does or is intended to strengthen students' pro-environmental values, attitudes, behaviors, or relationships (pro-EVABRs)? Do environmental educators have the right to influence students in this way or are they obligated to avoid pro-environmental advocacy in their teaching?

One under-researched issue concerns whether thinking philosophically about the ethics of environmental education can yield novel insights for environmental philosophy and ethics more generally. This entry focuses on the aims and legitimacy problems and closes with some thoughts on this and on the practice of environmental education.

DOI: 10.4324/9781315768090-72

The Aims Problem

As any education that is in, about, or for the environment in some way, environmental education can have many different purposes. Traditional classroom or lab-based education in the natural sciences, and as well as science education in the field (or other form of outdoor education) are forms of environmental education. Nominally, they aim to impart scientific knowledge and understanding. "Facts and information" campaigns (e.g. as carried out by special interest groups, like a hunting and fishing club circulating information *about* the ecological benefits of a strong base of hunters) often share this aim while taking a completely different form and seeking to influence learner EVABRs. As a component of formal schooling however, "environmental education" traditionally refers to education that is all of in, about, and for the environment involving some combination of the following: teaching some basic concepts of ecology or ecocentric ethics emphasizing notions of ecological stability and the interconnectedness and intrinsic value of all life, e.g. Aldo Leopold's Land Ethic (1949); imparting facts, knowledge, and information about the impact of human activities on ecosystems, such as species extinction rates; and local or outdoor experiences intended to inculcate attachment to particular places or broad appreciation and love of nature (Carson 1965; Hungerford & Volk 1990; Leopold 1949: 214; Van Matre 1990).

Environmental education in this traditional sense is widely regarded as an *activist view* of its purposes and methods (Jickling & Spork 1998; Johnson & Mappin 2005). It aims to strengthen the student's pro-EVABRs and is sometimes criticized as constituting more a kind of moral training than education (Jickling & Spork 1998). I return to this criticism below, but it is worth first noting that the *activist* ideal is a broad category that can take many different forms depending upon the ethics, epistemic focus, and experiential activities deployed.

In addition to the *traditional ecocentrist* approach just outlined, for example, canonical versions of *education for sustainable development* (ESD), such as the United Nations Decade of Education for Sustainable Development, broaden the ethical focus to include influencing student EVABRs relevant to the goals of sustainable development, or (by one common definition) meeting the needs of the present without compromising the ability of future generations to meet their own needs (UNESCO n.d.). The epistemic focus is broadened to include things like critical thinking and systems thinking. Forms of service learning (like volunteering with a state PIRG) or building and maintaining an aquaponic food production system might be common experiential features of ESD curricula. Other more recent ideals like *ecojustice education* merge the *traditional ecocentrist* aims and curriculum to a concern for environmental, food, or gender justice *via* an ecofeminist social analysis, i.e. the idea that systems of hierarchy are the common cause of social injustice and environmental destruction (Martusewicz et al. 2011). The ecojustice educator's epistemic focus shifts to include critical social theory and skepticism of science or science-driven or technocratic environmental decision-making. The site of her experiential curriculum broadens to include places like urban farmer's markets and school gardens.

Whatever particular form they may take, activist ideals have received a lot of criticism. As previously mentioned, one important objection claims that they are forms of moral training more so than teaching or education, which involves imparting knowledge, skills, and understanding. This criticism comes in two varieties. One is simply that, to the extent that they are indoctrinating, activist approaches fail to respect the autonomy of students to make up their own minds and thus lack legitimacy (Bell 2004; Jickling 2003; Mappin & Johnson 2005; Schinkel 2009). Teaching that is intentionally designed to prevent students from considering the weaknesses of a teacher's favored view or from considering alternatives to it can be indoctrinating in this way.

A related but different worry focuses on the consequences of advocacy for students' ability to form well-reasoned views of their own. By advocating for a particular ethical perspective, advocacy approaches may threaten to short-circuit critical questioning of the favored ethical perspectives being advanced, or worse, undermine acquisition of the critical capacities students need in order to become independent critical thinkers in the world outside of the classroom. Students will eventually become citizens who will have to decide for themselves what to believe about potentially very different problems from those they encountered in school. From anti-environmental special interests, they will encounter persuasive-looking arguments deploying rhetorical strategies like proof surrogate, scare tactics, red herring, and the like. If they are unprepared to detect the error in these, what reason is there to expect that they will form and firmly adhere to reasoned environmental perspectives in the face of the "strongly represented but weakly supported" anti-environmental perspectives, such as those espoused by organized climate change denial (Palmer 2006: 10)? If little, activist environmental educators are committed to methods that are ineffective means to achieving their own purposes.

Another related problem is that available empirical evidence suggests that "facts and information" approaches to influencing people's EVABRs are ineffective anyway (Heberlein 2012). The attitudes people have about the environment have a complex structure making them sometimes weak and easy to change, not always for the better, and at other times recalcitrant and almost impossible to change. Robust anti-environmental attitudes, however, are often easily whipped up when particular pro-environmental policies can be linked to things people deeply oppose. For example, very few people have sufficient connection to obscure wild species—like the humpback chub (a species of freshwater fish native to the Colorado river system and endangered by habitat loss from dam construction)—to form particularly robust pro-attitudes toward them, but many people strongly believe that economic growth is good and "big government" is bad. Opponents of efforts to preserve the humpback chub can thus easily solicit anti-preservation sentiment by linking preservation to economic stagnation and big government, *even while admitting* that preservation is the morally best policy. But even supposing that traditional environmental educators can succeed in fostering robust pro-environmental attitudes, other research indicates that the link between pro-environmental attitudes and behaviors (or pro-anything attitudes and behaviors) is weak or highly susceptible to the influence of numerous factors (e.g. the availability of local supports like bike lanes, recycling centers) beyond the educator's control (Kollmuss & Agyeman 2002).

A last criticism applies to traditional ecocentrist environmental education specifically. By some analyses, environmental issues, such as climate change, are crucially collective action problems that cannot be (easily) solved simply through individual behavior change. In making a priority of influencing individual students' environmental attitudes and behaviors, traditional approaches threaten to miss this point (Jickling & Wals 2013).

From here it is natural to move toward either a *science literacy* or *environmental civics* ideal of environmental education's purposes and methods. *Science literacy* is the view that scientific knowledge and understanding (of ecology, environmental science, or the human-environment relationship) should be the environmental educator's primary focus (Jickling & Spork 1998; Mappin & Johnson 2005; NRC 2012). *Environmental civics* is the view that environmental educators should focus on building citizens' capacity to participate in the environmental decision-making of a democratic society (Curren 2010; Krasny & Bonney 2005). These views have different advantages, but seem to share the advantage that they are not so obviously vulnerable to the legitimacy problem. Science literacy and environmental civics approaches make a priority of imparting scientific understanding or civic decision-making capacity, and so appear not to threaten the student's right or capacity to make up her own mind about environmental ethics.

Advocates of *science literacy* generally assume that the natural sciences are a trustworthy source of crucial environmental knowledge and understanding. They quite rightly point out that ecologists no longer regard ecosystems as essentially unified, stable, organism-like systems and that the notions of ecological stability, succession, or health and the like—often at the center of activist approaches—are now more at home in traditional knowledge systems or various moral or spiritual outlooks (Kolasa & Pickett 2005). In the American context, there is also evidence that citizens across widely different cultural groups already care a great deal about the environment, but also are relatively ignorant of important basic insights in environmental sciences (Kempton et al. 1996). They are consequently prone to draw erroneous conclusions about what sorts of policies will realize their environmental values. For example, many people confuse the science of climate change and atmospheric ozone and mistakenly believe that ozone depletion is a significant cause of global warming (Leiserowitz & Smith 2010: 3). If so, correcting citizens' scientific misunderstandings would seem to be the most pressing educational priority.

Advocates of *environmental civics*, however, argue that scientific knowledge and understanding are not sufficient for wise, just, or fair environmental decision-making. These kinds of decisions also require the moral and civic forms of knowledge and understanding imparted by the social sciences and humanities. One problem is that even highly science literate citizens will have limited capacity to assess scientific information for themselves and will have to make decisions based on scientific testimony (Anderson 2011). A crucial but potentially achievable skill for them is the ability to correctly discern which sources of scientific information are credible and which are not. But setting even that aside, environmental decisions are ultimately practical, or decisions about what we should do, which cannot be made without taking a position on questions of values, or what's right, good, worthwhile, just, and the like (Des Jardins 2005). Scientific information is certainly important to doing this well; if we do not know the likely consequences of the various courses of action open to us, we cannot make an informed decision about which ones are ethically preferable. It is helpful to know, for example, that option A involves an X% risk of exposure to some toxin, Y, whereas option B involves a Z%. But once this information is in we still must decide which level of risk would be *tolerable, best, right, good, fair, tolerable, just*, and the like. Does one of these options create *undue* risk for one community compared to another, or more *needed* wealth than the other, or involve any *unacceptable* costs to nonhuman species? The scientific information is ultimately factual information, or information about what *is* the case, and does not tell us how we should weigh these other considerations in our decision about what *should* be the case, i.e. what we ought to do. The judgments of scientists on these dimensions of environmental matters are not privileged above the judgments of ordinary well-informed, rational citizens. For these sorts of reasons environmental problems are widely believed to fall into a class of practical problems particularly ill-suited for technocratic resolution and thus require decision-makers having civic decision-making capacities.

The various ideals of environmental education's aims are not mutually exclusive and many (perhaps most) scholars agree that all of these aims are important. What they disagree about is the relative priority of the different goals. This disagreement is not trivial. Different curricula and pedagogical methods are suited to the different goals. Endorsing the primacy of one aim over another entails endorsing a potentially very contentious view of the structure and place of environmental education. The natural and some social sciences, not the arts and humanities, will be the proper place for environmental education if science literacy is the most important goal, but the arts and humanities will be at least as important if the aim is to instill pro-EVBARs. However, an infusion approach (in which environmental content is spread throughout the curriculum) might be most appropriate if environmental civics is the highest priority.

For all its problems, many lines of defense are open to advocates of activism. The complexity of attitudes and the attitude–behavior link is certainly one explanation of why traditional environmental education has failed to generate the green revolution its advocates hoped for. But another might be that traditional environmental education has been only weakly deployed. The charge that advocacy is indoctrinating is also dubious if so. Indoctrination generally seems to presuppose a mass effort (whether coordinated or uncoordinated) that imparts ignorance rather than knowledge and understanding of new ways of thinking (Taylor 2012). Against a backdrop in which a majority of environmental education fails to interrogate an environmentally destructive or unjust *status quo*, activist efforts might be precisely what civic environmental progress requires.

The criticism that imparting facts and information is ineffective is also a straw man inasmuch as traditional ecocentrist environmental education in formal settings is deeper and involves imparting skill, knowledge, and understanding, e.g., of ecological science or ethical reasoning. In addition, the Kempton et al. evidence that Americans generally place high value on the environment is dated and precedes some of the more radical steps the political right has taken against environmental causes in the past 15 or so years (Bailey 1996). By some measures, pro-environmental attitudes on the whole have declined since 1990 and more recent studies find wide variation in American's environmental attitudes that track differences in political orientation, gender, education, race, and the like (Franzen & Vogl 2013; McCright & Dunlap 2011). One demographic—conservative white males—accounts for a significant proportion of all American climate change deniers (Mc-Cright & Dunlap 2011). Worse, McCright and Dunlap also found that the most fervent deniers report the highest levels of self-reported understanding of the basic science of climate change, i.e. the more climate change deniers think they know about climate science, the more they fervently deny the climate change problem. This finding, according to McCright and Dunlap, is probably the result of the worst deniers reporting to know more than they actually do about climate change (2011). But the finding jibes with other evidence that people's perceptions of climate change risk are shaped as much or more by "cultural cognition," or the extent to which they perceive that action on climate change threatens their worldview, not their understanding of climate science (Kahan et al. 2011; McCright & Dunlap 2011). Apparently a majority of people most concerned about climate change *do* understand its basic science, while very few of those who are dismissive or unconcerned about it do (Leiserowitz & Smith 2010). Still, Americans are sharply divided about climate change by political orientation (Pew Research Center 2014) and general science literacy does not predict greater concern and may even negatively correlate with it among those who self-identify with more conservative ideological orientations (Kahan et al. 2012). It is open to advocacy environmental educators to argue that while science literacy may be absolutely crucial to making good environmental decisions, without sound environmental values, science literacy is either inaccessible or insufficient for sound environmental decision-making. If so, values and attitudes education still matter a great deal.

The Legitimacy Problem

That it aims to strengthen students' pro-EVABRs—that it is education *for* the environment—is the central reason why some question activism's legitimacy. Setting aside issues of efficacy (i.e. whether activism produces competent and active environmental citizens), the putative wrong is a wrong of disrespect for student freedom. Students as (prospective) citizens have a fundamental right to make up their own minds about ethical matters. Activist environmental education al-legedly violates this right by either tilting the curriculum more strongly toward one (or a limited range) of environmental perspectives over others or taking debate about (certain views of) the value of the environment off the table altogether, as well as using techniques that appeal to our desires and emotions, such as outdoor experiential or place-based learning.

Because science literacy and environmental civics do not aim to (intentionally) strengthen student's pro-EVABRs—they do not constitute forms of education *for* the environment—they do not seem to threaten this right, at least not in the same way or degree. Insofar as these approaches impart the kinds of critical knowledge and skills individuals need in order to make up their own minds, they actually support students' autonomy, or capacity for individual self-governance. This however is certainly not to say that they are uncontroversial, educate students *for* nothing in particular, or involve no moral training. Autonomy and critical thinking more broadly are controversial aims of education, preparation for which involves training in certain skills and moral and intellectual virtues, such as honesty, integrity, and a kind of humility (Brighouse 1998; Siegel 1988). That there is an element of moral training in this has been part of the canon of moral education since Aristotle, for whom the purpose of the study of ethics was so that we could learn to be good (1985). According to Aristotle, learning *via* habituation has to precede learning *via* teaching and reasoning because appeals to reason cannot move us the way that they move the virtuous unless we first come to see things as the virtuous see them. For example, unless I see that and why friendship is a good, I cannot be moved by mere appeals to my reason to treat my friend as another self. Nowadays we also understand that the capacity for moral cognition develops over time along certain broad pathways such that certain kinds of moral reasoning are simply inaccessible to those in earlier stages of development (Gibbs 2010). But Aristotle's point applies even to fully developed moral agents, who are not necessarily thereby virtuous but only capable of being virtuous. A fully developed moral agent can fail to see things as the virtuous see them and be in need of this training if he is to become fully good.

The need for training in science literacy and environmental civics can be brought out by revisiting the putative education/training contrast through one prominent account of educational *versus* noneducational activities. Among the most famous of these is R.S. Peters's education as initiation, according to which educational activities initiate the learner into the practices of inquiry in worthwhile fields of study, particularly those important to thinking critically about fundamental ethical questions (Peters 1964). Education using this model is crucially as much or more about imparting the skills, values, and attitudes of scholars in knowledge disciplines, such as critical thinking, open-mindedness, and a love of knowledge, as about imparting the received knowledge of those disciplines. This idea is similar to the central thrust of environmental civics education. But putting it this way reveals the similarity between traditional ecocentrist education and environmental civics, or even science literacy insofar as it aims to enable people to be able to think like scientists (NRC 2012). The latter approaches certainly do not aim simply to inculcate students unthinkingly into some worldview or set of desired behaviors. But neither can they render them scientifically or civically literate without initiating them into the attitudes, values, and behaviors of practitioners of those activities. The contrast between education and training is thus a red herring; even scholarly education involves a kind of moral training.

Notably scholarly aims of education are hardly uncontroversial these days. Some conservatives in the U.S. view academic science, humanistic scholarship, and critical thinking generally as just another religion or ideology opposed to their own faith (Stolzenberg 1993). However, many liberal educational researchers maintain that the focal purpose of education is advancing social justice, not imparting knowledge and understanding (Schiro 2012). This controversy does not by itself delegitimize teaching these subjects. But it does show that education inevitably has a moral dimension and involves taking a stance on fundamental questions of value. The question educators must answer is not whether to teach values or not, but which values they may legitimately teach and why.

Science literacy and environmental civics are putatively legitimate, whereas any form of activism is not, because they support the autonomy of individuals to make up their own minds. This value is the underlying source of the legitimacy problem. One line of defense for advocates of activism then is to reject this value, or liberal political morality generally. Some do so, advancing one or another of a few common arguments against liberalism. One argument targeting liberal neutrality (i.e. the idea that liberal governments should not intentionally favor some ways of life over others on grounds of their superiority) is that environmental education embodies controversial ideals about the best ways of life (e.g. recreation in nature is good for you, or nature is sacred). Since neutrality therefore forbids liberal governments to take a position on such ideals, it forbids them to take a position on environmental education. But environmental education is urgently needed (so the argument goes), so we should abandon liberal democracy for a "green" communitarian one in which citizens share the political goal of living environmentally (Postma 2006). Another argument is that because future generations cannot participate in mutually self-interested bargaining in setting the terms for social cooperation, contractarian ideals of justice like those of Hobbes, Locke, or John Rawls cannot justify principles of intergenerational justice, or principles requiring us to pass a liveable planet on to future generations (Postma 2002). A third is that liberal democracies are committed to ideals at the root of contemporary environmental problems (Martusewicz et al. 2011, chap.2). These include that individuals are "essentially self-interested creatures," that "private property and the accumulation of resources are…a primary right," and that liberal democracy involves a "hierarchized way of thinking" leading to a

> mindset where various aspects of the natural world [are] defined not as an interdependent set of relationships among living things, but rather as so many commodities to be harvested and used in the pursuit of both imperial and individual profit
>
> *(2011: p.49)*

Defenses of activism of these sorts have certain limitations. Insofar as being educated partly consists in a capacity to make up one's own mind, being educated and being (at least potentially) autonomous are linked. In addition, autonomy in this sense is perhaps one of the more widely valued and least controversial freedoms. While popularity is not by itself legitimating, the burden of proof seems to lie with critics rather than advocates of the liberal legitimacy constraints designed to protect that freedom in education. If so, it is difficult to see how any critique of these constraints can succeed without recognizing the importance of the autonomy aim in some sense. But it is difficult to see what sense this could be other than the liberal one in which individuals have a fundamental interest and right in the freedom to make up their own minds, a right that may not in general be subordinated in the service of collective ends (Rawls 2001: 57). This paradigm sense of liberal freedom is nowhere to be found in Martusewicz and company's description of liberal democracy and their critique of it is consequently straw man. In addition, no undue hierarchical thinking has any place in paradigmatic 20th-century articulations of liberalism, such as Rawls's liberal egalitarianism. Neither does the idea that individuals are egoistic and incapable of recognizing a common good, nor that property and accumulation of resources is a fundamental right. These ideas may characterize *neoliberalism*, but that is an altogether different animal from liberal egalitarianism.

Critiques from neutrality and future generations may also be moot in defense of advocacy (Ferkany & Whyte 2013). Both may be moot partly because liberal approaches to justice need not be neutralist or contractarian. The "perfectionist" and utilitarian liberalism of John Stuart Mill is neither (1978). The neutrality argument is moot because neutrality is compatible with state-mandated environmental education. The neutrality principle only forbids liberals to *intentionally* oppose any *permissible* way of life from condemnation of that way of life *qua* way of life, e.g. from condemnation of whaling ways of life because they are whaling ways. If critics are right that environmental education is urgently needed, anti-environmental ways of life may no longer be permitted (Michael 2000). But even if they are, liberals are free to intentionally oppose them on grounds that they violate the rights of others, such as future generations, or otherwise threaten the survival of liberal institutions; opposition on these grounds can be silent about the value of anti-environment ways of life for those who value them, and no disrespect is shown to those who value them.

Rejecting liberalism may also be unnecessary because activist environmental education, at least in some forms, may be compatible with support for student autonomy after all. Activist environmental education (as defined here) involves some intention to persuade students to come to see the environment as having value in some sense, to endorse some pro-environmental outlook or other, and to make better environmental lifestyle choices. Because such learning involves emotional and behavioral as much as cognitive change, strategic activist educators will also use methods that appeal to more than their students' reason. These aims and techniques certainly steer students away from positions teachers perceive to be anti-environment. But they also leave considerable room for variation in precisely how much space is also left for open, critical discussion and questioning (Ferkany & Whyte 2013). If so, there is no obvious reason why educators cannot ardently advocate independent student critical thinking and autonomy *and* certain controversial positions in environmental ethics. The latter may influence the beliefs students come to adopt, but it is not clear that influenced beliefs cannot be autonomously adopted; on a common account, if on due reflection we come to endorse any belief—influenced or not—then we endorse it autonomously (Dworkin 1988; Frankfurt 1988). The threat to our autonomy of any influence on our beliefs may also depend quite a bit on the content of the belief, whether it is false or in some way discourages being autonomous, and also on the spirit in which we are influenced to adopt it, e.g. whether dogmatically or by means of careful rational persuasion. On certain plausible assumptions, it may even be that strategies supporting student autonomy are very closely aligned with those for soliciting pro-EVABRs, such as outdoor experiential learning, selective reading lists, or more favorably grading pro-environmental student work. Suppose, for example, that the student's broader social world provides relatively little exposure to a diversity of conflicting, but newer ways of thinking in environmental ethics, alongside correspondingly little opportunity to think critically about the environmental *status quo*. In that context, students will quite naturally learn the environmental *status quo's* system of values, and few or no alternatives, unless teachers complement that learning by exposing them to alternative systems and enabling them to understand their underlying rationale. If so, just teaching in ways that enable students to access and understand new ways of thinking may radicalize their EVABRs, while necessarily enhancing their capacity to make up their own minds about environmental matters. In general, because autonomy involves choice from among a range of worthy options upon which one has been able to reflect, it is not obvious that activist environmental educators cannot both respect and support the autonomy of their students and plump in certain ways for what they take to be more pro-environmental positions.

This defense has some limitations, too. One is that it is not available to activist educators with certain radical agendas. Because it requires advocacy of open, critical inquiry, it is incompatible with efforts to inculcate very specific pro-environmental ideologies rather than a more generalized pro-environmental mindset, one open to interpretation and a range of differences of opinion. It is also incompatible with approaches in which students are persuaded primarily by non-rational means to adopt very specific or false beliefs, e.g. by taking critical debate about them entirely off the table. While it is quite possible to autonomously adopt false beliefs, persuading another to adopt false beliefs by non-rational means is paradigmatically manipulative. These features might also entail that traditional ecocentrist environmental education (e.g. Van Matre's earth education) is illegitimate insofar as the ideas of ecological interdependence and stability usually at the heart of it are quite specific and no longer supported in ecological science (Kolasa & Pickett 2005). For many, this might constitute a *reductio ad absurdum* of the view. More broadly the defense is also incompatible with neo-Nietzschean or constructivist approaches in which the very ideas of truth, knowledge, or justification—the essential components of the practice of rational persuasion—are questioned; outside of the possibility of this practice, it is difficult to make sense of the difference between manipulative and non-manipulative teaching methods, or cognitive and non-cognitive methods generally. If so, this *rational persuasion* defense may fail to constitute a defense of activism at all insofar as activism necessarily involves a preference for non-cognitive methods of persuasion.

If the rational persuasion defense is sound, however, some forms of activist environmental education can go beyond imparting scientific knowledge and understanding or environmental decision-making capacity while remaining of a kind with them. Advocacy approaches consistent with the use of rational persuasion and critical thinking thereby have more in common with science literacy and environmental civics than is ordinarily recognized. To that extent they share the features that have led scholars to regard science literacy and environmental civics as legitimate forms of environmental education.

Conclusion

Because it has traditionally aimed to strengthen pro-EVABRs, developments in environmental education have mostly tracked developments in environmental ethics or politics. If environmental ethics has historically been the handmaiden of the environmental movement, environmental education has been the handmaiden of environmental ethics. What novel insights for environmental ethics, if any, might come from philosophical study of the ethics of environmental education is an under-researched question. Kevin de LaPlante has proposed that broadening the focus of environmental philosophy teaching and research (from a focus on ethics to a focus on "the difference environment makes in understanding some phenomenon") both helps solve the legitimacy problem and usefully opens new horizons for scholarly inquiry and interdisciplinary work (2006: 52). Matthew Stichter has suggested a principle for the ethical use of animals from considering appropriate limits to their use in undergraduate environmental education (2012).

Speculatively however, this review suggests that philosophical investigation of the ethics of environmental education can yield novel insights for two major recent debates in environmental ethics. The first concerns whether the concept of intrinsic value—the idea that some things, potentially including nonhumans, have, or are worth regarding as having, value in themselves—is especially important for environmental ethics. This review suggests that it is and is not. It is insofar as serious engagement with this idea can advance the knowledge, understanding, and autonomy of those previously unfamiliar with it, and also contribute to a general civic willingness to interrogate the environmental *status quo*. It is not (or perhaps is again, in a different way) insofar as citizens of a liberal democracy can reasonably disagree on whether or what in the environment has intrinsic value, such that appeal to intrinsic value is unlikely ever to legitimize the use of coercion in environmental politics.

A second recent debate concerns the merits of thinking about environmental ethics in terms of the virtues instead of (or in addition to) general principles. Insofar as moral agency and intellectual excellence both grow along Aristotelian lines, involving various sorts of habituation and values training, this review suggests that environmental virtue ethicists are on to something in moving beyond principles to think about the qualities of environmentally good persons and citizens. If fully realized environmental agency involves suites of integrated habits of head, heart, and action, a deeper understanding of the moral and intellectual character virtues embodying these is essential to becoming ethical environmental agents ourselves, and helping others to become such agents.

These ideas are speculative, however. Much more work needs to be done to establish what insights for environmental ethics might be derived from philosophical study of the ethics of environmental education.

The discussion of this entry does suggest a few conclusions concerning the practice of environmental education, however. From respect for student autonomy—the primary limit on how far teachers can go in influencing students' EVABRs—teachers should enable students to critically reflect on the values they are being taught. Teachers should avoid disclosing their own ethical opinions where this will pressure students inappropriately and disclose when students might otherwise fail to note alternative perspectives. They should also teach the fundamental critical thinking and media literacy skills needed to detect fallacious arguments or mass campaigns of environmental misinformation.

However, all education influences students' values on some level and it is legitimate, within certain limits, for environmental educators to influence students' EVABRs. Experiences designed to do this, such as outdoor wilderness experiences or hydroponic food production, all may be legitimate ways of teaching, e.g., the values of ecocentrism or sustainability. Insofar as some students' values also impair their literacy of basic environmental science or their ability to participate in democratic environmental decision-making, experiences giving students access to alternative value perspectives are not only legitimate, but important components of a complete environmental education. In contexts where anti-environmental and anti-reason forces are increasingly aligned, the distance between activist, science literacy, and environmental civics forms of environmental education shrinks.

Acknowledgments

The author wishes to thank Ian Werkheiser, Zach Piso, Hannah Miller, Allison Freed, and Matthew Deroo for helpful discussion of an earlier draft of this chapter.

Related Topics

25. Skepticism and Denialism
33. Urban Sustainability
42. Community Gardens
50. Constitutional Rights
62. Community Participation

References

Anderson, E., 2011. Democracy, public policy, and lay assessments of scientific testimony. *Episteme*, 8(2), pp.144–164.

Aristotle, 1985. *Nichomachean Ethics*, Indianapolis: Hackett Publishing Co.

Bailey, C., 1996. The uncertain state of American environmentalism. *Environmental Politics*, 5(3), pp.551–554.

Bell, D., 2004. Creating green citizens? Political liberalism and environmental education. *Journal of Philosophy of Education*, 38(4), pp.37–53.

Brighouse, H., 1998. Civic education and liberal legitimacy. *Ethics: An International Journal of Social, Political, and Legal Philosophy*, 108(4), pp.719–745.

Carson, R., 1965. *The Sense of Wonder*, New York: Harper & Row.

Curren, R., 2010. Education for global citizenship and survival. In Y. Raley & G. Preyer, eds. *Philosophy of Education in the Era of Globalization*. London: Routledge, pp.67–87.

De LaPlante, K., 2006. Can you teach environmental philosophy without being an environmentalist? In C. Palmer, ed. *Teaching Environmental Ethics*. Leiden: Brill, pp.48–62.

Des Jardins, J.R., 2005. Scientific ecology and ecological ethics: the challenges of drawing ethical conclusions from scientific facts. In E. Johnson & M. Mappin, eds. *Environmental Education and Advocacy: Changing Perspectives Ecology and Education*. Cambridge: Cambridge University Press, pp.31–49.

Dworkin, G., 1988. *The Theory and Practice of Autonomy*, Cambridge: Cambridge University Press.

Ferkany, M. & Whyte, K.P., 2013. The compatibility of liberalism and mandatory environmental education. *Theory and Research in Education*, 11(1), pp.5–21.

Frankfurt, H.G., 1988. *The Importance of What We Care About: Philosophical Essays*, Cambridge: Cambridge University Press.

Franzen, A. & Vogl, D., 2013. Two decades of measuring environmental attitudes: a comparative analysis of 33 countries. *Global Environmental Change Part A: Human & Policy Dimensions*, 23(5), pp.1001–1008.

Gibbs, J., 2010. *Moral Development and Reality: Beyond the Theories of Kohlberg and Hoffman*, Boston: Pearson Allyn & Bacon.

Heberlein, T., 2012. *Navigating Environmental Attitudes*, Oxford: Oxford University Press.

Hungerford, H.R. & Volk, T., 1990. Changing learner behavior through environmental education. *Journal of Environmental Education*, 21(3), pp.8–21.Jickling, B., 2003. Environmental education and environmental advocacy: revisited. *Journal of Environmental Education*, 34(2), p.20(8).

Jickling, B. & Spork, H., 1998. Education for the environment: a critique. *Environmental Education Research*, 4(3), pp.309–327.

Jickling, B. & Wals, A.E.J., 2013. Probing normative research in environmental education: ideas about education and ethics. In R. B. Stevenson et al., eds. *International Handbook of Research on Environmental Education*. New York: Routledge.

Johnson, E. & Mappin, M. eds., 2005. *Environmental Education and Advocacy: Changing Perspectives Ecology and Education*, Cambridge: Cambridge University Press.

Kahan, D., Jenkins-Smith, H. & Braman, D., 2011. Cultural cognition of scientific consensus. *Journal of Risk Research*, 14(2), pp.147–174.

Kahan, D.M. et al., 2012. The polarizing impact of science literacy and numeracy on perceived climate change risks. *Nature Climate Change*, 2(10), pp.732–735.

Kempton, W.M., Boster, J.S. & Hartley, J.A., 1996. *Environmental Values in American Culture*, Cambridge, MA: MIT Press.

Kolasa, J. & Pickett, S.T., 2005. Changing academic perspectives of ecology: a view from within. In E. Johnson & M. Mappin, eds. *Environmental Education and Advocacy: Changing Perspectives Ecology and Education*. Cambridge: Cambridge University Press, pp.50–71.

Kollmuss, A. & Agyeman, J., 2002. Mind the gap: why do people act environmentally and what are the barriers to pro-environmental behavior? *Environmental Education Research*, 8(3), pp.239–60.

Krasny, M. & Bonney, R., 2005. A framework for integrating ecological literacy, civics literacy and environmental citizenship in environmental education. In E. Johnson & M. Mappin, eds. *Environmental Education and Advocacy: Changing Perspectives Ecology and Education*. Cambridge: Cambridge University Press.Leiserowitz, A. & Smith, N., 2010. *Knowledge of Climate Change Across Global Warming's Six Americas*, New Haven, Conn.: Yale University: Yale Project on Climate Change Communication.

Leopold, A., 1949. *A Sand County Almanac*, Oxford: Oxford University Press.

Lucas, A.M., 1980. Science and environmental education: pious hopes, self praise and disciplinary chauvinism. *Studies in Science Education*, 7, pp.1–26.

Mappin, M. & Johnson, E., 2005. Changing perspectives of ecology and education in environmental education. In E. Johnson & M. Mappin, eds. *Environmental Education and Advocacy: Changing Perspectives Ecology and Education*. Cambridge: Cambridge University Press, pp.1–27.

Martusewicz, R.A., Edmundson, J. & Lupinacci, J., 2011. *EcoJustice Education*, New York: Routledge.

McCright, A.M. & Dunlap, R.E., 2011. Cool dudes: the denial of climate change among conservative white males in the United States [electronic resource]. *Global Environmental Change*, 21(4), pp.1163–1172.

Michael, M.A., 2000. Liberalism, environmentalism, and the principle of neutrality. *Public Affairs Quarterly*, 14(1), pp.39–56.

Mill, J.S., 1978. *On Liberty*, Indianapolis: Hackett.

NRC, 2012. *A Framework for K-12 Science Education: Practices, Crosscutting Concepts, and Core Ideas*, Washington, DC: The National Academies Press.

Palmer, C. ed., 2006. *Teaching Environmental Ethics*, Leiden: Brill.

Peters, R.S., 1964. *Education as Initiation: An Inaugural Lecture Delivered at the University of London, Institute of Education, 9 December, 1963*, Evans.

Pew Research Center, 2014. Climate change: key data points from pew research. *Pew Research Center*. Available at: http://www.pewresearch.org/key-data-points/climate-change-key-data-points-from-pew-research/ [Accessed May 9, 2014].

Postma, D., 2002. Taking the future seriously: on the inadequacies of the framework of liberalism for environmental education. *Journal of Philosophy of Education*, 36(1), pp.41–56.

Postma, D., 2006. *Why Care for Nature?*, Dordrecht: Springer.

Rawls, J., 2001. *Justice as Fairness: A Restatement*, Cambridge, MA: Harvard University Press.

Schinkel, A., 2009. Justifying compulsory environmental education in liberal democracies. *Journal of Philosophy of Education*, 43(4), pp.507–536.

Schiro, M.S., 2012. *Curriculum Theory: Conflicting Visions and Enduring Concerns*, 2nd edition, Thousand Oaks, CA: SAGE Publications, Inc.

Siegel, H., 1988. *Educating Reason*, New York: Routledge.

Stichter, M., 2012. Justifying animal use in education. *Environmental Ethics: An Interdisciplinary Journal Dedicated to the Philosophical Aspects of Environmental Problems*, 34(2), pp.199–209.

Stolzenberg, N., 1993. "He drew a circle that shut me out": assimilation, indoctrination, and the paradox of a liberal education. *Harvard Law Review*, 106(3), pp.581–667.

Taylor, R., 2012. Indoctrination: a renewed threat to autonomy in today's educational environment. In Philosophy of Education Society of Great Britain. Oxford. Available at: http://www.philosophy-of-education.org/uploads/papers2012/Taylor.pdf [Accessed May 8, 2014].

Van Matre, S., 1990. *Earth Education...: A New Beginning*, Warrenville, IL: The Institute for Earth Education.

UNESCO, n.d., UN decade for sustainable development: The DESD at a glance. Available at: http://unesdoc.unesco.org/images/0014/001416/141629e.pdf [Accessed March 14, 2011].

Further Reading

Curren, R. and Ellen Metzger, 2017. *Living Well Now and in the Future: Why Sustainability Matters*, Cambridge, MA: MIT Press.

Dobson, A., 2003. *Citizenship and the Environment*, Oxford: Oxford University Press.

Palmer, J., 1998. *Environmental Education in the 21st Century: Theory, Practice, Progress and Promise*, London: Routledge.

Sagoff, M., 1988. Can environmentalists be liberals? In *The Economy of the Earth*. Cambridge: Cambridge University Press, pp.146–170.

Stevenson, R.B. et al. eds., 2013. *International Handbook of Research on Environmental Education*, New York: Routledge.

61

EVERYDAY AESTHETICS

Yuriko Saito

Introduction: The Battle of the Green

A green lawn is a symbol of American domesticity. Constant vigilance and work are required to achieve the aesthetic ideal of weeds-free, velvety-smooth, and green, not brown, lawn sporting uniform crew cut. Its cultivation and maintenance are thought to reflect the homeowner's industriousness, work ethic, orderliness, and civic-mindedness. The prototype of this aesthetic ideal goes back to Thomas Jefferson's Monticello, but it came to be promoted most heavily during the post-war period, thanks in no small measure to the chemical industry that created the market for lawn maintenance products. Americans spend an inordinate amount of time, energy, resources, and money to continue this practice, sometimes even risking their own health and safety (Steinberg 2006). In addition, what one observer calls this American "lawnoholic" obsession exacts a considerable environmental cost: heavy use of water, fertilizer, herbicide, insecticide, and gas-guzzling lawn-mower (Teyssot 1999: 30). For example, it is estimated that 30% of urban water is used for lawn care on the East Coast and 60% on the West Coast (Bormann 1993: 75).

Enter a challenger, the recent project of "Edible Estates" spearheaded by an American artist, Fritz Haeg. Specifically dubbed as "attack on the front lawn," this project replaces green lawn in residential front yard with gardens with fruits and vegetables (Haeg 2010). This project challenges the uniformity and monoculture of green lawn as "an icon of beauty," as well as the assumption that "plants that produce food are ugly and should not be seen" (Haeg 2010: 160, 17). Haeg instead calls for a paradigm shift in American domestic aesthetics, advocating fecundity, productivity, and "chaotic abundance of biodiversity" (Haeg 2010: 22). Furthermore, in addition to the literal fruits of labor harvested from such gardens, there are a number of other benefits, ranging from environmental stewardship to promoting neighborliness by prompting conversations among neighbors about how the crops are doing and by sharing bumper crops.

This battle of the green highlights the generally unrecognized fact that aesthetic considerations play a significant role in directing people's attitudes, decisions, and actions regarding their everyday life. Even less recognized is how such aesthetically motivated decisions and actions often lead to serious environmental consequences. Despite the recent establishment of environmental ethics and environmental aesthetics, the environmental ramifications of the aesthetics involved in people's daily life have not received adequate attention. This is because Western aesthetic discourse is primarily concerned with the aesthetic judgments people make purely as a spectator uninvolved with life affairs. A newly emerging discourse on everyday aesthetics challenges this mode of inquiry by calling attention to the fact that we are active agents, rather than disengaged spectators,

DOI: 10.4324/9781315768090-73

in our daily life. Everyday aesthetics sheds light on the way in which people's aesthetic tastes and judgments lead to decisions and actions, whether they be supporting a certain cause or literally engaging in actions such as maintaining a green lawn. Many of us tend to regard ourselves as passive receivers of the world fashioned by professionals like architects, designers, manufacturers, and policy-makers. However, whether we are aware or not, all of us participate in the collective and cumulative world-making project as citizens and consumers, as indicated by how we create and care for our yard. The following discussion highlights our contribution to this project motivated by aesthetic considerations, their environmental ramifications, and how to direct our everyday aesthetics toward better world-making.

Examples of Everyday Aesthetic Tastes

Let me list some more examples to illustrate how our aesthetic preferences and decisions lead to problematic environmental consequences. First, take our common attraction to scenic landscapes, particularly in the United States, typified by many national parks with wondrous, gorgeous, and exquisite beauty. Aldo Leopold characterizes this scenic aesthetics "an under-aged brand of esthetics which limits the definition of 'scenery' to lakes and pine trees" and "proper mountains with waterfalls, cliffs, and lakes" (Leopold 1966: 268, 179–80). In comparison, we tend to judge that "the Kansas plain is tedious" (Leopold 1966: 268) because of its monotonous appearance. The consequence of this scenic landscape aesthetics is that, while we loudly protest the destruction of scenic beauty, we tend to neglect protecting unscenic lands. This results in the decimation of unscenic lands, such as wetlands in the United States, by rendering them more "productive" through filling and paving (Vileisis 1997).

The same problem plagues nondescript-looking or unattractive creatures, such as fish, invertebrates, and insects. They do not garner the same kind of publicity and support when endangered, compared to creatures that are cute, cuddly, graceful, or awesome, such as seal pup, crane, whale, and bald eagle. Stephen Jay Gould points out the consequences of this aesthetic preference: "environmentalists continually face the political reality that support and funding can be won for soft, cuddly, and 'attractive' animals, but not for slimy, grubby, and ugly creatures (of potentially greater evolutionary interest and practical significance) or for habitats" (Gould 1993: 312).

Our environmental awareness is also influenced by the power of the aesthetic. We often refer to belching smoke stacks from factories and massive oil spill, such as the Exxon Valdez and BP disasters, as the quintessential examples of air and water pollution. Their aesthetic impact consists of dramatic images and an effective narrative structure of "an event" with a beginning, middle, and end, accompanied by an identifiable villain and hapless victims. Such aesthetically powerful events tend to eclipse our daily individual actions which are equally, if not more, serious as a source of pollution, because they lack comparable aesthetic effects (Tenner 1996: 88–94).

Aesthetics also plays a significant role in consumers' purchasing decisions and their attitudes toward and handling of their possessions. In the present-day United States, more often than not, the aesthetic interests seem to work against environmental concerns. A prime example is "perceived" obsolescence, a version of planned obsolescence, employed by the manufacturing industry to encourage consumers to throw away perfectly functional products that are no longer considered fashionable. As Virginia Postrel illustrates with many examples (Postrel 2003), we as consumers are constantly "updating" our appearance by buying new clothes, household items, automobiles, and other gadgets that are "in style," although their functional value may not differ from the older models. A 2007 animated film "Story of Stuff" (accessible online) presents a clear picture of the environmental cost of this phenomenon, including resource extraction, energy consumption in manufacture and transport, and disposal of waste.

Aesthetic considerations also direct how we care for our possessions. For example, our penchant for bright white shirts is responsible for many manufactures' inclusion of bleach and "optical brightener," a fluorescent blue dye harmful to the environment, in their laundry detergent. The environmentally conscientious consumers, however, have resigned themselves to "the reduction in standards from the 'whiter-than-white' effect we have come to expect from conventional washing powders to the noticeably less-than-white we get from bleach-free, environmentally friendly ones" (Whiteley 1993: 92).

Furthermore, it is no accident that the color, shape, and size of fresh produce we find at the supermarket are uniform. One commentator on America's food points out that "the apparent perfection" of "strangely uniform and incredibly shiny red tomatoes and picture-perfect peaches" reflects "the fact that perhaps one-third of the farm's fruits and vegetables have been discarded by the farmer or the supermarket for *aesthetic* reasons" (Blatt 2008: vii, emphasis added).

This wasteful practice motivated primarily by aesthetic considerations is not limited to the United States. The 1988 European Economic Community law stipulates the shape and size of fruits and vegetables that are fit to be sold directly to consumers. It includes, for example, "a mathematical definition for the acceptable curvature in the highest class of cucumber," although the law was superseded by the 2009 new legislation that "allows for many nonstandard fruits and vegetables to be sold directly to consumers as long as they are labeled 'intended for processing'" (Borasi 2010: 10, 12). In addition, without the use of chemical insecticides and pesticides, organic fruits may have holes created by worms, which may turn off the consumers not only because of the possible contamination by worms but also by the blemished surface.

If the green lawn discussed at the beginning exemplifies an aesthetically desirable but environmentally harmful phenomenon, the case of wind turbines offers the opposite example. It is best illustrated by the vociferous objection to the alleged eyesore qualities of the Cape Wind project in the Nantucket Sound, the largest off-shore wind farm in the world which received federal approval in 2011, after ten years of debate, although ultimately abandoned in 2017. Unlike the lawn, the Cape Wind project does not concern people's immediate surrounding, but it was primarily the area residents' aesthetic opposition to the ruinous effect on their vista that halted the project for many years. A similar debate is occurring over the plan to build wind turbines and solar plants in the Mojave Desert. The same aesthetic objection to eyesores underlies the prohibition of outdoor laundry hanging, despite its undisputed environmental benefit, in roughly 300,000 homeowners' associations in American suburban communities, affecting 60 million people.

One may challenge the environmental values and disvalues I am assigning to the forgoing examples. For example, one could take issue with the presumed environmental benefit of wind turbines by citing the structures' negative effects on birds and other wildlife, as well as their ineffectiveness in producing sufficient volume of electricity and the possible impact on human health caused by the buzzing sounds of whirling blades. In this discussion, however, the point I want to derive from these examples is not affected by such disputes. The moral of these examples is that seemingly innocent attitudes, choices, and actions we make as consumers and citizens are often guided by aesthetic considerations and have serious environmental consequences, whatever they may be. The first step toward developing an environmentally sound aesthetic choice, therefore, is to cultivate what may be called "aesthetic literacy": the awareness of how our aesthetic tastes and preferences inform our everyday decisions and actions, which in turn lead to environmental consequences.

What to Do with the Power of the Aesthetic

Is recognizing this power of the aesthetic and developing aesthetic literacy enough? Once recognized, what should we do with this power of the aesthetic? We have two options regarding this potent power of the aesthetic. One is to regard it sufficient to simply expose its potency. The other is to go further by engaging in a normative discourse to guide this power toward an environmentally sound direction.

The first option is to separate the aesthetic from other life values, in this case environmental, and train ourselves to act only on the basis of the latter, without letting ourselves be affected by any aesthetic considerations. So, for example, we should decide on the issues regarding preservation of unscenic landscape and unattractive creatures by their environmental values, *despite* their lack of aesthetic values. We should accept wind turbines and laundry hanging, *disregarding* their eyesore-like appearance. We should also *give up* our ideal green lawn and instead train ourselves to *tolerate* brownish spots and weeds on our lawn.

This way of promoting environmental good may be supported by the advocates of Kantian ethics who would want to appeal only to one's rationality as a moral decision-making compass. The separation of the aesthetic and other life values will also be supported by those who object to a kind of social engineering or "nudging" our aesthetic life to conform to what Marcia Eaton calls "aesthetic ought," that is, what we "ought to" appreciate aesthetically (Eaton 2001: 176). Furthermore, after considering the preceding examples where the powerful effects of aesthetics tend to lead us away from an environmentally sound future, we may be inclined to choose this option and sever the tie between the aesthetic dimensions and other life values.

However, as Friedrich Schiller argued in his vision of the aesthetic education of man, humans are creatures who are affected by and operate on the sensible as well as on the rational level, and what really moves us to act is that which appeals to the sensible part (Schiller 1965). I believe this is recognized by psychologists, educators, propagandists, and advertising agents, but curiously not sufficiently by aestheticians. Arnold Berleant is one of the few exceptions in this regard. He points out that aesthetics' significance "lies not only in the ability… to serve as a critical tool for probing social practice but as a beacon for illuminating the direction of social betterment" (Berleant 2010: 193). Those who have been promoting a sustainable future also recognize the potential of aesthetics to serve this cause and argue for its utilization. To cite only one example, David Orr holds that "we are moved to act more often, more consistently, and more profoundly by *the experience of beauty* in all of its forms than by intellectual arguments, abstract appeals to duty or even by fear" (Orr 2002: 178–179, emphasis added). Therefore, he continues, "we must be inspired to act by examples that we can see, touch, and experience," toward which we can develop an "emotional attachment" and a "deep affection" (Orr 2002: 25, 26, 185).

Given the potent power of the aesthetic, not utilizing it and steering it toward better world-making seems like a missed opportunity. Those professional world-makers cognizant of this power of the aesthetic have been advocating uniting the aesthetic appeal of the design with other values, such as environmental and social ones. For example, one landscape architect argues for the need to align aesthetics with ecology by making environmentally healthy landscape design attractive and appealing, so that people cherish, maintain, care for, and protect it, rendering it "culturally sustainable" (Nassauer 1997: 68). Another landscape architect observes that "this separation of art, ethics, utility and nature can leave aesthetics with an atrophied, and indeed, frivolous role in landscape education" and calls for the need to "make explicit a developing aesthetic criteria related to both ethics and utility" (Dee 2010: 21).

But what about those of us non-professionals who are nonetheless engaged in world-making project through purchasing goods, caring for our environs, and supporting certain causes? The current situation in which we are affected by the power of the aesthetic can be characterized as *laissez faire*. We are letting the power of the aesthetic be used for any purposes or agenda irrespective of its cumulative and collective consequences. For example, we continue laundering our shirts with detergent with bleach and optical brightener and drying them in a dryer, while also joining the frenzy of pursuing the newest and most fashionable model of clothes and gadgets. There is a compelling reason for supporting this *laissez faire* attitude: when it comes to aesthetic matters, we favor complete freedom and reject any attempt to regulate aesthetic taste, even if such legislation of aesthetic taste were possible.

However, the problem is that the power of the aesthetic has already been co-opted by those who seek to guide our aesthetic life toward a certain direction. We have looked at examples of coercion, such as the prohibition of hanging laundry and the pressure to keep up with the Joneses' green grass. Aesthetic strategies, such as branding of goods, food styling, and orchestration of a specific multi-sensory ambience in a store, also "nudge" us toward certain choices in every corner of commercial enterprise today. Perhaps the most problematic use of the power of the aesthetic in today's commercial industry is the aforementioned "perceived" obsolescence regarding the products' fashionableness. It is because satisfying such an artificially created "need" of the consumers not only requires more resource extraction and energy consumption but also creates an increasing amount of "waste," often toxic, that invariably ends up in developing nations. If we continue to endorse this *laissez faire* attitude, we are in effect supporting these existing "aesthetic ought" and "nudge" by default. As Richard Thaler and Cass Sunstein point out, "there is ... no way of avoiding nudging in some direction, and whether intended or not, these nudges will affect what people choose" (Thaler and Sunstein 2009: 10). In light of these considerations, there is a need to cultivate an environmentally informed everyday aesthetics that helps direct our decisions and actions toward a sustainable future.

Objections to Environmentally Informed Everyday Aesthetics

However, there is a persistent resistance particularly among aestheticians to connect the aesthetic and other life values. The primary reason is that bringing in values such as environmental concerns will compromise the core of the aesthetic, namely the sensuous appeal of the objects and the free play of the imagination. This resistance is understandable when considering the development of modern Western aesthetics. Since the 18th century, aesthetic discourse developed by declaring its independence from other considerations, in particular the moral. It culminated in the aestheticism of the late 19th century promoted notably by Oscar Wilde and J. A McNeill Whistler as a response to the prevailing moralistic view of art, followed by the aesthetic formalism of the early 20th century proposed by Clive Bell and Roger Fry as a defense for the then emerging abstract art. The concern is still alive and well today and generally takes the following two formulations: (1) the sensuous appeal of an object *is not* affected by the cognitive, such as moral and environmental associations with the object; (2) the sensuous appeal of an object *should not* be affected by such associations.

For the first of these claims, take the following questions posed by David E. Cooper: "Can the look of a lawn really change according to ecological savvy? Or wind farms begin to look beautiful when their benefits are explained at a consultation meeting?" (Cooper 2009: 23). There are two responses to these questions. First, consider the two already established discourses: art criticism and nature aesthetics. The mainstream aesthetic discourse regarding art presupposes that the interpretation and judgment regarding art is not merely a matter of subjective opinion but should be amenable to inter-subjective, reasoned, and critical discussion. Although there may not be *one* correct interpretation and evaluation of a work of art, within all-too-familiar disagreements, we do disregard those appreciations which are derived from highly idiosyncratic personal associations or not based upon sufficient or correct information. In a critical discourse on art, therefore, we expect not only the relevance but indeed the necessity of connecting the sensuous with other considerations, such as the object's art-historical context, the technique used in production, the artist's *oeuvre*, and the like.

These extra-sensory considerations are then expected to modify the expressive qualities of a work of art. The clearest example may be the change in our perception of a painting after we discover that it is a forgery. What we appreciated as a brilliant composition of Vermeer no longer appears to be brilliant, although we may see it now as being cleverly deceptive. Or, as pointed out by Arthur Danto in his theory of the artworld, the exact same writing that constitutes Cervantes' *Don Quixote* changes its meaning and expressive quality if it was written in the 20th century by Jorge Luis Borges' fictional writer, Pierre Menard (Danto 1978). Similarly, as Kendall Walton argues, the category in which an art object is experienced determines its expressive property, so that Picasso's *Guernica* will appear placid and peaceful if it is experienced under a fictional artistic category consisting of the same monochrome pattern of this painting with differing degrees of three-dimensional protrusion and recession (Walton 1978).

Nature aesthetics is following suit by developing the possibility of engaging in a critical discourse and educating one's aesthetic sensibility through scientific study, nature walks, nature writings, and works of art that represent or comment on nature. Despite considerable debates about the relevance and relative importance of scientific, historical, mythological, poetic, and imaginative associations in nature aesthetics, there is a sense in which some of these associations in certain contexts do render our aesthetic experience of nature richer, possibly more appropriate, and less trivial. Even those who advocate the imaginative aesthetic appreciation of nature seem to distinguish between "serious" and "trivial" associations (Hepburn 1993) as well as "imagining well" and the undisciplined "imagination let loose [which] can lead to the manipulation of the aesthetic object for one's own pleasure-seeking ends" (Brady 2004: 164). The seemingly monotonous and boring appearance of a salt marsh becomes richer and more complex when we associate its diverse environmental functions as well as its rather complex structure that negotiates varied saline content of the surrounding water. Furthermore, after discovering that the brilliant sunset is caused by air pollution, our perception of its crimson color is modified, because we experience what Cheryl Foster calls "aesthetic disillusionment" (Foster 1992).

When it comes to everyday artifacts and activities, we have not developed an equivalent discourse yet in which to analyze the appropriateness of our response or a strategy for educating and improving it. If everyday aesthetic responses are considered trivial because they lack a critical discourse, it is not clear whether the absence of such a discourse is endemic to everyday aesthetics or rather a lacuna that needs to be filled. My proposal is to pursue the latter possibility by leading everyday aesthetics to explore what sort of considerations are relevant and necessary to modify our aesthetic response so as to guide it toward better world-making.

There is another response to the objection that the cognitive considerations do not alter the aesthetic quality of an object. Admitting that the cognitive can modify or transform the sensuous does not necessarily commit us to allowing the cognitive considerations to *determine* its aesthetic value by *nullifying* the sensuous. Specifically, after learning the environmental cost of a green lawn and the benefit of wind turbines and laundry hanging, it is not the case that the lawn automatically becomes ugly while any kind of wind turbines and any method of laundry hanging become aesthetically positive. The green lawn still maintains its luscious appearance, but it no longer looks innocently and benignly gorgeous after we discover its environmental harm; its appearance is modified to become somewhat morbidly gorgeous or garishly beautiful.

At the same time, even if the environmental benefit remains the same, there should be an aesthetic difference between differently designed and arranged wind turbines. For example, early wind turbines were aesthetically problematic with their bare concrete surface, angular rather than round shape of the base column, disproportionate size and insensitivity to the site, uneven spacing, and the like. More recent models are aesthetically superior with their sleeker structure and sensitive arrangement and size to fit with the site; indeed, some can even be appreciated as a kind of kinetic sculpture.

Similarly, the unabashed, in-your-face parading of laundered underwear is aesthetically different from more discreet hanging which hides underwear behind less objectionable items such as linens and towels. There is also an aesthetic difference between arbitrarily hung laundry and thoughtfully arranged laundry according to the kind of items, colors, or sizes. The latter method indicates the consideration for its visual impression to the neighbors and passersby. Thus, making connections to other life values does not necessarily compromise the perceptual aspects of the aesthetic experience; instead, it enriches the experience.

The preceding discussion demonstrates that the cognitive considerations do in fact alter, though not determine, the aesthetic qualities of an object. The next challenge is to consider whether they *should*; that is, shouldn't the aesthetic realm be protected from what may amount to a kind of moral censure? Jane Forsey presents a typical view regarding the aesthetics of designed objects: the fact that some objects are made "in a third-world factory under dismal conditions … will certainly affect our moral judgements of the objects depending on our view on international trade and labour laws but not our aesthetic judgment of their beauty" (Forsey 2013: 186). Another version of the same point is suggested by Thomas Leddy who questions whether there is anything objectionable to aesthetically appreciating junkyards and roadside clutter. He points out that artists are particularly "sensitive observers of our world and capture aesthetic features in their works that we might not normally notice" and argues that it is important "to clear a space for a form of aesthetic appreciation that is freer, more imaginative, and more in tune with important discoveries of modernist art than is allowed by current morally-centered views in aesthetics" (Leddy 2008: sec. 5).

Leddy, however, recognizes the practical importance of morally based aesthetics and proposes the "toggle between interested and disinterested perception" of an object when its other life values are problematic (Leddy 2012: 114). The junkyard *as* a junkyard should be experienced with all of its life values, particularly when practical concerns are at stake, for example, in deciding whether or not to clean up the environment. However, there may be no compelling reason to *always* experience a junkyard in such a way. It will certainly impoverish our aesthetic life if we never experience things like junkyard and roadside clutter for their interesting colors and textures. Thus, there is a need to protect aesthetic experience free of concerns with other life values.

However, we should also note the urgency for cultivating environmentally informed everyday aesthetics today, precisely because our aesthetic judgments direct certain decisions and actions in the most literal sense, as demonstrated above. It is one thing to make an aesthetic judgment, such as that my wardrobe from last year is no longer fashionable, weeds-free green carpet is gorgeous, and the junkyard features intriguing artistic texture and composition. It is quite another to *act* on these judgments by buying a whole new wardrobe and discarding the "out-of-style" clothes, cultivating a green lawn with chemical fertilizer, herbicide, and pesticide, and remaining a mere spectator of the aesthetic appeal of a junkyard. The mainstream aesthetic discourse is not adequately equipped to address this latter issue because it characterizes an agent of an aesthetic experience as a disengaged, distanced, pure spectator who simply forms certain aesthetic tastes and judgments. In contrast, everyday aesthetics aims at addressing the reality that we are actively engaged in our everyday life by *acting* on aesthetic preferences and judgments.

Cultivation of Environmentally Informed Everyday Aesthetics

So, how do we cultivate an environmentally informed everyday aesthetic sensibility? I suggest that we begin by questioning the prevalent notion of beauty underlying our common aesthetic responses. David Orr proposes a new standard of beauty "as that which causes no ugliness somewhere else or at some later time" (Orr 2002: 134). It contrasts with the current situation where ugliness is rendered largely invisible:

> The problem is that we do not often see the true ugliness of the consumer economy and so are not compelled to do much about it. The distance between shopping malls and the mines, wells, corporate farms, factories, toxic dumps, and landfills, sometimes half a world away, dampens our perceptions that something is fundamentally wrong.
>
> *(Orr 2002: 179).*

Jonathan Maskit also calls for "the aesthetics of elsewhere" as "an environmentalist everyday aesthetics" that needs to be cultivated along with environmentalist everyday ethics (Maskit 2011). For example, in addition to the environmental harm caused by logging and mining, the aesthetics of elsewhere will expose the *aesthetic* devastation caused by them:

> mines, particularly open pit mines or mountain-top removal mines are seen by many as ugly. So too with logging sites as well as the various industrial facilities from which the raw materials for even the most beautiful buildings and artworks are sourced.
>
> *(Maskit 2011: 99)*

He argues for the role art can play in this regard, such as photographs of Edward Burtynsky and others that document various forms of ugliness caused by resource extraction and after-life of manufactured goods such as discarded electronics and automobiles.

One may claim that Orr's definition of beauty is impossible to put into practice because, lacking omniscience, we can never predict the future aesthetic consequences of our activities and products. The future ugliness may be unforeseen, just as environmentally harmful consequences can be unexpected. After all, a silent spring was not predicted by the use of DDT, nor was fish kill by the application of fertilizer. While this skepticism is justified and we may be taken by surprises in the future, we already know the negative aesthetic effects of mining, logging, chemical fertilizers, and disposal of electronic devices and automobiles. Disfigured landscapes, the visual blight and foul stench of fish kill, and ever-expanding junkyards already exist. The fact that we are not fully cognizant of future aesthetic consequences is no argument for ignoring the current aesthetic harm.

Orr's definition of beauty can also be put into a more positive manner. That is, even if something may not strike us immediately as beautiful in the conventional sense, its contribution to producing aesthetically positive values elsewhere or at later time can shed a different light on the object's aesthetics. Consider William James's anecdote regarding a cleared field he encountered in North Carolina. What at first appeared to him an "unmitigated squalor" and "a mere ugly picture on the retina" because of charred tree stumps and irregularly planted corn came to symbolize for him "a very paean of duty, struggle, and success" because of the residents' honest sweat and labor (James 1915: 231–234). Although I believe he goes too far by discounting altogether "the spectator's judgment," his story is instructive in calling attention to how the aesthetic value of something can be more than skin deep.

We can apply the same sensibility to many community gardens that are recently sprouting in American urban areas. They often appear to be crudely created and messily arranged, in comparison with the gorgeous gardens in suburban estates with well-manicured lawn and neatly planted exotic flowers. However, in addition to embodying the communal pride and collaborative spirit, such gardens often feature produces that feed the area residents, giving rise to the notion of rich fertility, particularly in comparison with sterile green lawn. They also attract bees and butterflies, providing liveliness, which constitutes a positive aesthetic value.

Wildflower gardens featuring indigenous plants can also be appreciated in the similar manner. Because they don't need extra water, fertilizer, pesticide, and herbicide, they also attract birds, butterflies, and other wildlife, contributing to a vibrant atmosphere. If environmentally problematic green lawn falls short of Orr's definition of beauty, community gardens and wildflower gardens earn a new sense of aesthetic value based upon their contribution to enlivening the area community for both human and nature.

Finally, let us return to "Edible Estates" with which I began. Their seemingly messy, chaotic, and disorderly appearance can begin to appear aesthetically positive when we consider their fecundity, productivity, and contribution to biodiversity and neighborliness. At the same time, the green lawn's beauty becomes compromised by its sterility, inhospitability toward living creatures, and downright environmental harm and health hazards. As Diana Balmori comments on Edible Estates:

Beauty has many dimensions, and … is a rather … complex concept that has cultural and moral dimensions. Will you look at this established icon deemed beautiful for generations with the same eyes once you know the effects it has on our environment? Ecological thinking has transformed how we see the lawn, and our concept of beauty has been transformed with it.

(Balmori 2010: 13)

In short, it is not possible to surgically remove life values from aesthetic values.

Conclusion

The most important reason for cultivating environmentally informed everyday aesthetics is that we are not merely disengaged spectators making aesthetic judgments, as characterized in the traditional aesthetic discourse. In our everyday life, we are also actively engaged in the world-making project in the most literal way. We have seen that, in our daily life, we often purchase, maintain, and dispose of objects, shape our immediate environment, and support or reject a public project, guided by aesthetic tastes and preferences. In light of this power of the aesthetic, it is crucial that we subject our everyday aesthetic judgments to a critical scrutiny. Sometimes such a critical examination suggests a paradigm shift. Following Haeg's justifications for the radical departure from the conventional American Dream, we should "question other antiquated conventions of home, street, neighborhood, city, and global networks that we take for granted," because

no matter what has been handed to us, each of us should be given license to be an active part in the creation of the cities that we share, and in the process, our private land can be a public model for the world in which we would like to live.

(Haeg 2010: 8)

Whether professional world-makers or not, we all share a responsibility in shaping the world and the future, and aesthetics has a surprisingly important role to play in this joint project.

Acknowledgments

I would like to thank Lydia Dixon and editors for their insightful comments on the first draft of this piece.

Related Topics

11. National Parks
12. Landscape
15. Wetlands
34. Urban Parks and Open Space
37. Waste and Consumption
42. Community Gardens
60. Education

References

Balmori, D. (2010) "Beauty and the Lawn: A Break with Tradition," in F. Haeg (ed.) *Edible Estates: Attack on the Front Lawn*. New York: Metropolis Books.

Berleant, A. (2010) *Sensibility and Sense: The Aesthetic Transformation of the Human World*. Exeter: Imprint Academic.

Blatt, H. (2008) *America's Food: What You Don't Know about What You Eat*. Cambridge: MIT Press.

Borasi, G. (ed.) (2010) *Journeys: How Travelling Fruit, Ideas and Buildings Rearrange Our Environment*. Montréal: Canadian Centre for Architecture.

Bormann, F. H., Balmori, D. and Gebelle, G. T. (1993) *Redesigning the American Lawn: A Search for Environmental Harmony*. New Haven: Yale University Press.

Brady, E. (2004) "Imagination and the Aesthetic Appreciation of Nature," in A. Carlson and A. Berleant (eds.) *The Aesthetics of Natural Environments*. Peterborough: Broadview Press.

Cooper, D. E. (2009) "Look of Lawns," *Times Literary Supplement* 5525 22–23.

Danto, A. (1978) "The Artworld" in J. Margolis (ed.) *Philosophy Looks at the Arts*. Philadelphia: Temple University Press.

Dee, C. (2010) "Form, Utility, and the Aesthetics of Thrift in Design Education," *Landscape Journal* 29 21–35.

Eaton, M. M. (2001) *Merit, Aesthetic and Ethical*. Oxford: Oxford University Press.

Forsey, J. (2013) *The Aesthetics of Design*. Oxford: Oxford University Press.

Foster, C. (1992) "Aesthetic Disillusionment: Environment, Ethics, Art," *Environmental Values* 1 205–215.

Gould, S. J. (1993) "The Golden Rule – A Proper Scale for Our Environmental Crisis." in S. J. Armstrong and R. G. Botzler (eds.) *Environmental Ethics: Divergence and Convergence*. New York: McGraw Hill.

Haeg, F. (2010) "Full-Frontal Gardening," in F. Haeg (ed.) *Edible Estates: Attack on the Front Lawn*. New York: Metropolis Books.

Hepburn, R. (1993) "Trivial and Serious in Aesthetic Appreciation of Nature," in S. Kemal and I. Gaskell (eds.) *Landscape, Natural Beauty, and the Arts*. Cambridge: Cambridge University Press.

James, W. (1915) *Talks to Teachers*. New York: Henry Holt and Company.

Leddy, T. (2008) "The Aesthetics of Junkyards and Roadside Clutter," *Contemporary Aesthetics* 6, sec. 5.

Leddy, T. (2012) *The Extraordinary in the Ordinary: The Aesthetics of Everyday Life*. Peterborough: Broadview Press.

Leopold, A. (1966) *A Sand County Almanac*. New York: Ballantine Books.

Maskit, J. (2011) "The Aesthetics of Elsewhere: An Environmentalist Everyday Aesthetics," *Aesthetic Pathways* 1:2 92–107.

Nassauer, J. I. (1997) "Cultural Sustainability: Aligning Aesthetics with Ecology," in J. I. Nassauer (ed.) *Placing Nature: Culture and Landscape Ecology*. Washington, DC: Island Press.

Orr, D. (2002) *The Nature of Design: Ecology, Culture, and Human Intention*. Oxford: Oxford University Press.

Postrel, V. (2003) *The Substance of Style: How the Rise of Aesthetic Value is Remaking Commerce, Culture, and Consciousness*. New York: HarperCollins Publishers.

Schiller, F. (1965) *On the Aesthetic Education of Man: In a Series of Letters*, trans. R. Snell. New York: Frederick Ungar Publishing.

Steinberg, T. (2006) *American Green; The Obsessive Quest for the Perfect Lawn*. New York: W.W. Norton.

Tenner, E. (1996) *Why Things Bite Back: Technology and the Revenge of Unintended Consequences*. New York: Alfred A. Knopf.

Teyssot, G. (1999) "The American Lawn: Surface of Everyday Life," in G. Teyssot (ed.) *The American Lawn*. New York: Princeton Architectural Press.

Thaler, R. H. and Sunstein, C. R. (2009) *Nudge: Improving Decisions about Health, Wealth, and Happiness*. New York: Penguin Books.

Vileisis, A. (1997) *Discovering the Unknown Landscape: A History of America's Wetlands*. Washington, DC: Island Press.

Walton, K. L. (1978) "Categories of Art," in J. Margolis (ed.) *Philosophy Looks at the Arts: Contemporary Readings in Aesthetics*. Philadelphia: Temple University Press.

Whiteley, N. (1993) *Design for Society*. London: Reaktion Books.

Further Reading

Berleant, A. and Carlson, A. (eds.) (2007) *The Aesthetics of Human Environments*. Peterborough: Broadview Press. (A collection of essays on the aesthetics of built environments.)

Carlson, A. and Lintott, S. (eds.) (2007) *Nature, Aesthetics, and Environmentalism: From Beauty to Duty*. New York: Columbia University Press. (A collection of essays on the aesthetics of nature and its ethical ramifications.)

Crouch, C., Kaye, N. and Crouch, J. (eds.) (2015) *An Introduction to Sustainability and Aesthetics: The Arts and Design for the Environment*. Boca Raton: Brown Walker Press. (A collection of essays on sustainable design.)

Drenthen, M. and Keulartz, J. (eds.) (2014) *Environmental Aesthetics: Crossing Divides and Breaking Ground.* New York: Fordham University Press. (A collection of essays on environmental aesthetics, its history and future, cultural dimensions, and relationship to art.)

Hosey, L. (2012) *The Shape of Green: Aesthetics, Ecology, and Design.* Washington, DC: Island Press. (An architect's discussion of the aesthetics of sustainable design.)

Minney, S. (2016) *Slow Fashion: Aesthetics Meets Ethics.* Oxford: New Internationalist. (Collection of examples of designers and companies producing sustainable products that are aesthetically pleasing.)

Saito, Y. (2007) *Everyday Aesthetics.* Oxford: Oxford University Press. (A discussion of prevailing aesthetics in everyday life and its impact on the quality of life.)

Saito, Y. (2017) *Aesthetics of the Familiar: Everyday Life and World-Making.* Oxford: Oxford University Press. (A discussion of the power of the aesthetic in everyday life and everyday aesthetics as a normative discourse.)

Saito, Y. (2018) "Consumer Aesthetic sand Environmental Ethics: Problems and Possibilities," *The Journal of Aesthetics of Art Criticism* 76:4 429–439. (A presentation of environmental problems associated with the aesthetics of today's consumerism and a proposal for possible aesthetic solutions.)

Soper, K. (2008) "Alternative Hedonism, Cultural Theory and the Role of Aesthetic Revisioning," *Cultural Studies* 22 567–587. (An argument for developing 'alternative hedonism' to support sustainable lifestyle.)

Steiner, W. (2009) "The Joy of Less," *Harvard Design Magazine* 30 8–9 (An argument against cultivating sustainable lifestyle by denying joy.)

Walker, S. (2006) *Sustainable by Design: Explorations in Theory and Practice.* London: Earthscan (A theory and practice of sustainable design written by a design educator).

62

COMMUNITY PARTICIPATION

W. S. K. "Scott" Cameron[1]

Introduction

From the 1950s through the 1990s, individual attitudes toward overpopulation, littering, recycling, and simplicity underwent dramatic change, but the most decisive environmental developments resulted from national and international political initiatives—in the US, the founding of the EPA and the passage of the Clean Air, Clean Water, and Endangered Species Acts; and internationally, the Montreal Accord and the Kyoto Protocols. Since then, however, the pendulum has swung so far back that several countries, most notably Australia, Canada, and the US, are bucking or backpedaling on the legislation urgently needed given the overwhelming scientific consensus that climate change threatens civilization itself.[2] Since gridlock has postponed effective action for 20 years, some hope that widely adopted voluntary efforts—individual lifestyle changes, or adaptations pioneered by small groups working on critical issues—can reverse the trend. After all, individual decisions determine many carbon emissions. Could and should we not modify them? Yet recently, several influential environmental philosophers have expressed skepticism.

Elsewhere, of course, such skepticism might appear obvious: those who regard climate change as illusory or who doubt its anthropogenesis must regard individual efforts to stop it as quixotic and fruitless. Yet in an important collection on climate change, co-editor Walter Sinnott-Armstrong concedes the reality, significance, anthropogenesis, and severe consequences of climate change especially for the young and poor, and then argues *against* any individual obligation to forego optional emissions such as a Sunday drive undertaken merely for the pleasures of the view and the satisfying roar of a borrowed gas-guzzler.[3] Though he shares the intuition that such a drive would be wrong, he lacks confidence in it:

> I would probably have different moral intuitions about this case if I had been raised differently or if I now lived in a different culture. My moral intuition might be distorted by overgeneralization from the other cases where I think that other entities (large governments) do have moral obligations to fight global warming. I also worry that my moral intuition might be distorted by my desire to avoid conflicts with my environmentalist friends. The issue of global warming …. [i]s also a peculiarly modern case, especially because it operates on a much grander scale than my moral intuitions evolved to handle …. In such circumstances, I doubt that we are justified in trusting our moral intuitions alone. We-need some kind of confirmation.
>
> *(288–289)*

DOI: 10.4324/9781315768090-74

Sinnott-Armstrong seeks a moral principle to ground his intuition; and having assessed many candidates and concluded that no general principle survives, he denies any individual obligation to mitigate one's contributions to climate change. Refraining from wasteful driving may be praiseworthy, and we should certainly pass laws that limit emissions. But in advance of such laws, I have, in his view, no individual obligation to forgo even trivial carbon-fueled pleasures.

A lone philosopher might just be wrong, but Sinnott-Armstrong leads an influential chorus. He develops themes first made famous in Garrett Hardin's "tragedy of the commons" argument and recently applied to climate change by Baylor Johnson.[4] A related argument for causal impotence features in Joakim Sandberg's prize-winning defense of Sinnott-Armstrong's position.[5] Together, these arguments suggest that individual and voluntary group initiatives are at best pointless, and at worst a distraction from effective political engagement.

In this chapter, I defend the apparently naive intuition that we *are* responsible to mitigate our contributions to climate change as well as to volunteer in groups identifying, educating others about, or cooperating to mitigate excessive emissions *in addition* to our responsibility to seek political change. Moreover, defending this view is important not only for practical, climate-related reasons. The arguments made by Sinnott-Armstrong and others have several weaknesses, but they are not daft. They are valuable as symptoms that reveal some critical, but highly contestable presuppositions of contemporary ethical theory more broadly.

And so our itinerary: I will outline Sinnott-Armstrong's argument and indicate how it is buttressed by others. I will then note the argument's methodological difficulties. Finally, I will defend five reasons to reject skepticism about the effects of voluntary action. My conclusions are given as follows:

a   The "moral mathematics" Sinnott-Armstrong and Sandberg presume is problematic: they discount individual effects but count collective ones. Vague, ambiguous, and context-insensitive descriptions blind them to the ways that collective effects grow from individual contributions.

b   Tragedy of the commons-style arguments, similarly, presumes far too narrow a view of historical structure and context of action.

c   The skeptics' argument for causal impotence presumes a weak and clearly inadequate model of causality. I cannot hope to elaborate the more sophisticated model we need, but I can show the limits of their model.

d   Sinnott-Armstrong doubts the effectiveness of voluntary individual or small group initiatives, but his deepest worry is that they distract us from advocating the collective solutions we *should* undertake. Yet he cannot consistently downplay individual, while emphasizing political, action. The two stand or fall together.

e   Finally, to forestall the worry that both individual and corporate action fail, I argue that individual action is far more politically potent than we usually recognize. In her exchange with Baylor Johnson, Marion Hourdequin thematizes the legitimating demand of integrity.[6] I concur, and add that the commitment to integrity is politically productive not only in individual but also in collective contexts. Most individual action follows and reinforces traditional—and in the climate case, problematic—patterns. But for that very reason, those bucking cultural defaults can have more influence than they may dare hope or imagine.

We must thus resist skepticism concerning voluntary individual and small group initiatives. Corporate changes are undoubtedly necessary to ensure a sustainable future. Yet far from being ineffective or distracting, voluntary initiatives are not only cumulatively significant, but are the most direct way to destabilize the learned helplessness and denial that has stymied political change both nationally and internationally.

The Argument for Skepticism, and Three Methodological Worries

When considering the putative obligation to forego a joyride, Sinnott-Armstrong assumes that it will have no measurable effect on atmospheric carbon. And since his drive "is neither necessary nor sufficient for global warming [in general]" (289) and its "exhaust on that Sunday does not cause any climate change [effects—e.g. storms or floods] at all" (291), he bears no blame for recreational driving. Of course *global warming* causes harm, yet he is only culpable if *his drive* does, and that is in fact far too trivial to harm any determinate individual (291, 293–294, 302).[7] One day's drive does not constitute a broader, harm-causing habit, and since his influence is limited, driving will not undermine anyone else's commitment to environmental goals. Sinnott-Armstrong concludes that "global warming is such a large problem that it is not individuals who cause it or who need to fix it" (304). Added together, millions of individual decisions *have* caused corporate effects that demand corporate regulation, but one Sunday's drive is both physically and socially insignificant—and thus blameless. On Sinnott-Armstrong's view, acknowledging this reduces the temptation perfectionistically to minimize one's own contributions to the problem and focuses attention on the far more important task of achieving political change.

Sinnott-Armstrong accepts the best current climate science—and yet appears to follow reasonable premises to absurd conclusions. When acts are rare and have negligible effects, it is unreasonably perfectionistic to condemn them. Yet Sinnott-Armstrong repeatedly shifts from this plausible view to denying any *general* obligation to avoid contributing to climate change—and this in a context that, by his own admission, demands immediate, effective, alternative patterns of living. Should we accept his skepticism regarding our individual obligations? Perhaps not, for methodological weaknesses undermine his case.

(a) Sinnott-Armstrong's Method of Elimination

First, Sinnott-Armstrong contests his initial intuition that a joyride would be wrong by evaluating and then eliminating the principles that purportedly justify it. But even if every critique succeeds, his argument invites suspicion that his enumeration of principles is incomplete. To be sure, this is merely an argument from silence—but it undermines confidence that his initial intuition is wrong, especially if, as he concedes, that intuition is widely shared.

(b) Sinnott-Armstrong's Appeals to Practical and Linguistic Intuitions

By evaluating baldly stated principles based on immediate practical and linguistic intuitions about briefly described cases, Sinnott-Armstrong's method is triply problematic. First, some candidate principles are very crudely formulated. Sinnott-Armstrong rejects any general obligation not to emit greenhouse gases, since then I would "have a moral obligation not to boil water (since water vapor is a greenhouse gas) or to exercise (since I expel carbon dioxide when I breathe heavily)" (293). Certainly that general principle would exclude those practices—but what thoughtful person would defend so ham-fisted a claim?

Second, Sinnott-Armstrong regularly relies on contestable intuitions to reject the principles he considers. He challenges the principle "not to increase the risk of harms to other people" by contrasting global warming (where it does not apply) with drunk driving (where it does), since "there is no way to identify any particular victim of my wasteful driving" whereas we can identify a drunk driver's victim(s) (293–294). But drunk driving is criminal even before it claims any determinate victim; and besides, does the mere difficulty of *identifying* a victim mean that there is *no* victim? During the Holocaust, many died without a trace—but far from erasing, the killers' callous disregard for their victims' personhood arguably increased their offense. Here I too deploy an immediate and perhaps contestable intuition about a quickly sketched case; my point is that the multiplying intuitions lead to contradiction and confusion, not clarity.[8]

Finally, Sinnott-Armstrong never explains why he accepts some, but rejects other common moral principles and intuitions—including his own initial intuition that the joyride was wrong. He notes the linguistic intuition that we identify exceptional factors as causes, thus blaming a struck match, rather than the oxygen always present, as the cause of a fire (290). But since this linguistic intuition is a rhetorically motivated shortcut (since listing *all* causes would often be tedious), it's not a strong guide to comprehensive practical judgment. Indeed, though Sandberg critiques Sinnott-Armstrong's appeals to intuition, he too regards "the difference between what happens as a result of [an] action and what would have happened otherwise" as "an immensely in-tuitive notion of harm which we should not give up very easily" (232, n.10, 231) in order to justify the conclusion that acts with negligible effects can be ignored. Yet the question is not whether this intuitive concept of harm is useful, but whether it needs qualification when individually negligi-ble acts have significant cumulative effects. Appeals to practical and linguistic intuitions can be helpful, but they demand a more circumspect method like John Rawls' reflective equilibrium;[9] without that, readers are abandoned on a slag heap of conflicting intuitions and principles.[10] The skeptics' method invites ethical paralysis—the last thing we can afford when facing the great moral challenge of our time.

(c) *Sinnott-Armstrong Misinterprets Some of the Principles He Considers*

Finally, Sinnott-Armstrong's interpretations of the principles he rejects are often contentious and in some cases outrageous. He interprets virtue ethics as a principle—that we must avoid vice—and asks "[h]ow can we tell whether driving a gas-guzzler for fun 'expresses a vice'?" (295–296). Similarly Sandberg:

> there is a deeper problem with the virtue oriented approach, namely, that it is not obvious that a fully virtuous person never would drive or fly. ... [Consequently] a virtuously green person cannot only have his or her head in the sky, but must also attend to the practical consequences of his or her choices.
>
> *(246–247)*

Adequately responding to such characterizations of virtue ethics would take far more space than I can afford, but even a quick sketch reveals problems. Aristotle developed virtue ethics in opposi-tion to Plato's aspiration to make ethics a science of principles.[11] Aristotle objected that ethics was more like medicine and navigation: areas irreducible to principles because the key question is not whether a treatment *usually* works or a navigational technique is *usually* effective, but whether it works *in this patient* or facing the challenges posed by *this trip*. Rather than vague principles, we need the careful, contextually sensitive intuition of a practically experienced person. Virtue ethics faces several reasonable challenges: e.g., who counts as a wise person, how does she gain her wis-dom, and what contextual features are relevant? But *pace* Sinnott-Armstrong and Sandberg, Aris-totle does not rule actions in or out deontologically; his goal is precisely context- and case-specific adjudication. Sinnott-Armstrong and Sandberg criticize him for failing to meet a standard he would reject, and in addition, Sandberg criticizes him for ignoring what is in fact his main goal.

The objections above reveal the methodological limitations of Sinnott-Armstrong and Sandberg's skepticism about individual obligations. Their skepticism must collapse, however, under the weight of the following four problems.

Deeper Problems with Skepticism about Individual and Small Group Initiatives

(a) *The Skeptics Rely on Vague, Ambiguous, and Context-Insensitive Descriptions*

Let's start with an obvious, if less significant, problem: Sandberg's appeal to the empirical difficulties of measurement to argue that a joyride is insignificant. Beginning from the truth that a Sunday's drive emissions are "extremely small," he concludes that they are, "[i]n any relevant practical sense … undetectable and negligible," and make climate change "not noticeably more likely to happen" (231). Yet his assertion that tiny emissions are "negligible" is contestable, as we will see; and even if they were "undetectable" or would "not noticeably" make climate change more likely, harms may be real, yet unnoticed or even undetectable, at least for a time.[12] Moreover the ride's emissions are *not* "undetectable" but, as Sandberg himself notes, can be measured easily and precisely by tracking the gas I burn. So back to the primary question: do such emissions cause harm? In fact, the right answer is neither yes nor no, for the concept of "a joyride" is ambiguous.

Sinnott-Armstrong and Sandberg justify their optimistic conclusion by observing that a single joyride produces an extremely tiny impulse toward climate catastrophe. All the drives *we* take may cause real, measurable harm as individually negligible impulses are multiplied. But no individual can stop all other drivers; and if all their drives *multiply* the effect of my drive by millions to produce a cumulatively disastrous effect, can my responsibility for that effect not be *divided* by the same factor? In Sinnott-Armstrong and Sandberg's terms, if climate change will happen because of millions of other decisions I cannot control, isn't it pointless to sacrifice my tiny contribution to achieve no marginal benefit?

These questions seem plausible—except that the right operation of "moral mathematics" is neither simple addition nor simple division. Both Sinnott-Armstrong and Sandberg assess the effects of *one drive alone*—i.e., the drive's "marginal effects," or "the difference between what happens as a result of the action and what would have happened otherwise" (Sandberg 231; Sinnott-Armstrong 292 and 305, n. 16). Yet this tight a focus so atomizes significance that it can offer no agent-orienting guidance. The skeptics rightly presume that the marginal effects of a single action may be so negative that we'd condemn it (e.g., even the murder of the "parasite" Raskolnikov considers his victim to be in *Crime and Punishment*); and they are also right that the marginal effects may be so small as to render an act insignificant despite any realistically likely multiplication (e.g. throwing pebbles into the sea, which merely speeds up normal geological processes).[13] But they fail to acknowledge a third class of action critical here. Derek Parfit famously identifies our tendency to ignore the accumulation of tiny effects into a cumulatively unfortunate result as one of "five mistakes in moral mathematics." Yet while I find one of his arguments very plausible,[14] Parfit often relies on contestable intuitions about science fiction examples. My objection is more direct. Sinnott-Armstrong's explicit question is whether one joyride is morally problematic. Yet he uses facts about a single *instance* to draw moral conclusions about a *type*—a subtle but critical ambiguity.

Since Aristotle, a fundamental principle of metaphysics and epistemology has been that "definition is of the universal and the form."[15] This insight, shared not only by ancient and medieval philosophers but by modern metaphysicians as various as Kant, Hegel, and Whitehead, recognizes that we cannot grasp anything as *unique*, but only as sharing common and knowable—i.e. *re*-cognizable—features. If we treat a joyride as entirely unconnected with other instances—i.e., as a unique event—it appears permissible because it is genuinely morally insignificant. But in that case, why report one's evaluation at all, since no judgment about it can inform conclusions about the permissibility of any other instance? Conversely if we want guidance for similar cases in the future, we cannot merely consider an isolated instance; what matters are the characteristics of the *type* of act. But typification presumes the potential of multiplication, and here it presumes *actual* multiplication: for not only *can* the drive produce cumulative effects, but here we see its effects multiplied millions of times in fact. Only by evaluating the consequences of the type can we license conclusions about future instances.[16] Consider this parallel: a wise spouse might disregard her tired partner's uncharacteristically insensitive put-down. But if he began repeating such comments, their meaning would shift. One remark might be trivial, but a repeated *type* can easily be corrosive.[17] In assessing this third class of actions, judgments about individual cases are insignificant. Only judgments about the type can legitimate future moral decisions.

(b) "Tragedy of the Commons" Arguments for Skepticism also Presume Too Narrow a View

Another worry that apparently justifies skepticism about individual and voluntary initiatives derives from Garrett Hardin's tragedy of the commons argument. Who could blame the villager who defied tradition to bring a second goat to the commons, reasoning that if the first ensured subsistence, the second might pay for her child's schooling? A new goat doesn't exhaust the land any more quickly than the first, and there is yet room to spare. Nevertheless if enough others follow her example, tragedy will unfold as the commons is grazed to exhaustion and collapse. Sinnott-Armstrong's focus on a joyride during a carbon emission crisis undoubtedly adds some complexities: the number of "villagers"—now over 7 billion; the fact that we necessarily produce emissions by breathing; and the remoteness of causal connection between tiny, individual carbon contributions and one massive, Earth-altering effect. But the upshot remains the same: individual initiatives appear pointless; and the only solution is "mutual coercion mutually agreed."[18] Baylor Johnson summarizes:

> The situation is nicely summed up by the phrase 'use it or lose it.' Resources foregone by the individual today are almost certain to be lost to some less enlightened herder tomorrow. Thus … personal sacrifice to preserve the commons tends to be self-eliminating, as the scrupulous users lose their livelihood to the ignorant, the unscrupulous, or those who reasonably doubt that all will voluntarily reduce their use. … [T]herefore there can be no reasonable expectation that unilateral reductions in use of the commons will be mirrored by enough other users to protect the commons.
>
> *(Johnson 274)*

Commons-style arguments, in addition, raise the Hotelling problem recently highlighted by Benjamin Hale: partial successes motivate global failure. It's not just that the resources uncaptured by the more enlightened are likely to be appropriated by the less; economic incentives push in that direction as well. If individual people or nations moderate their consumption, oil companies will strive to recover sunk costs by lowering prices to increase demand among less scrupulous individuals or countries.[19] And here too, legitimate worries about freeloading among the scrupulous will incentivize more self-interested, short-term choices.

Though the challenges here are more complex, the answer begins with the insight above: contemporary defaults toward individualism, strategic decision-making, and context-free description mislead much moral philosophy. It's tempting to see an innovating villager's second goat as harmless: it provides the innovator a small gain at what appears to be a bearable cost. But each two-goat owner produces two effects: a direct new pressure on the commons and the indirect effect of modeling collectively riskier behavior. Adding a goat has no serious consequences *if few enough follow one's model*—but without that caveat, the act is *not* harmless; it's merely ambiguous. Everything hangs on how many follow her lead.

As Johnson rightly insists, worries about freeloading—and especially its actual prevalence—will undermine incentives to act well in the long run. But the question is whether we can shift individual or international actors' reasoning from a short-term economic, to a long-term moral, framework. Doing so requires problematizing the public costs of private gains—a challenge to which we will return below.

(c) *Sinnott-Armstrong and Sandberg Presume an Inadequate Model of Causality*

A third problem hobbles the skeptics' case: assuming a narrow model of causality that presumes an identifiable individual using a predictable means to harm a determinate victim. This model is clearly useful and often adequate, so I rightly apologize after stepping on your toe.[20] Yet cases like the ones considered above prompt Derek Parfit to suggest an evolutionary explanation of our blindness to other possibilities:

> Until this century, most of mankind lived in small communities. What each did could affect only a few others. … [But e]ach of us can now … have real though small effects on thousands or millions of people. When these effects are widely dispersed, they may be either trivial, or imperceptible. It now makes a great difference whether we continue to believe that we cannot have greatly harmed or benefited others unless there are people with obvious grounds for resentment or gratitude.
>
> *(Parfit 87)*

If the cumulative consequences of apparently insignificant actions can now collectively benefit or harm others, he argues, we must acknowledge a new type of responsibility.[21]

But this explodes the simple agent—causal process—victim model. Consider Sandberg's claim that foregoing an unnecessary flight to Paris does not affect emissions, since the plane will fly despite my cancellation. Sandberg concludes that my decision is neither necessary nor sufficient to produce climate change, since "[i]f everyone else had acted differently, the threat of climate change would namely have been avoided; so my contribution is entirely dispensable" (Sandberg 240). Yet if appealing to necessary and sufficient conditions works in many cases, it is misleading here.

Above, I suggested the problem was ambiguity: my contribution is dispensable *only if others act differently than they are acting* (here, by not flying themselves); yet there is no reason to expect this, and if everyone's contribution is *singly* dispensable, everyone's *together* are not. Markets are not usually sensitive to individual decisions, but they are sensitive to stochastic effects, so a flight's cancellation requires many decisions to forego flying. Yet markets do follow the law of supply and demand, and demand shifts one mind at a time. Thus while my decision to forego a trip will not ground a plane, the airline's risk rises with every such decision, and at some point it will reduce trips. Individual decisions, considered singly, are insignificant; what counts is the larger pattern. But my decision, along with others', constitutes that pattern.[22] Here we need a more sophisticated causal model.

George Lakoff has introduced the term "systematic causation" to highlight this dynamic.[23] Some pollutants have identifiable effects on specific victims, thus following the simple agent—causal process—victim model. But often pollution's effects are far harder to pinpoint because they occur over long periods of exposure to very small doses distributed among many confounding variables, and harm is manifest only in adverse changes in the distribution of risk of disease in a population. Not all those exposed to pollutants develop diseases; and diseases caused by exposure *might* have arisen from other causes. Epidemiologists can identify "cancer blooms," but industry statisticians will remind us that correlations do not entail causation and that random distributions may include seemingly non-random clusters. Yet all these problems conceded, *no one* wants to live downstream from a major industrial polluter: that's the motive for NIMBY movements everywhere, and often the activists' worries are reasonable. The link between pollution and some diseases may defy our traditional causal model, but no honest person would deny it. Factory managers rarely live downstream or downwind of their factories.

Similarly, virtually all now admit (after a 40-year fight!) that smoking causes lung cancer, though not in a direct, linear way. No single cigarette, nor any specific number, is necessary or demonstrably sufficient to cause the disease: most smokers don't get lung cancer, and some who do have never smoked. The direct damage or even increased risk caused by one cigarette is tiny, perhaps immeasurable—and yet parents are distressed when their children experiment with tobacco, and deeply disappointed if they are hooked. On Lakoff's view, smoking "systematically causes" cancer since cancer risk rises with the amount smoked. This pattern cannot reasonably be divided into a determinate risk for each cigarette, yet the smoker's overall risk grows one cigarette at a time. And thus the case's relevance here, twice over: smoking cigarettes causes cancer, but in a broader, more diffuse way than Sinnott-Armstrong or Sandberg can acknowledge on their simple, standard model; and while we cannot specify a specific harm per cigarette, the growing cumulative risk is statistically obvious, as is its development through apparently insignificant individual decisions to light up. Shrugging off a cigarette as neither necessary nor sufficient for disease captures a small truth and obscures a larger one: smoking is risky, and smoking more is worse.[24] While no one cigarette causes cancer, smoking "systematically causes" cancer. By the same token, one joyride may have no individually measurable effects—but that does not mean that it has no cumulative effect.

(d) *Sinnott-Armstrong and Sandberg Inconsistently Recommend Legislative over Individual Action*

Sinnott-Armstrong is no climate skeptic; he agrees that climate change demands immediate, decisive action. He wants us to focus on collective change, for "global warming is such a large problem that it is not individuals who cause it or who need to fix it. Instead, governments need to fix it, and quickly" (SA 304). Individual solutions are insufficient: we cannot do enough

> simply by buying fuel-efficient cars, insulating our houses, and setting up a windmill to make our own electricity. That is all wonderful, but it does little or nothing to stop global warming and also does not fulfill our real moral obligation, which are to get governments to do their job to prevent the disaster of excessive global warming.
>
> (SA 304)

To begin with agreement: political action is unquestionably essential. And Sinnott-Armstrong concedes the necessity of individual action in one respect: "If I have an obligation to encourage the government to fulfill its obligation, then the government's obligation does impose some obligation on me" (SA 304, n. 9). He is clearly right that "I do not have an obligation to do what the government has an obligation to do" (SA 304, n. 9); the question is whether I have an obligation to forego the joyride. My worry is that his case against that obligation—relying as it does on ambiguity and an inadequate causal model—forces him to the general conclusion that individual actions are ineffective—and that general argument undermines the political advocacy he rightly affirms.

For just as no individual act will bring about or prevent climate change, no one's individual initiative will bring about political change. In both cases, constructive change arises as (and only as) the result of accumulating individual and small group efforts. Even the most politically potent American (whoever that might be) cannot control the economy; and the rest of us are far, far less powerful than that. Yet just as the economy grows or wanes as the result of individual decisions, political movements effect change only by means of accumulating individual and small group efforts. Sinnott-Armstrong must choose: either such actions *can* produce great cumulative effects—in which case political change is possible, and each of us must minimize his or her carbon footprint; or individual efforts cannot have cumulative effects, in which case minimizing one's carbon footprint may be ineffective—but so, also, is striving for political change.[25]

(e) The Skeptics Fail to Recognize the Political Significance of Individual and Small Group Initiatives

Sinnott-Armstrong and Sandberg concede the virtue, if not the obligation, of minimizing one's carbon footprint; and Sandberg and Johnson also recognize the virtue of integrity and the possibility that innovators modeling new, more adaptive lifestyles may influence others to follow. Regarding the too-general argument against the efficacy of individual efforts, I have argued that the potential effectiveness of minimizing one's carbon footprint and of political advocacy stand or fall together. Yet one could reply that individual and small group efforts fail in both contexts. And even if the skeptics concede the value of individual efforts, Sinnott-Armstrong, Sandberg, and Johnson all argue that the energy we spend on controlling emissions individually would better be spent on advocacy of a collective, political solution.

With Hourdequin, I contend that individual efforts do not compete with but rather complement the implementation of a collective goal. Whereas she draws on Buddhist philosophical anthropology to show that humans are more socially and other-directed than we usually recognize in the West, I will support her conclusion using more familiar categories. Foregoing trivial emissions like the joyride is a duty not only because of its direct effect on atmospheric carbon but because of indirect effects on others that the skeptics do not adequately acknowledge. A closer look at the category of action reveals that we are *always* reinforcing or challenging norms, and thus every action is politically potent. To be sure, the skeptics recognize that *some* acts may be influential; they simply doubt that many are. And I agree that many actions *appear* uninfluential. Yet in fact, the situation is far more complex. Even when following norms, we wield more influence than most of us recognize; and norm violation, while not always influential, can have disproportionate effects even if we cannot anticipate what or how extensive they will be.

To be sure, the skeptics offer compelling reasons to doubt our individual influence (Sinnott-Armstrong 291–292; Sandberg 243; Johnson, "Environmental Obligations" 285–286 and "Joint Communiqué" 149). Pride inclines us to self-dramatization and -congratulation, and our narrative orientation often discovers meaning where it may not actually exist. Quirks of personality or self-righteousness may discolor or undermine our initiatives; and without further explanation our counter-cultural efforts may appear quixotic or bizarre—assuming others trouble to notice them at all. Even when others *do* notice, it's hard to measure the effects of our actions, since those effects may take time to develop and may occur after we've gone—though as I'll suggest, we may as regularly under- as overestimate them. And so to the critical question: does individual action complement political change—and if so, how?

I take it as obvious that we're all members of many different sorts of groups. Sinnott-Armstrong worries about this, since he fears the possibility that we'll be assigned moral responsibility for arbitrarily assigned group identities—e.g., being held responsible for the group of "all terrorists plus me" (Sinnott-Armstrong 298). But who would make, much less take such claims seriously? The very words speakers use to identify themselves and others bear witness to countless *non*-arbitrary group memberships. We are hippies, soldiers, tax-and-spend–liberals, skateboarders, Birkenstock moms, or millennials—unsurprisingly, for "it is not good that man should live alone," [26] so we flourish in groups and continually take cues from one another. Economic cycles, architectural styles, trends in literature, education, and fashion all reflect our affiliation with others, as creators, interpreters, and fans. Indeed the tragedy of feral children reveals that human flourishing *requires* socialization into a linguistic community—i.e., into a shared way of grouping things and people.

Yet grouping ourselves and others is not merely an epistemic task, but a moral one. Take "adulterer," "murderer," "bachelor," "spinster," "teacher," "socialist," "employee," or "boss": many group concepts are constituted by widely shared moral evaluations that determine status, coordinate expectations and actions, and relieve us of some—though critically, not all—of the responsibility of independent judgment. Moreover we both give and deny moral cover through such group memberships. When Michael Jackson was photographed inadvertently dangling his child three stories above a Berlin street while greeting adulatory fans, virtually everyone who saw the pictures thought "that's flat out crazy"; yet we understand Native Americans sending their children on dangerous vision quests as participating in important cultural rituals even if we might not want our children to do likewise (though we let them drive cars, at greater risk, for reasons neither as heroic or meaningful).

Within a given culture, such forms of group identification and evaluation are relatively stable over time; yet both the groups to which we belong and the activities accepted as normal within them are not entirely static, as fashions, technologies, and political and moral views develop through adaptation or reinterpretation over time. And as above in the general causal story, here too in the historical one, individual decisions constitute the historical process of developing or repudiating forms of group identification and evaluation. When we act as others in our group do, we give ourselves moral cover and reinforce current norms; and by subtly or pointedly avoiding common activities, we deny their apparent inevitability and make explicit their belonging in the realm of responsible choice. Thus our everyday participation in the normalizing and de-normalizing of common activities is not merely of marginal, but of central importance. But how does this happen, more concretely?

When we're following patterns typical in our group, our actions are largely unconscious—the result of our prior socialization. I had never been explicitly aware of my knowledge of suburban property lines until my two-year-old, who didn't recognize them at all, ran through gardens to jump on swings hanging in the yards of unknown neighbors. Her obliviousness both made me aware of what I'd internalized long ago, and set me the task of teaching her these basic cultural norms that she (now 18) has long since internalized. Yet even if many actions and inhibitions are unselfconsciously familiar, they are meaningful, not just potentially, but always.

Aristotle's *Nicomachean Ethics* defines action as something I *do*, not merely an accident I undergo, like falling after being tripped. In action "the moving principle is in the agent himself, he being aware of the particular circumstances of the action."[27] The agent's awareness of and responsiveness to her circumstances distinguishes actions that appear identical but are different—offering a friendly wave, swatting away blackflies,[28] or involuntarily flailing one's hand when falling. As actors, we often explicitly intend our action's meaning, but even when our behavior is unreflectively habitual, we can identify, explain, and if need be justify intentions that were not fully conscious.[29]

Since actions are always meaningful, they are subject to the demand for justification. To be sure, explicit justification is rarely demanded, for most act in accord with well-accepted patterns and thus avoid question or rebuke. Yet when someone does something unexpected, she reveals the taken-for-granted response as optional—and thus as demanding justification. To be sure, we lack direct access to her intention, and it's possible that we have merely misinterpreted her action, or failed to recognize an established variation of the current pattern. But when it becomes clear that she has repudiated the default, we must wonder whether the default choice is the best one. When we know the actor well enough, we ask, "why did you do that?"

Both as interpreters and actors, then, we're continually monitoring ourselves and others. Typically our actions follow culturally accepted defaults, but even simple variations, like riding a bike to work in Canada or the US, for instance, reveal that other options are possible, and perhaps preferable. Our actions are thus more influential than we typically realize—both when we're conforming and thus quietly reinforcing norms, and when we break away as witnesses to other, perhaps better possibilities. We can reinforce the significance of unusual choices by drawing attention to and explaining them—and so protest marches, consciousness raising, and teach-ins are potentially threatening. But we need not step on a soapbox. Responding to the objection that new behaviors are too opaque to influence others, Hourdequin concedes that they may be, but often are not. Often we've already heard arguments for others' atypical choices, and we need less persuasion than a model of successful adaptation to inspire imitation (Hourdequin 2011, 160).[30] Seeing the odd cyclist riding to work will not wake many from complacency, but as more people join, they can reasonably demand reasonable accommodations at work,[31] bike paths in cities, and more sensible urban design—all of which will make it easier for others to follow their lead.

Typically, norms evolve slowly over time, but every new adopter reinforces the viability of new models, and at a certain point, the default shifts. My father rode a bike to work decades before it was cool, but seeing gray-haired professors cycling to the University of Bonn or grandmas and aunties riding home with their groceries in Nijmegen permanently changed my image of biking. The contemporary interest in and demand for "green" buildings, livable downtown cores, native gardens, and local, organic food, or the increasing support for gay marriage and the decriminalization of marijuana are among the most visible contemporary examples of cresting trends. But such evolution happens not in some nebulous social "space," but in concrete interactions among small knots of friends, colleagues, and neighbors. And whether a particular choice supports or destabilizes the status quo, our responsibility extends not only to its direct consequences, but to the role I play—whether self-consciously or not—in building or breaking down a current consensus. Even dictators respond surprisingly often to the letter writing campaigns of Amnesty International, despite the fact that the letters come from non-citizens who pose no greater threat to a sitting tyrant than the power of moral disapproval. Talk is helpful, and political action is necessary—of these there can be no doubt—but if we had waited for the government to encourage organic farming, community-supported agriculture, or farmer's markets, they'd have started far later and wouldn't now be among the fastest-growing segments of the food market. And if talk is helpful,[32] innovative action is critical. Self-consciously experimental communities provide the problem-solving models that can inspire citizens to demand government action.

So far, I have offered a superficial description of a topic that deserves fine-grained attention. But even without all the details, we can multiply examples of such individual influence—both for good and for ill. To acknowledge a few bright lights during a very dark time, those celebrated in the Garden of the Righteous among the Gentiles at Yad Vashem—the Israeli Holocaust memorial—forced and still force us to acknowledge that what was thought "normal" or at least "sensible" was not the only way possible way to act; and under more fortunate circumstances even a seamstress unwilling to relinquish her bus seat could spark the long-gathered tinder of injustice into a blazing beacon for the civil rights movement. Norm-following behavior can also be more paralyzingly powerful than we recognize, for norm-following obscures the degree of freedom others have to reevaluate.[33] The most moving Holocaust memorial I saw while living in Germany was a massive iron sculpture on the site of the destroyed Levetzowstrasse Synagogue portraying train departures to concentration camps with their "lading" of Jewish deportees: a reminder not just of the camps, but more poignantly of all the other pairs of hands—at least a million by one estimate—that played some small administrative role in carrying out the Holocaust by tracking, routing, driving, and braking the trains, building and provisioning the camps, designing progressively more efficient means of killing, knowingly passing on disinformation, and so on. No one of these people accomplished the Holocaust; indeed the vast majority did not directly kill a single person. But without all the colluding pairs of hands doing their little part, this mad "masterwork" could never have occurred.[34]

Both the positive and negative examples illustrate the same point: all our actions have a historical meaning insofar as they contribute to or interfere with the standing waves of current social norms. While the skeptics fear that our efforts must be divided between individual efforts and political actions that compete for attention and energy in a zero sum game, I contend that individual actions are always implicitly political. Though effective challenges to dysfunctional patterns do not arise often and cannot be sustained unless others join in, the first person to advocate a new response may have effects far broader and more surprising than she anticipates—and none of us knows ahead of time which acts will or will not bear fruit.[35]

Conclusion

After a long journey, I'll conclude with a brief look back. We began with Sinnott-Armstrong's case, buttressed by Sandberg, that we had no individual moral obligation to refrain from emitting carbon in trivial ways. Though they accept the reality and significance of the threat of climate change, they take individual action to be, at best, a distraction from the obligation they do recognize: to work for a collective, political solution. While their method, as I argued, had methodological weaknesses, its conclusions appeared to be reinforced both by the tragedy of the commons argument and by worries about the causal impotence of individual efforts.

In my view, however, we must address these challenges by adopting a wider point of view that recognizes the historical structure of action and the subtle but significant ways in which we influence one another's perception of "normal." The skeptical view presumes vague, ambiguous, and context-insensitive descriptions; it relies on far too simple a model of causal responsibility; its argument against individual initiative is inconsistent with the call for political action; and the assumption that individual and political efforts compete for attention fails to recognize the implicit political potential of all action as meaningful.

Before I close, however, I want to widen the view once more. I have been arguing for the significance of individual and small group initiatives, for they have already and can yet bring about significant collective effects. But the same arguments that justify the political significance of individual initiatives hold for the parallel claim that individual states ought to work toward a sustainable carbon economy even though no individual state caused or alone can solve the problem. Just as individual initiatives can challenge the perception of normal within a culture, the initiatives of individual states can challenge what is taken for granted in the community of nations. As Canada and Australia fail to attain (or even strive toward) their Kyoto commitments, they encourage backsliding by others; and as Denmark and Germany push above and beyond their original plans, they reveal another way forward. Only time will tell whether our better angels will win. But the more of us push in the right direction, the more likely we are to get there. The treaty process has not as yet produced a realistic, enforceable plan, despite the growing scientific consensus that we must do something, and soon. Fortunately we need not wait till everyone agrees. Again, no individual personal or governmental decision brought about or can forestall the climate crisis, so collective action is necessary both within and between nations. Yet we can only initiate such actions as individuals, small groups, and governments. Fortunately every time we step out, we make it more likely that others will eventually gain the conviction and courage to do more than the law currently demands. May we do so soon enough.

Notes

1 Editors' Note: Some may observe that this chapter is a bit longer than some of the others in the volume, which we've edited for brevity. Unfortunately, Professor Cameron passed away in 2016, as the essays for this volume were being collected and revised. Given that we are publishing this chapter posthumously, Professor Cameron was unable to make substantial revisions to his first draft. We felt it better to leave his words as he wrote them. Rest in peace, Scott. You are missed.

2 I do not say this lightly: climate change undermines the stable weather patterns critical to agriculture, and the agricultural revolution of 8,000–12,000 years ago was and remains the material foundation of civilization.

3 Walter Sinnott-Armstrong, "It's Not My Fault: Global Warming and Individual Moral Obligations," in Perspectives on Climate Change: Science, Economics, Politics, Ethics, ed. Walter Sinnott-Armstrong and Richard B. Howarth (Amsterdam: JAI Press, 2006) 285–307.

4 Garrett Hardin, "The Tragedy of the Commons," Science 162: 3859 (Dec. 13, 1968): 1243–1248; and Baylor L. Johnson, "Ethical Obligations in a Tragedy of the Commons," Environmental Values 12 (2003): 271–287.

5 Joakim Sandberg, "'My Emissions Make No Difference': Climate Change and the Argument from Inconsequentialism," Environmental Ethics 33 (Fall 2011): 229–248. This paper won the Holmes Rolston III Early Career Essay Prize sponsored by the International Society for Environmental Ethics.

6 Baylor and Hourdequin's remarkably productive exchange with Hourdequin's response to Baylor's 2003 article (above note 3), "Climate, Collective Action and Individual Ethical Obligations," Environmental Values Environmental Values 19 (2010): 443–464. Johnson responded with "The Possibility of a Joint Communiqué: My Response to Hourdequin," Environmental Values 20 (2011): 147–156; and Hourdequin responded in "Climate Change and Individual Responsibility: A Reply to Johnson," Environmental Values 20 (2011): 157–162.

7 I will use "climate change," for the process Sinnott-Armstrong calls "global warming." As has become clearer since he wrote, his term highlights a worrisome symptom, but misses the wide variety of effects, captured by the more inclusive term "climate change."

8 For a much more adequate argument revealing the problems inherent in this method, see Martha Nussbaum, Love's Knowledge: Essays on Philosophy and Literature (Oxford: Oxford University Press, 1990).

9 Rawls seeks "reflective equilibrium," i.e., the harmonization of apparently inconsistent practical intuitions and principles with the core of our most confident and consistent judgments. Rawls's appeals to intuition are productive because they are not "pot-luck": he presumes that we should give reasons for accepting some and rejecting other principles and intuitions in order to produce a consistent theory. See A Theory of Justice (Cambridge, MA: Harvard University Press, [1971] 1999), Chapter One.

10 Of course, to criticize Sinnott-Armstrong and Sandberg's skepticism by appealing to still other intuitions, as both Derek Parfit and Elizabeth Cripps do, for example, is no more effective a strategy—especially when those intuitions are based on science-fiction scenarios remote from—and thus only loosely informed by—our everyday moral sensibilities. See Derek Parfit, "Five Mistakes of Moral Mathematics," in Reasons and Persons (Oxford: Oxford University Press, 1986) 67–86, and Elizabeth Cripps, "Climate Change, Collective Harm and Legitimate Coercion," Critical Review of International Social and Political Philosophy, 14:2 (2011): 171–193.

11 Some contest Aristotle's characterization of Plato, but that interpretation motivated Aristotle's objection that ethical insights can only be stated "roughly and in outline," for we can only expect "precision in each class of things just so far as the nature of the subject admits." Aristotle, Nicomachean Ethics, trans. W. D. Ross, rev. J. O. Urmson, in Aristotle: The Complete Works, Vol. 2, ed. Jonathan Barnes (Princeton: Princeton University Press, 1984) 1094b12–26, cf. 1103b27–1104a9.

12 Of course, we call unnoticed or undetectable harms "real" because they later produce detectible effects. No one feels a flu virus beginning to multiply or a precancerous cell eluding the body's last defenses, but the resulting diseases reveal that those at-the-time-imperceptible harms were real.

13 I say "realistically likely" to avoid both the words "infinite," which is clearly impossible, and "indefinitely large," which is merely vague. But given the costs and constraints of action at a particular stage of technological development, we will stop repetition somewhere, constrained either by boredom or opportunity costs—and despite repetition to that level, the effects may be negligible.

14 Parfit appeals to the paradox of sorites (78): the puzzle that at some difficult-to-specify point during a long process of adding individual grains of sand, we end up with a heap. Identifying the transition is problematic, since it would be odd to suggest that any one grain made the difference. Yet the transition is undeniable: we move from not-a-heap to a heap, and our perplexities about localizing it assume the reality of change. So also with climate-changing emissions: each one is individually insignificant. But if a miniscule contribution is repeatable, and especially when it is actually repeated billions of times, we can neither assume that what's true of each contribution will be true of all together, nor that contributions that appear to make no change are different from those that eventually do. If we stop adding grains anywhere short of a heap, we will not get one; and even if we already have a heap we regret, stopping will prevent its getting any bigger—both possibilities relevant to the parallel case of joyriding emissions in a time of climate change.

15 Aristotle, Metaphysics, trans. W. D. Ross, in Aristotle: The Complete Works Vol. 2, ed. Jonathan Barnes (Princeton: Princeton University Press, 1984) 1036 b7–8.

16 Strictly speaking, Sinnott-Armstrong is right that "[t]he fact that it is morally wrong for me to do all of a hundred acts together does not imply that it is morally wrong for me to do one of those hundred acts." The first claim does not entail the other, as noted above (note 12), for we sometimes know that the consequences will not produce a negative outcome despite any realistic repetition. But in this case it's false, or at least hasty, to conclude that "[e]ven if it would be morally wrong for me to pick all of the flowers in a park, it need not be morally wrong for me to pick one flower in that park" (305, n. 16). It need not be, but often is: for a lone visitor may be blameless for taking one common flower from a rarely visited, richly blooming meadow, but in well-frequented parks, managers rightly insist that we take nothing home but pictures and stay on the hiking paths. My plucking a flower or stepping off the trail won't kill many plants, but our doing so thousands or millions of times a summer may destroy what the park was founded to protect.

17 In a course I regularly teach on hermeneutics, I demonstrate the power of repetition by saying: "Repeating a sentence exactly the same way changes its meaning." The first time I say it, a few people note it; the second time I say it, about half the class does; and after the third time, everyone has written it down. I need only note the accumulation of responses to show that repeating a sentence exactly subtly changes its meaning.

18 Garrett Hardin, "The Tragedy of the Commons" 1247; see also Baylor Johnson, who argues that political (and not individual) action is the only effective means to address this challenge in "Ethical Obligations in a Tragedy of the Commons," 283–284.

19 Benjamin Hale, The Monist 94 (3) (2011): 379–381.

20 Actually, as a Canadian I must apologize for stepping near your toe.

21 Parfit may too hastily conclude that such situations are new, even if he's right that we only now perceive them. On one influential account, the original inhabitants of North America wiped out its megafauna fairly shortly after their arrival. Likely they lacked any grasp of the disparity between their weapons—developed over millennia of hunter-prey co-adaptation in the Old World—and the short-term (less than 1,000 year) adaptive capacities of New World mammals. Yet if they did not anticipate the result of many entirely uncoordinated acts of slaughter, the aftermath of extinction is obvious. For this admittedly controversial case, see Jared Diamond, Guns, Germs and Steel (New York: W. W. Norton, 1997) Chapter 9; and for other premodern and indigenous examples of cumulative pressure on the natural world, see Diamond's Collapse: How Societies Choose to Fail or Succeed (New York: Viking, 2004) and Thomas Mann's 1491: New Revelations of the Americas Before Columbus (New York: Knopf, 2005).

22 I cannot resist two final examples. Our digestive system breaks food into nutrients and waste without distinguishing their source in fruit, vegetable, or donut, so balanced diets can include chocolates (whew!). Yet if no sugar is inherently dangerous, too high a total will harm, so we cannot evaluate calories without considering the context of total consumption. Sinnott-Armstrong ignores this when asserting that "wasteful driving is not habit-forming. ... I do not get addicted" (292). First, whether destructive or constructive, habits emerge through a historical process of developing or extinguishing behavior patterns. No one act constitutes (though it may exemplify) a habit; and single acts are usually insignificant: one cigarette won't kill me, nor will one run make me well. Thus we cannot identify a behavior as non-habitual without considering a historical trajectory. Second, we know that even if I never develop a joyriding habit, we already have, as marked by countess songs, stories, and movies about

roadtrips. It's misleading to ignore this larger context—like an alcoholic who crows about the night he didn't get drunk.

23 Lakoff has written several op-ed style introductions to this idea. For a quick introduction and several examples, see "Global Warming Systemically Caused Hurricane Sandy," posted on 10/30/2012 at http://www.huffingtonpost.com/george-lakoff/sandy-climate-change_b_2042871.html

24 Sinnott-Armstrong briefly addresses the question of risk when discussing drunk driving, which "is immoral, because it risks harm to others, even if the drunk driver gets home safely" (293). As we saw when discussing his method above, he contests any parallel between drunk driving and joyriding, since we cannot "identify any particular victim of my wasteful driving," and since "[i]f the risk principle were true, it would be unbelievably restrictive. Exercising and boiling water also expel greenhouse gases" (293–294).

25 Though he too favors collective solutions over individual initiatives, Baylor Johnson is not inconsistent in this way. Under Hourdequin's critique, he concedes an individual obligation to minimize emissions, though he regards that as less critical than the competing obligation to advocate political change. And Johnson denies "that individuals are powerless to fight climate change. Who, but individuals, will promote and participate in the collective schemes that we need?" ("Joint Communiqué," 155). Johnson's worry is motivational: without a collective solution, freeloading will undermine any gains. I take up that worry in the last section below.

26 Gen. 2:18—but of course this insight is common to many wisdom traditions and philosophers.

27 See Aristotle, Nicomachean Ethics 1111a22–1111b3.

28 Yes, I grew up in Northern Ontario.

29 Of course, there are limits to our own self-understanding. Even those skeptical of the possibility of "unconscious" intentions have likely had the humbling experience of being confronted with evidence of that our actions belied our acknowledged intentions.

30 In my own case, I needed not more arguments to justify biking to work, but the example of a colleague, Brian Treanor, whose own bike commute (much further, often carrying two children, usually twice a day) helped me persist as parenthood and laziness tempted me to rationalize using a car.

31 In North America, many businesses provide covered, secure places to park at significant expense. To save that expense, why not also offer sheltered bike parking, and change rooms with showers? As more riders participate, businesses might even negotiate lower health-care packages for their fitter employee group.

32 And note: some, but not all talk, will take the form of argument. Recall Plato's interest in parents and nurses as storytellers (Republic, Books 2–3), since the stories they tell children provide a template that orients them in the world. Changing the stories we tell may thus be more effective in changing fundamental attitudes than any arguments we produce, though arguments can help justify the need for new stories. Plato grew up in a privileged family that surely took most Athenian attitudes for granted, but his experience of Socrates as both moral innovator and intellectual challenge to the status quo transformed his own quest for meaning—and thus the arc of Western history as a whole.

33 The examples so far have revealed the indirect impact of individuals doing the right thing. But the same dynamic can achieve terribly regrettable effects, as when the silence of Germans after Kristalnacht licensed still more outrageous attacks on Jews, or when the first participants in the Rwandan and Burundian genocides "inspired" others to join what became an orgy of genocide—in Rwanda, 800,000 deaths, mostly by machete, in less than 100 days.

34 The memorial was designed by architects Jürgen Wenzel and Theseus Bappert; Peter Herbrich was the sculptor.

35 Johnson could legitimately distinguish his approach from Sinnott-Armstrong's, for he not only acknowledges that individual initiatives may (if rarely) bear fruit, but suggests that if individual actions are taken up by others, they are coordinated—and thus, in Johnson's sense, the type of collective action he enjoins (Joint Communiqué, 148). Against this, I'd only suggest that even if they are not coordinated, the mere fact that others can usually observe and understand our actions means that they may be more significant than we can imagine, both for better and for ill.

63

ENVIRONMENTAL JUSTICE

Robert Melchior Figueroa

Introduction

In this chapter, I provide an account that explores the role the Western philosophy has played in bringing useful theoretical inroads to environmental justice. Following some introduction to concepts and subject matter, I use the first section to bring the discussion of environmental justice to current crises our world is facing. In the second section, I briefly sketch the ways in which the Mainstream Environmental Movement and the Environmental Justice Movement have confronted and addressed some of their differences in racial, ethnic, class, and gender representations as well as examples where the movements have coalesced, especially at the grassroots. The third section surveys a few of the major theories that philosophers have brought to current interpretations of environmental justice. In the fourth section, I broaden the discourse to Environmental Justice Studies, to demonstrate the ways that frameworks and cases test the praxis of theory. The final section of this chapter focuses on an environment justice framework for a moral ecology that is guided by many of the insights explored in the previous sections.

Environmental justice (EJ) is a process that simultaneously reveals and confronts injustices and systemic harms in environmental relations. It significantly pertains to every moral dimension of socio-ecological relationships by continuing to grow new insights for relational environmental ethics and underscoring intersections of justice, humans, and nonhumans unlike other environmental philosophies. The intersectional relations draw from the experiences of underrepresented people of color, minorities, women, Indigenous people, disabled people, and peoples throughout the globe struggling for transformative justice that includes direct concern for nonhumans, multiple generations, and the ever-expanding socio–ecological connections. EJ aims towards direct inclusion in environmental decision-making, equity in environmental values, laws, and policies, as well as recognition in framing environmental institutions and knowledge. Western environmental philosophy maintained such deep historical assumptions on the established paths of particularly male, able-bodied, white, affluent, and colonial habits of hegemonic dualism that the very idea that the Environmental Justice Movement (EJM) would bolster itself around where people live, work, play, and pray was initially regarded by the mainstream as ecologically naïve and ethically shallow anthropocentrism. The tendency to demarcate EJ as an anthropocentric, and hence shallow environmentalism that is morally limited, if not intellectually lazy, would ultimately consider it anti-environmental (or anti-ecological) creating obstacles for inclusion in environmental philosophy and the mainstream environmental industrial complex (Figueroa, 2014; Mills, 2003; Brune, 2020)).

DOI: 10.4324/9781315768090-75

The Environmental Justice Movement (EJM) is a historical amalgamation of many grassroots efforts that are collectively intent on confronting the uses of state, social, corporate, and colonial power of vested in systemic practices around the world that marginalize, disenfranchise, and systemically impose environmental violence by corrupting the systems of relations between humans and the more-than-human world. "We Speak for Ourselves" is the EJM demand for self-determination and agency over environmental relations and ecological connections. Localized grassroots groups mobilize to tackle an environmental threat situated in a particular geographical, cultural, and historical experiences that are embodied and material, corporeal and vulnerable to the same misrecognition, systemic violence, and environmental marginalization indicated by disparities and gross inequities in environmental benefits and burdens. Embodied and material means justice pertains to situated material existence in environments that have ontological relations of justice. This means it is nearly impossible to use the terms or concepts of environment, ecology, or nature, without also requiring a critical analysis from Environmental Justice Studies (EJS) – the interdisciplinary, praxis paradigm combining grassroots activism, critical scholarship, and transformative governance (Figueroa, 2002; Pellow, 2018; Whyte, 2018; McGregor, 2009). Every localized mobilization has implications at multiple scales; thus, localized struggles are no less in complexity for matters of EJ.

Historically, the movements and the context of environmental justice have taken different forms, such as actions to avert and halt environmental violence, as in the Standing Rock Sioux mobilization against the Dakota Access Pipeline (Whyte, 2019). There are also EJ collectives honoring the fallen victims and assisting recovery in the aftermath of environmental violence. Compensation for these environmental injustices is vast, complicated, and slow, allowing cases to languish before compensation or real action is taken to address the harms for multiple victims. In the United States, students have mobilized against the violence of school shootings and the environmental trauma that follows. Compensation and responsibility for school shootings, like police violence upon Black and Brown citizens, are difficult to come by even when the video shows the obvious culprit and connections. In other cases, the aftermath of technological disasters involves similar problems with compensation and responsibility, especially as the devastation continues to take lives and dramatically impair others. Technological disasters of this magnitude often involve transnational corporations with pathetic environmental track records. The 1984 Union Carbide chemical gas leak in Bhopal, India, killed thousands and continually leaves severe health impacts for tens of thousands; similarly, the catastrophic 2020 Beirut Explosion of 2,750 tons of ammonium nitrate (fertilizer) that had been unsafely stored for six years killed over 1,500 people, dramatically impacted the health of tens of thousands in ways still under-determined, and left over 300,000 residents homeless (Hubbard & Abi-Habib, 2020). In the United States, the first responders, volunteers, and survivors of the 9/11 terrorist attack upon the World Trade Center suffered debilitating health conditions. Without culprits and responsible agents who could compensate the disaster, it took nearly a decade for Congress to enact the James Zadroga 9/11 Health and Compensation Act of 2010, and without constant mobilization, the Act would have lapsed in 2015. Instead, it was reauthorized for until 2090, but the Victim Compensation Fund in the reauthorization would lapse in five years, ending support for thousands who have suffered serious medical conditions since 2001. Again, without mobilization for the permanent reauthorization of the 9/11 Victim Compensation Fund, the Congress would have ignored the lapse without responsibility for compensation (U.S. Congress, 2019).

The diversity of climate change mobilization further exemplifies the scope of the EJM. Climate action mobilizes at multiple scales, from varied groups, sometimes working collectively at a distance, to address localized environmental threats. Indigenous and non-Indigenous youth have mobilized and connected with notable figures in tribal affiliations, Western and Indigenous sciences, governments, religious communities, and academia to be included as self-determined agents and leaders deliberating the environmental fate of the Earth's systems, the more-than-human world, and future generations that include world's young people, in the adaptation and mitigation decision-making and strategic management on the ground. These examples demonstrate the breadth and depth of the EJM, and the bewildering diversity of environmental justice struggles.

"We Speak for Ourselves," but "I Can't Breathe"

The COVID-19 novel corona virus pandemic presents a global environmental justice event of such proportions that it has exposed environmental violence and trauma lasting over the legacy of colonialism, and systematic discrimination against women, workers, minorities, people with disabilities, and the poor. Ecological conditions for nonhuman entities and human survival are entwined by severe weather events and fires brought on by climate change. The Australian fires decimating species and human communities, deadly fires throughout the Western United States and the unstable geology transforming livable landscapes, and incomparable flooding in Central America during the pandemic exposed the constancy systemic inequities and the institutionalized historical patterns of discrimination. Meeting with the Amazon Women's Indigenous Conference online provided the opportunity to hear directly from women leading environmental justice coalitions of Indigenous peoples in Brazil and Ecuador, who shared the history of pandemics threatening their peoples' survival since 15th century colonial invasions (WECAN, 2020). Under lockdown of the pandemic, and a legacy of pandemic devastation and resilience, women gave the account of keeping track of urban relatives away from their community for work and unable to return home, the recovery from extreme flooding cutting off communities in need of assistance, and continued violence from mining operations taking advantage of these vulnerabilities, while the state is away.

From the voices of the U.S. EJM, the pandemic made environmental racism and environmental justice popular terms for describing the continued disparate and compounding burdens upon Indigenous communities, communities of color around the world, ethnic and religious minorities, the poor, the elderly, women, Black and Latinx communities, frontline workers, first responders, LGBTQ communities, trans-communities, immigrants, mental healthcare workers, and the suffocation by state violence, brutal racist policing, and the prison industrial complex which have combined to forge urban violence, saturate incarceration with people of color, and separate immigrant children from their parents by caging them, if only to remind the world of the cruelty that white supremacy can impose from the federal level. The ineptitude and downright inhumanity represented across the spectrum of industrial development unleashed existing disparities of global and local economies. These distributive injustices compounded with simultaneous economic collapses occurring at multiple scales, from transnational corporate entities to national economies, and down to billions of households across the globe. In the United States, due to food shortages people who have never faced hunger or food scarcity waited the full day in food banks. These pandemic experiences glimpse the daily struggles of communities of color, blue-collar communities, women across the globe, Indigenous peoples, and the expansive range of marginalized communities, including allies from mainstream environmental groups. Magnifying the historical depth of these inequities were stark comparisons with ways that white communities were treated, especially regarding health, environmental, educational, and workplace disparities. These illuminate how much greater the inequities of environmental discrimination by distributive injustices and the inequalities of discriminatory environmentalism by failures in recognition justice have widened since the beginning of the pandemic in January 2020.

The world could no longer look away from systemic racism and environmental violence after May 25, 2020, when George Floyd was murdered by a Minnesota police officer kneeling on

Floyd's neck over nine minutes, as Floyd cried out, "I can't breathe," and then died before us all. The ensuing global scale of Black Lives Matter organizing to address police violence is how to confront environmental racism in 2020. The removal of State flags representing the white supremist Confederacy, statues honoring Confederate figures who fought for white supremacy and Black slavery, building names in universities across the world, street names, and heretofore unwavering organizations and private businesses could not look away. Protestors against police violence were continually gassed to ensure that the murdered Black lives would not be named. The white supremacy is shameless, as Chief Julian of the Great Odgalla Sioux Nation announced the President's Fourth of July Event at Mount Rushmore, the site of the Six Grandfathers in Sioux, is criminal trespassing and exclusionary to citizens of the United States under the 1980 U.S. Supreme Court decision in *US Government v. The Great Sioux Nation*, which determined the site illegally taken (Helmore, 2020). The posterchild of mainstream environmentalism, the Sierra Club, publicly shared the stain of white supremacy and disdain against the urban poor expressed by their founder and patron saint John Muir. American environmentalism could not look away from the history of relocation and environmental violence against Indigenous peoples and the programmatic morality Muir put on who was worthy of Yosemite National Park. President Wilson echoed the sentiment establishing Jim Crow regulations in the National Parks. Racial, gendered, and environmental violence were called out by cell phone over a Black man birdwatching and politely concerned about the white woman's unleashed dog. At the same time, during the COVID-19 pandemic, several states within the United States declared that systemic racism, especially regarding Indigenous, Black, Latinx, Asian, and Pacific Islanders in the United States, constitutes a fundamental health risk (CDC, 2021). However, this very same admission of the dangers that compounded injustices and denial of basic safety imposed upon people of color is also linked back through insidious loops of settler colonial narratives that these are inferior bodies, weaker against risks and environmental threats, and are therefore the more efficient data points for the sacrifice zones of modern industrial burdens. With the staggering number of COVID cases on Native American Reservations, the ratio between Black and white deaths, the subjugation of frontline workers from medical facilities to meat processing plants, and evidence of disparate impacts of climate change, the "greening" psychology immediately seized upon clear skies and wildlife roaming the empty streets, another attempt to look away (Garcia, 2020).

Environments of Justice

Environmental justice provides a conceptual paradigm that covers the shared philosophical space where the environmental relationships are recognized as historically situated and attached to the understanding of "nature" as a socially constructed institution of power introduced upon the advent of environmental colonialism. Unfortunately, environmental colonialism exists in the very environmental values, philosophies, and policies that emerged as mainstream environmentalism. From the perspective of EJ, every environment is an inherited space of justice. All environments, especially those naturalized, colonized, and racialized, are environments of justice. All environments, ecologically bound, materially interdependent, and more-than-human, are environments of justice. To this extent, the EJ discloses the colonial power that has depended upon a specific and powerful concept of nature for imposing continuous environmental trauma upon generations of communities made vulnerable by systematic marginalization and environmental violence. Naturalizing and colonizing are two in the same action; mainstream environmentalism, including the predominant work of Western environmental philosophy up to the end of the 20th century, emerges from slippery equivocations between environment and nature, or ecology and nature, to the extent that environment and ecology are often equivocated and then contrasted to social environments designated as human-centered. Regardless of socio-environmental shifts toward social justice as a fundamental and necessary feature of any environmental philosophy, the mainstream maintained a litmus of non-anthropocentrism that rigidly became devoted to anti-anthropocentrism, which is the very kind of hegemonic dualism like (pro)nature vs. (pro)human that creates oppressive conditions (Plumwood, 2002). Anti-anthropocentrism, as I call it, is the over-extension of the moral position of non-anthropocentrism to despise the standpoint of "other" environmentalists perceived to be anthropocentric according to the mainstream discourse. When mixing intrinsic values of nature to anti-anthropocentrism epistemological, cultural, and moral blind spots failed to notice the biological determinism required of slavery, sexism, and devaluing groups according to their relationship to "nature." As Charles Mills argues, non-white people and women were "naturalized" to be outside of the social contract of civil society – indeed, the sexual and racial contracts that prefigured the body politic upon which the mainstream environmental movement was shaped, protecting white parts of the "natural" environment while avoiding nature's waste, including society's excrement of industrial waste and minority communities (Mills, 2003).

Prior to the emergence of environmental philosophy and modern conservation settler colonial invasions existed centuries prior, establishing displacement and threats to every environmental relationship of colonized peoples. Non-Indigenous thinkers seem only now able to grasp the extent of Indigenous struggles by describing their description of the fearsome implications of the Anthropocene and the crisis of climate change. Environmental justice, as social movement and as an interdisciplinary academic venture, critically addresses failures and prejudices of the modern environmental movement, or Mainstream Environmental Movement (MEM). EJ discloses forms of settler colonialism from which the MEM is fundamentally borne, from the National Parks Movement at the turn of the 19th the century to inverting the use of "nature" to exclude minorities and women and the poor from the MEM throughout the 20th century. The MEM still has a record in its own defense for working with marginalized communities and collaborating to achieve environmental justice. For example, the collaboration of the United Farm Workers with the California Environmental Fund outlawing DDT spraying upon farmworkers, crops, and every environment within the United States. Likewise, the swamps, the suburbs, the forests, the beaches, the cities, and every place suffering to the point of cascading species extinction. The United Farm Workers collaborative campaign with the California Environmental Fund was also expressed in two languages of an environmentalism that would take decades for mainstream environmentalism to fully comprehend and accept as the language of the Environmental Justice Movement (EJM). UFW and CA Environmental Fund succeeded in abolishing the use of DDT, which accomplishes the first mission of the second wave of U.S. environmentalism. Rachel Carson arguing in *Silent Spring* (1962) that war was unleased upon decimates biodiversity and ecologies, human and nonhuman interactions, by the industrial chain of DDT production and saturation by a chemical pesticide. In this instance, the contribution to the cornerstone of the EJM was in the language of farmworkers – the Spanish language, the language of a people from centuries of farming tradition in the region, and a long environmental identity that dignifies the connection between work, land, and life. English expressed the CA Environmental Fund, but the obligation to see social justice in every environment linked to the food chain is indicative of the multi-lingual parlance of the environmental justice discourse.

In San Luis, Colorado, Ancient Forest Rescue a radical MEM joined the Chicano/a community in battling the deforestation of the La Sierra, the mountain supplying the watershed for traditional acequia farming lifeways. Numerous collaborations around the globe have been vital for limiting the discriminatory environmentalism found in governments, municipalities, and transnational industrial chains that manipulate policies, production, extraction, consumption, emissions, and waste disposal. Evidence that the MEM and EJM work together in a world of issues as deep as climate change and Indigenous climate justice, as far as forward as intergenerational justice, and as diverse as climate relocation that continues to require advocacy vital to envisioning the inherent human/nonhuman relations. Additionally, EJM continues its critical analysis of canonical Western environmental philosophy and MEM approaches that have privileged white, affluent membership while disregarding equitable membership and recognition of environmental struggles affecting marginalized people of color and the poor.. Indeed, the critical features of the EJM allow insights into environmental philosophy in ways other approaches can barely reach.

Recognition in Environmental Justice: A Philosophical Turn

In this section, I connect the early contributions of philosophers and justice scholars who exemplify some philosophical examples that form extended theories for the purpose of serving the implications and expectations of environmental justice. The ways that philosophy forges environmental justice theory have become dependent upon some expectations, such as the requirement that theories envision the critical attitudes and processes toward transformative environmental justice, while simultaneously revealing ineffective theories and raising alternative concepts of moral relations, sometimes by redefining the meanings of "environment" and "justice," in order to address issues that previous philosophical approaches have failed to consider and then evaluate those inadequacies for a better set of philosophical approaches.

Of the many formulations of Western environmental ethics and environmental philosophy only environmental justice, ecofeminism, and social ecology are among the philosophies designed to confront social injustices in environmental contexts. Social ecology championed by Murray Bookchin centers on the goal of socio-environmental equality, but environmental justice requires an approach that is broader than distributive conceptions of social justice making the goal to achieve a system of human and nonhuman interactions that are unoppressive broader than conceptions of justice underlying social ecology. Like social ecology, ecofeminism also provides an environmental philosophy that allows for the meaning of ecology, environment, and nature to be further interrogated for the systemic oppression against classes of people and the more-than-human world. However, since their arrival in scholarship, both environmental justice and ecofeminism have explicitly confronted environmental racism in their literature by detailing local struggles of systemic racism against Black and Brown communities. Facing a myriad of distributive environmental injustices of toxic landfills, point pollution sources, hazardous waste site, and participatory injustices in environmental decision-making and determining the epistemological power of sciences, academic environmental studies, and environmental policy at every political scale. At no time did the broader account of environmental racism fail to consider the role of women in the grassroots leadership of environmental justice struggles, and this kind of intersectionality is at the root of both environmental justice and ecofeminism. Similarly, the intersection of the environmental experiences pertaining to both environmental philosophies, EJ and ecofeminism predate their namesake. The history of settler colonialism, the environmental identity imposed for Latinx and African peoples upon colonial invasion, and the environmental heritage of these compounding historical conditions and continued environmental violence upon Indigenous peoples, all constitute historical origins of environmental justice (Figueroa, 2003; Whyte, 2018a). Intersected are the gendered marginalization forcing further environmental trauma upon the bodies of oppressed by environmental violence (Gaard, 2019). The historical reach environmental justice and ecofeminism provide foregrounds current environmental injustices and the intersections of movements, histories, localities, and experiences of grassroots struggles.

Philosophers Iris Marion Young (1983), Christian Hunold (1998), and Val Plumwood (2002), were already far along in their communicative democratic approaches to EJ. Hunold and Young's detail of substantive and procedural justice from a communicative democratic approach required direct participation, epistemological priority to communities in deciding locations of environmental burdens, and conditions for recognition of difference and procedures for conflict resolution

(1998). Val Plumwood demonstrated ways in which nonhumans are subjected to the same hegemonic dualism of "othering" as humans, providing a way that recognition between human and nonhuman agents would promote an ecological democracy (2002). Both approaches, Hunold and Young's, and Plumwood's, made for compelling ways to reconstruct the socio-environmental rationality before engaging in forms of distributive justice.

Another notable EJ participative philosophy includes Kristin Shrader-Frechette's conditions of consent, prima facie equality, and her philosophical influence on exposing fallacious science in risk assessments (Shrader-Frechette, 2002). Working away from the distributive justice paradigm that revealed too many problems regarding who decides and determines the disparity between environmental benefits and environmental burdens. Who decides what, where, and who (human and nonhuman) receive the environmental burdens? What causal connections to historical inequities in environmental burdens are maintained by a flawed system that perpetuates environmental injustices? It is necessary that participatory justice is as direct as possible to avoid the exploitation of forced options common in distributive and weaker participatory approaches (Whyte, K.P., 2010). The same challenges existed on the scale of global EJ, following the sustainable development wave of international reports and conferences, the 1992 Rio "Earth Summit," received criticism for continuing the elitism, and lack of Indigenous peoples among the representation of stakeholders. This was both an insult by exclusion and a gross reduction in the epistemological range of options. Dale Jamieson (1996), argued the concept of global environmental justice was a rebranding of distributive justice and sustainable development policies that fail any needed participatory or inclusive strategies. Expecting the participatory justice demands to be dramatically impaired by global institutions, Jamieson (1994), appealed to our moral vision of global environmental virtues that can be performed at all scales, and especially, more likely to be effective for local levels of participation and global environmental change.

In another challenge to distributive accounts for global environmental justice, Michael Glantz and I scaled global environmental justice with a proposal to award World Heritage status to the Aral Sea in Central Asia (Glantz and Figueroa, 1997). Our proposal followed the connective tissue of participatory and distributive justice in the Aral Sea Region, especially the Central Asian Republics (CARs), but we found we needed to construct a philosophy of heritage that could recognize various legators and legatees of the worse environmental disaster of the former Soviet Union. The dominant legators holding powerful offices previously under the Soviet Union became the powerbrokers and political leaders of the newly independent CARs. Immediately many of the strict distributive infrastructure previously centralized in Moscow collapsed for the peoples throughout the CARs and few would suffer the inequities and ethnic prejudices than the Karakalpak People at the Sea's quickly receding shores. Efforts to defend participatory requirements to recognize the plight of Karakalpak by the CARs seemed futile. We were involving the dominant legatees of the Cold War, the U.S. and NATO nations. What obligations do these global partners have as the dominant legatees to the Ara Sea Region, especially the Karakalpak? Would any other site of World Heritage status be better able to demonstrate where the Sea goes, so goes the culture and its people? The opportunity to confront this imposing quest for global participatory environmental justice in light reconciling competing environmental identity and heritage anticipated my introduction of recognition justice to EJ in 1997.

My reframing of Nancy Fraser's two-dimensional theory of justice (previous named bivalent) into a theory of environmental justice created a horizon for recognition to be transformed by the voices of the EJM. From Fraser's account, recognition justice and distributive justice are incorrectly posed from adversarial and opposing paradigms of justice (Fraser, 1997). She argues they are interdependent, coexisting, co-original paradigms (or dimensions) that should be operationalized from a perspectival dualism – a standpoint where both dimensions are active at multiple levels of analysis. One of my main targets for demonstrating the bivalent environmental justice theory was the early EJ debates around environmental racism. I argued that the language of EJ scholarship had already conceded in referencing "communities of color and the poor" that racial and socio-economic discrimination were historically intertwined. The very concept of environmental racism already assumed the inseparability of racial and economic discrimination; but the former is constitutionally illegal, the latter is not. Second, cases of environmental racism documented the struggles of a variety of communities of color. While one foundational study would reveal environmental racism against African Americans in Detroit, and many other studies will recount the 1981 Warren County, NC mass protest against the toxic landfill in Afton, other foundational studies such as *Toxic Wastes and Race in the United States* (1987, 1992, 2007) would identify "minority communities" more inclusively by noting Native Americans, African Americans, Latino Americans, Asian Americans, and Pacific Island Americans. My observation is intended to indicate that this variety in communities of color is precisely what makes environmental racism a fundamentally two-dimensional injustice. From my own Puerto Rican identity and experiences of other Latinx scholars and activists from Mexican American and Chicanx communities, the race/class division quickly falls out due to the overwhelmingly obvious history of environmental colonialism that portends identities of Latinx. From the United Farm Workers, to the Young Lords confronting urban colonialism of poor sanitation and lead poisoning in New York City's Puerto Rican neighborhoods, and from San Luis, Colorado, where acequia farming remains protected by its Chicano descendants and respect for La Sierra, to Mothers of East LA, and from Vieques, Puerto Rico's 50-year military occupation by the U.S. Navy to engage in live ammunition operations that would inevitably end in the errant missile death for a civilian sparking enormous protests to demand the Navy's removal. These all amount to the kind of environmental colonialism each Latinx, Chicanx, and Mexicanx environmental identity has as part of its environmental heritage. Generating a conception of environmental identity that would account for the Latinx cases, it was made clear that my bivalent account included nonhuman relations and recognition of different ecologies (Figueroa, 2003).

Although David Schlosberg attributes the introduction of recognition justice to my bivalent environmental justice approach, he misconstrues how far I extend the scope of recognition (Schlosberg, 207; Whyte, 2010). With Schlosberg, I agree with his characterization of Fraser's weaknesses in failing to account for nonhumans and ecologies, but those do not extend to the bivalent environmental justice I have advanced. Rather, my account includes the most thorough-going account of Fraser and environmental justice, including the original theory's failures of including nonhuman agents (Glantz and Figueroa, 1997; Figueroa, 2001). The Aral Sea case above demonstrates my attempt to include the agency and death of the Sea, the people who have closest relations to the Sea, the climate change of desertification, and the recognition struggles over a long epoch of environmental heritage. Similarly in my approach to Minamata disease in Japan (2007), I also argue for restorative justice to resolve the environmental trauma between differentially impacted environmental identities. The Latinx cases demonstrated environmental identities with nonhuman relations. The Indigenous environmental heritages embedded in relations with the more-than-human world were already conceded as the origins of environmental justice when I introduced recognition to environmental justice theory.

Restorative justice and responsibility are also absent from Fraser, but I argue these are vital for communities combating environmental violence on the job, at the home, and via alienation. *Adelante Mujeres* is a Mexican American organization which was formed under the guidance of restorative justice principles, to address environmental violence in the community of families of agricultural workers in Oregon's Willamette Valley. First, addressing conflicts that the women confront at home, workshops in restorative and reconciliation strategies provided families with the conditions for addressing issues of food, education, and opportunities. Then, addressing food issues for accessing healthier options by creating a farmer's market in the City of Forest Grove that runs Wednesday evenings for those who work during weekends or are unable to reach the typical farmer's market way out in Portland. The merchant's tables of the farmer's market are rented out by *Adelante Mujeres*, ultimately making farmers who hire seasonal agricultural workers pay members of the community to participate. Additionally, this community venture assured recognition for the community of families of agricultural workers, invested in accessible healthy foods, and provided opportunity through *Adelante Chicas* to establish business skills and educational opportunities. Additionally, *Adelante Mujeres* established an active afterschool tutoring program for the community's youth, training for women to establish women-run businesses, and a team of experts to oversee food production and assist local farmers and seasonal agricultural workers. All while continuing to promote restorative justice through workshops and training to continue the circle of recognition.

The general project of extending prominent justice theories moved to the Capability Theory of Martha Nussbaum and Amartya Zen. Most notably, Schlosberg has championed a version of Nussbaum's approach, largely accounting for recognition and distribution dimensions of justice, as well as advancing ecological justice (Schlosberg, 2007). Philosophers using capabilities can account for baselines in human and nonhuman flourishing, and the breadth of environmental justice issues fit many of the broad capabilities listed by Nussbaum. Schlosberg has explored numerous EJ cases from a capability approach and addresses relations where other justice theories are silent. Schlosberg found environmental justice to be lacking in its moral regard for nonhumans – that is true of many EJ scholars who misrepresented or completely left out recognition justice. Throughout my work I argue the distinction between non-anthropocentric and anthropocentric approaches common in Western traditions has radically separated humans from nonhumans and imposed hegemonic dualism for many non-white environmental identities. In these communities, this radical separation is much more difficult to allow because present and historical obligations toward nonhuman others are epistemologically woven into the social-ontology that recognizes the agency in nonhuman others and indeed, the reciprocal obligations nonhuman others may have to their human relatives are already evident in Indigenous Environmental Justice (Plumwood, 2002; Whyte, 2018b; McGregor, 2009). I also want to ensure that ecological justice does not reintroduce problematic dualisms of human and nonhuman spheres by distinguishing it from environmental justice. People of color and the poor do apprehend and articulate their relationships between the human community and more-than-human community. Schlossberg appears to agree by pivoting back to the Principles of Environmental Justice from the First National People of Color Environmental Summit in Washington, D. C., 1992; he observes the more-than-human world is front-and-center of the Preamble and considered multiple times throughout the 17 principles. Schlossberg and Collins (2014) add to this the influence the EJM has on climate justice citing the Bali Principles for Climate Justice modeled from the Principles of Environmental Justice, and the multiple establishment of rights for nonhuman entities, i.e., mountains, rivers, and lakes. Schlosberg's capabilities approach brings recognition justice to the more-than-human environmental justice, but distinguishing the ecological and environmental forms of justice must keep clear of Plumwood's warnings of dualistic hegemony imposed by the anthropocentric vs. non-anthropocentric split which alienated people of color, women, and the poor from the mainstream environmental movement and academic environmental philosophy (Plumwood, 2002).

Among the philosophical approaches to EJ, I return to the role that ecofeminist theory has contributed to recognition justice in a more-than-human world. In Victoria Davion's 2001 survey of Ecofeminism, she demonstrates that ecofeminist philosophers like Chris Cuomo, Lori Gruen, and Plumwood provided accounts of environmental racism early in the scholarship. Greta Gaard's more recent account of feminism and environmental justice (2019), points to the extensive connections with feminist theory, especially non-Western accounts, providing a better understanding of women in the EJM. Within Gaard's account, trans-identities in EJ theory challenge and advance sexuality studies where feminist theories may have been limited. Nancy Tuana's feminist philosophy of embodiment recognizes our existential and moral vulnerabilities by describing the viscous porosity of human and nonhuman connections (Tuana, 2015). She introduces embodied scales of justice that are especially fit for the connections of the pandemic and climate justice. As I mention further below, the countless non-Western feminist positions on EJ extend the theoretical and practical range of feminist influence well into the future. Lastly, a reminder that both the recognition and the capability EJ theories stem from feminist theories of Young's communicative democratic process, Plumwood's communicative democratic ecofeminist theory, Fraser's feminist pragmatist account of recognition and two-dimensional justice, and most Nussbaum's feminist philosophy of capabilities. Although these theories may have significant differences, the feminist base from which these theories have emerged cannot be understated.

Indigenous Environmental Justice scholars have emphasized relational ethics and recognition justice in ways that bring Western environmental philosophy new dimensions of nonhuman agency, reciprocity between humans and nonhumans, Indigenous science and traditional ecological knowledge that has been undergone empirical tests and observations, as well as adaptive strategies for resilience and what Kyle Powys Whyte has coined "collective continuance (2018a)." The ability to respond and adapt to environmental changes is the feature Whyte identifies as collective continuance, which is heavily invested in revealing the concretized settler colonial epistemologies that obstruct Indigenous self-determination. The conceptual devices of environmental identity and environmental heritage are part of the capacity for collective continuance, as Whyte contends, but these require many more specific details for understanding how other capacities serve collective self-determination and sovereignty over environmental lifeways, food systems, land, and cultural relations (Whyte 2018a). Whyte redefines *ecology* by the concept of collective continuance which,

> aims to describe an ecology (i.e., an ecological system) of interacting humans, nonhuman beings (animals, plants, etc.) and entities (spiritual, inanimate, etc.), and landscapes that are conceptualized and operate purposefully to facilities a collective's (such as an Indigenous people) adaptation to changes (Whyte 2018a, 354).

Vital for understanding the impacts of climate change upon Indigenous peoples is that these impacts are part of a long history of settler colonial attacks upon the collective capacities of Indigenous people to adapt to the environmental changes. As Whyte explains, "settler colonialism can be understood as a form of domination that directly targets the relationships that create collective capacities (the ecologies) that make up collective continuance" (Whyte 2018a, 358). Collective capacities and ecologies are synonymous for Whyte, as they are reciprocally "ecosystems but also calculated stewardship of them (hence, the -logy)" (Whyte 2018a, 354). Collective continuance then is the "overall adaptive capacity" that depends upon collective capacities as described above. In many ways, environmental identity and environmental heritage make up positive collective capacities, but there are also environmental identities and heritages that are dramatically changed by sea-level rise, desertification, forced relocation, and settler colonial violence which compete as the inherited environmental identity, a scar upon collective capacities struggling for collective continuance, or what can be saved and restored if not completely lost. It largely depends upon the relationships, the responsibility to others, and reciprocal interactions in order to reestablish collective capacities.

These EJ theories stem primarily from environmental philosophers and justice scholars who have extended the dimensions of EJ and constructed new theoretical frameworks for interdisciplinary and transdisciplinary perspectives across scales and levels of analysis. At this point in the history of environmental philosophy, and indeed all environmental endeavors, environmental justice is integral because it is intended to critically investigate the oppressive environmental conditions for all those impacted, starting with the historically marginalized humans and nonhumans, their relations, and ecologies. Environmental justice is now a primary dimension for many disciplines, practices, social policies, and socio-environmental movements. In the next section, I enlarge the scope of scholarly praxis to Environmental Justice Studies, a more inclusive representation of the many connections between EJ theorists, activists, and collectivities.

Framing Environmental Justice Studies

Environmental Justice Studies (EJS) is a transdisciplinary paradigm for praxis, where theory and practice combine grassroots activism and critical scholarship in a multiracial, multi-scalar, intersectional, interspecies, inter-ecological, anti-centrist/anti-dualistic, and transformative environmental movement. Environmental Justice Studies challenges us to rethinking the starting points and implications of theoretical worldviews. With these challenges, EJS and EJM are vital to rethinking environmental regulations, international agreements, every angle of climate justice, all forms of environmental activism, scholarship, and policy at multiple scales of analysis to be influenced by grassroots environmental struggles, environmental lifeways, environmental identities, and the continuous attention to environmental racism, environmental colonialism, environmental classism, environmental discrimination, discriminatory environmentalism, and environmental trauma in the face of environmental violence. In this section, I discuss the ways that framing devices in EJS have become preferable to extending theories of justice, given the scope and diversity of EJ scholarship, policies, local resistance, and community recovery that can now be tracked over centuries of environmental injustices.

EJS is akin to a rhizomatic connection that can grow in and around multiple environmental philosophies, environmental sciences, environmental arts, environmental humanities, and environmental activism that shapes justice around moral ecologies. Every theoretically significant approach to environmental justice is tested by its ability to address cases that exist outside of ideal theory, beyond academia, where it is possible to interpret and provide options driven by grassroots experiences. The perspective of activism better informs, modifies, and critiques theories. Each philosophical approach that has contributed to environmental justice discourse and its direct action may sustain its integrity in some version of "an Environmental Justice Frame," which provides the perspectival boundaries for what is needed to address the EJ case at hand. A single, grand unified theory of environmental justice would certainly alienate the very communities who compose the world's EJM. Thus, the merit of environmental justice theories will depend upon case-by-case basis as demonstrated through an EJ frame and the extent to which environmental injustices can be addressed at various levels of analysis at multi-scalar and on multiple dimensions of environmental relations. I use environmental justice paradigm in the weakest sense, meaning the set of theories of environmental justice that can be used in framing approaches and strategies for transformative justice.

A framework sets the parameters of the study partly by presenting specific examples and cases that exemplify the interpretive and practical implications of a theoretical approach. Given the vast and growing transdisciplinary character of EJS, a multitude of frameworks can be derived from multi-theoretical, or multi-dimensional approaches. Plumwood has argued that a broader range of framework choices opens greater options and better reasons for communicative relations between humans and nonhumans, especially by avoiding reductionist theoretical approaches (Plumwood, 2002:175). She regards reductionistic approaches as philosophical and scientific forms of colonial hegemony, usually expressed through professional jargon, vocabularies, and expert methodologies that historically have worked against non-white peoples. These have limited the framework choices and shut down the narrative, communicative ethics required for environmental justice between humans and nonhumans (Plumwood, 2002). The framework choices that EJS has presented over the past three decades have been astounding. Nearly every scholarly writing in EJS talks about "an Environmental Justice Frame," and the multi-dimensional, multi-scalar, multi-agent arrangements present greater choices. The extent of these frameworks in EJS reaches so many transdisciplinary strategies that it has become difficult to meet Plumwood's challenge to develop stances and framing options that can concretely communicate to the grassroots without new terminologies, jargons, and corners of expertise that may add to further reductionistic hegemony.

Among the many frames of environmental justice, I survey a few examples that highlight theoretical connections and fundamental norms that inform us of historical perspectives, move us to the standpoints expressed in grassroots experiences and agendas, and invite us to rethink points of emphasis in method, theory, and objectives. The few examples discussed below respect Plumwood's requirement to avoid the alienating jargon of expertise, even though they may appear quite jargoned at first glance. Giovanna DiChiro has provided a frame for "looking both ways" to better apprehend the bidirectional influence of the local and global, including large and small, and in-between-sized communities. Elsewhere DiChiro adopts the conceptual frame of "contact zones" to describe the translocal network building of activists in different parts of the world (Di-Chiro, 1997). Drawing upon Mary Louise Prat's notion of "contact zones" where colonial and neocolonial people separated historically and geographically connect establishing new, ongoing relationships . Contact zones are also where many assimilating, colonial, and interethnic conflicts can occur, and there are also mixed relations as in the new contact zones of populations repatriated by the forced relocation of climate impacts. As in the example of Tuvalu populations who have relocated to New Zealand and elsewhere to escape the sea level rise engulfing their nation (Figueroa, 2006). In DiChiro's work, she uses contact zones to work through democratizing citizen science and localized expertise represented in women environmental justice activists in the United States and India (DiChiro, 1997).

Kyle Whyte's theoretical contributions operate on multiple levels by several framing devices. Like DiChiro, Whyte uses directional frames such as "justice forward" to avoid the trappings of "backward-looking" justice that fails to recognize environmental heritage is that which legatees bring forward, including transformative environmental justice strategies for collective continuance. More recently his multi-scalar frame has involved "zooming in and zooming out" to capture the levels of analysis and scale-specific particulars of any given case (Whyte, 2018a). For example, food sovereignty is historical, existential, local, regional, and global requiring perspectival shifts for scales of knowledge and relationships. Thus, if we resume to a regional consideration for indigenous food sovereignty, we may also zoom back in for a specific indigenous community and zoom back out for the wider perspective of indigenous struggles across the globe. Local patterns of environmental justice and broader levels of analysis upon zooming out our frame will respectively alter some of the questions and critically disclose the patterns of oppression working through environmental dimensions of human life and relational impacts upon nonhumans. Thus, from the US-EJM, we have frames for specific historical, political, and policy orientations and the relationships between a variety of coalition members at the grassroots and community levels as well as at state and transnational levels. However, if we attempt to swipe to different historical and geographical locations around the globe, we should be responsible and zoom back to understand how contact zones will be approached. Failing to understand the local conditions, peoples, and histories could generate contact zones undermining localized interactions and strategies for environmental justice; in essence, assimilation and environmental colonialism may be the consequence. For this reason, we can appreciate the EJM emphasis upon self-determination for communities to "Speak for Ourselves."

A Moral Ecology Frame for Environmental Justice

In this section, I conclude by framing a moral ecology that Gordon Waitt and I developed for studying places with multiple conflicting environmental identities and environmental heritages. What we call moral terrains has been used in framing several cases of ecotourism, cosmological violence, settler colonial injustices, interdisciplinary reflections, scientific practices, and complex environmental identities. Here I provide the outline of the frame and improvements that help to clarify its scope for future use.

The use of moral terrains to contextualize EJ cases establishes the material conditions of embodiment and affective responsiveness to place and space. Allowing for these material conditions to be normatively perceived from multiple locations entails a dynamic, multi-layered, framework. The relational rethinking of identity and moral terrains depends upon the concept of environmental identity. Understood as the amalgamation of cultural identities, ways of life, and self perceptions that are connected to a given group's material environment, environmental identity provides a richer context for recognition justice relation to the moral terrains of the community in space and place. Environmental heritage is where the meanings and symbols of the past frame values, practices, and places we wish to preserve for ourselves as members of a community viewed over time (Figueroa, 2006, 373). Exploring the many theoretical transformations implied by providing an account of the moral terrains of our environmental identities and the related environmental heritages we inhabit has been a vital process of demonstrating the malleability of recognition justice – inclusive of place, affect, relations, space, time by heritage, and environmental identity superseding egoist normative ontologies.

Moral terrains are actually one level of a larger triadic moral ecology. The first level is this web of values layered over places through discourses that establish normative practices and socio-environmental belonging (Figueroa and Waitt, 2008:328). The second level marks convergences called moral gateways, where different environmental heritages and identities exist on the same moral terrain and must engage from embodied affectivity, in corporeal fashion, to the convergences and web of values (moral gateways and moral terrains). The triadic moral ecology has opened opportunities for non-Indigenous tourists to face the consequences of their actions on Indigenous lands, as in the case of Uluru-Kata Tjuta National Park in Australia, where tourists continually climbed Uluru despite constant request from the Anangu People who have 50,000 years of environmental heritage on the land. The climb of Uluru was permanently closed by the park at the demand of the Anangu on October 26, 2019.

Science tourism, voluntourism, the multiple legacies of Mesa Verde National Park, and science activism in environmental justice have all been studied on the framework of this triadic moral ecology of different webs of values (moral terrains), converging environmental identities and heritages (moral gateway that open or close to convergences) that are then corporeally expressed (bodily affective responses). The process has reflexivity as the bodily affects are inscribed by the moral terrains and moral terrains are carried by embodied agents. As Whyte and Dotson read it,

> This triadic moral ecology offers an ecological hermeneutics of moral sensibilities. That is to say, moral terrains are detectable through bodily affectivity, but also inform bodily affectivity. Moral gateways are open or closed through bodily affectivity, but also detected by and can inform bodily affectivity.
>
> *(Dotson and Whyte, 2013, 64)*

The place-based approach to moral terrains was an intended effort to study bodily affects at the site of the convergences and to provide different methodologies where engaged, field philosophy could engage others from the EJS. In the Mesa Verde Project, my colleagues included a member of the Santa Clara Pueblo, a filmmaker, an environmental literature author, a digital landscape artist, and an archaeologist of the Mesa Verde Region (Wolverton and Taylor, 2015). The triadic moral ecology was in constant consideration during site research in the region, meeting different tribal group members, visiting the Ute Mountain Park, meeting archaeologists at the Crow Canyon Archaeological Research Center, being led by ranger behind the chains that separated ordinary tourists from sections marked, "Archaeologist at Work."

The Mesa Verde group was able to return and devise a week-long workshop on Mesa Verde Environmental Heritage and Environmental Justice at Crow Canyon. However, these sites, like Uluru and those I visited in Aboriginal Arnhem Land, required a bodily affect into unknowability that opened to moral terrains that hugged the horizon of convergence. These were places that unknown others, ancestors that required respect, still exist in the cosmological moral terrains of the Pueblo People, the Imajaluk Community, and the Anangu Community. Thus, the heuristic device must break out to unknowable convergences, unknowable bodily affects, and unknown distant others who have moral terrains that we also need to be open to. The agency of the Rock, Uluru, and the ancestors of Pueblo First World, and the Dreamtime Ancestors who teach the people and the land, can be found by the fourth level in our moral ecology. Likewise, the knowledge and environmental heritage of agents outside of our purview must be included to address the "complex moral terrains," which push our limits of bodily affect. Again, Dotson and Whyte suggest the improvement:

> In many ways, one must begin to think beyond affectability as embodied emotion to also include affectability as brute, or unqualified, interdependence when considering a global community. In turn, considering ethics in a global community can aid in identifying strategies to counter current absence and the abjection of difference.
>
> *(Dotson and Whyte, 2013, 72)*

Indeed, this fourth level must be built into the moral ecology in order for nonhuman relations to be more open as Plumwood advised from her frame of the intentional recognition stance, where different embodiment and acts of mind can achieve communicative relational ethics. Likewise, our transitional climate change strategies will require us to gain comfort and insights on the unqualified affects of future others and nonhuman entities to whom we have obligations and for whom we call for environmental justice.

References

Brune, M. (2020) "Pulling Down Our Monuments," *Sierra Club* https://www.sierraclub.org/michael-brune/2020/07/john-muir-early-history-sierra-club#.XxkFnUGsfIc.facebook

Carson, R. (1962) *Silent Spring*, Boston: Houghton Mifflin.

DiChiro, G. (1997) "Local Actions, Global Visions: Remaking Environmental Expertise," *Frontiers: A Journal Women Studies* 18:203–231.

Dotson, K. and Whyte, K. (2013) "Environmental Justice, Unknowability and Unqualified Affectability," *Ethics and the Environment* 18:55–79.

Figueroa, R. (2002) "Teaching for Transformation: Lessons from Environmental Justice," in Adamson, J., Evans, M., and Stein, R. (eds.) *The Environmental Justice Reader: Politics, Poetics, and Pedagogy,* Tucson: Arizona University Press, 311–330.

Figueroa, R.M. (2003a) "Bivalent Environmental Justice and the Culture of Poverty," *Rutgers Journal of Law and Public Policy* 1:27–42.

Figueroa, R.M. (2003b) "Other Faces: Latinos and Environmental Justice," in Westra, L. and Lawson, B.E. (eds.) *Faces of Environmental Racism: Confronting Global Justice (2nd ed)*, Landham, MA: Rowan & Littlefield, Inc., 167–184.

Figueroa, R.M. (2006) "Evaluating Environmental Justice Claims," in Bauer, J. (ed.) *Forging Environmentalism: Justice, Livelihood, and Contested Environments*, Armonk, NY: M.E. Sharpe, 360–376.

Figueroa, R.M. and Waitt, G. (2008) "Cracks in the Mirror: (Un)covering the Moral Terrains of Environmental Justice at Uluru – Kata Tjuta National Park," *Ethics, Place, and Environment* 11:327–349.

Fraser, N. (1997) *Justice Interruptus: Critical Reflections on the "Postscocialist Condition,* New York: Routledge.

Gaard, G. (2019) "Feminism and Environmental Justice," in Holifield, R., Chakraborty, J., and Walker, G. (eds.) *Handbook of Environmental Justice,* New York: Routledge.

Garcia, S. (2020) "'We're the virus': The pandemic is bringing out environmentalism's dark side: Celebrating COVID-19's 'silver lining' is dangerous territory," *Grist Magazine, Inc.* Published 30 March 2020 https://grist.org/climate/were-the-virus-the-pandemic-is-bringing-out-environmentalisms-dark-side/

Glantz, M.H. and Figueroa, R.M. (1997) "Does the Aral Sea Merit Heritage Status?" *Global Environmental Change* 17:357–379.

Helmore, E. (2020) "Donald Trump should stay away from Mount Rushmore, Sioux leader says", *The Guardian,* July 2020. https://www.theguardian.com/us-news/2020/jul/01/mount-rushmore-donald-trump-sioux

Hubbard, B. and Abi-Habib, M. (2020) *"Deadly Explosions Shatter Beirut, Lebanon". The New York Times, Archived* from the original on 4 August 2020. *Retrieved 5 August 2020* https://www.nytimes.com/2020/08/04/world/middleeast/beirut-explosion-blast.html

Hunold, C. and Young, I.M. (1998) "Justice, Democracy, and Hazardous Siting," *Political Studies* 46: 82–95.

Jamieson, D. (1994) "Global Environmental Justice," *Royal Institute of Philosophy Supplement* 36:199–210.

McGregor, D. (2009) "Honouring Our Relations: An Anishnaabe Perspective on Environmental Justice," in Agyeman, J., Cole, P., and Haluza-Daley, R. (eds.) *Speaking for Ourselves: Environmental Justice in Canada,* Vancouver: University of British Columbia Press, 27–41.

Mills, C. (2003) "Black Trash," in Westra, L. and Lawson, B.E. (eds.) *Faces of Environmental Racism: Confronting Global Justice (2nd ed.),* Landham, MA: Rowan & Littlefield, Inc., 73–91.

Pellow, D.N. (2016) "Towards a Critical Environmental Justice Studies: Black Lives Matter as an Environmental Justice Challenge," *DuBois Review* 13:1–16. doi: 10.1017/S1742058X1600014X

Plumwood, V. (2002) *Environmental Culture: The Ecological Crisis of Reason,* New York: Routledge.

Schlosberg, D. (2007) *Defining Environmental Justice: Theories, Movements, and Nature,* New York: Oxford University Press.

Schlosberg, D. and Collins, L.B. (2014) "From Environmental to Climate Justice: Climate Change and the Discourse of Environmental Justice," *WIREs Climate Change.* doi: 10.1002/wcc.275.

Shrader-Frechette, K. (2002) *Environmental Justice: Creating Equality, Reclaiming Democracy,* New York: Oxford University Press.

Taylor, D. and Wolverton, S. (2015) (eds.) *Sushi in Cortez: Interdisciplinary Essays on Mesa Verde,* Salt Lake City, UT: The University of Utah Press.

Tuana, N. (2015) "Being Affected by Climate Change: The Anthropocene and the Body of Ethics," The Department of Philosophy Colloquium Series, University of Oregon, 3 March 2015. Permission for use by author 4 August 2022.

United Church of Christ Commission on Racial Justice (1987, 1992, 2007) *Toxic Wastes and Race in the United States.*

United States Center for Disease Control and Prevention, Office of Minority Health & Equity (2021) "Impact of Racism on our Nation's Health." Last reviewed April 8, 2021. https://www.cdc.gov/healthequity/racism-disparities/index.html

United States Congress 116th Session, Never forget the heroes: James Zadroga, Ray Pfeifer, And Luis Alvarez Permanent Authorization of the September 11th Victim Compensation Fund Act, Public Law 116–34—July 29, 2019. https://www.congress.gov/bill/116th-congress/house-bill/1327

Women's Earth and Climate Action Network (WECAN) (2020) "Indigenous Women of Brazil and Ecuador on the Frontlines: COVID-19 and Defending Communities and the Amazon" https://www.facebook.com/WECAN.Intl/videos/1089209014798381

Whyte, K.P. (2010) "An Environmental Justice Framework for Indigenous Tourism," *Environmental Philosophy* 7: 75–92.

Whyte, K.P. (2017) "Is it Colonial Déjà Vu? Indigenous Peoples and Climate Injustice," in Adamson, J. and Davis, M. (eds.) *Humanities for the Environment: Integrating Knowledges, Forging New Constellations of Practice,* New York: Routledge, 88–104.

Whyte, K.P. (2018a) "Food Sovereignty, Justice, and Indigenous Peoples: An Essay on Settler Colonialism and Collective Continuance," in Barnhill, A., Doggett, T., and Egan, A. (eds.) *Oxford Handbook on Food Ethics,* Oxford: Oxford University Press.

——— (2018b) "Settler Colonialism, Ecology, and Environmental Justice," *Environment and Society: Advances in Research* 9:125–144. doi:10.3167/ares.2018.090109.

———— (2019) "The Dakota Access Pipeline, Environmental Injustice, and US Settler Colonialism," in *The Nature of Hope: Grassroots Organizing, Environmental Justice, and Political Change*, Boulder: University of Colorado Press, 320–337.

Young, I.M. (1983) "Justice and hazardous waste," *The Applied Turn in Contemporary Philosophy: Bowling Green Studies in Applied Philosophy*, 5:171–183.

64

ENVIRONMENTAL CIVIL DISOBEDIENCE

Jennifer Welchman

Introduction

Most organized protests of environmental policies are both legal and peaceful. But some have been neither. In recent years, environmental activists have engaged in mass trespass, blockaded logging roads, 'vandalized' buildings with graffiti, scaled iconic structures to hang protest banners, destroyed genetically modified (GM) crops, and, in rare cases, committed arson. Detractors have condemned such protests, sometimes characterizing them as instances of militant activism or even terrorism. Those responsible have often denied this, claiming they were engaged in conscientious civil disobedience from justifiable concerns about important civic issues. To make sense of these counter-claims, we must try to get clear about what civil disobedience is, what distinguishes it from other forms of illegal action, and what circumstances (if any) can justify its employment.

The term 'civil disobedience' comes from the title of an 1866 essay in which Henry David Thoreau discussed his refusal to pay a poll tax as a means of protesting government policies he thought unjust (i.e., slavery and the Mexican-American War). Thoreau was arrested but released after a benefactor paid the tax. This annoyed Thoreau because he was not, as his benefactor seemed to think, merely refusing to become personally implicated by paying his poll tax. On the contrary, Thoreau believed that slavery and the American invasion of Mexico were social injustices no person of conscience could agree to support, financially or otherwise. Thoreau called for others to join his non-violent resistance to (or, in his words, "peaceable revolution against") slavery and militarism, arguing that "if the alternative is to keep all just men in prison, or give up war and slavery, the State will not hesitate which to choose" (Bedau 1991: 35–48).

Thoreau's essay was influential in three ways. First, it gave a name to a distinct form of political action. Second, it argued persuasively that civil disobedience provided ordinary citizens an effective means of influencing civil authorities and institutions. Third, it defended civil disobedience as morally justifiable – at least when practiced, as Thoreau had, publicly and non-violently. Since 1866, civil disobedience has been employed by reform movements around the world to champion women's suffrage, freedom from colonial oppression, elimination of racial segregation, nuclear disarmament, labor rights, and both the legalization and criminalization of abortion, among other causes.

By the 1960s, groups concerned about human degradation of the natural environment were also engaging in civil disobedience. Most early campaigns were primarily motivated by "anthropocentric" concerns that the depletion of natural resources would damage our own and future generations' rights or welfare. For example, the 'Greenpeace Expeditions,' launched by

DOI: 10.4324/9781315768090-76

the Canadian Don't Make a Wave Committee (parent to Greenpeace), protested nuclear weapons testing by sailing into maritime testing zones (Zelko 2013). The 1970s Indian Chipko Movement resorted to mass "tree hugging" to block government-approved logging from fear that deforestation would harm local communities (Shepard 1987). Later campaigns to defend natural species or environments for their own sakes emerged, grounded in the beliefs that all living things possess an inherent moral value worthy of our respect (biocentrism) and/or that all communities of which we are interdependent members, ecological and social, are morally due some measure of gratitude, loyalty, and protection (ecocentrism). Some of the groups who adopted biocentric or ecocentric outlooks – such as Friends of the Earth, Greenpeace, EarthFirst!, and the Environmental Liberation Front – would become notorious for incorporating subterfuge and/or sabotage in their campaigns for strengthening environmental protections – e.g., vandalizing earthmoving equipment, removing survey markers, and arson attacks (Scarce 2006). Indigenous protests are often motivated by all three outlooks, such as the Standing Rock Sioux protest of the Dakota Access pipeline project (USA), the Wet'suwet'en First Nation's Hereditary Chiefs' blockade of the Coastal GasLink pipeline project (Canada), the Lenca blockage of the Aqua Zarca Dam (Honduras), and the Ifugao protest of Oceanagold's Didipio Gold-Copper Mine (Philippines) (Whyte 2016).

Whatever its form, civil disobedience is always controversial. This is particularly true in democracies, whose citizens are thought to agree (implicitly) to abide by collectively established civil institutions, laws, and policies, even those they do not themselves endorse. However, no civil society treats the duty to obey the law as absolute. For example, if you parked your car illegally in order to help an accident victim escape a burning car, no society would demand punishment for your offense against the traffic code. Duties to obey the law are thus "*prima facie,*" non-absolute duties that can be overridden in special cases.

Threats to human life, rights, or welfare can justify failures to follow civil law in some cases, when the threat is serious and cannot be averted through purely legal action. But normally, civil disobedience is not the only tactic available to conscientious social reformers, especially in democratic polities where citizens have the rights to free expression, to lobby their public officials, to vote, and to initiate referendums. In these circumstances, it may be argued, civil disobedience is a breach of fidelity to the collaborative process of civil governance itself. As such, civil disobedience, even non-violent civil disobedience, is a form of treason against one's civil order that only direct, immediate, and serious threats to human life, welfare, or rights could possibly justify. (The practice of civil disobedience to avert threats to non-human species and communities for their own sake would never be justifiable on such views.)

Is this a fair charge? Does all civil disobedience necessarily involve a breach of fidelity to our social commitments that can rarely be justified, and least of all when the interests at stake are non-human? To answer these questions, we must first determine what the defining features of civil disobedience are.

What Civil Disobedience Is

Contemporary views of civil disobedience have been shaped partly by Thoreau and partly by influential proponents of civil disobedience from the 20th century, such as Martin Luther King, Jr., and M.K. Gandhi. What seemed to distinguish their disobedience from ordinary criminality on the one hand and from militant revolutionary action on the other was that their disobedience was (i) conscientious, (ii) 'public,' (iii) non-violent, (iv) limited to particular laws or policies, and that (v) they demonstrated fidelity to the civil order by their willingness to accept arrest and punishment for their disobedience. If we were to treat all five features as necessary (essential) conditions for acts to be classified as civil disobedience, then civil disobedience would be 'civil' in two distinct senses. They would be contributions to civil governance performed in the civil or public sphere and civil in the manner of their performance, i.e., polite, deferential, 'civilized.'

John Rawls' influential theory of civil disobedience and its justification draws on these models. Rawls argues that citizens of societies that closely approximate liberal ideals of justice generally have reason to respect democratic procedures and their results, even when these conflict with citizen's personal moral or religious views. But every society will make mistakes. When mistakes lead to the adoption of unjust policies, justice demands that citizens speak out. When words do not suffice, it may be necessary to bolster verbal protests with communicative actions. Thus Rawls likens civil disobedience to political speech, treating it as "a form of address, an expression of profound and conscientious political conviction … in a public forum" (Rawls 1999: 321). Concerns about unjust laws or policies may be communicated by either *direct* or *indirect* action. Dissenters may violate the offending law directly, as American Civil Rights activists did by disobeying laws segregating public facilities by race. Or if this is impossible, they may act indirectly, communicating their concerns by disobeying otherwise innocuous laws, as Thoreau did in violating tax laws to protest US government policies.

But whether direct or indirect, Rawls holds, for disobedience to qualify as 'civil' in the first or political sense, it must also be 'civil' in the second. What distinguishes civil disobedience from revolt is the dissenter's continuing fidelity to the principles of justice which order her society. Through her disobedience, she gives public notice of her rejection of particular injustices, while at the same time, her ongoing "fidelity to law is expressed by the public and nonviolent nature of the act, by [her] willingness to accept the legal consequences of [her] conduct." So Rawls defines

> civil disobedience as a nonviolent, conscientious yet political act contrary to law… [by which] one addresses the sense of justice of the majority of the community and declares… the principles of social cooperation among free and equal men are not being respected.
>
> *(Rawls 1999: 320)*

Critics have complained that Rawls' definition is unnecessarily restrictive in two respects. First, it is narrowly restrictive regarding the political issues to which civil disobedience can be a legitimate response. Second, Rawls appears to conflate questions of the definition of civil disobedience with questions of its justification, so that only justifiable illegal dissent will count as civil disobedience.

Rawls' definition may seem too narrow regarding the flaws in civil laws or policies to which it may be directed because he restricts them to violations of the principles of rights and justice underlying the majority's sense of justice. Peter Singer argues that this overlooks the facts that a majority's sense of justice may be flawed or that the majority may fail to realize what justice entails in particular cases (Singer 1973). Even in nearly just liberal societies we must presume errors of judgment will occur. When they do, we should acknowledge that illegal dissent may be warranted as a form of address calling for revisions to a majority's flawed sense of justice.

First, Rawls objects to viewing disobedience of this sort as 'civil' in the political sense, because it is addressed to the majority's moral *conscience* rather than its political convictions. For example, disobedient action intended to persuade a majority to expand its sense of justice to include animals, species, or ecological communities as rights bearers would be relying on moral rather than political principles, to which members of liberal polities possessing freedom of conscience cannot be required to assent (Bedau 1991: 122–129, List 1994). But as Singer points out, dissenters who called upon majorities to reconsider exclusions based on race, sex, and religious affiliation were once in precisely the same situation (Singer 1973). So we should not be quick to exclude disobedience that appeals to moral convictions, provided the appeal is ultimately directed to correcting or improving civil laws and policies.

Related objections have been raised against Rawls' restriction of the targets of civil disobedience to *present* violations of rights or standing principles of justice. As Alan Carter points out, this entails that civil disobedience targeting our societies' failure to regulate greenhouse gas emissions could not count as civil disobedience because the full impact of current emissions will not be felt for several generations (Carter 1998). Yet, we expect our civil institutions to promote rather than hinder the life prospects of our descendants. Thus it seems arbitrary to exclude individuals' concerns about the effect of civil institutions on future generations from the range of issues that civil disobedience can target.

We also expect our institutions to be open to citizens' views about the projects they should undertake. Absence of this civic virtue undermines realization of a second – democratic legitimacy. Civil institutions change so slowly that for the most part, citizens inherit their institutions, rather than creating them for themselves. To maintain democratic legitimacy, citizen engagement in reviewing and revising inherited institutions is essential. Daniel Markovits and William Smith point out that the inertia of civil institutions tends to reduce their openness to engagement by individuals and minorities, undermining their democratic legitimacy. For example, the failure of a government's institutions to respond to its citizens' concerns about rising rates of greenhouse gas emissions might be seen as evidence of a deterioration of its openness and the democratic legitimacy of its policies (Markovits 2005, Smith 2011). In such cases, civil disobedience can trigger a public discussion that ultimately enhances openness and restores legitimacy. Thus it seems arbitrary to exclude openness and democratic legitimacy from the category of issues that civil disobedience may legitimately target.

Second, Rawls' argument that acts are not truly 'civil' in the first sense unless they are also 'civil' in the second relies on a remarkably strict distinction between communicative and coercive action. According to Rawls, communicative acts, as 'forms of address,' are acts performed openly, without threats of harms to persons or property or evasion of punishment. Acts that are covert or threatening are attempts to override the majority's opinion rather than to persuade the majority to change its views. Thus he concludes they are not acts of *civil* disobedience.

Rawls' approach overlooks the fact that legitimate political practices in liberal democracies often exhibit these features. For example, anonymous political engagement is common place: elections are conducted by secret ballots, aliases are used by citizens tweeting their officials, and masks may be worn by participants in legal protest marches. Similarly, threats of political reprisal are also frequent in political practice: elected officials are threatened with recall campaigns, government agencies by strikes, businesses by boycotts, and so forth. Furthermore, in most liberal societies, no one is obligated to incriminate themselves before the courts. So it is difficult to see why anonymous action, employment of threats, or refusal to incriminate oneself should always be incompatible with fidelity to civil law in the case of illegal dissent.

No doubt 'civility' is important, but what it is important *for* is the justification of acts of civil disobedience (discussed below), not their classification *as* civil disobedience. For the purpose of classification, that disobedient acts are 'civil' in the first sense is sufficient to distinguish civil disobedience from either militant revolution or noncompliance from purely personal convictions (conscientious objection). Acts of civil disobedience are illegal acts, conscientiously performed from the belief that civil laws or policies exhibit flaws warranting immediate correction. Whether a particular act of civil disobedience is in fact justified in the circumstances is a separate matter.

If we go by common usage, illegal protests motivated by concerns about the effects of government policies on future generations, wild species, and nature itself are regularly described as civil disobedience. And even those who consider particular acts to be misguided and unjustifiable regularly classify them as 'civil disobedience' nonetheless. Consequently a broader definition of civil disobedience is desirable – one that aligns with contemporary conceptions, but which still allows for civil disobedience to be distinguished from other illegal conscientious dissent. To this end, we shall define 'civil disobedience' as *conscientious, illegal activity, intended to draw attention to and to correct perceived failings of civil laws, policies, or practices.* While these defining features of contemporary civil disobedience may be considered jointly 'sufficient' for an act meeting them to be classified as civil disobedience, none should be considered 'necessary' or essential. Civil disobedience is a social practice and no social practice is set in stone for all time. Thus it would be unreasonable to refuse to classify an act as civil disobedience simply because it lacked one or another of these features.

What Civil Disobedience Is Not

Legal Protest: Efforts to achieve social reform by boycotts, marches, petitions, and public meetings, if legal, are not instances of civil disobedience, no matter how controversial their messages may be.

Self-Interested Criminality: Law-breaking for personal benefit is not civil disobedience. For example, street artists illegally commandeering others' property to display their art are not engaged in civil disobedience when their prevailing motivations are personal satisfaction and/or career advancement. By contrast, street artists who illegally commandeer others' property to publicize their concerns about the adequacy of environment laws or policies could be engaged in (indirect) civil disobedience.

Conscientious Objection: Conscientious objectors are individuals who disobey specific laws or policies when these conflict with 'higher laws' (religious or moral) with which they feel bound to comply. But unless their disobedience is also motivated by a desire to influence public officials to change those laws or policies, it is not civil disobedience. Conscientious objectors who feel bound to disobey particular laws or policies often have no objection to obedience by others whose personal moral or religious convictions differ from their own. Moreover, as Kimberly Brownlee points out, conscientious objection is not always illegal, as many societies treat religious or moral conscientious objection as legally excusing many kinds of noncompliance with civil ordinances (Brownlee 2012).

Radical and Revolutionary Action: Like civil disobedience, radical and revolutionary dissent may be conscientiously motivated. However, they differ in the scope of the objections to civil laws and policies targeted and their respective overall fidelity to the existing civil order. Radical and revolutionary groups are less concerned with acting within the bounds of fidelity to the prevailing civil order, which they tend to view as deeply flawed, illegitimate, or corrupt. Thus radical and revolutionary activists are more likely to feel justified in coercing civil authorities and processes and in adopting violent means to do so. These differences are matters of degree, however. Individuals generally faithful to civil law may become persuaded that a failing in a particular law or policy is so serious that immediate and coercive measures are warranted. In special cases, their actions may be outwardly indistinguishable from those of more radical political activists, though still sufficiently narrowly targeted to be distinguishable from wholesale revolutionary dissent (Hettinger, this volume).

Terrorism: Historically, 'terrorism' was often defined as a form of guerilla warfare in which victory over an opposing force is sought by terrorizing a civilian populace into demanding capitulation. Radical environmental groups have not mounted terror campaigns against the general public, thus calling radical environmental activists 'eco-terrorists' may be viewed as a slur rather than a description. But should there be a campaign of environmental terrorism, the campaign could not be classified as civil disobedience (Hettinger, this volume).

Justifying Civil Disobedience

Most political theorists agree with Rawls that "civil disobedience used with due restraint and sound judgment helps to maintain and strengthen just institutions," as it serves to "inhibit departures from justice and to correct them when they occur" (Rawls 1999: 336). As noted above, civil disobedience also serves civil societies by repairing "democratic deficits" in collaborative decision-making, stimulating debate that overcomes institutional "inertia," and identifying and correcting other civic vices besides injustice in a community's laws or policies. As such, acts of civil disobedience, though illegal, can be justifiable on other grounds. Justification in practice will depend on what we take "due restraint and sound judgment" to involve.

But civil disobedience is never socially innocuous. Not only do disobedient dissenters violate their prima facie obligation to obey the law, their actions disrupt normal patterns of civic life. Public resources may be diverted to policing and prosecutions. Debate over the appropriateness of such actions may become socially divisive. And civil disobedience that involves damage to others' property impacts their rights and/or welfare. Even peaceful, non-violent disobedience can harm innocent parties. While illegal public rallies, sit-ins, and blockades are often highly effective means of attracting media attention to an issue, they can also delay innocent parties' abilities to conduct business, keep important appointments, or obtain help in emergencies.

Sound Judgment: To justify civil disobedience in light of these considerations, sound judgment is crucial. Prior to resorting to civil disobedience, one should gather all the relevant facts, carefully weigh the seriousness of the wrongs or harms one hopes to correct against those one would cause, and consider whether irreparable harm would occur if corrective measures were delayed as well as the likelihood of one's disobedience having beneficial results.

One can be conscientious but mistaken in one's beliefs about what civil laws or policies require or allow. Generally speaking, failing to check one's sources and confirm one's information prior to deciding upon civil disobedience is a serious error of judgment. This is especially true regarding disobedience targeted at environmental policies about which scientific information is often unavailable, incomplete, or evolving. For example, the worrying results of a single study of the effects of GM corn pollen on the larva of monarch butterflies prompted many to protest GM corn production, sometimes illegally. Subsequent research soon showed this particular concern about GM corn was misplaced. Public attention would have better directed to other aspects of the genetic modification of corn and other crops. Patience as well as thoroughness in gathering information is crucial for sound judgment about responses to concerns about environmental issues.

It is equally important to consider the risks and benefits of disobedience, whether direct or indirect. One must weigh the inconveniences others might suffer against the benefits one could bring about if the protest achieved its aims. One must also consider the possibility that the protest may achieve none of its aims. If there is little reason to think an act of civil disobedience will foster fruitful public discussion (let alone practical results), proceeding might be unjustifiable, especially if significant harm to others' welfare or rights might result. For example, when the Canadian street artist, Roadsworth, began painting whimsical bike lanes around the city of Montreal to protest automobile emissions, many Montrealers were delighted. But as city officials pointed out, his guerilla bike lanes might easily have confused motorists and caused traffic accidents. It was a happy accident that no harm resulted (van Toorn 2008). To put others at risk without taking steps to prevent loss and injury to innocent parties is difficult to justify morally or socially, even when, as in Roadsworth's case, one's environmental concerns are well founded.

This is not to say that one could never justifiably engage in civil disobedience with only incomplete information or without reasonable assurance that one's tactics will be both efficacious and non-injurious. But to do so, one should have reason to believe that inaction would permit serious and irreversible wrongs or harms to occur. Some of the 19th-century American abolitionists who harbored escaping slaves and the 20th-century citizens of Nazi Germany who helped Jewish compatriots evade arrest and detention had to make decisions quickly, with limited information and little time to evaluate their options, yet their acts were clearly justifiable. Swedish teenager Greta Thunberg's regular Friday 'School Strike for Climate,' begun in 2015, would eventually spark a global student movement, *Fridays for Future*, whose 2019 climate strikes drew worldwide attention to government inaction on climate change (Haynes 2019). Thunberg was soon widely praised even though when she began, she had no reason to suppose her conscientious truancy would have any positive impact at all.

Evidently some threats to the integrity of environmental systems or communities of life can justify decisions to disobey the law even when dissenters lack full information or reasonable assurance that their methods will prove efficacious and beneficial overall. For example, deciding to blockade a road to halt the clear-cutting of an old-growth forest in order to gain time for fuller public review of the logging plan could be justifiable, even if based on incomplete assessments of the situation, assuming there was little time to identify or pursue other options before irreversible damage was done. Many considered Timothy DeChristopher's disruption of an American government auction of oil and gas leases justifiable for these reasons. By bidding for leases he could not afford, Christopher intended to delay completion of the auction until an incoming presidential administration could review it (van der Zee 2011). No such justification could be offered for a campaign by the Swedish group, Asfaltsdjungelns indianern (Asphalt Indians), who protested city dwellers' use of sports utility vehicles by slashing their tires (Bynert 2007). The vehicles they damaged did not pose so serious or irreversible a threat as to justify a campaign so unlikely to win support for reforming Swedish transit policies.

Due Restraint: As Rawls rightly remarks, "due restraint" in the conduct of civil disobedience is also widely considered crucial for civil disobedience to be justifiable morally, socially, and politically. As noted above, 'due restraint' is usually taken to involve publicity, the absence of violence or of threats of reprisals, and submission to arrest and public prosecution.

Ideally civil disobedience should be public in several senses. The action should be "public" with regard to its subject (public policy), its target audience (public decision makers), and its performance. As a public act, civil disobedience is always (though not necessarily only) a means of communicating concerns about civil policies, laws, institutions, and practices, to others in the community. As Carl Cohen remarks, to be fully effective in this last respect, civil disobedience must be presented "before the public eye" (Cohen 1971: 16). Though acting openly puts disobedient campaigners at risk of arrest, prosecution, and punishment, overall this can increase the effectiveness of disobedient action, as these events provide disobedient campaigners further opportunities to testify to their conscientiousness and communicate their concerns.

Ideally, civil disobedience should be non-violent and non-threatening, i.e., eschew coercive use of force or threats of force against persons or property, for several reasons. First, violence and threats of violence against persons or their property are always at least *prima facie* morally wrong, as well as illegal. Thus disobedience dissent which employs violence will often, though not always, be impossible to justify. Second, as Andrew Sabl has argued, even in cases where violent protest might seem well justified, violence tends to inflame rather than heal social divisions. Thus those genuinely faithful to shared public governance have a further reason to avoid violence. Sabl notes that the appalling injustice of 20th-century American racial segregation policies "would have justified violent resistance" by civil rights campaigners (Sabl 2001: 314). Yet civil rights leaders, such as Martin Luther King, rightly refrained from violent protests, "based not on current obligation but on the desire not to foreclose future cooperation" with the majority whose sense of justice, though seriously flawed, was not so disordered as to be irremediable through sustained, peaceful protest (Sabl 2001: 309–310).

Finally, it is also widely believed that to be justified, disobedient dissenters should accept arrest and prosecution by public authorities. The reasons usually given overlap with those offered for believing that justified civil disobedience will be public and non-violent. To deliberately evade arrest or prosecution for one's disobedience is to invite doubts about one's conscientiousness, fidelity to civil law, and willingness to cooperate in shared processes of governance. To deliberately evade arrest may be taken as a sign that one's primary motivation was a self-interested desire to flaunt public law and public opinion, rather than conscientious concern about flaws in civil policies. It can also raise doubts about one's conscientiousness, as individuals genuinely convinced about serious flaws in civil policies would presumably use every forum available to draw public attention to them. Refusal to cooperate with authorities may also raise doubts about whether one's intent was merely civil disobedience, consistent with overall fidelity to the law, or was in fact radical or revolutionary revolt against the civil order. Finally, refusal to cooperate with authorities will raise doubts about the protestors' willingness to cooperate in future governance on fair and equal terms, as parties whose interests were injured by one's disobedient actions are denied the opportunity to appeal to courts for just compensation.

That full publicity, avoidance of coercive force (threatened or actual), and submissiveness to arrest are *generally* required for civil disobedience to be justifiable does not entail that absence of any one of these features necessarily renders an act of civil disobedience morally unjustifiable. The 19th-century American 'underground railroad,' which helped escaping slaves to flee north to Canada, could not have operated effectively if its members had revealed their identities or submitted themselves to arrest. Anonymity might be essential in other cases, such as delaying construction of a logging road by repeatedly removing surveyors' stakes, until public debate about the appropriateness of logging intended can take place. If conscientious individuals believe the risks of direct and immediate harm are high and anonymity temporarily required for effective protest, their covert action may be justifiable, especially if they take steps to meet the expectation of publicity in other ways, e.g., by issuing press releases or a subsequent public confession. Ironically in some cases, anonymity might actually enhance public confidence in a disobedient dissenter's good faith. A street artist who refuses to reveal her identity cannot benefit financially from public interest in her artwork. Thus her disobedience is less open to challenge as mere self-seeking criminality.

Since violence, actual or threatened, is generally immoral as well as illegal, disobedience involving violence will be more difficult to justify. Violence and threats of violent reprisals offend against individuals' rights and diminish their welfare. Moreover, violent coercive tactics tend to engender anger, fear, and resentment in others which can reduce the willingness of the majority to trust or cooperate with protesting minorities in future. Thus their costs – moral, social, and political – are likely to outweigh their potential benefits. That said, this does not entail that some forms of 'violent' or coercive action could not be potentially justified as a form of civil disobedience, if the damages caused were trivial in comparison with the seriousness of the flaws protestors perceived in civil policies they wished to correct. Joseph Raz notes cost-benefit accounting does not always work out in favor of the most peaceful or even legal methods of protest (Bedau 1991: 161,162). Legal methods of protest such as strikes or walk-outs, voluntary boycotts, and legal protest marches can cause much more inconvenience to those affected than illegal acts involving trivial damage to property. However significant property damage or injury to human beings would not readily be justifiable, absent direct and pressing threats to human or sentient animal life, at least in democratic polities, where other less coercive methods of communicating concerns about civil policies are usually available.

As a rule then, acts of civil disobedience are most likely to be seen as morally justifiable if 'civil' in both the two senses discussed above. Consider the results of the following two American pipeline protests. One was the student-organized XL Dissent rally held in Washington, D.C. (Stephenson 2014). The rally was organized to forcefully convey university students' disapproval of the Keystone XL pipeline project (intended to transport crude oil from Canada's oil sands to Texas refineries). Of the roughly 1,000 people attending, 398 participated in an illegal sit-in that blocked side-walks outside the White House. This action was publicly announced by the organizers who offered participants training in non-violent resistance and also recommended preparation for paying resulting fines. In this case, no one questioned whether the students' actions should be classified as civil disobedience nor were they widely condemned as unjustifiable. In fact, the local police department agreed to a 50% reduction in the fines imposed (Smith 2014).

Yet in the same month, a Michigan court sentenced three participants in an equally peaceful pipeline protest to 13 months of probation following felony convictions for obstructing police (Palmer 2014). This protest, organized by the Michigan Coalition Against Tar Sands (MICATS) against an Enbridge Incorporated's pipeline, involved participants trespassing on a worksite to halt operations, during which three protestors chained themselves to Enbridge equipment and refused police orders to release themselves. Enbridge's operations were delayed several hours until the protestors were cut free. This protest was just as public, conscientious, non-violent, narrowly targeted, and communicative as was the XL Dissent sit in, yet the individuals involved were much more harshly treated. This is partly explained by the more 'coercive' tactics employed by the MICATs protestors and the financial cost of the construction delay they achieved. Again, no one doubted that they were engaged in civil disobedience. But the relatively minor deviations from paradigm cases of justified civil disobedience reduced Michigan authorities' willingness to display leniency toward them.

The greater the deviation from paradigm cases of justifiable civil disobedience, such as the XL Dissent protest, the greater the potential for it to be morally condemned. Australian activist, Jonathan Moylan, was widely criticized for the methods he used against the Whitehaven Coal's mining operations. Moylan issued a fake press release claiming that the company's bankers had denied it a loan, causing Whitehaven Coal's share value to fall by over 300 million dollars (Bonacci 2014). No one doubted the sincerity of Moylan's motives, but his hoax involved "identity theft" and risked causing serious harm to investors (Spence 2013). The fire-bombing and destruction of a University of Washington scientist's offices by members of the Earth Liberation Front in the mistaken belief that he was developing GM trees more markedly deviate from much more markedly from paradigm cases than did Moylan's hoax (Bernton & Clarridge 2006). Not surprisingly, this act was roundly condemned in the public press, the perpetrators frequently being characterized as 'eco-terrorists,' rather than civilly disobedient. The conscientiousness of their convictions earned them no leniency from public authorities.

Conclusion

As these and other historical cases of environmental civil disobedience reveal, that an act of civil disobedience may be widely viewed as morally justifiable does not entail that it is legally justifiable. In exceeding rare cases, courts have accepted "necessity" defenses (the defense that an otherwise illegal action was excusable as necessary to prevent more serious harm or rights violations), the only clearly precedent-setting instance being the acquittal of Greenpeace activists on charges arising from an attempt to shut down a coal-fired power station in England in 2008 (Kaminski 2019, Vidal 2008). Civil authorities and courts in liberal democracies where free speech is protected are often but not always more lenient in their responses to civil disobedience than to other forms of law-breaking, particularly when the disobedience is restrained in such a way as to be more 'communicative' than coercive overall. But though contemporary liberal democracies do recognize rights of freedom of speech and conscience, they do not recognize rights to civil disobedience.

There are, as noted above, good reasons why this is so. Civil disobedience can play an important role in promoting the long-term stability and political legitimacy of civil societies. As Thoreau pointed out, civil disobedience is means by which ordinary people can bring general public attention to issues of concern, a power otherwise normally restricted to the wealthy and socially powerful. Civil disobedience can be an effective means for denouncing unjust policies, correcting democratic deficits in public institutions, or reforming a majority's flawed sense of justice.

When used to promote justice, openness, democratic legitimacy, and environmental responsibility, civil disobedience will often be justifiable – and even commendable. Well informed, soundly judged, and dully restrained, civil disobedience can be morally preferable to legal means, in cases where legal means are unlikely to avert serious wrongs or harms. But when ill-considered, poorly judged, or unrestrained, civil disobedience can also be harmful, ineffective, and socially divisive. Consequently, anyone contemplating civil disobedience must remember that even sympathetic civil authorities will always be motivated to try to deter the use of civil disobedience in favor of legal alternatives.

Related Topics

50. Constitutional Rights
63. Environmental Justice
65. Lawbreaking and Ecoterrorism

References

Bedau, Hugo Adam (1991) *Civil Disobedience in Focus*. New York: Routledge.

Bernton, H. and Clarridge, C. (2006) 'Earth Liberation Front Members Plead Guilty in 2001 Firebombing,' *Seattle Times*, 5 October. Available at: http://community.seattletimes.nwsource.com/archive/?date=20061005&slug=uwfire05m

Bonacci, M. (2014) 'Coal Prank Activist Jono Moylan Avoids Jail,' *Green Left Weekly*, 26 July. Available at: https://www.greenleft.org.au/node/56943

Brownlee, Kimberley (2012) *Conscience and Conviction: The Case for Civil Disobedience*. Oxford: Oxford University Press.

Bynert, S. (2007) '"Indiane" i kniv- attack på bildäck,' *Aftonbladet*, 26 July. Available at: http://www.aftonbladet.se/nyheter/article11147980.ab

Carter, Alan (1998) 'In Defense of Radical Disobedience,' *The Journal of Applied Philosophy*, 15: 29–47.

Cohen, Carl (1971) *Civil Disobedience Conscience Tactics and the Law*. New York: Columbia University Press.

Haynes, Suyin (2019) 'Students From 1,600 Cities Just Walked Out of School to Protest Climate Change. It Could Be Greta Thunberg's Biggest Strike Yet,' *Time*, 24 May. Available at: https://time.com/5595365/global-climate-strikes-greta-thunberg/

Kaminski, Isabella (2019) 'Climate Activists Win Necessity Defense Case in London,' *The Climate Docket*, 4 May. Available at: https://www.climatedocket.com/2019/05/10/necessity-defense-climate-change-activists/

List, Peter (1994) 'Some Philosophical Assessments of Environmental Disobedience' in Robin Attfied and Andrew Belsey (eds.) *Philosophy and the Natural Environment*, Cambridge: Cambridge University Press, 68–84.

Markovits, Daniel (2005) 'Democratic Disobedience,' *Yale Law Journal*, 114: 1897–1952.

Palmer, K. (2014) '3 Enbridge Protesters Get Time Served,' *Coloradoan*, 6 March. Available at: https://www.coloradoan.com/story/news/2014/03/05/3-enbridge-protesters-sentenced-to-time-served-probation/6085793/

Rawls, John (1999) *A Theory of Justice* (rev. ed.). Cambridge: Harvard University Press.

Sabl, Andrew (2001) 'Looking Forward to Justice: Rawlsian Civil Disobedience and its Non-Rawlsian Lessons,' *The Journal of Political Philosophy*, 9 (3): 307–330.

Scarce, Rik (2006) *Eco-warriors: Understanding the Radical Environmental Movement* (updated ed.). Walnut Creek, CA: Left Coast Press.

Shepard, Mark (1987) *Gandhi Today: A Report on Mahatma Gandhi's Successors*. Washington, DC: Seven Locks Press.

Singer, Peter (1973) *Democracy and Disobedience*. Oxford: Clarendon Press.

Smith, Heather (2014) 'Tied to the Rusty White House Fence: Two Keystone Protesters' Arrest Odyssey,' *Grist*, 4 March. Available at: https://grist.org/politics/tied-to-the-rusty-white-house-fence-two-keystone-protesters-arrest-odyssey/

Smith, William (2011) 'Civil Disobedience and the Public Sphere,' *The Journal of Political Philosophy*, 19: 145–166.

Spence, E. (2013) 'Whitehaven Hoax was an Unethical Act that Was Harmful to All,' *The Conversation*, 11 January. Available at: http://theconversation.com/whitehaven-hoax-was-an-unethical-act-that-was-harmful-to-all-11571

Stephenson, W. (2014) 'Nearly 400 Arrested at XL Dissent—the Largest White House Civil Disobedience Action in a Generation,' *The Nation*, 7 March. Available at: http://www.thenation.com/article/178739/nearly-400-arrested-xl-dissent-largest-white-house-civil-disobedience-action-generati

van der Zee, Bibi (2011). 'Tim DeChristopher on Trial for Sabotaging Oil and Gas Land Auction,' *The Guardian*, 28 February. Available at: http://www.theguardian.com/environment/2011/feb/28/tim-dechristopher-trial-oil-gas

van Toorn, Tai. (2008) 'The Rules of the Road: News Media, Street Art, and Crime,' *WRECK: Graduate Journal of Art History, Visual Art, and Theory*, 2: 7–30.

Vidal, J. (2008) 'Kingsnorth Trial: Coal Protesters Cleared of Criminal Damage to Chimney,' *The Guardian*, 10 September. Available at: http://www.theguardian.com/environment/2008/sep/10/activists.carbonemissions>

Whyte, Kyle Powys (2016) 'Indigenous Experience, Environmental Justice and Settler Colonialism in Nature and Experience' in Bryan E. Bannon, (ed.) *Phenomenology and the Environment*, London: Rowman & Littlefield, 157–174.

Zelko, Frank (2013) *Make it a Green Peace! The Rise of Countercultural Environmentalism*. New York: Oxford University Press.

65

LAWBREAKING AND ECOTERRORISM

Ned Hettinger

Those who make peaceful change impossible make violent revolution inevitable.

John F. Kennedy (1962)

I do not believe in violence as in any sense a solution to any problem.

Wendell Berry (2008)

Introduction

Imagine a future where climate change, mass extinction, and planetary poisoning are brought to an end by a campaign of political violence carried out by a coalition of radical environmental groups. Hackers manage to shut down computer systems that run coal-fired power plants. World-wide, targeted attacks on machinery stymie new human encroachment into natural areas. Industrial-scale fishing is hamstrung as vessels are sunk. Environmentally harmful consumption is seriously discouraged when the tires of gas-guzzling vehicles are routinely slashed, trophy houses in pristine areas are as likely to be burned to the ground as successfully built, and those who eat meat in public are frequently spray painted with a red X. Would such a campaign count as terrorism? Would it be morally justified?

The term "ecoterrorism" was likely first used by anti-environmentalists trying to discredit the environmental movement (Arnold 1983). They applied the label to acts of sabotage (e.g., tree spiking, bulldozer siltation, and more – see Foreman and Haywood, 1985) carried out in the late 1970s and 1980s by members of the group Earth First! In the 1990s and the following decade, activists who identified with the Earth Liberation Front employed tactics such as arson to thwart perceived environmental harm. In response, the Federal Bureau of Investigations (FBI) has classified groups and acts as ecoterrorist, and it has prosecuted environmental activists as terrorists, often imprisoning them with longer sentences known as "terrorism enhancements" (Harris et al. 2007). Some in the radical environmental movement have advocated tactics that seem to fit this rubric and there have been concrete instances of acts that likely warrant the label.

For many, "terrorism" is a word used to condemn more than describe. If something counts as terrorism, then it is morally odious. If something is morally permissible, then it cannot be terrorism. With this usage, the test of terrorism is not simply what is done but whether one believes it was justified. This leads to arbitrary and hypocritical applications of the term, as suggested by the cliché "one person's terrorist is another person's freedom fighter." Even though we use similar

DOI: 10.4324/9781315768090-77

tactics, it is only those who fight against us that are the terrorists. Consider that many of the thousands killed by U.S. drone strikes in the "War on Terror" are innocent civilians (Cavallaro et al. 2012) and yet few in the United States would consider this practice to involve terrorism.

When most people in the United States think of terrorism, they think of the September 11, 2001 al-Qaeda attack that killed almost 3,000 people in New York and Washington. They think of foreign suicide bombers exploding themselves in crowded public spaces. But many examples of terrorism are domestic, not international. Consider the thousands of lynchings of African Americans by the Ku Klux Klan in the late 19th and 20th centuries or the 1995 bombing of the Oklahoma City Federal Building by anti-government fanatics that killed and injured hundreds, including infants in a day-care center inside the building. Ted Kaczynski – known as the Unabomber for his targeting of university personnel – is perhaps the best example of an environmental terrorist. Between 1978 and 1995, he sent over a dozen bombs in the mail, killing a computer store owner, an advertising executive, and timber industry lobbyist. He also maimed a number of university professors. His goal was to help bring an end to the industrial-technological system that, he believed, was destroying the natural world and technologically enslaving humanity.

Terrorism seems the epitome of pure, unadulterated evil. Nevertheless, the assumption that terrorism is an absolute moral wrong is contentious. It would have us automatically condemn the allied "terror bombing" of German cities during WWII, a campaign that severely injured or killed over 1 million Germans, most of them civilians (Walzer 1977: 255–256). Its goal was to inflict sufficient death and destruction on civilians in order to break the German will to pursue the war. Or consider that Nelson Mandela, South Africa's first black President, was awarded the Nobel Prize for Peace despite having advocated violence in the early days of the campaign to overthrow the apartheid regime.

Although we should not assume that all terrorism (including ecoterrorism) is necessarily wrong, such acts face a seriously high burden of justification. While there can be good reasons for breaking the law, doing so is no trivial matter. In relatively just societies, it violates norms of fairness, promise keeping, and the duty to uphold just institutions (Hettinger 2001). If environmentalists expect those who would degrade the environment to obey laws with which they disagree, then environmentalists can also be expected to act in a law-abiding manner. The complexity of arguments for civil disobedience (Welchman, this volume) shows that even non-violent, submissive lawbreaking must overcome substantial moral objections. Lawbreaking that is coercive, intimidating, and violent is thus exceedingly difficult to defend. Only in extreme circumstances might it be justified. Are we now facing such a situation? Humans are destabilizing the earth's climate, wiping out other forms of life in massive numbers, acidifying and strip-mining the oceans, appropriating and homogenizing the earth's ecosystems, and generally taking over and poisoning the planet. Many environmentalists believe that we confront an extreme environmental emergency that requires drastic action. Might terrorism be an appropriate tool to bring about the significant changes needed?

What Is Terrorism?

Although there is considerable disagreement about what should count as terrorism, one fairly uncontroversial feature is that terrorism is motivated by political ideology, rather than self-interest, personal vendetta, or psychosis. Drug lords who massacre and behead people, though they intend to intimidate and instill fear, are not typically thought to engage in terrorism because their goals are financial, not ideological. Those who took advantage of the 2020 Black Lives Matter protests to loot stores solely in order to acquire goods are criminals not ideologically motivated terrorists. Nor are the U.S. school shootings typically terrorist activities, for, by and large, the perpetrators act out of revenge or serious mental instability rather than ideals about how the world should be.

While it is arguable that terrorists will often show evidence of psychological problems, ideological motivations dominate their thinking.

It is sometimes claimed that only individuals or sub-national groups can engage in terrorism and that states cannot (Hoffman 2006: Chapter One). Although one could insist that this is how the term "terrorism" should be used, such a restriction is problematic. Perhaps the resistance to allowing state terrorism comes from the idea that the state's use of violence and intimidation is typically part of a war against other states and terrorism is not possible in war. But this ignores the well-accepted idea that even in war there is an important moral distinction between targeting civilians and targeting combatants. Intentionally killing civilians as a mechanism to cow one's wartime opponent certainly seems like a terrorist tactic. Such a restriction also ignores that sometimes states use violence against their own citizens, aiming to intimidate them in order to achieve political objectives. The Nazi treatment of European Jews seems like terrorism, and yet it was performed by a state. Environmental activists have sometimes been the target of such state terrorism: Ken Saro-Wiwa, the leader of a campaign to protect the native homeland of the Ogoni people from oil-production pollution by Royal Dutch Shell, was executed by a Nigerian dictatorship in a terror campaign aimed at undermining the Ogoni drive for autonomy (Center for Constitutional Rights 2016).

At its core terrorism is violence that aims to intimidate for public purposes; it uses violence, or the threat of violence, to coerce by means of fear in order to achieve ideological objectives. Such acts can be performed by individuals, groups, and even governments.

Must Terrorism Target Innocents?

One important feature typically associated with terrorism is that the violence is directed at non-combatants or innocents. Targeting innocents seems essential if we are to distinguish terrorism from typical acts of war. When two countries go to war against each other, they are using violence in an attempt to intimidate and coerce the opposing nation and they are doing so for a political objective. If the violence is directed at military personnel and there is serious concern to avoid civilian casualties, then the terrorism label seems inappropriate. Perhaps this justifies the belief that U.S. drone attacks are not terrorism because the innocent civilians killed are not intentionally targeted and major steps are taken to avoid their deaths. Terrorism is most clearly indicated when violence is aimed at non-combatants, civilians, or those who are innocent in the sense that they are clearly not responsible for the injustice being opposed.

However, several examples challenge the requirement that the target of terrorism be innocents or non-combatants. Part of the trouble is that it is not always clear who the combatants are or who is responsible for the perceived injustice. If an al-Qaeda suicide bomber exploded himself inside the Pentagon, most would consider this terrorism. But because the Pentagon is central to the U.S. war-waging mechanism, it is arguable that people who work there are combatants. If we require that terrorism target innocent civilians (and assume that pentagon employees are combatants in the war on terror), then, implausibly, this attack would not count as terrorism.

The case of the Unabomber is also instructive. Kaczynski's targets were not random innocent people but those involved in perpetrating the "techno-fascism" he railed against. It turns out that one of his bombs detonated on an airplane and this suggests that he was also willing to kill innocent people, that is, those not directly involved in producing techno-fascism. Is it only because of this that Kaczynski is properly labeled a terrorist, or would he have been a terrorist without this targeting of "non-combatants?"

Consider certain tactics used in the campaign to end the production of foie gras. Almost ten years before California banned the production of fat-engorged liver from force-fed ducks and geese, animal activists targeted a chef and part owner of a restaurant that sold what plausibly can be considered a product of animal cruelty. In addition to flooding his restaurant, they vandalized his home and car and left a video-tape of him and his family with the message "stop or be stopped." This is clearly an attempt to intimidate for political goals. It seems clear that a chef who uses foie gras is not innocent of what the animal activists are trying to prevent. (Such activists might also consider those who consume this "delicacy of despair" as guilty parties.) If targeting non-combatants is required, then this might not be what the FBI called "an act of domestic terrorism." Was this terrorism only because the activists threatened not just the chef but also his family who are arguably innocent of the alleged evil the activists are trying to stop (despite benefiting from it)?

Those who advocate terrorism often stretch the notion of combatants to try to avoid our natural moral revulsion toward violence against innocent civilians. Militant jihadists, including Osama Bin Laden, have made the following pronouncement: "To kill the Americans and their allies—civil and military—is an individual duty for every Muslim" (Lewis 1998). This appears to accept the killing of innocents as a legitimate tactic. But after the September 11 attacks, Bin Laden suggested that American civilians are not innocent:

> The American people should remember that they pay taxes to their government and that they voted for their president. Their government makes weapons and provides them to Israel, which they use to kill Palestinian Muslims. Given that the American Congress is a committee that represents the people, the fact that it agrees with the actions of the American government proves that America in its entirety is responsible for the atrocities that it is committing against Muslims.
> *(Bin Laden 2005: 140–141, quoted in Primoratz 2018)*

Such a pronouncement stretches the notion of responsibility too far and is a thinly disguised attempt to avoid the natural moral horror we have at the targeting of innocent civilians.

Some in the environmental movement may be accused of a similar stretching of the notion of responsibility. It has been suggested that Derrick Jensen's *Endgame* (Jensen 2006) makes the following argument: Because civilization depends on widespread violence, all civilized people (even dogmatic pacifists) are complicit in violence simply by their own participation in the industrial economy. Thus to violently bring down industrial civilization, as Jensen clearly seems to advocate (see below), even if it were to involve the death of millions of ordinary people, would not count as terrorism (under the restriction that innocents must be targeted) because these ordinary people share in the responsibility for the destruction of nature.

As can be seen from these examples, those who support violence for political causes have a strong desire to avoid being seen as advocating violence against innocent parties. One reason to include the targeting of innocents in our conception of terrorism is that it explains and justifies the special moral horror that is associated with this tactic. Note, however, that we need not categorize an act of ideologically inspired, violent intimidation as terrorist in order to condemn it. Such violence might not be terrorism and yet still be severely wrong (perhaps even worse than an act of terrorism). We might say that the shooting of armed Jews by Nazi soldiers was not terrorism (as the Jewish resisters are obviously combatants) but insist it was clearly wrong (Coady 2001). So we could strongly object to the al-Qaeda bombing of the Pentagon, Kaczynski's bombing of technocrats, and the animal activists' attack on the chef while insisting that because they attack combatants, they are not terrorism.

Might Property Destruction Count as Terrorism?

Our working understanding is that terrorism involves intimidating violence that is motivated by ideology and that paradigmatically targets those not responsible for the evil being resisted, that is, innocents or non-combatants. What kind of violence does terrorism involve? Must the violence in terrorism be aimed at persons, or might violence aimed at property (sabotage) also count as terrorism? Virtually all violence perpetrated by radical environmentalists and animal activists has been aimed at property, not persons. Those who destroy property for environmental or animal protection reasons repeatedly say that their goal is to protect all life, including humans, from the forces destroying the earth. Here is Edward Abbey's articulation of the distinction between destroying property for political goals and terrorism:

> Sabotage is an act of force or violence against material objects, machinery, in which life is not endangered, or should not be. Terrorism, on the other hand, is violence against living things–human beings and other living things. That kind of terrorism is generally practiced by governments against their own peoples.
>
> *(Loeffler 2010: 8)*

The use of arson by members of the Earth Liberation Front (ELF) provides a useful test case. In 1998, ELF members burned $12 million worth of buildings and ski lifts at Vail Colorado because of plans to expand into endangered lynx habitat. ELF has claimed responsibility for burning sport utility vehicles (SUVs) and hummers at dealerships around the country (while others were spray painted with "Lazy Fat Americans" or "I love to pollute"). In 2001, a research laboratory at the University of Washington was burned down because it was believed (mistakenly) to be genetically engineering Popular trees. As recently as 2008, ELF members opposed to urban sprawl have set ablaze luxury homes and condos in various stages of construction. In one case, they mocked claims the homes were environmentally friendly by spray painting "Built Green? Nope black! McMansions… not green. ELF." These cases of arson involve targeting property, not people, but they also seem violent because they involve destructive force that harms.

Some of the activities of the Sea Shepard Conservation Society also provide examples of violent sabotage. This group has been fighting to protect marine mammals for decades using tactics such as cutting drift nets to prevent dolphin deaths – a tactic that does not seem violent – but also ramming whaling ships at sea and sinking one at a dock for allegedly violating international treaties regulating whaling. Paul Watson, who founded the group, specifically advocates violence, but never against life, only against property (Best 2005: 1720). One analysis of sabotage in defense of the environment argues that – with the notable exception of the Unabomber – "in more than twenty-five years there have been no cases where environmental activists have sought to kill or maim anyone" (Taylor and LeVasseur 2009: 290).

The distinction between violence to property and violence toward persons seems clear at a superficial level. Because of this, some theorists have argued that ecological sabotage should not count as terrorism:

> Acts that maintain the principled distinction between persons and property – neither harming nor threatening to harm persons – must be distinguished from genuine terrorism in theory and ought also to be distinguished from those more objectionable tactics under law… ecotage must be understood as a categorically different offence than terrorism.
>
> *(Vanderheiden 2008: 304)*

Nevertheless, the assumption that there is a bright line between violence directed at property and violence directed at people is problematic. While the distinction is important, we must not let it delude us into thinking that violence against property has no harmful effects on people, as if only things are being hurt and not people (Morreall 1976). Violence that harms people need not cause physical injury to them. Sabotage that destroys people's property, puts them out of work, or deprives them of income or profits harms them. The owners of the SUV dealerships were harmed by the arson campaign against them as were the researchers at the university who lost their labs and data. The owners of the fishing and whaling vessels were harmed as well by the sabotage directed at their property. Note too that violence that destroys a person's significant property can be far worse for her than a minor physical injury.

One important definition of terrorism explicitly includes property destruction. Coady (2001: 1697) suggests we might define terrorism as the tactic of intentionally targeting non-combatants (or non-combatant property when significantly related to life and security) with lethal or severe violence meant to produce political results via the creation of fear. Despite arguing for the importance of the distinction between harming persons and property, Vanderheiden also acknowledges that destroying property can terrorize:

> If a person or people can be "terrorized" – that is, illegitimately intimidated by the calculated use of force and implication of further violence – by random killing, then surely they can also be similarly terrorized by the significant destruction of certain kinds or quantities of property.
> *(Vanderheiden 2005: 430)*

If we allow that acts aimed at destroying property are not immune from the charge of terrorism simply because they don't directly aim at physical injury to a person, then the question remains whether ecological sabotage can be terrorism. Ecological sabotage seems to fit our working definition of terrorism fairly well. It is politically motivated, coercive, and illegal. Frequently it aims to change people's behavior by instilling fear of being the target of harmful illegal activity. ELF members who burned SUVs probably hoped to dissuade people from buying them and animal activists acts of sabotage often seem aimed at intimidating those who harmfully use animals (much as anti-abortion sabotage aims to intimidate abortion providers and those considering abortion). In so far as people lose their jobs or homes as the result of ecological sabotage, these losses seem "significantly related to people's life and security" (as Coady's definition requires). Note, however, that even non-violent, civil disobedience can intimidate and cause harm to people in the ways I'm suggesting sabotage can.

However, Coady's requirement that the property targeted must be that of "non-combatants" may exempt most acts of ecological sabotage from the category of terrorism. Ecological sabotage targets property used in what the saboteurs believe to be environmentally harmful ways and thus, from their perspective, those who own or use that property are not "innocent bystanders." The debate over who counts as combatants discussed above comes into play at this point: Perhaps the developers of Vail Resorts can legitimately be looked at as combatants but can those who own or build homes that constitute suburban sprawl be so considered? Note that whether or not we consider them to be terrorism, tactics like arson and ramming ships (compared with tactics like pulling up survey stakes or cutting drift nets) are more morally problematic as they present significant risks of physical injury to people.

Expanding the notion of terrorism to include some types of property destruction may contribute to an unfortunate tendency to overuse and extend the term. The FBI has an expansive definition of terrorism that both includes property destruction and does not require the violence be directed at non-combatants:

> Domestic terrorism is the unlawful use, or threatened use, of violence by a group or individual based and operating entirely within the United States (or its territories) without foreign direction, committed against persons or property to intimidate or coerce a government, the civilian population, or any segment thereof, in furtherance of political or social objectives.
> *(Jarboe 2002)*

The 2006 "Animal Enterprise Terrorism Act" criminalizes behavior which intentionally "damages or interferes with the operations of an animal enterprise," whether or not it involves "physical disruption." This act suggests not only that non-violent, civil disobedience might count as "terrorism" but even otherwise legal divestiture campaigns (ACLU 2013). Derrick Turner points out that if we call environmental and animal saboteurs terrorists we might "create the mistaken impression that they are no different, morally speaking, than the Unabomber" and that

> if we are too liberal in applying the term terrorism, we will no longer be able to use that term
> to capture what is distinctively bad about actions such as the September 11, 2001, attacks, or
> the 1995 Oklahoma City bombing.
>
> *(Turner 2009: 283)*

So, is ecological (and animal) sabotage terrorism or not? I think the best approach is to not assume there is some fact of the matter here and also to refuse to stipulate an answer by insisting on a precise definition. We should also not assume a strict all-inclusive dichotomy between two types of politically motivated lawbreaking: Either civil disobedience or terrorism. If our account of civil disobedience rules out property destruction (and it may or may not), it does not entail that such politically motivated property destruction is tantamount to terrorism. We have paradigms of environmental civil disobedience and also paradigms of environmental terrorism. We also have cases that fall on a spectrum in between. Some are closer to civil disobedience, such as when, in the late 1960s before the Clean Water Act, the middle-school science teacher and eco-activist known as "The Fox" "collected 50 pounds of sewage that a company had spewed into Lake Michigan and dumped it in the company's reception room" (Martin 2001). Others are closer to terrorism such as the burning of five luxury "green" homes outside Seattle in 2008 by eco-activists rejecting the claim that these high-end homes were environmentally friendly (Yardley 2008). The more environmentally motivated property destruction instantiates the core features of terrorism – namely, violence that harms (especially non-combatants), attempts to intimidate and coerce, and creates fear as a tool for ideological objectives – the more like terrorism it is. On this account, acts can be more or less terrorist-like, depending to what extend they embody the core features and resemble the paradigm cases of terrorism.

The Morality of Ecoterrorism

Some might think that there is an absolute ethical prohibition against ecoterrorism, because terrorism is necessarily morally wrong. Terrorism, they might think, is a unique evil that can never be justified. To consider this position in its strongest form, let's assume we are talking about terrorism in its paradigmatic instantiation: Killing and terrorizing innocent non-combatants in order to coerce a politically motivated change in policy. An absolute moral prohibition on terrorism entails rejecting as immoral the allied terror campaign in WWII aimed at forcing the Germans to surrender (as well as Truman's bombing of the Japanese to end WWII).

One way to defend this absolutist position is with the idea that no ends can ever justify a sufficiently evil means. In defending her objection to giving President Truman ("Two-Bomb Harry") an honorary degree at Oxford because he authorized the atomic bombing of Japan, G.E.M. Anscombe argued that "If you had to choose between boiling one baby and letting some frightful disaster befall a thousand people—or a million people, if a thousand is not enough—what would you do?" (Anscombe 1958: 3). The idea is that some acts by their nature are so wrong that they should never be done, no matter what the consequences.

Such a position rejects the consequentialist position in ethics that asserts that what makes actions right or wrong are solely the consequences – so that a sufficiently good consequence (preventing far more innocent deaths than one causes) could justify any act whatsoever (including terrorism). A consequentialist justification for ecoterrorism might look like this: Suppose a campaign of terrorism aimed at fossil fuel industry workers, executives, and users could quickly shift world economies away from fossil fuels to renewables. If such a shift prevented far more deaths than would have occurred from rapid climate change than the deaths this terrorism causes, then, other things being equal, it is morally justified.

Consequentialist moral reasoning is suspect in the eyes of many and so one might be tempted to reject this ethical approach and with it its defense of terrorism. But even non-consequentialist, rights-based ethics can be used to justify acts of terrorism. Virginia Held argues that non-consequentialist considerations of distributive justice can support a limited use of terrorism. Fairness in the distribution of rights violations is an important moral goal. "If we must have rights violations, a more equitable distribution of such violations is better than a less equitable one" (Held 2008: 88). Imagine a defective society (like 20th century apartheid South Africa) where one group's rights to personal safety are respected while another's are not. If limited terrorism against the privileged group were the only way to transition that society to one where everyone's fundamental rights were respected, then fairness in sharing the burden of rights violation would justify such terrorism.

Arguments justifying terrorism only begin to make sense in extreme circumstances. The presumption against using violence to bring about change, especially violence against innocents, is extremely high. But many environmentalists – and not just the few who advocate direct or (more strongly) violent action – articulate our environmental situation in such a way that it does constitute a supreme emergency, thus opening the door to the justification of extreme tactics.

Pulitzer Prize-winning author and award-winning scientist Jared Diamond has said that unless the world changes its environmental actions and policies, the world of 2050 will be "a world not worth living in" (Diamond 2014). On this account, our environmentally unfriendly, unsustainable lifestyles are not just harming other species, they are not just dramatically lowering the quality of our lives today and in the future, but they will also bring about a world so horrible that it would be better that our children not be born into it. Such a threat presents an emergency situation that could justify the most extreme tactics in order to prevent it.

One of those advocating such tactics is Derrick Jensen, author of *End Game* (Jensen 2006) and an advocate of what he (and others) calls "Deep Green Resistance" (McBay et al. 2011). Jensen believes that "all the world is at stake" and advocates "acting decisively to stop the industrial economy" (Jensen 2009). According to Jensen, industrial civilization is inherently violent, unsustainable, and irredeemable and we must therefore bring it to an end:

> This culture will not undergo any sort of voluntary transformation to a sane and sustainable way of living. If we do not put halt to it, civilization will continue to immiserate the vast majority of humans and to degrade the planet until it (civilization, and probably the planet) collapses... The longer we wait for civilization to crash – or the longer we wait before we ourselves bring it down – the messier the crash will be and the worse things will be for those humans and nonhumans who live during it, and for those who come after.
>
> *(Jensen 2006: ix–x)*

In response to the worry that dismantling civilization "will kill millions of people in cities," he argues that in our current predicament, no one has clean hands:

> No matter what you do, your hands will be blood red. If you participate in the global economy, your hands are blood red because the global economy is murdering humans and nonhumans the planet over.
>
> *(DeepGreenResistance.Org)*

Furthermore, "any option is better than a dead planet" (Jensen 2009). Morality requires that we act: "To fail to *effectively* stop the grotesque and ultimately absolute violence of civilization is by far the most immoral path any of us can choose. We are, after all, talking about killing the planet" (Jensen 2006: 253). He thus advocates "blowing up dams" (Jensen 2006: 252) and that we consider "assassination" (Jensen 2006: 252). "Some things, including a living planet… are worth fighting for, dying for, and killing for when other means of stopping the abuses have been exhausted" (Jensen 2006: 79).

Jensen – and other Deep Green Resistance advocates like Aric McBay who argues for "Decisive Ecological Warfare" (McBay et al. 2011: 462–463) – seem to be advocating ecoterrorism. Clearly Jensen is a "revolutionary" who wants to take down and replace our culture rather than to simply reform it, but much of what he claims about the current state and future of the environment coincides with the dire pronouncements of many mainstream reformist environmentalists who would not dream of advocating such tactics. But if we are making the planet uninhabitable for ourselves and most other life forms (as both revolutionaries like Jensen and mainstream environmentalists like Diamond claim) and if terrorism is not ruled out in principle as absolutely immoral, then ecoterrorism would seem to be a live option.

At this point, practical questions of political strategy become central. Would terrorism be effective at "saving the earth?" Is it the only way to do so? Clearly less violent and less harmful means should be used if they would be effective; at best, terrorism is a last resort. Because a variety of activities might quality as "terrorism," how effective terrorism might be depends on what kind we talking about (damaging property that hurts people, physically injuring people, harming only combatants or also non-combatants, and so on).

Might ecoterrorism function like environmental civil disobedience and provide a communicative tool aimed at persuading society about the severity of environmental problems and the importance of strong environmental policies? Because it fails to renounce violence and does not take legal responsibility for its actions, terrorism "compromises much of the sympathetic appeal characteristic of civil disobedience as a form of address and so is likely to be less persuasive" (Vanderheiden 2008: 309). However, we should not rule out this "appeal to the public" function of terrorism in principle. Although attempting to coerce people to change their behavior by causing fear is very different from trying to persuade them to change, it is possible it might also have that effect (and even have that as a secondary aim). For example, seeing the neighbor's Hummer burn might make a person realize the importance of taking fuel economy seriously. As Robert Young (1995: 206) argues:

> Sometimes the use of violence serves to highlight an injustice in a way no other form of protest can match… it is not until there is violent protest that any meaningful response to wrongs is likely to be made in many a society.

As an impetus for change, consider the civil unrest (including arson) occurring in 2020 in response to police brutality toward African Americans. If people really are willing to act coercively and violently to try to solve environmental problems, then perhaps we will take environmental problems more seriously.

Nonetheless terrorism, even less severe terrorism, seems unlikely to be an effective form of public appeal:

> As many radical activists have acknowledged in interviews—even those who have supported sabotage—the more an action risks or intends to hurt people, the more the media and public focus on the tactics rather than the concerns that gave rise to the actions. This means that the most radical tactics tend to be counterproductive to the goal of increasing awareness and concern in the general public.
>
> *(Taylor 2013: 311)*

It might even be argued that any form of lawbreaking, including civil disobedience, will cost environmentalism the public support it desperately needs. Still, it seems plausible that sometimes extremists can – in bad cop, good cop fashion – succeed in making ordinary environmental proposals and advocates look more modest and thus move the policy debate in the right direction.

Might ecoterrorism in one of its forms be successful in directly ending environmental exploitation? Interestingly, both the revolutionaries and sober analysts agree that ELF's ecological sabotage campaign has been ineffective. The eco-revolutionary Aric McBay claims that "It's hard to find a case in which a construction project has actually been given up because of ELF activity" (McBay et al. 2011: 418). Political science professor Steven Vanderheiden argues that "Despite causing over $100 million in damage over the past decade, the ELF has not significantly affected the underlying profit structure of targeted businesses" (Vanderheiden 2008: 307). Still another analyst argues that the prospect of ecoterrorists overthrowing industrial culture is "fanciful." He concludes that "It makes little sense to base strategy and tactics on such an unlikely possibility that communities of resistance will ever be able to mount a sustained campaign to bring down industrial civilization, even if that were a desirable objective" (Taylor 2013: 312). Of course, some would pessimistically claim that not only would terrorism be ineffective in bringing about the needed changes, but no tactic – civil or otherwise – will be successful.

If we leave open the possibility that ecoterrorism might be morally and practically justified, this presents the dangerous suggestion that environmental activists will have to decide for themselves whether their goals and their practical circumstances warrant this extreme behavior. The worry becomes clear when we reflect on cases where those with whom we disagree have sincerely believed their terrorism was justified. Consider the terrorist campaign against abortion which has included dozens of cases of arson, bombing, and assassination (resulting in almost a dozen deaths). Or consider al-Qaeda terrorists who seek to avenge innocents killed in the U.S. war on terror. Nevertheless, to allow the possibility that sometimes terrorism might be justified is not to commit to the view that any sincere act or campaign of terrorism is justified. Such justification depends on the righteousness of the cause and the effectiveness and necessity of the action. But how are activists to know if their possible terrorist acts are justified?

Those considering such radical tactics must begin their vetting process with supreme humility. Although overly absolutist, Gandhi's position on violence is instructive. He argued that because our beliefs are subject to error, no one should be so presumptuous as to inflict harm on others to further those beliefs (Gandhi 1957). Those who engage in violent activity are often arrogant, macho, and/or deluded and they are likely to have an irresponsibly high level of confidence in the rectitude of their cause and the effectiveness of their tactics. Would-be ecoterrorists have a solemn duty to vet their ideas with many thoughtful others, including those with whom they disagree, and to make sure they respect the humanity of those they oppose (and plan to harm).

For example, the ELF members who burned the research lab they believed was engaged in genetically engineering popular trees needed to ask if doing so would actually help stop the use of genetically modified crops. They should have been extraordinarily familiar with the debate over whether or not genetically modified organisms (GMOs) really are a significant threat to the environment. And since the scientists experimenting with the GMOs are not moral monsters who can be ignored, they needed to consider what will happen to the researchers and their livelihoods if the arson plan succeeds.

One useful constraint is to employ John Rawls requirement that those considering breaking the law must limit their justifications to conceptions of justice that have widespread acceptance and that they not rely simply on their own narrow political allegiances, religious conviction, or private moralities (Rawls 1971: 365). This would arguably rule out terrorism by abortion-opponents as well as terrorism in support of non-human life forms. Neither the right to life of human fetuses nor that of other species is a "commonly shared conception of justice that underlies the political order" (Rawls 1971: 365). However, this requirement seems to be met when eco-terrorism is justified by the fear that our industrial civilization is making the earth uninhabitable for humans. It is widely (if not universally) accepted that it would be a terrible injustice to destroy the human habitability of earth. Would-be ecoterrorists share this conception of justice both with their anti-environmentalists opponents and their less extreme environmentalist ones. What the advocates of ecoterrorism as a general strategy for the environmental movement do not share with their opponents is a sober, level-headed assessment of the likely success of using extreme violence to bring about a better environmental future and about the prospects for alternative, less extreme strategies to bring it about. It is hard to see how ecoterrorism could be justified when it is based on such rash and farfetched empirical assumptions.

Conclusion

Terrorism is the act or threat of violence that aims to intimidate and coerce by means of fear in or-der to achieve ideological objectives. In its paradigm form it targets innocents and attempts bodily injury to them. Because the distinction between innocent parties and those responsible for the alleged injustice is contentious and because one can harm people by destroying their property, the concept of terrorism can be expanded to include attacks on those indirectly involved in the alleged injustice, as well as to attacks on property. Given this expanded definition, the environmental movement has seen acts of ecoterrorism and some in the movement attempt to defend it. An abso-lute prohibition on terrorism (including ecoterrorism) as a supreme evil is not plausible. As Virginia Held has argued: "Terrorism is not uniquely atrocious but on a continuum with many other forms of political violence… wars, even 'good wars', are often morally far worse" (Held 2008: 9). The language of extreme environmental emergency used by many in the environmental movement – that we are creating a world not worth living in and that we need to "save the earth" – provides support for the radicals who seek to justify ecoterrorism. Nonetheless, it is highly dubious that ecoterrorism, at least in its extreme manifestations and as a general strategy for the environmental movement, could be effective in achieving the changes needed. Those contemplating this tactic should worry about their ability to judge wisely, and they have a solemn duty to vet their rationales with their opponents and to seriously consider the humanity of those they intend to harm.

Related Topics

20. Moral Bases of Responses to Climate Change
43. Remediation
44. Restoration
49. Pollution and Polluter Pays
51. Libertarianism
63. Environmental Justice
64. Environmental Civil Disobedience

References

ACLU. (2013) Brief of Amici Curiae before the United States court of appeals for the first circuit (on behalf of animal rights activists who had been charged with violating the Animal Enterprise Terrorism Act). Available at: https://www.aclum.org/sites/all/files/legal/blum_v_holder/20130729_aclum_amicus_brief_first_circuit.pdf (Accessed: 15 November 2016).

Anscombe, G. E. M. (1958) *Mr. Truman's degree,* a pamphlet published by the author. Available at: http://theahi.org/wp-content/uploads/2013/10/Anscombe-Mr.-Trumans-Degree-1958.pdf (Accessed: 15 June 2020).

Arnold, R. (1983) 'Eco-terrorism', *Reason*, February, pp. 31–35.

Berry, W. (2008) 'Speech against the state government', first delivered on February 14, 2008 at the "I Love Mountains" protest in Frankfort, Kentucky. Available at: http://bethwellington.wordpress.com/2008/10/19/wendell-berry-speech-against-the-state-government-and-mtr/ (Accessed: 15 June 2020).

Best, S. (2005) 'Watson, Paul (1950–) and the sea shepherd conservation society', in B. Taylor (ed.) *Encyclopedia of Religion and Nature*. New York: Continuum.

Bin Laden, O. (2005) 'The example of Vietnam', in Bruce Lawrence (ed.), James Howarth (trans.) *Messages to the World: The Statements of Osama Bin Laden*. London and New York: Verso, pp. 139–144.

Cavallaro, J., Sonnenberg, S., and Knuckey, S. (2012) 'Living under drones: death, injury, and trauma to civilians from US drone practices in Pakistan', Stanford, CA: International Human Rights and Conflict Resolution Clinic, Stanford Law School; New York: NYU School of Law, Global Justice Clinic. Available at: https://law.stanford.edu/publications/living-under-drones-death-injury-and-trauma-to-civilians-from-us-drone-practices-in-pakistan/ (Accessed: 15 June 2020).

Center for Constitutional Rights. (2016) *Wiwa et al v. Royal Dutch Petroleum et al.* Available at: https://ccrjustice.org/home/what-we-do/our-cases/wiwa-et-al-v-royal-dutch petroleum-et-al (Accessed: 15 June 2020).

Coady, C. (2001) 'Terrorism', in L. Becker and C. Becker (eds.) *Encyclopedia of Ethics*. 2nd edition. New York and London: Routledge, vol. 3, pp. 1696–1699.

DeepGreenResistance.Org (n.d.) *Frequently Asked Questions: If we Dismantle Civilization, Won't that Kill Millions of People in Cities? What About Them?* Available at: http://deepgreenresistance.org/en/who-we-are/deep-green-resistance-faqs (Accessed: 15 November, 2016).

Diamond, J. (2014) 'Jared Diamond speaks to the young on environmental challenges', Interviewed by Tom Ashbrook for *On Point*, 3 April. Available at: https://www.wbur.org/onpoint/2014/04/03/jared-diamond-climate-change (Accessed: 15 June 2020).

Foreman, D. and Haywood, B. (1985) *Ecodefense: A Field Guide to Monkeywrenching*. Tucson, AZ: Ned Ludd Book.

Gandhi, M. (1957/1971) 'Non-violence', in J. Murphy (ed.) *Civil Disobedience and Violence*. Belmont, CA: Wadsworth Publishing Company, pp. 93–102.

Harris, S. et al. (2007) 'The terrorism enhancement', *National Journal*, 39(28), pp. 34–40.

Held, V. (2008) *How Terrorism Is Wrong: Morality and Political Violence*. Oxford: Oxford University Press.

Hettinger, N. (2001) 'Environmental disobedience', in D. Jamieson (ed.) *Companion to Environmental Philosophy*, Malden, MA: Blackwell, pp. 498–509.

Hoffman, B. (2006) *Inside Terrorism*. 2nd edition. New York: Columbia University Press.

Jarboe, J. (2002) 'The threat of eco-terrorism', *Testimony Before the House Resources Committee, Subcommittee on Forests and Forest Health*, 12 February. Available at: https://archives.fbi.gov/archives/news/testimony/the-threat-of-eco-terrorism (Accessed: 15 June 2020).

Jensen, D. (2006) *Endgame: The Problem of Civilization*, volume 1. New York: Seven Stories Press.

Jensen, D. (2009) 'Forget shorter showers', *Orion*, July/August. Available at: https://orionmagazine.org/article/forget-shorter-showers/ (Accessed: 15 June 2020).

Kennedy, J. (1962) 'Address on the first anniversary of the alliance for progress', speech, delivered at the White House, 13 March. Available at: https://www.presidency.ucsb.edu/documents/address-the-first-anniversary-the-alliance-for-progress (Accessed: 15 June 2020).

Lewis, B. (1998) 'License to kill: Usama bin Ladin's declaration of Jihad', *Foreign Affairs,* 77(6). Available at: http://www.foreignaffairs.com/articles/54594/bernard-lewis/license-to-kill-usama-bin-ladins-declaration-of-jihad (Accessed: 15 June 2020).

Loeffler, J. (2010) *Headed Upstream: Interviews with Iconoclasts*. Santa Fe, NM: Sunstone Press.

Martin, D. (2001) 'James Phillips, 70, environmentalist who was called the Fox', *New York Times*, obituaries, 22 October [online]. Available at: http://www.nytimes.com/2001/10/22/us/james-phillips-70-environmentalist-who-was-called-the-fox.html (Accessed: 15 June 2020).

McBay, A., Keith, L., and Jensen, D. (2011) *Deep Green Resistance: Strategy to Save the Planet*. New York: Seven Bridges Press.

Morreall, J. (1976) 'The justifiability of violent civil disobedience', *Canadian Journal of Philosophy*, 6, pp. 35–47.

Primoratz, I. (2018) 'Terrorism', in Zalta, E. (ed.) *The Stanford Encyclopedia of Philosophy,* Winter 2018 edition. Available at: https://plato.stanford.edu/entries/terrorism (Accessed: 15 June 2020).

Rawls, J. (1971) *A Theory of Justice*. Cambridge, MA: Harvard University Press.

Taylor, B. (2013) 'Resistance: do the ends justify the means?', in Stark, L. (ed.) *State of the World 2013*. Washington, DC: Island Press, pp. 304–316.

Taylor, B. and LeVasseur, T. (2009) 'Ecotage and ecoterrorism', in Callicott, J. and Frodeman, R. (eds.) *Encyclopedia of Environmental Ethics and Philosophy*. Detroit: MacMillan Reference, pp. 286–291.

Turner, D. (2009) 'Ecosabotage', in Callicott, J. and Frodeman, R. (eds.) *Encyclopedia of Environmental Ethics and Philosophy*. Detroit: MacMillan Reference, pp. 281–284.

Vanderheiden, S. (2005) 'Eco-terrorism or justified resistance? Radical environmentalism and the "War on Terror"', *Politics & Society*, 33(3), pp. 425–447.

Vanderheiden, S. (2008) 'Radical environmentalism in an age of antiterrorism', *Environmental Politics*, 17(2), pp. 299–318.

Walzer, M. (1977) *Just and Unjust Wars: A Moral Argument with Historical Illustrations*. 3rd edition. New York: Basic Books.

Welchman, J. (2022). 'Environmental Civil Disobedience', *this volume*.

Yardley, W. (2008) 'Ecoterrorism suspected in house fires in Seattle suburb', *New York Times*, 4 March. Available at: https://www.nytimes.com/2008/03/04/us/04homes.html (Accessed: 15 June 2020).

Young, R. (1995) 'Monkeywrenching' and the processes of democracy', *Environmental Politics*, 4, pp. 199–214.

Additional Reference

Curry, Michael. (2011) 'If a tree falls: A story of the earth liberation front', Film Documentary, Oscilloscope Laboratories. (Excellent film about the Earth Liberation Front activities and prosecution.)

INDEX

Note: **Bold** page numbers refer to tables; *italic* page numbers refer to figures.

Abbate, C. E. 48
Abbey, E. 114, 116, 118, 119, 798
ableism 8, 9
acidification of ocean *see* ocean acidification
Ackerman, B. A. 667
Ackland, L. 522; *Making a Real Killing* 522
Adams, A. 136
Adams, P. 251, 637
adaptation 236, 239–240
adaptive management (AM): collaborative AM 718;
    command-and-control management 718; history
    of 717–718; intentional trial-and-error 717;
    leadership practices 718–723
Addison, J. 389
advocacy and activism 1; civil disobedience (*see*
    environmental civil disobedience); community
    participation 752–766; education (*see*
    environmental education); everyday aesthetics
    741–749; justice (*see* environmental justice (EJ));
    lawbreaking and ecoterrorism (*see* terrorism)
aesthetic literacy 743
agenda-setting 619
agrarian narrative 463–464
agriculture 1; animal 14, 21; biotechnology
    481–490; community gardens 503–512;
    industrial (*see* industrial agriculture); necessity
    and sport 79–80; sustainable (*see* sustainable
    agriculture); *see also* food
Agyeman, J. 493
Ahteensuu, M. 708
aims problem 729–733
Alberti, Leon Battista 417
Aldy, J. 610
Alison, A. 135, 414
Allen, P. 493, 494
Allen, Will 505, 508

American Public Opinion 304
Anarchy, State, and Utopia (Nozick) 624
Anderegg, W. R. 299, 300
Anderson, E. 691
Anderson, J. L. 146
Anderson, K. 442, 451
Anderson, M. 625, 626, 669
"angel hair" 126
Anhang, J. 21
"Animal Enterprise Terrorism Act" (2006) 800
animal liberation: *vs.* environmentalism 14–16;
    episodic memory 16; invertebrates 13–15;
    rights-based account of 12; self-awareness 15–16
Animal Liberation (Singer) 75
Animal Machines: The New Factory Farming Industry
    (Harrison) 476
animals 1; agriculture 14, 21; cognition 5, 9–11;
    cognition-heavy accounts 9; companion
    (*see* companion animals); eating (*see*
    eating animals); and environment 28–29;
    experimentation 34–37; and hunting (*see*
    hunting); moral significance 6; and moral status
    (*see* moral status); pain or distress 35–36; for
    research 39; sentience 5, 9–12, 16, 32, 34–35,
    64; species and wildlife 51–60; wild (*see* wild
    animals)
Animals (Scientific Procedures) Act (1986) 12
Animal Welfare Act 34
Animal Welfare Argument 120
Anscombe, G. E. M. 800
Anthropocene 57–58, 65–67, 79
anthropocentric account of moral status: ableism
    9; animal cognition-heavy accounts 9; being
    human 7–8; disability rights 8; distinctively
    human properties 7; full membership 9; *Homo
    sapiens* 7–8; neurotypicalism 8; secure full moral

status for humans 7–8; speciesism 8; species membership 7; species typicalism version 9
anthropocentric agricultural ethics 482–484
anthropogenic novel ecosystems 587–589
Appel, M. 12
appurtenance 160
Arab Spring 417
Arena Argument 113
Arendt, H. 416, 453
arguments for skepticism 757–758
Aristotle 144, 174, 416, 417, 734, 755, 756, 761; *Nicomachean Ethics* 761; *Politics* 144, 416
Aronson, J. 539, 540
Arras, J. D. 659
Art Gallery Argument 114
artificial trees 282
Assadourian, E. 450
assisted migration: background 546–548; or assisted colonization 239; or assisted colonization or managed relocation 547; Pleistocene rewilding 547–548; population reinforcement or augmentation 547; rapid ecological change 548–552; species reintroductions 546–547, 548–549; species translocations 546, 547, 549–552
Association of Zoos and Aquariums (AZA) 554, 556
Attfield, R. 536, 537
Auden, W. H. 215
authoritative voices, role of 353–355
auto: investing in infrastructure alternatives to 435–436; mass transit 435; oil, consumption of 431; "second class" citizenship 436; *see also* transportation
autonoetic (self-knowing) awareness 13, 16
AWA 34–36

Bacon, Francis 472
Baer, P. 272
"balance-of-nature" paradigm 586
Balantine, Bill 212
Bali Principles for Climate Justice 775
Baliunas, Sallie 303
Balmford, A. 559
Balmori, D. 748
band-aid objections 285
"Baptists and Bootleggers" 681
barrens 172
Barron, A. B. 11
Barr, S. 12
Basal, J. 595
Basile, B. M. 14
Battin, M. P. 658
Battle of Green 741
'the battle of the Vindel River' 351
Bayesianism 310n1
Beavan, C. 450–452; *No Impact Man* 450–451
being human 7–8

Bekoff, M. 72, 73
Bell, C. 136, 745
Bell, D. 367
beneficiaries 701
beneficiary-pays principle (BPP) 275
benefit-cost analysis (BCA) *see* cost-benefit analysis (CBA)
Bennett, N. J. 221
Bentham, J. 75
Berleant, A. 744
Bern Convention (1979) 352
Bernhardt, E. S. 186
Berry, W. 108, 450, 463, 471, 476
Betz, G. 252
Biden, Joe 177
biocentric consequentialism 388
biocentrism 388–389
biological species concept (BSC) 57
biomedical research 14, 31, 658
biotechnology: anthropocentric agricultural ethics 482–484; Deep Ecology 485–486; ecocentric environmental ethics 484–486; GM crops 481–484, 486, 489; Green Revolution 483; nature-polluting technologies 486; nature-replacing technologies 486; pragmatic environmental ethic 487–489
bleaching 196
Bly, R. 119
Boone, D. 121
Borges, J. L. 746
Borgmann, A. 105, 108
Borlaug, N. 482, 483
Botkin, D. B. 532
Botsman, R. 449
Boundary Waters Canoe Area Wilderness (BWCAW) 532
Bows, A. 442
Boym, S. 568, 569, 571, 575, 578
Bradshaw, A. D. 539
Brady, E. 571
Bramwell, B. 631, 634
Brandel, M. 350
Bratman, M. 505
Braverman, I. 555
"the breakthrough strategy" 201
BREEAM 407
Broome, J. 264, 318, 319
Brower, D. 122
Brownfields Law (2002) 516
Brown, L. 201
Brownlee, K. 787
Brubaker, E. 148, 149
Bryant, W. C. 115
Bugbee, H. G., Jr. 106; *The Inward Morning* 106
Bullard, R. D. 608
Burbank, Luther 471
Burgess-Jackson, K. 44
Burke, Edmund 135

Burnham, D. 398
Burtynsky, E. 136, 138, 141, 748
Busch, L. 495
Bush, George H. W. 177, 295

Cafaro, P. 451
Caldeira, K. 287
Callicott, J. B. 15, 16, 75, 90, 100, 110–122, 450, 487, 572
Camazine, S. 11
"Cambridge Declaration on Consciousness" 13
Caney, S. 261, 272
capabilities approach to rights 640
capacity building: and capabilities approach 645; data sharing and 645
Capper, J. L. 477
captivity 14, 41, 64, 65, 551, 554, 555, 558–560, 572
carbon capture and storage (CCS) 360
carbon dioxide removal (CDR) 236, 281
"Carbon Engineering" 280
carbon majors 317
Carbon Tracker 320
Carlisle, L. 497; *Lentil Underground* 497
Carmichael, Franklin 136
cars: affecting urbanization scale and scope 434; carbon-based fuels 433; carbon monoxide 434; environmental and urban issues of 433–435; life-cycle costs of oil 433; oil, consumption of 431; ownership 434–435; sulfur dioxide 434; *see also* transportation
Carson, R. 113, 470, 474, 771; *Silent Spring* 470, 471
Cartagena Protocol on Biosafety (2003) 707
Carter, A. 786
The Case for Animal Rights 10
Cathedral Argument 114–115
Cat Wars: The Devastating Consequences of a Cuddly Killer (Marra & Santella) 47
CBDR principle 274–276, 278
Cervero, R. 436
Cézanne, Paul 97, 136
Chamberlin, S. 449
Chan, K. M. A. 223, 596
Chapin, S. III 534
Charles, D. 481
Chittka, L. 11
Choi, Y. D. 533
Church, Frederic 135
cities 1; "anti-city ethos" 398–399; anti-nature, unsustainable 398–399; as dynamically balanced ecosociotechnical systems 400–401; industrial economy 401–402; as inhabited infrastructure 401–404; modern nature/city dualism 398–400; and nature, relationship between 400–404; preserving nature, paragon of sustainability 399–400; suburbs and exurbs 422–429; sustainable houses 400; transportation 431–439; urban parks and open space 412–420; urban

sustainability 397–409; waste and consumption 442–453; *see also Individual entries*
civic practice 652, 657, 659–661
civil-society groups 407
Clark, J. R. 218
Classroom Argument 116
Clayton, N. S. 14
"clean-coal" technologies 360
Clean Trade Act 321
Clean Water Act 162, 177, 182, 187, 188, 531, 609, 800
Clements, F. E. 172, 586, 594, 597; *Plant Succession: an Analysis of the Development of Vegetation* 172
Clewell, A. F. 534, 539
Clifford, W. K. 296
climate change 195; adaptation **238**, 239–240; CO_2 emissions 233, 237; and development of "just" climate services 638–639; geoengineering **238**, 240–241; mitigation (*see* mitigation of climate change); moral and prudential reasons 234–235; non-identity problem 235; ocean policy 195; precautionary approach 235; rectification 241
climate engineering 280
Climategate 303–304
climate justice and equity: in participation and development opportunity 276–278; in remedial liability for climate-related harm 272–274; responsibility and CBDR alternatives 274–276
climate modeling: climate knowledge, values and limits of 250–255; detection and attribution 245; earth system models of intermediate complexity 246; energy balance models 246; GCMs/ESMs 246–249; integrated assessment models 245; projection 245; regional climate models 246, 249; uses of 247–248; values in 248–250
Climate Research Unit (CRU) 305
climate services: Coupled Model Inter-comparison Project 5 (CMIP5) 252; data sharing and capacity building (ethical issue 2) 643–645; defined 251, 636, 638; economic arguments 641–642; ensemble climate modeling 251–252; ethics of climate services 253; funding (ethical issue 1) 641–643; integrity, transparency, humility, and collaboration 251; limitations of economic approaches 642–643; stakeholder deliberation, justice, and coproduction (ethical issue 3) 645–647
climate wager 296–297, **297**
Clinton, Bill 177, 680
cloning 59, 576–577
coal mining: climate change impacts 328; health conditions of workers 328–329; moral dilemmas of 327–329; overburden 328; proponents 328; U.S. electricity market 327–328; *see also* mining
Coase, R. H. 607, 624, 633
Cochrane, A. 43

cognitive biases: anchor and adjustment heuristic concerns 308; availability heuristic concerns 308; cultural cognition 309; framing effect heuristic concerns 308; loss aversion heuristic concerns 308; moral psychology 308; temporal discounting heuristic concerns 308
cognitivism: cultural 138–140; landscape 137–138; scientific 138, 140
co-harms (and co-benefits) 321
Cole, Glen F. 124
Cole, Thomas 114, 135
collective duties 262–263
command and control 667–675
Commoner, Barry 113
common water: complexity and 163–164; notion of "community" 164; politics and ethics 163
communitiveness 418
community gardens: collective or "we" intentions 505–506; community activism 508–509; education 506–507; gardens and community 503–506; individual growth and enhancement 509–510; related practices 511–512; tragedy of the commons 506; variety of functions 506–512
community participation *see* skepticism
community-supported agriculture (CSA) 498–499
companion animals: as citizens 44–45; condition of dependence 42; dependent agency 44; environmental impacts 45–48; ethical frameworks 43–45; excretion 47; food consumption 45–47; as living property 43–44; a "pet" 41; practice of keeping 41–43; predation 47–48; reflective agency 44
companionship 40–41, 43
compensatory mitigation (CM): "malicious" restoration projects 188; mitigation banking 187–188; "restoration thesis" 188
Comprehensive Environmental Response, Compensation, and Liability Act (CERCLA) 516, 603
Comprehensive Everglades Restoration Plan (CERP) 529
concentrated animal feeding operations (CAFOs) 474, 477
Conferences of the Parties (COPs) 276–277
conscientious objection 787
consensus, skepticism: Climategate 303–304; dissent in science 300–301; Doran and Zimmerman's 2009 Study 298–299, *299*; *Merchants of Doubt* 301–302; "Mike's Nature trick" 303; Oreskes' 2004 Consensus Study 298, *298*; scientific consensus 297–300
consequentialism: biocentric 388; and wild animals 67–68
conservation breeding and reintroduction 558–560
"Conservation Centers for Species Survival" (C2S2) 559
conservation responses 58–60
conservative and progressive ethics 350–353

constitutional rights: in centralized system 614; in decentralized system 614; mechanisms of constitutional amendment 614; procedural 613, 617–621; substantive 613, 614–617
Constructing Social Reality (Searle) 505
continental conservation 531
Convention on Biological Diversity (CBD) 216; Aichi Target 11 220
Convention on International Trade in Endangered Species (CITES) 194; of Wild Fauna and Flora 557
Conway, E. 302
Conway, W. G. 559
Cook, J. 299, 300
Cooper, D. E. 745
co-pollutant emissions 361
co-production 645
coralline algae 198
corals and sponges, new diseases 200
CORDEX program 249
corporate social responsibility (CSR): controversies in 330–333; "dark greens" 335; definition of sustainable development 331–332; expansion of neoliberal governance 331–332; impact on environmental and labor performance 333–334; "light greens" 335; models of participation 332–333; "New CSR" reforms 330; "old CSR" 330; public criticism 330–331; responsibility for regulation 332; "social license to operate" 333
cost-benefit analysis (CBA): anthropocentric 687; case of *Peevyhouse vs. Garland Coal* 685, 688, 692; critics of 686, 689; environmental quality 689; future generations 693; Multiple Accounts Analysis or Life Cycle Analysis 688; priceless 690–691; problem of external cost 688; qualitative values 690; utilitarian moral theory 687–688; willingness to pay 692–693
Coupled Model Inter-comparison Project 5 (CMIP5) 252
COVID-19 pandemic 89, 769
Cranor, C. 710, 711
Crease, R. P. 524
CRISPR technology 58
Crist, E. 16
Crocker, D. 387
Crompton, J. 414
Cronon, W. 425, 484, 486, 572, 590
Crusoe, Robinson 143, 144, 151
Cultural Diversity Argument 116–117
cultural landscapes 138
Cuomo, Chris 775

Dagger, R. 416
Daines, Steve 100
dam: channel modification 180, 183; construction 180, 183; flow diversion 180; removal, challenge of 187
Danto, A. 746

Darwin, C. 53, 172; *Origin* 172; paradigm-shattering tome 172
data sharing 645
Davidson, D. 168, 171
Davis, H. 408
DeChristopher, Timothy 789
decision-makers 701
de-extinction 59, 575–578; animal welfare and resource allocation 577; biological species 576–577; ecosystem health 577; genetic engineering 576; *Jurassic Park* 576; "the restorationist argument" 577; using living animals 575–576; woolly mammoths 576
Defense of Democracy Argument 119–120
Defoe, Daniel 143
Deforesting the Earth (Williams) 153
Degrazia, D. 9, 33
De LaPlante, K. 737
deliberation processes 619–620
deliberative justice 640–641
Demsetz, H. 143, 670
Denver, John 114
deontological approaches 53–54
De-Shalit, A. 415
Despommier, D. 511
detection and attribution 245
Diamond, J. 539, 720, 801
DiChiro, G. 778
Dickinson, A. 14
Diesendorf, M. 359, 360
Dilling, L. 309
direct air capture methods 282
Direction, Alignment, and Commitment (DAC) 720–721
disability rights 8
disaster response: civic practice 659–661; disaster preparedness planning 661–662; ecological being 658–659; ethical engineering 657–658; ethical situation of 654–656; ethics of 656–657; spectrum of disaster 652–654
'dis-climax' 586
discounting future costs 390
Disease Sequestration Argument 118
DiSilvestro, Roger L. 130
distributed carbon scrubbing 282
distributional issues (global poor) 287
divergence problem 303
Donaldson, S. 42–45, 49
Donlan, J. 574, 575
Donner, S. 643
Doran, P. T. 299, 300
Douglas, H. 252
Downing, A. J. 419
Doyle, M. 188
Drenthen, M. 595
Drew, C. A. 188
Dreyfus, H. 105, 106
Drosophila 11

Drury, R. T. 680
Dryzek: *The Politics of Earth* 720
Dryzek, J. S. 185, 720
Dublin Principles 158
due restraint 789–791
Dunlap, R. E. 733
DuPuis, E. M. 500
Dust Bowl 653

Earth Day 98
EarthFirst! 784
Earth Liberation Front (ELF) 798
Earth Summit's Rio Declaration (1992) 707
earth system models (ESMs) 246
earth system models of intermediate complexity (EMICs) 246
eating and environment: environmentalism and veganism 22–23; "imperfect" environmental duties 23–24; reasons to reduce 21–22
eating animals: consumption of 20; "factory farming" *vs.* "humane" farming 25; farmed animal facts 24–25; harm of death 25–26; human activities 20; moral justifications 26–27; objections to veganism 27–28
ecocentric environmental ethics 484–486
"eco footprints" 45–46
ecojustice education 730
ecological being 658–659
ecological council model 617–618
ecological restoration (ER): attributes of restored ecosystems 534–535; BWCAW 532; climate change 533; describing acts of 530–531; Elliot-Katz challenge for 535–537; goals and values of 531–535; historical fidelity 184, 533, 534; issue of disturbances paradigm 532; and people 540–541; philosophical issues 529–530; problem of mixed cultural/natural landscapes 532; rehabilitation 531; remediation 515; and restoration ecology 539–540; revegetation and reintroduction 530–531; river management 185; water ethics or water rights 185; wilderness preservation, and environmental paradigms 537–539
economic instruments: "Baptists and Bootleggers" 681; hot spots and environmental justice 679–681; polluter pays principle 677–678; saving planet 681–682; selling right to pollute 678–679
ecosociotechnical system 404–406
ecosystem health 535
ecosystem services 175
ecoterrorism: arguments 801; atomic bombing of Japan 800; consequentialist position in ethics 800–801; "Deep Green Resistance" 801–802; environmental exploitation 802–804; genetically modified organisms (GMOs) 803; morality of 800–804
education: environmental (*see* environmental education); outdoor or place-based education programs 729

education for sustainable development (ESD) 730
Ehrlich, A. H. 113, 473
Ehrlich, P. R. 113, 473
Eisenhower, Dwight D. 404–406
Eisner, T. 11
Elder, J. 570
Elefante, C. 400
Elliott, R. 188, 534–537, 541; challenge for ER 535–537
El Nino – Southern Oscillation (ENSO) phenomenon 247
Elwood, R. W. 12
The Emerald Necklace 402
emerging ecosystem 588
Emerson, Ralph Waldo 115, 135, 170, 478
emissions abatement 236
emotion: arousal of disgust 307; Haidt's cannibal story 307–308; moral and conventional transgressions 307; moral judgment 306–308; self-directed emotion 308
Endangered Species Act (ESA) 34, 51, 55, 88, 90, 98, 177, 557
Endres, D. 524
energy and extraction 1; energy poverty 387–393; fossil fuels 317–324; hydropower 349–358; mining 327–335; natural gas and fracking 374–383; nuclear power 338–346; renewable energy 359–368; *see also Individual entries*
energy balance models (EBMs) 246
Energy Policy Act (2005) 381
energy poverty: biocentric consequentialism 388; current 391–392; discounting future costs 390; energy security and energy for all 393; erosion of forests 387; 'fuel poverty' 387; importance of integrated policies of development 387–388; justice and biocentrism 388–389; present and future 389–390; sustainable solutions 390–391
energy security 393
Engels, F. 328
entertainment 14, 97, 103, 106, 555, 556, 561–563
environmental and labor performance, impact on 333–334
environmental civics 731, 732
environmental civil disobedience: civil institutions change 786; conscientious 783, 784, 785; due restraint 789–791; fidelity to civil order 784; 'Greenpeace Expeditions' 783–784; influential theory of 785; justifying 787–791; legal protest 787; limited to particular laws or policies 784; non-violent 784; public 784; 1970s Indian Chipko Movement 784; sound judgment 788–789; violations of rights or standing principles of justice 786
environmental education: aims problem 729, 730–733; legitimacy problem 729, 733–737
environmental ethics 4, 133–135, 221–223, 339–341, 525–526; criteria for protecting nature 340; criteria for protecting people 340; criteria

for sustainability 340–341; and environmental remediation 525–526
environmentalism 22–23; animal liberation *vs.* 14–16
environmental justice (EJ) 276; *Adelante Mujeres* 774–775; Aral Sea case 774; Capability Theory 775; "collective continuance (2018a)" 776; concept of collective capacities 776; DDT production 771; distributive accounts 773; ecofeminism 772; EJM 767–768; EJS 768, 777–778; MEM 771–772; Mesa Verde National Park 779–780; moral ecology frame 779–780; perspective of 770–771; philosophical approaches 775–776; prima facie equality 773; recognition in 772–777; 1992 Rio "Earth Summit" 773; social ecology 772; two-dimensional theory of justice 773–774
Environmental Justice Movement (EJM) 768–769
Environmental Justice Studies (EJS) 777–778
Environmental Liberation Front 784
Environmental Modification (ENMOD) Convention 289
environmental modification techniques 289
environmental nostalgia 568–569
Environmental Protection Agency (EPA) 183, 376
environmental quality 689
environmental radicalism: Mack and "Live and Let Live" 630–632; Nozick and attenuation of rights 627–628; Rothbard and reformulation of torts 628–630
environmental remediation 525–526
environmental transformation 1; assisted migration and reintroduction 546–552; ecosystems (*see* novel ecosystems); remediation 515–526; restoration (*see* ecological restoration (ER)); rewilding 568–579; zoos and conservation 554–563
environment narratives and food: agrarian narrative 463–464; developing world narrative 464–465; ethical consumerism narrative 465–466; meat narratives 466–467; political-economy narrative 464; religious narratives 466; romantic narrative 463; scientific narrative 461; techno-dystopian narrative 462–463; techno-utopian narrative 461–462; travelogue 465
episodic memory 15, 16; in humans 14; in nonhuman animals 14
Epstein, R. A. 634
equity and rights 219–221
eruptive disasters 653
An Essay on the Picturesque (Price) 418
ethical considerations for forest management in the 21st century 90–93; responsibilities 92–93; restoration baselines 91; restoration reparations 91; restoration scale 91–92
ethical consumerism narrative 465–466
ethical deficit 699; distribution of risks 700; exposed to risks 700; intentionally

expose 700; involuntary risk-taking 700; sensitive individuals 700

ethical risk assessment method: beneficiaries 701; combinations of roles (*see* roles, combinations of); decision-makers 701; risk-exposed 701

Ethics and Animals (Gruen) 476

ethics of coping with disaster 656

ethics of preparing for disaster 656

Ethics, Policy and Environment 2

ethnocentrism 32

European Union's Maastricht Treaty (1992) 707

Everett, J. 77

everyday aesthetics: aesthetic disillusionment 746; art criticism 745; Battle of Green 741–742; cultivation of environmentally informed everyday aesthetics 747–749; examples of everyday aesthetic tastes 742–743; nature aesthetics 745, 746; objections to environmentally informed everyday aesthetics 745–747; power of aesthetic 743–745; project of "Edible Estates" 741, 748

experimentation: animal research in medicine 32; animal suffering 31; applied ethics in animal experimentation 34–37; duty of beneficence 34; duty of nonmaleficence 34, 36; harm to animals 33–34; human and nonhuman animals 32–33; moral concern 32; speciesism condition 32

extended producer responsibility (EPR) 448–449

extensionism, pluralism, and nonanthropocentrism 74–76

extinction 57

extreme energy 375

"factory farming" *vs.* "humane" farming 25

"facts and information" campaigns 730–731

Fair Labor and Standards Act 177

fair risk exchanges 706

Family and Medical Leave Act 177

farming methods 494; agroecology 495–497; CRISPR-Cas9 technology 496; farmed animal facts 24–25; Food and Agricultural Organization 495; generations of GM crops 496; protecting ecosystems 496; sustainable intensification 495–496

Farr, D. 399; *Sustainable Urbanism: Design with Nature* 399

Faulkes, Z. 12

fault and wrong-doing in causally complex cases 604–605

fault tree analysis 697

Favre, D. 43–45

Finley, R. 505, 507, 508

fishing: collapse of global fisheries 208–209; domesticated stocks 210; fish-sourcing practices 211; globalization 205, 208; illegal fishing 193; and importance of harvesting 211–212; ITQ system 210–211; lives and livelihoods of fisherfolk and communities 198–199; migration 194; policy of "maximum sustainable yield" 193; productivity 198; setting limits to catch 210–211; state of world's fisheries 206–208; sustainable business approach 209–213; United Nations Fish Stocks Agreement 193; and warming 198; Western fisheries management 194

Florini, A. 364

Floyd, George 769, 770

food: and environment narratives (*see* environment narratives and food); localizing food production 499–500; narrative knowing 459–460

Food and Agriculture Organization (FAO) 205

Food Biotechnology in Ethical Perspective (Thompson) 475

food democracy 493

"foodies" and "locavores" 72

Foreman, D. 486

Forest Reserves 88

forests: Black Forest Fire 87; High Park Fire 87; land ethics 89–90; Lower North Fork Fire 87; management 89–90; recreation 88; threat of wildfires 87; timber 88; 21st century, ethical considerations for management in 90–93; US historical context 88–89; and wildfire management, equity considerations 93

Forest Service Organic Administration Act 88

formalism 136

Forsey, J. 747

fossil fuels: average utilitarian 323; carbon majors 317; co-harms and co-benefits 321; costs of 317–318, 322; discounting problem 323; double dividend hypothesis 323; international relationship 317, 321–323; national relationship 317, 319–321; personal relationship 318–319; potential harms of 342–343; social cost of carbon 322–323; total utilitarianism 323; usage of 317

Foster, C. 746

Fourier, J. 295

Fowlie, M. 680

Fox, J. 380

Fox, Warwick 115

"Frac Focus" 381

fracking: definition of 374; fluids 375; hydraulic fracturing 375; impact on 374; nonacid hydraulic fracturing (Hydrafrac) 375; at oil and gas pad sites 378; and technological wager 375–377; *see also* natural gas

Francione, G. 41, 42, 44, 45

free-market environmentalism 632–633

"free to grow" stage 537

Friedman, D. 626

Friedman, M. 668, 669

Friends of the Earth 784

Fry, Roger 745

Full Body Burden (Iverson) 522

Full Cost Accounting 686

Future Generations Argument 121
future-oriented system of checks and balances
620–621

Gaard, G. 775, 776
Gaia Hypothesis Argument 120–121
Galt, R. E. 498
Gandhi, M. K. 784, 803
Gardiner, S. M. 265, 320
Garreau, J. 422
Gates, Bill 482, 483, 496
Geddes, P. 400
Gelbspan, R. 301, 302
general circulation models 246
genetically modified (GM) crops 481–484, 486,
489, 496
genetic editing 59
geoengineering 280; alternative conceptions
283–284; band-aid objections 285; benefits,
costs, and risks 284–286; carbon dioxide
removal 281–283; Carbon Engineering 280,
281; climate engineering 280; DICE model
285; direct air capture methods 282; distributed
carbon scrubbing 282; *Dual Use Criterion* 284;
geo-steering and geo-remediation technologies
284; governance 289–290; justice considerations
286–288; mitigation proposals 281; moral hazard
286; ocean fertilization proposals 282; social
240–241; solar radiation management 281, 283;
stratospheric sulfate injection 281–282; technical
distinctions 281–283
Getz, C. 498
Gibbs, L. M. 608
Gilpin, William 135
Gleason, H. A. 586, 587, 594, 597
global climate models (GCMs) *see* general
circulation models
global warming 754
Goldschmidt, W. 476
good governance 221
Goodland, R. 21
Goodman, D. 500
good water: "bloodstream of the biosphere"
158; policy and management challenges 158;
requirements 158; water availability and quality
157, 159; *see also* water
Gore, Al 295
Gorelick, S. M. 360
Graber, David M. 125
Great Acceleration 594
Great Barrier Reef 199
Greater Perfections: The Practice of Garden Theory
(Hunt) 504
Greaves, H. 323
Greene, J. D. 307
green electricity 365–366
Green Globes 407
greenhouse effect 195

greenhouse gas (GHG) emissions 195, 256;
consumption per person 359; population 359;
technology choice 359
Greenpeace 784
"Grey Towers Protocol" 89
Gruen, L. 476, 775; *Ethics and Animals* 476
Grunebaum, J. O. 147
Guernica (Picasso) 133
Gunn, A. S. 80, 536, 537
Guthman, J. 498, 500

Hagenback, Carl 557
Haidt, J. 307; cannibal story 307–308
Hale, B. 517, 518, 694
Half-Earth Project 58
"Halliburton Loophole" 381
Hamilton, H. 723
Hampton, R. R. 14
Hansen, J. 449
Hardin G. 194, 473, 506, 670, 753, 757
harmful and toxic algal blooms 200
Harper, J. L. 538
Harris, Lawren 136
Harrison, Ruth 476; *Animal Machines: The New
Factory Farming Industry* 476
harvesting 211–212
Harvey, D. 350
Harvey, Paul 128
Hawken, Paul 98
Hayek, F. A. 624
Hayward, T. 615
Health Research Extension Act (1985) 34
Heck cattle 575
Hedahl, M. 281
Heede, R. 317
Hegel, Georg Wilhelm Friedrich 756
Heifetz, R. A. 720, 723
Held, V. 801, 804
heritage landscapes 138
Herring, R. J. 477
heterosexism 32
Hettinger, N. 76, 541
Heyward, C. 281
Higgs, E. 107, 184, 533, 540, 541, 595
Hinrichs, C. C. 500
Hippodamus 417
Hirsch, J. 510
Hobbes, Thomas 147, 735
Hobbs, R. J. 533, 588, 591, 597
Hoffman, M. L. 14, 558
Holden, M. 408
Holland, A. 181
Holling, C. S. 589
Homo sapiens 7–8, 15, 16
Hopkins, R. 449
Hornaday, William T. 556
Horowitz, A. 13
The Horticulturist magazine 419

hot spots and environmental justice 679–681
Hourdequin, M. 287, 318, 533, 534, 753, 762
HREA 34–35
Hughes, Thomas 427
Hull, D. L. 576
Hull, R. B. 718
"human chauvinism" 74
human moral patients 7
Humboldt, Alexander von 171
hummocks 172
Hunold, Christian 772, 773
hunting: agricultural subsistence 80;
    anthropocentrism 76; anti-hunting 77, 80;
    conservation management 73; environmental
    ethics 73; environmental fascism 75; ethical
    nonanthropocentrism 74–75; ethic of "fair
    chase" 81; "foodies" and "locavores" 72;
    human-centered sustainability 74; human
    culture and nonhuman nature 78–79;
    instrumental valuing 77; moral extensionism 75;
    nature-culture dualism 78; necessity and sport
    79–80; North American Wildlife Conservation
    Model 72; participatory affirmation and intrinsic
    value 76–78; plurality of ethics 76; predation
    problem 76; science and policy 78–79; "Social
    Carrying Capacity" 81; "solutions-based"
    approach 81; subsistence hunting 79–80; vision
    of ethical hunting 72–73; wilderness and
    technology 80–82
Hunting Argument 111–112
Hunt, J.: *Greater Perfections: The Practice of Garden
    Theory* 504
Hunt, John Dixon 504
hydropower: authoritative voices, role of 353–355;
    'the battle of the Vindel River' 351–352;
    conservative and progressive ethics 350–353;
    critique against expansion 350–351; effectiveness
    of environmental resistance 356; elite-based
    opposition 354; fatalistic debate climate 353;
    green electricity in exchange for environmental
    destruction 365–366; indigenous Sami
    population 354; industrial fatalism 352; oil crisis
    (1973) 352; oppositions 352–354; parliamentary
    venue 353; resource mobilization 355; Sarek
    Peace (1961) 351; Social Democrats 354; Swedish
    hydropower resistance 349–350; Swedish Society
    for Nature Conservation (SSNC) 350, 353;
    Swedish State Power Board (SSPB) 350; World
    Commission of Dams' landmark declaration 356
Hyson, J. 557

Individual Transferrable Quota (ITQ) system
    210–211
industrial agriculture: agrarian alternative 478–479;
    The Agrarian Vision 474; concentrated animal
    feeding operations (CAFOs) 474, 477; critics
    of industrial food system 475–477; defending
    477–478; early industrial revolution (1760–1840)

471; industrial food systems 472; industrial
    philosophy of agriculture: part I 472–473;
    industrial philosophy of agriculture: part II
    474–475; pesticide use in agriculture 470;
    rights-based or deontological approach 474–475;
    traditional agriculture 471
industrial food systems 472; critics of 475–477;
    environmental impact 476; human health
    475–476; non-human animals 476; social impact
    476–477
Industrial Revolution 195
informed consent 378–379
inherent value 10
initial condition ensemble (ICE) 251–252
"inside the Perimeter" (ITP) 422–423, 425
Inspiration Argument 114
institutional animal care and use committees
    (IACUCs) 35
"institutional control" 520
integrated assessment models (IAMs) 245
integrity 221
Intemann, K. 250
intergenerational issues (future generations) 288
Intergovernmental Panel on Climate Change
    (IPCC) 234, 546
Intergovernmental Science-Policy Platform on
    Biodiversity and Ecosystem Services (IPBES)
    206
International Union for the Conservation of
    Nature (IUCN) 126, 215
"the interpretation trap" 616
Intrinsic Value Argument 121–122
invasive species 15
"invasive" species 15
invertebrates 13–15
The Inward Morning (Bugbee) 106
Irwin, A. 711
Iverson, K. 522; *Full Body Burden* 522

Jackson, Michael 761
Jackson, S. T. 533, 586, 587
Jackson, T. 449
Jackson, W. 471, 487
Jacobson, J. A. 658
James, W. 294, 748
James Zadroga 9/11 Health and Compensation Act
    (2010) 768
Jamieson, D. 319, 415, 773
Jastrow, Robert 302
Jefferson, Thomas 415, 418, 478, 523, 741
Jeffers, Robinson 114
Jeffreys, Kent 302
Jensen, D. 797, 801, 802
Johnson, Bea 451; *Zero Waste Home* 451
Johnson, Baylor 753, 757, 758, 760
Jones, Phil 303, 304
Jordan, A. 708
Jordan, W. R., III 538, 539, 541, 595

Jozet-Alves, C. 14
justice considerations, geoengineering 286–288; distributional issues (global poor) 287; intergenerational issues (future generations) 288; participation issues (informed, democratic participation in decision-making) 288; recognition issues (tribal claims) 287

Kahan, D. M. 304, 305; Scientific Literacy Study 305–306, *306*
Kahneman, D. 308
Kain, J. 434
Kane, G. S. 538
Kania, J. 723
Kant, Immanuel 23, 77, 135, 756
Kaswan, A. 361, 362
Katz, E. 188, 485, 486, 530, 535, 536, 537, 541; challenge for ER 535–537
Keeling, C. D. 296
Keith, David W. 280, 281
Kelly, S. D. 105, 106
Kelman, S. 690, 692
Keyes, Robert 668, 669
King, M. L., Jr. 265, 784, 789
King, R. 81
Kingsolver, B. 450
Kitcher, P. 248
Klein, C. 11
Knight, Richard Payne 135
Korsgaard, C. M. 44
Kover, T. R. 80
Kutz, C. 452
Kymlicka, W. 42–45, 49
Kyoto Protocol 302
Kyoto treaty 295

Laboratory Argument 115
Ladkin, D. 537
Lakoff, George 758, 759
land 1; and forest management (*see* forests); meeting mountains (*see* mountains); and parks (*see* national parks); and property (*see* property); and wilderness (*see* wilderness)
land ethics 487, 489, 490, 497
landscape: art and landscapes 134–135; attitudes and actions 133–134; Bitterroot Range 141; cognitivism 137–138; cultural cognitivism 138–140; and ethics 140; formalism 136; heritage landscapes 138, 139; of human habitation 138; of human production 138; objects of aesthetic appreciation 133–135; and Picturesque 135–136; postmodern landscape 136–137; preservation 140; scientific cognitivism 139; significant form 136; urbanism 398
land-use changes 59
Lappé, F. 449
Larding the Lean Earth (Stoll) 471
lawbreaking and ecoterrorism: morality of ecoterrorism 800–804; terrorism 795–796

Lawford-Smith, H. 364
Lawhon, Lydia A. 694
Law of the Sea 193, 209
Lawson, L. J. 506, 508
leadership practices for AM: collaborative and disruptive 720–722; Direction, Alignment, and Commitment (DAC) 720–721; disruptive double-loop learning 722; humble and courageous 718–720; influence without authority 721; overcoming differences 721–722; reflective and action-oriented 722–723; stakeholder uncertainty 718–719; strategy uncertainty 719–720; successful AM 721; system uncertainty 718–719
"Leave No Trace" principles 90
Le Corbusier: *Plan Voisin* for Paris *404*
Leddy, T. 747
Lee, K. 486
legal and social contexts 516–518
legal protest, of civil disobedience 787; conscientious objection 787; radical and revolutionary action 787; self-interested criminality 787; terrorism 787
Legates, D. R. 300
legitimacy problem 729; defense of advocacy 735; defenses of activism 735; learning *via* habituation 734; limitations 736; rational persuasion defense 736–737; rejecting liberalism 735–736; strengthen students' pro-EVABRs 733–734, 737; training in science literacy and environmental civics 734–735
Lentil Underground (Carlisle) 497
Leonardo: *Mona Lisa* 133
Leopold, A. 15, 52, 76, 89, 90, 94, 98–108, 113, 115, 119–121, 133, 147, 148, 169, 172–177, 182, 424, 442, 487, 488, 490, 497, 539, 556, 719, 730, 742; on Wetlands 172–175
"The Leopold Report" 532
libertarianism: backing away from environmental radicalism 627–632; ending pollution 626–627; free-market environmentalism 623, 632–633; property rights 624–626; rights-based libertarianism 624
Life Cycle Analysis *see* Full Cost Accounting
"lifecycle assessments" (LCAs) 361
Life-Support Argument 113
Light, A. 181, 398–400, 424, 533, 537, 540, 541, 595
Lindzen, Richard 300
Lintott, S. 538
Living Building Challenge 407
'local foods' 500
Lockean proviso 671
Locke, J. 147, 149, 473, 624, 626, 671, 735
Loftin, R. 74
Loos, J. 497
Lorrain, Claude 135
Lovelock, James 120
Lozada, M. 12

Lubick, G. M. 539
Luke, B. 81

MacBride, S. 447, 448, 451, 452; *Recycling Reconsidered* 447, 451
Machiavelli, Niccolò 416
Mack, E. 624, 628, 630–632, 634; absolutism of libertarian property rights 631; case of *Bamford v. Turnley* 631; "elbow room postulate" 631; environmental problems 631; and "Live and Let Live" 630–632; moral paralysis or "hog-tying" 630; problem of interconnectedness 631; ur-claim 630
"Madagascar periwinkle argument" *see* Pharmacopoeia Argument
Mainstream Environmental Movement (MEM) 771
Making a Real Killing (Ackland) 522
Mallory, C. 76
Malthus, Thomas 473
"mammoth steppe" 575
"management issues" 127
Mandela, Nelson 795
Mander, J. 485
Maniates, M. 451
Mann, Michael 303, 304
Mansourian, S. 92
Marine Mammal Protection Act (1972) 557
marine protected areas (MPAs): benefits for nature and humans 217; consequentialist, deontological, and virtue-based ethics 222–223, 224; defined 216; deontological ethics 223; environmental ethics 221–223; environmental outcomes 220; environmental protection 221–222; environmental values of 218; equity and rights 219–221; ethical schools of thought and implications **216**; fish nurseries 217; integrity 221; key ethical considerations 223, **224**; Locally Managed Marine Areas 220; marine and estuarine systems 218; "no-take" MPAs 220; objectives of Thai **217**; potential ecological benefits 217–218, 221, **222**; relational values 223; utility 217–219; wilderness 219
Marine Stewardship Council (MSC) 210
market-based instruments 675
Markovits, D. 786
Marris, E. 59, 533
Marshall, Bob 118, 125
Marshall, George C. 302
Marshall, Robert 113
Martin-Ordas, G. 14
Martin, P. 574
Mascaro, J. 583, 587, 592
Maskit, J. 748
Mason, J. 477
Massachusetts Toxic Use Reduction Act (TURA) (1989) 713
Mathews, F. 569, 571
Matisse, Henri 136
Matthews, H. D. 287

Mayr, E. 55
McBay, A. 802, 803
McCool, D. 184
McCright, A. M. 733
McKibben, B. 67, 96, 99, 100, 102, 103, 108, 320, 449
McLean, D. 390
McMahan, J. 9
McNeill, J. A. 745
Meadows, Donella 112
Measurement While Drilling (MWD) 376
meat narratives 466–467
Menard, Pierre 746
Mendelsohn, R. 678
"Men's Movement" 119
Mental Therapy Argument 114
Merchants of Doubt 301–302
Messiah (Handel) 133
Michaels, Patrick 301, 302
Midgley, M. 143
Miller, D. 272
Miller, M. 504
Mill, J. S. 81, 149, 472, 702, 735
Mills, C. 771
mining: coal, moral dilemmas of 327–329; controversies in CSR (*see* corporate social responsibility (CSR)); uranium, moral dilemmas of 329–330
Mintz-Woo, K. 323
Mitchell, G. P. 376
mitigation of climate change: burdens of 239; CO_2 emissions 236–237; collective duties 262–263; "command-and-control" policies 238–239; future impacts of climate change 257–258, *258*; GHG emissions 256, *257*; individual duties 263–265; moral reasons 259–261; optimal pathway approach 237, *238*; risk management approach 237; scientific and economic reasons 256–259; social cost of carbon 237–238
mixoecology 587
Mona Lisa (Leonardo) 133
monetary value 641
Monfort, S. 559
monitoring 379–381
Montreal Accord 752
Montreal Protocol (1987) 707
Moore, S. A. 406
moral and prudential reasons 234–235
moral corruption 265
moral ecology frame 779–780
moral hazard 286
moral patients 7
moral status 6; anthropocentric account of 7–9; "human" and "person" 6; moral agents 6–7; moral patients 7; sentientist account of 9–10
Moriarty, P. V. 78
Morrow, David R. 281
Moser, S. C. 309
Moss, D. A. 607

mountains: Absaroka-Beartooth Wilderness
100; animating nature 105; challenge 107;
disclosure of 107; "Forest Reserve Revenue
Areas" 100; hyper-anthropocentric flattening
104; interanimating symmetry 97, 104–108;
intrinsic value 102; "the key-log" 103; living
by mountains 108; Livingston Peak or Mount
Baldy 97–98, 100, 102; Montana's Paradise
Valley 97; mountain-choking developments
104; New Hampshire's White Mountains 106;
ontology of technology 104–105; Pipestone Pass
near Butte, Montana 97; revising "Thinking
like a Mountain" 100–104; steep slopes and deep
powder of 104; transformation of Earth 98–100;
as valuable 106–107; Wilderness areas 103–104;
Yellowstone Park 97
Moylan, Jonathan 791
Muir, J. 114, 150, 151, 169–174, 176, 177, 217, 770;
love of wetlands 170–172
Muller, N. Z. 678
multi-model ensemble (MME) 251
Multiple Accounts Analysis *see* Full Cost
Accounting
Multiple Use Sustained Yield Act (MUSYA) 88
Mumford, L. 400, 401, 404, 405, 409; biotechnic
era 401, 404; eotechnic era or "water and
wood" complex 400, 401; neotechnic era 401,
405; paleotechnic era 401, 405; *Technics and
Civilization* 400
Murchison, K. M. 610
Murcia, C. 540
Murray, B. 322
Mythopoetic Argument 119

narratives, food: stories depict scenarios 460; stories
humanize characters and make events relatable
460; stories interpret events 459–460; stories
make arguments 460; stories portray characters
460
Nash, J. R. 677
Nash, R. F. 111, 120
National Aeronautics and Space Administration
(NASA) 36
National Building Institute (NBI) 407
National Character Argument 117
National Environmental Policy Act (NEPA) 91
National Forest Management Act (1976) 88
National Institutes of Health (NIH) 37
national parks: Glacier National Park 128; Great
Smoky Mountains National Park 125; Hopi
tribe, native Americans 126–127; Mammoth
Cave National Park 126; management issues 127;
National Park Service 124; and natural parks
124–125; people in parks/people *vs.* parks
125–127; recreation and creation 129–131;
Rocky Mountain National Park 127;
Shenandoah National Park 127; Theodore
Roosevelt National Park 128; Tower National

Monument, Wyoming 127; Wildlife: "Let
Nature Take Its Course" 128–129; "working
landscapes" 124; Yellowstone National Park 217;
Yosemite Park Service 126
Natural Channel Design (NCD) approach 186
natural gas: compulsory integration or forced
pooling 378; definition, history, and overview
374–376; and fracking (*see* fracking); informed
consent 378–379; monitoring 379–381;
proactionary principle 377; renovation 381–383;
and technological wager 376–377; therapy and
research 377
"naturalistic fallacy" 54
Natural Resources Argument 111
Natural Resources Law 353, 354
Necessity Argument 119
Neely, M. 569
Nelson, M. P. 158, 572
neoliberal governance, expansion of 331–332
neurotypicalism 8
For a New Liberty (Rothbard) 626
Nierenberg, William 302
NIMBY (not in my backyard) cases 703
Niven, J. 11
nociceptors 13
No Impact Man (Beavan) 450–451
nonacid hydraulic fracturing (Hydrafrac) 375
Non-Aggression Principle 628
non-anthropocentric nuclear power ethics 345–346
non-consequentialism and wild animals 68–70
non-governmental organizations (NGOs) 407
Nordhaus, W. 234, 285
Norgaard, R. B. 364
Northern Virginia Stream Restoration Bank
(NVRRB) 188
Norton, B. G. 179, 185, 426, 429, 487
Noss, R. 572–574, 572–575, 577, 578
novel ecosystems 588, *588*; "the Anthropocene"
583–584; anthropogenic 587–589; appearance
of coywolves 584, 585; "balance-of-nature"
view 586; characterization of 586; classic
restoration 595; community-based succession
586; composition 597n1; definition of ecosystem
584–585, 588; eco-centric theory of value
586–587; environmental values for new nature
593–597; "Franken-earth" 584; function
597n1; Great Acceleration 594; historic and
wild ecosystem 588, *589*, 594–595; historic
naturalness in conservation and restoration
590–591; "human-dominated" landscapes 593,
595; hybrid ecosystem 588, *589*; individualistic
conception of ecosystem 587; mixoecology 587;
modeling space of ecosystem states and pathways
of change 591–593; natural ecosystems 591, *592*;
novelty and human agency 592, *592*; orientation
by analogy 583–586; recombinant ecologies 587;
resilience 584, 589; structure 597n1; theories of
ecological change and natural ecosystem novelty

586–587; value pluralism 596–597; wide range of variation 595

Nozick, R. 276, 624–628; *Anarchy, State, and Utopia* 624; and attenuation of rights 627–628

nuclear legacy of rocky flats 522–525

nuclear power: debate 338–339; environmental ethics 339–341; non-anthropocentric nuclear power ethics 345–346; opponents of 339; policy implications 344–345; potential harms of power sources 341–344; proponents 338–339

nuclear power, potential harms of 341–342

Nussbaum, M. C. 9, 640, 775, 776

Obama, Barak 127, 177

Obama, M. 507

Occupational Safety and Health Act 177

Occupy Wall Street 417

ocean acidification 192, 197–198, 199

ocean fertilization 282

ocean policy: acidification of ocean (*see* ocean acidification); climate changes 195; climate warming 192, 199; decline of seagrass meadows and oyster reefs 200; ethics and morality 194–195; fisheries (*see* fishing); harmful and toxic algal blooms 200; new diseases ravaging corals and sponges 200; oceans warming 195–196; plasticization of sea 200; sea levels 196–197; spreading dead zones 199–200; storms 197

oceans warming 195–196

Octopus vulgaris 12

Oil Pollution Act (OPA) 603

O'Keefe, Georgia 136

Okin, G. S. 48

Olmsted, F. L. 412, 413, 415, 417–420; barbarisms 418; communitiveness 418; farming 417–418; father of American Landscape Architecture 413; legacy 417–419; Republicanism and history of public parks 417–419; social freedom 419; traditional park values 417

O'Neill, J. 179, 181, 185

Ontogeny Argument 116

"optimal pathway" approach 237, 238

"Oregon Petition" 313n10

Oreskes, N. 297, 298, 300, 302

Origin (Darwin) 172

O'Riordan, T. 708

Orr, D. W. 408, 449, 744, 747

Ortega y Gasset, J. 81, 82

"outside the Perimeter" (OTP) 422, 425

Owens, D. 398

ownership: classification of 144; institution of 147; property and 143–144; rules of 147

Palamar, C. R. 538

Palmer, C. 42

Palmer, M. A. 185, 186

"pan *situ*" conservation 560

Parfit, D. 161, 235, 389, 756, 758

Parker, W. S. 252

Parsons, G. 506

participation: issues (informed, democratic participation in decision-making) 288; models of 332–333; participatory affirmation and intrinsic value 76–78

Pascal, B. 297

passivity of preservation 538

"pasture tree" 570

Pauly D. 206

Peeveyhouse vs. Garland Coal, case of 685, 688, 692

Peirce, C. S. 405, 406

Pellow, D. N. 445

Pence, G. 483, 484

permafrost 195

perturbed-physics or perturbed-parameter ensemble (PPE) 251

Peters, R. S. 734

Pharmacopoeia Argument 112

phylogenetic species concept (PSC) 57

Physical Therapy Argument 113

Picasso: *Guernica* 133, 746

Pigou, A. C. 677

Plant Succession: an Analysis of the Development of Vegetation (Clements) 172

plasticization of sea 200

Plato: *Republic* 416; social and ecological criticism 145; *The State* 144

Pleistocene overkill hypothesis 574, 578

"Pleistocene Park" 575

Pleistocene Rewilding 573–575

Pletsch, C. 538

Plumwood, Val 772, 773, 775–778, 780

policy frameworks and response measures 1; constitutional rights 613–621; disaster response 652–662; libertarianism 623–634; pollution and polluter pays 603–611; prediction and forecasting 636–649; *see also Individual entries*

political-economy narrative 464

Politics (Aristotle) 144, 416

The Politics of Earth (Dryzek) 720

polluter pays principle 603, 677–678

pollution: consequences of strict liability 609–610; "hot spots" 678; impossible and implausible demand 626–627; Malloy Corporation, case of 604–605; and polluter pays 603; right of recourse to government action 606–607; right to self-defense 606; risk imposition 605–606; risk reduction 606, 607; risk shifting 606, 608; risk shifting in context 610–611; risk spreading 606, 607–608; Russian roulette 605; strict liability in environmental law 608–609

Positive Aesthetics 138

Postel, S. 180, 183, 189

postmodern landscape 136–137

poverty 389–390

Powell, J. 710

Powell, M. 710

power sources: potential harms of fossil fuels
342–343; potential harms of nuclear power
341–342; potential harms of renewables 343–344
pragmatic environmental ethic 487–489
Prat, Mary Louise 778
precautionary principles (PP): alternatives
assessment 713; asymmetries of knowledge,
power, and policy making 711, 713–716; concept
of sustainability 709; decision rules 709, 711;
deliberative and participatory approaches 713;
epistemic rules 709; ethical norms 377, 378;
formulations of 709; interpretation of 712;
opposition 382–383; procedural requirements
709; proponents of 708, 711, 712–714, 714;
resolving ignorance and uncertainty 710–711;
selective ignorance 710, 714
predation 47–48, 76
prediction and forecasting: background 637;
capabilities approach to rights 640; climate
change and development of "just" climate
services 638–639; climate services 638;
deliberative justice 640–641; examining current
climate services discourses (*see* climate services);
rights-based justice 639–640
Preston, C. 486
priceless 690–691
Price, U. 135, 418, 419; *An Essay on the
Picturesque* 418
pricing approaches 437
principle of "cross and compensate" 628
principle of equal consideration of interests 9, 52
principle of maximal equal influence 706
Prior, J. 571
proactionary principle 377, 379, 382
probabilistic risk assessment (PRA) 697, 698
probabilistic safety assessment (PSA) 697, 698
procedural constitutional rights: decision-making
procedures 617; ecological council model
617–618; interpersonal deliberation 619–620,
622n7; intrapersonal deliberation 619–620,
622n7; submajority rule model 618–621
procedural memory 14, 16
projection 245
property: (regulated) common property 145;
contents 147–149; European Court of Human
Rights (ECHR) 146–147; "high-tech property,"
emergence of 145; incompatibilism 148; open-
access or non-property 145; and ownership
143–144; private property 145; "Private
property" 149; real property 143; rights
624–626; state property 145; subjects 149–151;
types 144–147
protected areas (PA): defined 215; marine
(*see* marine protected areas (MPAs))
psychology, skepticism: American Public Opinion
304, *305*; cognitive biases 308–309; cultural
cognition thesis (CCT) 304–306; egalitarian
communitarians (EC) 305; emotion and moral

psychology 306–308; hierarchical individualists
(HI) 305; science comprehension thesis (SCT)
304–306; Scientific Literacy Study 305–306, *306*
public water 161
Purdom, R. 540
Puri, S. 12

qualitative values 690

Rachels, J. 9
racism 32
radical and revolutionary action 787
Railton, P. 632
rapid ecological change: and species reintroductions
548–549; and species translocations 549–552
rational method 405
Rawls, J. A. 482, 639, 735, 755, 785–787, 789, 803
Reagan, Ronald 302, 460, 625, 669
reclamation 531
recognition issues (tribal claims) 287
rectification **238**, 241
Recycling Reconsidered (MacBride) 447, 451
Rees, J. H. 308
Rees, W. E. 397
Regan, T. 10, 13–15
regional climate models (RCMs) 246
regulatory tools 1; adaptive management 717–724;
CBA (*see* Cost-Benefit Analysis (CBA));
command and control 667–673; economic
instruments 675–682; PP (*see* precautionary
principles (PP)); risk assessment 696–705
rehabilitation 531
religious narratives 466
remediation: environmental ethics 525–526;
environmental remediation 515–516, 525–526;
ethical concerns 517; legal and social contexts
516–518; nuclear legacy of rocky flats 522–525;
Rocky Flats National Wildlife Refuge 518,
522–524; rocky mountain arsenal 519–521;
Rocky Mountain Arsenal National Wildlife
Refuge 518–521; superfund remediation
518–525
renewable energy (RE): carbon capture and storage
360–361; "clean-coal" technologies 360; *vs.*
clean fossil-fuel-based technologies 360–361;
CO_2 emissions 359–360; ethical problems
365–368; feasibility of zero-carbon economy
362–365; greenhouse gas emissions 359, 362;
hydropower 365–366; lifecycle assessments
361; long-term energy security 362; solar
photovoltaic 366; wider benefits of renewable
technologies 361–362; wind power 366–368
renewables 343; *vs.* clean fossil-fuel-based
technologies 360–361; potential harms of
343–344; technologies, benefits of
361–362
renovation 381–383
Republic (Plato) 416

resilience 654
Resource Conservation and Recovery Act (RCRA) 516
Respect Principle 10
responsibility for regulation 332
restoration 185; Comprehensive Everglades Restoration Plan (CERP) 529; ecological (*see* ecological restoration (ER)); "the restorationist argument" 577; "restoration thesis" 188
restorative nostalgia 578–579
resurrection science 576
rewilding: active rewilding 578; as conservation strategy 572–573; Cs: cores, carnivores, and corridors 572–573; de-extinction 575–578; environmental nostalgia 568–569; "The Irony of Eastern Wilderness" 570–571; passive rewilding 578; Pleistocene rewilding 573–575; reflective nostalgia 568, **569**, 571; restorative nostalgia 568, **569**, 578–579; unplanned rewilding 570–571; wolf tree in Cockaponsett State Forest *570*
Reynolds, R. J. 302
Richter, B. 180, 183
Rieux, Bernard 655, 661
right of minorities to require delays 619
right of minorities to require referendums: agenda-setting 619; future-oriented system of checks and balances 620–621; processes of deliberation 619–620
rights-based account of animal liberation: inherent value 10; subject-of-a-life 10
rights-based justice 639–640
Rio Declaration (1992) 708
risk assessment 697; acceptable risk 704; dichotomous model of decision-making 696–697; ethical (*see* ethical risk assessment method); ethical deficit 699–700; expectation value 697, 698; fair risk exchanges 706; principle of maximal equal influence 706; probabilistic analysis 698; scenario development 697–698; standard methodology 697–699
risk-exposed 701
risk imposition and right to self-defense 605–606
risk management approach 237, 697
risk reduction 607
risk shifting 608; in context 610–611
risk spreading 607–608
river(s): challenge of dam removal 187; channelization of 180; compensatory mitigation 187–188; European settlement 180; Kissimmee River Restoration Project 179; restoration (*see* river restoration (RR)); valuing 181; and watershed management (*see* watersheds)
river restoration (RR): adaptive management 185; challenges 189; discursive democracy 185; ecological value of 182–183; economic value of 181–182; ethical reflection and reformation 179; flow characteristics 183; function and processes 183; healthy river 182; historical approaches 183–184, 185; importance of scale 185–187; integrity of rivers 182; proliferation of 180; resolving conflicts 184–185; stakeholder or social concerns 184
Rivers, N. 322
Robinson, M. 262
Rodman, J. 75
Rogers, R. 449
roles, combinations of: risk role 1, only beneficiary 701; risk role 2, beneficiary and decision-maker 701–702; risk role 3, only decision-maker 702; risk role 4, risk-exposed and beneficiary 702; risk role 5, risk-exposed, beneficiary and decision-maker 702–703; risk role 6, risk-exposed and decision-maker 703; risk role 7, only risk-exposed 703–704
Rollin, B. E. 9, 34, 476
Rolston, H., III 75–78, 80, 98, 105, 113, 115, 424, 482, 483
romantic narrative 463
Roosevelt, Eleanor 507
Roosevelt, Franklin D. 507
Roosevelt, Theodore 112, 128, 556
Rosa, Salvator 135
Ross, M. 433
Rothbard, M. N. 624–626; distinction between trespass and nuisance 630; homestead easements 629; *For a New Liberty* 626; Non-Aggression Principle 628; and reformulation of torts 628–630; strict causal connection 629; Trespass/Nuisance distinction 629
Routley, R. 143
Royce, Josiah 415
ruralism 415–416
Russow, L. 57

Sachs, C. 493, 494
Sachs, J. 483
Safe Drinking Water Act 381
Sagoff, M. 158, 689
Salazar, Ken 127
Salm, R. V. 218
Salomon, D. 8
Salvation of Freedom Argument 118–119
Sanbonmatsu, J. 16
Sandberg, J. 753, 755–756, 758–760, 763; inconsistently recommend legislative over individual action 761; presume an inadequate model of causality 758–759
A Sand County Almanac [SCA] (Leopold) 98, 101–102, 147, 173–174
Sandin, P. 708
Sarek Peace (1961) 351
savannas 172
"Saving Animals From Extinction" (SAFE) program 554
Schiller, F. 744
Schlosberg, D. 276, 774, 775

Schmidtz, D. 149
Schon, D. 722
Schwartz, B. 656
Schwartz, D. 452
science literacy 731–732
scientific cognitivism 138
scientific consensus 297–300
scientific literacy and cultural cognition 304–306
scientific narrative 461
Scoville, J. M. 537
sea levels 196–197
Searle, J. 505; *Constructing Social Reality* 505
second-generation rights to food, shelter, and
    healthcare 260
Seitz, Frederick 302
self-awareness 15–16
self-interested criminality 787
Self-Realization Argument 117–118
semantic memory 14, 16
Senge, P. 723
sentientist account of moral status 9–10
Service Argument 112–113
Seuss, H. 296
sexism 32
sexual predators 76
shadow prices 176
Shapiro, M. 616
Sharpe, K. 656
Shelley, M. 583
"shellfish poisoning" 200
Shiva, V. 476, 477
Short, L. L. 55
Should Trees Have Standing? (Stone) 75
Shrader-Frechette, K. 773
Shue, H. 393
"significant form" 136
Silent Spring (Carson) 470, 471
Simon, H. 308, 404
Singer, J. W. 143, 146, 147
Singer, P. 9, 10, 14, 32, 34, 52, 53, 75, 77, 476, 477,
    785; *Animal Liberation* 75
Sinnott-Armstrong, W. 752–760, 763; appeals
    to practical and linguistic intuitions 754–755;
    drunk driving 754, 766n24; inconsistently
    recommend legislative over individual action
    761; method of elimination 754; misinterprets
    some of principles 755–756; presume an
    inadequate model of causality 758–759
skepticism: about individual and small group
    initiatives 756–763; argument for 754–756;
    climate wager 296–297; communicative
    strategies 309–310; consensus 297–304; evidence
    294–296; fail to recognize political significance
    of individual and small group initiatives
    760–763; human greenhouse gas emissions
    293; increasing global temperatures 293;
    psychology 304–309; rely on vague, ambiguous,
    and context-insensitive descriptions 756–757;

terminology 293–294; and three methodological
    worries 754–756; "Tragedy of the Commons"
    arguments 757–758
skiing 104
Smith, K. K. 44
Smith, Wake 285
Sneddon, L. U. 12
Snow, C. P. 540
Social Bonding Argument 120
Social Darwinism 76
social distancing plans 659
social imaginary 656
socioeconomic benefit studies (SEB) *see* cost-benefit
    analysis (CBA)
soft constraints 364
solar photovoltaic 366
solar radiation management (SRM) 281
solar radiation management or solar
    geoengineering 236
Soon, W. 303, 304
Soule, M. 572–575, 578
sound judgment 788–789
species and wildlife: aesthetic debates, wildness
    56; Anthropocene 57–58; axiological *vs.*
    deontological approaches 53–54; conservation
    responses 58–60; contemporary debates 56–58;
    decline and change of common wildlife 56–57;
    environmental ethics 51–52; epistemic debates
    54; extinction 57; individualism *vs.* holism
    52–53; ontological debates, species 55;
    species-centered arguments 57; wildlife
    ethics *vs.* animal ethics 52
species extinctions 59
species introductions 59
speciesism 8
species membership 7
species typicalism version 9
stakeholder or social concerns 184
Standard of Land Health Argument 115–116
The State (Plato) 144
Steel, D. 250, 709, 712
Stewart, Potter 470
Stewart, R. B. 667
Stichter, M. 737
Stoll, S. 471; *Larding the Lean Earth* 471
Stone, C. 42, 75, 90, 91
Storage Silo Argument 115
Storch, Hans von 303, 304
storms 197
Story of Stuff (Leonard) 443, 446
stratospheric sulfate injection (SSI) 281–282
subject-of-a-life 10
submajority rule model 617; right of minorities to
    require delays 619; right of minorities to require
    referendums 619–621
substantive constitutional rights: *The Constitution of
    Norway* 615; *The Constitution of South Africa* 615;
    democratic problem 615–616; environmental

quality 614–615; judicial review and legal
indeterminacy 615–617; rule of law problem
616–617
substitution problem 530
suburbs and exurbs: hope 428–429; ITP 422–423,
425; metropolitan system (second challenge)
426–428; myth of suburbs and wild (first
challenge) 423–426; OTP 422, 425
"Sundowner Principles" 36
Sunstein, C. R. 707, 745
Superfund law *see* Comprehensive Environmental
Response, Compensation, and Liability Act
(CERCLA)
sustainability: concept of 709; conservation
management, and anthropocentrism 73–74; food
democracy 493–494
sustainable agriculture: community-supported
agriculture 498–499; ecosystem services 495;
ethical concerns 492; farming methods 494,
495–497; integrating ethics 493–495; making
food local and fair 499–500; organic agriculture
497–498; place of labor 497–499; science and
technology 494–495; sustainable intensification
and agroecology 495–497
Sustainable Fisheries Act (1996) 199
sustainable houses 400
sustainable intensification 495
Sustainable Sites Initiative (SSI) 407
sustainable solutions 390–391
sustainable urbanism *see* urban
sustainability
Sustainable Urbanism: Design with Nature
(Farr) 399
Sylvan, R. 531, 536
synthetic biology 59
Szasz, A. 608

Tansley, A. G. 586
Taylor, P. 52, 53
Taylor, S. 8
Technics and Civilization (Mumford) 400
techno-dystopian narrative 462–463
technological wager 376–377
techno-utopian narrative 461–462
terrorism: 2006 "Animal Enterprise Terrorism Act"
800; core features of 800; domestic terrorism
799; enhancements 794; Nazi treatment of
European Jews 796; property destruction
797–800; September 11, 2001 al-Qaeda attack
795; state's use of violence and intimidation 796;
targeting non-combatants or innocents 796–797;
"terror bombing" of German cities 795; "War
on Terror" 795; *see also* ecoterrorism
Thaler, R. H. 745
Thompson, P. B. 338, 339, 487–489, 492; *Food
Biotechnology in Ethical Perspective* 475
Thompson, R. P. 483, 484, 487
Thomson, J. J. 606

Thoreau, H. D. 98, 107, 115, 116, 135, 150, 151,
169–177, 478, 783–785, 791; love of wetlands
169–170
"thought experiment" 168
A Thousand Mile Walk to the Gulf (Muir) 170
three Cs: cores, carnivores, and corridors 572–573
"3Rs" framework 37; reduce harmful animal use
35; refinement of experiments 35; replacement
of sentient animals 35
Throop, W. 536, 537, 540
Tickner, J. 708, 712
"timber wars" 88
Time to eat the dog? The real guide to sustainable living
(Vale and Vale) 45
Torreya Guardians 547
Toxic Substances Control Act (TSCA) 713
Trachtenberg, Z. 350
traditional park values 413–415
"Tragedy of the Commons" 506, 670, 757–758
transformation of Earth 98–100
transit-oriented development (TOD) 434–437
translocation 530, 560
transportation: conclusions for environmental
ethics 438–439; consequences for individuals
432–433; environmental and urban issues of
car use 433–435; investing in infrastructure
alternatives to auto 435–436; mode 432;
passenger mobility 432; pricing approaches 437;
proposed policy and planning solutions 435–439;
technology solutions 437–438; transit-oriented
development 436–437
travelogue 465
tree lawn 508
Trenberth, Kevin E. 303
Truman, Harry S. 800
Trump, Donald 177, 178, 328, 380, 383
Tuana, N. 776
Turner, D. 800
Turner, F. 538
Turner, J. 106, 107
Tversky, A. 308
"two cultures problem" 540
Tyndall, J. 295, 451

Unilever 210
United Kingdom Animals Act 35
United Nations Convention on Biodiversity 201
United Nations Convention on the Law of
the Sea 209
United Nations Framework Convention on
Climate Change (UNFCCC) 233, 237,
271–272, 302, 639
Unknown and Indirect Benefits Argument 121
uranium mining: history of 329; moral dilemmas
of 329–330; nuclear waste disposal 330;
occupational safety for 329; problems of workers'
exposure to radiation and radon gas 329; *see also*
mining

urban parks and open space: civic values 413; economic development 414–415; isonomia 416, 417; "leisure services" 413; as "the lungs of the city" 414; *Modus vivendi* agreements 417; nature 413–414; public values 415; Republicanism and history of public parks 417–419; role in politics 417; ruralism 415–416; traditional park values 413–415; and urban values 416–417; from values to design 419–420; zoning and transit 412
urban sustainability: achieving dynamically balanced ecosociotechnical system 404–406; assessing and regulating urban sustainability 406–407; Bioswale Detail for Elmer Avenue Retrofit *406*; changing neighborhoods, Social Reform from 7 Chicago Settlement Houses *399*; concept of rational planning 405; context-dependent design intentions 407; context-independent standard 407; definition of sustainable development 397; eco-districts 403–404; familiar knowledge 405; Fenway and Simmons College, Boston, Mass *403*; Le Corbusier's *Plan Voisin* for Paris *404*; modern nature/city dualism 398–400; *Oberlin Project* in Oberlin *408*, 408–409; public talk and social learning 408; rational method 405; relationship between nature and city 400–404; *see also* cities
urban values and urban parks 416–417
US Environmental Protection Agency (EPA) 374
US Forest Service (USFS) 88
US Green Building Council (USGBC) 407
US historical context 88–89
US Surface Mining Control and Reclamation Act (1977) 531
US Wetland Policy 177–178
utilitarian moral theory 687–688

Vale, B. 45
Vale, R. 45
Van Den Berg, A. E. 398
Vanderheiden, S. 799, 803
Varner, G. 41, 74
Vedung, E. 350
veganism 22–23; moral argument for 20–21; objections to 27–28
virtue-based ethics 222–224
Vitali, T. 82
Von Wehrden, H. 342

Wagner, Gernot 285
Waldheim, C. 398
Walker, C. 414
Walmart 498
Walton, K. L. 746
warming 199
waste and consumption: culture of consumerism 450; "Earth First"-type resistance 449;

electronic waste 445; environmental ethics of stuff 443; ethical consumption 450; ethical nonconsumption 450–451; ethics of "being the change" 450–453; extended producer responsibility 448–449; gross domestic product (GDP) 446; *Keep America Beautiful* "crying Indian" campaign 452; moral structure 452–453; quantity of waste 444; role in fueling climate change 442, 445; and stuff upstream 444–447; systemic, transformative change 448–450; TerraCycle 448; zero waste 448
waste lands 175
water 1; availability 159–161; community 159–160; ecosystem services 163; fishing and harvesting 206–209; good water 157–159; MPA (*see* marine protected areas (MPAs)); object-given *vs.* subject-given forms of ethical reasoning 161; and ocean (*see* ocean policy); preferences via market mechanisms 160; prior appropriation rights 160; quality 162–163; and rivers (*see* rivers); sustainable business approach 209–213; value of wetlands (*see* wetlands)
Water Quality Improvement Act (1970) 609
watersheds: management 180–181; valuing 181; *see also* rivers
Watson, P. 798
Webber, S. 643
welfare 64–65
Wells, M. 219
Wenar, L. 321, 322
Weston, Edward 136
wetlands: intrinsic and instrumental values 176–177; Leopold's environmental philosophy 172–175; "Marshland Elegy" features 173–175; Millennium Ecosystem Assessment 175; Muir, love of wetlands 170–172; Thoreau, love of wetlands 169–170; tidal marsh vegetation 176; US Wetland Policy 177–178; value and valuation 175–177; and Western imaginary 168–169
Wheatley, T. 307
Whitehead, Alfred North 756
Whitman, Walt 116
Whyte, K. P. 524, 776, 778–780
Wiener, J. B. 682
wild animals: Anthropocene 65–67; consequentialism and wild animals 67–68; non-consequentialism and wild animals 68–70; suffering 63; welfare 63, 64–65
Wilde, Oscar 745
wilderness 114; areas 111, 112; *The Great New Wilderness Debate* 111; psychological health benefits 114; technology and fair chase 80–82; 2020 introduction 110–124
Wilderness Act 78, 90, 98, 177, 574
wilderness preservation arguments: Animal Welfare Argument 120; Arena Argument 113;

Art Gallery Argument 114; Cathedral
Argument 114–115; Classroom Argument
116; Cultural Diversity Argument 116–117;
Defense of Democracy Argument 119–120;
Disease Sequestration Argument 118; Future
Generations Argument 121; Gaia Hypothesis
Argument 120–121; Hunting Argument
111–112; Inspiration Argument 114; Intrinsic
Value Argument 121–122; Laboratory Argument
115; Life-Support Argument 113; Mental
Therapy Argument 114; Mythopoetic Argument
119; National Character Argument 117; Natural
Resources Argument 111; Necessity Argument
119; Ontogeny Argument 116; Pharmacopoeia
Argument 112, Physical Therapy Argument
113; Salvation of Freedom Argument 118–119;
Self-Realization Argument 117–118; Service
Argument 112–113; Social Bonding Argument
120; Standard of Land Health Argument
115–116; Storage Silo Argument 115; Unknown
and Indirect Benefits Argument 121
wildlife ethics *vs.* animal ethics 52
Williams, M. 150
willingness to pay (WTP) 641, 692–693
Wilson, B. B. 406
Wilson, E. O. 58, 78, 115
wind power 366–368
Winsberg, E. 252
Woodley, S. 534
Woods, M. 78
world narrative 464–465
World Wildlife Federation (WWF) 210

Worm, B. 206
Wright, Frank Lloyd 415
Wynne, B. 711

Yandle, B. 681
Young, I. M. 416, 772, 773

Zavaleta, E. S. 534
Zen, Amartya 775
zero-carbon economy, feasibility of 362–365
Zero Waste Home (Johnson) 451
Zimmerman, M. E. 118
Zimmerman, M. K. 299, 300
Zoback, M. D. 360
zoos: "Amphibian Ark" 561; "assisted colonization"
and "Pleistocene rewilding" 562; AZA 554, 556;
Bronx Zoo 556, 557, 560; challenges 554–555;
Cincinnati Zoo 556; conservation agenda 556;
conservation breeding and reintroduction
558–560; "Conservation Centers for Species
Survival" 559–560; Copenhagen Zoo 561;
extinction 556; London Zoological Garden
555; New York Zoological Park 555–557;
People for the Ethical Treatment of Animals
561; Philadelphia Zoological Society 555;
SAFE program 554; science and growth of
modern zoo 555–558; Species Survival Plans
557–558; Wildlife Conservation Society 557;
World Association of Zoos and Aquariums
556; zoological ethics on changing landscape
560–562
Zwolinski, M. 669